Fundamental Principles of Bacteriology

Antoni van Leeuwenhoek (1632–1723)

van Leeuwenhoek is generally given credit as being the first to see and describe bacteria. He is known as the father of microbiology. van Leeuwenhoek worked with a simple microscope, that is, with one lens, ground by himself, and fixed between two copper or silver plates. He may have made as many as 543 such microscopes. There are now only a few in existence. The focal distance is generally not more than about 1 to 2.5 mm. The best instrument that still remains is in Utrecht, the Netherlands. It magnifies approximately 270 times and has a resolving power of 1.4 μ. On the strength of calculations from his letter of November 2, 1680, the maximum magnification is about 480 times with a resolving power of 1 μ or a little less.

Taken from
"ANTONI VAN LEEUWENHOEK,"
"ON THE CIRCULATION OF THE BLOOD,"
NETHERLANDS SOCIETY FOR
THE HISTORY OF MEDICINE,
MATHEMATICS AND EXACT SCIENCES,
1962. (*Courtesy of De GRAAF Antiquarian Booksellers/Publishers. Nieuwkoop, the Netherlands.*)

FUNDAMENTAL PRINCIPLES OF BACTERIOLOGY

Seventh Edition

A. J. SALLE, B.S.,M.S.,Ph.D.
Professor of Bacteriology, Emeritus
University of California, Los Angeles

McGraw-Hill Book Company

New York
St. Louis
San Francisco
Düsseldorf
Johannesburg
Kuala Lumpur
London
Mexico
Montreal
New Delhi
Panama
Rio de Janeiro
Singapore
Sydney
Toronto

Fundamental Principles of Bacteriology

2 3 4 5 6 7 8 9 0 KPKP *7 9 8 7 6 5 4*

This book was set in Press Roman. The editors were James R. Young, Jr., and M. E. Margolies; the designer was Wladislaw Finne; and the production supervisor was Sally Ellyson. New drawings were done by John Cordes, J & R Technical Services, Inc. The printer and binder was Kingsport Press, Inc.

Library of Congress Cataloging in Publication Data

Salle, Anthony Joseph, 1895–
Fundamental principles of bacteriology.

Includes bibliographies.
1. Bacteriology. I. Title.
QR41.2.S22 1973 589.9 73-4251
ISBN 0-07-054480-8

To the memory of
Cecelia Daverso Salle
this book is affectionately dedicated

Contents

Preface

The seventh edition of this textbook has been thoroughly revised and completely rewritten to bring the contents up to date. Every chapter has been included in the revision. To name the significant changes which have been made would mean the inclusion of almost the entire book.

The author has attempted to include only sound fundamental material to give the beginner a solid foundation for more advanced work in the subject. Explanations of all phenomena are included insofar as it was possible to do so.

Some of the material appears for the first time and may require proper evaluation by others before it can be accepted as fact. The student should be warned against uncritical acceptance of views that may not be solidly established. It is believed that the training afforded by contemporary science and education should prove adequate to develop critical, inquiring students.

Emphasis is placed on the use of chemistry for a clearer understanding of the composition of bacteria and the reactions they produce. A student may conceivably pursue bacteriology without much of the chemistry contained in this book, and the instructor may wish to minimize some or most of it in a beginning course. For some phases of bacteriology, emphasis on chemistry is perhaps not necessary. However, it is included for those inquiring students who may desire it for a clearer understanding of the subject and especially for proper development of certain or most phases of bacteriology where it is absolutely indispensable.

The classification and names of enzymes follow the recommendations of the International Union of Biochemistry ("Report of Commission on Enzymes," 1965). The new names of enzymes are in most instances quite different from the previously accepted terminology. Because of this fact, some of the older names have been included with the new in order to prevent confusion as much as possible.

Many new illustrations have been added. Some have been prepared and photographed by the author. Others have been selected from the literature and reproduced from original prints furnished upon request. The author is greatly indebted to the large number of investigators who have kindly cooperated with his request for illustrative materials.

The outline classification of bacteria and the names of the organisms used throughout the text are based almost entirely on the seventh edition of "Bergey's Manual of Determinative Bacteriology," which is generally accepted as standard throughout the world. The author is grateful to the editor-in-chief, the late Dr. Robert S. Breed, for permission to quote from that book.

The new eighth edition of Bergey's manual has not yet appeared in print and could not be reviewed in this textbook. This was an unfortunate happening, as many new genera and species of bacteria are included in the new classification. Also the names of some of the species in the seventh edition of the manual have been changed. However, the author trusts that the user of this textbook will consider this an unavoidable situation and hopes that any confusion which may result will not prove too serious.

Most of the references to the literature contain the names of more than one author. If the publication is authored by one or not more than two investigators, all names are mentioned; if more than two names appear on the publication, only the first author is mentioned followed by the Latin phrase et al. This procedure was followed solely for the purpose of saving space. A few exceptions may appear in the chapters prepared by the contributors. Exceptions may also be noted in the captions of illustrations, where all names of contributors appear regardless of the number on the publications.

The chapter on the bacteriology of the sea, by Dr. C. E. ZoBell, has been revised by the same investigator and brought up to date. Likewise the chapters on bacterial genetics and bacterial viruses, by Dr. W. R. Romig, have been upgraded to reflect the latest information on those subjects. The chapter on immunity and the immune response has been revised by Dr. E. E. Sercarz. The author is greatly indebted to the above-named investigators for their aid in the preparation of those chapters.

Thanks are also due to all who have offered valuable suggestions and criticisms during the preparation of the manuscript.

Lastly, the author has attempted to acknowledge the sources of the text materials and illustrations. He alone accepts full responsibility for any defects that may be inherent in the plan and scope of the book, and for any errors or omissions which may have escaped detection.

A. J. Salle

Introduction 1

Biology is that branch of knowledge that treats of living organisms. The two major divisions of biology are botany, the science of plant life, and zoology, the science of animal life. Bacteriology deals with the study of organisms known as bacteria (singular, bacterium). The term includes a large group of typically unicellular microscopic organisms widely distributed in air, water, soil, the bodies of living plants and animals, and dead organic matter. Microbiology in its broadest meaning is the science that deals with the study of all kinds of microorganisms, both plant and animal, such as bacteria, yeasts, molds, algae, and protozoa. The modern tendency is to confine the term to the study of organisms classified with the lowest groups of botanical or plant life, i.e., bacteria, yeasts, and molds. The term *microbe* is taken from the French and means a microscopic organism or microorganism, being usually applied to the pathogenic forms. The term *germ*, in popular usage, refers to any microorganism but especially to one of the pathogenic or disease-producing bacteria. The terms microbe and germ are probably synonymous with bacterium. Although this book will include a discussion of the lowest groups of plant life, the major portion of the material is devoted to the study of organisms classified with the bacteria.

An organism is a complex, highly organized specific system that is capable of self-reproduction. A new organism never appears except from a preexisting identical organism. Each organism synthesizes all its necessary components if provided with food and energy. Under these conditions the organism grows, metabolizes, and multiplies.

A living organism, even one that is typically unicellular, is very complex in composition. It is composed of macromolecules such as proteins, nucleic acids, carbohydrates, and lipides. The smallest living cell may contain several thousand different kinds of macromolecules. Only systems that are living are able to reproduce their complex macromolecular makeup. This is a characteristic of living systems.

Bacteria belong to a group of living organisms known as microorganisms, without distinction regarding whether they are plants or animals. They are so called because of their small size. All cells are small, but bacteria do not exist as parts of organisms. They exist as single cells.

Because bacteria are small, they have a very high ratio of surface area to volume, which is approximately 200,000 times greater than a similar ratio for an adult human being. Since bacteria absorb their nutrients through their cell membranes, they exhibit very high metabolic rates. This, coupled with their rapid rate of multiplication, explains why they are able to produce such great changes in the environment in a short period of time.

Man, who is forever classifying things, has placed living organisms into either the plant or the animal kingdom. Most living organisms possess the characteristics of one kingdom or the other and may be sharply differentiated. However, bacteria are exceptions. Some higher bacteria border on the plants and are more related to them; other higher bacteria border on the animals. The so-called true bacteria are in a class by themselves. They are neither plants nor animals. Therefore, it is no longer possible to classify all living organisms into either the plant or the animal kingdom.

For this reason a number of newer classifications have been offered for consideration. The one outlined in Table 1 was proposed by Whittaker (1969) and divides the living world into five kingdoms as follows: (1) *Monera*, (2) *Protista*, (3) *Plantae*, (4) *Fungi*, and (5) *Animalia*.

Table 1 A Classification of the Living World from Kingdoms through Phyla

KINGDOM	*Monera.** Procaryotic cells, lacking nuclear membranes, plastids, mitochondria, and advanced (9 + 2-strand) flagella; solitary unicellular or colonial-unicellular organization (but in one group mycelial). Predominant nutritive mode absorption, but some groups are photosynthetic or chemosynthetic. Reproduction primarily asexual by fission or budding; protosexual phenomena also occur. Motile by simple flagella or gliding, or nonmotile.
Branch	*Myxomonera.* Without flagella, motility (if present) by gliding.
PHYLUM	*Cyanophyta.* Blue-green algae.
PHYLUM	*Myxobacteriae.* Gliding bacteria.
Branch	*Mastigomonera.* Motile by simple flagella (and related nonmotile forms).
PHYLUM	*Eubacteriae.* True bacteria.

* A more detailed classification of the bacteria is found in Chap. 15, page 507.

PHYLUM *Actinomycota*. Mycelial bacteria.

PHYLUM *Spirochaetae*. Spirochetes.

KINGDOM *Protista*. Primarily unicellular or colonial-unicellular organisms (but simple multinucleate organisms or stages of life cycles occur in a number of groups), with eucaryotic cells [possessing nuclear membranes, mitochondria, and in many forms plastids, (9 + 2-strand) flagella, and other organelles]. Nutritive modes diverse—photosynthesis, absorption, ingestion, and combinations of these. Reproductive cycles varied, but typically including both asexual division at the haploid level and true sexual processes with karyogamy and meiosis. Motile by advanced flagella or other means, or nonmotile.

PHYLUM *Euglenophyta*. Euglenoid organisms.

PHYLUM *Chrysophyta*. Golden algae.

PHYLUM *Pyrrophyta*. Dinoflagellates and cryptomonads.

PHYLUM *Hyphochytridiomycota*. Hyphochytrids.

PHYLUM *Plasmodiophoromycota*. Plasmodiophores.

PHYLUM *Sporozoa*. Sporozoans.

PHYLUM *Cnidosporidia*. Cnidosporidians.

PHYLUM *Zoomastigina*. Animal flagellates.

PHYLUM *Sarcodina*. Rhizopods.

PHYLUM *Ciliophora*. Ciliates and suctorians.

KINGDOM *Plantae*. Multicellular organisms with walled and frequently vacuolate eucaryotic cells and with photosynthetic pigments in plastids (together with closely related organisms which lack pigments or are unicellular or syncytial). Principal nutritive mode photosynthesis, but a number of lines have become absorptive. Primarily nonmotile, living anchored to a substrate. Structural differentiation, leading toward organs of photosynthesis, anchorage, and support, and in higher forms toward specialized photosynthetic, vascular, and covering tissues. Reproduction primarily sexual with cycles of alternating haploid and diploid generations, the former being progressively reduced toward the higher members of the kingdom.

Table 1 (*continued*)

SUBKINGDOM	*Rhodophycophyta.* Chlorophyll a and (in some) d, with *r*-phycocyanin and *r*-phycoerythrin also present, food storage as floridean starch, flagella lacking.
PHYLUM	*Rhodophyta.* Red algae.
SUBKINGDOM	*Phaeophycophyta.* Chlorophyll a and c, with fucoxanthin also present, food storage as laminarin and mannitol, zoospores with two lateral flagella, one of whiplash and one of tinsel type.
PHYLUM	*Phaeophyta.* Brown algae.
SUBKINGDOM	*Euchlorophyta.* Chlorophyll a and b, food storage as starch within plastids, ancestral flagellation two or more anterior whiplash flagella.
Branch	*Chlorophycophyta.* Primarily aquatic, without marked somatic cell differentiation.
PHYLUM	*Chlorophyta.* Green algae.
PHYLUM	*Charophyta.* Stonewarts.
Branch	*Metaphyta.* Primarily terrestrial, with somatic cell and tissue differentiation.
PHYLUM	*Bryophyta.* Liverworts, hornworts, and mosses.
PHYLUM	*Tracheophyta.* Vascular plants.
KINGDOM	*Fungi.* Primarily (except subkingdom *Gymnomycota*) multinucleate organisms with eucaryotic nuclei dispersed in a walled and often septate mycelial syncytium, plastids and photosynthetic pigments lacking. Nutrition absorptive. Somatic tissue differentiation absent or limited, reproductive tissue differentiation and life-cycle elaboration marked in higher forms. Primarily nonmotile (but with protoplasmic flow in the mycelium), living embedded in a medium or food supply. Reproductive cycles typically including both sexual and asexual processes; mycelia mostly haploid in lower forms but dicaryotic in many higher forms.

† More complete classifications of yeasts and molds are found in Chap. 5, page 143, and Chap. 6, page 185, respectively.

Table 1 (continued)

SUBKINGDOM	*Gymnomycota.* Deviant organizations including in life cycles separate cells, aggregations of cells, and sporulation stages.
PHYLUM	*Myxomycota.* Syncytial or plasmodial slime molds.
PHYLUM	*Acrasiomycota.* Cellular or pseudoplasmodial slime molds.
PHYLUM	*Labyrinthulomycota.* Cell-net slime molds.
SUBKINGDOM	*Dimastigomycota.* Biflagellate (heterokont) zoospores present, chytrid to simply mycelial organization, cellulose walls.
PHYLUM	*Oomycota.* Oospore fungi.
SUBKINGDOM	*Eumycota.* Predominantly mycelial organization, zoospores uniflagellate if present, chitin walls, other characters as stated for kingdom.
Branch	*Opisthomastigomycota.* Uniflagellate (opisthokont) zoospores present, chytrid to simply mycelial organization, mainly aquatic.
PHYLUM	*Chytridiomycota.* True chytrids and related fungi.
Branch	*Amastigomycota.* Flagellated zoospores absent, simple to advanced mycelial organization (but secondarily unicellular in yeasts). Mainly terrestrial.
PHYLUM	*Zygomycota.* Conjugation fungi.
PHYLUM	*Ascomycota.* Sac fungi.
PHYLUM	*Basidiomycota.* Club fungi.
KINGDOM	*Animalia.* Multicellular organisms with wall-less eucaryotic cells lacking plastids and photosynthetic pigments. Nutrition primarily ingestive with digestion in an internal cavity, but some forms are absorptive and a number of groups lack an internal digestive cavity. Level of organization and tissue differentiation in higher forms far exceeding that of other kingdoms, with evolution of sensorineuromotor systems and motility of the organism (or in sessile forms of its parts) based on contractile fibrils. Reproduction predominantly sexual, haploid stages other than the gametes almost lacking above the lowest phyla.

Table 1 (*continued*)

SUBKINGDOM	*Agnotozoa.* Nutrition absorptive and ingestive by surface cells, internal digestive cavity and tissue differentiation lacking. Minute, motile by cilia.
PHYLUM	*Mesozoa.* Mesozoans.
SUBKINGDOM	*Parazoa.* Nutrition primarily ingestive by individual cells lining internal water canals. Cell differentiation present but tissue differentiation lacking or very limited; cells with some motility but the organism nonmotile.
PHYLUM	*Porifera.* Sponges.
PHYLUM	*Archaeocyatha.* (Extinct.)
SUBKINGDOM	*Eumetazoa.* Advanced multicellular organization with tissue differentiation, other characteristics of the kingdom.
Branch	*Radiata.* Animals of radiate or biradiate symmetry.
PHYLUM	*Cnidaria*, coelenterates.
PHYLUM	*Ctenophora*, comb jellies.
Branch	*Bilateria.* Animals of bilateral symmetry.
Grade	*Acoelomata.*
PHYLUM	*Platyhelminthes,* flatworms.
PHYLUM	*Nemertea* or *Rhynchocoela,* ribbon worms.
Grade	*Pseudocoelomata.*
PHYLUM	*Acanthocephala*, spiny-headed worms.
PHYLUM	*Aschelminthes,* diverse pseudo-coelomate worms.
PHYLUM	*Entoprocta* or *Kamptozoa*, pseudo-coelomate polyzoans.
Grade	*Coelomata.*
SUBGRADE	*Schizocoela.*
PHYLUM	*Bryozoa* or *Ectoprocta*, coelomate, ectoproct polyzoans.
PHYLUM	*Brachiopoda*, lamp shells.

PHYLUM *Phoronida,* lophophorate, phoronid worms.

PHYLUM *Mollusca*, molluscs.

PHYLUM *Sipunculoidea,* peanut worms.

PHYLUM *Echiuroidea*, spoon worms.

PHYLUM *Annelida*, segmented or annelid worms.

PHYLUM *Arthropoda*, arthropods.

SUBGRADE *Enterocoela*

PHYLUM *Brachiata* or *Pogonophora*, beard worms.

PHYLUM *Chaetognatha,* arrow worms.

PHYLUM *Echinodermata*, echinoderms.

PHYLUM *Hemichordata*, acorn worms.

PHYLUM *Chordata*, chordates.

For more information: Schopf (1967, 1968); Schopf and Barghoorn (1967).

Distribution of bacteria

Bacteria are widely distributed in nature, being found almost everywhere. They are present in still ponds and ditches, in running streams and rivers, in seawater, in soil, in air, in foods, in petroleum oil from deeply seated regions, in rubbish and manure heaps, in decaying organic matter of all kinds, on the body surface, in body cavities, and in the intestinal tracts of man and animals. The kinds and numbers vary from one locality to another, depending upon the environmental conditions.

Some bacteria are always present in certain places. The common occurrence of one or more species in a particular environment is spoken of as the natural flora of that environment. For example, the normal souring of milk is caused by *Streptococcus lactis*. This organism is a normal inhabitant of the soil and is present on grains and hay. Since these are consumed by cows, the organism appears in manure and on the coat of the animals, from where it gains entrance to milk at the time of collection. Practically every sample of raw or pasteurized milk contains the organism.

Bacteria in soil The numbers and kinds of organisms present in soil depend upon the type of soil, quantity of plant and animal debris (humus), acidity or alkalinity, depth, degree of aeration, moisture content, and treatment. The great majority of soil organisms are found in the surface layer. The numbers decrease with depth, owing to lack of oxygen and food materials. The bacterial population of rich garden soil is considerably greater than that of poor, uncultivated soil.

Bacteria in air Bacteria are found in air, being carried there principally by wind currents. Organisms do not grow and multiply in air because conditions are not favorable for this to occur. There is no such thing as a normal atmospheric flora. The numbers and kinds depend upon location, amount of moisture, dust particles, wind currents, and the presence of toxic gases. The air over the oceans, far removed from shore, also shows the presence of microorganisms. In general, marine air contains fewer microorganisms than terrestrial air. The air over high mountains is usually free from organisms. The air of the city and country differs in the kinds of species and numbers present. Dusty rooms usually have considerably more organisms than rooms kept free from dust. Bacteria are found usually adhering to particles of dust, which means that the more particles suspended in the air, the greater will be the extent of microbial contamination. Viable spores of bacteria, yeasts, and molds are commonly found in air because these bodies are more resistant to the ultraviolet rays of the sun than are the vegetative cells producing them. These bodies are a frequent cause of air contamination in bacteriological laboratories and, because of their great resistance to heat, require high temperatures for their destruction.

Bacteria in water Most waters contain large numbers of bacteria. The numbers may vary considerably depending upon the source of the water, e.g., deep or shallow wells, springs, rivers, lakes, ponds, or streams. Water polluted with sewage may contain thousands or even millions of organisms per milliliter. Under some conditions disease organisms may also be present. Practically all bacterial species found in soil may at times be present in water. Some species are always present and constitute the natural flora of that water. Usually fewer bacteria occur in seawater than in soil. This is probably because of its poorer qualities as a culture medium.

Bacteria in food Foods are rarely free from living organisms. Some organisms are of benefit in producing desirable fermentations such as occur in the oxidation of alcohol to vinegar, the lactic fermentation of cabbage to sauerkraut, etc. Frequently, the undesirable organisms may gain access to foods and produce abnormal changes. Some diseases or intoxications may be produced by the consumption of foods contaminated with certain organisms or their growth products.

Bacteria in milk Normal udders of cows are probably never free from bacteria, which means that freshly drawn milk is not sterile. The first milk drawn always contains more organisms than milk collected at the close of the operation, because most of the bacteria are washed away from the udders early in the process. However, most of the bacteria present in milk are not those which are present in the udders but forms which gain entrance after the milk has been collected. Unless the milk is properly stored immediately after collection, the organisms present may be capable of producing undesirable changes, making the milk unfit for human consumption.

Bacteria of the body The outer surface of the skin of the body always contains bacteria. The same applies to the respiratory passages and the alimentary and intestinal tracts. These environments contain normal floras which are for the most part harmless. Occasionally a species may invade the body by penetrating the broken skin or mucous membrane. Unless the body is invaded by massive numbers, the organisms are usually destroyed by the defense mechanisms of the host. Sometimes the body cannot destroy the invaders. Under these conditions, a disease process may be established.

It has been said that as much as one-third of the dry weight of the intestinal contents of man is composed of bacterial cells.

Escherichia coli is found in the large intestine of man. Other organisms are present, but in an adult on a mixed diet this organism predominates. *E. coli*, then, is largely responsible for the natural flora of the large intestine. Changes in the environment produce changes in the bacterial flora. If the diet of an adult is changed from a high-protein to a high-carbohydrate diet, the *E. coli* organisms will gradually disappear, only to be replaced by a much larger rod-shaped organism known as *Lactobacillus acidophilus*. If this particular diet is maintained, *L. acidophilus* will become the predominating organism of the large intestine.

For more information: Marples (1969).

Functions of bacteria

Those who are not familiar with the elementary rudiments of bacteriology have an erroneous concept of the role of bacteria in nature. Since the early development of bacteriology was concerned with a study of disease-producing organisms, the impression is generally held that the sole purpose of bacteria on this earth is to cause human ills. This belief is entirely erroneous. Only a few bacterial species are harmful to man. The great majority are not only harmless but absolutely necessary for the existence of living things. Life could not exist in the complete absence of bacteria.

Plants and animals owe their existence to the fertility of the soil, and this, in turn, depends upon the activity of the soil population. Plants absorb their nutrients from the soil in the form of minerals. They cannot utilize organic compounds such as fats, carbohydrates, and proteins. The soil organisms attack human and animal carcasses and mineralize the organic constituents, making them available to plants. The same is true for the remains of plant crops such as plant stubble and leaves. In the absence of soil organisms plants could not live or grow, and in the absence of plants there would be no animal life on this earth. Man, of course, is dependent upon both plants and animals for food.

A few species of soil bacteria are capable of invading the roots of certain plants, where they take free nitrogen from the air and convert it into organic compounds which are utilized by the plants. The bacteria, in turn, derive their nutrients from the sap of the plants. This may be cited as an excellent example of a symbiotic relationship occurring in nature between two different species. In the absence of fertilizers such as animal manures, nitrates, and ammonium salts, there would be no available nitrogen in the soil were it not for the activities of these organisms. Sulfur and phosphorus, two elements necessary for plant growth, are converted into soluble inorganic salts by bacteria and absorbed by plant roots.

Fertile soils may always be distinguished from poor soils in containing greater numbers of viable organisms. If the soil is rich in plant remains, is well aerated, contains sufficient moisture, shows the right temperature and hydrogen-ion concentration (reaction), many organisms will be present to attack the plant residues, converting the insoluble and indiffusible constituents into soluble, inorganic compounds utilizable by plants.

Bacteria are necessary for the disposal of sewage. They convert the insoluble proteins, fats, and carbohydrates (cellulose) into soluble, odorless compounds which may be disposed of in an inoffensive manner.

The souring of milk is the result of bacterial action. This is the first step in the preparation of butter. The separation of butterfat is more easily accomplished and the yield improved if the milk or cream is first permitted to sour. Also, bacteria are selected which improve the aroma and flavor of the butter.

Various types of industrial fermentations are produced by the action of bacteria, yeasts, and molds on carbohydrates. Butyric acid, acetone, butyl alcohol, lactic acid, and propionic acid are produced by bacterial action. Ethyl alcohol is formed by yeasts. Gluconic acid and citric acid are the result of the activities of molds.

These are only a few examples of the activities of organisms in nature. Many other useful purposes will be discussed in the various chapters of this book.

For more information: Baker and Allen (1968); Benton and Werner (1966); Biological Sciences Curriculum Study (1963); Cockrum et al. (1966); Curtis (1968); Hale (1967); Hardin (1966); Johnson et al. (1969); Keeton (1967); Marsland (1964); Milne and Milne (1965); Nason (1965); Nelson (1967); Orians (1969); Platt and Reid (1967); Ramsey (1965); Simpson and Beck (1965); Villee (1967); Weisz (1967).

References

Baker, J. J. W., and G. E. Allen: "A Course in Biology," Reading, Mass., Addison-Wesley Publishing Company, Inc., 1968.

Benton, A. H., and W. E. Werner, Jr.: "Field Biology and Ecology," New York, McGraw-Hill Book Company, 1966.

Biological Sciences Curriculum Study, American Institute of Biological Sciences: "Biological Science: Molecules to Man," Boston, Houghton Mifflin Company, 1963.

Cockrum, E. L., et al.: "Biology," Philadelphia, W. B. Saunders Company, 1966.

Curtis, H.: "Biology," New York, Worth Publishers, Inc., 1968.

Devons, S., ed.: "Biology and the Physical Sciences," New York, Columbia University Press, 1969.

Hale, M. E.: "The Biology of Lichens," London, Edward Arnold (Publishers) Ltd., 1967.

Hardin, G.: "Biology: Its Principles and Implications," San Francisco, W. H. Freeman and Company, 1966.

Johnson, W. H., et al.: "Essentials of Biology," New York, Holt, Rinehart and Winston, Inc., 1969.

Keeton, W. T.: "Biological Science," New York, W. W. Norton and Company, Inc., 1967.

Marples, M. J.: Life on the human skin, *Sci. Am.*, **220**:108, 1969.

Marsland, D.: "Principles of Modern Biology," New York, Holt, Rinehart and Winston, Inc., 1964.

Milne, L. J., and M. J. Milne: "The Biotic World and Man," Englewood Cliffs, N.J., Prentice-Hall, Inc., 1965.

Nason, A.: "Textbook of Modern Biology," New York, John Wiley & Sons, Inc., 1965.

Nelson, G. E.: "Fundamental Concepts of Biology," New York, John Wiley & Sons, Inc., 1967.

Orians, G. H.: "The Study of Life," Boston, Allyn and Bacon, Inc., 1969.

Platt, R. B., and G. K. Reid: "Bioscience," New York, Reinhold Publishing Corporation, 1967.

Ramsay, J. A.: "The Experimental Basis of Modern Biology," Cambridge, England, The University Press, 1965.

Schopf, J. W.: Antiquity and evolution of precambrian life. From "McGraw-Hill Yearbook of Science and Technology," New York, McGraw-Hill Book Company, 1967.

——: Microflora of the bitter springs formation, late precambrian, *Central Australia J. Paleontol.*, **42**:651, 1968.

—— and E. S. Barghoorn: Alga-like fossils from the early precambrian of South Africa, *Science*, **156**:508, 1967.

Simpson, G. G., and W. S. Beck: "Life," New York, Harcourt, Brace & World, Inc., 1965.

Villee, C. A.: "Biology," Philadelphia, W. B. Saunders Company, 1967.

Weisz, P. B.: "The Science of Biology," New York, McGraw-Hill Book Company, 1967.

Whittaker, R. H.: New concepts of kingdoms of organisms, *Science*, **163**:150, 1969.

2 *The microscope*

Bacteria are too small to be seen by the unaided eye. They must be greatly magnified to be clearly seen and studied. The microscope is absolutely indispensable to the biologist in general and to the bacteriologist in particular.

A microscope may be defined as an optical instrument, consisting of a lens, or combination of lenses, for making enlarged or magnified images of minute objects.

A simple microscope, or single microscope, consists of only one lens or magnifying glass held in a frame, usually adjustable, and often provided with a stand for conveniently holding the object to be viewed and a mirror for reflecting the light.

A compound microscope differs from a simple microscope in having two sets of lenses, one known as the objective and the other as the eyepiece, mounted in a holder commonly known as a body tube (Fig. 1). The lens system nearest the specimen, called the objective, magnifies the specimen a definite number of times. The second lens system, called the eyepiece, further magnifies the image formed by the objective. The image seen by the eye has a magnification equal to the product of the magnifications of the two systems. The individual or initial magnifications of the objective and eyepiece are engraved on each such part. Accurate focusing is attained by a special screw appliance known as a fine adjustment. Compound microscopes give much greater magnifications than simple microscopes and are necessary for viewing and examining minute objects such as bacteria.

Every user of the microscope should first understand the principles involved in order that the instrument may be employed to the greatest advantage. As Sir A. E. Wright stated:

> Every one who has to use the microscope must decide for himself the question as to whether he will do so in accordance with a system of rule of thumb, or whether he will seek to supersede this by a system of reasoned action based upon a study of his instrument and a consideration of the scientific principle of microscopical technique.

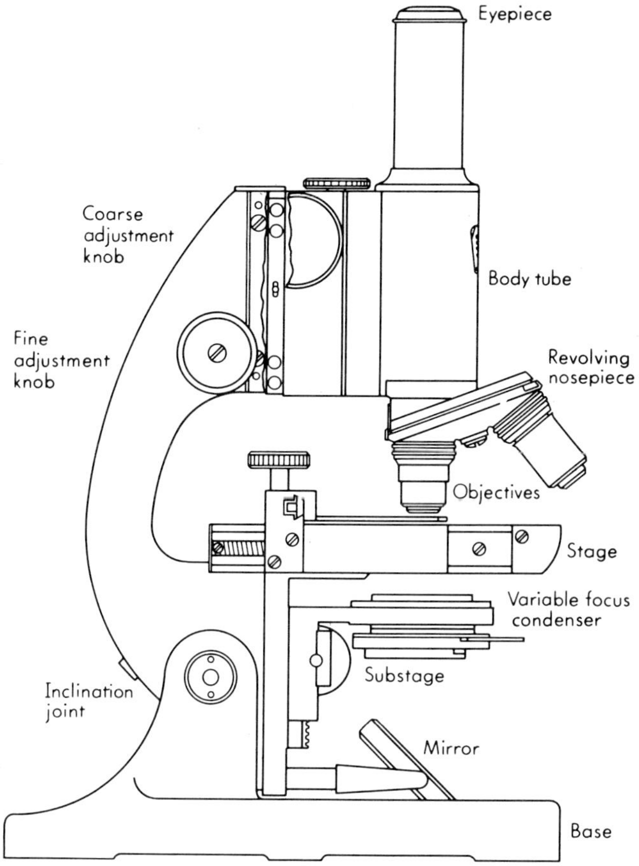

Fig. 1 Compound microscope and its parts. (Courtesy of Bausch & Lomb Optical Company.)

General principles of optics

The path of light through a compound microscope is illustrated in Fig. 2. The light, in passing through the condenser, object in plane I, and objective lens, would form a real and inverted image in plane II if the ocular or eyepiece were removed. In the presence of the ocular *F*, the rays are intercepted, forming the image in plane III. The real image is then examined, with eye lens *E* of the ocular acting as a single magnifier and forming a virtual image in plane IV. The distance between the virtual image (plane IV) and the eyepoint is known as the projection distance. The object is magnified first by the objective lens and second by the ocular, or eyepiece. With a tube length of 160 mm (most microscope manufacturers have adopted 160 mm as the standard tube length), the total magnification of the microscope is equal to the magnifying power of the objective lens multiplied by the magnifying power of the ocular.

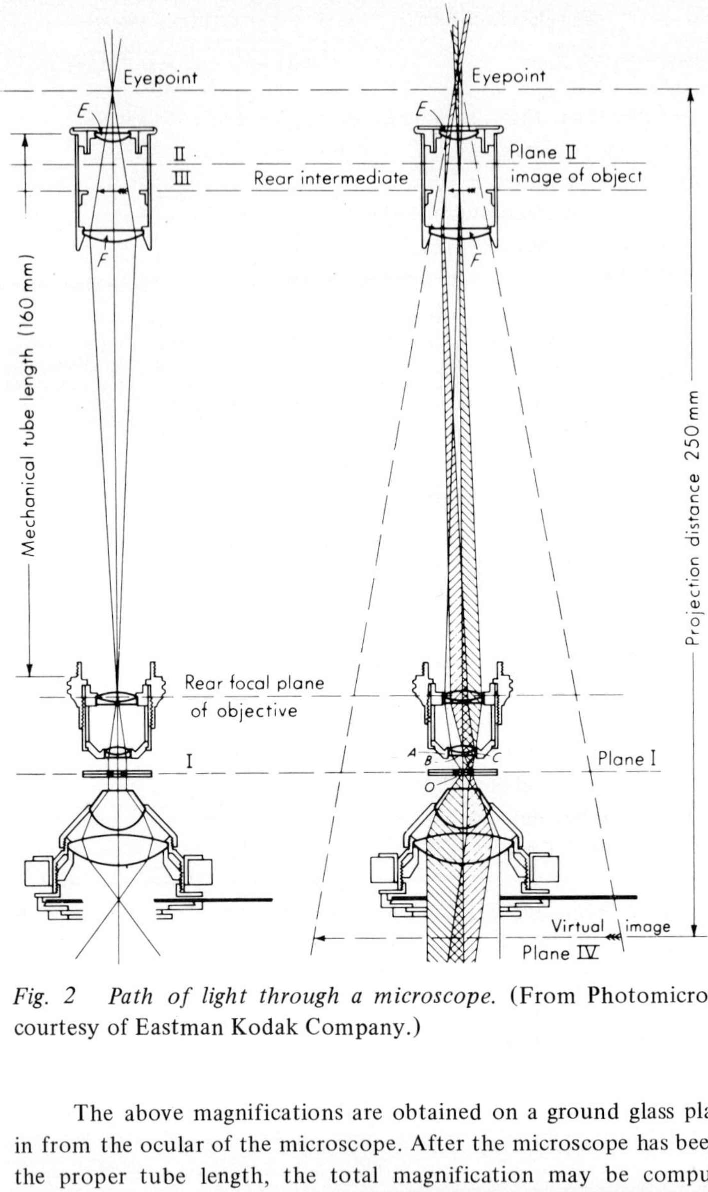

Fig. 2 Path of light through a microscope. (From Photomicrography, courtesy of Eastman Kodak Company.)

The above magnifications are obtained on a ground glass placed 10 in from the ocular of the microscope. After the microscope has been set at the proper tube length, the total magnification may be computed by multiplying the magnifying power of the objective by that of the eyepiece and by one-tenth of the distance from the eyepiece to the ground glass, measured in inches. For example, if the ground glass is placed 10 in from the eyepiece of the microscope, the total magnification will be as given on the ocular and objective. If placed 20 in from the eyepiece, the magnification will be twice as great. If placed 5 in from the eyepiece, the magnification will be one-half as great. To take a specific example:

Magnification of objective	97x
Magnification of ocular	10x

Distance of ground glass from ocular	7 in
Total magnification	97 x 10 x 0.7 (0.1 x 7) = 679x

It may be seen that almost any degree of magnification could be obtained by using oculars of different magnifying powers or by varying the length of the draw tube. Even though the magnifying powers of the microscope could be greatly increased in this manner, the amount of detail that can be seen is not improved since this is strictly limited by the structure of light.

Structure of light According to the undulatory, or wave, theory, light is transmitted from luminous bodies to the eye and other objects by an undulatory or vibrational movement. In its interaction into matter, light behaves as though it were composed of many individual bodies called photons, which carry such particle-like properties as energy and momentum. The velocity of this transmission is about 186,300 miles per second, and the vibrations of the ether are transverse to the direction of propagation of the wave motion. The waves vary in length from 3850 to 7600 angstroms (Å) approximately. The color evoked when the energy impinges on the retina varies with the wavelength, the amplitude of vibration, and various other factors and conditions. Waves of a similar character whose lengths fall above or below the limits mentioned are not perceptible to the average eye under normal conditions. Those between 1000 and 3850 Å constitute ultraviolet light and are manifested by their photographic or other chemical action. Those exceeding 7600 Å are the infrared waves and are detected by their thermal effects.

When a beam of white light is passed through a prism, a spectrum is obtained in which several colors form a series from deep red through orange, yellow, green, blue, and indigo to deepest violet. The wavelengths of the various colors are different; red shows the longest and violet the shortest waves of the visible spectrum.

The length of a light wave is the distance from the crest of one wave to the crest of the next (Fig. 3). The unit of measurement is the angstrom

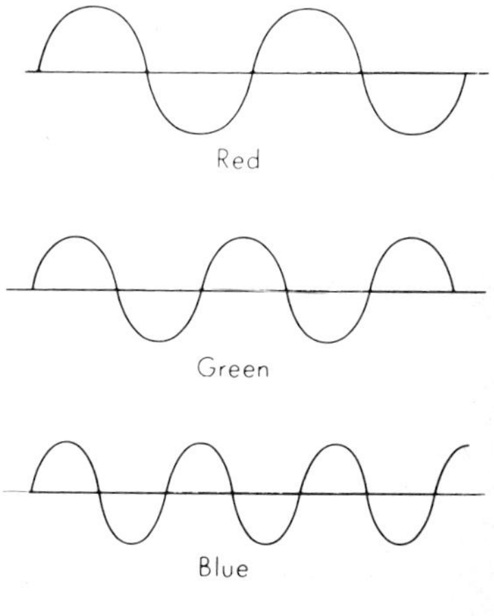

Fig. 3 Wavelengths of light of different colors.

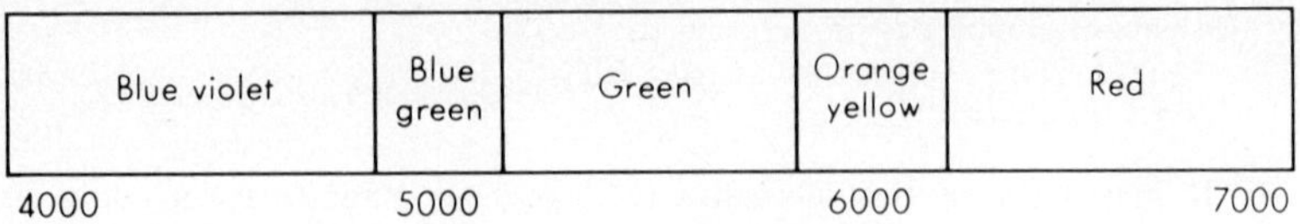

Fig. 4 Light rays of the visible spectrum and their corresponding wavelengths in angstroms.

(Å), which is equal to 1/10,000,000 mm, or to approximately 1/250,000,000 in. The visible spectrum, together with the corresponding wavelengths of the light rays in angstroms, may be represented as shown in Fig. 4. Visible light waves, ranging in length from 4000 to 7000 Å, may be roughly divided into three portions: blue-violet, from 4000 to 5000 Å; green, from 5000 to 6000 Å; red, from 6000 to 7000 Å.

Objectives

The objective is the most important lens on a microscope because its properties may make or mar the final image. An objective capable of utilizing a large angular cone of light coming from the specimen will have better resolving power than an objective limited to a smaller cone of light. The chief functions of the objective lens are (1) to gather the light rays coming from any point of the object, (2) to unite the light in a point of the image, and (3) to magnify the image.

There are three major types of objectives, namely, achromatic, fluorite, and apochromatic (Fig. 5). The achromats are the simplest in construction and the least expensive. They are adequate for most purposes. Correction for both color and spherical aberration is quite good in the lower-power objectives, but the control of aberrations becomes more difficult as the power is increased. Aberrations are largely eliminated

Fig. 5 General construction of the 4-mm achromatic objective, the 4-mm fluorite objective, and the 4-mm apochromatic objective, in that order.

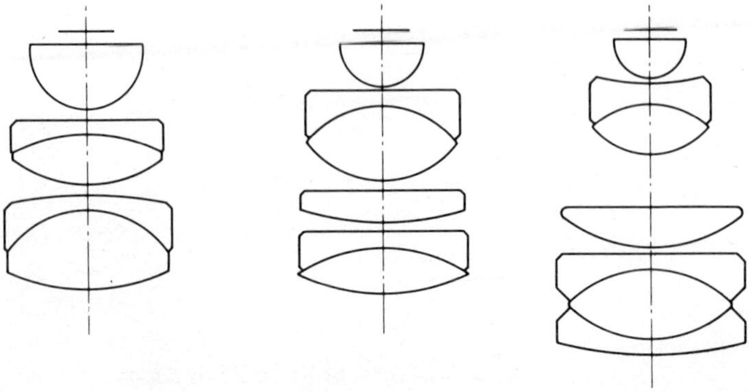

by the use of fluorite (semiapochromatic) objectives and, especially, the apochromats. The latter are more highly corrected with respect to aberrations than any other type of objective and are preferred for the most critical work.

Numerical aperture The resolving power of an objective may be defined as its ability to separate distinctly two small elements in the structure of an object that are a short distance apart. The measure for the resolving power of an objective is the numerical aperture (NA). The larger the NA, the greater the resolving power of the objective and the finer the detail it can reveal.

Since the limit of detail or resolving power of an objective is fixed by the structure of light, objects smaller than the smallest wavelength of visible light cannot be seen. In order to see such minute objects, it would be necessary to use rays of shorter wavelengths. Invisible rays, such as ultraviolet light, are shorter than visible rays but since they cannot be used for visual observation (but only for photography) their usefulness is limited.

The image of an object formed by the passage of light through a microscope will not be a point but, in consequence of the diffraction of the light at the diaphragm, will take the form of a bright disk surrounded by concentric dark and light rings (Fig. 6). The brightness of the central disk will be greatest in the center, diminishing rapidly toward the edge. The image cone of light composed of a bright disk surrounded by concentric dark and light rings is spoken of as the antipoint. If two independent points in the object are equidistant from the microscope lens,

Fig. 6 The Airy disk. Photomicrograph of a pinhole in an aluminum mirror taken with a high-power dry objective. The radius of the first dark ring, h', is a measure of the resolving power. (Courtesy of Bausch & Lomb Optical Company.)

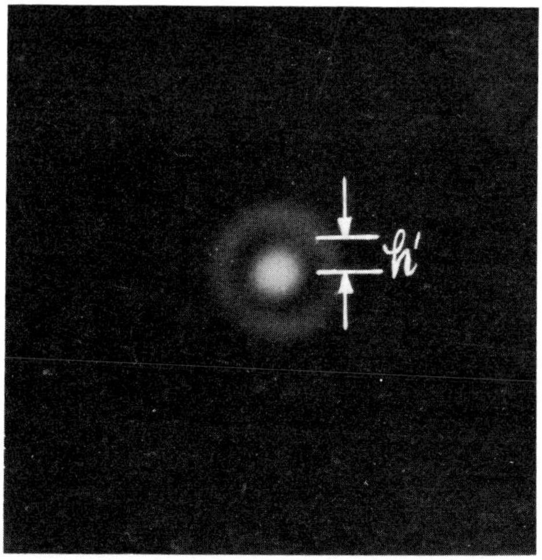

each will produce a disk image with its surrounding series of concentric dark and light rings. The disks will be clearly visible if completely separated, but if the images overlap they will merge into a single bright area, the central portion of which appears quite uniform. The two disks will not, therefore, be seen as separate images. It is not known how close the centers of the images can be and still be seen as separate antipoints.

The minimum distance between the images of two distinct object points decreases as the angle of light AOC (Fig. 2), coming from the object O, increases. The angle formed by the extreme rays is known as the aperture of the objective. The ability of the objective lens system to form distinct images of two separate object points is proportional to the trigonometric sine of the angle. The latter, then, is a measure of the resolving power of the objective. Actually, however, the sine of angle AOB is used, which is just one-half of angle AOC. This is usually referred to as $\sin \mu$. Since the sine of an angle may be defined as the ratio of the side opposite the angle in a right-angled triangle to the hypotenuse, then

$$\sin \mu = \frac{AB}{AO}$$

The light in passing through the objective is influenced by the refractive index n of the space directly in front of the lens. This is another factor that affects the resolving power of an objective. The two factors, refractive index n and $\sin \mu$, may be combined into a single expression, the NA, which may be expressed as follows:

$$\text{NA} = n \sin \mu$$

Importance of NA If a very narrow pencil of light is used for illumination, the finest detail that can be revealed by a microscope with sufficient magnification is equal to

$$\frac{\text{wl}}{\text{NA}}$$

where wl is the wavelength of the light used for illumination and NA is the numerical aperture of the objective. The resolving power of the objective is proportional to the width of the pencil of light used for illumination. This means that the wider the pencil of light, the greater the resolving power. The maximum is reached when the whole aperture of the objective is filled with light. In this instance, the resolving power is twice as great. The finest detail that the objective can reveal is now equal to

$$\frac{\text{wl}}{2\ \text{NA}}$$

For example, the brightest part of the spectrum shows a wavelength of 5300 Å. An objective having a numerical aperture equal to 1.00 will resolve two lines separated by a distance of 5300 Å/1.00 = 5300 Å (48,000 lines to the inch) if a very narrow pencil of light is used, and 5300 Å/(2 x 1.00) = 2650 Å (95,000 lines to the inch) if the whole aperture of the objective is filled with light.

From the above, it is evident that the maximum efficiency of an objective is not reached unless the back lens is completely filled with light. This may be ascertained by removing the eyepiece from the microscope and viewing the back lens of the objective with the naked eye. If the back lens is completely filled with light, the efficiency will then be in accordance with the numbers engraved on the objective.

Resolving power The relationship between wavelength and resolving power is illustrated in Fig. 7. The shorter the wavelength of light, the finer the detail revealed by the objective. With an objective having an NA of 1.00 and a yellow filter (light transmission of 5790 to 5770 Å), it is possible to see about 88,000 lines to an inch; with a green filter (light transmission of 5460 Å), about 95,000 lines to an inch; with a violet filter (light transmission of 4360 Å), about 115,000 lines to an inch; and with ultraviolet light (light transmission of 3650 Å), about 140,000 lines to an inch.

Fig. 7 Amphipleura pellucida, a diatom. Effect of light of different wavelengths on the resolving power of the objective. (A) Yellow filter, 5790 Å; (B) green filter, 5460 Å; (C) violet filter, 4360 Å; (D) ultraviolet filter 3650 Å. (From Photomicrography, courtesy of Eastman Kodak Company.)

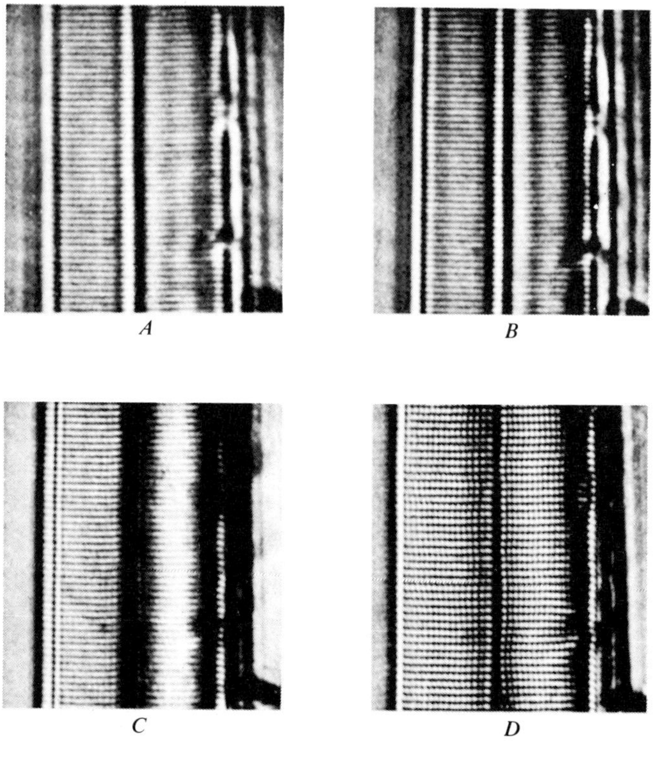

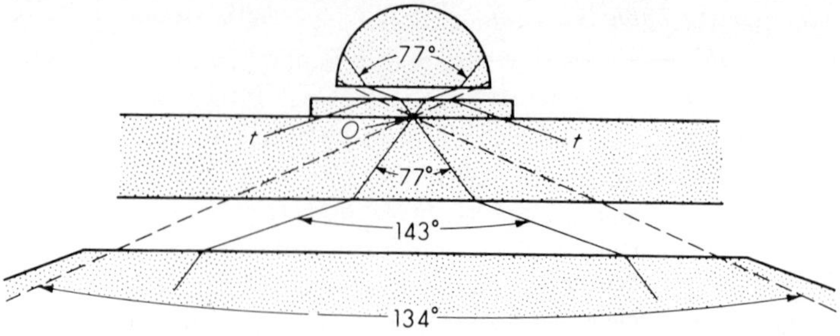

Fig. 8 Passage of light through an object on a glass slide using dry and immersion objectives. See text for details. (Redrawn from Photomicrography, courtesy of Eastman Kodak Company.)

Immersion objectives When a dry objective is used, an air space is present on both sides of the microscope slide and cover slip. The largest cone of light coming from *O* (Fig. 8) that could possibly be used is 180° in air, which is equal to an angle of about 82° in the glass. This corresponds to an NA of 1.0. In actual practice, however, these figures become 143° and 77°, respectively, owing to the fact that the air space must be wide enough to correspond to a practical working distance of the objective. Rays of greater angular aperture than 82° in glass, which originate at the object point *O* by diffraction, will be completely reflected at the upper surface of the cover slip *t*.

The refractive index *n* of the air is equal to 1.0. If the air space between the cover slip and the objective is filled with a fluid having a higher refractive index, such as water ($n = 1.33$), or, what is still better, a liquid having a refractive index approaching that of glass, such as cedarwood oil ($n = 1.51$), angles greater than 82° are obtained. Numerical apertures greater than 1.0 are realized by this method. Cedarwood oil causes the light ray to pass right through the homogeneous medium, with the result that a cone of light of about 134° is obtained, which corresponds to an NA of 1.4. Finer detail can, therefore, be resolved by this procedure. With an oil-immersion objective and an NA of 1.4, two lines as close together as $^1/_{100,000}$ in (0.25 μ) can be separated. The greater the NA of the objective, the greater will be its resolving power or ability to record fine detail.

The refractive indexes of a number of media which are employed for immersion objectives are given in Table 2.

Depth of focus The depth of focus is known also as the depth of sharpness or penetration. The depth of focus of an objective depends upon the NA and the magnification, and is inversely proportional to both. This means that the higher the NA and the magnification, the lower the depth of focus. Therefore, high-power objectives must be more carefully focused than low-power objectives. These conditions cannot be changed by the optician.

Table 2

Medium	*Refractive index at 25°C*
Water	1.33
Mineral (paraffin) oil	1.47
Cedarwood oil	1.51
Sandalwood oil	1.51
Shillaber's immersion oil	1.52
Balsam	1.53
Crown oil	1.55

Equivalent focus Objectives are sometimes designated by their equivalent focal lengths measured in either inches or millimeters. An objective designated by an equivalent focus of $^1/_{12}$ in, or 2 mm, means that the lens system produces a real image of the object of the same size as is produced by a simple biconvex or converging lens having a focal distance of $^1/_{12}$ in, or 2 mm.

Working distance of uncovered objects

If the object on a glass slide is not covered with a cover slip, the working distance may be defined as the distance between the front lens of the objective and the object on the slide when in sharp focus. The working distance is always less than the equivalent focus of the objective. This is illustrated in Fig. 9*A*.

The working distance may be easily determined by noting the number of complete turns of the micrometer screw (fine adjustment)

Fig. 9 Working distance of an objective. A, object not covered with a cover slip; B, C, object covered with a cover slip. (Redrawn from Gage, "The Microscope," Comstock Publishing Associates, Inc.)

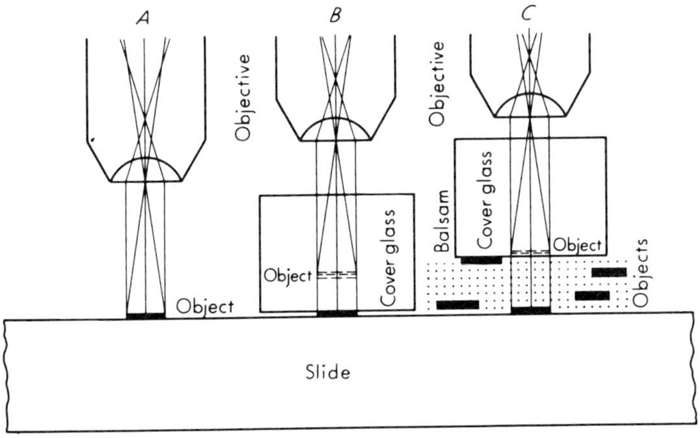

required to raise the objective from the surface of the slide, where the object is located, to a point where the microscope is in sharp focus.

To take a specific example:

Each turn of the micrometer screw = 0.1 mm

Number of turns required to bring object in sharp focus = 6

Working distance = 6 x 0.1 = 0.6 mm

Working distance of covered objects

If the object is covered with a cover slip, the free distance from the upper surface of the cover slip to the front of the objective will be less than in the case of an uncovered object. It is obvious from this that if the cover glass is thicker than the working distance of the objective, it will be impossible to get the object in focus. On the other hand, if the glass is thin it will be possible to get the object in focus, but the focus of the microscope on a covered object will be different from that on an uncovered object. It follows from this that an object covered with a glass cover slip or other highly refractive body will appear as if raised, and the amount of elevation will depend upon the refractive index of the glass or other medium covering the object. Also, the greater the refraction of the covering body, the more will be the apparent elevation. This is shown in Fig. 9 (*B* and *C*). The apparent depth of the object below the surface of the covering medium may be calculated by taking the reciprocal of its index of refraction. For example, if a glass cover slip is used, it will have an index of refraction of 1.52. The reciprocal of this figure is $1/1.52 = \frac{2}{3}$, approximately. This means that the apparent depth of the object is only two-thirds its actual depth.

The working distance of covered objects may be determined by noting the number of complete turns of the micrometer screw (fine adjustment) required to raise the objective from the surface of the cover slip to a point where the objective is in sharp focus.

To take a specific example:

Each turn of micrometer screw = 0.1 mm

Number of turns required to bring object in sharp focus = 3.5

Working distance = 3.5 x 0.1 = 0.35 mm

Aberrations in objectives Perfect lens systems have not yet been designed. All lens systems have aberrations to a greater or lesser degree, depending upon the skill of the designer and the magnitude of the design problem. Lens systems are made up of lenses having spherical surfaces, and such surfaces do not form perfect images. This defect may be largely counteracted by combining lens shapes and different glasses.

The principal defects in the image are the result of chromatic aberration, spherical aberration, distortion, curvature of field, astigmatism, coma, and lateral color.

Chromatic aberration White light in passing through a prism is broken up into its constituent colors, the wavelengths of which are different (page 15). A simple or compound lens, composed of only one material, will exhibit different focal lengths for the various constituents of white light. This is due to the dispersive power of the lens. Every wavelength is differently refracted, the shortest waves most and the longest waves least. The blue-violet rays cross the lens axis first and the red rays last. There will be a series of colored foci of the various constituents of white light extending along the axis (Fig. 10). As a result, the lens will not produce a sharp image with white light. Instead, the image will be surrounded by colored zones or halos which interfere with the visual observation of its true color. This is spoken of as chromatic aberration. It may be lessened by reducing the aperture of the lens or, better still, by using a lens composed of more than one material (compound lens). Two or more different glasses or minerals are necessary for correcting the chromatic aberration of an objective, and the amount of correction depends upon the dispersive powers of the components of the objective.

If two optical glasses are carefully selected to image light of two different wavelengths at the same focal point, the lens is said to be achromatic, and an objective containing such a lens system is spoken of as an achromatic objective. The remaining rays of the white light will be imaged at approximately the same point. An achromatic objective will yield images free from pronounced color halos. If the focus is shifted slightly, faint green and pink halos may be observed. The slight residual color will not prove objectionable for the usual microscopic work. Achromats are the universal objectives for visual work and are very satisfactory in photomicrography when used with monochromatic light (obtained by the use of filters).

Lens systems corrected for light of three different wavelengths are called apochromatic objectives. These objectives are composed of fluorite in combination with lenses of optical glass. The images produced by

Fig. 10 Chromatic aberration with white light. White light, in passing through a lens, is dispersed into its constituent colors. The red or long waves are refracted less than the blue or short waves. The blue rays (fb) cross the optical axis of the lens before the red rays (fr). The blue light will focus nearer the lens than the red light.

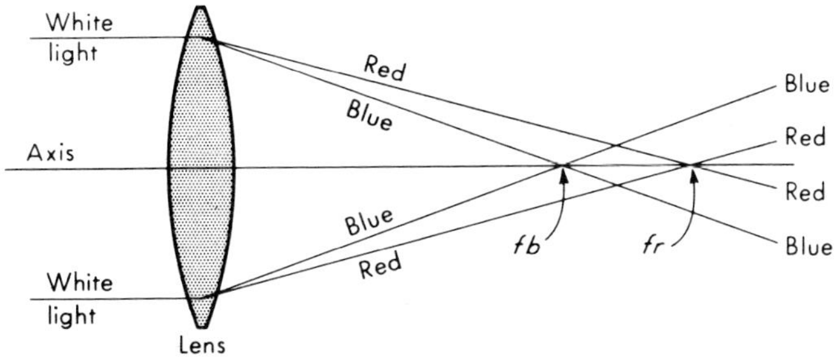

objectives in this group exhibit only a faint blue or yellow residual color. Since these objectives are corrected for three colors instead of for two, they are superior to the achromats. Their finer color correction makes possible a greater usable numerical aperture. The violet rays are brought to the same focus as visual rays. This fact makes these objectives excellent for photographic use for both white and monochromatic light.

Another group of objectives exhibits qualities intermediate between the achromats and the apochromats. They are called *semiapochromats*. If the mineral fluorite is used in their construction, they are termed fluorite objectives. These objectives also yield excellent results when used for photomicrography.

Spherical aberration This refers to the greater power in the outer portion of a spherical surface than in the inner portion. This is overcome by judicious combinations of convergent and divergent lens elements, properly shaped to minimize the variation of focal power with aperture. Spherical aberration causes some of the light which should be in the central spot to diffuse out into the ring structure (Fig. 11). This causes a loss in contrast in the normal microscope preparation.

Distortion This type of aberration renders a square object as an image with curved sides. If the rulings near the edge appear curved inward, the aberration is known as cushion distortion. If the opposite effect occurs, where the rulings appear curved outward, it is known as barrel distortion. Distortion is caused by the lens surface having different magnifications at the marginal and central portions of the image.

Curvature of field This aberration is caused by a spherical lens

Fig. 11 Spherical aberration. This causes some of the light which should be in the central spot to diffuse out into the ring structure, resulting in a loss in contrast in the normal microscope preparation. (Courtesy of Bausch & Lomb Optical Company.)

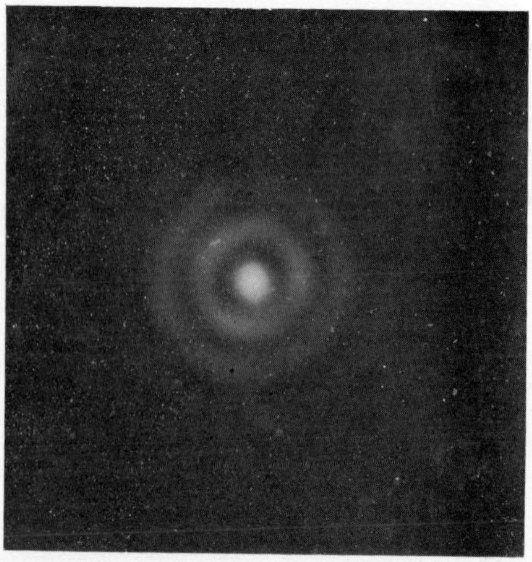

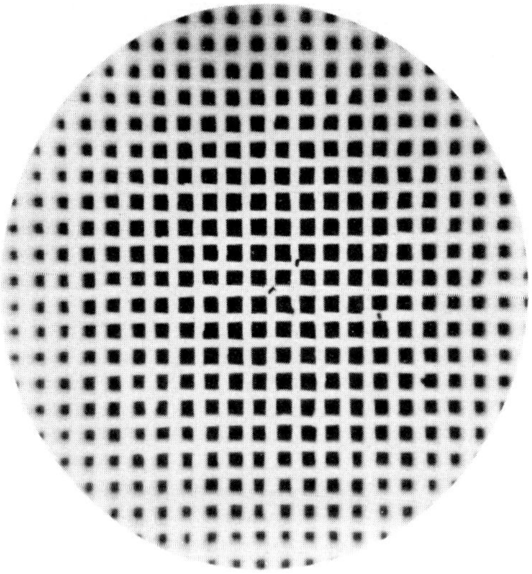

Fig. 12 Curvature of field. The center is sharply focused; the periphery is out of focus. (Courtesy of Bausch & Lomb Optical Company.)

surface, which produces a curved image of a flat object because the marginal portions of the image come to focus at a different distance from those of the central portions of the image (Fig. 12).

Astigmatism If a marginal point object is drawn out into two separate-line images lying at different distances from the lens surface, the result is called astigmatism (Fig. 13). It causes a general deterioration of the off-axis image. An astigmatic image can never be focused sharply except for detail parallel or perpendicular to a radius of the field.

Fig. 13 The appearance of a point object due to the presence of astigmatism. (Courtesy of Bausch & Lomb Optical Company.)

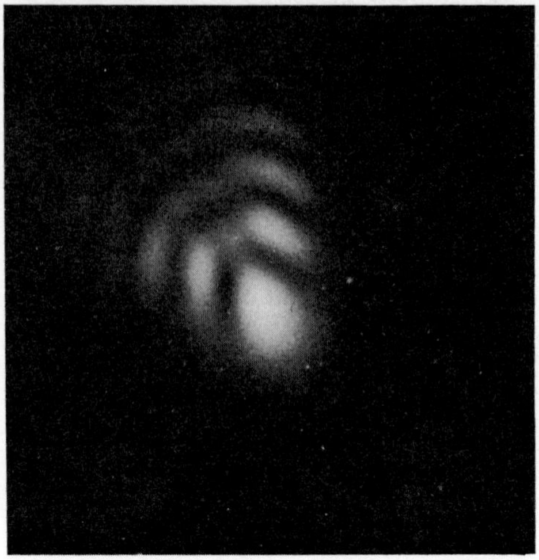

Fig. 14 The appearance of a point object due to the presence of coma. (Courtesy of Bausch & Lomb Optical Company.)

Coma This name is given to the defect in which different circular concentric zones of the lens surface give different magnifications to an off-axis image. This results in a point object being imaged as a comet-shaped image (Fig. 14). Coma in the center of the field is an indication of damage to the objective.

Lateral color The presence of this defect results in light of one color being imaged at a greater magnification than light of another color. This causes an off-axis image of a point object to be spread out into a tiny spectrum or spread of color.

Oculars

The chief functions of the ocular, or eyepiece, are the following:

1. It magnifies the real image of the object as formed by the objective.
2. It corrects some of the defects of the objective.
3. It images cross hairs, scales, or other objects located in the eyepiece.

Several types of eyepieces are employed, depending upon the kind of objective located on the microscope. Those most commonly used are known as Huygenian, hyperplane, and compensating oculars.

Huygenian eyepiece In this type of eyepiece, two simple plano-convex lenses are employed, one of which is below the image plane (Fig. 15). The convex surfaces of both lenses face downward. Oculars in this group are sometimes spoken of as negative eyepieces. This type of ocular is made with a large field lens which bends the pencils of light coming from the

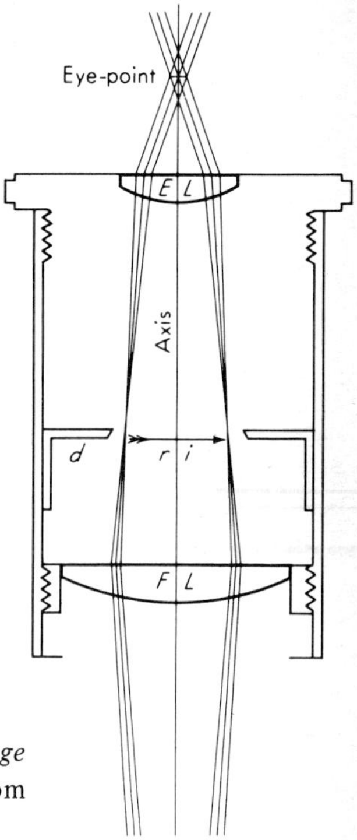

Fig. 15 Huygenian eyepiece. EL, eye lens; FL, field lens; ri, real image formed between the ocular lenses and the diaphragm d. (Redrawn from Gage, "The Microscope," Comstock Publishing Associates, Inc.)

objective toward the axis without altering to any great extent the convergence or divergence of the rays in the individual pencils. Above the field lens, and at some distance from it, is a smaller lens known as the eye lens, the function of which is to convert each pencil of light into a parallel or only slightly diverging ray system capable of being focused by the eye. The rays, after emerging through this lens, then pass through a small circular area known as the Ramsden disk, or eyepoint. It may be seen that the real image of the object is formed between the two eyepiece lenses. In an eyepiece of this type, the distance separating the two lenses is always a little greater than the focal length of the eye lens. This is to prevent any dirt on the field lens from being seen sharply focused by the eye. An image should be viewed with the eye placed at the Ramsden disk in order to obtain the largest field of view and also to obtain the maximum brightness over the field.

The Huygenian eyepiece works well with the low-power achromats but gives undercorrected curvature of field and lateral color with the intermediate and higher-power objectives. The degree of compensation required increases with the objective power, making it highly desirable to have a graded series of eyepieces. Therefore, the Huygenian eyepiece should be used to cover the low powers; the hyperplane eyepiece, the intermediate powers; and the compensating eyepiece, the higher powers.

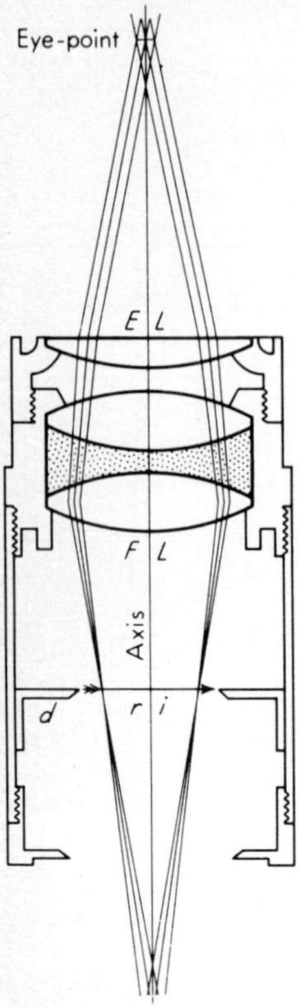

Fig. 16 Compensating eyepiece. EL, eye lens; FL, field lens composed of three components; ri, real image formed below the lenses at the diaphragm d. (Redrawn from Gage, "The Microscope," Comstock Publishing Associates, Inc.)

Hyperplane eyepiece Apochromatic objectives, when used with compensating eyepieces, give fields that are not flat. Flat-field eyepieces have been designed to correct this defect. They give much flatter fields than do the other two types but they are less perfectly corrected chromatically. Oculars of this type are referred to as hyperplane, planoscopic, periplane, etc. They may be employed with the higher-power achromatic, fluorite, and apochromatic objectives without introducing chromatic aberrations in the image. Their color compensation falls about midway between the Huygenian and the compensating eyepieces.

Compensating eyepiece Oculars of this type consist of an achromatic triplet combination of lenses (Fig. 16). These eyepieces are more perfectly corrected than are those of the Huygenian and hyperplane types. A compensating eyepiece is corrected to neutralize the chromatic difference of magnification of the apochromatic objectives. Such eyepieces are intended, therefore, to be used primarily with apochromatic objectives, although they may be employed with the higher-power achromatic and fluorite objectives with good results.

Condensers

Several methods are employed for illuminating the object under examination. In bacteriology, the two methods commonly used are (1) illumination by transmitted light and (2) dark-field illumination.

Illumination by transmitted light A condenser may be defined as a series of lenses for illuminating, with transmitted light, an object to be studied on the stage of the microscope. It is located under the stage of the microscope between the mirror and the object, whereas the objective and ocular lenses are located above the stage. It is sometimes referred to as a substage condenser. The most popular substage optical system is known as the Abbe condenser (Fig. 17*A*).

A condenser is necessary for the examination of an object with an oil-immersion objective to obtain adequate illumination. A condenser is also preferable when working with high-power dry objectives. Probably the most commonly employed condenser has a 1.25 NA.

A good condenser sends light through the object under an angle sufficiently large to fill the aperture of the back lens of the objective. When this is accomplished, the objective will show its highest NA. This may be determined by first focusing the oil-immersion objective on the object. The eyepiece is then removed from the ocular tube. The back lens of the objective is observed by looking down the microscope tube, care

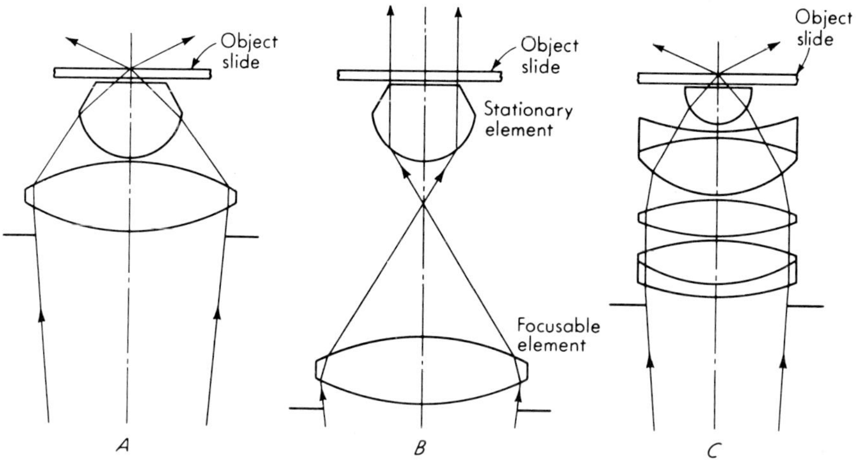

Fig. 17 Substage condensers. (A) Abbe condenser; (B) variable-focus condenser; (C) achromatic condenser. (Courtesy of Bausch & Lomb Optical Company.)

being taken not to disturb the focus. The back lens of the objective should be evenly illuminated. If it is not, the mirror should be properly centered. If the condenser has a smaller numerical aperture than the objective, the peripheral portion of the back lens of the objective will not be illuminated, even though the condenser iris diaphragm is wide open. If the condenser has a greater NA than the objective, the back lens of the objective may receive too much light, resulting in a decrease in contrast. The smaller the aperture, the greater the depth of focus and the greater the contrast of the components of the image. The lowest-permissible aperture is reached when diffraction bands become evident about the border of the object imaged. This difficulty may be largely overcome by closing the iris diaphragm of the condenser until the leaves of the iris appear around the edges of the back lens of the objective. The diaphragm is then said to be properly set. The setting of the iris diaphragm will vary with different objectives.

The Abbe condenser (NA, 1.25) utilizes only two lenses. Because of its simplicity and good light-gathering ability, it is extensively used for general microscopy. It is not corrected for spherical and chromatic aberration but for general visual observation it serves very well.

The variable-focus condenser (Fig. 17*B*) is a two-lens system (NA, 1.25 maximum) in which the upper lens element is fixed and the lower element focusable. By this means it is possible to fill the field of low-power objectives without the necessity of removing the top element. This condenser is basically similar to the Abbe condenser when the lower lens is raised to its top position. When the focusable lens is lowered, the light is focused in between the elements, and when this focus is at the point indicated in the diagram, the light emerges as a large-diameter parallel bundle.

The achromatic condenser (NA, 1.40) is corrected for both chromatic and spherical aberrations (Fig. 17*C*). Because of its high degree of correction, it is recommended for research microscopy and for color photomicrography where the highest degree of perfection in the image is desired.

Dark-field illumination The microscope is most commonly employed by allowing light to pass through the object. This is called microscopy by transmitted light or bright-field microscopy.

Many transparent and semitransparent objects are not readily visible in a bright field. Visibility is dependent upon contrast between the object and its background and can be improved by using a dark background.

The cone of light normally illuminating an object must not enter the objective; only the light scattered or reflected by the specimen should be seen by the objective.

If the aperture of the condenser is opened completely and a dark-field stop inserted below the condenser, the light rays reaching the object form a hollow cone. If a stop of suitable size is selected, all the direct rays from the condenser can be made to pass outside the objective. Any object within this beam of light will reflect some light into the objective and be visible. This method of illuminating an object, where the object appears self-luminous against a dark field, is known as dark-field illumination.

Three types of condensers are employed for dark-field illumination: (1) the Abbe, (2) the paraboloid, and (3) the cardioid.

The Abbe condenser is probably more commonly employed than the other two because it is especially suitable for objects that do not require the highest magnifications to make them visible. It may be employed either by inserting a dark-field stop below the condenser (Fig. 18) or by

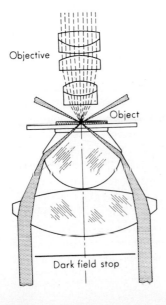

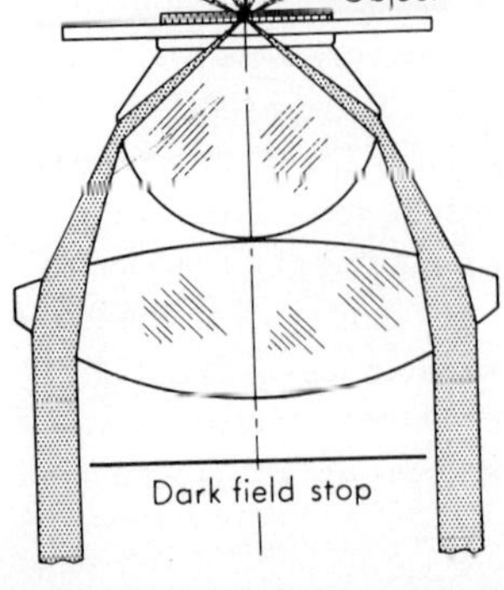

Fig. 18 Abbe condenser with dark-field stop inserted below the condenser. (From Dark Field Optical Systems, courtesy of Bausch & Lomb Optical Company.)

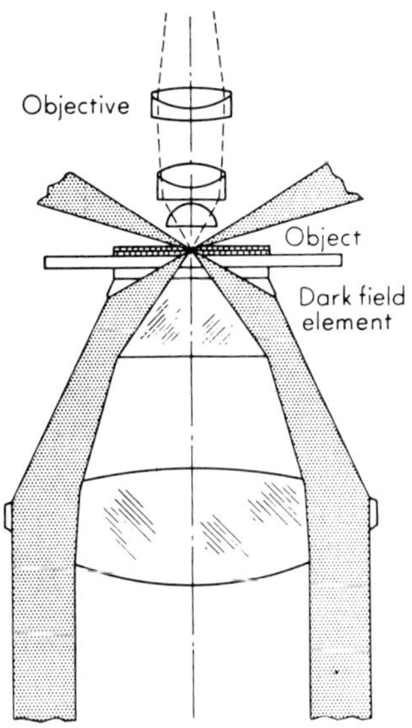

Fig. 19 Abbe condenser with top part of condenser replaced by a dark-field element. (From Dark Field Optical Systems, courtesy of Bausch & Lomb Optical Company.)

unscrewing the top part of the condenser and substituting for it a dark-field element (Fig. 19).

The paraboloid condenser is designed to be used with high-power oil-immersion objectives and an intense source of light (Fig. 20). In using this condenser, it is necessary to place cedar oil or glycerin between the condenser and the slide. Also, the specimen must be mounted in a liquid or cement and protected with a cover slip. The numerical aperture of the objective must not be greater than that of the condenser.

The cardioid condenser is the most refined type of dark-field illuminator (Fig. 21). It is especially designed to be used for the examination of colloidal solutions or suspensions, i.e., particles measuring less than 0.25 μ in diameter.

The cardioid condenser is best employed with a strong arc lamp. Since the concentration of light is so great, ordinary glass slides and cover slips should not be used. Visible defects and the difficulty of removing foreign objects from the glass ruin the visibility of ultramicroscopic particles. It is better to employ fused quartz object slides and fused quartz cover slips. These are available and are highly recommended for use. They resist corrosion and abrasion better than conventional glass slides and cover slips.

For more information: Barer and Cosslett (1968); Barron (1965); Bradbury (1967, 1968); Busacca (1966); Connes (1968); Cosslett (1966); Eastman Kodak Company (1966); Feinberg (1968); Françon (1961); Gage (1947); Gray (1967); Jones (1968); Malies (1959); Martin (1966); Michel

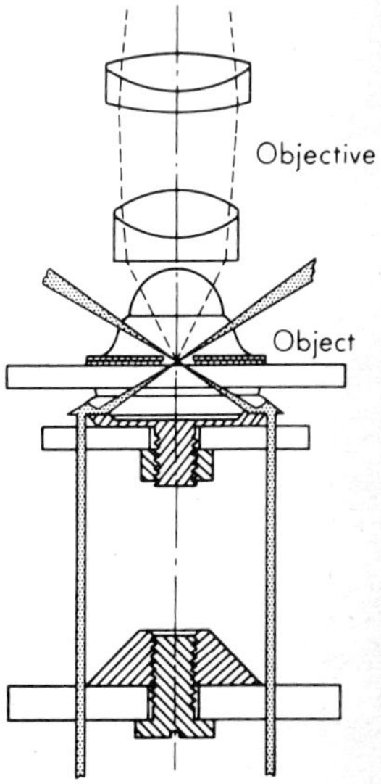

Fig. 20 Paraboloid condenser. (From Dark Field Optical Systems, courtesy of Bausch & Lomb Optical Company.)

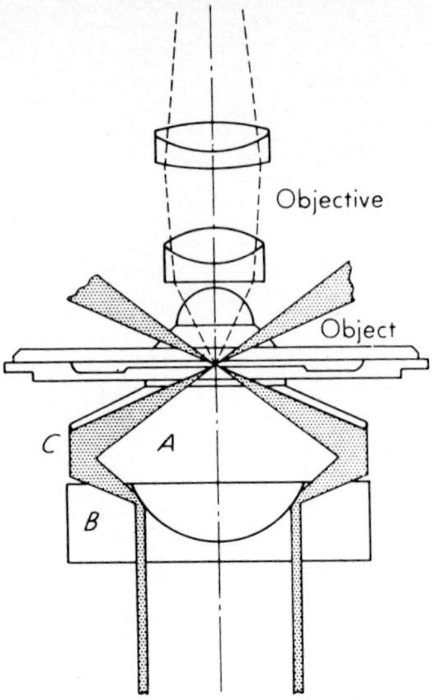

Fig. 21 Cardioid condenser. (From Dark Field Optical Systems, courtesy of Bausch & Lomb Optical Company.)

(1964); Needham (1968); Schenk and Kistler (1963); Smith (1968); Weisskopf (1968).

Phase microscope

The principle of the phase microscope is not new, having been discovered as early as 1892, but only recently has it been developed to practical use. It complements rather than replaces existing methods of microscopy.

The commonly employed light microscope reproduces very well the structure of objects which are examined in fixed and stained preparations. In the absence of staining, very little if anything can be seen since details are only distinguishable by their refractive power. However, there are many unknown facts of growth, motion, propagation, and exchange of cellular constituents which can only be revealed by a study of living preparations in the unstained state.

Phase microscopy is a method for controlling the contrast in the image so that unstained living cells and cytological details within them become visible. Phase microscopy also greatly enhances visibility of stained material of low contrast.

The method employed for fixing and staining bacteria may make a difference in their size. The bacterial cell usually shrinks considerably under such treatment. Since unstained microorganisms show sharp outlines under the phase microscope, accurate measurement of living cells is now possible.

A schematic diagram of the general optical arrangement of a phase microscope is shown in Fig. 22. An annular aperture in the diaphragm, placed in the focal plane of the substage condenser, controls the illumination on the object. The aperture is imaged by the condenser and objective at the rear focal plane, or exit pupil, of the objective. In this plane a phase shifting element, or phase plate, is placed.

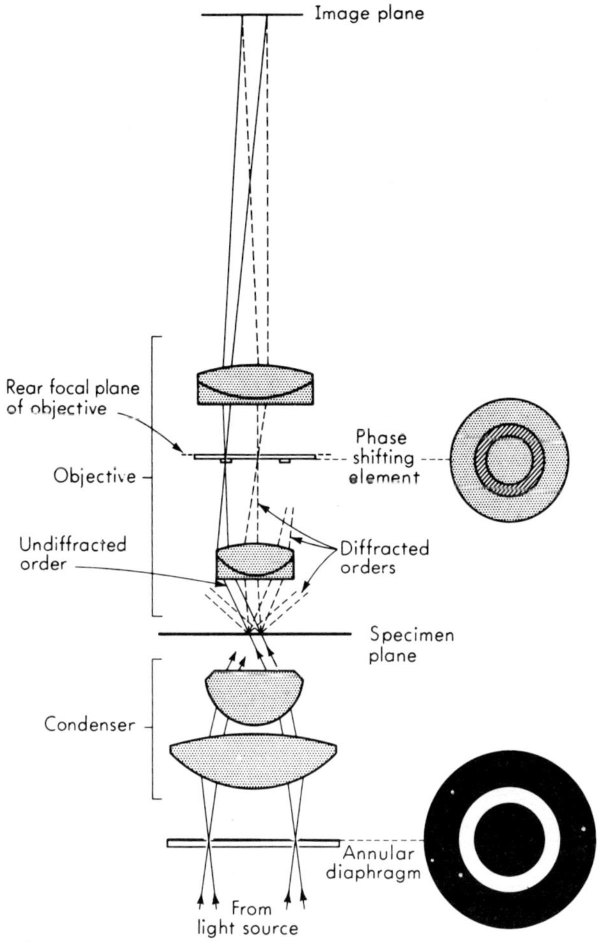

Fig. 22 Schematic diagram of the general optical arrangement of a phase microscope. (Courtesy of Bausch & Lomb Optical Company.)

Light, shown by the solid lines and undeviated by the object structure, in passing through the phase altering pattern, acquires a one-quarter wavelength of green light advance over that diffracted by the object structure (broken lines) and passing through that region of the phase plate not covered by the altering pattern. The resultant interference effects of the two portions of light form the final image. Altered phase relations in the illuminating rays, induced by otherwise invisible elements in the specimen, are translated into brightness differences by the phase altering plate.[1]

For more information: Bausch & Lomb (1966), Clark et al. (1961), Ross (1967).

[1] Taken from J. R. Benford, "The Theory of the Microscope," Bausch & Lomb Optical Company, Rochester, N.Y., 1951.

Fluorescence microscope[2]

In fluorescence microscopy the preparation becomes self-luminous while the radiation exciting the luminosity does not contribute to the image formation but is eliminated by barrier filters. The fluorescing part of the preparation appears bright, usually colored, against a dark background. For excitation, light of shorter wavelength than that emitted by the preparation is used. Thus blue and green fluorescence can only be excited by ultraviolet irradiation, whereas yellow and red fluorescence may also be excited by intense blue-violet.

Since only a small part of the incident radiant energy is converted into fluorescent light, it is necessary to employ the most intense sources of light. Besides the excitation radiations of short wavelengths, such lamps also emit light of greater wavelengths which completely flood the relatively weak fluorescences.

For this reason two kinds of filters are a part of every fluorescence microscope: (1) excitation filters which transmit in the illuminating beam only the excitation radiations of the total radiations emitted by the light source; and (2) barrier filters which bar the further passage of excitation radiations in the imaging beam (Fig. 23).

Just as the inherent color of a substance is due to its transmission or reflection of the nonabsorbed light falling upon it, so a primary fluorescence, where it occurs, is mainly a function of the chemical constitution. On this basis, guided by the presence or absence of fluorescence of definite quality, conclusions can be drawn concerning basic chemical composition (fluorescence analysis). Besides the inherent color, affinity to certain stains may be characteristic. Fluorochromes do not necessarily have a pronounced color. In fact, some are practically colorless. But they fluoresce brightly in a characteristic color when exposed to appropriate exciting radiations.

Suitable fluorochromes include: acridine orange NO, acridine yellow, acriflavine, auramine O, berberine, coriphosphine O, fluorescein, primuline, rheonine A, rhodamine B, rivanol, rose bengale, thioflavine S, thioflavine T, and titan yellow G.

Various cellular constituents show a specific affinity for the fluorochromes. A peculiar phenomenon occurring more frequently with the fluorochromes than with ordinary stains is the property of staining in a different color, e.g., the appearance of a cellular element in one color (red) when treated with a stain of a different color (blue).

Cellular constituents clearly show fluorescence after treatment with high dilutions of fluorochromes in contrast to vital stains which are effective only in relatively high concentrations. Hence it is possible to fluorochrome a cell in full function without producing any damage.

[2] Taken from J. R. Benford, "The Theory of the Microscope," Bausch & Lomb Optical Company, Rochester, N.Y., 1951.

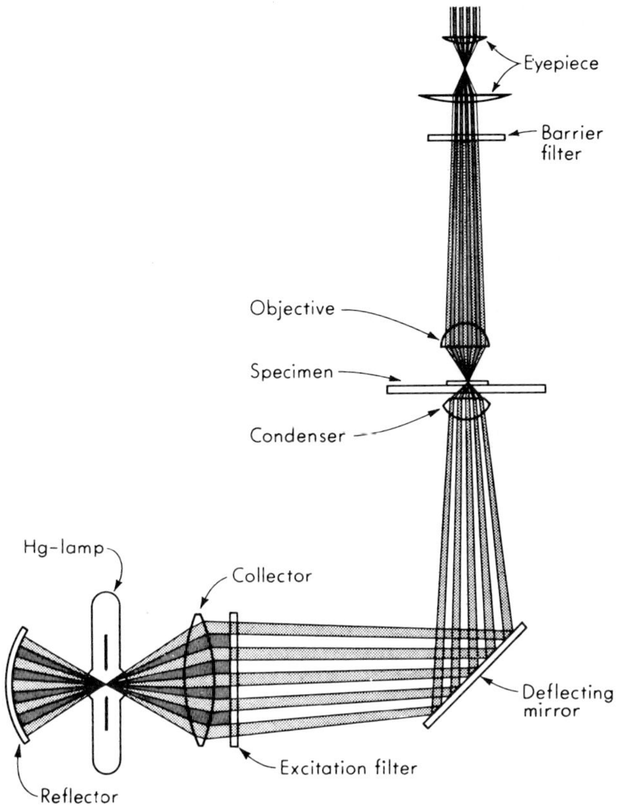

Fig. 23 Fluorescence microscope for transmitted light. Diagrammatic representation of the path of rays and the action of excitation and barrier filters. (Redrawn from Fluorescence Microscopes, courtesy of Carl Zeiss.)

For more information: Berger (1967); Beutner (1961); Coons (1961); Darken (1961*a, b*); De Repentigny (1961); Drawert (1968); Goldman (1968); Jutte and Lemke (1968); Nairn (1969).

Electron microscope

A new type of instrument has been developed having a higher resolving power than the ordinary light microscope. This instrument is known as the electron microscope (Figs. 24, 25, and 26).

The electron microscope uses a beam of high-speed electrons having an equivalent wavelength of x-ray dimension (about 0.05 Å or one-fifth of a billionth of an inch). It is this extremely short wavelength that gives the electron microscope its fundamental superiority over the light microscope. A close analogy exists between the action of a magnetic or electric field of rotational symmetry or an electron beam and the action of a glass lens on a light beam (Fig. 27).

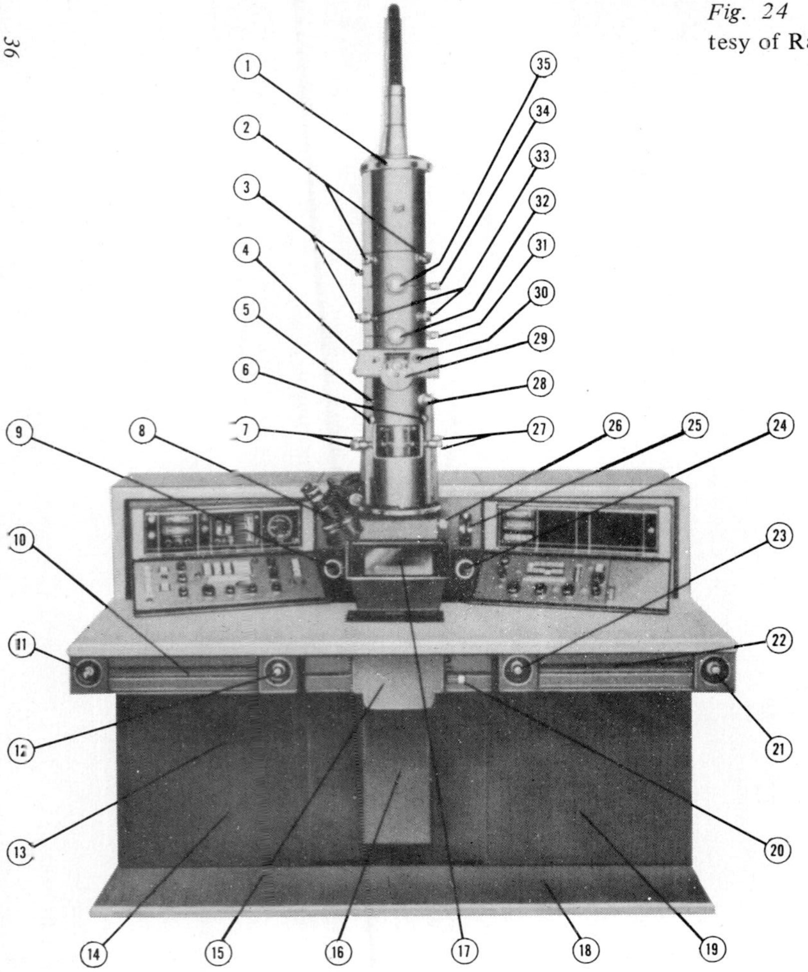

Fig. 24 Electron microscope. Center-of-interest console design. (Courtesy of Radio Corporation of America.)

1. Knurled-ring electron-gun access
2. Gun traverse controls
3. Condenser stigmator controls
4. Chamber feedthrough connections and port (side)
5. Objective aperture centering
6. Objective lens traverse
7. Intermediate-lens traverse (micrometer)
8. Wray 10 x 30 binoculars
9. Left-hand stage traverse
10. Ball-bearing-mounted storage drawer
11. Key-lock master power switch
12. Console panel-light dimmer
13. Lens-current regulators (behind panel)
14. 22-liter ballast tank (behind panel)
15. Photo chamber access door
16. 720-liter/sec. diffusion pump (inside)
17. Panoramic viewing window
18. Comfort-slanted toe board
19. Space reserved for image intensifier (inside)
20. Photo-magazine and cassette controls
21. Small parts tray
22. Ball-bearing-mounted accessory drawer
23. Room-light dimmer control
24. Right-hand stage traverse
25. Exposure-size selector lever
26. Photo-frame mask selector
27. Intermediate-lens traverse (micrometer)
28. Objective aperture centering
29. Specimen airlock
30. Specimen Chamber Window
31. Lower condenser aperture centering
32. Lower condenser aperture selector
33. Double condenser traverse controls
34. Upper condenser aperture centering
35. Upper condenser aperture selector

Fig. 25 Electron microscope. left-hand control group of console;

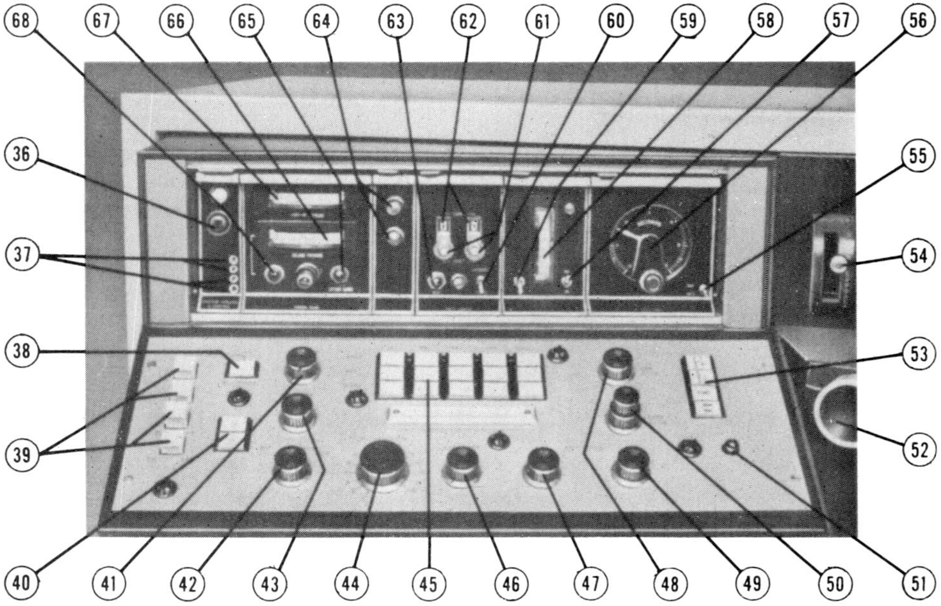

Left-hand control group

36. Vacuum sensing
37. Maintenance test points
38. Lens power master control
39. Pushbutton vacuum-valving control
40. Microscopy/diffraction mode switches
41. High-voltage modulator control
42. Diffraction focus/magnification vernier
43. Diffraction magnification control
44. 15-step magnification control (root-2 steps)
45. Lens current modulate/reverse/normalize switches
46. Auxiliary fine focus control
47. Lower condenser lens control
48. Left-hand gun alignment control
49. Upper condenser lens control
50. Left-hand condenser lens alignment control (coarse and fine)
51. Photo-plate in-transit indicator light
52. Left-hand stage traverse
53. Photo-cassette position control
54. Image screen tilt lever/shutter control
55. MANual/AUTOmatic timer reset switch
56. 0.4 to 6.0-second exposure timer
57. Photometer sensitivity selector switch
58. Photometer read-out scale
59. Photometer on/off switch
60. Exposure/plates counter selector switch
61. Exposure-count preset controls
62. 0-99 exposure/plate count read-out
63. Counter reset button
64. Indicator lights for aperture heaters
65. Vacuum-gauge range-extender button
66. Column-pressure read-out
67. Foreline pressure read-out
68. Plate-load column-pressure bypass pushbutton

Fig. 25 (continued) right-hand control group of console. (Courtesy of Radio Corporation of America.)

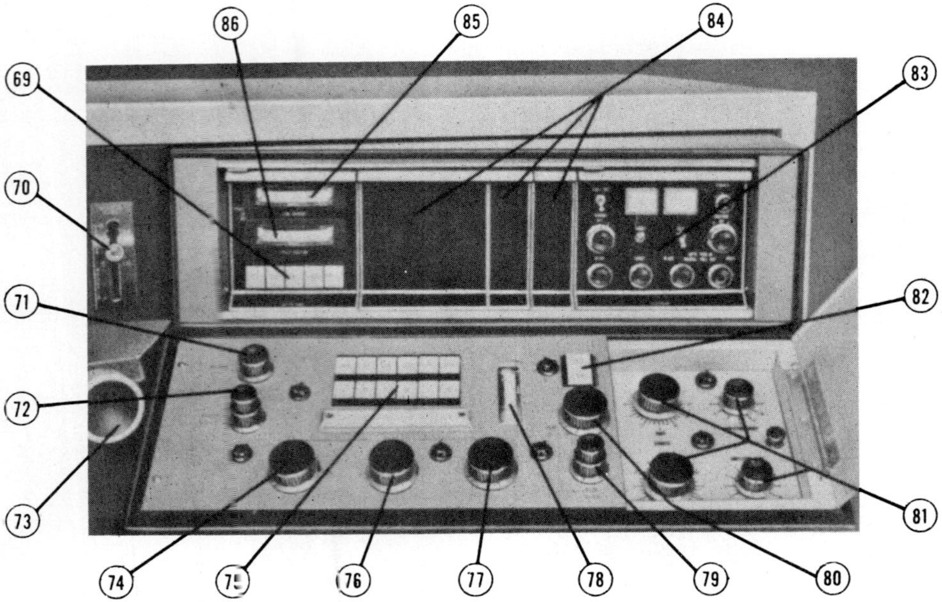

Right-hand
control group

69. Lens-current meter selector switches
70. Photo-size selector lever
71. Right-hand gun-alignment control
72. Right-hand condenser-lens alignment control (coarse and fine)
73. Right-hand stage traverse control
74. Fine-focus control
75. Step-focus pushbuttons
76. Medium-fine focus control
77. Coarse focus control
78. Beam current read-out
79. Electron-gun bias control
80. Gun emission/HV on-off control
81. Objective lens stigmator controls (behind door)
82. 50/100-kV HV selector switches
83. Image-intensifier TV-camera control (optional equipment)
84. Reserved module space for accessories
85. Lens-current multimeter read-out
86. Photo-plate position read-out

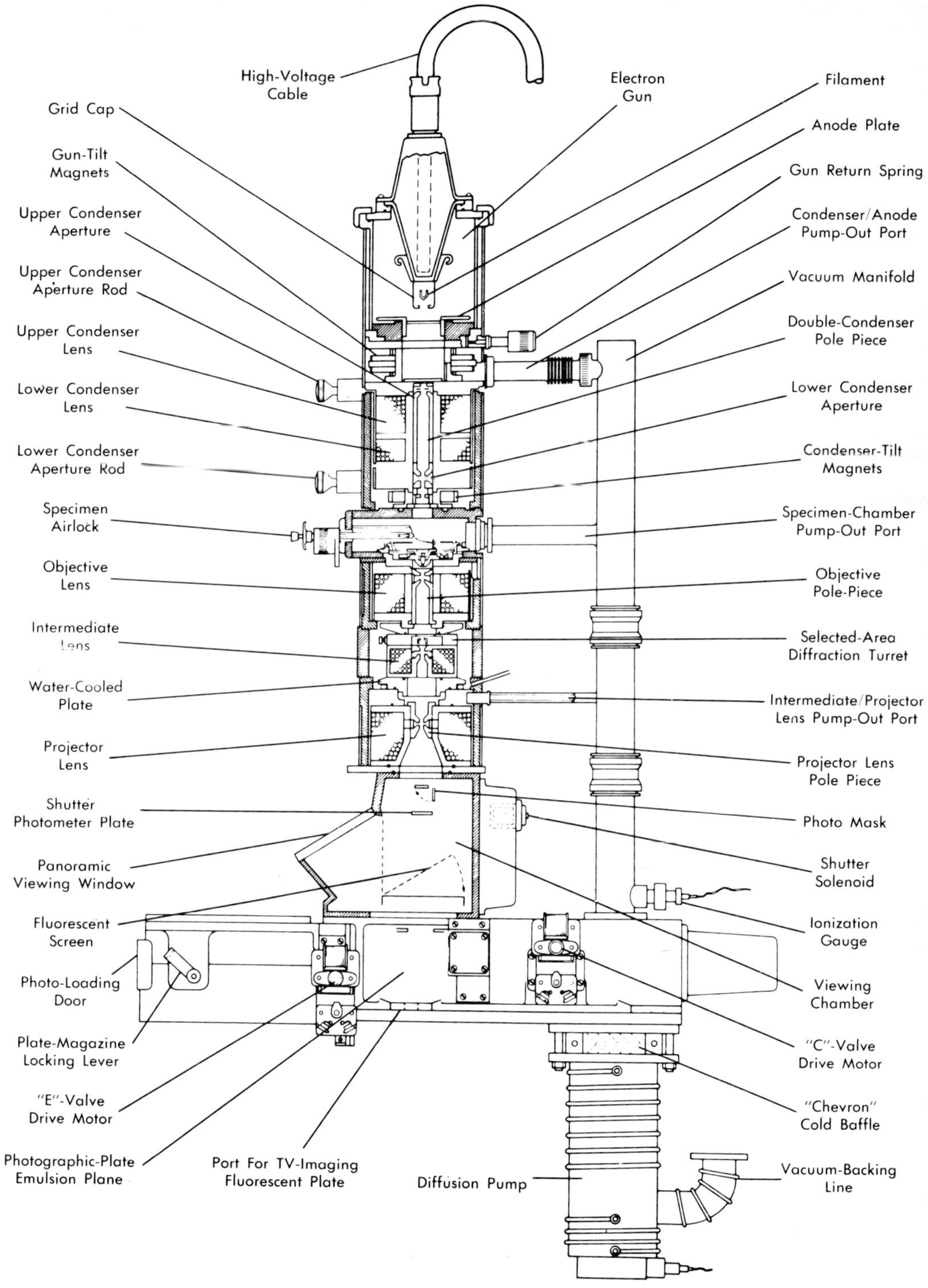

Fig. 26 Cross section of electron microscope column. Note straight-line vacuum manifold specimen chamber evacuation, large diffusion pump, and barrel-like design of column. (Courtesy of Radio Corporation of America.)

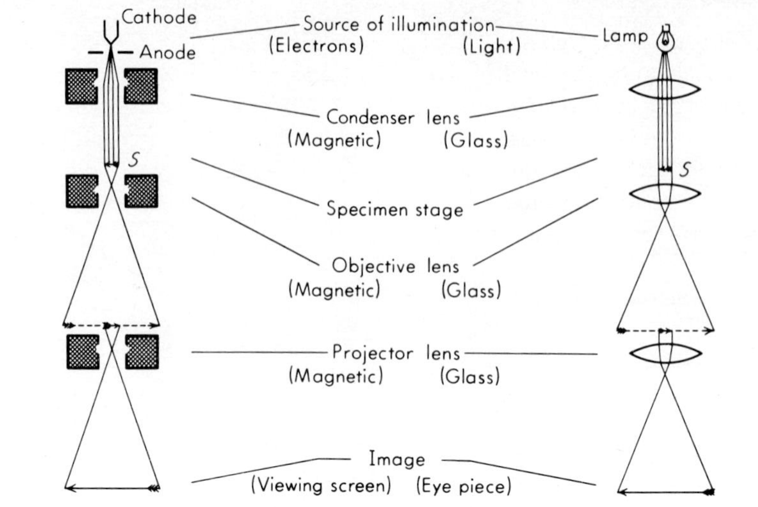

Fig. 27 Similarity of light and electron microscopes. (Courtesy of Radio Corporation of America.)

The electrons obtain their high velocity and low wavelength as a result of their acceleration through both 50 and 100 kV with instant changeover, thereby providing maximum flexibility of operation. After reaching this high velocity, the electrons penetrate the specially prepared specimen and are then focused by means of magnetic fields to form an enlarged image on a fluorescent screen or a photographic plate.

An optical-lens microscope magnifies a maximum of about 2000 diameters. If ultraviolet light is used, this magnification can be increased to about 3000 diameters. With the electron microscope, the maximum magnification for sharpness of details is about 250,000 diameters. However, under the best conditions this 250,000x image will contain much more detail than the unaided eye can see. By exposing a suitable photoplate to this image and through later optical enlargement of this image by 10 diameters, one can obtain a final magnification of about 2.5 million diameters. The resolving power is so much greater than that of the ordinary optical microscope that it is now possible to obtain images of protein molecules, viruses, bacterial viruses (bacteriophages), unstained flagella, internal structures of cells, etc.

For more information: Barer and Cosslett (1968); Bartlett (1967); Causey (1962); Greenwood and O'Grady (1969); Grimstone (1968); Haggis (1967); Hall (1966); Hayat (1970); Kay (1965); Meek (1970); Mercer and Birbeck (1966); Parsons (1970); Reimer (1967); Siegel (1964); Swift (1970); Thornton (1968); Toner and Carr (1968); Wischnitzer (1970); Everhart and Hayes (1972).

Damage to lenses by molds

Surfaces of optical glass are susceptible to mold growth, especially when the temperature and relative humidity are high for appreciable periods. Molds from a number of genera have been isolated and identified.

The etching of glass is due chiefly to (1) moisture associated with the mold leaching the surface, and (2) the organic acids produced as end products of mold metabolism. Cleanliness, lowered humidity, and fungicides have been found useful in protecting lens surfaces from mold damage.

For more information: Baker (1967, 1968).

References

Baker, P. W.: An evaluation of some fungicides for optical instruments, *Int. Biodeterioration Bull.,* **3**:59, 1967.

——: Possible adverse effects of ethyl-mercury chloride and meta-cresyl-acetate if used as fungicides for optical/electronic equipment, *Int. Biodeterioration Bull.,* **4**:59, 1968.

Barer, R., and V. E. Cosslett, eds.: "Advances in Optical and Electron Microscopy," New York, Academic Press, Inc., 1968.

Barron, A. L. E.: "Using the Microscope," London, Chapman and Hall, Ltd., 1965.

Bartlett, G. A.: Scanning electron microscope: Potentials in the morphology of microorganisms, *Science,* **158**:1318, 1967.

Bausch & Lomb Optical Company: "Dynazoom and Dynoptic Laboratory and Laboratory Research Microscopes," Rochester, N.Y., 1966.

Berger, J.: "Die Fluoreszenzmikroscopie in der Früherfassung der Weiblichen Genital-karzinome," Basel, Switzerland, Karger, 1967.

Beutner, E. H.: Immunofluorescent staining: The fluorescent antibody method, *Bact. Rev.,* **25**:49, 1961.

Bradbury, S.: "The Evolution of the Microscope," Oxford, England, Pergamon Press, Ltd., 1967.

——: "The Microscope; Past and Present," Oxford, England, Pergamon Press, Ltd., 1968.

Busacca, A.: "Manuel de Biomicroscopie Oculaire," Paris, Doin, 1966.

Causey, G.: "Electron Microscopy," Edinburgh, E. and S. Livingstone, 1962.

Clark, G. L., et al.: "The Encyclopedia of Microscopy," New York, Reinhold Publishing Corporation, 1961.

Connes, P.: How light is analyzed, *Sci. Am.,* **219**:72, 1968.

Coons, A. H.: The beginnings of immunofluorescence, *J. Immunol.,* **87**:499, 1961.

Cosslett, V. E.: "Modern Microscopy," Ithaca, N.Y., Cornell University Press, 1966.

Darken, M. A.: Natural and induced fluorescence in microscopic organisms, *Appl. Microbiol.,* **9**:354, 1961*a*.

——: Applications of fluorescent brighteners in biological techniques, *Science,* **133**:1704, 1961*b*.

De Repentigny, J.: La microscopie a fluorescence en microbiologie, *Trans. Roy. Soc. Can.,* **55**:5, 1961.

Drawert, H.: "Vitalfärbung und Vitalfluorochromierung Pflanzenlicher Zellen und Gewebe," New York, Springer-Verlag Inc., 1968.

Eastman Kodak Company: "Photography through the Microscope," Rochester, N.Y., 1966.

Everhart, T. E., and T. L. Hayes: The scanning electron microscope, *Sci. Am.,* **226**:55, 1972.

Feinberg, G.: Light, *Sci. Am.,* **219**:50, 1968.

Françon, M.: "Progress in Microscopy," New York, Harper & Row, Publishers, Incorporated, 1961.

Gage, S. H.: "The Microscope," Ithaca, N.Y., Comstock Publishing Associates, a division of Cornell University Press, 1947.

Goldman, M.: "Fluorescent Antibody Methods," New York, Academic Press, Inc., 1968.

Gray, P.: "The Use of the Microscope," New York, McGraw-Hill Book Company, 1967.

Greenwood, D., and F. O'Grady: Antibiotic-induced surface changes in microorganisms demonstrated by scanning electron microscopy, *Science,* **163**:1076, 1969.

Grimstone, A. V.: "The Electron Microscope in Biology," New York, St. Martin's Press, Inc., 1968.

Haggis, G. H.: "The Electron Microscope in Molecular Biology," New York, John Wiley & Sons, Inc., 1967.

Hall, C. E.: "Introduction to Electron Microscopy," New York, McGraw-Hill Book Company, 1966.

Hayat, M. A.: "Principles and Techniques for Electron Microscopy," New York, Van Nostrand Reinhold Company, 1970.

Jones, R. C.: How images are detected, *Sci. Am.,* **219**:110, 1968.

Jütte, A., and L. Lemke: "Intravitalfärbung am Augenhintergrund mit Fluoreszein-Natrium," Stuttgart, Germany, Enke, 1968.

Kay, D.: "Techniques for Electron Microscopy," Oxford, Blackwell Scientific Publications, Ltd., 1965.

Malies, H. M.: "Applied Microscopy and Photomicrography," London, Fountain Press, 1959.

Martin, L. C.: "The Theory of the Microscope," New York, American Elsevier Publishing Company, Inc., 1966.

Meek, G. A.: "Practical Electron Microscopy for Biologists," New York, Wiley-Interscience, 1970.

Mercer, E. H., and M. S. C. Birbeck: "Electron Microscopy," Oxford, Blackwell Scientific Publications, Ltd., 1966.

Michel, K.: "Die Grundzüge der Theorie des Mikroskops in Elementarer Darstellung," Stuttgart, Germany, Wissenschaftliche Verlagsgesellschaft M.B.H., 1964.

Nairn, R. C., ed.: "Fluorescent Protein Tracing," Edinburgh, E. and S. Livingstone, 1969.

Needham, G. H.: "The Microscope," Springfield, Ill., Charles C Thomas, Publisher, 1968.

Parsons, D. F., ed.: "Some Biological Techniques in Electron Microscopy," New York, Academic Press, Inc., 1970.

Reimer, L.: "Elektronenmikroskopische Untersuchungsund Präparationsmethoden," Berlin, Springer-Verlag OHG, 1967.

Ross, K. F. A.: "Phase Contrast and Interference Microscopy for Cell Biologists," London, Edward Arnold (Publishers) Ltd., 1967.

Schenk, R., and G. Kistler: "Photomicrography," Palisade, N.J., Franklin Publishing Company, Inc., 1963.

Siegel, B. M.: "Modern Developments in Electron Microscopy," New York, Academic Press, Inc., 1964.

Smith, F. D.: How images are formed, *Sci. Am.,* **219**:96, 1968.

Swift, J. A.: "Electron Microscopes," London, Kogan Page, 1970.

Thornton, P.: "Scanning Electron Microscopy," London, Chapman and Hall, Ltd., 1968.

Toner, P. G., and K. E. Carr: "Cell Structure: An Introduction to Biological Electron Microscopy," Edinburgh, E. and S. Livingstone, 1968.

Weisskopf, V. F.: How light interacts with matter, *Sci. Am.,* **219**:60, 1968.

Wischnitzer, S.: "Introduction to Electron Microscopy," New York, Pergamon Press, 1970.

3 *Stains and staining solutions*

Bacteria are semitransparent and difficult to see in the unstained state. Stains are used (1) to render microscopic and semitransparent objects visible, (2) to reveal their shape and size, (3) to show the presence of various internal and external structures, and (4) to produce specific physical or chemical reactions. The presence of certain structures and staining reactions aids in their identification and classification.

The terms stains and dyes generally are used interchangeably by the biologist, but they are not the same. A coloring agent that is to be used for general purposes is called a dye; one that is to be used as a biological is called a stain. Biological coloring agents are manufactured with greater care under more rigid specifications so that they will be satisfactory for the procedures in which they are employed. Textile coloring agents which are not so exacting in their characteristics are called dyes. Stains also may be used for textile dyes, although less purified preparations are satisfactory for such purposes.

Natural dyes predominated during the early years of bacteriology but at present only a few are being used. They have been gradually discarded in favor of the artificial or synthetic dyes. Since the first artificial dyes were produced from aniline, they are generally referred to as aniline dyes. However, many are not derived from aniline and bear no relation to the compound. Since all of them are derived from one or more substances found in coal tar, they are more correctly referred to as coal-tar dyes.

The coal-tar dyes may be considered as derivatives of the cyclic compound benzene, or benzole:

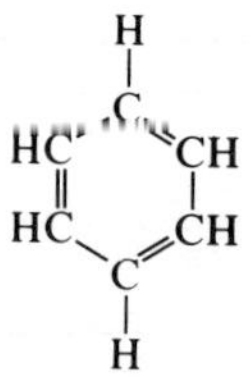

The empirical formula is C_6H_6. It is customary to write the structural formula by omitting the double bonds and the hydrogen atoms, abbreviating it to a hexagon, each corner of which represents an atom of carbon and one of hydrogen:

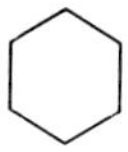

This hexagon is known as the benzene ring.

One or more hydrogen atoms may be replaced by some element or radical. For example, if one hydrogen atom is replaced by a hydroxyl (OH) group, the compound phenol, or carbolic acid, is produced:

If another hydrogen atom is replaced by a methyl group (CH_3), the compound known as cresol is produced. Three different cresols are possible, depending upon which hydrogen atoms are substituted:

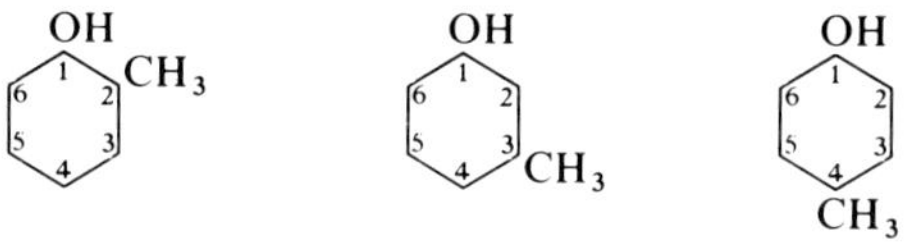

The substituted radicals are in the 1-2 (ortho), 1-3 (meta), and 1-4 (para) positions. The compounds are named orthocresol, metacresol, and paracresol, respectively. The prefixes are usually abbreviated to the letters *o*-, *m*-, and *p*-.

The quinones are compounds produced by the elimination of two hydroxyl-hydrogen atoms from aromatic dihydroxy derivatives. The simplest quinone is benzoquinone. It is also called quinone, the formula of which is

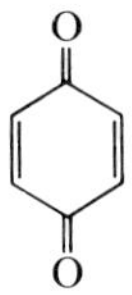

The benzene ring of the quinones contains two double bonds instead of three as in cresol. The formula of benzoquinone shows that it is not a true benzene derivative but the diketone of a *p*-hydrobenzene. Substances containing the quinone ring are called *quinonoid compounds*. The double bonds in the quinonoid compounds are supposed to be fixed, not mobile as in benzene. A large number of dyes contain the quinone ring.

Dyes

Definition of a dye A dye may be defined as an organic compound containing both chromophore and auxochrome groups linked to benzene rings. A chromophore group imparts to the compound the property of color. Compounds of benzene containing chromophore radicals have been called *chromogens.* Such a compound, even though colored, is not a dye. It possesses no affinity for, nor ability to unite with, fibers and tissues. The color may be easily removed by mechanical methods. To be a dye, a color must contain not only a chromophore group but also another group that imparts to the compound the property of electrolytic dissociation. Such auxiliary groups are known as *auxochromes.* In some instances they may alter the shade of the dye but are not the cause of the color. The function of auxochrome groups is to furnish salt-forming properties to the compound.

This may be illustrated by the following example: The nitro group (NO_2) may be considered a chromophore. When three hydrogen atoms in the benzene molecule are replaced by three nitro groups, the compound trinitrobenzene is formed:

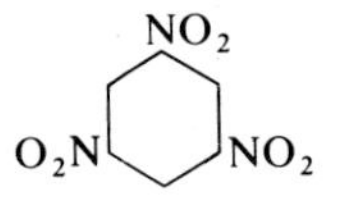

This yellow-colored compound is a chromogen but not a dye. It does not dissociate electrolytically and is unable, therefore, to form salts with either acids or bases. If, however, another hydrogen atom is replaced by an auxochrome group, such as (OH), the compound known as picric acid is formed:

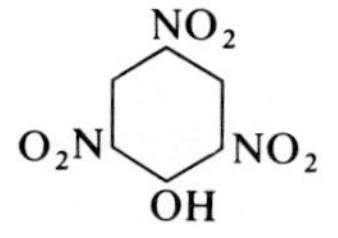

This compound is capable of dissociating as follows:

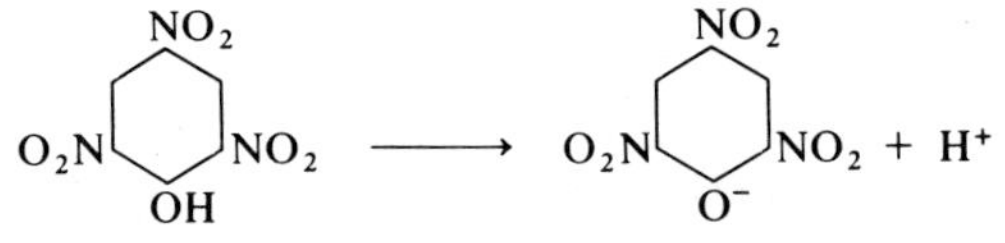

Picric acid also is yellow in color and is a dye. The dye portion of the molecule has a negative electrical charge. It is, therefore, an acid dye, being capable of forming salts with bases. The color of picric acid is due to the chromophoric nitro groups (NO_2), and its dyeing properties to the auxochromic hydroxyl group (OH), which imparts to the molecule the property of electrolytic dissociation.

Acidic and basic dyes The dyes of commerce are not acids or bases in the true sense. The terms do not refer to the hydrogen-ion concentrations of their solutions. The distinction depends on whether the dye portion of the molecule has a positive or negative electrical charge. Acidic dyes ionize to give the dye portion of the molecule a negative electrical charge. They are salts of color acids, usually the sodium salts, sometimes the potassium, calcium, or ammonium salts. Basic dyes ionize to give the dye portion of the molecule a positive electrical charge. They are salts of color bases, usually the chloride, sometimes the sulfate or acetate.

Some auxochromic groups are acidic (e.g., OH), whereas others are basic (e.g., NH_2). The amino group is basic by virtue of the ability of its nitrogen atom to become pentavalent on the addition of water or acid:

With water:

$$R-N\begin{array}{c}H\\H\end{array} + H_2O \longrightarrow R-N\begin{array}{c}H\ H\\OH\\H\end{array}$$

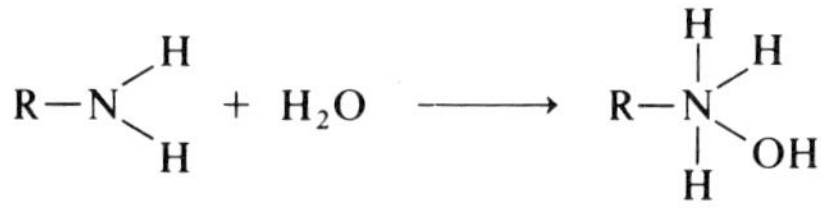

| Amine | Organic ammonium base |

With an acid:

$$R-N\begin{array}{c}H\\H\end{array} + HCl \longrightarrow R-N\begin{array}{c}H\ H\\Cl\\H\end{array}$$

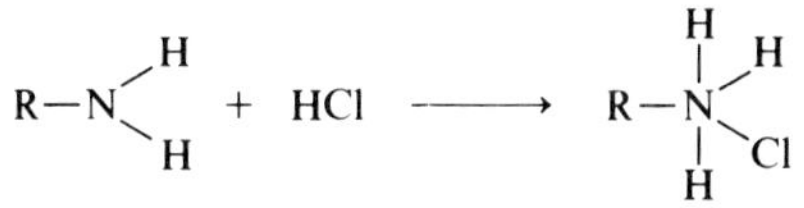

| Amine | Amine hydrochloride |

The hyroxyl group is weakly acidic by virtue of its power to furnish hydrogen ions by dissociation. The more of either of these two groups (i.e., OH or NH_2) in a compound, the more strongly acidic or basic it becomes. The amino group is more strongly basic than the hydroxyl group is acidic. If one of each of these two radicals is present, the basic character of the amino group predominates.

Some dyes have the sulfonic group (SO_2OH) attached to a benzene ring. It is a strongly acidic group possessing salt-forming properties. The radical is only weakly auxochromic. It serves two very important purposes in the dye molecule: (1) It renders a dye soluble in water, and (2) it changes a basic dye to one acidic in character by the introduction of the sulfonic group into the benzene ring. Since the radical is only weakly auxochromic, a compound containing a chromophore and a sulfonic group is not a dye unless a true auxochrome radical is also present.

Both acidic and basic stains are used in bacteriology. The acidic stains are used chiefly to stain cytoplasm. The basic stains color acid constituents (nuclei, metachromatic granules) more intensely than cytoplasmic material.

Chromophores   In order that a compound be a dye, it must contain at least one group that imparts the property of color to the substance. This is known as a chromophore group. Some chromophores are basic in reaction; others are acidic.

Basic chromophores   The basic chromophores include (1) the azo group, (2) the azine group, and (3) the indamine group.

1. The azo group

   —N=N—

   is found in all azo dyes. In these compounds a benzene ring is attached to each nitrogen atom. The dyes in this group may be considered as derivatives of azobenzene:

   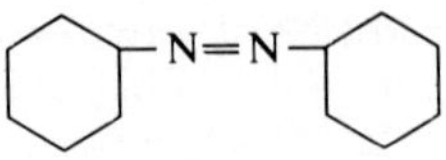

   Examples of biological stains containing this chromophore are bismarck brown, methyl red, and methyl orange.
2. The azine group

   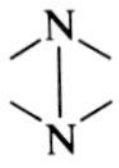

   is found in the phenazines. Neutral red and the safranines are examples of azine stains.
3. The indamine group

   —N=

   is found in the indamines, the thiazines, and other dyes. Many of the dyes have two benzene rings attached to a nitrogen atom. One of the rings shows the quinonoid structure:

   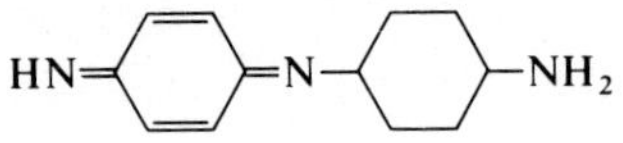

   The thiazines have the two benzene rings further joined together by an atom of sulfur. The simplest thiazine nucleus has the following structure:

   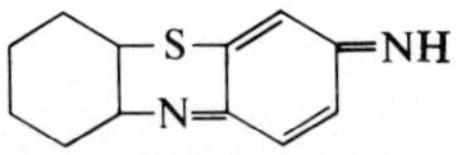

The best-known biological stain having the thiazine base is methylene blue.

Acid chromophores The acid chromophores include (1) the nitro group and (2) the quinonoid ring.

1. The nitro group (NO_2) is found in many compounds, an example of which is picric acid.
2. The quinonoid ring

   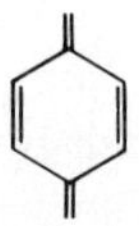

occurs in many dyes such as the indamines, the xanthenes, and the di- and triphenyl methanes. Some of the well-known stains in this group are rosolic acid, fuchsin, the methyl violets, methyl green, crystal violet, and pararosaniline.

Leuco compounds

The chromophores all have unsatisfied affinities and are easily reduced by combining with hydrogen at the double bonds. The nitro group may be reduced to an amino radical; the bond between the nitrogen atoms of the azine group may break and be replaced by two atoms of hydrogen; the double bonds of the quinonoid ring may break and one atom of hydrogen be taken up by each valence set free. A reduction of the chromophore group results in a loss of color. These decolorized stains are known as leuco compounds. Stains often are used as indicators of oxidation and reduction. The decolorization of pararosaniline may be represented by the following equation:

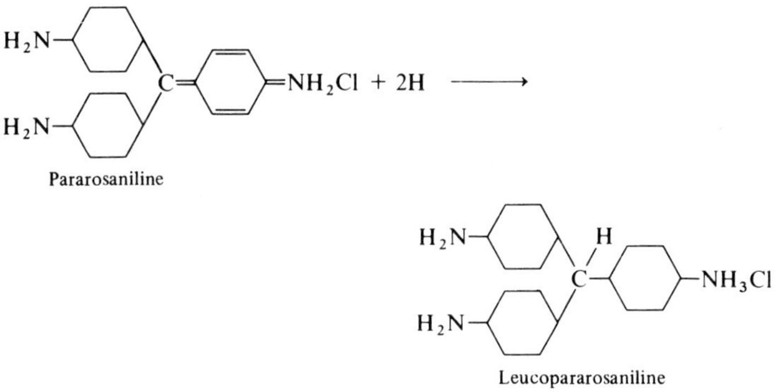

For another type of leuco compound, see page 67 under Feulgen Stain.

Classification of biological stains

The important stains used in bacteriology are given below. Some are acidic; others are basic. The basic stains, for reasons already given, are the most important to the bacteriologist. The stains are classified according to the chromophore groups present.

I. The nitro stains. The chromophore is $-NO_2$. The stains are all acidic.

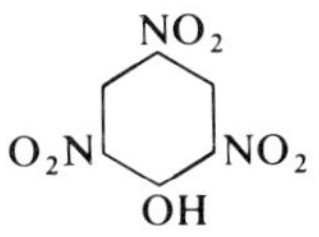

Picric acid

Group includes: *aurantia, Martius yellow, picric acid.*

II. The azo group. The azo group may be subdivided into the (1) monoazo stains and (2) disazo and polyazo stains.

A. Monoazo stains. The chromophore –N=N– joins together benzene or naphthalene rings. In the monoazo stains the group occurs only once:

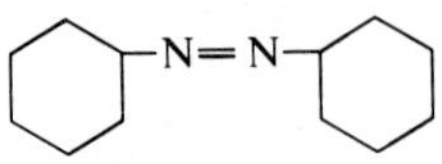

A hydroxyl or amino group on a benzene ring is usually in the para position in relation to the azo group. The azo chromophore is distinctly acid. The addition of OH groups makes the stains acidic; the addition of NH_2 groups makes them basic.

Group includes: *Bordeaux red, brilliant yellow S, chrysoidin Y, fast yellow, Janus green B, methyl orange, methyl red, orange G, orange II, Sudan R, Sudan II.*

B. Disazo and polyazo stains. Sometimes the chromophore occurs more than once in a molecule:

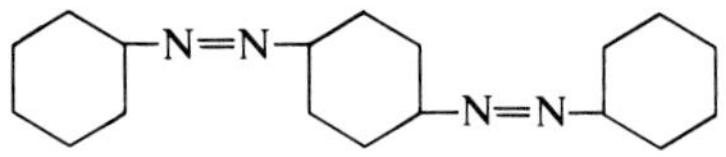

Group includes: *azo blue, Biebrich scarlet* (water soluble), *bismark brown Y, brilliant purpurin R, chlorazol black E, Congo red, Evans blue, ponceau S, Sudan III, Sudan IV, Sudan black B, Sudan red 7B, trypan blue, vital red.*

III. The anthraquinone group. The anthraquinone stains include derivatives of anthracene through its oxidation product anthraquinone:

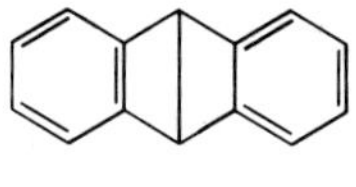

Anthracene

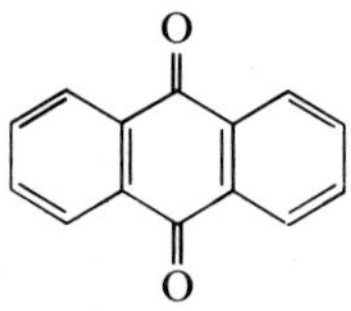

Anthraquinone

The quinonoid ring is the chromophore and anthraquinone the chromogen.

Group includes: *alizarin, alizarin red S, purpurin.*

IV. The thiazole stains. The stains in this group contain the thiazole ring

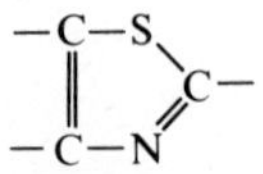

in which the indamine group is the chromophore.

Group includes: *geranine G, primuline, titan yellow G, thioflavine S.*

V. The quinonimine stains. The stains in this group contain two chromophores, the indamine group (–N=) and the quinonoid ring:

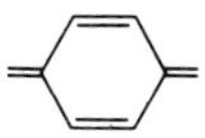

They are derivatives of the theoretical compound para-quinone-di-imine:

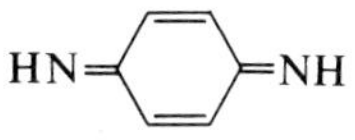

In a typical formula one of the imine hydrogen atoms is replaced by a phenyl group:

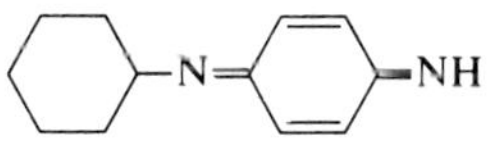

A. The indamines. These stains consist of methylated amino derivatives of indamine. There are no members of any biological importance.
B. The indophenols. The indophenols are closely related to the indamines. The simplest member of this group is indophenol blue or indophenol:

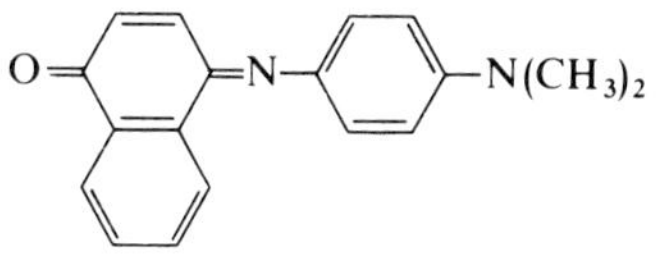

C. The thiazines. The thiazines have a phenyl and a quinonoid ring joined together by an atom of sulfur and one of nitrogen to form a third closed ring:

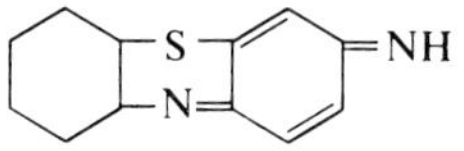

This group contains some of the most important biological stains.

Group includes: *azure A, azure B, methylene azure, methylene blue, methylene green, methylene violet, thionine, toluidine blue O.*

D. The oxazines. In these stains the sulfur of the thiazines is replaced by an atom of oxygen:

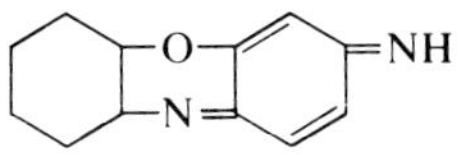

Oxazine

Group includes: *brilliant cresyl blue, cresyl violet acetate, gallocyanin, new blue R, Nile blue sulfate, resazurin.*

E. The azines. The azines are derivatives of phenazine, a compound consisting of two benzene rings, or one benzene ring and one quinonoid ring, joined together through two nitrogen atoms to form a third ring:

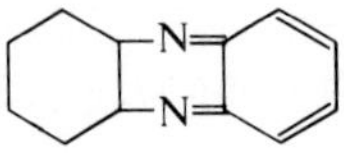

and

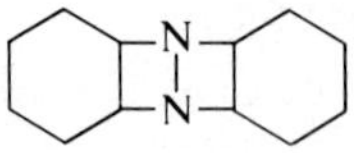

In the first formula the quinonoid ring is the chromophore; in the second formula the azine group=N–N=is the chromophore.

1. The aminoazines. The aminoazines are produced by the introduction of one or more amino groups into the phenazine molecule. They are useful chiefly as indicators.

   Group includes: *neutral red, neutral violet.*

2. The safranines. In the safranines, one of the nitrogen atoms of the azine group is pentavalent and another benzene ring is attached to it.

   Group includes: *amethyst violet, azocarmine G, phenosafranine, safranine O.*

3. The indulines. The indulines are highly phenylated amino derivatives of the safranines.

   Group includes: *induline* (alcohol soluble), *induline* (water soluble), *nigrosine* (water soluble).

VI. The phenylmethane stains. This group includes the most important stains used in bacteriology. The compounds are substituted methanes. One or more hydrogen atoms of methane may be replaced by methyl, ethyl, or phenyl groups. If three hydrogen atoms are replaced by ethyl groups, the compound triethylmethane is formed:

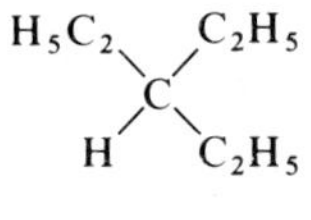

If two hydrogens of methane are replaced by phenyl groups, diphenylmethane is formed:

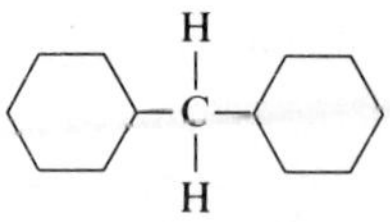

If three hydrogens are replaced, triphenylmethane is produced:

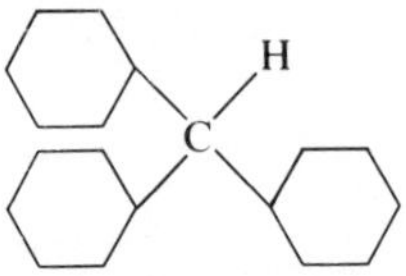

The introduction of amino and other groups, and substituted

amino groups, account for the large number of possible compounds.

A. The diamino triphenylmethane stains. With the exception of the sulfonated derivatives, these are strongly basic dyes. They are derivatives of diamino triphenylmethane:

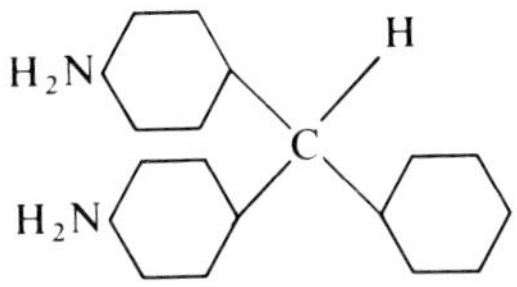

Group includes: *brilliant green, fast green FCF, light green SF* (yellowish), *malachite green.*

B. The triamino triphenylmethane stains. These are also strongly basic stains, except the sulfonated derivatives which are acidic. They are derivatives of triamino triphenylmethane:

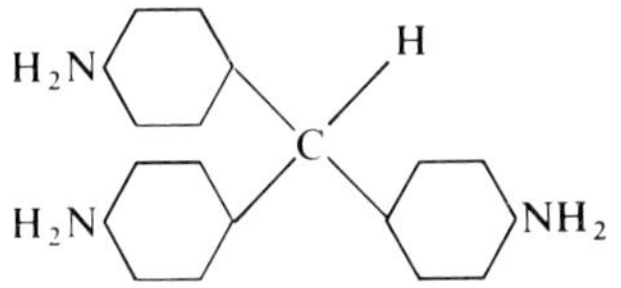

Group includes: *acid fuchsin, aniline blue* (water soluble), *basic fuchsin, crystal violet, ethyl green, ethyl violet, Hofmann's violet, methyl blue, methyl green, methyl violet 2B, new fuchsin, pararosaniline, rosaniline, spirit blue, Victoria blue 4R.*

C. The hydroxy triphenylmethane stains. These are triphenylmethane derivatives in which the amino groups of the rosanilines are replaced with hydroxyl groups giving the stains acidic rather than basic properties:

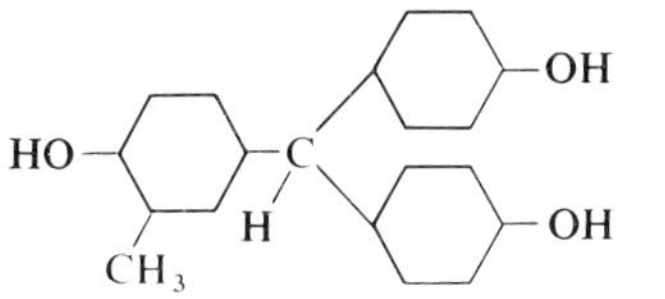

Leucorosolic acid

The members of this group are used chiefly as indicators.

Group includes: *rosolic acid.*

D. Diphenyl-naphthyl methane stains. These are naphthyl derivatives of the diphenylmethane stains.

Group includes: *Victoria blue B, Victoria blue R, night blue, wool green S.*

VII. The xanthene stains. The xanthenes are derivatives of xanthene:

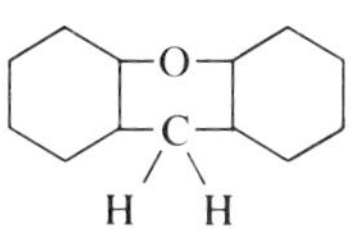

Some of the stains are basic; others are acidic. The most useful indicators employed in bacteriology fall into this group.

A. The pyronine stains. The pyronines are methylated diamino derivatives of xanthene. They are closely related to the diphenylmethane stains. Their formula is similar to the oxazines except that the nitrogen atom of the central ring is replaced by a —CH= radical. Three formulas are possible:

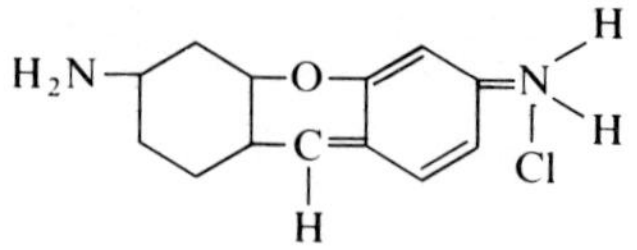

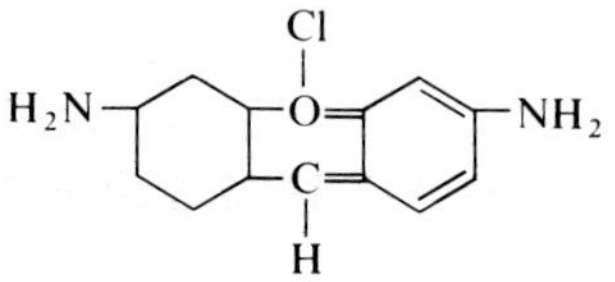

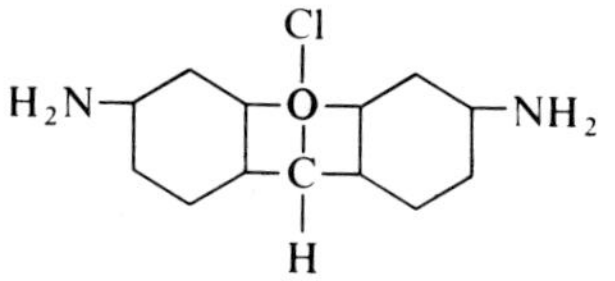

Group includes: *pyronine B, pyronine Y.*

B. The rhodamine stains. The rhodamines are similar to the pyronines except that they have another benzene ring to which is attached a carboxyl group in the ortho position:

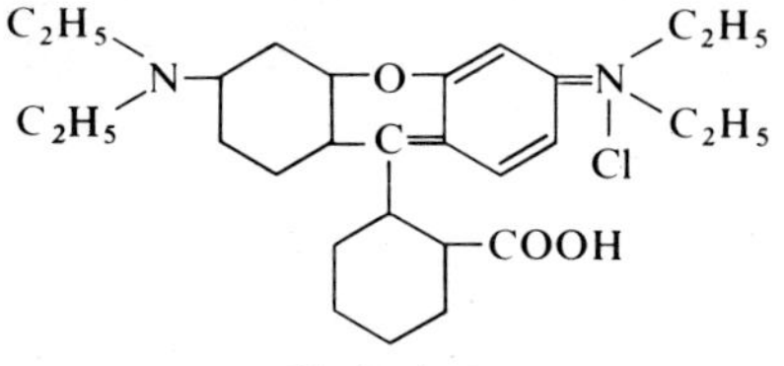

Rhodamine B

Since the stains contain two amino groups to one carboxyl, they are basic in character.

Group includes: *fast acid blue R, rhodamine B.*

C. The fluorane stains. The fluoranes are derivatives of fluorane:

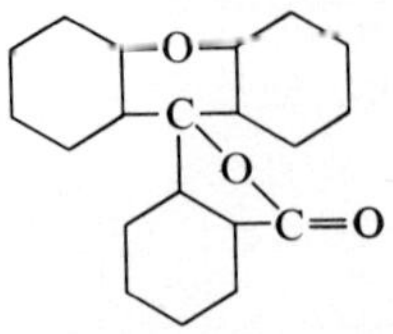

They are sometimes considered as derivatives of fluorescein, a salt of dihydroxyfluorane:

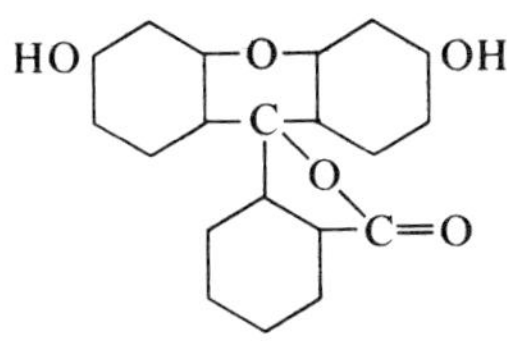

Group includes: *eosine B* (bluish), *eosine Y* (yellowish), *ethyl eosine, erythrosine* (bluish), *erythrosine* (yellowish), *fluorescein, Mercurochrome 220, phloxine B, rose bengale.*

D. The phenolphthalein and the sulfonephthalein stains. A phthalein is a compound of phthalic anhydride,

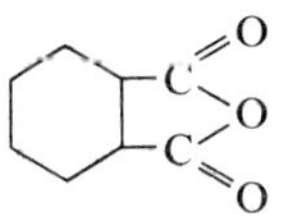

with phenol or a phenol derivative. A phenolphthalein is a compound of phthalic acid with two molecules of phenol:

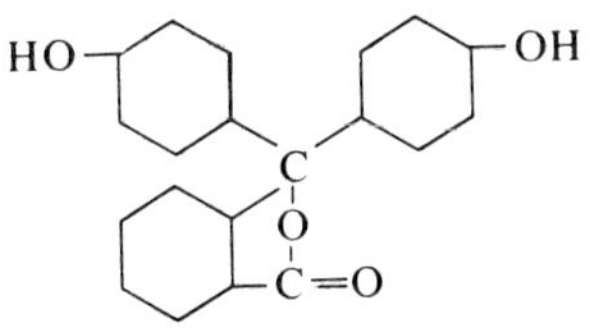

Phenolphthalein

A sulfonephthalein is a compound of orthosulfobenzoic acid,

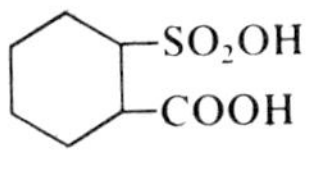

with phenol or a phenol derivative:

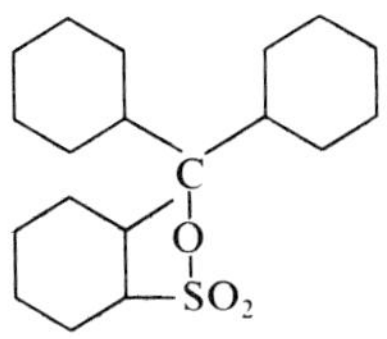

Sulfonephthalein

Group includes: *bromochlorophenol blue, bromocresol green, bromocresol purple, bromophenol blue, bromophenol red, bromothymol blue, chlorocresol green, chlorophenol red, cresolphthalein, cresol red, metacresol purple, phenolphthalein, phenol red, thymol blue.*

E. The acridine stains. The acridines are derivatives of acridine,

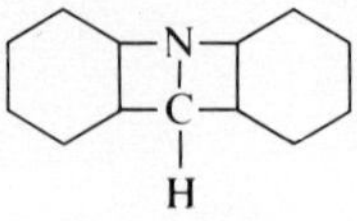

a compound closely related to xanthene. The stains are used chiefly as disinfectants against bacteria and protozoa.

Group includes: *acridine orange NO, acridine yellow, acriflavine, atabrine, neutral acriflavine, phosphine, proflavine, rivanol.*

VIII. The natural stains. Natural stains predominated during the early years of bacteriology, but they have been almost completely replaced by the synthetic stains. The natural stains still found useful are those which have not yet been synthesized. Only a few are of any importance in bacteriology.

A. Indigo. Several species of plants of the genus *Indigofera* contain a glucoside, indican, which on fermentation yields the stain indigo:

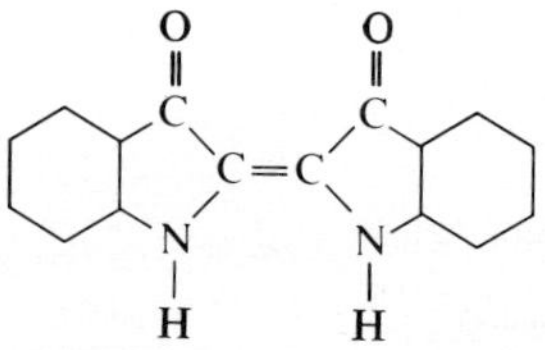

B. Indigo carmine. This stain is the sodium salt of indigo disulfonic acid. It is a blue-colored stain having acidic properties:

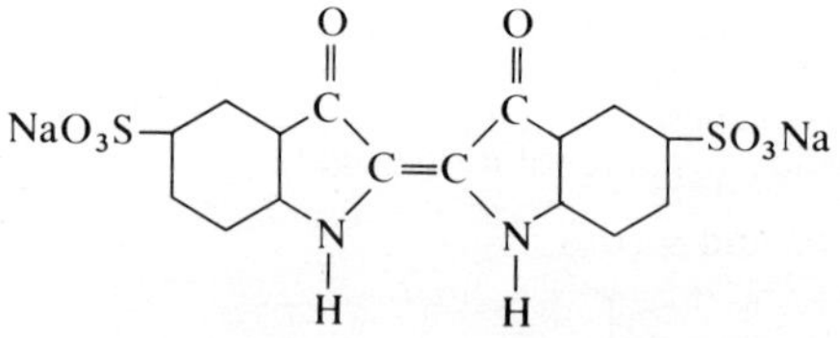

C. Cochineal. This principle is obtained by grinding the dried bodies of the female insect *Coccus cacti* and extracting with water. Cochineal possesses no affinity for tissues and is of limited value when used alone. However, it becomes a useful stain when converted into carmine.

D. Carmine. This stain is prepared by treating cochineal with alum or other metal salts. Carmine is a valuable nuclear stain. Its exact formula is not known.

E. Orcein. The precursor of orcein is obtained from the lichens *Lecanora tinctoria* and *Roccella tinctoria.* The plants contain certain colorless, crystalline, phenolic compounds. One of these

is orcinol,

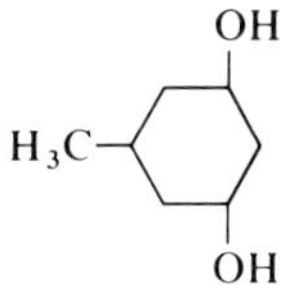

which changes to orcein, a violet-colored compound, on the addition of ammonia and oxygen (air). It is a weakly acidic stain. The formula is not known.

F. Litmus. Litmus is obtained from the same lichens as orcein by treating the plants first with lime and soda, followed by ammonia and air. Litmus is believed to be a mixture of colored compounds, but the primary principle is known as azolitmin. Its exact formula is not known.

G. Brazilin. This principle is obtained by extracting the bark from brazilwood. It is colorless when freshly extracted but becomes oxidized to the red stain brazilein on exposure to air:

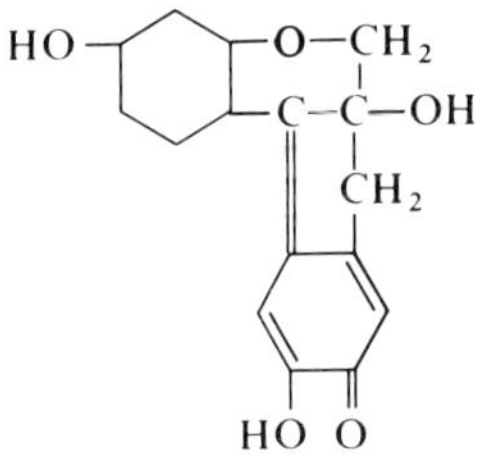

In combination with alum or iron, it is employed as a nuclear stain.

H. Hematoxylin. Hematoxylin is obtained from logwood, a legume growing in South America and other tropical areas. It is prepared by extracting the wood with water, evaporating the extract to dryness, extracting the residue with ether, evaporating the ether extract to dryness, dissolving the residue in water, filtering it, and setting it aside for crystals to separate from the aqueous extract. Like brazilin, hematoxylin is not a stain but on standing in the presence of air it is oxidized to the stain hematein, which is homologous with brazilein. It is believed to have the following formula:

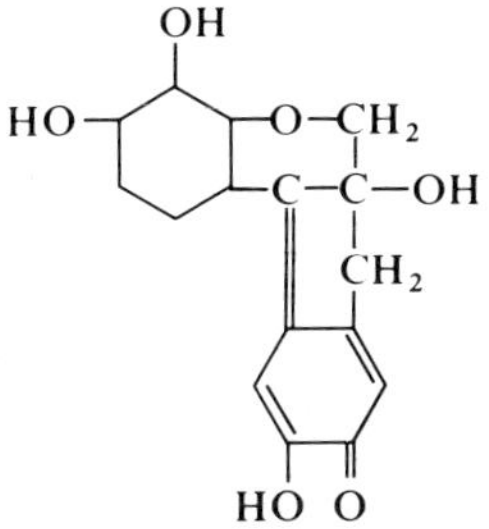

Theories of staining

Many theories have been advanced to explain the phenomenon of staining. All attempt to explain the process on a purely physical or chemical basis.

Physical A physical process may be defined as a reaction between two substances without the formation of a new compound. When bacteria are stained, there is no evidence that the stain has been changed chemically to form a new compound. It is usually possible to extract all or nearly all the stain from the cells by sufficiently long immersion in water, alcohol, or other solvent. The bacterial protoplasm never completely removes all the stain from solution. This is generally believed to be contrary to a chemical reaction, in which a new compound is formed having properties different from either component entering into its formation.

The proponents of the physical theory claim that all staining reactions can be explained on the basis of capillarity, osmosis, adsorption, and absorption. There does not appear to be any general agreement on the amount of weight that should be given to each force, although all authorities agree that each occurs in the staining process.

Chemical Some parts of a cell are acidic in reaction whereas other parts are basic. This fact led chemists to explain staining on a purely chemical basis. The synthetic stains are either anionic (acidic) or cationic (basic); i.e., the color portion of the molecule is either the negative or the positive ion. The proponents of the theory stated that the acidic constituents of the cell (nucleus) reacted with basic stains, and the basic constituents (cytoplasm) reacted with acidic stains. The process is not as simple as this, however, and the theory probably does not explain all the facts.

McCalla and Clark (1941) showed that basic stains were adsorbed at pH values higher than the isoelectric point (page 62), and acidic stains at pH values lower than the isoelectric point. Under normal conditions bacteria possess negative electrical charges and, therefore, attract positively charged particles. In the McCalla and Clark experiments, if bacteria were placed in a solution having a pH value lower than their isoelectric points, they became positively charged (Fig. 28). Under these conditions, they were capable of attracting negatively charged stains such as acid fuchsin.

McCalla and Clark concluded that the reaction of stains with bacteria was an adsorption exchange process reaching stoichiometrical proportions. On the basic side of the isoelectric point, basic stains acted as cations, replacing similarly charged ions from the bacterial system; on the acidic side of the isoelectric point, acidic stains acted as anions, replacing similarly charged ions from the bacterial system. Stains appeared to react with the bacterial cell at the same positions as do inorganic cations and anions.

Summarizing, it may be stated that available evidence seems to point to the fact that staining is neither entirely physical nor entirely chemical but probably a combination of both.

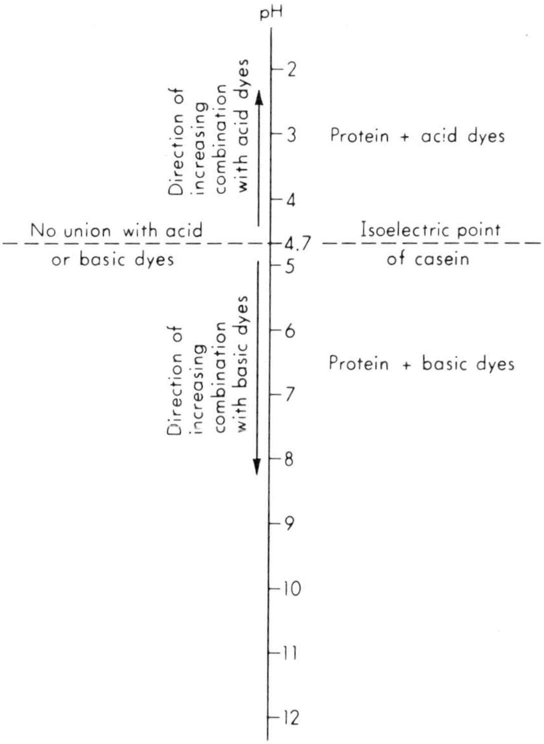

Fig. 28 Combination of acid and basic dyes with casein, an amphoteric compound.

For more information: Finkelstein and Bartholomew (1953); Harris (1951); Lillie (1969).

Staining solutions

Preparations employed for staining bacteria are largely aqueous solutions. In some cases concentrated stock solutions in alcohol are prepared; then the stock solutions are diluted with water as needed. Since alcohol removes stains from cells, pure alcoholic solutions should not be employed.

Staining solutions generally contain low concentrations of stains, rarely exceeding 1 percent. A very dilute staining solution acting for a relatively long period of time will, in general, produce much better results than a more concentrated solution acting for a shorter interval. This is the method followed where it is desired to reveal internal structures in bacteria. In routine practice, concentrated staining solutions generally are used because of the saving in time, but where time is not a serious factor, more dilute preparations may be employed to greater advantage.

Mordant A mordant may be defined as any substance that forms an insoluble compound with a stain and serves to fix the color to the bacterial cell.

Mordants generally are applied first to the bacterial smear, followed by the addition of the staining solution. In some procedures the mordant is added to the staining solution, and the preparation applied to the bacteria in one application.

Compounds which function as mordants are tannic acid and salts of aluminum, iron, tin, zinc, copper, and chromium.

Examples of staining procedures employing mordants are the various flagella stains.

Simple stains Many different kinds of staining solutions are employed in the various bacteriological procedures. Some are for general use; others are designed for special purposes. A simple staining solution contains only one stain dissolved in a solvent. It is applied to the bacteria in one application. The bacteria are given the color characteristic of the staining solution. The purpose of a simple stain generally is to color the bacteria so that they may be more easily seen and to reveal their size and shape. The simple staining solutions commonly employed by the bacteriologist for routine purposes are dilute carbolfuchsin, crystal violet, and methylene blue.

Carbolfuchsin staining solution This is prepared by dissolving about 0.3 percent of basic fuchsin, a triamino triphenylmethane or rosaniline stain, in a 5 percent solution of phenol. For use as a simple stain, it is advisable to dilute the solution about 10 times with distilled water.

For information on the composition of the various fuchsins of commerce, see pages 66 and 67.

Crystal violet staining solution Crystal violet also is a member of the triamino triphenylmethane stains. Chemically it is hexamethyl pararosaniline (see page 61). Other names for the stain are methyl violet 10B, gentian violet, hexamethyl violet, and violet C, G, or 7B. It produces the deepest shade of the pararosanilines and is considered the most satisfactory of all the violet compounds as a simple bacterial stain.

Methylene blue staining solution Chemically, methylene blue is tetramethyl thionine chloride, a basic stain, having the following formula:

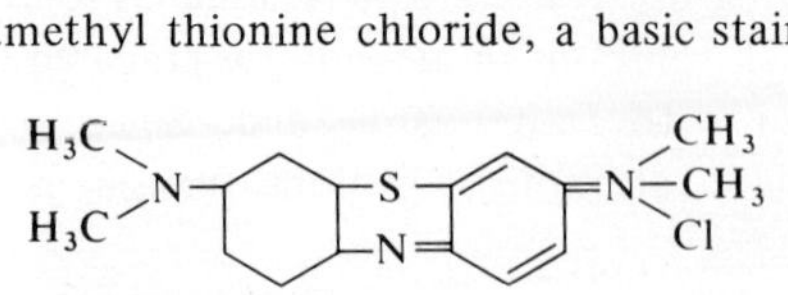

Methylene blue is used more than any other stain in bacteriology. Because of its strongly basic nature, it stains nuclei and nucleic acid granules very intensely. It is very useful in making a rapid survey of the bacterial population of milk (page 719). Methylene blue is usually preferred in staining smears for the diagnosis of diphtheria. It is combined with eosin for staining blood films. The stain is incorporated with eosin in a lactose agar base for distinguishing typical *Escherichia coli* from typical

Aerobacter aerogenes (page 693). These are only a few of its many uses in bacteriology.

Differential staining solutions Differential staining procedures utilize more than one stain. In some techniques the stains are applied separately; in others they are mixed and applied in one solution. The two most important differential stains used by the bacteriologist are the Gram stain and the acid-fast stain.

Gram stain In this method of staining, the bacterial film is covered with a solution of one of the methyl violet stains and allowed to act for a definite period of time. The stain is washed off with water, and a dilute solution of iodine is added and allowed to remain for the same period of time. Next the slide is treated with alcohol or a mixture of alcohol and acetone until almost all the color is removed from the film. Finally a counterstain such as safranine, dilute carbolfuchsin, bismarck brown, or pyronine B is applied.

Some organisms retain the violet stain even after treatment with the decolorizing agent, and the color is not modified by the application of the counterstain. Other organisms readily lose the primary stain and take the counterstain. Those organisms that retain the first stain are called gram-positive; those that fail to retain the primary stain but take the counterstain are called gram-negative. Organisms may be placed into one group or the other on the basis of the Gram stain.

The pararosaniline stains give the best results in the Gram technique. The two most important members are methyl violet and crystal violet.

Methyl violet is a mixture of the more highly methylated fuchsins, consisting principally of the chloride of pentamethyl pararosaniline. Commercially it is designated methyl violet 2B. Crystal violet is the chloride of hexamethyl pararosaniline. Their structural formulas are as follows:

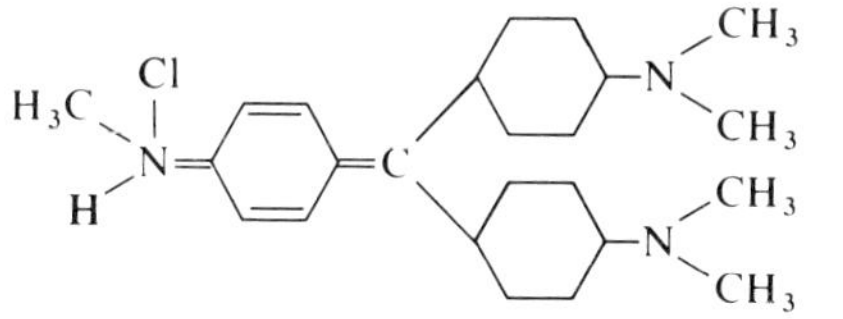

Pentamethyl pararosaniline

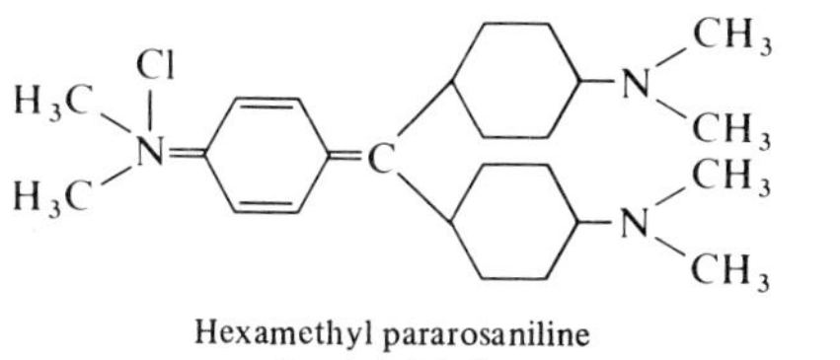

Hexamethyl pararosaniline
(crystal violet)

The shade of color of pararosaniline is deepened by increasing the number of methyl groups in the molecule. Hence, hexamethyl pararosaniline is deeper in shade than the pentamethyl compound. The designations methyl violet 3R, 2R, R, B, 2B, 3B, etc., refer to the number of methyl groups in the molecule. The letter R denotes the red shades and B, the blue shades. Hexamethyl pararosaniline (crystal violet) contains six methyl groups and is considered the most satisfactory primary stain in the Gram technique.

The ability of some organisms to retain the primary stain is not characteristic of all living cells but is confined almost entirely to the bacteria and yeasts. The cells of higher plants and animals do not retain the primary stain. Molds stain somewhat irregularly; granules present in the mycelia tend to retain the stain. The Gram reaction is not a hard-and-fast one. It may vary with the age of the culture, with the pH of the medium, with the choice of stains, with the viability of the organisms, and perhaps in other ways.

Dead gram-positive organisms are said to stain gram-negatively. Gram-negative cells usually are observed in a smear of a gram-positive organism, the numbers increasing with the age of the culture.

The common assumption seems to be that gram-positive organisms give the strongest reaction when they are very young, with a tendency to become negative as they grow older. This assumption appears to be correct in examining the results on spherical bacteria but not necessarily true on the rod-shaped cells. Hucker (1923, 1927) examined a large number of rod-shaped species from the soil and found that a greater number of positive reactions was obtained on the fourth and seventh days. The results showed such great variation that he was unable to determine a definite time when an organism gained or lost its power to retain the violet stain.

Gianni (1952) found that the gram-positive organisms *Bacillus subtilis* and *B. anthracis* stained gram-negatively when cultures were 2 to 3 hr old; then a substance developed inside the cell wall to reverse the reaction.

A number of theories have been advanced to explain the mechanism of the Gram stain. Stearn and Stearn (1923, 1924*a*, *b*) based their theory on a chemical combination between stain and bacterial protein. Proteins and amino acids are amphoteric compounds, i.e., have the power to react with both acids and bases by virtue of their amino (NH_2) and carboxyl (COOH) groups. In acid solutions they react with acids; in alkaline solutions they react with bases.

Isoelectric point According to the classical theory, the isoelectric point is defined as that pH where an amphoteric compound shows the least amount of dissociation. In other words, the maximum amount of the compound is present in the un-ionized or molecular state.

Opposed to this theory is the newer concept which defines the isoelectric point as that pH where the acidic and basic groups of the amphoteric compound are completely ionized (page 268). On the acidic side of the isoelectric point, the compound behaves as a base; on the basic side, it behaves as an acid.

Acidic and basic stains also combine with proteins. The acidic stains react on the acidic side of the isoelectric points; the basic stains react on the basic side. The amount of combination in either case is proportional to the degree of acidity or alkalinity of the solutions. At the isoelectric points, proteins do not combine with either acidic or basic stains. Using the protein casein as an example, the action of acidic and basic stains may be represented as shown in Fig. 28.

Stearn and Stearn found that the staining reactions of bacteria were due largely to their protein content. Bacteria behaved as amphoteric compounds, combining with acidic stains in acid solutions and with basic stains in basic solutions. Combination did not occur at the isoelectric point. Since organisms contain more than one protein, the isoelectric point becomes an isoelectric range or zone. The gram-positive organisms exhibited an isoelectric range at a lower pH than the gram-negative bacteria.

On the basis of their data, Stearn and Stearn concluded as follows:

1. Gram-positive organisms can be rendered gram-negative by increasing acidity.
2. Gram-negative organisms can be rendered gram-positive by increasing alkalinity.
3. Acidic stain-positive organisms can be rendered gram-negative by increasing alkalinity.
4. Basic stain-positive organisms can be rendered gram-negative by increasing acidity.
5. At the isoelectric range, there is little tendency for any stain to be retained. This range is characteristic of each species.
6. There appears to be good evidence that the proteins of bacteria are not simple proteins but a loose combination of proteins with lipoidal or fatty substances. An example of such a substance is lecithin, and a combination of lecithin and protein is known as lecithoprotein.
7. The lipoidal material extracted from gram-positive organisms differs from that extracted from gram-negative organisms in that the former contains a much larger proportion of unsaturated acids that have a great affinity for oxidizing agents. All mordants (such as iodine) used in the Gram stain are oxidizing agents. Their effect is in general to render the substance oxidized more acid in character. This increases the affinity of an organism for basic stains.
8. The change of Gram character with age is especially true of those organisms which are only weakly gram-positive and are cultivated in media containing fermentable substances that become acid in reaction as growth proceeds.

For more information: Harden and Harris (1953), Kennedy and Woodhour (1956).

Another approach to the explanation of the Gram reaction depends upon evidence suggesting the existence of an outer layer surrounding a gram-negative core.

Henry and Stacey (1943) treated heat-killed gram-positive organisms with a 2 percent bile solution and were able to dissolve away an outer surface layer. The gram-positive organisms so treated became gram-negative. The extracted material was composed of polysaccharides, protein, and the magnesium salt of ribonucleic acid. The extracted material could be "replated" back on the gram-negative "skeleton" forms, restoring their Gram reaction, provided the skeleton was maintained in a suitable state of reduction by the presence of formalin. Neither the skeleton nor the extracted material was gram-positive except in combination, and recombination of these materials was not possible unless the skeleton was reduced by the presence of formalin.

Bartholomew and Umbreit (1944) used a 2 percent solution of sodium choleate (purified bile extract) instead of whole bile and obtained similar results. In addition, they found that purified ribonuclease (page 449) was also capable of destroying the gram-positive character of heat-killed cells.

Baker and Bloom (1948) found that gram-positiveness could be imposed on normal gram-negative *E. coli* by the addition of a highly viscous deoxyribonucleic acid. The gram-positiveness could be removed by washing the treated cells with distilled water but not with 0.85 percent sodium chloride solution. Deoxyribonucleic acid is soluble in distilled water but not in saline.

It appears that both ribonucleic and deoxyribonucleic acids are capable of imposing gram-positiveness on gram-negative bacteria and that both types might be present in the outer shells of gram-positive bacteria (page 844). Neither type of acid has been shown to be present in the outer shells of gram-negative bacteria.

Libenson and McIlroy (1955) and Gerhardt et al. (1956) reported that if the gram-positive reaction were dependent upon the formation of a complex combination between the components of the Gram stain and the cell-wall proteins, it should be expected that bacteria disintegrated by physical means would still retain the Gram stain since such treatment could not change the chemical nature of the cell-wall constituents. On the contrary, disintegrated gram-positive organisms lost the ability to retain the primary color and stained gram-negatively. The investigators supported the view that the permeability of the cell wall to crystal violet, the low solubility of the stain-iodine complex in alcohol and acetone, and the free access of the solvent to the complex constituted the main factors involved in the mechanism of the Gram stain.

Chelton and Jones (1959) found untenable the hypothesis that the gram-positive reaction was due to the inability of the stain-iodine complex to diffuse through the cell membranes. They employed a modification of the Gram technique and found that distintegrated *Clostridium perfringens* stained in part gram-positive. Using the same procedure, disintegrated gram-negative bacteria always stained gram-negatively. They concluded that the gram-staining reaction was due in part to a specific staining component and in part to a factor dependent upon the intact nature of the cells. The latter might be associated with the stability or orientation of the staining component or with some membrane effect.

Barbaro et al. (1956) noted that the degree of gram-positivity could be altered by chemical treatment of the cells before staining. The gram-negative *E. coli* increased in gram-positivity after treatment with crystal violet, sodium bisulfite, or sodium hydroxide. The gram-positive *Staphylococcus aureus* decreased in gram-positivity after treatment with picric acid. The omission of Gram's iodine from the staining procedure in converted *E. coli* resulted in complete removal of crystal violet by decolorization. Microscopically, crystal violet was still retained when iodine was omitted from the procedure. The counterstain, safranine, was capable of replacing or masking the residential crystal violet.

The cell walls of microorganisms have little affinity for stains. However, the walls of gram-positive species can be stained with certain stains such as basic fuchsin and the methyl violets. On the other hand, the cell walls of gram-negative bacteria are not colored by the above stains.

Lamanna and Mallette (1954) reported that the specificity of the triphenyl methanes as primary stains in the Gram technique rested on the ability of solutions of these stains to color the cell walls of gram-positive bacteria. The cell walls of gram-negative bacteria were unable to sorb the primary basic stain.

Basu et al. (1968*a*) found that crystal violet could be extracted completely from gram-stained *S. aureus* by 95 percent ethanol if the stained bacteria were pretreated with dilute sodium thiosulfate solution. Thiosulfate removed the iodine from the cell component–stain–iodine complex instantly and rendered the stain extractable by the differentiating medium. The findings indicated that the stability of the cell component–stain–iodine complex determined the Gram character of the cell.

In a later communication (1968*b*), Basu et al. reported that gram-negative bacteria contained more lipides than gram-positive cells. Extraction of the lipides from *E. coli* induced the cells to become gram-positive. They concluded that the more lipides present in the cell wall or cell membranes, the less retention of the stain was exhibited by the bacteria.

About the only conclusion that can be drawn from the above reports is that no single theory gives a satisfactory interpretation of all of the known facts of the Gram technique. More work is necessary for a proper explanation of the mechanism of action of this very important differential stain.

Some characteristics of gram-positive and gram-negative bacteria are given in Table 3.

For more information: Bartholomew (1962); Bartholomew and Cromwell (1965); Bartholomew et al. (1965); Bartholomew and Finkelstein (1958); Fischer and Larose (1952); Gianni et al. (1954); Lamanna and Mallette (1964); Mastroeni and Contadini (1966); Tucker and Bartholomew (1962).

Acid-fast stain The great majority of bacteria are easily stained by the usual simple procedures. However, there are exceptions. Some bacteria are surrounded by a covering composed of fatty or waxy substances. These organisms are not readily stained, but when once stained are able to retain the color even after treatment with drastic decolorizing agents. They are called acid-fast because the stained bacteria are resistant to decolorization with acid alcohol. Two well-known members of this group are *Mycobacterium tuberculosis*, the etiological agent of tuberculosis in man, and *M. leprae*, believed to be the cause of leprosy or Hansen's disease.

In the Ziehl-Neelsen method, the primary stain is carbolfuchsin (basic fuchsin dissolved in a dilute solution of carbolic acid or phenol). This is applied to a smear, and then the slide is steamed for 3 to 5 min. More stain is applied as evaporation occurs. The slide is washed in water and the film decolorized with acid alcohol until only a suggestion of pink remains. The slide is again washed in water and the film counterstained with methylene blue. The acid-fast organisms stain deep red; everything else on the slide stains blue.

The basic fuchsin of commerce is a mixture of pararosaniline, rosaniline, and magenta II. Another stain, new fuchsin (magenta III), may be purchased in pure form and is frequently employed in the acid-fast staining method. The formulas of these stains are as follows:

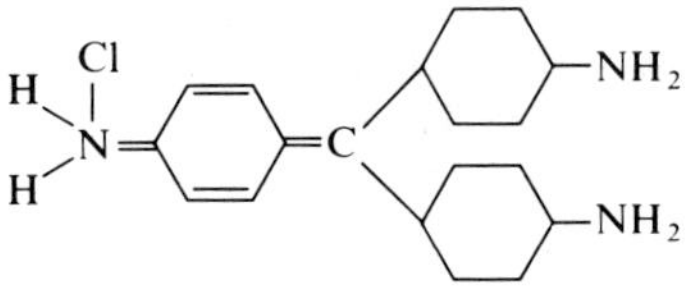

Pararosaniline

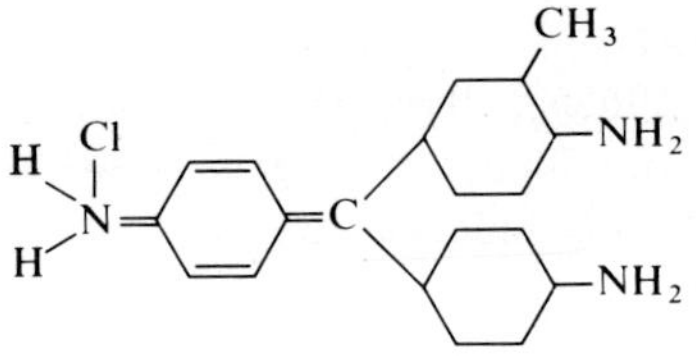

Rosaniline

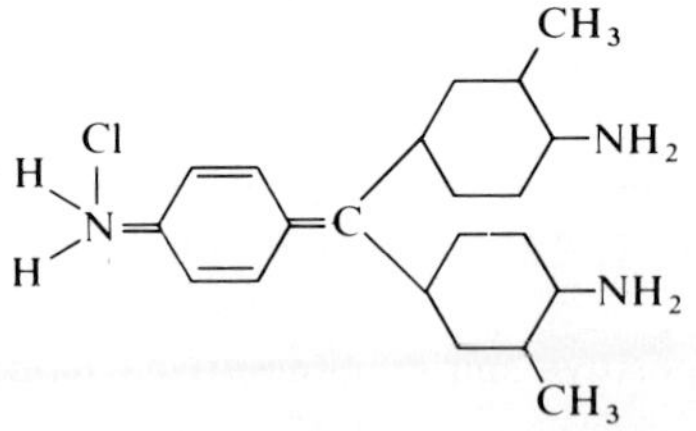

Magenta II

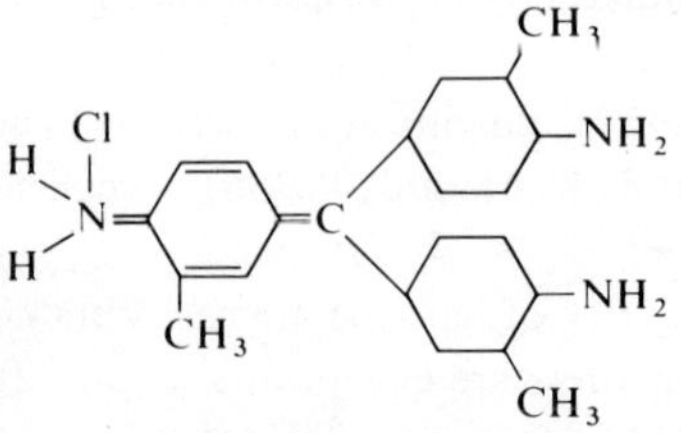

New fuchsin

Table 3

Gram-positive bacteria	*Gram-negative bacteria*
Contain magnesium ribonucleate	Do not contain magnesium ribonucleate
Very sensitive to triphenylmethane dyes	Less sensitive to triphenylmethane dyes
Sensitive to penicillin	Sensitive to streptomycin
Resistant to alkalies; not dissolved by 1 percent KOH	Sensitive to alkalies; dissolved by 1 percent KOH
Isoelectric range pH 2.5–4	Isoelectric range pH 4.5–5.5
Usually cocci or spore-forming rods (exceptions, *Lactobacillus, Corynebacterium*)	Usually non-spore-forming rods (exception, *Neisseria*, which are cocci)
May be acid-fast	Probably never acid-fast
Contain a lower content of lipides than gram-negative bacteria, ranging from 4.8 percent in *S. citreus* to 8.9 percent in *S. aureus*	Contain a higher content of lipides than gram-positive bacteria, ranging from 13.2 percent in *E. coli* to 27.7 percent in *K. pneumoniae*

The more methyl groups present in the molecule, the deeper will be the shade of red. New fuchsin with three methyl groups is the deepest in shade and pararosaniline with no methyl groups the lightest in shade of the above compounds. Generally, the deeper the shade of stain used in the acid-fast stain, the better will be the degree of differentiation.

M. tuberculosis can be made to stain solid or beaded by varying the staining procedure. The presence of beads depends to a large extent on how the staining reaction is carried out. Addition of small amounts of electrolytes to the stain solution increases the number of beads. Washing the stained smears with alcohol, after decolorization with acid alcohol, removes practically all beads and leaves most cells evenly stained with a pink tinge.

The composition of the phenylmethane stain also affects the appearance of the stained organisms. Smears stained with rosaniline or pararosaniline acetate show 50 to 100 percent beaded organisms; smears stained with rosaniline or pararosaniline chloride show solidly stained rods. For best results, the chloride salts should be used.

Woodhour (1956) found virulent *M. tuberculosis* to be more strongly acid-fast than nonvirulent forms. The data indicated the existence of quantitative degrees of acid-fastness whether stained with carbolfuchsin or crystal violet. The virulent strains could be distinguished from the nonvirulent by the amount of stain taken up by the cells.

Feulgen stain Chromosomes or chromatinic bodies are composed of deoxyribonucleoprotein, a compound of deoxyribonucleic acid (DNA) and a basic protein.

The Feulgen reaction is almost a specific test for DNA. It employs acid hydrolysis of the DNA with the liberation of aldehyde groups. The aldehyde groups form an addition compound with the Feulgen reagent to give a deep color reaction. On the other hand, ribonucleic acid (RNA) treated in the above manner does not give a color reaction. It is generally believed that a positive Feulgen reaction indicates the presence of nuclear material in cells.

The Feulgen stain is prepared by decolorizing basic fuchsin with sulfurous acid. The decolorized stain is then known as Schiff's reagent. Feulgen and Rossenbeck (1924) were the first to apply Schiff's reagent to tissue cells to detect the presence of aldehyde-like substances in nuclei. A positive Feulgen reaction is indicated by the reappearance of the chromatic form of the stain. However, the restored stain is not the same as the original compound, being violet rather than red in color. The Feulgen technique reveals certain structures in bacteria which are not stained in the usual methods of staining.

Several modifications of the original Feulgen stain have been recommended, one being as follows:

Dissolve 0.5 gm basic fuchsin in 100 ml boiling distilled water; cool to 50°C; filter, and add 10 ml 1 *N* HCl to the filtrate; add 0.5 g potassium metabisulfite ($K_2S_2O_5$) and allow solution to stand overnight in the dark; if solution is not completely decolorized, add 500 mg charcoal, shake for 2 min, and filter.

DeLamater (1948) postulated the following linkage between Feulgen's reagent and DNA:

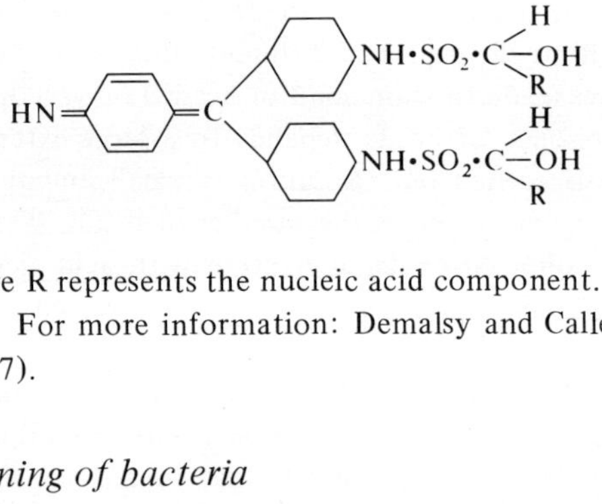

where R represents the nucleic acid component.

For more information: Demalsy and Callebaut (1967), Pérez-Zapata (1967).

Staining of bacteria

Preparation of smears To obtain good stained smears for microscopic examination, it is necessary to take certain well-recognized precautions in their preparation.

The slides must be free from grease. A drop of water should spread evenly over the surface without forming globules. Clean slides should be handled by grasping only at their edges to prevent redeposition of grease by the fingers.

It is frequently difficult to prepare smears from liquid cultures. Lipoids present in such cultures prevent the fixing of the organisms to the slides. It is better to use a solid culture and to emulsify some of the surface growth in a droplet of neutral distilled water on the slide. In this way excessive lipoidal material is avoided and the organisms become more firmly fixed to the slide.

For best results, the distilled water should be neutral or very slightly alkaline in reaction. Freshly distilled water is most likely satisfactory but on standing it absorbs CO_2 from the air and becomes acid in reaction. An acid reaction interferes in many staining procedures. Acid water may be corrected by boiling to expel CO_2 or by adding a small amount of a neutral phosphate buffer.

A bacterial smear is prepared by removing a loopful of a liquid culture with a sterile wire loop and spreading it on a glass slide over an area of about 15 x 30 mm. If a solid culture is used, a minute amount of the growth is emulsified in a droplet of neutral distilled water, previously placed in the center of the slide, and spread out over an area of about 15 x 30 mm. The smear is carefully dried by holding the slide high over a low gas flame to avoid steaming. The dried smear is fixed by quickly passing the slide five or six times through the upper portion of the bunsen flame. This should prevent the film from washing away during the staining process. The dried and fixed smear is then ready to be stained.

Viability of fixed and stained organisms Dried and heat-fixed smears of bacteria, even after being stained by one of the simple staining procedures, may still contain viable cells. With the possible exception of some of the spore-forming organisms, vegetative cells fail to survive after being treated by the more drastic Gram procedure of staining. To be on the safe side, care should be observed in the handling of stained preparations of pathogenic organisms to prevent the possibility of laboratory infections.

For more information: Committee on Bacteriological Technic (1957); Conn et al. (1960); Gurr (1960, 1965); Lillie (1969).

References

Baker, H., and W. L. Bloom: Further studies on the Gram stain, *J. Bact.*, **56**:387, 1948.

Barbaro, J. F., et al.: II. The effect of the fixative and of chemical treatment of fixed bacteria on the adsorption and retention of dye, *J. Bact.*, **72**:451, 1956.

Bartholomew, J. W.: Variables influencing results, and the precise definition of steps in Gram staining as a means of standardizing the results obtained, *Stain Technol.*, **37**:139, 1962.

——, and T. Cromwell: Relative contribution of the cell wall, cytoplasmic membrane, and cytoplasm to the gram-positive characteristic of *Bacillus megaterium*, *J. Bact.*, **90**:643, 1965.

—— et al.: Analysis of the mechanism of Gram differentiation by use of a filter-paper chromatographic technique, *J. Bact.*, **90**:766, 1965.

—— and H. Finkelstein: Relationship of cell wall staining to Gram differentiation, *J. Bact.*, **75**:77, 1958.

—— and W. W. Umbreit: Ribonucleic acid and the Gram stain, *J. Bact.*, **48**:567, 1944.

Basu, P. S., et al.: Extraction of crystal violet-iodine complex from gram-positive bacteria by different solvents and its implication on Gram differentiation, *Histochemie,* **14**:221, 1968*a*.

—— Gram positivity induced in *Escherichia coli* by chloroform-methanol treatment, *Histochemie,* **16**:150, 1968*b*.

Chelton, E. T. J., and A. S. Jones: The gram-staining reaction of disintegrated microorganisms, *J. Gen. Microbiol.,* **21**:652, 1959.

Committee on Bacteriological Technic (American Society for Microbiology): "Manual of Microbiological Methods," New York, McGraw-Hill Book Company, 1957.

Conn, H. J., et al.: "Staining Procedures," Baltimore, The Williams & Wilkins Company, 1960.

DeLamater, E. D.: Basic fuchsin as a nuclear stain, *Stain Technol.,* **23**:161, 1948.

Demalsy, P., and M. Callebaut: Plain water as a rinsing agent preferable to sulfurous acid after the Feulgen nucleal reaction, *Stain Technol.,* **42**:133, 1967.

Feulgen, R., and H. Rossenbeck: Mikroskopisch-chemischer Nachweis einer Nucleinsaure vom Typus der Thymonucleinsaure und die darauf beruhende elektive Farbung von Zellkernen in mikroskopischen Praparaten, *Z. Physik. Chem.,* **135**:203, 1924.

Finkelstein, H., and J. W. Bartholomew: Quantitative determination of dye uptake by bacterial cells, *Stain Technol.,* **28**:177, 1953.

Fischer, R., and P. Larose: Mechanism of Gram stain reversal, *J. Bact.,* **64**:435, 1952.

Gerhardt, P., et al.: Gram reaction of isolated protoplasts and surface membranes of *Bacillus megaterium, J. Bact.,* **72**:721, 1956.

Gianni, A.: Sulle fasi di sviluppo della gram-positività nei germi, *Boll. Ist. Sieroterap. Milan.,* **31**:427, 1952.

—— et al.: Il polipeptide dell' acido glutamico estraibile dal *B. subtilis* e suo rapporto con la gram-positività, *Bol 1st Sieroterapico Milanese,* **33**:129, 1954.

Gurr, E.: "Encyclopaedia of Microscopic Stains," Baltimore, The Williams & Wilkins Company, 1960.

——: "The Rational Use of Dyes in Biology," London, Leonard Hill, 1965.

Harden, V. P., and J. O. Harris: The isoelectric point of bacterial cells, *J. Bact.,* **65**:198, 1953.

Harris, J. O.: A study of the relationship between the surface charge and the adsorption of acid dyes by bacterial cells, *J. Bact.,* **61**:649, 1951.

Henry, H., and M. Stacey: Histochemistry of the gram-staining reaction for microorganisms, *Nature,* **151**:671, 1943.

Hucker, G. J.: Methods of Gram staining, *N.Y. Agr. Exp. Sta. Tech. Bull.,* **93**, 1923.

——: Further studies on the methods of Gram staining, *N.Y. Agr. Exp. Sta. Tech. Bull.,* **128**, 1927.

Kennedy, E. R., and A. F. Woodhour: I. The effect of pH on the quantity of dye retained by bacteria and the apparent isoelectric point, *J. Bact.,* **72**:447, 1956.

Lamanna, C., and M. F. Mallette: The cytological basis for the role of the primary dye in the Gram stain, *J. Bact.*, **68**:509, 1954.

—— and ——: Chromatographic analysis of the state of association of the dye-iodine complex in decolorization solvents of the Gram stain, *J. Bact.*, **87**:965, 1964.

Libenson, L., and A. P. McIlroy: On the mechanism of the Gram stain, *J. Infectious Diseases*, **97**:22, 1955.

Lillie, R. D.: "H. J. Conn's Biological Stains," Baltimore, The Williams & Wilkins Company, 1969.

Mastroeni, P., and V. Contadini: The relationship between lipid and taxonomic classification of microorganisms. I. Difference between gram-positive and gram-negative bacteria, *Giorn. Microbiol.*, **14**:99, 1966.

McCalla, T. M., and F. E. Clark: Dye adsorption by bacteria at varying H-ion concentrations, *Stain Technol.*, **16**:95, 1941.

Pérez-Zapata, A. J.: Técnica para la coloración de material nuclear en bacterias, *Rev. Latinoamer. Microbiol. Parasitol.*, **9**:47, 1967.

Stearn, E. W., and A. E. Stearn: The mechanical behavior of dyes, especially gentian violet, in bacteriological media, *J. Bact.*, **8**:567, 1923.

—— and ——: The chemical mechanism of bacterial behavior. I. Behavior toward dyes – factors controlling the Gram reaction, *J. Bact.*, **9**:463, 1924*a*; II. A new theory of the Gram reaction, *ibid.*, **9**:479, 1924*b*.

Tucker, F. L., and J. W. Bartholomew: Variations in the Gram staining results caused by air moisture, *Stain Technol.*, **37**:157, 1962.

Woodhour, A. F.: III. A quantitative acid-fast stain, *J. Wash. Acad. Sci.*, **46**:344, 1956.

4 *Morphology of bacteria*

Bacteria are placed in the class of organisms known as the *Schizomycetes* (*schizo*, fission, and *mycetes*, fungi). They are single-celled organisms and reproduce normally by transverse or binary fission (page 509).

The class Schizomycetes is divided into ten orders. The largest order, the *Eubacteriales*, contains most of the common bacterial species. The other orders are: *Pseudomonadales, Chlamydobacteriales, Hyphomicrobiales, Caryophanales, Actinomycetales, Beggiatoales, Myxobacterales, Spirochaetales,* and *Mycoplasmatales.*

Bacteria are typically unicellular plants, the cells being usually small, sometimes ultramicroscopic. They are frequently motile. Using modern cytological techniques, a true nucleus has been demonstrated in bacteria. Cells may be spherical or straight, curved, or spiral rods. They may occur in regular or irregular masses, or even in cysts. Where they remain attached to each other after cell division, they may form chains or even definite trichomes. The latter may show some differentiation into holdfast cells and into motile or nonmotile reproductive cells. Some grow as branching mycelial threads whose diameter is not greater than that of ordinary bacterial cells, i.e., about 1 μ. Some species produce pigments. The true purple and green bacteria possess photosynthetic pigments much like or related to the true chlorophylls of higher plants. The phycocyanin found in the blue-green algae does not occur in the Schizomycetes. Multiplication is typically by cell division. Endospores are formed by some species of Eubacteriales. Sporocysts are found in Myxobacterales. Bacteria are free-living, saprophytic, parasitic, or even pathogenic. The latter types cause diseases of either plants or animals.

Shape of bacteria Bacteria exhibit three fundamental shapes: (1) spherical, (2) rod, and (3) spiral or curved rod. All bacteria exhibit pleomorphism in more or less degree under normal or other conditions, but a bacterial species is still generally associated with a definite cell form when grown on a standard medium under controlled conditions.

The spherical bacteria (singular, coccus; plural, cocci) divide in one, two, or three planes. Division in one plane produces single cells, pairs, or chains; division in two planes produces clusters; and division in three planes produces packets. Some are apparently perfect spheres; others are slightly elongated or ellipsoidal in shape.

The streptococci divide in only one plane. They grow normally in pairs or chains. Depending upon the species, the distal ends of each pair may be lancet-shaped or flattened at the adjacent sides to resemble a coffee bean.

The staphylococci divide in two planes, producing pairs, tetrads, or clusters of bacteria, the latter resembling bunches of grapes.

The sarcinae divide in three planes, producing regular packets. These are cubical masses with one layer of bacteria atop another.

The rod forms also show considerable variation. A rod is usually considered to be a cylinder with the ends more or less rounded. Some rod forms are definitely ellipsoidal in shape. The ends of the rods are not the same in all species. Some are markedly rounded; others exhibit flat ends perpendicular to the sides. Gradations between these two extremes may be seen.

Rods may show marked variation in their length/width ratio. Some rods are very long in comparison to their width; others are so short they may be confused with the spherical forms.

The shape and other characteristics of bacteria depend to a large extent upon certain environmental factors such as temperature of incubation, age of the culture, concentration of substrate, and composition of the medium. As these change, the morphological and tinctorial properties may show some changes. Bacteria usually exhibit their characteristic morphology in young cultures and on media possessing favorable conditions for growth. The classical textbook pictures of bacteria are usually made on 24-hr cultures.

Some workers have described changes in morphology as orderly stages in the life cycles of bacteria. However, there is little evidence that bacteria pass through a series of well-defined stages, but rather such forms result from changes in the media or environments in which they are growing.

Young cells are, in general, larger than old organisms of the same species. As a culture ages, the cells become progressively larger until a maximum is reached, after which the reverse effect occurs. Bacterial variations resulting from changes in age are only temporary; the original forms reappear when the organisms are transferred to fresh medium.

Variations in shape, Brzin (1963) observed striking changes in the size and shape of the rod-shaped organism *Bacterium anitratum* under the influence of temperature. Incubation at room temperature produced small, single-celled rods; at 37°C, cell division was inhibited and the bacteria became considerably larger with fusiform and club-shaped forms appearing in the culture.

Felter et al. (1969) found coccoid or round bodies frequently present in early and late stages of stationary growth cultures of *Vibrio marinus* (Fig. 29). The round bodies contained at least one, and often three or four, cell units.

For more information: Adler et al. (1968).

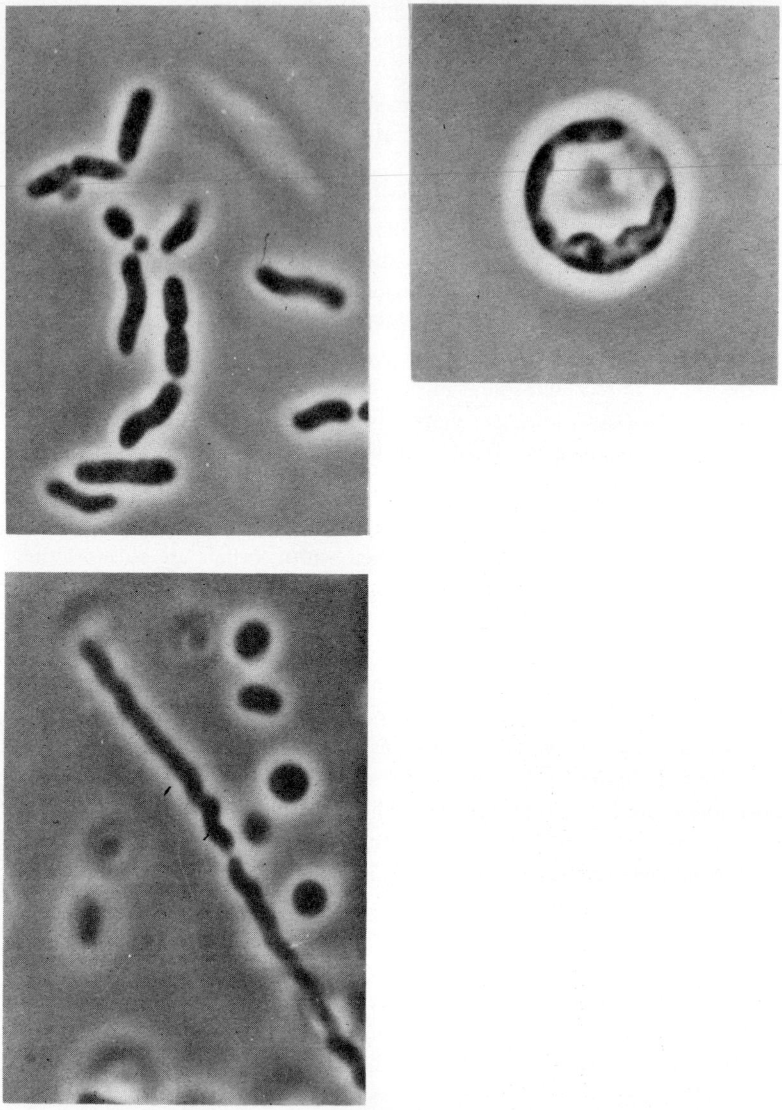

Fig. 29 Phase-contrast micrographs of Vibrio marinus. Upper left, 12-hr culture, x3000; lower left, 48-hr culture showing numerous round bodies, x3000; upper right, enlarged round body showing internal structure, x5000. (Courtesy of Felter, Colwell, and Chapman.)

Size of bacteria Bacteria show great variation in size. Some species are so small they approach the limit of visibility of a light microscope; others are so large they are almost visible to the naked eye. The majority of bacteria occupy a range intermediate between these two extremes. Regardless of size, none can be clearly seen without the aid of a microscope.

A spherical form is measured by its diameter; a rod or spiral form by its length and width. Calculation of the length of a spiral organism by this method gives only the apparent length, not the true length. The true length may be computed by actually measuring the length of each turn of the spiral. Mathematical expressions have been formulated for making such computations.

The units of length used in bacteriology are the following:

Centimeter (cm) = 10^{-2} meter(m)
Millimeter (mm) = 10^{-1} cm = 10^{-3} m
Micron (μ) = 10^{-3} mm = 10^{-4} cm = 10^{-6} m
Millimicron (mμ) = 10^{-3} μ = 10^{-6} mm = 10^{-7} cm = 10^{-9} m
Nanometer (nm) = mμ (This term is preferred to mμ.)
Angstrom unit (Å) = 10^{-1} nm = 10^{-4} μ = 10^{-7} mm = 10^{-10} m

The term micron (μ) is firmly established in expressing size in bacteria and is used throughout this book.

Some bacteria measure as large as 80 μ in length; others as small as 0.2 μ. However, the majority of the commonly encountered bacteria, including the disease producers, measure about 0.5 μ in diameter for the spherical cells and 0.5 by 2 to 3 μ for the rod forms. Spore-producing bacteria are generally larger than the non-spore-producing species. The sizes of some common species in dried and stained smears are as follows: *Escherichia coli*, 0.5 by 1 to 3 μ; *Proteus vulgaris*, 0.5 to 1 by 1 to 3 μ; *Salmonella typhosa*, 0.6 to 0.7 by 2 to 3 μ; *Streptococcus lactis*, 0.5 to 1 μ in diameter; *Streptococcus pyogenes*, 0.6 to 1 μ in diameter; *Staphylococcus aureus*, 0.8 to 1 μ in diameter; *Lactobacillus acidophilus*, 0.6 to 0.9 by 1.5 to 6 μ; *Bacillus subtilis* rods, 0.7 to 0.8 by 2 to 3 μ, spores, 0.6 to 0.9 by 1 to 1.5 μ; *B. megaterium* rods, 0.9 to 2.2 by 1 to 5 μ, spores, 1 to 1.2 by 1.5 to 2 μ; *B. anthracis* rods, 1 to 1.3 by 3 to 10 μ, spores, 0.8 to 1 by 1.3 to 1.5 μ.

The most commonly employed method for measuring bacteria is by means of an ocular micrometer. Measurements may also be made by using a camera-lucida attachment and drawing oculars, or by projecting the real image on a screen and measuring the bacteria.

The same factors that cause variations in the shape of bacteria also affect their size. With few exceptions, young cells are much larger than old or mature forms. Cells of *B. subtilis* from a 4-hr culture measure five to seven times longer than cells from a 24-hr culture. Variations in width are less pronounced. The organism *Corynebacterium diphtheriae* is a notable exception to the rule of decreasing cell size with age.

Variations in cell size with age are due to a variety of factors. The major causes appear to be changes in the environment with the accumulation of waste products. An increase in the osmotic pressure of the medium will also cause a decrease in cell size and may very well be the most important factor.

The method employed for fixing and staining bacteria may make a difference in their size. The bacterial cell shrinks considerably during drying and fixing. This will vary somewhat, depending upon the type of

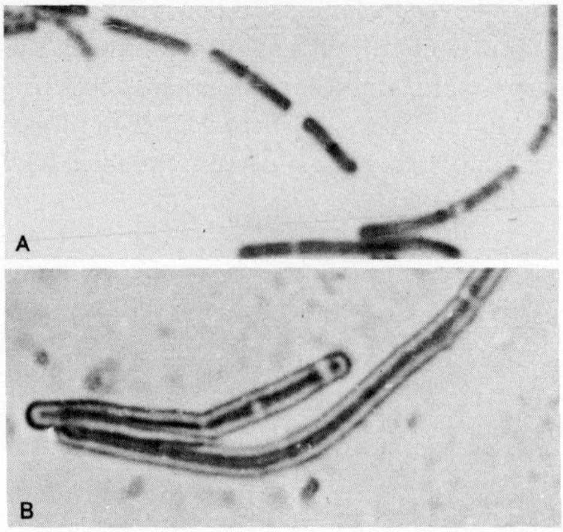

Fig. 30 Bacillus cereus. (A) Cells from a 6-hr-old agar slant culture incubated at 33°C and stained with methylene blue; (B) cells from the same culture but stained by a different technique to reveal the cell wall. (Courtesy of Knaysi.)

medium employed for their cultivation. Shrinkage generally averages about one-third of the length of the cell as compared to an unstained hanging-drop preparation. Young cells of *B. megaterium* may shrink from 15 to 25 percent when transferred from nutrient broth to the same medium containing sodium chloride in 2 *M* concentration.

Measurements show some variation, depending upon the staining solution used and the method of application. In dried and fixed smears, the cell wall and slime layer do not stain with weakly staining dyes such as methylene blue but do stain with the intensely staining pararosaniline, new fuchsin, crystal violet, and methyl violet (Fig. 30).

Hoeniger (1966) inoculated an agar medium with *Proteus mirabilis* and described the difference in cell size depending upon the staining procedure. Cells stained to reveal the cell walls were much larger than those treated by a different procedure to demonstrate the presence of nuclei (Fig. 31).

The great majority of bacteria have been measured in fixed and stained preparations. In some instances dried, negatively stained smears have been used. Therefore the method employed should be specified when measurements of bacteria are reported; otherwise the results will be of doubtful value.

Filament formation Bacteria, especially the rod-shaped forms, may be induced to elongate into filaments. This may be accomplished by changing the conditions of growth (for example, by adding chemicals, antibiotics, or minimal concentrations of bacteriostatic agents), by varying the incubation temperature, and by varying the types of radiations.

Hoffman and Frank (1963) obtained filaments of *E. coli* by incubating a culture at room temperature overnight (Fig. 32).

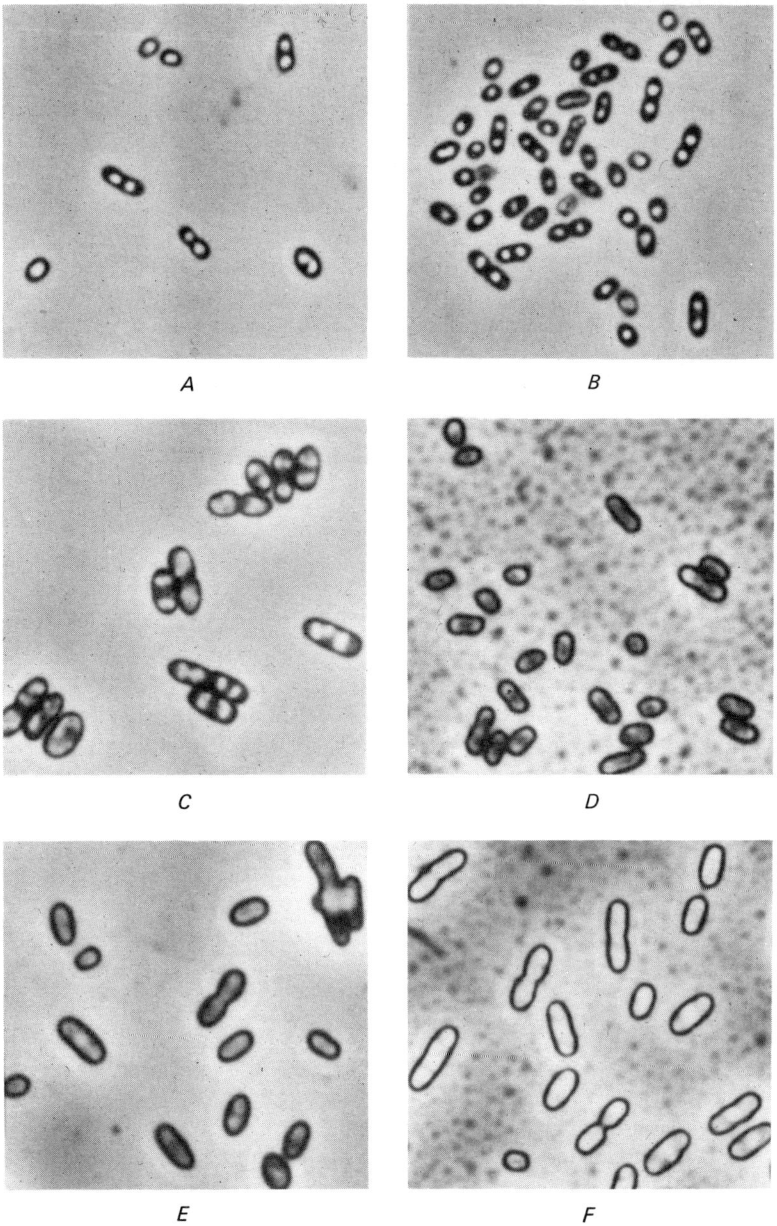

Fig. 31 Proteus mirabilis. (A, B, C) nuclear staining of ½-hr-, 1-hr-, and 1½-hr-old cultures respectively. Phase-contrast microscopy. x2900. (D, E, F) staining of cell wall of ½-hr-, 1-hr-, and 1½-hr-old cultures respectively. x2900. Note the increase in size when cell wall is stained. (Courtesy of Hoeniger.)

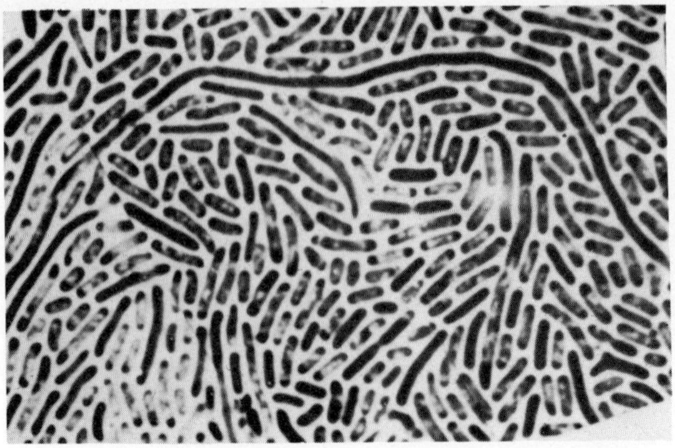

Fig. 32 Escherichia coli microculture, incubated at room temperature overnight. Note extremely long filament. (Courtesy of Hoffman and Frank.)

Grula (1960*a, b*) found cells of an *Erwinia* species grew to a length of 10 to 20 μ when cultured on one type of medium, and 1 to 2 μ when cultured on another type of medium. Division was inhibited by growing the cells in the presence of the D forms of serine, methionine, phenylalanine, threonine, tryptophan, or histidine. Division inhibition could be reversed by the addition of inorganic ammonium salts, D- or L-alanine or *p*-aminobenzoic acid.

Variation in the magnesium (Mg) content of the medium may exert a marked effect on cell division of some bacteria. In a Mg-deficient medium, gram-positive rods grow in the form of long filaments. Such filaments revert to normal forms when transferred to the same medium supplemented with suitable concentrations of Mg. Filament formation is enhanced by the addition of zinc and cobalt and also in media containing an excess of Mg.

Hoeniger (1966) cultivated *P. mirabilis* on agar and reported the morphological changes which took place ranging from short rods to long filamentous swarmers. As the culture aged, the filamentous forms appeared followed by swarming on the surface of the agar (Fig. 33).

With the exception of the radiations, such changes are usually temporary, the cells reverting to their original forms by reestablishing the initial conditions. Radiations, on the other hand, may give rise to temporary or permanent induction of filamentous cells.

In the light of the above results, cell mass or volume and cell division may be considered to some extent as separate and independent processes, at least insofar as growth may occur either with or without the operation of the cell-division mechanism.

For more information: Deibel et al. (1956); Epstein and Weiss (1960); Hughes (1956); Rosenberg et al. (1967); Shaw (1968); Terry et al. (1966); Wahren et al. (1967).

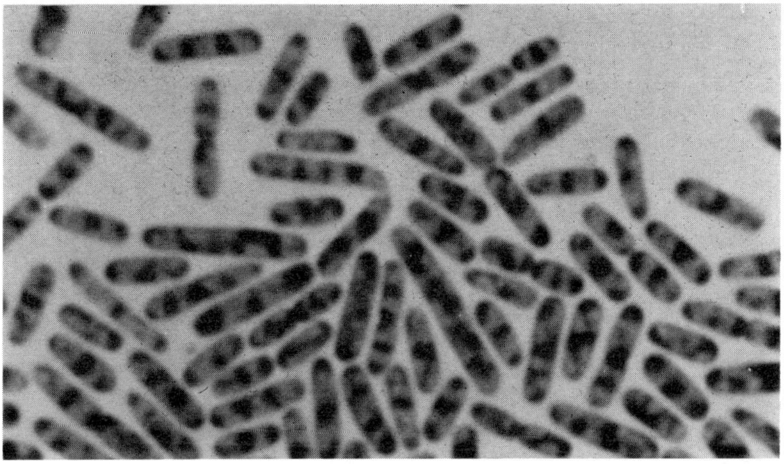

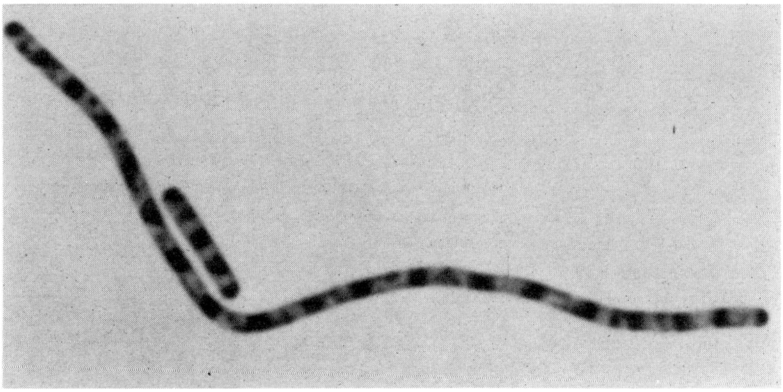

Fig. 33 Proteus mirabilis. Upper, staining of cytoplasm in a 2½-hr-old culture. x3500. Lower, staining of cytoplasmic bands in swarming cells. x3500. (Courtesy of Hoeniger.)

The bacterial cell

A bacterial cell consists of a compound membrane enclosing protoplasm. Protoplasm exists in many different modifications but ordinarily it is a thick viscous semifluid or almost jelly-like, colorless, transparent material which makes up the essential substance of both the cell body and the nucleus, including the cytoplasmic membrane but not the cell wall. It contains a high percentage of water and holds fine granules in suspension.

Structures visible in the protoplasm by appropriate examination include: nucleus, vacuoles, mesosomes, mitochondria, ribosomes, polysaccharides, lipides, metachromatic granules, and spores. Externally, bacterial cells may show capsules, fimbriae (pili), and organs of locomotion called flagella.

Cell membranes The cell membranes consist of an inner cytoplasmic membrane, the cell wall proper, and an outer slime layer or capsule.

Cytoplasmic membrane This membrane appears in young cells as an interfacial fluid film, and becomes thicker and denser as surface-active material accumulates. It is finally converted into a firm structure (Fig. 34).

The membrane comprises about 10 percent of the total dry weight of the cell and contains up to 75 percent protein, 20 to 30 percent lipide, and about 2 percent carbohydrate. The protein consists of phosphoprotein as well as a number of enzymes and carriers.

The membrane takes a deep stain with basic and neutral dyes over a wide range of pH. It is a semipermeable membrane and is principally responsible for the Gram and acid-fast reactions. The membrane stains gram-positive in gram-positive bacteria and acid-fast in acid-fast organisms.

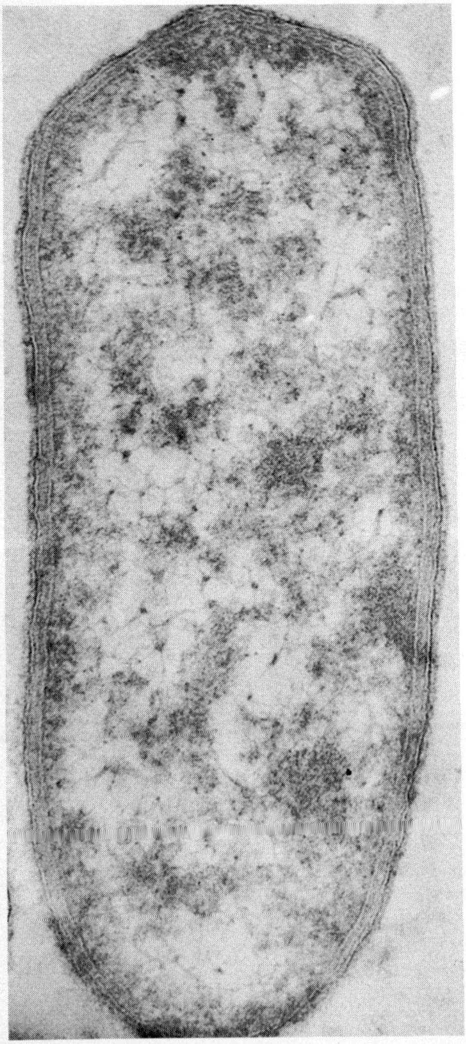

Fig. 34 Electron micrograph of the cell membranes of a strain of Bacteroides. The light-colored innermost membrane is the cytoplasmic membrane. This is surrounded by the darker-appearing cell wall. A dense crust appears around the cell wall. (Courtesy of Bladen and Waters.)

When a cell is plasmolyzed by immersion in a hypertonic solution, this membrane is drawn in with the cytoplasmic constituents. The thickness of the membrane varies even in a single cell. Measurements on a strain of *B. cereus* at various stages of development ranged from 5 to 10 nm in thickness.

The cytoplasmic membrane possesses the property of selective permeability. It permits solutes from the environment to enter the cell. The concentration of metabolites inside the cell may reach an osmotic pressure as high as 20 atmospheres or about 300 lb per sq in. It is able to withstand this high pressure only because it is supported by the rigid cell wall. The absence of the wall would result in the disruption of the membrane.

In addition to the cytoplasmic membrane, many bacteria possess other intracellular membrane systems such as mesosomes or chondrioids. The mesosome structure is most likely formed by an invagination of the cytoplasmic membrane. This is discussed on page 87.

The cytoplasmic and other internal membranes in gram-positive bacteria can be readily isolated intact by employing appropriate cell-wall degrading enzymes. On the other hand, the membranes of gram-negative bacteria can be obtained only as particulate fractions from the fragmented structures. The membranes of the gram-negative bacteria are, in general, quite complex as compared to those of the gram-positive organisms.

Cell wall The cell wall is a rigid structure and is responsible for the form of the bacterial body. It behaves as a selectively permeable membrane and apparently plays a fundamental role in the life activities of the cell.

The cell wall has a low affinity for stains and is not stained in most of the usual staining procedures. It is lightly stained by certain basic stains such as basic fuchsin and the methyl violets. Where deep staining of the wall is desired, the use of a mordant such as tannic acid is necessary. The mordant increases the affinity of the cell for stain and also increases the thickness of the wall (Fig. 35).

The cell wall accounts for an average of about 20 percent of the dry weight of bacteria and represents the major structural component. In thickness, it ranges from 10 to 23 nm, depending upon the species.

Bacteria do not change in their outer appearance when placed in solutions of high or low osmotic pressures. In this respect they differ markedly from higher animal and plant cells. The thick and tough bacterial cell wall is unchanged in such environments. Removal of the wall by appropriate means does not kill the cell but does cause the "naked" cell to change shape by rounding up into a spherical form (see section on protoplasts).

Electron micrographs of the cell wall and protoplasm of *Rickettsia typhi* are shown in Fig. 36.

For more information: Abram (1965); Bayer (1968); Brostrom and Binkley (1969); Costerton (1970); Fiil and Branton (1969); Gilleland et al. (1971); Hurst and Stubbs (1969); Imaeda et al. (1968); Martin and

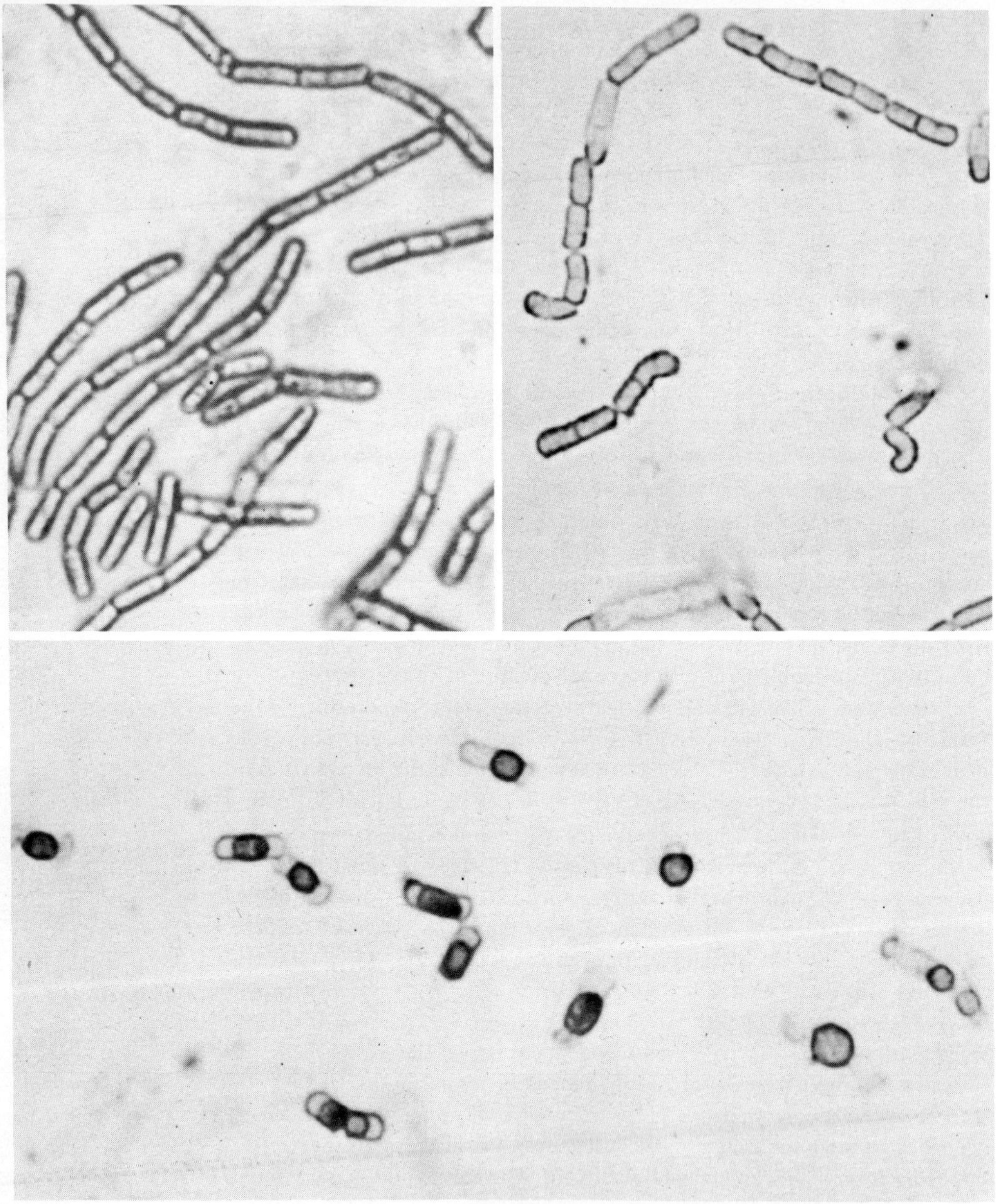

Fig. 35 Cell-wall stain of Bacillus M. Upper left, stained with tannic acid violet; upper right, stained with alcian blue; lower, treated with egg albumin and stained with Hale's cell-wall stain (Courtesy of Tomcsik and Grace.)

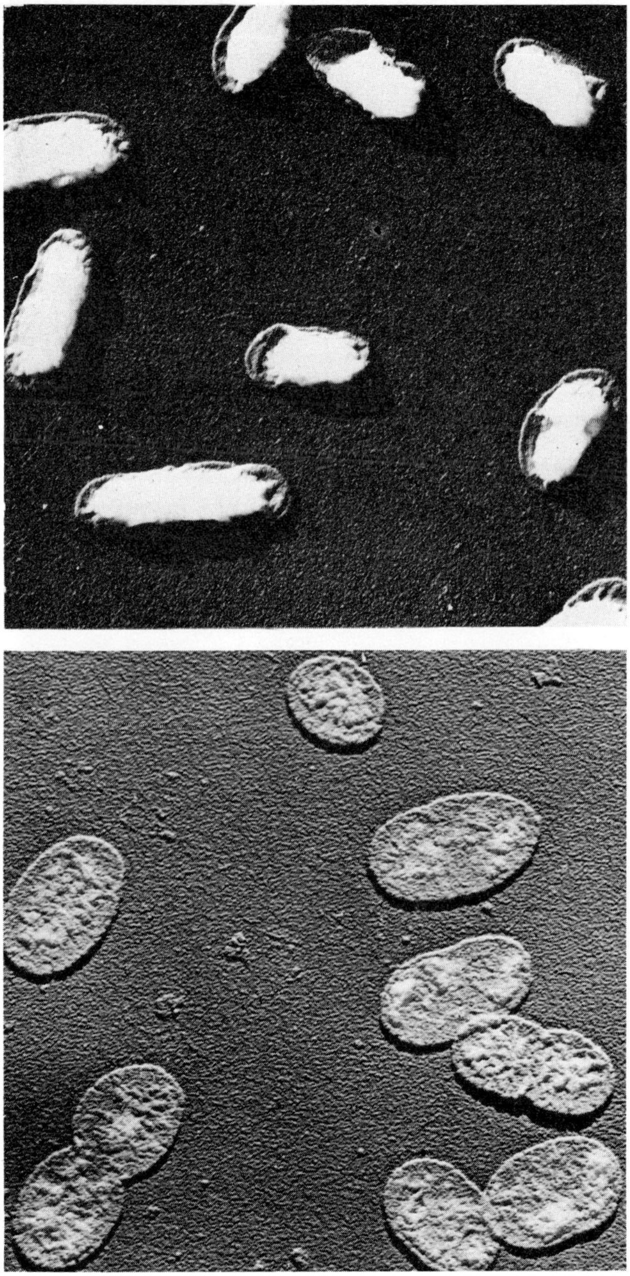

Fig. 36 Rickettsia typhi. Upper, air-dried and shadowed preparation showing cell walls surrounding the central, shrunken protoplasm; lower, purified preparation of cell walls. Sample taken 4 hr after addition of 1 percent sodium deoxycholate. (Courtesy of Schaechter, Tousimis, Cohn, Rosen, Campbell, and Hahn.)

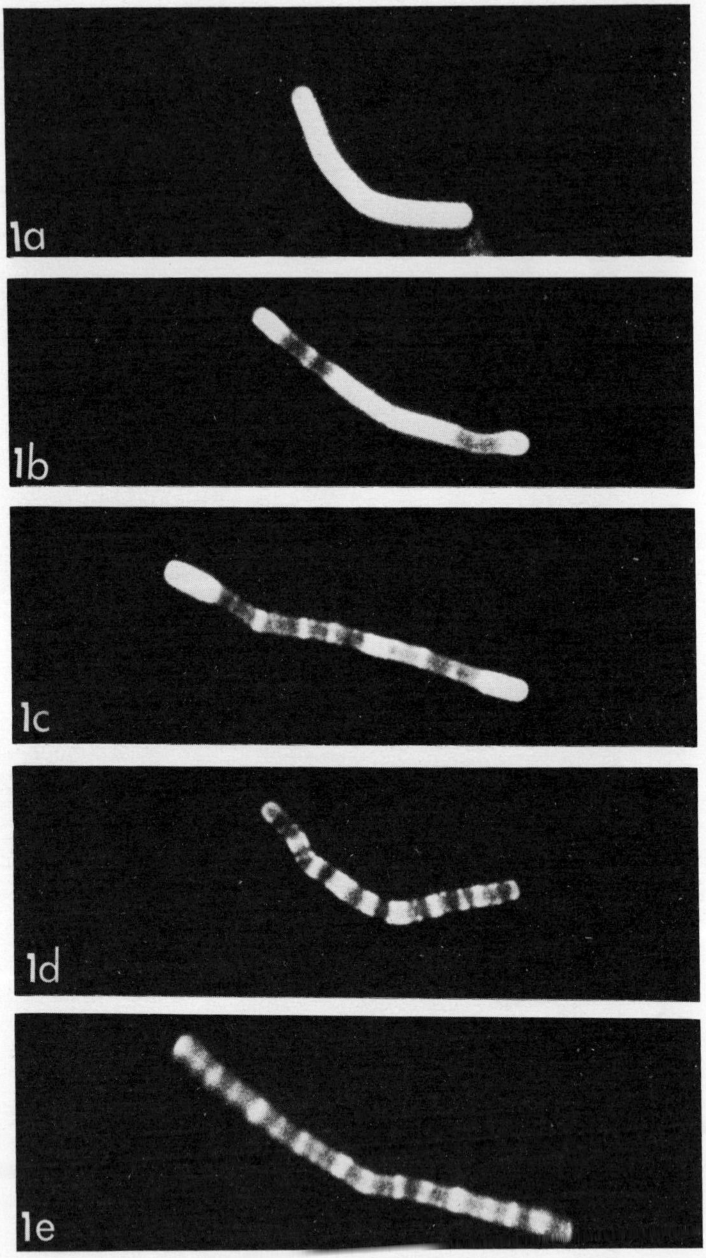

Fig. 37 Cell-wall replication in Bacillus cereus. Cells incubated and examined after the following intervals: 1a, no incubation; 1b, 10 min; 1c,

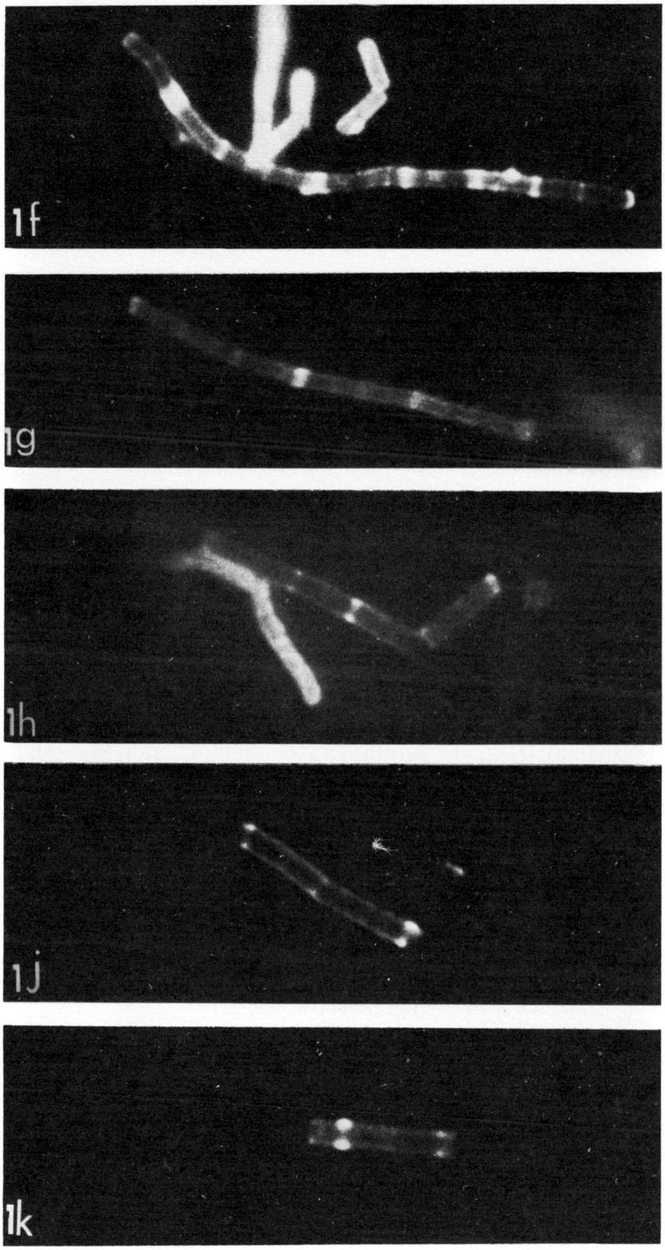

20 min; 1d, 30 min; 1e, 40 min; 1f, 50 min; 1g, 60 min; 1h, j, k, 70 min. See text for details. (Courtesy of Chung, Hawirko, and Isaac.)

MacLeod (1971); Nanninga (1970); Schnaitman (1971); Sharon (1969); van Iterson (1960).

Cell-wall replication Chung et al. (1964) studied cell-wall replication by differential labeling with fluorescent and nonfluorescent antibody. In *B. cereus*, growth of new cell wall was initiated near the poles (Fig. 37). In the old wall, additional new wall segments gradually developed to form an alternating pattern of new and old wall segments. Further growth elongated the new wall and pushed the old segments apart. Separation of daughter cells appeared to involve splitting of the transverse septa laid down at or near the old wall segments.

In *B. megaterium*, growth of new cell wall was initiated either at one of the poles or at the central area of the cell. Multiple segments of new and old wall appeared along the cell length. Further elongation was followed by formation of transverse septa and separation of daughter cells incorporating either old or new wall segments.

For more information: Cole (1964); Cole and Hahn (1962).

Cell-wall composition The cell wall of higher plant cells is composed largely of cellulose, a polymer of glucose. The supporting structure in molds is largely chitin, a polymerized acetylated glucosamine. The cell wall of true bacteria does not contain either cellulose or chitin as the major component.

Thin sections of gram-positive bacteria show a cell wall 20 to 80 nm (mμ) in thickness. The wall covers an alternating electron-dense, electron-transparent layer, the so-called cytoplasmic membrane of about 7.5 nm in thickness. The cell wall is composed of a peptidoglycan (also called mucopeptide, glycopeptide, or murein). The peptidoglycan is composed of six different components, namely, *N*-acetylglucosamine, *N*-acetylmuramic acid, L-alanine, D-alanine, D-glutamic acid, and either L-lysine or *meso*-diaminopimelic acid. In addition, the peptidoglycan in some species of gram-positive bacteria may contain glycine, L-serine, L-threonine, D-aspartic acid, and amide ammonia.

The cell wall of gram-negative bacteria is considerably more complex in organization and composition. The rigid peptidoglycan layer (about 2 to 3 nm in thickness) is located between the cytoplasmic membrane and an outer multiple-track layer. The latter layer is believed to be composed of lipoprotein and lipopolysaccharide complexes, which means that the gram-negative bacteria contain a much higher fat content than the gram-positive bacteria. The three layers, namely, the external multiple-track layer, the rigid peptidoglycan layer, and the cytoplasmic membrane, constitute what is sometimes called the bacterial envelope.

In some instances cell-wall synthesis can proceed independently from that of other parts of the cell. For example, both wall and cell protein may require an essential amino acid for growth. After depletion of this essential acid in the medium, synthesis of both components of the cell ceases. However, if the amino acid is required only for cell synthesis, wall synthesis will continue with the result that wall substance continues to increase and in some instances may even double in amount.

For more information: Adams et al. (1970); De Petris (1965); Ellwood (1970); Ghuysen (1968); Herzberg and Green (1964); Hungerer and Tipper (1969); Hungerer et al. (1969); Jeanloz (1967); Kane et al. (1969); Kanetsuna (1968); Kingan and Ensign (1968); Lopes and Inniss (1970); Manasse and Corpe (1967); Meynell and Lawn (1965); Rogers (1965); Salton (1964, 1967*a*); Šharon (1969); Sutow and Welker (1967); Takumi and Kawata (1970); Tipper and Pratt (1970); Welker (1971); Wheat and Ghuysen (1971).

Mesosomes Thin-section preparations of many bacteria reveal the presence of other intracytoplasmic membranes which have been named mesosomes or chondrioids. The mesosome structure is apparently formed by an invagination of the cytoplasmic membrane. It appears to be most conspicuous in the gram-positive bacteria although the membrane has been observed in some gram-negative cells.

The mesosome membranes may appear in two forms: (1) As a lamellar type of structure resulting from the coiled-up membranes, as seen in *Lactobacillus plantarum* (Figs. 38 and 39); or (2) as the vesicular type as

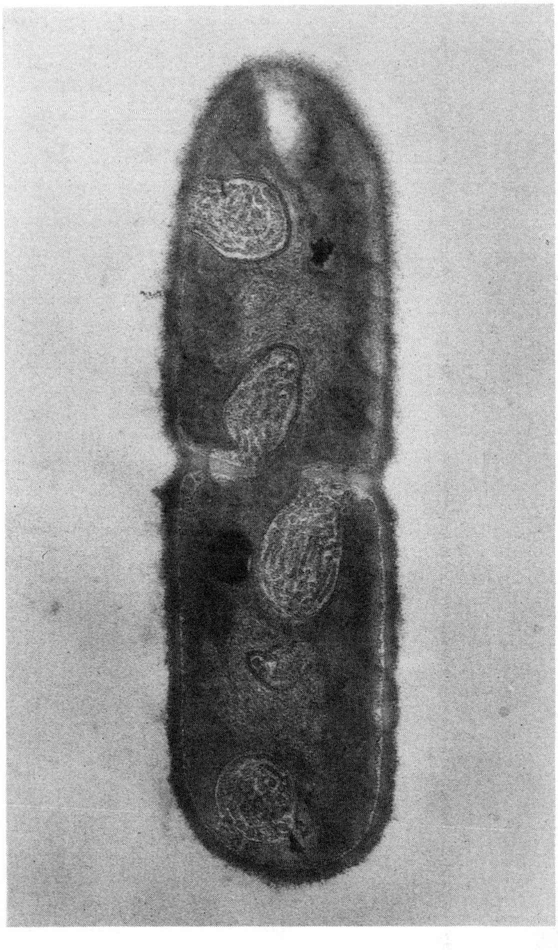

Fig. 38 Dividing cell of Lactobacillus plantarum showing prominent mesosomes associated with a newly forming cross wall. The black bodies are believed to be polyphosphate granules. x94,300. (Courtesy of Kakefuda, Holden, and Utech.)

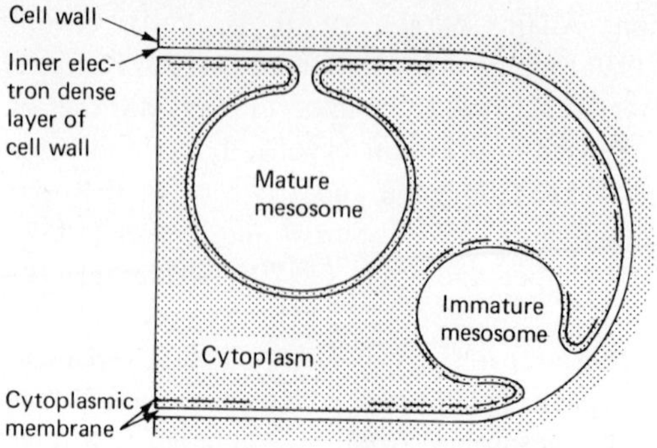

Fig. 39 Schematic drawing of the cytoplasmic membrane, cell wall, and mesosomes of Lactobacillus plantarum. The wall-associated electron-dense layer (inner electron-dense layer of the cell wall) and the cytoplasmic membrane in which the inner layer of the unit membrane is almost invisible in an intact cell is shown. The cytoplasmic membrane is continuous with the boundary membrane of the mesosome. The cytoplasmic side layer of an immature mesosome boundary is faint and discontinuous in appearance as it is in the cell membrane. However, a mature mesosome is surrounded by a distinct, triple-layer unit membrane. (Courtesy of Kakefuda, Holden, and Utech.)

Fig. 40 Cross-fractured cell of Bacillus subtilis showing cell content, outer surface of the cytoplasmic membrane, cross-fractured cell wall, and a mesosome composed of numerous vesicles. x63,000. (Courtesy of Nanninga.)

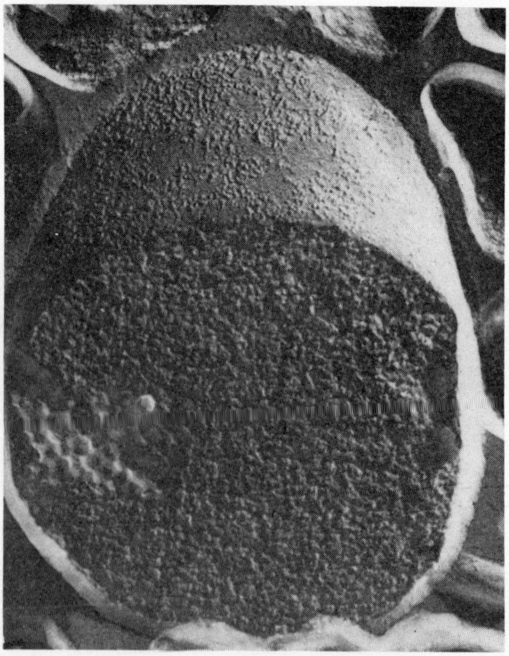

present in *B. subtilis* (Fig. 40). In the lamellar type of structure it appears that the removal of the cell wall by appropriate means causes the mesosome to uncoil and become a part of the surface cell membrane. In the vesicular type of structure, the mesosome usually appears as a sac of membrane vesicles surrounded by the invaginated cytoplasmic membrane. The vesicles are actually situated external to the cytoplasmic membrane and are most likely released after removal of the cell wall.

The exact function of the mesosome is not clearly understood. Pontefract et al. (1969), from their observations, concluded that the mature mesosome appeared to form a link between the cytoplasmic membrane and the nucleus of the cell. This led to the hypothesis that division of the nucleus was accomplished by two separate polar mesosomes. One mesosome was derived from the parent cell and was present at one pole of the daughter cell. The other was freshly synthesized at or near the newly forming pole of the daughter cell. Whereas the old mesosome remained attached to the chromosome received from the parent cell, the newly synthesized mesosome became attached to and initiated replication of the new chromosome. As the cell increased in size and matured, the two mesosomes separated from the chromosomes resulting in the division of the nucleus.

For more information: Kakefuda et al. (1967); Nanninga (1968); Salton (1967*b*); Schnaitman and Greenawalt (1966); Weigand et al. (1970).

Slime layer and capsule An extracellular, slimy or gelatinous, polysaccharide-like material is formed by bacteria, especially those producing mucoid growths. This material may remain firmly adherent as a discrete covering layer of each cell, or it may part freely from the cell as it is formed. In the former case, it is called a capsule; in the latter, free slime or gum.

Capsules and slime are believed to be distinct from the morphological and biochemical point of view. The capsule is a part of the cell, the slime a secretion. Capsules are of definite shape, of more or less definite density throughout, and of definite outline (Figs. 41 and 42). On the other

Fig. 41 Capsules of Klebsiella pneumoniae. (After Klieneberger-Nobel.)

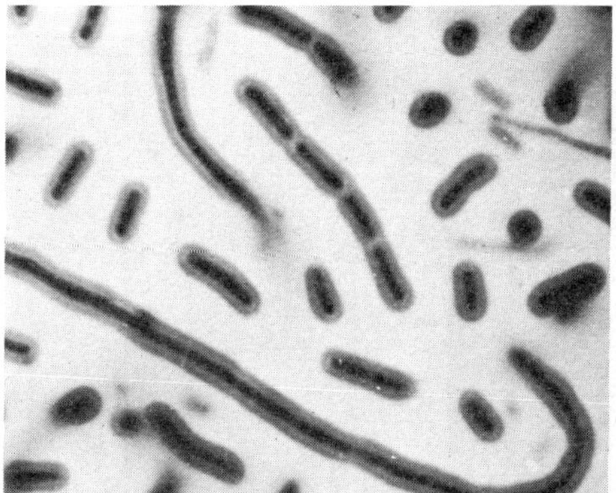

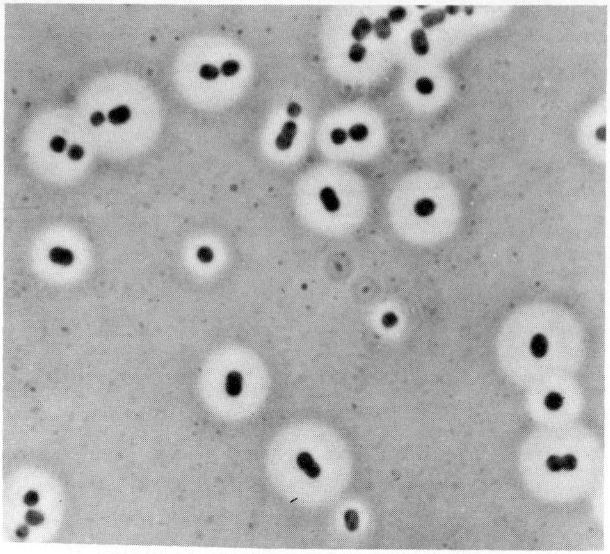

Fig. 42 Capsules produced by a gram-negative coccus. (Courtesy of Juni and Heym.)

hand, slime envelopes are amorphous and can be drawn out into manifold structures, are most concentrated in the vicinity of the bacterial cells, and decrease in density with increasing distance from the cell.

Capsules are not believed to be homogeneous accumulations of amorphous viscous substances around the cell surface but rather physically and chemically heterogeneous structures. According to Salton (1961), capsules may show (1) a continuous layer around the cell, (2) banded fibrils, (3) localized patches of polysaccharide and polypeptide, or (4) discontinuities in the capsular surface (Fig. 43).

Broth cultures of capsule-producing organisms are usually stringy in texture, and agar colonies exhibit a very moist, glistening surface that is described as mucoid. Capsule formation may be dependent to some extent upon the composition of the medium but especially upon the variant phase of the organism. Some disease-producing species form large capsules in culture media rich in animal fluids; others produce prominent capsules when cultures are incubated at low temperatures (4 to 20°C).

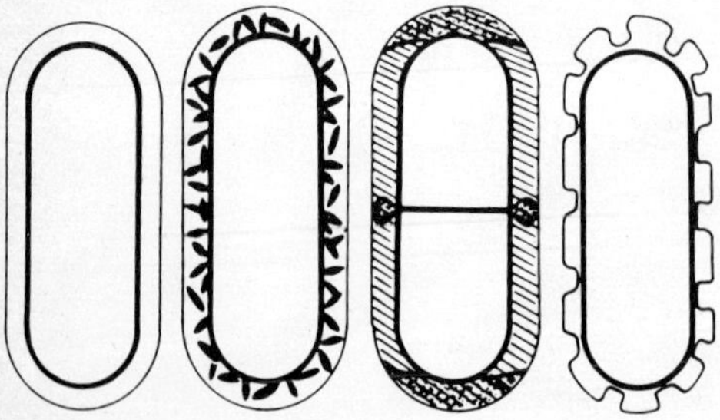

Fig. 43 Diagrammatic representation of type of capsular structures. From left to right: (1) Cell surrounded with continuous layer of capsular substance; (2) capsular layer with banded fibrils; (3) complex capsule with localized patches of polysaccharide and polypeptide; (4) discontinuities in capsular surface. (Courtesy of Salton.)

Gums and capsules show wide differences in composition. Gums or slime appear, in most cases, to be pure polysaccharides. On hydrolysis they yield their constituent sugars. Some appear to be composed of only one sugar, such as glucose, galactose, and levulose; others contain more than one sugar. Apparently no two are alike in composition. Capsules, on the other hand, may show the presence of nitrogenous compounds as well as polysaccharides. The compounds may be (1) glycoproteins, (2) protein-polysaccharide complexes, or (3) a framework of polysaccharides with the spaces filled in with polypeptides. Here again, apparently no two are alike in composition.

Capsular material is difficult to distinguish from those gums which flow away from the cells as they are formed. Organisms producing gums do so when grown in sugar solutions. Some organisms produce gums only in the presence of a specific sugar; others produce gums in the presence of any one of several sugars. In the absence of sugar, usually very little, if any, gum is formed. Organisms producing gums of this type are the cause of considerable losses in the sugar industry. The increased viscosity produced by the gum interferes with the filtration of the sugar solution.

The species commonly encountered in sugarcane juice is *Leuconostoc mesenteroides* (Fig. 44). The cells are surrounded by a thick, gelatinous, colorless polysaccharide consisting of dextran (glucose polymer).

The formation of gums is of common occurrence by soil bacteria. From 5 to 16 percent of such forms have been shown to be capable of synthesizing gums from sugars.

For more information: Anderson (1961); Cadmus et al. (1963); Eagon (1962); Gaudy and Wolfe (1962); Izumi (1962); Jones et al. (1969); Jones et al. (1962); Juni and Heym (1964); Lindberg and Svensson (1968); Meynell and Meynell (1966); Walker and Short (1969); Wang et al. (1963).

Anatomy of the cell surface The outer surface of bacterial cells is

Fig. 44 Leuconostoc mesenteroides. Three-day culture on 10 percent raw sugar agar incubated at room temperature. (After McCleskey, Faville, and Barnett.)

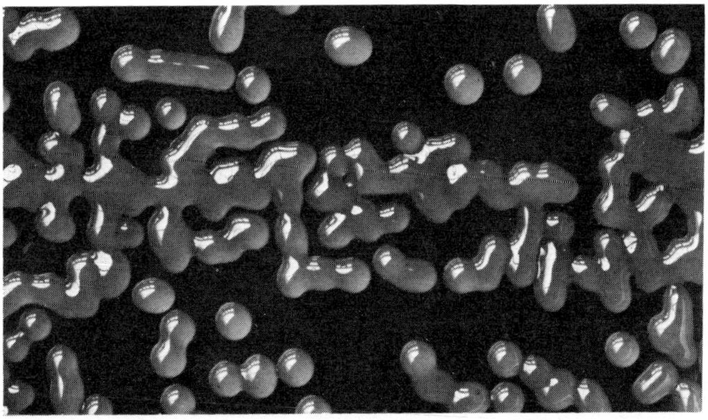

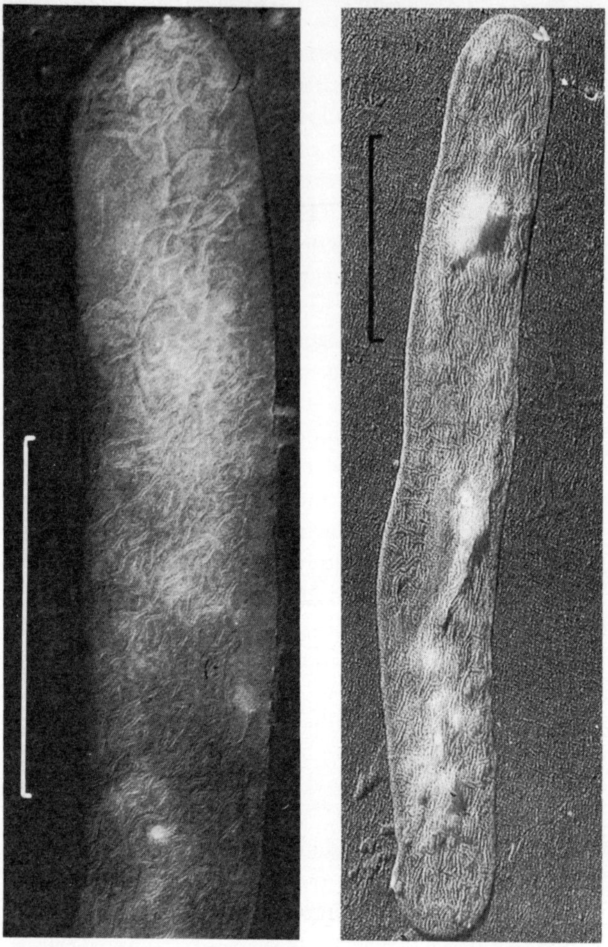

Fig. 45 Left, electron micrograph of a "ghost cell' of Mycobacterium tuberculosis showing the paired fibrous structure of the cell envelope; right, same of M. lepraemurium. (Courtesy of Takeya, Mori, Tokunaga, Koike, and Kazuhito.)

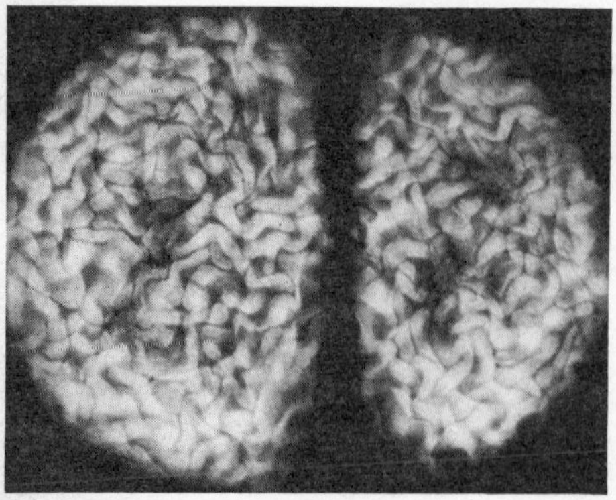

Fig. 46 Electron micrograph of the surface structure of a species of Veillonella. One-half of the diplococcus appears larger than the other. (Courtesy of Bladen and Mergenhagen.)

not a smooth uniform structure but possesses a definite sculptured pattern which shows variation depending upon the species.

Takeya et al. (1961) cultivated mycobacteria on streptomycin agar and observed a paired fibrous structure over the cell envelope when viewed under the electron microscope (Fig. 45).

Bladen and Mergenhagen (1964) found the surface structure of cells of a species of *Veillonella* to be extremely convoluted and uneven (Fig. 46). The diplococcal shape can be seen with one-half of the cell appearing larger than the other.

DeBoer and Spit (1964) examined a gram-negative motile bacterium by replica technique. Electron micrographs showed cell walls with a striated structure, with areas of different sizes and shapes, each with its own direction of striation (Fig. 47).

Nermut and Murray (1967) revealed the macromolecular structure of the cell surface of *B. polymyxa* by shadowing with tungsten oxide. It

Fig. 47 Electron micrographs of a photo-organotroph showing cell walls having a striated structure with areas of different sizes and shapes, each with its own direction of striation. (Courtesy of deBoer and Spit.)

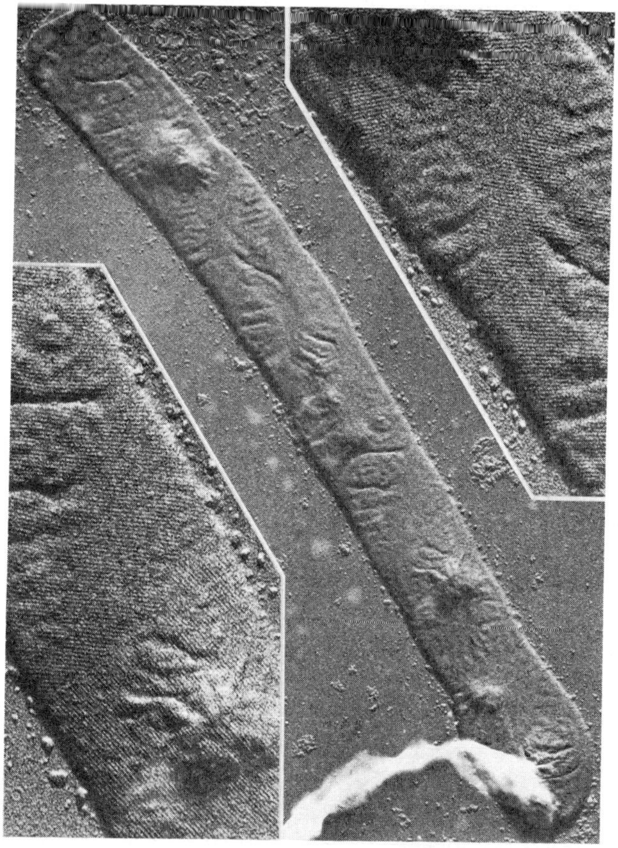

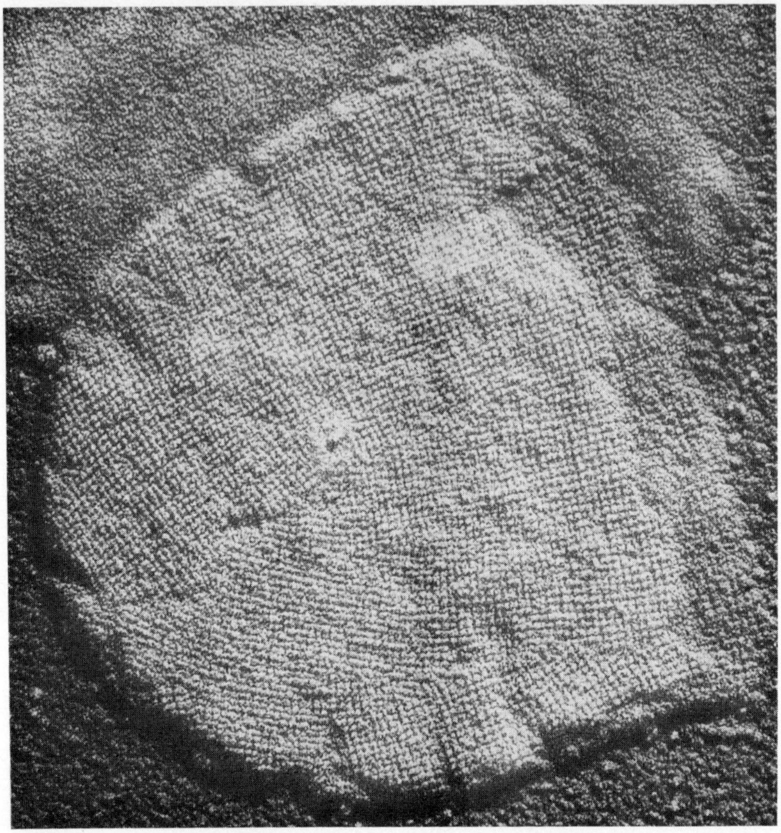

Fig. 48 Electron micrograph of a fragment of a cell wall from a 5-day-old culture of Bacillus polymyxa. (Courtesy of Nermut and Murray.)

consisted of a rectangular array of 70-Å globules with a repeating interval of 100 Å (Fig. 48).

For more information: Bayer and Remsen (1970); Hageage and Gherna (1971); Homma et al. (1963); Salton (1961, 1964).

Flagella True bacterial motion, with few exceptions, is associated with organs of locomotion called flagella (singular, flagellum). The exceptions are some of the higher bacteria which may exhibit a creeping or jerky motion not associated with any external appendages. The presence of flagella does not mean necessarily that the bacteria are always motile, but it indicates a potential power to move.

Independent bacterial motion is a true movement of translation and must be distinguished from the quivering or back-and-forth motion exhibited by very small particles suspended in a liquid. The latter type of motion is called brownian movement and is caused by the bombardment of the bacteria by the molecules of the suspending fluid.

Properties of flagella Flagella are very delicate organs and easily detached from the cell. In the stained condition they are long, slender, undulating organs. They are directed backward to the direction of motion at an angle of about 45°. Reversal of direction occurs by swinging the flagella through an angle of about 90°. Turning movements take place by swinging the flagella forward on one side only. Flagella propel the organism by a spiral or corkscrew motion.

The thickness of flagella may show wide variation from species to species. In *P. vulgaris* they measure about 12 nm. Seidler and Starr (1968) found the flagella of *Bdellovibrio bacteriovorus* to measure about 28 nm. These figures are considerably below the shortest wavelength of visible light and explain why flagella cannot be seen in hanging-drop preparations or in smears stained by the usual simple procedures. When special staining methods are employed, sufficient stain becomes deposited on the flagella to make their diameters greater than the smallest wavelength of visible light. They may then be seen under a light microscope.

Chemistry of flagella Flagella and bacterial bodies differ in composition. Flagella consist largely of protein with small amounts of carbohydrate and possibly some lipide and in some cases surrounded by a sheath of unknown chemical composition. When treated with heat, acid, and other agents, flagella disintegrate into individual protein molecules (monomers) called flagellins. Flagellins have molecular weights ranging from 15,000 to 40,000.

Flagellins produced by acid disaggregation are resistant to digestion by pepsin, trypsin, and papain. The same is true for whole flagella. As many as 36 separable peptides have been detected in flagellins. Flagellin prepared from the flagella of *P. vulgaris* is an incomplete protein containing only 16 known amino acids.

Abram and Koffler (1964) reported that flagellin monomers prepared from the flagella of *Bacillus pumilus* by acid disintegration reassembled in vitro in at least three forms, two of which were highly organized in a linear fashion, and appeared as straight structures and flagella-like filaments. The latter closely resembled native flagella in morphology and in other properties. The rate and extent of reaggregation, and the structure of the aggregates produced, were affected by the concentration of flagellin, pH, salt, and temperature.

Wakabayashi et al. (1969) found flagellin from *Salmonella* to polymerize spontaneously into flagellar filaments in the presence of moderately high concentrations of certain salts at neutral pH. Rapid and complete polymerization of flagellin occurred after the addition of F, CO_3, SO_4, HPO_4, and citrate ions in final concentrations higher than 0.3 *M*. The reformed filaments were indistinguishable from the original intact flagella. On the other hand, high concentrations of Mg and Ca ions produced the reverse effect, namely, depolymerization.

The *H* antigens of bacteria are associated with the flagella, and the *O* antigens with the bodies (see page 939). Purified flagella are agglutinated by *H* antiserum but not by *O* antiserum; *O* antigens are not agglutinated by *H* antiserum. This is another indication that flagella and bacterial bodies differ in composition.

X-ray diffraction studies show flagellins to have a helical structure. Flagellins prepared by acid treatment of flagella at pH 4 consist of 20 to 40 percent α-helices. In stained preparation of partially disintegrated flagella of *S. typhimurium*, the monomers distinguishable within the flagella appear to be spheroid or ovoid in shape.

Some workers believe that flagellin molecules are linked into fibers. Flagella of *S. typhimurium* have been reported to contain three coiled and five parallel strands of subunits; flagella of *B. pumilus* consist of six coiled fibers made up of spheroid or ovoid subunits with hollow central cores. Other workers have not been able to substantiate such findings.

An electron micrograph demonstrating the presence of a helical structure in the flagella of *Bordetella bronchiseptica* is shown in Fig. 49.

For more information: Aamodt and Eisenstadt (1968); Asakura and Eguchi (1964); Joys (1968); Klein et al. (1968); Lowy and Hanson (1964); Lowy and McDonough (1964); McDonough (1965); Asakura and Iino (1972).

Origin of flagella Flagella were at one time believed to have their origin in the cell wall but since the introduction of the electron microscope they have now been shown to originate from the interior of

Fig. 49 Electron micrographs of flagella of Bordetella bronchiseptica. Left, cell and attached flagella, x13,000; right, three sections of flagella showing a helical structure, x77,000. Palladium shadowed. (Courtesy of Labaw and Mosley.)

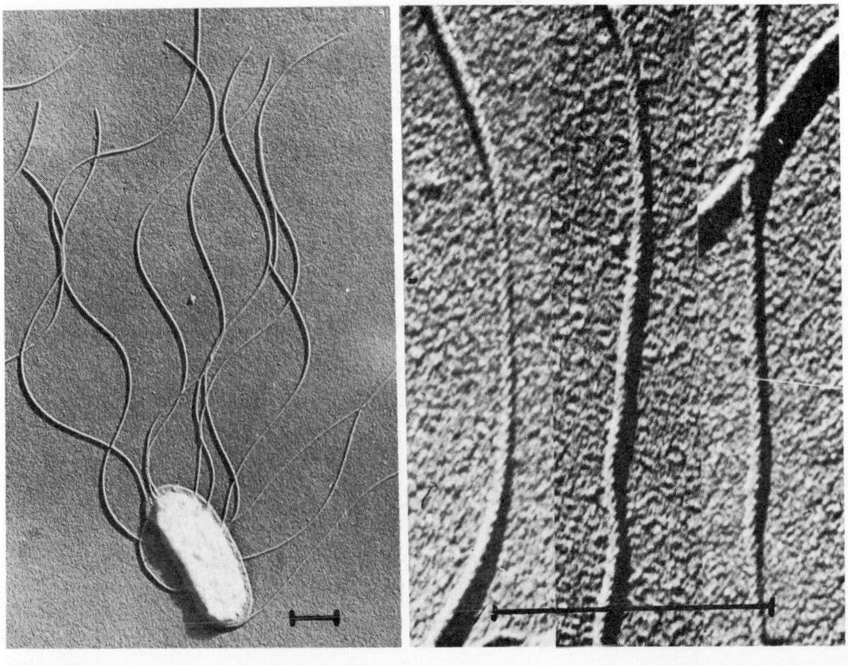

the cell, most likely the cytoplasm. De Pamphilis and Adler (1971*a*) studied the hook-basal body complex comprising the basal end of purified flagella from *Escherichia coli* and *B. subtilis* by electron microscopy (Fig. 50). A model of the basal end of a flagellum from *E. coli* is shown in Fig. 51. The *E. coli* hook showed five or six concentric helical coils. The basal body consisted of four rings arranged in pairs and mounted on a rod. The top pair of rings was connected near their periphery, resembling a closed cylinder. In *B. subtilis*, the top pair of rings was not observed. They proposed the existence of both structures as exemplified by *E. coli* (gram-negative) and *B. subtilis* (gram-positive). A schematic representation of the attachment of the hook-basal body complex of a flagellum of *Rhodospirillum rubrum*, as proposed by Cohen-Bazire and London (1967), is shown in Fig. 52.

For more information: Abram et al. (1965, 1966, 1970); Betz (1969); De Pamphilis and Adler (1971*b, c*); Dimmitt and Simon (1971); Hoeniger et al. (1966); Tawara (1965); Vaituzis and Doetsch (1969).

Number and arrangement of flagella The number and arrangement of flagella vary with different bacteria, but they are generally constant for each species. Some have only one flagellum; others have two or more flagella.

In rod-shaped cells the flagella arise either at one or both poles, or are distributed laterally with the poles being generally bare. In some species flagella are located both laterally and at the poles. A species may show considerable variation in the number and arrangement of flagella. Single *Alcaligenes* cultures may contain forms with a polar flagellum only; some with several lateral flagella; and some with both lateral and polar flagella. Sometimes a species may show cells which are flagellated in one environment and nonflagellated in another.

Leifson et al. (1955) reported the presence of four definite types of curvature in the flagella of *P. vulgaris* (Fig. 53). Individual organisms were seen with more than one type of flagella, and individual flagella with more than one type of curve. Environmental factors such as pH and formalin (Leifson, 1961) may change the curvature of the flagella on some strains. In acid media the curly curvature tends to predominate; in alkaline media the normal form predominates.

Iino and Mitani (1967) reported the presence of straight flagella on the cells of a mutant strain of *S. typhimurium* (Fig. 54). The original nonmutated culture, containing cells with curly flagella, exhibited active motility whereas cells of the mutant strain, with straight flagella, were nonmotile.

Martinez et al. (1968) reported similar results in a mutant culture of *B. subtilis*. They found that the conversion of the curly motile flagella to the straight nonmotile form resulted from the substitution of the amino acid valine for alanine in the altered dipeptide of flagellin. They concluded that a single amino acid substitution at a critical position in the primary sequence of flagellin, a protein having a molecular weight of about 40,000, led to a loss of the long-period helix of the flagella and thus to an

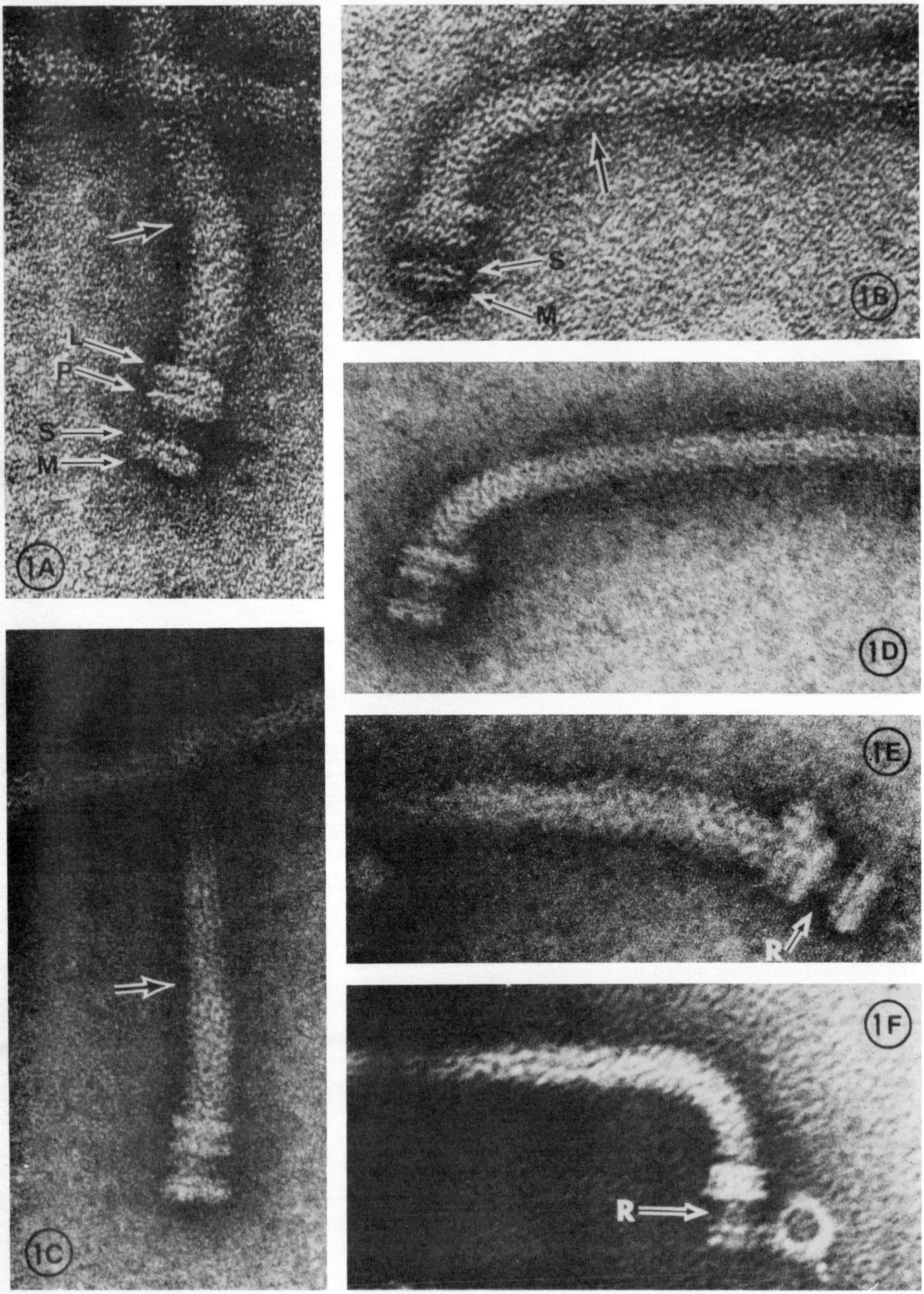
L
P
S
M
1A
S
M
1B
1C
1D
1E
R
1F
R

Fig. 50 Basal end of intact flagellum from Escherichia coli. (A) Both top and bottom edges of L and P rings are seen. Arrow marks junction between hook and filament. The bar in this and subsequent figures represents 30 nm. (B) Top edge of S ring and top and bottom edges of M ring can be seen giving a circular appearance. Arrow marks junction between hook and filament. (C) Arrow marks junction between hook and filament. (D) L ring appears extended in D and E probably because fragment of outer membrane remained attached. (E) R marks rod connecting top and bottom rings. (F) R marks rod connecting top and bottom rings. Detached ring has associated with bottom rings. (Courtesy of DePamphilis and Adler.)

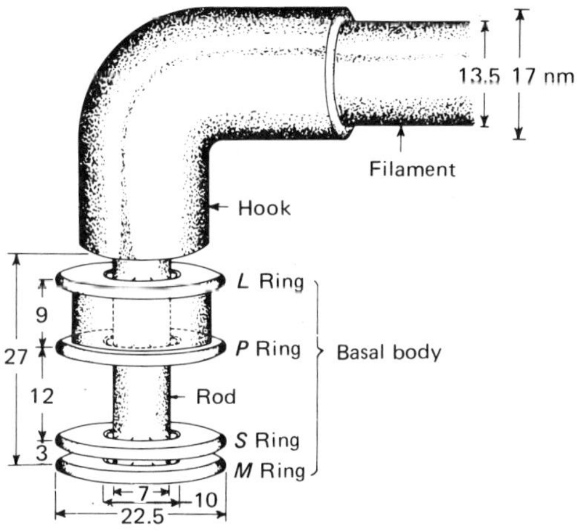

Fig. 51 Model of the basal end of a flagellum from Escherichia coli. Dimensions are expressed in nanometers. (Courtesy of DePamphilis and Adler.)

Fig. 52 Schematic representation of the structural relationship between the cortical layers of the polar region of Rhodospirillum rubrum and the basal region of the flagella. (After Cohen-Bazire and London.)

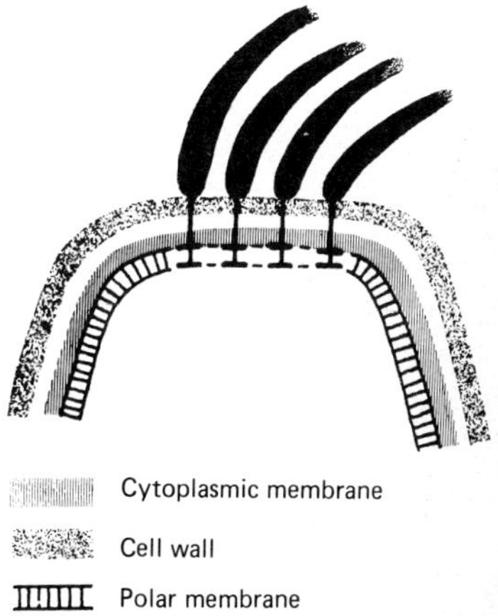

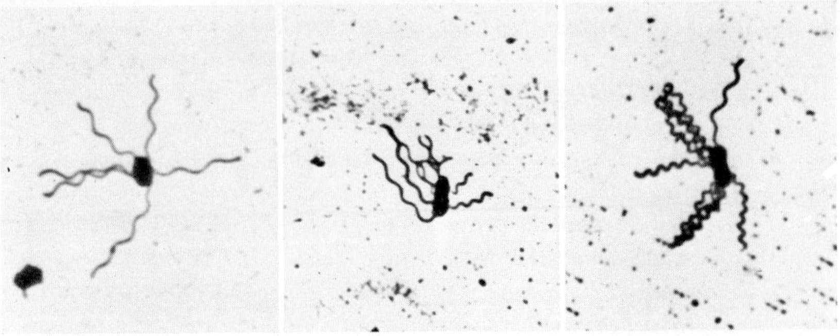

Fig. 53 Flagella of Proteus species. Left, P. mirabilis showing normal flagella; center, P. morganii with six normal and two curly flagella; right, P. mirabilis with one normal and several curly flagella. (Courtesy of Leifson, Carhart, and Fulton.)

Fig. 54 Salmonella typhimurium. Upper, photomicrographs of motile cells with curly normal flagella, and nonmotile cells with straight flagella. Lower, electron micrograph of a greatly enlarged nonmotile cell with straight flagella. (Courtesy of Iino and Mitani.)

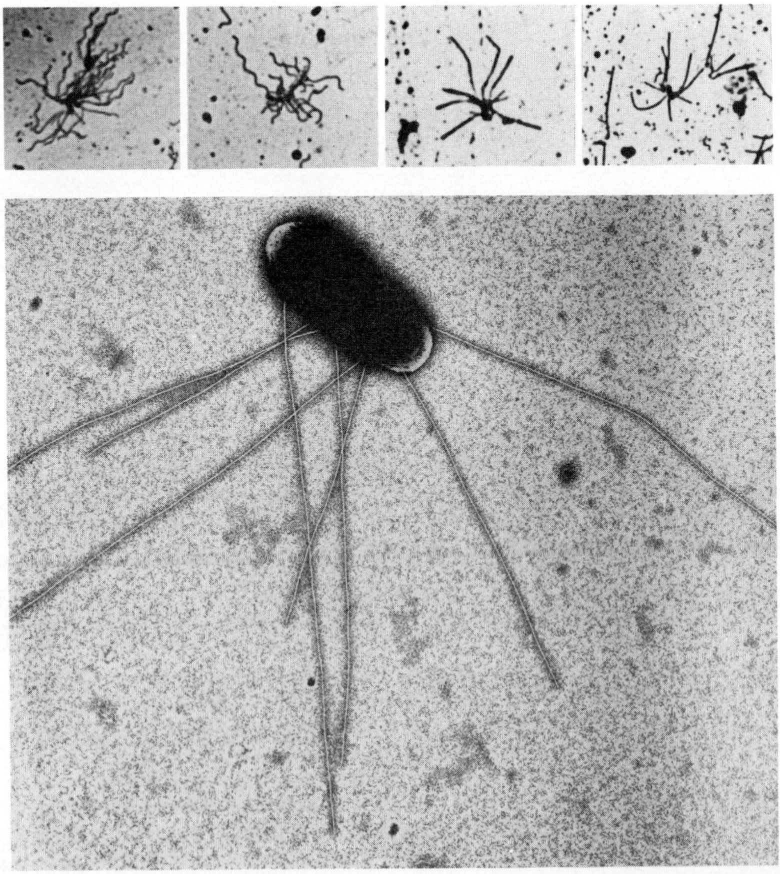

alteration in the gross morphology of the structure. This alteration appeared to be responsible for the loss of function of the flagella.

Organisms have been classified on the basis of the number and arrangement of flagella as follows:

Monotrichous – A single flagellum at one end of the cell
Lophotrichous – Two or more flagella at one end or both ends of the cell
Amphitrichous – One flagellum at each end
Peritrichous – Flagella surrounding the cell

Staining of flagella The staining of flagella is a difficult technique, especially in the hands of the beginner. For this reason many methods have been proposed. Regardless of the method employed, the film must first be treated with a mordant to make the flagella take the stain heavily. Mordants consist usually of a mixture of tannic acid and some metallic salt. In some methods the mordant and stain are applied separately; in others they are combined in one solution.

The most important factors which influence the staining results are (1) the skill of the investigator, (2) the organisms under study, (3) the culture medium on which the organisms are grown, and (4) the staining method. The most important appears to be 1, and the least important, 4.

In stained preparations not all parts of the slide are the same. Some areas are better stained than others. This means that the slide should be examined thoroughly for the possible presence of satisfactory areas. Since flagella are easily torn away from the cells, it is a common observation to find bacteria without flagella, and clumps of stained flagella without bacteria.

Photomicrographs of the arrangement of flagella on bacteria, according to Leifson (1951), are shown in Fig. 55.

For more information: Hoeniger (1965); Holwill and Burge (1963); Iino (1969); Iino and Mitani (1967); Lautrop and Jessen (1964); Leifson (1963); Mitani and Iino (1965); Morrison and McCapra (1961); Roberts and Doetsch (1966); Weinberg and Brooks (1963).

Fimbriae The term fimbriae (singular, fimbria) was first proposed by Duguid (1955) to designate certain structures that were observed as surface appendages on some bacteria (Figs. 56 and 57). Other names which have been proposed to describe the same structures are threads, fibers, filaments, and pili (singular, pilus).

Fimbriae on *E. coli* measure 0.01 μ in width by 0.3 to 1.0 μ in length, rarely as long as 4 μ. They are not organs of locomotion and should not be confused with flagella.

Several different types of fimbriae, differing in dimensions, number, distribution, adhesive properties, and genetic control, have been described in species of *Enterobacteriaceae* and *Pseudomonadaceae*. They have been designated types 1, 2, 3, 4, and 5, and the sex fimbriae types F and 1b. The commonest type, type 1, confers adhesive properties on bacteria enabling them to stick firmly to mold, plant, and animal cells, including red blood corpuscles, and the epithelial cells of the alimentary, respiratory, and urinary tracts.

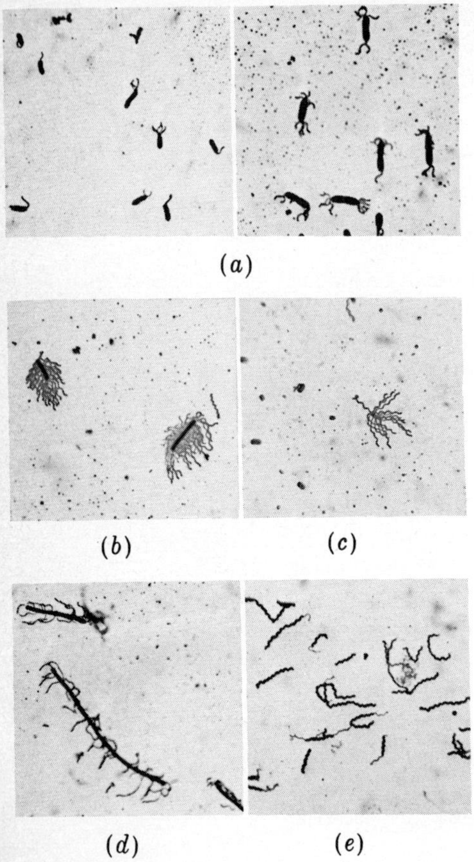

Fig. 55 Shape and arrangement of bacterial flagella. (a) Alcaligenes species; (b) Salmonella typhosa; (c) Proteus vulgaris; (d) mutant of a nitrogen-fixing organism; (e) Treponema pallidum. (After Leifson.)

Because bacteria with type 1 fimbriae have a tendency to adhere to one another, they show a strong capacity for growing in the form of pellicles on the surface of aerobic static liquid media. This property may be of value to saprophytic fimbriate bacteria that are found in poorly aerated, stagnant, waters.

For more information: Duguid (1968); Hashimoto et al. (1963*a*, *b*, 1965); Mayer (1971); Novotny and Lavin (1971); Novotny et al. (1969); Old and Duguid (1970); Old et al. (1968); Yanagawa and Otsuki (1970).

Motion of colonies Motile organisms have been described which exhibit colonial movement when grown on solid media. The first such organism isolated capable of exhibiting this characteristic is *Bacillus alvei*. Agar colonies measuring 0.2 to 0.5 mm in diameter exhibited an average linear motion of about 14 mm per hr. Comparing this figure with the speed of individual cells of other species of motile bacteria gave the following results:

S. typhosa	65 mm per hr
B. megaterium	27 mm per hr
B. alvei colonies	14 mm per hr

Fig. 56 Fimbriae on Escherichia coli. (Courtesy of Duguid.)

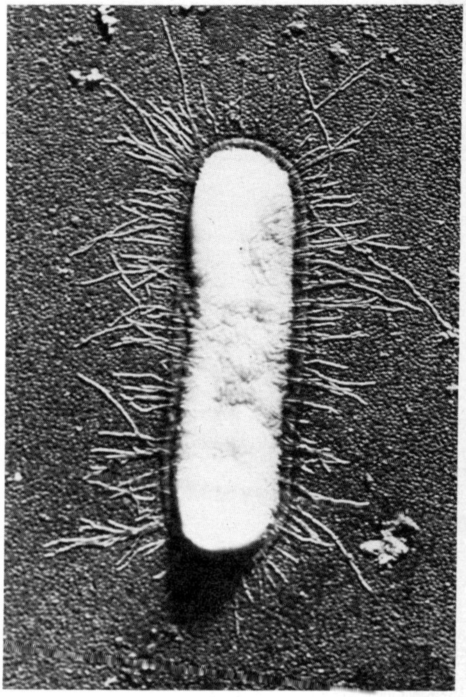

Motile colonies exhibit both a linear and a slow rotary motion. The direction of rotation may be clockwise or counterclockwise, depending upon the species. Both types may be observed in the same species although one type predominates.

Turner and Eales (1941*a, b*) streaked the organism *Clostridium oedematiens* on agar plates. Colonies developed and reached a certain size before migrating. When a colony migrated, it left a peculiar track on the agar surface. A small number of cells were left behind, mostly at the edges of the track, which formed two parallel lines separated by the width of the moving colony (Fig. 58).

Typical migrating colonies pursued curved or spiral paths which were often very elaborate and of relatively great length, even 2 or 3 cm (Fig. 59). The direction of rotation was either clockwise or counterclockwise. After wandering for a variable distance, a colony approached the center of its spiral path with rapidly shortening radius, ceased to migrate, began to

Fig. 57 Electron micrograph of Escherichia coli, showing protoplast, cell wall, typical fimbriae, and one lateral flagellum. (Courtesy of Hashimoto et al.)

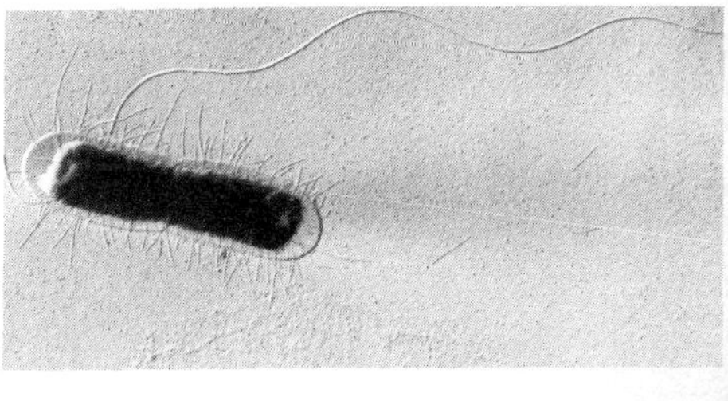

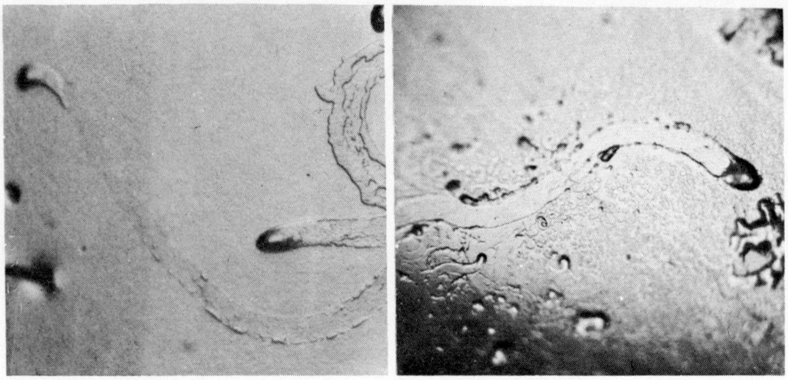

Fig. 58 Rotating and migrating colonies of Clostridium oedematiens. Left, showing tracks and moving colonies; right, showing a large wandering cell group pushing through the swarm film. (Courtesy of Turner and Eales.)

rotate around its center, lost its elongated shape, and increased in size to several times the width of the track at the end of which it was formed.

For more information: Johnston (1965).

Nucleus The nucleus (plural, nuclei) is a structure present in almost all plant and animal cells and is considered to be an essential agent in metabolism, growth, multiplication, and the hereditary transmission of characters. It encloses material known as chromatin, which is regarded as the physical basis of heredity. Chromatin exists in small granules in the resting stage of the nucleus and becomes aggregated in a definite manner into chromosomes before mitotic division.

The search for a nucleus in a bacterial cell has been a lengthy one for three reasons: (1) limitations of the light microscope, (2) lack of a satisfactory staining procedure, and (3) lack of appreciation of the importance of the age of the culture. With the introduction of the electron microscope, appropriate methods of staining, and control of the time and

Fig. 59 Sketch of convoluted track of a wandering colony, showing two series of clockwise spirals followed by a final counterclockwise spiral. The colony had increased considerably in size after coming to rest and showed curved radial markings indicating rotation. The total length of the track was about 2 cm. (After Turner and Eales.)

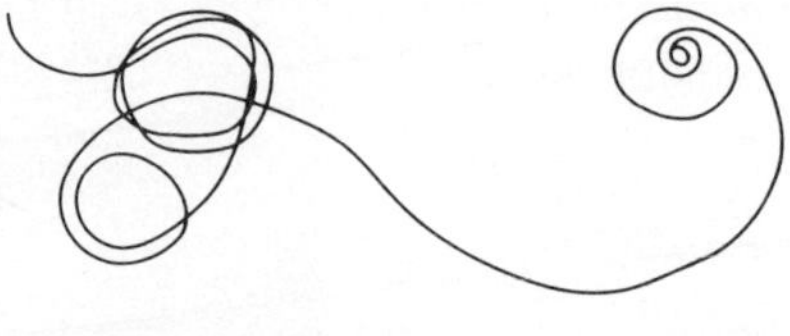

temperature of incubation of the culture, the above difficulties have been largely overcome.

The staining procedure developed to demonstrate the presence of nuclear material in bacteria is the HCl-Giemsa technique of Piekarski and its many modifications. A satisfactory modification by Robinow is as follows: A wet smear of the bacteria is prepared on a glass slide, fixed in osmic acid vapor, dried, and immersed in 1 *N* HCl for about 9 min at 53 to 55°C, then washed and stained in 1:20 Giemsa solution for 10 to 60 min, depending upon the staining properties of the specimen. The deeper colors (blue and violet) tend to localize in the chromatin material comprising the nuclear structures.

Bacterial cells lack a well-organized nucleus. However, structures of varying size, shape, and configuration are demonstrable in bacteria. These bodies take up acid-Giemsa and Feulgen stains, which are routinely employed to demonstrate nuclear apparatus in other cells, and are variously called chromatin bodies, nucleoid, and nuclear material.

The chromatin bodies, according to Robinow (1956), normally lie separately in the cytoplasm, and all those in one bacterium are homologous. Growth and division of chromatin bodies are attended by changes of form only, not by visible changes of texture (Fig. 60). In the simplest type of body, that which in profile looks like a bar or dumbbell, division begins at one end and causes the successive appearance of V-, U-, and H-shaped phases. The chromatin structures of certain bacteria are net-like or sponge-like, and their mode of division is not easily imagined.

Direct division of the chromatin bodies of *E. coli* has been demonstrated by Mason and Powelson (1956) in a series of phase-contrast photomicrographs (Fig. 61). These observations were made on living bacteria. The nuclear areas in the dividing cells appeared to be as clearly defined as the areas in fixed, hydrolyzed, and stained cells.

With the aid of the electron microscope, nuclei may be clearly visible in bacterial cells (Fig. 62). In this electron micrograph by Pontefract et al. (1969), the mesosome appears to form a link between the cytoplasmic membrane and the naked nucleus.

For more information: Cairns (1966); Ryter (1968).

Composition of the nucleus The nucleus consists largely of deoxyribonucleic acid (DNA) combined with protein to produce a deoxyribonucleoprotein. Another nucleic acid, ribonucleic acid (RNA), involved in protein synthesis, is also present to some extent in the nucleus but it is found largely in the cytoplasm (see pages 845 and 846).

A bacterial nucleus contains a single, very long molecule of DNA. It has a molecular weight of about 5×10^9. The length has been calculated as being about 3000 μ. This figure is approximately 1000 times the length of the cell in which it was produced. Electron micrographs of such molecules are shown in Figs. 90 and 317.

When dissolved in a solution of sodium hydroxide, both nucleic acids are hydrolyzed into a mixture of smaller units called nucleotides.

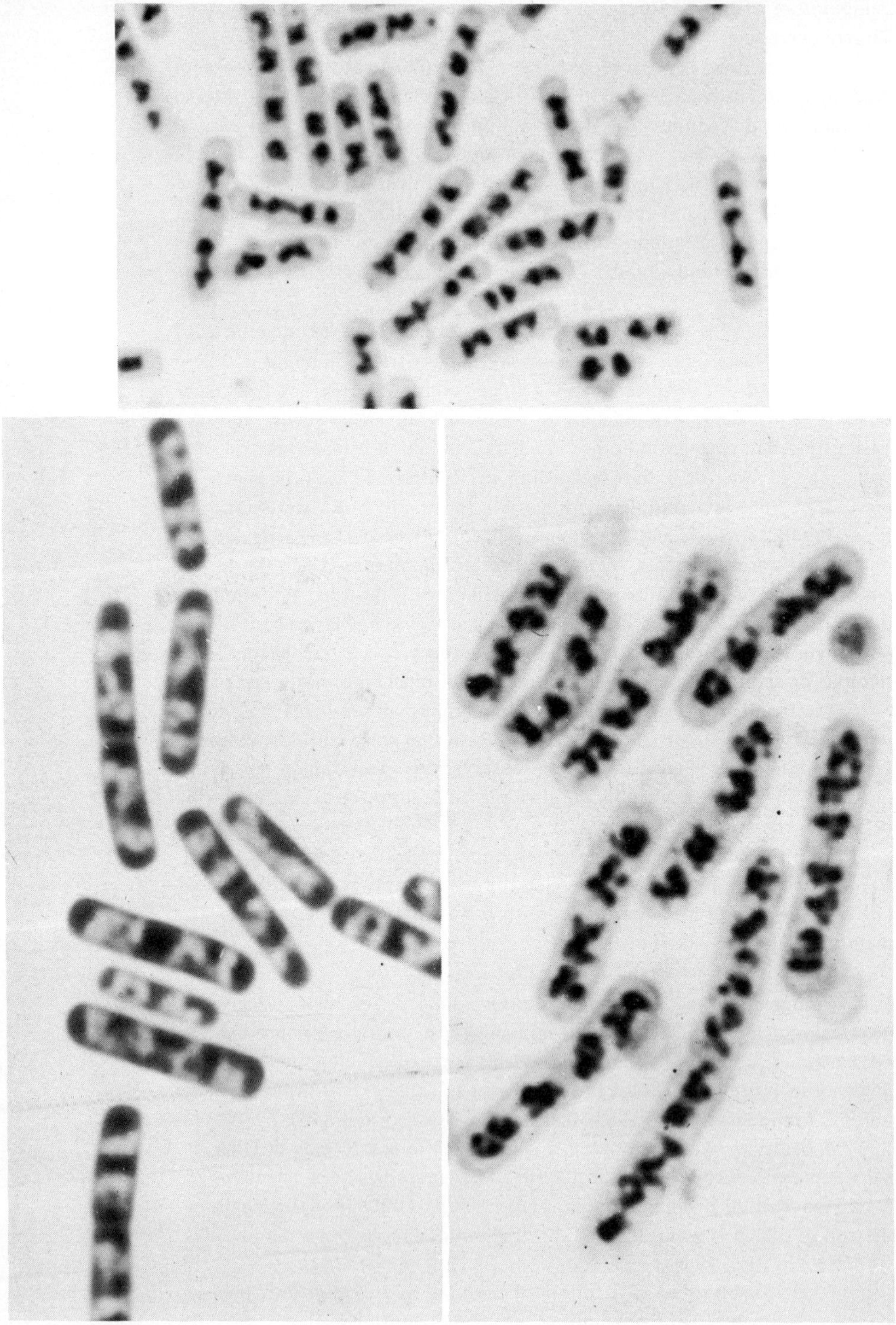

Fig. 60 Chromatinic bodies in bacteria. Upper, Bacillus cereus, successive stages of growth and direct division of the chromatin bodies in bacteria of increasing length; lower left, B. cereus, chromatin bodies stand out, unstained, from the basophile cytoplasm; lower right, B. megaterium, successive stages of the third division of the chromatin bodies in a group of cells growing out of recently germinated spores. (Courtesy of Robinow.)

Fig. 61 Nuclear division in live Escherichia coli. Age of cells in minutes indicated in lower left-hand corner of each photograph. (Courtesy of Mason and Powelson.)

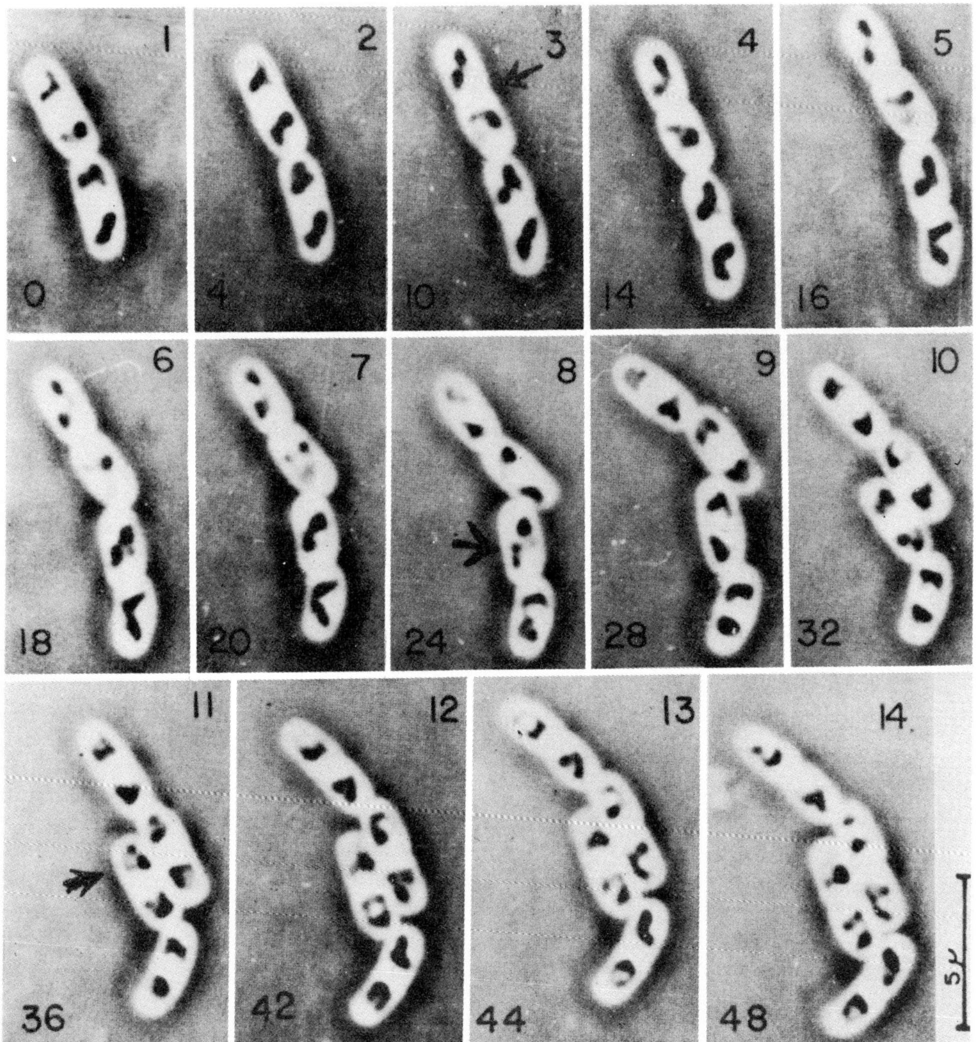

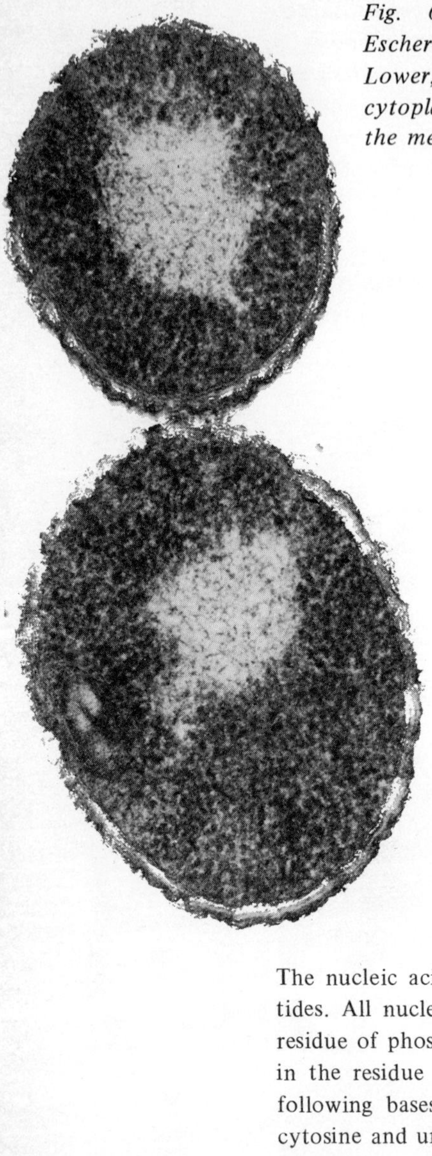

Fig. 62 Electron micrograph of cross sections of two Escherichia coli cells. Nuclei appear near the center of the cells. Lower, a coiled mesosome can be seen in close proximity to the cytoplasmic membrane. An extension of the nucleus is close to the mesosome. (Courtesy of Pontefract, Bergeron, and Thacher.)

The nucleic acid molecule contains normally only four kinds of nucleotides. All nucleotides from RNA contain the sugar ribose, and all have a residue of phosphoric acid attached to carbon 5 of the ribose. They differ in the residue attached to carbon 1 of the sugar. It can be one of the following bases: The purines adenine and guanine, and the pyrimidines cytosine and uracil. The four nucleotides are linked together as shown on page 846.

DNA is similar to RNA but differs in having the sugar deoxyribose instead of ribose, and the base thymine instead of uracil. DNA has been isolated from a few bacteria having a nucleotide containing cytosine with a methyl or hydroxymethyl group on carbon 5. Both RNA and DNA molecules are best visualized as chains of alternating units of ribose or deoxyribose and phosphoric acid, and one of the four bases normally linked to carbon 1 of each sugar unit.

A more detailed discussion on the nucleic acids may be found on page 844.

Function of the nucleus DNA is the genetic material of bacterial cells and some viruses. In other viruses RNA is the genetic material. Therefore both DNA and RNA carry genetic information. Both single-stranded and double-stranded DNA and RNA have been isolated from bacteria and viruses. In general DNA occurs more often as double-stranded and RNA as single-stranded.

For more information: Barry (1964); Lwoff (1962).

Vacuoles Vacuoles have been identified in bacteria. They are cavities in the cytoplasm and contain a fluid called cell sap. As the cell approaches maturity, some of the water-soluble reserve food materials manufactured by the cell dissolve in the sap. Insoluble constituents precipitate out as cytoplasmic inclusion bodies.

Gas vacuoles have also been identified in some bacteria and blue-green algae. The exact function of such structures is not clearly understood. It has been suggested that the gas vacuoles absorb air, after which the oxygen component is metabolized. Gas vacuoles are prominent in species of *Halobacterium* which are obligately aerobic bacteria found in saturated salt brines. Since brine solutions are poor in oxygen, the gas vacuoles aid the cells to rise to the surface to insure greater access to air.

For more information: Cohen-Bazire et al. (1969); Larsen et al. (1967), Van Ert and Staley (1971).

Lipides A number of phospholipides have been isolated from bacterial cells. The cytoplasmic membrane contains from 20 to 30 percent lipide, probably as phospholipides. The cell wall also contains lipides, as lipoprotein and lipopolysaccharide complexes. The cell wall of gram-negative bacteria is richer in lipides than the cell wall of gram-positive bacteria.

Lipides are also found in the cytoplasm of bacteria in the form of fat globules. Fat globules begin to appear in 24-hr cultures and reach a maximum usually in about 48 hr (Fig. 63). Some cells may contain only one large globule; others may show the presence of several small, scattered globules.

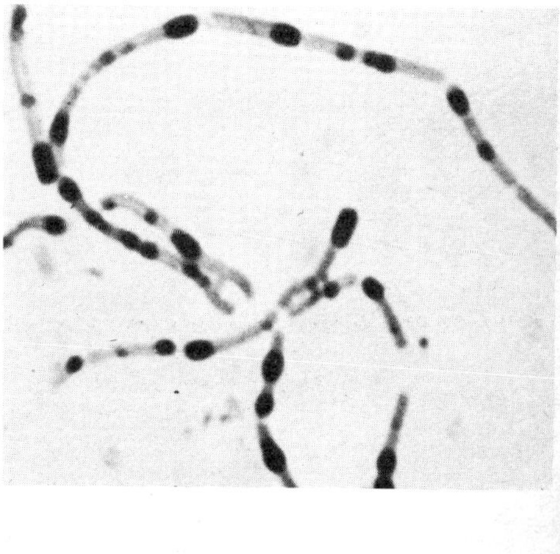

Fig. 63 Bacillus cereus, showing fat globules. From a 48-hr glycerin agar culture stained with Sudan black B and safranine. Fat globules appear bluish black; cytoplasm stains pink. (Courtesy of Burdon, Stokes, and Kimbrough.)

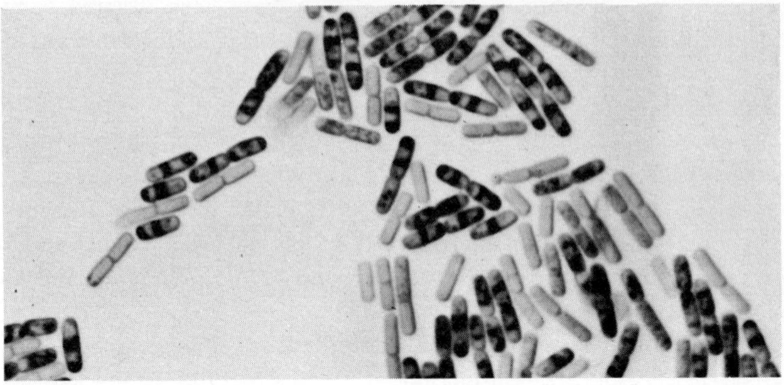

Fig. 64 Bacillus cereus (grown from spores) oxidized with sodium metaperiodate and stained with sulfite-decolorized basic fuchsin. Dark areas indicate the presence of polysaccharide. Cells 10 hr old. (After Pennington.)

Lipides are believed to function as a storage form of carbon atoms. Globules usually cannot be demonstrated in young, vigorously growing cells. As the cells age and slow down in activity, fat globules accumulate in the cytoplasm and may be recognized by appropriate staining.

Polysaccharides Polysaccharides occur (1) in cell membranes, (2) extracellularly in capsules and gums, and (3) intracellularly in the cytoplasm. The first two have already been discussed.

Polysaccharides may be demonstrated by treating bacteria with sodium metaperiodate followed by staining with sulfite-decolorized basic fuchsin (Fig. 64). In *B. cereus* the polysaccharide is concentrated in the cytoplasmic membrane as well as in the cell wall. Polysaccharide areas are colored red by the stain.

In *E. coli* polysaccharides may account for 4 to 25 percent of the dry weight of the cell and there is evidence that it consists of granules of glycogen. The glycogen is formed in media containing a rich source of

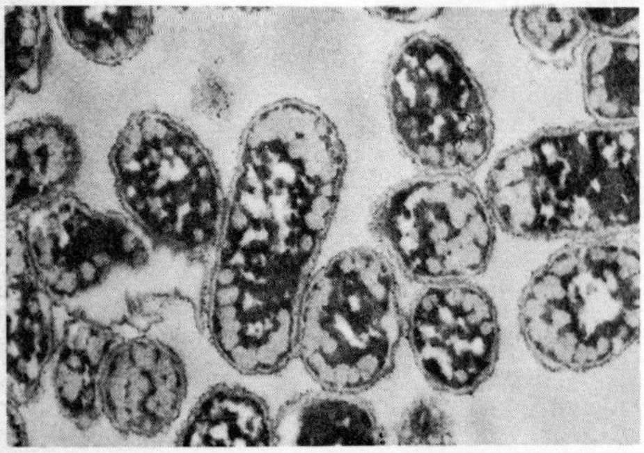

Fig. 65 Electron micrograph of glycogen granules in the cytoplasm of Escherichia coli. The granules are predominantly located at the ends of the cells. x30,000. (Courtesy of Holme and Cedergren.)

Fig. 66 Electron micrographs of sections through cells of Sarcina ventriculi showing intercellular cellulose with fibrous structure. (Courtesy of Canale-Parola, Borasky, and Wolfe.)

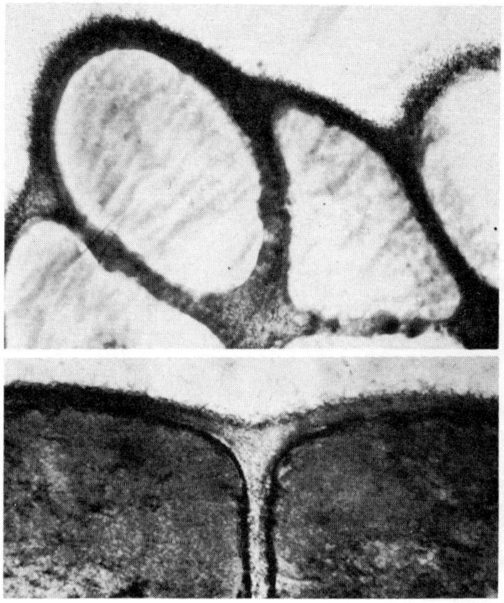

carbon and a minimum concentration of nitrogen. The glycogen accumulates in the cytoplasm at the ends of the cell in the form of granules having diameters of 50 to 100 nm (Fig. 65).

Glycogen synthesis appears to bear an inverse ratio to growth rate. The growth rate may be decreased by supplying the medium with a limiting concentration of nitrogen. When nitrogen-starved cells are supplied with adequate nitrogen, the glycogen granules are rapidly degraded. The result would indicate that glycogen functions as a reserve source of food, being produced during periods of inactivity and utilized by the cell when rapid conditions for growth are reestablished.

Cellulose The presence of the polysaccharide cellulose has been reported in the cells of *Sarcina ventriculi*. Electron micrographs indicate that cellulose constitutes either totally or in great part the cementing material that binds the cells together into large packets. Cellulose was found to have a fibrous structure. It is not present in the cell wall (Fig. 66).

For more information: Canale-Parola et al. (1961); Carrier and McCleskey (1962); Holme and Cedergren (1961).

Metachromatic granules Bacteria contain certain inclusion bodies known as volutin or metachromatic granules (Fig. 67). The granules are small in young cells and become larger with age. They are believed to originate in the cytoplasm of young cells and to localize in the vacuoles of mature forms. The granules show a strong affinity for basic dyes indicating that they are acid in character.

The exact composition of metachromatic granules is not known. They are said to contain inorganic polyphosphate, lipoprotein, RNA, and magnesium. Their presence and size in cells is related to the phosphate

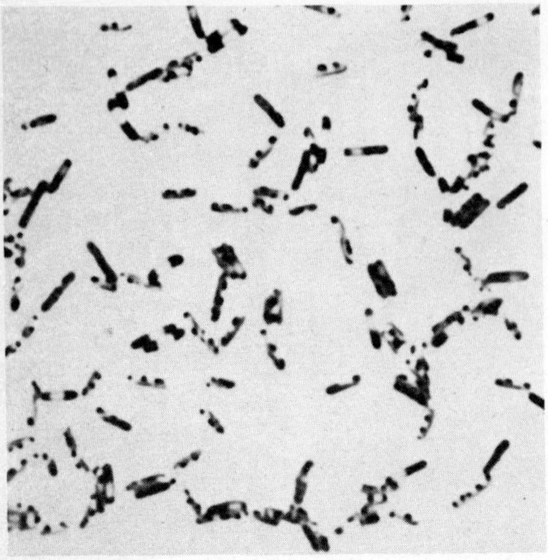

Fig. 67 Metachromatic granules in Bacillus subtilis. Culture 18 hr old and stained with a 1:5000 solution of crystal violet.

concentration of the medium in the presence of an energy source and certain specific divalent ions. Older cells contain larger granules. Their basophilic nature apparently does not depend on the presence of RNA or DNA.

Harold (1963) and Harold and Sylvan (1963) reported that growing cells of *Aerobacter aerogenes* contained traces of inorganic polyphosphate, but large amounts accumulated when growth ceased as a result of a nutritional deficiency. The reciprocal relationship between growth and phosphate accumulation was traced to competition between nucleic acid synthesis and polyphosphate for intracellular phosphorus. Polyphosphate accumulated only after nucleic acid synthesis ceased. This explains why young, rapidly growing cells contain very few, if any, metachromatic granules whereas old cells show an abundance of such granules.

Several functions have been attributed to metachromatic granules: (1) reserve source of food, (2) phosphate storage structures, (3) linkage with energy metabolism, and (4) involvement in nucleic acid synthesis and cell division.

It was shown many years ago that the addition of inorganic phosphate to a suspension of yeast cells, previously subjected to phosphate starvation, caused a massive accumulation of the compound in the cells in the form of granules. The granules were subsequently shown to be composed largely of polyphosphate. Friedberg and Avigad (1968) demonstrated the presence of polyphosphate granules in the bacterium *Micrococcus lysodeikticus* (Fig. 68). Such particles may or may not be the same as metachromatic granules although they are generally regarded as distinct entities. Both metachromatic and polyphosphate particles appear to accumulate when growth ceases for physiological reasons or because of the presence of metabolic inhibitors. The organism *M. lysodeikticus* appears to be an exception to the rule, in that polyphosphate particles accumulate during the logarithmic growth phase and gradually disappear during the stationary phase.

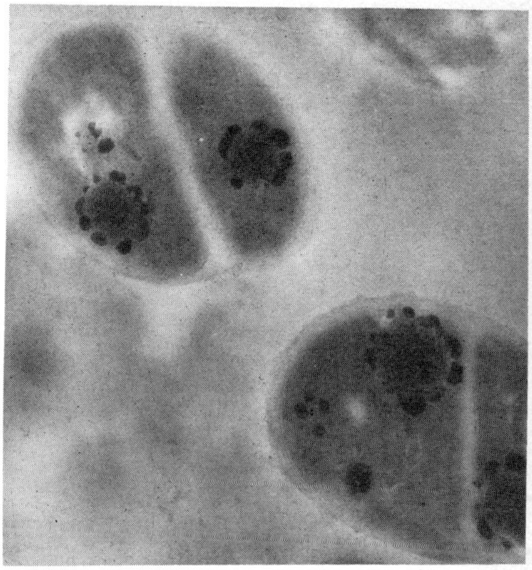

Fig. 68 Electron micrograph of thin section of Micrococcus lysodeikticus showing the presence of polyphosphate particles. x50,000. (Courtesy of Friedberg and Avigad.)

Polyphosphate is a linear phosphate compound with a chain length of from four to almost any number of phosphate residues linked together as follows:

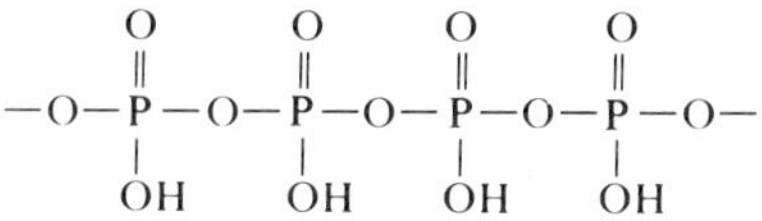

The exact function of polyphosphate in bacteria and other living cells is not clearly understood.

For more information: Friedberg and Avigad (1970); Harold (1966); Widra (1959).

Mitochondria Respiratory enzymes in plant and animal cells are confined to the membranes of organelles known as mitochondria (singular, mitochondrion). The mitochondrial membranes contain the enzymes which produce oxidative phosphorylations, the reactions by which the cells obtain the energy required to sustain life.

Examination of bacterial cells does not reveal structures resembling mitochondria. However, bacteria are able to perform the same chemical reactions with similar enzymes differently arranged. The bacterial respiratory enzymes are also bound to membranes. The concentration of such enzymes is determined by the area of the membranes. The smaller the cell, the larger will be the ratio of surface area to volume. Since bacterial cells are smaller than plant and animal cells and about the same size as mitochondria, they can dispense with such structures.

As stated above, the mitochondrion supplies most of the energy of cell metabolism by virtue of its ability to carry out oxidative phosphorylation. Because of this fact, it has been called the power house of the cell.

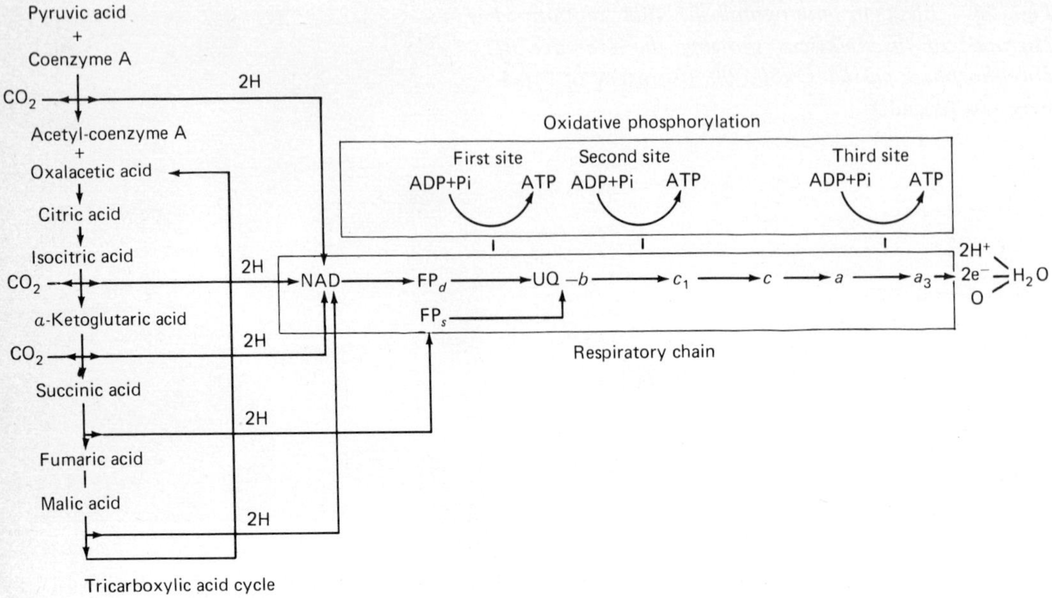

Fig. 69 Oxidative phosphorylation. See text for description. (After Racker with modifications.)

The universal carrier of energy of the cell is adenosine triphosphate (ATP). The ATP molecule functions by transferring the terminal high-energy phosphate group to another molecule converting itself to adenosine diphosphate (ADP). The ADP is then regenerated to ATP by energy-generating systems in the cell after which it can transfer more energy-rich phosphate to another molecule.

The mitochondrion in an animal cell produces three molecules of ATP for each pair of electrons released by oxidation. The same is most likely true of the chloroplasts of plant cells. This makes the chloroplast and the mitochondrion the favored power organelles for generating energy in living cells.

Racker (1968) advanced a scheme, similar to the one that appears in Fig. 69, to show how the process of oxidative phosphorylation harnesses the energy from foodstuffs to produce ATP, the energy carrier of the cell. Fats, proteins, and carbohydrates are partially metabolized and then, in mitochondria, enter the tricarboxylic acid cycle where they are broken down to carbon dioxide. In the process, hydrogen atoms are accepted by the coenzyme NAD. The chain of respiratory enzymes including flavoproteins and cytochromes catalyzes the transfer of electrons to produce water. Phosphorylation is coupled to electron transfer at three sites.

Structure of the mitochondrion A schematic cross section of a mitochondrion, according to Racker (1968), is shown in Fig. 70. The organelle has two membranes (*a*) each about 60 Å (six-millionths of a

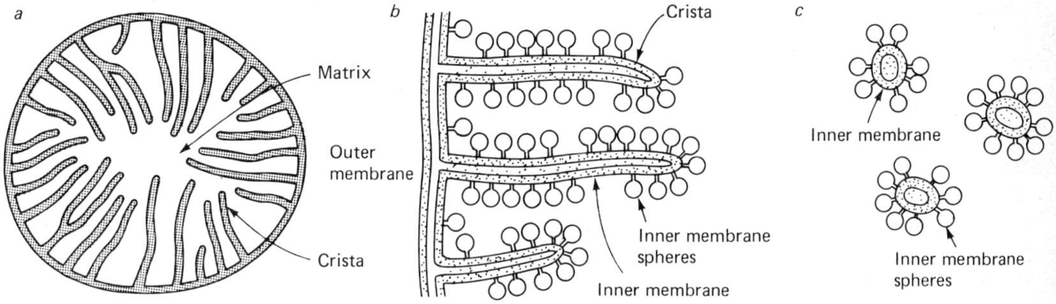

Fig. 70 Schematic cross section of a mitochondrion and its fragmented particles. See text for descriptions. (Courtesy of Racker.)

millimeter) in thickness. The inner membrane is deeply folded into cristae (singular, crista) covered with the inner-membrane spheres, each about 85 Å in diameter (*b*). The inner membrane, with its spheres, is the site of oxidative phosphorylation. Mitochondria exposed to sonic vibrations become fragmented into small submitochondrial particles (*c*), which are still capable of oxidative phosphorylation.

The fluid between the two membranes provides communication between the two layers and supplies the enzymes in the membranes with the necessary coenzymes.

The membranes are composed of two principal structural materials: (1) a protein, making up about 80 percent of the weight of the membranes; and (2) a lipide, making up the remaining 20 percent. Both protein and lipide by themselves are insoluble in water. However in a structural combination known as a micelle, the lipide becomes soluble in water. This is because one end of the lipide molecule is hydrophobic (water-repelling) and the other end is hydrophilic (water-attracting). Lining up the lipide molecules one against another and all facing the same way, the hydrophilic ends face the water, and the hydrophobic ends are anchored to tissue. Therefore, the membranes consist of a combination of structural protein and phospholipide micelles.

For more information: Lehninger (1964); Meyer and Simpson (1968); Nass (1969).

Ribosomes Ribosomes are particles located in the cytoplasm of bacteria and are the sites of protein synthesis. They are discussed under Protein Synthesis in Chap. 13, page 466.

Protoplasts and spheroplasts The cell wall may be removed from bacteria without destroying the vital nature of protoplasm. In the absence of the wall, the cell loses its rigidity and rounds up into osmotically sensitive, spherical forms (Fig. 71).

The walls of gram-positive and gram-negative bacteria differ in composition. The latter contain lipoproteins and lipopolysaccharide complexes which are not present in the former. By appropriate treatment, cell

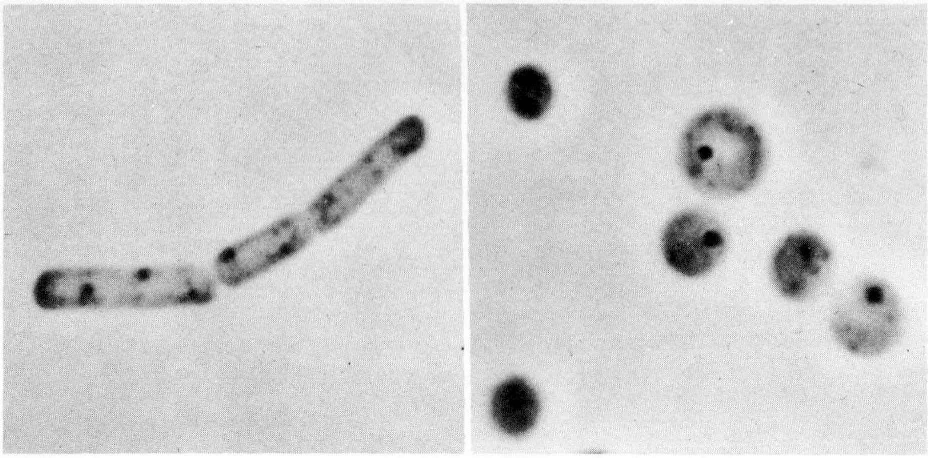

Fig. 71 Photomicrographs of Bacillus megaterium. Left, cells before lysis with lysozyme; right, cells after lysis with lysozyme in 0.02 M sucrose and 0.03 M phosphate. (Courtesy of Weibull.)

walls may be removed from bacteria without destroying the cells. The walls of gram-positive bacteria are more easily removed than those of gram-negative bacteria. Therefore the term protoplast should be used to indicate living protoplasm surrounded by the cytoplasmic membrane but with the cell wall completely removed. If parts of the cell wall still remain, as is likely after treatment of gram-negative bacteria, the term spheroplast should be used.

Methods for removal of cell wall A number of methods have been employed for the removal of the cell walls from bacteria. A good method for one organism may not be good for another. Generally a method that is satisfactory for one member of a genus may be employed on all the other species in that genus. Methods which have been employed include: (1) treatment with lysozyme at pH 9; (2) treatment with a mixture of lysozyme and phospholipase or acid phosphatase; (3) treatment with lysozyme followed by the addition of ethylenediaminetetraäcetic acid (EDTA); (4) action of appropriate bacteriophage; (5) growth in the presence of penicillin; (6) growth in sucrose broth at appropriate pH; deprival of diaminopimelic acid; and (7) growth in the presence of 1 to 5 percent glycine.

Properties of protoplasts The difficulty in handling and studying protoplasts is their extreme fragility and sensitivity to osmotic shock, shaking, centrifugation, and aeration. Removal of the cell wall does not change the structure and capabilities of the protoplasm. Permeability and respiration appear to be the same for protoplasts and normal intact cells.

Kawata et al. (1961) prepared protoplasts of *B. megaterium* and found that the spherical cells still retained their flagella, another indication that such appendages are not a part of the cell wall (Fig. 72).

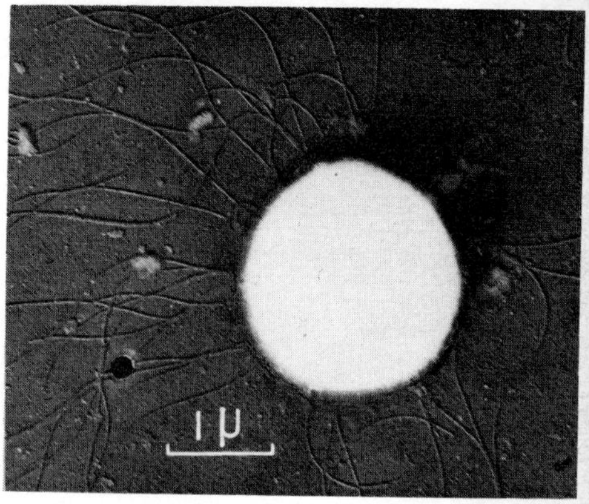

Fig. 72 Electron micrograph of a spheroplast of Bacillus megaterium apparently devoid of cell wall but still possessing flagella. (Courtesy of Kawata, Asaki, and Takagi.)

Op den Kamp et al. (1969) found the shape and osmotic susceptibility of protoplasts of *B. subtilis* to depend on the growth conditions. Cells harvested from a neutral growth medium gave spherical protoplasts which lysed rapidly, whereas cells grown in an acid medium maintained their normal rod-shaped form to a great extent after removal of the cell wall, even after being suspended in a hypotonic medium. The latter observation suggested the presence of a more rigid membranous structure in cells which have been exposed to a low pH during growth.

Kusaka (1967) reported that protoplasts of *B. megaterium* grew well and divided in nutrient broth containing 0.5 *M* NaCl as the stabilizer. Protoplasts enlarged from 1.8 to 3.5 μ in diameter, then a small protuberance formed in the enlarged cells. The swellings enlarged until eventually symmetrical, dumbbell-shaped bodies were formed, which then separated into two daughter protoplasts having a diameter of about 2.5 μ. Membrane-bound amino sugar content was considerably reduced when the cells were grown in the presence of penicillin (Fig. 73).

Prepared protoplasts of *B. megaterium*, followed by incubation, are capable of sporulation. The morphological properties, germination, resistance to ultraviolet irradiation, and thermal resistance of the protoplast spores are the same as the spores produced by normal cells.

Under special conditions protoplasts grow and probably divide like normal cells. However, there is no evidence that protoplasts form colonies directly. The metabolism of protoplasts and normal cells appears to be very similar but probably not identical.

Reversion of spheroplasts to rods Altenbern (1963), Hirokawa (1962), and Nermut and Svoboda (1962) found that spheroplasts of *P. vulgaris*, *P. mirabilis* and *E. coli*, produced by penicillin, reverted to the bacillary forms when cultivated in the absence of the antibiotic.

Landman and Forman (1969) and Landman et al. (1968) reported that protoplasts of *B. subtilis*, placed in a 25 percent gelatin medium and incubated at 26°C, reverted back to the normal walled forms.

The process appears to be a reversible one, being dependent upon the conditions of the environment.

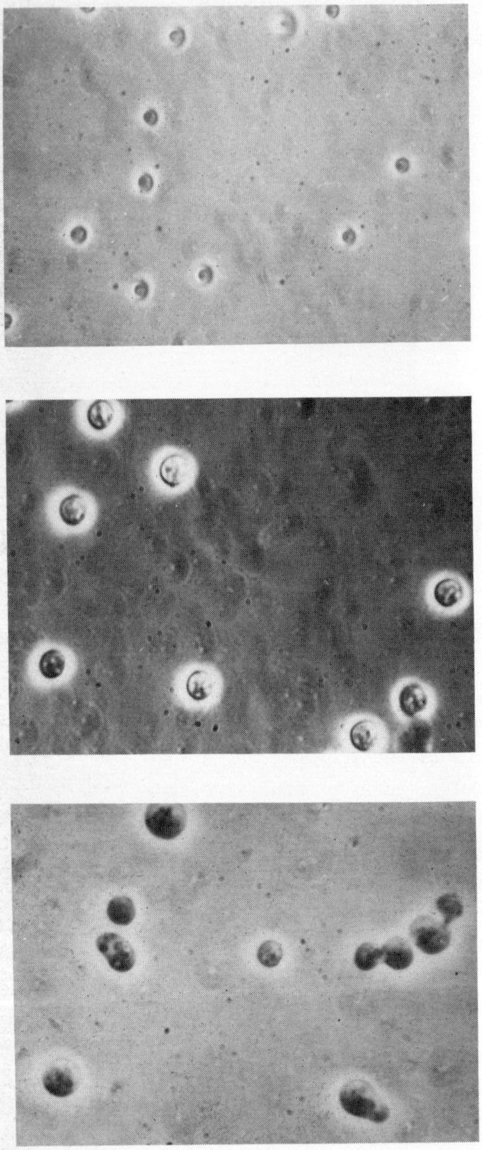

Fig. 73 Electron micrographs of Bacillus megaterium protoplasts. Upper, initial protoplasts; center, enlarged protoplasts 2½ hr after incubation; lower, dividing protoplasts 3½ hr after incubation. See text for description. (Courtesy of Kusaka.)

Filterability of protoplasts Sinkovics (1958) reported the spontaneous occurrence of units in aged cultures of *E. coli* which conformed to the description of artificially induced bacterial protoplasts. The disintegration of cells from aged cultures was preceded by swelling of the cell and rupture of the rigid cell wall. Centrifugation of the culture gave a supernate which, after filtration, contained units capable of regeneration when placed in fresh medium. The smallest units capable of regeneration measured about 350 nm in diameter. The units underwent fusion before cell-wall formation occurred.

The results supported the assumption that aged *E. coli* cultures could survive in the form of units having no cell walls and which, under adequate conditions, regenerated into vegetative forms.

Immunogenicity of spheroplasts Diena et al. (1968) prepared spheroplasts of various serotypes of *E. coli* by cultivation in the presence of glycine. Then the spheroplasts were inactivated by the addition of thimerosal. The killed suspensions were adjusted to the proper pH and lysed by the addition of pronase (a protease). The lysed spheroplasts proved to be good vaccines when used in experiments on protection of mice against challenge by all serotypes. The lysed spheroplast vaccines were considerably less toxic than killed whole-cell vaccines.

For more information: Birdsell and Cota-Robles (1968); Boatman (1964); Corner and Marquis (1969); Cundliffe (1968); DeVoe et al. (1970); Diena et al. (1965, 1968); Fitz-James (1964, 1971); Guze (1968); Hines et al. (1964); Iannetta and Wedgwood (1967); Kawata and Inoue (1963); Kusaka (1971); Marquis (1967); Mohan et al. (1965); Roth et al. (1971); Rubenstein et al. (1970); Sud (1967); Vaituzis and Doetsch (1966); Willett and Thacore (1967).

Endospores Endospores are bodies produced within the cells of a considerable number of bacterial species. They are more resistant to unfavorable environmental conditions, such as heat, cold, desiccation, osmosis, and chemicals, than the vegetative cells producing them. However, it is debatable whether such extreme conditions actually occur in nature. For instance, the resistance of spores to high temperatures is a laboratory phenomenon and probably never occurs in a natural environment.

A number of theories have been advanced to explain the greater resistance of spores to adverse conditions when compared to the vegetative cells producing them. From available information, the theories do not offer a satisfactory explanation and the problem remains unsolved.

The bulk of evidence indicates the existence of a close relationship between spore formation and the exhaustion of nutrients essential for continued vegetative growth. Sporulation is a defense mechanism to protect the cell when the occasion arises.

Desser and Broda (1969) measured the endogenous respiration of spores of *B. cereus* through the CO_2 produced. Dry spores exhibited no measurable activity. Wet spores showed slight activity, being greatest at a temperature of 40°C.

Spore formation is limited almost entirely to two genera of rod-shaped bacteria: *Bacillus* (aerobic or facultatively anaerobic) and *Clostridium* (anaerobic or aerotolerant). With few exceptions, the spherical bacteria do not sporulate. Some spore-bearing species can be made to lose their ability to produce spores. When the ability to produce spores is once lost, it is seldom regained. Sporulation is not a process to increase bacterial numbers because a cell rarely produces more than one spore.

For more information: Falcone et al. (1966); Grecz et al. (1970); Maeda et al. (1968).

Properties of spores Spores may be spherical, ellipsoidal, or cylindrical in shape. The position of the spore in the cell may be central, eccentric, subterminal, or terminal. A fully grown spore may have a diameter the same size as, or greater than, that of the vegetative cell. This large a spore causes a bulging of the cell into a clostridium, if central, and a plectridium, if terminal. As a rule, each species has its own characteristic size, shape, and position of the spore but this is subject to variation under different environmental conditions.

Fine structure of spores Franklin and Bradley (1957), by means of carbon replicas, reported that the spores of a majority of species of

Fig. 74 Electron micrographs of carbon replicas of spores. Upper left, Bacillus circulans; upper right, B. subtilis; lower left, B. alvei; lower right, Clostridium perfringens. (Courtesy of Franklin and Bradley.)

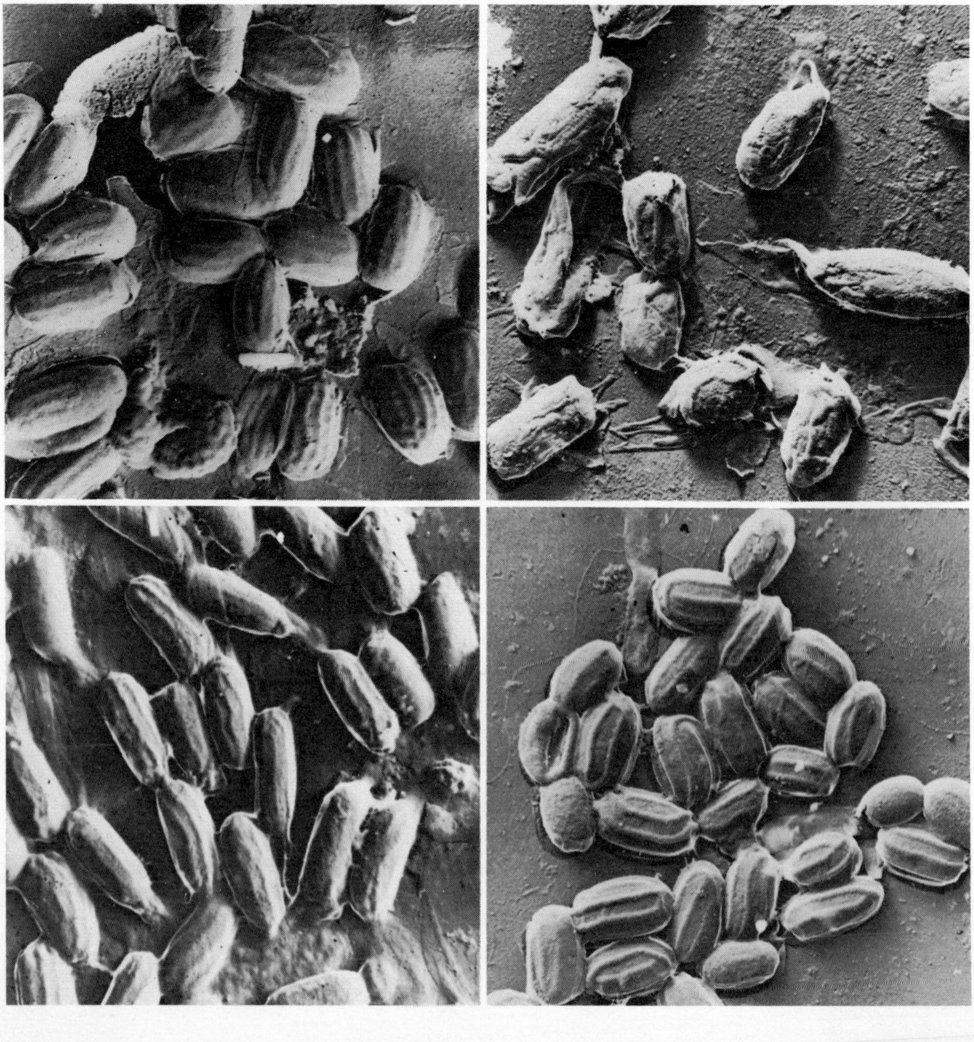

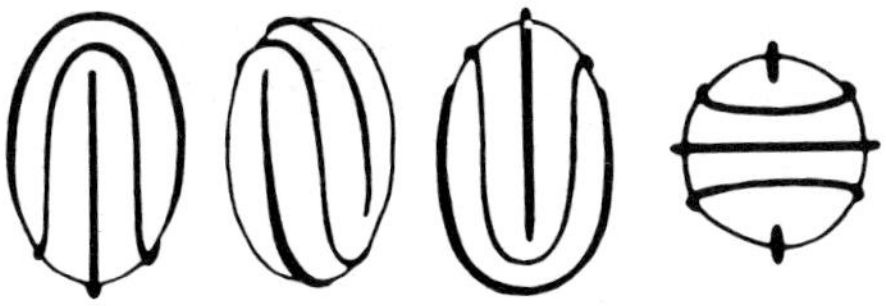

Fig. 75 Surface structure of a spore of Bacillus polymyxa. From left to right: side view; same rotated a quarter turn from right to left; same rotated a further quarter turn; view of a pole. (From Franklin and Bradley.)

Bacillus and *Clostridium* were readily distinguished by surface patterns. The surface may be smooth or ribbed, with the ribs usually longitudinal (Fig. 74).

Drawings of the surface sculpturing of spores of *B. polymyxa* are shown in Fig. 75. The sculpturing consists of a single endless ridge in the form of two loops similar to the markings on a tennis ball, together with two separate ridges terminating within the loops. The markings resemble those on the spores of *C. perfringens* (Fig. 74, lower right).

Ultrathin sections of *B. polymyxa* spores, by van den Hooff and Aninga (1956), are shown in Fig. 76. The spore coat consists of an outer and inner layer separated by a space. The outer layer is called the exine and the inner layer the intine. The intine faintly follows the surface relief. The central core is separated from the intine by a regular nonosmophilic space. A peripheral spot is present in the core which probably represents nuclear material.

Rode (1968) examined the spore structure of six strains of *B. megaterium.* Three were designated AL types and three GN types on the basis of their nutritional requirements. AL spores were morphologically

Fig. 76 Electron micrograph of ultrathin sections of spores of Bacillus polymyxa. See text for details. (Courtesy of van den Hooff and Aninga.)

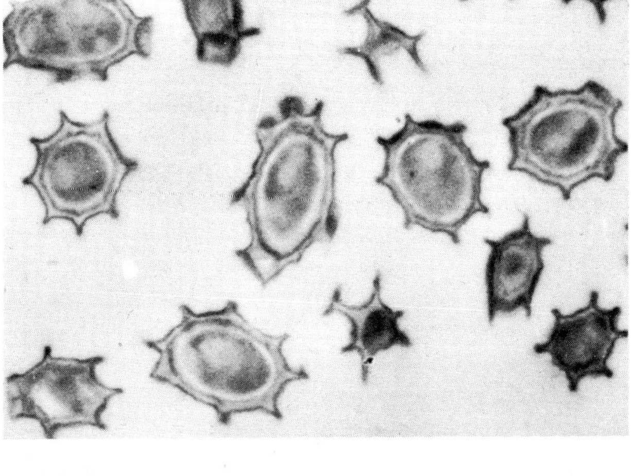

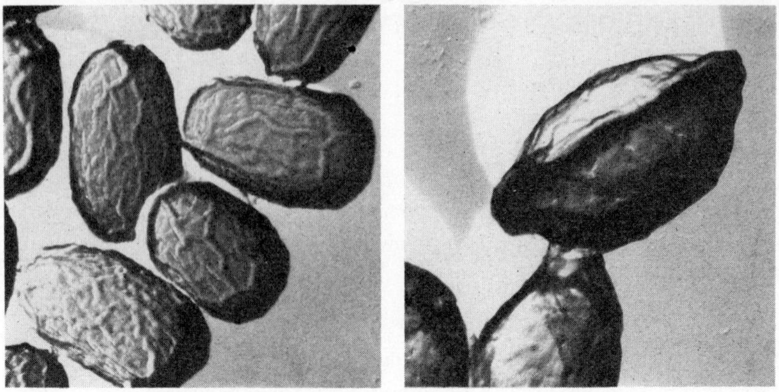

Fig. 77 Carbon replicas of spores of two strains of Bacillus megaterium. Left, spores germinated in a mixture of L-alanine and inosine; right, spores germinated in a mixture of glucose and potassium nitrate. Note the contrasting appearance of the two types. (Courtesy of Rode.)

distinct from the GN spores (Fig. 77). The two spore groups differed also in germination inhibition characteristics, dipicolinic acid content, hexosamine content, phosphorus and magnesium content, spore coat features, ion exchange properties, and heat resistance.

In other investigations, Rode et al. (1967) and Pope et al. (1967) described numerous broad ribbon-like appendages attached to one end of spores of *Clostridium* species (Figs. 78 and 79). The appendages are two or three times the length of the spore and, at their maximal dimension, may be two-thirds the width of the spore. They are attached to the spore body by a common trunk which is continuous with the outer spore coat. Each appendage is a multilayered structure and is enclosed in an amorphous material. The function of the appendage is not known.

For more information: Bulla et al. (1969); Cole et al. (1970); Gerhardt and Ribi (1964); Murphy and Campbell (1969); Pope and Rode (1969); Rode et al. (1971); Samsonoff et al. (1970, 1971); Santo et al. (1969); Beaman et al. (1972); Yolton et al. (1972).

Fig. 78 Replica series showing sequential steps in development of the free spore of a Clostridium species. Age of culture varied from 6 days (A through E) to 7 days (F through J). (A) Vegetative cell with peritrichous flagella; (B) early indication of spore formation; (C) sporangium with maturing spore; (D) spore body is free and has a rough surface; (E, F) first evidence of spirally arranged ribbon-like appendages in the disintegrating vegetative cell; (G) spore body with exposed spiral of attached appendages; (H) free spore with some appendages still spirally arranged; (I) free spore with appendages appearing now as a tuft of parallel ribbons; (J) free spore with ribbon-like appendages separated and flared. (Courtesy of Rode, Crawford, and Williams.)

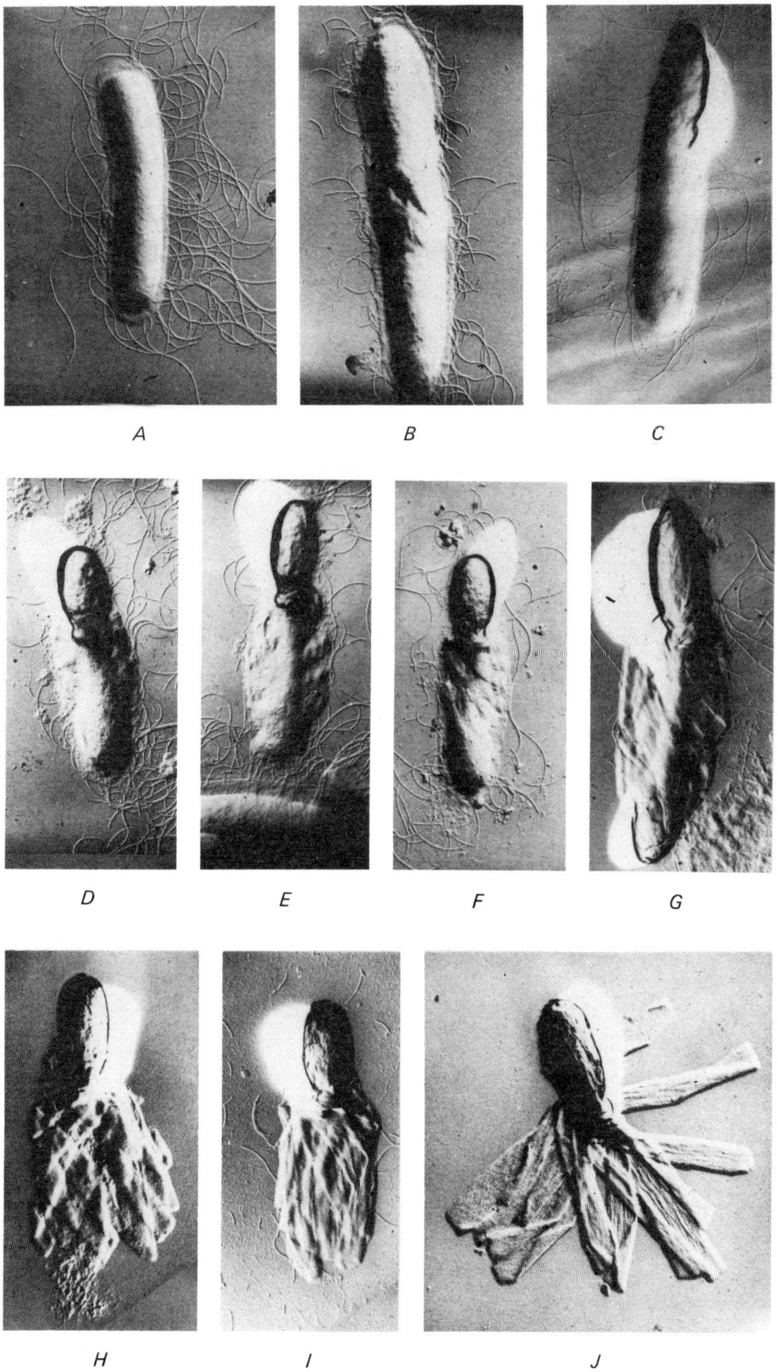
A
B
C
D
E
F
G
H
I
J

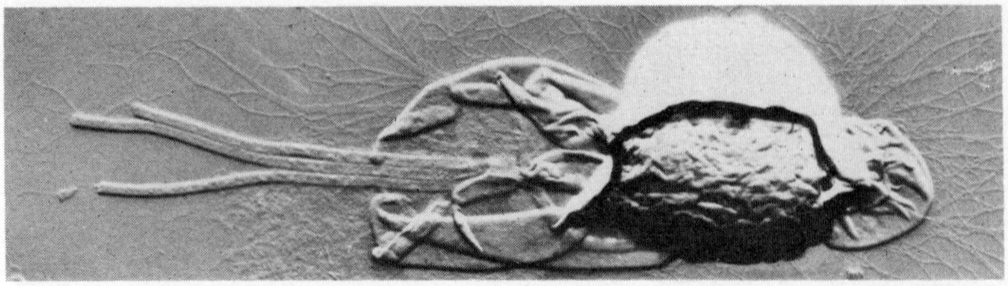

Fig. 79 Appendages of Clostridium bifermentans spore. Typical free spore. The irregular surface of the spore body, the folded and collapsed exosporium, and three appendages which project well beyond the exosporium margin are apparent. (Courtesy of Pope, Yolton, and Rode.)

Enzymes of spores The presence of enzyme systems in *Bacillus* spores has been reported. Some of these are inorganic pyrophosphate phosphohydrolase that requires manganese for activation, adenine ribosidase that hydrolyzes adenosine, L-alanine racemase that converts L-alanine to D-alanine, several glucose oxidoreductases, keto-1-phosphate aldehydelyase, proteinases, esterases, and adenylate kinase.

The L-alanine racemase is an interesting enzyme in that it not only converts L-alanine into a racemic mixture of L- and D-alanine but is active in the intact spore, is heat-resistant to temperatures as high as 100°C, and remains heat-resistant even after the spore has germinated. Furthermore, its resistance to heat depends upon attachment to some colloidal material. Upon separation from such material, the enzyme becomes heat-sensitive.

Other heat-resistant enzymes which have been demonstrated in spores are catalase and ribosidase (hydrolyzes adenosine to adenine and ribose). Doubtless other enzymes in addition to the above are also present.

For more information: Bach and Sadoff (1962); Sierra (1967).

Sporulation process Conditions necessary for sporulation in one species do not necessarily apply to another. The subject appears to be in such a state of confusion that it is impossible to discuss sporulation in terms of generalities.

Conditions which have been reported as favoring sporulation include: the addition of certain metals to the medium such as Cu^{2+}, $MoO_4{}^{2+}$, Fe^{2+}, Zn^{2+}, Mn^{2+}, and Ca^{2+}; addition of glucose; addition of glutamic acid; addition of tomato juice; addition of $CaCO_3$ to a carbohydrate medium to prevent accumulation of acid; shaking a culture of vegetative cells with distilled water at 37°C; incubating the culture at an appropriate temperature; the presence of O_2 (air); the addition of certain amino acids.

As already stated, spores are generally produced when essential nutrients for continued vegetative growth become exhausted. Kerravala et al. (1964) found *B. subtilis* to be an exception in that it could sporulate when cultivated in the presence of essential nutrients.

Ellar and Lundgren (1966) diagrammed the morphological changes, duration, and time of occurrence, during sporulation of *B. cereus*

Fig. 80 Diagrammatic representation of the morphological changes, their duration and time of occurrence, during sporulation of Bacillus cereus. The duration of each stage is indicated. (Courtesy of Ellar and Lundgren.)

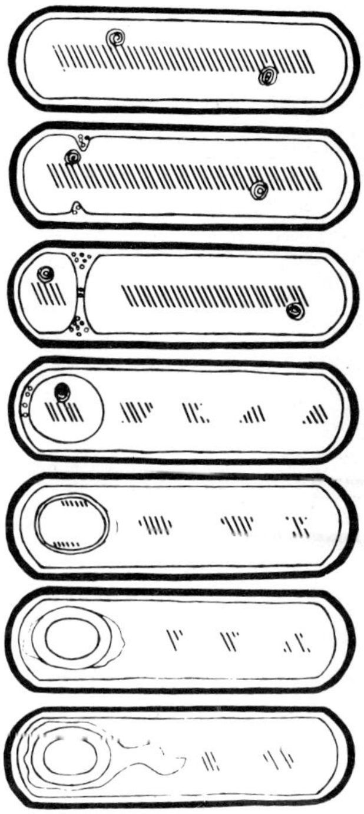

(Fig. 80). The developmental stages of sporulation occurred in a fairly synchronous manner and were complete after 14 hr. Sporulation could be separated into six morphological stages. The initiation of the spore (forespore) septum took the form of an inward folding of the cytoplasmic membrane toward the pole of the cell. The inward folding formed a characteristic Y-shaped membrane structure enclosing an area within which vesicles were found. The vesicles comprised the perisporal mesosome of the cell. The membranes on opposite sides of the cell progressed toward the cell center where they fused to form the double unit membrane of the spore septum. As proliferation of the spore septum continued, the vesicular areas moved toward the pole. The end result was a double forespore membrane that completely enclosed a part of the chromatin of the vegetative cell. Sporal mesosomes, as well as membrane vesicles, were involved in the proliferation of the forespore.

For more information: Bonsen et al. (1969); Buono et al. (1966); Haynes and Rhodes (1966); Hoeniger et al. (1968); Kay and Warren (1968); Nakata (1964, 1966); Somerville (1971); Spudich and Kornberg (1969).

Microcycle sporogenesis The phenomenon of microcycle sporogenesis, or the formation of a spore from a spore without intervening cell division, was first reported by Vinter and Slepecky (1965). Primary cells resulting from the germination of spores of *B. cereus* and *B. megaterium* were capable of elongation and sporulation without intermediate cell division. Apparently second-stage spores of the same species were unable to complete microcycle sporogenesis in media which permitted microcycle sporogenesis of primary spores.

For more information: Freer and Levinson (1967); MacKechnie and Hanson (1968).

Parasporal body Some spore formers produce a nonviable parasporal body during the process of sporulation. It is globular and may be as large as or larger than the spore. In *B. cereus*, according to Hannay (1961), the spore is terminal and the parasporal body paracentral, lying between the spore and the remaining vegetative cell chromatin bodies (Fig. 81). On completion of sporulation, both spore and parasporal body are contained

Fig. 81 Fowler's bacillus. Top, 4-hr culture stained to reveal the cell wall; center, 16-hr culture showing unstained spore and parasporal body surrounded by heavily stained cytoplasm; bottom, 48-hr culture after completion of sporulation, showing disintegrated cytoplasm, leaving spore and parasporal body in the exosporium. (Courtesy of Hannay.)

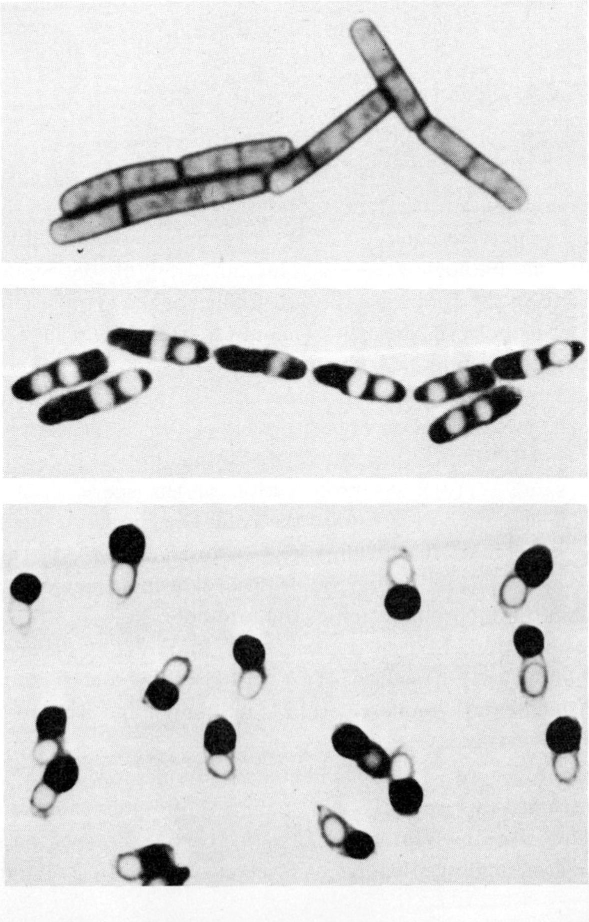

within an exosporium. When sporulation is well advanced the body begins to grow from a single crystal, then develops as a multicrystalline body with the lattices oriented at different angles. When the parasporal body approximates the spore in size, a coat is formed and an exosporium develops which eventually encircles the parasporal body and the spore.

Germination of spores The first change in a spore prior to germination is a loss in refractility accompanied by an imbibition of water. This stage is associated with a loss in heat-resistance and dry weight, and a gain in stainability. Spore germination may be regarded as the change from a heat-resistant spore to a labile entity which may not necessarily be a true vegetative cell. Later the spore coat breaks, followed by emergence from the spore case of a new germ cell, which eventually matures into a vegetative cell.

Some conditions which have been shown to stimulate spore germination are the following:

1. Heat shock at sublethal temperatures for varying periods of time, from 1 min to an hour or more. The time of heating depends upon species, age of spores, and conditions of storage, i.e., frozen or refrigerated.
2. Incubation of spores at 30°C for 5 min in water-ethanol mixtures. Ethanol was effective over a broad range of concentrations. Pure water alone or ethanol alone at 30°C did not stimulate germination.
3. Addition of yeast extract. The active components were glucose, a few amino acids, with or without nucleosides.
4. Addition of L-alanine. L-alanine alone could initiate germination of spores of some strains of *B. subtilis*. L-alanine and adenosine proved to be sufficient for germination of most species of spores. The effect of L-alanine could be reversed by the addition of D-alanine. L-asparagine (or L-glutamine), fructose, glucose, and K^+ initiated germination of *B. subtilis* spores in the absence of L-alanine.
5. Presence of various metal ions. Some ions such as Co^{2+} and Ni^{2+} inhibited spore germination whereas Ca^{2+} and Mg^{2+} stimulated germination.
6. Presence of dipicolinic acid. Dipicolinic acid (DPA) is an organic compound having the following structure:

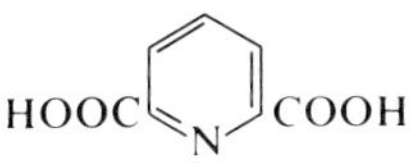

Dipicolinic acid

The calcium salt of DPA is a major constituent of bacterial spores. The compound constitutes from 5 to 15 percent of the dry weight of spores of aerobic and anaerobic bacteria. DPA is not detectable in vegetative cells but is rapidly synthesized during the process of sporulation. Spores become heat-resistant only after development of DPA, which occurs approximately 1 hr after synthesis of the compound. The acid is released into the medium when spores germinate.

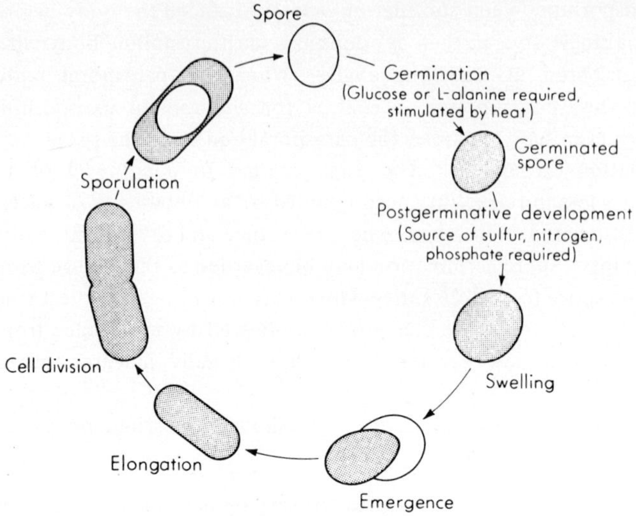

Fig. 82 Diagrammatic representation of the stages of germination and postgerminative development in Bacillus megaterium. (Courtesy of Levinson and Hyatt.)

Spores contain from 2 to 10 times more Ca^{2+} than the corresponding vegetative cells. Levinson et al. (1961) reported a relationship between the Ca/DPA ratio and heat resistance of spores. In general, heat resistance increased as the ratio increased. The loss of DPA during germination correlated with the drop in heat resistance, loss of refractility, and gain in stainability.

A diagrammatic representation of the stages of germination and post-germinative development, according to Levinson and Hyatt (1963), is shown in Fig. 82.

Spores germinate in a variety of ways. A classification based on the modes of germination is as follows:

I. Spore germination by shedding of spore coat. Characteristics of this method are
   A. Spore does not expand greatly in volume previous to the germ cell breaking through the spore coat. The limit of volume increase of the spore may be considered to be twice its original volume.
   B. Spore coat does not lose all its refractive property previous to germination.
   C. After the second division of the germ cell, giving a chain of three organisms, the original spore coat, remaining attached to the cells, is visible for a long time after germination.
      1. Equatorial germination (Fig. 83).
      2. Polar germination (Fig. 83).
      3. Comma-shaped expansion (Fig. 83).

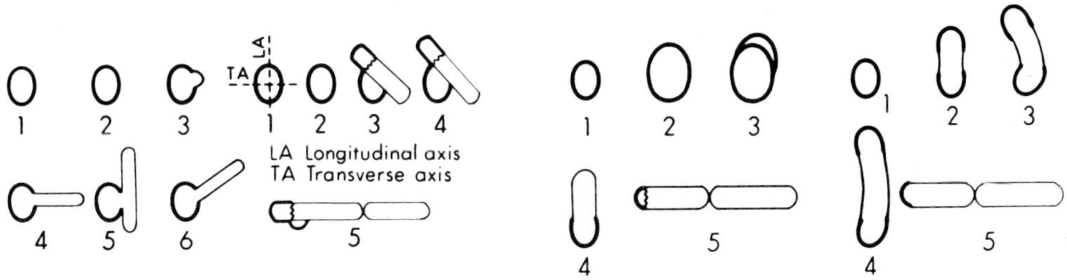

Fig. 83 Methods of spore germination. From left to right: equatorial germination without splitting along transverse axis; equatorial germination with splitting along transverse axis; polar germination; spore germination by comma-shaped expansion. (After Lamanna.)

II. Spore germination by absorption of the spore coat. Characteristics of the method are
   A. The spore expands greatly during germination. A tripling or greater increase of the original volume occurs (Fig. 84).
   B. The spore loses its characteristic refractiveness during germination, so that it is difficult to say when the spore has disappeared and the germ cell appeared.
   C. After the second division of the germ cell, even if a thin capsule originally remains, all traces of the spore coat are gone.

   Some strains germinating by absorption regularly show a thin capsule remaining about one end of the growing cell. This would appear as a polar germination (Fig. 83). In other cases, equatorial capsules are seen (Fig. 83). Yet, in all instances, the spore is considered to germinate by absorption inasmuch as the three characteristics of the method are still adhered to.

For more information: Ando and Iida (1970); Buono et al. (1966); Fields and Frank (1969); Gould and Hurst (1969); Hashimoto et al. (1969); Haynes and Rhodes (1966); Hermier et al. (1970); Hoeniger and Headley (1968); Holmes and Levinson (1967); Holmes et al. (1965); Hyatt

Fig. 84 Spore germination by absorption. (After Lamanna.)

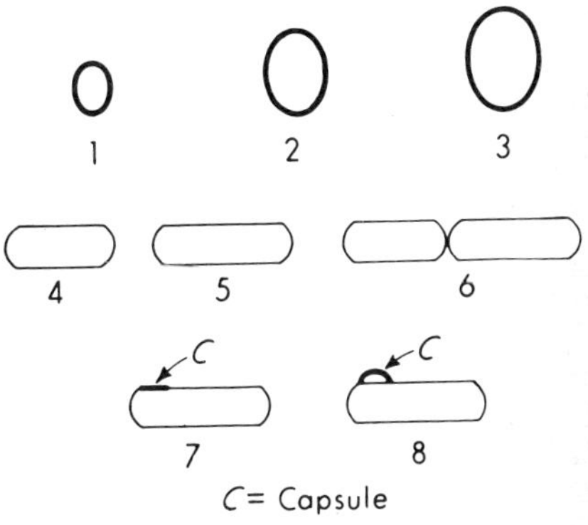

and Levinson (1964, 1968); Jaye and Ordal (1966); Keynan and Halvorson (1962); Knaysi (1964); Lee and Ordal (1963); Levinson and Hyatt (1964, 1966, 1969, 1970*a*, *b*); Murakami (1970); Riemann and Ordal (1961); Rode and Foster (1966); Šťastná and Vinter (1970); Vinter and Slepecky (1965); Wax and Freese (1968).

References

Aamodt, L. W., and J. M. Eisenstadt: Flagellar synthesis in *Salmonella typhimurium*: Requirement for ribonucleic acid synthesis, *J. Bact.*, **96**:1079, 1968.

Abram, D.: Electron microscope observations on intact cells, protoplasts, and the cytoplasmic membrane of *Bacillus stearothermophilus, J. Bact.,* **89**:855, 1965.

—— and H. Koffler: In vitro formation of flagella-like filaments and other structures from flagellin, *J. Mol. Biol.,* **9**:168, 1964.

—— et al.: Basal structure and attachment of flagella in cells of *Proteus vulgaris, J. Bact.,* **90**:1337, 1965.

—— et al.: Attachment and structural features of flagella of certain bacilli, *J. Bact.,* **91**:2045, 1966.

—— et al.: Differentiation within the bacterial flagellum and isolation of the proximal hook, *J. Bact.,* **101**:250, 1970.

Adams, G. A., et al.: The chemical composition of cell-wall lipopolysaccharides from *Moraxella duplex* and *Micrococcus calco-aceticus, Can. J. Microbiol.,* **16**:1, 1970.

Adler, H. I., et al.: Giant cells of *Escherichia coli, J. Bact.,* **95**:139, 1968.

Altenbern, R. A.: Reversion of L forms and spheroplasts of *Proteus mirabilis, J. Bact.,* **85**:269, 1963.

Anderson, E. S.: Slime-wall formation in the salmonellae, *Nature,* **190**:284, 1961.

Ando, Y., and H. Iida: Factors affecting the germination of spores of *Clostridium botulinum* type E, *Japan. J. Microbiol.,* **14**:361, 1970.

Asakura, S., and G. Eguchi: Reconstitution of bacterial flagella *in vitro, J. Mol. Biol.,* **10**:42, 1964.

Asakura, S., and T. Iino: Polymorphism of *Salmonella* flagella as investigated by means of *in vitro* copolymerization of flagellins derived from various strains, *J. Mol. Biol.,* **4**:251, 1972.

Bach, J. A., and H. L. Sadoff: I. Glucose dehydrogenase of *Bacillus cereus, J. Bact.,* **83**:699, 1962.

Barry, J. M.: "Molecular Biology: Genes and the Chemical Control of Living Cells," Englewood Cliffs, N.J., Prentice-Hall, Inc., 1964.

Bayer, M. E.: Areas of adhesion between wall and membrane of *Escherichia coli, J. Gen. Microbiol.,* **53**:395, 1968.

—— and C. C. Remsen: Structure of *Escherichia coli* after freeze-etching, *J. Bact.,* **101**:304, 1970.

Beaman, T. C., et al.: Ultrastructure of the exosporium and underlying inclusions in spores of *Bacillus megaterium* strains, *J. Bact.,* **109**:1198, 1972.

Betz, J. V.: Structure of the proximal ends of flagella of *Clostridium sporogenes, Can. J. Microbiol.,* **15**:761, 1969.

Birdsell, D. C., and E. H. Cota-Robles: Lysis of spheroplasts of *Escherichia coli* by nonionic detergent, *Biochem. Biophys. Res. Commun.,* **31**:438, 1968.

Bladen, H. A., and S. E. Mergenhagen: Ultrastructure of *Veillonella* and morphological correlation of an outer membrane with particles associated with endotoxic activity, *J. Bact.,* **88**:1482, 1964.

Boatman, E. S.: Observations on the fine structure of spheroplasts of *Rhodospirillum rubrum, J. Cell Biol.,* **20**:297, 1964.

Bonsen, P. P. M., et al.: Biochemical studies of bacterial sporulation and germination. XII. A sulfonic acid as a major sulfur compound of *Bacillus subtilis* spores, *J. Bact.,* **98**:62, 1969.

Brostrom, M. A., and S. B. Binkley: Membrane alteration and the formation of metachromatic granules in *Escherichia coli* treated with *p*-fluorophenylalanine, *J. Bact.,* **98**:62, 1969.

Brzin, B.: The influence of temperature on the size and shape of *Bacterium anitratum, Acta Pathol. Microbiol. Scand.,* **57**:188, 1963.

Bulla, L. A., et al.: Scanning electron and phase-contrast microscopy of bacterial spores, *Appl. Microbiol.,* **18**:490, 1969.

Buono, F., et al.: Physiology of growth and sporulation in *Bacillus cereus.* I. Effect of glutamic and other amino acids, *J. Bact.,* **91**:2291, 1966.

Cadmus, M. C., et al.: New bacterial polysaccharide from *Arthrobacter, Appl. Microbiol.,* **11**:488, 1963.

Cairns, J.: The bacterial chromosome, *Sci. Am.,* **214**:37, 1966.

Canale-Parola, E., et al.: Studies on *Sarcina ventriculi.* III. Localization of cellulose, *J. Bact.,* **81**:311, 1961.

Carrier, E. B., and C. S. McCleskey: Intracellular starch formation in corynebacteria, *J. Bact.,* **83**:1029, 1962.

Chung, K. L., et al.: Cell wall replication. I. Cell wall growth of *Bacillus cereus* and *Bacillus megaterium, Can. J. Microbiol.,* **10**:43, 1964.

Cohen-Bazire, G., and J. London: Basal organelles of bacterial flagella, *J. Bact.,* **94**:458, 1967.

——— et al.: Comparative study of the structure of gas vacuoles, *J. Bact.,* **100**:1049, 1969.

Cole, R. M.: Cell wall replication in *Salmonella typhosa, Science,* **143**:820, 1964.

——— and J. J. Hahn: Cell wall replication in *Streptococcus pyogenes, Science,* **135**:722, 1962.

——— et al.: Ultrastructure of a temperature-sensitive rod mutant of *Bacillus subtilis, J. Bact.,* **103**:793, 1970.

Corner, T. R., and R. E. Marquis: Why do bacterial protoplasts burst in hypotonic solutions? *Biochim. Biophys. Acta,* **183**:544, 1969.

Costerton, J. W.: The structure and function of the cell envelope of gram-negative bacteria, *Rev. Can. Biol.,* **29**:299, 1970.

Cundliffe, E.: Preparation and some properties of active protoplasts of *Bacillus megaterium, J. Gen. Microbiol.,* **53**:425, 1968.

De Boer, W. E., and B. J. Spit: A new type of bacterial cell wall structure revealed by replica technique, *Antonie van Leeuwenhoek, J. Microbiol. Serol.*, **30**:239, 1964.

Deibel, R. H., et al.: Filament formation by *Lactobacillus leichmannii* when desoxyribosides replace vitamin B_{12} in the growth medium, *J. Bact.*, **71**:255, 1956.

De Pamphilis, M. L., and J. Adler: Fine structure and isolation of the hook-basal body complex of flagella from *Escherichia coli* and *Bacillus subtilis*, *J. Bact.*, **105**:384, 1971*a*.

—— and ——: Purification of intact flagella from *Escherichia coli* and *Bacillus subtilis*, *J. Bact.*, **105**:376, 1971*b*.

—— and ——: Attachment of flagellar basal bodies to the cell envelope: Specific attachment to the outer, lipopolysaccharide membrane and the cytoplasmic membrane, *J. Bact.*, **105**:396, 1971*c*.

De Petris, S.: Ultrastructure of the cell wall of *Escherichia coli*, *J. Ultrastruct. Res.*, **12**:247, 1965.

Desser, H., and E. Broda: Radiochemical investigation of the respiration of spores of *Bacillus cereus*, *Arch. Mikrobiol.*, **65**:76, 1969.

De Voe, I. W., et al.: Stability and comparative transport capacity of cells, mureinoplasts, and true protoplasts of a gram-negative bacterium, *J. Bact.*, **101**:1014, 1970.

Diena, B. B., et al.: Reversion and biological characteristics of *Salmonella typhi* spheroplasts, *Can. J. Microbiol.*, **11**:427, 1965.

—— et al.: Spheroplasts of enteropathogenic *Escherichia coli*. II. Immunogenicity of spheroplast vaccines, *Can. J. Microbiol.*, **14**:679, 1968.

Dimmitt, K., and M. I. Simon: Purification and partial characterization of *Bacillus subtilis* flagellar hooks, *J. Bact.*, **108**:282, 1971.

Duguid, J. P.: Non-flagellar filamentous appendages ("fimbriae") and haemagglutinating activity in *Bacterium coli*, *J. Pathol. Bacteriol.*, **70**:335, 1955.

——: The function of bacterial fimbriae, *Arch. Immunol. Therapiae Exp.*, **16**:173, 1968.

Eagon, R. G.: Composition of an extracellular slime produced by *Pseudomonas aeruginosa*, *Can. J. Microbiol.*, **8**:585, 1962.

Ellar, D. J., and D. G. Lundgren: Fine structure of sporulation in *Bacillus cereus* grown in a chemically defined medium, *J. Bact.*, **92**:1748, 1966.

Ellwood, D. C.: The wall content and composition of *Bacillus subtilis* var. *niger* grown in a chemostat, *Biochem. J.*, **118**:367, 1970.

Epstein, S. S., and J. B. Weiss: Measuring the size of isolated cells, *Nature*, **187**:461, 1960.

Falcone, G., et al.: Temperatura di sporulazione e termoresistenza delle spore di *B. stearothermophilus*, *Boll, Ist. Sieroterap. Milan.* **45**:11, 1966.

Felter, R. A., et al.: Morphology and round body formation in *Vibrio marinus*, *J. Bact.*, **99**:326, 1969.

Fields, M. L., and H. A. Frank: Dipicolinate-induced germination of *Bacillus stearothermophilus* spores, *J. Bact.*, **97**:464, 1969.

Fiil, A., and D. Branton: Changes in the plasma membrane of *Escherichia coli* during magnesium starvation, *J. Bact.*, **98**:1320, 1969.

Fitz-James, P. C.: Sporulation in protoplasts and its dependence on prior forespore development, *J. Bact.*, **87**:667, 1964.

——: Formation of protoplasts from resting spores, *J. Bact.*, **105**:1119, 1971.

Franklin, J. G., and D. E. Bradley: A further study of the spores of species of the genus *Bacillus* in the electron microscope using carbon replicas, and some preliminary observations on *Clostridium welchii*, *J. Appl. Bact.*, **20**:467, 1957.

Freer, J. H., and H. S. Levinson: Fine structure of *Bacillus megaterium* during microcycle sporogenesis, *J. Bact.*, **94**:441, 1967.

Friedberg, I., and G. Avigad: Structures containing polyphosphate in *Micrococcus lysodeikticus*, *J. Bact.*, **96**:544, 1968.

Gaudy, E., and R. S. Wolfe: Composition of an extracellular polysaccharide produced by *Sphaerotilus natans*, *Appl. Microbiol.*, **10**:200, 1962.

Gerhardt, P., and E. Ribi: Ultrastructure of the exosporium enveloping spores of *Bacillus cereus*, *J. Bact.*, **88**:1774, 1964.

Ghuysen, J.-M.: Use of bacteriolytic enzymes in determination of wall structure and their role in cell metabolism, *Bact. Rev.*, **32**:425, 1968.

Gilleland, H. E., Jr., et al.: Ultrastructure of the cell wall and the mechanism of cellular division of a gram-variable coccus, *Can. J. Microbiol.*, **17**:421, 1971.

Gould, G. W., and A. Hurst, eds.: "The Bacterial Spore," New York, Academic Press, Inc., 1969.

Grecz, N., et al.: Sorption of water by spores, heat-killed spores, and vegetative cells, *Can. J. Microbiol.*, **16**:573, 1970.

Grula, E. A.: Cell division in a species of *Erwinia*. I. Initial observations relating to nutritional dependency, *J. Bact.*, **80**:369, 1960*a*; II. Inhibition of division by D-amino acids, *ibid.*, **80**:375, 1960*b*.

Guze, L. B., ed.: "Microbial Protoplasts, Spheroplasts and L-forms," Baltimore, The Williams & Wilkins Company, 1968.

Hageage, G. J., Jr., and R. L. Gherna: Surface structure of *Chromatium okenii* and *Chromatium weissei*, *J. Bact.*, **106**:687, 1971.

Hannay, C. L.: Fowler's bacillus and its parasporal body, *J. Biophys. Biochem. Cytol.*, **9**:285, 1961.

Harold, F. M.: Accumulation of inorganic polyphosphate in *Aerobacter aerogenes*. I. Relationship to growth and nucleic acid synthesis, *J. Bact.*, **86**:216, 1963.

——: Inorganic polyphosphates in biology: Structure, metabolism, and function, *Bact. Rev.*, **30**:772, 1966.

—— and S. Sylvan: II. Environmental control and the role of sulfur compounds, *J. Bact.*, **86**:222, 1963.

Hashimoto, M., et al.: Report I. Demonstration of fimbriae in gram-negative enteric bacteria, *Bull. Tokyo Med. Dental Univ.*, **10**:181, 1963*a*; Report II. Adhesive properties of the fimbriae, *ibid.*, **10**:493, 1963*b*; Report III. Fimbriation and Acid Agglutination, *ibid.*, **12**:283, 1965.

Hashimoto, T., et al.: Microgermination of *Bacillus cereus* spores, *J. Bact.*, **100**:1385, 1969.

Haynes, W. C., and L. J. Rhodes: Spore formation by *Bacillus popilliae* in liquid medium containing activated carbon, *J. Bact.*, **91**:2270, 1966.

Hermier, J., et al.: Role des déshydrogénases a nicotinamide adénine dinucléotide dans la phase initiale de la germination chez la spore de *Bacillus subtilis*, *Ann. Inst. Pasteur*, **118**:611, 1970.

Herzberg, M., and J. H. Green: Composition and characteristics of cell walls of smooth strains of *Salmonella typhimurium* and derived rough variants, *J. Gen. Microbiol.*, **35**:421, 1964.

Hines, W. D., et al.: Fine structure of *Brucella suis* spheroplasts, *J. Bact.*, **87**:1492, 1964.

Hirano, T.: Observation of nuclear fusion in living cells of *Bacillus megaterium*, *Antonie van Leeuwenhoek, J. Microbiol. Serol.*, **27**:457, 1961.

Hirokawa, H.: Biochemical and cytological observations during the reversing process from spheroplasts to rod-form cells in *Escherichia coli*, *J. Bact.*, **84**:1161, 1962.

Hoeniger, J. F. M.: Development of flagella by *Proteus mirabilis*, *J. Gen. Microbiol.*, **40**:29, 1965.

——: Cellular changes accompanying the swarming of *Proteus mirabilis*, *Can. J. Microbiol.*, **12**:113, 1966.

—— and C. L. Headley: Cytology of spore germination in *Clostridium pectinovorum*, *J. Bact.*, **96**:1835, 1968.

—— et al.: Basal bodies of bacterial flagella in *Proteus mirabilis*, *J. Cell Biol.*, **31**:603, 1966.

—— et al.: Cytology of spore formation in *Clostridium perfringens*, *J. Bact.*, **96**:1818, 1968.

Hoffman, H., and M. E. Frank: Temperature limits, genealogical origin, developmental course, and ultimate fate of heat-induced filaments in *Escherichia coli* microcultures, *J. Bact.*, **85**:1221, 1963.

Holme, T., and B. Cedergren: Demonstration of intracellular polysaccharide in *Escherichia coli* by electron microscopy and by cytochemical methods, *Acta Pathol. Microbiol. Scand.*, **51**:179, 1961.

Holmes, P. K., and H. S. Levinson: Activation of *Bacillus megaterium* spores with aqueous ethyl alcohol: their deactivation and reactivation, *Currents Mod. Biol.*, **1**:256, 1967.

—— et al.: Concurrent heat activation and suppression of *Bacillus megaterium* spore germination, *J. Bact.*, **90**:827, 1965.

Holwill, M. E. J., and R. E. Burge: A hydrodynamic study of the motility of flagellated bacteria, *Arch. Biochem. Biophys.*, **101**:249, 1963.

Homma, J. Y., et al.: The surface structure of *Pseudomonas aeruginosa, J. Immunol.*, **90**:819, 1963.

Hughes, W. H.: The structure and development of the induced long forms of bacteria. In "Bacterial Anatomy," edited by E. T. C. Spooner and B. A. D. Stocker, London, Cambridge University Press, 1956.

Hungerer, K. D., and D. J. Tipper: Cell wall polymers of *Bacillus sphaericus* 9602. I. Structure of the vegetative cell wall peptidoglycan, *Biochemistry,* **8**:3577, 1969.

—— et al.: Structure of the cell wall peptidoglycan of *Lactobacillus casei* RO94, *Biochemistry,* **8**:3567, 1969.

Hurst, A., and J. M. Stubbs: Electron microscopic study of membranes and walls of bacteria and changes occurring during growth initiation, *J. Bact.*, **97**:1466, 1969.

Hyatt, M. T., and H. S. Levinson: Effect of sugars and other carbon compounds on germination and postgerminative development of *Bacillus megaterium* spores, *J. Bact.,* **88**:1403, 1964.

—— and ——: Water vapor, aqueous ethyl alcohol, and heat activation of *Bacillus megaterium* spore germination, *J. Bact.,* **95**:2090, 1968.

Iannetta, A., and R. J. Wedgwood: Culture of serum-induced spheroplasts from *Vibrio cholerae, J. Bact.,* **93**:1688, 1967.

Iino, T.: Polarity of flagellar growth in *Salmonella, J. Gen. Microbiol.,* **56**:227, 1969.

—— and M. Mitani: A mutant of *Salmonella* possessing straight flagella, *J. Gen. Microbiol.,* **49**:81, 1967.

Imaeda, T., et al.: Ultrastructure of cell walls of genus *Mycobacterium, J. Ultrastruct. Res.,* **25**:46, 1968.

Izumi, K.: Mucopolysaccharides produced by a strain of *Clostridium perfringens, J. Bact.,* **83**:956, 1962.

Jaye, M., and Z. J. Ordal: Germination response of spores of *Bacillus megaterium* after exposure to calcium dipicolinate at 60°C., *Can. J. Microbiol.,* **12**:199, 1966.

Jeanloz, R. W.: The chemical structure of the cell wall of gram-positive bacteria, *Pure Appl. Chem.,* **14**:57, 1967.

Johnston, D. E.: Motile daughter colonies of *Clostridium capitovale, Can. J. Microbiol.,* **11**:119, 1965.

Jones, H. C., et al.: Electron microscopic study of a slime layer, *J. Bact.,* **99**:316, 1969.

Jones, J. K. N., et al.: Biosynthesis of sugars found in bacterial polysaccharides. II, *Can. J. Chem.,* **40**:1798, 1962.

Joys, T. M.: The structure of flagella and the genetic control of flagellation in *Eubacteriales*: A review, *Antonie van Leeuwenhoek, Microbiol. Serol.,* **34**:205, 1968.

Juni, E., and G. A. Heym: IV. Capsule resynthesis by decapsulated resting-cell suspensions, *J. Bact.,* **87**:461, 1964.

Kakefuda, T., et al.: Ultrastructure of the membrane system in *Lactobacillus plantarum, J. Bact.,* **93**:472, 1967.

Kane, J., et al.: Chemical studies on the structure of mucopeptide isolated from *Streptococcus bovis, J. Bact.,* **99**:175, 1969.

Kanetsuna, F., Chemical analyses of mycobacterial cell walls, *Biochim. Biophys. Acta,* **158**:130, 1968.

Kawata, T., and T. Inoue: Spheroplast formation of clostridia by penicillin and glycine, *Japan. J. Bact.,* **18**:354, 1963.

—— et al.: Autolytic formation of spheroplasts of *Bacillus megaterium* after cessation of aeration, *J. Bact.,* **81**:160, 1961.

Kay, D., and S. C. Warren: Sporulation in *Bacillus subtilis, Biochem. J.,* **109**:819, 1968.

Kerravala, Z. J., et al.: II. Growth and sporulation in *Bacillus subtilis, J. Bact.,* **88**:374, 1964.

Keynan, A., and H. O. Halvorson: Calcium dipicolinic acid-induced germination of *Bacillus cereus* spores, *J. Bact.,* **83**:100, 1962.

Kingan, S. L., and J. C. Ensign: Chemical composition of the cell wall of *Bacillus thuringiensis* var. *thuringiensis, J. Bact.,* **95**:724, 1968.

Klein, D., et al.: Conformational transitions in flagellins. I. Hydrogen ion dependency, *J. Biol. Chem.,* **243**:4931, 1968.

Knaysi, G.: Effects of temperatures above the maximum for germination on the endospore of *Bacillus cereus, J. Bact.,* **87**:1129, 1964.

Kusaka, I.: Growth and division of protoplasts of *Bacillus megaterium* and inhibition of division by penicillin, *J. Bact.,* **94**:884, 1967.

——: Electron microscopic observations on growing and dividing protoplasts of *Bacillus megaterium, J. Gen. Microbiol.,* **63**:199, 1971.

Landman, O. E., and A. Forman: Gelatin-induced reversion of protoplasts of *Bacillus subtilis* to the bacillary form: Biosynthesis of macromolecules and wall during successive steps, *J. Bact.,* **99**:576, 1969.

—— et al.: Gelatin-induced reversion of protoplasts of *Bacillus subtilis* to the bacillary form: Electron-microscopic and physical study, *J. Bact.,* **96**:2154, 1968.

Larsen, H., et al.: On the gas vacuoles of the halobacteria, *Arch. Mikrobiol.,* **59**:197, 1967.

Lautrop, H., and O. Jessen: On the distinction between polar monotrichous and lophotrichous flagellation in green fluorescent pseudomonads, *Acta Pathol. Microbiol. Scand.,* **60**:588, 1964.

Lee, W. H., and Z. J. Ordal: Reversible activation for germination and subsequent changes in bacterial spores, *J. Bact.,* **85**:207, 1963.

Lehninger, A. L.: "The Mitochondrion: Molecular Basis of Structure and Function," New York, W. A. Benjamin, Inc., 1964.

Leifson, E.: Staining, shape, and arrangement of bacterial flagella, *J. Bact.,* **62**:377, 1951.

——: The effect of formaldehyde on the shape of bacterial flagella, *J. Gen. Microbiol.,* **25**:131, 1961.

——: Mixed polar and peritrichous flagellation of marine bacteria, *J. Bact.,* **86**:166, 1963.

—— et al.: Morphological characteristics of flagella of *Proteus* and related bacteria, *J. Bact.,* **69**:73, 1955.

Levinson, H. S., and M. T. Hyatt: Bacterial spore germination with special reference to *Bacillus megaterium, Ann. N.Y. Acad. Sci.,* **102**:773, 1963.

—— and ——: Effect of sporulation medium on heat resistance, chemical composition, and germination of *Bacillus megaterium* spores, *J. Bact.,* **87**:876, 1964.

—— and ——: Sequence of events during *Bacillus megaterium* spore germination, *J. Bact.,* **91**:1811, 1966.

—— and ——: Activation of *Bacillus megaterium* spore germination. In "Spores IV," American Society for Microbiology, 1969, 1913 I St., N. W., Washington, D.C. 20006.

—— and ——: Effects of temperature on activation, germination, and outgrowth of *Bacillus megaterium* spores, *J. Bact.,* **101**:58, 1970*a*.

—— and ——: Activation energy for glucose-induced germination of *Bacillus megaterium spores, J. Bact.,* **103**:270, 1970*b*.

—— et al.: Dependence of the heat resistance of bacterial spores on the calcium:dipicolinic acid ratio, *Biochem. Biophys. Res. Commun.,* **5**:417, 1961.

Lindberg, B., and S. Svensson: Structural studies on dextran from *Leuconostoc mesenteroides* NRRL B-512, *Acta Chem. Scand.,* **22**:1907, 1968.

Lopes, J., and W. E. Inniss: Electron microscopic study of lipopolysaccharide from an avian strain of *Escherichia coli* 018, *J. Bact.,* **103**:238, 1970.

Lowy, J., and J. Hanson: The structure of bacterial flagella, *3d Europ. Reg. Conf. Elec. Microscop.,* 1964.

—— and M. W. McDonough: Structure of filaments produced by reaggregation of *Salmonella* flagellin, *Nature,* **204**:125, 1964.

Lwoff, A.: "Biological Order," Cambridge, Mass., The M.I.T. Press, 1962.

MacKechnie, I., and R. S. Hanson: Microcycle sporogenesis of *Bacillus cereus* in a chemically defined medium, *J. Bact.,* **95**:355, 1968.

Maeda, Y., et al.: Physical properties of water in spores of *Bacillus megaterium, J. Gen. Appl. Microbiol.,* **14**:217, 1968.

Manasse, R. J., and W. A. Corpe: Chemical composition of cell envelopes from *Agrobacterium tumefaciens, Can. J. Microbiol.,* **13**:1591, 1967.

Marquis, R. E.: Osmotic sensitivity of bacterial protoplasts and the response of their limiting membrane to stretching, *Arch. Biochem. Biophys.,* **118**:323, 1967.

Martin, E. L., and R. A. MacLeod: Isolation and chemical composition of the cytoplasmic membrane of a gram-negative bacterium, *J. Bact.,* **105**:1160, 1971.

Martinez, R. J., et al.: A single amino acid substitution responsible for altered flagellar morphology, *J. Mol. Biol.,* **34**:559, 1968.

Mason, D. J., and D. M. Powelson: Nuclear division as observed in live bacteria by a new technique, *J. Bact.,* **71**:474, 1956.

Mayer, F.: Elektronenmikroskopische untersuchung der fimbrienkontraktion bei dem sternbildenden bodenbakterium *Pseudomonas echinoides, Arch. Mikrobiol.*, **76**:166, 1971.

McDonough, M. W.: Amino acid composition of antigenically distinct *Salmonella* flagellar proteins, *J. Mol. Biol.*, **12**:342, 1965.

Meyer, R. R., and M. V. Simpson: DNA biosynthesis in mitochondria: Partial purification of a distinct DNA polymerase from isolated rat liver mitochondria, *Proc. Natl. Acad. Sci.*, **61**:130, 1968.

Meynell, G. G., and A. M. Lawn: Inheritance of capsule and the manner of cell-wall formation in *Bacillus anthracis, J. Gen. Microbiol.*, **39**:423, 1965.

—— and E. Meynell: The biosynthesis of poly D-glutamic acid, the capsular material of *Bacillus anthracis, J. Gen. Microbiol.*, **43**:119, 1966.

Mitani, M., and T. Iino: Electron microscopy of bundled flagella of the curly mutant of *Salmonella abortivoequina, J. Bact.*, **90**:1096, 1965.

Mohan, R. R., et al.: Autolytic mechanism for spheroplast formation in *Bacillus cereus* and *Escherichia coli, J. Bact.*, **90**:1355, 1965.

Morrison, R. B., and J. McCapra: Flagellar changes in *Escherichia coli* induced by temperature of the environment, *Nature*, **192**:774, 1961.

Murakami, H.: Studies on the germination mechanism of *Bacillus subtilis* PCI 219 spores. Part I. Basic studies of germination conditions and influences of cations and anions on the germination, *J. Agr. Chem. Soc. Japan*, **44**:539, 1970.

Murphy, J. A., and L. L. Campbell: Surface features of *Bacillus polymyxa* spores as revealed by scanning electron microscopy, *J. Bact.*, **98**:737, 1969.

Nakata, H. M.: Organic nutrients required for growth and sporulation of *Bacillus cereus, J. Bact.*, **88**:1523, 1964.

——: Role of acetate in sporogenesis of *Bacillus cereus, J. Bact.*, **91**:784, 1966.

Nanninga, N.: Structural features of mesosomes (chondrioids) of *Bacillus subtilis* after freeze-etching, *J. Cell. Biol.*, **39**:251, 1968.

——: Ultrastructure of the cell envelope of *Escherichia coli* B after freeze-etching, *J. Bact.*, **101**:297, 1970.

Nass, M. M. K.: Mitochondrial DNA: Advances, problems, and goals, *Science*, **165**:25, 1969.

Nermut, M. V., and R. G. E. Murray: Ultrastructure of the cell wall of *Bacillus polymyxa, J. Bact.*, **93**:1949, 1967.

—— and A. Svoboda: Reversion of spheroplasts produced by lysozyme into rods in *Proteus vulgaris, Nature*, **193**:396, 1962.

Novotny, C. P., and K. Lavin: Some effects of temperature on the growth of F pili, *J. Bact.*, **107**:671, 1971.

Novotny, C., et al.: Mechanical removal of F pili, type I pili, and flagella from Hfr and RTF donor cells and the kinetics of their reappearance, *J. Bact.*, **98**:1294, 1969.

Old, D. C., and J. P. Duguid: Selective outgrowth of fimbriate bacteria in static liquid medium, *J. Bact.*, **103**:447, 1970.

—— et al.: Fimbriation, pellicle formation and the amount of growth of salmonellas in broth, *J. Gen. Microbiol.*, **51**:1, 1968.

Op den Kamp, J. A. F., et al.: Phospholipid composition of *Bacillus subtilis*, *J. Bact.*, **99**:298, 1969.

Pontefract, R. D., et al.: Mesosomes in *Escherichia coli*, *J. Bact.*, **97**:367, 1969.

Pope, L., and L. J. Rode: Spore fine structure in *Clostridium cochlearium*, *J. Bact.*, **100**:994, 1969.

—— et al.: Appendages of *Clostridium bifermentans* spores, *J. Bact.*, **94**:1206, 1967.

Racker, E.: The membrane of the mitochondrion, *Sci. Am.*, **218**:32, 1968.

Riemann, H., and Z. J. Ordal: Germination of bacterial endospores with calcium and dipicolinic acid, *Science*, **133**:1703, 1961.

Roberts, F. F., Jr., and R. N. Doetsch: Some singular properties of bacterial flagella, with special reference to monotrichous forms, *J. Bact.*, **91**:414, 1966.

Robinow, C. F.: The chromatin bodies of bacteria. In "Bacterial Anatomy," edited by E. T. C. Spooner and B. A. D. Stocker, London, Cambridge University Press, 1956.

Rode, L. J.: Correlation between spore structure and spore properties in *Bacillus megaterium*, *J. Bact.*, **95**:1979, 1968.

—— and J. W. Foster: Influence of exchangeable ions on germinability of bacterial spores, *J. Bact.*, **91**:1582, 1966.

—— et al.: *Clostridium* spores with ribbon-like appendages, *J. Bact.*, **93**:1160, 1967.

—— et al.: Spore appendages and taxonomy of *Clostridium sordellii*, *J. Bact.*, **108**:1384, 1971.

Rogers, H. J.: The outer layers of bacteria: The biosynthesis of structure. In "Function and Structure in Micro-organisms," London, Society for General Microbiology, 1965.

Rosenberg, B., et al.: Platinum-induced filamentous growth in *Escherichia coli*, *J. Bact.*, **93**:716, 1967.

Roth, G. S., et al.: Balanced macromolecular biosynthesis in "protoplasts" of *Streptococcus faecalis*, *J. Bact.*, **105**:710, 1971.

Rubenstein, K. E., et al.: Synthetic capabilities of plasmolyzed cells and spheroplasts of *Escherichia coli*, *J. Bact.*, **104**:443, 1970.

Ryter, A.: Association of the nucleus and the membrane of bacteria: a morphological study, *Bact. Rev.*, **32**:39, 1968.

Salton, M. R. J.: The anatomy of the bacterial surface, *Bact. Rev.*, **25**:77, 1961.

——: "The Bacterial Cell Wall," New York, American Elsevier Publishing Company, Inc., 1964.

——: Structure and composition of bacterial membranes, *Protides Biol. Fluids, Proc. Colloq.*, **15**:279, 1967*a*.

——: Bacterial membranes. In "The Specificity of Cell Surfaces," edited by B. D. Davis and L. Warren, Englewood Cliffs, N.J., Prentice-Hall, Inc., 1967*b*.

Samsonoff, W. A., et al.: Ultrastructural changes associated with germination and outgrowth of an appendage-bearing clostridial spore, *J. Bact.*, **101**:1038, 1970.

—— et al.: Appendage development in *Clostridium bifermentans, J. Bact.*, **106**:269, 1971.

Santo, L. M., et al.: Ultrastructure of putrefactive anaerobe 3679h during sporulation, *J. Bact.*, **99**:824, 1969.

Schnaitman, C. A.: Effect of ethylenediamine tetraacetic acid, Triton X-100 and lysozyme on the morphology and chemical composition of isolated cell walls of *Escherichia coli, J. Bact.*, **108**:553, 1971.

—— and J. W. Greenawalt: Intracytoplasmic membranes in *Escherichia coli, J. Bact.*, **92**:780, 1966.

Seidler, R. J., and M. P. Starr: Structure of the flagellum of *Bdellovibrio bacteriovorus, J. Bact.*, **95**:1952, 1968.

Sharon, N.: The bacterial cell wall, *Sci. Am.*, **220**:92, 1969.

Shaw, M. K.: Formation of filaments and synthesis of macromolecules at temperatures below the minimum for growth of *Escherichia coli, J. Bact.*, **95**:221, 1968.

Sierra, G.: Dissociation of esterase from proteinase activity of *Bacillus subtilis* spores, *Can. J. Microbiol.*, **13**:673, 1967.

Sinkovics, J.: Occurrence and filterability of protoplast-like elements in aged bacterial cultures, *Nature*, **181**:566, 1958.

Somerville, H. J.: Formation of the parasporal inclusion of *Bacillus thuringiensis, Europ. J. Biochem.*, **18**:226, 1971.

Spudich, J. A., and A. Kornberg: Biochemical studies of bacterial sporulation and germination. XIII. Adenylate kinase of vegetative cells and spores of *Bacillus subtilis, J. Bact.*, **98**:69, 1969.

Šťastná, J., and V. Vintner: Spores of microorganisms. XXIII. Interdependence of intra- and extra-cellular levels of calcium: Its effect on the germination of bacterial spores in different media, *Folia Microbiol. (Prague)*, **15**:103, 1970.

Sud, I. J.: Spheroplasts of vibrios: Formation in hypertonic sugar media, *Indian J. Path. Bacteriol.*, **10**:226, 1967.

Sutow, A. B., and N. E. Welker: Chemical composition of the cell walls of *Bacillus stearothermophilus, J. Bacteriol.*, **93**:1452, 1967.

Takeya, K., et al.: Further studies on the paired fibrous structure of mycobacterial cell wall, *J. Biophys. Biochem. Cytol.*, **9**:496, 1961.

Takumi, K., and T. Kawata: Chemical composition of the cell walls of *Clostridium botulinum* type A, *Japan. J. Microbiol.*, **14**:57, 1970.

Tawara, J.: The root of flagella of *Vibrio cholerae, Japan. J. Microbiol.*, **9**:49, 1965.

Terry, D. R., et al.: Filament formation in *Clostridium acidiurici* under conditions of elevated temperatures, *J. Bacteriol.*, **91**:1625, 1966.

Tipper, D. J., and I. Pratt: Cell wall polymers of *Bacillus sphaericus* 9602. II. Synthesis of the first enzyme unique to cortex synthesis during sporulation, *J. Bacteriol.*, **103**:305, 1970.

Turner, A. W., and C. E. Eales: An aerobic, sporulating bacillus that forms rotating and migrating colonies, *Australian J. Exp. Biol. Med. Sci.*, **19**:161, 1941*a*.

—— and ——: Motile daughter colonies in the *Cl. oedematiens* group and some other clostridia (*Cl. botulinum* C, *Cl. tetani, Cl. septicum*), *Australian J. Exp. Biol. Med. Sci.*, **19**:167, 1941*b*.

Vaituzis, Z., and R. N. Doetsch: Flagella of *Escherichia coli* spheroplasts, *J. Bact.*, **91**:2103, 1966.

—— and ——: Relationship between cell wall, cytoplasmic membrane, and bacterial motility, *J. Bact.*, **100**:512, 1969.

Van den Hooff, A., and S. Aninga: An electron microscope study on the shape of the spores of *Bacillus polymyxa, Antonie van Leeuwenhoek, J. Microbiol. Serol.*, **22**:327, 1956.

Van Ert, M., and J. T. Staley: Gas-vacuolated strains of *Microcyclus aquaticus, J. Bact.*, **108**:236, 1971.

Van Iterson, W.: Membranes, particular organelles and peripheral bodies in bacteria, *Proc. European Regional Conf. Electron Microscopy*, Vol. II, 1960.

Vinter, V., and R. A. Slepecky: Direct transition of outgrowing bacterial spores to new sporangia without intermediate cell division, *J. Bact.*, **90**:803, 1965.

Wahren, A., et al.: Studies on filamentous forms of *Bacillus cereus* strain T, *J. Gen. Microbiol.*, **49**:59, 1967.

Wakabayashi, K., et al.: Polymerization of *Salmonella* flagellin in the presence of high concentrations of salts, *Biochim. Biophys. Acta*, **175**:195, 1969.

Walker, P. D., and J. Short: Location of bacterial polysaccharide during various phases of growth, *J. Bact.*, **98**:1342, 1969.

Wang, M., et al.: Extracellular polysaccharide of mucoid *Lactobacillus bifidus, J. Bact.*, **86**:898, 1963.

Wax, R., and E. Freese: Initiation of the germination of *Bacillus subtilis* spores by a combination of compounds in place of L-alanine, *J. Bact.*, **95**:433, 1968.

Weigand, R. A., et al.: Formation and ultrastructure of extra membranes in *Escherichia coli, J. Bact.*, **102**:240, 1970.

Weinberg, E. D., and J. I. Brooks: Trace metal control of bacterial flagellation, *Nature*, **199**:717, 1963.

Welker, N. E.: Structure of the cell wall of *Bacillus stearothermophilus*: Mode of action of a thermophilic bacteriophage lytic enzyme, *J. Bact.*, **107**:697, 1971.

Wheat, R. W., and J. M. Ghuysen: Occurrence of glucomuramic acid in gram-positive bacteria, *J. Bact.*, **105**:1219, 1971.

Widra, A.: Metachromatic granules of microorganisms, *J. Bact.*, **78**:664, 1959.

Willett, H. P., and H. Thacore: Formation of spheroplasts of *Mycobacterium tuberculosis* by lysozyme in combination with certain enzymes of rabbit peritoneal monocytes, *Can. J. Microbiol.*, **13**:481, 1967.

Yanagawa, R., and K. Otsuki: Some properties of the pili of *Corynebacterium renale, J. Bact.*, **101**:1063, 1970.

Yolton, D. P., et al.: Isolation and partial chemical characterization of the spore appendages of *Clostridium taeniosporum, J. Bact.*, **109**:881, 1972.

5 Yeasts

The yeasts are spherical, ovoid, or rod-shaped ascomycetous fungi in which the usual and dominant growth form is unicellular. They are widely distributed in nature, being found in soil, in dust, and on fruits and leaves of many plants. The organisms are particularly numerous in the soils of orchards and vineyards. Yeasts appear as a surface froth or as a thick sediment in fruit juices, malt worts, and other saccharine liquids.

Numerous species of yeasts and yeast-like fungi will, under certain conditions, grow in the form of filaments or mycelia, whereas under other conditions they will grow in yeast phases (Fig. 115). It is common to find occasional short chains of yeast phase cells in young, actively growing yeast cultures, but the designation of "filamentous" should not be made unless such growth becomes the usual morphology and is accompanied by cell elongation. There are thus two basic morphological types to be considered: the ellipsoidal or ovoid yeast-like cell and the elongated cell, the latter being elongated beyond that which is the usual morphology. In this respect the yeasts differ from the molds, which are typically filamentous. On the other hand, some molds may grow temporarily as single cells and take on the appearance of yeasts.

Some of the mucors grow as single cells capable of budding under reduced oxygen tensions. The organisms causing sporotrichosis, coccidioidal granuloma, and blastomycosis may appear as single cells in the tissues of the host, and as mycelia on artificial culture media.

Filament formation in yeasts may be enhanced by controlling a number of environmental factors such as temperature, presence of certain nutritional substances, effect of carcinogenic chemicals, age of the culture, and irradiation.

The fact that molds display dimorphism has led some investigators to believe that the yeasts were at one time mold-like but have permanently lost the ability to produce mycelia. For example, it has been shown that *Aspergillus oryzae*, when grown under reduced oxygen, gave rise to yeast-like cells from the conidia that were submerged in the medium. The cells failed to revert to the mycelial stage but grew permanently as yeasts, even under aerobic conditions.

For more information: Brown (1970); Scherr and Weaver (1953).

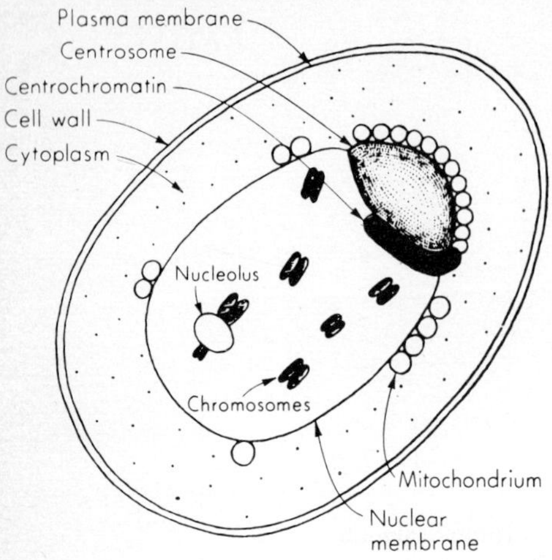

Fig. 85 Diagram of a yeast cell. (After Lindegren.)

Cytology of yeasts

The structure of the yeast cell is shown in Fig. 85.

Cell membranes Special staining methods reveal the presence of a cytoplasmic (plasma) membrane surrounded by the cell wall.

Cell wall The cell wall of yeasts is not a rigid structure as in the bacteria but flaccid and readily deformed by light centrifugation (Fig. 86). The cell wall of intact yeast has three zones of different electron densities with the intermediate zone having a lower electron density than the outer and inner zones (Fig. 87). These three layers are not believed to be independent but differ only in chemical composition. The cell wall occupies about 40 percent of the volume of the cell. In thickness it is fairly constant in all cells, measuring about 1.2 μ.

The cell wall is composed largely of a polysaccharide known as yeast cellulose although it is somewhat different from the cellulose present in the cell walls of higher plants.

Chitin also has been reported as being present in the cell wall of yeasts. It is a nitrogenous compound closely related to the polysaccharides. Chemically it is a polymerized acetylated glucosamine. In plants it

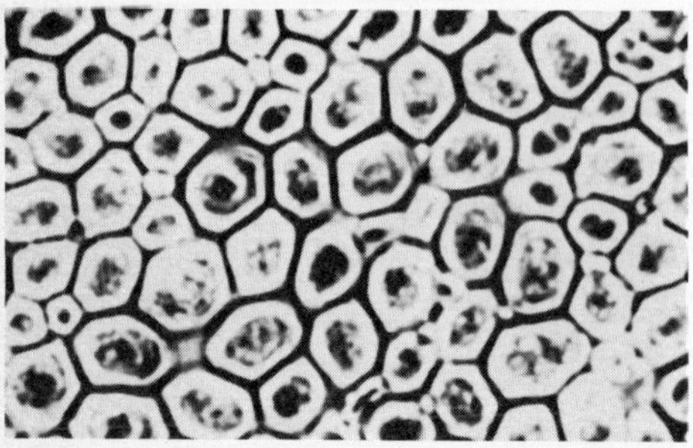

Fig. 86 Saccharomyces. Hanging-drop preparation showing flaccid cell wall. (Courtesy of C. C. Lindegren.)

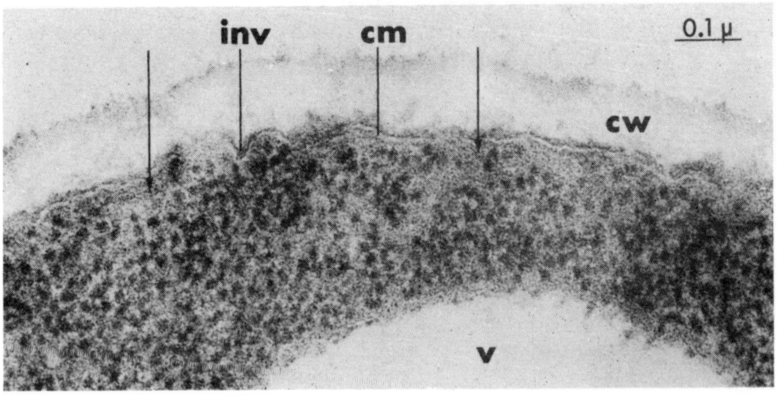

Fig. 87 Part of a cell wall (cw) of Saccharomyces cerevisiae which appears to consist of three zones of different electron densities. The inner boundary of the cell wall appears to be in intimate contact with the cytoplasmic membrane (cm). The cytoplasm is surrounded by a triple-layered cytoplasmic membrane showing many invaginations (inv). Membranous profiles are seen adjacent to the cytoplasmic membrane in the cytoplasm (arrow), which is densely packed with ribosomes. Part of a large membrane-bound vacuole (v) is seen. x118,000. (Courtesy of Darling, Theilade, and Birch-Andersen.)

frequently occupies a corresponding position to cellulose in cell structure. Chitin is present in the cell wall of molds, in the shells of crabs and lobsters, and in the shards of beetles, but there is still some question as to its presence in bacteria.

For more information: MacWilliam (1970).

Cytoplasmic membrane The cytoplasmic membrane is a triple-layered structure showing many invaginations (Fig. 87). It is very delicate and relatively thin in young cells and becomes thicker as the cell ages (Fig. 88). In higher plant cells subjected to high osmotic pressures, the membrane may be withdrawn smoothly, without rupture, from the inner surface of the cell wall. Lindegren (1963*a*) found that when yeast cells were placed in solutions of high osmotic pressures, the cytoplasmic membrane almost never pulled away from the cell wall. He concluded that either the cell wall was very elastic or the cell wall and cytoplasmic membrane were integrated.

The cytoplasmic membrane in yeasts is believed to function in a manner similar to that of the same membrane in bacteria, i.e., by determining which substances may enter and leave the cell (see page 81).

Capsule The cell walls of some yeasts have been shown to be surrounded by a polysaccharide material similar to the capsular substance of bacteria. Mager (1947) isolated two polysaccharides from the capsulated yeasts *Torulopsis rotundata* and *T. neoformans*. One was a zymohexose (starch), and the other a pentosan. The hexose in the

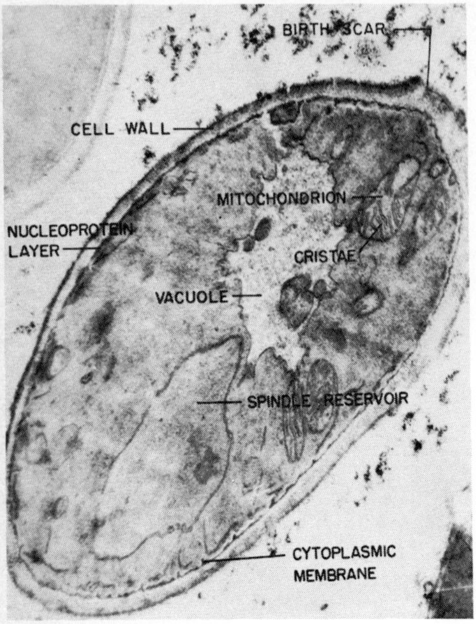

Fig. 88 Electron micrograph of a thin section of a vegetative cell of Saccharomyces cerevisiae. (Courtesy of C. C. Lindegren.)

former was shown to be glucose, and the pentose in the latter to be D-xylose.

Protoplasts and spheroplasts Cell walls of yeasts may be completely or partially removed without destroying the vital nature of the protoplasm. A protoplast by definition retains no remnants of the original cell wall; otherwise it should be called a spheroplast.

One method of removing the cell wall is by enzymatic digestion. An enzyme preparation from the snail *Helix pomata*, mixed with a sulfur-containing amino acid such as cysteine or methionine, is very effective in digesting yeast cell walls without disturbing the intact cytoplasmic membrane. Cells so treated are known as protoplasts or spheroplasts depending upon the extent of removal of cell wall material (Fig. 89).

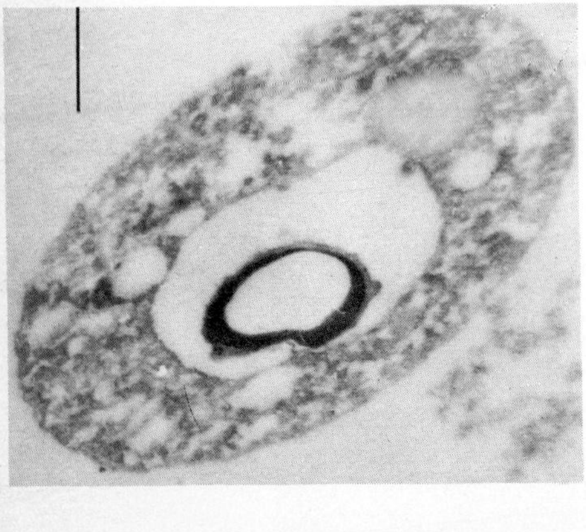

Fig. 89 Electron micrograph of a thin section of a protoplast of Saccharomyces cerevisiae. (Courtesy of C. C. Lindegren.)

Protoplasts may be maintained in such condition on appropriate culture media.

For more information: Cirillo (1962); Darling et al. (1969); Hutchison and Hartwell (1967); Matile et al. (1969); Nickerson (1963); Streiblová (1968); Svihla et al. (1963); Darling et al. (1972); Mann et al. (1972).

The cytoplasm and its contents The cytoplasm is a clear fluid rich in ribonucleic acid (RNA). It contains the mitochondria, centrosome, centrochromatin, nucleus, vacuoles, granules, and fat globules.

Mitochondria The mitochondria are spherical, rod-shaped, or thread-like and branched structures, measuring about 0.3 to 1μ in diameter and up to 3 μ in length. They are usually found adhering to the surface of the centrosome or the nuclear membrane.

The mitochondria have a high lipid content. In baker's yeast the lipid content may be as high as 25 percent and the phospholipid up to 13 percent of the mitochondrial dry mass. The lipids are present in the mitochondrial membranes.

Yeast mitochondria contain a small amount of deoxyribonucleic acid (DNA) and considerably higher amounts of RNA. A large amount of the RNA is most likely ribosomal. Sinclair et al. (1967) reported mitochondrial DNA isolated from purified *Saccharomyces cerevisiae* to contain a satellite DNA with a density of 1.682. It appears to exist as open-ended filaments at least 5 μ in length (Fig. 90). DNA from intact cells contains circular filaments whose lengths vary from 0.5 to 7 μ with a great majority at 1.95 μ. The circular DNA has a density of 1.697 which is similar to that of the major nuclear peak.

Hirano and Lindegren (1963) observed lamellated mitochondria in old cultures held under partially anaerobic conditions, which appeared to result from changes in the inner membranes of certain mitochondria (Fig. 88).

The mitochondria may function as storage reservoirs in the insulation of essential substances inside an oily layer and may assist in establishing dormancy. Granular or fibrous deposits of glycogen may be present which sometimes spread through the entire cytoplasm. Mitochondria are also said to be the seat of cellular respiration, the energy-producing process by which cells convert oxygen and sugar to carbon dioxide and water.

For more information: Avers (1967); Carnevali et al. (1969).

Centrosome This is a rigid structure that stains readily with acidic stains but not with basic stains. It probably contains largely basic proteins. The centrosome is always attached to the nuclear membrane and plays a leading role in budding, copulation, and meiosis.

Centrochromatin This structure is attached to the external surface of the centrosome, with some portion of it always in contact with the nucleus. It is acid in reaction. In the resting cell, it may cover most of the centrosome; in division, it usually is present in the form of a long strand.

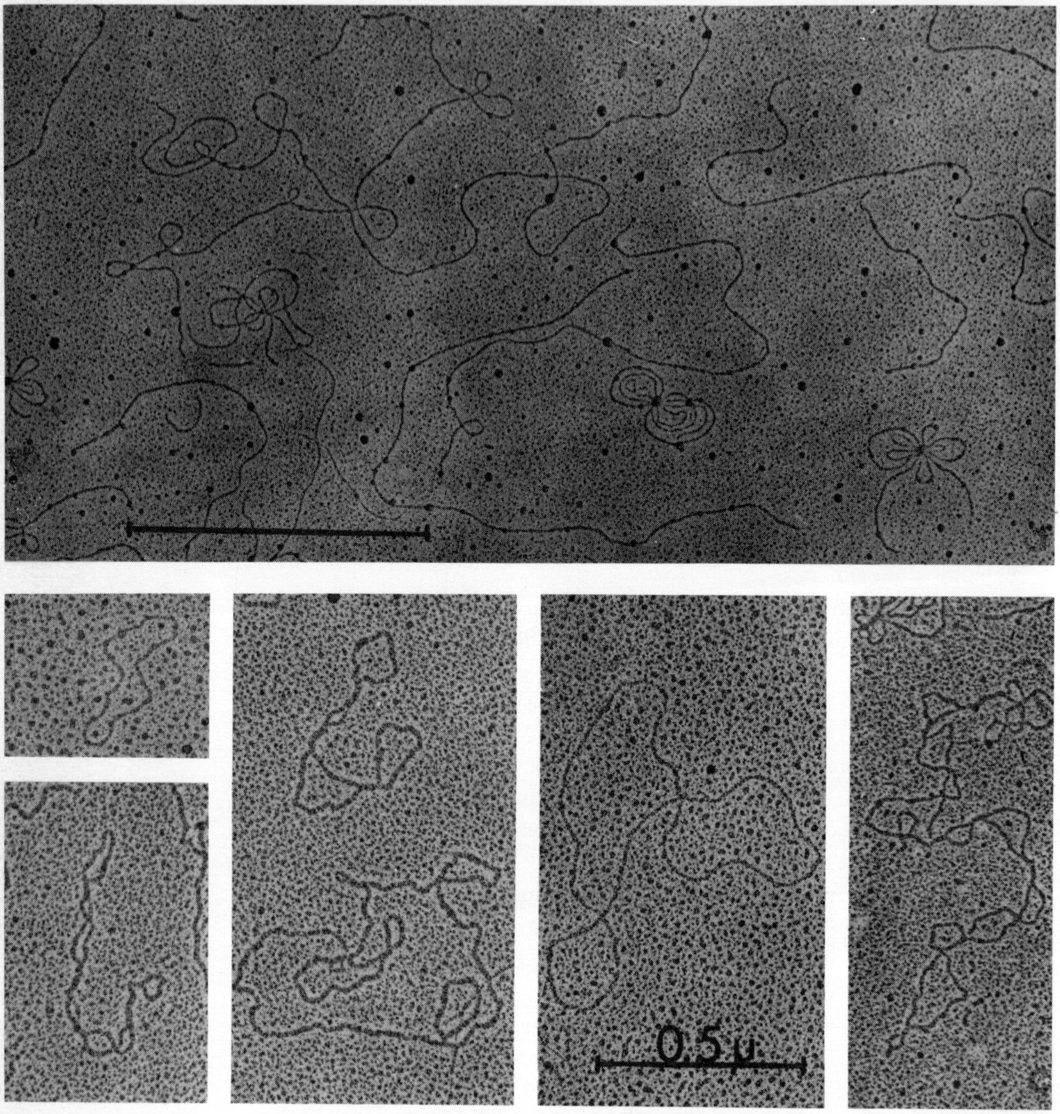

Fig. 90 Surface-spread yeast DNA shadowed with uranium oxide. Upper, native mitochondrial-satellite DNA purified by subfractionation with cesium chloride; lower, various examples of circular DNA filaments from whole-yeast-cell DNA. (Courtesy of Sinclair, Stevens, Sanghavi, and Rabinowitz.)

Nucleus and its contents Yeast cells differ from the true bacteria in being usually much larger and in possessing well-defined nuclei. The nucleus is difficult to see in hanging-drop preparations but may be seen in stained smears. It is spherical to oval and is bounded by a double membrane (Fig. 85).

The nucleus contains the chromosomes and the nucleolus. The chromosomes are partitioned between mother cell and bud in a precise and orderly manner without recourse to a spindle. The wall of the nucleus does not break down at any time in the life cycle. It is a permanent cellular structure.

The nucleus consists largely of DNA and protein. Another nucleic acid, RNA, involved in protein synthesis, is present to some extent in the nucleus but it is found largely in the cytoplasm.

The nucleic acid molecule consists of a mixture of smaller molecules called nucleotides. There are normally four kinds of nucleotides. All nucleotides from RNA contain the sugar ribose, and all have a residue of phosphoric acid attached to carbon 5 of the ribose. They differ in the residue attached to carbon 1 of the sugar. It can be one of the following bases: the purines adenine and guanine, and the pyrimidines cytosine and uracil. The four nucleotides are linked together as shown on page 844.

DNA is similar to RNA but differs in having deoxyribose instead of ribose, and one of the four bases is thymine instead of uracil. Both RNA and DNA molecules are best visualized as chains of alternating units of ribose or deoxyribose and phosphoric acid, and one of the four bases is normally linked to carbon 1 of each sugar unit.

DNA is a double-stranded polynucleotide chain having the structure of a helix. The two chains are held together by hydrogen bonds between the bases (see page 306). A pyrimidine base is bound to a purine base, i.e., thymine to adenine, and cytosine to guanine. In a double-stranded DNA helix the ratio of thymine to adenine and cytosine to guanine is equal to 1. The two chains are coiled around each other and run in opposite directions.

Some viruses consist entirely of RNA nucleoprotein (tobacco mosaic virus, polio virus); others consist almost entirely of DNA nucleoprotein (bacterial viruses or bacteriophages).

Vacuoles Vacuoles are cavities in the protoplasm. They are especially characteristic of plant cells and protozoa but occur also in the cells of higher animals. In plants, the vacuoles contain a fluid known as cell sap, that is commonly an aqueous solution of various organic acids and their salts. In protozoa, vacuoles may contain secretions of the protoplasm or substances about to be excreted, or food in various stages of digestion and assimilation.

Svihla and Schlenk (1959, 1960) cultivated *T. utilis* and *S. cerevisiae* in a culture medium containing the amino acid L-methionine and reported the accumulation of large amounts of *S*-adenosylmethionine in the vacuoles (Fig. 91). The compound is formed by the interaction of

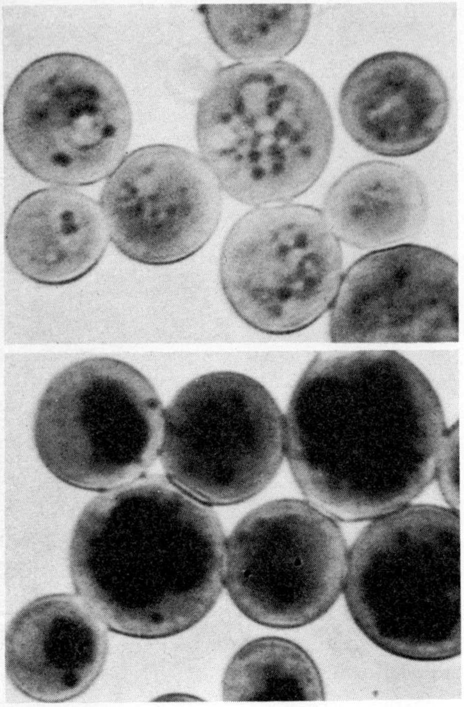

Fig. 91 Ultraviolet photomicrographs of Saccharomyces cerevisiae. Upper, control cells showing vacuoles; lower, cells showing localization of S-adenosylmethionine in vacuoles. (Courtesy of G. Svihla.)

L-methionine and adenosine:

$$\text{Adenine-ribose-}\underset{\displaystyle CH_3}{\underset{|}{S^+}}\text{—}CH_2\cdot CH_2\cdot CHNH_2\cdot COOH$$

A part of the material is transferred from the vacuoles of the mother cells into the vacuoles of the buds and daughter cells.

Balish and Svihla (1968) employed ultraviolet microscopy to study the capacity of yeast and filamentous strains of *Candida albicans* to accumulate UV-absorbing materials from a medium supplemented with purines, pyrimidines, amino acids or related compounds as the main nitrogen source. All strains accumulated UV-absorbing compounds when adenine, adenosine, isoguanine, xanthine, or uric acid was supplied but did not accumulate UV-absorbing compounds when pyrimidines were supplied. The filamentous strain accumulated UV-absorbing material from medium supplemented with hypoxanthine but the yeast strains did not. Since significant differences occurred among various species, the investigators hesitated to formulate a general theory on the vacuole as a storage compartment at this time.

For more information: Botsford and Parks (1967); Knudsen et al. (1969); Svihla et al. (1963).

Granules The protoplasm of yeast cells is more clearly differentiated than that of bacteria. Young cells have a very thin cell wall and a relatively homogeneous cytoplasm. As the cells become older, granules and vacuoles appear. The granules consist of metachromatic granules (volutin) and glycogen.

Metachromatic granules are widely distributed in fungi and constitute important cellular elements present in yeasts. The granules originate most likely in the cytoplasm and localize almost exclusively in the vacuoles later. In old cells, the metachromatic granules may be present in relatively large masses. The granules appear as refractive bodies in unstained preparations and show a great affinity for basic stains, revealing their acid character.

The granules are said to be composed of metaphosphate or another form of inorganic phosphate, some fat, and possibly small amounts of protein and ribonucleic acid. The granules can be extracted by perchloric and trichloracetic acids. Their presence and size in cells are related to the phosphate concentration of the growth medium in the presence of an energy source and specific divalent ions (Mn and Zn). Older cells possess larger granules.

Glycogen is a white, amorphous carbohydrate related to starch and dextrin. On boiling with dilute mineral acids, or by the action of amylolytic enzymes, it is hydrolyzed to glucose. Glycogen is easily recognized by the brown color produced on the addition of a dilute solution of iodine. The color disappears when the solution is heated to 60°C and reappears on cooling.

Glycogen has been shown to be abundant in well-nourished yeast cells and to disappear during starvation. The concentration increases with age and reaches a maximum after 48 hr. Glycogen is usually localized in the vacuoles distinct from those which contain the metachromatic granules. It accumulates in the asci during sporulation and is absorbed by the spores during their maturity.

Glycogen is believed to be a reserve food material that accumulates in old cells after activity ceases. It disappears when old cells are placed in a new medium and reappears again as the cells increase in age.

For more information: Roth (1970).

Fat globules Fat globules of variable size are distributed throughout the cytoplasm of yeast cells. They stain brown with osmic acid. Fat globules are prominent in yeast cells, especially during sporulation, and serve as food for the ascospores. They are considered to be reserve food material.

White and Werkman (1948) reported the following conditions under which relatively large and reproducible increases in the fat content of cells of *S. cerevisiae* could be obtained.

1. Cells were grown in an appropriate medium at 30°C for 24 hr.
2. Cells were harvested by means of a Sharples centrifuge and suspended in a phosphate buffer of pH 7.0 containing acetate. Under such conditions the cells no longer proliferated.
3. The suspension of nonproliferating cells was aerated for 24 hr. During this period the acetate was converted into fat and stored in the cells.

For more information: Greenfield and Klein (1960), Klein (1955), Klein and Booher (1955), Starr and Parks (1962).

Reproduction in yeasts

Yeasts reproduce by budding, fission, asexual spore formation, sexual reproduction or copulation, and parthenogenesis. However, the usual method of reproduction is by budding. Sometimes thick-walled, resistant cells known as chlamydospores are produced. Since only one spore is produced from each cell, this is not a means of increasing numbers but a method for perpetuating the species.

Budding Reproduction by budding is observed in almost all the yeasts (Fig. 92). Budding occurs after the organism acquires a certain size. At the first step in budding (Fig. 92*D*), the centrosome produces a small conical process which forces its way through the cytoplasm and erupts into a new bud (Fig. 92*E*). The nuclear vacuole puts out a process which enters the bud (Fig. 92*F*). The bud vacuole finally receives the chromosome complex, after which the connection between the bud and mother vacuole is closed off (Fig. 92*G*). Cytoplasm passes into the bud, followed by the centrochromatin, which travels along the elongated centrosome (Fig. 92*H*). The centrochromatin divides transversely, and the mother and daughter cells are held together by a plug of centrosomal material which

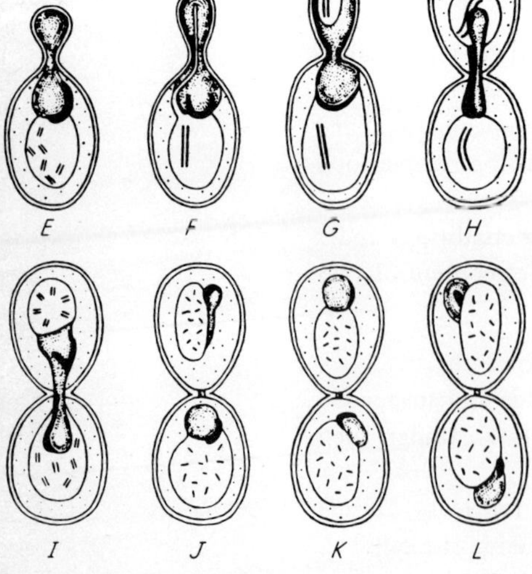

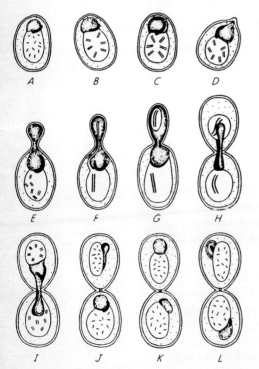

Fig. 92 Multiplication by budding. See text for descriptions. (After Lindegren.)

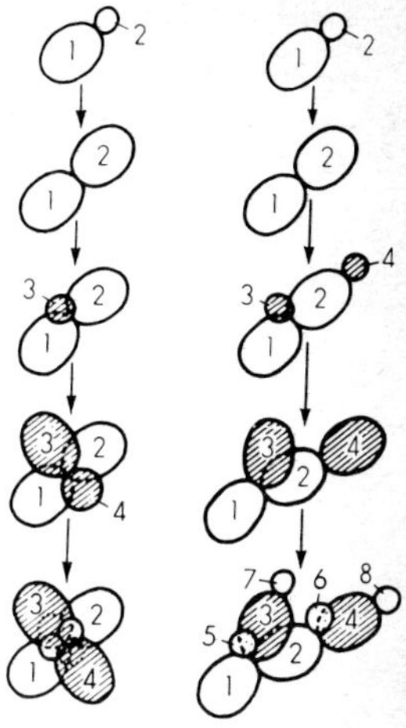

Fig. 93 Saccharomyces cerevisiae. Left, haploid mode of budding; right, polyploid mode of budding. Shaded cells represent second generation. In both diagrams, 3 and 4 have budded from 1 and 2, respectively. In the next generation, 5, 6, 7, and 8 are progeny of 1, 2, 3, and 4 in that order. (After Freifelder.)

makes up the bud scar (Fig. 92*I*). This plug remains at the junction of the bud and mother cell after the bud and mother centrosome have separated (Fig. 92*J*, *K*, and *L*). After budding is completed, the nuclei in bud and mother cell reorient themselves so that the centrosome of each is distal to the point of connection between the two cells. The young bud receives from the mother cell a complement of cytoplasm rich in mitochondria and ribosomes essential for protein synthesis, and a full nuclear apparatus comprising both the mechanisms for genetic control and subsequent budding.

During periods of rapid division, buds may be formed at different points on the surface of the cell. The daughter cells likewise may bud at different points before separation from the mother cell. This results in the formation of a small colony or chain of yeast cells.

Freifelder (1960) observed the growth of *S. cerevisiae* by means of time-lapse photography. He found the position of the new cell to be dependent upon ploidy, being at the birth end in haploid and the opposite end in polyploid cells (Fig. 93). The first few buds tended to arise from the same point of the mother cell.

For more information: Hashimoto et al. (1959); McClary et al. (1962); Matile et al. (1969).

Transverse fission In a few species, multiplication occurs by transverse fission. These yeasts resemble the bacteria in their mode of division.

In *Schizosaccharomyces octosporus*, the spherical or ovoid-shaped cells elongate to a certain size, then form cross walls in the middle (Fig. 108). The two cells pull apart, and the ends become rounded. When the two new cells reach maturity, they elongate and repeat the cycle. During periods of rapid multiplication, cells may divide without separating. When this occurs, chains of cells are produced resembling a mycelium. The cells eventually break apart.

Streiblová et al. (1966) studied growing and dividing cells of *S. pombe* by means of fluorescence and electron microscopy with the use of metal-shadowed isolated walls, replicas, and ultrathin sections. Vegetative cells contained division scars (six at most). Structural changes of cell surfaces and lateral walls during fission are represented schematically and described in Figs. 94 and 95.

In the genus *Saccharomyces*, a form of division intermediate between budding and fission occurs. Buds are generally produced at the extremities of the cells. The cells first elongate, then a tube puffs out at one end. This enlarges and is slowly transformed into a bud, which remains

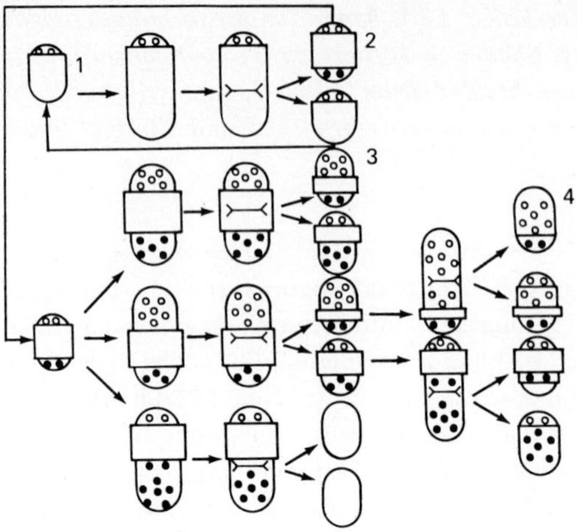

Fig. 94 Schematic representation of fission in Schizosaccharomyces pombe (partially up to the fourth generation). (1) Cells with a single scar are doubled by longitudinal growth and a septum is formed. (2) Cleavage of the septum gives rise to two cells; the single-scar cell undergoes the same process as the original cell, whereas three alternatives of terminal growth exist for a two-scar cell. (3) Cells of the third generation have two scars; the original wall ring has a different width and localization. The third alternative of the terminal growth has not as yet been sufficiently studied, and therefore the cells of the third generation are drafted without any structures. (4) After growth of the arthrospore from one pole, fourth generation contains three-scar cells in addition to single-scar cells. Original wall ring is shown as white area; scar plugs, polar cell walls, and additional wall rings by small circles. (Courtesy of Streiblová, Málek, and Beran.)

attached to the cell by a collar. Finally a wall is formed which separates the cell from the bud.

Asexual spore formation Although budding is the usual process of multiplication in yeasts, such a method does not perpetuate the species. The usual process for perpetuating the species is by spore formation, which is a form of resistance that permits yeast cells to remain viable after budding has stopped.

Sporulation may be induced by growing yeast on a special agar medium, then transferring the cells to the surface of a plaster-of-paris block moistened with dilute acetic acid. On the presporulation medium yeasts accumulate enormous reserves of fat and glycogen. The fat is stored on the mitochondria, and the glycogen in the cytoplasm. The nucleus and centrosome assume a central position in the cell in contrast to the eccentric arrangement of the centrosome in the budding cell.

There are at least four pairs of chromosomes which segregate regularly at meiosis. During meiosis, it is believed that first the chromatin

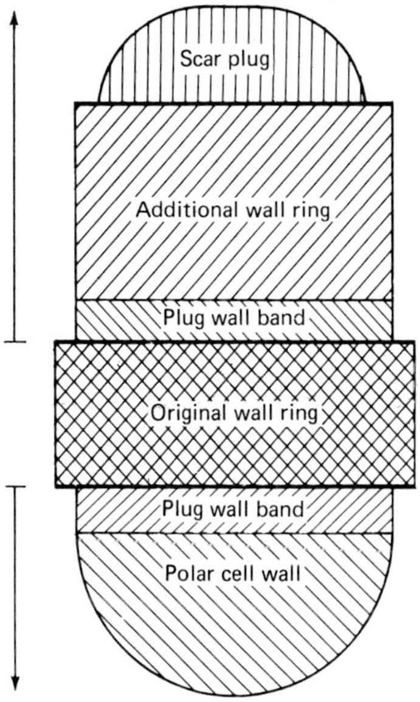

Fig. 95 Diagram of a three-scar cell. Arrows indicate the direction of the terminal growth (terminal growth continues on the lower pole). The scar plug is the region of cell wall on the cell pole bounded by the scar margin. This region has the same properties as the plug wall band. The original wall ring is the oldest part of the cell wall; it is bounded on both sides by scar margins. The plug wall band is a belt of the scar plug wall; it preserves the properties of the scar plug and is a part of the polar cell wall and of the additional wall ring. The polar cell wall is a region of the younger cell wall of the cylindrical cell pole; it is bounded by one scar margin, and has a plug wall band determining the direction of wall extension. The additional wall ring of the younger wall, bounded on both sides by scar margins, has a plug wall band determining the direction of wall extension. (Courtesy of Streiblová, Málek, and Beran.)

divides into two long bands, and then these divide again to form four structures. At this point neither centrosome nor nuclear vacuole is visible.

At the onset of sporulation the nuclear vacuole becomes indistinct and is not observed clearly again until the spores are fully formed. The most prominent structure in the cell is the chromatin, which occupies a central position in the cell and may be clumped in various patterns around the centrosome. At this stage the chromosomes cannot be distinguished from the chromatin of the centrosome. The chromatin may be arranged over the surface of the centrosome, or it may be at one side, or may surround it completely. Neither the centrosome nor the nuclear vacuole is clearly distinguishable, and the chromatin lies in two or four long granular strands, with the granules of the different strands apparently paired (Fig. 96).

A definite number of spores is usually produced in the cells of each species, although under certain conditions this may show wide variation.

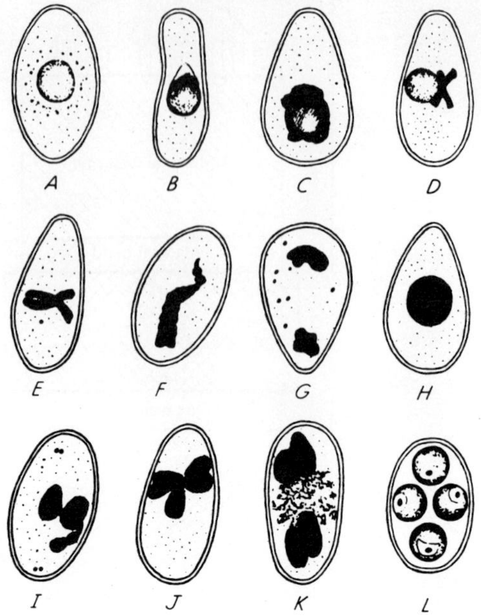

Fig. 96 Sporulation. (A) Unstained cell showing centrally located nuclear vacuole with enlarged centrosome above the vacuole. The latter is surrounded by mitochrondria. (B) Stained cell showing shrunken nuclear vacuole and large centrosome with closely applied chromatin. (C) Cell with large centrosome and conspicuous chromatin complex closely applied to centrosome. Nuclear vacuole not visible. (D) Centrosome with two associated double chromatin rods. Vacuole not visible. (E, F) Paired chromatic strands, probably representing two chromosome complexes. (G) Chromosome complexes separated. (H) Centrosome completely covered with chromatin. (I, J) Division of the chromatin. (K) Division of chromatin completed. (L) Four spores in an ascus, each with a vacuole and darkly stained centrosome. (After Lindegren.)

The cells bearing spores are called asci (singular, ascus), and the spores are called ascospores. In the majority of species four spores are produced in a cell. In other species the number may be one, two, or less frequently, eight. *S. cerevisiae*, the common baker's yeast, usually shows asci containing four spores. Santa María (1957, 1959*a*, *b*) studied a large number of strains of this species and found asci containing eight and even nine spores.

Fowell and Moorse (1960) found that yeasts tended to dissociate into mixtures of cell types with different powers of sporulation. They recommended the use of single-cell isolates for sporulation studies.

For more information: Conti and Naylor (1960*a*, *b*); Fowell (1969).

Sexual reproduction or copulation Sexual multiplication occurs by (1) isogamic copulation, (2) heterogamic copulation, (3) a form of copula-

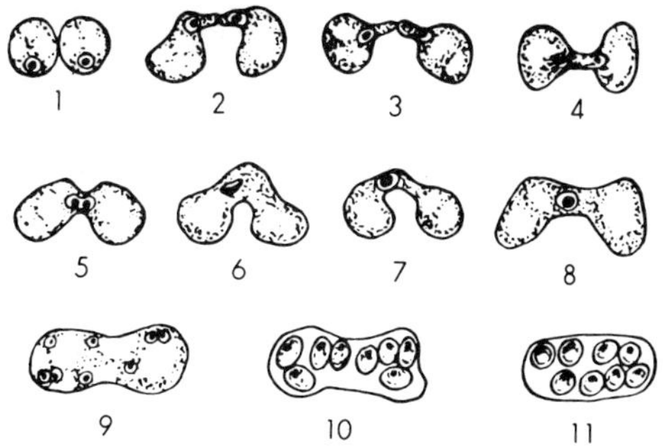

Fig. 97 Schizosaccharomyces octosporus. Isogamic copulation of two cells to form an ascus containing eight ascospores. (From Guilliermond-Tanner, "The Yeasts," John Wiley & Sons, Inc.)

tion intermediate between isogamy and heterogamy, and (4) copulation of ascospores.

Isogamic copulation Isogamic copulation may be defined as the fusion of two similar gametes. *Schizosaccharomyces octosporus* is an example of a yeast that reproduces in this manner. Two cells lying adjacent in a colony are joined by a copulation canal (Fig. 97). The two cells are now known as gametes. The wall that separates the two cells quickly disappears and the nucleus of each passes into the copulation canal. This results in the formation of a single cell or zygospore. The zygospore increases in size, followed by a division of the nucleus. The nuclei become surrounded with cytoplasm, around which are formed the spore walls. The zygospore now becomes an ascus.

Fusion of nuclear material occurs first between the centrosomes (Fig. 98*A*), which apparently project through the cell walls at the point of contact of the copulating gametes. Next, there is an interchange of cytoplasm between the copulants (Fig. 98*B*). The strands of centrochromatin from each cell travel along the centrosome to fuse, each

Fig. 98 Isogamic copulation. (A) a pair of copulating cells with the centrosomes fused. The centrochromatin elongated preparatory to union. (B) incomplete zygote. Centrochromatin fused. Nuclear vacuole putting out a process preparatory to fusion. (C) fusion of vacuoles complete. (D) completed zygote with chromosomes paired in common nuclear vacuole. (After Lindegren.)

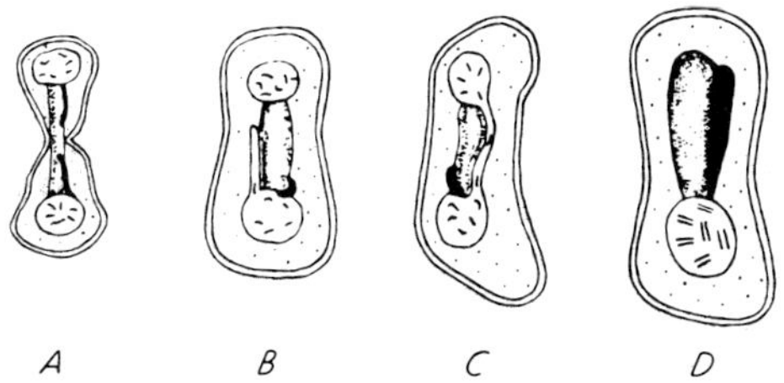

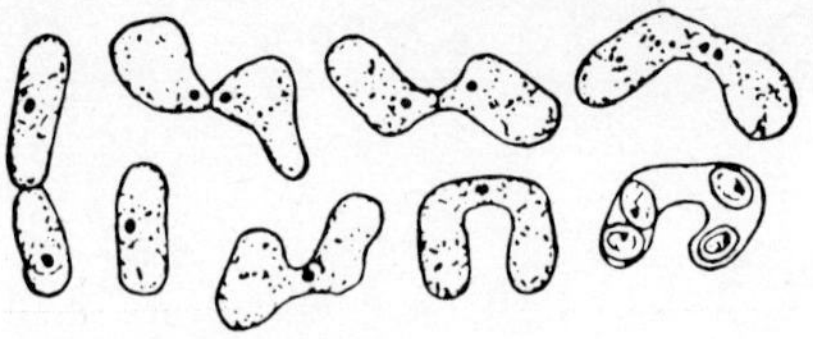

Fig. 99 Schizosaccharomyces pombe. Copulation and incomplete fusion of two cells to form an ascus containing four ascospores. (Reprinted from Guilliermond-Tanner, "The Yeasts," John Wiley & Sons, Inc.)

retaining contact with its respective nuclear vacuole. Simultaneously, one or both nuclear vacuoles send out processes which also fuse, joining both vacuoles into a single one. The chromosomes pass into the vacuole, which is in contact with fused centrochromatins (Fig. 98*C*) and pair up (Fig. 98*D*). The zygospore produces a hybrid diploid bud.

In *S. pombe*, copulation occurs as in *S. octosporus*, except that fusion usually remains incomplete. Copulation occurs between two adjacent cells in the same colony. The gametes are joined by a canal through which nuclear and protoplasmic fusion occurs. The nucleus resulting from the fusion quickly divides, and the two nuclei migrate to both enlargements of the zygospore. The nuclei undergo a second division, resulting in the formation of four spores. The zygospore becomes an ascus, and the spores are known as ascospores (Fig. 99).

Heterogamic copulation Heterogamic copulation may be defined as the fusion of two unlike gametes (Fig. 100). The ascus that develops is composed of two unequal enlargements, the larger representing the mother cell and the other the bud. Because of lack of space in the bud, the spores develop in the mother cell.

In the genus *Nadsonia*, copulation occurs by heterogamy between an adult cell and one of its buds (Fig. 101). After the two cells fuse, the contents of the male gamete or bud pass into the female gamete or mother cell. A new cell then forms by budding, and the contents of the mother cell pass into it. This new cell now becomes an ascus, and it usually contains a single ascospore.

Intermediate form of copulation A rare form of copulation, intermediate between isogamy and heterogamy, has been observed. In these yeasts the two cells or gametes are of the same size and do not show

Fig. 100 Saccharomyces. Ascus produced by heterogamic copulation. (From Guilliermond's Clef dichotomique pour la détermination des levures, courtesy of Librairie le Francois, Paris.)

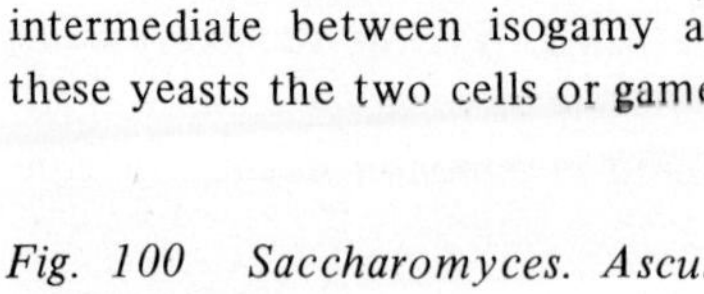

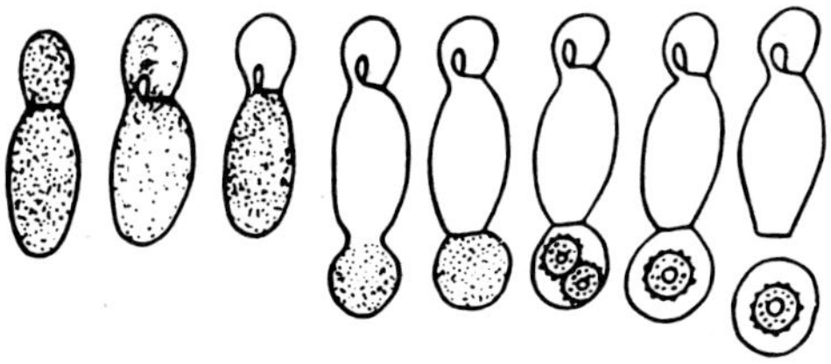

Fig. 101 Nadsonia. Formation of an ascus by heterogamic copulation. (From Guilliermond's Clef dichotomique pour la détermination des levures, courtesy of Librairie le Francois, Paris.)

any sexual differentiation. After fusion takes place, the contents of one cell pass into the other. The former may be regarded as the male cell and the latter as the female cell. The ascospores originate from the female cell and are usually two in number (Fig. 102).

Copulation of ascospores In some species, such as in *Saccharomycodes ludwigii, S. johannisbergensis,* and *Hansenula saturnus,* an isogamic copulation occurs between the ascospores produced in an ascus (Fig. 103). In the species *S. ludwigii*, an ascus usually contains four spores. The spores copulate in pairs and are joined by a canal. The nuclear material fuses, resulting in the formation of a zygospore. However, the fusion remains incomplete. The zygospore elongates into a germination tube, from which develop numerous vegetative cells by budding.

Copulation occurs normally between two spores in the same ascus. It has also been observed between ascospores from different asci and even those more distantly related. This may be observed in old asci where many of the spores are dead. Some of the spores can germinate alone; others must fuse in pairs before germination can occur. This means that some of the spores in the latter group are forced to fuse with spores from different asci. The fusion of ascospores is not regarded as a true copulation but as a new process that takes the place of normal sexual fertilization.

Fig. 102 Schwanniomyces. Asci with ascospores. (From Guilliermond's Clef dichotomique pour la détermination des levures, courtesy of Librairie le Francois, Paris.)

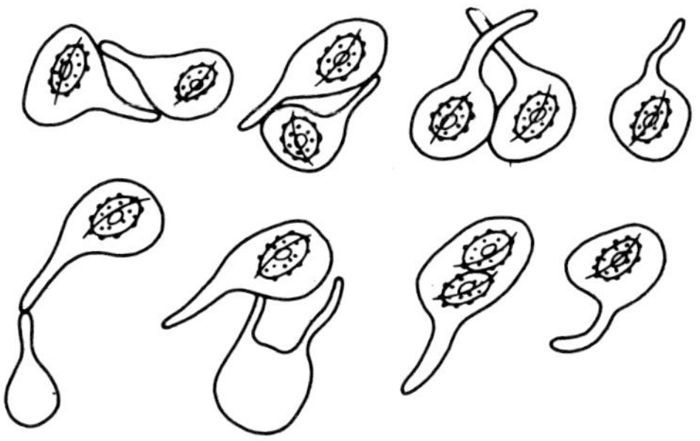

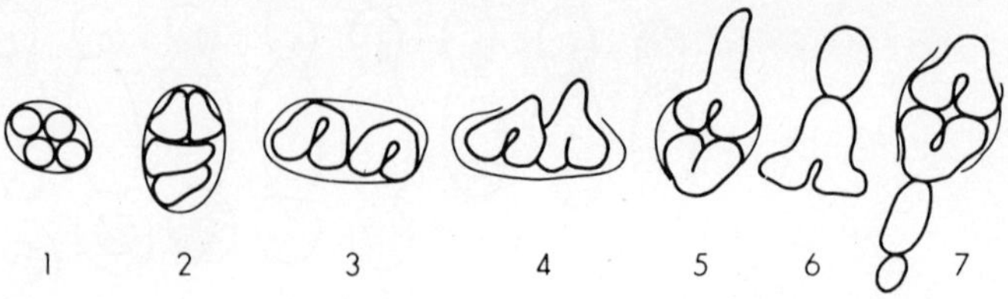

Fig. 103 Saccharomycodes ludwigii. Conjugation of ascospores inside an ascus to form zygospores. The zygospores germinate by budding. (From Guilliermond's Clef dichotomique pour la détermination des levures, courtesy of Librairie le Francois, Paris.)

Shapes of spores Yeast spores assume various shapes (Fig. 104), and some species may be easily recognized by this character. Yeast spores are usually spherical or ovoid. Such spores are found in *Saccharomyces cerevisiae,* the common bread or beer yeast, and in other lesser-known species. The spores in *Hansenula anomala* and in the genus *Hanseniaspora* are hemispherical, and their adjacent surfaces are provided with a projecting border, giving them the appearance of a hat (Fig. 103*b*). In the species *Pichia membranaefaciens,* the spores are irregularly shaped into ovoid, elongated, triangular, kidney-shaped, or hemispherical forms (Fig. 103*c*). Cells of *H. saturnus* produce spores that are lemon-shaped and surrounded by a projecting ring (Fig. 103*d*). In the species *Schwanniomyces occidentalis,* the spores are surrounded by a projecting ring and the membrane is covered with stiff, erect protuberances (Fig. 103*e*). The spores of *Debaryomyces* are globular and also covered with protuberances (Fig. 103*f*). Other shapes are shown in Fig. 103*g, h*, and *i*.

Parthenogenesis This term may be defined as the development of an organism from an unfertilized cell. Parthenogenesis is a modification of ordinary sexual reproduction and is not to be confused with asexual multiplication.

Fig. 104 Various types of ascospores. (From Guilliermond's Clef dichotomique pour la détermination des levures, courtesy of Librairie le Francois, Paris.)

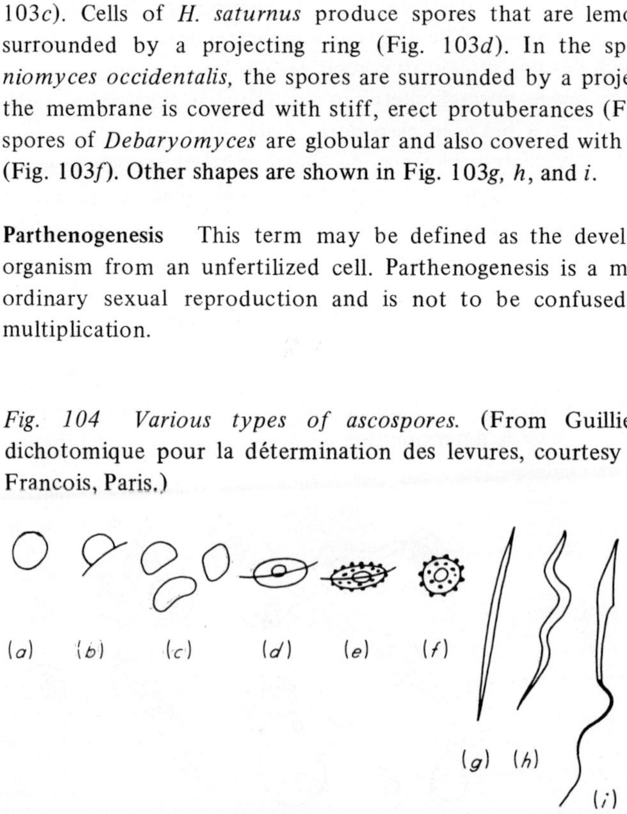

In some yeasts, such as in the *Schwanniomyces*, sexuality has not been observed. The cells forming the asci produce projections of different lengths which attempt to fuse together as in true copulation (Fig. 102). However, fusion fails to occur, and it appears that the cells have retained only a portion of their sexual characteristics.

Chlamydospores Under unfavorable conditions, yeast cells stop reproducing. When this occurs, some yeasts may become filled with reserve food materials, such as fat and glycogen, and enclose themselves in a thick wall. These cells are known as durable cells or chlamydospores. When a chlamydospore is placed in a favorable environment, it germinates into a cell that reproduces by budding. However, since only one chlamydospore is formed in a cell, it should not be considered a true method of reproduction but rather a method for perpetuating the species.

Life cycle of yeasts

Genes are factors in the nucleus concerned with the transmission and development or determination of hereditary characters. Genes consist of deoxyribonucleic acid, and chromosomes are largely DNA combined with basic protein as deoxyribonucleoprotein.

In most organisms, cells can exist with either one, two, three, or more sets of chromosomes. Yeast cells may exist in haploid, diploid, triploid, tetraploid, and even pentaploid and hexaploid forms. In each case, the degree of ploidy of a culture remains constant during growth by budding. Changes in ploidy, as well as sexual differentiation, are an integral part of the life cycle of yeasts.

Mating types Many yeasts exhibit heterothallism. Thus strains of *Saccharomyces cerevisiae* fall into two classes: α-mating type and a-mating type. When haploid cells of type α are mixed with cells of type a, cells become connected in pairs by a tubular structure, and from two fused cells a single cell or zygote is produced. The zygote is diploid and is not capable of copulating with a similar cell.

Cultivation on an appropriate culture medium causes the diploid zygote to sporulate. The cell nucleus divides twice to produce four nuclei, internal walls are formed around each of the four nuclei, and the cell becomes an ascus containing four ascospores. The spores are smaller and more resistant to adverse environmental conditions than normal, vegetative cells. When the spores are placed in a favorable medium, they enlarge and become normal, budding, haploid vegetative cells. The cells have now completed the life cycle, starting from haploid cells, to copulating cells, to diploid zygotes, to four-spored asci, to free-germinating ascospores, and finally back to vegetative haploid cells again (Fig. 105). The life cycle is completed in 3 days. The final four cultures from each ascus consist of two α-mating and two a-mating types.

Certain tetraploid strains can also be made to sporulate forming an

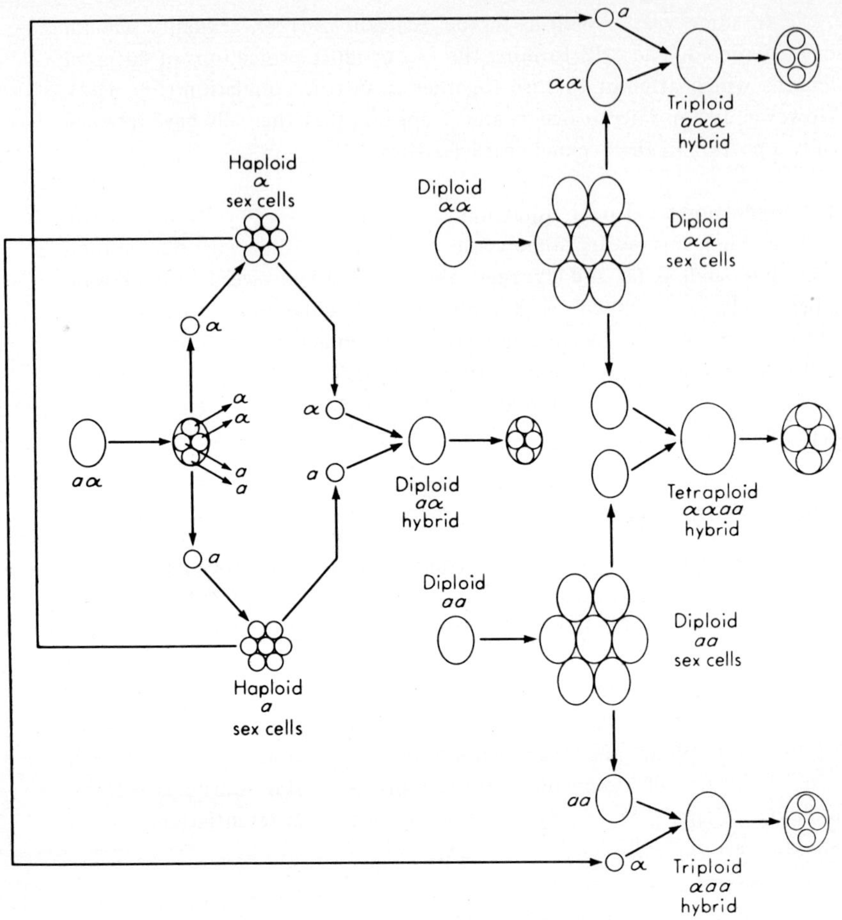

Fig. 105 Diagram of the life cycle of Saccharomyces. (After Lindegren.)

ascus containing four diploid spores. Cultures derived from these spores follow a simple rule. If they are able to sporulate, they will not mate; if they can mate, they cannot be made to sporulate. Diploid segregant cultures from tetraploids are of three types: αα, aa, and aα. The diploid mating cultures αα and aa show mating reactions similar to α-mating and a-mating type cultures, respectively, whereas aα cultures do not mate but sporulate like normal diploid zygotes. Thus each of the diploid products produced by sporulation of a tetraploid cell can be assigned to one of the above three classes.

For more information: Fowell (1969); Lindegren (1962, 1963*c*); Lindegren et al. (1962); Ouchi and Lindegren (1963); Windisch (1961).

Hybridization

A hybrid may be defined as a new organism developed from two cells of different species or different genetic makeup.

Winge and Laustsen (1938, 1939) showed that it is possible to breed new varieties of yeasts by hybridization. They placed two spores of opposite sexes from different species in a drop of culture solution to enable them to copulate. A zygospore was formed from which the hybrid yeast germinated. The investigators succeeded in producing 14 new yeast types from 8 different species and strains of *Saccharomyces* and one of *Zygosaccharomyces*.

Lindegren (1944, 1945, 1949) and Lindegren and Lindegren (1943, 1945) also succeeded in hybridizing yeast spores from different species or strains. They developed a number of new yeast strains strong in characteristics required for special needs or purposes.

While this procedure is in itself very intricate, the even greater problem is to determine which spore cultures have the specific characteristics required to meet a certain need. The outline drawings in Fig. 106 illustrate four steps in the process of hybridization.

Step 1. Yeast cells from two strains (strains A and B) are shown after they have been induced to form spores. The spores are removed by microdissection by means of a micromanipulator.

Step 2. Each individual spore is planted into a nutrient medium and allowed to multiply to produce separate spore cultures. After careful testing for the desirable characteristics, two spore cultures of different sexes are selected for mating or crossbreeding.

Step 3. When these two spore cultures are brought together, instead of budding they fuse in pairs and produce a completely new yet stable combination of inherited qualities.

Step 4. The new hybrid strain shown is a combination of the best qualities of strains A and B. The new hybrid grows and reproduces by budding. As it grows, the new inherited qualities are equally transmitted to all cells reproduced.

In Fig. 107, five stages are shown in the use of the microtool for cutting the cell wall of an ascus and removing the individual spores. It consists of a high-powered microscope combined with a manipulator that reduces hand movements several thousandfold. When the operator moves a handle 1/4 in, the microneedle moves less than 1/10,000 in.

For more information: Winge (1952*a, b*).

Classification of yeasts

Yeasts have been classified with difficulty and confusion still exists. The abridged classification given here has been taken largely from the works of Lodder and Kreger-van Rij (1953) and Kreger-van Rij (1969).

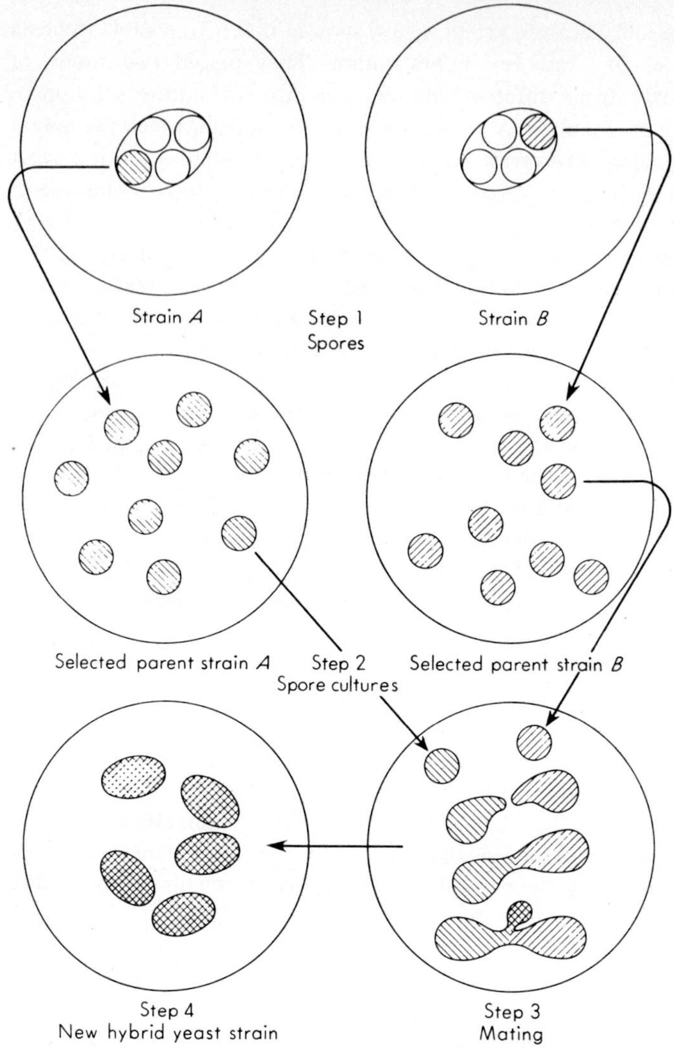

Fig. 106 Four steps in the process of hybridization. See text for description. (Courtesy of Anheuser-Busch, Inc.)

ASCOSPOROGENOUS YEASTS

FAMILY I *Saccharomycetaceae*

GROUP I

Genus I *Schizosaccharomyces.* Cells cylindrical, rectangular, oval, or spherical. Vegetative reproduction by fission only. True mycelium which breaks up into arthrospores may be

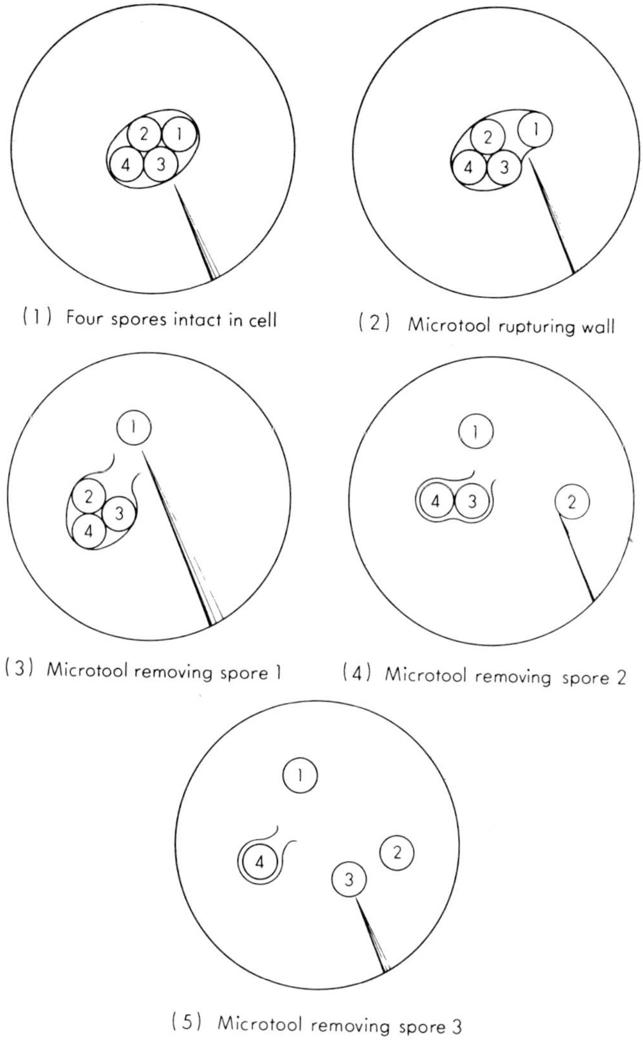

Fig. 107 Five stages in the microdissection of yeast spores. See text for descriptions. (Courtesy of Anheuser-Busch, Inc.)

formed. Asci formed by isogamic copulation and contain four or eight spherical, oval, or kidney-shaped spores (Fig. 108).

GROUP II

Genus I *Endomycopsis.* Growth form true mycelial with arthrospores; also pseudomycelial and budding cells. Vegetative reproduction by budding. Ascospores hemispherical, oval, sickle-shaped, smooth, or warty (Fig. 109).

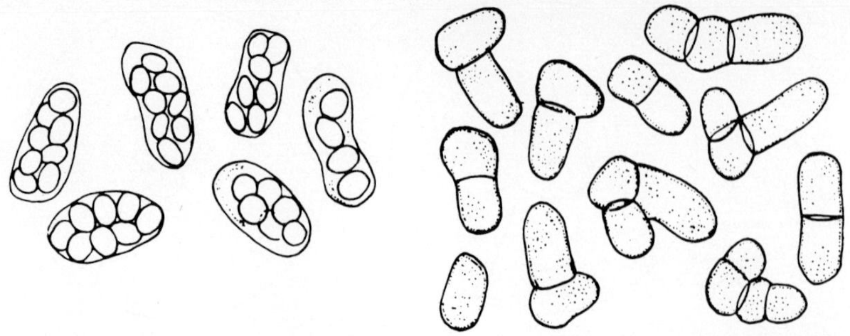

Fig. 108 Schizosaccharomyces octosporus. Right, round or cylindrical yeast cells; left, asci containing eight or fewer round spores. (From Lodder and Kreger-van Rij's "The Yeasts," courtesy of North Holland Publishing Company, Amsterdam.)

Genus II *Hansenula.* Cells spherical, oval, or elongate. Pseudomycelium formation frequent. Vegetative reproduction by multipolar budding. Isogamic or heterogamic conjugation may or may not immediately precede ascus formation. Asci contain one to four hat-shaped, Saturn-shaped, spherical, or hemispherical spores (Fig. 110).

For more information: Crandall and Brock (1968); Mills and Blackwood (1967).

Genus III *Dekkera.* Vegetative reproduction by multilateral budding. Mycelium not formed. Ascospores hat-shaped.

Fig. 109 Endomycopsis fibuliger. Left, budding yeast cells and septate, branched hyphae; right, asci containing mostly four hat-shaped ascospores. (From Lodder and Kreger-van Rij's "The Yeasts," courtesy of North Holland Publishing Company, Amsterdam.)

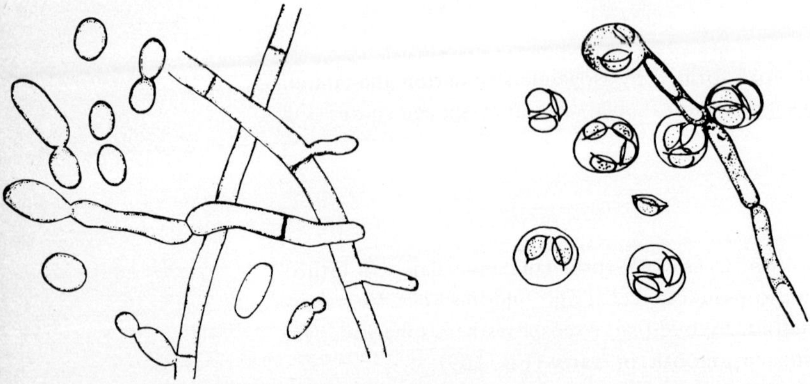

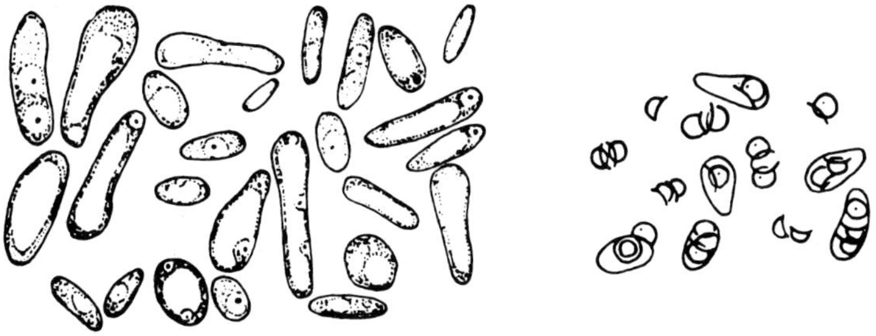

Fig. 110 Hansenula anomala. Left, oval and cylindrical yeast cells; right, asci containing one to four hat-shaped ascospores. (From Lodder and Kreger-van Rij's "The Yeasts," courtesy of North Holland Publishing Company, Amsterdam.)

Genus IV *Pichia.* Growth form pseudomycelial. Cells ovoid to long cylindrical. Vegetative reproduction by multipolar budding. Heterogamic conjugation may or may not precede ascus formation. Asci contain one to four spherical, hat- or Saturn-shaped, smooth or warty spores (Fig. 111).

For more information: Poncet (1967).

Genus V *Debaryomyces.* Cells spherical or oval. Pseudomycelium sometimes present. Vegetative reproduction by multipolar budding. Asci formed by conjugation between mother cell and bud. Asci contain usually one, sometimes up to four, round or oval warty spores (Fig. 112).

Genus VI *Schwanniomyces.* Cells oval or egg-shaped. Pseudomycelium not formed. Vegetative reproduction by multipolar budding. Asci formed by conjugation between mother cell and bud. Asci contain one, sometimes two, spherical, or oval-shaped warty spores with a projecting border in the center (Fig. 113).

Genus VII *Saccharomyces.* Cells spherical, ovoid, or elongate. Reproduction by multipolar budding. Pseudomycelium may be formed. Isogamic or heterogamic conjugation may or may not precede ascus formation. Asci contain one to

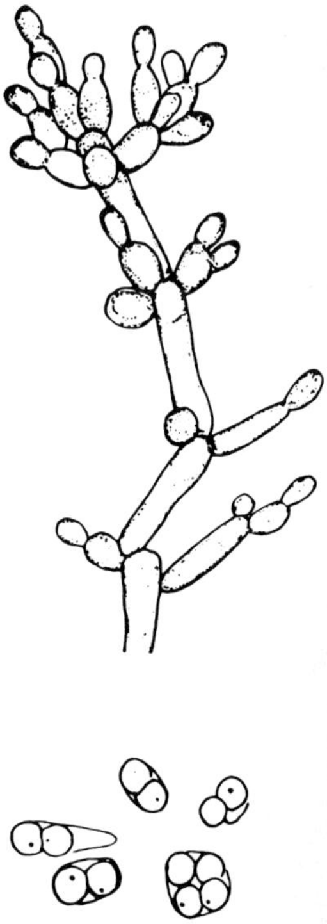

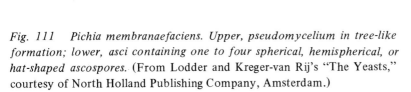

Fig. 111 Pichia membranaefaciens. Upper, pseudomycelium in tree-like formation; lower, asci containing one to four spherical, hemispherical, or hat-shaped ascospores. (From Lodder and Kreger-van Rij's "The Yeasts," courtesy of North Holland Publishing Company, Amsterdam.)

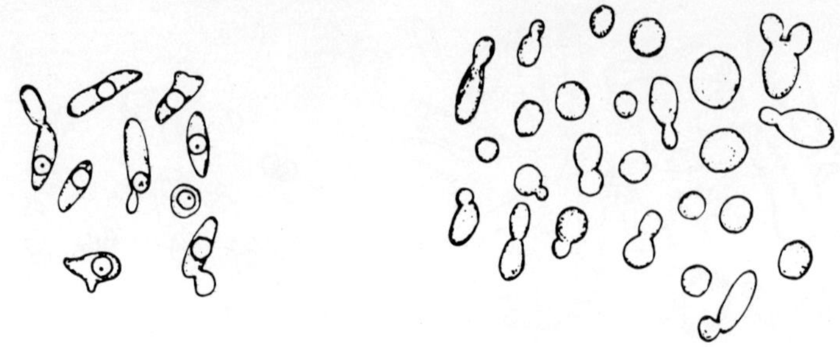

Fig. 112 Debaryomyces kloeckeri. Right, budding yeast cells; left, asci containing usually one, sometimes two spherical ascospores. (From Lodder and Kreger-van Rij's "The Yeasts," courtesy of North Holland Publishing Company, Amsterdam.)

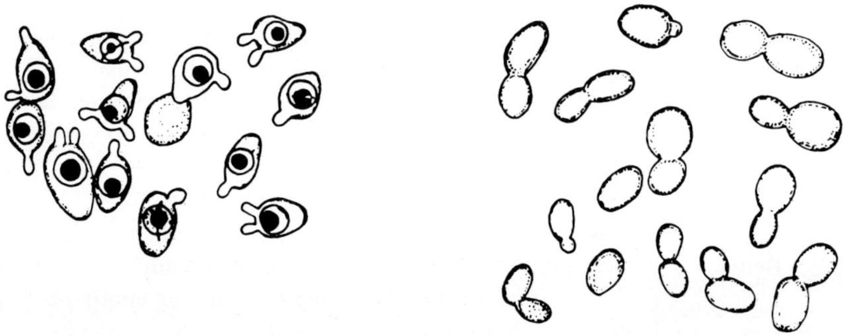

Fig. 113 Schwanniomyces occidentalis. Right, budding yeast cells; left, asci containing usually one round, or oval, warty ascospore with a projecting border in the center. (From Lodder and Kreger-van Rij's "The Yeasts," courtesy of North Holland Publishing Company, Amsterdam.)

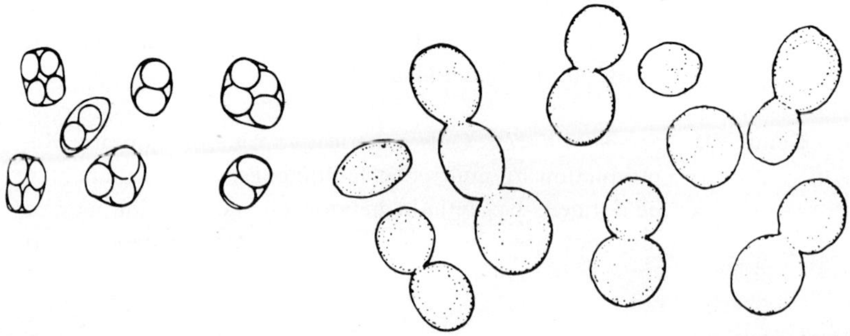

Fig. 114 Saccharomyces cerevisiae. Right, budding yeast cells; left, asci containing four round or slightly oval ascospores. (From Lodder and Kreger-van Rij's "The Yeasts," courtesy of North Holland Publishing Company, Amsterdam.)

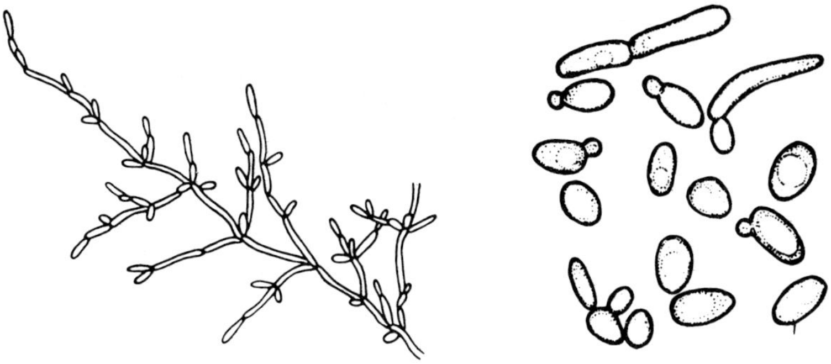

Fig. 115 Saccharomyces fragilis. Left, well-developed pseudomycelium produced on potato agar; right, budding yeast cells grown on malt extract agar. (From Lodder and Kreger-van Rij's "The Yeasts," courtesy of North Holland Publishing Company, Amsterdam.)

four spherical, or oval, smooth or warty spores (Figs. 114 and 115).

For more information: Klein et al. (1967); Santa María and Sanchez (1970).

Genus VIII *Saccharomycopsis.* Cells long oval or cylindrical. Vegetative reproduction by multipolar budding. Asci contain oval spores with two membranes. Spores germinate by budding.

Other and more recently created genera in this group are *Citeromyces, Kluyveromyces, Pachysolen,* and *Wingea.*

GROUP III

Genus I *Lipomyces.* Cells spherical or oval, surrounded by a slimy capsule. Pseudomycelium absent. Vegetative reproduction by multipolar budding. Vegetative cells produce saclike protuberances which become asci. Ascus usually remains attached to mother cell. In some cases more than one ascus arises from a mother cell. Asci contain one to sixteen ascospores. Spores are oval, amber-colored, smooth or warty, or with ridges (Fig. 116).

For more information: Slooff et al. (1969).

GROUP IV

Genus I *Nadsonia.* Cells oval, elongate, or lemon-shaped. Pseudomycelium not formed. Vegetative reproduction by bipolar budding. Asci produced by heterogamic conjugation

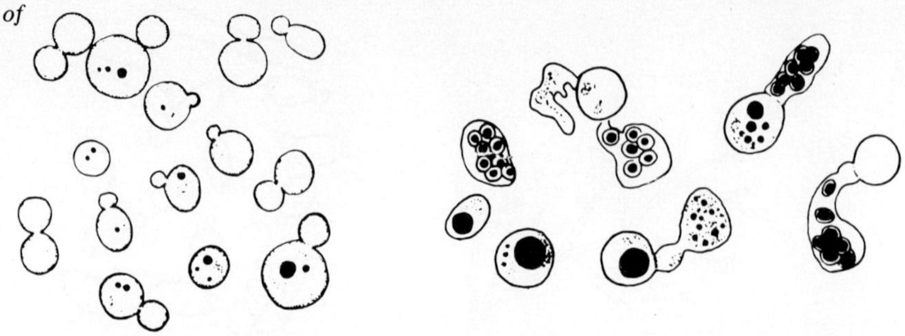

Fig. 116 Lipomyces lipoferus. Left, round to short-oval yeast cells with buds; right, asci containing usually eight smooth, oval ascospores. (From Lodder and Kreger-van Rij's "The Yeasts," courtesy of North Holland Publishing Company, Amsterdam.)

between bud and mother cell, and contain usually one, sometimes two, spherical, warty, brown spores (Fig. 101).

For more information: Spencer and Gorin (1968).

Genus II *Saccharomycodes.* Cells lemon-shaped or sausage-shaped. Pseudomycelium poorly developed. Vegetative reproduction by bipolar budding. Asci contain two to four spherical, smooth spores with a ledge. Spores conjugate in pairs (Fig. 117).

Genus III *Hanseniaspora.* Cells lemon-shaped or oval. Pseudomycelium not formed. Vegetative reproduction by bipolar budding. Asci produced by parthenogenesis and

Fig. 117 Saccharomycodes ludwigii. Right, lemon-shaped or sausage-shaped yeast cells with buds; left, asci containing four round and smooth ascospores. (From Lodder and Kreger-van Rij's "The Yeasts," courtesy of North Holland Publishing Company, Amsterdam.)

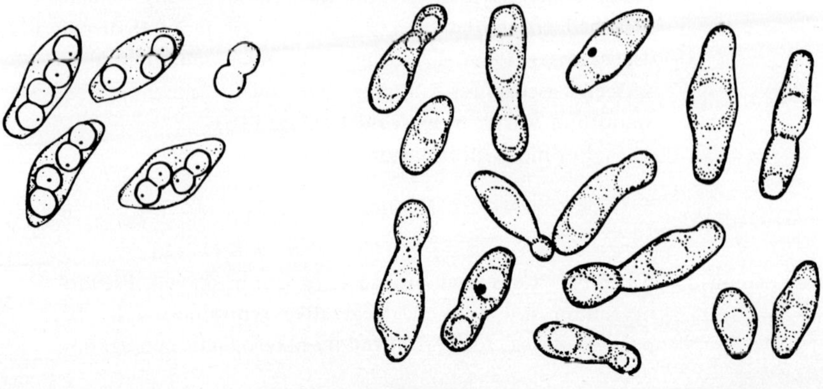

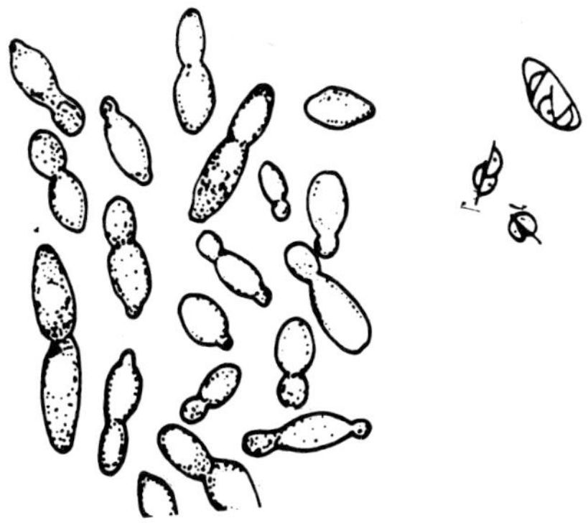

Fig. 118 Hanseniaspora valbyensis. Left, lemon-shaped, oval, or long-oval yeast cells with buds; right, asci containing one to four ascospores. The spores are round when young, later becoming hat-shaped. (From Lodder and Kreger-van Rij's "The Yeasts," courtesy of North Holland Publishing Company, Amsterdam.)

contain one to four round, hat-, helmet-, or Saturn-shaped, smooth or warty spores (Fig. 118).

Another and more recently created genus in this group is *Wickerhamia*.

FAMILY II *Spermophthoraceae*

Genus I *Nematospora.* Cells spherical, oval, elongate, or irregularly shaped. Vegetative reproduction by multipolar budding. True mycelium and pseudomycelium always present. Asci produced by parthenogenesis and are much larger than vegetative cells. Asci contain four to eight spindle-shaped ascospores, each with a whip (Fig. 119).

Genus II *Coccidiascus.* Cells spherical or elongate, slightly curved. Mycelium not formed. Vegetative reproduction by budding. Asci large, banana-shaped, and produced by isogamic conjugation. Asci contain eight fusiform, nonflagellate spores (Fig. 120).

Another and more recently created genus in this group is *Metschnikowia*.

For more information: Bos and de Boer (1968); Donkersloot (1966); Gilliland (1971); Makula and Finnerty (1968); Meyer and Phaff (1969); Scheda and Bos (1966).

Saccharomyces This genus contains almost all of the species of industrial importance and is by far the most important group of yeasts.

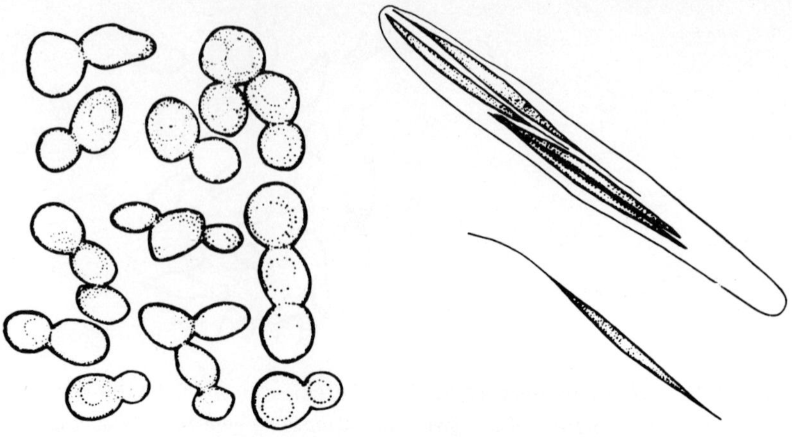

Fig. 119 Nematospora coryli. Left, round or oval-shaped yeast cells with buds; right, ascus containing four fusiform-shaped spores, each provided with a whip. Usually asci contain eight spores in two bundles of four each. (From Lodder and Kreger-van Rij's "The Yeasts," courtesy of North Holland Publishing Company, Amsterdam.)

The cells are spherical, ovoid, or elongate. Reproduction occurs by multiple budding. Asci are formed by isogamic or heterogamic copulation, or by parthenogenesis, and contain one to four spherical or oval, smooth or warty, spores.

The species of greatest importance is *S. cerevisiae*, the common yeast employed by bakers, brewers, and distillers. It is the same species that is extensively used therapeutically as a source of vitamins of the B complex.

BALLISTOSPOROGENOUS YEASTS

FAMILY I *Sporobolomycetaceae*

Genus I *Sporidiobolus.* Cells reproduce by budding and by asymmetrical ballistospore formation. True mycelium formed with clamp connections.

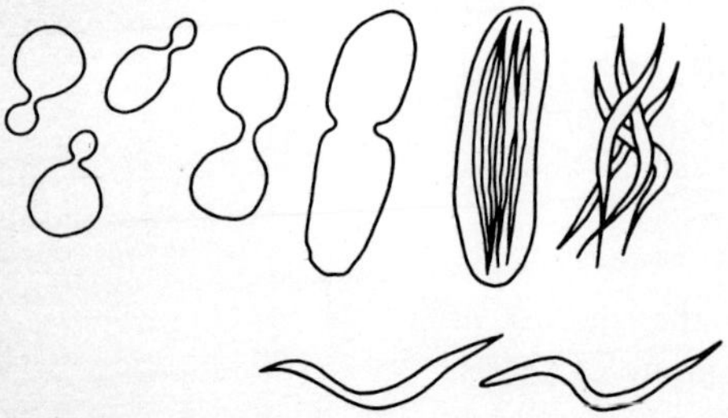

Fig. 120 Coccidiascus. Ascus produced by isogamic copulation. Four ascospores in each ascus. (From Guilliermond's Clef dichotomique pour la détermination des levures, courtesy of Librairie le Francois, Paris.)

Genus II *Sporobolomyces.* Cells reproduce by budding and by asymmetrical ballistospore formation. True mycelium formed.

Genus III *Bullera.* Cells reproduce by budding and by symmetrical ballistospore formation.

Genus IV *Rhodosporidium.* Cells reproduce by budding. True mycelium with or without clamp connections produced. Chlamydospores and sporidia may also be present.

For more information: Newell and Hunter (1970).

Genus V *Leucosporidium.* Cells reproduce by budding. True mycelium with or without clamp connections produced. Chlamydospores and sporidia may also be present.

FAMILY II *Cryptococcaceae.* Yeasts that do not produce ballistospores or ascospores.

Genus I *Torulopsis.* Pseudomycelium absent or very primitive. Cells spherical or oval, rarely elongate. Reproduction by multipolar budding. Cells occasionally capsulated. Starch-like polysaccharide absent (Fig. 121).

For more information: Yarrow (1968).

Fig. 121 Asporogenous yeasts. Upper row: left, Trigonopsis sp.; center, Trichosporon margaritiferum; right, Kloeckera sp. Lower row: left, Pityrosporum ovale; center, Torulopsis sp.; right, Cryptococcus neoformans. (From Mrak, Phaff, and Stadtman, "Taxonomy and Morphology of Yeasts," University of California, Berkeley.)

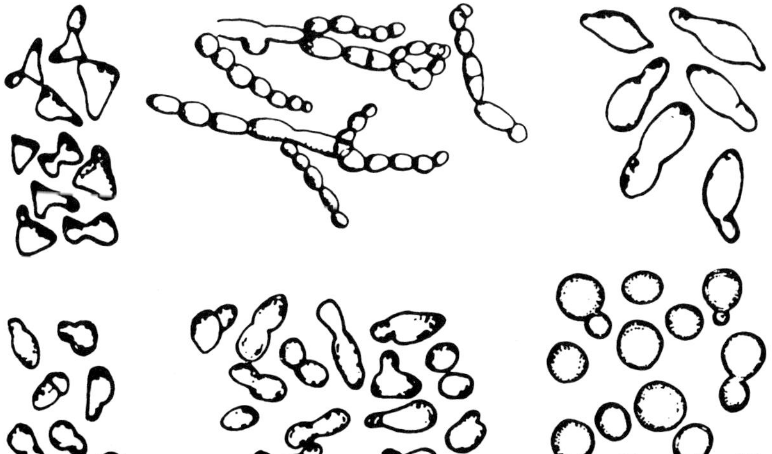

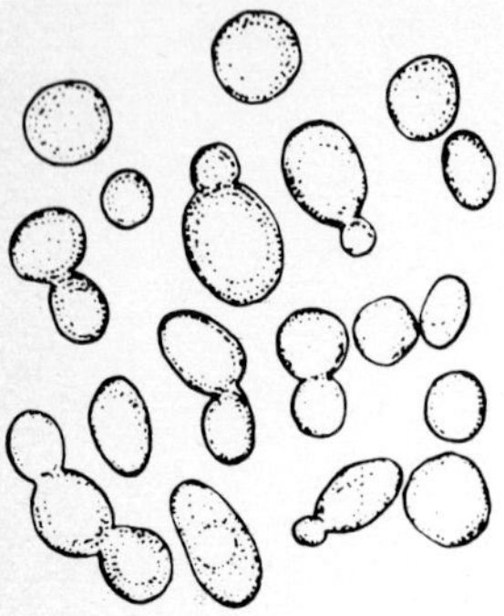

Fig. 122 Rhodotorula glutinis. Budding yeast cells. (From Lodder and Kreger-van Rij's "The Yeasts," courtesy of North Holland Publishing Company, Amsterdam.)

Genus II *Rhodotorula.* Cells spherical, oval, or elongate. Occasionally a primitive pseudomycelium observed. Reproduction by multipolar budding. Yellow or red pigment produced (Fig. 122).

Genus III *Cryptococcus.* Pseudomycelium absent or rudimentary. Cells spherical or oval, occasionally long oval, amoeboid, or polymorph. Reproduction by multipolar budding. Cells surrounded by a capsule composed of a starchlike polysaccharide. Cultures on solid media have a mucoid appearance (Fig. 121).

Genus IV *Candida.* Pseudomycelium more or less abundantly developed; also true mycelium. Blastospores may be attached to pseudomycelium in a characteristic manner. Cells of varying shapes. Reproduction by multipolar budding (Fig. 123). Chlamydospores may be present. The accompany-

Fig. 123 Asporogenous yeasts. Various types of pseudomycelia formed in the genus Candida. (From Mrak, Phaff, and Stadtman, "Taxonomy and Morphology of Yeasts," University of California, Berkeley.)

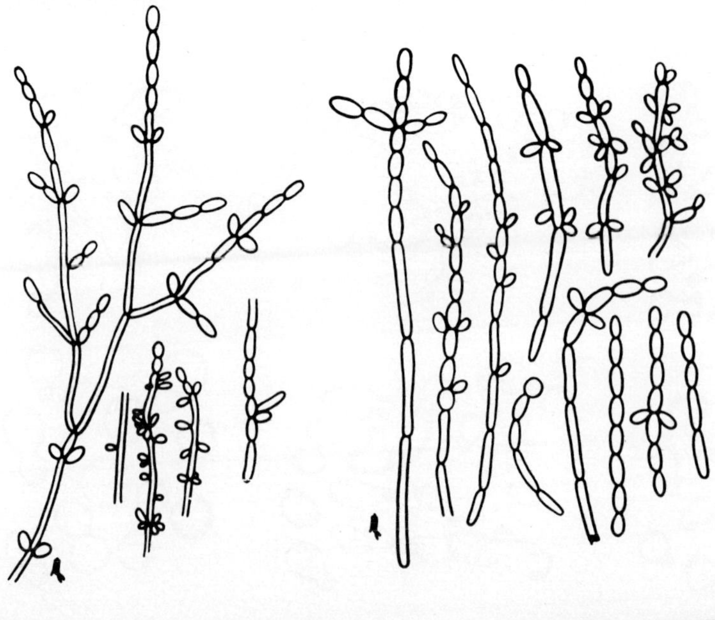

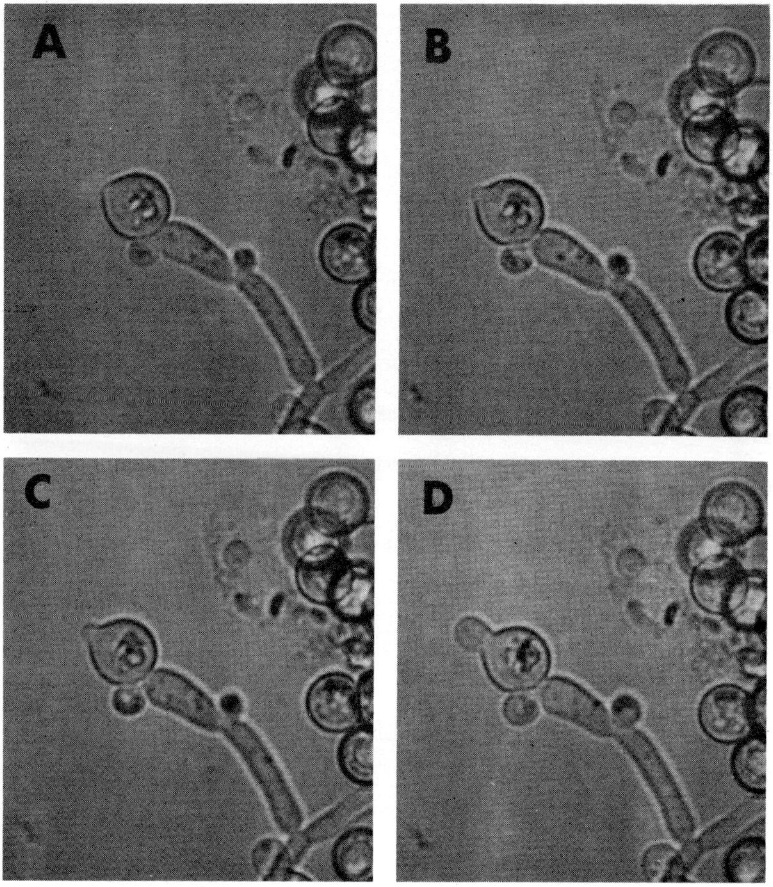

Fig. 124 Candida albicans. Budding of chlamydospores. (A to C) three min elapsed time; (D) thirty min after C. (Courtesy of Jansons and Nickerson.)

ing micrographs in Figs. 124 and 125 are by Jansons and Nickerson (1970*a*) and by Barnes et al. (1971).

For more information: Jansons and Nickerson (1970*b*); Kocková-Kratochvilová et al. (1967).

Genus V *Trigonopsis*. Pseudomycelium absent. Cells triangular or ellipsoidal. Reproduction by budding, at the angles in the triangular cells, and multipolar in the ellipsoidal cells (Fig. 121).

Genus VI *Kloeckera*. Primitive, rarely well-developed pseudomycelium occasionally observed; true mycelium absent. Cells lemon-shaped, oval, or sausage-shaped. Reproduction by bipolar budding (Fig. 121).

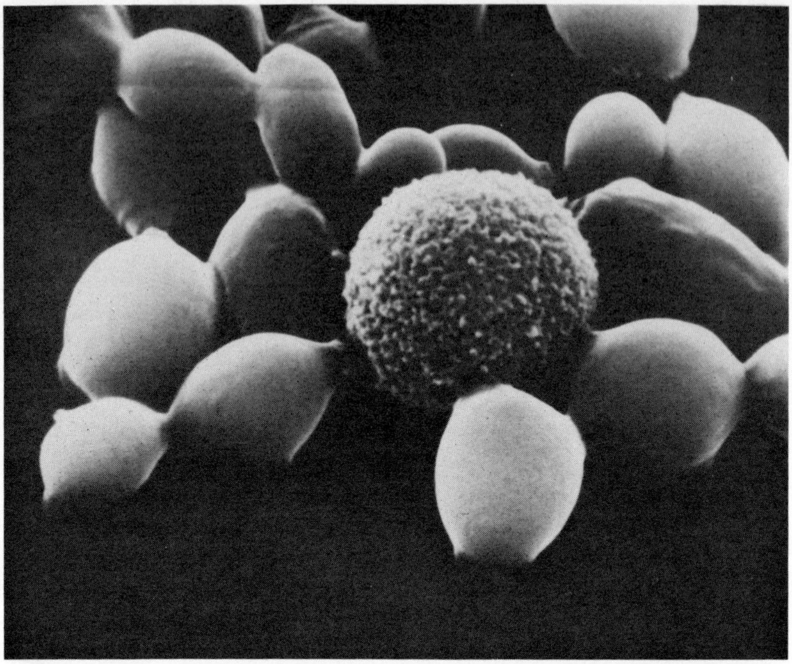

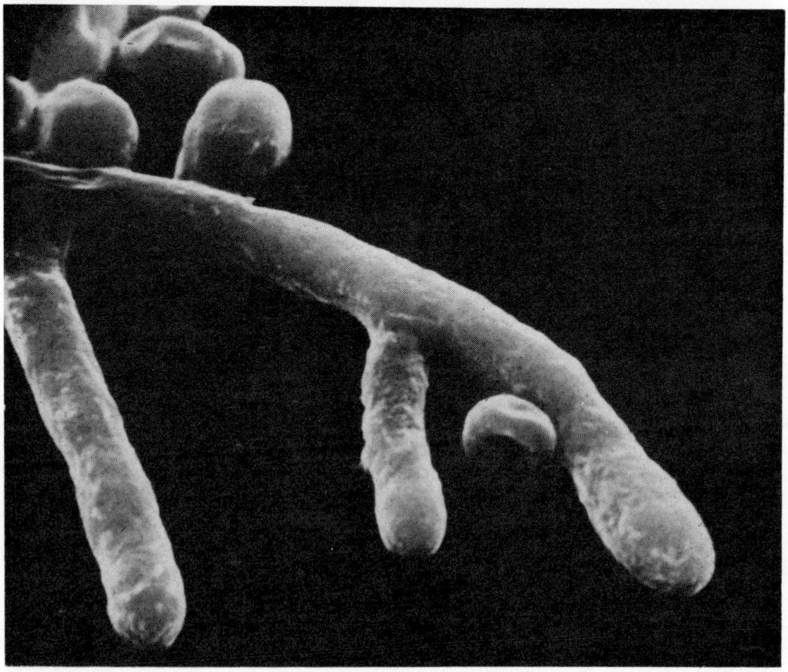

◀ *Fig. 125 Upper, budding yeast cells of Candida albicans and a large "rough" cell which exhibits convolutions on its surface. The smooth surface of the budding cells is evident. Lower, filamentous stage after 3 days of growth showing mycelial-like processes and accompanying blastospore formation.* (Courtesy of Barnes, Flesher, Berger, and Arnold.)

Genus VII *Pityrosporum.* Mycelium or pseudomycelium absent. Cells oval or bottle-shaped. Reproduction by polar budding (Fig. 121).

Genus VIII *Brettanomyces.* Primitive pseudomycelium may be formed. Cells spherical, oval, elongate, frequently ogive-shaped. Reproduction by multipolar budding, leading to formation of irregular chains of cells (Fig. 126).

Genus IX *Trichosporon.* True mycelium, pseudomycelium, blastospores, and arthrospores present. Blastospores develop in small chains or in wreaths. Cells of varying shapes. Chlamydospores may occur (Fig. 121).

Genus X *Oosporidium.* Cells reproduce by budding on a broad base. True mycelium present.

Genus XI *Sterigmatomyces.* Budding cells formed on sterigmata.

Pityrosporum A number of species of the genus *Pityrosporum* have been isolated from skin, but they all appear to be sufficiently alike to be

Fig. 126 Asporogenous yeasts. Left and right, Brettanomyces lambicus. (From Mrak, Phaff, and Stadtman, "Taxonomy and Morphology of Yeasts," University of California, Berkeley.)

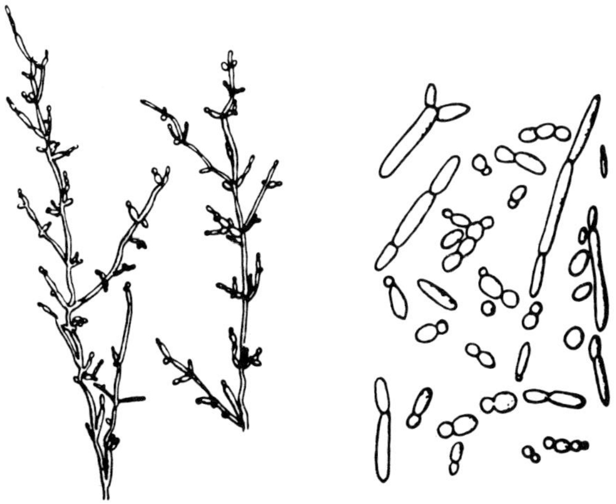

considered identical with *P. ovale*. The organism is usually referred to as the bottle bacillus because its shape resembles a bottle. Since the organism is present on skin, especially of the scalp, it is claimed by some to be the etiological agent of dandruff. However, there appears to be no evidence that it has anything to do with the formation of dandruff or any pathological condition of the skin or scalp.

Brettanomyces This genus includes a number of species of industrial importance. They are concerned with the afterfermentation of certain English and Belgian beers and ales. The afterfermentation is a secondary fermentation taking place in bottles in which the residual oxygen is utilized and replaced by carbon dioxide. Fermentation under anaerobic conditions is very slow, requiring several months for the reaction to go to completion. Under aerobic conditions, the organisms oxidize alcohol to acetic acid.

Laboratory technique

The methods used for the isolation and study of yeasts in pure culture are, in general, the same as those employed for bacteria.

Most species of yeasts grow best at a temperature of about 25°C.

Induced sporulation The conditions responsible for sporulation to occur are not clearly understood. Numerous factors are involved, and these seem to vary from species to species. In other words, a set of conditions cannot be laid down that will apply to all species of yeasts.

The following factors are generally believed to be of importance in causing vegetative cells to sporulate:

1. The yeast cells must be vigorous, well nourished, and young.
2. A temperature of about 25°C is most favorable for sporulation to occur. This generally corresponds to the optimum growth temperature of yeasts. Temperatures somewhat below and above 25°C are less favorable for sporulation to occur.
3. An abundance of oxygen must be present. Strictly anaerobic suspensions of ascosporogenous yeasts fail to sporulate. Cross sections of yeast colonies reveal sporulating cells only on the extreme outer layer. Stored yeast cakes contain spores only on the outside where oxygen is present.
4. Reports on the effect of pH on sporulation are contradictory. Some maintain that acid or neutral conditions are necessary for sporulation to occur; others claim that slightly alkaline conditions are more favorable. Welten (1914) found that yeasts readily sporulated on prune-extract agar. More spores were produced in concentrated than in dilute prune extract. Since no spores were observed in alkaline media, he concluded that a high acidity was necessary for sporulation to occur. It is likely that sporulation is independent of reaction and that a pH of 6 or 7 will give satisfactory results, provided other conditions are

satisfied. In this connection it may be mentioned that the pH of expressed yeast juice is about 5.6 to 6.0 for brewer's yeasts, and somewhat lower for baker's yeasts.

5. An abundance of moisture is necessary for sporulation. Porous materials composed of plaster of paris, clay, wood, filter paper, agar, gelatin, etc., become soaked up with water or nutrient solution and are satisfactory as carriers of moisture.

   A plaster-of-paris block is placed in a petri dish with the lower portion of the block immersed in water or liquid culture medium. The petri dish is sterilized in an autoclave. Then the upper surface of the block is heavily inoculated with a young liquid culture and incubated at 25°C for at least 1 week. This method generally produces good sporulation.

   On the other hand, Hartelius and Ditlevsen (1953) found plaster-of-paris blocks to be unsatisfactory because of their inability to withstand sterilization. They recommended blocks cast from a mixture of diatomaceous earth and portland cement. Such blocks were found to be heat-resistant and superior for spore formation. *Saccharomyces cerevisiae* sporulated profusely after 48 hr at 26°C.

6. An appropriate carbon compound in the medium may promote sporulation. The type added varies with the species under study. Mannose and maltose appear to produce the best results. Salts of organic acids, especially acetates, may stimulate sporulation.

   Gorodkowa (1908) employed an agar medium containing 1 percent peptone and 0.25 percent glucose. Sporulation occurred after 4 days. A similar medium containing 5 percent glucose failed to induce sporulation, which indicated that conditions essential to sporulation involved primarily the sugar in the medium.

7. Earlier methods made use of vegetable wedges prepared from carrots, beets, potatoes, cucumbers, or turnips. Of these, carrot wedges gave superior results.

   Also agar slants prepared with an aqueous extract of carrots, beets, cucumbers, and potatoes induce sporulation and serve as an excellent stock culture medium.

   Lindegren and Lindegren (1944) recommended a medium containing extracts of beet leaves and roots, apricot juice, grape juice, yeast, glycerol, calcium carbonate, and agar. Most species of yeasts studied produced spores on slants in a few weeks.

8. Sporulation may be observed in old cultures where the environment is less favorable for growth. Nickerson and Thimann (1943) observed that sporulation occurred more abundantly when many dead cells were present, and postulated that some stimulating substance was derived from the dead and autolyzed cells.

It may be concluded that a specific nutrient is essential for abundant sporulation to occur. If this nutrient is satisfactory and the sugar content of the medium is not too high, sporulation may occur, even on agar slants. If gypsum blocks are used, the water should be acid in reaction and contain by-products of yeast growth, and possibly some substances produced on the death and disintegration of the yeast cells.

For more information: Esposito et al. (1969); Phaff and Mrak (1948, 1949); Roth and Halvorson (1969).

Isolation of pure cultures Many methods have been recommended for the isolation of yeasts in pure culture. Most of the methods are concerned with the isolation of single cells and their propagation in a suitable culture medium. Such methods require considerable skill and patience. Yeasts may also be isolated in pure culture by the same methods employed for the separation of bacterial species (see page 273).

For more information: Cook (1958); Fowell (1969); Ingram (1955); Kreger-van Rij (1969); Kudrjawzew (1960); Matile et al. (1969); Phaff et al. (1966); Reiff et al. (1960); Rose and Harrison (1969).

References

Avers, C. J.: Heterogeneous length distribution of circular DNA filaments from yeast mitochondria, *Proc. Natl. Acad. Sci.,* **58**:620, 1967.

Balish, E., and G. Svihla: Ultraviolet microscopy of purines and amino acids in the vacuole of *Candida albicans, J. Bact.,* **96**:259, 1968.

Barnes, W. G., et al.: Scanning electron microscopic studies of *Candida albicans, J. Bact.,* **106**:276, 1971.

Bos, P., and W. E. de Boer: Some aspects of the utilization of hydrocarbons by yeasts, *Antonie van Leeuwenhoek, J. Microbiol. Serol.,* **34**:241, 1968.

Botsford, J. L., and L. W. Parks: Role of S-adenosylmethionine biosynthesis in yeast, *J. Bact.* **94**:966, 1967.

Brown, M. L.: Development of chains of cells of *Saccharomyces cerevisiae* during fermentation, *J. Inst. Brewing,* **76**:61, 1970.

Carnevali, F., et al.: Cytoplasmic DNA from petite colonies of *Saccharomyces cerevisiae*: A hypothesis on the nature of the mutation, *Science,* **163**:1331, 1969.

Cirillo, V. P.: Sugar transport by *Saccharomyces cerevisiae* protoplasts, *J. Bact.,* **84**:1251, 1962.

Conti, S. F., and H. B. Naylor: Electron microscopy of ultrathin sections of *Schizosaccharomyces octosporus.* II. Morphological and cytological changes preceding ascospore formation, *J. Bact.,* **79**:331, 1960*a*; III. Ascosporogenesis, ascospore structure, and germination, *ibid.,* **79**:417, 1960*b*.

Cook, A. H.: "The Chemistry and Biology of Yeasts," New York, Academic Press, Inc., 1958.

Crandall, M. A., and T. D. Brock: Molecular basis of mating in the yeast *Hansenula wingei, Bact. Rev.,* **32**:139, 1968.

Darling, S., et al.: Kinetic and morphological observations on *Saccharomyces cerevisiae* during spheroplast formation, *J. Bact.,* **98**:797, 1969.

Darling, S., et al.: Structure and chemical composition of prospheroplast envelopes of *Saccharomyces cerevisiae* and *Hansenula anomala, J. Bact.,* **110**:336, 1972.

Donkersloot, J. A.: Utilization of mixtures of carbon compounds by yeasts, *Antonie van Leeuwenhoek, J. Microbiol. Serol.,* **32**:419, 1966.

Esposito, M. S., et al.: Acetate utilization and macromolecular synthesis during sporulation of yeast, *J. Bact.,* **100**:180, 1969.

Fowell, R. R.: Sporulation and hybridization of yeasts. In "The Yeasts," edited by A. H. Rose and J. S. Harrison, New York, Academic Press, Inc., 1969.

—— and M. E. Moorse: Factors controlling the sporulation of yeasts. I. The presporulation phase, *J. Appl. Bact.,* **23**:53, 1960.

Freifelder, D.: Bud position in *Saccharomyces cerevisiae, J. Bact.,* **80**:567, 1960.

Gilliland, R. B.: Yeast classification, *J. Inst. Brewing,* **77**:276, 1971.

Gorodkowa, A. A.: Ueber das Verfahren rasch die Sporen von Hefepilzen zu gewinnen, *Bull. jard. imp. bot. St. Peters.,* **8**:163, 1908.

Greenfield, S., and H. P. Klein: Lipid synthesis in low protein homogenates of yeast, *J. Bact.,* **79**:691, 1960.

Hartelius, V., and E. Ditlevsen: Cement blocks, heat-stable blocks, for ascospore-formation in yeast, *Compt. rend. trav. lab. Carlsberg, Série physiol.,* **25**:213, 1953.

Hashimoto, T., et al.: IV. Observations on budding *Saccharomyces cerevisiae* by light and electron microscopy, *J. Bact.,* **77**:344, 1959.

Hirano, T., and C. C. Lindegren: Electron microscopy of mitochrondrial changes in *Saccharomyces, J. Ultrastruct. Res.,* **8**:322, 1963.

Hutchison, H. T., and L. H. Hartwell: Macromolecule synthesis in yeast spheroplasts, *J. Bact.,* **94**:1697, 1967.

Ingram, M.: "An Introduction to the Biology of Yeasts," London, Sir Isaac Pitman & Sons, Ltd., 1955.

Jansons, V. K., and W. J. Nickerson: Induction, morphogenesis, and germination of the chlamydospore of *Candida albicans, J. Bact.,* **104**:910, 1970*a*.

—— and ——: Chemical composition of chlamydospores of *Candida albicans, J. Bact.,* **104**:922, 1970*b*.

Klein, H. P.: Synthesis of lipids in resting cells of *Saccharomyces cerevisiae, J. Bact.,* **69**:620, 1955.

—— and Z. K. Booher: Synthesis of lipids in cell-free extracts of yeast, *Proc. Soc. Exp. Biol. Med.,* **89**:43, 1955.

—— et al.: Membranes of *Saccharomyces cerevisiae, J. Bact.,* **94**:475, 1967.

Knudsen, R. C., et al.: Uptake and utilization of S-adenosyl-L-methionine and S-adenosyl-L-homocysteine in an adenine mutant of *Saccharomyces cerevisiae, J. Bact.,* **98**:629, 1969.

Kocková-Kratochvílová, A., et al.: The genus *Candida* Berkhout. VIII. Fermentation type II species, *Folia Microbiol. (Prague),* **12**:327, 1967.

Kreger-van Rij, N. J. W.: Taxonomy and systematics of yeasts. In "The Yeasts," edited by A. H. Rose and J. S. Harrison, New York, Academic Press, Inc., 1969.

Kudrjawzew, W. I.: "Die Systematik der Hefen," Berlin, Akademie-Verlag GmbH, 1960.

Lindegren, C. C.: The improvement of industrial yeasts by selection and hybridization, *Wallerstein Lab. Commun.*, **7**:153, 1944.

——: Yeast genetics: Life cycles, cytology, hybridization, vitamin synthesis, and adaptive enzymes, *Bact. Rev.*, **9**:111, 1945.

——: "Yeast of Tomorrow," Anheuser-Busch, Inc., St. Louis, Mo., 1949.

——: "Yeast genetics–1962," Biol. Res. Lab., Southern Illinois University, Carbondale, Ill., 1962.

——: Nucleoprotein layer of the yeast cell, *Nature*, **198**:1325, 1963*a*.

——: The flaccid cell wall of *Saccharomyces*, *Can. J. Genet. Cytol.*, **5**:254, 1963*b*.

——: Directed mutations in yeasts and bacteria, *Bull. Res. Council Israel*, **11A**:363, 1963*c*.

—— and G. Lindegren: A new method for hybridizing yeast, *Proc. Nat. Acad. Sci. U.S.*, **29**:306, 1943.

—— and ——: Sporulation in *Saccharomyces cerevisiae*, *Botan. Gaz.*, **105**:304, 1944.

—— and ——: Vitamin-synthesizing deficiencies in yeasts supplied by hybridization, *Science*, **102**:33, 1945.

—— et al.: Centromeres, sites of affinity and gene loci on the chromosomes of *Saccharomyces*, *Nature*, **194**:260, 1962.

Lodder, J., and N. J. W. Kreger-van Rij: "The Yeasts–A Taxonomic Study," New York, Interscience Publishers, Inc., 1953.

MacWilliam, I. C.: The structure, synthesis and functions of the yeast cell wall–A review, *J. Inst. Brewing*, **76**:524, 1970.

Mager, J.: Studies on the polysaccharides of capsulated yeasts, *Biochem. J.*, **41**:603, 1947.

Makula, R., and W. R. Finnerty: Microbial assimilation of hydrocarbons. II. Fatty acids derived from 1-alkenes, *J. Bact.*, **95**:2108, 1968.

Mann, J. W., et al.: Yeast spheroplasts formed by cell wall-degrading enzymes from *Oerskovia* sp., *J. Bact.*, 821, 1972.

Matile, Ph., et al. Yeast cytology. In "The Yeasts," edited by A. H. Rose and J. S. Harrison, New York, Academic Press, Inc., 1969.

McClary, D. O., et al.: Ultraviolet microscopy of budding *Saccharomyces*, *J. Bact.*, **83**:276, 1962.

Meyer, S. A., and H. J. Phaff: Deoxyribonucleic acid base composition in yeasts, *J. Bact.*, **97**:52, 1969.

Mills, C., and A. C. Blackwood: Pellicle formation and ester production by *Hansenula anomala*, *Can. J. Microbiol.*, **13**:1259, 1967.

Newell, S. Y., and I. L. Hunter: *Rhodosporidium diobovatum* sp. n., the perfect form of an asporogenous yeast (*Rhodotorula* sp.), *J. Bact.*, **104**:503, 1970.

Nickerson, W. J.: Symposium on biochemical bases of morphogenesis in fungi. IV. Molecular bases of form in yeasts, *Bact. Rev.*, **27**:305, 1963.

—— and K. V. Thimann: The chemical control of conjugation in *Zygosaccharomyces, Am. J. Botany,* **28**:617, 1941; II. *ibid.*, **30**:94, 1943.

Ouchi, S., and C. C. Lindegren: Genic interaction in *Saccharomyces, Can. J. Genet. Cytol.*, **5**:257, 1963.

Phaff, H. J., and E. M. Mrak: Sporulation in yeasts, Part I, *Wallerstein Lab. Commun.*, **11**:261, 1948; Part II, *ibid.*, **12**:29, 1949.

—— et al.: "The Life of Yeasts," Cambridge, Mass., Harvard University Press, 1966.

Poncet, S.: A numerical classification of yeasts of the genus *Pichia* Hansen by a factor analysis method, *Antonie van Leeuwenhoek, J. Microbiol. Serol.*, **33**:345, 1967.

Reiff, F., et al.: "Die Hefen," vols. 1 and 2, Verlag Hans Carl Nürnberg, 1960.

Rose, A. H., and J. S. Harrison, eds.: "The Yeasts," New York, Academic Press, Inc., 1969.

Roth, R.: Carbohydrate accumulation during the sporulation of yeast, *J. Bact.*, **101**:53, 1970.

—— and H. O. Halvorson: Sporulation of yeast harvested during logarithmic growth, *J. Bact.*, **98**:831, 1969.

Santa María, J.: Formation by *Saccharomyces cerevisiae* of asci with more than four spores, *J. Bact.*, **74**:692, 1957.

——: Poliploidia en *Saccharomyces, Anales Inst. Nac. Invest. Agron. (Madrid),* **8**:679, 1959*a*.

——: Sobre especies pluriesporuladas del genero *Saccharomyces. Anales Inst. Nac. Invest. Agron. (Madrid),* **8**:737, 1959*b*.

—— and C. Sanchez: Significación taxónomica de las propiedades fisiológicas de las especies incluidas en el género *"Kluyveromyces," Inst. Nac. Invest. Agron.*, 62, June 1970.

Scheda, R., and P. Bos: Hydrocarbons as substrates for yeasts, *Nature,* **211**:660, 1966.

Scherr, G. H., and R. H. Weaver: The dimorphism phenomenon in yeasts, *Bact. Rev.*, **17**:51, 1953.

Sinclair, J. H., et al.: Mitochondrial-satellite and circular DNA filaments in yeast, *Science,* **156**:1234, 1967.

Slooff, W. Ch., et al.: Polysaccharides, ascospore walls and classification of *Lipomyces, Antonie van Leeuwenhoek, J. Microbiol. Serol.,* **35**:A23, *Supplement: Yeast Symposium,* 1969.

Spencer, J. F. T., and P. A. J. Gorin: Mannose-containing polysaccharides of the apiculate yeasts, *Nadsonia, Hanseniaspora, Kloeckera,* and *Saccharomycodes,* and their use as an aid in classification, *J. Bact.,* **96**:180, 1968.

Streiblová, E.: Surface structure of yeast protoplasts, *J. Bact.,* **95**:700, 1968.

—— et al.: Structural changes in the cell wall of *Schizosaccharomyces pombe* during cell division, *J. Bact.,* **91**:428, 1966.

Svihla, G.: Spheroplasts of the yeast *Candida utilis, J. Bact.*, **82**:808, 1961.

——and F. Schlenk: Localization of S-adenosylmethionine in *Candida utilis* by ultraviolet microscopy, *J. Bact.*, **78**:500, 1959.

——and ——: S-adenosylmethionine in the vacuole of *Candida utilis, J. Bact.*, **79**:841, 1960.

——et al.: Ultraviolet microscopy of purine compounds in the yeast vacuole, *J. Bact.*, **85**:399, 1963.

Welten, H.: Wann bildet die hefe sporen? *Mikrokosmos*, **8**:3, 41, 1914.

White, A. G. C., and C. H. Werkman: Fat synthesis in yeast, *Arch. Biochem.*, **17**:475, 1948.

Windisch, S.: Genetic yeast research: Methods and some new results, *Wallerstein Lab. Commun.*, **24**:316, 1961.

Winge, Ö.: The basis for the present position of yeast genetics, *Wallerstein Lab. Commun.*, **15**:21, 1952*a*.

——: The genetic situation concerning fermentation in yeasts, *Heredity*, **6**:263, 1952*b*.

——and O. Laustsen: Artificial species hybridization in yeast, *Compt. rend. trav. lab. Carlsberg, Série physiol.*, **22**:235, 1938.

——and ——: On 14 new yeast types, produced by hybridization, *Compt. rend. trav. lab. Carlsberg, Série physiol.*, **22**:337, 1939.

Yarrow, D.: *Torulopsis peltata*, sp. n., *Antonie van Leeuwenhoek, J. Microbiol. Serol.*, **34**:81, 1968.

Molds 6

Molds are minute saprophytic or parasitic filamentous fungi which reproduce by means of asexual and sexual spores. They constitute a large heterogeneous group of plant-like organisms which form a subdivision of the *Thallophyta*, one of the five divisions of the plant kingdom (Fig. 127). The *Thallophyta* grow in irregular plant masses not differentiated into roots, stems, and leaves like higher plants. Such a mass of plant tissue is known as a thallus.

A thallus consists of branching, filamentous, thread-like growths called hyphae (singular, hypha). A mass of threads taken collectively is spoken of as a mycelium. The filaments of hyphae usually are colorless. Hyphae which are concerned in the production of spores are the fertile hyphae; those which serve to secure nutrients are the vegetative hyphae. The fertile hyphae extend into the air, forming and discharging spores. The vegetative hyphae burrow into the substrate, digesting and absorbing nutrients.

Molds differ from algae in not containing chlorophyll, the green pigment that enables plants to synthesize carbohydrates from carbon dioxide and water in the presence of sunlight as the source of energy.

A lichen is a composite organism consisting of a mold living symbiotically with an alga. The mold obtains food materials from the alga, while the alga is in turn protected from external injury or exposure. The fructifications of the lichens are produced by the molds, not by the algae, which are always vegetative. The natural indicator litmus is derived from a lichen.

Typical plants are chemosynthetic, being capable of utilizing simple substances and building them up into compounds of greater complexity. They utilize carbon dioxide to form carbohydrates and fats and eliminate oxygen. Generally speaking, animals are chemoanalytic, being capable of breaking down complex organic matter into simple compounds. They cannot utilize inorganic compounds of nitrogen but require this element in the form of proteins or of the constituent amino acids. Animals differ from plants in that they take in oxygen and eliminate carbon dioxide. Molds resemble plants in structure and generally in being able to utilize

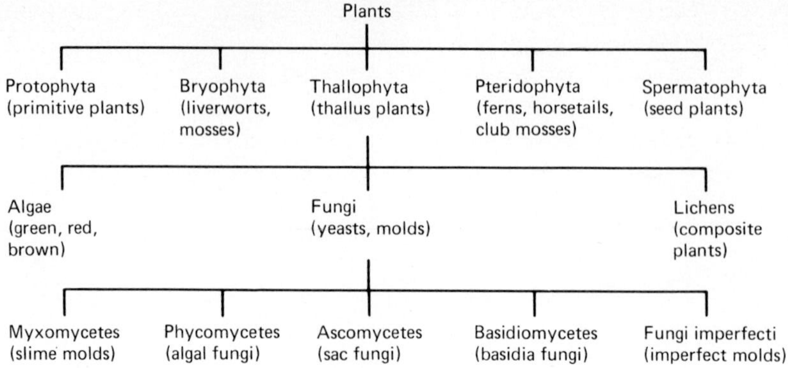

Fig. 127 A schematic representation of the Thallophyta in relation to other plants.

inorganic nitrogen compounds but appear to be more related to animals in requiring oxygen in their metabolism and eliminating carbon dioxide.

Structure of hyphae Hyphae are more or less branched, continuous tubes filled with protoplasm (Fig. 128). They may be single-celled (nonseptate) or many-celled (septate). The transverse walls in the septate hyphae are known as septa (singular, septum). Longitudinal or oblique septa are very rare. In most molds, increase in length of hyphae occurs by apical growth. Cells lying back from the tips of hyphae may start to grow and develop into branches. In a few species, all cells may continue to grow and divide. The nonseptate hyphae form one large cell containing many nuclei. This absence of septa makes possible a flowing of protoplasm. The septate hyphae may contain one, two, or more nuclei in each cell. Septa are rare in the Phycomycetes, except in the fruiting bodies, though occasionally they may be present in old cultures.

Fig. 128 Mold hyphae. Left, nonseptate; right, septate. (After Kurung.)

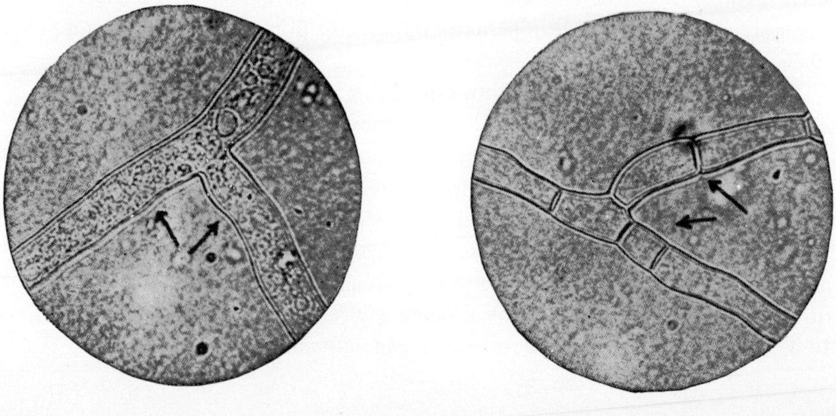

The nuclei usually are small and, in most fungi, seen only by special staining methods. In young organisms the cytoplasm may nearly fill the cell, whereas in old cells it is often limited to a thin layer lying next to the cell wall. The central portion may show the presence of a large vacuole filled with colorless cell sap. Various reserve food materials such as fat globules, glycogen, and metachromatic granules may also appear (page 151).

Cell wall In some molds, the cell wall appears to be composed largely of cellulose. In others, particularly the higher forms, the cell wall is composed largely of a polymerized acetylated glucosamine known as chitin, together with some cellulose and other substances of an obscure nature. Chitin is also present in the shells of crabs and lobsters and in the shards of beetles (see page 144).

The composition of the cell wall of any species is not the same under all circumstances. Substances which may be present in young hyphae may be absent or almost so in old filaments. External factors such as composition of the substrate, pH, and temperature greatly influence the composition of the cell wall of molds.

For more information: Baxby and Gray (1968); Gray and Baxby (1968).

Classes of molds The molds are placed in that division of the plant kingdom known as the Thallophyta. The Thallophyta contains the yeasts and molds but not the bacteria. The Thallophyta is subdivided into the Myxomycetes (slime molds), Phycomycetes (algal fungi), Ascomycetes (sac fungi, including all yeasts and some molds), Basidiomycetes (basidia fungi), and the Fungi Imperfecti (imperfect fungi).

The species of Myxomycetes consist of naked masses of protoplasm which creep very slowly over the surface and ingest solid food. Except for a few parasitic species, they are found on damp earth and decaying vegetable matter. The members of the Phycomycetes form sexual spores which are in some cases single and in others double. In the Ascomycetes the sexual spores are endogenous and typically eight in number. In the Basidiomycetes the sexual spores or basidiospores are borne exogenously and are typically four in number. This class includes in part the large fleshy fungi such as the mushrooms, puffballs, and bracket fungi which grow on trees. The Fungi Imperfecti includes all molds in which a sexual or perfect stage has not been observed.

Reproduction in molds

Most of the common molds may be cultivated by transferring any part of the plant to fresh medium, but the normal process of development begins with the germination of a spore. Spores are of different shapes and sizes and may be composed of one cell or more than one cell.

A spore consists of an outer wall, the epispore, and an inner wall, the

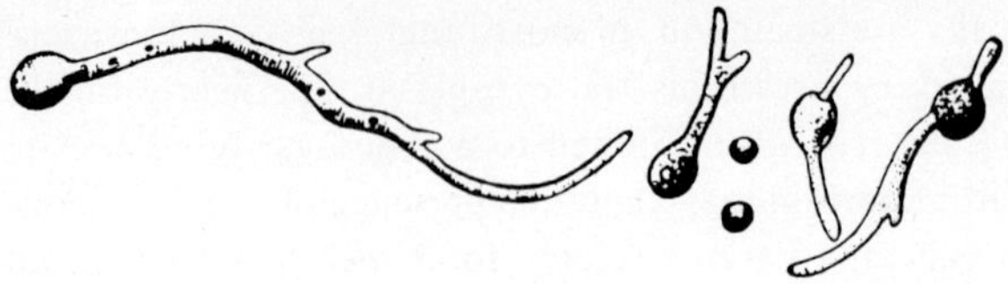

Fig. 129 Germination of conidia. (After Wehmer.)

endospore. The epispore may be smooth, pitted, or roughened by small projections; the endospore encloses the protoplasm in which may be seen droplets of oily or fatty material and one or more nuclei. Under favorable conditions the spore first swells, then throws out one or more germ tubes (Fig. 129). Each germ tube elongates and becomes branched, forming a network of hyphae, or a mycelium. Later spore-bearing bodies develop on the fertile hyphae, or some of the hyphae show the presence of special fruiting bodies in which spores are formed.

Two types of spores are produced by molds: asexual and sexual. Asexual reproduction does not involve the union of sex cells or sex organs (Fig. 130). On the other hand, sexual reproduction is characterized by the union of two nuclei. A few molds produce several kinds of spores corresponding to different stages in their development. Practically all the commonly encountered molds produce asexual spores. Some produce both asexual and sexual spores. Asexual reproduction appears to be more important for the propagation of the species because it may occur several times during a season, whereas sexual multiplication generally occurs only once a year.

Types of multiplication in molds

These may be classified as follows:

1. Fragmentation of the thallus.
   a. Reproduction by budding as occurs in yeasts and yeast-like organisms (page 152).
   b. Budding of mycelium to form spores known as blastospores (Fig. 131).
   c. Segmentation of terminal hyphae followed by a rounding up and separation of the segments to form arthrospores (Fig. 132). Segments capable of giving rise to new hyphae.
   d. Rounding up and thickening of cell walls of a vegetative thallus to form chlamydospores (Fig. 132). Spores may be intercalary or terminal. They occur either simply or in chains in ordinary vegetative hyphae or in special branches. They are resistant bodies and may be considered as resting spores.
2. Asexual spore formation. This represents the imperfect stage of a mold.
   a. Fungi Imperfecti are septate molds which reproduce only by formation of asexual spores (Fig. 130). A sexual stage in unknown.

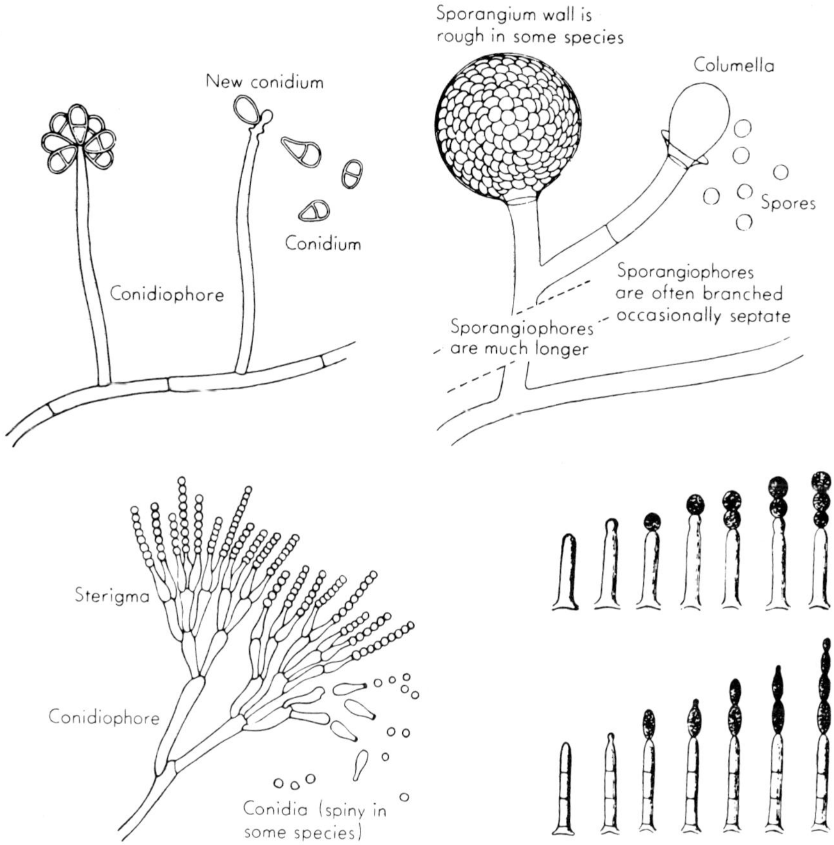

Fig. 130 Asexual spore formation. Upper left, Trichothecium. Septate mold that reproduces only by formation of asexual spores known as conidia. Upper right, Mucor. Development of a closed structure or sporangium on the tip of a sporangiophore in which are produced sporangiospores. Lower left, Penicillium. Development of spores at tip of hyphal branch or conidiophore known as conidia. (Courtesy of Plunkett and Wilson.) *Lower right, formation of conidia. Upper half, the topmost spore is the oldest, lower half, the topmost spore is the youngest.*

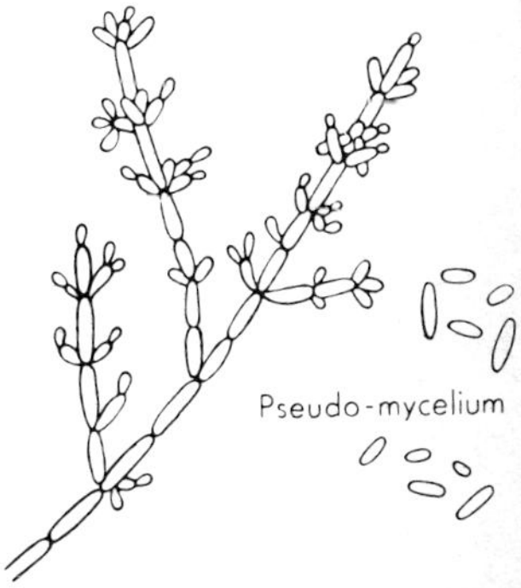

Fig. 131 Candida. Fragmentation of the thallus. Budding of mycelium to form blastospores. (After Plunkett and Wilson.)

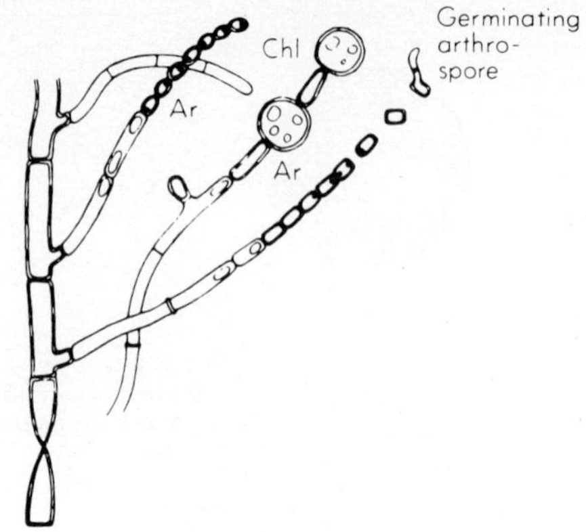

Fig. 132 Geotrichum. Segmentation of terminal hyphae to form arthrospores. Rounding up and thickening of cell walls of a vegetative thallus to form chlamydospores. (Courtesy of Plunkett and Wilson.)

Reproduction occurs by spores developed in various ways. The molds are classified according to types of spores and manner in which they are produced.

b. Phycomycetes develop asexually by the formation of a closed structure or sporangium on the tip of a sporangiophore in which are produced sporangiospores (Fig. 130).

c. Ascomycetes develop thin-walled spores at tip of hyphal branch or conidiophore known as conidiospores or conidia (Fig. 130). They may be (1) pinched off from tip of conidiophore or (2) budded off from existing conidiospores instead of directly from conidiophore (Fig. 130). In the former case the topmost spore is the oldest and largest; in the latter case the topmost spore is the youngest and smallest.

3. Sexual spore formation. This represents the perfect stage of a fungus. Spores are the result of nuclear fusion and are formed in or on specialized cells. Hyphal fusion frequently occurs but is not necessarily indicative of sexuality.

a. Sexual spores of Phycomycetes. Fusion of (1) unlike gametes in large cell to form oöspores or (2) like gametes to form zygospores (Fig. 133). Spores are in some cases single and in others multiple.

b. Sexual spores of Ascomycetes. Ascus is a spore mother cell which arises as a result of nuclear fusion to form ascospores by reduction division (Fig. 133). Ascospores may be formed (1) by direct conjugation as in yeasts, (2) by fusion of two mycelial cells, (3) by fusion of definite sex organs in which oögonium becomes an ascus, or (4) from ascogenous hyphae budded off a binucleate sex organ. In all cases the young ascus is binucleate. The nuclei fuse, and the diploid nucleus undergoes reduction division to yield four haploid nuclei. Haploid nuclei divide to yield typically eight spores.

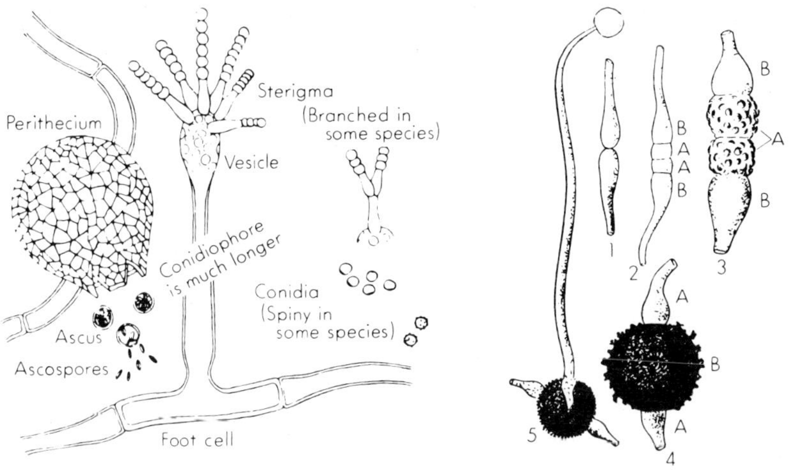

Fig. 133 Sexual spore formation. Left, Aspergillus. Ascus is a spore mother cell which arises as a result of nuclear fusion to form ascospores. (Courtesy of Plunkett and Wilson.) *Right, Mucor. Formation of zygospore. (1) two hyphae in terminal contact; (2) articulation into gamete a and suspensor b; (3) fusion of gametes a; (4) ripe zygospore b supported by the suspensors aa; (5) germination of zygospore.* (After Brefeld.)

c. Sexual spores of Basidiomycetes. Sexual organs absent. Basidium typically club-shaped, nonsegmented, and bears exogenously four basidiospores. Initially basidia are ends of hyphae composed of binucleate cells. As each terminal cell enlarges, the two nuclei also enlarge. Nuclei fuse to form a diploid fusion nucleus which then divides twice to form four daughter nuclei. Each nucleus is haploid. Four small projections (sterigmata) appear at apex of basidium. These projections expand terminally, and a nucleus passes into each inflated tip. When ripe, the basidiospores are abstricted. Basidium is similar to an ascus except that spores are produced outside rather than inside.

Resistance of mold spores

Molds have been responsible for enormous losses in the home and the industries. These losses are due largely to the fact that molds produce spores in great masses which may be spread widely by wind. They are abundant in the atmosphere.

Mold spores are very resistant to unfavorable environmental conditions such as heat, cold, desiccation, ultraviolet light, high osmotic pressure, and deficient food supply. They are more resistant to heat than mycelium and generally less resistant than bacterial spores. Spores are easily disseminated by wind and air currents. They are commonly present in the air of laboratories and are the frequent cause of contaminations of cultures and culture media. Therefore, laboratory windows should be kept closed to prevent the wind from stirring up the dust and spores in the laboratory air.

Molds occur particularly in damp places. Spores will not germinate in a dry environment. Many industrial products, such as paper, leather, textiles, and foods, readily absorb moisture from the atmosphere and are susceptible to attack by molds. In order to decrease mold contaminations, laboratories and rooms should be kept as free as possible from excessive moisture.

In general, molds grow more slowly than bacteria. Consequently they are not found growing to any extent in environments where they must compete with bacteria, but are found growing under conditions which are unfavorable to the growth of bacteria. Environments unfavorable to bacterial growth but favorable to molds include starchy foods, high osmotic salt and sugar solutions, acid materials such as fruit juices and sour milk, salted and smoked meats, butter, and cheese.

Kerner-Gang (1968) reported the development of etching patterns on optical lenses produced by mold growth. The extent of action on glass depended upon the presence of moisture, acids, alkalies, temperature, and species of mold. Damage to lenses is especially evident in the tropics where the humidity is high. Unless mold growth is removed from the glass, the moisture will damage the surface and leave it etched or raised where the growth occurred.

The genera of molds isolated from glass surfaces included *Alternaria, Aspergillus, Dactylium, Paecilomyces, Penicillium,* and *Trichoderma*.

Generally speaking, molds do not multiply in the absence of oxygen. They are, therefore, unable to grow in commercially canned foods. However, a few species may grow to some extent under reduced oxygen conditions. An example is the mold *Penicillium roqueforti*, which is responsible for ripening of Roquefort cheese.

Morphology of the common molds

Several hundred genera and thousands of species of molds (at least 80,000) have been described. Only a few genera are of common occurrence, and these may be easily recognized. They are *Mucor, Rhizopus, Cephalothecium* (*Trichothecium*), *Geotrichum, Candida, Aspergillus, Penicillium, Cladosporium,* and *Alternaria.*

Their characteristics are as follows:

Mucor This is the largest genus of the order Mucorales. The mucors are found in soil and manure, and on fruits, vegetables, bread, and other starchy foodstuffs. The vegetative mycelium penetrates the food materials and sends out long, slender threads known as aerial hyphae. The mycelium generally is white in color. A septum forms near the apex of each hypha. The tip cell of the hypha swells into a round, cylindrical, or pear-shaped structure called a columella. Around the columella a globular structure or sporangium then forms, within which develop numerous oval, asexual spores known as sporangiospores. The sporangia are almost black in color. The wall of the ripe sporangium easily breaks, discharging the enclosed spores. Each spore is capable of repeating the cycle (Fig. 130).

Under certain conditions conjugation of two cells from different hyphae precedes spore formation, resulting in the development of a zygospore. This is sexual reproduction. A germ tube arises from the matured zygospore, developing a sporangium at the apex (Fig. 133).

Separation of the various species is based on the length and diameter of the sporangiophores, the type of branching if any, the size and color of the sporangia, the character of the sporangial wall, the characteristics of zygospores and chlamydospores if any, the size and shape of the columellae, the size and shape of spores, and general colony characteristics such as color and height of aerial growth.

Bartnicki-García and Nickerson (1962) reported that vegetative development of *M. rouxii* may follow either of two patterns of morphogenesis (mold-yeast dimorphism) depending upon the atmosphere of incubation. Under air, a filamentous mold-like growth developed, commonly followed by fragmentation of hyphae into spherical cells or arthrospores. Cultivation under an atmosphere of carbon dioxide (CO_2) induced development of spherical, budding, yeast-like cells (Fig. 134). Anaerobically, a pCO_2 of at least 0.3 atmosphere produced purely yeast-like development; presence of oxygen annulled the effect of CO_2.

For more information: Bartnicki-García (1963, 1968); Bartnicki-García and Lipman (1969); Elmer and Nickerson (1970*a*, *b*); Terenzi and Storck (1969).

Rhizopus Members of this genus are of common occurrence and the frequent cause of laboratory contaminations. Growth on the usual laboratory media is very rapid. The molds spread widely by means of stolons or runners. Culture tubes and petri dishes soon become filled with a dense, cottony mycelium. Species of this genus are easily distinguished from the mucors by the presence of stolons. Stolons often reach a length of several centimeters and bear tufts of root-like hyphae or rhizoids, which emerge from the points where the stolons come in contact with the medium or the surface of the glass. The columella is hemispherical, not round, cylindrical, or pear-shaped as in the mucors, and rests in a cup-shaped expansion of the sporangiophore called the apophysis. Spores may be ovoid, polygonal, or striated. The members are usually grayer in color and produce a more luxuriant growth than the mucors (Fig. 135).

Cephalothecium (Trichothecium) This genus contains several species, but only one, *C. roseum*, is of common occurrence. The colonies are thin, spreading, floccose, at first white in color, then becoming slowly pale pink. The conidiophores bear clusters of spores attached to the tip. The spores are ovoid, with a nipple-like projection at the point of attachment, and are composed of two cells. The cell closest to the conidiophore is the smallest

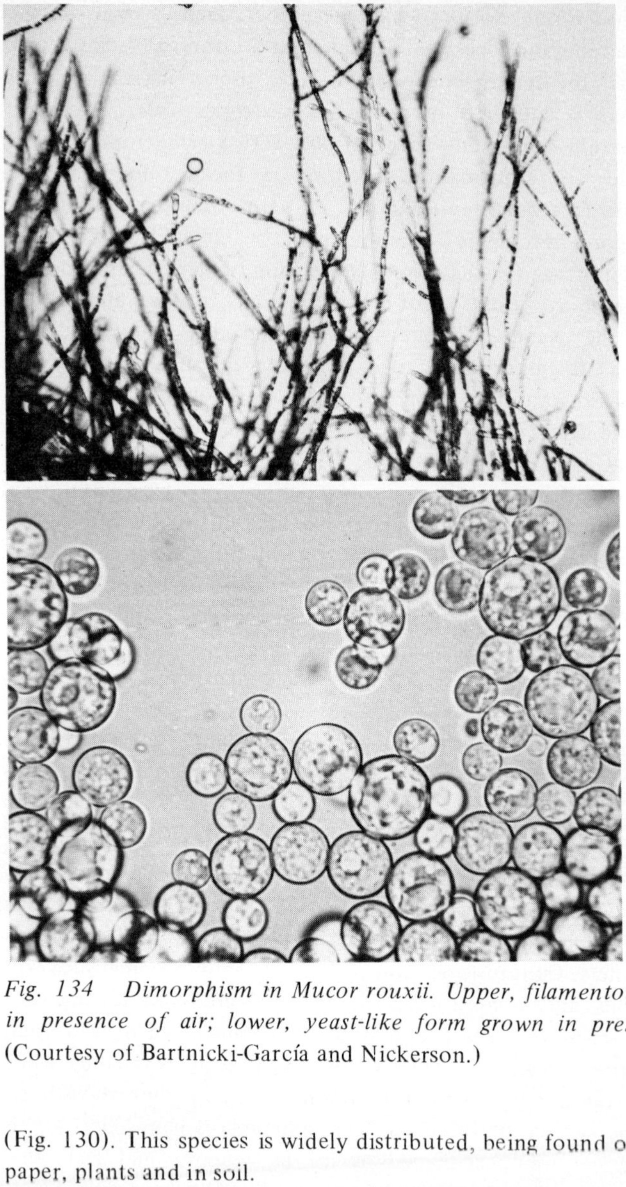

Fig. 134 Dimorphism in Mucor rouxii. Upper, filamentous form grown in presence of air; lower, yeast-like form grown in presence of CO_2. (Courtesy of Bartnicki-García and Nickerson.)

(Fig. 130). This species is widely distributed, being found on fruits, wood, paper, plants and in soil.

Geotrichum The best-known member is *G. candidum*. It is found in various milk products. It grows readily on milk or wort agar, producing a thin, spreading, slimy growth. Colonies on agar are creamy white in color. In young cultures, long hyphal threads are present; in old cultures, the hyphae break up into short single cells, more or less cylindrical in shape, known as arthrospores. Each spore germinates into a new plant (Fig. 132).

Candida Members of this genus are essentially yeast-like organisms which have a tendency to form mycelium on culture media low in

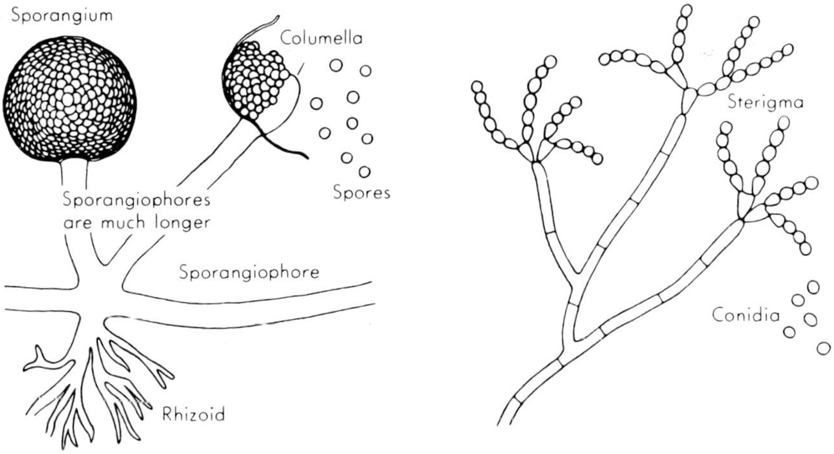

Fig. 135 Left, Rhizopus; right, Cladosporium. See text for description. (Courtesy of Plunkett and Wilson.)

nutrients. Young glucose agar cultures may show a white, pasty growth consisting entirely of round or oval budding yeast cells. In old agar cultures there may be present fine filaments penetrating into the agar. Microscopic examination shows the presence of filaments and budding yeast cells. In a nutrient-gelatin stab culture, yeast-like cells may be present on the surface, whereas tufts of mycelium that radiate into the gelatin may be observed along the line of puncture. A few species are of some industrial importance.

For more information: Balish and Svihla (1966).

Aspergillus Species of this genus are relatively common in air. They are found almost everywhere on nearly all types of substrates, such as decaying fruits, vegetables, grains, bread, and other articles of food. Aspergilli are commonly present in incompletely sterilized culture media. They may appear in various colors, such as green, yellow, orange, brown, and black. Colonies have a powdery appearance. In marked contrast to the mucors, the hyphae of the aspergilli are branched and septate. The hyphae enlarge at the apices to form conidiophores. The conidiophores are not branched. Numerous short stalks or sterigmata develop from the apical or swollen ends of the conidiophores. Chains of spores or conidia are produced from the tips of the sterigmata, sometimes developing to a considerable length (Fig. 136). A few species produce perithecia. These are spherical, cylindrical, or flask-shaped hollow structures which contain the asci and are usually open by a terminal pore. The asci contain the ascospores (Fig. 133).

Species of aspergilli are used commercially for the hydrolysis of starch as a preliminary to the fermentation of the released sugar to alcohol. Some are used in the manufacture of soy sauce. The black aspergilli are employed for the production of oxalic and citric acids.

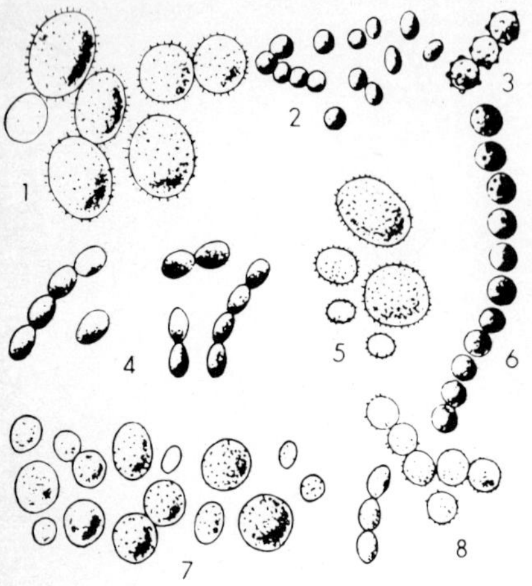

Fig. 136 Conidia of various species of Aspergillus. (1) A. glaucus; (2) A. fumigatus; (3) A. niger; (4) A. clavatus; (5) A. tokelau; (6) A. varians; (7) A. oryzae; (8) A. wentii. (After Lafar.)

For more information: Raper and Fennell (1965); Ruiz-Herrera and Starkey (1969); Takebe (1960); Takebe and Shimizu (1959); Thom (1954).

Penicillium The penicillia are closely related to the aspergilli and are also widely distributed in nature. The genus includes the characteristic blue-green-colored mold so often observed on citrus and other fruits, vegetables, grains, hay, organic infusions, cheeses, and other food materials (Fig. 130). The vegetative mycelium penetrates the food substances, after which aerial hyphae or conidiophores appear. The conidiophores branch one or more times from the same joint, giving rise to a terminal cluster of parallel hyphae known as sterigmata. A chain of conidia develops from each sterigma (Fig. 137).

Some species are destructive, whereas others are beneficial. Two beneficial species are *P. roqueforti* and *P. camemberti*, which are responsible for the desirable changes occurring in Roquefort, Camembert, Gorgonzola, and similar cheeses. The penicillia are employed for the manufacture of a number of substances of commercial importance, one being penicillin.

The penicillia are more difficult to classify than the aspergilli. The conidia show less variation in color. Most species show some shade of green during the period of active growth. The colors vary in shade under different environmental conditions and with age of the cultures. Also, there are more species of *Penicillium* than of *Aspergillus*.

For more information: Raper and Thom (1949); Scott and Stolk (1967); Thom (1954).

Cladosporium The common species *C. herbarum* is widely distributed in nature, being found on rubber, textiles, leather, paper, foodstuffs, decaying leaves, straw, and in soil and air. It grows over a wide range of temperature, and has been isolated from meat in cold storage.

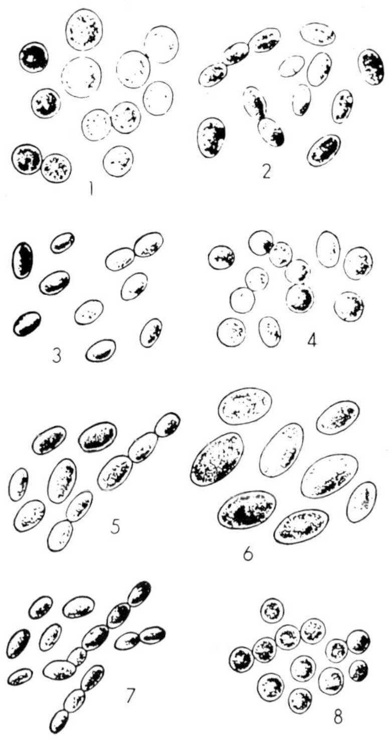

Fig. 137 Conidia of various species of Penicillium. (1) P. camemberti; (2) P. purpurogenum; (3) P. claviforme; (4) P. rubrum; (5) P. italicum; (6) P. olivaceum; (7) P. luteum; (8) P. glaucum. (After Lafar.)

On culture media it produces a thick velvety growth, and the color varies from deep green to dark gray-green. The first spores are formed directly from the tips of the conidiophores. Additional spores are produced by budding of existing conidia. In this manner a chain or mass of spores is formed in which the terminal spore is the youngest. Young spores are usually one-celled; old spores are mostly two-celled or even three-celled.

Alternaria Species of this genus are commonly found on organic matter. Parasitic forms have been isolated from cultivated plants.

Spores are not conidiospores but thallospores, and may vary considerably in size, shape, and degree of septation. Spores are produced in chains.

The best-known species is *A. tenuis.* On cultivation, it grows rapidly and produces a thin velvety colony, almost black, with aerial growth consisting almost entirely of chains of spores. Spores are yellow-brown to dark brown in color, irregular in shape and size, and most have a definite beak (Fig. 138).

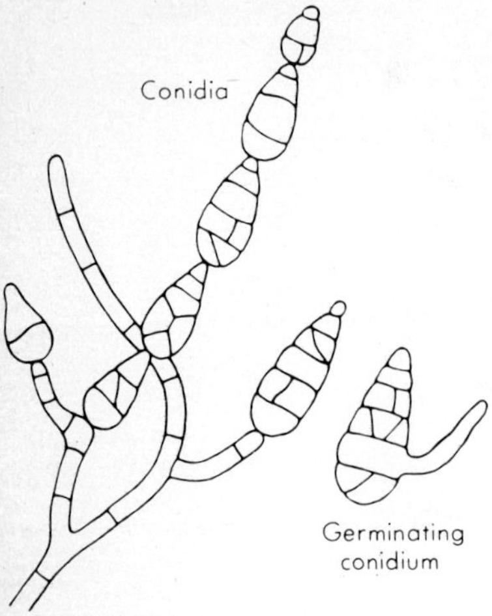

Fig. 138 *Alternaria. See text for description.* (Courtesy of Plunkett and Wilson.)

Classification of molds

A classification is of value for several reasons: (1) to show natural relationships; (2) to show how the different forms have evolved; and (3) to facilitate identification of the large number of species.

The general principles followed for the classification of molds are similar to those employed for bacteria (page 505).

A classification of the above described genera, including some important species is as follows:

CLASS I *Phycomycetes.* Vegetative mycelium typically nonseptate. Septations commonly formed in connection with the development of reproductive structures. They may also appear in old hyphae. Sporangia developed which contain motile or nonmotile sporangiospores, or conidia. Oöspores or zygospores formed, sometimes after fertilization, sometimes without any preliminary fusion of gametes or gametangia. Some are parasitic; others are saprophytic.

ORDER *Mucorales.* Accessory multiplication by sporangiospores. Sporangia globose to ovoid, usually contain numerous spores, sometimes one or a few. Zygospores formed from the whole of the two gametangia. Columella present. Zygospores naked or invested by outgrowths from its own wall or from those of the suspensors. Principal sporangia contain numerous spores.

Family *Mucoraceae.* Sporangiola, if any, developed on lateral branches of principal sporangiophores. Sporangia of one kind. Sporangiophore simple or branched but not repeatedly dichotomous. Suspensors without appendages at maturity.

GENUS 1 *Mucor.* Three main types of branching recognized: the monomucors show unbranched sporangiophores; the racemomucors show a main stem with lateral branches; the cymomucors show sporangiophores typically branched. Stolons absent. Sporangia single and terminal.

Sporangiophores rarely or never branched (monomucors):

M. mucedo. Sporangiophores erect, with grayish sporangia encrusted with crystals, probably calcium oxalate. Terminal sporangia 100 to 200 μ in diameter. Columella pear-shaped to cylindrical. Spores elliptical, 6 to 12 μ long. Arthrospores and chlamydospores absent.

M. hiemalis. Sporangiophores simple or sympodially branched. Sporangia 50 to 80 μ in diameter, olive to grayish-brown when ripe, with diffluent walls. Columellae spherical or oval, with collarette, up to 50 μ long. Spores irregular but mostly oval and rough. Chlamydospores formed in mycelium.

M. ramannianus. Sporangiophores mostly unbranched, 2 to 6 μ in diameter. Sporangia reddish, 20 to 40 μ in diameter, with diffluent walls. Columellae spherical, 5 to 10 μ in diameter. Spores spherical to oval, 2 to 3 μ long. Chlamydospores numerous and variously shaped. Giant cells formed in submerged mycelium.

M. piriformis. Sporangiophores mostly unbranched, up to 50 μ in diameter. Sporangia spherical, 250 to 350 μ in diameter, with diffluent walls. Columellae pear-shaped, 200 to 300 μ long. Spores elliptical, 5 to 13 μ by 4 to 8 μ.

Sporangiophores with main stem and secondary lateral branches, racemose (racemomucors):

M. racemosus. Sporangiophores simple at first, later becoming branched. Sporangia spherical and small, 20 to 70 μ in diameter. Columellae spherical or oval, with collarette. Spores mostly elliptical, 6 to 10 μ by 5 to 8 μ. Black chlamydospores formed in aerial hyphae, about 20 μ in diameter. Submerged mycelium breaks up into yeast-like cells or arthrospores, capable of fermenting sugar to alcohol.

M. erectus. Closely related to *M. racemosus* except that columellae spherical. Sporangia grayish-yellow.

M. fragilis. Closely related to *M. racemosus*, except that columellae spherical. Sporangia black.

Sporangiophores typically branched (cymomucors):

M. spinosus (plumbeus). Sporangiophores branched in sympodial cymes, about 1000 μ long and 10 μ in diameter. Sporangia usually flattened, about 65 μ in diameter, encrusted with fine needle-shaped crystals. Columellae oval or pear-shaped with spiny projections at top. Spores round, 7 to 8 μ in diameter.

M. circinelloides. Exhibits true cymose branching. Sporangia in two rows, alternating. Spores spherical or ellipsoidal. Submerged mycelium breaks up into yeast-like cells or arthrospores which germinate by budding.

M. alternans. Sporangia in two rows, alternating. Spores ellipsoidal.

M. rouxii. Sporangiophores sparingly branched. Sporangia 20 to 30 μ in diameter. Spores oval, about 5 μ long. Columellae spherical. Black chlamydospores appear in aerial mycelium and budding cells in submerged portions. Contains necessary enzymes to convert starch to alcohol. Used in manufacture of alcohol.

GENUS 2 *Rhizopus.* Readily distinguished from *Mucor* by the presence of stolons (runners) and of tufts of rhizoids (root-like hyphae) emerging from points where stolons touch medium.

R. stolonifer. Common black bread mold. Stolons clearly differentiated. Sporangiophores erect, 2500 μ long and 20 μ in diameter. Sporangia round, white at first, then black as spores ripen, up to 200 μ in diameter. Spores of various shapes, about 10 to 15 μ in long axis.

R. oryzae. Similar to *R. stolonifer* but spores smaller. Employed in fermentation of starch to alcohol.

R. japonicus. Also similar to *R. stolonifer*, and employed in fermentation of starch to alcohol.

CLASS II *Ascomycetes.* With exception of yeasts, all possess well-developed mycelium of branched and septate hyphae. Cells of mycelium may be uninucleate or contain several nuclei. Multiplication by conidia and chlamydospores, but characteristic method by means of ascospores. Asci contain usually eight spores, more rarely a smaller or larger number. Some of the *Aspergillus, Penicillium,* and *Alternaria* produce ascospores and should be included here.

However, asci have not been identified in the great majority of species, and for that reason all are grouped under the Fungi Imperfecti for convenience.

CLASS III *Basidiomycetes.* All possess well-developed mycelium. Basidiospores produced and borne externally on the mother cell or basidium. Young basidium contains two nuclei that fuse; the fused nucleus then divides to provide the nuclei of the spores. Spore formed on a sterigma through which nucleus passes from basidium to enter the developing spore. Basidiospores unicellular, round, or oval, asymmetrically attached to their sterigmata, usually with a smooth, rather thin wall. Echinulate spores occur in a few species.

CLASS IV *Fungi Imperfecti.* Characteristic method of reproduction by conidia. In some species, arthrospores and chlamydospores may be present. Sporangiospores, ascospores, and basidiospores absent.

ORDER *Moniliales.* Conidiophores free, arising irregularly from mycelium.

Family 1 *Moniliaceae.* Hyphae colorless or in pale or bright colors.

GENUS 1 *Trichothecium* (*Cephalothecium*). Spores two-celled, in small clusters on ends of erect conidiophores.

T. roseum. Commonly occurring pink mold found on decaying fruit. At times causes contamination of laboratory media.

GENUS 2 *Geotrichum.* Reproduction regularly by fragmentation of mycelium into arthrospores. Blastospores rarely, if ever, produced.

G. candidum. Commonly found in milk and cheese. Imparts flavor and aroma to many types of cheese.

GENUS 3 *Candida* (*Monilia*). Reproduction by fragmentation of mycelium into blastospores. Arthrospores absent. Colonies dry, flat, and wrinkled on Sabouraud agar; heavy pellicle on liquid media.

C. krusei (*M. krusei*). Colonies moist and creamy on Sabouraud agar; slight or no pellicle on liquid media.

C. parakrusei (*M. parapsilosis*). Glucose only carbohydrate fermented.

C. tropicalis (*M. candida*). Glucose, maltose, sucrose fermented; lactose, raffinose not fermented.

C. pseudotropicalis (*M. mortifera*). Glucose, lactose, sucrose fermented; maltose, raffinose not fermented.

C. guilliermondi (*M. guilliermondi*). Glucose, sucrose, raffinose fermented.

C. albicans (*M. albicans*). Blastospores in dense clusters. Glucose fermented; sucrose not fermented.

GENUS 4 *Aspergillus.* Vegetative mycelium consisting of septate hyphae, branching, colorless, or colored. Conidiophores arising from specialized foot cells, usually nonseptate, terminating in a swelling which bears the sterigmata. Conidia borne in chains formed by abscission from sterigmata. Conidia vary greatly in color, size, shape, and markings. Perithecia found in some groups (unknown in most species), producing asci and ascospores within a few weeks.

A. clavatus group. Conidiophores generally coarse, smooth-walled, uncolored. Conidial heads clavate, pale blue-green. Sterigmata in one series. Conidia elliptical, smooth, thick-walled.

A. glaucus group. Conidiophores smooth-walled, terminating in dome-like vesicles. Perithecia yellow, globose to subglobose, thin-walled, suspended in network of red or yellow hyphae. Asci contain eight lenticular, smooth- or rough-walled ascospores. Sterigmata in one series. Conidial heads abundant, radiate to somewhat columnar, typically green. Conidia elliptical to subglobose, uniformly roughened.

A. fumigatus group. Conidiophores smooth-walled, usually green. Sterigmata in one series, crowded. Vesicles flask-shaped. Conidial heads columnar, green to dark green. Conidia globose, echinulate, green.

A. nidulans group. Conidiophores smooth-walled, terminating in dome-like or hemispherical vesicles. Perithecia usually present. Ascospores purple-red. Conidial heads short columnar, usually dark green. Conidia globose, echinulate. Large, thick-walled, globose bodies form an irregular layer around perithecia.

A. ustus group. Conidiophores yellow-brown, smooth. Conidial heads irregular, ranging from radiate to hemispherical to loosely columnar. Vesicles hemispherical.

Sterigmata in two series. Conidia roughened, ranging in color from pale blue-green to olive-green to deep brown. Characteristic thick-walled, globose to subglobose, Hülle cells present.

A. flavipes group. Conidiophores smooth, yellow, with color often confined to outer layer. Conidial heads barrel-shaped to columnar. Vesicles subglobose to elliptical. Conidia colorless, smooth, thin-walled. Hülle cells generally present, helicoid or variously twisted.

A. versicolor group. Conidiophores smooth, colorless, more or less sinuous. Vesicles globose to ovate or elliptical, with radiate sterigmata borne over upper half to three-fourths of surface. Sterigmata in two series. Conidial heads hemispherical to globose, usually green or blue-green. Conidia globose to subglobose, echinulate.

A. terreus group. Conidiophores smooth, colorless. Conidial heads columnar, in cinnamon, pale buff, or light flesh colors. Vesicles hemispherical, upper half to two-thirds covered by sterigmata. Sterigmata in two series, generally crowded. Conidia small, smooth, globose to slightly elliptical.

A. candidus group. Conidiophores smooth, colorless to slightly yellow. Conidial heads white or becoming yellowish with age, typically globose but approaching columnar in small heads. Sterigmata in two series. Conidia smooth, globose or subglobose. Sclerotia present in some strains.

A. niger group. Conidiophores smooth, colorless or tinged with yellow-brown colors in upper third or less. Conidial heads black, brownish black or purple-brown, typically large and globose. Vesicles globose in large heads, fertile over entire surface; in small heads, often reduced to dome-like apices of short conidiophores. Conidia rough, showing bands of brown-black coloring matter. Sclerotia characteristic of many strains.

A. wentii group. Conidiophores smooth-walled or nearly so. Conidial heads large, typically globose, varying in color from dull yellowish to ecru-olive, from light brown to dark brown, depending on species and strain. Vesicles globose. Sterigmata in two series. Conidia commonly elliptical, smooth or roughened.

A. tamarii group. Conidiophores colorless, typically roughened throughout a part or entire length. Conidial heads radiate, hemispherical to globose, yellow-brown to olive-brown. Vesicles subglobose to globose, fertile over upper half to entire surface. Sterigmata in one or two series depending upon species and strain, sometimes both conditions in same head. Conidia heavy-walled, rough, elliptical, pyriform, or subglobose. Sclerotia commonly present.

A. flavus-oryzae group. Conidiophores rough or pitted, colorless. Conidial heads hemispherical to columnar to subglobose. Sterigmata in one or two series, often varying in same head. Vesicles variable in form, from hemispherical to dome-shaped in small heads; globose in large heads. Conidia roughened, varying in color. Sclerotia present in many strains.

A. ochraceus group. Conidiophores normally yellow in outer layers of wall, rough or pitted. Conidial heads globose or radiate with conidial chains commonly adhering to divergent columns, sulfur yellow to varying shades of ochraceous. Sterigmata in two series with the primary often large and septate. Conidia in some species thin-walled and smooth; in others double-walled, roughened or echinulate. Sclerotia present in most species and strains.

GENUS 5 *Penicillium.* Conidia produced from sterigmata which in turn are produced from short branches called metulae given off from tips of conidiophores. Penicillia divided into four groups, division being based on nature of branching of spore heads. Branching may be symmetrical about the axis of conidiophore or unsymmetrical. Symmetrical types separated into (1) monoverticillata, with a single whorl or sterigmata at tip of conidiophore; (2) biverticillata, in

Fig. 139 Branching of spore heads in Penicillium. Left, monoverticillate; center, biverticillate; right, polyverticillate. See text for description.

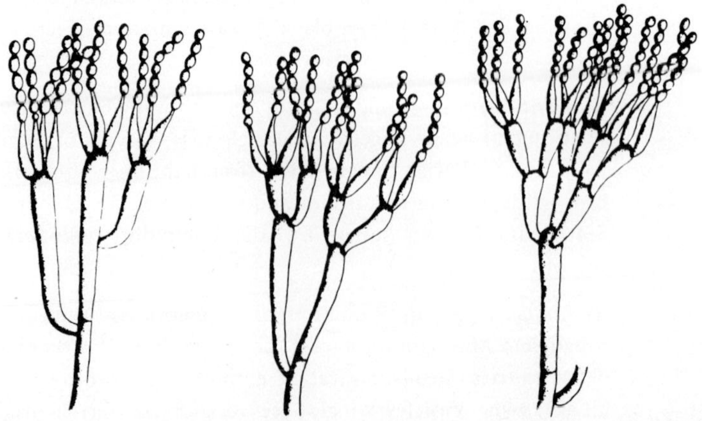

which the verticils of sterigmata are produced from two symmetrical branches; and (3) polyverticillata, in which verticils of sterigmata arise from three or more symmetrical branches (Fig. 139).

P. camemberti. Found in Camembert and Brie cheese. Colonies floccose, white or grayish-green in color.

P. brevicaule. Found in Camembert and Brie cheese. Yellow-brown areas formed. Spores rough.

P. roqueforti. Found in Roquefort cheese. Forms green streaks inside cheese.

P. italicum. Produces a soft rot of citrus fruits. Colonies blue-green.

P. digitatum. Produces a shriveling and drying of infected citrus fruits. Colonies dull yellow-green or olive-green.

P. expansum. Produces a soft rot of apples and pears. Colonies blue-green.

P. notatum. Produces penicillin. Used commercially.

P. chrysogenum. Produces penicillin. Used commercially.

Family 2 *Dematiaceae.* Conidiophores detached, not compacted. Mycelium, spores, or both, dark brown to black.

GENUS 1 *Cladosporium.* Spores increase by budding, forming branched chains. Spores one-celled but become two-celled with age.

C. herbarum. Found on decaying paper, straw, and similar materials.

GENUS 2 *Alternaria.* Spores many-celled, club-shaped, and in chains.

A. tenuis. Found on moldy grains and in soil. Frequently present in laboratory air.

Laboratory technique

In order that accurate studies may be made on molds, it is necessary to grow them in pure cultures. The methods employed for isolating and studying molds in pure cultures are, in general, similar to those used for bacteria (page 272).

Culture media Many types of solid and liquid media are employed for the cultivation of molds but only a few are for general use.

Vegetables and vegetable extracts are commonly employed as culture media. These are suitable with or without the addition of sugar. The solid media may be composed of solid substances such as potatoes, carrots, and beans, or liquid media made solid by the addition of agar or gelatin.

Molds generally grow more slowly than bacteria. If both are present in the material to be cultivated, the bacteria are likely to grow at a faster rate, making it difficult, if not impossible, to obtain pure culture isolations of the molds. This may be overcome by making the media unfavorable to the growth of the bacteria but favorable to the molds. Since molds can tolerate acid conditions better than bacteria, the procedure generally followed is to make the media highly acid in reaction (pH 2 to 5).

A useful solid medium for general use is Sabouraud agar. It has the following composition: glucose, 2 g; peptone, 1 g; agar, 2 g; distilled water, 100 ml. The pH is adjusted to 5.6. Some prefer a medium composed of pure inorganic salts. Such a medium is constant in composition and strictly comparable when prepared in different laboratories. A simple medium of this type for the cultivation of molds is Czapek's solution. A modification of the original formula is as follows: sodium nitrate, 2.0 g; potassium chloride, 0.5 g; magnesium sulfate, 0.5 g; ferrous sulfate, 0.01 g; potassium acid phosphate, 1.0 g; distilled water, 1000 ml. If a solid medium is desired, add 20 g agar. The medium is acid (pH 4.2), which is unfavorable to the growth of almost all bacteria. Various carbohydrates may be added, usually 50 g glucose or 30 g sucrose per liter. This is a very useful liquid or solid medium for the cultivation of molds.

Isolation and purification of molds Molds may be easily and satisfactorily purified by the pour-plate method, as used for the purification of bacterial species (page 274).

Where molds are heavily contaminated with bacteria, it is advantageous to add an antibacterial agent to the plating medium. The stain, rose bengale, in a concentration of about 1:15,000, generally prevents bacterial growth without exhibiting any toxic effect on molds.

Microscopical examination Considerable information may be obtained by first examining dry, living cultures under the low-power objective. With the lids removed, petri-dish cultures are placed on the stage of the microscope and examined by transmitted or reflected light. Aerial mycelium, conidiophores, fruiting heads, chains of spores, and other structures may be seen and examined. This gives a preliminary indication of what to look for when slide preparations are examined, since mold structures are easily broken when disturbed.

For high-power examination, slide preparations are necessary. Mold specimens are very difficult to remove from culture media without being greatly broken. Therefore, great care must be exercised in preparing satisfactory mounts. Water should not be used for the mounting fluid since it rapidly evaporates, produces a shrinkage of the hyphae by osmosis, and causes the various parts to adhere together as a tangled mass. Obviously, such preparations are unsatisfactory for accurate observations.

A useful mounting fluid is known as lactophenol. It has the following composition: phenol, cp, crystals, 20 g; lactic acid, cp, 20 g; glycerol, cp, 40 g; and distilled water, 20 ml. The solution is prepared by first dissolving the phenol in the water, then adding the lactic acid and glycerol.

This fluid does not cause shrinkage of the cells and does not evaporate, thus permitting permanent preparations to be prepared. A stain may be added to the fluid to stain the various mold structures. This is especially desirable for mounting molds that are to be photographed.

Molds are mounted by first placing a drop of lactophenol in the center of a clean glass slide. A small portion of the mold growth is removed from the culture and placed in the drop of fluid. It is gently teased out with a pair of needles until the various parts are well separated and wetted by the fluid. It is then carefully covered with a cover slip to avoid, as much as possible, air bubbles being entrapped.

For more information: Gray et al. (1963); Smith (1969).

Biochemistry of the molds

The biochemical activities of molds are of great importance in the industrial world. They are probably not so important in this respect as the yeasts and bacteria, but they do, nevertheless, produce certain changes not carried out by the other two members of the fungi group. A few of the more important reactions produced by molds are the following:

Alcoholic fermentation Alcohol is produced industrially by the fermentation of various sugars by yeasts (page 476). The fermentable materials include sucrose, glucose, molasses, pineapple juice and other fruit juices, various starches (corn, potato, wheat), whey, and celluloses.

Saccharine materials usually require no special treatment other than proper concentration. Starches, celluloses, and other polysaccharides must be first hydrolyzed to soluble sugars before utilization by yeasts.

In general, the methods of saccharifying complex carbohydrates involve the use of enzymes, or dilute acids, or a combination of both. A number of molds, bacteria, and plants elaborate the necessary enzyme or enzymes that effect the saccharification of the various starches and other polysaccharides, after which certain yeasts are capable of fermenting the sugars to alcohol. Takamine (1914) advocated the use of mold bran (Takakoji), an enzyme preparation obtained by growing *Aspergillus oryzae* on moist, sterilized bran (see Underkofler and Fulmer, 1943).

Some species of *Mucor* are capable of fermenting starch to alcohol. The molds secrete both amylase (diastase) and zymase. Amylase is an enzyme that hydrolyzes starch to sugar. Zymase is the name sometimes used for a complex of enzymes and coenzymes that converts sugar to alcohol. *M. rouxii* was the first species of the Mucoraceae to be used in alcoholic fermentation by the "amylo process." The mold is used in the Orient for preparing alcoholic beverages from rice. It is mixed with rice meal and marketed as Chinese yeast.

For more information: Johnson et al. (1968).

Citric acid Citric acid is a natural constituent of citrus and other fruits. Scheele (1784) first isolated and crystallized the acid from lemon juice.

Many species of molds are capable of fermenting sugar to citric acid but only a few can produce the compound in sufficient concentration to enable the process to compete with the natural source. Currie (1917) found that the yield could be greatly increased by employing *A. niger*. This mold is employed commercially, and in some localities the process competes quite favorably with the extraction of citric acid from natural sources.

The two most important controlling factors in the production of citric acid are the supply of oxygen and the composition of the medium. The mold is strongly aerobic and requires a plentiful supply of oxygen. Sufficient nitrogen and minerals must be present in the medium for the mold to grow satisfactorily. The highest yield of citric acid is obtained when the medium contains about 15 percent sucrose.

Currie found the following medium to be most favorable for citric acid production: sucrose, 150 g; ammonium nitrate, 2.5 g; potassium acid phosphate, 1.0 g; magnesium sulfate, 0.25 g; distilled water, to make 1000 ml. The medium was adjusted with hydrochloric acid to give a pH of about 3.5.

The fermentation yields oxalic acid and carbon dioxide in addition to citric acid. The fermented medium is concentrated by evaporation, which permits the less soluble oxalic acid to crystallize out, leaving the citric acid in solution.

Foster et al. (1941) found carbon dioxide to be necessary in the production of citric acid through the Wood-Werkman reaction (pyruvic acid + CO_2 → oxaloacetic acid). Their reaction scheme is shown in Fig. 140.

The overall result may be represented by the equation:

$$\underset{\text{Glucose}}{C_6H_{12}O_6} + 3O \longrightarrow \underset{\text{Citric acid}}{H_3C_6H_5O_7} + 2H_2O$$

According to this scheme: (1) pyruvic acid is probably formed by the same series of reactions as in alcoholic fermentation (page 476); (2) the formation of citric acid is dependent upon the presence of carbon dioxide; (3) the carbon dioxide liberated in the decarboxylation of some pyruvic acid is fixed in the Wood-Werkman reaction, converting other molecules of pyruvic acid to oxaloacetic acid; (4) in unaerated cultures the carbon dioxide arising from respiration accumulates in the vicinity of the mycelium, giving considerably more citric than oxalic acid; (5) in aerated

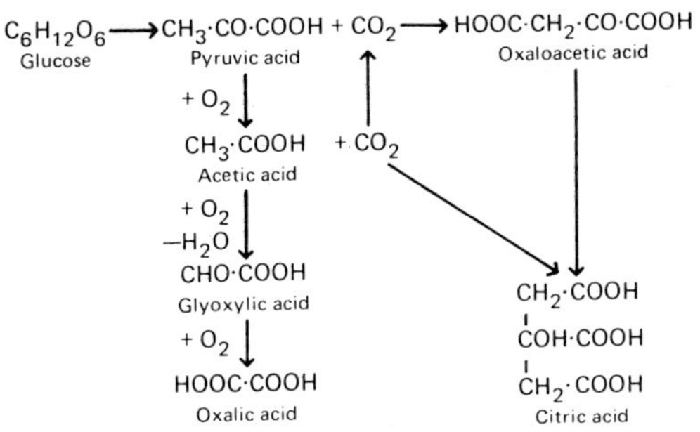

Fig. 140 Reaction scheme for the production of citric acid from glucose. (After Foster et al.)

cultures the carbon dioxide is carried away by the air, giving considerably more oxalic acid than citric acid.

For more information: Clark (1962); Clark and Lentz (1963); Foster (1949); Karow and Waksman (1947); Meyrath (1967); Shu and Johnson (1947).

D-gluconic acid This acid is produced by several molds, especially by species of *Aspergillus* and *Penicillium*. Wells et al. (1937) selected a strain of *A. niger* that sporulated readily and produced good yields of the acid. Takao and Sasaki (1964) found several strains of the mold *Pullularia pullulans* to be excellent gluconic acid producers from glucose.

A satisfactory medium for gluconic acid production has the following formula: glucose, 200 g; magnesium sulfate ($MgSO_4 \cdot 7H_2O$), 0.25 g; disodium phosphate ($Na_2HPO_4 \cdot 7H_2O$), 0.10 g; potassium chloride, 0.05 g; sodium nitrate, 1.00 g; distilled water, to make 1000 ml. The medium produces a good yield in about 10 days at 25 to 30°C.

The overall reaction for the oxidation of glucose to gluconic acid is:

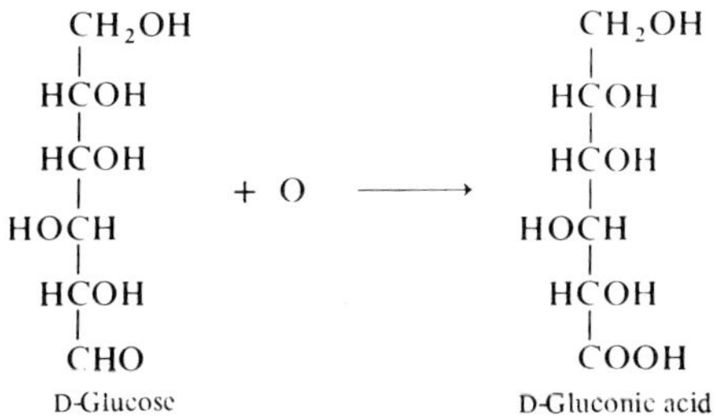

Fumaric acid Fumaric acid is an unsaturated acid produced by a number of molds, principally by species of the genus *Rhizopus*. It has the formula $COOH \cdot CH = CH \cdot COOH$.

The acid is produced from various monosaccharides, and from sucrose, maltose, molasses, and starches. The following medium may be used for good fumaric acid production: glucose, 150 g; ammonium sulfate,

5 g; magnesium sulfate ($MgSO_4 \cdot 7H_2O$), 0.5 g; dipotassium phosphate (K_2HPO_4), 0.5 g; zinc sulfate ($ZnSO_4 \cdot 7H_2O$), 0.01 g; ferric sulfate [$Fe_2(SO_4)_3$], 0.02 g; water, to make 1000 ml. Calcium carbonate is added to the medium to neutralize the fumaric acid as it is formed.

Foster and Davis (1948) proposed the following scheme for the formation of fumaric acid from glucose:

$C_6H_{12}O_6$
Glucose

↓

$CH_3{\cdot}CHOH{\cdot}COOH$ ← $CH_3{\cdot}CO{\cdot}COOH$ → $CH_3{\cdot}CHO + CO_2$
Lactic acid ← Pyruvic acid → Acetaldehyde

Pyruvic acid ↓ $HOOC{\cdot}CH_2{\cdot}CO{\cdot}COOH$ (Oxaloacetic acid)

Acetaldehyde ↓ $CH_3{\cdot}CH_2OH$ (Ethyl alcohol)

↓

$HOOC{\cdot}CH_2{\cdot}CHOH{\cdot}COOH$
Malic acid

↓

$HOOC{\cdot}CH{=}CH{\cdot}COOH$
Fumaric acid

According to this scheme: (1) pyruvic acid, lactic acid, acetaldehyde, and alcohol probably are formed by the same series of reactions as in alcoholic fermentation (page 476); (2) the decarboxylation of some pyruvic acid furnishes CO_2, which reacts with the remainder of the pyruvic acid to give oxaloacetic acid, from which (3) malic acid, and finally (4) fumaric acid are produced.

For more information: Rhodes et al. (1959); Romano et al. (1967).

D-Mannitol Smiley et al. (1967) reported that nearly all strains of white and many black *Aspergillus* produced significant amounts of the hexahydric alcohol mannitol from glucose. One strain, grown under optimal conditions, converted about 50 percent of the consumed glucose to mannitol.

The overall reaction for the conversion is:

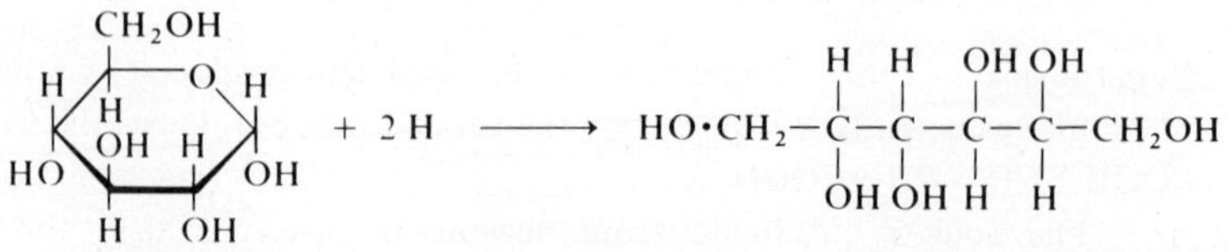

α-D-Glucose D-Mannitol

For more information: Nelson et al. (1971).

Cheeses A number of well-known cheeses are ripened by molds: (1) soft cheeses of the Camembert and Brie types; and (2) green-streaked cheeses of the Roquefort, Gorgonzola, and Stilton types.

The cheeses in the first group are ripened by the mold *Penicillium camemberti.* The curd is produced by the action of rennase, then allowed to drain without the application of pressure. The soft curd is shaped into cakes, salted on the surface, and inoculated with the spores of the mold. The cakes are placed in a damp room where the mold multiplies rapidly on the surface, then gradually penetrates and softens the entire mass of curd. When complete, about 80 percent of the nitrogenous matter becomes water-soluble. The process requires about 4 weeks.

The cheeses in the second group are prepared by first inoculating the curd with a pure culture of *P. roqueforti.* The curd is then pressed so as to leave irregular cracks in the cake. The cake is aerated from time to time during the ripening process by piercing it with wires. The mold produces a dense growth along the cracks, giving the finished product a streaked appearance. Salt is periodically applied to the surface of the cheese during the ripening process to decrease the population of undesirable organisms. The ripening process requires 5 to 6 months.

A more extensive treatment of cheese preparation is given in Chap. 22, page 748.

Antibiotics A large number of molds elaborate certain specific toxic substances which interfer with the metabolism of other organisms to such an extent that they are either killed or prevented from multiplying. These specific toxic compounds are known as antibiotics. The best-known representative of this group is penicillin, produced by the mold *P. notatum*, and other species. This important subject is discussed under Antibiosis (page 604).

Miscellaneous metabolic products of molds In addition to those already discussed, molds produce a large number of compounds, which, with few exceptions, are of minor industrial importance. They are, for the most part, nonnitrogenous metabolic products and probably the result of the action of the molds on carbohydrates or carbohydrate-like compounds. It should be noted that, on the whole, molds produce compounds of greater complexity than do bacteria. The compounds which have been isolated and characterized include organic acids, pigments, polysaccharides, antibiotics, aldehydes, esters, higher alcohols, sterols, etc.

For more information: Ainsworth and Sussman (1966); Barron (1968); Bracken et al. (1954); Burnett (1968); Gould (1947); Hawker (1966); Hesseltine (1965); Morton and Broadbent (1955); Müller and Loeffler (1968); Sasaki and Takao (1967*a, b, c*); Smith (1969); Sommer et al. (1963); Takao (1964); Tsao (1963); Galun (1971).

References

Ainsworth, G. C., and A. S. Sussman, eds.: "The Fungi," vol. 2, New York, Academic Press, Inc., 1966.

Balish, E., and G. Svihla: Ultraviolet microscopy of *Candida albicans, J. Bact.,* **92**:1812, 1966.

Barron, G. L.: "The Genera of Hyphomycetes from Soil," Baltimore, The Williams & Wilkins Company, 1968.

Bartnicki-García, S.: III. Mold-yeast dimorphism of *Mucor, Bact. Rev.,* **27**:293, 1963.

———: Control of dimorphism in *Mucor* by hexoses: Inhibition of hyphal morphogenesis, *J. Bact.,* **96**:1586, 1968.

——— and E. Lippman: Fungal morphogenesis: Cell wall construction in *Mucor rouxii, Science,* **165**:302, 1969.

——— and W. J. Nickerson: Induction of yeast-like development in *Mucor* by carbon dioxide, *J. Bact.,* **84**:829, 1962.

Baxby, P., and T. R. G. Gray: Chitin decomposition in the soil. I. Media for isolation of chitinoclastic micro-organisms from soil, *Trans. Brit. Mycol. Soc.,* **51**:287, 1968.

Bracken, A., et al.: Studies in the biochemistry of micro-organisms, *Biochem. J.,* **57**:587, 1954.

Burnett, J. H.: "Fundamentals of Mycology," London, Edward Arnold (Publishers) Ltd., 1968.

Clark, D. S.: Submerged citric acid fermentation of ferrocyanide-treated beet molasses: Morphology of pellets of *Aspergillus niger, Can. J. Microbiol.,* **8**:133, 1962.

——— and C. P. Lentz: Submerged citric acid fermentation of beet molasses in tank-type fermenters, *Biotechnol. Bioengineer.,* **5**:193, 1963..

Currie, J. N.: The Citric Acid Fermentation of *Aspergillus niger, J. Biol. Chem.,* **31**:15, 1917.

Elmer, G. W., and W. J. Nickerson: Filamentous growth of *Mucor rouxii* under nitrogen, *J. Bact.,* **101**:592, 1970*a.*

——— and ———: Nutritional requirements for growth and yeast-like development of *Mucor rouxii* under carbon dioxide, *J. Bact.,* **101**:595, 1970*b.*

Foster, J. W.: "Chemical Activities of Fungi," New York, Academic Press, Inc., 1949.

——— and J. B. Davis: Anaerobic formation of fumaric acid by the mold *Rhizopus nigricans, J. Bact.,* **56**:329, 1948.

——— et al.: Radioactive carbon as an indicator of carbon dioxide utilization. VII, *Proc. Natl. Acad. Sci. U.S.,* **27**:590, 1941.

Galun, E.: Scanning electron microscopy of intact *Trichoderma* colonies, *J. Bact.,* **108**:938, 1971.

Gould, B. S.: Chemical compounds formed from sugars by molds, Scientific Report Series No. 7, Sugar Research Foundation, Inc., New York, 1947.

Gray, T. R. G., and P. Baxby: Chitin decomposition in soil. II. The ecology of chitinoclastic micro-organisms in forest soil, *Trans. Brit. Mycol. Soc.,* **51**:293, 1968.

Gray, W. D., et al.: Growth of fungi in sea water medium, *Appl. Microbiol.,* **11**:501, 1963.

Hawker, L. E.: "Fungi: An Introduction," London, Hutchinson & Co. (Publishers), Ltd., 1966.

Hesseltine, C. W.: A millennium of fungi, food, and fermentation, *Mycologia,* **57**:149, 1965.

Johnson, D. E., et al.: Starch hydrolysis by conidia of *Aspergillus wentii, Appl. Microbiol.,* **16**:1678, 1968.

Karow, E. O., and S. A. Waksman: Production of citric acid in submerged culture, *Ind. Eng. Chem.,* **39**:821, 1947.

Kerner-Gang, W.: Zur frage der entstehung von schimmelpilzspuren auf optischen gläsern, *Mater. Organismen,* **3**:1, 1968.

Meyrath, J.: Citric acid production, *Process Biochem.,* October 1967.

Morton, A. G., and D. Broadbent: The formation of extracellular nitrogen compounds by fungi, *J. Gen. Microbiol.,* **12**:248, 1955.

Müller, E., and W. Loeffler: "Mykologie," Stuttgart, Georg Thieme Verlag, 1968.

Nelson, G. E. N., et al.: Production of D-mannitol by conidia of *Aspergillus candidus, Appl. Microbiol.,* **22**:484, 1971.

Raper, K. B., and D. I. Fennell: "The Genus Aspergillus," Baltimore, The Williams & Wilkins Company, 1965.

—— and C. Thom: "A Manual of the Penicillia," Baltimore, The Williams & Wilkins Company, 1949.

Rhodes, R. A., et al.: Production of fumaric acid by *Rhizopus arrhizus, Appl. Microbiol.,* **7**:74, 1959.

Romano, A. H., et al.: Mechanism of fumaric acid accumulation in *Rhizopus nigricans, J. Bact.,* **93**:600, 1967.

Ruiz-Herrera, J., and R. L. Starkey: Dissimilation of methionine by fungi, *J. Bact.,* **99**:544, 1969.

Sasaki, Y., and S. Takao: Organic acid production by basidiomycetes. II. Acid production from xylose, *J. Fac. Agr., Hokkaido Univ.,* **55**:174, 1967*a*.

—— and ——: Organic acid fermentation using the genus *Rhizopus, J. Ferment. Technol.,* **45**:211, 1967*b*.

—— and ——: Cultural conditions for L-malic acid production, *Appl. Microbiol.,* **15**:373, 1967*c*.

Scott, D. B., and A. C. Stolk: Studies on the genus *Eupenicillium Ludwig.* II. Perfect states of some penicillia, *Antonie van Leeuwenhoek, J. Microbiol., Serol.,* **33**:297, 1967.

Shu, P., and M. J. Johnson: Effect of the composition of the sporulation medium on citric acid production by *Aspergillus niger* in submerged culture, *J. Bact.,* **54**:161, 1947.

Smiley, K. L., et al.: Biosynthesis of D-mannitol from D-glucose by *Aspergillus candidus, Biotechnol. Bioengineer.,* **9**:365, 1967.

Smith, G.: "An Introduction to Industrial Mycology," London, Edward Arnold (Publishers) Ltd., 1969.

Sommer, N. F., et al.: Production pf pectolytic enzymes by *Rhizopus stolonifer* sporangiospores after "lethal" gamma irradiation, *Appl. Microbiol.*, **11**:463, 1963.

Takamine, J.: Enzymes of *Aspergillus oryzae* and the application of its amyloclastic enzyme to the fermentation industry, *Chem. News*, **110**:215, 1914.

Takao, S.: Riboflavin biosynthesis by *Candida robusta*. Part I, *Agr. Biol. Chem. (Tokyo)*, **28**:559, 1964.

——— and Y. Sasaki: Gluconic acid fermentation by *Pullularia pullulans*, Part I, *Agr. Biol. Chem. (Tokyo)*, **28**:752, 1964.

Takebe, I.: Choline sulfate as a major soluble sulfur component of conidiospores of *Aspergillus niger*, *J. Gen. Appl. Microbiol.*, **6**:83, 1960.

——— and N. Shimizu: Some aspects of the nutritional requirement of conidiospores of *Aspergillus niger* for the initiation of germination, *Annual Rept. Inst. Food Microbiol.*, **12**:29, 1959.

Terenzi, H. F., and R. Storck: Stimulation of fermentation and yeast-like morphogenesis in *Mucor rouxii* by phenethyl alcohol, *J. Bact.*, **97**:1248, 1969.

Thom, C.: The evolution of species concepts in *Aspergillus* and *Penicillium*, *Ann. N.Y. Acad. Sci.*, **60**:24, 1954.

Tsao, G. T.-N.: Production of oxalic acid by a wood-rotting fungus, *App. Microbiol.*, **11**:249, 1963.

Underkofler, L. A., and E. I. Fulmer: Microbial amylases for saccharification of starch in the alcoholic fermentation, *Chronica Botan.*, **7**:420, 1943.

Wells, P. A., et al.: Gluconic acid production: Effect of pressure, air flow, and agitation on gluconic acid production by submerged mold growths, *Ind. Eng. Chem.*, **29**:653, 1937.

Nutrition of bacteria 7

Culture media (singular, medium) are solid, semisolid, and liquid preparations employed for the cultivation and identification of bacteria. Because of the great variation in the nutrient requirements of bacteria and the large number of tests employed for their differentiation, many kinds of media must be made available.

Strict autotrophic bacteria The strict autotrophs cannot utilize organic matter and may even be harmed by its presence. These organisms are able to synthesize complex compounds composing their protoplasm from simple inorganic salts. They obtain their carbon from carbon dioxide and their energy from the oxidation of certain inorganic compounds or even elements. Because of this fact, they are independent of vegetable and animal life.

Strict heterotrophic bacteria The strict heterotrophs cannot synthesize their complex protoplasm from simple inorganic salts but must have organic compounds, such as proteins, peptones, amino acids, and vitamins, for growth.

Facultative organisms The facultative heterotrophic forms show characteristics intermediate between the two, being able to utilize both inorganic and organic compounds. They comprise the great majority of bacteria that have been studied and classified.

At one end of the scale, the organisms exhibit complete independence; at the other end, they show complete parasitism. Fildes (1934) advanced the theory that parasitism involved the loss of enzymes essential for the synthesis of bacterial protoplasm, making it necessary to add certain complex ingredients to culture media.

Common ingredients of culture media

The ingredients commonly added to culture media and their uses are as follows:

Water Water is absolutely necessary for the existence of living cells. Tap water may show considerable variation in composition from one locality to another, with the result that uniform culture media cannot always be prepared. The calcium and magnesium in tap water react with the phosphates present in peptones, beef extract, and other ingredients of culture media to give insoluble phosphates. The insoluble phosphates may not form in the cold, but during sterilization such media throw down considerable precipitate, which usually proves objectionable. Since distilled water is of definite composition, it should be used in preference to tap water for the preparation of culture media. However, tap water may be preferable for the cultivation of certain autotrophic forms that do not use organic matter in their nutrition.

Peptones Peptones are intermediate products of hydrolysis formed by the action of certain proteolytic enzymes (trypsin, pancreatin, papain, etc.) on native proteins. As hydrolysis proceeds, the large colloidal protein molecules are broken up into a series of smaller fragments which are designated, respectively, as proteoses, peptones, peptides, and finally amino acids. The proteoses still exhibit colloidal properties, and it is customary to consider them as the last hydrolytic product still possessing truc protein characteristics. In other words, the protein nature of the molecule disappears on further hydrolysis.

The commercial peptones are not the same as the peptone of the chemist, who uses the term in its narrow, chemical sense. The commercial preparations employed by the bacteriologist are composed of proteoses or albumoses, peptones, peptides, and amino acids. The proportions vary, depending upon the type of peptone. The usual commercial preparations contain high percentages of peptones and amino acids, and only negligible quantities of proteoses and other nitrogenous constituents. Others contain a higher content of proteoses with smaller amounts of peptones and amino acids. Still others contain all fractions in more or less balanced proportions. Some organisms prefer one type of peptone; others grow best in another type.

A mixture of pure amino acids as a source of nitrogen is not satisfactory for the growth of almost all bacteria even though culture media are optima in all other respects. Daniels (1966), in his studies on *Pseudomonas denitrificans,* reported that D-amino acids were more toxic than the corresponding L-forms: neutral amino acids were more toxic than acidic amino acids; and the basic amino acids were the least inhibitory of all.

A whole protein, such as casein or egg albumin, is probably not attacked by bacteria when incorporated in a medium as the only source of nitrogen and carbon. The molecules are believed to be too large to enter a bacterial cell. If a trace of peptone is added to the medium, it will enter the cell and stimulate the elaboration of an extracellular proteolytic enzyme capable of attacking the whole protein. The protein fractions can then enter the bacterial cell, where they are acted upon by the intracellular enzymes.

Svihla et al. (1969), by means of ultraviolet micrographs, showed that cytochrome c, a small compact protein molecule with a molecular weight of about 12,000, can enter the yeast cell. However, it has not been demonstrated that the larger protein molecules can enter the cell unless first reduced to smaller fragments.

The most important function of peptones in culture media is to furnish an available source of nitrogen. Since amino acids are amphoteric compounds, peptones are also excellent buffers.

For more information: Hishinuma et al. (1969).

Meat extract Meat extract is prepared by cutting fresh, lean beef into pieces, placing them in a vessel with an appropriate amount of water, and heating for several hours with occasional stirring. The liquid portion is poured off and the solid material subjected to gentle pressure to separate the remaining liquid. The extract is cooled to remove fat, and strained. The clear liquid is evaporated in vacuo to the consistency of a pasty mass.

The constituents removed from muscle by boiling it in water are known as extractives and amount to about 2 percent of the weight of the tissue. Some extractives contain nitrogen; others do not. The nitrogenous extractives include creatine, xanthine, hypoxanthine, uric acid, adenylic acid, inosinic acid, carnosine, carnitine, glycocoll, urea, glutamine, β-alanine, etc. The nonnitrogenous extractives include glycogen, hexosephosphate, lactic acid, succinic acid, inositol, fat, inorganic salts, etc.

The use of beef extract in culture media was introduced by Loeffler (1881) and has been a routine procedure in bacteriology ever since. Meat extract is added to media to supply certain substances that stimulate bacterial activity. It contains enzyme exciters, which cause accelerated growth of microorganisms. The following vitamins of the B complex have been shown to be present in beef extract: glutamine, β-alanine, thiamine, riboflavin, pantothenic acid, nicotinic acid, biotin, pyridoxin, folic acid, and *p*-aminobenzoic acid.

Yeast extract Brewer's yeast is a by-product of the brewing industry, and constitutes the basic raw material for the manufacture of yeast extract. Yeast extract consists of protein and other intracellular constituents of the cell extracted by a process known as autolysis.

Yeast cells are commonly autolyzed by heating to 45°C in distilled water adjusted to a pH of 6.5, with occasional stirring for a period of about 14 hr. The autolysis is carefully controlled to avoid destruction of the natural-occurring vitamins of the B complex.

Yeast extract is an excellent stimulator of bacterial growth and is frequently used in culture media in place of meat extract. It is a rich source of the B vitamins and is used to supply these factors in culture media. For this reason it is superior to meat extract in most culture media.

For more information: Hough and Maddox (1970).

Gelatin Gelatin is a protein obtained by boiling skin, tendons, ligaments, bones, etc., with water. It is an incomplete protein, lacking tryptophan and containing only small amounts of other important amino acids.

Gelatin is not soluble in cold water but swells up and absorbs 5 to 10 times its weight when immersed in it. It is quite soluble in boiling water. On cooling, it solidifies to form a transparent gel.

Gelatin varies in composition from one batch to another and from manufacturer to manufacturer. According to Rossmoore and Goehler (1963):

The term gelatin is merely a descriptive one including a whole family of macromolecular substances of very similar chemical composition, but with wide differences in both molecular weight, distribution, and structure. The feature common to all of these substances is their ability to form a gel under certain conditions. The differences are given by the source from which they are derived, the extraction procedure by either acid or alkaline treatment, and the purification manipulations.

Gelatin is rarely used as a substitute for agar in the preparation of solid media because (1) it is attacked and decomposed by many bacteria and (2) it melts at 37°C. Gelatin is added to media principally to test the ability of organisms to liquefy it. Some organisms can liquefy it; others cannot. It is of importance in the identification and classification of bacteria.

Agar Agar is the dried mucilaginous substance extracted from *Gelidium corneum* and other species of *Gelidium* and closely related algae (class, *Rhodophyceae*). The plants are found growing chiefly off the coasts of Japan, China, Ceylon, Malaya, and southern California.

Agar is the sulfuric acid ester of a linear galactan. It is insoluble in cold water but slowly soluble in hot water to give a viscid solution. A 1 percent solution melts at 80 to 100°C and sets at 35 to 50°C to a firm gel. Since agar is attacked by relatively few bacteria, it is the most satisfactory solidifying agent for the growth and isolation of bacterial species.

The species of bacteria capable of digesting agar are members of the genera *Agarbacterium, Cytophaga, Pseudomonas*, and *Vibrio*. Some species of *Alginomonas* produce extensive softening of agar rather than complete liquefaction. The enzyme responsible for agar liquefaction is extracellular.

For more information: Girard et al. (1968).

Sodium chloride Sodium chloride is commonly added to culture media although it is not necessary for growth.

Sometimes media containing blood are required for (1) cultivation of bacteria or (2) the recognition of a hemolytic reaction on agar. Red blood cells are hemolyzed when added to water or to media having low osmotic pressures. Obviously such media cannot be used for the recognition of a hemolytic reaction. This may be prevented by the addition of sodium chloride in a concentration approximating that of an isotonic solution.

Sodium chloride does not act as a buffer. Salts such as phosphates and carbonates do possess a strong buffering action and are frequently added to culture media for this purpose (page 243).

Inorganic requirements The inorganic requirements of bacteria are not well understood. The chief obstacle to work of this nature is the difficulty encountered in obtaining media sufficiently free of inorganic contaminants to permit accurate observations to be made. In the absence of accurate information, the following elements are usually supplied: Na, K, Mg, Fe, S, and P. In addition, Cl, C, N, and H are usually obtained from organic matter.

Shankar and Bard (1952) reported the necessity of Ca, Mg, Fe, Na, and K for the growth of *Clostridium perfringens* but not Zn, Mn, Co, or Cu. In the absence of Ca, the bacteria grew in an aggregated state, whereas Mg and K deficiency resulted in the appearance of filaments. Similar results were reported by Webb (1948), MacLeod (1951), and Rochford and Mandle (1953) on the same and other organisms.

MacLeod and Snell (1948) found K to be necessary for the growth of five lactic acid bacteria. The necessity of K for other organisms was shown by Brown and Gibbons (1955); Haynes et al. (1954); Hems and Krebs (1962); Lester (1958); Shooter and Wyatt (1956); Wyatt (1963*a, b*).

Sulfur appears to be a universal constituent of living cells. Cowie et al. (1950) found that the sulfate ion readily passed through cells of *Escherichia coli* and that the uptake of the element was directly proportional to cellular growth.

Spitznagel (1961) cultivated *Mycobacterium bovis* in a sulfate-deficient medium and reported the appearance of long forms without division. Development of the long forms was accompanied by changes in the patterns of macromolecule synthesis.

Iron also is a growth requirement for bacteria. Winder and O'Hara (1962) reported that *M. smegmatis* required the element for full growth. Many investigators have reported the necessity of iron for other bacteria.

Certain chelating agents in low concentrations produce a stimulating action on the growth of bacteria. Mayer and Traxler (1962) found that if the concentration is increased beyond the optimum, the compounds become inhibitory. *Bacillus subtilis* requires a critical concentration of Mn for growth initiation. If the Mn concentration is increased, there is a corresponding increase in the lag time and, therefore, in the growth time of the culture. The Mn inhibition can be reversed by the addition of a chelate.

Petty (1961) showed the need of Cl and other halogens for a large number of higher bacteria and molds.

For more information: Eagon and Asbell (1969); Heinen (1963); Johnson and Kyker (1961); Kempner (1967); Kennell and Katoulas (1967*a, b, c*); Lusk et al. (1968); Marchesi and Kennell (1967); Nicholas et al. (1962); Frieden (1972).

Fermentable compounds Fermentable compounds serve two functions in media: (1) they furnish readily available sources of energy, provided the organisms are able to utilize the compounds; and (2) fermentation reactions are helpful in identifying and classifying organisms.

The fermentable compounds commonly added to culture media include:

Monosaccharides

Pentoses: arabinose, rhamnose, ribose, xylose
Hexoses: galactose, glucose, levulose, mannose

Disaccharides

Lactose, maltose, melibiose, sucrose, trehalose

Trisaccharides

Gentianose, melezitose, raffinose

Polysaccharides

Cellulose, dextrin, glycogen, inulin, starch

Polyhydric alcohols

Tritol: glycerol
Tetritol: erythritol
Pentitols: adonitol, arabitol, ribitol
Hexitols: dulcitol, mannitol, sorbitol

Glucosides

Amygdalin, salicin

Noncarbohydrate compound

Inositol

Types of culture media

Culture media may be divided into two groups on the basis of the nature of the ingredients entering into their composition: (1) synthetic media and (2) nonsynthetic media.

Synthetic media Synthetic media are composed of compounds of known chemical composition. They may be composed entirely of inorganic salts; or mixtures of inorganic salts and organic compounds such as amino acids, lower fatty acids, hydroxy acids, alcohols, and carbohydrates; or inorganic and organic compounds with added vitamins. The exact chemical makeup of all ingredients is known so that two batches of the same medium can be duplicated to a high degree of accuracy. Synthetic media are employed where it is desired to ascertain what effect an organism will have on a certain compound. The nutritional requirements of bacteria may be accurately determined only by the use of synthetic culture media.

The literature shows an increasing number of synthetic media being recommended for the cultivation of all types of organisms, including the fastidious disease producers. Since the nutritional requirements of organisms are becoming better understood, it appears to be only a question of time when most, or possibly all, culture media will be of the synthetic type.

For more information: Erlandson and Mackey (1958); Friedman and Roessler (1961); Sergeant et al. (1957).

Nonsynthetic media Nonsynthetic media are composed of ingredients of unknown chemical composition. Some of these are beef extract, yeast extract, various peptones, meat infusion, blood, serum, and casein hydrolysate. It is practically impossible to prepare two identical lots of the same medium from different batches of the ingredients.

The usual culture media employed by the bacteriologist are of the nonsynthetic type.

Growth factors in culture media are absolutely necessary for the successful cultivation of most bacteria (page 224). The failure of an organism to grow on a certain medium is probably due to the absence of one or more of the essential growth accessory substances. Media are usually selected for their ability to produce results rather than because they are known to contain the necessary growth substance. It seems highly probable that few media containing all the necessary accessory substances could be successfully employed for the cultivation of bacteria. Until such investigations are made and reported, the bacteriologist will continue to employ many kinds of media, each more or less specific for a particular purpose.

Nitrogen sources The strict autotrophic bacteria are able to utilize inorganic ammonium salts as the only source of nitrogen. They cannot utilize exogenous organic nitrogen compounds.

The strict heterotrophic bacteria do not utilize inorganic ammonium salts but must have organic nitrogen such as is present in amino acids.

The facultative bacteria are able to utilize both inorganic and organic sources of nitrogen. Since it is believed that the bulk of the amino acids in a medium are dissimilated with the liberation of ammonia, compounds furnishing nitrogen, whether amino acids or inorganic compounds, are generally of equal value as available sources of the element.

Bacteria show wide differences in their amino acid requirements. Amino acids essential for one organism may not be necessary for another. No rule can be formulated that will apply to all bacteria. This can be determined only by actual experimentation.

Hunt and Pittillo (1968) employed a chemically defined medium to determine the nitrogen requirement of a single cell of *Escherichia coli*. Ammonium chloride and glucose were added to a nutritionally deficient culture of the organisms. Immediately thereafter, samples were removed at 3-min intervals and counts of viable cells made. Appropriate calculations revealed that approximately 10^{-13} g of ammonium chloride was required per cell.

Fildes et al. (1933) found the amino acid tryptophan to be necessary for the growth of *B. typhosa*. However, some strains which initially required tryptophan could be trained to grow without the compound whereas others could not. They concluded that tryptophan was an essential constituent of protoplasm and if the organism could not synthesize the compound it must be added to the culture medium.

Curcho (1948) produced mutant strains of *Eberthella typhosa* capable of growing in a medium without tryptophan. The mutants retained their tryptophan independence through daily transfers for over a year. The mutants, no doubt, were able to synthesize their own requirements of this essential amino acid.

Carlton (1967), on the basis of experimental evidence, proposed a scheme for the synthesis of tryptophan in *E. coli, Salmonella typhimurium*, and several species of molds (Fig. 141).

Fig. 141 Pathway of tryptophan biosynthesis as determined for Escherichia coli, Salmonella typhimurium, and several molds. Mutant classes having the following properties have been isolated in all of these organisms: (B) mutants grow only on tryptophan and accumulate indole in culture filtrates; (A) mutants grow on either indole or tryptophan and accumulate indoleglycerol. Both classes are defective in the final step in tryptophan biosynthesis; (C) mutants accumulate 1-(O-carboxyphenylamino)-1-deoxyriboluse, the dephosphorylated form of 1-(O-carboxyphenylamino)-1-deoxyribulose-5-phosphate (CDRP), and are defective in the enzyme indole-3-glycerol phosphate (IGP) synthetase; (D) mutants grow on either indole or tryptophan, accumulate an acid-labile intermediate believed to be the coupling product [N-5′-phosphoribosyl anthranilic acid (PRA)] of anthranilic acid and 5′-phosphoribosyl-1-pyrophosphate (PRPP). They are defective in the enzyme activity PRA-isomerase; (E) mutants also grow on indole or tryptophan but are defective in enzymatic function for coupling PRPP and anthranilic acid (PR-transferase). (After Carlton.)

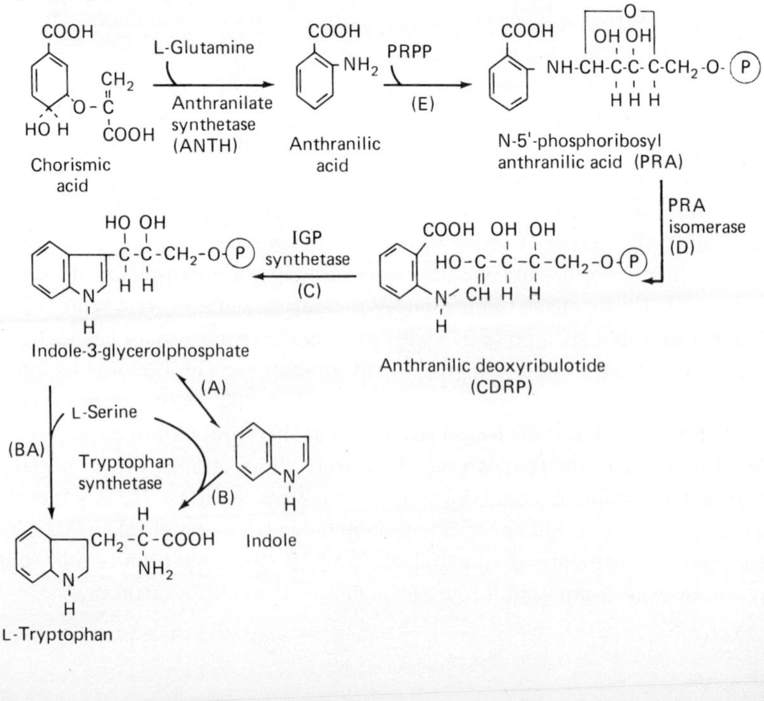

Fildes (1956) showed that strains of *Salmonella typhi* in which the indole → tryptophan enzyme was unimpaired synthesized tryptophan in excess during growth. The amount produced might be as small as to be undetectable by the methods used, but when indole in excess was added to the culture, the production of tryptophan was increased. Experiments with *E. coli* also showed an accumulation of tryptophan in cultures containing added indole (Fig. 142). The accumulation of tryptophan in this case was due to inhibition of the enzyme tryptophanase by the indole.

Fig. 142 Two plates of casein agar were flooded with a tryptophan-requiring mutant of Salmonella typhosa, and the surfaces dried. Then a few drops of a tryptophan-producing mutant were spotted on the plates. Into the lid of the upper plate was inserted a piece of filter paper carrying a few drops of dilute indole. The S. typhosa spotted on the surface synthesized tryptophan from the indole vapors. The tryptophan was utilized by the organisms flooded over the surface of the agar, causing the formation of halos of growth. In the absence of indole (lower plate), the spotted organisms failed to synthesize sufficient tryptophan to produce halos of the organisms flooded on the plate. (Courtesy of P. Fildes.)

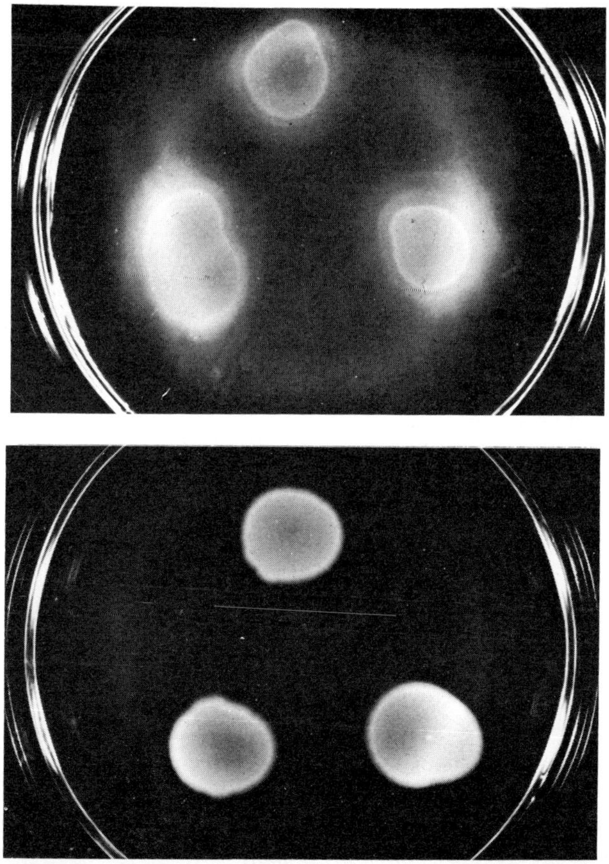

For more information: Brenner (1955); Doy and Pittard (1960); Goldberg et al. (1968); Morton and Macmillan (1954); Schlessinger and Ben-Hamida (1966); Yanofsky (1956).

Carbon requirements Organisms show great variation in their carbon requirements. The strict autotrophic bacteria obtain their carbon from carbon dioxide or carbonates. The strict heterotrophic bacteria must have organic carbon compounds. The facultative bacteria can utilize both inorganic and organic sources of carbon. The nature of the carbon compounds available to bacteria varies from one species to another. Variation in the nature of the carbon source required by living cells probably accounts for the flora of specific environments.

The heterotrophic forms show wide differences in their ability to utilize organic sources of carbon. Carbohydrates and related compounds are most generally utilized by bacteria. These are followed by certain organic acids such as malic, citric, succinic, and lactic; next come the lower fatty acids; and lastly the monohydric alcohols.

Formation of lipides The formation of lipides by bacteria is dependent upon the nature of the carbon compounds added to media.

Stephenson and Whetham (1922) reported that in a medium containing acetate, lactate, and glucose, singly or in various combinations, *Mycobacterium phlei* synthesized sufficient lipide to become acid-fast; whereas in the same medium without the carbon source the cells stained non-acid-fast.

Larson and Larson (1922) found that lipide synthesis occurred only if the organisms utilized the carbon source (glycerol or glucose) without producing a fermentation. Glycerol was found to be superior to glucose.

Kaneda (1963*a, b*) reported the synthesis of six branched-chain and two straight-chain fatty acids by *B. subtilis* when grown on a glucose yeast-extract peptone medium. The addition to the medium of short-chain fatty acids, such as propionic, butyric, isobutyric, valeric, isovaleric, and α-methylbutyric produced an increased formation of long-chain fatty acids.

For more information: Kates et al. (1961).

Vitamins

The term *growth factor* was originally applied to any substance which, in minute amount, produced a stimulatory effect on the growth of microorganisms. In the same sense, the term *vitamin* was used in connection with the nutrition of animals. Williams coined the term *nutrilite*. Since growth factors, vitamins, and nutrilites have been shown to be the same, the terms are used interchangeably, but vitamin is now used almost exclusively.

A large number of vitamins have been studied and their chemical makeup determined. Most of these have been synthesized. Some are simple in structure; others are quite complex.

Response of organisms to vitamins That an organism will grow in the absence of a particular vitamin in the medium does not mean necessarily that the factor is not required. The strict autotrophic bacteria can synthesize all their vitamins; the strict heterotrophic bacteria probably are unable to synthesize any vitamin; the facultative forms fall between and show great variation in their ability to synthesize vitamins. A further complication is that some organisms can synthesize a certain vitamin when grown aerobically but cannot do so anaerobically; with other organisms the reverse is true. Therefore it is erroneous to conclude that an organism does not require a certain vitamin because it is absent from the medium in which the cells are growing.

Some bacteria have been adapted to dispense with certain vitamins by repeated subculture to fresh medium. For example, strains of *Propionibacterium* species have been successfully cultivated without riboflavin and thiamine added to the medium.

For more information: Goodwin (1963); Koser (1968); Ortega and Brown (1960); Reeves (1963).

Vitamins required by bacteria The vitamins most frequently reported as being necessary for bacteria include: thiamine chloride, biotin, pantothenic acid, riboflavin, pyridoxin, nicotinic acid, and *p*-aminobenzoic acid. In addition, a number of miscellaneous compounds have been shown to be indispensable for some, but not necessarily for all, bacteria. Some of these are nicotinamide, folic acid, inositol, vitamin B_{12} (cyanocobalamin), glutamic acid, glutamine, asparagine, purines, pyrimidines, codehydrogenase I (NAD), codehydrogenase II (NADP), hemin, putrescine, and glutathione.

Thiamine hydrochloride (vitamin B_1) Chemically, thiamine hydrochloride is a pyrimidine-thiazole compound having the structural formula:

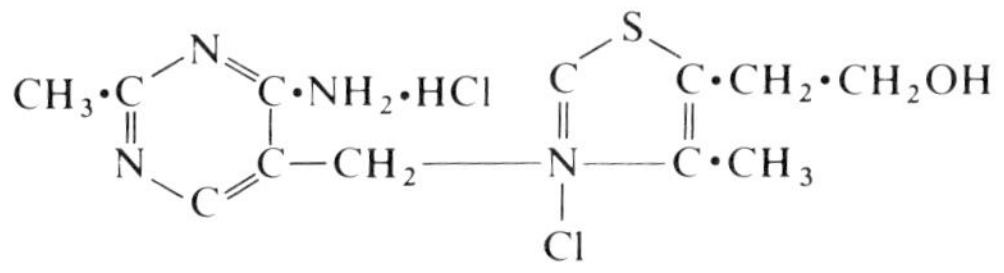

Thiamine hydrochloride

This vitamin is probably necessary for the growth of all bacteria. A number of heterotrophic organisms have been shown to be capable of synthesizing the compound.

The pyrophosphate ester (thiamine pyrophosphate) is known as cocarboxylase (page 410). It participates in the decarboxylation of many α-keto acids with the formation of aldehydes and carbon dioxide. The coenzyme can replace thiamine hydrochloride in the nutrition of yeasts. It is probable that thiamine hydrochloride serves as a precursor of cocarboxylase.

For more information: Bánhidi (1960*a, b*).

Biotin This factor is identical with vitamin H. Chemically it is 2′keto-3,4-imidazolido-2-tetrahydrothiophene-*n*-valeric acid:

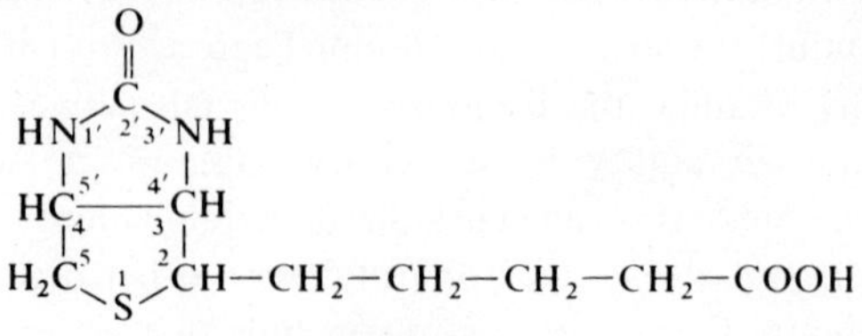

Biotin

Biotin is present in all living cells in minute amounts. Bacteria, yeasts, and molds are the best sources of the vitamin. An amount as small as 1 part in 50 billion parts of medium is sufficient to produce a 100 percent increase in yeast growth.

The exact function of biotin is not clearly understood. It has been suggested that it may participate in the deamination of aspartate, threonine, and serine; the carboxylation of pyruvate, adenine, and guanine; the decarboxylation of oxaloacetate and succinate; and the oxidation of pyruvate and lactate. The carboxylation of pyruvate to oxaloacetate is as follows:

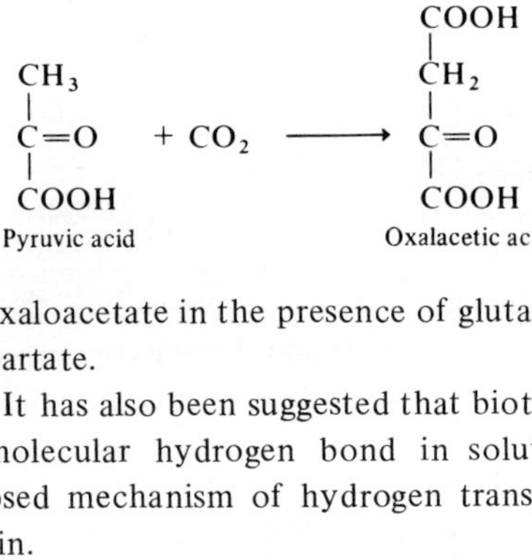

Pyruvic acid   Oxalacetic acid

The oxaloacetate in the presence of glutamate or alanine may be converted to aspartate.

It has also been suggested that biotin might be capable of forming an intramolecular hydrogen bond in solution. This would strengthen the proposed mechanism of hydrogen transport as a mode of action of the vitamin.

For more information: Birnbaum et al. (1967); Pai and Lichstein (1967).

Pantothenic acid Pantothenic acid is a growth determinant of universal occurrence in nature. Rice bran and molasses are good vegetable sources of the vitamin.

Chemically, the vitamin consists of β-alanine united to a saturated dihydroxy acid by a peptide-like combination:

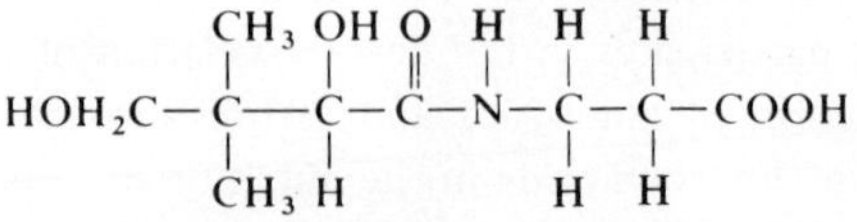

Pantothenic acid

Pantothenic acid is a component of coenzyme A which is involved in the acetylation of aromatic amines and choline (page 410). It is also related to the utilization of other vitamins, especially riboflavin.

Brown (1959) believed certain bacteria converted pantothenic acid to coenzyme A through the following series of reactions:

Pantothenic acid ⟶ 4′phosphopantothenic acid
⟶ 4′phosphopantothenylcysteine
⟶ 4′phosphopantetheine ⟶ coenzyme A.

For more information: Reeves (1963).

Riboflavin This factor is also known as vitamin B_2, vitamin G, and lactoflavin. Chemically it is 6,7-dimethyl-9-(D-1′-ribityl) isoalloxazine:

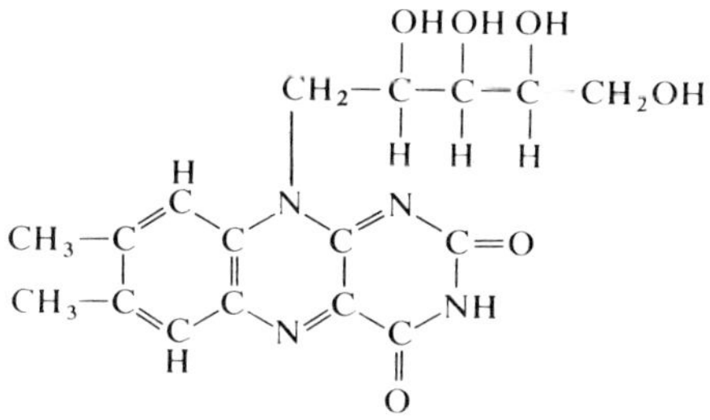

D-Riboflavin

In cells riboflavin occurs also as riboflavin-phosphoric acid and as riboflavin-phosphoric acid-adenine dinucleotide.

Riboflavin plays an important function in many enzyme systems. It is a component of several enzymes known as flavoproteins such as Warburg's yellow enzyme, $NADH_2$:lipoamide oxidoreductase, cytochrome c:oxidoreductase, xanthine:O_2 oxidoreductase, D-amino acid:O_2 oxidoreductase, etc.

Pyridoxin This factor is known also as vitamin B_6. It is 2-methyl-3-hydroxy-4,5-bis(hydroxymethyl)-pyridine. Two other naturally occurring substances with vitamin B_6 activity are pyridoxal (2-methyl-3-hydroxy-4-formyl-5-hydroxymethylpyridine), and pyridoxamine (2-methyl-3-hydroxy-4-aminomethyl-5-hydroxymethylpyridine):

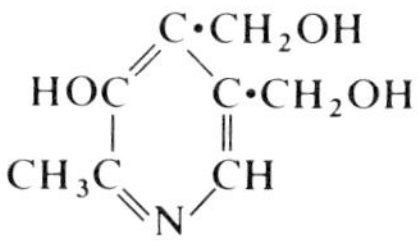

Pyridoxin

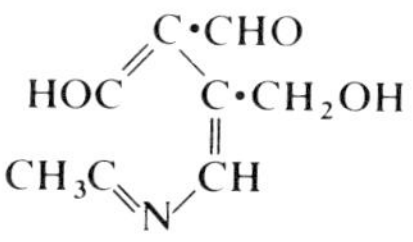

Pyridoxal

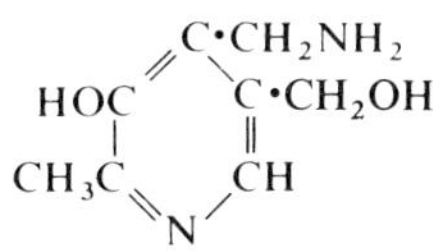

Pyridoxamine

These three vitamins function in the form of a coenzyme, pyridoxal-5-phosphate:

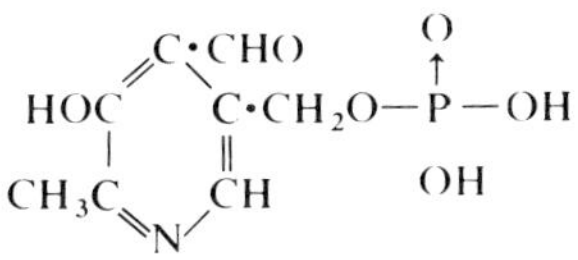

Pyridoxal-5-phosphate

Other names for the coenzyme are codecarboxylase, cotransaminase, etc.

Pyridoxin, pyridoxal, and pyridoxamine owe their vitamin activity to the ability of the organism to convert them into the active pyridoxal-5-phosphate. It functions as a transaminase for the synthesis of amino acids from their keto analogs. Pyridoxal-5-phosphate has also been reported to function as a glutamate-aspartate transaminase. When combined with protein (apoenzyme), pyridoxal-5-phosphate catalyzes the decarboxylation of amino acids.

Nicotinic acid (niacin) and nicotinamide (niacinamide) Chemically nicotinic acid is 3-pyridine carboxylic acid, and nicotinamide is 3-pyridine carboxylic acid amide:

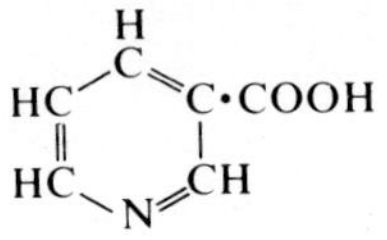

Nicotinic acid

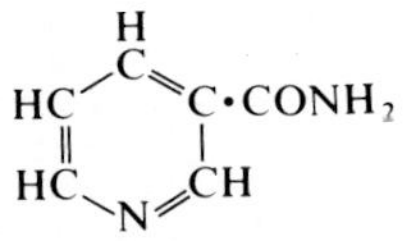

Nicotinamide

Nicotinic acid or its amide is required by all living cells. It is a component of nicotine adenine dinucleotide (NAD) and nicotine adenine dinucleotide phosphate (NADP). As far as known, this is the only necessity for the vitamin.

For more information: Greenberg et al. (1960); Ortega and Brown (1960).

***p*-Aminobenzoic acid (PABA)** PABA has the following structural formula:

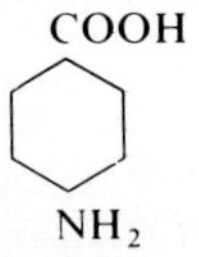

p-Aminobenzoic acid

PABA is highly active in reversing the bacteriostatic action of the sulfonamides (page 585). Sulfonamides compete with PABA for an enzyme and thus interfere with some essential metabolic reaction. Sulfonamide-resistant strains of *Staphylococcus aureus* produce greater amounts of PABA than their parent strains. The quantity of PABA synthesized by resistant strains is sufficient to account for their resistance to the sulfonamides.

PABA is a unit in the structure of another vitamin, namely, pteroylglutamic acid or folic acid. The sulfonamides act by blocking the incorporation of PABA into the folic acid molecule.

Folic acid Chemically folic acid is pteroylglutamic acid having the following formula:

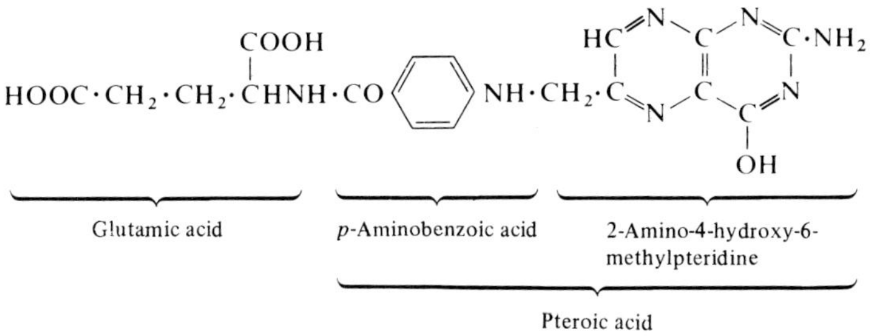

p-Aminobenzoic acid and glutamic acid are components of folic acid. Folic acid appears to be primarily concerned with the synthesis of purines and pyrimidines. It is also believed to be a requirement for the synthesis of certain amino acids which involve the incorporation of a single carbon fragment. It appears that carbon dioxide may serve as the source of the carbon fragment in this experimental system.

For more information: Brown et al. (1961); Dyke (1965); Goodwin (1963).

Inositol Nine stereoisomeric forms of inositol are possible, seven being optically inactive or *meso* forms, and two asymmetric enantiomorphs. The form that is most widely distributed in nature and that is of importance in the nutrition of organisms is optically inactive. It is sometimes called *meso*-inositol or *myo*-inositol, and has the following structure:

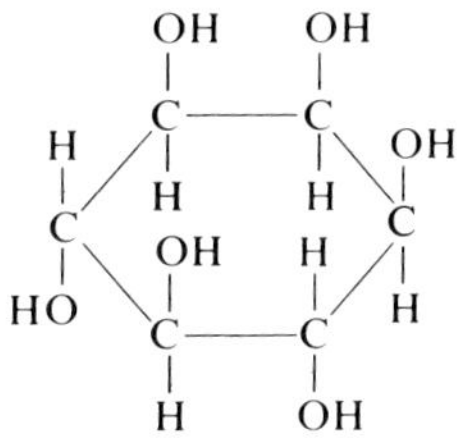

Inositol

Inositol is of widespread occurrence and great abundance in nature. It is believed to act both as a vitamin and an energy-yielding nutrient. Its exact function in living cells in terms of enzymes and coenzymes is not clearly understood. It is a constituent of certain phospholipides isolated from bacteria. It combines with phosphate, proteins, fatty acids, glycerol, and galactose.

For more information: Ridgeway and Douglas (1958*a*, *b*).

Vitamin B_{12} This is a cobalt- and phosphorus-containing vitamin of great complexity. Chemically it is 5,6-dimethylbenzimidazolyl cyanocobamide and is known commercially as cyanocobalamin. It is produced by the growth of suitable organisms in appropriate culture media.

Vitamin B_{12} appears to be involved in protein biosynthesis. The

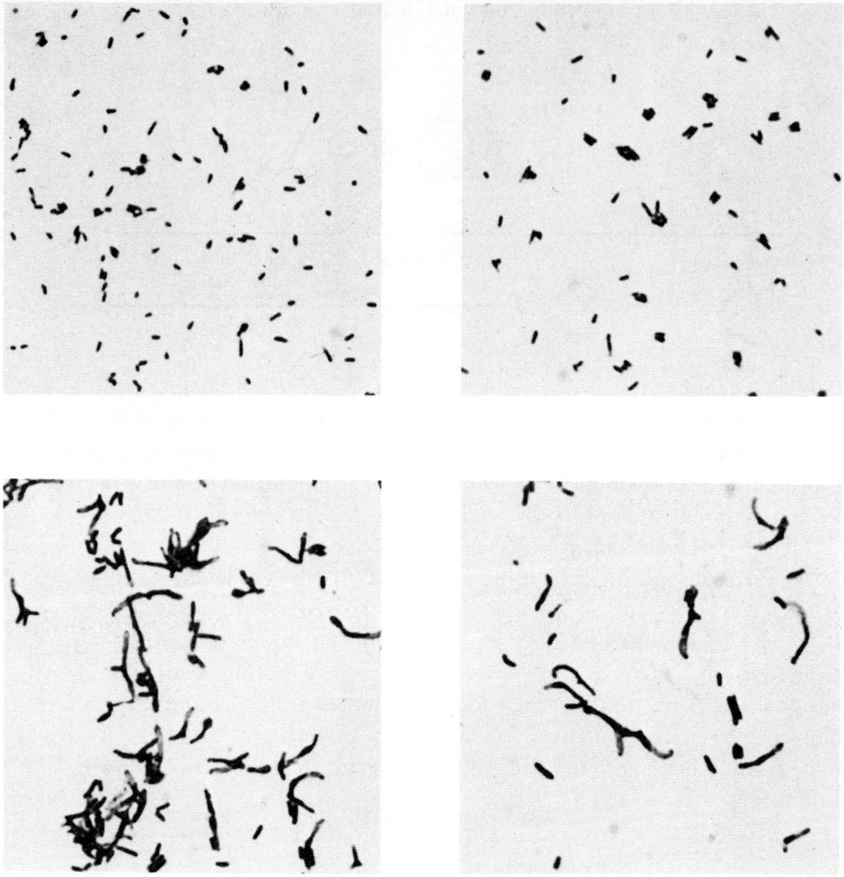

Fig. 143 Effect of various concentrations of vitamin B_{12} on the growth of an Arthrobacter species. Gram stain. Upper left, 2 μg/ml. Normal cells. Upper right, 1 μg/ml. Normal cells slightly larger. Lower left, 0.075 μg/ml. Cells swollen and elongated with slight branching. Lower right, 0.05 μg/ml. More pronounced branching. Large gram-negative areas. (After Chaplin and Lochhead.)

vitamin has also been reported as increasing the uptake of phosphorus in the synthesis of deoxyribonucleic acid by *Lactobacillus leichmannii.*

Chaplin and Lochhead (1956) studied the B_{12} requirements of a species of the genus *Arthrobacter*. They showed that with suboptimal concentrations of the vitamin, growth in liquid medium was flocculent, whereas cultures with adequate amounts were uniformly turbid. Floccula-tion was associated with abnormal cell morphology. The cells were noticeably swollen, elongated, and irregularly bent and showed rudimen-tary branching (Fig. 143).

For more information: Dyke (1965); Goodwin (1963); Smith (1965).

Pimelic acid and β-alanine These two factors have been shown to be necessary for the growth of the diphtheria bacillus in a synthetic medium. Their structural formulas are as follows:

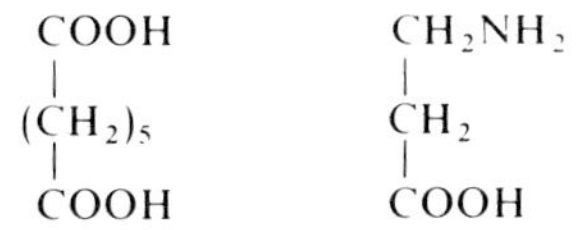

Pimelic acid β-Alanine

Pimelic acid is one of the degradation products of biotin and has been shown to be capable of serving as a biotin precursor for some microorganisms.

β-Alanine is a component of pantothenic acid, being united to a saturated dihydroxy acid by a peptide-like linkage (page 226).

Glutamic acid and glutamine This acid or its amide is required by most bacteria for the synthesis of cell proteins. Glutamine is a constituent of beef extract. Also, glutamic acid is a component of another vitamin, namely, folic acid.

The structural formulas for glutamic acid and glutamine are as follows:

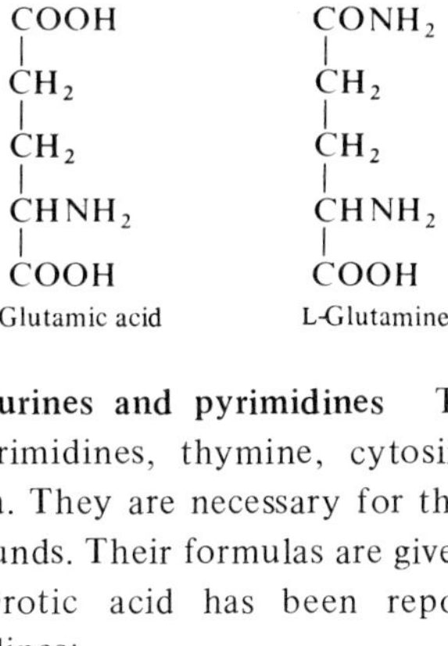

L-Glutamic acid L-Glutamine

Purines and pyrimidines The purines, adenine and guanine, and the pyrimidines, thymine, cytosine, and uracil are required by most bacteria. They are necessary for the synthesis of nucleic acids and related compounds. Their formulas are given on page 845.

Orotic acid has been reported to act as a precursor of the pyrimidines:

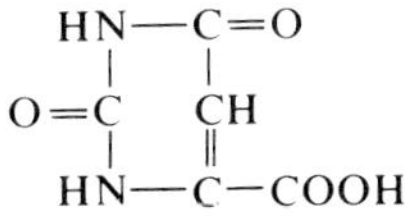

Orotic acid
(uracil-4-carboxylic acid)

Since folic acid contains a purine-like component, probably small amounts of purines are utilized in the synthesis of that vitamin.

For more information: Reynolds et al. (1955); Sen (1960).

***V* and *X* factors** *Haemophilus influenzae* will not grow unless supplied with two factors (*V* and *X*), both of which are present in blood. The *V* factor is also present in many plant extracts and in a large number of bacteria. It is thermolabile, being destroyed in 15 min at 90°C, is very sensitive to alkali but not acid, diffuses through parchment membranes, and is not readily destroyed by atmospheric oxygen. The *X* factor is present in potatoes and some bacteria. It is thermostable, resisting a temperature of 120°C for 45 min.

H. influenzae is unable to grow on media containing only the *X* factor. However, it will grow on media in association with an organism, such as *S. aureus*, which is capable of producing the *V* factor. The characteristic arrangement of colonies of *H. influenzae* in such an association is sometimes referred to as the satellite phenomenon. The hemophilic organisms grow as satellites in isolated colonies at some distance from the colonies of *S. aureus*.

H. parainfluenzae requires only the *V* factor for growth. The *V* factor has been isolated from yeast extract and shown to consist of nicotinamide-adenine dinucleotide (NAD) or nicotinamide-adenine dinucleotide phosphate (NADP). Either coenzyme can substitute for the *V* factor of blood (page 406).

The *X*-factor requirements are supplied largely by the addition of hemin. Since the addition of small amounts of blood enhances activity of the organisms, doubtless other factors are also involved.

For more information: Gibbons and Macdonald (1960), Gingrich and Schlenk (1944), White and Granick (1963).

Glutathione This compound has been shown to act as a growth factor for several organisms. It is found in meat infusion, yeast infusion, and red blood cells.

Glutathione possesses several functions: (1) It acts as a carrier of hydrogen; (2) it prevents inactivation of sulfhydryl groups; and (3) it is believed to maintain ascorbic acid in the reduced form.

Its structural formula is given on page 418.

Putrescine This putrefactive compound has been reported as being necessary for the growth of *H. parainfluenzae* and *Neisseria perflava*.

Chemically it is tetramethylene diamine and is produced by the decarboxylation of ornithine (page 459). Ornithine itself is produced by the hydrolysis of the amino acid arginine.

Asparagine Garvie et al. (1961) reported the necessity of this compound for the growth of a species of *Pediococcus*.

pH and growth-factor requirements The growth-factor requirements of bacteria may be influenced by the pH of the medium.

The amount of nicotinic acid required by *S. aureus* is 15 times greater at pH 8.0 than at pH 6.0. *Shigella paradysenteriae* must be supplied nicotinic acid at pH 7.0 and 8.0 but not at pH 6.0. At pH 6.0 the organisms can synthesize the factor. *L. casei* must be supplied with pyridoxin at pH 6.0 to 7.0 but not at pH 5.0. Folic acid, riboflavin, biotin, nicotinic acid, and pantothenic acid are most effective at pH 6.0.

Destruction of vitamins by microorganisms Vitamin C (ascorbic acid) is destroyed by the enteric bacteria. The vitamin is utilized as a carbon source. However, the factor is protected in the presence of an easily fermentable carbohydrate.

Pseudomonas riboflavina oxidizes riboflavin to lumichrome

$$\underset{\text{Riboflavin}}{C_{17}H_{20}O_6N_4} + 5\tfrac{1}{2}O_2 \longrightarrow \underset{\text{Lumichrome}}{C_{12}H_{10}O_2N_4} + 5CO_2 + 5H_2O$$

which accumulates in the culture as lemon-yellow crystals.

Goodhue and Snell (1966) reported that a species of *Pseudomonas* attacked pantothenic acid with the formation of β-alanine, pantoic acid, α-ketoisovaleric acid, valine, and several unidentified products.

Yang et al. (1969) found strains of molds and yeasts of the genera *Penicillium, Rhodotorula,* and *Endomycopsis*, which were capable of degrading biotin.

For more information: Brady et al. (1965); Burg et al. (1960); Harkness et al. (1964); Koser and Baird (1944); Tsai et al. (1963).

Vitamins in ingredients of culture media Information on the vitamin content of ingredients of culture media should prove of value in the cultivation of organisms. Such information could be used to decide whether a particular medium satisfies the vitamin requirements of an organism.

Stokes et al. (1944) assayed a number of culture media for their content of eight members of the B complex: thiamine, riboflavin, pantothenic acid, nicotinic acid, biotin, pyridoxin, folic acid, and *p*-aminobenzoic acid, and compared the results with the vitamin requirements of various microorganisms that were unable to synthesize these factors. They arranged the various culture-media ingredients in order of descending value on the basis of their content of vitamins of the B complex:

1. Yeast extract
2. Meat extract, brain infusion, heart infusion
3. Various peptones

With the possible exception of thiamine, yeast extract is an excellent source of all vitamins of the B complex. This explains why yeast extract is held in such high favor as an ingredient of culture media.

They concluded that if peptone, meat extract, etc., were used in concentrations of 1 or 2 percent, singly or in combinations, the resultant media were likely to be deficient in thiamine, riboflavin, pantothenic acid, pyridoxin, and *p*-aminobenzoic acid, but not in nicotinic acid, biotin, and folic acid. They suggested that such growth-factor deficiencies in culture media could be remedied by proper combination of ingredients or by the addition of yeast concentrates or synthetic vitamins.

For more information: Dyke (1965); Gibson and Pittard (1968); Goodwin (1963); Kasai (1953); Kihara et al. (1961); Marks (1968); Robinson (1966); Sebrell and Harris (1967).

Metabolite antagonists (antimetabolites) A metabolite may be defined as any substance essential to growth and reproduction of cells. It may be synthesized by the cells or obtained from the environment. A metabolite antagonist is any substance that blocks or inhibits the normal function of the metabolite. The best examples of antimetabolites involve substances structurally related to the substrate.

The enzyme succinic dehydrogenase is inhibited by malonic acid:

$$COOH{\cdot}CH_2{\cdot}COOH$$

Malonic acid

$$COOH{\cdot}CH_2{\cdot}CH_2{\cdot}COOH$$

Succinic acid

The homologous dibasic malonic acid is believed to inhibit the enzyme by competing with the normal metabolite succinic acid. The competitive inhibition between the two compounds results almost in a complete blocking of the enzymic reaction.

p-Aminobenzoic acid (PABA) antagonizes the action of sulfanilamide. The close structural relationship between the two compounds is shown in the following formulas:

p-Aminobenzoic acid

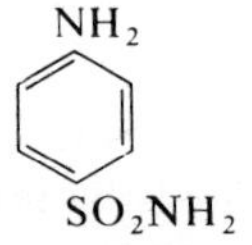

Sulfanilamide

PABA is an essential vitamin normally associated with an enzyme. Sulfanilamide displaces PABA from its enzyme, thereby stopping this essential line of metabolism. Other sulfa drugs (sulfonamides) behave in a similar manner.

Arginine is an essential metabolite for several species of lactic acid bacteria. None of the species tested could substitute canavanine for the arginine.

$$NH{=}C\begin{cases} NH_2 \\ \underset{H}{N}{\cdot}CH_2(CH_2)_2{\cdot}CHNH_2{\cdot}COOH \end{cases}$$

Arginine

$$NH{=}C\begin{cases} NH_2 \\ \underset{H}{N}{-}O{-}(CH_2)_2{\cdot}CHNH_2{\cdot}COOH \end{cases}$$

Canavanine

Canavanine functions as an effective growth inhibitor. This is true for organisms that synthesize arginine as well as for those that require the amino acid preformed.

For more information: Knight (1949); Pittillo and Foster (1954); Roblin (1949); Woolley (1959, 1960).

Growth phases in a culture

Most bacteria multiply at a very rapid rate and produce pronounced changes in culture media in a short period of time. Under favorable conditions a single cell of *E. coli* divides into two about every 20 min. If this rate is maintained, a single organism will give 1 billion new cells after a period of about 10 hr. However, this rate of multiplication is not maintained indefinitely, owing to the exhaustion of the nutrients, to the accumulation of toxic metabolic waste products, and to the fact that many of the cells die. The rate of death increases as the culture ages. The more vulnerable cells die first, leaving the most resistant forms at the end of the cycle.

When an organism is inoculated into a medium, such as nutrient broth, multiplication does not take place in a regular manner. A number of distinctive growth phases (generally seven) are recognized and these are designated as follows:

1. Initial stationary phase. During this phase, the number of bacteria remains constant. Plotting the results on graph paper gives a straight horizontal line (1*a*) in Fig. 144.
2. Lag phase or phase of positive growth acceleration. During this phase, the rate of multiplication increases with time (*ab*).
3. Logarithmic growth phase. During this phase, the rate of multiplication remains constant (*bc*). This means that the generation time is the same throughout.
4. Phase of negative growth acceleration. During this phase, the rate of multiplication decreases (*cd*). The average generation time increases. The organisms continue to increase in numbers but at a slower rate than during the logarithmic growth phase.
5. Maximum stationary phase. During this phase, the number of living organisms remains constant, i.e., the death rate equals the rate of increase (*de*).

Fig. 144 Growth phases in a culture. (After Buchanan.)

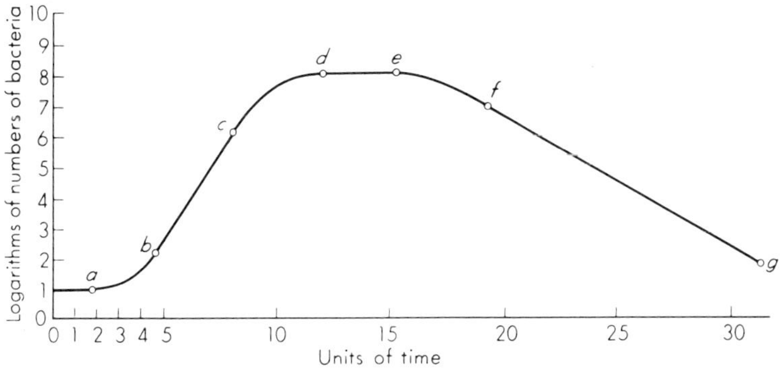

6. Phase of accelerated death. During this phase, the numbers fall off with increasing rapidity (*ef*). The average rate of death per organism increases to a maximum.
7. Logarithmic death phase. During this phase, the rate of death is constant (*fg*).

It is generally believed that the phases in a culture result from changes in the environment, such as alteration in pH, exhaustion of food supply, and accumulation of waste products.

For more information: Sinclair and Stokes (1962).

Growth phase and multiplication rate When bacteria are transferred from an old culture (24 hr) to fresh medium, they exhibit a period of delayed multiplication or lag. The lag period may be increased or decreased depending upon the age of the transfer culture. If the inoculum is taken from a culture in the logarithmic death phase, the lag period will be greatly extended; if taken from a culture in the logarithmic growth phase, the lag period will be greatly reduced or eliminated completely. Organisms in the logarithmic growth phase are multiplying in a regular manner at the maximum rate and continue to do so when transferred to fresh medium. Generally, transfer from any phase of a bacterial culture cycle to fresh medium is followed by a continuance of the phase of the parent culture.

Organisms show considerable variation to harmful influences in the various growth phases. Cultures in the lag and early logarithmic growth phases exhibit greater sensitivity to heat than those in the ensuing phases. In other words, young bacteria are more sensitive than old cells. Then resistance shows a steady increase as the culture passes through the subsequent phases.

Lag phase Organisms show a marked increase in size in the lag phase. The average length of *B. megaterium* may show a sixfold increase over the same organism taken from an old culture. Generally, this increase manifests itself after 2 hr and reaches a maximum between 4 and 6 hr. In the lag phase, the cells show considerable fluctuation in form. On passing from the lag phase to the logarithmic death phase, the organisms gradually decrease in size and exhibit a more constant cell form.

It is well known that the transfer of a parent culture into a fresh medium is followed by an initially slow rate of increase in numbers. This slow rate of multiplication cannot be interpreted as indicating a period of lag in the sense of decreased viability and activity. What actually happens is that the rate of multiplication decreases but the individual cells increase in mass. During the first 2 or 3 hr, fresh cultures inoculated from 3-hr-old parent cultures show a lower multiplication rate than fresh cultures prepared from 24-hr parent cultures but the increase in protoplasmic growth remains the same. The cells from a young parent culture show the same increase in cell mass as the cells from an old parent culture even though they multiply at a lower rate. The rate of increase in cell mass is nearly constant from the time growth first begins until the maximum population is reached.

This suggests that conditions in fresh medium favor an increase in cell size rather than cell division with the result that a majority of the cells attain an abnormal size before fission occurs. Transfer of a large number of cells to a fresh medium tends to produce the reverse effect, namely, the average size is smaller and attained sooner than with small inoculations. If cells are transferred to fresh medium before attaining maximum size, the organisms become larger and the critical point takes place later than in the original culture. This would indicate that size is dependent upon the density of the culture. The concentration of nutrients in the medium is another factor. Organisms in a dilute medium are smaller and the critical point is reached earlier. A dilute medium shows less growth than a more concentrated preparation. This means that a concentrated medium showing a heavy growth produces the same effect on cell size as a dilute medium showing a light growth.

Effect of carbon dioxide on lag phase Organisms require carbon dioxide (CO_2) for growth. If a fresh culture is aerated with CO_2-free air, growth does not occur. Some believe that the phenomenon of lag is due largely, if not entirely, to the time required for the culture to build up the CO_2 content of the medium or of the cells themselves to a value essential for growth. The amount of CO_2 present in a new culture is quite likely an important factor in controlling the length of the lag phase but probably not the only one.

For more information: Duncan and Nickerson (1963); Hoffman and Frank (1965); Lockhart (1960); Powell and Errington (1963); Quesnel (1960).

Effect of temperature on growth The rate of multiplication of bacteria increases with temperature until a maximum is reached. Then the reverse effect occurs, finally ending in the death of the bacteria.

The generation times of *E. coli* at different temperatures of incubation are:

°C	*Time*	*°C*	*Time*
10	14 hr, 20 min	35	22 min
15	120 min	40	17½ min
20	90 min	45	20 min
25	40 min	47½	77 min
30	29 min		

For more information: Ng (1969).

Effect of medium on growth The composition of the medium also affects the rates of growth of bacteria. The generation time of *S. typhosa* in a 1 percent solution of peptone is about 40 min at 37°C. If the content of the peptone is less than 0.2 percent, the generation time is almost inversely proportional to its concentration. If the peptone content is above 0.4 percent, an increase in its concentration is almost without effect on the growth rate. The addition of 0.175 percent glucose to a medium containing only 0.1 percent peptone lowers the generation time from 111 to 50 min. The addition of the same concentration of glucose to a 1 percent peptone solution lowers the generation time only from 39 to 34 min.

Cell density plays an important role on survival in the absence of nutrients. Harrison (1960) reported that suspensions of logarithmic phase cells of *Aerobacter aerogenes* at various cell densities held at growth temperature in the absence of added nutrients showed a survival which was maximum at a particular cell density. At densities above and below this critical cell density, survival was proportionately less. At a density of 10^5 or 10^6 cells per ml the death rate was very great and the suspension became sterile within several hours. At a density of 10^7 cells per ml the death rate showed fluctuations. At the critical density, 10^8 cells per ml, an initially rapid death gradually diminished until it eventually became negligible. Harrison concluded that logarithm phase cells in the absence of food died as a result of loss of cell substance (effluent). The concentration of effluent increased with time. At low densities the effluent never became sufficiently concentrated to be recoverable, so none of the cells survived. But at higher densities the effluent eventually became sufficiently concentrated to sustain those cells which were still viable at that time.

Strange et al. (1963) found the effluent to consist of ammonia, nucleic acid bases, and inorganic phosphate, which would suggest its origin from macromolecules of ribonucleic acid (RNA) and protein. Since most of the RNA and a substantial fraction of the total protein are present in the ribosomes, it appeared likely that the latter suffered these losses.

Galdiero et al. (1967) reported that (1) in the logarithmic growth phase, protein, polysaccharide, and RNA increased, especially the latter; (2) in the maximum stationary phase, protein and polysaccharide increased while RNA decreased; and (3) in the phase of accelerated death, protein remained unchanged while polysaccharide and RNA decreased.

For more information: Baarda and Lockhart (1962); Bohinski and Mallette (1967); Ecker and Lockhart (1961); Ecker and Schaechter (1963); Nishi et al. (1967); Postgate and Hunter (1964).

Effect of age on morphology Bacteria exhibit their largest size in the initial stationary and early lag phases. Also, the cell protoplasm becomes more hyaline and stains homogeneously without exhibiting a granular structure. As the cells age, they become increasingly smaller and more granular. Old cells are, in general, smaller and very granular, whereas young cells are larger and do not exhibit a granular appearance (Fig. 145). When old granular cells are transferred to a fresh medium, the above cycle is repeated. An exception to this rule is *Corynebacterium diphtheriae* which appears to be smaller in young cultures and to exhibit the presence of granules in both young and old cells.

For more information: Clifton (1967).

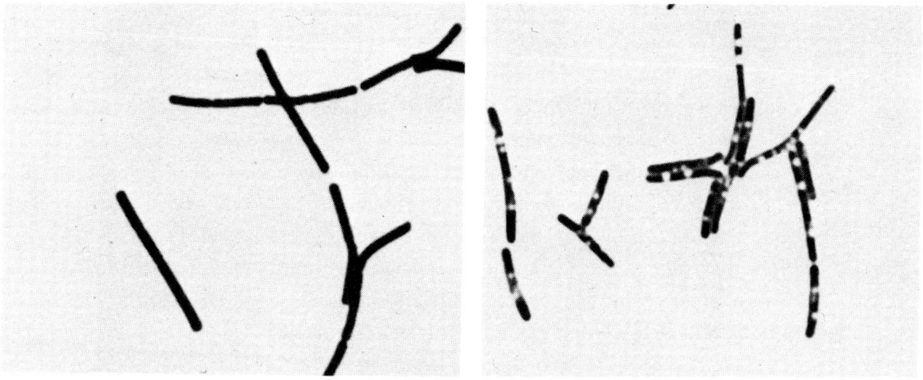

Fig. 145 Bacillus mycoides. Left, 6-hr culture; right, 24-hr culture. Note the development of granules with age.

Effect of continuous culture on growth Devices have been evolved for the purpose of providing growing organisms with frequent replenishments of nutrient media. The devices permit automatic addition of nutrient and the attainment of a steady state at a relatively uniform population density, or growth rate. Under these conditions it is possible to maintain the concentration of various nutrients and the generation time at predetermined values, and to study the effects of changes in the variables on the culture.

The desired product of a fermentation is frequently not the cell material but rather a substance secreted into the medium. The optimum conditions for synthesis of the latter may be quite different from the conditions required for cell growth. For example, in penicillin production, the optimum pH for growth of the mold is 7.0, whereas for antibiotic production it is 7.4.

For more information: Baschnagel-de Pamphilis and Hanson (1969); Eisler (1968); Eisler and Webb (1968); Hedén and Holme (1960); Johnson (1967); Johnson and Larkin (1963); Kono (1968); Mitchell and Plummer (1954); Mitchison (1969); Petropulos (1964); Taub and Dollar (1968).

Hydrogen-ion concentration of culture media

Culture media are adjusted to varying degrees of acidity or alkalinity, depending upon the requirements of the organisms under cultivation. Some organisms grow best in acid media; others grow best in alkaline media; still others prefer media which are approximately neutral in reaction. This last group includes the majority of bacteria which have been studied. So for maximum growth, it is necessary to adjust the reaction of media to the requirements of the organisms under examination.

Two methods are employed for adjusting the reaction of culture media: (1) the determination of the actual numbers of free hydrogen ions and (2) the determination of the net amount of acid- or base-binding groups present. The former is spoken of as the hydrogen-ion (H^+) concentration; the latter as the titratable acidity or alkalinity.

The hydrogen-ion concentration may be determined either colorimetrically or electrometrically. The titratable acidity is determined by titration of a known volume of medium with a standard solution of NaOH to the predetermined end point as shown by a glass electrode or by the color of a suitable indicator. Both methods serve useful purposes in bacteriology. The adjustment of media is more accurately carried out by the hydrogen-ion method. The titratable acidity determinations are of value in learning the buffer content of media, i.e., their ability to resist changes in reaction on the addition of acid or alkali.

Measuring the concentration of hydrogen (H^+) ions Pure water is neutral in reaction because it ionizes into equal numbers of hydrogen and hydroxyl ions.

$$H_2O \rightleftharpoons H^+ + OH^-$$

However, it is dissociated to an extremely small extent.

One liter of 1 *N* HCl contains approximately 1 g of H^+ ions. One liter of pure water contains approximately 0.0000001 g of hydrogen ions. This may be written 10^{-7} g per liter. For each H^+ ion, there is a corresponding and neutralizing OH^- ion.

According to the law of mass action,

$$\frac{(H^+)(OH^-)}{HOH} = K$$

Since the concentration of undissociated water is very great in comparison to the concentration of free H^+ and OH^- ions, the equation may be written

$$(H^+)(OH^-) = K$$

The numbers of H^+ and OH^- ions being equal, each must have a concentration of 1×10^{-7}. The product of the concentrations of H^+ and OH^- is equal to 1×10^{-14}. The equation now becomes

$$(H^+)(OH^-) = 1 \times 10^{-14}$$

Pure water has a H^+ concentration of 1×10^{-7} and is neutral in reaction. Since the product of the H^+ and OH^- is always equal to 1×10^{-14}, an increase in H^+ produces a decrease in OH^- and vice versa. If the H^+ concentration of a medium is less than 1×10^{-7}, it will be alkaline in reaction; if more than 1×10^{-7}, it will be acid in reaction.

The symbol pH is used to express the acidity or alkalinity of a medium. It is defined as the logarithm of the reciprocal of the H^+ concentration. For convenience only the exponent is used in expressing pH. If the pH of a medium is less than 7.0, it is acid, and as the figure approaches 0, the medium becomes progressively more acid; if the pH is more than 7.0, it is alkaline, and as the figure approaches 14, the medium

Table 4 The relation of pH to H^+

	pH	Normality in terms of hydrogen ions	Normality in terms of hydroxyl ions
Acid	0	1	10^{-14}
	1	10^{-1}	10^{-13}
	2	10^{-2}	10^{-12}
	3	10^{-3}	10^{-11}
	4	10^{-4}	10^{-10}
	5	10^{-5}	10^{-9}
	6	10^{-6}	10^{-8}
Neutral point	7	10^{-7}	10^{-7}
Alkaline	8	10^{-8}	10^{-6}
	9	10^{-9}	10^{-5}
	10	10^{-10}	10^{-4}
	11	10^{-11}	10^{-3}
	12	10^{-12}	10^{-2}
	13	10^{-13}	10^{-1}
	14	10^{-14}	1

becomes progressively more alkaline. Most organisms which have been studied prefer media having a pH of about 7.0.

The relation of pH to H^+ is shown in Table 4.

Colorimetric method The determination of the hydrogen-ion concentration by the colorimetric method depends upon the color changes produced in certain weakly acidic or weakly basic dyes by varying the reaction of the medium. Such dyes are called indicators. An indicator changes in color within a short distance each side of that point on the pH scale at which it is 50 percent dissociated. At that point, one-half of the dye is undissociated and the other half is dissociated. The pH at which this occurs is denoted by the symbol pK.

A short distance each side of the pK point gives a zone that is referred to as the sensitive range of the indicator (Fig. 146). Every shade of color of the indicator in this sensitive range corresponds to a definite pH value so that by comparing the color with standards of known reaction, the H^+ concentration may be determined. Indicators can be selected to cover the range from pH 1.0 to 11.

Sometimes it is difficult to know which indicator to select in making a pH determination. In such instances an indicator paper may be used (Fig. 147) that covers the range from pH 1 to 11. The color change gives an approximate determination and aids in the proper selection of one of the indicators given in Fig. 146.

Potentiometric method Most pH determinations by the electrometric method are now made with the glass electrode.

Under suitable conditions, a thin glass membrane separating two solutions of different pH values exhibits an electric potential that is proportional to the difference in the pH of the solutions. For the construction of the glass electrode, a special type of glass is employed.

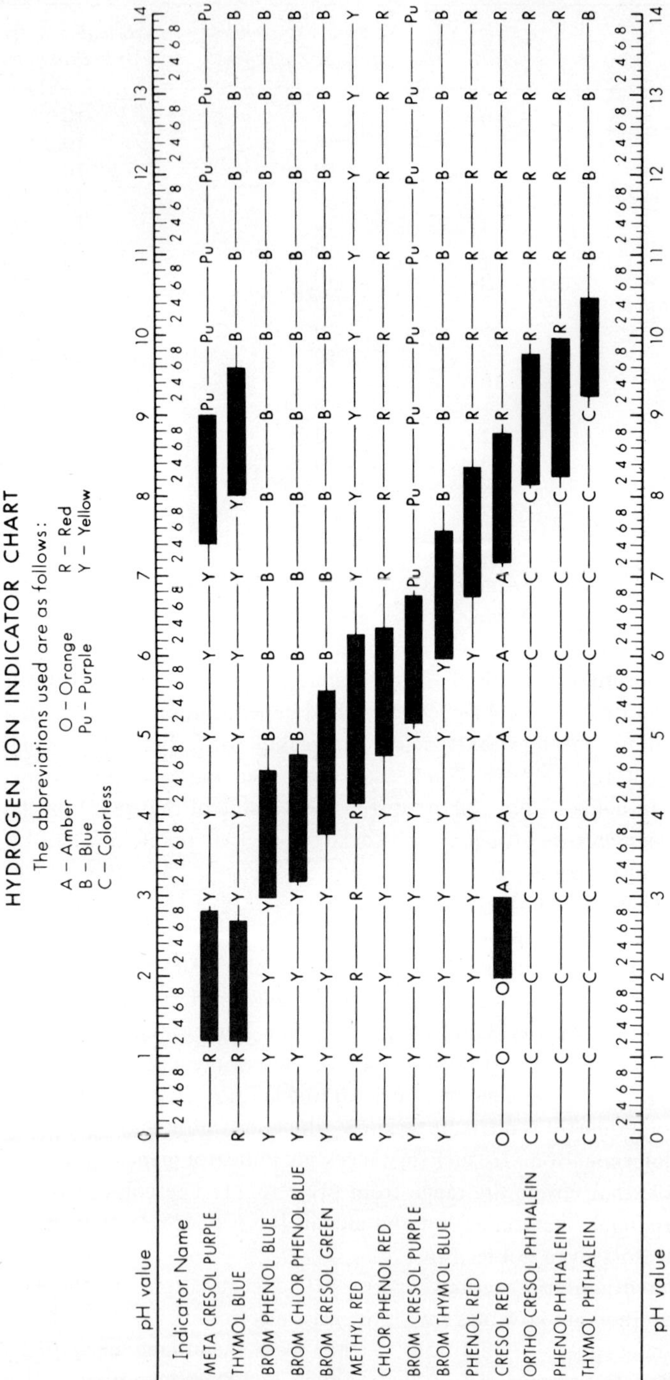

Fig. 146 Selector chart for widely used pH indicators.

Fig. 147 Universal indicator paper, pH 1 to 11. (Courtesy of Fluka AG, Buchs SG, Switzerland.)

The potentiometric method is more accurate than the colorimetric procedure. Also, it can be employed for the determination of the pH values of highly colored solutions which cannot be tested satisfactorily by the use of indicators.

For extensive discussions on the theoretical and practical aspects of the potentiometric method, see Bates (1964); Christensen (1964); Clark (1928).

Buffers The salts of weak acids have the power of preventing pronounced changes in the reactions of solutions on the addition of acids and alkalies. Substances of this nature are called buffers.

The titration curve of a weak acid is S-shaped. Each end of the curve has a steep slope and the central portion a gentle slope. This central, almost horizontal, portion of the curve expresses the buffer action of the system or its ability to resist changes in pH on the addition of acids or bases.

The addition of 1 ml of *N*/10 hydrochloric acid to 1 liter of neutral distilled water (pH 7.0) gives a solution having a pH of about 4.0. The addition of 1 ml of *N*/10 sodium hydroxide to 1 liter of neutral distilled water gives a solution having a pH of about 10.0. The addition of the same amount of acid or alkali to 1 liter of distilled water in which are dissolved a few grams of sodium phosphate produces very little change in the reaction. Sodium phosphate is classed as a buffer. This may be shown in the following reactions:

The addition of a strong acid:

$$Na_2HPO_4 + HCl \rightarrow NaH_2PO_4 + NaCl$$

The strong acid (HCl) reacts with the weak alkali (Na_2HPO_4) to give the weak acid (NaH_2PO_4) and sodium chloride. In other words, the strong HCl is replaced by the weak acid phosphate, resulting in a relatively small change in the final hydrogen-ion concentration.

The addition of a strong alkali:

$$NaH_2PO_4 + NaOH \rightarrow Na_2HPO_4 + H_2O$$

The strong alkali (NaOH) reacts with the weak acid (NaH_2PO_4) to give the weak alkali (Na_2HPO_4) and water. The strong NaOH is replaced by the weak basic phosphate, resulting in a relatively small change in the final hydrogen-ion concentration.

The important salts commonly added to nutrient media for their buffering action include phosphates and carbonates. These compounds are particularly valuable because they are relatively nontoxic.

Bacteriological peptones contain such substances as proteoses, peptones, peptides, and amino acids, all of which are buffers. These possess both acidic and basic properties, i.e., have the power of uniting with both bases and acids. Therefore, all culture media containing peptone are well buffered, the degree of buffering being dependent upon the amount of peptone added.

Buffers are of special importance in carbohydrate media that are vigorously fermented. In the various fermentations, organic acids are produced. The acidity soon builds up to a concentration that prevents further multiplication of the organisms. In the presence of a buffer, this takes place usually in 24 to 48 hr. In the absence of a buffer, the activity of the organisms may cease after a few hours. A good culture medium, besides containing the necessary nutrients, should also be well buffered.

Buffer standards Clark and Lubs (1917) proposed a series of buffer standards covering the range from pH 1.2 to 10 at intervals of 0.2 pH. By selection of the appropriate indicators (Fig. 146), the buffer standards may be used for the adjustment of the reactions of culture media.

For more information: Bates (1964); Christensen (1964); Clark (1928); Cohen (1957).

References

Baarda, I. F., and W. R. Lockhart: Characteristics and significance of the physiological patterns accompanying growth limitation, *J. Bact.*, **84**:1085, 1962.

Bánhidi, Z. G.: Contributions to the molecular physiology of thiamine, *Acta Physiol. Scand. Suppl.* 174, **50**:2, 1960*a*.

——: Growth promotion of some heterofermentative lactobacilli with different derivatives of thiamine, *Arkiv Kemi*, **15**:229, 1960*b*.

Baschnagel-De Pamphilis, J., and R. S. Hanson: Relationship between glucose utilization and growth rate in *Bacillus subtilis, J. Bact.*, **98**:222, 1969.

Bates, R. G.: "Determination of pH: Theory and Practice," New York, John Wiley & Sons, Inc., 1964.

Birnbaum, J., et al.: Biosynthesis of biotin in microorganisms. V. Control of vitamer production, *J. Bact.*, **94**:1846, 1967.

Bohinski, R. C., and M. F. Mallette: Response of logphase cells of *Escherichia coli* to medium limited in both sulfate and phosphate, *J. Bact.*, **93**:1316, 1967.

Brady, R. N., et al.: Bacterial and enzymatic degradation of biotin, *Biochem. Biophys. Res. Commun.*, **19**:777, 1965.

Brenner, S.: Tryptophan biosynthesis in *Salmonella typhimurium, Proc. Nat. Acad. Sci. U.S.*, **41**:862, 1955.

Brown, G. M.: The metabolism of pantothenic acid, *J. Biol. Chem.*, **234**:370, 1959.

—— et al.: The biosynthesis of folic acid. I, *J. Biol. Chem.*, **236**:2534, 1961.

Brown, H. J., and N. E. Gibbons: The effect of magnesium, potassium, and iron on the growth and morphology of red halophilic bacteria, *Can. J. Microbiol.*, **1**:486, 1955.

Burg, R. W., et al.: Bacterial oxidation of vitamin B_6. III, *J. Biol. Chem.*, **235**:1164, 1960.

Carlton, B. C.: Transformation mapping of the genes controlling tryptophan biosynthesis in *Bacillus subtilis, J. Bact.*, **94**:660, 1967.

Chaplin, C. E., and A. G. Lochhead: Abnormal morphology of a bacterium resulting from vitamin B_{12} deficiency, *Can. J. Microbiol.*, **2**:340, 1956.

Christensen, H. N.: "pH and Dissociation," Philadelphia, W. B. Saunders Company, 1964.

Clark, W. M.: "The Determination of Hydrogen Ions," Baltimore, The Williams & Wilkins Company, 1928.

—— and H. A. Lubs: The colorimetric determination of hydrogen-ion concentration, *J. Bact.*, **2**:1, 109, 191, 1917.

Clifton, C. E. Aging of *Pseudomonas aeruginosa, J. Bact.*, **94**:2077, 1967.

Cohen, B.: The measurement of pH, titratable acidity, and oxidation-reduction potentials. In Committee on Bacteriological Technic (Society of American Bacteriologists), "Manual of Microbiological Methods," New York, McGraw-Hill Book Company, 1957.

Cowie, D. B., et al.: Sulfur metabolism in *Escherichia coli*. I, *J. Bact.*, **60**:233, 1950.

Curcho, M. de la G.: Mutation of tryptophan independence in *Eberthella typhosa, J. Bact.*, **56**:374, 1948.

Daniels, H. J.: Inhibition of growth of *Pseudomonas denitrificans* by amino acids, *Can. J. Microbiol.*, **12**:1095, 1966.

Doy, C. H., and A. J. Pittard: Feedback control of tryptophan biosynthesis, *Nature*, **185**:941, 1960.

Duncan, D. W., Jr., and J. T. R. Nickerson: Effect of environmental and physiological conditions on the phase of adjustment of *Pseudomonas fragi*, *Appl. Microbiol.*, **11**:179, 1963.

Dyke, S. F.: "The Chemistry of the Vitamins," New York, Interscience Publishers, a division of John Wiley & Sons, Inc., 1965.

Eagon, R. G., and M. A. Asbell: Effect of divalent cations on the uptake and oxidation of substrates by *Pseudomonas aeruginosa*, *J. Bact.*, **97**:812, 1969.

Ecker, R. E., and W. R. Lockhart: Specific effect of limiting nutrient on physiological events during culture growth, *J. Bact.*, **82**:511, 1961.

—— and M. Schaechter: Bacterial growth under conditions of limited nutrition, *Ann. N.Y. Acad. Sci.*, **102**:549, 1963.

Eisler, W. J., Jr.: Automatic sample collector for continuous cultures, *Appl. Microbiol.*, **16**:1381, 1968.

—— and R. B. Webb: Electronically controlled continuous culture device, *Appl. Microbiol.*, **16**:1375, 1968.

Erlandson, A. L., Jr., and W. H. Mackey: Nutrition of *Shigella*: growth of *Shigella flexneri* in a simple, chemically defined medium, *J. Bact.*, **75**:253, 1958.

Fildes, P.: Some medical and other aspects of bacterial chemistry, *Proc. Roy. Soc. Med.*, **28**:79, 1934.

——: Production of tryptophan by *Salmonella typhi* and *Escherichia coli*, *J. Gen. Microbiol.*, **15**:636, 1956.

—— et al.: The nitrogen and vitamin requirements of *B. typhosus*, *Brit. J. Exp. Pathol.*, **14**:189, 1933.

Frieden, E.: The chemical elements of life, *Sci. Amer.*, **227**:52, 1972.

Friedman, M. E., and W. G. Roessler: Growth of *Listeria monocytogenes* in defined media, *J. Bact.*, **82**:528, 1961.

Galdiero, F., et al.: Modificazioni biochimiche in cellule di *Escherichia coli* in sviluppo ed in condizioni resting, *Riv. Ist. Sieroterap. Ital.*, **42**:130, 1967.

Garvie, E. I., et al.: The effect of asparagine on the growth of a gram-positive coccus, *J. Gen. Microbiol.*, **24**:25, 1961.

Gibbons, R. J., and J. B. Macdonald: Hemin and vitamin K compounds as required factors for the cultivation of certain strains of *Bacteroides melaninogenicus*, *J. Bact.*, **80**:164, 1960.

Gibson, F., and J. Pittard: Pathways of biosynthesis of aromatic amino acids and vitamins and their control in microorganisms, *Bact. Rev.*, **32**:465, 1968.

Gingrich, W., and F. Schlenk: Codehydrogenase I and other pyridinium compounds as V-factor for *Haemophilus influenzae* and *H. parainfluenzae*, *J. Bact.*, **47**:535, 1944.

Girard, A. E., et al.: A nutritional study of some agarolytic marine bacteria, *Can. J. Microbiol.*, **14**:1193, 1968.

Goldberg, M. E., et al.: Fluorescence studies of substrate and subunit interactions of the B_2 protein of *Escherichia coli* tryptophan synthetase, *Biochemistry*, **7**:3662, 1968.

Goodhue, C. T., and E. E. Snell: The bacterial degradation of pantothenic acid. I. Over-all nature of the reaction, *Biochemistry,* **5**:393, 1966.

Goodwin, T. W.: "The Biosynthesis of Vitamins and Related Compounds," New York, Academic Press, Inc., 1963.

Greenberg, L., et al.: The production of niacin by mycobacteria, *Can. J. Comp. Med. Vet. Sci.,* **24**:187, 1960.

Harkness, D. R., et al.: Bacterial degradation of riboflavin. V, *Arch. Biochem. Biophys.,* **108**:323, 1964.

Harrison, A. P., Jr.: The response of *Bacterium lactis aerogenes* when held at growth temperature in the absence of nutriment: An analysis of survival curves, *Proc. Roy. Soc., Ser. B.,* **152**:418, 1960.

Haynes, W. C., et al.: The effect of potassium upon the growth of *Micrococcus pyogenes, Appl. Microbiol.,* **2**:339, 1954.

Hedén, C.-G., and T. Holme: Studies on a simple multistage system for continuous culture under pulsating aeration, *Soc. Chem. Ind. Monograph,* no. 12, 1960.

Heinen, W.: Silicium-stoffwechsel bei mikroorganismen. IV, *Arch. Mikrobiol.,* **45**:162, 1963.

Hems, R., and H. A. Krebs: Further experiments on the potassium uptake by *Alcaligenes faecalis, Biochem. J.,* **82**:80, 1962.

Hishinuma, F., et al.: Effects of glycine and D-amino acids on growth of various microorganisms, *Agr. Biol. Chem.,* **33**:1577, 1969.

Hoffman, H., and M. E. Frank: Time-lapse photomicrography of cell growth and division in *Escherichia coli, J. Bact.,* **89**:212, 1965.

Hough, J. S., and I. S. Maddox: Yeast autolysis, *Process Biochem.,* May 1970.

Hunt, D. E., and R. F. Pittillo: The calculated nitrogen requirements of a single cell of *Escherichia coli, Can. J. Microbiol.,* **14**:480, 1968.

Johnson, G. T., and G. C. Kyker: Fission-product and cerium uptake by bacteria, yeasts and molds, *J. Bact.,* **81**:733, 1961.

Johnson, M. J.: Liquid-metering device for continuous culture, *Biotechnol. Bioengineer.,* **9**:630, 1967.

Johnson, R. M., and J. M. Larkin: An apparatus for continuous synchronization of growing cultures, *Can. J. Microbiol.,* **9**:907, 1963.

Kaneda, T.: Biosynthesis of branched chain fatty acids. I. Isolation and identification of fatty acids from *Bacillus subtilis* (ATCC 7059), *J. Biol. Chem.,* **238**:1222, 1963*a*: II. Microbial synthesis of branched long chain fatty acids from certain short chain fatty acid substrates, *ibid.,* **238**:1229, 1963*b*.

Kasai, G. J.: Growth response of microorganisms to vitamins at different temperatures, *J. Infectious Diseases,* **92**:58, 1953.

Kates, M., et al.: The lipid composition of *Micrococcus halodenitrificans* as influenced by salt concentration, *Can. J. Microbiol.,* **7**:427, 1961.

Kempner, E. S.: Trace metal analysis of nutrient broth, *Appl. Microbiol.,* **15**:1525, 1967.

Kennell, D., and A. Kotoulas: Magnesium starvation of *Aerobacter aerogenes.* I. Changes in nucleic acid composition, *J. Bact.,* **93**:334, 1967*a*; II. Rates of nucleic acid synthesis and methods for their measurement, *ibid.;* **93**:345; 1967*b*; IV. Cytochemical changes, *ibid.,* **93**:367, 1967*c*.

Kihara, H., et al.: Peptides and bacterial growth. X, *J. Biol. Chem.,* **236**:172, 1961.

Knight, B. C. J. G.: Essential metabolites and antimetabolites, *J. Mt. Sinai Hosp. N.Y.,* **15**:281, 1949.

Kono, T.: Kinetics of microbial cell growth, *Biotechnol. Bioengineer.,* **10**:105, 1968.

Koser, S. A.: "Vitamin Requirements of Bacteria and Yeasts," Springfield, Ill., Charles C Thomas, Publisher, 1968.

—— and G. R. Baird: Bacterial destruction of nicotinic acid, *J. Infectious Diseases,* **75**:250, 1944.

Larson, L. W., and W. P. Larson: Factors governing the fat content of bacteria and the influence of fat on pellicle formation, *J. Infectious Diseases,* **31**:407, 1922.

Lester, G.: Requirement for potassium by bacteria, *J. Bact.,* **75**:426, 1958.

Lockhart, W. R.: Measurement of generation lag, *Can. J. Microbiol.,* **6**:381, 1960.

Lusk, J. E., et al.: Magnesium and the growth of *Escherichia coli, J. Biol. Chem.,* **243**:2618, 1968.

MacLeod, R. A.: Further mineral requirements of *Streptococcus faecalis, J. Bact.,* **62**:337, 1951.

MacLeod, R. A., and E. E. Snell: The effect of related ions on the potassium requirement of lactic acid bacteria, *J. Biol. Chem.,* **176**:39, 1948.

Marchesi, S. L., and D. Kennell: Magnesium starvation of *Aerobacter aerogenes, J. Bact.,* **93**:357, 1967.

Marks, J.: "The Vitamins in Health and Disease," London, Churchill, 1968.

Mayer, G. D., and R. W. Traxler: Action of metal chelates on growth initiation of *Bacillus subtilis, J. Bact.,* **83**:1281, 1962.

Mitchell, C. A., and P. J. G. Plummer: A continuous method for the cultivation of microorganisms in liquid medium, *Can. J. Comp. Med. Vet. Sci.,* **18**:1, 1954.

Mitchison, J. M.: Enzyme synthesis in synchronous cultures, *Science,* **165**:657, 1969.

Morton, A. G., and A. Macmillan: The assimilation of nitrogen from ammonium salts and nitrate by fungi, *J. Exp. Botany,* **5**:232, 1954.

Ng, H.: Effect of decreasing growth temperature on cell yield of *Escherichia coli, J. Bact.,* **98**:232, 1969.

Nicholas, D. J. D., et al.: Cobalt requirement for inorganic nitrogen metabolism in microorganisms, *Proc. Natl. Acad. Sci.,* **48**:1537, 1962.

Nishi, A., et al.: Shift of cell-age distribution pattern in the later phases of *Escherichia coli* culture, *J. Gen. Microbiol.,* **13**:103, 1967.

Ortega, M. V., and G. M. Brown: Precursors of nicotinic acid in *Escherichia coli*, *J. Biol. Chem.*, **235**:2939, 1960.

Pai, C. H., and H. C. Lichstein: VI. Further evidence for desthiobiotin as a precursor in *Escherichia coli*, *J. Bact.*, **94**:1930, 1967.

Petropulos, S. F.: Automatic sampling device for study of synchronized cultures of microorganisms, *Science*, **145**:268, 1964.

Petty, M. A.: An introduction to the origin and biochemistry of microbial halometabolites, *Bact. Rev.*, **25**:111, 1961.

Pittillo, R. F., and J. W. Foster: Potentiation of inhibitor action through determination of reversing metabolites, *J. Bact.*, **67**:53, 1954.

Postgate, J. R., and J. R. Hunter: Accelerated death of *Aerobacter aerogenes* starved in the presence of growth-limiting substrates, *J. Gen. Microbiol.*, **34**:459, 1964.

Powell, E. O., and F. P. Errington: Generation times of individual bacteria: Some corroborative measurements, *J. Gen. Microbiol.*, **31**:315, 1963.

Quesnel, L. B.: The behaviour of individual organisms in the lag phase and the formation of small populations of *Escherichia coli*, *J. Appl. Bact.*, **23**:99, 1960.

Reeves, R. E.: Pantethine-requiring *Bacteroides*, *J. Bact.*, **85**:1197, 1963.

Reynolds, E. S., et al.: The metabolism of orotic acid in aerobic bacteria, *J. Bact.*, **69**:250, 1955.

Ridgeway, G. J., and H. C. Douglas: Distribution of inositol in subcellular fractions of yeast cells, *J. Bact.*, **75**:85, 1958*a*.

—— and ——: Unbalanced growth of yeast due to inositol deficiency, *J. Bact.*, **76**:163, 1958*b*.

Robinson, F. A.: "The Vitamin Co-factors of Enzyme Systems," New York, Pergamon Press, 1966.

Roblin, R. O., Jr.: Metabolite antagonists, *Chem. Eng. News*, **27**:3624, 1949.

Rochford, E. J., and R. J. Mandle: The production of chains of *Diplococcus pneumoniae* in magnesium deficient media, *J. Bact.*, **66**:554, 1953.

Rossmoore, H. W., and B. Goehler: Utilization of gelatin by *Bacillus* species, *Develop. Ind. Microbiol.*, **3**:150, 1963.

Schlessinger, D., and F. Ben-Hamida: Turnover of protein in *Escherichia coli* starving for nitrogen, *Biochim. Biophys. Acta*, **119**:171, 1966.

Sebrell, W. H., Jr., and R. S. Harris, eds.: "The Vitamins: Chemistry, Physiology, Pathology, Methods," New York, Academic Press, Inc., 1967.

Sen, R.: Purine requiring strains of shigellae, *J. Bact.*, **80**:585, 1960.

Sergeant, T. P., et al.: Initiation of growth of *Bacillus* species in a chemically defined medium, *J. Bact.*, **74**:728, 1957.

Shankar, K., and R. C. Bard: The effect of metallic ions on the growth and morphology of *Clostridium perfringens*, *J. Bact.*, **63**:279, 1952.

Shooter, R. A., and H. V. Wyatt: Mineral requirements for growth of *Staphylococcus pyogenes*: Effect of potassium ions, *Brit. J. Exp. Pathol.*, **37**:311, 1956.

Sinclair, N A., and J. L. Stokes: Factors which control maximal growth of bacteria, *J. Bact.,* **83**:1147, 1962.

Smith, E. L.: "Vitamin B_{12}," London, Methuen & Co., Ltd., 1965.

Spitznagel, J. K.: Macromolecule synthesis in long forms of *Mycobacterium bovis* induced by sulfate deficiency, *J. Bact.,* **81**:933, 1961.

Stephenson, M., and M. D. Whetham: Studies in the fat metabolism of the timothy grass bacillus, *Proc. Roy. Soc.* (*London*), *Ser. B,* **93**:262, 1922.

Stokes, J. L., et al.: Vitamin content of ingredients of microbiological culture media, *J. Bact.,* **47**:293, 1944.

Strange, R. E., et al.: The catabolism of proteins and nucleic acids in starved *Aerobacter aerogenes, Biochem. J.,* **86**:197, 1963.

Svihla, G., et al.: Ultraviolet micrography of penetration of exogenous cytochrome c into the yeast cell, *J. Bact.,* **100**:498, 1969.

Taub, F. B., and A. M. Dollar: Improvement of a continuous-culture apparatus for long-term use, *Appl. Microbiol.,* **16**:232, 1968.

Tsai, L., et al.: Bacterial degradation products of riboflavin. IV, *Biochem. Z.,* **338**:561, 1963.

Webb, M.: The influcnce of magnesium on cell division. I. The growth of *Clostridium welchii* in complex media deficient in magnesium, *J. Gen. Microbiol.,* **2**:275, 1948.

White, D. C., and S. Granick: Hemin biosynthesis in *Haemophilus, J. Bact.,* **85**:842, 1963.

Winder, F. G., and C. O'Hara: Effects of iron deficiency and of zinc deficiency on the composition of *Mycobacterium smegmatis, Biochem. J.,* **82**:98, 1962.

Woolley, D. W.: Antimetabolites, *Science,* **129**:615, 1959.

——: 3-Methylaspartic acid as a potent antimetabolite of aspartic acid in pyrimidine biosynthesis, *J. Biol. Chem.,* **235**:3238, 1960.

Wyatt, H. V.: The effect of alkali metals on the growth of *Staphylococcus pyogenes, Exp. Cell Res.,* **30**:56, 1963*a*.

——: The uptake of potassium and rubidium by *Staphylococcus pyogenes, Exp. Cell Res.,* **30**:62, 1963*b*.

Yang, H.-C., et al.: Studies on degradation of *d*-biotin by microorganisms. Part I. *d*-Biotin degradation by yeasts and molds, *Agr. Biol. Chem.* (*Tokyo*), **33**:1730, 1969.

Yanofsky, C.: The enzymatic conversion of anthranilic acid to indole, *J. Biol. Chem.,* **223**:171, 1956.

Sterilization 8

Sterilization may be defined as the complete destruction of all living organisms by physical or chemical agents. The term may also include the removal of organisms from liquids by means of filtration.

Plugged glassware and petri dishes must be sterilized before use to destroy all living organisms adhering to the inner surfaces. Pipettes are placed in containers and heated to sterilize both inner and outer surfaces. Likewise, culture media must be sterilized prior to use to destroy all contaminating organisms present. Studies on pure cultures could not be made if the glassware and culture media were contaminated before use. When once sterilized, glassware may be kept in a sterile condition indefinitely if protected from outside contamination. The same applies to culture media if, in addition, evaporation can be prevented.

The usual methods employed for the sterilization of laboratory materials involve the use of dry or moist heat. Three types of sterilizers are generally used in bacteriology: (1) the hot-air sterilizer; (2) the Arnold sterilizer; and (3) the autoclave.

Hot-air sterilizer This is a dry-air type of sterilizer (Fig. 148). It is constructed with three walls and two air spaces. The outer walls are covered with thick asbestos to reduce the radiation of heat. A burner manifold runs along both sides and rear between the outside and intermediate walls. Convection currents travel a complete circuit through the wall space and interior of the oven, and the products of combustion escape through an opening in the top.

The hot-air sterilizer is operated at a temperature of 160 to 180°C (320 to 356°F) for 1 to 1 1/2 hr. If the temperature goes above 180°C, there will be danger of the cotton stoppers charring. The above temperature may be increased for glassware, such as petri dishes in metal containers where cotton stoppers are not involved. This will shorten the sterilization period resulting in a saving of time.

The hot-air sterilizer is used for sterilizing all kinds of glassware such as test tubes, pipettes, petri dishes, and flasks. In addition, it may be used to treat other laboratory materials and equipment, which are not burned or injured by the high temperature of the sterilizer. Under no conditions

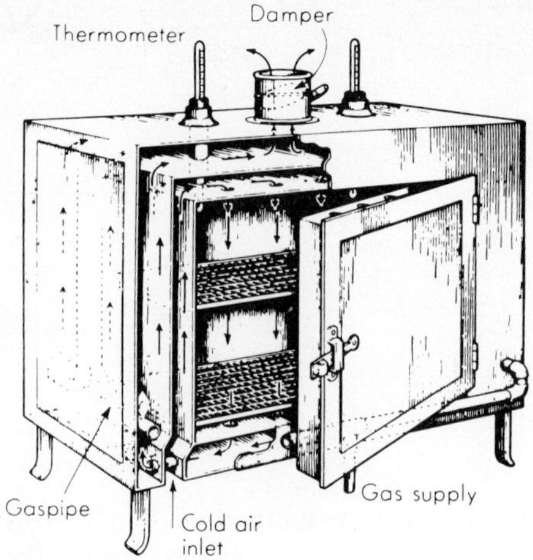

Fig. 148 Hot-air sterilizer. (From Belding and Marston, "A Textbook of Medical Bacteriology," Appleton-Century-Crofts, Inc.)

must the sterilizer be used to sterilize rubber goods and laboratory coats, as such materials would burn. The same applies to culture media, as the liquids would boil to dryness.

Arnold sterilizer It is well known that moist heat is more effective as a sterilizing agent than dry heat. This is believed to be true for two reasons: (1) moist heat has greater penetrating power; and (2) death is believed to result from the coagulation of the proteins of the protoplasm. An increase in the water content of the protoplasm decreases the temperature required to coagulate the proteins.

The Arnold (Fig. 149) is built with a quick-steaming base that is automatically supplied with water from an open reservoir. The water passes through small apertures into the steaming base to which the heat is applied. Since the base contains only a thin layer of water, steam is produced very rapidly. The steam rises through a funnel in the center of the equipment and passes into the sterilizing chamber.

The sterilizer employs streaming steam at a temperature of approximately 100°C (212°F) for a period of 20 min or more on three consecutive days. The length of the exposure period will depend on the nature of the materials being treated and the size of the containers. Agar, for example, must be completely melted before recording the beginning of the exposure period.

A temperature of 100°C for 20 min is not sufficient to destroy spores. A much higher temperature is required to effect a complete sterilization in one operation over a relatively short exposure period. However, sterilization in the Arnold may be realized by employing the intermittent method. The first exposure period kills all vegetative cells

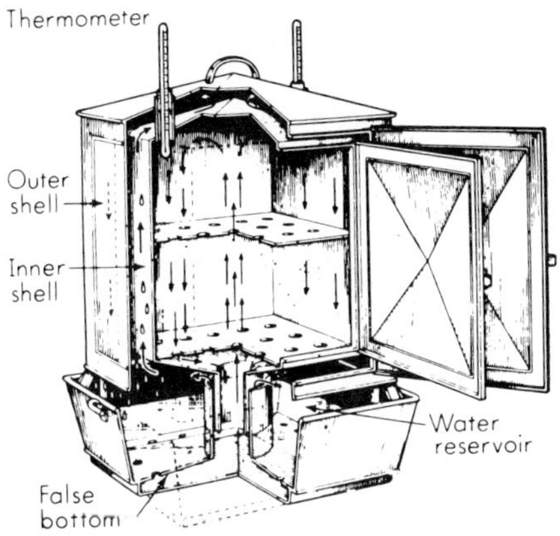

Fig. 149 *Arnold sterilizer.* (From Belding and Marston, "A Textbook of Medical Bacteriology," Appleton-Century-Crofts, Inc.)

present. Then the materials are kept at a warm temperature for 24 hr. During this holding period, any spores present should germinate into vegetative cells. The second heating period again will destroy all vegetative cells. It sometimes happens that all spores do not germinate into vegetative cells before the second exposure period. Therefore, an additional 24-hr period is allowed to elapse to make sure all spores have germinated into vegetative cells.

It may be seen that unless the spores germinate, the method will fail to sterilize. Failure may be due to the following: (1) The medium may be unsuited for the germination of the spores. Distilled water, for example, is not a favorable environment for the growth of bacteria and will not permit spores to germinate into vegetative cells. (2) Spores of anaerobic bacteria, if present, will not germinate in a medium in contact with air.

The Arnold is used principally for the sterilization of gelatin, milk, and carbohydrate media. Higher temperatures in the autoclave, or longer single exposures in the Arnold, may hydrolyze or decompose carbohydrates and prevent gelatin from solidifying. Obviously, such media would then be unsuited for use.

Autoclave The autoclave is a cylindrical metal vessel having double walls except at the opening (Fig. 150). It is built to withstand a steam pressure of at least 60 lb per sq in.

The principle of the autoclave is as follows; Water boils at about 100°C, depending upon the vapor pressure of the atmosphere. If the vapor pressure is increased, the temperature will be increased. If the steam pressure inside the autoclave is increased to 15 lb per sq in (2 atm absolute), the temperature will rise to 121°C. The relationship between pressure and temperature is shown in Table 5.

Table 5 Relationship between pressure and temperature

Pressure, lb per sq in	Corresponding temperature	
	°C	°F
5	109	228
10	115	240
15	121	250
20	126	259
25	130	267
30	135	275

The autoclave is operated usually at 15 lb steam pressure (121°C) for a period of 15 min. This temperature is sufficient to destroy vegetative cells and spores in one operation.

Certain precautions must be observed to prevent sterilization failures. The most important single cause is incomplete evacuation of air from the chamber. Observation of the pressure gauge alone is not sufficient. The proper degree of temperature must also be taken into consideration. The temperature figures given in Table 5 are true only if all air is evacuated from the sterilizing chamber.

The temperature of a mixture of steam and air at a given pressure is less than that of pure steam alone. This means that although the autoclave is kept at the desired pressure, the temperature may not be sufficient to

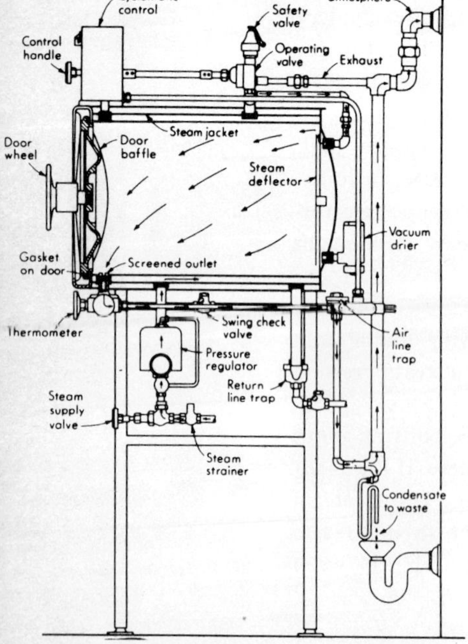

Fig. 150 Cross section of autoclave. Arrows indicate flow of steam from source through various operating parts of sterilizer. (Courtesy of American Sterilizer Company.)

Table 6 Temperature with various degrees of air discharge

Gauge pressure, lb	Pure steam, complete air discharge		Two-thirds air discharge, 20-in vacuum		One-half air discharge, 15-in vacuum		One-third air discharge, 10-in vacuum		No air discharge	
	°C	°F	°C	°F	°C	°F	°C	°F	°C	°F
5	109	228	100	212	94	202	90	193	72	162
10	115	240	109	228	105	220	100	212	90	193
15	121	250	115	240	112	234	109	228	100	212
20	126	259	121	250	118	245	115	240	109	228
25	130	267	126	259	124	254	121	250	115	240
30	135	275	130	267	128	263	126	259	121	250

give complete sterilization. The actual temperatures attained in the autoclave under ordinary conditions of proper and improper usage are given in Table 6.

Another important precaution to be observed is that the steam must have access to the materials to be treated. If the steam is prevented from penetrating the materials, the method will be of doubtful value. For example, suppose it is desired to sterilize some cotton contained in a bottle. If the bottle is closed with a rubber stopper, the steam cannot enter. The autoclave will be no more effective than a hot-air sterilizer kept at 121°C for 15 min. It has already been shown that treatment at such a temperature and for such a time period is not sufficient to destroy spores in a dry-air sterilizer. On the other hand, if the mouth of the bottle is covered with one or more thicknesses of muslin, permitting free access of steam, the cotton should be sterilized.

The autoclave is used to sterilize anything that is not injured by steam and high temperatures. This includes most types of solid and liquid media with and without carbohydrate, gelatin media, distilled water, saline solutions, rubber tubing and stoppers, discarded cultures and contaminated media prior to washing, laboratory coats and aprons, etc. This is the type of sterilizer commercially used for processing canned foods.

For more information: Ernst (1968); Perkins (1963); Wilson (1968).

Sterilization by filtration

Some solutions cannot be sterilized by heat without being greatly altered in their physical and chemical properties. Serum in culture media is easily coagulated by heat. If the serum content is high enough, the medium becomes changed from a liquid to a solid preparation. Certain physiological salt solutions containing the unstable compound sodium bicarbonate are ruined if heated. The bicarbonate easily loses carbon dioxide and is converted into the more alkaline sodium carbonate. Enzymes and bacterial toxins in solution are easily destroyed by heat. These are only a few examples.

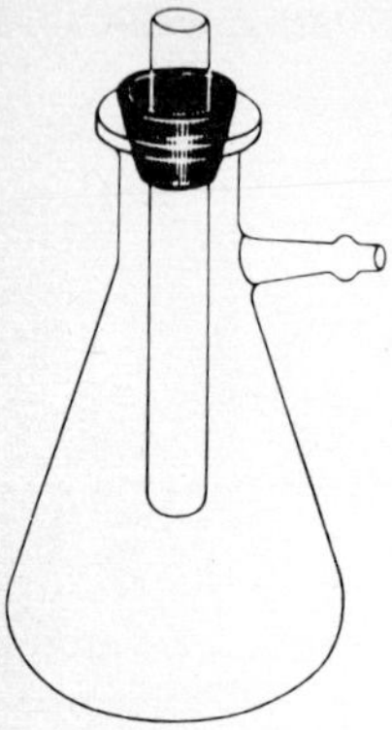

Fig. 151 Porcelain filter assembled in a filter flask and ready for filtration.

Solutions containing heat-sensitive compounds are best sterilized by filtration. The types of filters employed for this purpose include porcelain filters, diatomaceous earth filters, fritted-glass filters, asbestos filters, collodion membranes or ultrafilters, Polypore filters, Millipore filters, and Triacetate Metricel, Type GA filters.

Porcelain filters Filters of this type are hollow cylinders and unglazed except for a short distance at the open end. They are composed of hydrous aluminum silicate or kaolin and are baked at a temperature as high as possible without sintering the clay. The filters are prepared in five graduated degrees of porosity (no. 1 porosity, from 13.5 to 15 μ; no. 5, from 0.65 to 0.8 μ; with gradations between these two extremes). Cylinders numbered 1 and 2 are prefilters for the removal of coarse particles; the others are for the removal of bacteria. Porcelain filters may be purchased in varying lengths and diameters.

The filters arc assembled as shown in Fig. 151. The rubber stopper fits over the glazed portion of the filter.

Another filter of the porcelain type is known as the Jenkins filter (Fig. 152). It consists of a metal mantle holding a soft rubber sleeve and a porcelain filter block. The porcelain block is held in the rubber sleeve and made watertight by screwing together two metal parts. The mantle is fitted with a rubber stopper, wrapped in paper, and sterilized in an autoclave. For use, the filter is removed from the paper wrapping, attached to a filter flask, and the liquid is drawn through by vacuum. After each use, the filter block is washed, dried, and reassembled in the mantle.

The Jenkins filter is designed for the filtration of small quantities of liquids.

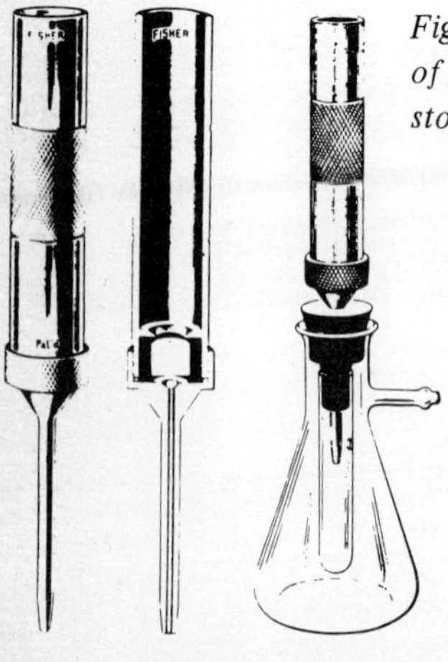

Fig. 152 Jenkins filter. Left, completely assembled; center, cross section of filter; right, filter assembled and attached to a filter flask by a rubber stopper. Small tube inside of flask is used to collect the filtrate.

Diatomaceous earth filters Diatomaceous earth is a fine, usually white, siliceous powder composed chiefly or wholly of the remains of diatoms. It is also called infusorial earth and kieselguhr.

Filters of this type were first manufactured in Germany and are known as Berkefeld filters. They are prepared by mixing carefully purified diatomaceous earth with asbestos and organic matter, pressing into cylinder form, and drying. The dried cylinders are heated in an oven to about 2000°C to bind the materials together. Then the cylinders are machined to the desired sizes and each filter fitted with a metal ferrule and nipple to the open end.

The cylinders are graded as W (dense), N (normal), and V (coarse), depending upon the sizes of the pores. The grading depends upon the rate of flow of pure filtered water under a constant pressure.

Filters of a similar type prepared in the United States are known as Mandler filters. They are composed of 60 to 80 percent diatomaceous earth, 10 to 30 percent asbestos, and 10 to 15 percent plaster of paris. The proportions vary depending upon the sizes of pores desired. The ingredients are mixed with water, subjected to high pressure, then baked in ovens at 980 to 1650°C to bind the materials together.

The finished cylinders are tested by connecting a rubber tube to the nipple of the filter, submerging in water, and passing compressed air to the inside. A gauge records the pressure when air bubbles first appear on the outside of the cylinder in the water. Each cylinder is marked with the air pressure obtained in actual test.

The filters are assembled in glass mantles as shown in Fig. 153. The filter is held in place by a rubber washer and screw nut, and provided with a rubber stopper to fit a filter flask. The assembled filter is wrapped in

Fig. 153 Center, Mandler filter; right, glass mantle for use with either Mandler or Berkefeld filter; left, cross section of filter assembled in a mantle.

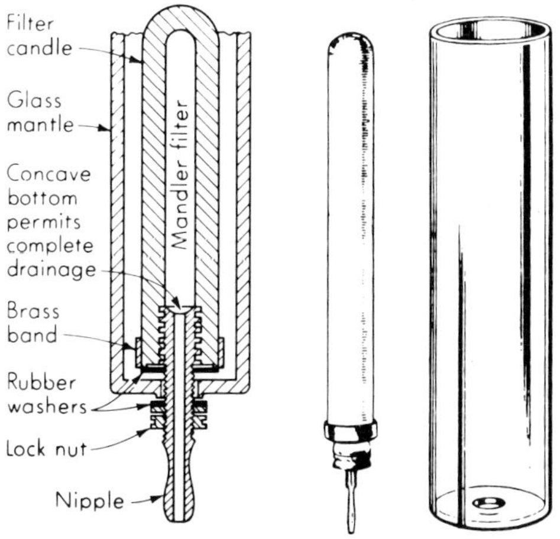

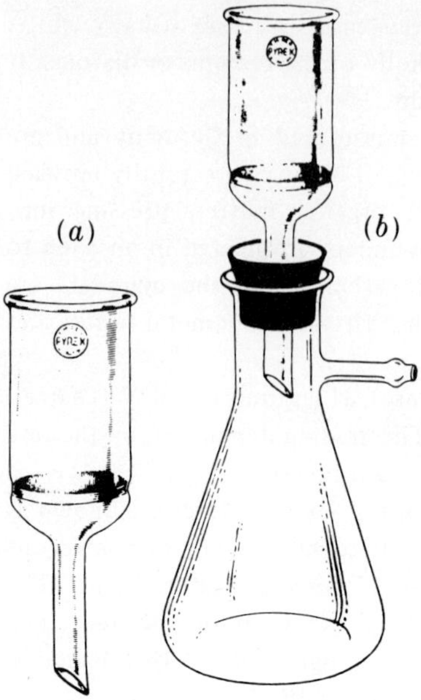

Fig. 154 (a) Fritted-glass filter; (b) filter assembled in a filter flask and ready for filtration.

paper and autoclaved. The flask opening and tubulation are both plugged with cotton and sterilized in the autoclave or dry-air sterilizer. For use, the assembled filter is removed from the paper wrapping and attached aseptically to the flask. The liquid to be filtered is poured into the mantle where it passes through the filter by vacuum applied to the side tubulation of the flask. After completion of filtration, the vacuum is released, the filter is disengaged from the flask, and the filtrate is transferred aseptically to a sterile container.

Fritted-glass filters Filters of this type are prepared by molding finely pulverized glass in disk form, and heating just high enough to cause the particles to become a coherent solid mass, without destroying the porosity of the disk. Then the disk is carefully fused in a glass funnel and the whole attached to a filter flask by a rubber stopper (Fig. 154). A better arrangement is the coupling of the filter to the flask through a ground-glass joint (Fig. 155).

Fritted-glass filters are marketed in five degrees of porosity: EC (extra coarse), C (coarse), M (medium), F (fine), and UF (ultrafine).

Filters are employed generally under conditions of reduced pressure. Bush (1946) recommended filtration through glass disks by the use of

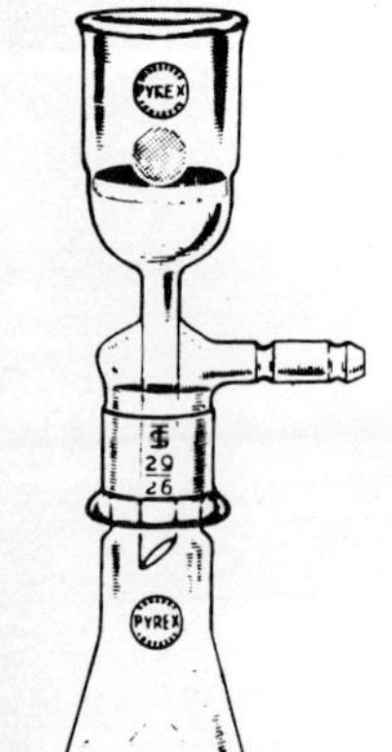

Fig. 155 Fritted-glass filter coupled to a flask through a ground-glass joint. (Courtesy of Corning Glass Works.)

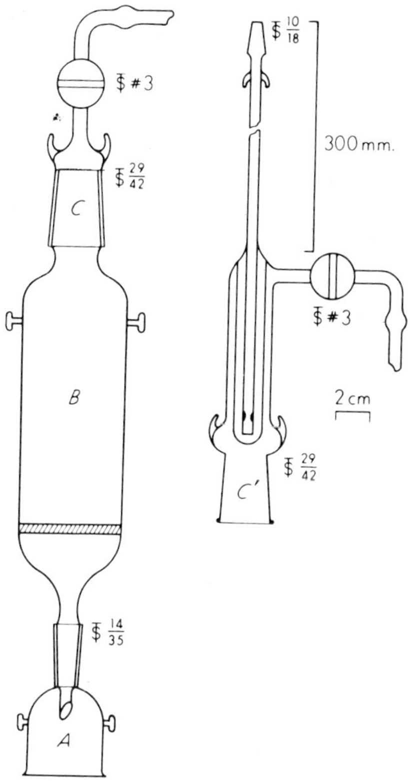

Fig. 156 Arrangement of bacteria filter for positive-pressure filtration. (After Bush.)

positive pressure because it reduces or eliminates evaporation of the filtrate and facilitates the interchange of receivers. This is particularly important in filtration which must be handled aseptically.

A convenient arrangement is shown in Fig. 156. The main body of the filter *B* contains a fritted-glass disk. A shield *A* protects the receiver from dust, and a pressure head *C* carries a stopcock. An alternate pressure head *C'* contains a built-in mercury manometer. The stopcock on *C* (*C'*) permits the retention of pressure after the apparatus is detached from the source of compressed air. An ordinary rubber pressure bulb is satisfactory for producing pressures up to at least 450 mm mercury. If the ground-glass joints are well lubricated and the parts held together with strong rubber bands or springs, the apparatus should hold this pressure for days.

Asbestos filters The best known asbestos filter is the Seitz filter (Fig. 157).

Purified asbestos is pressed together in disk form, and clamped tightly between two smooth metal rims by screw clamps. The liquid to be filtered is poured into the funnel-like structure and drawn through the asbestos disk by vacuum. The disks are capable of effectively retaining

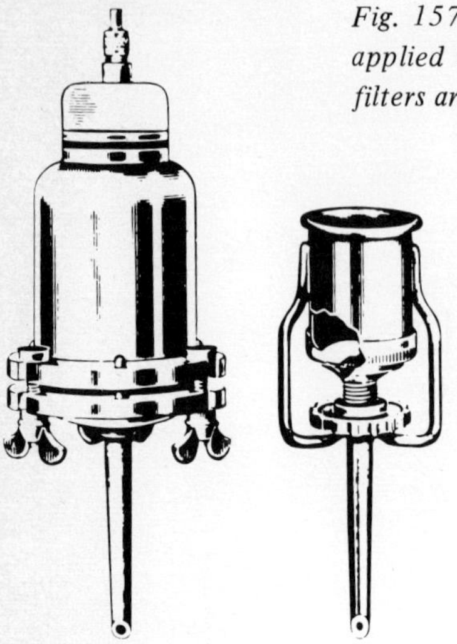

Fig. 157 Seitz filter. Left, filter holder for pressure filtration. Pressure is applied to upper tubulation. Right, filter holder for vacuum filtration. The filters are assembled as shown in Fig. 154.

bacteria and other suspended matter. After completion of filtration, the disk is removed from the holder, a new one inserted, and the assembled filter resterilized. Since preliminary cleaning is not necessary, the Seitz filter is very convenient to use.

A modification of the Seitz filter, utilizing centrifugal force instead of suction or pressure, has been recommended by Boerner (Fig. 158*B*). The filter consists of a cylinder and a funnel-shaped part with stem, which holds the filter pad supported on a wire gauze disk. The cylinder screws into the funnel with the filter disk pressed between them. The assembled filter fits closely into the top of a 15-ml metal centrifuge tube, with the knurled collar of the funnel portion resting on the top of the metal tube.

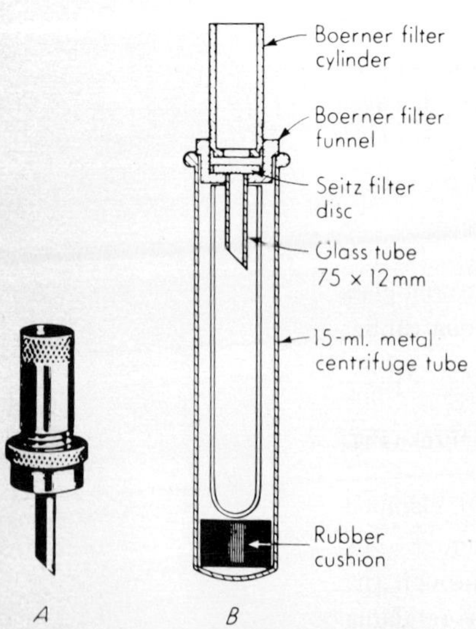

Fig. 158 Left, Boerner centrifugal filter; right, Boerner filter assembled in a 15-ml metal centrifuge tube with glass collecting tube inside.

The filtrate is collected in a glass tube inside the cup. The filter can also be used for vacuum filtration in the conventional manner by inserting the stem through a rubber stopper fitted to a filter flask (Fig. 158*A*).

For more information on the Seitz filter disks: Webb (1946).

Ultrafilters Ultrafiltration may be defined as the separation of colloidal particles from their solvents and from crystalloids by means of special dense filters known as ultrafilters.

The early ultrafilters were composed of gelatin and of silicic acid, but these have been replaced by collodion in membrane or sac form, or collodion deposited in a porous supporting structure. This may be filter paper in sheet or thimble form, unglazed porcelain, dishes and crucibles, Büchner funnels, filter cylinders, etc.

Collodion Several kinds of collodion are employed for ultrafiltration. The earliest kind was prepared by dissolving pyroxylin or soluble guncotton in a mixture of 1 part of alcohol and 3 parts of ether. A more popular type is prepared by dissolving pyroxylin in glacial acetic acid. Pore size may be controlled by increasing or decreasing the pyroxylin content, or by adding various liquids such as glycerol or ethylene glycol to alcohol-ether collodion. Elford (1931) recommended a type of collodion for the preparation of a graded series of filters which he called *Gradocol* membranes. The filters are prepared by incorporating a definite amount of amyl alcohol with an alcohol-ether collodion, then adding graded amounts of water or acetic acid to control pore size. Since membranes prepared in this manner are quite strong, it is not necessary to deposit the collodion in a porous supporting structure.

Sacs may be prepared by pouring the collodion into a test tube or beaker, inverting and twirling continuously until a thin even coating remains. The tube or beaker is then plunged into cold water to gel the collodion. After the solvent has been washed away, the sac may be loosened from the glass and removed. Collodion membranes in sheet form may be prepared by cutting the sacs at right angles at the closed ends, and then lengthwise.

Membranes employing filter paper as the porous supporting structure are usually prepared with acetic collodion. Pore size depends upon the strength of the collodion. A strong collodion gives finer pores than a weak one. It is a simple matter to prepare a graded series of membranes. The paper foundation gives the membranes a strength capable of withstanding a pressure as high as 20 atm or even higher.

Collodion filters are used for the isolation of viruses and, before the invention of the electron microscope, were extensively employed for their measurement. Such measurements proved to be largely inaccurate because at that time it was assumed that all virus particles were round. Now it is known that they assume other shapes.

A type of filter known commercially as cellulose dialyzer tubing may be purchased in varying diameters and wall thicknesses; it functions similarly to the prepared membranes. The material is marketed in rolls,

Fig. 159 Structure of a Polypore filter reconstructed from a photomicrograph. Structure resembles a kitchen sponge. (Courtesy of Gelman Instrument Company, Chelsea, Mich.)

and pieces of the desired length may be cut as needed. The lower open end may be closed by tying tightly with string. The filter has an average pore diameter of 4.8 nm which is small enough to hold back bacteria, large protein molecules, and practically all viruses which have been isolated and studied. The membrane is stronger than the prepared type and does not require a porous supporting structure.

For more information: Polson and Madsen (1953).

Polypore filters Polypore is a secret material composed of a microporous cellulose ester.[1] It is noncontaminating, chemically inert, and will not interact with the solution being filtered.

The filters are available in a range of pore sizes, from 4 to 100 nm (0.1 μ). Particles above the filter's rated pore size are trapped on the surface, and even smaller particles are impinged on the interlocked internal structure of the membrane.

Polypore filters look like a kitchen sponge when the structure is reconstructed from photomicrographs (Fig. 159). Its filtering action is very much like a sieve at the surface but it has depth of filtering capacity. Polypore is consistent in pore characteristics, strong, flexible, and easy to handle.

Polypore filters are assembled as shown in Fig. 160. At the left, a filter is inserted into the metal funnel base and supported by a fritted stainless-steel disk. The top is screwed into the base to hold the filter in place and form a leaktight seal. The complete filter is attached to a sterile

[1] Manufactured by Gelman Instrument Company, P.O. Box 1448, Ann Arbor, Mich. 48106.

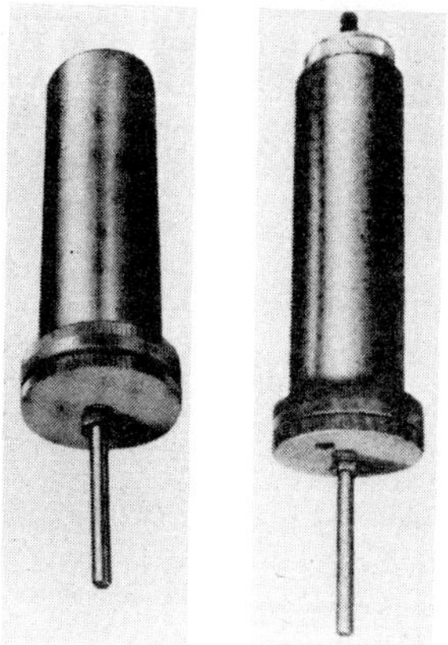

Fig. 160 Polypore filter. Left, assembly for vacuum filtration; right, assembly for pressure filtration. Pressure is applied to upper tubulation. Filters are assembled as shown in Fig. 154. (Courtesy of Gelman Instrument Company, Chelsea, Mich.)

filter flask, and the liquid drawn through by vacuum. At the right, the filter holder is the same except that filtration is produced by pressure applied to the upper tubulation. Pressure can increase the speed of filtration 5 to 10 times over that obtainable by vacuum. After each use the membrane is removed from the holder and replaced with a new one.

Another filter by the same manufacturer, and used for miniaturized filtration, is known as the *Triacetate Metricel*, Type GA Membrane. This membrane is assembled in a holder fastened to a hypodermic syringe (Fig. 161) and recommended for the processing of 1 to 5 ml quantities of liquids. Since the liquid is forced through the membrane by pressure, no vacuum is necessary. The membranes measure 13 mm in diameter and are available in several pore sizes, ranging from 5 to 0.2 μ.

The filter holder with membrane is attached to a syringe and the complete unit sterilized in the autoclave.

Millipore filters Filters of this type also are composed of pure and biologically inert cellulose esters.[2] They are prepared in thin, porous, circular membranes of about 150 μ in thickness, and contain millions of capillary pores of uniform dimension per square centimeter of filter surface.

Millipore filters (MF) are availabe in 12 different porosity grades ranging from 10 nm to 8 μ. The pore sizes of the various types of filters,

[2] Manufactured by Millipore Filter Corporation, Bedford, Mass. 01730.

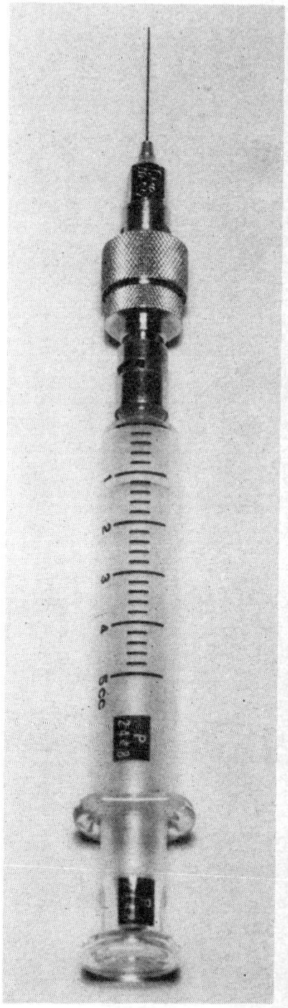

Fig. 161 Chrome-plated Swinny syringe filter holder for filtering very small volumes of fluids. (Courtesy of Gelman Instrument Company, Chelsea, Mich.)

Table 7 Pore sizes and flow rates of Millipore filters

MF Type	Pore size	Rate of flow Water*	Air†
SC	8.0 μ ± 1.4 μ	850	55
SM	5.0 μ ± 1.2 μ	540	35
SS	3.0 μ ± 0.9 μ	400	20
RA	1.2 μ ± 0.3 μ	300	15
AA	0.80 μ ± 0.05 μ	212	11
DA	0.65 μ ± 0.03 μ	150	10
HA	0.45 μ ± 0.02 μ	64	4.5
PH	0.30 μ ± 0.02 μ	40	3.7
GS	0.22 μ ± 0.02 μ	21	2.5
VC	100 nm ± 8 nm	2.0	0.49
VM	50 nm ± 3 nm	1.0	0.31
VF	10 nm ± 2 nm	0.2	0.22

*Water flow rates are given in approximate milliliters per minute per square centimeter of filter area at 25°C (77°F) with a pressure differential of 70 centimeters of mercury (13.5 psi).

†Air flow rates are given in approximate liters per minute per square centimeter of filter area at 25°C (77°F) with a pressure differential of 70 centimeters of mercury (13.5 psi).

together with rates of flow of water and air, are listed in Table 7. A comparison of pore sizes of some typical Millipore filters with size references is given in Table 8. The sizes of the pores are controlled to an extraordinary degree of precision.

The two most popular porosity grades for bacteriological sterility applications are:

Type HA with a pore size of 0.45 μ ± 0.02 μ.
Type GS with a pore size of 0.22 μ ± 0.02 μ.

The pores which pass through the MF occupy 80 to 85 percent of the total filter volume while the cellulose matrix which defines the pores occupies only 15 to 20 percent. This unique porosity characteristic permits extremely high flow rates for liquids.

The pore sizes of some typical MFs suitable for filtration of various living organisms and nonliving particles are listed in Table 8.

The MF is not attacked by water, dilute acids and alkalies, aliphatic and aromatic hydrocarbons, halogenated hydrocarbons, or nonpolar liquids. It is soluble in ketones, esters, ether alcohols, and nitroparaffins.

The MF is assembled as shown in Fig. 277. The assembled filter is wrapped in paper and sterilized in the autoclave. For use, the filter is removed from the wrapping and attached aseptically to a sterile filter

Table 8 A comparison of pore sizes of some typical Millipore filters with size references

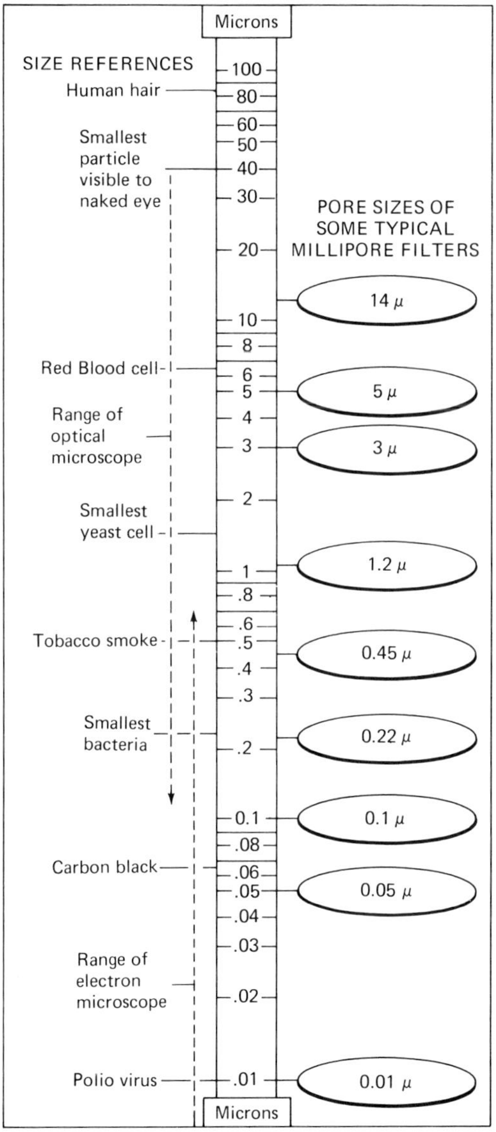

flask. Each membrane is used once, then discarded and replaced with a new one.

The filters find wide use in water examinations (see page 697).

For more information: Cliver (1968); Ver et al. (1968).

Electrical charge of filters The filtration of solutions containing enzymes, toxins, antibodies, viruses, etc., may result in a loss of active material. If the active material is present in low concentrations, it may be completely removed from the solution.

Filters composed of porcelain, diatomaceous earth, fritted glass, and asbestos consist chiefly of metal silicates and carry negative electrical charges.

The metal cations (Mg^{2+}, Al^{3+}, Ca^{2+}), or positive ions, are more soluble than the silicate anions, or negative ions, and show a greater tendency to pass into solution. When a liquid containing one of the above agents is passed through a silicate filter, positively charged particles react with negative silicate ions, and negatively charged particles react with positive metal ions. Since the metal ions are soluble, they react with negatively charged particles and pass through the filter into the filtrate. On the other hand, the insoluble silicate ions react with the positively charged particles and remain fixed to the walls of the filter.

Adsorption of compounds from solution can be demonstrated by the use of basic and acidic stains such as toluidine blue and picric acid, respectively.

Toluidine blue contains thiazine as the chromophore, and it ionizes as follows:

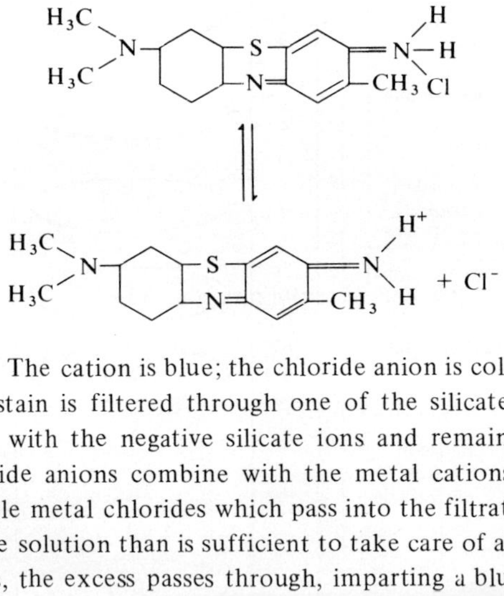

The cation is blue; the chloride anion is colorless. When a solution of this stain is filtered through one of the silicate filters, the blue cations react with the negative silicate ions and remain fixed to the filter. The chloride anions combine with the metal cations of the silicates to form soluble metal chlorides which pass into the filtrate. If more stain is present in the solution than is sufficient to take care of all silicate ions of the filter pores, the excess passes through, imparting a blue color to the filtrate. On the other hand, if the amount of stain is insufficient to take care of all silicate ions, the toluidine blue is completely removed from solution and the filtrate is colorless. The reaction is reversible, however, since the passage of distilled water through the filter saturated with stain gives a blue-colored filtrate.

Picric acid or trinitrophenol is an acid stain and it ionizes as follows:

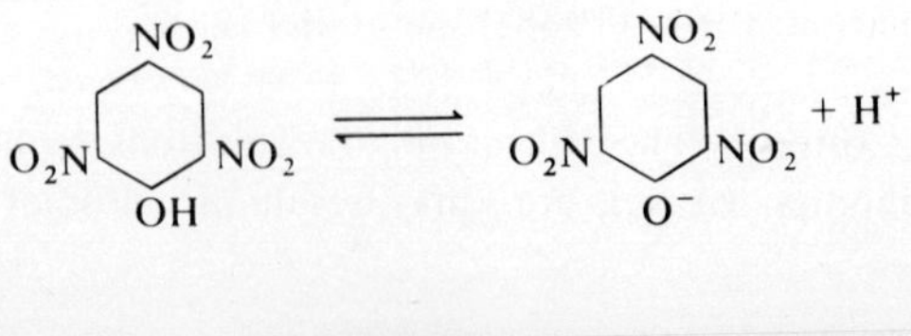

In this case, the cation is colorless and the anion is yellow. When a solution of this stain is filtered through a silicate filter, not a trace of it is adsorbed because an exchange of ions results in the formation of soluble picrates, which pass through in the filtrate.

Michaelis (1925) found collodion membranes to be nonionogenic but that they also carried a negative charge. He believed the negative charge was the result of the adsorption of negative ions. Elford (1933) reported that proteins in solution adjusted to different pH values with HCl and NaOH were most strongly adsorbed in the isoelectric zone (see below). On the other hand, proteins in solution buffered with $M/15$ phosphate were adsorbed on the acid side of the isoelectric zone. The negatively charged collodion now preferentially adsorbed the positively charged proteins. Elford concluded that the effect probably was associated with some specific influence of the phosphate ion.

Amphoteric nature of proteins and amino acids   An important characteristic of proteins and amino acids is that they contain both acidic (COOH) and basic (NH_2) groups. In acidic solutions, the compounds act as bases; in basic solutions, they act as acids. Representing the formula of an amino acid as $R \cdot CHNH_2 \cdot COOH$, the reactions with acids and bases are as follows:

With an acid

$$\underset{\underset{\text{COOH}}{|}}{\overset{\overset{\text{R}}{|}}{H-C-NH_2}} + HCl \longrightarrow \underset{\underset{\text{COOH}}{|}}{\overset{\overset{\text{R}}{|}}{H-C-NH_3Cl}}$$

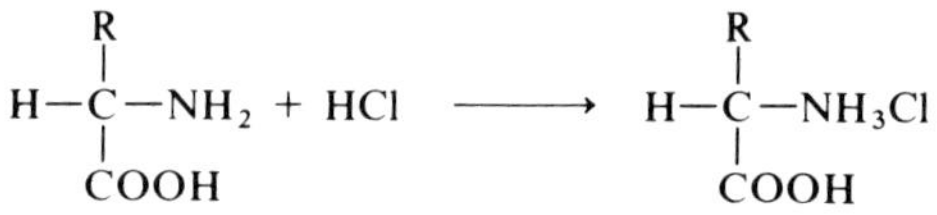

On ionization, this gives

$$\underset{\underset{\text{COOH}}{|}}{\overset{\overset{\text{R}}{|}}{H-C-NH_3Cl}} \rightleftharpoons \underset{\underset{\text{COOH}}{|}}{\overset{\overset{\text{R}}{|}}{H-C-NH_3^+}} + Cl^-$$

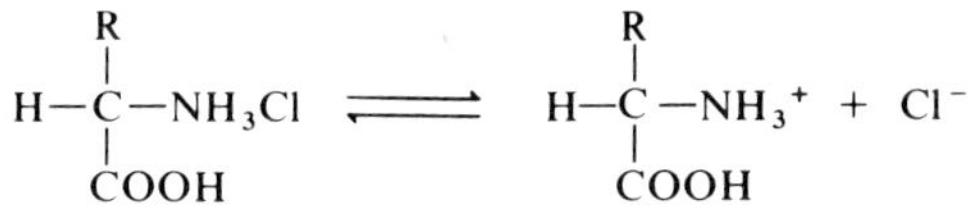

The acid reacts with the basic amino group. The amino acid molecule has a positive charge and, therefore, behaves as a base.

With a base,

$$\underset{\underset{\text{COOH}}{|}}{\overset{\overset{\text{R}}{|}}{H-C-NH_2}} + NaOH \longrightarrow \underset{\underset{\text{COONa}}{|}}{\overset{\overset{\text{R}}{|}}{H-C-NH_2}} + H_2O$$

On ionization, this gives

$$\underset{\underset{\text{COONa}}{|}}{\overset{\overset{\text{R}}{|}}{H-C-NH_2}} \rightleftharpoons \underset{\underset{\text{COO}^-}{|}}{\overset{\overset{\text{R}}{|}}{H-C-NH_2}} + Na^+$$

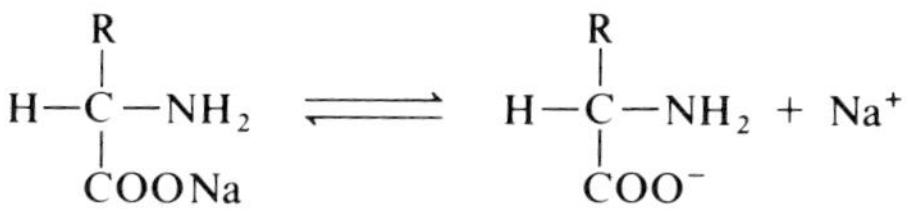

The base reacts with the acidic carboxyl group. The amino acid molecule now has a negative charge, and, therefore, behaves as an acid. Compounds of this nature which are capable of reacting with both acids and bases are all said to be amphoteric.

Isoelectric point   According to the zwitterion hypothesis, the isoelectric point is defined as that pH where ionization of an amphoteric compound is at a maximum:

$$
\begin{array}{c}
CH_3 \\
| \\
H-C-NH_3{}^+ \\
| \\
COO^-
\end{array}
$$

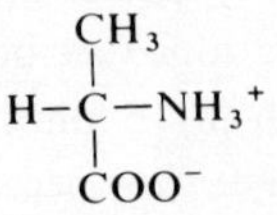

The molecule is neutral and is assumed to be due to complete ionization of the acidic and basic groups. The zwitterion reacts with acids and bases as follows:

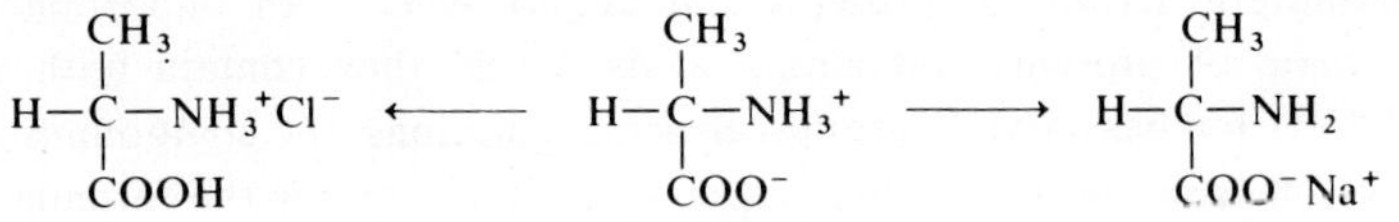

$$
\begin{array}{ccccc}
CH_3 & & CH_3 & & CH_3 \\
| & & | & & | \\
H-C-NH_3{}^+Cl^- & \longleftarrow & H-C-NH_3{}^+ & \longrightarrow & H-C-NH_2 \\
| & & | & & | \\
COOH & & COO^- & & COO^-\,Na^+
\end{array}
$$
$$\text{Zwitterion}$$

The isoelectric point of a protein is not necessarily the neutral point (pH 7.0). As a matter of fact, most proteins which have been studied have isoelectric points on the acid side of neutrality. The isoelectric points of a few of the common proteins are as follows:

Common proteins	Isoelectric point (pH)
Casein (milk protein)	4.7
Egg albumin	4.6
Gelatin	4.7
Hemoglobin	6.8
Serum albumin	4.8
Serum globulin	5.6

A knowledge of the isoelectric points is of considerable value in the filtration of solutions containing proteins, amino acids, bacterial toxins, enzymes, viruses, antitoxins, etc. If a solution is acid with respect to the isoelectric point, the active constituent will behave as a base and possess a positive electrical charge. The filtration of such a solution through a silicate filter, which has a negative charge, will result in the complete or partial adsorption of the active constituent on the walls of the filter pores. To avoid this, it would be necessary to use a filter having a positive charge, or to change the reaction of the active constituent to the alkaline side of its isoelectric point.

The adjustment of a solution to correspond to the acidic or basic side of the isoelectric point can be carried out only provided the change in pH will not result in a destruction of the active material. After filtration, the pH of the filtrate should be readjusted to correspond to the optimum pH range of the active component. If the active material is very sensitive to slight changes in pH, a filter having an appropriate electrical charge should be selected instead.

Cleaning filters Some filters, such as the asbestos, collodion, Polypore, and Millipore filters, are used only once, then discarded. Others, such as the porcelain, diatomaceous earth, and fritted-glass filters, are too expensive to be used only once, but are easily cleaned.

Porcelain filters are cleaned by heating to a red heat in a muffle furnace. This burns off the organic matter in the pores and restores the filters to their original condition.

Diatomaceous earth filters are cleaned by placing the cylinders in a special metal holder connected to a faucet. The filter is reversed in the holder so that the water passes from within outward. This should be continued until all foreign matter has been washed from the pores; otherwise any albuminous substances remaining in the pores are coagulated by heat during sterilization, rendering the filter useless for further work unless given special treatment.

Clogged filters may be cleaned in various ways but probably most conveniently by continuous suction of full-strength Clorox, or similar solution, for 5 to 15 min (Vaisberg, 1938). This treatment quickly dissolves the coagulated material and restores the usefulness of the filter. Thorough washing is necessary to remove the last traces of the oxidizing solution.

Fritted-glass filters may be cleaned by treatment with concentrated sulfuric acid containing sodium nitrate. The strong acid quickly oxidizes and dissolves the organic matter. Thorough washing is necessary to remove the last traces of acid.

Sterilization of filters Collodion membranes cannot be sterilized by heat. Since acetic collodion is sterile, it is not necessary to heat sterilize membranes made of filter paper impregnated with this material if aseptic precautions are observed.

The other filters already discussed are assembled in their appropriate holders, wrapped in paper, and sterilized in the autoclave.

References

Bush, M. T.: Glass bacteriological filters arranged for positive pressure, *J. Bact.*, **51**:531, 1946.

Cliver, D. O.: Virus interactions with membrane filters, *Biotechnol. Bioengineer.*, **10**:877, 1968.

Elford, W. J.: A new series of graded collodion membranes suitable for general bacteriological use, especially in filterable virus studies, *J. Pathol. Bacteriol.*, **34**:505, 1931.

——: The principles of ultra-filtration as applied to biological studies, *Proc. Roy. Soc. (London), Ser. B,* **112**:384, 1933.

Ernst, R. R.: Sterilization by heat. From "Disinfection, Sterilization, and Preservation," edited by C. A. Lawrence and S. S. Block, Philadelphia, Lea & Febiger, 1968.

Michaelis, L.: Contribution to the theory of permeability of membranes for electrolytes, *J. Gen. Physiol.,* **8**:33, 1925.

Perkins, J. J.: "Principles and Methods of Sterilization," Springfield, Ill., Charles C Thomas, Publisher, 1963.

Polson, A., and T. I. Madsen: Optical control of graded collodion membranes, *Biochim. Biophys. Acta,* **12**:584, 1953.

Vaisberg, M.: Method for clearing coagulated serum blocked Berkefeld filters, *J. Lab. Clin. Med.,* **23**:542, 1938.

Ver, B. A., et al.: Efficient filtration and sizing of viruses with membrane filters, *J. Virology,* **2**:21, 1968.

Webb, H. B.: Composition of Seitz filter pads, *Am. J. Clin. Pathol.,* **16**:442, 1946.

Wilson, D. A.: A study of the factors influencing sterilization by heat and radiation, *J. Med. Lab. Technol.,* **25**:301, 1968.

Techniques of pure cultures 9

A pure culture consists of a nutrient medium containing the growth of a single species of organism. Pure cultures are required for studying the morphology and physiology of organisms. All laboratory studies, with few exceptions, are based on the use of pure cultures. Occasionally two species are grown together in making studies of the various types of bacterial associations.

A mixed culture consists of a nutrient medium containing the growth of two or more species of organisms.

A plate culture consists of an organism growing on a solid medium contained in a petri dish.

A slant culture consists of an organism growing on the inclined surface of a solid medium such as nutrient agar. This is referred to as a nutrient agar slant culture. Other types of solid media include coagulated blood serum, potato wedges, and coagulated egg. Slant cultures are sometimes referred to as streak cultures. A nutrient agar slant culture may be called a nutrient agar streak culture; a coagulated blood serum slant culture may be called a coagulated blood serum streak culture; etc.

A solid medium in a slanted position greatly increases the surface area and gives a much greater growth of organisms. In general, solid media give much heavier growths of bacteria than liquid media and are preferred where great numbers of bacteria are required.

A stab culture is prepared by stabbing a solid medium, such as nutrient gelatin or agar, to almost the complete depth of the column with a previously inoculated straight wire needle. A gelatin medium is used to study the character of liquefaction produced by bacteria. An agar medium containing a carbohydrate and indicator is used to detect the production of acid and gas. Acid is indicated by a change in the color of the medium; gas is detected by the appearance of gas bubbles in the agar. Under some conditions, the agar may be split into disks with a layer of gas separating each disk.

A liquid culture consists of a liquid medium, such as nutrient broth or milk, containing the growth of organisms.

A shake culture is prepared by inoculating a liquefied agar medium and rotating the tube to obtain a uniform suspension of organisms before solidification occurs. A shake culture is valuable for indicating the oxygen requirement of an organism. Anaerobic organisms grow in the deeper parts of the medium; aerobic organisms grow at or near the surface exposed to air. All gradations of oxygen tensions occur between these two extremes.

Enrichment cultures Enrichment cultures involve the development of environmental conditions to guarantee growth and survival of the desired cells in a mixed population. To quote from Schlegel and Jannasch (1967),

. . . The mixed microbial population of soil, water, etc., is considered to be a community of competing species or metabolic types. Changes of external conditions will favor growth of one or a few particular species and will result in the establishment of their predominant population at the expense of other species. It is the aim and the art of the enrichment culture technique to control those selective conditions which quickly and reproducibly lead to the predominant population of one special organism, thereby facilitating its isolation. Successful enrichment of an unknown organism provides basic information for the subsequent study of its growth requirements in pure culture. After the physiology of an isolate is investigated, the originally empirical enrichment process may be fully understood and, occasionally, improved. Conversely, some groups of morphologically characteristic microorganisms, known only from microscopic observations, have been inaccessible for detailed studies in pure culture until enrichment techniques were developed.

Employing the enrichment principle in an ingenious manner, the most important microbial transformations in soil and water and the major metabolic types of microorganisms involved were discovered in a relatively short period of time by a small group of workers.

Pure culture study of bacteria

A study of pure cultures of bacteria is limited here to a determination of their characteristics and identity. Included are: methods of inoculation of media; methods of isolation; microscopic study of pure species; determination of morphological, cultural, and physiological characters; animal inoculation; and antigenic structure (serology).

Methods employed for inoculation of media The following procedures are recommended:

Agar deep cultures Sterilize the wire needle in the flame and allow it to cool for a few moments. Remove the cotton stopper or screw cap from an agar slant culture, by grasping it with the small finger of the right hand, and flame the neck of the tube. Hold the tube slanted, not upright, to minimize aerial contamination. Remove a small amount of the growth with the sterilized wire needle. Again flame the neck of the agar slant culture, replace the cotton stopper or screw cap, and set the tube in the test-tube block. Remove the cotton stopper or screw cap from the tube to be inoculated by grasping it with the small finger of the right hand, and flame the neck of the tube. Stab the straight wire containing the inoculum to the bottom of the tube. Carefully withdraw the needle. Again flame the neck of the tube, and replace the cotton stopper or screw cap. Flame the wire needle before setting it down on the table. Mark the tube with the name of the organism and the date. Use a china-marking pencil. Incubate the culture at the proper temperature.

If a transfer is to be made from a liquid culture, use the wire loop instead of the needle. Remove a loopful of the culture and force the wire loop to the bottom of the tube of agar. Carefully withdraw the wire loop to avoid tearing the agar. The procedure in every other detail is the same as the foregoing.

Agar slant cultures Sterilize the wire needle or wire loop in the flame, depending upon whether a solid or a liquid culture is to be used. Allow the wire to cool for a few moments. Remove the cotton stopper or screw cap from the culture, by grasping it with the small finger of the right hand, and flame the neck of the tube. Hold the tube slanted, not upright, to minimize aerial contamination. Remove a small amount of the growth with the sterilized wire needle, or a loopful of the liquid culture with the wire loop. Again flame the neck of the culture, replace the cotton stopper or screw cap, and set the tube aside in the test-tube block. Remove the cotton stopper or screw cap from the agar slant to be inoculated by grasping it with the small finger of the right hand. Flame the neck of the tube. Spread the inoculum over the surface of the agar slant by making streaks back and forth a few millimeters apart. Start at the butt of the slant and work up to the top. Withdraw the needle or loop from the tube. Again flame the neck of the tube, and replace the cotton stopper or screw cap. Flame the wire needle or loop before setting it down on the table. Mark the tube with the name of the organism and the date. Use a china-marking pencil. Incubate the culture in an upright position at the proper temperature.

Broth cultures The same procedure as given under Agar Slant Cultures is followed except that the inoculum is transferred to broth. If growth from an agar slant culture is transferred to broth, vigorous shaking may be necessary to dislodge the inoculum from the wire. If a liquid inoculum is used, only gentle shaking is necessary to remove the culture from the wire loop before withdrawing from the tube.

Isolation of species in pure culture

Bacteria are rarely found in nature in pure culture. Mixed species is the rule. Before accurate studies can be made on an organism, it first must be isolated in pure culture. Two different species growing together frequently produce reactions quite different from those given by each organism when studied separately.

A number of methods have been employed for the propagation of cultures from single cells. Most of these are too difficult and time-consuming to be of practical value, except in certain special instances.

Plate cultures offer a means of isolating pure species of organisms in a comparatively simple manner. Two methods are generally followed: (1) the streak-plate method and (2) the pour-plate method.

Streak-plate method Melt two tubes of appropriate agar medium in boiling water or in an Arnold sterilizer. Allow the agar to cool to about 50°C to minimize steaming; otherwise wetting of the agar surface may occur. A wet surface causes colonies to run together.

Remove the stopper from one of the tubes and flame the neck of the tube. Lift the lid of a sterile petri dish on one side just high enough to insert the opening of the tube and pour the melted agar into the plate. In like manner, pour the second tube of melted agar into another sterile petri dish. Care must be observed in pouring from one container into another to avoid external contamination. Set both plates aside until the agar is firm.

Sterilize the wire loop in the flame and allow it to cool for about 5 sec. Remove the stopper from the mixed broth culture by grasping it with the small finger of the right hand, and flame the neck of the tube. Remove a loopful of the culture with the wire loop. Again flame the neck of the tube, replace the stopper, and set aside in the test-tube block. Raise the lid of the petri dish at the nearest side just high enough to insert the wire loop. With a free arm movement from the elbow, spread the loopful of culture at the upper end of the plate to thin it out; then make steaks back and forth over the surface of the agar about ½ cm apart. The first streak will contain more of the culture than the second, the second more than the third, etc. The last streaks should thin out the culture to give isolated colonies. Without reinoculating the wire loop, streak a second plate. This gives greater certainty in securing well-isolated colonies. Each colony usually represents the growth from a single organism. Mark the plates and incubate at the proper temperature.

The colonies appear only on the surface of the agar. A pure culture may be obtained by transferring, with the wire needle, a portion of a well-isolated colony to an appropriate medium (Fig. 162). It is advisable to make a Gram stain from a portion of the colony first, to ensure greater certainty in the selection of a pure colony.

Pour-plate method Place three tubes, each containing 10 ml sterile distilled water, in a test-tube block, and number them. Transfer to the first tube 0.1 ml of the mixed broth culture and shake to obtain a uniform suspension. Transfer 0.1 ml from the first to the second tube, and mix thoroughly. Transfer 0.1 ml from the second to the third tube and mix as before. Observe aseptic precautions.

Mark three petri dishes to correspond to the numbers appearing on the dilution tubes. Transfer 0.1 ml from each dilution tube to the petri dishes.

Melt three tubes of appropriate agar medium. Allow the agar to cool to about 50°C. Pour the melted and cooled agar into the petri dishes, and

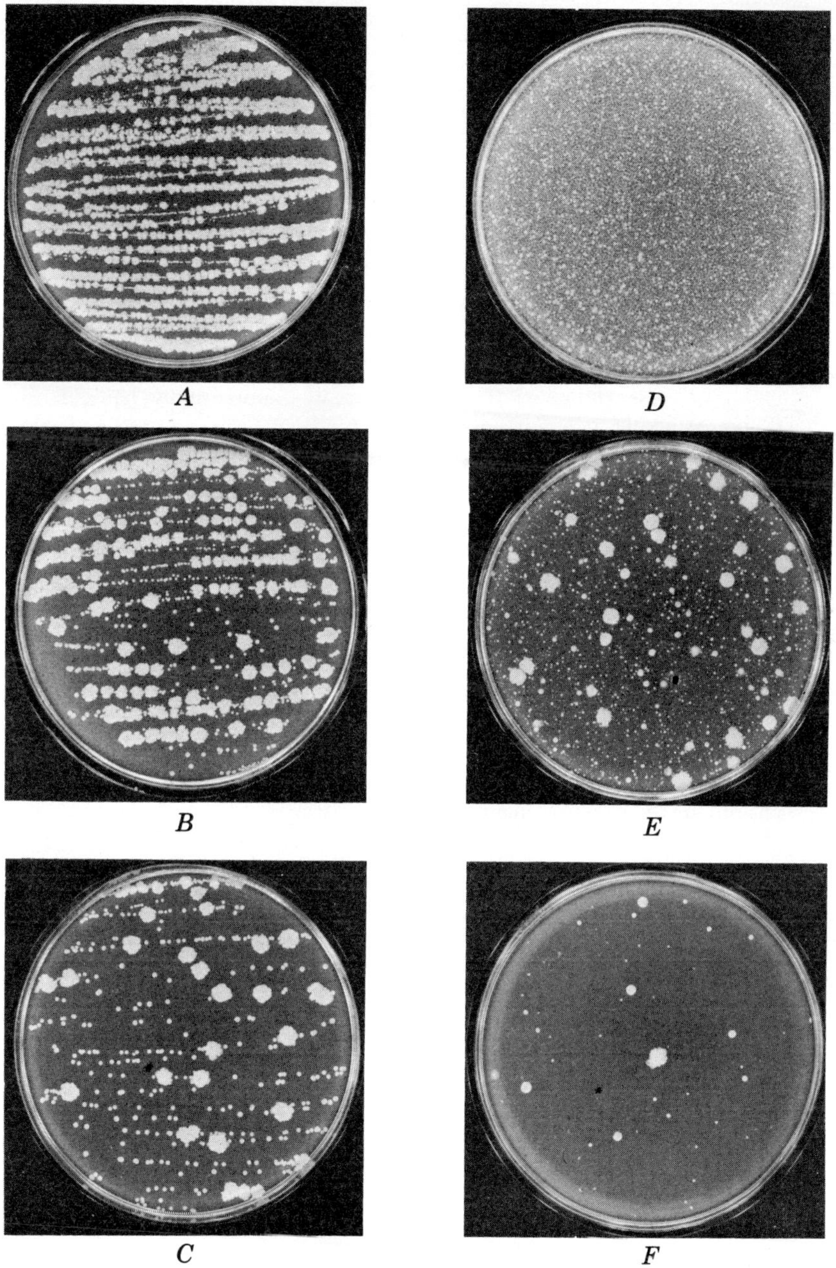

Fig. 162 Isolation of pure species from a mixed culture of Bacillus subtilis and Staphylococcus aureus. Streak-plate method: (A) one loopful of culture streaked over the surface of agar. Without recharging loop, a second and third plate (B, C) were streaked. Pour-plate method: D, tube of agar mixed with one loopful of culture and poured into a petri dish; E, agar mixed with one loopful from tube D and poured into a petri dish; F, agar mixed with one loopful from tube E and poured into a petri dish.

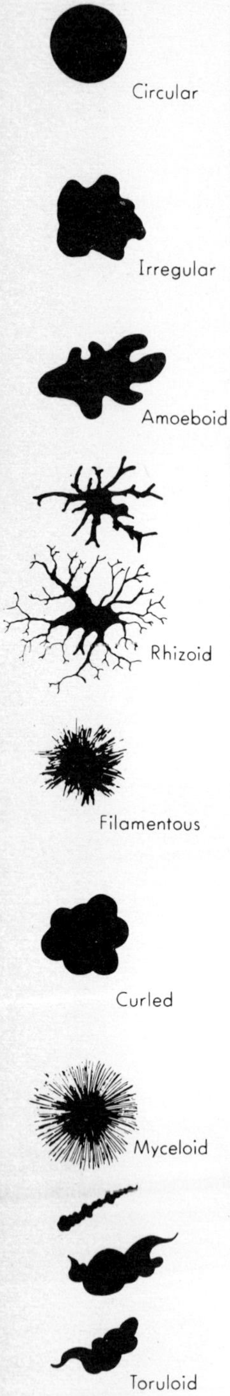

Fig. 163 Shape or form of colonies. (After Thomas.)

tilt from side to side to obtain a uniform distribution of organisms. When firm, invert the plates and incubate at the proper temperature.

In this procedure, most of the colonies are embedded in the agar (Fig. 162). The first plate usually contains colonies so crowded together that good separation is not possible. Also, colonies very close together do not develop to normal size. The second or third plate should show well-separated colonies of normal size. A pure culture may be obtained by removing a portion of a well-separated and embedded colony with the wire needle and transferring it to an appropriate medium. It is advisable to make a Gram stain from a portion of the colony first, to ensure greater certainty in the selection of a pure colony. Well-isolated surface colonies usually are round and quite characteristic for each species. On the other hand, colonies embedded in agar are smaller and usually lenticular in shape. As a rule, it is not possible to distinguish one species from another by the appearance of their subsurface colonies.

It is usually easier to obtain pure cultures by the pour-plate method because the organisms separate better when mixed with melted agar. Bacteria that produce mucoid colonies are very difficult to separate from nonmucoid organisms by the streak-plate method.

Clark (1967) made a comparison of the bacterial counts by the two methods and reported that the streak-plate technique gave results from 70 to 80 percent higher than by the pour-plate procedure. This finding may be of importance where the objective is to isolate various species or strains of bacteria from a mixture for identification and estimation of their relative proportions. Certain species or strains could conceivably be missed entirely or their relative significance underestimated.

Identification of bacterial species A descriptive chart prepared by the Committee on Bacteriological Technic of the American Society for Microbiology will be found on pages 278 to 281. The chart is used for the identification and classification of bacteria.

Tests required to complete the chart include: morphological, cultural, and physiological characteristics of the organism under examination; fermentation reactions using a variety of carbohydrates; animal inoculation; and antigenic structure. Numbers may be recorded to indicate the results of each test. By glancing at the numbers in the right-hand margin and at the bottom edge, a large number of charts may be quickly processed and compared.

Before proceeding to the study of an unknown species, it is necessary to know (1) the nutritional requirements for optimum growth, i.e., composition of the enrichment medium, such as presence or absence of organic matter; (2) the firmness of the medium, i.e., solid, semisolid, or liquid, for best growth; (3) the relation to free oxygen, i.e., strict aerobe, strict anaerobe, or facultative; and (4) the temperature requirement for optimum growth, i.e., 55, 37, or 21 to 25°C.

Colony formation Generally speaking, each bacterial species, when grown on a standard solid medium, forms a characteristic type of

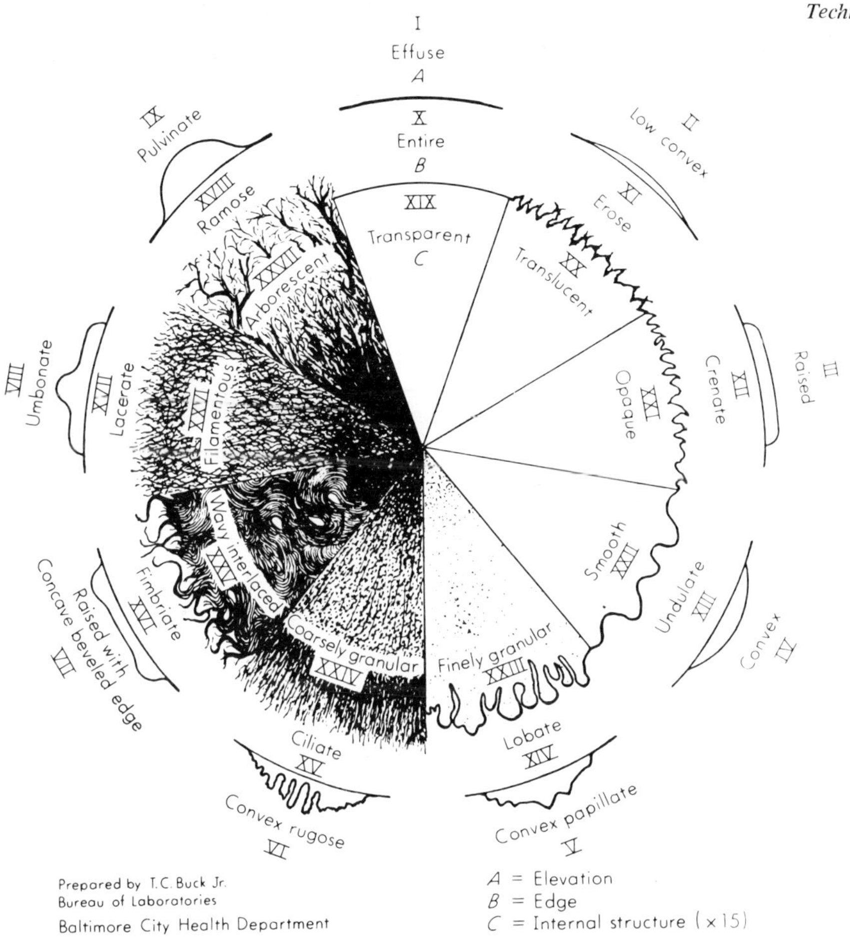

Fig. 164 Bacterial colony formations.

colony. The colonies differ in size, shape, edge, elevation, internal structure, color, and in other ways (Figs. 163 and 164).

Whittaker and Drucker (1970) examined intact colonies of *Streptococcus mutans* OMZ61 by scanning electron microscopy (Fig. 165). The magnification of the raised periphery of the same colony (Fig. 166) was approximately 65 times greater than that of the colony. In a later report Drucker and Whittaker (1971) extended the investigations to include *Bacillus cereus, B. megaterium, B. mycoides, Corynebacterium hoffmanii,* and *Lactobacillus acidophilus* (Figs. 167 to 171).

The terms used to describe the various characteristics are defined in the glossary at the end of this chapter.

For more information: Cooper et al. (1968); Finkelstein and Punyashthiti (1967); Springer and Roth (1972).

Stab and streak cultures The types of growth on the surface and in the depth of agar are characteristic of bacterial species (Figs. 172 and

Name of organism ____________ Source ____________
Date of isolation ____________ Habitat ____________
Is phase variation observed? ______ Phase on this Chart: S, R, M, G (smooth, rough, mucoid, gonidial) ___

Underscore required terms	SKETCHES
VEGETATIVE CELLS: Medium used ____________ Reaction (pH) ______ Temp. ______ Age ______ d. Size of Majority ____________ Ends, *rounded, truncate, concave, tapering* MOTILITY: In broth ______ On agar ______ SPORANGIA and ENDOSPORES: *present, absent.* Medium used ____ pH ____ Temp. ______ Age ____ d. Endospore Form: *spherical, ellipsoid, cylindrical* IRREGULAR FORMS: Present on ______ in ____ days at ______ °C	

	Surface Colonies	Deep Colonies
AGAR COLONIES: Temperature ______ °C. Age ______ d. Form, *punctiform (i.e. under 1 mm. diam.), circular (i.e. over 1 mm diam.), filamentous, irregular, rhizoid.* Surface, *smooth, rough, concentrically ringed, radiately ridged.* Edge, *entire, undulate, lobate, erose, filamentous, curled.* Elevation of growth, *effuse, flat, raised, convex.* Optical Characters, *opaque, translucent, opalescent, iridescent.*		
GELATIN COLONIES: Temperature ______ °C. Age ______ d. Form, *punctiform, circular, irregular, filamentous.* Elevation, *flat, raised, convex, pulvinate, crateriform (liquefying).* Edge, *entire, undulate, lobate, erose, filamentous, curled.* Liquefaction, *cup, saucer, spreading.* Surface, *smooth, contoured, rugose.* Optical Characters, *opaque, translucent, opalescent, iridescent.*	Surface Colonies	Deep Colonies

AGAR STROKE: Temperature ______ °C. Age ______ d. Growth, *scanty, moderate, abundant, none.* Form of growth, *filiform, echinulate, beaded, spreading, arborescent, rhizoid.* Lustre, *glistening, dull.* Chromogenesis ______ *photogenic, fluorescent.* Odor, *absent, decided, resembling* ______ Consistency, *butyrous, viscid, membranous, brittle.* Medium, *grayed, browned, reddened, blued, greened, unchanged.*		______ Medium Temperature ______ °C Age ______ d.	
NUTRIENT BROTH: Temperature ______ °C. Age ______ d. Surface growth, *ring, pellicle, flocculent, membranous, none.* Clouding, *slight, moderate, strong, transient, persistent, none, fluid turbid, granular growth.* Odor, *absent, decided, resembling* ______ Sediment, *compact, flocculent, granular, flaky, viscid.* Amount of sediment, *abundant, scanty, none.*		______ Medium: Temperature ______ °C Age ______ d.	
GELATIN STAB: Temperature ______ °C. Age ______ d. Growth, *uniform, best at top, best at bottom.* Line of puncture, *filiform, beaded, papillate, villous, arborescent.* Liquefaction, *none, crateriform, infundibuliform, napiform, saccate, stratiform:* begins in ____ d. complete in ____ d. Degree of liquefaction in ______ days ______ Method used ____________ Medium, *fluorescent, browned, unchanged.*		______ Medium: Temperature ______ °C Age ______ d.	

Fermentation — Temperature ______ °C

Medium ______ containing ______ ______ and:	Monosaccharides							Disaccharides						Polysaccharides						Alcohols								Glucosides				
	Arabinose	Rhamnose	Xylose	Glucose	Fructose	Galactose	Mannose	Lactose	Sucrose	Maltose	Trehalose	Melibiose	Cellobiose	Raffinose	Melezitose	Starch	Inulin	Dextrin	Glycogen	Glycerol	Erythritol	Arabitol	Adonitol	Mannitol	Sorbitol	Dulcitol		Salicin	Aesculin	Coniferin	a-Methyl Gluc.	
Gas in ______ fermentation tube																																
Amt. Co_2 *in Eldredge tube*																																
Reaction (pH) after ______ d.																																
Titrable acidity in ml. of N/ ______ NaOH																																

Studied by ____________________ Culture No. ____________

Optimum conditions: Media ____________________ Temp. ____________ °C.

Phases recorded on other charts: ____________________

Brief Characterization

As each of the following characteristics is determined, indicate in proper marginal square by means of figure, as designated below. In case any of these characteristics is doubtful or has not been determined, indicate with the letters U, V, and X according to the following code:

U, undetermined; V, variable; X, doubtful.

Morphological	VEGETATIVE CELLS	Form & arrangement: 1, streptococci; 2, diplococci; 3, micrococci; 4, sarcinae; 5, rods; 6, commas; 7, spirals; 8, branched rods; 9, filamentous	
		Diameter: 1, under 0.5 μ; 2, between 0.5 μ and 1 u; 3, over 1 μ	
		Gram stain: 0, negative; 1, positive	
		Flagella: 0, absent; 1, peritrichic; 2, polar; 3, present but undetermined	
		Capsules: 0, absent; 1, present	
		Chains (4 or more cells): 0, absent; 1, present	
	SPORANGIA: 0, absent; 1, elliptical; 2, short rods; 3, spindled; 4, clavate; 5, drumsticks		
	ENDOSPORES: 0, absent; 1, central to excentric; 2, subterminal; 3, terminal		
Cultural	AGAR STROKE	Growth: 0, absent; 1, abundant; 2, moderate; 3, scanty	
		Lustre: 1, glistening; 2, dull	
	AGAR COLONIES	Form: 1, punctiform; 2, circular (over 1 mm diameter); 3, rhizoid; 4, filamentous; 5, curled; 6, irregular	
		Surface: 1, smooth; 2, contoured; 3, rugose	
	GELATIN COLONIES	Form: 1, punctiform; 2, circular (over 1 mm), 3, irregular; 4, filamentous	
		Surface: 1, smooth; 2, contoured; 3, rugose	
Physiological	Biologic relationships: 1, pathogenic for man; 2, for animals but not for man; 3, for plants; 4, parasitic but not pathogenic; 5, saprophytic; 6, autotrophic		
	Relation to free oxygen: 1, strict aerobe; 2, facultative anaerobe; 3, strict anaerobe; 4, microaerophile		
	In nitrate media: 0, neither nitrite nor gas; 1, both nitrite and gas; 2, nitrite but no gas; 3, gas but no nitrite		
	Chromogenesis: 0, none; 1, pink; 2, violet; 3, blue; 4, green; 5, yellow; 6, orange; 7, red; 8, brown; 9, black		
	Other photic characters: 0, none; 1, photogenic; 2, fluorescent; 3, iridescent		
	Indole: 0, negative; 1, positive		
	Hydrogen sulfide: 0, negative; 1, positive		
	Hemolysis: 0, negative; 1, positive		
	Methemoglobin: 0, negative; 1, positive		
	PROTEIN LIQUEFAC-TION OR DIGESTION	Gelatin: 0, negative; 1, positive	
		Casein: 0, negative; 1, positive	
		Egg albumin: 0, negative; 1, positive	
		Blood serum: 0, negative; 1, positive	
	INDICATOR REDUC-TION	Litmus: 0, negative; 1, positive	
		Methylene blue: 0, negative; 1, positive	
		Janus green: 0, negative; 1, positive	
	Rennet production: 0, negative; 1, positive		

Temperature Relations

Medium ________________ pH ________
Optimum temperature for growth ________ °C
Maximum temperature for growth ________ °C
Minimum temperature for growth ________ °C

THERMAL DEATH POINT: Time 10 minutes: °C
Medium ________________ pH ________

THERMAL DEATH TIME:
Medium ________________ pH ________

Temp.	Time	Temp.	Time
____ °C	____ min	____ °C	____ min
____ °C	____ min	____ °C	____ min
____ °C	____ min	____ °C	____ min
____ °C	____ min	____ °C	____ min
____ °C	____ min	____ °C	____ min

Chromogenesis

Gelatin ________________
Agar ________________
Potato ________________

Other Photic Characters

Photogenesis on ________________
Iridescence on ________________
Fluorescence in ________________

Relation to Reaction (pH) of Medium

Medium ________________
Optimum for growth: *about pH* ________
Limits for growth: *from pH* ________ *to* ________

Relation to Free Oxygen

Method ________________
Medium ________________ Temp. ____ °C
Aerobic growth: *absent, present, better than anaerobic growth, micro-aerophilic*
Anaerobic growth: *absent, occurs in presence of glucose, of sucrose, of lactose, of nitrate; better than aerobic growth*
Additional data: ________________

Milk

Temperature ________ °C

Reaction: ____ d. ____ ; ____ d. ____ ; ____ d. ____
Acid curd: ____ d. ____ ; ____ d. ____ ; ____ d. ____
Rennet curd: ____ d. ____ ; ____ d. ____ ; ____ d. ____
Peptonization: ____ d. ____ ; ____ d. ____ ; ____ d. ____

Litmus Milk

Temperature ________ °C

Reaction: ____ d. ____ ; ____ d. ____ ; ____ d. ____
Acid curd: ____ d. ____ ; ____ d. ____ ; ____ d. ____
Rennet curd: ____ d. ____ ; ____ d. ____ ; ____ d. ____
Peptonization: ____ d. ____ ; ____ d. ____ ; ____ d. ____
Reduction of litmus *begins in* ____ *days, ends in* ____ *days*

PATHOLOGY

Animal Inoculation

Medium used ________________ Age of culture ________ Amount ________ Incubation period ________

		Whole culture			Cells			Filtrate		
Animal		*								
Type of Injection	Subcutaneous									
	Intraperitoneal									
	Intravenous									
	Per os									

* In each instance where pathogenicity is observed, indicate location of lesion, and type, *e.g.* edema, histolysis, gas, hemorrhage, ulcer, diphtheritic, etc.

Antigenic Action

Animal ________________ Medium used ________________ Age of culture ________
Type injection ________________ Number of injections ________
Culture causes production of *cytolysins, agglutinins, precipitins, antitoxin*
Specificity: Antibodies produced effective against other antigens as follows ________________

Immune sera from ________________ effective against this organism as antigen

This DESCRIPTIVE CHART presented at the annual meeting of the SOCIETY OF AMERICAN
Prepared by a sub-committee consisting

TARY DATA

Action on Erythrocytes

Cells: ____________________
Method: *plate, broth, filtrate*
Hemolysis: *negative, positive*
Methemoglobin: *negative, positive*

Production of Indole

Medium ____________________
Test used ____________________
Indole *absent, present in* ____________________ *days*

Production of Hydrogen Sulfide

Medium ____________________
Test used ____________________
H_2S *absent, present in* ____________________ *days*

Action on Nitrates

Medium ____________________ Temp. ____ °C
Nitrite ____ d. ____ ; ____ d. ____ ; ____ d. ____ ; ____ d. ____
Gas (N_2): ____ d. ____ ; ____ d. ____ ; ____ d. ____ ; ____ d. ____
Medium ____________________ Temp. ____ °C
Nitrite: ____ d. ____ ; ____ d. ____ ; ____ d. ____ ; ____ d. ____
Gas (N_2): ____ d. ____ ; ____ d. ____ ; ____ d. ____ ; ____ d. ____
Ammonia production (in amino-N-free nitrate medium):
negative, positive
Complete disappearance of nitrate in ____________ medium:
negative, positive
Disappearance of 2 ppm nitrite in ____________ medium:
negative, positive

Reduction of Indicators

Medium ____________ pH ______ Temp. ______ °C.

Indicator	Conc.	Reduction:
________	____ % ____ hr	______ ; ______ hr ____
________	____ % ____ hr	______ ; ______ hr ____
________	____ % ____ hr	______ ; ______ hr ____
________	____ % ____ hr	______ ; ______ hr ____

Staining Reactions

Gram ______ d. ____ ; ____ d. ____ ; ____ d. ____ ; ____ d. ____
Method ____________________
Spores: Method ____________________
Capsules: Method ____________________
Medium ____________________
Flagella: Method ____________________
Special Stains ____________________

Additional Tests

Methyl red: *negative, positive*
Voges-Proskauer: *negative, positive*
Growth in sodium citrate: *absent, present*
Hydrolysis of starch: *complete (iodine colorless); partial (iodine reddish-brown); none (iodine blue)*
Nitrogen obtained from the following compounds: ____________________

SPECIAL TESTS

BACTERIOLOGISTS, Dec. 28, 1934, by the Committee on Bacteriological Technic
of M. W. Jennison and H. J. Conn.

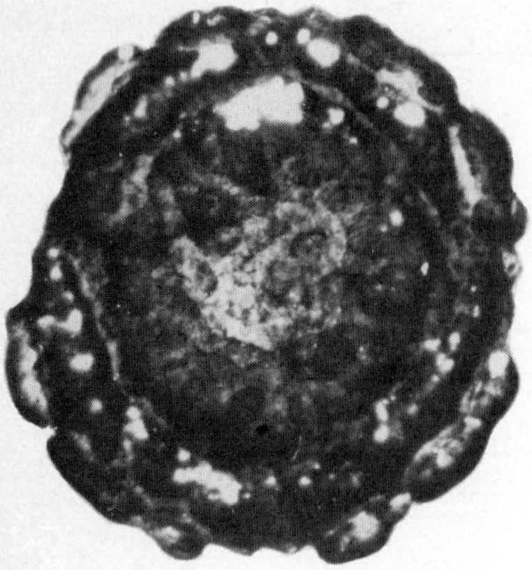

Fig. 165 Scanning electron microscopy of an intact colony of Streptococcus mutans OMZ61. (Courtesy of Whittaker and Drucker.)

173). The various characteristics are made use of in the identification and classification of bacteria.

Pitfalls in the use of the charts In making routine tests for the identification of bacterial species, the beginner is likely to encounter a number of pitfalls, the most important being (1) danger of contaminated cultures, (2) variation of species into more than one phase, and (3) differences in methods of study.

Fig. 166 Scanning electron microscopy of organisms in the raised periphery of the Streptococcus mutans OMZ61 colony. Note the banded appearance and the tightly packed arrangement of the cells. (Courtesy of Whittaker and Drucker.)

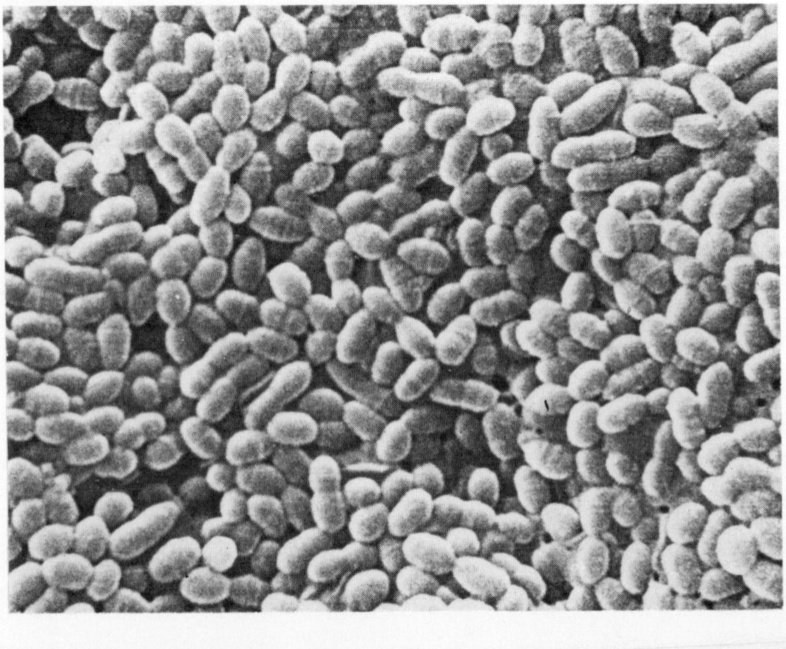

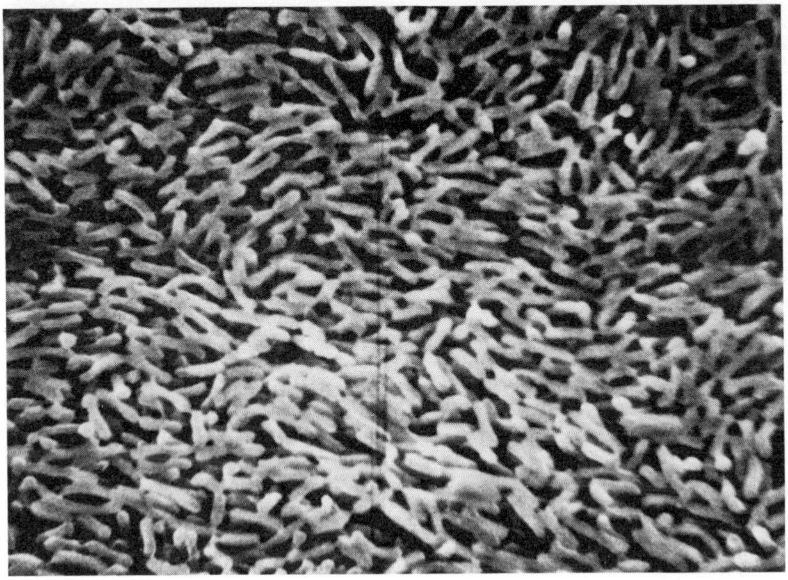

Fig. 167 Bacillus cereus. Scanning electron micrograph of organisms in central area of a colony showing localized orientation. There is little extracellular material. (Courtesy of Drucker and Whittaker.)

Contaminated cultures Needless to say, the beginner cannot be too careful in handling cultures and making transfers to fresh media and at the same time preventing external contaminations. Unless a culture is kept pure, the results are certain to be misleading. As may be seen in the following section, it is sometimes very difficult to know when a culture is contaminated with another species or the original organism has changed in one or more morphological characteristics.

Bacterial variation During the early years of bacteriology most bacteriologists believed that a bacterial species could exist in more than one cell form.

Some years later this concept was altered in favor of a fixed cell form for each species. Forms that departed more or less widely from the normal types were usually dismissed as being involution forms, degenerate cells, or different species present as contaminants.

At the present time sufficient evidence has accumulated to support the original hypothesis of variability in the morphological characters of an organism. As a result of this change in viewpoint, it is easy for a careless beginner to believe that he is observing two phases in a culture when in reality one of the phases observed is a contaminant. Conversely, it is also easy for a beginner to consider a culture composed of two different organisms when actually they are different phases of the same species.

Differences in methods of study In order that results of tests be significant, it is necessary to include the procedure employed. For example, it is not sufficient to state that an organism does or does not

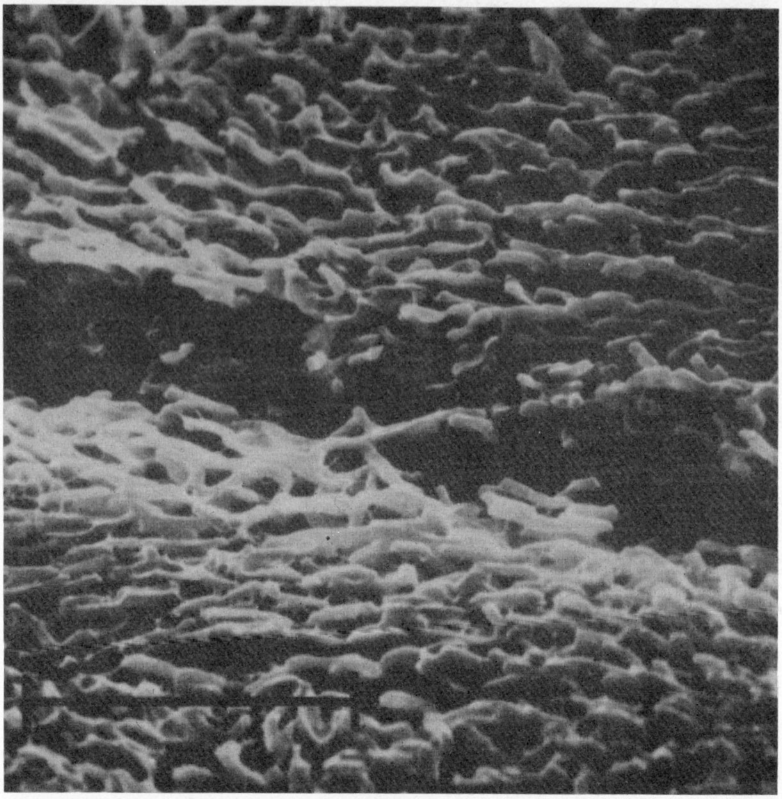

Fig. 168 Bacillus megaterium. Scanning electron micrograph of organisms in the center of a colony. Note the incomplete separation of cells. Bar = 2 nm. (Courtesy of Drucker and Whittaker.)

produce hydrogen sulfide without including the conditions under which it was investigated. It is necessary to mention the composition of the culture medium, the indicator incorporated in the medium to test for hydrogen sulfide, the temperature of incubation, and the length of the incubation period. Under one set of conditions the test may be positive; under another the result may be negative. Unless the conditions are mentioned, disagreements in results may occur.

Glossary of terms used on the descriptive chart[1]

A number of scientific terms are used on the descriptive chart to describe the various characteristics of organisms growing on different media.

[1] Taken from "Manual of Microbiological Methods," by the Society of American Bacteriologists (name now changed to American Society for Microbiology). Courtesy of McGraw-Hill Book Company, 1957.

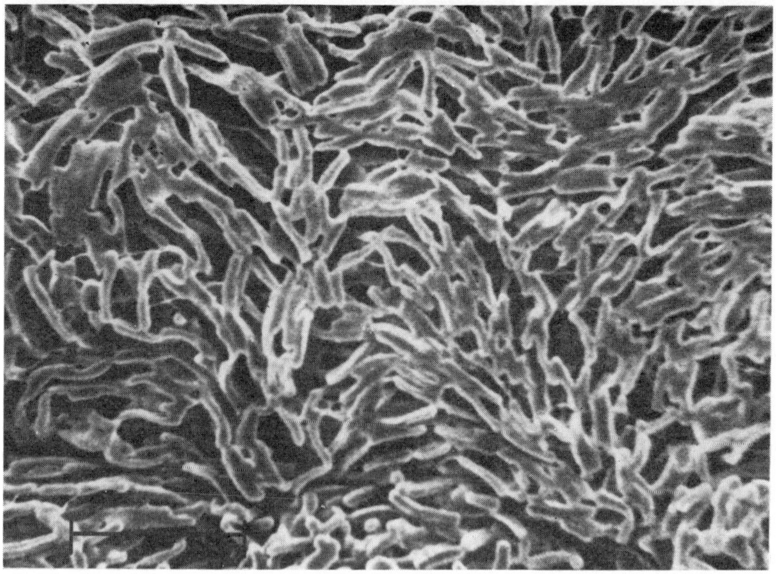

Fig. 169 Bacillus mycoides. Scanning electron micrograph of chains of cells in a colony near the agar surface. Note the lack of overall orientation of bacteria. Bar = 10 nm. (Courtesy of Drucker and Whittaker.)

Fig. 170 Corynebacterium hoffmanii. Scanning electron micrograph shows tightly packed cells in the center of a colony without any covering film. Bar = 2 nm. (Courtesy of Drucker and Whittaker.)

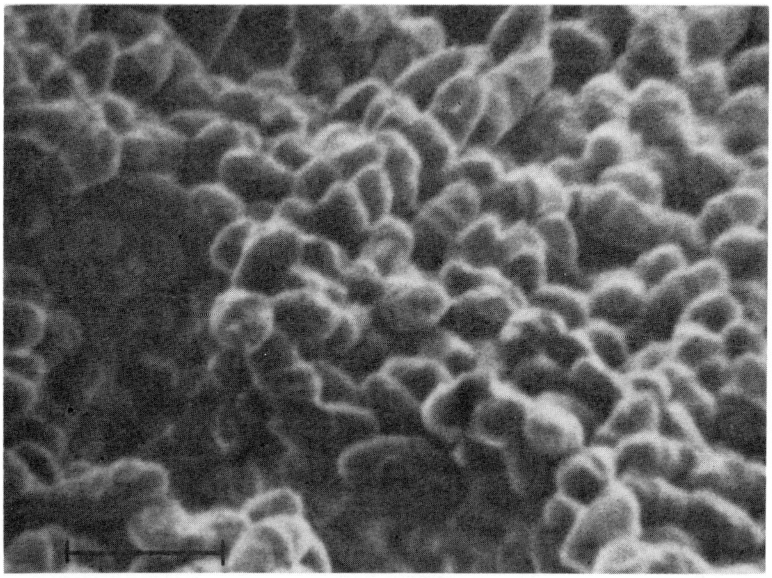

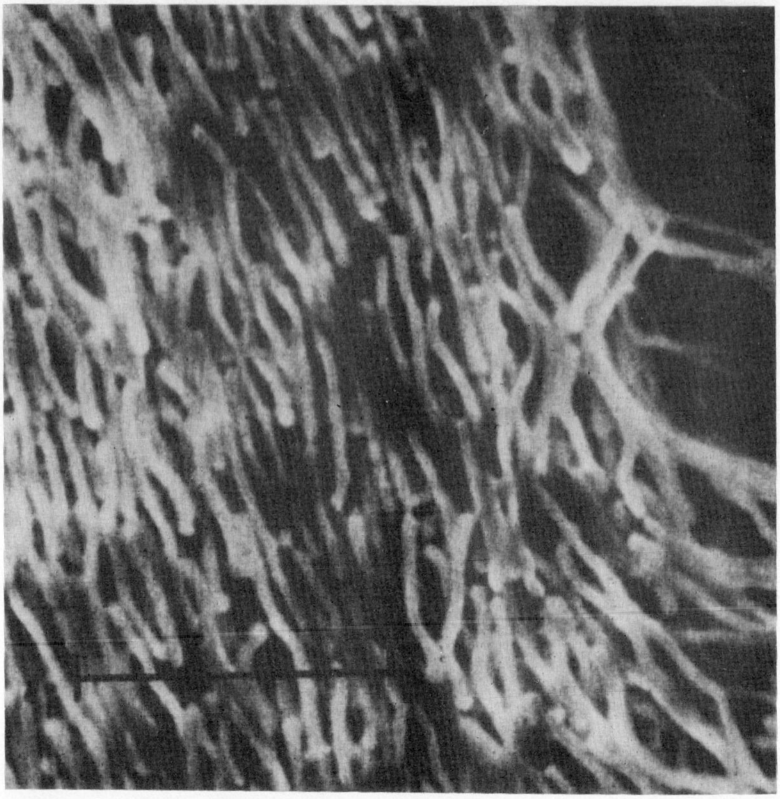

Fig. 171 Lactobacillus acidophilus. Scanning electron micrograph of a colony containing closely packed strongly oriented rods. Bar = 10 nm. (Courtesy of Drucker and Whittaker.)

Fig. 172 Growth in agar stab cultures. (After Thomas.)

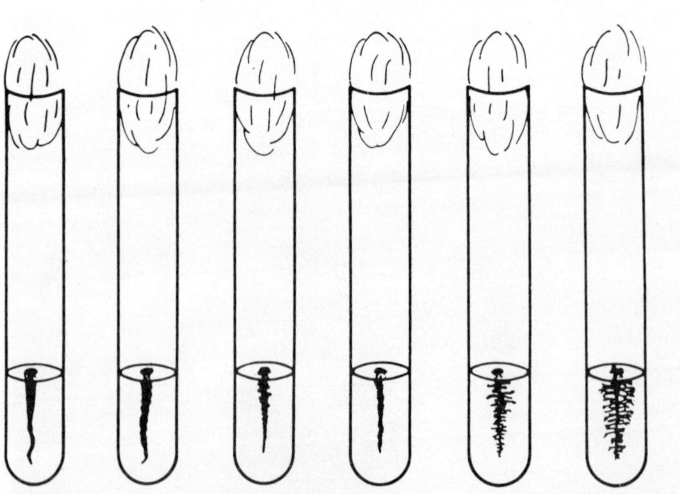

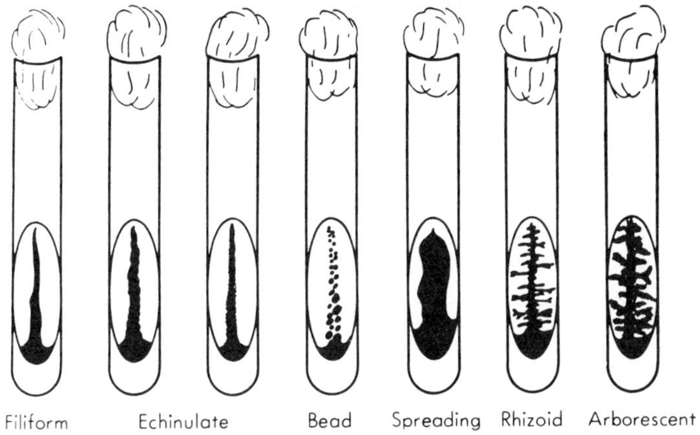

Fig. 173 *Growth on agar streak cultures.* (After Thomas.)

These terms, together with a number of others, are defined as follows:

Acid curd, coagulation of milk due to acid production.
Adherent, applied to sporangium wall, indicates that remnants of sporangium remain attached to endospore for some time.
Aerobic, growing in the presence of free oxygen; strictly aerobic, growing only in the presence of free oxygen.
Agglutinin, an antibody having the power of clumping suspensions of bacteria.
Anaerobic, growing in the absence of free oxygen; strictly anaerobic, growing only in the absence of free oxygen; facultative anaerobic, growing both in presence and in absence of oxygen.
Antibody, a specific substance produced by an animal in response to the introduction of an antigen.
Antigen, a substance which, when introduced into an animal body, stimulates the animal to produce specific bodies that react or unite with the substance introduced.
Antigenic action, behavior as an antigen.
Antiserum, serum containing antibody.
Antitoxin, an antibody having the power of uniting with or destroying a toxic substance.
Arborescent, branched, tree-like in growth.
Aseptically, without permitting microbial contamination.
Autotrophic, able to grow in absence of organic matter.
Bacteremia, the presence of bacteria in the blood stream.
Bactericidal, capable of killing bacteria.
Bacteriostasis, prevention of bacterial growth, but without killing the bacteria.
Beaded (in stab or stroke culture), separate or semiconfluent colonies along the line of inoculation.
Bipolar, at both poles or ends of the bacterial cell.

Bleb, vesicle or blister-like swelling.
Brittle, growth dry, friable under the platinum needle.
Butyrous, showing growth of butter-like consistency.
Capsule, an envelope surrounding the cell membrane of some kinds of bacteria.
Chains, four or more bacterial cells attached end to end.
Chromogenesis, the production of color.
Clavate, club-shaped.
Coagulation, formation of a clot or curd.
Coccus (plural, cocci), a spherical organism.
Comma, a short curved rod; comma-shaped.
Compact, referring to sediment in the form of a single, fairly tenacious mass.
Complement, a nonspecific enzyme-like substance, destroyed if subjected to heat (56°C or over for 30 min), which occurs in blood serum and is necessary, in conjunction with a specific antibody, in order to bring about cytolysis.
Concave, presenting a depressed or hollow surface.
Concentrically ringed, marked with rings, one inside the other.
Contoured, having an irregular, smoothly undulating surface, like that of a relief map.
Crateriform, referring to a saucer-shaped liquefaction of the medium.
Cuneate, wedge-shaped.
Curd, precipitated milk protein consisting largely of casein.
Curled, composed of parallel chains in wavy strands, as in anthrax colonies.
Cytolysin, an antibody causing cytolysis.
Cytolysis, a dissolving action on cells.
Diastatic action, conversion of starch into simpler carbohydrates, such as dextrins or sugars, by means of diastase.
Diphtheritic, diphtheria-like
Dissociation, separation of characters, usually referring to phase variation (*q.v.*).
Echinulate, showing a growth along the line of inoculation with toothed or pointed margins.
Edema, intercellular accumulation of fluid in a part of an animal body.
Effuse, of thin growth, veily, unusually spreading.
Ellipsoid, ovate or ovoid (egg-shaped).
Elliptical, same as ellipsoid.
Endospore, thick-walled spore formed within bacteria, i.e., typical bacterial spores like those of *B. subtilis.*
Endotoxin, a toxic substance produced within a microorganism and not excreted.
Entire, applied to colonies, indicates a smooth margin.
Enzyme, a chemical ferment produced by living cells.
Erose, irregularly notched.

Excentric, slightly to one side of the center, between the positions denoted central and subterminal.
Exogenous, originating outside the organism.
Exotoxin, a toxic substance excreted by a microorganism and hence found outside the cell body.
Facultative anaerobe, see anaerobic.
Fibrinolysin, an enzyme that dissolves fibrin.
Filamentous, denoting growth composed of long, irregularly placed or interwoven threads.
Filaments, as applied to morphology of bacteria, refers to thread-like forms, generally unsegmented; if segmented the organisms are enclosed in a sheath.
Filiform, in stroke or stab cultures, a uniform growth along line of inoculation.
Flagellum (plural, flagella), a motile, whip-like attachment; an organ of locomotion.
Flaky, refers to sediment in the form of numerous separate flakes.
Flocculent, containing small adherent masses of various shapes floating in the field.
Fluorescent, having one color by transmitted light and another by reflected light.
Gonidia, asexual spores.
Gonidial, referring specifically to a bacterial phase producing gonidia-like bodies.
Granular, composed of small granules.
Habitat, the place where an organism normally lives under natural conditions.
Hemolysin, a substance causing hemolysis either alone or in the presence of complement.
Hemolysis, a dissolving action on red blood corpuscles.
Hemorrhage, an escape of blood from the vessels.
Histolysis, breaking down of tissues.
Hydrolysis of starch, destruction of starch by the formation of a chemical union with water; includes diastatic action, but is a more general term.
Immune serum, an animal fluid containing an antibody.
Inactivate, to destroy complement by heat (at 56°C for 30 min).
Infundibuliform, in form of a funnel or inverted cone.
Intraperitoneal, within the peritoneum.
Intravenous, within a vein.
Iridescent, exhibiting changing rainbow colors in reflected light.
Lesion, a local injury or morbid structural change.
Lobate, having lobes, or rounded projections.
Maximum temperature, temperature above which growth does not take place.
Membranous, of thin growth, coherent, like a membrane.
Metabolite, a substance produced by metabolism.

Microaerophilic, growing best in presence of small quantities of oxygen.

Minimum temperature, temperature below which growth does not take place.

Morphology, form, shape, structure.

Motility, ability to move spontaneously.

Mucoid, mucus-like, referring specifically to a bacterial phase producing slimy growth.

Mycelioid, colonies having the radiately filamentous appearance of mold colonies.

Napiform, denoting liquefaction in form of a turnip.

Opalescent, milky white with tints of color as in an opal.

Opaque, not allowing light to pass through.

Optimum temperature, temperature at which most growth occurs.

Papillate, denoting growth beset with small nipple-like processes.

Parasitic, deriving its nourishment from some living animal or plant upon which it lives and which acts as host; not necessarily pathogenic.

Pathogenic, not only parasitic (*q.v.*) but also causing disease in the host.

Pellicle, bacterial growth forming either a continuous or an interrupted sheet over the culture fluid.

Peptonization, rendering curdled milk soluble by the action of peptonizing enzymes.

Peritrichiate, applied to the arrangement of flagella, indicates that they are distributed over the entire surface of an organism.

Peritrichic, having flagella in peritrichiate arrangement.

Per os, through the mouth.

Persistent, lasting many weeks or months.

Phase variation, separation of a species into strains having somewhat different characters.

Photogenic, glowing in the dark, phosphorescent.

Polar, at the end or pole of the bacterial cell.

Precipitin, an antibody having the power of precipitating soluble proteins.

Pulvinate, cushion-shaped.

Punctiform, very small, but visible to naked eye; under 1 mm in diameter.

Radiate, to diverge or spread from a common point.

Raised, denoting thick growth, with abrupt or terraced edges.

Reduction, removal of oxygen or its equivalent from a chemical compound; or addition of hydrogen or its equivalent. Refers to the conversion of nitrate to nitrite, ammonia, or free nitrogen; also to the decolorization of litmus.

Rennet curd, coagulation of milk due to rennet or rennet-like enzymes, distinguished from acid curd by the absence of acid.

Rhizoid, growth of an irregular branched or root-like character, as colonies of *Bacillus mycoides*.

Ring, growth at the upper margin of a liquid culture, adhering to the glass.

Rough, colonies with an irregular, nonsmooth surface.

Rugose, wrinkled.

Saccate, liquefying in the form of an elongated sac, tubular, cylindrical.

Saprophytic, living on dead organic matter.
Saucer, a concave liquefaction of the medium, shallower than crateriform.
Sensitize, to render sensitive, usually to a foreign protein.
Sepsis, a state of infection.
Sheath, an envelope similar to a capsule (*q.v.*), but surrounding a filamentous organism.
Smooth, colonies with an even surface.
Spindled, larger at the middle than at the ends. Applied to sporangia, refers to the forms frequently called clostridia.
Sporangium (plural, sporangia), cells containing endospores.
Spreading, denoting growth extending much beyond the line of inoculation, i.e., several millimeters or more.
Stratiform, liquefying to the walls of the tube at the top and then proceeding downward horizontally.
Strict aerobe, see aerobic.
Strict anaerobe, see anaerobic.
Subcutaneous, under the skin.
Subterminal, situated toward the end of the cell but not at the extreme end, i.e., between the positions denoted excentric (*q.v.*) and terminal.
Synergism, cooperative action of two organisms, resulting in an end product which neither could produce alone.
Thermophilic, growing best at high temperatures, i.e., 50°C or over.
Toxic, poisonous.
Transient, lasting a few days.
Translucent, allowing light to pass through without allowing complete visibility of objects seen through the substance in question.
Trituration, thorough grinding in a mortar.
Truncate, with ends abrupt, square.
Turbid, cloudy with flocculent particles, i.e., cloudy plus flocculent.
Ulcer, an open sore.
Umbonate, having a raised center; knob-like.
Undulate, wavy.
Vegetative cells, bacterial cells not containing spores; cells primarily concerned with nutrition and growth.
Villous, having short, thick, hair-like processes on the surface intermediate in meaning between papillate and filamentous (*q.v.*).
Virulence, degree of pathogenicity (referring to infectiousness).
Virus, a self-propagating cause of disease, often referring to one too small to be seen with a microscope.
Viscid, denoting growth on a culture medium that follows the needle when touched and withdrawn; or referring to sediment that on shaking rises as a coherent swirl.

References

Breed, R. S., et al.: "Bergey's Manual of Determinative Bacteriology," Baltimore, The Williams & Wilkins Company, 1957.

Buchanan, R. E., et al., eds.: "Index Bergeyana," Baltimore, The Williams & Wilkins Company, 1966.

Clark, D. S.: Comparison of pour and surface plate methods for determination of bacterial counts, *Can. J. Microbiol.,* **13**:1409, 1967.

Committee on Bacteriological Technic (Society of American Bacteriologists): "Manual of Microbiological Methods," New York, McGraw-Hill Book Company, 1957.

Cooper, A. L., et al.: Factors affecting the growth of bacterial colonies on agar plates, *Proc. Roy. Soc. B.,* **171**:175, 1968.

Drucker, D. B., and D. K. Whittaker: Microstructure of colonies of rod-shaped bacteria, *J. Bact.,* **108**:515, 1971.

Finkelstein, R. A., and K. Punyashthite: Colonial recognition, a "new" approach for rapid diagnostic enteric bacteriology, *J. Bact.,* **93**:1897, 1967.

Schlegel, H. G., and H. W. Jannasch: Enrichment cultures. In "Annual Review of Microbiology," **21**:49, 1967.

Skerman, V. B. D.: "The Genera of Bacteria," Baltimore, The Williams & Wilkins Company, 1967.

Springer, E. L., and I. L. Roth: Scanning electron microscopy of bacterial colonies. I. *Diplococcus pneumoniae* and *Streptococcus pyogenes, Can. J. Microbiol.,* **18**:219, 1972.

Whittaker, D. K., and D. B. Drucker: Scanning electron microscopy of intact colonies of microorganisms, *J. Bact.,* **104**:902, 1970.

10 Effect of environment upon bacteria

It is well known that the life activities of organisms are conditioned by their environment. Any marked change in the environment produces a corresponding change in the morphological and physiological characters of organisms. Bacteria are able to withstand great changes in the environment and quickly adapt themselves to the new conditions. In this respect, they differ markedly from higher plant and animal cells. By understanding the various physical factors controlling survival and multiplication, bacterial activity may be increased, decreased, or destroyed completely.

Bacteria multiply normally by binary or transverse fission. The rate at which division takes place can be made to vary widely. Any alteration of the time between consecutive cell divisions (generation time) indicates that one or more environmental factors have changed.

The death time of bacteria by physical agents is related logarithmically to the number of surviving bacteria. This means that the disinfection process does not take place suddenly but is a gradual operation in which the number of organisms killed in unit time is greater at the beginning and becomes less and less as action continues. If the numbers of survivors in unit time are plotted against time and lines drawn, the points lie on smooth curves. On the other hand, if the logarithms of the numbers of survivors are plotted against time, the points fall on a straight line (see page 564). This is a general rule applicable to all agents employed for the destruction of monocellular organisms.

Effect of temperature

Bacteria are able to survive wide limits of temperature, but the range in which they can grow and carry on their activities generally falls between 0 and 90°C. Bott and Brock (1969) immersed microscope slides to a depth of at least one meter in one of the hot springs of Yellowstone National Park. The slides were removed at intervals and counts made. Conclusive evidence was obtained that growth of some bacterial species occurred at a temperature as high as 95°C.

Low temperature A decrease in temperature decreases the kinetic energy of all molecules in the system. The decrease in kinetic energy in turn gives rise to secondary effects, including a reduction in vapor pressure and density of water, an increase in the viscosity of water, and a decrease in the rates of chemical reactions. The magnitude of the changes is proportional to the degree to which the temperature is lowered.

When the water surrounding a suspension of cells changes to ice, the intracellular water freezes also. Because of the presence of electrolytes in the cell, it will require a temperature somewhat lower than 0°C to convert the intracellular water to ice. In the frozen state, all metabolic activity ceases.

Most likely little death occurs in the absence of freezing. Hence low temperatures are harmless in themselves. Freezing water is necessary for death to occur although not sufficient in itself. The evidence seems to point to the fact that death by freezing is related directly to the increase in volume of ice over the original water, resulting in cell injury. In other words, death by freezing is probably due to injury by ice crystals.

Survival appears to be related to the temperature to which cells are cooled. It remains relatively high until a certain critical zone of temperature is reached, then survival drops sharply. For example, the survival of cells of *Pasteurella tularensis* (causative agent of tularemia) remains over 60 percent down to −10°C, then drops sharply between −30 and −45°C (Fig. 174). If cells are cooled slowly instead of rapidly, the damage at temperatures below the critical zone is less extensive although the reverse of this may be true with other species.

Nei et al. (1967) found a correlation between (1) preliminary salt

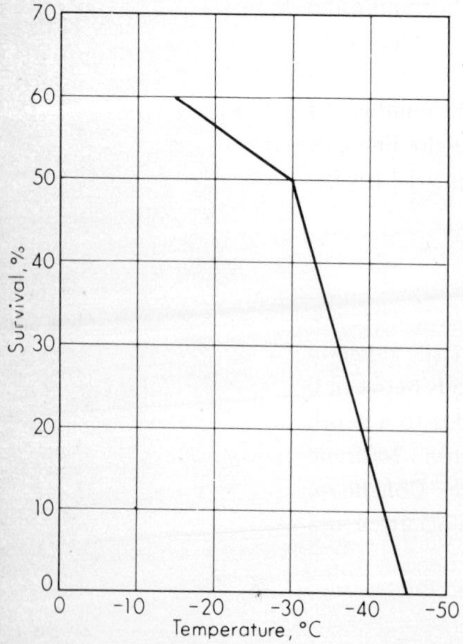

Fig. 174 Percentage of cells of Pasteurella tularensis surviving rapid cooling to indicated temperatures and subsequent slow warming. (After Mazur.)

dehydration of cells of *Escherichia coli* and (2) rates of cooling and the permeability of the cell membrane to water. An increase in cell dehydration, by exposure to a hypertonic salt solution, was accompanied by a decrease in intracellular freezing and resulted in an increase in the survival rate. They concluded that the formation of intracellular ice crystals depended upon the relationship between the velocity of dehydration due to cooling and the permeability of the cell membrane to water.

Exposure of cells to subzero temperatures alters their viability, composition, and nutritional requirements. Bacteria become more dependent upon certain nutritional factors after freezing. Frozen and thawed cells of certain bacteria can no longer grow on a restricted agar but can develop on a more complete medium. This suggests that cells exposed to subzero temperatures are injured in such a way that their nutritional requirements are altered.

The warming phase of low-temperature treatment is also very important on survival rates of microorganisms. When cells of *P. tularensis* are warmed rapidly, a relatively high proportion survive, regardless of whether initial cooling is rapid or slow. On the other hand, when cells are warmed slowly, the percentage of survivors is much lower.

The sensitivity of organisms to freezing damage varies during the growth cycle. The maximum susceptibility to freezing damage occurs during the logarithmic growth phase (see page 235).

Increasing the time of storage in the frozen state results in a progressive decrease in the number of viable cells. However, they may survive for very long periods, and it is not safe to attempt to sterilize foods, water, and other liquids by freezing.

There are three requirements for successful preservation by the technique of freezing: (1) very rapid freezing, (2) low-temperature storage, and (3) very rapid thawing. Most failures to obtain survival after rapid freezing can be attributed to insufficient recognition of these requirements.

During rapid freezing the intracellular water is not changed to ice crystals but to a glass-like or vitreous, amorphous substance. If cells are vitrified without the formation of ice crystals, they can be held in a viable form at low temperature for long periods. During slow thawing, the water may change to ice crystals and cause death of organisms. On the other hand, rapid thawing will most likely prevent the formation of ice crystals and prove harmless to bacteria.

Young bacteria are more sensitive to sudden changes in temperature than old forms. If a suspension of young cells is quickly cooled from 45 to 10°C, as many as 95 percent of the cells may be killed. Gradual cooling produces very little, if any, effect. On the other hand, mature cells are not affected by either an initial cold shock or a prolonged holding at 0°C. The sensitivity of young cells to cold appears to be related in some manner to cell division and to changes within the individual cell.

For more information: Bretz and Ambrosini (1966); De Siervo (1969); Doebbler and Rinfret (1963); Efthimion and Corpe (1969);

Harrison and Cerroni (1956); Klein et al. (1969); Kocka and Bretz (1969); Mazur (1960); Packer et al. (1965); Strange and Dark (1962).

High temperature Bacteria which grow best at high temperatures are called thermophilic organisms; those which live best at low temperatures are called psychrophilic bacteria; and those which fall between the two extremes are the mesophilic forms.

There exists for every organism a maximum, a minimum, and an optimum growth temperature. The temperatures vary somewhat with changes in the environment. Therefore, the values are significant only if the experimental conditions are specified.

Maximum growth temperature The maximum growth temperature may be defined as that temperature above which growth does not take place. The psychrophilic organisms, i.e., those which grow at low temperature, do not develop well even at room temperature. Saprophytic mesophilic bacteria, i.e., those which live best on dead organic matter, show a maximum at about 30°C. A majority of the pathogenic forms for man fall between 30 and 50°C. The thermophilic or heat-loving bacteria may show growth at 60 to 75°C or higher.

These figures may vary, depending upon the composition of the environment. For example, Sie et al. (1961) found that a certain mesophilic bacterium was able to grow at 55°C when yeast autolysate was added to the medium. Dul and McDonald (1971) subjected vegetative forms of *Bacillus subtilis* to a temperature of 53°C and reported the cells exhibited an increase in their thermal resistance.

Optimal growth temperature The optimum temperature is the most favorable temperature for growth. The psychrophilic bacteria have an optimum temperature below 20°C. This group includes many of the pigment-producing bacteria. The mesophilic organisms have optimum temperatures of 18 to 45°C. The saprophytic mesophiles grow best at temperatures of 18 to 25°C; the parasitic mesophiles grow best at the temperature of the host. The thermophilic bacteria vary greatly in their temperature optima. Many grow best at about 55°C. They occur in soil, manure, and other extreta, decaying organic matter, etc. Owing to their great resistance to heat, they are the source of considerable trouble in the canning industry.

Minimum growth temperature The minimum growth temperature is that temperature below which growth does not take place.

The psychrophilic bacteria have a minimum temperature of about 0°C; the mesophilic of 5 to 25°C; and the thermophilic of 25 to 45°C.

A drop in temperature from the optimum to the minimum produces a corresponding drop in the multiplication rate of an organism. At the minimum temperature, growth is exceedingly slow; and below the minimum, it ceases entirely.

Growth temperature range This is defined as the number of degrees between the minimum and maximum growth temperatures. With some organisms the range is very wide; with others it is quite narrow.

The effect of different temperatures on the growth of several organisms is shown in Fig. 175.

The temperature ranges of the psychrophilic, mesophilic, and thermophilic bacteria are summarized as follows:

Bacteria	*Minimum, °C*	*Optimum, °C*	*Maximum, °C*
Psychrophilic	0	15	30
Mesophilic	5–25	18–45	30–50
Thermophilic	25–45	55	60–90

Adaptation of bacteria to growth at elevated temperatures Dowben and Weidenmüller (1968) adapted several strains of mesophilic *B. subtilis* to growth at elevated temperatures, as high as 72°C, by slowly increasing the temperature of incubation. When the adapted bacteria were subsequently grown at 37°C, the ability to grow at elevated temperatures was lost and it was necessary to readjust the bacteria. Similar results have been reported by others working with different bacterial species.

For more information: Brock et al. (1971); Christophersen (1967).

Fig. 175 Effect of temperature of incubation on growth. Organisms from left to right in each illustration: Bacillus viridulus, B. subtilis, Escherichia coli, and Flavobacterium brunneum. Each set of tubes incubated at the following temperatures: (A) 15°C; (B) 25°C; (C) 37°C; (D) 55°C.

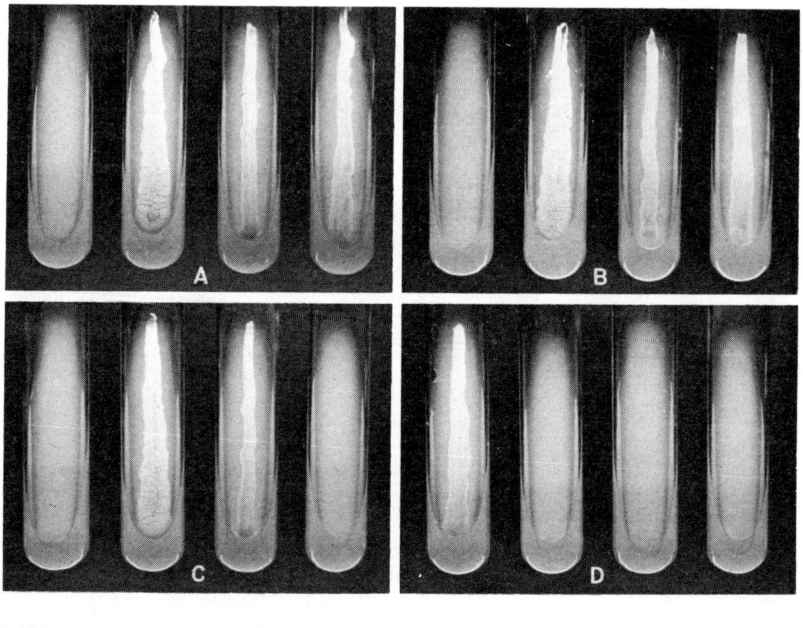

Psychrophilic bacteria The term psychrophilic is defined in various ways, but as used here it includes only those organisms which are capable of growing at 5°C and below, regardless of their optimum temperatures. Eddy (1960) suggested a definition to include growth at 5°C and below, and a maximum temperature below 35°C. As is true with most definitions, exceptions may be found.

Straka and Stokes (1959) isolated psychrophilic bacteria from soil, skua gull and seal feces, and other materials obtained from Antarctica. Thousands per gram were present in some of these.

The organisms were isolated by cultivation at 0°C. They consisted of rods, cocci, and coccobacillary forms and were virtually all gram-negative and nonmotile. The bacteria grew well at 0°C and some at −7°C. The maximum growth temperature was 35°C or lower, with the exception of two isolates which grew at 37°C.

In a later report, Upadhyay and Stokes (1962) found psychrophiles to be widely distributed in nature and could be readily isolated from a variety of foods, soil, sewage, and other habitats. However, obligate anaerobic species were not isolated. Maximum populations were obtained at 5°C aerobically and at 25°C anaerobically. Growth was most rapid in the range 20 to 35°C, with maximum temperatures of 25 to 45°C.

Morita (1968) stated that since 90 percent by volume of the oceans is colder than 5°C, the activities of the psychrophiles below the thermocline become very important in the mineralization of the various organic compounds found in such waters. The Antarctic and Arctic waters are high in productivity and obligately psychrophilic bacteria have been demonstrated in both waters.

For more information: Kenis and Morita (1968); Larkin and Stokes (1966, 1968); McDonald et al. (1963); Sinclair and Stokes (1963, 1964); Stokes and Redmond (1966). (See also page 730).

Thermal death rate The thermal death rate is defined as that temperature at which an organism is killed after an exposure period of 10 min under controlled conditions. This temperature is generally called the *thermal death point*, but since organisms subjected to unfavorable conditions are not all killed in the same period of time, the term *rate* is more appropriate than *point*.

Conditions producing variations in the thermal death rates include (1) water content of the medium, (2) water content of the organisms, (3) hydrogen-ion concentration of the medium, (4) composition of the medium, (5) age of the cells, (6) presence of spores in a culture of a spore-forming species, and (7) incubation temperature of recovery cultures. A change in one or more factors produces a corresponding change in the thermal death rate.

Water content of the medium Death of bacteria by heat is believed to be due to coagulation of the proteins of the protoplasm. Within limits, the greater the percentage of water in a medium, the lower will be the temperature required to kill bacteria. Moist heat is more

Table 9 Relation between moisture content and temperature of coagulation of egg albumin

Amount of water, percent	*Temperature of coagulation, °C*
50	56
25	74– 80
18	80– 90
6	145
0	160–170

effective as a sterilizing agent than dry heat. Dry egg albumin may be heated to a point at which it decomposes without showing any appreciable coagulation. As the percentage of moisture is increased, the temperature of coagulation becomes progressively less. This may be seen in Table 9.

Water content of the organisms Yesair et al. (1946) showed that moist or dry micrococci heated under moist conditions were low in resistance. When heated in moist fat, the resistance was higher. The resistance of dry micrococci in dry fat was exceptionally high and approached that obtained under conditions of dry sterilization. The investigators concluded that dry micrococci might survive sterilization under the usual conditions, and that the mechanism of fat protection in foods appeared to rest upon the localized absence of moisture.

Jensen (1954) reported similar results. He found that dried micrococci suspended in moist butterfat survived only 15 min at 100°C, but in dry butterfat a period of 50 min at 115°C was necessary to achieve the same result.

Hydrogen-ion concentration of the medium The pH of the medium plays an important role in the survival rate. Most organisms are more easily killed in acid or alkaline solutions than in a neutral environment. In general the greater the degree of acidity or alkalinity, the lower will be the temperature required to kill bacteria. Acid-fast organisms are exceptions in that they survive heating in acid media far better than in slightly alkaline solutions. Since the organisms are protected by a waxy capsule (see page 65), it seems reasonable to postulate that this covering is destroyed under alkaline conditions making the cells more vulnerable to attack.

For more information: Chislett and Kushner (1961); Walker (1964).

Composition of the medium The composition of the suspending medium may greatly affect the thermal death rate. Media containing high concentrations of albumins or other proteins usually increase the temperature required to destroy bacteria. The proteins form a film around the organisms, protecting them from unfavorable influences.

The composition of the medium used for the recovery cultures is also important. Variations in the composition of the subculture medium may produce variations in the final results. The same applies to the time required for the germination of spores of some species.

Age of cells The age of the cells influences the thermal death rate. Old cells are generally more resistant to adverse conditions than young cells. Cultures of definite age, usually 24 hr, should be used in the test.

For more information: Harris (1963); Lemcke and White (1959); Russell and Harries (1968).

Presence of spores Non-spore-forming bacteria and the vegetative forms of the spore bearers are generally killed by moist heat at temperatures of 60 to 70°C. Spores can withstand a temperature of 100°C or higher. In reporting the thermal death rate of a spore-forming organism, care should be taken to make sure spores are present in the culture.

Spores and vegetative cells of different species, and of different strains of the same species, exhibit marked differences in heat resistance. Vegetative cell resistance is not related to spore resistance. Williams and Zimmerman (1951) reported that cultures might contain vegetative cells of high resistance and spores of low resistance, or vegetative cells of low resistance might be associated with spores of high resistance.

According to Curran (1952), spore resistance varies widely from species to species, to a considerable extent within a species, and within a given spore population. It is independent of the luxuriance of growth and of the luxuriance of sporulation.

Spores of *Clostridium botulinum* are most resistant to heat when newly formed, whereas in other species maximum resistance is attained after a certain period of maturation. Among the aerobic sporing bacteria those species with the highest temperature of growth usually show the greatest resistance to heat.

The nature of the nutrients in the medium is important. *B. subtilis* spores produced in vegetable infusions show high heat resistance, whereas the spores formed in certain digest media exhibit low thermal resistance. Spores of *C. botulinum* produced in a medium containing certain fatty acids are more heat-resistant than spores formed in media from which the fatty acids are previously extracted. Media deficient in certain essential metallic ions, such as magnesium, calcium, and iron, yield spores of low thermal resistance.

Rao and Visweswaram (1961) incubated spores of *B. subtilis* in 1 percent peptone solution at 37°C and found that 90 percent of the spores lost their heat resistance within 60 min. When 1 percent glucose was substituted for the peptone, only 5 percent of the spores lost their heat resistance in the same period of time. In a mixture of 1 percent peptone and 1 percent glucose, the results were the same as in 1 percent peptone alone.

Spores show maximum heat resistance when produced at the optimum growth temperature. Any deviation from the optimum yields spores of lowered heat resistance. The pH of the medium affects spore resistance. Spores usually exhibit maximum resistance at or near the neutral point. In strongly acid or alkaline substrates, thermal resistance is materially reduced. Salt in low concentrations increases heat resistance of aerobic and anaerobic mesophilic species. Cook and Gilbert (1968) reported that 2, 4, and 8 percent sodium chloride added to the heating medium progressively reduced the heat resistance of spores of *B. stearothermophilus*.

The concentration of moisture in the medium influences the heat resistance of spores. Moist heat at 115°C for 30 min usually is sufficient to sterilize spores, whereas dry heat at 160 to 180°C requires 1 hr. The presence of organic matter in the medium increases the resistance of spores against heat. The presence of added vitamins may alter spore resistance. The addition of 100 ppm of ascorbic acid (vitamin C) to a neutral phosphate buffer increases the spore resistance of a *Clostridium* species. Spores that survive heat treatment are more readily killed by further application of the same treatment than are untreated spores.

The thermal resistance of spores suspended in sucrose solutions increases with the concentration of sucrose. With spores suspended in 50 percent sucrose, the increase in heat resistance is almost immediate and remains high for many hours.

Williams and Robertson (1954) cultivated several strains of *B. stearothermophilus* at increasing temperatures of incubation. The spores which developed increased in resistance to heat as the temperature of incubation was increased.

El-Bisi and Ordal (1956) determined the effect of sporulation temperatures of 30, 37, and 45°C on the thermal resistance of spores of *B. coagulans* and concluded that the higher growth temperature markedly enhanced their heat resistance.

Hoffman et al. (1968) investigated the effect of spore moisture content on the rate of death. Spores of *B. subtilis* were equilibrated to 11, 33, and 85 percent relative humidity (RH), then heated to several temperatures ranging from 108 to 164°C. They found that spores preequilibrated to 11 percent RH were least resistant and those preequilibrated to 85 percent RH were most resistant.

Levinson and Hyatt (1969) activated spores of *B. megaterium* by heating at 10 temperatures (50 to 75°C). Activation was measured by increase in germination rate over that of dormant spores. They reported the exposure time required to achieve a given germination rate increased with decreasing temperature.

For more information: Alderton and Snell (1969); Amaha and Sakaguchi (1954); Angelotti et al. (1968); Bruch and Smith (1968); El-Bisi et al. (1962); Evans and Curran (1960); Fox and Pflug (1968); Levinson and Hyatt (1971); Mullican and Hoffman (1968); Nath and Clegg (1969); Shull et al. (1963); Yokoya and York (1965).

Incubation temperature of recovery cultures Incubation temperatures of 24, 27, and 31°C have been shown to be more favorable for the development of heated spores of *C. botulinum* than 37°C. This was also found to be true with other species of bacteria.

These are probably the most important factors which influence the thermal resistance of bacteria and spores. Unless the conditions of the test are mentioned, the thermal death rates will be of doubtful value.

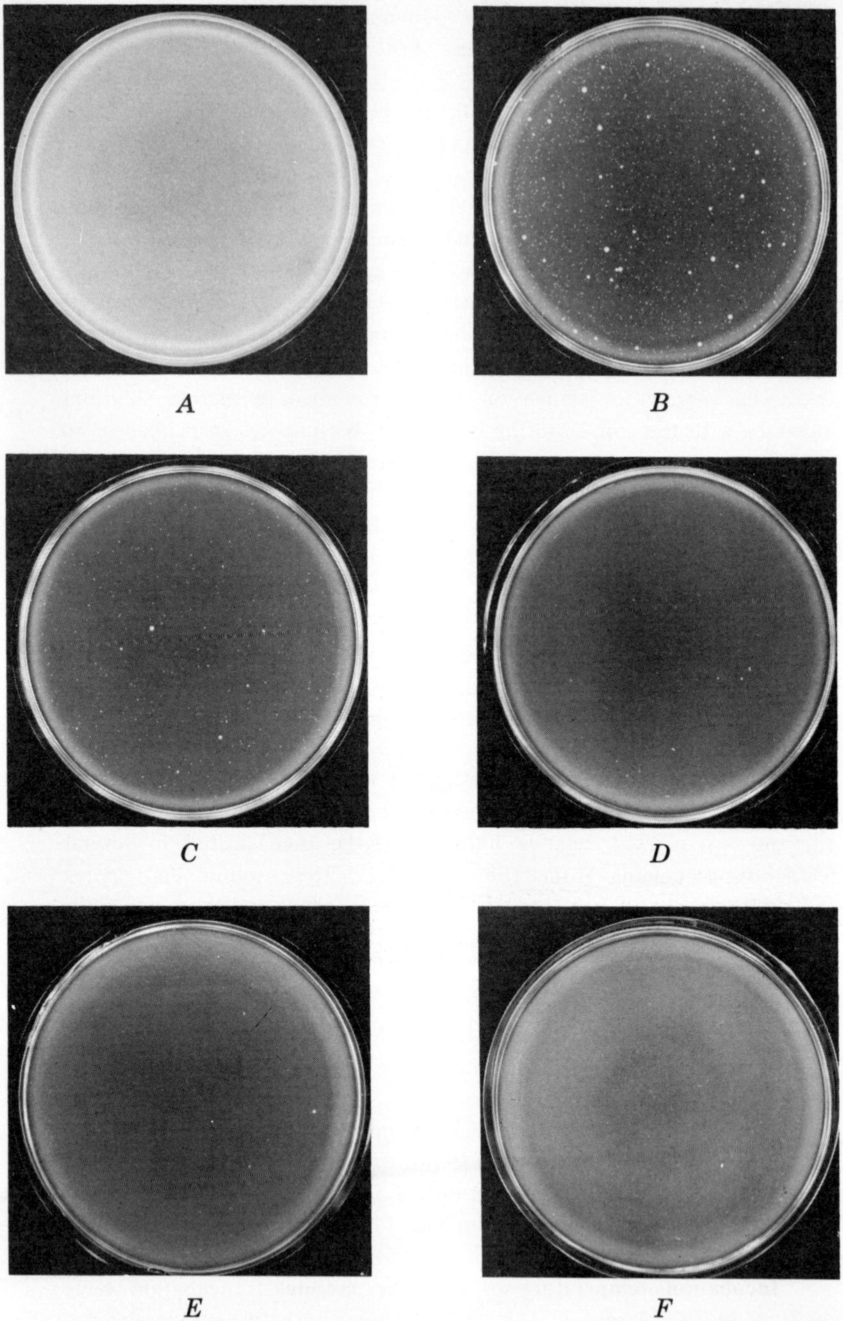

Fig. 176 Thermal death rate of Serratia marcescens. Six tubes, each containing 1 ml of culture, were heated for 10 min at the following temperatures: (A) 50°C; (B) 55°C; (C) 60°C; (D) 65°C; (E) 70°C; (F) 75°C. Then the contents of each tube were transferred to petri dishes and mixed with melted agar.

Thermal death time In testing for the thermal death rate, the time is kept constant and the temperature necessary to produce sterilization is determined. In finding the thermal death time, the temperature is kept constant and the time required to kill all cells is determined.

Results of thermal death rates and thermal death times are of great value in applied bacteriology and especially in the canning industry. They aid the canner in determining the temperatures and times required to process canned foods.

It is the usual practice to isolate the organisms causing the spoilage of a certain kind of food and to determine their thermal death rates or times under controlled environmental conditions. The results then may be used as a guide in determining the temperature and time required to process the food.

The thermal death rates of *Serratia marcescens* and *B. cereus* are shown in Figs. 176 and 177.

For more information: Hansen and Riemann (1963); Licciardello and Nickerson (1963).

Effect of ultraviolet irradiation

A few bacteria elaborate photosynthetic pigments which function in these organisms in a manner similar to chlorophyll in plants. The photosynthetic bacteria are benefited by light rays. However, the great majority of bacteria do not contain photosynthetic pigments and are unable to utilize sunlight as a source of energy. Furthermore, such organisms are harmed by exposure to ultraviolet (UV) rays.

UV rays are the invisible components of the sun's radiation. They are short rays and are measured in angstroms (Å). One millimeter is equal to 10 million Å, or 1 μ equals 10,000 Å. UV rays centering around 2537 Å, emitted by special lamps, exert a pronounced destructive action on bacteria and other microorganisms. However, the highly germicidal range is between 2000 and 2967 Å with a maximum at 2650 Å (Fig. 178). A wavelength of 2537 Å from germicidal tubes is about 10 times more bactericidal than a wavelength of 2967 Å from sun lamps, about 4000 times more germicidal than a wavelength of 3650 Å from photochemical lamps, and about 30,000 times more germicidal than the most visible wavelength from the sun and from artificial light sources. UV rays are able to kill cells, temporarily delay cell division and also the synthesis of certain substances by cells, change the manner in which substances pass across cellular membranes, cause abnormalities in chromosomes, and produce mutations. Such rays are extensively employed by experimenters for the study of living systems.

All living microorganisms, such as molds, yeasts, bacteria, viruses, and spores are sensitive to UV treatment. However, spores require approximately twice the exposure to such rays to effect the same percentage reduction in numbers as the vegetative cells producing them.

The penetrating power of UV light into water is dependent upon the

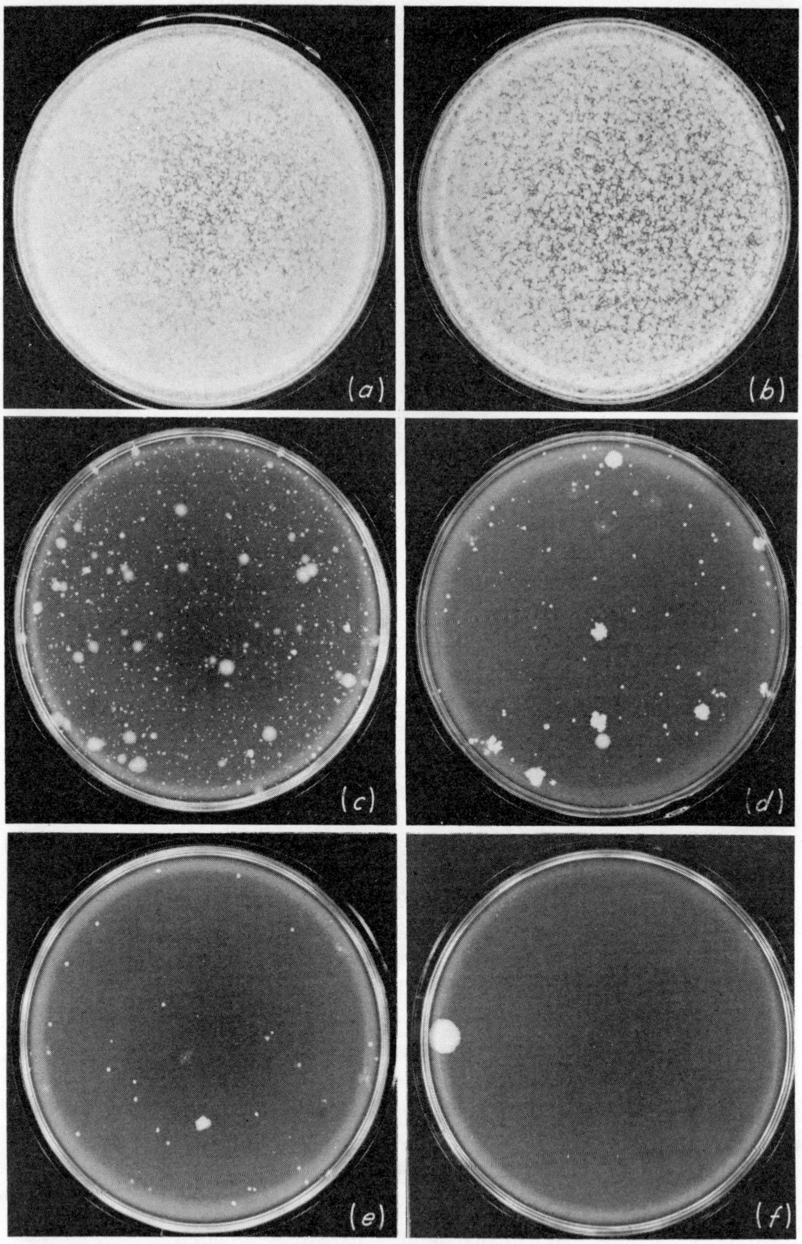

Fig. 177 Thermal death rate of Bacillus cereus. Six tubes, each containing 1 ml of culture, were heated for 10 min at the following temperatures: (A) 75°C; (B) 80°C; (C) 85°C; (D) 90°C; (E) 95°C; (F) 100°C. Then the contents of each tube were transferred to petri dishes and mixed with melted agar.

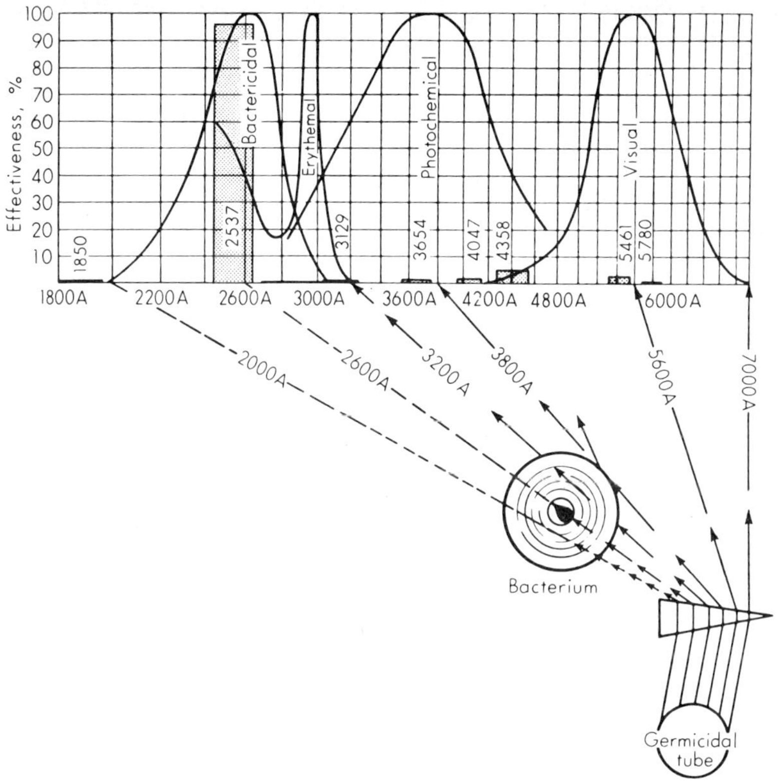

Fig. 178 Germicidal portion of the electromagnetic spectrum. Wavelengths are measured in angstroms. One angstrom is equal to 1/10,000,000 mm, or approximately 1/250,000,000 in. (Courtesy of General Electric Company.)

nature and quantity of suspended matter. Ladanyi and Morrison (1968) determined the effect of UV on (1) several species of organisms frozen in ice cubes and held at −20°C, and (2) buffer suspensions of the same organisms placed beneath an ice block 19 cm thick. Results showed that 1 min of UV treatment killed as many as 97 percent of the gram-negative and at least 60 percent of the gram-positive bacteria frozen in ice cubes 30 mm thick. UV transmitted through an ice block 19 cm thick inactivated 98 percent of the bacteria suspended in the buffer solutions.

Radiations are effective only if they are absorbed. The rays are absorbed by the proteins, by the deoxyribonucleic acid (DNA) of the nucleus, by the ribonucleic acid (RNA) of the cytoplasm, and by other organic compounds. Many of the effects of UV on living cells are exerted solely or chiefly by means of changes in the DNA. In the case of viruses, this may be the whole story. When a virus attacks a normal cell, only the nucleic acid is injected into the cell. Most proteins in solution show marked absorption bands in the UV region. DNA, the genetic material, absorbs most strongly at 2650 Å, the most potent wavelength in its effects

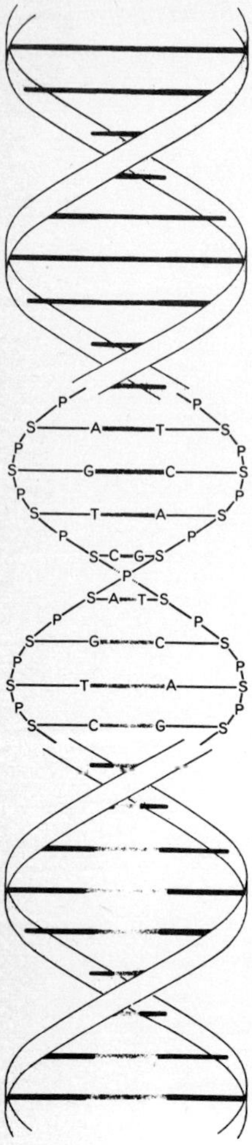

Fig. 179 Diagram of a double-helix DNA molecule. See text for description. (Courtesy of R. A. Deering and Scientific American.)

on living cells. UV rays exert a negligible bactericidal action on bacteria suspended in blood. The blood proteins absorb the rays, preventing them from reaching the bacteria.

Natural DNA consists of a double-strand helix (Fig. 179). The helices proper consist of an alternation of sugar (S) as deoxyribose and phosphate (P) groups. Attached to each of the sugars are the four nitrogenous bases, generally the purines adenine (A) and guanine (G), and the pyrimidines thymine (T) and cytosine (C). The bases on the two backbones are joined in pairs by hydrogen bonds (gray color), adenine on one chain always being paired with thymine on the other, and guanine with cytosine (Fig. 180). Genetic information is provided by the sequence of bases along a strand. The hydrogen bonds which join the pairs are easily broken by heating, and the backbones are partially or completely separated into two strands of so-called denatured DNA.

UV rays are absorbed primarily by the bases, especially the two pyrimidines thymine and cytosine which are considerably more sensitive than the purines adenine and guanine.

Available evidence indicates that UV alters DNA activity by causing two molecules of thymine to form a double molecule or dimer (Figs. 181 and 182). Irradiation with high doses of the longer wavelengths of UV

Fig. 180 The four nitrogenous bases of DNA. Thymine is always paired with adenine, and cytosine with guanine. See text for description.

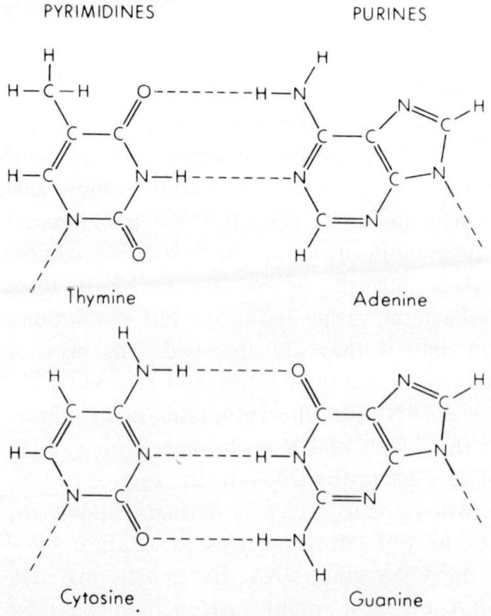

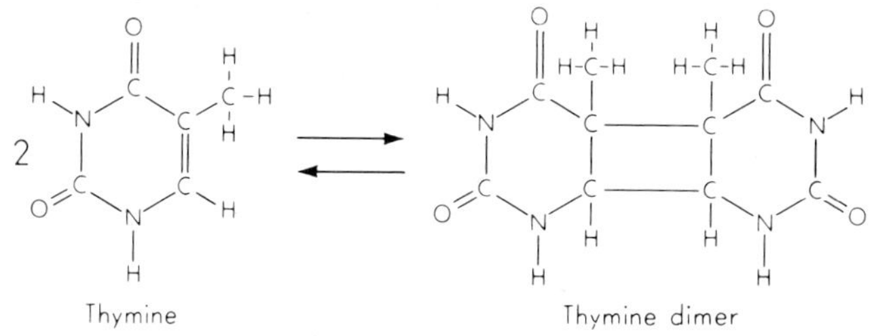

Fig. 181 Ultraviolet light alters DNA activity by causing two molecules of thymine to form a double molecule or dimer. See text for description.

(2600 to 2800 Å) leads to an equilibrium condition in which most of the thymine molecules are dimerized; high doses of shorter wavelengths of UV (2300 to 2500 Å) tend to reverse the reaction, partially restoring the original activity of the DNA.

For more information: Ashwood-Smith et al. (1968); Camerman and Camerman (1968); Deering (1962); Gallant and Suskind (1961); Ginoza (1967); Jennings et al. (1970); Kapp and Smith (1970); Kleczkowski and Govier (1969); Riklis (1965); Sawada and Suzuki (1961); Smith and Meun (1970).

Fig. 182 Diagrammatic representation of alterations in the DNA molecule by ultraviolet light. See text for description. (Courtesy of R. A. Deering and Scientific American.)

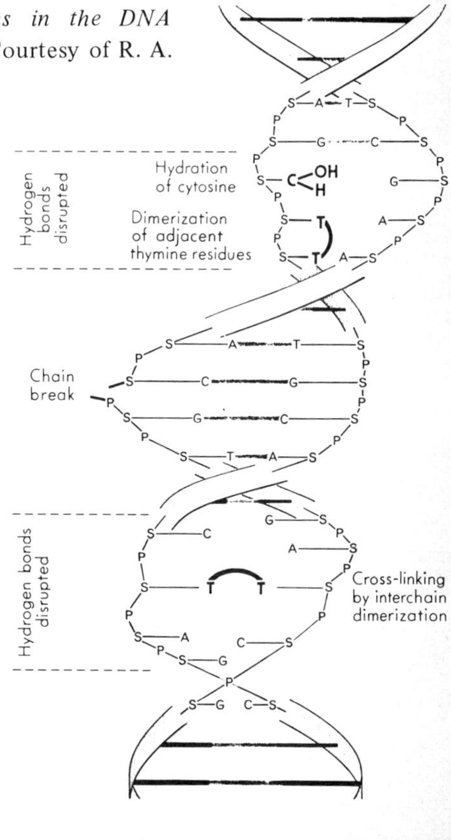

Protection of bacteria against ultraviolet

A number of substances have been shown to be capable of protecting bacteria against the lethal and mutagenic action of UV light.

Pyruvate added to bacterial suspensions prior to irradiation protects the cells against radiation damage. Prolonged exposure to UV may cause the pyruvate to become both toxic and mutagenic. The addition of pyruvate to cells after exposure does not modify the biological response to the radiations.

Sodium azide added to suspensions of some bacteria prior to irradiation reduces their sensitivity to the lethal action of the rays.

Intracellularly synthesized catalase and the cytochromes exhibit a protective action on *E. coli* exposed to UV.

Leif and Hebert (1960) reported that urocanic acid and dipicolinic acid, two UV-absorbing compounds, protected bacteria suspended in broth and in physiological sodium chloride solution.

Shimada and Takagi (1967) found that caffeine inhibited UV-induced DNA breakdown in bacteria.

Photoreactivation of irradiated bacteria

Kelner (1949) showed that UV-damaged bacteria could be recovered in viable form by exposure to daylight. To give a typical experiment: A suspension containing 8 million *E. coli* cells per milliliter was irradiated until the viable count was reduced to 100 cells per milliliter. A portion of this suspension was exposed to light of 3600 to 4900 Å for 30 min, then assayed again. The suspension then contained 2 million viable cells per milliliter, whereas a portion of the original irradiated culture kept in the dark still contained only 100 bacteria per milliliter.

The degree of recovery possible for irradiated cells was dependent upon the original dose of UV light. The greater the dose, the smaller the proportion of inactivated cells that could be revived. Given a sufficiently high dose of UV, probably not a single cell would recover, regardless of the amount of reactivating light.

A temperature effect was also involved in photoreactivation. The warmer the cells during exposure to visible light, the greater the amount of photoreactivation up to a temperature of 45 to 50°C. Beyond 50°C, the cells were subject to heat injury.

Low doses of UV had little immediate effect on the synthesis of ribonucleic acid by the cells but stopped immediately and completely the synthesis of DNA. After reactivating light, synthesis of DNA was resumed at an accelerated rate.

Kelner believed that specific inhibition of DNA synthesis was the basic immediate effect of UV light.

More recently, Wulff and Rupert (1962) identified one mechanism of reactivation. They found that a particular enzyme preparation in the presence of blue light could break up 90 percent of the thymine dimers in irradiated DNA.

Kantor and Deering (1967) reported that small doses of UV (2650 Å) caused *E. coli* B to grow into long, multinucleate, nonseptate, filamentous cells. Filament formation was prevented by irradiation with photoreactivating light (4060 Å) after UV treatment. They concluded that UV-induced division inhibition in *E. coli* B was initially induced by repairable lesions in the DNA, probably pyrimidine dimers.

For more information: Castellani et al. (1964); Ganesan and Smith (1969, 1970); Govier and Kleczkowski (1969); Hanawalt and Buehler (1960); Harm (1968, 1970); Jagger (1958); Kleczkowski (1968); Pootjes (1968); Town et al. (1971).

Reactivation by other methods Ellison et al. (1955) reversed the lethal and mutagenic effects of UV light by plating the irradiated bacteria in nutrient agar containing sodium acetate. The degree of reactivation was directly proportional to the acetate concentration.

Heinmetz and Lehman (1955) obtained high numbers of viable cells when irradiated bacteria were incubated with certain combinations of metabolites of the tricarboxylic acid cycle (page 412) and the coenzymes nicotinamide-adenine dinucleotide (NAD) and coenzyme A. The combined action of photoreactivation and metabolic reactivation was more effective than the action of a single process alone.

Weatherwax (1956) reported that survival of *E. coli* after irradiation was strongly dependent on the pH of the plating medium. With a given dose of radiation, the number of survivors producing colonies on agar was increased as much as a thousandfold when the pH of the plating medium was decreased from 8 to 5. The actual decision for or against survival occurred during the interval of incubation between 2 and 4 hr after irradiation. At any time prior to the end of 2 hr of incubation, survival could be determined by changing the pH of the medium. This opportunity was rapidly lost during the next 2 hr of incubation, after which time the radiation damage had become irreversible.

Irradiation of culture media The observations already reported are limited to the action of UV on microorganisms and their spores. It has been shown that culture media exposed to light become less suited to bacterial growth. Bedford (1927) reported that the irradiation of media caused the formation of hydrogen peroxide, a compound that is toxic to organisms. The concentration of peroxide that accumulated in media depended upon the wavelength of light and the period of exposure.

Similar results were reported by Wyss et al. (1948). They found the toxic effect of irradiated broth could be negated by the addition of the enzyme catalase, which is capable of decomposing hydrogen peroxide as it is formed.

For more information: Cohn and Middlebrook (1965).

Commercial uses Air sanitation is closely analogous to water sanitation. Its purpose is substantially the same, namely, to make the air in confined spaces safer under the particular circumstances of its use and to guard against the possibility that airborne organisms may cause clinical infections. This is especially true of the respiratory diseases, because the organisms involved are caught from respired air. Many hospital rooms and operating rooms are irradiated to sterilize or to decrease the number of organisms.

UV rays are used commercially for the destruction of various organisms in foods such as sugar, apple cider, maple sap, meats, and bakery products. Spores of *B. stearothermophilus* are commonly present in sugar and may cause serious losses in many food preparations. The spores may be destroyed by irradiation of the sugar crystals. Kissinger and Willits (1966) found UV light to be an effective agent for the destruction of

Fig. 183 Effect of ultraviolet light. (A) Bacillus subtilis exposed to an ultraviolet lamp for 1 hr; (B) same exposed for 2 hr. Since the culture contained many spores, a longer exposure was necessary to produce complete sterilization; (C) staphylococcus aureus exposed for 1 hr; (D) same exposed for 2 hr. Culture was completely sterilized after 2 hr.

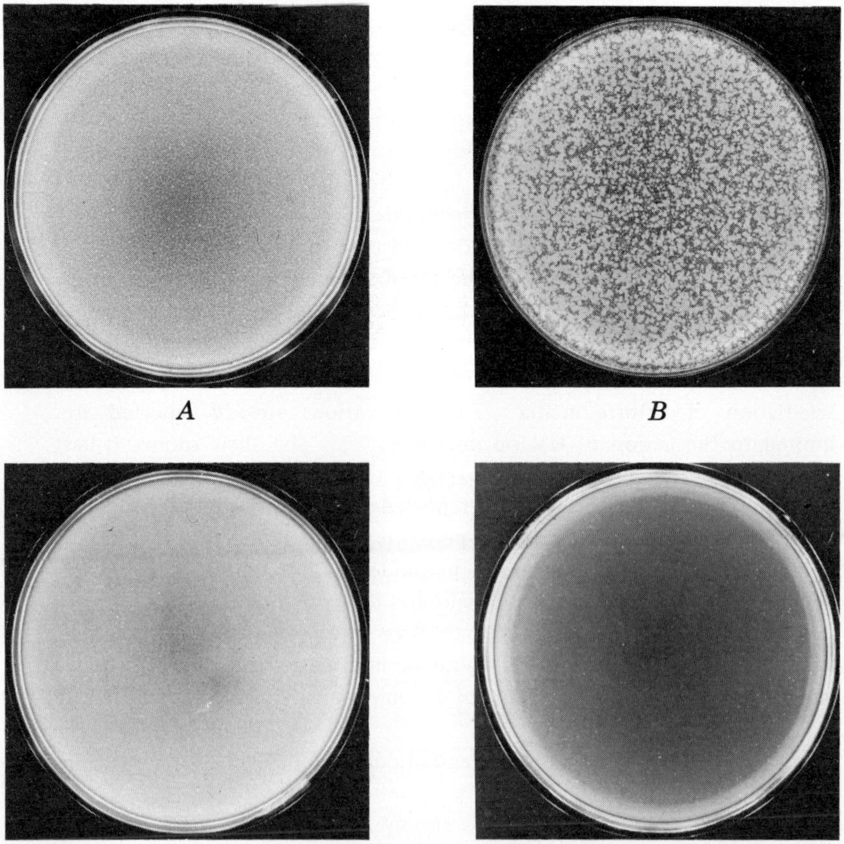

A B C D

organisms in maple sirup. Harrington and Hills (1968) employed specially designed UV lamps for reducing the bacterial population of fresh cider. UV light is effective in sterilizing surfaces of dairy equipment such as tinned dippers, cans, and pasteurizing vats, also milk bottles. Luckiesh et al. (1947) found low-pressure mercury-vapor lamps to be effective in destroying bacteria in water. They recommended the use of such lamps commercially for the treatment of drinking water.

These are only a few examples of the many commercial applications of UV rays. More information on UV light may be found in Chap. 19, page 663.

The effect of UV light on the growth of *B. subtilis* and *Staphylococcus aureus* is shown in Fig. 183.

Effect of osmotic pressure

Osmosis may be defined as the diffusion that takes place between two miscible liquids separated by a membrane, especially a semipermeable membrane, which tends to equalize their concentrations.

The term osmotic pressure refers to the unbalanced pressure that gives rise to the phenomena of diffusion and osmosis, as in a solution in which there are differences in concentration. It is proportional to the absolute temperature and also to the molecular concentration.

Plasmolysis The rate at which water passes into and out of cells is in part determined by the ratio that exists between the concentrations of electrolytes inside and outside the cell membranes. If an organism is immersed in a solution having a higher osmotic pressure, water will leave the cell. This will continue until an equilibrium is established between the osmotic pressures inside and outside the cell. If the initial difference in osmotic pressure between the inside and outside of the cell is sufficiently great, the cytoplasmic membrane will be drawn in with the cytoplasmic contents and collect in the center of the cell. The cell is then said to be plasmolyzed, and the process is called plasmolysis. The solution on the outside is hypertonic with respect to the solution on the inside of the cell.

Zimmerman (1969) reported that dilute solutions of salts (0.017 *N*) did not cause cells of *Serratia marcescens* to lose water whereas stronger solutions (1.71 *N*) markedly reduced the weight and water content of centrifugally packed cells. He concluded that: (1) the cells lost weight and water to concentrated salt solutions through a nonosmotic competitive dehydration, causing a shrinkage of the protoplasmic gel. The cell wall was too rigid to show any appreciable shrinkage. Therefore a space appeared between the cell wall and the cytoplasmic membrane; (2) cells may have equilibrated their water activity with that of their environment by two mechanisms, namely, the loss of water by plasmolysis or competitive dehydration, and alteration in cell permeability that admitted previously excluded solutes to the cell interior.

Scheie (1969) found cells of *E. coli* B/r to be plasmolyzed in sucrose solutions and observed under phase contrast microscopy. The degree of

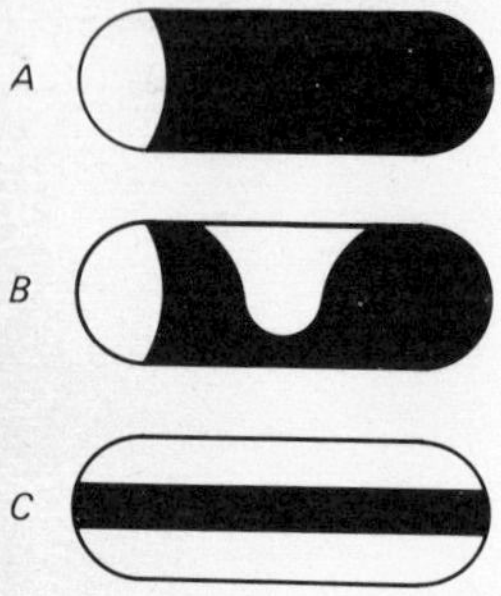

Fig. 184 Schematic diagrams of three degrees of plasmolysis. (a) Slight (b) extensive, and (c) severe. (After Scheie.)

plasmolysis was categorized as slight, extensive, or severe (Fig. 184). Survival curves indicated that extensive plasmolysis was not lethal to colony-forming ability.

Plasmoptysis If an organism is immersed in a solution having a lower osmotic pressure, water will enter the cell. This will continue until an equilibrium is established between the osmotic pressures inside and outside the cell. If the initial difference in osmotic pressure between the inside and the outside of the cell is of sufficient magnitude, the cell membranes will burst, releasing their contents. The cell is then said to be plasmoptyzed, and the process is called plasmoptysis. In this case, the solution on the outside is hypotonic with respect to the solution on the inside of the cell.

Isotonic solutions If the concentrations of ions and molecules on the inside and outside of the cell membranes are equal, there will be no difference in their osmotic pressures. The result will be neither shrinking nor swelling of the cell contents. The two solutions are then said to be isotonic with respect to each other.

Cellular changes during plasmolysis and plasmoptysis can easily be demonstrated in higher plant and animal cells. Until recently such changes were not reported as having been observed in the true bacteria. However, bacteria are sensitive to osmotic effects but to a lesser extent than plant and animal cells. A greater increase in the osmotic pressure of the surrounding solution is necessary before any destructive action on bacteria is observed.

Cota-Robles (1963) plasmolyzed cells of *E. coli* by immersion in 12 percent sucrose solution for 5 to 20 min prior to fixation. Then ultrathin sections were prepared and examined under an electron microscope. The illustrations leave no doubt that plasmolysis occurred.

The use of high osmotic pressures finds practical applications in the preservation of some foods from attack by fungi. This principle is employed in the preservation of jams, jellies, and condensed milk by means of sugar; and of meats, corned beef, fish, etc., by the use of salt (Fig. 185).

Iyer and Bhat (1954) found 13 strains and species of the genus *Bacillus* to tolerate salt and sugar concentrations of 10 to 15 percent and 30 to 60 percent, respectively. The concentrations of these two ingredients as natural preservatives in the food industry may often prove inadequate to prevent spoilage. Such high concentrations will easily destroy higher plant and animal cells.

For more information: Bateman and White (1963); Christian and Waltho (1961); Dombrowski (1963); Gossling (1959); Mohr and Larsen (1963); Anton (1972); Bernheim (1971); Scheie and Rehberg (1972).

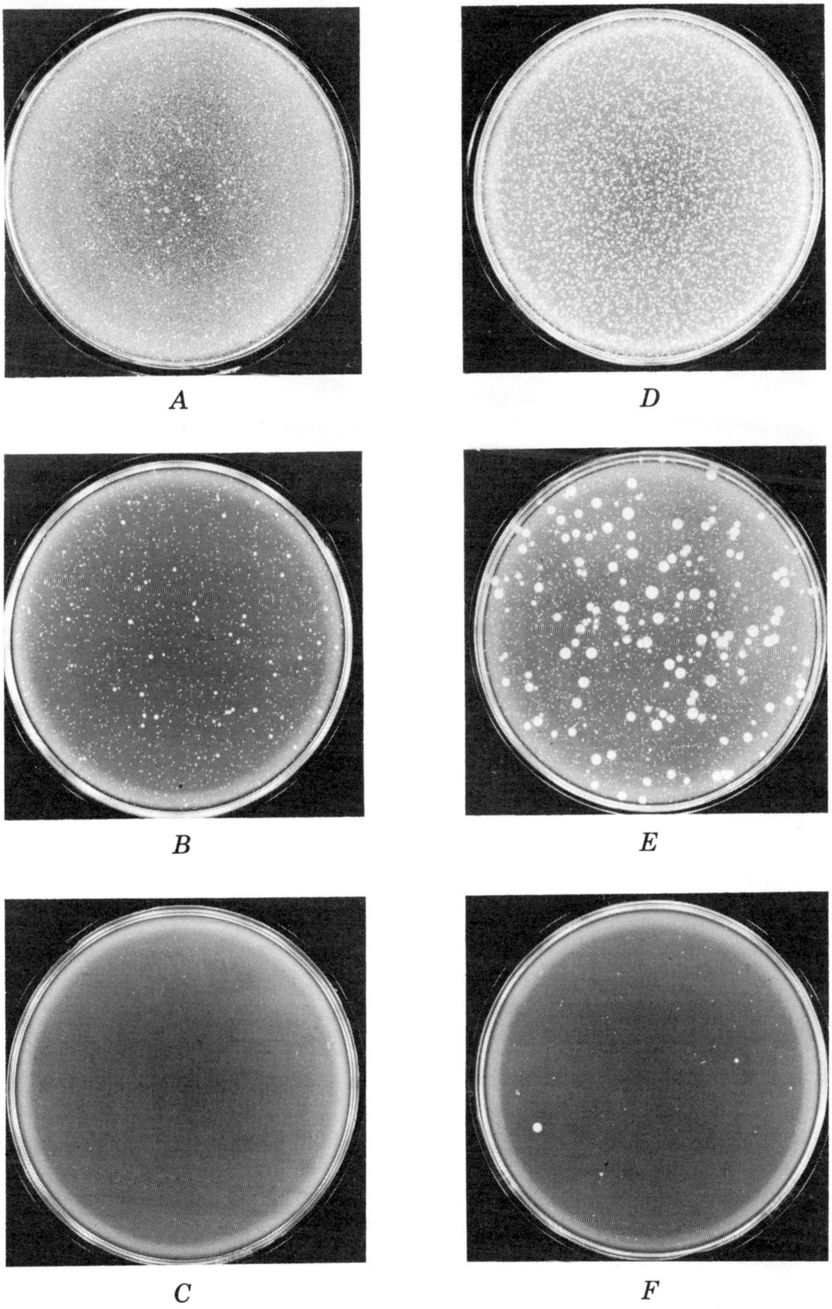

Fig. 185 Effect of osmotic pressure. Staphylococcus aureus and Escherichia coli exposed to 30 percent sodium chloride solution, then mixed with melted agar and poured into plates. (A) S. aureus poured immediately; (B) same after 24 hr; (C) same after 48 hr; (D) E. coli poured immediately; (E) same after 24 hr; (F) same after 48 hr.

Marine bacteria Most marine bacteria differ from freshwater forms in that they are able to tolerate much greater concentrations of salt. ZoBell and Feltham (1933) found that less than 10 percent of the bacteria isolated from seawater were able to multiply in nutrient freshwater media and that a smaller number of species isolated from freshwater could multiply in media prepared with undiluted seawater.

ZoBell and Michener (1938) acclimatized marine bacteria to hypotonic solutions by gradually diluting the seawater medium with each successive transfer of the cultures. The species could be acclimatized to 25 to 30 percent seawater. Below this concentration, considerable difficulty and delay were encountered in making the bacteria grow. Most of the original cultures which were kept in the refrigerator on undiluted seawater agar multiplied when transferred to freshwater medium. The old stock cultures adapted themselves better to hypotonic solutions than did cultures of the same organisms gradually acclimatized to decreasing concentrations of seawater medium.

Young bacteria are more susceptible to adverse environmental conditions than old cells. Cultures gradually acclimatized to decreasing concentrations of seawater tend to keep the organisms in a physiologically young condition. On the other hand, the parent stock cultures become physiologically old and senescent and less susceptible to environmental changes. These cells are better able to adapt themselves to adverse conditions.

Influence of hydrostatic pressure

Hydrostatic pressure has been shown to exert a marked effect on the growth of a number of representative species of bacteria. ZoBell and Johnson (1949) found that at 30°C terrestrial organisms developed abundantly within 48 hr at normal pressure but failed to multiply under 600 atm (9,000 lb per sq in) of pressure. In fact, some of the bacteria were killed by this pressure during the period of incubation. Growth of most of the species was visibly retarded by a hydrostatic pressure of 400 atm, and plate counts indicated that growth was slower and death faster at only 300 atm than at normal pressure. On the other hand, marine species, which were isolated from depths where the pressure approximated 500 atm grew readily under 600 atm at both 30 and 40°C. Other species isolated near the surface of the sea were intermediate, or more nearly resembled terrestrial forms in their sensitivity to pressure.

In general the retarding effect of pressure was more marked at low than at high temperatures. The net effect of high temperatures was to oppose the retarding action of hydrostatic pressure.

ZoBell and Cobet (1962) found pressures of 100 to 500 atm to retard the growth and reproduction of *E. coli* in nutrient medium. Although cell division was retarded, cell size (biomass) increased. When incubated near the threshold of pressure tolerance (about 475 atm at 30°C), some cells of *E. coli* grew into long slender filaments showing little evidence of cell division. The inhibitory effects of pressure on reproduction and growth were found to be less at 30 than at 20 or 40°C. Pressures of 400 to 1000 atm accelerated the death rate of *E. coli* in nutrient medium. The lethal effects of pressure were greater at 40 than at 20°C.

Bornside (1967*a, b*) exposed shallow stationary broth cultures of *Pseudomonas aeruginosa* to 100 percent oxygen at 3 atm of absolute pressure and reported a marked decrease in growth. Subsequent exposure of such cultures to air restored normal rate of growth. The minimal inhibitory concentration of the antibiotic polymyxin B for the same organism was determined after exposure periods of 3, 6, and 12 hr to high oxygen pressure. The amount of antibiotic required to inhibit growth of the organism was found to decrease with time.

Johnson and ZoBell (1949*a, b*) reported that spores of *B. subtilis*, suspended in a buffered salt solution of pH 7.0, slowly lost their viability at 25°C, and this decrease in viability was accelerated by a hydrostatic pressure of 600 atm. At 92.5 or 93.6°C corresponding suspensions rapidly lost their viability, but this loss was retarded by a hydrostatic pressure of 600 atm.

Clouston and Wills (1969) subjected phosphate-buffered spores of *B. pumilus* to pressures as high as 1700 atm at 25°C. Phosphate-buffered spores were more sensitive to compression than spores suspended in distilled water. Initiation of germination occurred at pressures exceeding 500 atm and was the prerequisite for inactivation by compression.

Effect of pressure on enzymes Vignais and Vignais (1954) reported that high hydrostatic pressures also inhibited the action of enzymes. Morita and ZoBell (1956) inactivated the succinic dehydrogenase system in *E. coli*. Some inactivation occurred at 200 atm and a temperature of 30°C. The degree of inactivation increased progressively with time of compression. All of the enzyme was irreversibly inactivated after 4 hr at 1000 atm.

Haight and Morita (1962) reported that the enzyme aspartase isolated from *E. coli* also was inactivated by high pressures.

For more information: Albright and Morita (1968); Clouston and Wills (1970); Hedén (1964); Morita and Mathemeier (1964); Paul and Morita (1971); Timson and Short (1965); ZoBell (1964); ZoBell and Budge (1965); Albright and Henigman (1971).

Preservation of bacteria

Organisms in the desiccated state are not capable of multiplication. Moisture is necessary for this to occur. When organisms are dried they gradually die, the rate of death being dependent upon several factors. Slow drying is more effective than rapid drying. Death may be due to a denaturation of the proteins, to a destruction of the essential enzymes, or to other causes.

The percentage of survivors after desiccation may show wide variation, depending upon the species, the age of the culture, the surface

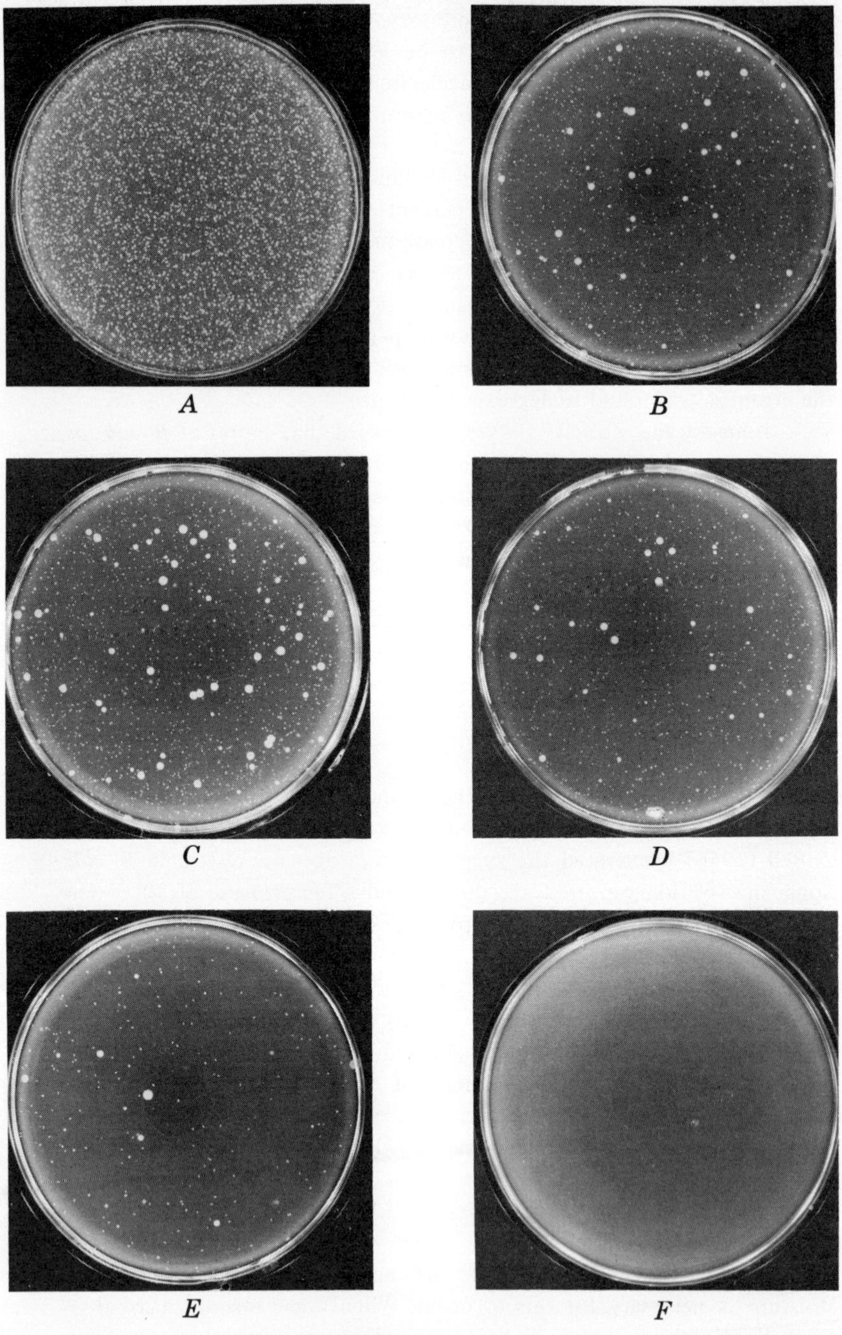

Fig. 186 Effect of desiccation on Escherichia coli. (A) Control, not dried; (B) dried for 24 hr at 37°C; (C) dried for 48 hr; (D) dried for 72 hr; (E) dried for 96 hr; (F) dried for 216 hr.

on which the cells are dried, the temperature of drying, and the composition of the suspending medium.

Young cells are, in general, less resistant to desiccation than old cells. Cells in the logarithmic growth phase are more easily destroyed by drying than cells in the later phases of growth.

Bacteria suspended in broth, milk, serum, and other albuminous fluids survive drying fairly well. The proteins of the medium act as protective colloids which make the drying process more gentle and less abrupt. On the other hand, bacteria suspended in water or saline are more easily destroyed by desiccation.

Spores are more resistant to desiccation than the vegetative cells producing them. They are better able to withstand adverse conditions. Likewise, capsulated organisms are more resistant to drying than non-capsulated bacteria. The mucilaginous deposit surrounding the organisms acts as a protective layer, decreasing the rate of desiccation.

Bagga (1967) dried cultures of 26 organisms at 0, 10, 20, 30, and 40°C over $CaCl_2$ in desiccators for 24 hr. The cultures were then removed from the desiccators and stored in sealed jars at laboratory and refrigerator temperatures. Cultures dried at 0, 10, and 20°C retained viability longer than cultures dried at 30 or 40°C. Relative humidities of 0 and 5 percent were the most suitable for maintaining viability.

Desiccated bacteria and spores are more resistant to destructive agencies than the same organisms in the moist state. Death of bacteria by moist heat is believed to result from the coagulation of the protoplasm. Much higher temperatures are required to coagulate the protoplasm of partially dried bacteria. Heat probably does not coagulate the protoplasm of completely dried bacteria. Apparently the same applies to enzymes. This explains why higher temperatures are required to sterilize glassware in the dry-air sterilizer than in the autoclave.

Death of bacteria by dry heat is believed to be the result of desiccation followed by oxidation. The death rate increases with temperature, owing to an increase in the rate of oxidation.

The effect of desiccation at 37°C on survival of *E. coli* and *B. subtilis* is shown in Figs. 186 and 187.

For more information: Leach et al. (1959); Monk and McCaffrey (1957); Webb et al. (1964).

Maintenance of stock cultures The maintenance of a large number of cultures of bacteria requires frequent attention to prevent loss of the organisms. Transfers must be made to suitable media at definite intervals before the organisms are killed by their own waste products. This requires not only a considerable amount of time but involves the possible loss of certain biological, immunological, and cultural characteristics of the organisms.

A number of methods are used for maintaining organisms in a viable condition over long periods.

Agar slant cultures For class use, bacteria are generally preserved on nutrient agar slants contained in screw-cap tubes to prevent drying.

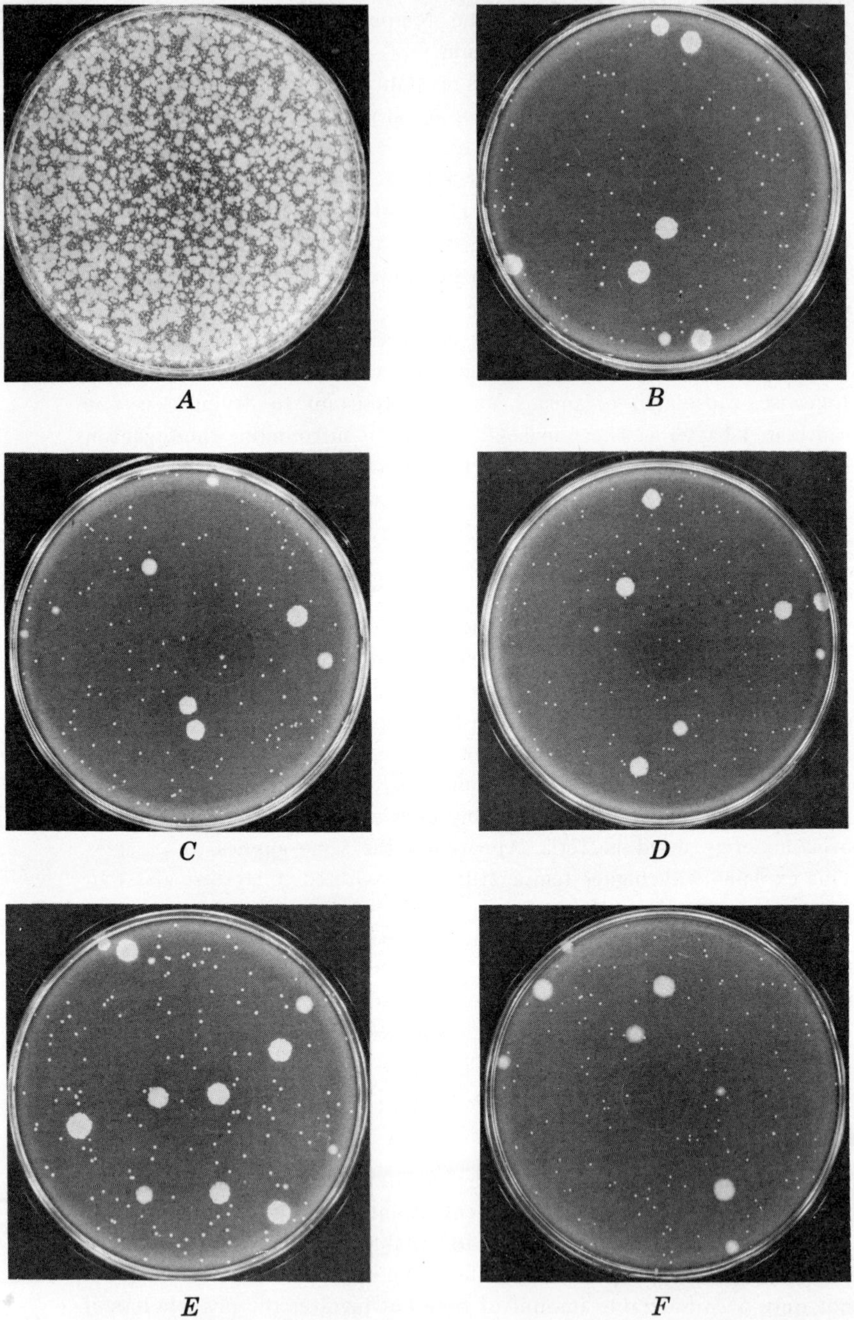

Fig. 187 Effect of desiccation on Bacillus subtilis. (A) control, not dried; (B) dried for 24 hr at 37°C; (C) dried for 48 hr; (D) dried for 72 hr; (E) dried for 96 hr; (F) dried for 216 hr. The organism produces spores which were not destroyed in that period of time.

Some exceptions to the use of nutrient agar are: certain pathogenic bacteria which require the addition of serum, blood, or other body fluids to the medium; anaerobic organisms which must be provided with a reducing environment; disease-producing streptococci which prefer semi-solid media; and some autotrophic species which grow only in mineral salts media.

The toxic metabolic waste products secreted by bacteria diffuse into the agar and away from the growth. Because of this property, bacteria are able to survive longer on agar than in liquid media. After preparation, cultures are incubated for 24 hr or more, stored in a cool dark place or refrigerator, and used as needed. They are referred to as stock cultures.

Agar slant cultures covered with oil Agar slants in screw-cap tubes are inoculated and incubated until good growth appears, then covered with sterile mineral oil to a depth of 1 cm above the agar slant. Transplants are made by removing a loopful of the growth, touching the loop to the glass surface to drain off excess oil, and transferring the inoculum to fresh medium.

This method possesses the following advantages:

1. Practically all species tested live longer under oil than in the control tubes without oil.
2. Transplants may be prepared when needed without affecting the preservation of the stock cultures.
3. The method is especially advantageous in working with unstable variants where occasional transfers to fresh media or growth in mass cultures results in changes in the developmental stages of the strains.
4. Special equipment such as a centrifuge, desiccator, or vacuum pump is not required.

Hartsell (1956) reported that 27 of 52 species of yeasts survived for 14 years and set the probable optimum survival time from 5 to 7 years.

For more information: Floodgate and Hayes (1961); Hartsell (1953, 1955); Little and Gordon (1967).

Saline suspensions Sodium chloride in high concentrations is frequently used as an inhibitor of bacterial growth. Chance (1963) suspended various bacteria in 1 percent salt solution (sublethal concentration) in screw-cap tubes to prevent evaporation. The tubes were stored at room temperature. After almost 4 years, transfers were made to agar slants and incubated. In every case vigorous growth occurred in less than 24 hr.

The procedure offers an easy method of storing bacterial cultures for 2 or 3 years, possibly longer. It appears to be particularly valuable in keeping stock cultures for class use in laboratories with limited equipment.

Freezing The preservation of organisms by freezing has been practiced for many years.

Floodgate and Hayes (1961) preserved 45 species of marine bacteria, representing 10 genera, by suspending the organisms in nutrient broth containing 15 percent glycerol, freezing the mixtures, and storing at $-29°C$. After 2 years, 86 percent of the suspensions were still viable.

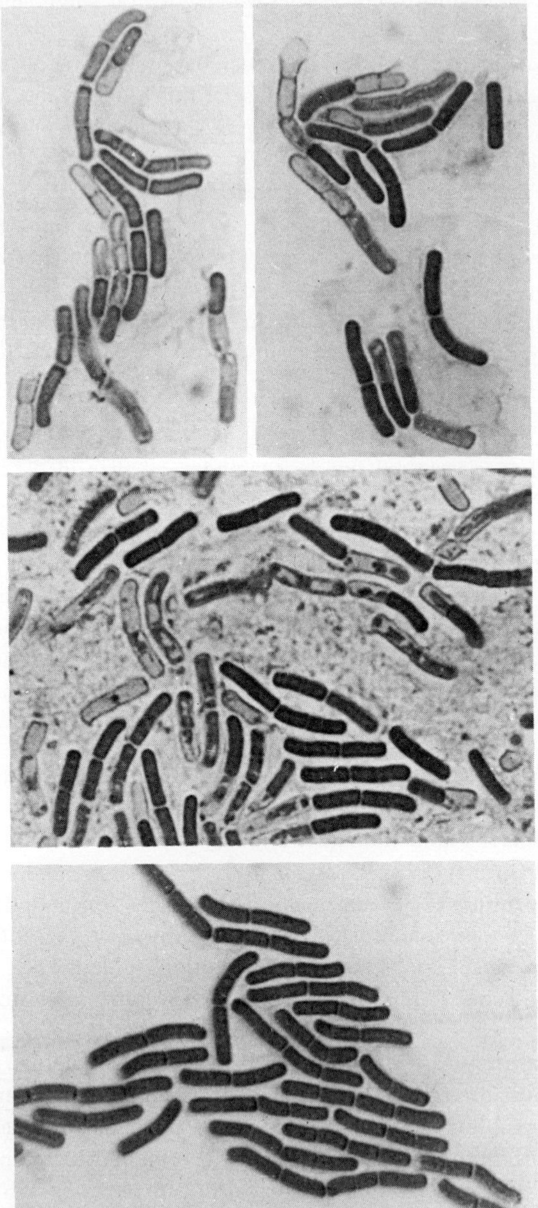

Fig. 188 Photomicrographs of megaterium stained with 1 percent fuchsin solution. Left, unfrozen control; center and right, repeatedly frozen cells. Note the destroyed and lightly stained cells. (After Nei.)

Schmittdiel (1961) suspended psittacosis virus in skim milk containing 7.5 percent glucose, and stored it at −15°C. He reported similar results.

Gibson et al. (1965) froze 16 cultures of lactic streptococci, followed by storage at −17.8 and −23.3°C. The numbers of surviving bacteria decreased as storage time was increased. For most species, the survival rate was greater at the lower temperature.

On the other hand, Nei (1960) showed that a high percentage of cells of *B. megaterium* was more or less damaged both morphologically and physiologically by repeated freezing and thawing (Fig. 188).

For more information: Asahina (1967); Ashwood-Smith et al. (1967); Bretz (1961); Clement (1961); Goldenberg et al. (1964); Gruft et al. (1968); Leibo and Mazur (1969); Tarshis (1961).

Storage in liquid nitrogen Hwang (1966) stored 162 species and strains of fastidious molds in liquid nitrogen and reported excellent results for long-term preservation of stock cultures.

The cultures were sealed in ampoules and cooled to −35°C at the rate of approximately 1°C per min. Further cooling to the storage temperature of −165 to −196°C was uncontrolled and took place at an accelerated rate. Viable cultures developed from specimens after a storage period of at least 5 years.

For more information: Jarvis et al. (1967).

Drying in vacuo Kitasato (1889) observed that the organism of cholera, *Vibrio comma*, survived longer when dried in a desiccator than when air-dried. Since that time many investigators have desiccator-dried other organisms with similar results.

Miller and Simons (1962) dried 202 cultures, representing 67 species of bacteria, over calcium chloride in a vacuum, then stored the organisms in the refrigerator at 10°C (Brown's method, 1940). After a period of 10 years, they found that only 13 cultures failed to grow. However, duplicates of 9 of the 13 cultures did show growth. The gram-positive species survived better than the gram-negative bacteria. This appears to be a simple and satisfactory procedure for the preservation of bacteria for long periods.

For more information: Annear (1964).

Freezing and drying Shackell (1909) recommended freezing as a preliminary step to rapid desiccation. The cultures may be frozen in a mixture of dry ice (carbon dioxide) and alcohol or acetone, then dried in a vacuum over sulfuric acid as the desiccant. This is the best method at present and, with its many modifications, is universally employed for the preservation of bacteria, viruses, sera, toxins, enzymes, and other biological materials. It is referred to as the freeze-drying technique or lyophilization.

Stein and Rogers (1950) designed a simple apparatus for rapid high-vacuum desiccation of frozen suspensions of biologic materials (Fig. 189). The apparatus consists of (1) glass ampoule containing frozen suspension, (2) rubber tubing connecting glass vials with multiple glass

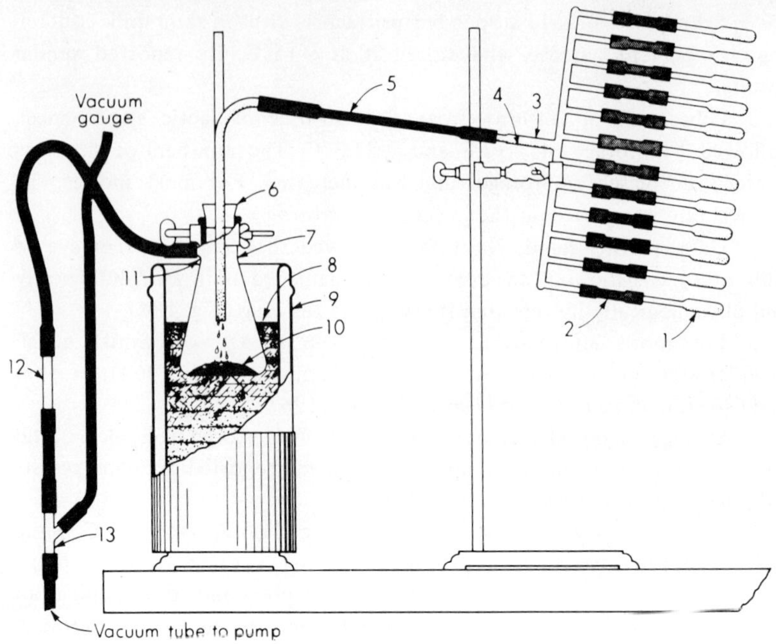

Fig. 189 Apparatus for rapid high-vacuum desiccation of frozen suspensions of biologic materials. See text for description. (After Stein and Rogers.)

pipette, (3) multiple glass pipette with 12 prongs, (4) vapor, (5) rubber tubing, (6) rubber stopper, (7) condenser (Pyrex filter flask), (8) refrigerant (dry ice and alcohol) −78°C, (9) vacuum jar, (10) frozen distillate from vapor, (11) outlet in filter flask, (12) absorbent-cotton filter in glass tubing, and (13) T-shaped glass tubing for attachment of vacuum gauge and vacuum pump.

Conditions affecting viability of freeze-dried organisms include: (1) composition of the suspending culture medium before lyophilization; (2) addition of certain substances to the bacterial suspensions prior to lyophilization to protect the organisms against the toxic effects of oxygen and moisture when stored in unsealed tubes; (3) sealing the tubes after lyophilization to prevent access of oxygen and moisture; (4) incomplete removal of oxygen and moisture on the survival rate of lyophilized organisms; and (5) storage temperature of lyophilized and sealed cultures.

In the freeze-drying procedure already described, the moisture is removed from the frozen state by high vacuum. However, organisms vary in their sensitivity to such treatment. With some species the percentage of survivors is quite high; with others it is too low to be satisfactory.

Dewald (1966) stressed the importance of the removal of water and air from lyophilized cultures prior to sealing of the tubes. Similar results were reported by Nei et al. (1966).

Wagman and Weneck (1963) obtained much higher yields of viable

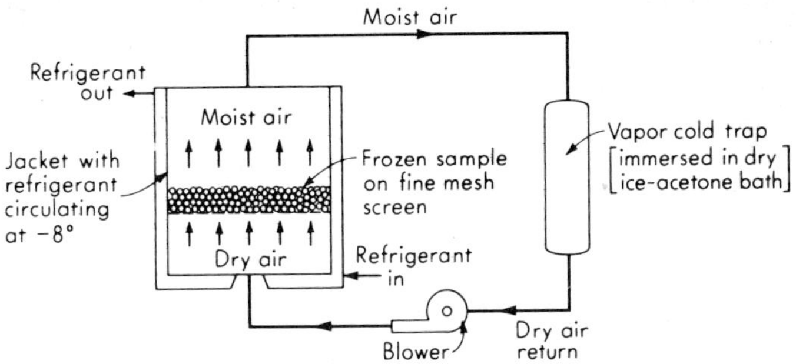

Fig. 190 Apparatus for convective freeze-drying procedure. (After Wagman and Weneck.)

organisms by employing the convective freeze-drying procedure (Fig. 190). In this method, the organisms are quickly frozen, then desiccated by dry air circulated by a blower at atmospheric pressure. Sublimed water vapor is continuously removed from the sample and condensed in the cold trap.

Bacteria may be preserved for longer periods by freeze-drying than is possible by any of the other methods described.

Rhoades (1970) stored duplicate sets of 52 lyophilized cultures of bacteria, molds, a yeast, and viruses at room temperature (approximately 22 to 32°C) and in the refrigerator (approximately 6 to 7°C); 47 cultures were stored at refrigerator temperature only. Of the 52 lyophilized cultures stored at room temperature, 40 were viable after 20 years. In the group of 52 cultures, the viruses of fowlpox and Newcastle disease survived better in the refrigerator than at room temperature.

Organisms stored in nonlyophilized cultures gradually change in their morphological, biochemical, and immunological properties. On the other hand, lyophilized organisms do not exhibit any appreciable differences in such properties compared to the same organisms prior to freeze-drying.

For more information: Berman et al. (1968); Chatelain (1963); Flosdorf (1949); Greiff and Rightsel (1967); Harrison and Pelczar (1963); Heckly and Dimmick (1968); Lion and Bergmann (1961*a*, *b*); MacLeod et al. (1966); Marshall et al. (1963); Nei (1969); Rey (1966); Rüger (1970); Sinskey and Silverman (1970); Watko and Heddleston (1966); Webb (1967); Zimmerman (1962).

Effect of hydrogen-ion concentration

The hydrogen-ion (H^+) concentration of culture media is of prime importance for the successful cultivation of bacteria. Some species grow best in acid environments; others grow best in alkaline media; still others prefer substrates approximately neutral in reaction.

There exists for each organism an optimum H^+ concentration where it grows best. The highest pH value that just shows growth is known as the

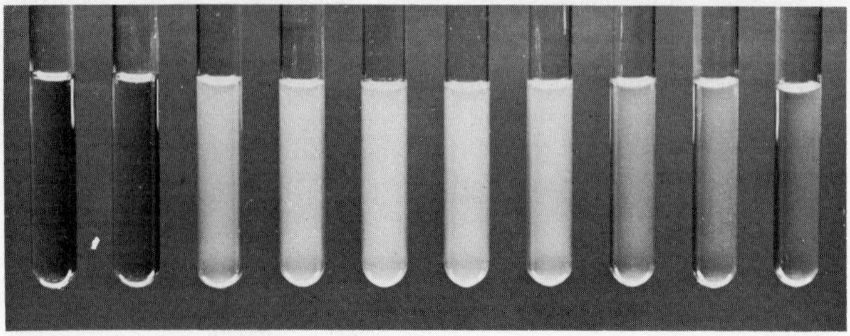

Fig. 191 Effect of pH in the growth of Escherichia coli. From left to right, pH 2.8 to 10.0 at intervals of 0.8 pH. Tubes 1 and 2, no growth; tubes 3 to 6 inclusive, increasing turbidity; tubes 7 to 10 inclusive, decreasing turbidity. Maximum turbidity occurred in tube 6 at pH 6.8.

minimum H^+ concentration; and the lowest pH value that just shows growth as the maximum H^+ concentration. The pH growth range of an organism is the difference between the maximum and minimum values. With some species the difference is wide; with others it is narrow.

The values are true only if other environmental factors are controlled. Variations in the composition of the medium, temperature of incubation, and osmotic pressure of the medium, even though slight, may produce significant changes in the minimum, optimum, and maximum H^+ values of an organism.

An organism that ferments a carbohydrate vigorously will be slowed in its growth as acids accumulate in the medium. This will continue until a certain critical pH is reached when growth will be stopped completely. This may require only a few hours during incubation of the culture. The final H^+ concentration is spoken of as the physiological constant and probably is different for every fermenting organism. It is, therefore, possible to separate some species on differences in their final H^+ concentrations (see page 701).

The effect of pH on the growth of *E. coli* is shown in Fig. 191.

Effect of surface tension

Surface tension may be defined as that property, due to molecular forces, which exists in the surface film of all liquids and tends to bring the contained volume into a form having the least superficial area.

Molecules attract their neighbors and are attracted by them. A molecule situated in a liquid will be in equilibrium by virtue of the equal attractions on all sides. On the other hand, a molecule situated on the surface of a liquid will have equal horizontal attractions but unequal vertical attractions. This results in an unbalanced attraction toward the interior of the liquid. This resultant force reaches a maximum at the surface, and the mass of liquid behaves as if surrounded by an elastic membrane, tending to compress the liquid into the smallest possible volume.

The composition of the surface layer of a culture medium inoculated with an organism may differ appreciably from the composition of the same medium taken as a whole. The surface tension of media may, therefore, play an important role in the growth of an organism. This is especially true of those organisms which have a tendency to grow on the surface of culture media in a film or pellicle.

Organisms growing on the surface of a medium in the form of a pellicle were at one time regarded as strict aerobes. The surface layer, exposed to air, gave the organisms more oxygen than could be obtained from the deeper portions of the medium. It is true that the pellicle-forming bacteria are aerobic but not obligately aerobic. They are capable of growth and multiplication under both aerobic and anaerobic conditions. Bacteria have a density slightly greater than that of the culture medium in which they are grown. If the pellicle produced by a pellicle-producing organism, such as *B. subtilis*, is sedimented, the pellicle will not rise to the surface again but will remain at the bottom of the tube. A new pellicle may develop on the surface of the medium. It is obvious that the surface film is supported in this position by some force in the medium. This force is spoken of as surface tension.

The unit of force in the cgs system of physical units is the dyne. It is such a force that, under its influence, a particle whose mass is 1 g would experience during each second an acceleration of 1 cm per sec. The dyne is approximately the force exerted by a milligram weight under the influence of gravity.

The surface tension of the usual laboratory media varies between 57 and 63 dynes. The surface tension of pure water is 73 dynes. Some substances may be added to culture media to raise the surface tension, among which may be mentioned charcoal and calcium chloride. Charcoal is effective by virtue of its ability to remove some surface-tension depressants from media. Also, the growth of bacteria may in some cases raise the surface tension of the media in which they are cultivated. On the other hand, many substances may be added to culture media to lower their surface tensions. Among these may be mentioned ethyl alcohol, glycerol, soaps, and synthetic detergents. The soaps have been almost the only efficient compounds, but they are being replaced by the synthetic detergents or wetting agents (see page 582).

Larson et al. (1919) found that if the surface tension of nutrient broth was depressed to some point below 40 dynes by means of soap, and then inoculated with *B. subtilis*, the organism grew in a diffuse manner rather than in the form of a pellicle on the surface. The same principle was found to apply to other pellicle-producing organisms. The formation of a pellicle apparently has nothing to do with the oxygen requirements of an organism, but is dependent upon the surface tension of the medium (Fig. 192).

Wetting is a function of surface tension. If bacteria are not wetted

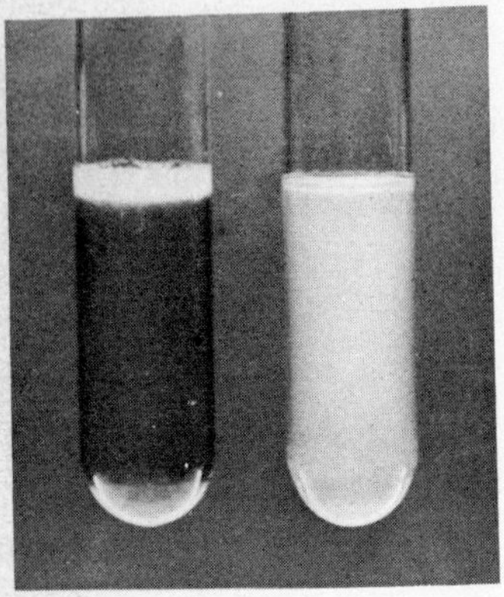

Fig. 192 Effect of surface tension on the growth of Bacillus subtilis. Left, growth in nutrient broth. The growth is confined almost entirely to the pellicle. Right, nutrient broth containing 0.05 percent sodium ricinoleate to lower the surface tension. The growth occurs as a uniform turbidity.

by the medium, they will grow on the surface in the form of a film; if they are wetted, they will produce a uniform clouding of the medium. It has been shown that bacteria growing on the surface of media contain a higher content of lipoidal substances than non-pellicle-producing organisms. The organism of tuberculosis has been found to contain as high as 40 percent lipoidal substances in contrast to nonpellicle producers, which contain about 7 percent.

Larson and Larson (1922) showed that organisms which ordinarily produced a uniform turbidity in the medium could be made to grow in the form of a pellicle if the lipoid content was increased. Accordingly, the fat content of *S. aureus* was increased by growing the organism in a medium containing a carbohydrate or glycerol, neither of which was fermented. The organism produced a pellicle on the surface that resembled the growth of the tubercle bacillus.

For more information: Larson (1928); Nakamura and Ramage (1963).

Antagonistic action of ions

Winslow and Falk (1923*a*) reported that *E. coli* maintained itself in distilled water at pH 6.0 without material reduction in numbers for almost 24 hr. At pH 5.0, the reduction was somewhat greater. The viability decreased as the solution was adjusted to more acid or alkaline conditions (Table 10). A heavy line is drawn to indicate the time and pH most nearly corresponding to a one-third reduction in bacterial numbers.

On the other hand, 0.0145 *M* NaCl (0.85 percent) exerted a distinctly favorable action on the viability of *E. coli*. Instead of a slight but definite decrease in numbers after 24 hr, as occurred in distilled water at pH 6.0, the bacteria maintained themselves in undiminished numbers. Above a concentration of 0.0145 *M* NaCl, the percentage of survivors decreased with increasing concentrations of salt.

Similar results were obtained when $CaCl_2$ was substituted for NaCl. The organisms maintained themselves better in the presence of this salt

Table 10 Viability of Escherichia coli *in distilled water*

	Percent alive at pH						
Hours	*4.0*	*5.0*	*6.0*	*6.5*	*7.0*	*7.5*	*8.0*
1	87	88	84	92	68	77	79
3	39	71	74	66	54	24	52
6	4	48	64	30	24	8	12
9	1	68	82	7	17	5	12
24	0	6	77	2	23	3	10
Number of experiments	2	2	4	2	10	2	4

than in distilled water. Here again, pH 6.0 appeared to be most favorable for growth. The most favorable concentration of $CaCl_2$ was 0.00145 *M*.

In another report, Winslow and Falk (1923*b*) found solutions of 0.725 *M* NaCl and over, and 0.435 *M* $CaCl_2$ and over exhibited distinctly toxic actions on *E. coli* at all pH values. However, in a solution containing a mixture of these two salts in appropriate proportions, an antagonistic effect was manifested which tended to protect the bacteria against the toxic action of either salt when present alone (Table 11). This phenomenon is spoken of as the antagonistic action of ions.

MacLeod and Snell (1950) found zinc to be toxic for *Lactobacillus arabinosus* and that its toxicity could be overcome by manganese. Magnesium, calcium, and strontium also counteracted the toxicity of zinc for this organism. On the other hand, the above ions failed to reverse the toxicity of zinc for *L. mesenteroides*. They believed that the toxicity of zinc for *L. arabinosus* resulted from the formation of a catalytically inactive zinc protein from this ion and some protein normally activated by manganese, magnesium, calcium, or strontium. The latter ions

Table 11 Viability of Escherichia coli *in solutions of NaCl and* $CaCl_2$*, singly and in combinations*

		Percentage of bacteria alive after 9 hr in:			
*Total isotonic concentration**	*Total molar concentration*	*Pure NaCl*	*Pure* $CaCl_2$	*NaCl +* $CaCl_2$	*Ratio, Na/Ca*
0	0	89	89	89	
1	0.145	82	22		
2	0.290			41	1 : 1
3	0.435	55	0+	28	2 : 1
4	0.580			40	3 : 1
5	0.725	46	0+	117	4 : 1
6	0.870	33		30	5 : 1

*1 tonicity = 0.145 *M* = 0.85 percent.

counteracted the toxicity of zinc by displacing zinc from the protein to form a catalytically active metalloprotein.

Abelson and Aldous (1950) found nickel, cobalt, zinc, cadmium, and manganese to be toxic to *E. coli*; and that the toxicity was reduced in the presence of much magnesium. If magnesium was not present, the elements were toxic in very low concentrations.

The addition of cobalt to a medium supporting growth of *C. perfringens* was found by Shankar and Bard (1955) to yield filamentous cells which fermented glucose chiefly to lactic acid. The inhibitory effect of cobalt was quantitative and reversible by magnesium. Metabolic studies indicated that cobalt caused a fermentative shift to the homolactic type by affecting enzyme synthesis during growth.

A practical application of this phenomenon was the development, by Ringer, of a physiological salt solution before ionic antagonism was understood. He showed that when a beating heart was perfused with a 0.75 percent sodium chloride solution, pulsation stopped completely. On the addition of 0.0125 percent calcium chloride to the solution, the heartbeat was restored, although not in a normal manner. On the further addition of 0.01 percent potassium chloride and a small amount of sodium bicarbonate to adjust the pH of the solution, the heartbeat became normal. This solution is known as Ringer's solution. Other preparations of this type are Tyrode's and Locke's solutions.

Oligodynamic action of heavy metals

Naegeli (1893) noted that silver in very high dilutions produced a toxic action on certain organisms. He found that 1 part of silver in 100 million parts of water killed algae of the genus *Spirogyra*. He believed that silver in such a high dilution could not produce a chemical action on living cells and attributed its toxic effect to an oligodynamic action. The term oligodynamic may be used to identify certain substances effective in exceedingly minute amounts.

Other metals also exert a toxic action on organisms. Copper in a dilution of 1 part in 77 million of water is toxic to certain algae. The spores of *Aspergillus niger* fail to germinate in the presence of 1 part of silver in 1.6 million parts of water. Water distilled from a copper still is toxic to bacteria. This is due to the presence of traces of dissolved copper in the water. Water distilled from stills made of other heavy metals also exhibits the same phenomenon, but to a lesser degree. This is not due to the greater toxicity of copper, as it is known that mercury is probably the most toxic metal, but to the fact that copper is more soluble in water than the other toxic metals. Therefore, metal stills should be avoided for the preparation of distilled water intended for biological use.

ZoBell (1941) found that containers made of copper, zinc, tin, or nickel alloys were not suitable for the collection of samples of seawater for bacteriological analysis owing to the inimical oligodynamic action of the metals. Under certain conditions, most of the bacteria in seawater were killed within a few minutes, and the seawater itself was rendered bacteriostatic by exposure to the metals.

Burrows and Hemmens (1943) reported that bacteria swabbed on the polished surface of a silver chalice died off rapidly. Experiments on the transmission of test organisms from one person to another by common use of the chalice showed that approximately 0.001 percent of the organisms were transferred, even under the most favorable conditions. When the conditions approximated those of actual use, no transmission could be detected.

On the other hand, Hobbs et al. (1967) found that the common communion cup, made of silver and containing wine (14.5 percent alcohol), could serve as a means of transmitting infection. However, the risk involved was believed to be small and probably much smaller than that of contracting infection by other methods in any gathering of people.

For more information: Burke and McVeigh (1967); Ronald and Morris (1967); Zimmerman (1966).

Demonstration of oligodynamic action

Silver exerts a marked bactericidal or oligodynamic action on bacteria. This may be demonstrated by placing a piece of metallic silver or a silver coin in a petri dish and pouring over it melted agar, previously inoculated with an organism such as *Sarcina lutea* or *Serratia marcescens.* After incubation for 24 hr at 37°C, a clear zone will be seen immediately surrounding the silver metal or coin. This is the oligodynamic zone (Fig. 193). Beyond this will appear a narrower zone in which growth is stimulated. Minute amounts of metallic ions stimulate growth, whereas

Fig. 193 Oligodynamic action of silver. (A) Agar inoculated with Sarcina lutea and poured into a dish containing a silver coin; (B) same inoculated with Serratia marcescens. The zone immediately surrounding the silver coin is the oligodynamic zone. This is followed by a narrow stimulating zone. Normal growth occurs in the remainder of the agar.

A B

greater concentrations produce an inhibitory effect. Normal growth occurs in the remainder of the agar. The same result is obtained if a piece of copper or a copper coin is substituted for the silver.

Gibbard (1937) came to the following general conclusions regarding the action of silver on bacteria:

1. The width of the oligodynamic zone is increased by treating the silver with nitric acid, and decreased by cleaning the metal mechanically.
2. Pure silver metal shows no bactericidal or oligodynamic action. However, if silver is first treated with nitric acid, a film of silver oxide forms on the surface from which silver ions are produced. If silver oxide is prevented from forming, no inhibitory action occurs. This may be shown by melting silver, allowing a portion to cool in hydrogen and another to cool in air. The silver cooled in hydrogen shows no bactericidal action, while that cooled in air exhibits a pronounced bactericidal effect.
3. Silver nitrate, silver oxide, and electrically dispersed colloidal silver, when properly diluted to contain the same concentration of silver, all possess a similar bactericidal action.
4. The bactericidal properties of silver nitrate and silver oxide are greatly reduced in the presence of proteins. Silver proteinates are formed, removing the metal from solution.

Application of oligodynamic action Silver has been recommended for the treatment of water, milk, vinegar, wine, cider, fruit juices, liquors, etc. Its greatest application appears to be in the treatment of water for drinking purposes. According to Gibbard, three methods are generally used commercially for this purpose:

1. The water is exposed to silver deposited on sand, porcelain, and other solid materials.
2. The silver is applied by electrolysis.
3. Use is made of the difference in emf that exists between nickel and silver electrodes kept in the material at different temperatures.

In all three methods the objective is the same, namely, to obtain a solution of silver.

Silver poisoning There is still considerable discussion about whether sufficient silver remains in treated water, foods, beverages, etc., to produce silver poisoning or argyria in man and animals. It is important to know how much of the silver ingested will be retained in the body. So far as is known, there are no available data on this point. More work is required before this controversy can be definitely settled.

Effect of shaking

Microorganisms can be killed by agitation with various substances. Curran and Evans (1942) agitated bacterial and spore suspensions with glass beads, sand, and carborundum, and obtained large reductions in viable counts. King and Alexander (1948) reported similar results.

The cells were killed by fragmentation. Recovery was not possible unless damage was slight. It is reasonable to suppose that recovery occurs only when the nucleus or nuclear material remains uninjured.

For more information: Eaton (1962); Lamanna et al. (1959); Sacks and Bailey (1963).

References

Abelson, P. H., and E. Aldous: Ion antagonisms in microorganisms: Interference of normal magnesium metabolism by nickel, cobalt, cadmium, zinc, and manganese, *J. Bact.*, **60**:401, 1950.

Albright, L. J., and R. Y. Morita: Effect of hydrostatic pressure on synthesis of protein, ribonucleic acid, and deoxyribonucleic acid by the psychrophilic marine bacterium, *Vibrio marinus, Limnol. Oceanog.*, **13**:637, 1968.

Albright, L. J., and J. F. Henigman: Seawater salts-hydrostatic pressure effects upon cell division of several bacteria, *Can. J. Microbiol.*, **17**:1246, 1971.

Alderton, G., and N. Snell: Chemical states of bacterial spores: Dry-heat resistance, *Appl. Microbiol.*, **17**:745, 1969.

Amaha, M., and K. Sakaguchi: Effects of carbohydrates, proteins, and bacterial cells in the heating media on the heat resistance of *Clostridium sporogenes, J. Bact.*, **68**:338, 1954.

Angelotti, R., et al.: Influence of spore moisture content on the dry-heat resistance of *Bacillus subtilis* var. *niger, Appl. Microbiol.*, **16**:735, 1968.

Annear, D. I.: Recoveries of bacteria after drying in glutamate and other substances, *Australian J. Exp. Biol. Med. Sci.*, **42**:717, 1964.

Anton, D. N.: Osmotic-sensitive mutant of *Salmonella typhimurium, J. Bact.*, **109**:1273, 1972.

Asahina, E., ed.: The mechanism of cellular injury by freezing in microorganisms. In "Cellular Injury and Resistance in Freezing Organisms," *Proc. International Conf. Low Temperature Sci.*, vol. 2, 1967.

Ashwood-Smith, M. J., et al.: Sunlight and frozen bacteria, *Nature*, **214**:33, 1967.

——: Response of bacterial spores and *Micrococcus radiodurans* to ultraviolet irradiation at low temperatures, *Nature*, **217**:337, 1968.

Bagga, H. S.: Effect of different drying temperatures and levels of relative humidity during storage on longevity of dried cultures of pathogenic and industrial microorganisms, *Plant Disease Reptr.*, **51**:1055, 1967.

Bateman, J. B., and F. E. White: Relative humidity and the killing of bacteria: The survival of *Serratia marcescens* dehydrated by concentrated glycerol and sucrose solutions, *J. Bact.*, **85**:918, 1963.

Bedford, T. H. B.: The nature of the action of ultraviolet light on microorganisms, *Brit. J. Exp. Pathol.*, **8**:437, 1927.

Berman, S., et al.: Freeze-drying various strains of *Shigella, Appl. Microbiol.,* **16**:1779, 1968.

Bernheim, F.: The effects of alcohols and sugars on the swelling rate of cells of *Pseudomonas aeruginosa* in various salts, *Microbios.,* **4**:49, 1971.

Bornside, G. H.: Exposure of *Pseudomonas aeruginosa* to hyperbaric oxygen: Inhibited growth and enhanced activity of polymyxin B., *Proc. Soc. Exp. Biol. Med.,* **125**:1152, 1967*a*.

——: Enhancement of antibiotic activity against *Staphylococcus aureus* by exposure to hyperbaric oxygen, *Appl. Microbiol.,* **15**:1020, 1967*b*.

Bott, T. L., and T. D. Brock: Bacterial growth rates above 90°C. in Yellowstone Hot Springs, *Science,* **164**:1411, 1969.

Bretz, H. W.: Death of *Escherichia coli* frozen on cellophane and on membrane filters, *Can. J. Microbiol.,* **7**:793, 1961.

—— and R. A. Ambrosini: Survival of *Escherichia coli* frozen in cell extracts, *Cryobiology,* **3**:40, 1966.

Brock, T. D., et al.: Microbial life at 90°C: The sulfur bacteria of Boulder Spring, *J. Bact.,* **107**:303, 1971.

Bruch, M. K., and F. W. Smith: Dry heat resistance of spores of *Bacillus subtilis* var. *niger* on kapton and teflon film at high temperatures, *Appl. Microbiol.,* **16**:1841, 1968.

Burke, C. M. W., and I. McVeigh: Toxicity of copper to *Escherichia coli* in relation to incubation temperature and method of sterilization of media, *Can. J. Microbiol.,* **13**:1299, 1967.

Burrows, W., and E. S. Hemmens: Survival of bacteria on the silver communion cup, *J. Infectious Diseases,* **73**:180, 1943.

Camerman, N., and A. Camerman: Photodimer of thymine in ultraviolet-irradiated DNA: Proof of structure by x-ray diffraction, *Science,* **160**:1451, 1968.

Castellani, A. J., et al.: Overlap of photoreactivation and liquid holding recovery in *Escherichia coli* B., *Science,* **143**:1170, 1964.

Chance, H. L.: Salt—a preservative for bacterial cultures, *J. Bact.,* **85**:719, 1963.

Chatelain, R.: A propos de l'utilisation du glutamate de sodium dans la lyophilisation des bactéries, *Ann. Inst. Pasteur,* **105**:111, 1963.

Chislett, M. E., and D. J. Kushner: A strain of *Bacillus circulans* capable of growing under highly alkaline conditions, *J. Gen. Microbiol.,* **24**:187, 1961.

Christian, J. H. B., and J. A. Waltho: The sodium and potassium content of nonhalophilic bacteria in relation to salt tolerance, *J. Gen. Microbiol.,* **25**:97, 1961.

Christophersen, J.: Adaptive temperature responses of microorganisms. In "Molecular Mechanisms of Temperature Adaptation," Washington, American Association for the Advancement of Science, 1967.

Clement, M. T.: Effects of freezing, freeze-drying, and storage in the freeze-dried and frozen state on viability of *Escherichia coli* cells, *Can. J. Microbiol.,* **7**:99, 1961.

Clouston, J. G., and P. A. Wills: Initiation of germination and inactivation of *Bacillus pumilus* spores by hydrostatic pressure, *J. Bact.,* **97**:684, 1969.

—— and ——: Kinetics of initiation of germination of *Bacillus pumilus* spores by hydrostatic pressure, *J. Bact.,* **103**:140, 1970.

Cohn, M. L., and G. Middlebrook: The effect of near-ultraviolet light on culture media, *Am. Rev. Respirat. Diseases,* **91**:929, 1965.

Cook, A. M., and R. J. Gilbert: The effect of sodium chloride on heat resistance and recovery of heated spores of *Bacillus stearothermophilus, J. Appl. Bact.,* **32**:96, 1968.

Cota-Robles, E. H.: Electron microscopy of plasmolysis in *Escherichia coli, J. Bact.,* **85**:499, 1963.

Curran, H. R.: Resistance in bacterial spores, *Bact. Rev.,* **16**:111, 1952.

—— and F. R. Evans: The killing of bacterial spores in fluids by agitation with small inert particles, *J. Bact.,* **43**:125, 1942.

Deering, R. A.: Ultraviolet radiation and nucleic acid, *Sci. Am.,* **207**:135, 1962.

De Siervo, A. J.: Alterations in the phospholipid composition of *Escherichia coli* B during growth at different temperatures, *J. Bact.,* **100**:1342, 1969.

Dewald, R. R.: Preservation of *Serratia marcescens* by high-vacuum lyophilization, *Appl. Microbiol.,* **14**:561, 1966.

Doebbler, G. F., and A. P. Rinfret: Survival of microorganisms after ultrarapid freezing and thawing, *J. Bact.,* **85**:485, 1963.

Dombrowski, H.: Bacteria from paleozoic salt deposits, *Ann. N.Y. Acad. Sci.,* **108**:453, 1963.

Dowben, R. M., and R. Weidenmüller: Adaptation of mesophilic bacteria to growth at elevated temperatures, *Biochim. Biophys. Acta,* **158**:255, 1968.

Dul, M. J., and W. C. McDonald: Morphological changes and antibiotic-induced thermal resistance in vegetative cells of *Bacillus subtilis, J. Bact.,* **106**:672, 1971.

Eaton, N. R.: New press for disruption of microorganisms, *J. Bact.,* **83**:1359, 1962.

Eddy, B. P.: The use and meaning of the term "psychrophilic," *J. Appl. Bact.,* **23**:189, 1960.

Efthimion, M. H., and W. A. Corpe: Effect of cold temperatures on the viability of *Chromobacterium violaceum, Appl. Microbiol.,* **17**:169, 1969.

El-Bisi, H. M., et al.: Chemical events during death of bacterial endospores by moist heat, *J. Food Sci.,* **27**:219, 1962.

—— and Z. J. Ordal: The effect of sporulation temperature on the thermal resistance of *Bacillus coagulans* var. *thermoacidurans, J. Bact.,* **71**:10, 1956.

Evans, F. R., and H. R. Curran: Influence of preheating, pH, and holding temperature upon viability of bacterial spores stored for long periods in buffer substrates, *J. Bact.*, **79**:361, 1960.

Floodgate, G. D., and P. R. Hayes: The preservation of marine bacteria, *J. Appl. Bact.*, **24**:87, 1961.

Flosdorf, E. W.: "Freeze-Drying," New York, Reinhold Publishing Corporation, 1949.

Fox, K., and I. J. Pflug: Effect of temperature and gas velocity on the dry-heat destruction rate of bacterial spores, *Appl. Microbiol.*, **16**:343, 1968.

Frank, H. A., and C. O. Willits: XIII. Sterilizing effect of sunlight on maple sap in transparent tubes, *Appl. Microbiol.*, **8**:141, 1960.

Gallant, J., and S. R. Suskind: Relationship between thymineless death and ultraviolet inactivation in *Escherichia coli, J. Bact.*, **82**:187, 1961.

Ganesan, A. K., and K. C. Smith: Dark recovery processes in *Escherichia coli* irradiated with ultraviolet light. II. Effect of *uvr* genes on liquid holding recovery, *J. Bact.*, **97**:1129, 1969; III. Effect of *rec* mutations on recovery of excision-deficient mutants of *Escherichia coli* K-12, *ibid.*, **102**:404, 1970.

Gibbard, J.: Public health aspects of the treatment of water and beverages with silver, *Am. J. Public Health*, **27**:122, 1937.

Gibson, C. A., et al.: Survival of strains of lactic streptococci during frozen storage, *J. Dairy Res.*, **32**:151, 1965.

Ginoza, W.: The effects of ionizing radiation on nucleic acids of bacteriophages and bacterial cells. In "Annual Review of Microbiology," **21**:325, 1967.

Goldenberg, M. I., et al.: The survival of *Pasteurella pestis* in materials preserved by solid carbon dioxide (dry-ice), *Bull. World Health Organ.*, **30**:741, 1964.

Gossling, B. S.: The loss of viability of bacteria in suspension due to changing the ionic environment, *J. Appl. Bact.*, **21**:220, 1959.

Govier, D. A., and A. Kleczkowski: Residual infectivity and the extent of photoreactivation in the different host plants of u.v.-irradiated potato virus X, *J. Gen. Virol.*, **4**:347, 1969.

Greiff, D., and W. A. Rightsel: Stabilities of suspensions of viruses after freezing or drying by vacuum sublimation and storage, *Cryobiology*, **3**:432, 1967.

Gruft, H., et al.: Preservation of mycobacterial cultures, *Appl. Microbiol.*, **16**:355, 1968.

Haight, R. D., and R. Y. Morita: Interaction between the parameters of hydrostatic pressure and temperature on aspartase of *Escherichia coli, J. Bact.*, **83**:112, 1962.

Hanawalt, P., and J. Buehler: Photoreactivation of macromolecular synthesis in *Escherichia coli, Biochim. Biophys. Acta*, **37**:141, 1960.

Hansen, N.-H., and H. Riemann: Factors affecting the heat resistance of nonsporing organisms, *J. Appl. Bact.*, **26**:314, 1963.

Harm, W.: Dark repair of photorepairable uv lesions in *Escherichia coli*, *Mutation Res.*, **6**:25, 1968.

——: VIII. Inhibition of photoenzymatic repair of UV lesions in *E. coli* DNA by caffeine, *Mutation Res.*, **10**:319, 1970.

Harrington, W. O., and C. H. Hills: Reduction of the microbial population of apple cider by ultraviolet irradiation, *Food Technol.*, **22**:1451, 1968.

Harris, N. D.: The influence of the recovery medium and the incubation temperature on the survival of damaged bacteria, *J. Appl. Bact.*, **26**:387, 1963.

Harrison, A. P., Jr., and R. E. Cerroni: Fallacy of "crushing death" in frozen bacterial suspensions, *Proc. Soc. Exp. Biol. Med.*, **91**:577, 1956.

—— and M. J. Pelczar, Jr.: Damage and survival of bacteria during freeze-drying and during storage over a ten-year period, *J. Gen. Microbiol.*, **30**:395, 1963.

Hartsell, S. E.: The preservation of bacterial cultures under paraffin oil, *Appl. Microbiol.*, **1**:36, 1953.

——: Maintenance of cultures under paraffin oil, annotated summary, *Lab. Bacteriol., Purdue University*, March 1955.

——: Maintenance of cultures under paraffin oil, *Appl. Microbiol.*, **4**:350, 1956.

Heckly, R. J., and R. L. Dimmick: Correlations between free radical production and viability of lyophilized bacteria, *Appl. Microbiol.*, **16**:1081, 1968.

Hedén, C.-G.: Effects of hydrostatic pressure on microbial systems, *Bact. Rev.*, **28**:14, 1964.

Heinmetz, F., and J. J. Lehman: Preliminary studies on the restoration of viability of ultraviolet-inactivated bacteria by metabolites and cofactors, *Arch. Biochem. Biophys.*, **59**:313, 1955.

Hobbs, B. C., et al.: Experiments on the communion cup, *J. Hyg.*, **65**:37, 1967.

Hoffman, R. K., et al.: Effect of cell moisture on the thermal inactivation rate of bacterial spores, *Appl. Microbiol.*, **16**:1240, 1968.

Hwang, S.-W.: Long-term preservation of fungus cultures with liquid nitrogen refrigeration, *Appl. Microbiol.*, **14**:784, 1966.

Iyer, V., and J. V. Bhat: Growth of spore-forming bacilli in relation to common salt and sugar, *J. Sci. Ind. Res.* (*India*), **13B**:336, 1954.

Jagger, J.: Photoreactivation, *Bact. Rev.*, **22**:99, 1958.

Jarvis, J. D., et al.: Storage of bacteria in liquid nitrogen, *J. Med. Lab. Tech.*, **24**:312, 1967.

Jennings, B. H., et al.: Photosensitized dimerization of thymine, *Photochem. Photobiol.*, **11**:215, 1970.

Jensen, L. B.: "Microbiology of Meats," Champaign, Ill., The Garrard Press, 1954.

Johnson, F. H., and C. E. ZoBell: The retardation of thermal disinfection of *Bacillus subtilis* spores by hydrostatic pressure, *J. Bact.*, **57**:353, 1949*a*.

—— and ——: The acceleration of spore disinfection by urethan and its retardation by hydrostatic pressure, *J. Bact.*, **57**:359, 1949*b*.

Kantor, G. J., and R. A. Deering: Recovery of division ability in ultraviolet-irradiated *Escherichia coli* induced by photoreactivation, photoprotection, and liquid holding treatment, *J. Bact.*, **94**:1946, 1967.

Kapp, D. S., and K. C. Smith: Repair of radiation-induced damage in *Escherichia coli* II, *J. Bact.*, **103**:49, 1970.

Kelner, A.: Photoreactivation of ultraviolet-irradiated *Escherichia coli* with special reference to the dose-reduction principle and to ultraviolet-induced mutation, *J. Bact.*, **58**:511, 1949.

Kenis, P. R., and R. Y. Morita: Thermally induced leakage of cellular material and viability in *Vibrio marinus*, a psychrophilic marine bacterium, *Can. J. Microbiol.*, **14**:1239, 1968.

King, H. K., and H. Alexander: The mechanical destruction of bacteria, *J. Gen. Microbiol.*, **2**:315, 1948.

Kissinger, J. C., and C. O. Willits: The control of bacterial contamination in maple sap stored in field storage tanks by ultraviolet irradiation, *J. Milk Food Technol.*, **29**:279, 1966.

Kitasato, S.: Die Widerstandfähigkeit der Cholerabacterien gegen das Eintrocknen und gegen Hitze, *Z. Hyg.*, **5**:134, 1889.

Kleczkowski, A.: Dark reactivation of ultraviolet-irradiated tobacco necrosis virus, *J. Gen. Virology*, **3**:19, 1968.

—— and D. A. Govier: Action spectrum for inactivation of the infectivity of potato virus X by u.v. radiation, *Photochem. Photobiol.*, **10**:53, 1969.

Klein, F., et al.: Interacting factors that influence long-term storage of live *Pasteurella tularensis* vaccine and Rift Valley fever virus, *Appl. Microbiol.*, **17**:427, 1969.

Kocka, F. E., and H. W. Bretz: Survival of *Escherichia coli* after storage in various frozen menstrua, *Antonie van Leeuwenhoek, J. Microbiol. Serol.*, **35**:65, 1969.

Ladanyi, P. A., and S. M. Morrison: Ultraviolet bactericidal irradiation of ice, *Appl. Microbiol.*, **16**:463, 1968.

Lamanna, C., et al.: Pumping as a means for the mechanical rupture of microorganisms mixed with glass beads, *J. Bact.*, **77**:104, 1959.

Larkin, J. M., and J. L. Stokes: Isolation of psychrophilic species of *Bacillus*, *J. Bact.*, **91**:1667, 1966.

—— and ——: Growth of psychrophilic microorganisms at subzero temperatures, *Can. J. Microbiol.*, **14**:97, 1968.

Larson, L. W., and W. P. Larson: Factors governing the fat content of bacteria and the influence of fat on pellicle formation, *J. Infectious Diseases*, **31**:407, 1922.

Larson, W. P.: The effect of the surface tension of the menstruum upon bacteria and toxins. In "The Newer Knowledge of Bacteriology and Immunology," edited by E. O. Jordan and I. S. Falk, Chicago, University of Chicago Press, 1928.

—— et al.: The influence of the surface tension of the culture medium on the growth of bacteria, *J. Infectious Diseases,* **25**:41, 1919.

Leach, R. H., et al.: The death of microorganisms during drying in relation to solute concentration and drying temperature, *J. Gen. Microbiol.,* **21**:658, 1959.

Leibo, S. P., and P. Mazur: Freezing of bacteriophage T_4: Temperature and rate effects as a function of salt concentration, *Virology,* **38**:558, 1969.

Leif, W. R., and J. E. Hebert: Effect of urocanic acid and dipicolinic acid on bacteria exposed to ultraviolet radiation, *Am. J. Hyg.,* **71**:285, 1960.

Lemcke, R. M., and H. R. White: The heat resistance of *Escherichia coli* cells from cultures of different ages, *J. Appl. Bact.,* **22**:193, 1959.

Levinson, H. S., and M. T. Hyatt: Heat activation kinetics of *Bacillus megaterium* spores, *Biochem. Biophys. Res. Commun.,* **37**:909, 1969.

—— and ——: Distribution and correlation of events during thermal inactivation of *Bacillus megaterium* spores, *J. Bact.,* **108**:111, 1971.

Licciardello, J. J., and J. T. R. Nickerson: Some observations on bacterial thermal death time curves, *Appl. Microbiol.,* **11**:476, 1963.

Lion, M. B., and E. D. Bergmann: The effect of oxygen on freeze-dried *Escherichia coli, J. Gen. Microbiol.,* **24**:191, 1961*a*.

—— and ——: Substances which protect lyophilized *Escherichia coli* against the lethal effect of oxygen, *J. Gen. Microbiol.,* **25**:291, 1961*b*.

Little, G. N., and M. A. Gordon: Survival of fungus cultures maintained under mineral oil for twelve years, *Mycologia,* **59**:733, 1967.

Luckiesh, M., et al.: Killing bacteria in water under pressure, *Gen. Elec. Rev.,* **50**:16, 1947.

MacLeod, R. A., and E. E. Snell: The relation of ion antagonism to the inorganic nutrition of lactic acid bacteria, *J. Bact.,* **59**:783, 1950.

—— et al.: I. Effect of freezing and storage on the requirements of *Aerobacter aerogenes* and *Escherichia coli* for growth, *Can. J. Microbiol.,* **12**:61, 1966.

McDonald, I. J., et al.: Proteolytic activity of some cold-tolerant bacteria from Arctic sediments, *Can. J. Microbiol.,* **9**:303, 1963.

Marshall, B. J., et al.: The effect of water activity, solutes and temperature on the viability and heat resistance of freeze-dried bacterial spores, *J. Gen. Microbiol.,* **31**:451, 1963.

Mazur, P.: Physical factors implicated in the death of microorganisms at subzero temperatures, *Ann. N.Y. Acad. Sci.,* **85**:610, 1960.

Miller, R. E., and L. A. Simons: Survival of bacteria after twenty-one years in the dried state, *J. Bact.,* **84**:1111, 1962.

Mohr, V., and H. Larsen: On the structural transformations and lysis of *Halobacterium salinarium* in hypotonic and isotonic solutions, *J. Gen. Microbiol.,* **31**:267, 1963.

Monk, G. W., and P. A. McCaffrey: Effect of sorbed water on the death rate of washed *Serratia marcescens, J. Bact.,* **73**:85, 1957.

Morita, R. Y.: The basic nature of marine psychrophilic bacteria, *Bull. Misaki Marine Biol. Inst. Kyoto Univ.*, February 1968.

—— and P. F. Mathemeier: Temperature-hydrostatic pressure studies on partially purified inorganic pyrophosphatase activity, *J. Bact.*, **88**:1667, 1964.

—— and C. E. ZoBell: Effect of hydrostatic pressure on the succinic dehydrogenase system in *Escherichia coli, J. Bact.*, **71**:668, 1956.

Mullican, C. L., and R. K. Hoffman: Dry heat or gaseous chemical resistance of *Bacillus subtilis* var. *niger* spores included within water-soluble crystals, *Appl. Microbiol.*, **16**:1110, 1968.

Nakamura, M., and C. M. Ramage: Increased survival of ultraviolet-irradiated *Shigella sonnei* by decreasing the surface tension, *Nature*, **197**:1028, 1963.

Nath, E. J., and L. F. L. Clegg: Studies on the activation of spores of some *Bacillus* species, *Can. Inst. Food Technol. J.*, **2**:86, 1969.

Nei, T.: Effects of freezing and freeze-drying on microorganisms. In "Recent Researches in Freezing and Drying," 1960.

——, ed.: Conference on "Mechanisms of Cellular Injury by Freezing and Drying in Microorganisms," Tokyo, Japan, University of Tokyo Press, 1969.

—— et al.: Effect of residual moisture content on the survival of freeze-dried bacteria during storage under various conditions, *Cryobiology*, **2**:276, 1966.

—— et al.: IV. The mechanism of formation of ice crystals in *E. coli* cells, *Low Temp. Science, Ser. B*, **25**:113, 1967.

Packer, E. L., et al. Factors affecting the rate of killing of *Escherichia coli* by repeated freezing and thawing, *J. Bact.*, **89**:718, 1965.

Paul, K. L., and R. Y. Morita: Effects of hydrostatic pressure and temperature on the uptake and respiration of amino acids by a facultatively psychrophilic marine bacterium, *J. Bact.*, **108**:835, 1971.

Pootjes, C. F.: Repair of ultraviolet light damage by a facultative autotroph: *Hydrogenomonas facilis, Can. J. Microbiol.*, **14**:1205, 1968.

Rao, V. S., and D. Visweswaram: I. Heat resistance of *Bacillus subtilis* spores during incubation in peptone and glucose solutions, *Indian J. Microbiol.*, **1**:51, 1961.

Rey, L.: "Advances in Freeze-Drying," Paris, Ed. Hermann, 1966.

Rhoades, H. E.: Effects of 20 years' storage on lyophilized cultures of bacteria, molds, viruses, and yeasts, *Am. J. Vet. Res.*, **31**:1867, 1970.

Riklis, E.: Studies on mechanism of repair of ultraviolet-irradiated viral and bacterial DNA in vivo and in vitro, *Can. J. Biochem.*, **43**:1207, 1965.

Ronald, G. W., and R. L. Morris: The effect of copper on distilled water quality for use in milk and water laboratories, *J. Milk Food Technol.*, **30**:305, 1967.

Rüger, H.-J.: Konservierung mariner bakterien durch trocknung im vakuum, *Veröff. Inst. Meeresforsch. Bremerh.*, **12**:297, 1970.

Russell, A. D., and D. Harries: Factors influencing the survival and revival of heat-treated *Escherichia coli, Appl. Microbiol.*, **16**:335, 1968.

Sacks, L. E., and G. F. Bailey: Dry rupture of bacterial spores, *J. Bact.*, **85**:720, 1963.

Sawada, F., and K. Suzuki: Recovery of *Escherichia coli* irradiated with ultraviolet light, *J. Bact.*, **81**:288, 1961.

Scheie, P. O.: Plasmolysis of *Escherichia coli* B/r with sucrose, *J. Bact.*, **98**:335, 1969.

Scheie, P. O., and R. Rehberg: Response of *Escherichia coli* B/r to high concentrations of sucrose in a nutrient medium, *J. Bact.*, **109**:229, 1972.

Schmittdiel, E.: Versuche zur Konservierung des Psittakose-Virus durch die Vakuumgefriertrocknung, *Zentr. Bakteriol., Orig.*, **181**:446, 1961.

Shackell, L. F.: An improved method of desiccation with some applications to biological problems, *Am. J. Physiol.*, **24**:325, 1909.

Shankar, K., and R. C. Bard: Effect of metallic ions on the growth, morphology, and metabolism of *Clostridium perfringens, J. Bact.*, **69**:444, 1955.

Shimada, K., and Y. Takagi: The effect of caffeine on the repair of ultraviolet-damaged DNA in bacteria, *Biochim. Biophys. Acta*, **145**:763, 1967.

Shull, J. J., et al.: Kinetics of heat activation and of thermal death of bacterial spores, *Appl. Microbiol.*, **11**:485, 1963.

Sie, E. H. C., et al.: Factors inducing mesophilic bacteria to grow at 55°C., *Biochem. Biophys. Res. Commun.*, **6**:206, 1961.

Sinclair, N. A., and J. L. Stokes: Role of oxygen in the high cell yields of psychrophiles and mesophiles at low temperatures, *J. Bact.*, **85**:164, 1963.

—— and ——: Isolation of obligately anaerobic psychrophilic bacteria, *J. Bact.*, **87**:562, 1964.

Sinskey, T. J., and G. J. Silverman: Characterization of injury incurred by *Escherichia coli* upon freeze-drying, *J. Bact.*, **101**:429, 1970.

Smith, K. C., and D. H. C. Meun: Repair of radiation-induced damage in *Escherichia coli*. I, *J. Molec. Biol.*, **51**:459, 1970.

Stein, C. D., and H. Rogers: Recovery of viable microorganisms and viruses from vapors removed from frozen suspensions of biologic material during lyophilization, *J. Vet. Research*, **11**:339, 1950.

Stokes, J. L., and M. L. Redmond: Quantitative ecology of psychrophilic microorganisms, *Appl. Microbiol.*, **14**:74, 1966.

Straka, R. P., and J. L. Stokes: Metabolic injury to bacteria at low temperatures, *J. Bact.*, **78**:181, 1959.

Strange, R. E., and F. A. Dark: Effect of chilling on *Aerobacter aerogenes* in aqueous suspension, *J. Gen. Microbiol.*, **29**:719, 1962.

Tarshis, M. S.: The preservation of mycobacteria by freezing in various diluents, *Am. Rev. Respirat. Diseases*, **83**:762, 1961.

Timson, W. J., and A. J. Short: Resistance of microorganisms to hydrostatic pressure, *Biotechnol. Bioengineer.*, **7**:139, 1965.

Town, C. D., et al.: Production and repair of radiochemical damage in *Escherichia coli* deoxyribonucleic acid; its modification by culture conditions and relation to survival, *J. Bact.*, **105**:127, 1971.

Upadhyay, J., and J. L. Stokes: Anaerobic growth of psychrophilic bacteria, *J. Bact.*, **83**:270, 1962.

Vignais, P., and P. Vignais: Inhibition sélective de quelques enzymes respiratoires chez *Escherichia coli*, *Experientia*, **10**:305, 1954.

Wagman, J., and E. J. Weneck: Preservation of bacteria by circulating-gas freeze drying, *Appl. Microbiol.*, **11**:244, 1963.

Walker, H. W.: Influence of buffers and pH on the thermal destruction of spores of *Bacillus megaterium* and *Bacillus polymyxa*, *J. Food Sci.*, **29**:360, 1964.

Watko, L. P., and K. L. Heddleston: Survival of shell-frozen, freeze-dried, and agar slant cultures of *Pasteurella multocida*, *Cryobiology*, **3**:53, 1966.

Weatherwax, R. S.: Reactivations of ultraviolet-irradiated *Escherichia coli*, *J. Bact.*, **72**:329, 1956.

Webb, S. J.: The influence of oxygen and inositol on the survival of semidried microorganisms, *Can. J. Microbiol.*, **13**:733, 1967.

—— et al.: Relative humidity, inositol and the effect of radiations on air-dried microorganisms, *Nature*, **201**:1103, 1964.

Williams, O. B., and W. J. Robertson: Studies on heat resistance. VI. Effect of temperature of incubation at which formed on heat resistance of aerobic thermophilic spores, *J. Bact.*, **67**:377, 1954.

—— and C. H. Zimmerman: Studies on heat resistance. III. The resistance of vegetative cells and spores of the same organism, *J. Bact.*, **61**:63, 1951.

Winslow, C.-E. A., and I. S. Falk: Studies on salt action. VIII. The influence of calcium and sodium salts at various hydrogen-ion concentrations upon the viability of *Bacterium coli*, *J. Bact.*, **8**:215; 1923*a*; IX. The additive and antagonistic effects of sodium and calcium chlorides upon the viability of *Bacterium coli*, *ibid.*, **8**:237, 1923*b*.

Wulff, D. L., and C. S. Rupert: Disappearance of thymine photodimer in ultraviolet irradiated DNA upon treatment with a photoreactivating enzyme from baker's yeast, *Biochem. Biophys. Res. Commun.*, **7**:237, 1962.

Wyss, O., et al.: The role of peroxide in the biological effects of irradiated broth, *J. Bact.*, **56**:51, 1948.

Yesair, J., et al.: Effect of certain environmental conditions on heat resistance of micrococci, *Food Res.*, **11**:327, 1946.

Yokoya, F., and G. K. York: Effect of several environmental conditions on the "thermal death rate" of endospores of aerobic thermophilic bacteria, *Appl. Microbiol.*, **13**:993, 1965.

Zimmerman, L.: Survival of *Serratia marcescens* after freeze-drying or aerosolization at unfavorable humidity. I. Effects of sugars, *J. Bact.*, **84**:1297, 1962.

——: Toxicity of copper and ascorbic acid to *Serratia marcescens, J. Bact.*, **91**:1537, 1966.

——: Permeability of *Serratia marcescens* to some inorganic salts, *J. Bact.*, **97**:749, 1969.

ZoBell, C. E.: Apparatus for collecting water samples from different depths for bacteriological analysis, *J. Marine Res.*, **4**:173, 1941.

——: Hydrostatic pressure as a factor affecting the activities of marine microbes. In "Recent Researches in the Fields of Hydrosphere, Atmosphere and Nuclear Geochemistry," Tokyo, Japan, Maruzen Co., Ltd., 1964.

—— and K. M. Budge: Nitrate reduction by marine bacteria at increased hydrostatic pressures, *Limnol. Oceanograph.*, **10**:207, 1965.

—— and A. B. Cobet: Growth, reproduction, and death rates of *Escherichia coli* at increased hydrostatic pressures, *J. Bact.*, **84**:1228, 1962.

—— and C. B. Feltham: Are there specific marine bacteria? *Proc. 5th Pacific Sci. Cong.*, **3**:2097, 1933.

—— and F. H. Johnson: The influence of hydrostatic pressure on the growth and viability of terrestrial and marine bacteria, *J. Bact.*, **57**:179, 1949.

—— and H. D. Michener: A paradox in the adaptation of marine bacteria to hypotonic solutions, *Science*, **87**:328, 1938.

11 *Enzymes of bacteria*

Catalysis may be defined as the acceleration of a reaction produced by the presence of a substance known as a catalyst.

A catalyst is an agent that accelerates a chemical reaction without itself being destroyed or used up. It may be recovered practically unchanged at the end of the reaction.

An enzyme or ferment may be defined as a thermolabile organic catalyst elaborated by a living cell and capable of functioning independently of the cell.

Every living cell may be considered a gross catalyst. It converts nutrients to waste products and, in so doing, obtains energy. The reaction

$$6CO_2 + 6H_2O + \text{energy} \rightleftharpoons C_6H_{12}O_6 + 6O_2$$

from left to right is the classical equation for the photosynthetic reaction in green plants; and from right to left in animals, plants, and many microorganisms.

An enzyme acts by catalysis; i.e., it increases the velocity of a chemical reaction without itself being permanently changed. The enzyme may be recovered in an active condition after completion of the reaction. There is a close analogy between inorganic catalysts and enzymes. For example, hydrogen peroxide slowly decomposes into water and oxygen as according to the reaction

$$2H_2O_2 \longrightarrow 2H_2O + O_2$$

In the presence of an inorganic catalyst, such as platinum, or the enzyme catalase, the decomposition of the peroxide is greatly accelerated and ceases only when the destruction of the compound is complete.

Nature of enzymes Enzymes possess the properties of proteins. They form colloidal solutions, dialyze through membranes either very slowly or not at all, are amphoteric, form opalescent solutions, are precipitated from solutions by the same agents which precipitate proteins, and have large molecular weights. When treated with acids, alkalies, or proteolytic enzymes, they yield mixtures of amino acids.

Over 700 enzymes are known and well over 100 have been crystallized. The crystalline enzymes appear to be pure molecules.

However, crystallinity in itself is not an index of purity as was at one time supposed. Some enzymes have been repeatedly recrystallized without yielding to separation from other components.

Photomicrographs of some crystalline enzyme preparations are shown in Fig. 194.

Great similarities exist in enzyme systems from bacteria to man, indicating common patterns of catalysis in living cells.

Union of enzyme and substrate Available evidence indicates that enzymes combine with the substrates upon which they act. Whether this union is physical or chemical is not clearly understood but certainly the evidence points to some type of unstable complex between catalyst and substrate. After the substrate is decomposed, free enzyme is regenerated and becomes available to react with more substrate.

The substrate is believed to bind itself to the enzyme at specific sites, usually one or a few per molecule of enzyme. These are referred to as active sites of the enzyme. They are highly specific, being able to distinguish between closely similar compounds.

In most cases the enzyme molecule is considerably larger than the substrate acted upon. The actual substrate for the enzyme action is usually a low molecular weight compound or a small portion of a high molecular weight molecule. Therefore, only a fraction of enzyme protein could be in contact with the substrate. This means that only a relatively few amino acids are directly involved in the process. If all amino acids in a region were essential, it would be possible to group these around the substrate so that a concerted action of many residues would be conceivable. That amino acid residues, adjacent in sequence to these essential residues, do not play a part makes it probable that the contact amino acids involved must be very few in number.

A schematic representation of an enzyme-substrate complex is shown in Fig. 195. It may be seen that some amino acids are in direct contact with the substrate, whereas others can play only an indirect role. Some may play no role whatsoever, at least as far as the enzyme action is concerned.

Polyribonucleotide 2-oligonucleotidotransferase (ribonuclease) is the first enzyme in which the complete amino acid sequence has been determined. This enzyme is obtained from pancreatic juice and it hydrolyzes ribonucleic acid. For a schematic drawing of the molecule in which active sites are indicated, see Hirs et al. (1960).

Structure and mechanism of action of enzymes

Enzymes are proteins having molecular weights from 10,000 to well up into the millions. The smallest enzymes are composed of about 100 amino acids; the largest ones of many more.

The amino acids in an enzyme follow a unique pattern that is precise. Each position is occupied by one of the known amino acids and

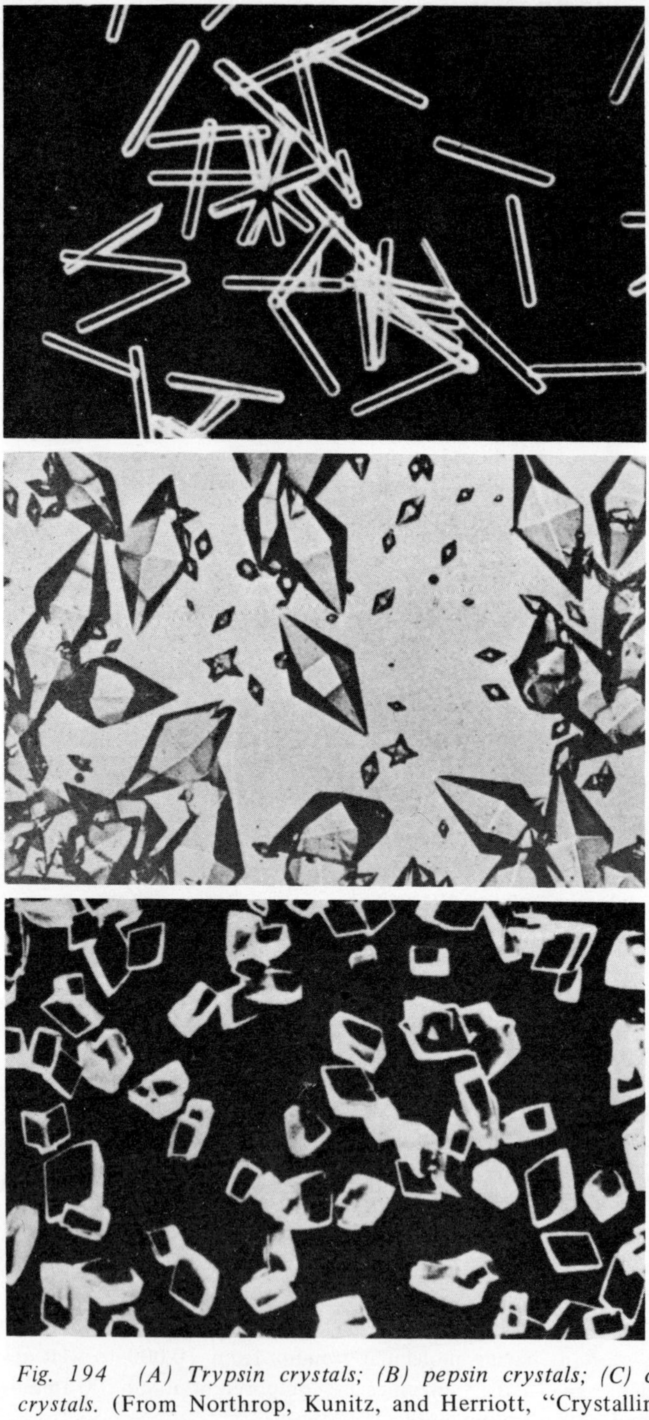

(a)

(b)

(c)

Fig. 194 (A) Trypsin crystals; (B) pepsin crystals; (C) chymotrypsin crystals. (From Northrop, Kunitz, and Herriott, "Crystalline Enzymes," Columbia University Press.)

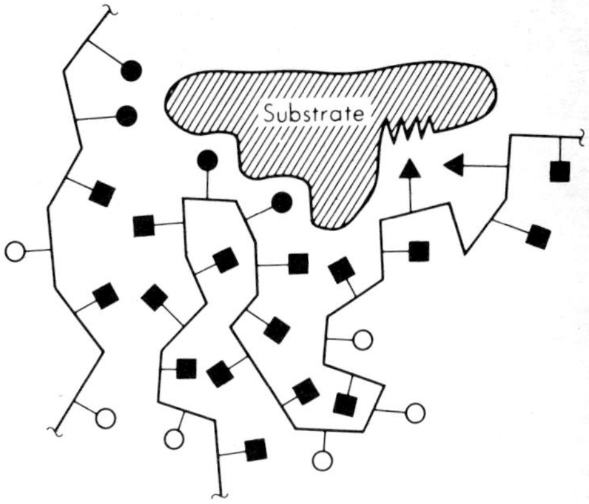

Fig. 195 Schematic representation of an active site. Solid circles represent contact amino acid residues whose fit with substrate determines specificity; triangles represent catalytic residues acting on substrate bond, indicated by a jagged line; empty circles represent nonessential residues on the surface; and squares represent residues whose interaction with each other maintains the three-dimensional structure of the protein. (Redrawn from Koshland.)

the same sequence is repeated as the enzyme is synthesized over and over by the cell.

An enzyme protein, in x-ray diffraction studies, looks like a piece of tangled rope. The chain of precisely joined amino acids is folded in a complex manner so that it possesses a reproducible three-dimensional structure. The interactions of the amino acid side chains of the protein are believed to determine the three-dimensional structure.

Lysozyme Lysozyme is an enzyme that is capable of dissolving many species of bacteria. Although the term is generally used in the singular, some 20 closely related lysozymes are now known.

They are found in the nasal mucus, in some molds, in egg white, in the latex of different plants, etc. The lysozyme present in egg white may easily be isolated and crystallized. The enzymes are stable, have relatively low molecular weights (14,500) and have been some of the most thoroughly investigated proteins. The lysozymes have been classified as muramidases or N-acetylmuramide glycanohydrolases. They are believed to be one of nature's defense mechanisms.

Susceptible bacteria exposed to lysozyme undergo a series of characteristic changes (Fig. 196). First the cells begin to swell; then they become transformed into large spheres. These soon lyse and lose their visible outline. In *Micrococcus lysodeikticus* it is the cell-wall mucopolysaccharide that is the substrate for the enzyme. It is a complex material containing both sugars and peptides. Subsequently similar insoluble mucopolysaccharides have been found in other bacteria studied and soluble saccharides and glycopeptides formed by action of lysozyme on cell walls have been isolated and identified. The egg white lysozyme also hydrolyzes chitin, a linear polymer of N-acetyl-D-glucosamine (GlcNac).

In addition to vegetative cells, Suzuki and Rode (1969) reported that resting spores from a strain of *Bacillus megaterium* were also markedly affected by the action of lysozyme (Fig. 197).

Phillips and his colleagues, by means of x-ray crystallographic studies, showed that egg white lysozyme consists of 129 amino acids

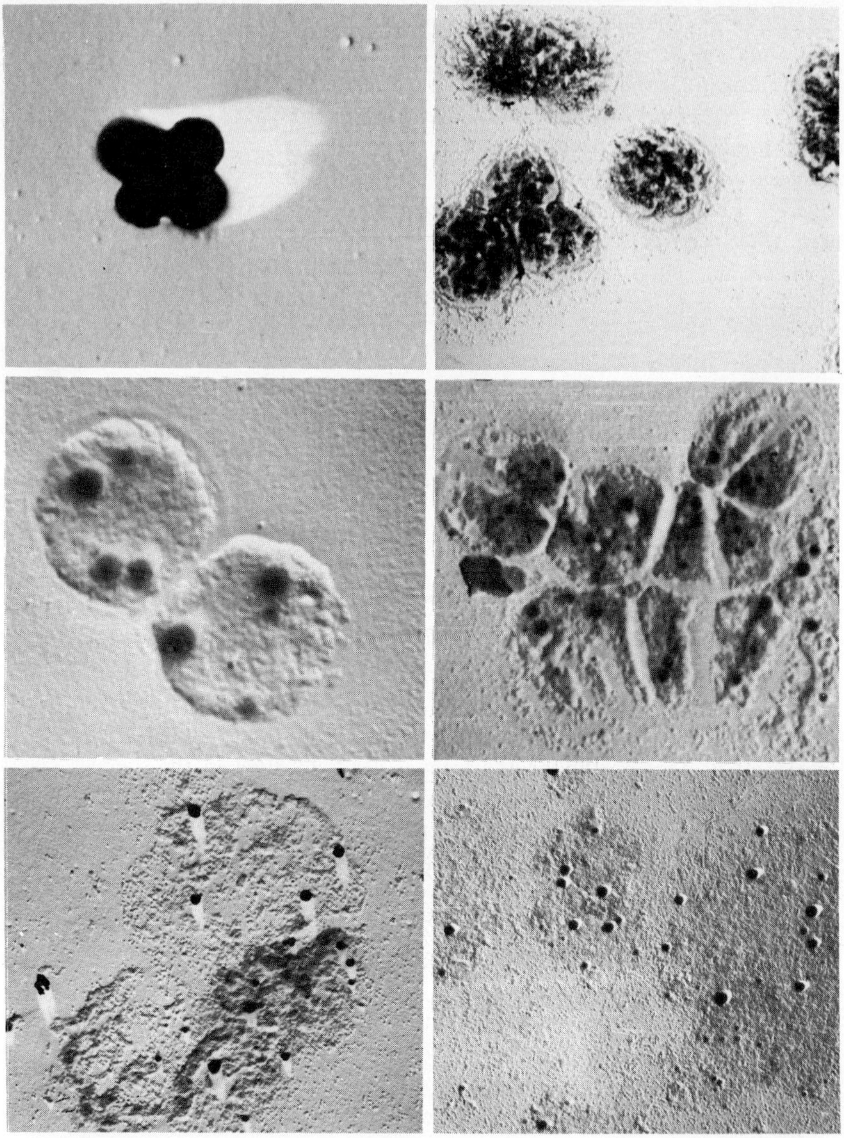

Fig. 196 Action of lysozyme on Micrococcus lysodeikticus. Upper left, normal cells after 1 hr in double-distilled water. Upper right, removal of cell wall immediately after addition of lysozyme. Cells appear larger owing to swelling and flattening. Center left, partially solubilized and disrupted protoplasts after cell wall has been removed by lysozyme. Granules exposed from this point on. Center right, protoplasts undergoing plasmoptysis. Lower, end points of lysis in double-distilled water. (Courtesy of Grula and Hartsell.)

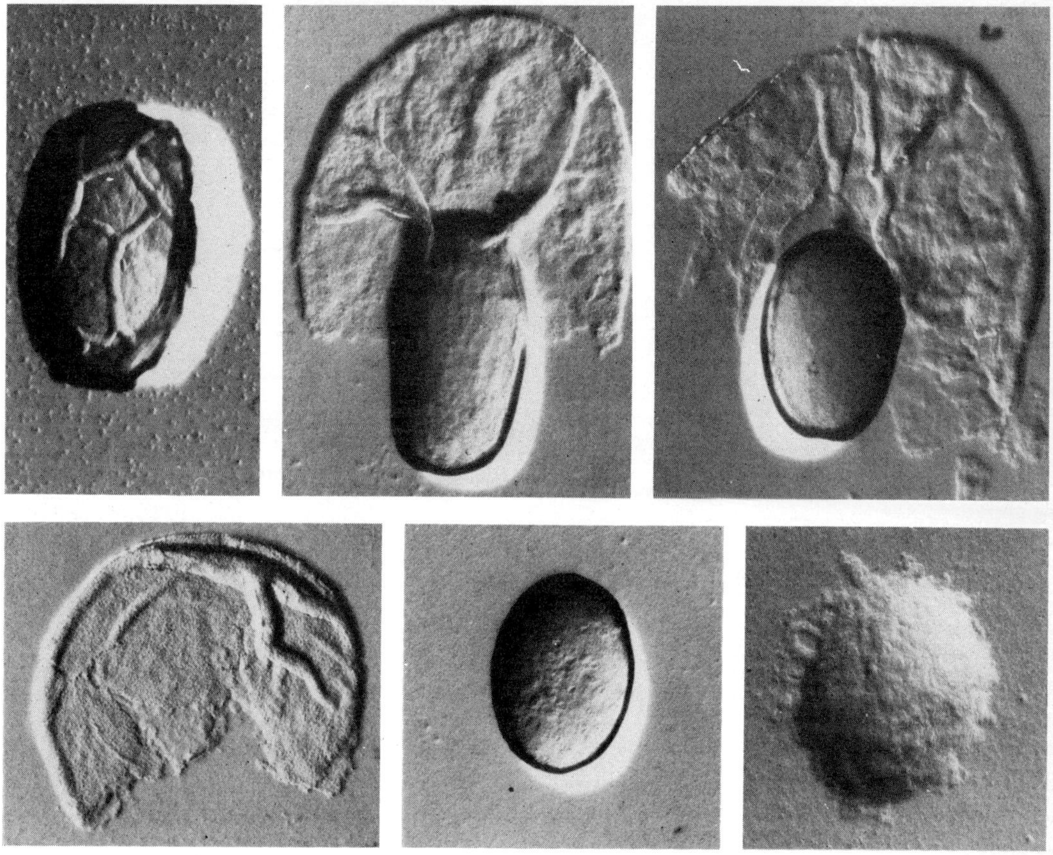

Fig. 197 Effect of lysozyme on spores of Bacillus megaterium. Carbon replicas. Upper left, spore not treated with lysozyme; upper center, spore central body with ruptured coat attached, treated 5 min with lysozyme; upper right, central body detached from adjacent coat, treated 5 min with lysozyme; lower left, isolated coat, treated 5 min with lysozyme; lower center, isolated central body, treated 5 min with lysozyme; lower right, disintegrating central body, treated 10 min with lysozyme. (Courtesy of Rode and Suzuki.)

arranged in definite sequence, and described its three-dimensional structure (Figs. 198 and 199).

For more information: Chaloupka and Veres (1961); Chipman and Sharon (1969); Hartsell and Acker (1961); Iwata and Eda (1968); Iwata et al. (1968); Phillips (1966); Smith and Liener (1967); Hendrich and Pospíšil (1971).

Comparison of enzyme and nonenzyme velocities The velocities at which enzymatic reactions occur are extraordinarily more rapid than their nonenzymatic analogs. According to Koshland (1963), a mixture of amino acids added to a solution of starch produces no detectable hydrolysis of the molecule. The enzyme α-1,4-glucan maltohydrolase

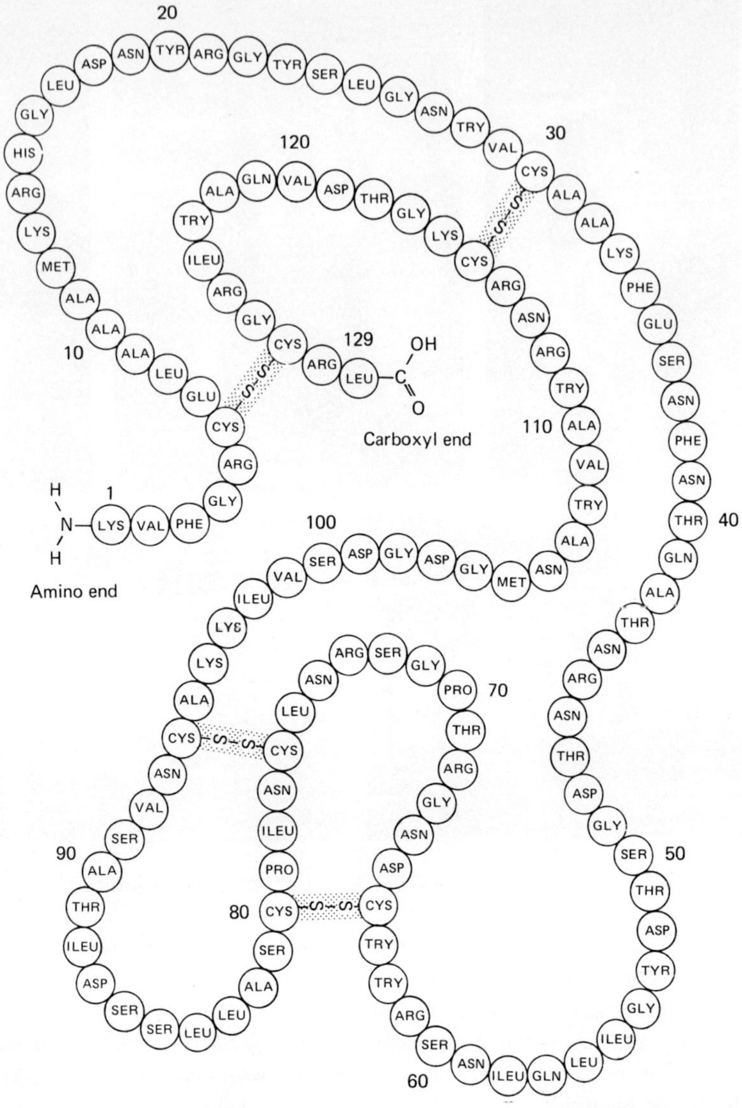

ALA	Alanine
ARG	Arginine
ASN	Asparagine
ASP	Aspartic acid
CYS	Cysteine
GLU	Glutamic acid
GLN	Glutamine
GLY	Glycine
HIS	Histidine
ILEU	Isoleucine
LEU	Leucine
LYS	Lysine
MET	Methionine
PHE	Phenylalanine
PRO	Proline
SER	Serine
THR	Threonine
TRY	Tryptophan
TYR	Tyrosine
VAL	Valine

Fig. 198 Two-dimensional model of the lysozyme molecule. Lysozyme is a protein containing 129 amino acid subunits, commonly called residues. These residues form a polypeptide chain that is cross-linked at four places by disulfide (–S–S–) bonds. The three-dimensional structure of the lysozyme molecule has now been established (see Fig. 199). The function of lysozyme is to split a particular long-chain molecule of a complex sugar found in the outer membrane of many living cells. The complex sugar fits into a cleft, or pocket, formed by the three-dimensional function of the enzyme molecule. (Courtesy of Phillips.)

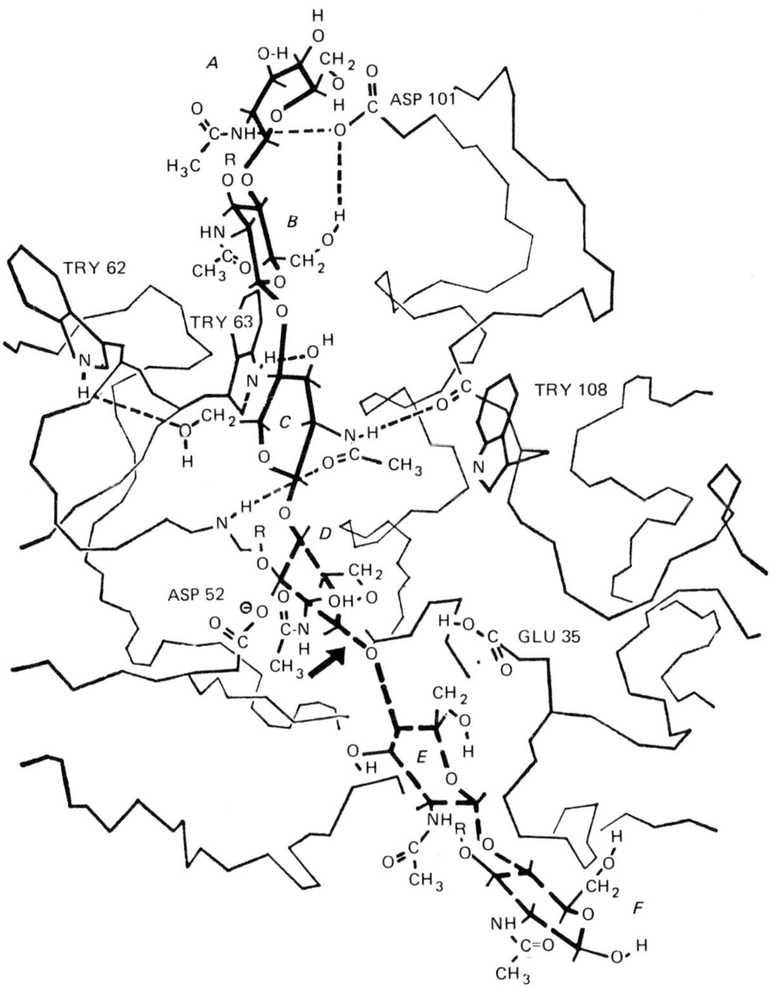

Fig. 199 The three-dimensional model of the area around the active site of the lysozyme-substrate complex. (After Phillips). *The pyranose rings of the substrate are shown in heavy lines, with the monosaccharide units of the lysozyme-$(GlcNAc)_3$ complex (A–C) solid and those placed by model building (D–F) dashed. Except for some groups of particular interest, only the peptide backbone of the protein is shown. The light dashed lines are the six hydrogen bonds between $(GlcNAc)_3$ and the enzyme. The groups R are hydrogen in the case of chitin substrates, and lactyl groups in cell-wall saccharides. The arrow indicates the bond cleaved.* (Courtesy of Chipman and Sharon.)

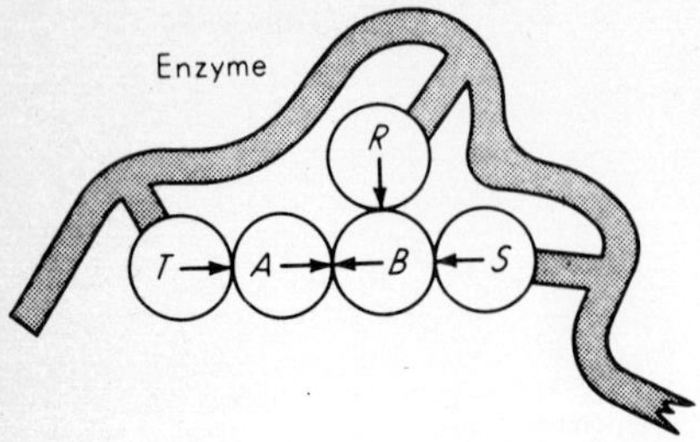

Fig. 200 Schematic representation of an active site, showing reaction between substrates A and B catalyzed by catalytic side chains R, S, and T. Orientation of side chains is indicated by arrows. Orientation and juxtaposition are possible because of the structure of the protein and because of the attraction of the substrate to the active site. (Redrawn from Koshland.)

(β-amylase), which contains the same amino acids, can hydrolyze 4000 glucosidic bonds per second per molecule of enzyme.

A possible explanation of this is given in Fig. 200, which illustrates a reaction between two substrates, A and B, catalyzed by three amino acid side chains, R, S, and T. To quote Koshland

It is apparent from the simple chemical analogues that the orientation of A and B relative to each other and of R, S, and T relative to A and B will be crucial. Yet, the probability of a five-body oriented collision in aqueous solutions is extremely remote. If, however, the three catalysts are held in a fixed and appropriate position because of the 3-dimensional structure of the enzyme, and if the substrates A and B are absorbed in a precise geometrical relationship to each other and to the catalytic groups, the probabilities of reaction are enormously increased.

Specificity of enzymes Inorganic catalysts such as platinum, palladium, and nickel are able to catalyze many reactions, and display a certain amount of specificity. Enzymes, on the other hand, display a greater degree of specificity with respect to the nature of the substrates acted upon.

Enzymes exhibit absolute specificity, group specificity, stereochemical specificity, and relative specificity.

Absolute specificity Enzymes in this group act on only a single substrate. Urea amidohydrolase (urease), for example, hydrolyzes urea to ammonia and carbon dioxide. This is the only substrate attacked by the enzyme.

Group specificity These enzymes are not as specific as those in the preceding group. Peptide hydrolases (proteases) act on the peptide bonds of proteins, not on fats; and glycerol ester hydrolase acts on true fats, not on proteins.

Stereochemical specificity α-D-Glucoside glucohydrolase (α-glucosidase) catalyzes the hydrolysis of α-D-glucosides but has no effect on β-D-glucosides; L-arginine ureohydrolase (arginase) acts on L-arginine but not on D-arginine.

Relative specificity These enzymes act preferentially on one class of compounds but will attack members of another class to a certain extent. The term is also used to describe different reaction rates within a given group or class.

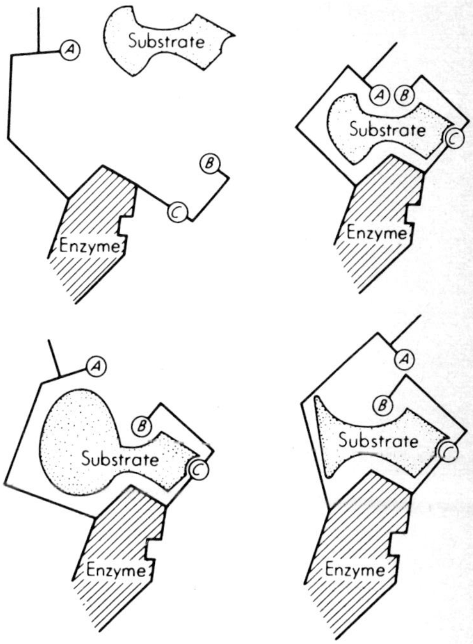

Fig. 201 Schematic representation of a flexible active site. Top row, substrate binding induces proper alignment of catalytic groups A to B so that reaction ensues. Bottom row, compounds which are either too large or too small are bound but fail to cause alignment of catalytic groups, and do not react. (Redrawn from Koshland.)

Koshland (1963) believes that the specificity properties of a number of enzymes demand a flexible active site in which the substrate induces a conformation change leading to an appropriate alignment of catalytic groups. This is illustrated in Fig. 201. This hypothesis predicts that changes in protein structure accompany the binding of the substrate to the enzyme surface.

For more information: Niemann (1964); Wang (1968).

Prosthetic groups Some enzymes consist entirely of protein. Examples are the digestive enzymes pepsin and trypsin. Other enzymes contain a nonprotein component in their molecules. Enzymes of this type are called conjugate enzymes. The nonprotein component is known as the prosthetic group, and the protein component as the apoenzyme. The complete enzyme, consisting of both apoenzyme and prosthetic group, is called the holoenzyme. The prosthetic group is firmly fixed to the apoenzyme, in contrast to the union of enzyme and substrate which is believed to be an unstable complex.

Location of enzymes in bacteria

Considerable interest has developed in attempts to associate enzyme activity with certain structures present in cells.

Fractionation The cell wall of bacteria may be removed by hydrolysis in hypertonic solution to form protoplasts which may be emptied of cytoplasmic contents by subsequent osmotic shock. The method has proved successful in isolating components of certain cells, the walls of which are completely hydrolyzed by lysozyme. Possible damage to internal cellular structures may be avoided by opening the cytoplasmic membrane with lipase.

The individual subcellular structures may be separated by centrifugation. By this procedure the extract may be resolved into three fractions: (1) structures which are large enough to be observed with a light microscope by sedimenting at 10,000 x *g* for 15 min; (2) submicroscopic particles which sediment at 100,000 x *g* for 60 to 120 min; and (3) the soluble fraction which contains substances that remain in solution in the supernate that is separated from the submicroscopic particles.

The bulk of the respiratory enzymes of bacteria are recovered in fraction 1. This fraction consists of cytoplasmic membrane and a network of internal membranes in the form of vesicles and tubules in section (Fig. 202). These envelopes are free of all cytoplasmic material and ribosomes. The intracytoplasmic membranes are the probable locus of the respiratory enzymes of the cell. The proteolytic enzymes appear to be about equally divided between cytoplasm and cytoplasmic membrane.

The ribosomes are found in fraction 2. They are rich in RNA and phospholipides and do not contain the respiratory enzymes.

Fig. 202 Section of envelopes of Azotobacter agilis prepared by brief ballistic disintegration. Numerous internal membranes appear within the envelope. Tears are apparent in the walls of several envelopes. One cell has escaped disruption and has retained its cytoplasm. X31,000. (Courtesy of Pangborn, Marr, and Robrish.)

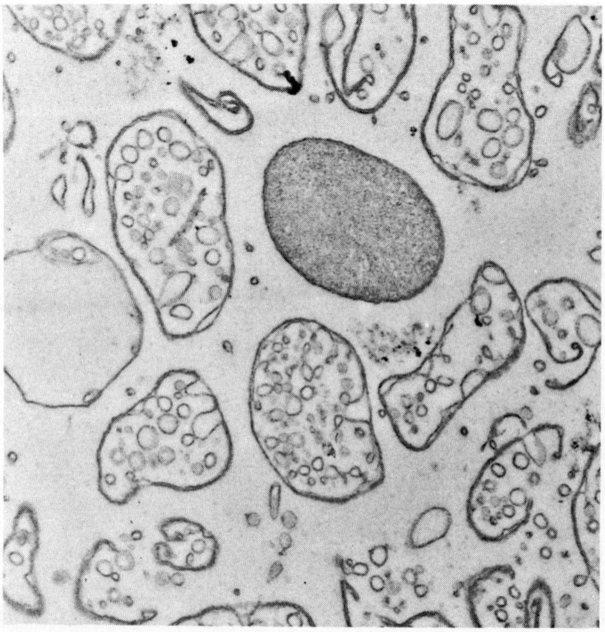

For more information: Chaloupka (1961); Duerre and Ribi (1963); Heppel (1967); Marr (1960); McDonald (1964); Munkres and Wachtel (1967); Neu and Chou (1967); Pangborn et al. (1962); Weibull et al. (1959).

Factors which influence enzyme reaction rates

The rates of enzyme reactions are influenced by a number of factors, the most important being (1) temperature, (2) pH of the solution, (3) ultraviolet light, (4) concentration of enzyme, (5) concentration of substrate, (6) presence of activators, and (7) presence of inhibitors. The same factors which influence the growth and multiplication of bacteria also affect the action of enzymes. However, enzymes are generally more resistant to unfavorable environmental conditions than the cells producing them. For example, dried yeast heated to 100°C for 6 hr loses the power of growth and multiplication but still retains the ability to ferment sugar to alcohol and carbon dioxide.

Effect of temperature The velocity of an enzyme reaction is accelerated by an increase in temperature. This continues until a maximum is reached, after which the velocity gradually decreases, resulting finally in the destruction of the enzyme. In general, a rise of 10°C doubles the velocity of an enzyme reaction.

Each enzyme has its own characteristic optimum, maximum, and minimum temperatures. An enzyme displays its greatest activity at the optimum temperature. That temperature above which an enzyme is no longer active is known as the maximum temperature; and that temperature below which it cannot function is known as the minimum temperature. The optimum, maximum, and minimum temperatures are influenced by a number of factors, such as concentration of enzyme, nature and concentration of substrate, pH, and presence of activating and paralyzing substances.

Most enzymes in solution are more or less stable at temperatures below 45°C, but above 50°C inactivation increases rapidly with a rise in temperature. The majority of enzymes are irreversibly destroyed at temperatures of 70 to 80°C. On the other hand, enzymes in the dry state are more resistant to high temperatures than the same enzymes in solution. For example, dry rennin is only slowly destroyed at 158°C.

In general, soluble enzymes may be preserved by freezing and kept at low temperatures, or by lyophilization to dry powders.

Effect of pH Enzymes are markedly influenced by the hydrogen-ion concentration of the solution. Some enzymes act best in acid solutions; others require alkaline solutions; still others do not function well unless their environments are neutral in reaction. There exists for every enzyme a maximum, an optimum, and a minimum pH. These pH values vary with temperature, type and concentration of substrate, type of buffer, presence or absence of inhibitors or activators, and period of action of enzyme.

Two different types of behavior may be distinguished in this connection: reversible and irreversible effect. An enzyme can tolerate slight changes in acidity or alkalinity without being destroyed. Readjustment of the reaction to the optimum pH restores fully the activity of the enzyme. On the other hand, if the solution is made strongly acid or strongly alkaline, the enzyme undergoes an irreversible loss of activity. Readjustment of the reaction back to the optimum pH does not restore the activity of the enzyme.

Effect of ultraviolet light Ultraviolet light either destroys or modifies the action of enzymes. The rate of destruction is practically independent of temperature but is affected by pH and other environmental conditions.

Purified enzymes are more easily destroyed by light rays than the same enzymes in less purified preparations. Impurities from the culture media, especially proteins and protein split products, may afford considerable protection. In general, the greater the degree of purification of an enzyme, the more susceptible it becomes to the toxic light rays.

Enzymes, like bacteria and other living cells, are most sensitive to the same wavelengths of ultraviolet, i.e., those in the vicinity of 2650 Å (page 303).

Effect of concentration of substrate and enzyme An increase in the substrate concentration may increase or decrease the velocity of an enzymatic reaction. If the substrate concentration is small in proportion to the amount of enzyme, an increase in the substrate content will increase the velocity of the reaction. For example, the enzyme sucrase hydrolyzes sucrose to glucose and levulose. An increase in the sucrose concentration, up to 4 or 5 percent, increases the velocity of hydrolysis. Above this concentration, the relative amount of sugar hydrolyzed becomes progressively less as the sugar concentration is increased. It may be concluded that a strong solution of cane sugar diluted to a concentration of about 5 percent will be more efficiently hydrolyzed by sucrase than one more concentrated.

The time required for sucrose to be hydrolyzed is proportional to the concentration of the enzyme. In general this is true for all enzymes when the amount of enzyme is much smaller than the concentration of substrate so that all of the enzyme can combine with the latter. When an excess of enzyme is used, the velocity appears to be proportional to the square root of the enzyme concentration.

Effect of activators and inhibitors Some substances restore the activity of enzymes, whereas others produce the reverse effect. The former are called activators; the latter, inhibitors. These substances may be either specific or nonspecific.

Heavy metals Many enzymes are inactivated by heavy metals such as silver, mercury, and copper. The inactivated enzymes usually may be reactivated by treatment with hydrogen sulfide to precipitate the metals.

Urea amidohydrolase (urease) is an example of an enzyme that is affected by such treatment. This would indicate that the enzyme contains a sulfhydryl group. The enzyme is also inactivated by fluorides, halogens, borates, quinones, formaldehyde, hydrogen peroxide, and certain basic dyes. The enzyme is most reactive at a pH of about 7.0.

Acids Pepsin is a proteolytic enzyme that hydrolyzes peptide linkages, but it attacks only those links which contain the tyrosyl or phenylalanyl group attached to the imino side of the peptide linkage (page 452).

The enzyme is active only in acid solutions. Its greatest activity occurs at about pH 1.6. If a solution of pepsin is adjusted to a pH higher than 6.8, the enzyme is slowly destroyed.

Alkalies Trypsin is another proteolytic enzyme that hydrolyzes peptide linkages, but it attacks only those links which contain the lysyl or arginyl group attached to the carbonyl side of the peptide linkage (page 452).

The enzyme is active only in alkaline solutions. Its greatest activity occurs in the pH range of 8 to 9.

Other chemicals Donor: H_2O_2 oxidoreductase (peroxidase) and H_2O_2:H_2O_2 oxidoreductase (catalase) are inactivated by hydrocyanic acid, sodium azide, hydrogen sulfide, and hydroxylamine. Removal of these substances restores activity to the enzymes.

Oxidation-reduction Papain (vegetable pepsin) is a proteolytic enzyme obtained from the juice of the green fruit of the papaya; also from the leaves and other parts of the tree.

Papain has an optimum pH range of 4 to 7. It is a common constituent of commercial meat tenderizers.

A sulfhydryl group is believed to be an essential part of the active enzyme. Oxidation inactivates the enzyme, and reduction restores its activity. Activation can be effected by hydrogen sulfide, cysteine, glutathione, or hydrocyanic acid.

Peptide hydrolases These are proteolytic enzymes which are destructive to other enzymes. This is another indication that enzymes are proteins.

Pepsin and papain, at pH 4.3, digest urease; trypsin at pH 8 to 9 digests maltase; pepsin, at pH 2.0, digests trypsin, while at pH 8.0, the reverse effect occurs.

For more information: Gutfreund (1968); Vogel et al. (1968).

Synthetic activities of enzymes

A great number of chemical reactions are termed reversible reactions. The characteristic of a reaction of this type is that it progresses in one direction or the other until an equilibrium specific to the reaction and to the concentration of the reacting materials is established. Any change in the concentration of the reacting components is immediately followed by a change to a new equilibrium. All chemical reactions strive to reach the equilibrium point, but the rate in many cases is infinitely slow.

Enzymes, on the other hand, speed up reactions so that equilibrium is established much sooner than otherwise. Like chemical reactions, all enzyme reactions are reversible. Some reactions, because of the nature of the end products, are only slightly reversible. Therefore, the term reversible is used only in a qualitative sense.

Classification of enzymes

The number of enzymes known has increased very rapidly in recent years. For this reason many difficulties have arisen in connection with their terminology. Because of the investigations in different laboratories, the same enzyme became known by more than one name. Also, many of the names gave no information on the nature of the reactions catalyzed. Therefore, a systematic classification, however imperfect, became necessary to bring order out of confusion.

With possibly a few exceptions the exact chemical structures of enzymes are not known. This has made it necessary to name and classify enzymes on the basis of what they do rather than on what they are. The latest classification recommended by the International Union of Biochemistry (1961, 1965) follows this system as closely as possible.

Enzymes which are known to be single enzymes have the ending *-ase*. Where more than one enzyme is involved, the word *system* is included in the name when referring to the overall reaction catalyzed by the enzymes. As is true with all classifications, there are a number of names of enzymes (hydrolases) which are exceptions and do not conform to this rule.

Enzymes are placed in six main groups as follows: (1) oxidoreductases, (2) transferases, (3) hydrolases, (4) lyases, (5) isomerases, and (6) ligases (synthetases).

The outline that follows is by no means complete but it shows how enzymes are named and classified. The chemical reactions are included wherever possible.

At the bottom of each enzyme reaction in the classification, and placed centrally on the page, are given the recommended trivial or common names of the catalysts by which they are better known.

I. OXIDOREDUCTASES

A. Acting on CH–OH group of donors

1. With NAD or NADP as acceptor

a. Alcohol: NAD oxidoreductase + primary or secondary alcohols or semiacetals + NAD $\rightleftharpoons$ aldehyde or ketone + reduced NAD

$$\underset{\text{Alcohol}}{R\cdot CH_2OH} + NAD \rightleftharpoons \underset{\text{Aldehyde}}{R\cdot CHO} + \text{reduced NAD}$$

Alcohol dehydrogenase

b. Alcohol: NADP oxidoreductase + primary alcohols only $\rightleftharpoons$ aldehyde + reduced NADP

$R{\cdot}CH_2OH + NADP \rightleftharpoons R{\cdot}CHO + \text{reduced NADP}$

Alcohol — Aldehyde

Alcohol dehydrogenase (NADP)

c. 2,3-Butylene glycol: NAD oxidoreductase + 2,3-butylene glycol + NAD $\rightleftharpoons$ acetoin + reduced NAD

$CH_3{\cdot}CHOH{\cdot}CHOH{\cdot}CH_3 + NAD \rightleftharpoons$

2,3-Butylene glycol

$CH_3{\cdot}CO{\cdot}CHOH{\cdot}CH_3 + \text{reduced NAD}$

Acetoin

Butylene glycol dehydrogenase

d. Glycerol: NAD oxidoreductase + glycerol + NAD $\rightleftharpoons$ dihydroxyacetone + reduced NAD

$CH_2OH{\cdot}CHOH{\cdot}CH_2OH + NAD \rightleftharpoons$

Glycerol

$CH_2OH{\cdot}CO{\cdot}CH_2OH + \text{reduced NAD}$

Dihydroxyacetone

Glycerol dehydrogenase

e. L-Glycerol-3-phosphate: NAD oxidoreductase + L-glycerol-3-phosphate + NAD $\rightleftharpoons$ dihydroxyacetone phosphate + reduced NAD

$CH_2O{\cdot}PO_3H_2{\cdot}CHOH{\cdot}CH_2OH + NAD \rightleftharpoons$

L-Glycerol-3-phosphate

$CH_2O{\cdot}PO_3H_2{\cdot}CO{\cdot}CH_2OH + \text{reduced NAD}$

Dihydroxyacetone phosphate

Glycerol-3-phosphate dehydrogenase

f. L-Lactate: NAD oxidoreductase + NAD + L-lactate $\rightleftharpoons$ pyruvate + reduced NAD

$CH_3{\cdot}CHOH{\cdot}COOH + NAD \rightleftharpoons$

$CH_3{\cdot}CO{\cdot}COOH + \text{reduced NAD}$

L-Lactic acid — Pyruvic acid

Lactate dehydrogenase

g. D-3-Hydroxybutyrate: NAD oxidoreductase + D-3-hydroxybutyrate + NAD $\rightleftharpoons$ acetoacetate + reduced NAD

$CH_3\cdot CHOH\cdot CH_2\cdot COOH$ + NAD $\rightleftharpoons$

D-3-Hydroxybutyric acid

$CH_3\cdot CO\cdot CH_2\cdot COOH$ + reduced NAD

Acetoacetic acid

3-Hydroxybutyrate dehydrogenase

h. L-Malate: NAD oxidoreductase + L-malate + NAD $\rightleftharpoons$ oxaloacetate + reduced NAD

$COOH\cdot CH_2\cdot CHOH\cdot COOH$ + NAD $\rightleftharpoons$

L-Malic acid

$COOH\cdot CH_2\cdot CO\cdot COOH$ + reduced NAD

Oxaloacetic acid

Malate dehydrogenase

i. β-D-Glucose: NAD(P) oxidoreductase + β-D-glucose + NAD(P) $\rightleftharpoons$ D-glucono-δ-lactone + reduced NAD(P)

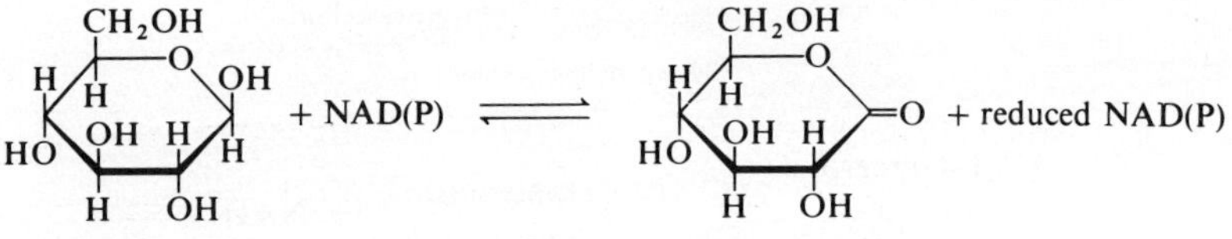

β-D-Glucose D-Glucono-δ-lactone

Glucose dehydrogenase

j. D-Glucose-6-phosphate: NADP oxidoreductase + D-glucose-6-phosphate + NADP $\rightleftharpoons$ D-glucono-δ-lactone-6-phosphate + reduced NADP

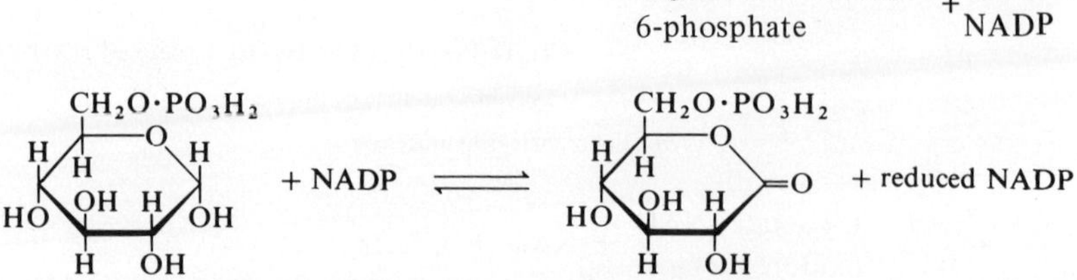

D-Glucose-6-phosphate D-Glucono-δ-lactone-6-phosphate

Glucose-6-phosphate dehydrogenase

2. With a cytochrome as acceptor

   a. L-Glycerol-3-phosphate: cytochrome c oxidoreductase + L-glycerol-3-phosphate + oxidized cytochrome c ⇌

   dihydroxyacetone phosphate + reduced cytochrome c

   $CH_2O{\cdot}PO_3H_2{\cdot}CHOH{\cdot}CH_2OH$ + oxidized cytochrome ⇌

   L-Glycerol-3-phosphate

   $CH_2O{\cdot}PO_3H_2{\cdot}CO{\cdot}CH_2OH$ + reduced cytochrome c

   Dihydroxyacetone phosphate

   Glycerophosphate dehydrogenase

3. With O_2 as acceptor

   a. β-D-Glucose: O_2 oxidoreductase + β-D-glucose + O_2 ⇌

   D-glucono-δ-lactone + H_2O_2

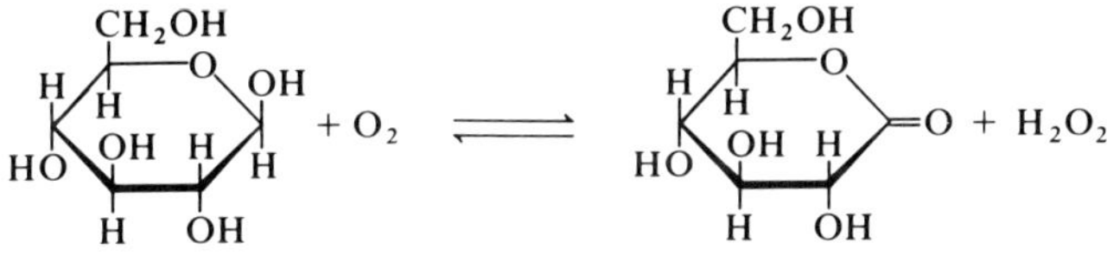

β-D-Glucose D-Glucono-δ-lactone

Glucose oxidase

4. With other acceptors

   a. Choline: (acceptor) oxidoreductase + choline + acceptor ⇌

   betaine aldehyde + reduced acceptor

   Choline dehydrogenase

B. Acting on aldehyde or keto group of donors

5. With NAD or NADP as acceptor

   a. Aldehyde: NAD(P) oxidoreductase + aldehyde + NAD(P) + H_2O ⇌

   acetic acid + reduced NAD(P)

   $CH_3{\cdot}CHO$ + NAD(P) + H_2O ⇌

   Acetaldehyde

   $CH_3{\cdot}COOH$ + reduced NAD(P)

   Acetic acid

   Aldehyde dehydrogenase

   b. D-Glyceraldehyde-3-phosphate: NADP oxidoreductase + D-glyceraldehyde-3-phosphate + NADP + H_2O ⇌

   D-3-phosphoglycerate + reduced NADP

$CH_2O{\cdot}PO_3H_2{\cdot}CHOH{\cdot}CHO + H_2O + NADP \rightleftharpoons$

D-Glyceraldehyde-3-phosphate

$CH_2O{\cdot}PO_3H_2{\cdot}CHOH{\cdot}COOH$ + reduced NADP

D-3-Phosphoglyceric acid

Glyceraldehydephosphate dehydrogenase

6. With a cytochrome as acceptor
   a. Pyruvate: cytochrome oxidoreductase + pyruvate + H_2O + oxidized cytochrome $\rightleftharpoons$ acetate + CO_2 + reduced cytochrome

$CH_3{\cdot}CO{\cdot}COOH + H_2O$ + oxidized cytochrome $\rightleftharpoons$

Pyruvic acid

$CH_3{\cdot}COOH + CO_2$ + reduced cytochrome

Acetic acid

Pyruvate dehydrogenase

7. With O_2 as acceptor
   a. Xanthine: O_2 oxidoreductase + xanthine + $H_2O + O_2 \rightleftharpoons$ urate + H_2O_2

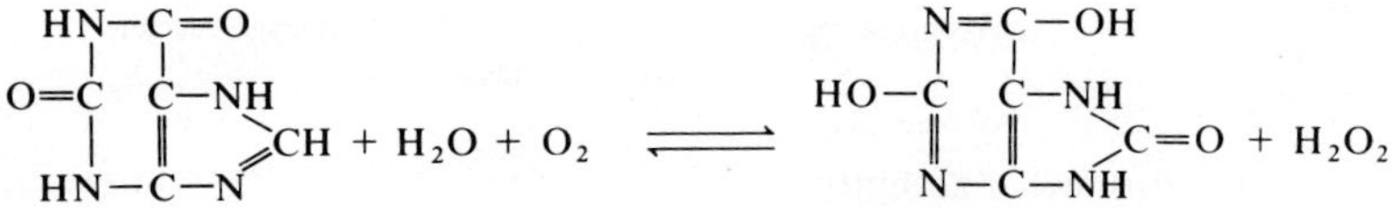

Xanthine   Uric acid

Xanthine oxidase

C. Acting on CH–CH group of donors
   8. With NAD or NADP as acceptor
      a. 4,5-Dihydrouracil: NAD oxidoreductase + 4,5-dihydrouracil + NAD $\rightleftharpoons$ uracil + reduced NAD

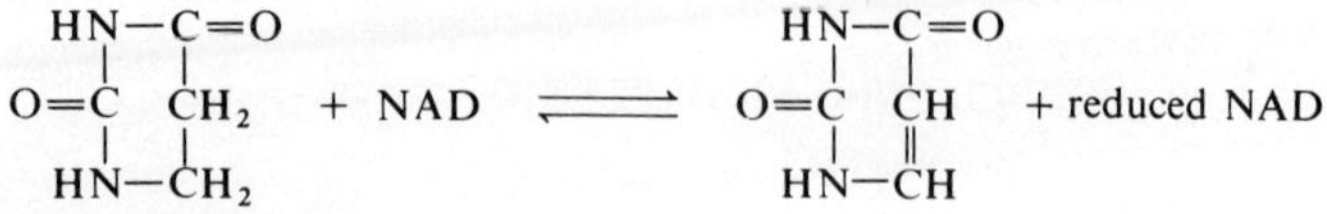

4,5-Dihydrouracil   Uracil

Dihydrouracil dehydrogenase

D. Acting on $CH{-}NH_2$ group of donors
   9. With NAD or NADP as acceptor
      a. L-Glutamate: NADP oxidoreductase + L-glutamate + NADP + H_2O $\rightleftharpoons$ 2-oxoglutarate + NH_3 + reduced NADP

$COOH \cdot CH_2 \cdot CH_2 \cdot CHNH_2 \cdot COOH + NADP + H_2O \rightleftharpoons$

L-Glutamic acid

$COOH \cdot CH_2 \cdot CH_2 \cdot CO \cdot COOH + NH_3 +$ reduced NADP

2-Oxoglutaric acid

Glutamate dehydrogenase (NADP)

10. With O_2 as acceptor

a. L-Amino acid: O_2 oxidoreductase + an L-amino acid + $H_2O + O_2 \rightleftharpoons$

2-oxo-acid + $NH_3 + H_2O_2$

$R \cdot CHNH_2 \cdot COOH + H_2O + O_2 \rightleftharpoons$

L-Amino acid

$R \cdot CO \cdot COOH + NH_3 + H_2O_2$

2-Oxo-acid

L-Amino acid oxidase

E. Acting on reduced NAD or NADP

11. With a cytochrome as acceptor

a. Reduced NAD: cytochrome c oxidoreductase + reduced NAD + oxidized cytochrome c $\rightleftharpoons$

NAD + reduced cytochrome c

Reduced NAD cytochrome c reductase

b. Reduced NADP: cytochrome c oxidoreductase + reduced NADP + oxidized cytochrome c $\rightleftharpoons$

NADP + reduced cytochrome c

Reduced NADP cytochrome c reductase

12. With a disulfide compound as acceptor:

a. Reduced NAD(P): glutathione oxidoreductase + reduced NADP + oxidized glutathione $\rightleftharpoons$

NAD(P) + 2 reduced glutathione

Glutathione reductase

13. With a quinone or related compound as acceptor

a. Reduced NAD(P): quinone oxidoreductase + quinone + reduced NAD(P) $\rightleftharpoons$

p-dihydroxy benzene + NAD(P)

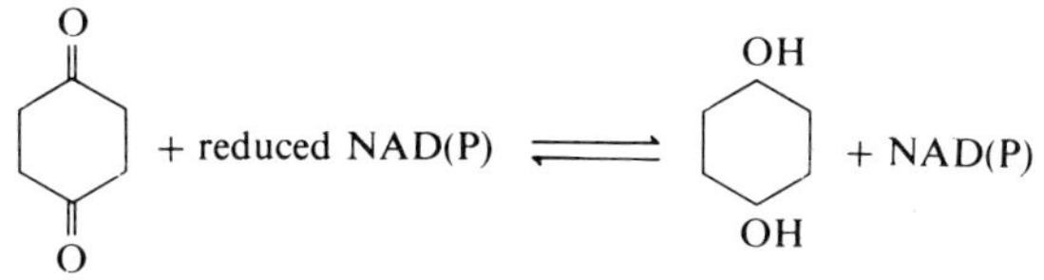

Quinone p-Dihydroxy benzene

Quinone reductase

b. Reduced NAD(P):
dehydroascorbate oxidoreductase + dehydroascorbate + reduced NAD(P) ⇌ ascorbate + NAD(P)

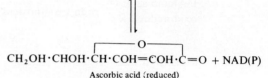

$CH_2OH{\cdot}CHOH{\cdot}CH{\cdot}CO{\cdot}CO{\cdot}C{=}O$ (ring through O) + reduced NAD(P)
Dehydroascorbic acid (oxidized)

⇅

$CH_2OH{\cdot}CHOH{\cdot}CH{\cdot}COH{=}COH{\cdot}C{=}O$ (ring through O) + NAD(P)
Ascorbic acid (reduced)

Oxidized ascorbate reductase

14. With a nitrogenous group as acceptor

a. Reduced NAD:
nitrate oxidoreductase + nitrate + reduced NAD ⇌ nitrite + H_2O + NAD

$NaNO_3$ + reduced NAD ⇌ $NaNO_2$ + H_2O + NAD
Sodium nitrate — Sodium nitrite

Nitrate reductase

b. Reduced NAD(P):
nitrite oxidoreductase + nitrite + reduced 2NAD(P) ⇌ NH_2OH + NaOH + 2NAD(P)

$NaNO_2$ + reduced 2NAD(P) ⇌ NH_2OH + NaOH + 2NAD(P)
Sodium nitrite — Hydroxylamine

Nitrite reductase

c. Reduced NAD(P):
nitrite oxidoreductase + 2 nitrite + reduced NAD(P) ⇌ 2NO + $2H_2O$ + NAD(P)

$2NaNO_2$ + reduced NAD(P) ⇌ 2NO + 2NaOH + NAD(P)
Sodium nitrite — Nitric oxide

Nitrite reductase

15. With other acceptors

a. Reduced NADP:
diaphorase oxidoreductase + acceptor + reduced NADP ⇌ reduced acceptor + NADP

Reduced NAD(P) dehydrogenase

F. Acting on haem groups of donors

16. With O_2 as acceptor

a. Cytochrome c: O_2 oxidoreductase + reduced cytochrome c + O_2 ⇌ oxidized cytochrome c + H_2O

4 Reduced cytochrome c + O_2 ⇌ 4 oxidized cytochrome c + $2H_2O$

Cytochrome oxidase

17. With a nitrogenous group as acceptor

a. Cytochrome: nitrate oxidoreductase + nitrate + reduced cytochrome ⇌ nitrite + oxidized cytochrome

$NaNO_3$ (Sodium nitrate) + reduced cytochrome ⇌ $NaNO_2$ (Sodium nitrite) + oxidized cytochrome

Nitrate reductase

G. Acting on diphenols and related substances as donors

18. With O_2 as acceptor

a. L-Ascorbate: O_2 oxidoreductase + L-ascorbate + O_2 ⇌ dehydroascorbate + H_2O

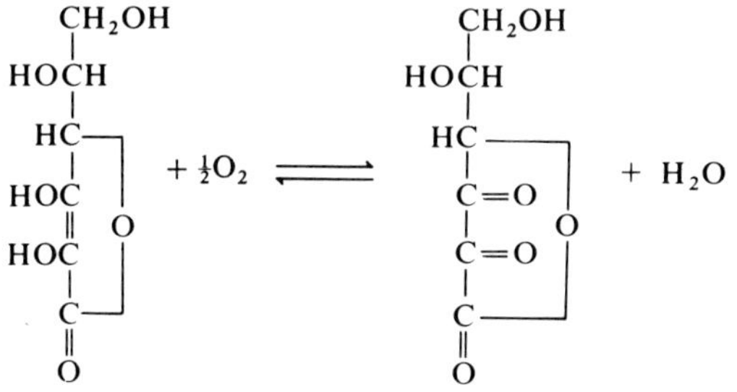

L-Ascorbic acid — Dehydroascorbic acid

Ascorbate oxidase

H. Acting on H_2O_2 as acceptor

a. Reduced NAD: H_2O_2 oxidoreductase + H_2O_2 + reduced NAD ⇌ H_2O + NAD

$2H_2O_2$ + reduced NAD ⇌ $2H_2O$ + NAD

NAD peroxidase

b. Reduced NADP: H_2O_2 oxidoreductase + H_2O_2 + reduced NADP ⇌ $2H_2O$ + NADP

$2H_2O_2$ + reduced NADP ⇌ $2H_2O$ + NADP

NADP peroxidase

c. $H_2O_2 : H_2O_2$ oxidoreductase + H_2O_2 ⇌ $H_2O + O_2$

$$H_2O_2 + H_2O_2 \rightleftharpoons 2H_2O + O_2$$

Hydrogen peroxide

Catalase

II. TRANSFERASES

A. Transferring one-carbon groups

1. Carboxyl- and carbamoyltransferases

a. Methylmalonyl-CoA: pyruvate carboxyltransferase + 2-methylmalonyl-CoA + pyruvate ⇌ propionyl-CoA + oxaloacetate

Methylmalonyl-CoA carboxyltransferase

B. Acyltransferases

2. Acyltransferases

a. Acetyl-CoA: imidazole N-acetyltransferase + imidazole + acetyl-CoA ⇌ N-acetylimidazole + CoA

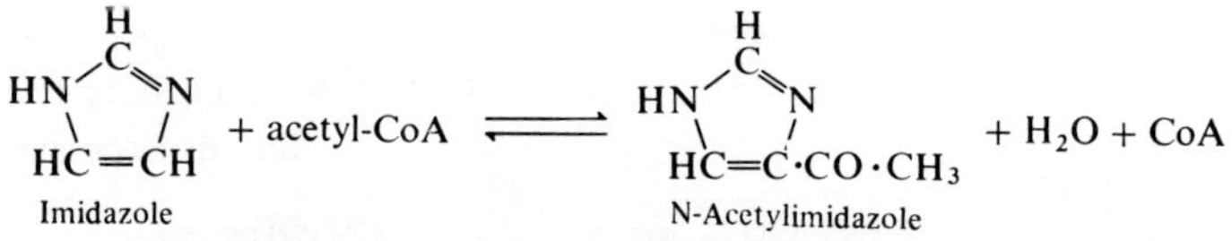

Imidazole acetyltransferase

b. Acetyl-CoA: orthophosphate acetyltransferase + orthophosphate + acetyl-CoA ⇌ acetylphosphate + CoA

Phosphate acetyltransferase

C. Transferring nitrogenous groups

3. Aminotransferases

a. L-Aspartate: 2-oxoglutarate aminotransferase + L-aspartate + 2-oxoglutarate ⇌ oxaloacetate + L-glutamate

$COOH{\cdot}CH_2{\cdot}CHNH_2{\cdot}COOH + COOH{\cdot}CH_2{\cdot}CH_2{\cdot}CO{\cdot}COOH \rightleftharpoons$

L-Aspartic acid — 2-Oxoglutaric acid

$COOH{\cdot}CH_2{\cdot}CO{\cdot}COOH + COOH{\cdot}CH_2{\cdot}CH_2{\cdot}CHNH_2{\cdot}COOH$

Oxaloacetic acid — L-Glutamic acid

Aspartate aminotransferase

b. L-Alanine: 2-oxoglutarate aminotransferase + L-alanine + 2-oxoglutarate ⇌ pyruvate + L-glutamate

$$CH_3 \cdot CHNH_2 \cdot COOH + COOH \cdot CH_2 \cdot CH_2 \cdot CO \cdot COOH \rightleftharpoons$$

L-Alanine — 2-Oxoglutaric acid

$$CH_3 \cdot CO \cdot COOH + COOH \cdot CH_2 \cdot CH_2 \cdot CHNH_2 \cdot COOH$$

Pyruvic acid — L-Glutamic acid

Alanine aminotransferase

c. L-Tyrosine:
2-oxoglutarate aminotransferase + L-tyrosine + 2-oxoglutarate ⇌ *p*-hydroxyphenyl-pyruvate + L-glutamate

$$HO \cdot C_6H_4 \cdot CH_2 \cdot CHNH_2 \cdot COOH + COOH \cdot CH_2 \cdot CH_2 \cdot CO \cdot COOH$$

L-Tyrosine — 2-Oxoglutaric acid

$$\rightleftharpoons$$

$$HO \cdot C_6H_4 \cdot CH_2 \cdot CO \cdot COOH + COOH \cdot CH_2 \cdot CH_2 \cdot CHNH_2 \cdot COOH$$

p-Hydroxyphenyl pyruvic acid — L-Glutamic acid

Tyrosine aminotransferase

D. Transferring phosphorus-containing groups
4. Phosphotransferases with an alcohol group as acceptor
a. ATP: D-hexose-6-phosphotransferase + D-hexose + ATP ⇌ D-hexose-6-phosphate + ADP

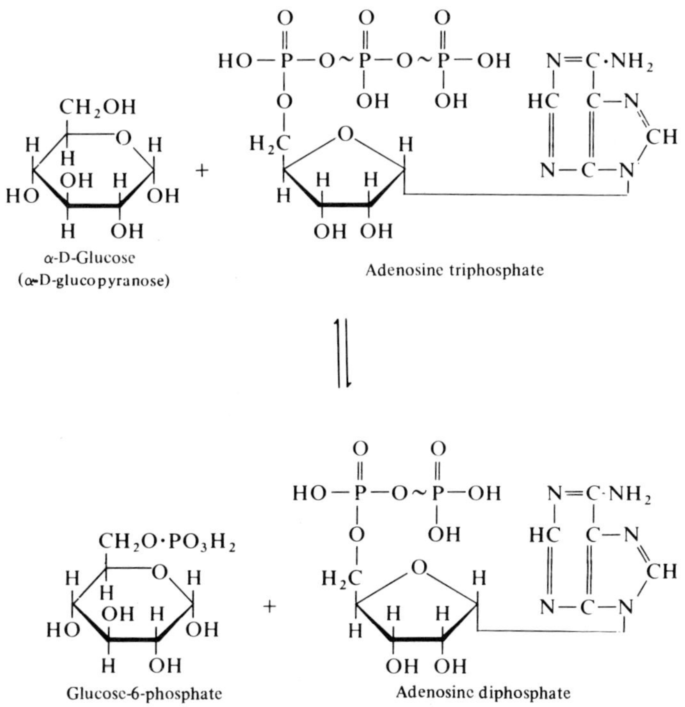

Hexokinase

b. ATP: D-glucose-1-phosphate 6-phosphotransferase + D-glucose-1-phosphate + ATP $\rightleftharpoons$ D-glucose-1,6-diphosphate + ADP

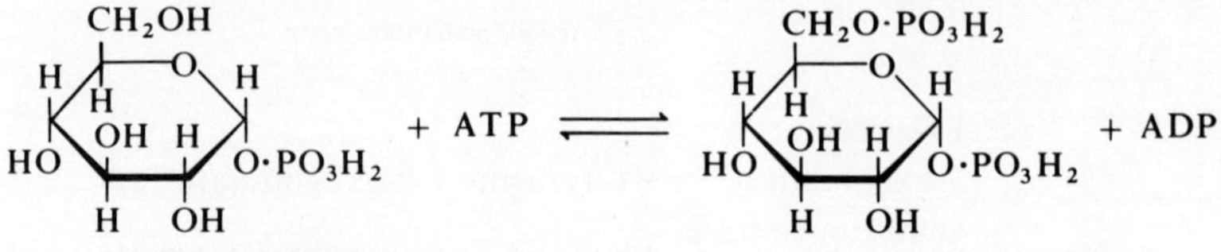

D-Glucose-1-phosphate D-Glucose-1,6-diphosphate

Phosphoglucokinase

c. ATP:NAD 2-phosphotransferase + ATP + NAD $\rightleftharpoons$ ADP + NADP

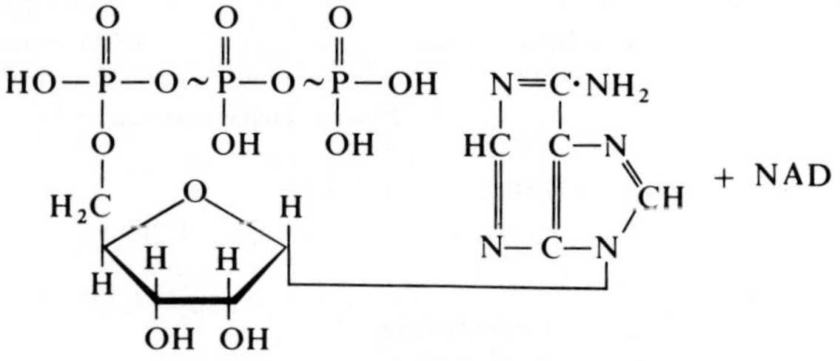

Adenosine triphosphate

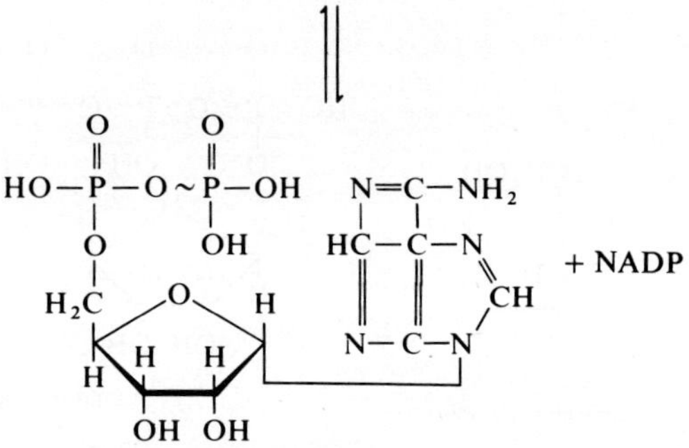

Adenosine diphosphate

NAD kinase

d. ATP: glycerol phosphotransferase + glycerol + ATP $\rightleftharpoons$ L-glycerol-3-phosphate + ADP

$CH_2OH{\cdot}CHOH{\cdot}CH_2OH$ + ATP $\rightleftharpoons$

Glycerol (above)

$CH_2O{\cdot}PO_3H_2{\cdot}CHOH{\cdot}CH_2OH$ + ADP

L-Glycerol-3-phosphate (above)

Glycerol kinase

e. ATP: pyruvate phosphotransferase + pyruvate + ATP ⇌ phosphoenolpyruvate + ADP

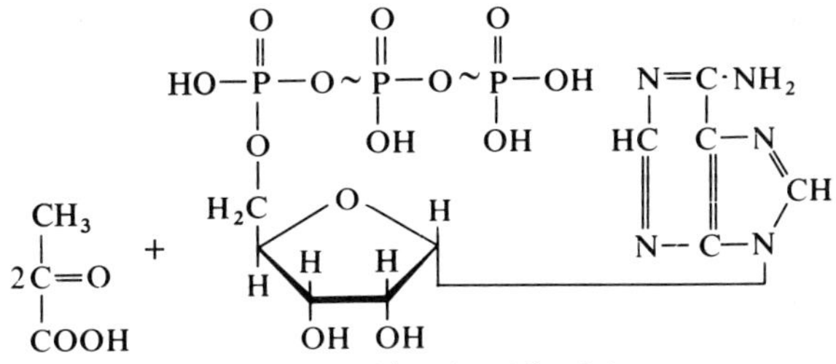

Pyruvic acid

Adenosine triphosphate

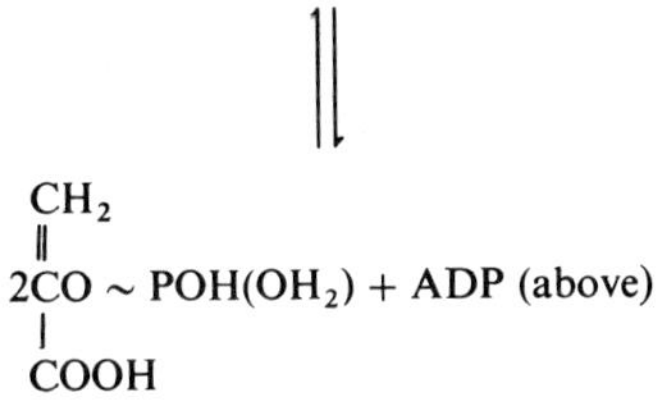

Phosphoenolpyruvate

Pyruvate kinase

f. ATP: adenosine-5-phosphotransferase + adenosine + ATP ⇌ adenylic acid + ADP

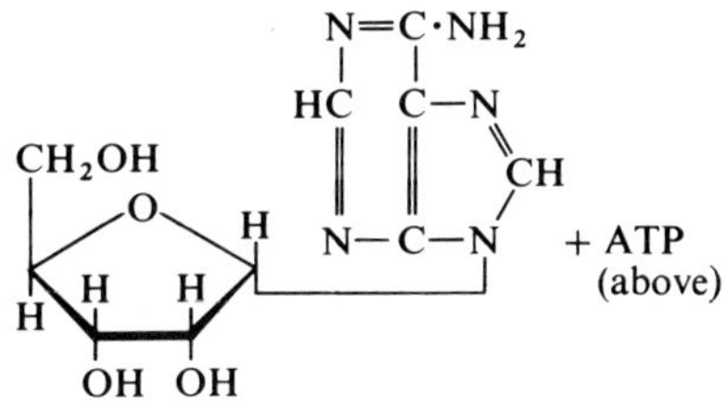

Adenosine

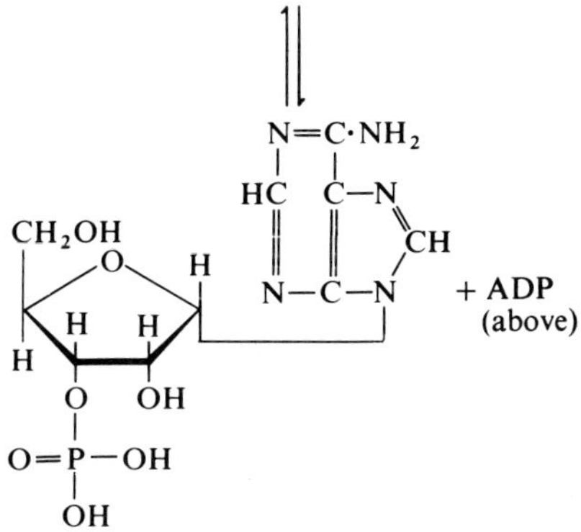

Adenylic acid

Adenosine kinase

g. ATP: L-arginine phosphotransferase + L-arginine + ATP ⇌ L-phosphoarginine + ADP

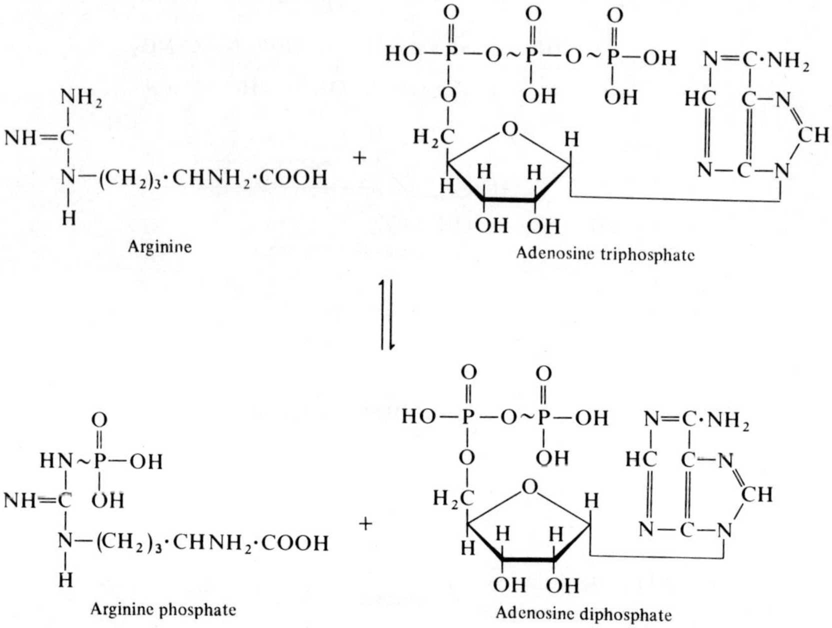

Arginine kinase

E. Transferring sulfur-containing groups

5. CoA-transferases

a. Acetyl-CoA: propionate CoA-transferase + acetyl-CoA + propionate ⇌ acetate + propionyl-CoA

$$\text{Acetyl-CoA} + \underset{\text{Propionic acid}}{CH_3 \cdot CH_2 \cdot COOH} \rightleftharpoons \underset{\text{Acetic acid}}{CH_3 \cdot COOH} + \text{propionyl-CoA}$$

Propionate CoA-transferase

III. HYDROLASES

A. Acting on ester bonds

1. Carboxylic ester hydrolases

a. Carboxylic ester hydrolase + a carboxylic ester + H_2O ⇌ carboxylic acid + an alcohol

$$\underset{\text{Carboxylic ester}}{R \cdot COOR'} + H_2O \rightleftharpoons \underset{\text{Carboxylic acid}}{R \cdot COOH} + \underset{\text{Alcohol}}{R'OH}$$

Carboxylesterase

b. Glycerol ester hydrolase + a triglyceride + H_2O $\rightleftharpoons$ a diglyceride + a fatty acid

$$\begin{array}{l} CH_2O\cdot OC\cdot C_{17}H_{33} \\ | \\ CHO\cdot OC\cdot C_{17}H_{33} \\ | \\ CH_2O\cdot OC\cdot C_{17}H_{33} \end{array} + H_2O \rightleftharpoons \begin{array}{l} CH_2O\cdot OC\cdot C_{17}H_{33} \\ | \\ CHO\cdot OC\cdot C_{17}H_{33} \\ | \\ CH_2OH \end{array} + C_{17}H_{33}COOH$$

Triolein Diolein Oleic acid

Lipase

c. Acetylcholine esterase + acetylcholine + H_2O $\rightleftharpoons$ choline + acetic acid

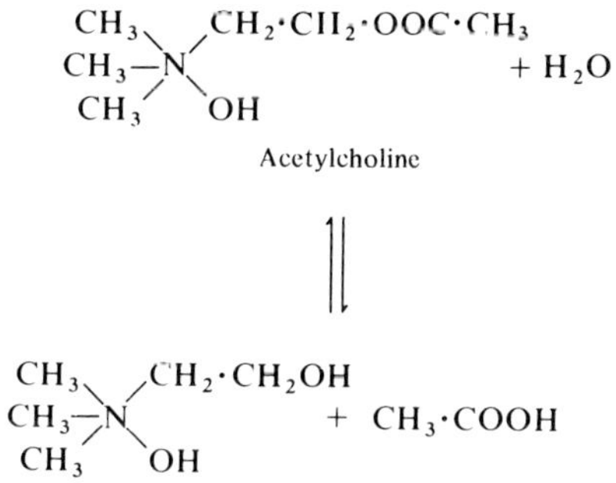

Acetylcholine

Choline Acetic acid

Acetylcholine esterase

d. Sterol ester hydrolase + a cholesterol ester + H_2O $\rightleftharpoons$ cholesterol + an acid

Cholesterol esterase

e. Chlorophyll chlorophyllido-hydrolase + chlorophyll + H_2O $\rightleftharpoons$ chlorophyllide + phytol

$$C_{32}H_{30}N_4MgO(COOCH_3)\cdot COO\cdot C_{20}H_{39} + H_2O \rightleftharpoons$$

Chlorophyll

$$C_{32}H_{30}N_4MgO(COOCH_3)\cdot COOH + C_{20}H_{39}OH$$

Chlorophyllide Phytol

Chlorophyllase

f. Tannin acyl-hydrolase + digallate + H_2O $\rightleftharpoons$ gallate

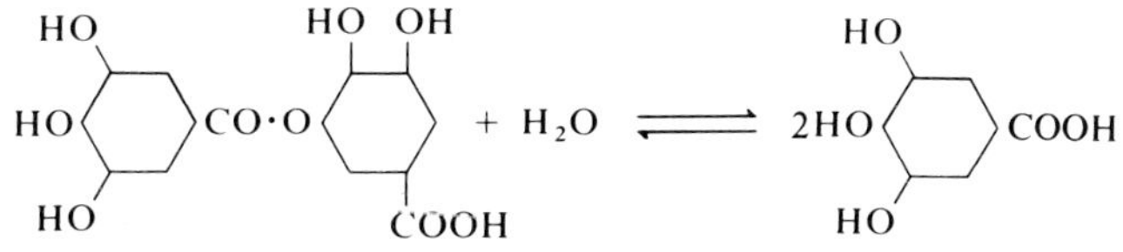

Digallic acid Gallic acid

Tannase

g. Phosphatide acyl-hydrolase + lecithin + H_2O $\rightleftharpoons$

lysolecithin + oleic acid

$$CH_2OOC \cdot C_{17}H_{35}$$
$$CHOOC \cdot C_{17}H_{33}$$

$$CH_2-O-\underset{\underset{O}{\parallel}}{\overset{\overset{OH}{|}}{P}}-O-CH_2 \cdot CH_2-\underset{\underset{OH}{|}}{\overset{\overset{CH_3}{\diagup}}{N}}\overset{CH_3}{\underset{CH_3}{\diagdown}} \quad + \ H_2O$$

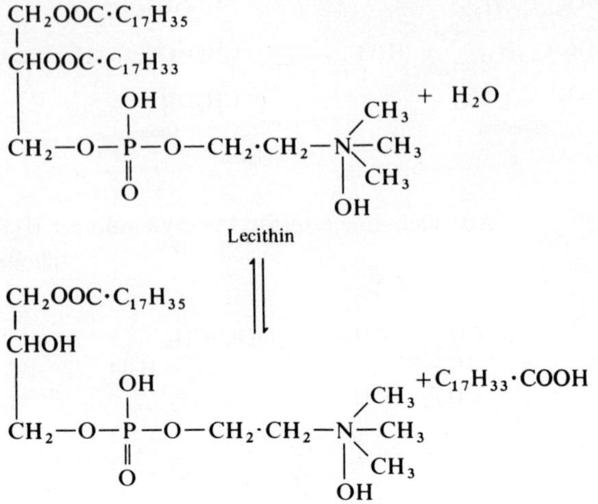

Lecithin

$$CH_2OOC \cdot C_{17}H_{35}$$
$$CHOH$$

$$CH_2-O-\underset{\underset{O}{\parallel}}{\overset{\overset{OH}{|}}{P}}-O-CH_2 \cdot CH_2-\underset{\underset{OH}{|}}{\overset{\overset{CH_3}{\diagup}}{N}}\overset{CH_3}{\underset{CH_3}{\diagdown}} \quad +C_{17}H_{33} \cdot COOH$$

Lysolecithin   Oleic acid

Phospholipase A

h. Lysolecithin acyl-hydrolase + lysolecithin + H_2O $\rightleftharpoons$

glycerophosphocholine + fatty acid

$$CH_2OOC \cdot C_{17}H_{35}$$
$$CHOH$$

$$CH_2-O-\underset{\underset{O}{\parallel}}{\overset{\overset{OH}{|}}{P}}-O-CH_2 \cdot CH_2-\underset{\underset{OH}{|}}{\overset{\overset{CH_3}{\diagup}}{N}}\overset{CH_3}{\underset{CH_3}{\diagdown}} \quad + \ H_2O$$

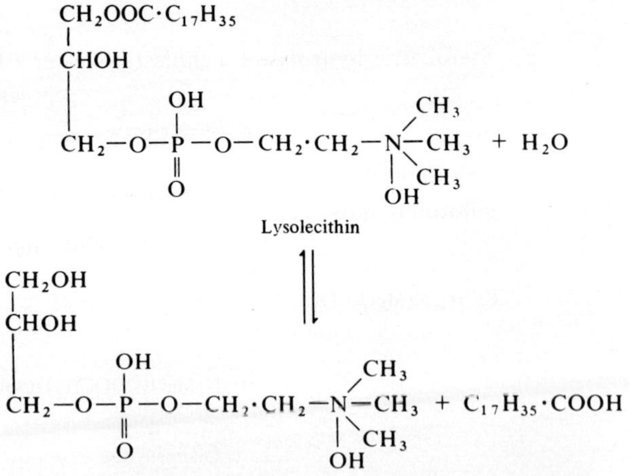

Lysolecithin

$$CH_2OH$$
$$CHOH$$

$$CH_2-O-\underset{\underset{O}{\parallel}}{\overset{\overset{OH}{|}}{P}}-O-CH_2 \cdot CH_2-\underset{\underset{OH}{|}}{\overset{\overset{CH_3}{\diagup}}{N}}\overset{CH_3}{\underset{CH_3}{\diagdown}} \quad + \ C_{17}H_{35} \cdot COOH$$

Glycerophosphocholine ester   Stearic acid

Phospholipase B

i. Pectin pectyl-hydrolase + pectin + H_2O $\rightleftharpoons$

pectic acid + methyl alcohol

$$(R \cdot COOCH_3)n + nH_2O \rightleftharpoons (R \cdot COOH)n + nCH_3OH$$

Pectin   Pectic   Methyl
                         acid   alcohol

Pectinesterase

j. Acylcholine acyl-hydrolase + an acylcholine + H_2O ⇌ an acid + choline

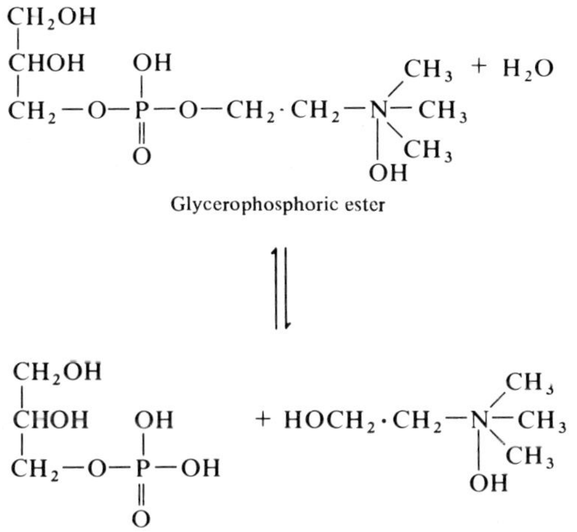

Glycerophosphoric ester

Glycerophosphoric acid

Choline

Cholinesterase

2. Phosphoric monoester hydrolases

a. Orthophosphoric monoester phosphohydrolase + glycerophosphate ⇌ glycerol + phosphate

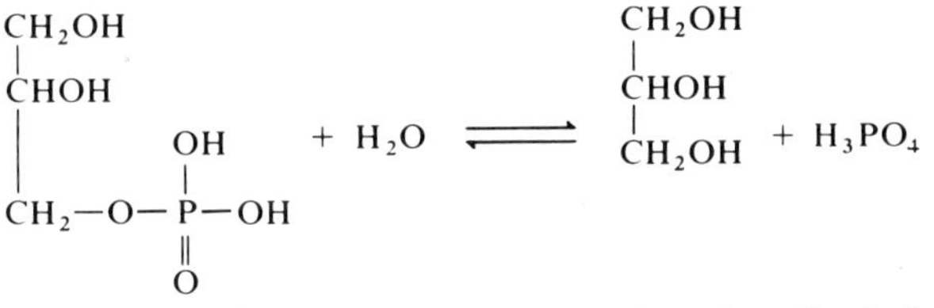

Glycerophosphoric acid

Glycerol

Phosphoric acid

Alkaline phosphatase

b. *Meso*-inositol-hexaphosphate phosphohydrolase + *Meso*-inositol hexaphosphate + $6H_2O$ ⇌ *meso*-inositol + 6 orthophosphate

Meso-inositol hexaphosphate + $6H_2O$ ⇌ 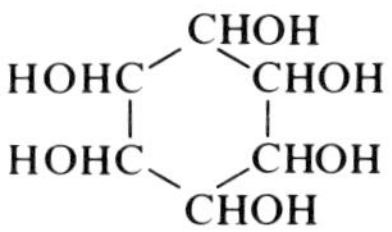 + $6\,H_3PO_4$

Inositol

Phosphoric acid

Phytase

c. D-Fructose-1,6-diphosphate 1-phosphohydrolase + D-fructose-1,6-diphosphate + H_2O ⇌ D-fructose-6-phosphate + phosphate

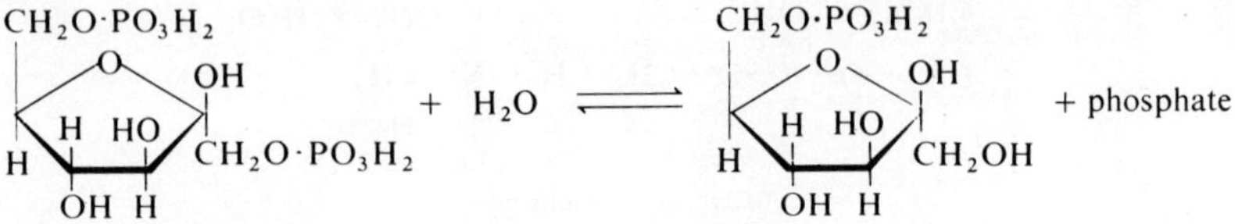

D-Fructose-1,6-diphosphate D-Fructose-6-phosphate

Hexosediphosphatase

d. D-Glucose-6-phosphate phosphohydrolase + D-glucose-6-phosphate + H_2O ⇌ D-glucose + phosphate

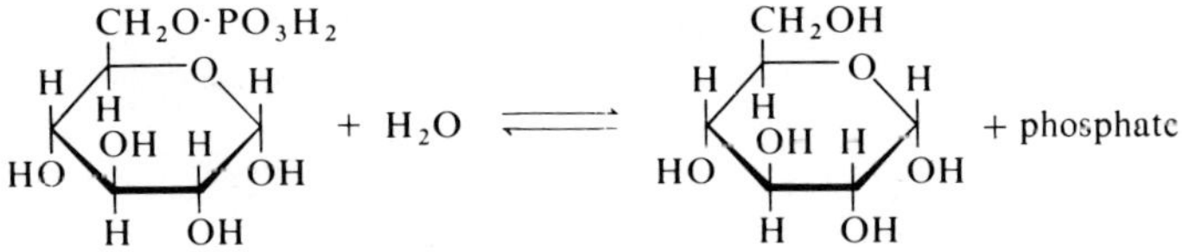

Glucose-6-phosphate α-D-Glucose (α-D-glucopyranose)

Glucose-6-phosphatase

3. Sulfuric ester hydrolases
   a. Aryl-sulfate sulfohydrolase + a phenol sulfate + H_2O ⇌ a phenol + sulfate

$$C_6H_5O{\cdot}SO_3K + H_2O \rightleftharpoons C_6H_5OH + KHSO_4$$

Potassium phenyl sulfate — Phenol — Potassium bisulfate

Arylsulphatase

B. Acting on glycosyl compounds
   1. Glycoside hydrolases
      a. α-1,4-Glucan-4-glucanohydrolase + starch, glycogen, etc. ⇌ dextrin + maltose

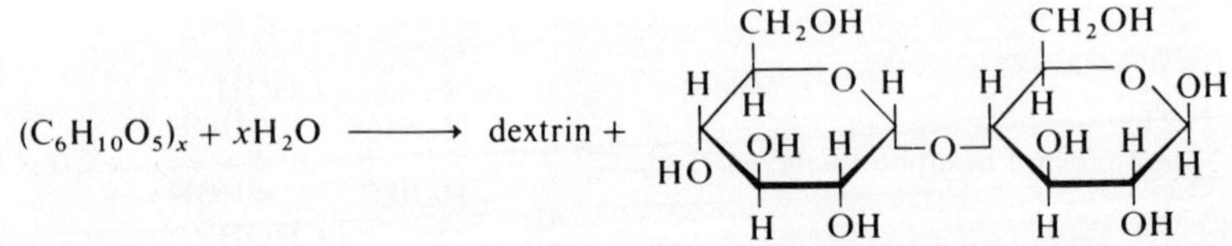

Starch — Maltose (4-*O*-α-D-glucopyranosyl-D-glucose)

α-Amylase

b. α-1,4-Glucan maltohydrolase + starch, glycogen, etc. ⇌ maltose

$(C_6H_{10}O_5)_x + xH_2O \rightleftharpoons$ 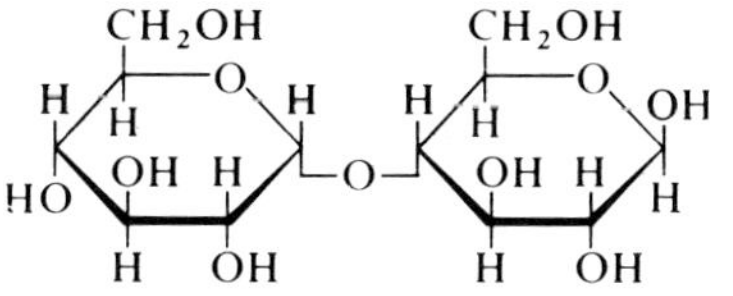

Maltose
(4-*O*-α-D-glucopyranosyl-D-glucose)

β-Amylase

c. α-1,4-Glucan 4-glucanohydrolase + starch, glycogen, etc. ⇌ α-D-glucose

$(C_6H_{10}O_5)_x + xH_2O \rightleftharpoons$ 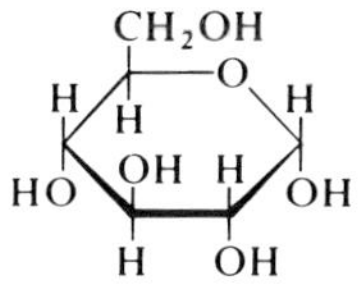

α-D-Glucose
(α-D-glucopyranose)

d. β-1,4-Glucan 4-glucanohydrolase + cellulose ⇌ β-cellobiose

$(C_6H_{10}O_5)_x + xH_2O \rightleftharpoons$ 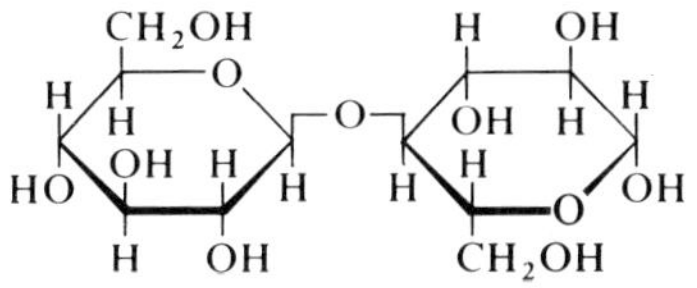

Cellulose

Cellobiose
(4-*O*-β-D-glucopyranosyl-D-glucose)

Cellobiose

e. Inulin 1-fructanohydrolase + inulin ⇌ β-D-fructose

$(C_6H_{10}O_5)_x + xH_2O \rightleftharpoons$ 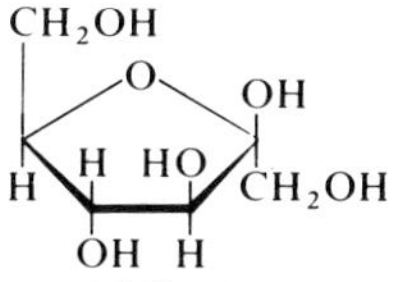

Inulin

β-D-Fructose
(β-D-fructofuranose)

Inulase

f. Polygalacturonide + pectin glycanohydrolase ⇌ D-galactose + α-D-galacturonic acid

Pectic acid ⇌

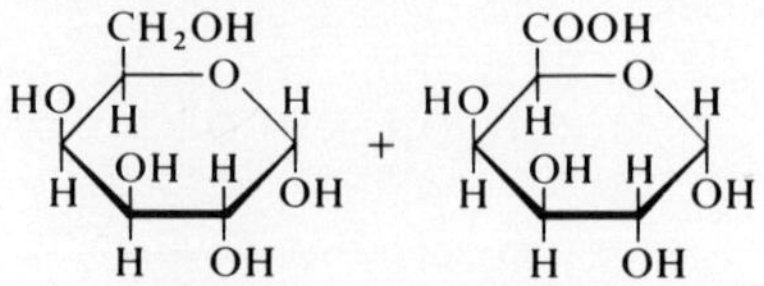

D-Galactose (α-D-galactopyranose) + α-D-Galacturonic acid

Polygalacturonase

g. α-D-Glucoside glucohydrolase + α-D-glucosides ⇌ α-D-glucose + alcohol or phenol residue

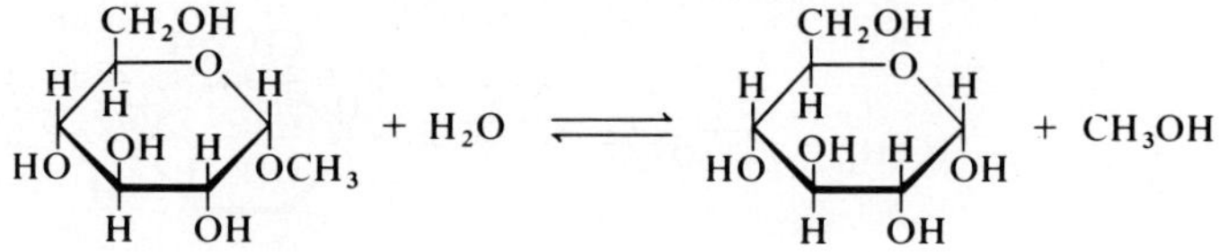

α-Methyl-D-glucoside (α-methyl-D-glucopyranoside) + H_2O ⇌ β-D-Glucose (α-D-glucopyranose) + Methyl alcohol

α-Glucosidase

h. β-D-Glucoside glucohydrolase + β-D-glucosides ⇌ β-D-glucose + alcohol or phenol residue

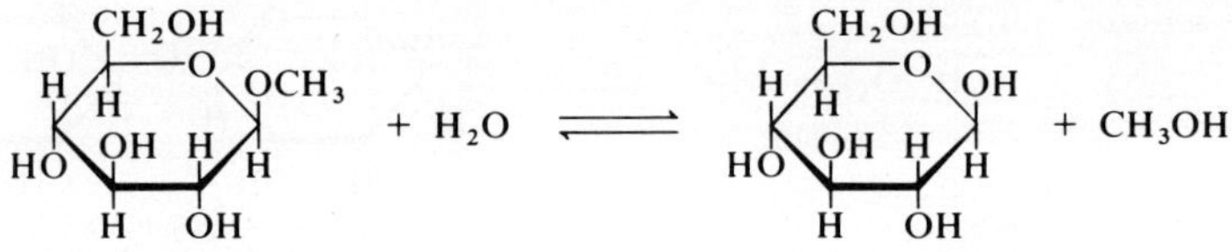

β-Methyl-D-glucoside (β-methyl-D-glycopyranoside) + H_2O ⇌ β-D-Glucose (β-D-glucopyranose) + Methyl alcohol

β-Glucosidase

i. β-D-Mannoside mannohydrolase + a β-D-mannoside + H_2O ⇌ β-D-mannose + an alcohol

β-D-Mannoside + H_2O ⇌

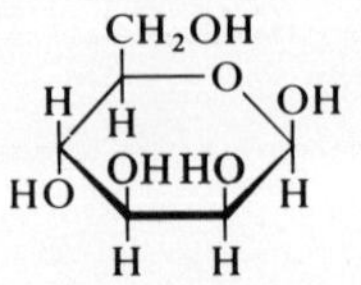

\+ an alcohol

β-D-Mannose

β-Mannosidase

j. α-D-Galactoside galactohydrolase + α-D-galactoside + H_2O ⇌ D-galactose + an alcohol

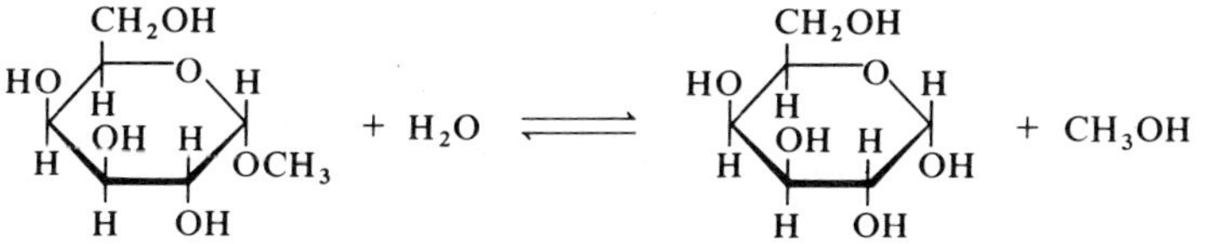

α-D-Methyl galactoside α-D-Galactose Methyl alcohol

α-Galactosidase

k. β-D-Galactoside galactohydrolase + β-D-galactoside + H_2O ⇌ D-galactose + an alcohol

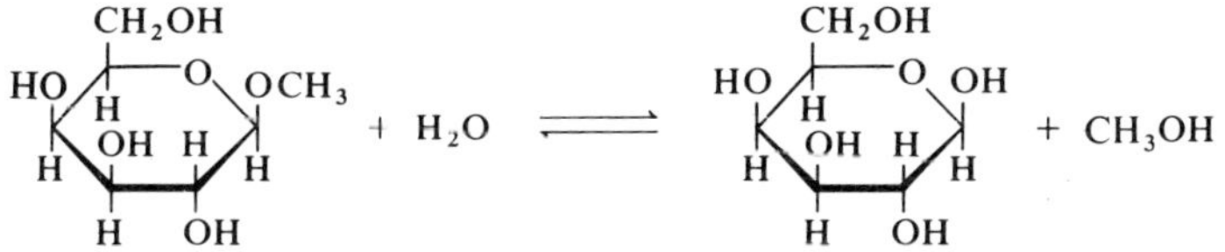

β-D-Methyl galactoside β-D-Galactose Methyl alcohol

β-Galactosidase

l. Trehalose 1-glucohydrolase + trehalose + H_2O ⇌ 2 D-glucose

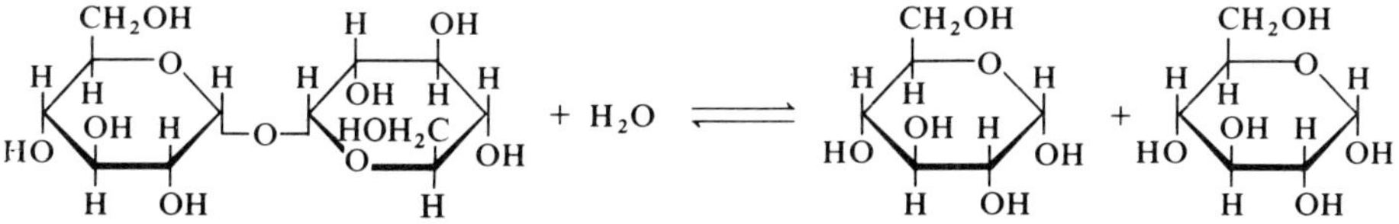

α,α-Trehalose (α-D-glucopyranosyl-α-D-glucopyranoside) α-D-Glucose (α-D-glucopyranose) α-D-Glucose (α-D-glucopyranose)

Trehalase

2. Hydrolyzing N-glycosyl compounds

a. Inosine ribohydrolase + inosine + H_2O ⇌ hypoxanthine + α-D-ribose

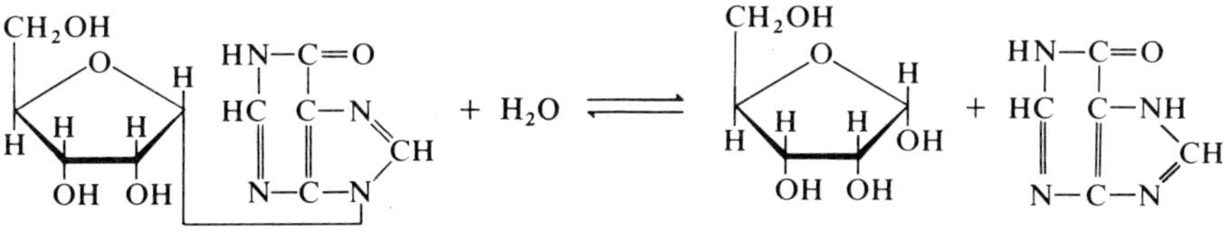

Inosine α-D-Ribose Hypoxanthine

Inosinase

b. Uridine ribohydrolase + uridine + H_2O ⇌ uracil + α-D-ribose

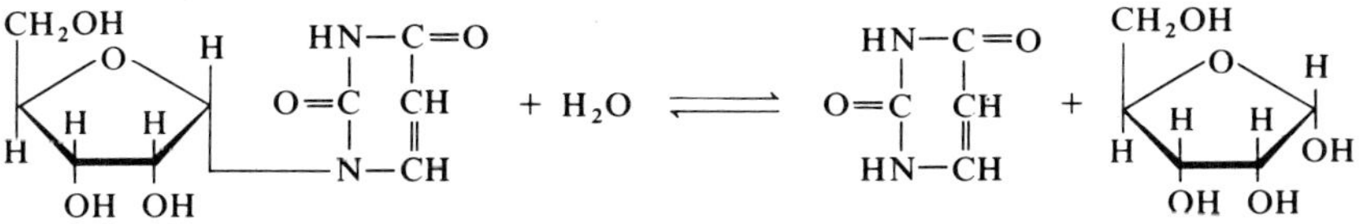

Uridine Uracil α-D-Ribose

Uridine nucleosidase

c. N-Ribosyl-purine ribohydrolase + nucleosides from ribonucleic acid ⇌ purines + D-ribose

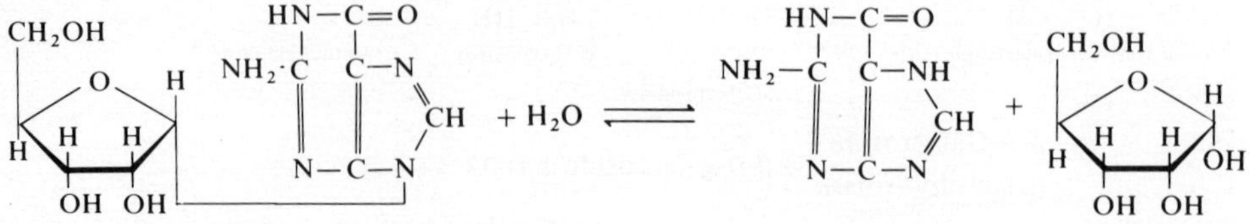

Guanosine Guanine α-D-Ribose

Nucleosidase

C. Acting on peptide bonds (peptide hydrolases)

3. α-Aminopeptide aminoacidohydrolases

   a. Leucine aminopeptidase hydrolyzes L-peptides, splitting off leucine from end of chain if it has a free amino group. For example, the amino acid leucine is liberated on hydrolysis.

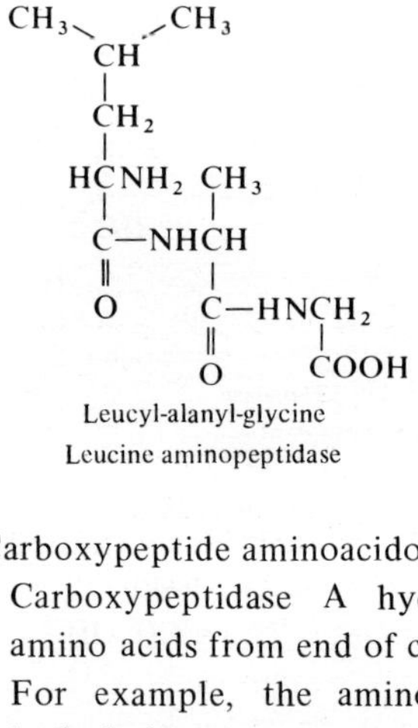

Leucyl-alanyl-glycine
Leucine aminopeptidase

4. α-Carboxypeptide aminoacidohydrolases

   a. Carboxypeptidase A hydrolyzes peptides, splitting off amino acids from end of chain having free carboxyl groups. For example, the amino acid tyrosine is liberated on hydrolysis.

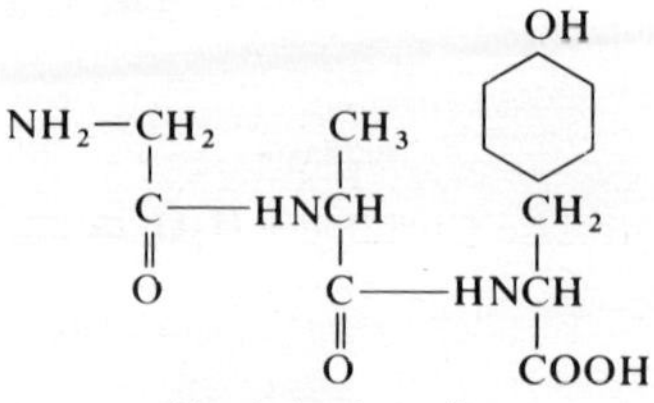

Glycyl-alanyl-tyrosine
Carboxypeptidase A

   b. Carboxypeptidase B hydrolyzes peptides having terminal arginine or lysine groups. For example, the amino acid lysine is liberated on hydrolysis.

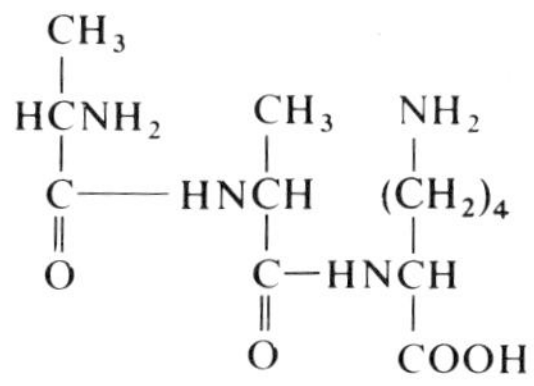

$$\begin{array}{c} CH_3 \\ | \\ HCNH_2 \qquad CH_3 \quad NH_2 \\ | \qquad\qquad | \qquad | \\ C \!\!-\!\!-\!\! HNCH \quad (CH_2)_4 \\ \| \qquad\qquad | \qquad | \\ O \qquad\qquad C\!-\!HNCH \\ \qquad\qquad \| \qquad | \\ \qquad\qquad O \qquad COOH \end{array}$$

Alanyl-alanyl-lysine

Carboxypeptidase B

5. Dipeptide hydrolases

    a. Glycylglycine hydrolase + glycylglycine $\rightleftharpoons$ 2 glycine

$$CH_2NH_2 \cdot CONH \cdot CH_2 \cdot COOH + H_2O \rightleftharpoons 2CH_2NH_2 \cdot COOH$$

Glycylglycine   Glycine

Glycyl-glycine dipeptidase

    b. Glycyl-L-leucine hydrolase + glycyl-L-leucine + H_2O

$$\rightleftharpoons glycine + L\text{-leucine}$$

$$\begin{array}{c} CH_3 \quad CH_3 \\ \diagdown \;\diagup \\ CH \\ | \\ CH_2 \\ | \\ H_2NCH_2 \cdot CONH \cdot CH \cdot COOH \end{array} + H_2O \rightleftharpoons CH_2NH_2 \cdot COOH + \begin{array}{c} CH_3 \quad CH_3 \\ \diagdown \;\diagup \\ CH \\ | \\ CH_2 \\ | \\ CHNH_2 \\ | \\ COOH \end{array}$$

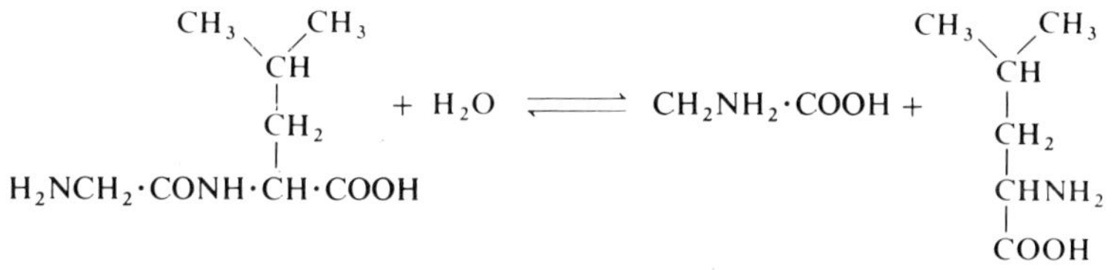

Glycyl-L-Leucine   Glycine   L-Leucine

Glycyl-leucine-dipeptidase

    c. Prolidase + glycyl-L-proline + H_2O $\rightleftharpoons$

$$glycine + L\text{-proline}$$

$$H_2N\!-\!CH_2\!-\!CO\!-\!N\!\!\begin{array}{c} \diagup CH_2\!-\!CH_2 \\ | \\ \diagdown CH\!-\!CH_2 \end{array} + H_2O$$
$$HOOC$$

Glycyl-L-proline

$\Updownarrow$

$$CH_2NH_2 \cdot COOH + \begin{array}{c} CH_2\!-\!CH_2 \\ | \qquad\quad \diagdown NH \\ CH_2\!-\!CH \\ \qquad\quad \diagdown COOH \end{array}$$

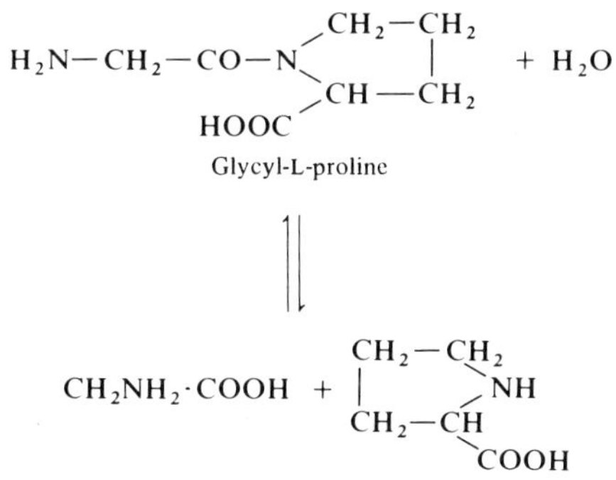

Glycine   L-Proline

Prolidase

6. Peptidyl peptide hydrolases

    a. Pepsin hydrolyzes peptides, including those with bonds adjacent to aromatic or dicarboxylic L-amino residues

Native proteins $\rightleftharpoons$ proteoses + peptones

Pepsin

b. Rennin hydrolyzes peptides. Specificity similar to that of pepsin.

$$\text{Casein} \underset{\text{Rennin}}{\rightleftharpoons} \text{paracasein}$$

c. Trypsin hydrolyzes peptides, amides, esters, etc., at bonds involving carboxyl groups of L-arginine or L-lysine.

Native proteins, proteoses, peptones, and peptides $\rightleftharpoons$

$$\underset{\text{Trypsin}}{} \text{amino acids and polypeptides}$$

d. Papain hydrolyzes peptides, amides, and esters, especially at bonds involving basic amino acids, or leucine, or glycine.

$$\text{Native proteins} \underset{\text{Papain}}{\rightleftharpoons} \text{polypeptides + dipeptides}$$

D. Acting on C–N bonds other than peptide bonds
   7. In linear amides
     a. L-Asparagine amidohydrolase + β-L-asparagine + H_2O

$$\rightleftharpoons \text{L-aspartate} + NH_3$$

$$\underset{\beta\text{-L-Asparagine}}{CONH_2 \cdot CH_2 \cdot CHNH_2 \cdot COOH} + H_2O \rightleftharpoons$$

$$\underset{\text{L-Aspartic acid}}{COOH \cdot CH_2 \cdot CHNH_2 \cdot COOH} + NH_3$$

$$\text{Asparaginase}$$

     b. L-Glutamine amidohydrolase + L-glutamine + H_2O $\rightleftharpoons$

$$\text{L-glutamate} + NH_3$$

$$\underset{\text{L-Glutamine}}{CONH_2 \cdot CH_2 \cdot CH_2 \cdot CHNH_2 \cdot COOH} + H_2O \rightleftharpoons$$

$$\underset{\text{L-Glutamic acid}}{COOH \cdot CH_2 \cdot CH_2 \cdot CHNH_2 \cdot COOH} + NH_3$$

$$\text{Glutaminase}$$

     c. Urea amidohydrolase + urea + H_2O $\rightleftharpoons$ $CO_2 + NH_3$

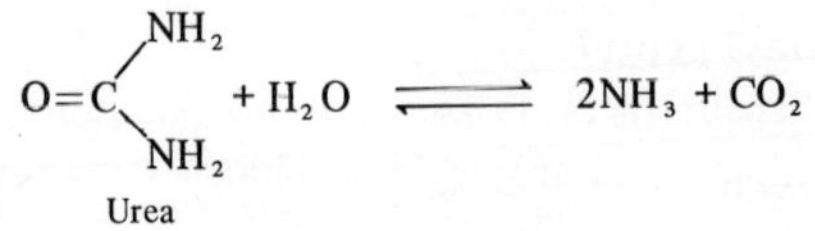

$$\underset{\text{Urea}}{O=C\begin{smallmatrix} \nearrow NH_2 \\ \searrow NH_2 \end{smallmatrix}} + H_2O \rightleftharpoons 2NH_3 + CO_2$$

$$\text{Urease}$$

8. In linear amidines

   a. L-Arginine ureohydrolase + L-arginine + H_2O $\rightleftharpoons$

$$\text{urea} + \text{L-ornithine}$$

$$\text{HN}=\text{C}\underset{\displaystyle\underset{H}{|}}{\overset{\displaystyle\text{NH}_2}{\big\langle}}\text{N(CH}_2)_3\cdot\text{CHNH}_2\cdot\text{COOH} + H_2O$$

L-Arginine

$\big\Downarrow$

$$\text{O}=\text{C}\overset{\displaystyle\text{NH}_2}{\underset{\displaystyle\text{NH}_2}{\big\langle}} + \text{NH}_2(\text{CH}_2)_3\cdot\text{CHNH}_2\cdot\text{COOH}$$

   Urea   L-Ornithine

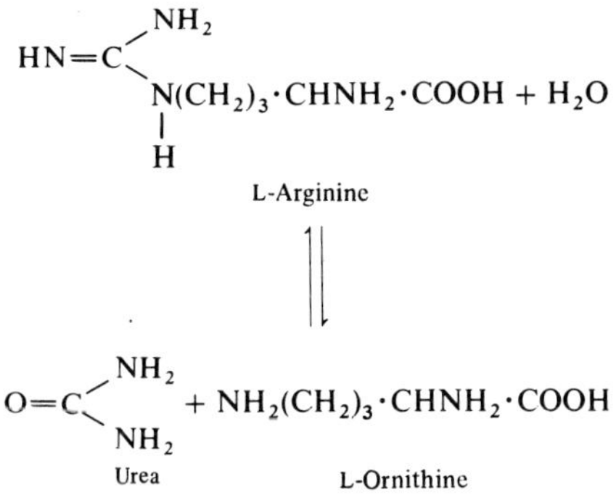

        Arginase

9. In cyclic amidines

   a. Cytosine aminohydrolase + cytosine + H_2O $\rightleftharpoons$

$$\text{uracil} + \text{NH}_3$$

   Cytosine   Uracil

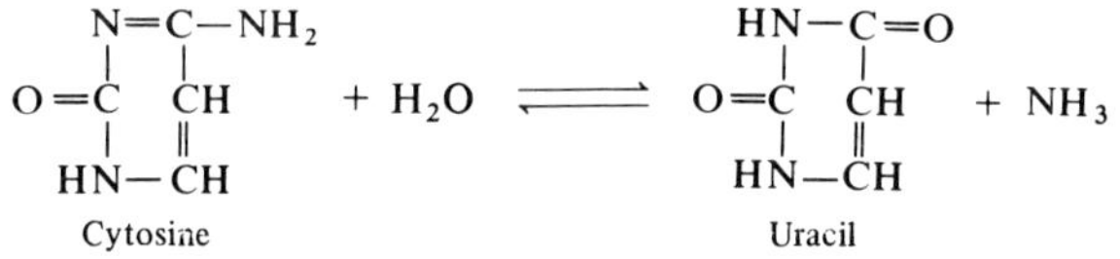

            Cytosine deaminase

   b. Adenine aminohydrolase + adenine + H_2O $\rightleftharpoons$

$$\text{hypoxanthine} + \text{NH}_3$$

   Adenine   Hypoxanthine

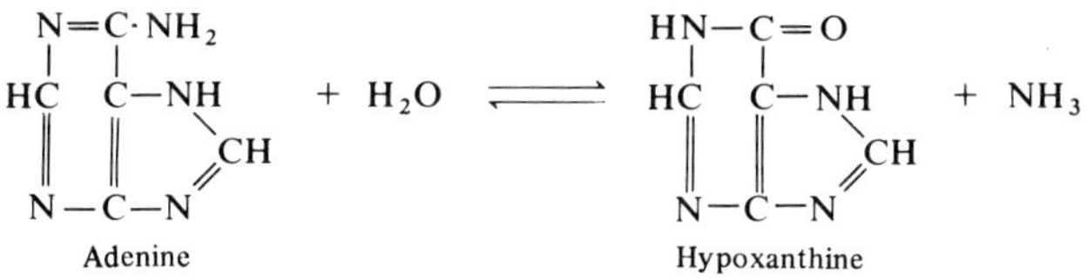

            Adenine deaminase

   c. Guanine aminohydrolase + guanine $\rightleftharpoons$

$$\text{xanthine} + \text{NH}_3$$

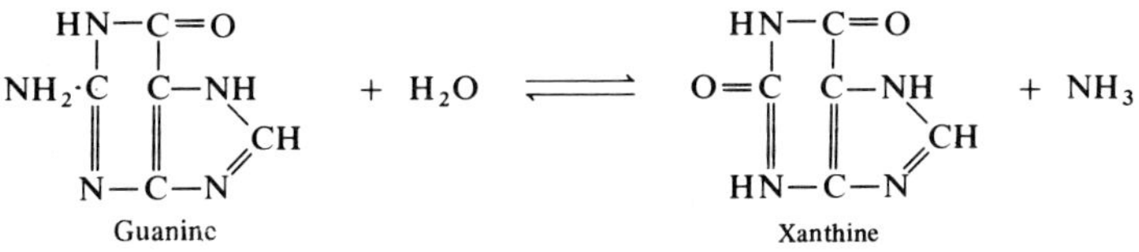

   Guanine   Xanthine

           Guanine deaminase

d. Adenosine aminohydrolase + adenosine + $H_2O \rightleftharpoons$ inosine + NH_3

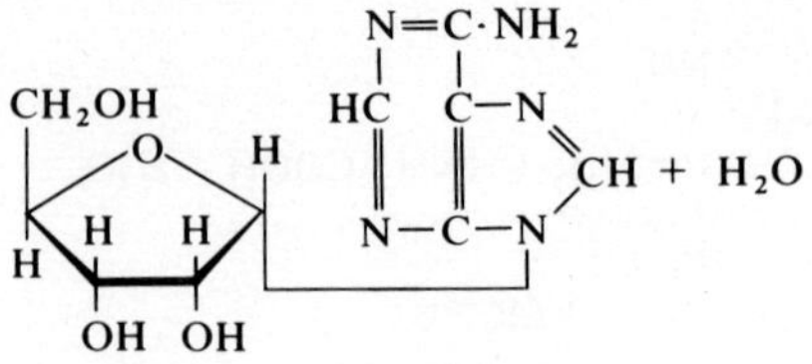

Adenosine

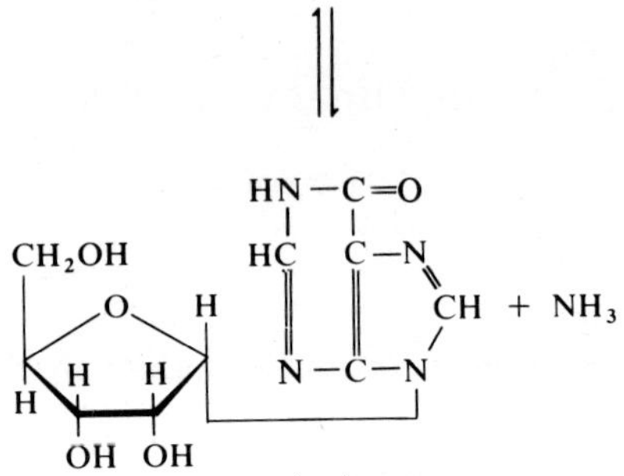

Inosine

Adenosine deaminase

e. Guanosine aminohydrolase + guanosine + $H_2O \rightleftharpoons$ xanthosine + NH_3

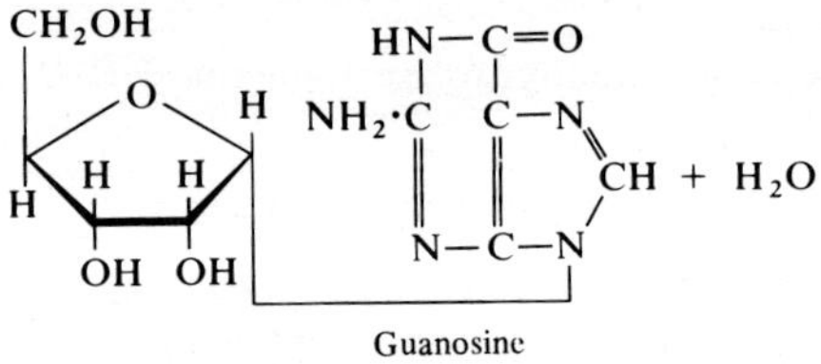

Guanosine

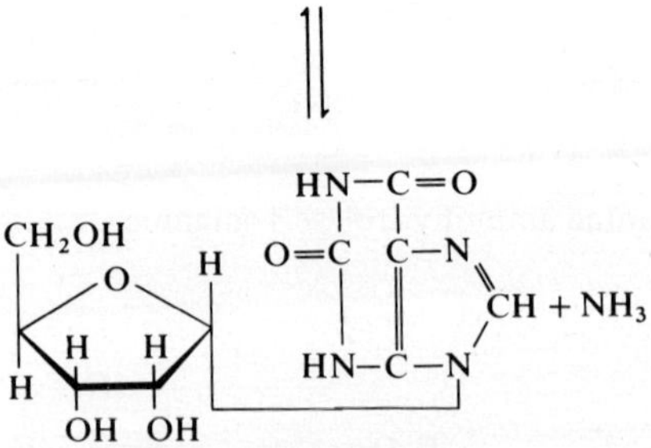

Xanthosine

Guanosine deaminase

f. Cytidine aminohydrolase + cytidine + $H_2O \rightleftharpoons$ uridine + NH_3

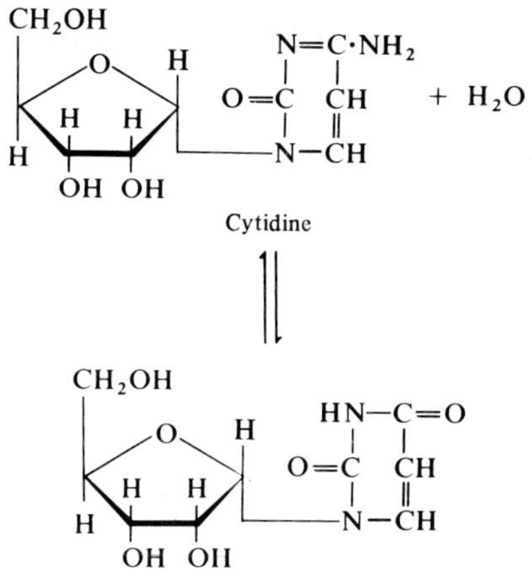

Cytidine

Uridine

Cytidine deaminase

E. Acting on C–C bonds

10. In ketonic substances

a. Oxaloacetate acetylhydrolase + oxaloacetate + H_2O ⇌ oxalate + acetate

$$COOH{\cdot}CH_2{\cdot}CO{\cdot}COOH + H_2O \rightleftharpoons COOH{\cdot}COOH + CH_3COOH$$

Oxaloacetic acid — Oxalic acid — Acetic acid

Oxaloacetase

IV. LYASES

A. Carbon-carbon lyases

1. Carboxy-lyases

a. 2-Oxo-acid carboxy-lyase + a 2-oxo-acid ⇌ an aldehyde + CO_2

$$CH_3{\cdot}CO{\cdot}COOH \rightleftharpoons CH_3{\cdot}CHO + CO_2$$

Pyruvic acid — Acetaldehyde

Pyruvate decarboxylase

b. Oxaloacetate carboxy-lyase + oxaloacetate ⇌ pyruvate + CO_2

$$COOH{\cdot}CH_2{\cdot}CO{\cdot}COOH \rightleftharpoons CH_3{\cdot}CO{\cdot}COOH + CO_2$$

Oxaloacetic acid — Pyruvic acid

Oxaloacetate decarboxylase

c. Acetoacetate carboxy-lyase + acetoacetate ⇌ acetone + CO_2

$$CH_3{\cdot}CO{\cdot}CH_2{\cdot}COOH \rightleftharpoons CH_3{\cdot}CO{\cdot}CH_3 + CO_2$$

Acetoacetic acid — Acetone

Acetoacetate decarboxylase

d. L-Aspartate 1-carboxy-lyase + L-aspartate ⇌ β-alanine + CO_2

$$COOH{\cdot}CH_2{\cdot}CHNH_2{\cdot}COOH \rightleftharpoons CH_2NH_2{\cdot}CH_2{\cdot}COOH + CO_2$$

L-Aspartic acid → β-Alanine

Aspartate 1-decarboxylase

e. L-Aspartate 4-carboxy-lyase + L-aspartate ⇌ L-alanine + CO_2

$$COOH{\cdot}CH_2{\cdot}CHNH_2{\cdot}COOH \rightleftharpoons CH_3{\cdot}CHNH_2{\cdot}COOH + CO_2$$

L-Aspartic Acid → L-Alanine

Aspartate 4-decarboxylase

f. L-Glutamate 1-carboxy-lyase + L-glutamate ⇌ 4-aminobutyrate + CO_2

$$COOH{\cdot}CH_2{\cdot}CH_2{\cdot}CHNH_2{\cdot}COOH \rightleftharpoons CH_2NH_2{\cdot}CH_2{\cdot}CH_2{\cdot}COOH + CO_2$$

L-Glutamic acid → 4-Aminobutyric acid

Glutamate decarboxylase

g. L-Valine carboxy-lyase + L-valine ⇌ isobutylamine + CO_2

$$(H_3C)_2{>}CH{-}CHNH_2{\cdot}COOH \rightleftharpoons (H_3C)_2{>}CH{\cdot}CH_2NH_2 + CO_2$$

L-Valine → Isobutylamine

Valine decarboxylase

h. L-Ornithine carboxy-lyase + L-ornithine ⇌ putrescine + CO_2

$$NH_2CH_2{\cdot}CH_2{\cdot}CH_2{\cdot}CHNH_2{\cdot}COOH \rightleftharpoons NH_2CH_2{\cdot}CH_2{\cdot}CH_2{\cdot}CH_2NH_2 + CO_2$$

L-Ornithine → Tetramethylenediamine (putrescine)

Ornithine decarboxylase

i. L-Lysine carboxy-lyase + L-lysine ⇌ cadaverine + CO_2

$$NH_2CH_2{\cdot}CH_2{\cdot}CH_2{\cdot}CH_2{\cdot}CHNH_2{\cdot}COOH \rightleftharpoons NH_2CH_2{\cdot}CH_2{\cdot}CH_2{\cdot}CH_2{\cdot}CH_2NH_2 + CO_2$$

L-Lysine → Pentamethylenediamine (cadaverine)

Lysine decarboxylase

j. L-Arginine carboxy-lyase + L-arginine

⇌ agmatine + CO_2

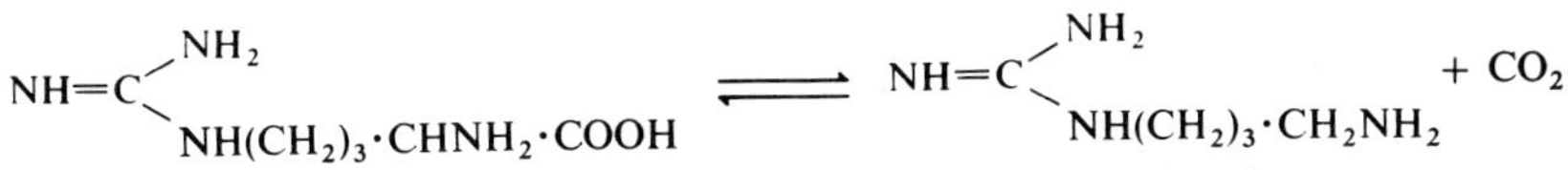

L-Arginine　　Agmatine

Arginine decarboxylase

k. L-Histidine carboxy-lyase + L-histidine ⇌

histamine + CO_2

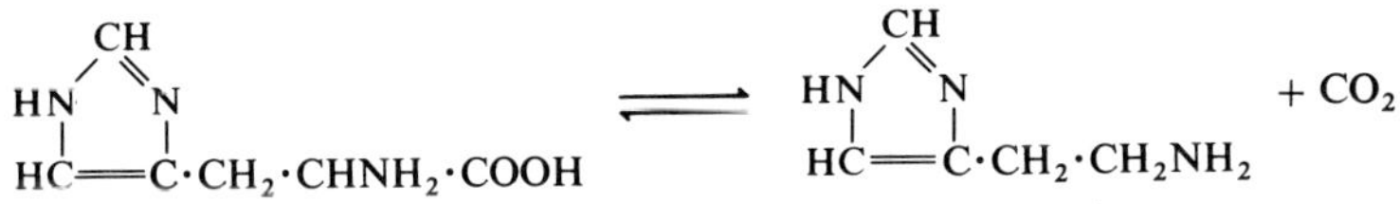

L-Histidine　　Histamine

Histidine decarboxylase

l. L-Tyrosine carboxy-lyase + L-tyrosine ⇌

tyramine + CO_2

L-Tyrosine　　Tyramine

Tyrosine decarboxylase

m. L-Tryptophan carboxy-lyase + L-tryptophan ⇌

tryptamine + CO_2

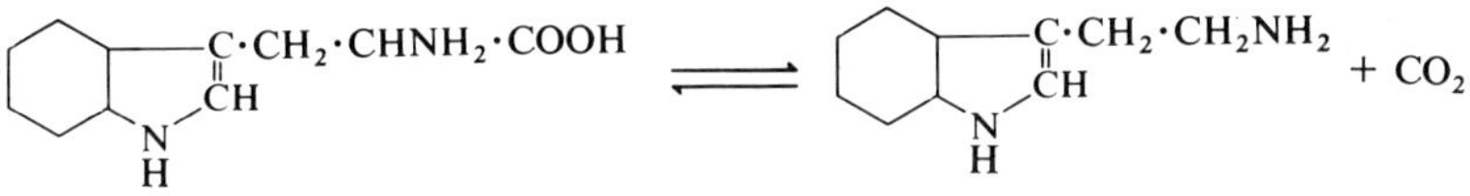

L-Tryptophan　　Tryptamine

Tryptophan decarboxylase

2. Aldehyde-lyases
   a. 2-Deoxy-D-ribose-5-phosphate acetaldehyde-lyase + 2-deoxy-D-ribose-5-phosphate ⇌

D-glyceraldehyde 3-phosphate + acetaldehyde

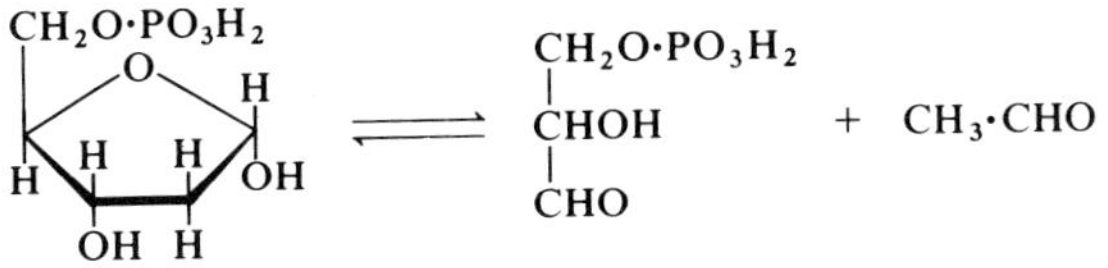

2-Deoxy-D-ribose-5-phosphate　　D-Glyceraldehyde 3-phosphate　　Acetaldehyde

Deoxyriboaldolase

b. L-Threonine acetaldehyde-lyase + L-threonine ⇌ glycine + acetaldehyde

$$CH_3\cdot CHOH\cdot CHNH_2\cdot COOH \rightleftharpoons CH_2NH_2\cdot COOH + CH_3\cdot CHO$$

L-Threonine — Glycine — Acetaldehyde

Threonine aldolase

c. Ketose-1-phosphate aldehyde-lyase + a ketose 1-phosphate ⇌ dihydroxyacetone phosphate + an aldehyde

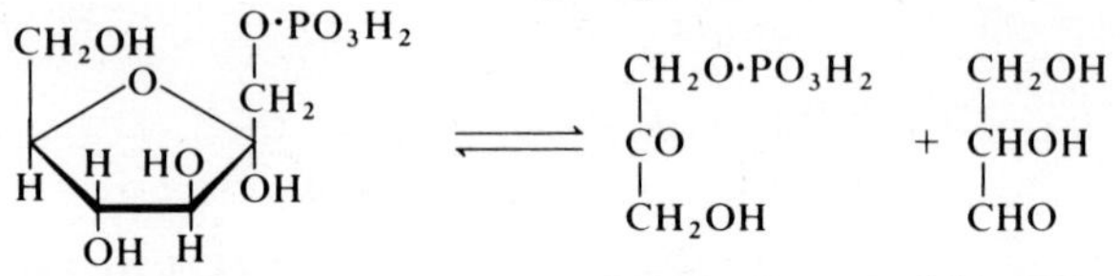

Fructofuranose-1-phosphate — Dihydroxyacetone phosphate — 3-Glyceraldehyde

Ketone-1-phosphate aldolase

3. Keto acid-lyases

a. Citrate oxaloacetate-lyase + citrate ⇌ oxaloacetate + acetate

$$\begin{array}{l} CH_2\cdot COOH \\ | \\ COH\cdot COOH \\ | \\ CH_2\cdot COOH \end{array} \rightleftharpoons COOH\cdot CO\cdot CH_2\cdot COOH + CH_3\cdot COOH$$

Citric acid — Oxaloacetic acid — Acetic acid

Citrate lyase

B. Carbon-oxygen lyases

4. Hydro-lyases

a. L-Malate hydro-lyase + L-malate ⇌ fumarate + H_2O

$$COOH\cdot CHOH\cdot CH_2\cdot COOH \rightleftharpoons COOH\cdot CH{=}CH\cdot COOH + H_2O$$

L-Malic acid — Fumaric acid

Fumarate hydratase

b. D-2-Phosphoglycerate hydro-lyase + D-2-phosphoglycerate ⇌ phospho-enolpyruvate + H_2O

$$\begin{array}{l} CH_2OH \\ | \\ CHO\cdot PO_3H_2 \\ | \\ COOH \end{array} \rightleftharpoons \begin{array}{l} CH_2 \\ \| \\ CO\cdot PO_3H_2 + H_2O \\ | \\ COOH \end{array}$$

2-Phosphoglyceric acid — Phosphopyruvic acid (enol)

Phosphopyruvate hydratase

c. D-Serine hydro-lyase + D-serine + H_2O $\rightleftharpoons$ pyruvate + NH_3 + H_2O

$CH_2OH \cdot CHNH_2 \cdot COOH + H_2O \rightleftharpoons CH_3 \cdot CO \cdot COOH + NH_3 + H_2O$

D-Serine → Pyruvic acid

D-Serine dehydratase

d. L-Threonine hydro-lyase + L-threonine + H_2O $\rightleftharpoons$ 2-oxobutyrate + NH_3 + H_2O

$CH_3 \cdot CHOH \cdot CHNH_2 \cdot COOH + H_2O \rightleftharpoons CH_3 \cdot CH_2 \cdot CO \cdot COOH + NH_3 + H_2O$

L-Threonine → 2-Oxobutyric acid

Threonine dehydratase

e. L-Serine hydro-lyase (adding indole) + L-serine + indole $\rightleftharpoons$ L-tryptophan + H_2O

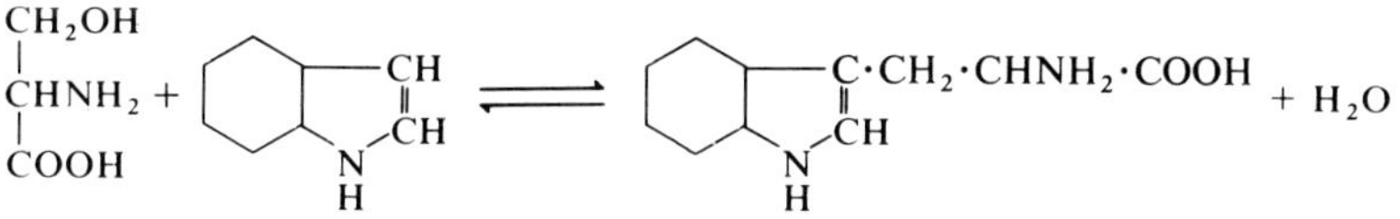

L-Serine Indole L-Tryptophan

Tryptophan synthase

f. L-Serine hydro-lyase (adding H_2S) + H_2S $\rightleftharpoons$ L-cysteine + H_2O

$CH_2OH{-}CHNH_2{-}COOH + H_2S \rightleftharpoons CH_2SH{-}CHNH_2{-}COOH + H_2O$

L-Serine → L-Cysteine

Cysteine synthase

C. Carbon-nitrogen lyases

5. Ammonia-lyases

a. L-Aspartate ammonia-lyase + L-aspartate $\rightleftharpoons$ fumarate + NH_3

$COOH \cdot CH_2 \cdot CHNH_2 \cdot COOH \rightleftharpoons COOH \cdot CH{=}CH \cdot COOH + NH_3$

L-Aspartic acid → Fumaric acid

Aspartate ammonia-lyase

V. ISOMERASES

A. Racemases and epimerases

1. Acting on aminoacids and derivatives

   a. Alanine racemase + L-alanine ⇌ D-alanine

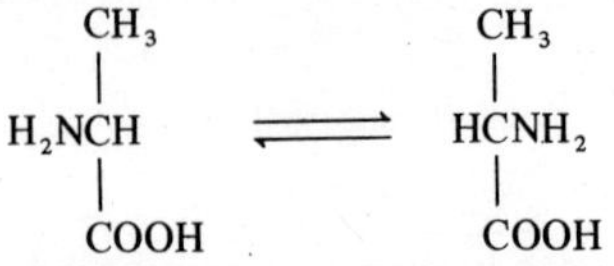

L-Alanine D-Alanine

Alanine racemase

   b. Lysine racemase + L-lysine ⇌ D-lysine

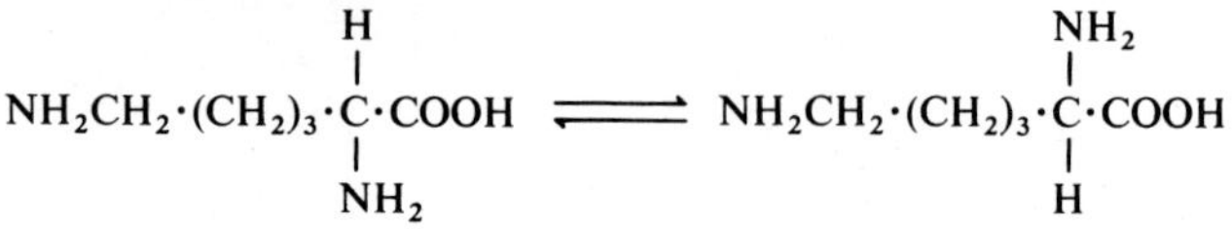

L-Lysine D-Lysine

Lysine racemase

2. Acting on hydroxyacids and derivatives

   a. Lactate racemase + L-lactate ⇌ D-lactate

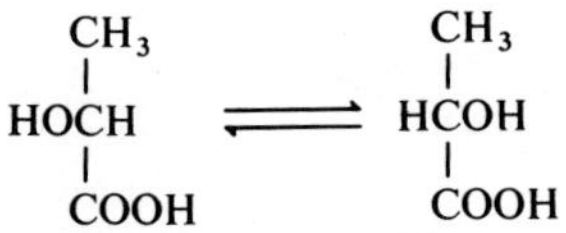

L-Lactic acid D-Lactic acid

Lactate racemase

B. Intramolecular oxidoreductases

3. Interconverting aldoses and ketoses

   a. D-glyceraldehyde-3-phosphate keto-isomerase + D-glyceraldehyde-3-phosphate ⇌ dihydroxyacetone phosphate

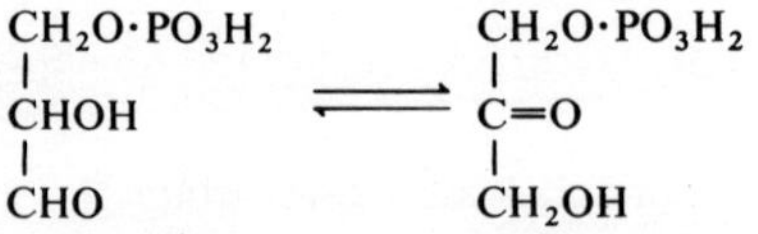

D-Glyceraldehyde-3-phosphate Dihydroxyacetone phosphate

Triosephosphate isomerase

   b. β-D-Mannose keto-isomerase + β-D-mannose ⇌ β-D-fructose

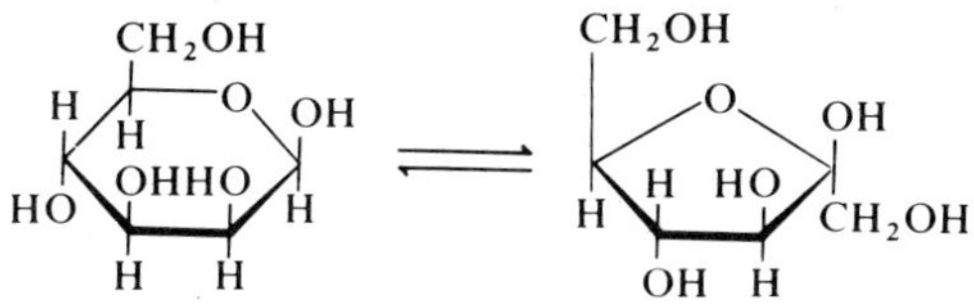

β-D-Mannose β-D-Fructose

Mannose isomerase

c. β-D-Glucose-6-phosphate keto-isomerase + β-D-glucose-6-phosphate ⇌ β-D-fructose-6-phosphate

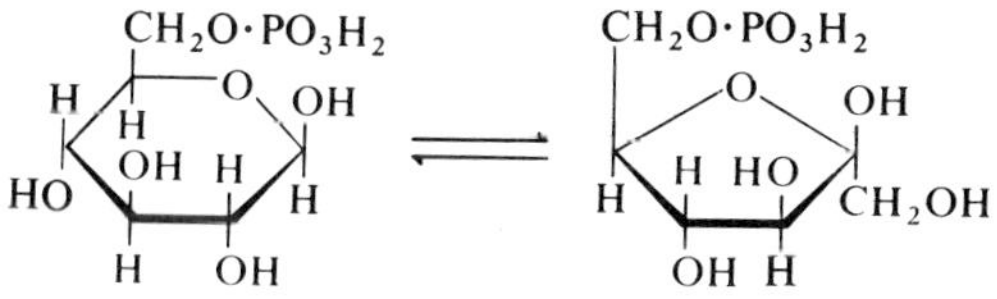

β-D-Glucose-6-phosphate β-D-Fructose-6-phosphate

Glucosephosphate isomerase

VI. LIGASES

A. Forming C–N bonds

1. Acid-ammonia ligases

a. L-aspartate:ammonia ligase (AMP) + L-aspartate + ATP + NH_3 ⇌ L-asparagine + pyrophosphate + AMP

$$COOH \cdot CH_2 \cdot CHNH_2 \cdot COOH + ATP + NH_3 \rightleftharpoons$$

L-Aspartic acid

$$CONH_2 \cdot CH_2 \cdot CHNH_2 \cdot COOH + H_3PO_4 + ADP$$

L-Asparagine

Asparagine synthetase

2. Acid-aminoacid ligases

a. D-Alanine: D-alanine ligase (ADP) + 2 D-alanine + (ATP) ⇌ D-alanyl-D-alanine + phosphate + ADP

$$2CH_3 \cdot CHNH_2 \cdot COOH + ATP \rightleftharpoons$$

D-Alanine

$$CH_3 \cdot CHNH_2 \cdot CONH \cdot CH(CH_3) \cdot COOH + H_3PO_4 + ADP$$

D-Alanyl-D-alanine

D-Alanylalanine synthetase

For more information: Barman (1969); Bernhard (1968); Boyer et al. (1970, 1971, 1972); Commission on Biochemical Nomenclature (1971); Dixon and Webb (1964); Gutfreund (1965); Hayashi et al. (1967); King et al. (1965); Locke (1969); Morino and Snell (1967*a, b*); Moss (1968); Webb (1963); Yamada et al. (1968).

Constitutive and induced enzymes

Most of the proteins synthesized by bacteria are enzymes which must be produced in a balanced array for proper functioning of the cell. Changes in the external environment may also involve changes in the constitution of enzymes. Bacteria grown under aerobic conditions require certain enzymes which may not be necessary when the cells are placed in an anaerobic environment; however, these enzymes must be replaced by other enzymes. Within limits, bacteria are able to adapt themselves to new environments. Such changes are accompanied by a resynthesis of the necessary enzyme systems.

Bacterial enzymes may be classed as (1) constitutive enzymes and (2) induced enzymes.

Constitutive enzymes Most bacterial enzymes are elaborated by the cell irrespective of the presence or absence of the homologous substrates. These fall into the first class and are called constitutive enzymes.

The constitutive enzymes are believed to be essential enzymes. Their levels show fluctuations with changes in the environment. Nevertheless, they are believed to be always produced, regardless of the conditions under which the bacteria are growing.

Another indication that constitutive enzymes are essential is that they maintain their activity very well even when the nitrogen content of the medium is reduced. Under these conditions the nonessential enzymes, i.e., those which expand the living conditions of the cell but are not necessarily required, lose most or all of their activity.

For more information: MacKelvie et al. (1968); Pardee and Beckwith (1962).

Induced enzymes A bacterium, being a single-celled organism, is in close contact with its environment. To survive, it must adapt itself to changes in the environment. This may involve not only the phenotype but also the genetic makeup (genotype). The importance of the environment in determining the phenotype is greater in bacteria than in more organized forms of life. Most of the phenotypic changes in a bacterium are due to its ability to manufacture new proteins, especially enzymes.

Escherichia coli can attack the disaccharide lactose, hydrolyzing it to the monosaccharides glucose and galactose. Since the enzyme is not secreted from the cell, the reaction is intracellular. If the cell is not permeable to lactose, the reaction cannot occur even though the organism produces the enzyme. If the cell is permeable to lactose but the sugar is absent, the enzyme is useless.

Occasionally a strain of *E. coli* is encountered that is unable to attack lactose [lactose (–)]. This is a question of genetic makeup. A lactose (–) strain is a mutant that cannot manufacture the necessary enzyme. In the absence of the enzyme, the organism is unable to grow in a medium containing lactose as the only source of carbon.

A lactose (+) strain must be permeable to lactose. A cell that forms the necessary enzyme but is not permeable to lactose is not capable of utilizing the sugar. On the basis of lactose permeability and lactose utilization, four types of *E. coli* are possible and have been isolated:

1. A^+B^+, the normal type that forms the enzyme and is permeable to lactose.
2. A^+B^-, the type that forms the enzyme but is not permeable to lactose.
3. A^-B^+, the type that does not form the enzyme but is permeable to lactose.
4. A^-B^-, the type that does not form the enzyme and is not permeable to lactose.

A lactose (−) or A^-B^+ strain can become lactose (+) or A^+B^+ by mutation. This can occur by growing the lactose (−) strain in the presence of lactose as the sole carbon source. There will be a lag period before growth begins. During the lag period the organism synthesizes the necessary enzyme protein. An enzyme produced under such conditions is called an induced enzyme.

E. coli utilizes lactose by synthesizing enzyme protein de novo. The utilization of lactose does not result from the activation of a previously present enzyme, but rather the cells synthesize a protein not previously manufactured.

Many enzymes which are normally not produced by an organism may be induced by adding the inducers to the medium. Some enzymes are induced rapidly; others require more time. The induced enzymes gradually disappear when the inducers are withheld from the medium because the new cells being reproduced do not synthesize the enzymes. Continued growth simply dilutes the enzymes until they disappear completely.

For more information: Bilezikian et al. (1967); Boos et al. (1967); Bradley (1963); Fukumoto et al. (1967); Kobayashi et al. (1968); Lwoff (1962); McClatchy and Rosenblum (1963); Pastan and Perlman (1970); Sistrom (1969); Welker and Campbell (1963*a, b*).

Secretion of enzymes

Enzymes of bacteria may be divided into two groups, depending upon whether they are secreted into the surrounding culture medium or remain confined within the cell. The enzymes that belong to the first group are known as extracellular enzymes or exoenzymes; those classified in the second group are called intracellular enzymes or endoenzymes.

Extracellular enzymes Enzymes are capable of producing their specific actions in the complete absence of the living cell. In the case of the extracellular enzymes, this may be shown by centrifugating a liquid culture, passing the clear supernate through a bacteriological filter, and demonstrating enzyme activity by adding the filtrate to the appropriate substrates.

Extracellular enzymes may also be demonstrated by incorporating insoluble compounds such as starch, inulin, fats, gelatin, and casein (milk) in agar, and streaking prepared plates with the test organisms. Clear zones around colonies of the organisms indicate the presence of extracellular enzymes capable of utilizing the compounds.

Nature of starch Starch is a polymer of glucose found widely distributed throughout the plant kingdom. It occurs as a reserve material in roots, seeds, fruits, and pith of plants. During periods of active growth, especially in the spring, starch is converted into sugar and transported to wherever needed for conversion into cellulose or other products.

Starch occurs in the form of granules having characteristic striations. The size, shape, and striations of the granules are typical of many plants and usually enable the skilled microscopist to determine the source of the starch.

Starch granules are insoluble in cold water even after long immersion. After the outer membranes are broken, the contents swell in cold water. Hot water produces a swelling of the intact granules, causing the membranes to burst, after which the contents form viscous solutions or gels. On cooling such solutions, rigid gels may be formed by the crystallization of the dissolved or dispersed starch contents as networks occluding much liquid.

When starch is heated with dilute mineral acids, it is hydrolyzed to dextrins, maltose, and D-glucose. These products are all soluble in water. Concentrated mineral acids hydrolyze the molecule completely to D-glucose. Starch is also hydrolyzed by the two enzymes α-1,4-glucan 4-glucanohydrolase and α-1,4-glucan maltohydrolase.

Starch can be separated into two fractions possessing different physical properties: amylose and amylopectin. Amylose is soluble in water and is colored blue with iodine. Amylopectin is insoluble in water and gives a red color with iodine.

Both amylose and amylopectin are composed of D-glucopyranose units. In amylose the units are linked by 1 → 4 α-glucosidic bonds which tend to create a spiraling of the molecule in the form of a helix:

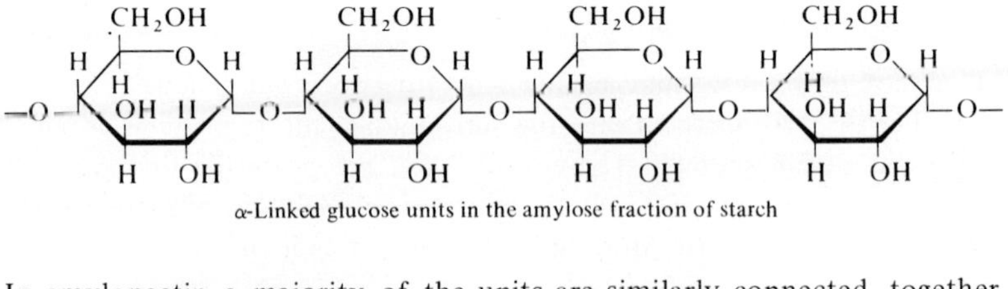

α-Linked glucose units in the amylose fraction of starch

In amylopectin a majority of the units are similarly connected, together with a few 1 → 6 α-glucosidic bonds, and the molecule shows considerable branching:

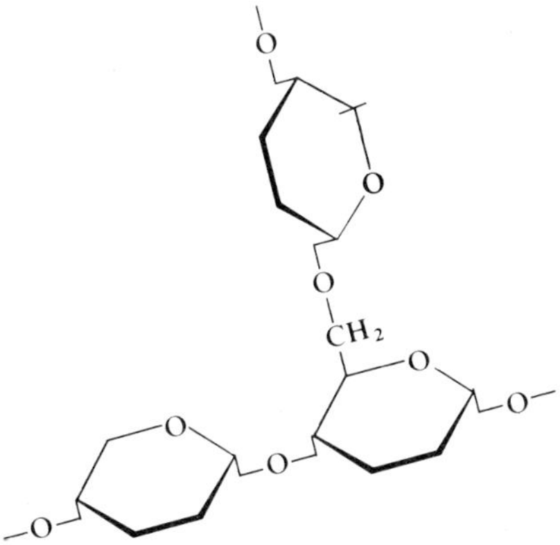

α-Linked glucose units in the amylopectin fraction of starch

Amylose has a smaller molecular weight than amylopectin and is hydrolyzed to a greater extent by α-1,4-glucan maltohydrolase. Amylose and amylopectin from different sources are heterogeneous both in regard to molecular size and degree of branching.

Soluble starch Raw starch on treatment with 7.5 percent HCl for 7 days at room temperature becomes soluble in water with the formation of a limpid solution. Soluble starch is more satisfactory than raw starch for use in culture media.

Hydrolysis of starch Many bacteria produce the enzymes which are capable of hydrolyzing starch. α-1,4-glucan 4-glucanohydrolase is known as the liquefying enzyme, being capable of hydrolyzing starch chiefly to dextrin with some maltose. α-1,4-glucan maltohydrolase is called the saccharifying enzyme, being capable of hydrolyzing starch to maltose.

The two enzymes are extracellular. They convert indiffusible starch into diffusible substances which then can enter the cell and be acted upon by the intracellular enzymes.

The presence of the enzymes may be demonstrated by filtering a broth culture and mixing the filtrate with starch. Disappearance of the starch indicates the presence of the enzymes. The addition of iodine may be used as an indicator of the reaction. A blue color indicates the presence of starch; a brown color indicates the absence or complete hydrolysis of starch (Fig. 203).

Another method is to streak a loopful of the test culture over the surface of a starch agar plate. After incubation, the plate is flooded with a dilute solution of iodine. The absence of a blue color at some distance from the bacterial growth indicates the extracellular hydrolysis of starch (Fig. 204).

Liquefaction of gelatin Gelatin is a protein that possesses the property of forming a gel when dissolved in warm water. Since it is a protein, it can be attacked by many bacteria, resulting in the loss of its property to gel.

The enzyme that is responsible for the hydrolysis of gelatin is known as gelatinase. In the presence of carbohydrates which are rapidly

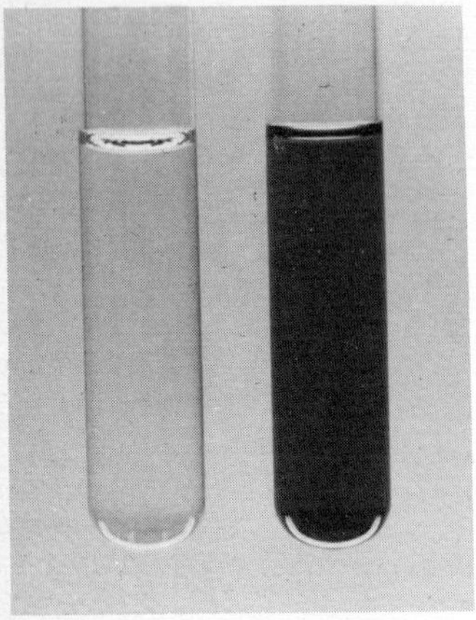

Fig. 203 Hydrolysis of starch in the absence of bacteria. Culture filtrates added to tubes of starch broth and incubated at 37°C for 24 hr. Left, Bacillus subtilis filtrate, no color with iodine, indicating hydrolysis. Right, Escherichia coli filtrate, blue color with iodine, indicating no hydrolysis.

fermented, gelatinase either is not produced or only in minute amounts. In general, noncarbohydrate media should be used to demonstrate the presence of the enzyme. The test is of value in identifying and classifying bacteria.

For more information: Fukushi et al. (1968*a, b*).

Hydrolysis of casein Casein is the chief protein present in milk. Some organisms are capable of hydrolyzing the protein to smaller units. This transformation is generally spoken of as peptonization. The enzyme involved is called casease.

The presence of casease may be demonstrated as follows: Agar is mixed with sufficient milk to give it an opalescent appearance. The organism to be tested is then streaked over the surface of the medium. If a culture of *B. subtilis* is used, clear zones will appear around each colony

Fig. 204 Hydrolysis of starch in the presence of bacteria. Left, Bacillus subtilis, no color with iodine surrounding the colonies, indicating hydrolysis of starch. Right, Escherichia coli, blue color with iodine, indicating absence of starch hydrolysis.

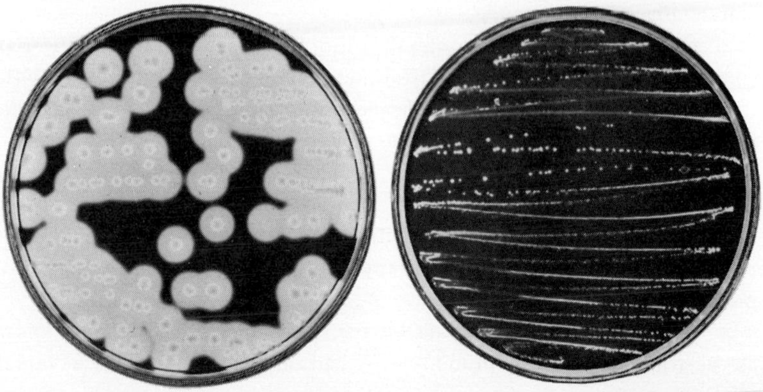

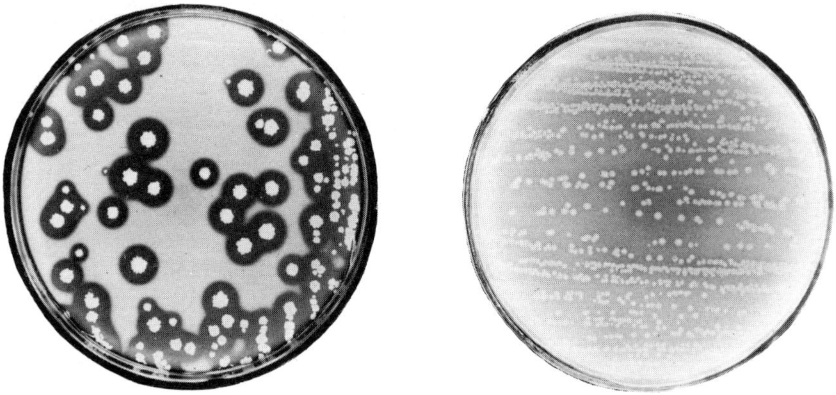

Fig. 205 Hydrolysis of casein. Left, milk agar plate streaked with Bacillus subtilis. Clear zones appear around the colonies, indicating hydrolysis of the casein. Right, same medium streaked with Escherichia coli. The casein is not hydrolyzed.

developing on the plate. This indicates that the casein has been digested to soluble compounds by the extracellular enzyme. On the other hand, *E. coli* does not produce an extracellular caseolytic enzyme and is unable to attack the protein (Fig. 205). The test is of value in identifying and classifying bacteria.

For more information: Johnson et al. (1967); May and Elliott (1968); Shieh and Ware (1968); Whiteside and Corpe (1969).

Commercial uses of enzymes

Enzymes are becoming increasingly more important in various industrial processes. They are obtained from all three divisions of the fungi, namely, molds, yeasts, and bacteria.

Commercially enzymes are produced by both surface and submerged fermentation methods (page 606). Strains of the same species vary considerably in the amounts of enzymes synthesized. For this reason strains are selected that have the capacity for the highest production of the particular enzymes desired.

Enzymes produced by microbial growth include one or several kinds of the following: amylase, catalase, cellulase, glucose oxidase, invertase, lactase, lipase, pectinase, protease, etc.

Lipases are used for (1) modifying milk fats for flavor, (2) improving whip of egg albumin, (3) waste disposal, (4) digestive aid. Cellulase is used for (1) tenderizing food products, (2) modifying cellulose wastes, (3) digestive aid. Catalase is used for (1) peroxide bleaching, (2) peroxide sterilization of milk for the manufacture of cheese. Glucose oxidase is used for (1) the removal of sugar from eggs, (2) the removal of oxygen from food products, (3) analytical reagent, (4) diagnostic aid in diabetes.

At the present time at least two enzymes produced by *B. subtilis*, namely protease and amylase, are being added to various soaps and detergents for their ability to digest proteins and carbohydrates in stains on soiled clothes. These enzymes can tolerate the high temperature of water used in washing machines.

For more information: Cayle et al. (1970); Underkofler (1966).

Intracellular enzymes The intracellular enzymes are concerned with the respiratory activities of bacteria. They are not secreted outside of cells and cannot, therefore, be demonstrated in culture filtrates. Such enzymes may be detected by employing living, disintegrated, or dissolved bacteria. Cells may be disintegrated by making use of various types of ball mills. Suspensions of pneumococci and other species, when mixed with bile, are dissolved. The disintegrated cells and lysates may be used for demonstrating the presence of various endoenzymes by incorporating such preparations with appropriate substrates.

Oxidizing and reducing enzymes Oxidations may occur by (1) the addition of oxygen, (2) the loss of hydrogen, or (3) the loss of electrons. Oxidation is the opposite of reduction. The oxidation of one compound is accompanied by the reduction of another.

Enzymes, coenzymes, and carriers involved in intracellular reactions are discussed in the chapter that follows.

References

Barman, T. E.: "Enzyme Handbook," New York, Springer-Verlag New York, Inc., 1969.

Bernhard, S. A.. "The Structure and Function of Enzymes," New York, W. A. Benjamin, Inc., 1968.

Bilezikian, J. P., et al.: Mechanism of tryptophanase induction in *Escherichia coli, J. Mol. Biol.*, **27**:495, 1967.

Boos, W., et al.: Untersuchungen zur induktion der lacenzyme. 1. Induktionswirkung und permeation, *Europ. J. Biochem.*, **1**:382, 1967.

Boyer, P. D., et al., eds.: "The Enzymes," New York, Academic Press, Inc., vols. I, II, 1970; vols. III, IV, V, 1971; VI, VII, 1972.

Bradley, S. G.: Loss of adaptive enzyme during storage. In "Culture Collections: Perspectives and Problems," Toronto, Canada, University of Toronto Press, 1963.

Cayle, T., et al.: Enzymes as detergent additives, *Wallerstein Lab. Commun.*, **33**:37, 1970.

Chaloupka, J.: Localization of proteases in cells of *Escherichia coli* and *Bacillus megaterium, Folia Microbiol.*, **6**:231, 1961.

——— and K. Veres: Formation of osmotically fragile rods by the action of lysozyme on *Bacillus megaterium* KM, *Folia Microbiol.*, **6**:379, 1961.

Chipman, D. M., and N. Sharon: Mechanism of lysozyme action, *Science*, **165**:454, 1969.

Dixon, M., and E. C. Webb: "Enzymes," New York, Academic Press, Inc., 1964.

Duerre, J. A., and E. Ribi: Enzymes released from *Escherichia coli* with the aid of a Servall Cell Fractionator, *Appl. Microbiol.,* **11**:467, 1963.

Fukumoto, J., et al.: Part I. Induced production of uricase by *Streptomyces* sp., *J. Agr. Chem. Soc. Japan,* **41**:540, 1967.

Fukushi, T., et al.: Changes in enzymatic activity and conformation during regeneration of native bacterial amylase from denatured form, *J. Biochem.,* **63**:409, 1968*a*; Regeneration of the native three-dimensional structure of *Bacillus subtilis* α-amylase and its formation in biological systems, *ibid.,* **64**:283, 1968*b*.

Gutfreund, H.: "An Introduction to the Study of Enzymes," Oxford, Blackwell Scientific Publications, Ltd., 1965.

——: Rapid reactions and the resolution of individual steps during enzyme catalysis, *Fractions,* no. 2, 1968.

Hartsell, S. E., and R. F. Acker: Lysozymic spheroplasts and the phage infectious process, *Proc. 2nd Intern. Symp. Fleming's Lysozyme,* Milan, Italy, April, 1961.

Hayashi, Shin-ichi, et al.: Tyrosinase aminotransferase, *J. Biol. Chem.,* **242**:3998, 1967.

Hendrich, F., and L. Pospíšil: Lysozyme, Czechoslovakia, *Acta Facultatis Medicae, Universitatis Brunensis,* 1971.

Heppel, L. A.: Selective release of enzymes from bacteria, *Science,* **156**:1451, 1967.

Hirs, C. H. W., et al.: The sequence of the amino acid residues in performic acid-oxidized ribonuclease, *J. Biol. Chem.,* **235**:633, 1960.

International Union of Biochemistry: "Commission on Enzymes," New York, Pergamon Press, 1961.

——: "Enzyme Nomenclature," Amsterdam, Elsevier Publishing Company, 1965.

IUPAC-IUB Commission on Biochemical Nomenclature (CBN): The nomenclature of multiple forms of enzymes, *Eur. J. Biochem.,* **24**:1, 1971.

Iwata, K., and T. Eda: Studies on the extracellular enzymes of staphylococci. 3, *Japan. J. Bact.,* **23**:465, 1968.

—— et al.: Studies on the extracellular enzymes of staphylococci, *Japan. J. Bact.,* **23**:700, 1968.

Johnson, G. G., et al.: The extracellular protease from *Pseudomonas aeruginosa* exhibiting elastase activity, *Can. J. Microbiol.,* **13**:711, 1967.

King, T. E., et al., eds.: "Oxidases and Related Redox Systems," vols. I, II, New York, John Wiley & Sons, Inc., 1965.

Kobayashi, T., et al.: II. Culture conditions in solid culture and some enzymatic properties of phytase of *Asp. terreus* Thom (No. 9A-1), *J. Ferment. Technol.,* **46**:863, 1968.

Koshland, D. E., Jr.: Correlation of structure and function in enzyme action, *Science,* **142**:1533, 1963.

Locke, D. M.: "Enzymes, the Agents of Life," New York, Crown Publishers, Inc., 1969.

Lwoff, A.: "Biological Order," Cambridge, Mass., The M.I.T. Press, 1962.

MacKelvie, R. M., et al.: Survival and intracellular changes of *Pseudomonas aeruginosa* during prolonged starvation, *Can. J. Microbiol.,* **14**:639, 1968.

Marr, A. G.: Enzyme localization in bacteria. In "Annual Review of Microbiology," 14:241, 1960.

May, B. K., and W. H. Elliott: Selective inhibition of extracellular enzyme synthesis by removal of cell wall from *Bacillus subtilis, Biochim. Biophys. Acta,* **166**: 532, 1968.

McClatchy, J. K., and E. D. Rosenblum: Induction of lactose utilization in *Staphylococcus aureus, J. Bact.,* **86**:1211, 1963.

McDonald, I. J.: Location of proteinase, peptidase, catalase, and NADH oxidase in cells of *Staphylococcus lactis, Can. J. Microbiol.,* **10**:197, 1964.

Morino, Y., and E. E. Snell: A kinetic study of the reaction mechanism of tryptophanase-catalyzed reactions, *J. Biol. Chem.,* **242**:2793, 1967*a.*

—— and ——: The relation of spectral changes and tritium exchange reactions to the mechanism of tryptophanase-catalyzed reactions, *J. Biol. Chem.,* **242**: 2800, 1967*b*.

Moss, D. W.: "Enzymes," London, Oliver & Boyd Ltd., 1968.

Munkres, M., and A. Wachtel: Histochemical localization of phosphatases in *Mycoplasma gallisepticum, J. Bact.,* **93**:1096, 1967.

Neu, H. C., and J. Chou: Release of surface enzymes in *Enterobacteriaceae* by osmotic shock, *J. Bact.,* **94**:1934, 1967.

Niemann, C.: Alpha-chymotrypsin and the nature of enzyme catalysis, *Science,* **143**:1287, 1964.

Pangborn, J., et al.: Localization of respiratory enzymes in intracytoplasmic membranes of *Azotobacter agilis, J. Bact.,* **84**:669, 1962.

Pardee, A. B., and J. R. Beckwith: Genetic determination of constitutive enzyme levels, *Biochim. Biophys. Acta,* **60**:452, 1962.

Pastan, I., and R. Perlman: Cyclic adenosine monophosphate in bacteria, *Science,* **169**:339, 1970.

Phillips, D. C.: The three-dimensional structure of an enzyme molecule, *Sci. A.,* **215**:78, 1966.

Shieh, T. R., and J. H. Ware: Survey of microorganisms for the production of extracellular phytase, *Appl. Microbiol.,* **16**:1348, 1968.

Sistrom, W. R.: "Microbial Life," New York, Holt, Rinehart and Winston, Inc., 1969.

Smith, R. A., and I. E. Liener: Amino acid sequences involving the histidine residues of porcine trypsin, *J. Biol. Chem.,* **242**:4039, 1967.

Suzuki, Y., and L. J. Rode: Effect of lysozyme on resting spores of *Bacillus megaterium, J. Bact.,* **98**:238, 1969.

Underkofler, L. A.: Manufacture and uses of industrial microbial enzymes, *Chem. Eng. Prog. Symp. Series,* **62**:11, 1966.

Vogel, R., et al.: "Natural Proteinase Inhibitors," New York, Academic Press, Inc., 1968.

Wang, J. H.: Facilitated proton transfer in enzyme catalysis, *Science,* **161**:328, 1968.

Webb, J. L.: "Enzyme and Metabolic Inhibitors," New York, Academic Press, Inc., 1963.

Weibull, C. H., et al.: Localization of enzymes in *Bacillus megaterium*, strain M, *J. Gen. Microbiol.,* **20**:519, 1959.

Welker, N. E., and L. L. Campbell: Induced biosynthesis of α-amylase by growing cultures of *Bacillus stearothermophilus, J. Bact.,* **86**:1196, 1963*a*.

—— and —— : De novo synthesis of α-amylase by *Bacillus stearothermophilus, J. Bact.,* **86**:1202, 1963*b*.

Whiteside, T. L., and W. A. Corpe: Extracellular enzymes produced by a *Pseudomonas* sp. and their effect on cell envelopes of *Chromobacterium violaceum, Can. J. Microbiol.,* **15**:81, 1969.

Yamada, K., et al.: Phytase from *Aspergillus terreus.* I., *Agr. Biol. Chem.,* **32**:1275, 1968.

12 *Respiration of bacteria*

In higher organisms, the term respiration is defined as the intake of oxygen and the release of carbon dioxide. Since there are bacteria which can live and multiply in the absence of oxygen, respiration cannot be defined in the above manner. Respiration in its broadest meaning is concerned with the oxidation of metabolites, resulting in the release of energy to the organisms. Only in this sense can the term be used in bacteriology.

During the early years of bacteriology it was believed that free oxygen was necessary for all organisms and that life was not possible in the complete absence of this element. Pasteur demonstrated very early in his studies that such a statement, which was then considered fundamental, had to be abandoned. He showed that there were bacteria that could not grow in the presence of free oxygen. He classified bacteria as aerobic, anaerobic, facultative aerobic, and facultative anaerobic, depending upon their reaction to free oxygen.

Pasteur considered all fermentations, i.e., the action of organisms on carbohydrates, to be anaerobic. The organisms utilized the free oxygen of the medium and released carbon dioxide (CO_2). The escaping stream of CO_2 aided in preventing the entrance of more free oxygen from the air. Even though the media were exposed to air, conditions well below the surface were anaerobic. Yet the cells produced oxidations in the absence of free oxygen.

Oxidizing-reducing activities of bacteria

Biological oxidations may be defined as the withdrawal of electrons from a substance with or without the addition of oxygen or elements analogous to oxygen, or as the withdrawal of electrons with or without the withdrawal of hydrogen or elements analogous to hydrogen. It is believed that oxygen acts as a hydrogen acceptor to form H_2O or H_2O_2. Oxygen as such does not enter the molecule of the substrate. Compounds which are active hydrogen acceptors take the place of oxygen in various oxidations, being themselves reduced in the reactions. Bacterial oxidations and reductions are associated phenomena and must be studied together.

Almost all oxidations in living cells can be explained on the basis of the removal of hydrogen rather than of the addition of oxygen. Very few cases are known in which oxygen is used directly as the oxidizing agent. Hydrogen acceptors are necessary for the reactions to take place. Oxidations that occur in this manner are called dehydrogenations. Exceptions include the oxidation of aldehydes and purine bases. In these instances, a preliminary hydration occurs prior to removal of hydrogen:

1. Oxidation without preliminary hydration:

$$\underset{\text{Alcohol}}{R\cdot CH_2OH} \longrightarrow \underset{\text{Aldehyde}}{R\cdot CHO} + 2H + \text{hydrogen acceptor}$$

2. Oxidation with preliminary hydration:

$$\underset{\text{Aldehyde}}{R\cdot CHO + H_2O} \longrightarrow \underset{\text{Aldehyde hydrate}}{R\cdot C(H)(OH)OH}$$

$$\underset{\text{Aldehyde hydrate}}{R\cdot C(H)(OH)OH} \longrightarrow \underset{\text{Acid}}{R\cdot COOH} + 2H + \text{hydrogen acceptor}$$

Bacteria are capable of reducing such stains as methylene blue, neutral red, and litmus to their colorless forms. The speed of decolorization is proportional to the bacterial population. A test based on this principle is the determination of the approximate bacterial count of milk by noting the methylene blue reduction time.

Reduction of methylene blue Methylene blue acts as a hydrogen acceptor and becomes reduced to the colorless form. If air or oxygen is bubbled through the medium containing the reduced compound, the blue color is restored. The reaction may be represented as follows:

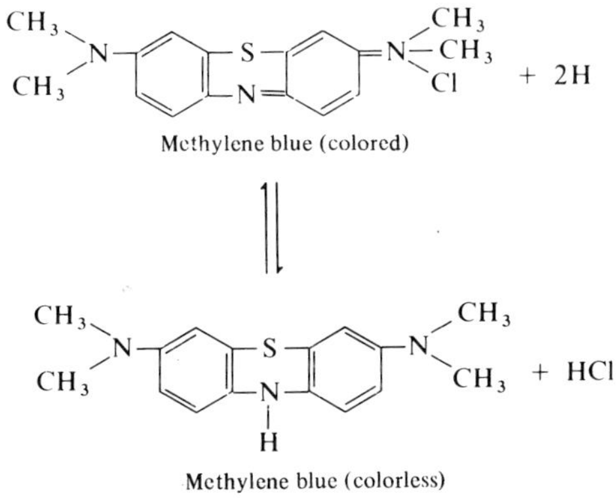

One atom of hydrogen is taken up by the double-bonded nitrogen, converting the stain into the reduced or colorless form. The reaction is easily reversible from one form to the other.

By means of the methylene blue technique, it is possible to demonstrate the presence of oxidoreductases which are capable of oxidizing compounds by the removal of hydrogen. The speed of reduction of the stain is an indication of the rate of oxidation taking place. The reaction occurs in an anaerobic environment. The presence of air results in the reoxidation of the reduced methylene blue to the colored form.

Tests for the presence of oxidoreductases in bacteria may be performed by (1) centrifugating a broth culture of the test organism, (2) removing the supernate by aspiration, and (3) resuspending the compacted bacteria in water or saline and aerating for 1 hr to remove oxidizable constituents. The bacteria are now ready for use.

The following components are added to a sterile vacuum tube:

1. Buffer solution of the proper reaction.
2. Sterile solution of the substrate under examination.
3. Suspension of bacteria prepared as above.
4. Solution of methylene blue.

The tube is evacuated of air and incubated at the appropriate temperature. Decolorization of the stain indicates the presence of the oxidoreductase capable of oxidizing the substrate.

A series of tubes may be assembled, each containing a different substrate. Decolorization of the stain indicates the presence of oxidoreductases capable of oxidizing the substrates. In this manner the presence of a large number of oxidoreductases may be demonstrated in bacteria.

Oxidoreductases

Some oxidoreductases produce oxidations by the removal of hydrogen; others by the addition of oxygen.

Hydrogen-transferring enzymes These enzymes function with the following hydrogen acceptors:

1. Acting on the CH–OH groups of donors, with NAD or NADP as acceptor. (*a*) Alcohol: NAD oxidoreductase, (*b*) alcohol: NADP oxidoreductase, (*c*) acetoin: NAD oxidoreductase, (*d*) glycerol: NAD oxidoreductase, (*e*) L-lactate: NAD oxidoreductase, (*f*) D-lactate: NAD oxidoreductase, (*g*) L-malate: NAD oxidoreductase (decarboxylating), (*h*) L-malate: NADP oxidoreductase (decarboxylating), (*i*) D-glucose-6-phosphate: NADP oxidoreductase, (*j*) 6-phospho-D-gluconate: NADP oxidoreductase (decarboxylating).
2. Acting on the CH–OH groups of donors, with a cytochrome as acceptor. (*a*) L-Lactate: ferricytochrome c oxidoreductase, (*b*) D-lactate: ferricytochrome c oxidoreductase, (*c*) D-mannitol: ferricytochrome oxidoreductase.
3. Acting on the aldehyde or keto-group of donors, with NAD or NADP as acceptor. (*a*) Formaldehyde: NAD oxidoreductase, (*b*) aldehyde: NAD oxidoreductase, (*c*) aldehyde: NADP oxidoreductase, (*d*) D-glyceraldehyde-3-phosphate: NADP oxidoreductase.

4. Acting on the aldehyde or keto-group of donors, with a cytochrome as an acceptor. (*a*) Pyruvate: ferricytochrome b_1 oxidoreductase.
5. Acting on the CH–CH group of donors, with NAD or NADP as acceptor. (*a*) 4,5-Dihydro-uracil: NAD oxidoreductase, (*b*) 4,5-dihydro-uracil: NADP oxidoreductase.
6. Acting on the $CH-NH_2$ group of donors, with NAD or NADP as acceptor. (*a*) L-Alanine: NAD oxidoreductase (deaminating), (*b*) L-glutamate: NADP oxidoreductase (deaminating), (*c*) L-amino acid: NAD oxidoreductase (deaminating).
7. Acting on the C–NH group of donors, with NAD or NADP as acceptor. (*a*) L-Proline: NAD(P) 2-oxidoreductase, (*b*) L-proline: NAD(P) 5-oxidoreductase.
8. Acting on reduced NAD or NADP, with a cytochrome as acceptor. (*a*) Reduced NAD; ferricytochrome b_5 oxidoreductase.
9. Acting on reduced NAD or NADP, with a disulfide compound as acceptor. (*a*) Reduced NAD: L-cystine oxidoreductase, (*b*) reduced NAD(P): oxidized glutathione oxidoreductase.
10. Acting on reduced NAD or NADP, with a nitrogenous group as acceptor. (*a*) Reduced NAD(P): nitrate oxidoreductase, (*b*) reduced NAD(P): nitrite oxidoreductase, (*c*) reduced NAD: hyponitrite oxidoreductase.
11. Acting on sulfur groups of donors, with NAD or NADP as acceptor. (*a*) Hydrogen sulfide: NADP oxidoreductase.
12. Acting on haem groups of donors, with a nitrogenous group as acceptor. (*a*) Ferrocytochrome: nitrate oxidoreductase.

The above oxidoreductases produce oxidations by the removal of hydrogen which is, in turn, transferred to reducible substances. Certain stains, e.g., methylene blue, may be used in place of naturally occurring hydrogen acceptors. Since the metabolites become oxidized by the removal of hydrogen, this type of oxidation is spoken of as a dehydrogenation.

Almost all oxidations occurring in living cells involve dehydrogenations. The hydrogen is transferred to appropriate acceptors. Eventually all reducible compounds will have their hydrogen-accepting bonds satisfied after which all cellular oxidations will cease. In actual practice, however, a cell contains a chain of acceptors, the hydrogen being passed from one to the other. Finally the last member of the chain passes its hydrogen to an intracellular pigment named cytochrome c. This pigment in the reduced form can be reoxidized by molecular oyxgen in the presence of the enzyme cytochrome c: O_2 oxidoreductase. Oxygen enters the picture only at the end of the chain and is reduced to hydrogen peroxide (H_2O_2). The purpose of the oxygen is to keep cytochrome c in the oxidized condition so that the oxidation of metabolites may be a continuous process.

This may be represented as follows:

Metabolite ⟶ oxidoreductase ⟶ NAD
⟶ flavoproteins ⟶ cytochromes ⟶ O_2

Oxygen-adding enzymes These oxidoreductases function with oxygen as the acceptor of hydrogen. The oxygen is reduced to H_2O_2. They differ from the hydrogen-transferring oxidoreductases in that they transfer hydrogen directly to oxygen to form H_2O_2 (a few form H_2O instead of H_2O_2), whereas the latter transfer hydrogen to acceptors other than oxygen.

Some oxygen-adding oxidoreductases are the following:

1. Those that act on the CH–OH group of donors, with oxygen as acceptor. (*a*) L-lactate: oxygen oxidoreductase, (*b*) L-malate: oxygen oxidoreductase, (*c*) β-D-glucose: oxygen oxidoreductase.
2. Acting on the aldehyde or keto-group of donors, with oxygen as acceptor. (*a*) Aldehyde: oxygen oxidoreductase, (*b*) xanthine: oxygen oxidoreductase.
3. Acting on the $CH{-}NH_2$ group of donors, with oxygen as acceptor. (*a*) L-amino acid: oxygen oxidoreductase (deaminating), (*b*) D-amino acid: oxygen oxidoreductase (deaminating), (*c*) D-aspartate: oxygen oxidoreductase (deaminating).
4. Those that act on sulfur groups of donors, with oxygen as acceptor. (*a*) Sulfite: oxygen oxidoreductase.
5. Acting on haem groups of donors, with oxygen as acceptor. (*a*) Ferrocytochrome c: oxygen oxidoreductase.
6. Acting on diphenols and related substances as donors, with oxygen as acceptor. (*a*) L-ascorbate: oxygen oxidoreductase.
7. Those that act on hydrogen peroxide as acceptor. (*a*) Reduced NAD: hydrogen peroxide oxidoreductase, (*b*) reduced NADP: hydrogen peroxide oxidoreductase, (*c*) hydrogen peroxide: hydrogen peroxide oxidoreductase, (*d*) glutathione: hydrogen peroxide oxidoreductase, (*e*) donor: hydrogen peroxide oxidoreductase.
8. Acting on single donors with incorporation of oxygen (oxygenases). (*a*) *Meso*-Inositol: oxygen oxidoreductase, (*b*) L-tryptophan: oxygen oxidoreductase.

H_2O_2: H_2O_2 oxidoreductase This is the new name for the enzyme catalase.

Most bacteria produce H_2O_2 in the presence of free oxygen. Since the compound is toxic to living cells, its destruction is of great importance. The accumulation of H_2O_2 in cultures is controlled by two factors: (1) bacterial catalase and (2) the degree of sensitiveness of the organisms to the compound.

Catalase is capable of decomposing H_2O_2 to water and molecular oxygen, according to the equation

$$2H_2O_2 + \text{catalase} \longrightarrow 2H_2O + O_2$$

This is sometimes referred to as the catalatic reaction.

The presence of H_2O_2 was first detected in cultures of the pneumococcus, an organism that does not produce catalase and is only moderately sensitive to peroxide. Organisms that do not produce catalase

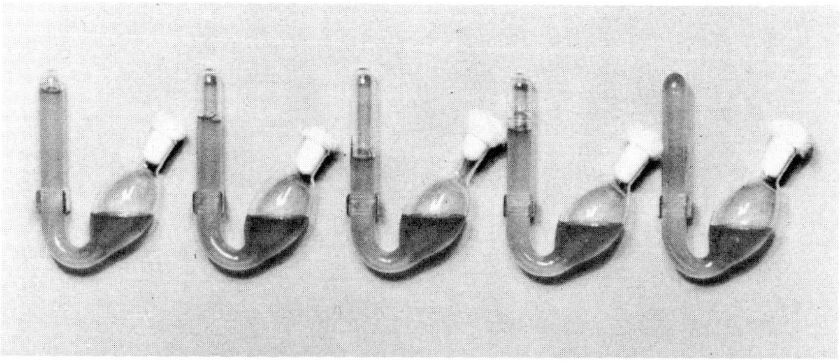

Fig. 206 Reduction of hydrogen peroxide with liberation of oxygen. From left to right: Bacillus subtilis, Staphylococcus aureus, Pseudomonas fluorescens, Escherichia coli, and control. B. subtilis shows slight decomposition, whereas P. fluorescens shows vigorous decomposition. Note absence of gas in the control.

may be protected by being cultivated with certain plant or animal tissues, or with other organisms capable of producing the enzyme.

Catalase is produced by many bacteria. Some produce more of it than others. It is present in largest amounts in the strictly aerobic bacteria. On the other hand, the enzyme is not produced by the obligately anaerobic bacteria because they would have no need for it.

The concentration of catalase may be determined by adding H_2O_2 to bacterial cultures and noting the amount decomposed, or by adding H_2O_2 to bacterial cultures in fermentation tubes and measuring the volumes of oxygen gas evolved (Fig. 206).

For more information: Clayton (1959).

Donor: H_2O_2 oxidoreductase This enzyme is commonly known as peroxidase.

The function of peroxidase is to transfer oxygen from peroxides to oxidizable substances according to the equation

$$H_2O_2 + \text{organic molecule} + \text{peroxidase} \longrightarrow H_2O + \text{oxidized organic molecule}$$

This is known as the peroxidatic reaction.

Peroxidase does not decompose a peroxide in the absence of an oxidizable substance. In this respect it differs from catalase which decomposes H_2O_2 in the absence of an oxygen acceptor. This would indicate that catalase should be classed as a splitting enzyme, and peroxidase as a transferring enzyme.

A large number of compounds may function as oxygen acceptors, thereby permitting the peroxidase to act. Among these may be mentioned glutathione and cytochrome c. Potassium cyanide completely inhibits the action of the enzyme, whereas carbon monoxide has no effect.

The oxidase is widely distributed in nature, being present in vegetables, fruits, many sprouts, sugar beets, wheat flour, bran, brewer's yeast, insects, birds, animals, aerobic bacteria, milk, potatoes, etc. Horseradish is one of the richest sources of peroxidase and is frequently employed for the preparation of the enzyme.

Peroxidase is very resistant to heat. It is destroyed at 98°C in 10 min, but on standing, as much as 25 percent of the original activity is restored.

L-Ascorbate: O_2 oxidoreductase This enzyme is better known as ascorbic acid oxidase.

The enzyme is capable of oxidizing ascorbic acid or vitamin C to dehydroascorbic acid (page 363):

$$2 \text{ L-Ascorbic acid} + O_2 \rightleftharpoons 2 \text{ dehydroascorbic acid} + 2H_2O$$

The oxidase is widely distributed in the plant kingdom, being present in a large number of fruits and vegetables. It is not present in animal tissues, where ascorbic acid may be oxidized through the cytochrome system.

Ascorbic acid is readily oxidized to dehydroascorbic acid. Hydrogen peroxide is not produced in the reaction. The oxidized form still displays some antiscorbutic activity which appears to parallel to a remarkable degree its reducing capacity.

It is generally believed that ascorbic acid functions as a hydrogen carrier. It is found in all animal tissues in amounts compatible with hydrogen transport or coenzyme function. Some believe that glutathione is the chief protective substance in cells for maintaining ascorbic acid in the reduced form. It guards the vitamin against irreversible oxidation with a loss of antiscorbutic activity. Glutathione added to a mixture of ascorbic acid and its oxidase maintains the vitamin in the reduced or active form.

Cytochromes Cytochromes are respiratory pigments which were discovered by MacMunn (1886) but were forgotten until 1925, when they were rediscovered by Keilin. The latter worker named the compounds cytochrome, which means cellular pigments. They are concerned in the later stages of many biological oxidations. They become reduced by $NADH_2$ under the action of flavoproteins and the reduced cytochromes are reoxidized in the presence of cytochrome oxidase.

The cytochromes are widely distributed in nature, being found in higher plants and animals, birds, insects, mollusks, crustacea, bacteria, yeasts, molds, etc. They are absent from most anaerobic bacteria. Exceptions appear to be certain anaerobic photosynthetic sulfur bacteria. It is generally believed that the more aerobic the bacteria, the more fully developed will be the cytochrome system.

Cytochrome is not a single compound but consists of many components. They fall into four groups which have been designated cytochromes a, b, c, and d. Separation of the cytochromes is based on (1) the position of the α-band of pyridine ferrohaemochrome in alkali, and (2) the ether solubility of the product formed after treatment of cytochrome with acetone-hydrochloric acid (Table 12).

Table 12 Classification of cytochromes

Group	*α-Band of pyridine ferrohaemochrome in alkali (nm)*	*Solubility of product in ether after treatment of cytochrome with acetone-HCl*
a	580–590	Soluble
b	556–558	Soluble
c	549–551	Insoluble
d	600–620	Soluble

Cytochrome c occurs more abundantly in nature than the other components. It is freely soluble in water and was the first to be extracted and purified. For this reason it has been studied in greater detail than any of the others. The present knowledge of cytochromes is based largely on this component.

Chemically cytochromes are haemoproteins having structures similar to haemoglobin. They act as electron and/or hydrogen carriers by virtue of a reversible valency change of their haem iron. All the cytochromes can undergo oxidation and reduction, and most of them serve as carriers of hydrogen. Under normal conditions the cytochromes which function as carriers are present in cells in the oxidized or partly reduced form. During periods of great activity the cytochromes are partly oxidized; during periods of inactivity they are again reduced.

Cytochrome c does not react directly with oxygen. The reduced form is oxidized only by ferrocytochrome c: oxygen oxidoreductase with which it is closely associated in cells.

Ferrocytochrome c: oxygen oxidoreductase The common or trivial name of this enzyme is cytochrome oxidase.

Most cytochromes are believed to function as carriers of hydrogen in the complicated bucket brigade which leads to the transfer of hydrogen from metabolites to free oxygen. Cytochrome oxidase plays the final role in that it brings about the oxidation, by molecular oxygen, of the reduced forms of cytochromes, which leads to the formation of water.

The autoxidizable flavoproteins reduce molecular oxygen to hydrogen peroxide, whereas the cytochrome system converts free oxygen to water. A satisfactory explanation for this difference has not been given.

Cytochrome oxidase has been called indophenolase and indophenol oxidase. The enzyme is widely distributed in nature, being present in higher plants and animals, yeasts, molds, algae, aerobic bacteria, etc.

Cytochrome oxidase is capable of oxidizing *p*-phenylenediamine to indophenol. The presence of the enzyme is easily detected by pouring a solution of *p*-phenylenediamine over the surface of an agar plate containing bacterial colonies. In the presence of the enzyme, the *p*-phenylenediamine is oxidized to indophenol, resulting in the appearance of a blue color. The oxidase is believed to oxidize phenols, amines, etc., not directly, but by oxidizing cytochrome c. The cytochrome c then oxidizes the phenols and amines.

The oxidation of reduced cytochrome c is the only property that has been definitely assigned to cytochrome oxidase. This may be illustrated in the following reaction:

Metabolite · 2H + oxidoreductase
+ cytochrome c ⟶ reduced cytochrome c

Reduced cytochrome c + cytochrome oxidase + O_2 ⟶
oxidized cytochrome c + H_2O

The flavoproteins are believed to be the reducing agents which bridge the gap between NAD or NADP and the cytochromes (see page 400). The reduced cytochromes are then reoxidized by cytochrome oxidase, which transfers the hydrogen to free oxygen to produce water.

For more information: Ariji et al. (1968); Baldwin (1967); Hayaishi (1966); Hayaishi and Nozaki (1969); International Union of Biochemistry (1965); Keilin (1966); Okunuki et al. (1968); Saunders et al. (1964); Wagner and Folkers (1964); Cusanovich (1971).

Coenzymes

Enzymes may be defined as organic catalysts elaborated by living cells and capable of functioning independently of the cells. They are heat-labile and nondialyzable and have high molecular weights. Coenzymes are also catalytic agents produced by living cells, and are necessary in many enzymatic reactions, but they differ from enzymes in being heat-stable, dialyzable, and having smaller molecular weights.

Nicotinamide-adenine dinucleotide This coenzyme is abbreviated to NAD. It is the new name of diphosphopyridine nucleotide (DPN), also known as codehydrogenase I and coenzyme I.

NAD consists of one molecule of β-nicotinamide, one of adenine, two of pentose, and two of phosphoric acid. The compound is capable of alternate oxidation and reduction and functions as a carrier of hydrogen. It has the following structural formula:

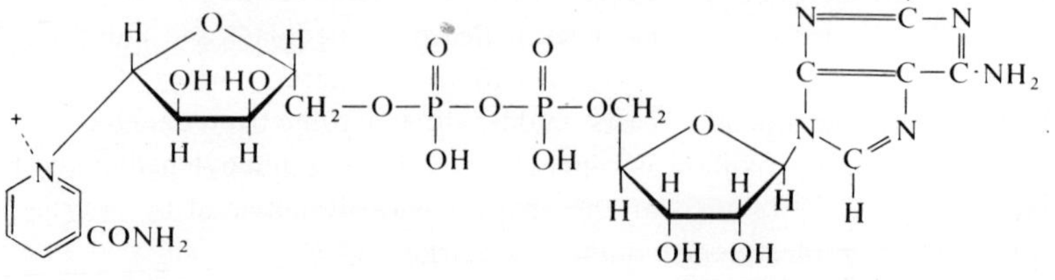

NAD is universally present in living cells. It is present in high concentration in baker's yeast. It functions as a coenzyme in many oxidoreductase reactions. The oxidoreductase transfers hydrogen from the metabolite to NAD. The pyridine ring of NAD accepts an atom of hydrogen, converting a strongly basic quaternary nitrogen to a weakly basic tertiary nitrogen atom:

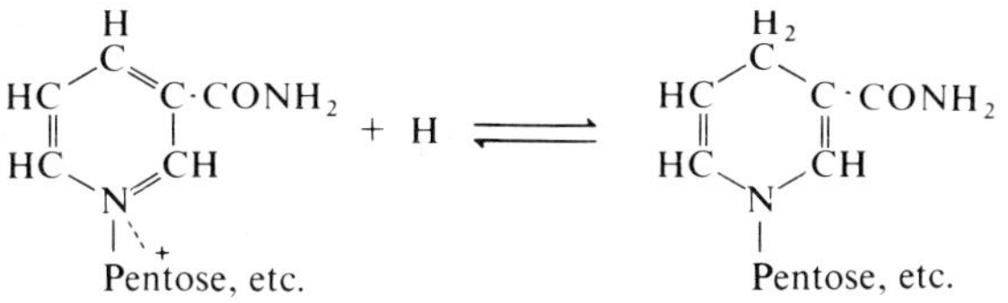

Nicotinamide-adenine dinucleotide phosphate This coenzyme is abbreviated to NADP. It is the new name of triphosphopyridine nucleotide (TPN), also known as codehydrogenase II or coenzyme II. Like NAD, it is widely distributed in living cells. It functions in the same manner as NAD and differs from it only in containing one more phosphate radical, which is joined to the D-ribose. NADP has the following structural formula:

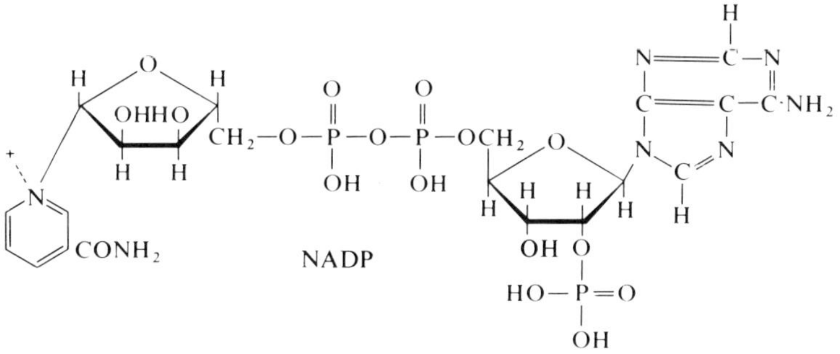

The two coenzymes are generally not interchangeable. Some oxidoreductases require NAD; others must have NADP. Their function is to transport hydrogen from metabolites to other carriers, usually the flavin nucleotides, and become reoxidized again to accept more hydrogen.

This may be illustrated in the following series of reactions in which glucose becomes oxidized to gluconic acid:

$$\text{Glucose} + H_3PO_4 \rightleftharpoons \text{glucose-6-phosphate}$$

$$\text{Glucose-6-phosphate} + \text{NADP (ox.)} \rightleftharpoons \text{gluconic acid-6-phosphate} + \text{NADP (red.)}$$

$$\text{NADP (red.)} + \text{flavin nucleotide (ox.)} \rightleftharpoons \text{NADP (ox.)} + \text{flavin nucleotide (red.)}$$

$$\text{Flavin nucleotide (red.)} + O_2 \rightleftharpoons \text{flavin nucleotide (ox.)} + H_2O_2$$

Flavin mononucleotide (FMN) This coenzyme is the phosphate ester of riboflavin (vitamin B_2). Chemically it is isoalloxazine ribitol phosphate (riboflavin-5-phosphate):

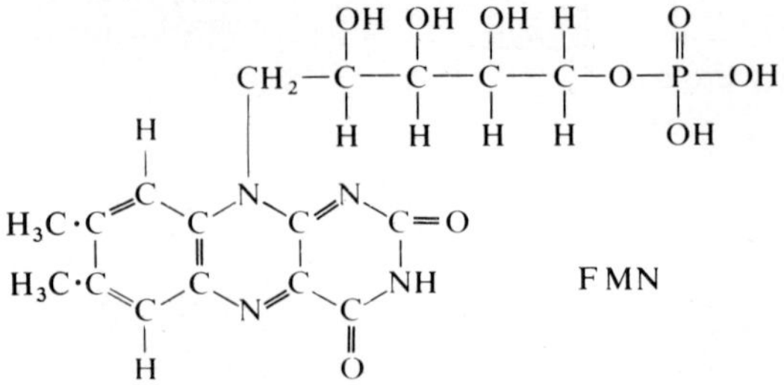

FMN is found in a large number of systems which function in carbohydrate metabolism and serves as an agent for the transfer of hydrogen from NAD or NADP to the cytochromes.

Flavin-adenine dinucleotide (FAD) Like FMN, FAD serves as an agent in the transfer of hydrogen from NAD or NADP to the cytochromes.

It has the following structure:

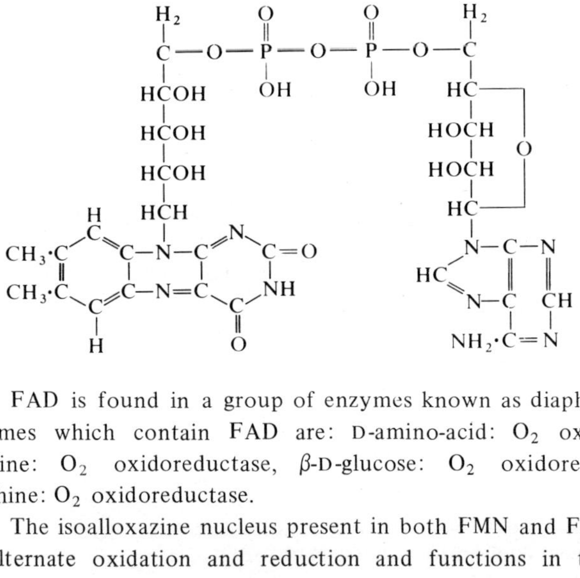

FAD is found in a group of enzymes known as diaphorases. Some enzymes which contain FAD are: D-amino-acid: O_2 oxidoreductase, diamine: O_2 oxidoreductase, β-D-glucose: O_2 oxidoreductase, and xanthine: O_2 oxidoreductase.

The isoalloxazine nucleus present in both FMN and FAD is capable of alternate oxidation and reduction and functions in the following manner in the transfer of hydrogen.

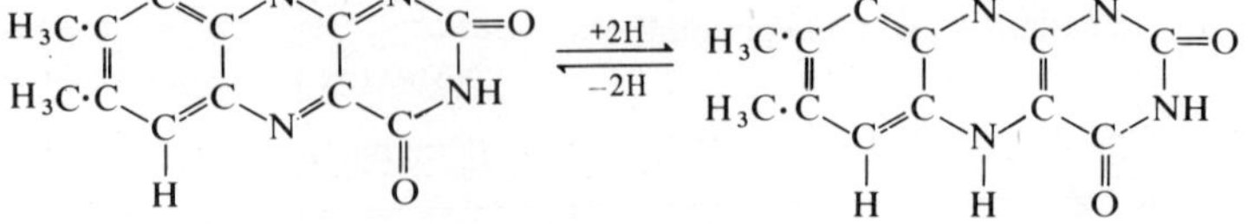

Some typical reactions catalyzed by FMN and FAD are given in Table 13.

Table 13 Some reactions catalyzed by FMN and FAD

Enzyme	*Coenzyme*	*Substrate*	*Products produced*
Type: $R\cdot CHNHR'\cdot COOH$ + flavoprotein (ox.) + $H_2O \rightleftharpoons R\cdot CO\cdot COOH + H_2NR'$ + flavoprotein (red.)			
D-*Amino acid* : O_2 *oxidoreductase*	FAD	D-Amino acids + H_2O + O_2	Keto acids + NH_3 + H_2O_2
L-*Amino acid* : O_2 *oxidoreductase*	FMN	L-Amino acids + H_2O + O_2	Keto acids + NH_3 + H_2O_2
Type: NAD (red.) / NADP (red.) + flavoprotein (ox.) $\rightleftharpoons$ NAD (ox.) / NADP (ox.) + flavoprotein (red.)			
NAD (red.): cytochrome c oxidoreductase	FAD	NAD (red.) + cytochrome c (ox.)	NAD (ox.) + cytochrome c (red.)
NADP (red.):cytochrome c oxidoreductase	FAD or FMN	NADP (red.) + cytochrome c (ox.)	NADP (ox.) + cytochrome c (red.)
Miscellaneous			
NAD (red.): nitrate oxidoreductase	FAD	NAD (red.) + nitrate	NAD (ox.) + nitrite + H_2O
NADP (red.): nitrate oxidoreductase	FAD	NADP (red.) + nitrate	NADP (ox.) + nitrite + H_2O
NAD (red.): nitrite oxidoreductase	FAD	NAD (red.) + nitrite	NAD (ox.) + 2NO + $2H_2O$
NADP (red.): nitrite oxidoreductase	FAD	NADP (red.) + nitrite	NADP (ox.) + 2NO + $2H_2O$

Thiamine pyrophosphate (cocarboxylase) The enzyme carboxylase catalyzes the decarboxylation of pyruvic acid to acetaldehyde and CO_2. The coenzyme thiamine pyrophosphate or cocarboxylase is necessary for the activity of the enzyme:

$$\underset{\text{Pyruvic acid}}{CH_3 \cdot CO \cdot COOH} + \text{carboxylase} + \text{cocarboxylase} \longrightarrow \underset{\text{Acetaldehyde}}{CH_3 \cdot CHO} + CO_2$$

Thiamine pyrophosphate is widely distributed in living cells. It is composed of one molecule of thiamine (vitamin B_1) and two of phosphoric acid:

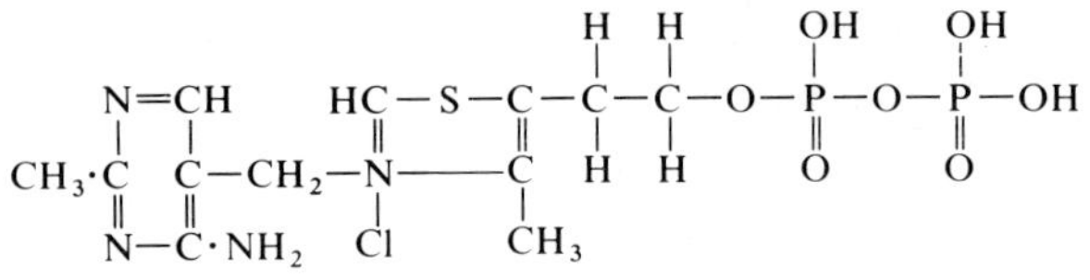

Coenzyme A (CoA) This coenzyme functions as a carrier of the acetyl group. It participates in the acetylation of choline and in the condensation of acetate with oxaloacetate to form citrate.

CoA is composed of adenylic acid, pantothenic acid, and thioethanolamine:

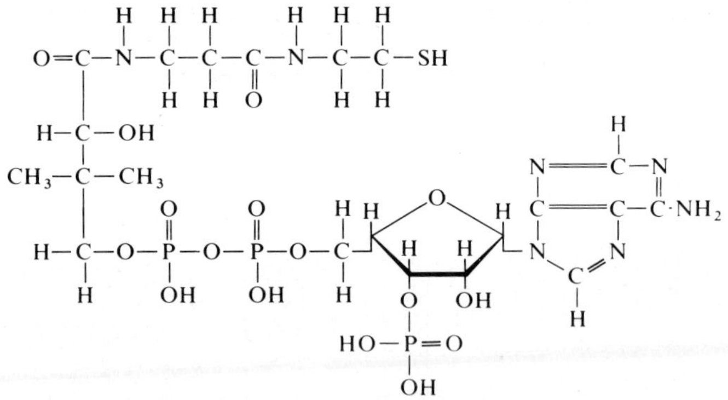

Methylmalonyl coenzyme A Methylmalonyl coenzyme A is an intermediate in the metabolism of propionic acid to succinic acid. The reaction sequence proceeds through the stages (1) propionyl coenzyme A → (2) methylmalonyl coenzyme A → (3) succinyl coenzyme A. The conversion of (1) through (2) is mediated by the vitamin biotin; the isomerization of (2) through (3) is mediated by dimethylbenzimidazolyl cobamide coenzyme.

It has the following formula:

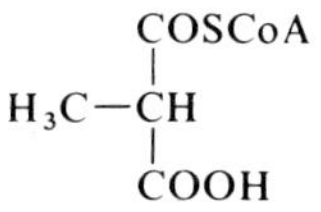

Methylmalonyl CoA

Coenzyme Q (ubiquinone) There exists a family of quinone coenzymes designated Q_6 through Q_{10}. Coenzyme Q_{10} was first isolated from beef heart mitochondria. Coenzymes Q_6 through Q_9 have been isolated from microorganisms. The number of isoprenoid units varies with the source of the coenzyme. The substance in beef heart has 10 units (Q_{10}) and coenzymes isolated from microorganisms have 9, 8, 7, and 6 isoprenoid units.

Coenzyme Q has this structure:

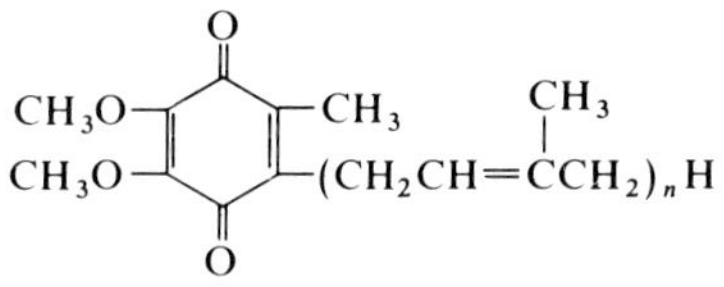

Coenzyme Q
n = 6, 7, 8, 9, or 10 isoprenoid units

Coenzyme Q is widely distributed in animal and plant cells. It is a constituent of mitochondria where it occupies a key position in the respiratory process. It functions as an electron carrier in citric acid cycle oxidations coupled to phosphorylation.

Adenosine diphosphate (ADP), adenosine triphosphate (ATP) These are phosphate-transferring coenzymes having the following structures:

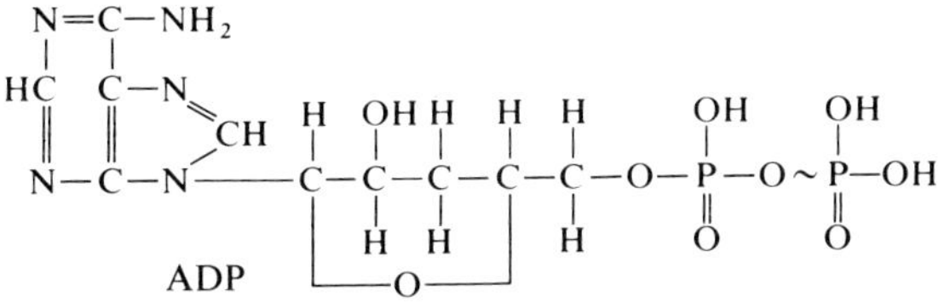

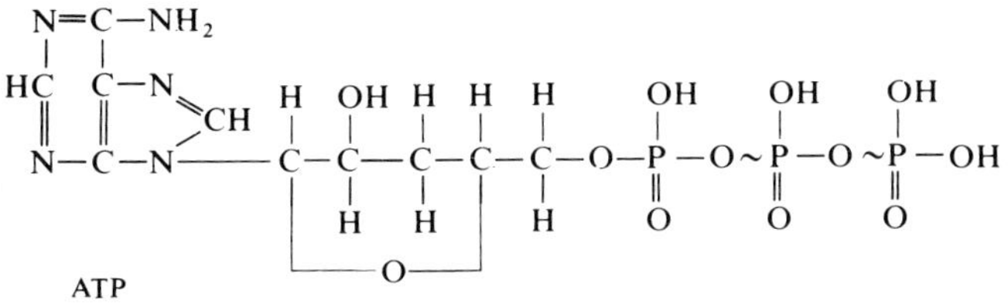

ATP

ATP is the universal fuel of the living cell. It contains two high-energy phosphate bonds ($\sim$) and each stores about 12,000 calories and releases about 7,500 calories when broken.

ATP is produced by two series of reactions: (1) an aldehyde reacts with an inorganic phosphate to give hydrogen and an acid phosphate; and (2) the acid phosphate reacts with ADP to give an organic acid and ATP:

$$R-CHO + HO-\overset{\overset{\displaystyle OH}{|}}{\underset{\underset{\displaystyle O}{\|}}{P}}-OH \longrightarrow 2H + R-\overset{\overset{\displaystyle }{}}{\underset{\underset{\displaystyle O}{\|}}{C}}-O-\overset{\overset{\displaystyle OH}{|}}{\underset{\underset{\displaystyle O}{\|}}{P}}-OH + H_2O$$

$$R-\overset{\overset{\displaystyle }{}}{\underset{\underset{\displaystyle O}{\|}}{C}}-O-\overset{\overset{\displaystyle OH}{|}}{\underset{\underset{\displaystyle O}{\|}}{P}}-OH + ADP + H_2O \longrightarrow R-\overset{\overset{\displaystyle OH}{|}}{\underset{\underset{\displaystyle O}{\|}}{C}}-OH + ATP$$

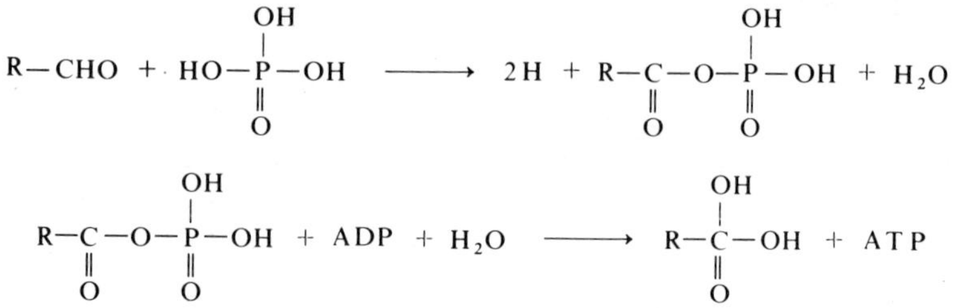

Acid phosphate

R denotes radicals, or side groups.

For more information: Baldwin (1967); Galivan and Allen (1968*a, b*); Gel'man et al. (1967); International Union of Biochemistry (1965); Knook and Planta (1971); Singer (1968); Thiele and Hoffmann (1968); Wagner and Folkers (1964); Yagi (1968).

Tricarboxylic acid cycle

Aerobic cellular respiration comprises two separate phases. The first phase includes a series of complex reactions whereby carbohydrates, fats, and other organic nonnitrogenous compounds are oxidized to CO_2 with successive removal of pairs of hydrogen atoms. This is known as the tricarboxylic acid cycle (also as the Krebs cycle or the citric acid cycle). Once the substrate enters the cycle, the sequence of steps is the same regardless of the type of food which is ingested by the organism. The second phase involves the oxidation of hydrogen atoms by oxygen, with the formation of ATP. The two phases working together lead to the eventual oxidation of the substrates to CO_2 and H_2O, and to the formation of biologically useful energy as ATP.

In the anaerobic phase, only a small amount of energy is extracted from the oxidation of the glucose molecule. Aerobic cells can obtain considerably more energy by breaking down pyruvic acid to H_2O and CO_2. Approximately 16 energy-rich bonds in the form of ATP are formed for each molecule of pyruvate oxidized. This means that 32 or more energy-rich bonds are produced during the aerobic oxidation of each molecule of glucose.

A very important cofactor entering into the cycle is coenzyme A. It has an active sulfhydryl group (−SH) and is generally represented as CoA·SH. When the coenzyme enters into a chemical combination, it does so through its −SH group. This blocks the active group and the coenzyme becomes totally inactive. However, the coenzyme becomes active again when the active group is unblocked.

The mechanism of the tricarboxylic acid cycle operates in minute intracellular bodies known as mitochondria (see page 113). They are believed to contain all the enzymes of the tricarboxylic acid cycle, in addition to CoA, NAD, NADP, the flavins (FMN, FAD), and the cytochromes. That is why the mitochondria are sometimes called the powerhouses of the cells.

Steps in the tricarboxylic acid cycle:

1. The first step involves a reaction between pyruvic acid and CoA·SH to give acetyl-coenzyme A:

$$\underset{\text{Pyruvic acid}}{CH_3\cdot CO\cdot COOH} + CoA\cdot SH \longrightarrow \underset{\text{Acetyl-coenzyme A}}{CH_3\cdot CO\cdot S\cdot CoA} + 2H + CO_2$$

2. The acetyl-coenzyme A reacts with oxaloacetic acid to give citric acid:

$$\underset{\text{Oxaloacetic acid}}{COOH\cdot CO\cdot CH_2\cdot COOH + CH_3\cdot CO\cdot S\cdot CoA + H_2O} \longrightarrow \underset{\text{Citric acid}}{\begin{array}{l} CH_2\cdot COOH \\ | \\ COH\cdot COOH \\ | \\ CH_2\cdot COOH \end{array}} + CoA\cdot SH$$

3. By a rearrangement of atoms, citric acid is converted to isocitric acid:

$$\begin{array}{l} CH_2\cdot COOH \\ | \\ COH\cdot COOH \\ | \\ CH_2\cdot COOH \end{array} \longrightarrow \underset{\text{Isocitric acid}}{\begin{array}{l} CH_2\cdot COOH \\ | \\ CH\cdot COOH \\ | \\ CHOH\cdot COOH \end{array}}$$

4. The isocitric acid loses CO_2 and 2H to form α-ketoglutaric acid:

$$\begin{array}{l} CH_2\cdot COOH \\ | \\ CH\cdot COOH \\ | \\ CHOH\cdot COOH \end{array} \longrightarrow \underset{\alpha\text{-Ketoglutaric acid}}{\begin{array}{l} CH_2\cdot COOH \\ | \\ CH_2 \\ | \\ C{=}O \\ | \\ COOH \end{array}}$$

5. The α-ketoglutaric acid is converted to succinic acid by the loss of CO_2 and 2H:

$$\begin{array}{l} CH_2\cdot COOH \\ | \\ CH_2 \\ | \\ C{=}O \\ | \\ COOH \end{array} + H_2O \longrightarrow \underset{\text{Succinic acid}}{\begin{array}{l} COOH \\ | \\ CH_2 \\ | \\ CH_2 \\ | \\ COOH \end{array}} + CO_2 + 2H$$

6. The succinic acid loses 2H and becomes converted to fumaric acid:

$$\begin{array}{l} COOH \\ | \\ CH_2 \\ | \\ CH_2 \\ | \\ COOH \end{array} \longrightarrow \underset{\text{Fumaric acid}}{\begin{array}{l} COOH \\ | \\ CH \\ \| \\ CH \\ | \\ COOH \end{array}} + 2H$$

7. Next, fumaric acid takes up water and is converted to malic acid:

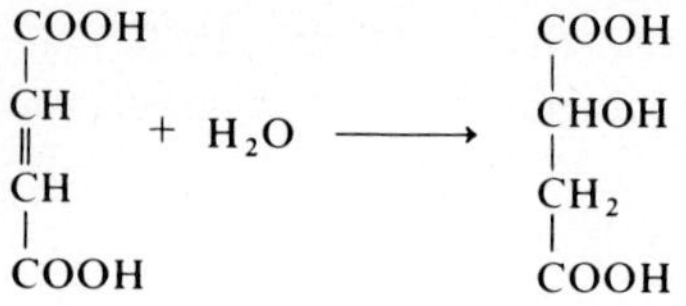

Malic acid

8. Finally, malic acid loses 2H to give oxaloacetic acid and the cycle is ready to repeat itself:

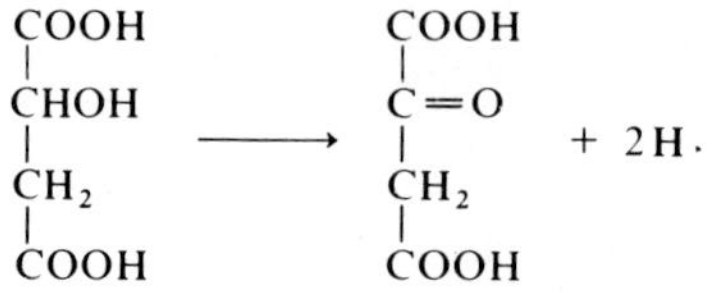

Oxaloacetic acid

Oxaloacetic acid is required to start the whole cycle. A constant supply of the compound must be formed unless, as so commonly happens in biochemical systems, it is regenerated.

Ignoring the CoA, which acts as a catalyst, the overall reaction in the cycle becomes:

$$CH_3 \cdot COOH + 2O_2 \longrightarrow 2CO_2 + 2H_2O$$

Acetic acid

A schematic drawing of the tricarboxylic acid cycle is shown in Fig. 207.

Generation of adenosine triphosphate

Most of the ATP is produced during the transfer of hydrogen atoms from the substrate to oxygen. The hydrogen atoms are produced mostly by dehydrogenations in the various steps of the tricarboxylic acid cycle.

Two hydrogen atoms (as two electrons and two protons) are transferred from acceptor to acceptor and the energy released is utilized to couple inorganic phosphate with ADP at three points. This results in the formation of three molecules of ATP containing energy-rich bonds. The capture of energy from the flow of electrons in this manner is commonly called oxidative phosphorylation. This may be diagrammed as on page 415.

Bioluminescence

The emission of light by living organisms is called bioluminescence. It is sometimes spoken of as phosphorescence.

Light of the sea, of fish, flesh, or wood is always caused by the presence of living organisms. They include certain bacteria, molds,

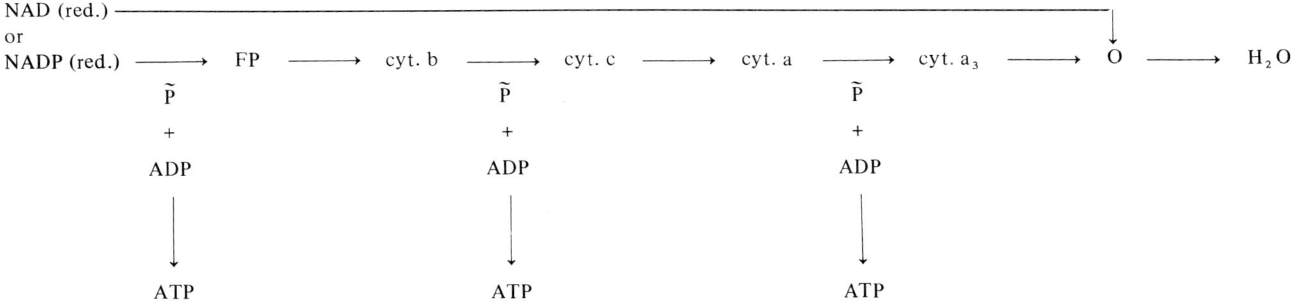

where FP = flavoprotein
cyt. = cytochrome
P = energy-rich phosphate

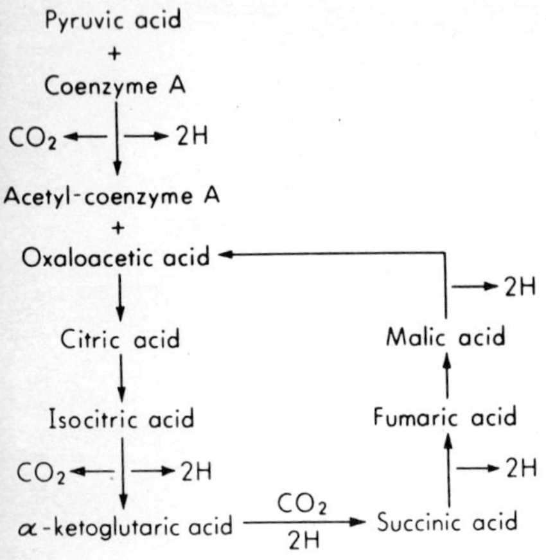

Fig. 207 The tricarboxylic acid cycle.

sponges, radiolarians, corals, flagellates, hydroids, nemerteans, ctenophores, crustaceans, clams, snails, squid, fishes, centipedes, millepedes, and insects such as fireflies or glowworms. Most, if not all, luminous bacteria are found in the genera *Photobacterium* and *Vibrio*. These are marine forms. With a few doubtful exceptions, luminous freshwater bacteria have not been encountered.

Cultivation of luminescent bacteria

In order to assure good growth and light emission of luminescent bacteria, the osmotic tension of inorganic salt solutions used for their cultivation should be equivalent to that produced in a 3 to 5 percent sodium chloride solution. The optimum temperature varies between 15 and 40°C, depending upon the species. The optimum temperature for growth is not necessarily the same as the optimum for luminescence.

Nature of light emitted Luminescent bacteria and molds may be distinguished from other light-emitting organisms by the uniform intensity of the light, shining day and night, independent of any stimulation. On the other hand, animals emit light only when disturbed or stimulated.

Since very little heat is produced by luminous organisms, bioluminescence is often spoken of as "cold light." The rise in temperature in some light-producing organisms is less than 0.001°C. The light emitted has properties similar to other kinds of light. It will affect photographic film and can induce chemical reactions. Infrared or ultraviolet radiations are not produced, which means that the luminous efficiency is nearly 100 percent.

Mechanism of bioluminescence Bioluminescence is an aerobic reaction and oxygen must be present for the emission of light.

Luciferin is the key compound that actually emits the light. In the reduced form it has the following structure:

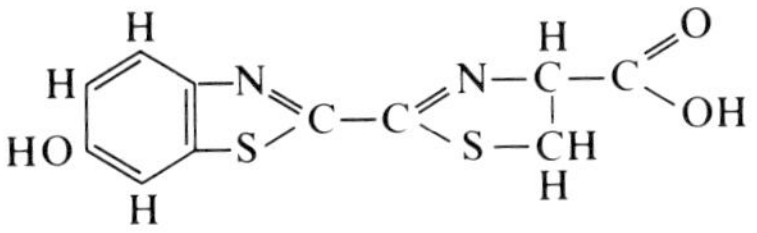

Luciferin (red.)

In the presence of oxygen, it loses two atoms of hydrogen to give oxidized luciferin and water:

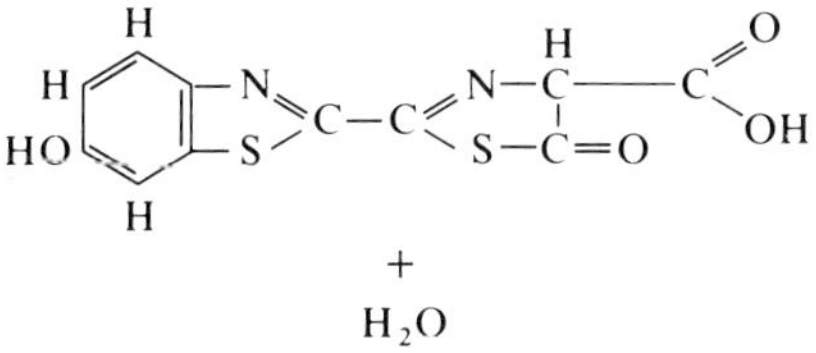

Luciferin (ox.)

The light-emitting reaction is intimately related to the oxidative, or electron transport, processes by which the cell extracts energy from nutrients. The requirements for luminescence are a reduced flavin such as FMN, a long-chain aldehyde, oxygen, and the enzyme luciferase. Bacterial luciferase catalyzes a bioluminescent oxidation of reduced FMN by molecular oxygen, the photon yield of the reaction being greatly stimulated by the presence and absolute necessity of a long-chain aldehyde. Luciferases from different bacterial species and strains apparently differ somewhat in the quantitative responses to different reduced flavins and different aldehydes.

The reaction scheme may be diagrammed as follows:

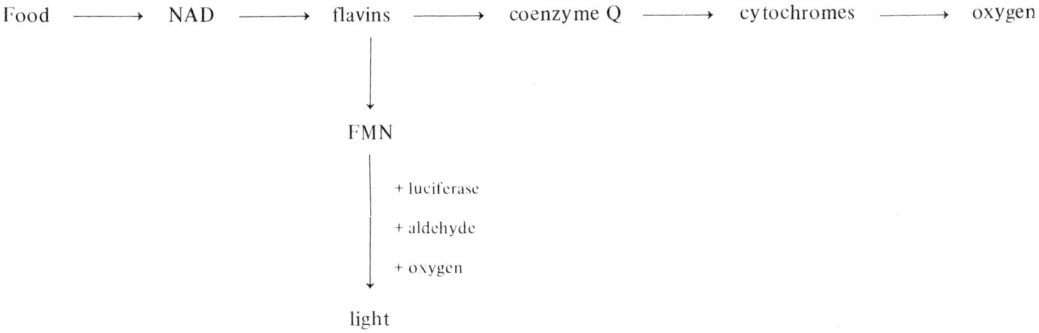

Hydrogen atoms are removed from the nutrient and passed along to a series of acceptors in the direction of the arrows. The final hydrogen acceptor is oxygen and the end product is H_2O. At certain steps in the sequence, energy is removed from the reactants and stored as ATP. Light is emitted when one of the reduced flavins (FMN) reacts with luciferase and oxygen in the presence of a long-chain aldehyde. In the above scheme, FMN plays the role of luciferin.

For more information: Coffey (1967); Cormier et al. (1969); Eley et al. (1970); Friedland and Hastings (1967); Hastings et al. (1969); Klein (1965); Krieger and Hastings (1968); Kuwabara et al. (1965); McElroy (1960); McElroy and Seliger (1961, 1962); Mitchell and Hastings (1970); Seliger et al. (1961); Tsuji et al. (1963).

Glutathione

Glutathione is a tripeptide present in living cells and functions as an oxidation-reduction (O-R) system. It is composed of the three amino acids glutamic acid, cysteine, and glycine joined together in a peptide linkage. It has the following structural formula:

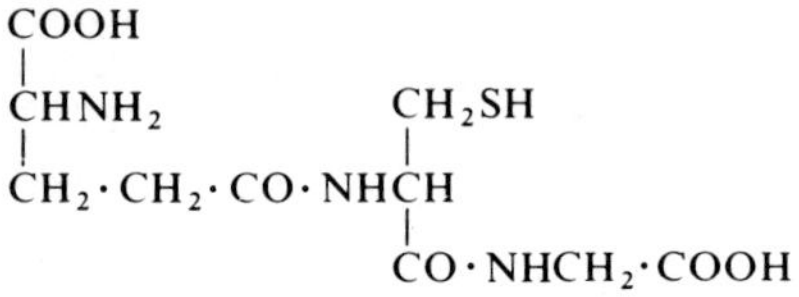

Glutathione (glutamyl cysteinyl glycine)

Since it is the sulfhydryl (–SH) group that is of importance in O-R reactions, the above formula may be abbreviated to GSH. Two molecules of GSH readily give up the hydrogen of their –SH groups and become oxidized to a disulfide:

$$2GSH + B \rightleftharpoons GSSG + BH_2$$

Reduced glutathione — Oxidized glutathione

The –S–S– form is readily reduced to the –SH compound by the addition of two atoms of hydrogen. Cells contain systems which rapidly reduce –S–S– and which quickly oxidize –SH by means of molecular oxygen.

Glutathione has been shown to be of almost universal occurrence in living tissue where the concentration roughly parallels the metabolic activity of the cells. The concentration is higher in rapidly growing cells than in older cells. Blood is said to contain 34 to 47 mg of glutathione per 100 g. It is also present in many species of aerobic and anaerobic bacteria, in yeasts, and in molds.

Some believe that glutathione is the chief protective substance in cells for maintaining ascorbic acid in the reduced form. It guards the vitamin against irreversible oxidation with a loss of antiscorbutic activity.

Since the O-R mechanism of glutathione is due to the cystine-cysteine combination, cystine also functions in a similar manner:

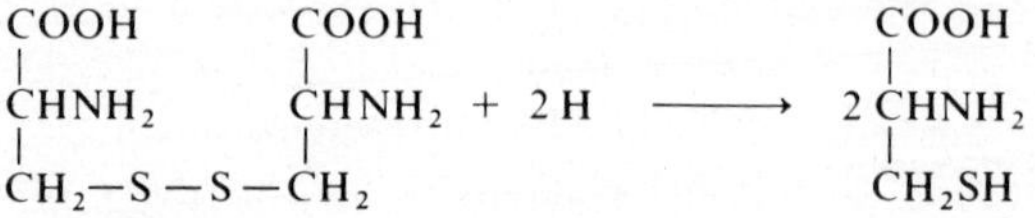

Cystine — Cysteine

Oxidation-reduction potentials

Oxidation-reduction potentials are of great importance in biology. Intracellular reactions which release energy to the organism involve a study of O-R potentials.

An oxidation occurs either by the addition of oxygen, or by the removal of hydrogen. The oxidation of one compound is accompanied by the reduction of another.

A typical oxidation by the addition of oxygen involves the formation of cupric oxide in the reaction:

$$2Cu + O_2 \longrightarrow 2CuO$$

The conversion of alcohol to aldehyde is an example of an oxidation by the removal of hydrogen:

$$CH_3 \cdot CH_2OH \longrightarrow CH_3 \cdot CHO + 2H$$

Oxidations occur also by the withdrawal of electrons, and reductions by the addition of electrons. An atom consists of a nucleus of positive electricity surrounded by a shell of electrons possessing negative electrical charges. The sum total of the negative charges must be equal to the positive charge of the nucleus. Some elements easily lose electrons, whereas others add electrons. An oxidation involves the loss of one or more electrons; a reduction involves a gain of electrons:

$$Cu^+ \underset{\text{Reduction}}{\overset{\text{Oxidation}}{\rightleftharpoons}} Cu^{++} + (e)$$

A substance that readily gives up electrons is a good reducing agent; conversely, a substance that readily takes up electrons is a good oxidizing agent.

The transfer of electrons from one compound to another sets up a potential difference between the reactants, which may be measured by an appropriate instrument. The magnitude of this potential difference depends upon the ease with which the electrons are lost or gained. The greater the oxidizing or reducing power of a substance, the greater will be the electrical potential on one side or the other of a zero point. The more highly oxidized a substance, the more positive will be its electrical potential and the more highly reduced a substance, the more negative will be the electrical potential. The direction in which a reaction proceeds is dependent upon the free electrons in the system. If the number of electrons is increased, the system will produce more of the reductant; if the number is reduced, the system will produce more of the oxidant. The electronic state of the system is a measure of its oxidizing or reducing power.

The O-R potential of a system is expressed by the symbol Eh. The greater the proportion of reduced substance present, the lower will be the Eh value; conversely, the greater the proportion of oxidized substance, the higher will be the Eh value. When the concentration of the oxidant is equal to that of the reductant, the term becomes zero and the observed potential is equal to E_0.

Measurement of O-R potentials Two methods may be followed for determining the O-R potentials of bacterial cultures: (1) the colorimetric method and (2) the potentiometric method. Each has its advantages and disadvantages, but the latter is generally preferred.

Litmus and methylene blue are used in bacterial cultures as indicators of reduction. For the determination of various degrees of reduction intensity, a selection of a series of indicators is necessary. It is now possible to cover the range from +0.300 to −0.450 V at pH 7.0 and a temperature of 30°C. Usually the oxidants of the oxidation-reduction systems are colored and the reductants are practically colorless.

For more information: Society of American Bacteriologists (1957), Zador (1961).

Anaerobic bacteria

Anaerobes are usually defined as organisms which can live and multiply only in the complete absence of oxygen. This statement is not correct as all obligate anaerobes can tolerate some free oxygen. However, they show considerable variation in the amount of oxygen they can tolerate. *Clostridium tetani* can grow in a liquid medium exposed to 5 to 15 mm air pressure. *C. perfringens* produces good growth in an atmosphere containing 200 mm air pressure and shows slight growth at 380 mm.

The most important anaerobic bacteria are placed in the genus *Clostridium*. Some members are obligately anaerobic; others are anaerobic, aerotolerant. They are large, gram-positive rods which produce spores that cause a bulging of the cell wall. H_2O_2:H_2O_2 oxidoreductase (catalase) is lacking except in small amounts in certain aerotolerant forms. Some species are strongly fermentative; others are actively putrefactive. A few species are obligately thermophilic. They are commonly found in soil and in human and animal feces.

Mechanism of oxygen inhibition The mechanism involved in the inhibition of growth of anaerobes when exposed to air has been the subject of many investigations. The most important theories are:

1. Oxygen is directly toxic to the cell.
2. H_2O_2 is produced and, since the organisms do not form catalase, the compound is toxic to the cell.
3. The growth of anaerobes is dependent upon a low oxidation-reduction potential that is not possible in the presence of air.

1. Free oxygen does prevent growth of anaerobic bacteria but it does not kill them. Growth is reestablished when cultures exposed to air are again placed in an anaerobic environment. Oxygen prevents growth without poisoning the bacteria.
2. Some anaerobes produce small amounts of H_2O_2 when exposed to air

and, since they are very sensitive to the compound, its presence prevents growth. This statement is based on certain observations made in connection with the growth of anaerobes in a blood medium. It was observed that anaerobes produced a zone of greenish discoloration about 1/8 in below the surface of tubes of cooked blood (chocolate) agar. The growth was very similar to that produced by *Diplococcus pneumoniae* on the same medium and which is known to be a producer of H_2O_2.

The function of catalase is to decompose H_2O_2 into water and molecular oxygen. Although catalase is unable to promote growth of anaerobes in contact with air, it raises the level of growth in deep tubes of agar almost to the surface. Also, the appearance of a green-colored ring in chocolate agar cultures is greatly delayed and decreased by the addition of catalase.

It is exceedingly difficult to demonstrate the presence of H_2O_2 in the cultures of anaerobes because their active life is inhibited by exposure to air before sufficient peroxide has accumulated to give a positive test. If colonies of *C. botulinum* on a blood agar plate are treated with benzidine, dark halos appear within an hour in the presence of air. The test indicates the production of H_2O_2. Since the organisms do not produce catalase, the H_2O_2 becomes toxic to the bacteria.

Gordon et al. (1953) showed that direct tests for H_2O_2 in oxygenated sediments from liquid cultures of anaerobes gave quite variable and usually negative results. This was due to the fact that the residual medium and metabolic products of the organisms masked the H_2O_2 produced or interfered with the reactions used for its detection. If the organisms were centrifugated in such a way that the residual medium and metabolites were almost completely removed, and the sediment resuspended in peroxide-free distilled water and then oxygenated, positive reactions for H_2O_2 were obtained in almost every case. It was suggested that indirect reactions were less readily obtained with proteolytic anaerobes because this activity increased the concentration of substances capable of masking H_2O_2.

3. Some believe that anaerobic growth is dependent entirely on a low oxidation-reduction potential that is not possible in the presence of air. The optimum O-R potential appears to be at an Eh value of about -0.2 V. Reducing substances which are of value in lowering the O-R potentials of culture media include sodium thioglycollate, cysteine, ascorbic acid, sodium formaldehyde sulfoxylate, and glucose. Sodium thioglycollate is extensively used as an addition to culture media for the cultivation of anaerobes in the presence of air.

   Regardless of which theory is correct, the fact remains that special methods, all based on the exclusion or elimination of free oxygen, must be employed for the successful cultivation of anaerobic bacteria.

For more information: Mateles and Zuber (1963).

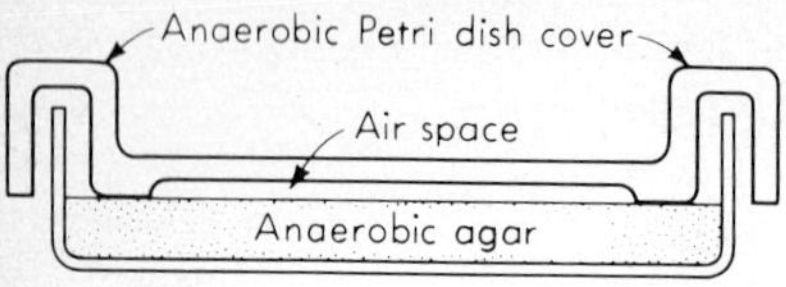

Fig. 208 Technique for using the Brewer petri dish cover and anaerobic agar.

Cultivation of anaerobic bacteria A large number of procedures have been devised for the cultivation of anaerobes. Some employ liquid media; others make use of solid preparations. All procedures aim to exclude atmospheric oxygen from the environment. In some procedures certain constituents are added to media to reduce the oxygen potential; in others, especially where solid media are employed, the cultures are incubated within closed containers from which the oxygen is either absorbed or removed mechanically.

Use of reducing compounds The addition of reducing compounds such as glucose, cysteine, sodium formaldehyde sulfoxylate, and sodium thioglycollate to liquid media usually permits growth of anaerobes under aerobic conditions. Heavy inocula are more effective than light inocula because of the transfer of reducing substances to the new media.

Brewer (1942) introduced a petri dish cover, together with an agar medium containing sodium thioglycollate and methylene blue, for the surface cultivation of anaerobic bacteria (Fig. 208). The anaerobic agar is melted, cooled to 50°C, poured into a petri dish, and allowed to harden. The surface of the agar is streaked with the organism. Then the petri dish cover is replaced by the Brewer anaerobic lid, which is designed to touch the agar at the periphery and trap a small amount of air (less than 1 mm in thickness) over the surface of the agar. The sodium thioglycollate in the medium uses up the oxygen in this small amount of air to create anaerobic conditions. The glass rim on the lid forms a seal with the moist agar, preventing the entrance of more atmospheric oxygen. The methylene blue in the agar acts as an indicator, and the center of the dish, which is anaerobic, becomes colorless; the oxygenated edge of the plate, about 5 mm in diameter, remains blue.

For more information: Gorini (1961).

Use of aerobe to absorb oxygen Microorganisms have been used to achieve anaerobiosis. This can be handled in two ways: (1) by keeping the aerobe and anaerobe separated and (2) by growing the aerobe and anaerobe together as a temporary mixed culture.

1. A tube of agar is melted and poured into a petri dish. When the agar has hardened, the plate is divided into two parts by making a line across the bottom with a china-marking pencil. One-half of the plate is streaked with a culture of the anaerobe; the other half is streaked with a culture of an aerobe. The cover is replaced, and the edges of the two halves are sealed with modeling clay. The plate is incubated in the inverted position to prevent accumulation of moisture on the surface of the agar. The moisture may cause the growth to spread and prevent the development of well-isolated colonies.

The aerobe utilizes the free oxygen and eliminates carbon dioxide. The oxygen concentration is soon reduced to a level that permits growth of the anaerobe.

2. Kneteman (1957) employed an obligate aerobic *Micrococcus* to create anaerobic conditions. The *Micrococcus* and the anaerobic organism to be cultivated were both inoculated into the same liquid medium. During incubation the *Micrococcus* gradually utilized the free oxygen creating conditions favorable for the growth of the anaerobe. After anaerobiosis was achieved, the *Micrococcus* died leaving the anaerobe in pure culture.

   If agar plates were used, it was necessary to prevent diffusion of oxygen from the air into the medium. This was effected by covering the agar surface with a very thin film of cellophane, which is impervious to oxygen. The plate method is carried out in ordinary petri dishes. Anaerobic jars are not used.

Use of semisolid agar The addition of 0.05 to 0.2 percent agar to an appropriate medium generally provides optimum conditions for the rapid development of anaerobes. The agar creates various degrees of oxygen tension, making possible the growth of anaerobes without special precautions. The medium should be inoculated deep, where the oxygen concentration is very low. The upper portion of the medium is well suited for aerobic growth. Below this the agar develops widely variable degrees of oxygen concentrations, providing a medium suitable for the cultivation of many types of bacteria.

Exclusion of atmospheric oxygen A tube of deep nutrient broth or an appropriate carbohydrate broth is inoculated and then covered with a 1/2-in layer of sterile melted vaspar (a mixture of equal parts of petroleum jelly and paraffin). The seal does not entirely prevent the entrance of atmospheric oxygen, but it is usually sufficient in establishing initial growth of the less fastidious anaerobes. After growth has once set in, the elimination of carbon dioxide by the organisms creates optimum conditions for multiplication. It is advisable to heat the medium in an Arnold sterilizer for 10 min previous to inoculation to drive out as much of the dissolved oxygen as possible.

Agar may be substituted for the broth. A tube of the medium is melted and cooled to about 50°C. The agar is inoculated with the anaerobe, thoroughly mixed by gently shaking and rotating the tube. The medium is allowed to solidify in a vertical position, after which the surface is covered with a 1/2-in layer of melted vaspar.

The value of this medium is that it affords a simple means of grading the oxygen tension in the medium. On the surface, the pressure is atmospheric; at the bottom, the conditions are anaerobic.

The agar cylinder is removed by rimming the tube in the center by means of a triangular file, pulling the two halves apart, and collecting the agar in a sterile petri dish. The colonies can then be fished from the agar and transferred to fresh media.

Use of living or dead tissue Pieces of kidney, liver, etc., removed aseptically from an animal, or minced and heat-sterilized beef heart or brain tissue have been used for the cultivation of obligate anaerobes.

Minced and sterilized brain medium has been used probably more than any other tissue preparation for the cultivation of anaerobes. It is prepared by suspending cooked and minced sheep brains in glucose broth. Previous to inoculation, the medium is heated for 10 min in an Arnold sterilizer to drive out as much of the dissolved oxygen as possible. The tube is allowed to cool to a temperature of about 50°C, then inoculated by loop or pipette plunged deep. A surface seal is not necessary because the minced brain tissue is very effective in increasing the reducing power of the medium. Anaerobic growth is first established near the bottom of the tube.

Use of anaerobic liquid culture tube Wilson (1950) described a simple device for the cultivation of anaerobic bacteria in a liquid medium without the addition of reducing compounds or pieces of tissue (Fig. 209). It consists of two test tubes, differing in size, which are fitted at the top with screw caps and joined at the bottom with thick-walled capillary tubing. The plastic cap of the small tube is perforated with a small hole, and both caps are supplied with solid, heavy rubber gaskets so that airtight seals are formed when screwed in place.

The tubes are prepared as follows: Broth of the desired composition is added to fill the small arm. It is prevented from running into the large arm by tightening the cap on the latter. After the small arm is filled, its cap is screwed tightly in place. Then the cap in the large arm is removed and about 1 in of broth added. The cap is replaced loosely on the large arm, and the device with its contained broth is sterilized in an upright position in an autoclave at 121°C for 20 min. When the tube is removed from the autoclave and allowed to cool, the small arm becomes filled with the medium. The medium in the small arm is maintained in the reduced state, whereas the medium in the large arm becomes oxidized shortly from contact with the air.

The inoculum may be added to the broth in the large arm by a pipette directed at the orifice of the capillary, or it may be introduced into the small arm through the rubber gasket by means of a syringe and needle.

Growth of the anaerobic organisms begins in the small arm, but when growth is particularly vigorous, the medium in both arms will show growth and will be reduced as a result.

Absorption of oxygen by activated iron Parker (1955) recommended the use of activated iron wool for the removal of oxygen from sealed containers. The iron was activated by immersion in an acid solution of copper sulfate, then drained. Since oxygenated iron absorbs CO_2, more of this gas must be added to the closed container; otherwise growth will be seriously affected.

Fig. 209 Anaerobic liquid culture tube. See text for description. (After Wilson.)

Absorption of oxygen by chromous salts Marshall (1960) recommended the use of chromous salts for the production of anaerobic conditions. Tubes and/or plates of cultures are placed in a suitable anaerobic jar, together with a dish containing zinc metal and calcium carbonate. A solution of chromic sulfate dissolved in sulfuric acid is introduced into the jar through a long-stemmed funnel, the jar closed, and evacuated immediately to a pressure of 80 to 100 mm mercury. Then the jar is sealed and incubated. Both hydrogen and chromous ions are produced. The chromous ions have a strong affinity for oxygen.

Absorption of oxygen by pyrogallol An agar slant is inoculated with the organism to be cultivated. The cotton stopper is cut off flush with the neck of the tube and pushed down about 1/2 in below the opening. The surface of the cotton stopper is covered with a layer of pyrogallol crystals, then moistened with a few drops of sodium hydroxide solution. A tight-fitting rubber stopper is immediately inserted into the opening of the tube to exclude atmospheric oxygen. The alkaline pyrogallol absorbs oxygen from the environment, creating conditions compatible with the growth requirements of the anaerobe. This procedure is usually satisfactory for the cultivation of those organisms which are not too exacting in their requirements.

A more convenient arrangement is to employ a Bray or a Spray anaerobic culture dish (Fig. 210). The dish is separated at the bottom by a raised center ridge. In one compartment is placed a solution of pyrogallol and in the other a solution of sodium hydroxide. Melted agar is poured into a petri dish and allowed to harden. The surface is streaked with the anaerobic organism. The agar plate is then inverted over the top of the culture dish, and the edges are sealed with a suitable material, such as plasticine, to prevent entrance of atmospheric oxygen. Finally, the solutions in the bottom are mixed by gently tilting the dish.

Replacement of atmospheric oxygen with hydrogen In this procedure the inoculated tubes are placed in a closed jar. The air is evacuated and replaced with hydrogen gas. The jar is then placed in an incubator.

This procedure does not always prove satisfactory, owing to the fact that sufficient oxygen usually remains in the medium to prevent growth. The method becomes considerably more efficient if an alkaline solution of pyrogallol is added to the jar just before the air is removed. This is best performed by placing some pyrogallol crystals on the bottom of the jar, followed by the addition of sufficient solution of sodium hydroxide to dissolve the compound. The lid is replaced immediately and the air

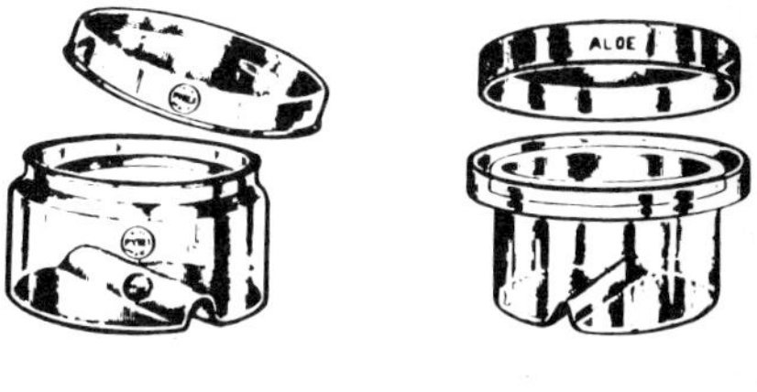

Fig. 210 Left, Bray anaerobic culture dish; right, Spray anaerobic culture dish.

removed as quickly as possible. The pyrogallol usually removes any oxygen still remaining in the medium, permitting growth of the strict anaerobes.

A solution of methylene blue is frequently placed in anaerobic jars as an indicator of anaerobiosis. Methylene blue is blue in the oxidized state and colorless when reduced.

A procedure for the preparation of a satisfactory indicator solution is as follows:

Solution A 6 ml $N/10$ NaOH solution diluted to 100 ml with distilled water.

Solution B 3 ml of a 0.5 percent aqueous solution of methylene blue diluted to 100 ml with distilled water.

Solution C 6 g of glucose dissolved in 100 ml of distilled water, to which is added a small crystal of thymol as a preservative.

Mix equal parts of the three solutions in a test tube, place the tube in boiling water, and heat until the color of the dye disappears. Transfer the tube to the anaerobic jar containing the organisms to be cultivated and immediately evacuate the container. If the solution remains colorless, it indicates that the container is satisfactorily deoxygenated; if the blue color returns, it indicates that conditions are not anaerobic. The mixture of the three solutions should be freshly prepared before use.

Fig. 211 Cross section of modified Brewer anaerobic jar showing construction. (After Evans, Carlquist, and Brewer.)

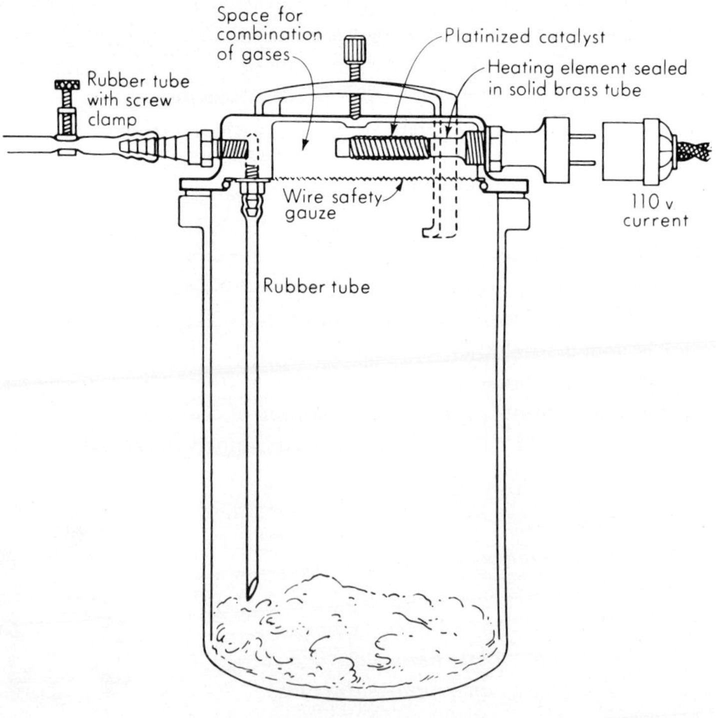

Combustion of oxygen with hydrogen Surface colonies are most essential for purifying and identifying anaerobes. The removal of oxygen by combustion with hydrogen is one of the most efficient methods for obtaining surface colonies of anaerobes.

The first practical development of a jar for this purpose was announced by Fildes (1931). A convenient modification by Evans et al. (1948) is shown in Fig. 211.

More recently Brewer and Allgeier (1968*a*) described a self-contained carbon dioxide–hydrogen anaerobic system that allows ease of operation not previously possible with other systems (Fig. 212). Gas cylinders, vacuum pumps, valves, and gauges have been eliminated. A new anaerobic lid was developed that is fitted only with a snap-in rubber gasket, a double stainless-steel gauze flash arrestor, and a catalyst holder. The holder contains a palladium pellet catalyst that is active at room temperature and requires no heat for activation. The system was made specifically for use with a disposable hydrogen–carbon dioxide generator (Fig. 213), and a disposable methylene blue anaerobic indicator. The jar incorporates unique safety features which eliminate the possibility of an explosion. The oxygen-free atmosphere, composed primarily of nitrogen and carbon dioxide, was quickly achieved within the jar to insure maximal growth.

For more information: Brewer and Allgeier (1968*b*); Brewer et al. (1966); Carroll and Ward (1965); Gordon (1963); Khairat (1964); Moore et al. (1964); Society of American Bacteriologists (1957).

Fig. 212 Brewer anaerobic jar. See text for description. (Courtesy Brewer and Allgeier.)

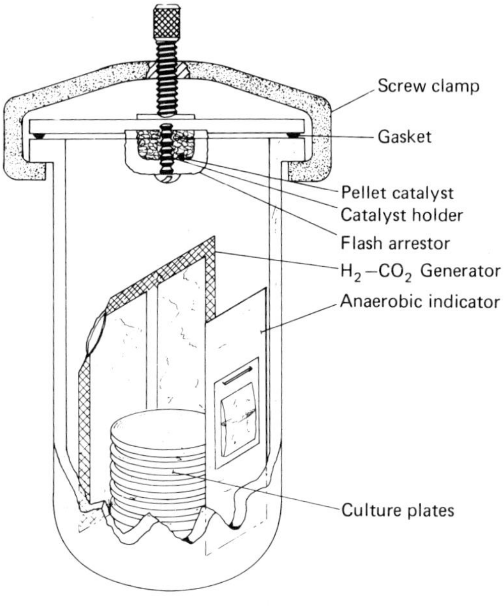

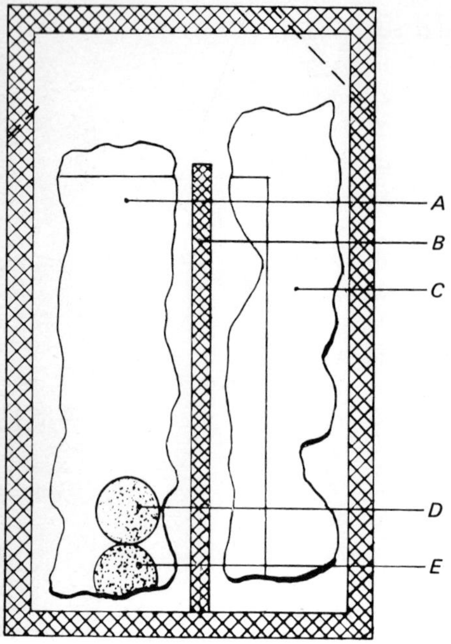

Fig. 213 Cutout view of the hydrogen–carbon dioxide generator. (A) Filter; (B) heat seal; (C) water added to this side; (D) carbon dioxide tablet; (E) hydrogen tablet. See text for description. (Courtesy of Brewer and Allgeier.)

Reduction of nitrate

The term *nitrate reduction* includes all processes in which nitrate disappears under the influence of bacterial action and reappears in some less oxidized state. In many cases the reduction does not proceed beyond the stage of nitrite. However, a few organisms are able to convert nitrate to molecular nitrogen.

An organism that takes up nitrate from the medium and utilizes it for the synthesis of its cell protein may also be considered a nitrate-reducing organism. However, there are good reasons for separating nitrate-assimilating organisms from those which have the property of reducing the compound and excreting most of the reduction products into the surrounding medium.

From the foregoing considerations, biological nitrate reductions may be classified into the following three types:

1. Assimilatory nitrate reduction, which may be termed *nitrate assimilation*.
2. Incidental dissimilatory nitrate reduction, in which the nitrate merely acts as a nonessential hydrogen acceptor.
3. True dissimilatory nitrate reduction, in which the nitrate acts as a hydrogen acceptor essential for growth.

In the third type, nitrate is reduced by bacteria only when growing under anaerobic or partial anaerobic conditions. The organisms substitute nitrate for oxygen as a hydrogen acceptor. On the other hand, if the cultures are well aerated from the start, the organisms prefer oxygen as the acceptor of hydrogen and nitrate is not reduced.

Some of the enzyme systems involved in the reduction of nitrate include: (1) Reduced-NAD or NADP: nitrate oxidoreductase; (2) ferrocytochrome: nitrate oxidoreductase; and (3) reduced-NAD(P): nitrite oxidoreductase:

$$(1)\ \text{Reduced-NAD or NADP} + NO_3^- \longrightarrow \text{oxidized-NAD or NADP} + NO_2^- + H_2O$$

$$(2)\ \text{Ferrocytochrome} + NO_3^- \longrightarrow \text{ferricytochrome} + NO_2^- + H_2O$$

$$(3)\ \text{Reduced-NAD(P)} + NO_2^- \longrightarrow \text{oxidized-NAD(P)} + NH_3 + H_2O$$

Verhoeven (1956*a*) proposed the following scheme for true dissimilatory nitrate reduction:

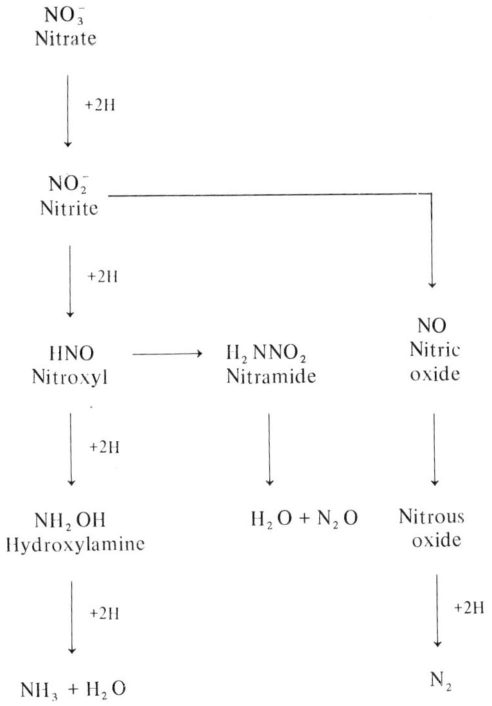

The end products include nitrite, ammonia, and nitrogen gas. Reduction of nitrate to nitrogen gas may be detected by dispensing nitrate medium in test tubes containing inverted vials. The nitrogen gas is trapped in the vial as it is formed and may be observed.

In making tests for nitrate reduction, observe for (1) the reduction of nitrate to nitrite, (2) the disappearance of nitrite to ammonia, and (3) the presence of nitrogen gas. It is highly important that tests be made for these three products.

The presence of nitrite indicates that nitrate has been reduced. The presence of gas shows that nitrite has been reduced to ammonia and finally to nitrogen. Negative results mean that (1) the organism under examination is unable to reduce nitrate, or (2) the medium is not satisfactory for

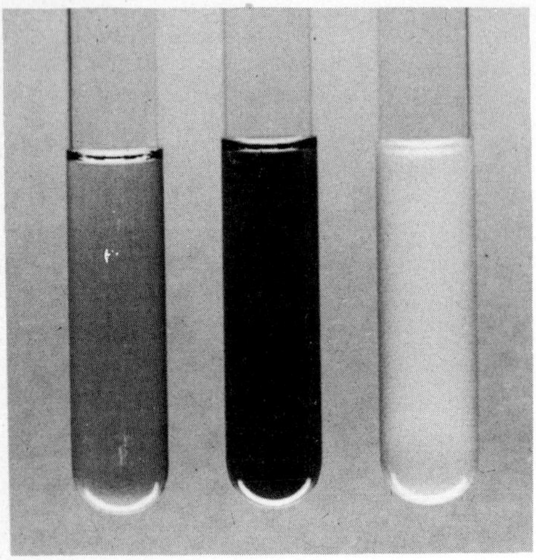

Fig. 214 Reduction of nitrate to nitrite. Left, control, uninoculated; center, culture of Proteus vulgaris showing reduction of nitrate to nitrite (red color with reagents); right, culture of Pseudomonas fluorescens showing no reduction.

growth. The medium may be improved by (1) increasing or decreasing the amount of peptone, (2) increasing the concentration of nitrate, (3) adjusting the reaction of the medium to a more favorable pH, (4) adding a readily available carbohydrate to stimulate growth, or (5) incorporating a small amount of agar to increase the viscosity of the medium.

The nitrite test may be negative even though good growth has occurred. This may indicate no action on the nitrate or complete reduction beyond the nitrite stage. Therefore, tests for nitrate should be made in every case where a negative nitrite test is obtained. If tests do not show nitrate reduction, it is probable that the organism is not capable of attacking the compound.

The reduction of nitrate has long been used as a test for separating bacteria (Fig. 214). The test is of value in identifying and classifying bacteria.

For more information: Hadjipetrou and Stouthamer (1965); Hooper (1968); Jacobs et al. (1964); Matsubara and Mori (1968); Miyata and Mori (1968); Nason (1962); Renner and Becker (1970); Van 'T Riet et al. (1968); Verhoeven (1956*b*).

Reduction of sulfate

Most bacteria can reduce sulfate as the sole source of sulfur. However, they do not produce detectable amounts of sulfide directly from the sulfate. This small-scale reduction of sulfate has been termed assimilatory sulfate reduction.

In contrast to the above organisms, there are a few species present in seawater, marine mud, freshwater, and soil which utilize sulfate as the terminal electron acceptor in anaerobic respiration and produce large amounts of sulfide. This large-scale reduction of sulfate to sulfide has been termed dissimilatory sulfate reduction. Organisms capable of producing dissimilatory sulfate reduction are placed in the genus *Desulfovibrio*. The type species is *D. desulfuricans*.

D. desulfuricans is the first nonphotosynthetic anaerobe in which the presence of a cytochrome was demonstrated. The cytochrome functions as an electron carrier for the production of hydrogen and CO_2 from formate, and for the reduction of sulfite and thiosulfate by hydrogen. It is also believed to be an electron carrier for the reduction of sulfate by hydrogen, since reduced cytochrome is oxidized in the presence of sulfate. The thiosulfate appears to be an intermediate in the reduction of sulfite to sulfide.

Desulfovibrio obtains its energy from the anaerobic reduction of sulfate accompanied by the simultaneous oxidation of organic matter. Using sodium lactate as an organic source, the reduction of sulfate to sulfide proceeds according to the equation:

$$2CH_3 \cdot CHOH \cdot COONA + 3MgSO_4 \longrightarrow 3MgCO_3 + Na_2CO_3 + 2CO_2 + 3H_2S + 2H_2O$$

A general equation for the reduction of sulfate may be represented as follows:

$$2C + MeSO_4 + H_2O \longrightarrow MeCO_3 + CO_2 + H_2S$$

in which C represents an organic substrate and Me a metal.

A large number of organic compounds may be substituted for the lactate, all of which are oxidized in an appropriate sulfate medium. The compounds are oxidized by the removal of hydrogen; i.e., they act as hydrogen donors. A few typical examples are the following:

$$\underset{\text{Formic acid}}{H \cdot COOH} \longrightarrow CO_2 + 2H$$

$$\underset{\text{Acetic acid}}{CH_3 \cdot COOH} + 2H_2O \longrightarrow 2CO_2 + 8H$$

$$\underset{\text{Lactic acid}}{CH_3 \cdot CHOH \cdot COOH} + 3H_2O \longrightarrow 3CO_2 + 12H$$

$$\underset{\text{Glycerol}}{CH_2OH \cdot CHOH \cdot CH_2OH} + 3H_2O \longrightarrow 3CO_2 + 14H$$

Each molecule of sulfate requires eight atoms of hydrogen for its reduction to hydrogen sulfide. The dehydrogenation of the organic substrate proceeds in steps, followed by a corresponding hydrogenation of the sulfate to sulfide:

$$H_2SO_4 + 8H \longrightarrow H_2S + 4H_2O$$

The organic material must be present in large excess for complete oxidation to occur.

Akagi (1967) found that *D. desulfuricans* oxidized pyruvate to acetate, CO_2, and H_2. The electron carriers were ferredoxin (a low-molecular-weight protein) and cytochrome c_3. Ferredoxin accepted the

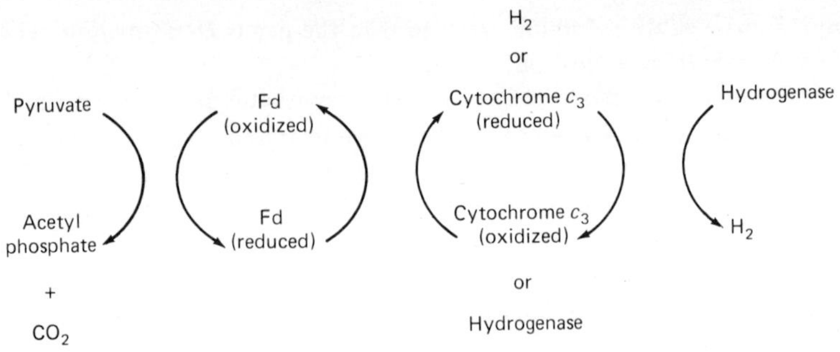

Fig. 215 Proposed scheme for electron transfer. See text for legend. (After Akagi.)

electrons from the pyruvate: ferricytochrome oxidoreductase complex whereas cytochrome did not. However, a combination of the two carriers resulted in maximum production of H_2 and acetylphosphate. Akagi proposed the sequential scheme shown in Fig. 215 for the transfer of electrons.

Organisms The cells of *Desulfovibrio* are curved rods of variable length, usually occurring singly but sometimes in short chains which have the appearance of spirilla. Swollen pleomorphic forms are common. They are actively motile by means of a single polar flagellum (Fig. 216). The cells are strictly anaerobic.

The organisms may be cultivated by inoculating soil or mud into an appropriate sulfate medium and incubating the culture at 25 to 30°C under anaerobic conditions. Isolated colonies may be obtained by streaking a liquid culture over the surface of a sulfate agar medium containing a trace of iron as an indicator. The hydrogen sulfide produced from the sulfate reacts with the iron to give iron sulfide, which imparts a black color to the colonies.

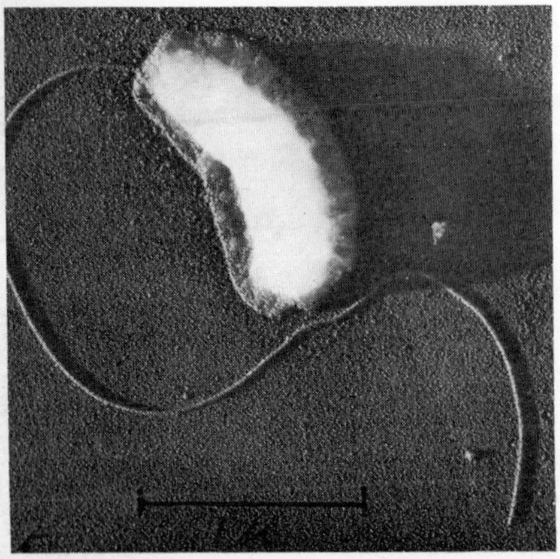

Fig. 216 Electron micrograph of Desulfovibrio desulfuricans. (Courtesy of L. Leon Campbell, Jr.)

Mechanism of sulfate reduction In the reduction of sulfate to sulfite at least two enzymes and most likely a third are involved.

According to Peck (1962*a*), the reactions are believed to be as follows:

1. Sulfate reacts with ATP in the presence of the enzyme ATP-sulfurylase to produce adenosine-5′-phosphosulfate (APS) and pyrophosphate (PPi):

$$ATP + SO_4^{++} \xrightarrow{\text{ATP-sulfurylase}} APS + PPi$$

APS has the following structure:

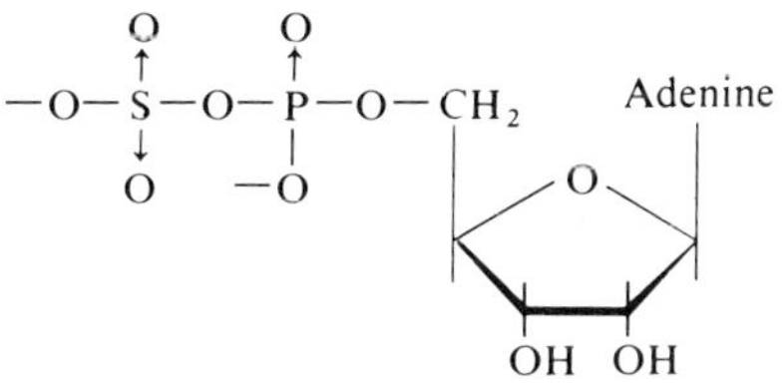

Adenosine-5′-phosphosulfate (APS)

The structure of ATP is given on page 477.

2. The PPi, in the presence of the enzyme pyrophosphatase, reacts with water to give two molecules of phosphate:

$$PPi + H_2O \xrightarrow{\text{Pyrophosphatase}} 2Pi$$

3. APS, in the presence of APS-reductase, accepts two electrons (2e) to give adenosine monophosphate (AMP) and sulfite:

$$APS + 2e \xrightarrow{\text{APS-reductase}} AMP + SO_3^{++}$$

The sulfite is further reduced to sulfide by hydrogen in the presence of reduced cytochrome. The structure of AMP is given on page 477.

These are the only bacteria which are able to effect a direct reduction of sulfate to sulfide. Because of this property, they are included among the nuisance organisms in water. The sulfide makes water unpalatable and odoriferous, and may cause corrosion of steel and cast iron in water-distribution systems.

The organisms are said to contribute to the formation of petroleum and to its modification after formation. It has been claimed that some strains are capable of releasing oil from oil-bearing sediments.

For more information: Alico and Liegey (1966); Deevey (1970); Drucker et al. (1970); Ehlert (1967); Findley and Akagi (1970); Iverson (1966, 1967, 1968); Laishley et al. (1969); la Rivière (1955); Peck (1962*b*); Suh and Akagi (1969); Tuttle et al. (1969*a, b*).

Production of methane

A few species of soil bacteria are capable of fermenting certain organic compounds with the production of methane (CH_4). The important methane producers are found in the genera *Methanobacterium*, *Methanococcus*, and *Sarcina*. They are characterized as follows:

Methanobacterium The methanobacteria are straight or slightly curved rods, sometimes united into bundles or long chains. They are strictly anaerobic. The cells may be chemoautotrophic or chemoheterotrophic, fermenting the simpler organic alcohols, fatty acids, and one-carbon compounds as substrates, and reducing CO_2 to CH_4. They are gram-negative. The organisms are found growing in canal mud and sewage where anaerobic conditions prevail.

Methanococcus The cells are spherical, occurring singly, in pairs, or in masses. They may be motile or nonmotile. The Gram stain is variable. They are chemoheterotrophic, fermenting various organic compounds with the production of CH_4. The cells are strictly anaerobic. The organisms are saprophytes, being found in garden soil, sewage sludge, and black mud.

Sarcina The cells are spherical, multiplying in three perpendicular planes to produce packets. They arc usually nonmotile. The Gram reaction is positive. The organisms are found in mud, swamp waters, and fermenting sewage sludge.

Methanobacterium omelianski catalyzes the oxidation of primary alcohols, such as ethanol and propanol, to the corresponding acids and methane; and secondary alcohols, such as isopropanol, to acetone and methane:

$$\underset{\text{Ethanol}}{2CH_3\cdot CH_2OH} + CO_2 \longrightarrow \underset{\text{Acetic acid}}{2CH_3\cdot COOH} + \underset{\text{Methane}}{CH_4}$$

$$\underset{\text{Isopropanol}}{4CH_3\cdot CHOH\cdot CH_3} + CO_2 \longrightarrow \underset{\text{Acetone}}{4CH_3\cdot CO\cdot CH_3} + CH_4 + 2H_2O$$

M. suboxydans ferments (1) saturated fatty acids, such as valeric acid, to give propionic and acetic acids, and methane, and (2) butyric acid to give acetic acid and methane:

$$\underset{\text{Valeric acid}}{2CH_3CH_2\cdot CH_2\cdot CH_2\cdot COOH} + CO_2 + 2H_2O \longrightarrow \underset{\text{Propionic acid}}{2CH_3\cdot CH_2\cdot COOH} + \underset{\text{Acetic acid}}{2CH_3\cdot COOH} + CH_4$$

$$\underset{\text{Butyric acid}}{2CH_3\cdot CH_2\cdot CH_2\cdot COOH} + CO_2 + 2H_2O \longrightarrow \underset{\text{Acetic acid}}{4CH_3\cdot COOH} + CH_4$$

The acetic acid may, in turn, be fermented to methane by species of *Methanococcus* and *Sarcina*, and the propionic acid may be fermented to methane by a species of *Methanobacterium*:

$$\underset{\text{Acetic acid}}{CH_3\cdot COOH} \longrightarrow CO_2 + CH_4$$

$$4CH_3 \cdot CH_2 \cdot COOH + 2H_2O \longrightarrow 4CH_3 \cdot COOH + CO_2 + 3CH_4$$

Propionic acid — Acetic acid

The methane-producing bacteria can convert mixtures of CO_2 and H_2 to methane:

$$CO_2 + 4H_2 \longrightarrow 2H_2O + CH_4$$

Any oxidizable compound may be substituted for the H_2. Therefore, a generalized reaction may be written for the methane fermentation wherein CO_2 acts as an oxidant or terminal electron acceptor for the oxidation of the organic alcohols, acids, fatty acids, and other acids entered in the above equations:

$$4H_2A + CO_2 \longrightarrow 4A + 2H_2O + CH_4$$

For more information: Breed et al. (1957); Bryant et al. (1968); Jeris and McCarty (1965); Langenberg et al. (1968); Pine (1958); Pine and Barker (1956); Pine and Vishniac (1957); Stadtman (1967); Zeikus and Wolfe (1972).

Hydrogen bacteria

The genus *Hydrogenomonas* contains a few species of organisms whose natural habitat is mud and soil. The cells are short rods with polar flagella if motile. The hydrogen bacteria are able to grow autotrophically in simple substrates consisting of H_2, O_2, CO_2, ammonium N, and various metal salts. They are also able to grow heterotrophically in organic media exposed to air. The cells can oxidize elemental H_2 as a source of energy for growth. Under these conditions CO_2 is used as a source of carbon, being itself reduced in the process. The cells are gram-negative.

The reduction of CO_2 by hydrogen is shown in the following reaction:

$$2H_2 + CO_2 \longrightarrow (CH_2O) + H_2O$$

The utilizable energy, as ATP, is generated from the oxidation of H_2 by oxidative phosphorylation.

For more information: Bongers (1970); DeCicco and Stukus (1968); MacElroy et al. (1969); McFadden and Tu (1967); Ramsey (1968); Repaske (1966); Stukus and DeCicco (1970).

Carbon dioxide requirements of bacteria

It is well established that bacteria do not grow in the complete absence of CO_2. An accumulation of a definite concentration of the gas is necessary to initiate growth. Many bacteria grow better in the presence of an increased concentration of CO_2.

Some organisms, when freshly isolated from their natural environments, either do not grow or grow very poorly unless the CO_2 concentration is increased to 5 to 10 percent. After growth on artificial media is once established, an increased concentration of the gas is no longer necessary.

It is now generally accepted that CO_2 plays an important role in the metabolism of a variety of heterotrophic bacteria. Tracer studies using radioactive carbon ($^{13}CO_2$, $^{14}CO_2$) have shown that the gas is incorporated in a number of metabolic products including acetate, lactate, pyruvate, succinate, aspartate, glutamate, oxalacetate, phosphoglycerate, purines, pyrimidines, nucleic acids, and proteins.

Eastin and Thorne (1963) found that virulent strains of *Bacillus anthracis* required a greater concentration of CO_2 than that in normal air for the production of capsular material. Fixation of $^{14}CO_2$ was detected in aspartate, glutamate, succinate, and glycocoll. Austrian and Collins (1966) reported that 8 percent of the strains of pneumococci isolated from man required the presence of CO_2 for detectable growth on solid media.

Wood and Werkman proposed a mechanism for the incorporation of CO_2 into C–C bonds of organic compounds before radioactive carbon became available. They showed the relationship between CO_2 fixation and succinate production by the following series of reactions:

$$\underset{\text{Pyruvic acid}}{CH_3 \cdot CO \cdot COOH} + CO_2 \longrightarrow \underset{\text{Oxaloacetic acid}}{COOH \cdot CO \cdot CH_2 \cdot COOH}$$

$$\underset{\text{Oxaloacetic acid}}{COOH \cdot CO \cdot CH_2 \cdot COOH} + 2H \longrightarrow \underset{\text{Malic acid}}{COOH \cdot CHOH \cdot CH_2 \cdot COOH}$$

$$\underset{\text{Malic acic}}{COOH \cdot CHOH \cdot CH_2 \cdot COOH} + 2H \longrightarrow \underset{\text{Succinic acid}}{COOH \cdot CH_2 \cdot CH_2 \cdot COOH}$$

Later when radioactive carbon became available, conclusive proof of CO_2 assimilation was made possible. Since that time, a considerable literature has developed leaving no doubt of the absolute necessity of CO_2 for the growth of all organisms.

For more information: Goff and Hartman (1970); Hartman (1970); Lachica (1968); Lachica and Hartman (1968, 1969); McFadden et al. (1967), Myoda and Werkman (1960); Whitcomb et al. (1962); Wood (1946); Schulman (1972).

Oxygen requirements of bacteria

All living animals and higher plants consume O_2 and eliminate CO_2. Bacteria also utilize O_2, either free or combined, and eliminate CO_2.

Organisms differ widely in their sensitiveness to free O_2. Some are depressed in growth by a slight increase in the concentration of O_2; others are stimulated in growth by concentrations up to 60 percent. In general, the percent yield of aerobic organisms varies directly with aeration of the cultures.

The O_2 supply for bacteria in the usual culture methods is generally far from adequate. As an example, *Mycobacterium tuberculosis* is said to require 1.22 percent O_2 for the complete oxidation of 1 percent glycerol.

The medium at 37°C contains only 0.00065 percent O_2. The disproportion between supply and demand is so great that a majority of the cells in a culture probably starve for lack of oxygen.

For more information: Gottlieb and Pakman (1968); Hernandez and Johnson (1967); Johnson (1967); Kaye (1967); Longmuir (1954).

Chromogenesis

A considerable number of bacteria produce nonphotosynthetic colored compounds known as pigments. This is especially true of the strictly aerobic species. Many colors are produced covering the entire range of the chromatic spectrum.

Types of bacterial pigments Most bacterial pigments may be classified as carotenoids, anthocyanins, melanins, tripyrrylmethenes, and phenazines.

Carotenoids Most of the pigments which have been isolated, purified, and studied are members of the carotenoid group. These are red, orange, or yellow nitrogen-free pigments which are insoluble in water but soluble in lipoid solvents such as ether, chloroform, and carbon disulfide. They derive their name from the unsaturated hydrocarbon carotene, $C_{40}H_{56}$, the same pigment found in butter, egg yolks, flour, and carrots, which is the precursor of vitamin A. Oxidized carotene, $C_{40}H_{56}O_2$, is named xanthophyll which is now known to be composed of a mixture of at least 12 pigments in some plant species. Carotene and xanthophyll are usually found together in nature.

Carotenoids may be hydrocarbons, alcohols, aldehydes, ketones, or acids. Their central structure is a methylated 18-carbon conjugated polyene radical, $C_{22}H_{26}$, having four methyl groups in the 1,5- or 1,6-position with respect to each other:

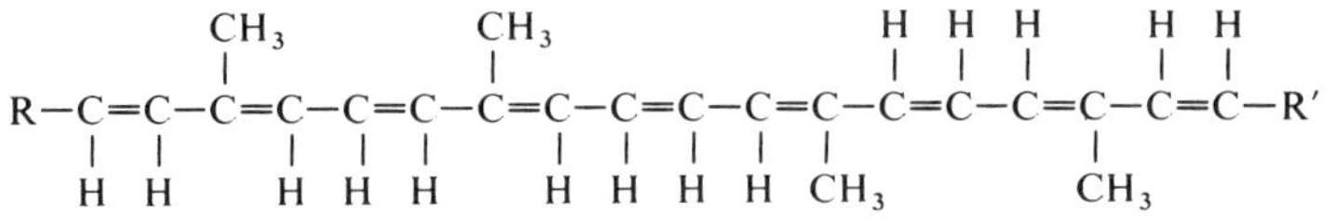

This central section never contains oxygen.

Stephens and Starr (1963) released the carotenoid from *Xanthomonas juglandis* by sonic and ballistic treatment and showed the pigment to be localized in the cytoplasmic membrane.

These results are consistent with other localization studies which have shown that respiratory enzymes and carotenoids are located in the cytoplasmic membrane portion of the bacterial cell envelope.

Anthocyanins The anthocyanins are red, blue, and intermediate shades of pigments found in the petals of many flowers and in some bacteria. They are soluble in water and alcohol but not in ether. Most, if not all, are natural indicators, changing usually from red in acid solutions to green in alkaline solutions. They are glucosides and on hydrolysis yield a sugar and some derivative of benzopyrylium:

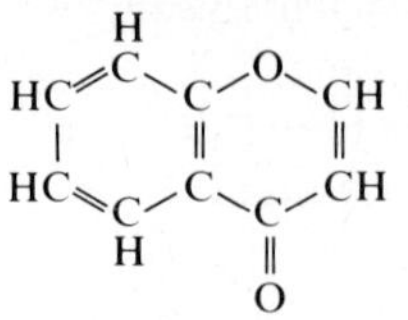

Benzopyrylium

Some species of *Streptomyces* have been reported as being capable of producing anthocyanins.

Melanins The melanins include certain brown, black, orange, and red pigments produced especially by higher organisms such as molds and streptomycetes. They are very complex substances of high molecular weight and are insoluble in most solvents. The pigments are produced by the decomposition of proteins by treatment with boiling concentrated mineral acids or by the action of an enzyme on the amino acid tyrosine. In addition to tyrosine, the enzyme attacks other compounds, including the amino acid tryptophan.

The colored zones surrounding colonies of streptomycetes growing on a peptone agar medium are usually the result of the action of the specific microbial enzyme acting on the tyrosine of the medium.

Tripyrrylmethenes *Serratia marcescens* produces the red pigment prodigiosin, which has been characterized as a derivative of tripyrrylmethene. It has been fractionated into four components, each showing a different shade of color. The pigment is generally insoluble in water. An occasional strain has been encountered in which one or more of the components are said to be water soluble.

Phenazines *Pseudomonas aureofaciens* and three other species of *Pseudomonas* are capable of producing phenazine α-carboxylic acid:

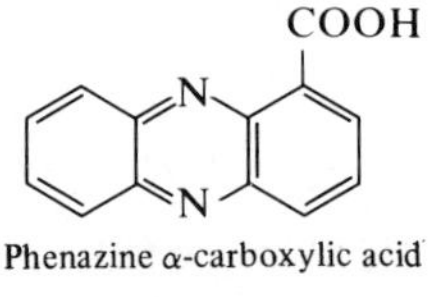

Phenazine α-carboxylic acid

The pigment is water-soluble and imparts first an orange-yellow, then a deep orange, and finally a reddish-orange color to the medium.

Conditions necessary for pigment production It is probable that bacterial pigments are produced only in the presence of oxygen. Pigmented cultures placed in an anaerobic atmosphere slowly lose their color. The color is gradually restored when such cultures are again exposed to air.

Media complete in all nutrients such as minerals, vitamins, proteins, certain amino acids, and carbohydrates or carbohydrate-like compounds appear to be necessary for strong pigment production. Yeast extract, glucose, casein hydrolysates, glycerol, and mineral salts are popular constituents of such media. Both solid and liquid media have been used with favorable results. The pH of media influences the color of those pigments which are natural indicators. Since most of the chromogenic bacteria are soil and water forms, a temperature of 20 to 30°C appears to be best for strong pigment production. Pigment formation decreases with a rise in temperature even though growth occurs. In general, pigment formation takes place best in the dark.

Some pigments are soluble in water and dissolve in culture media; others are not water soluble and remain confined within the bacterial cells. The water-insoluble pigments are soluble in fat solvents such as alcohol, acetone, ether, chloroform, and carbon tetrachloride.

Function of bacterial pigments Little is known concerning the physiological function of nonphotosynthetic pigments. They are colored in the oxidized forms and colorless when reduced. Since the colored forms are generally present only in the presence of free oxygen, some believe they act as respiratory carriers. Evidence available at present is not sufficient to support this view.

For more information: Aasen and Jensen (1967); Aasen et al. (1969); Conti and Benedict (1962); Eimhjellen and Jensen (1964); Hammond and White (1970); Hertzberg and Jensen (1966); Kelly et al. (1970); Knackmuss et al. (1969); Liaaen-Jensen et al. (1968); Norgard et al. (1970); Pfennig et al. (1968); Powers et al. (1960); Pratt et al. (1960); Purkayastha and Williams (1960); Starr et al. (1967); Ungers and Cooney (1968); Weeks and Andrewes (1970); Weeks and Garner (1967); Weeks et al. (1969).

References

Aasen, A. J., and S. L. Jensen: The carotenoids of *Thiorhodaceae*. 6. Total synthesis of okenone and related compounds, *Acta Chem. Scand.*, **21**:970, 1967.

——— et al.: XXIX. The carotenoids of two yellow halophilic cocci—including a new glycosidic methyl apo-lycopenoate, *Acta Chem. Scand.*, **23**:2605, 1969.

Akagi, J. M.: Electron carriers for the phosphoroclastic reaction of *Desulfovibrio desulfuricans, J. Biol. Chem.*, **242**:2478, 1967.

Alico, R. K., and F. W. Liegey: Growth of *Desulfovibrio desulfuricans* under heterotrophic and anaerobic conditions, *J. Bact.*, **91**:1112, 1966.

Ariji, F., et al.: Electron microscopy of the sites of respiratory enzymes in *Escherichia coli, J. Elec. Micros.*, **17**:139, 1968.

Austrian, R., and P. Collins: Importance of carbon dioxide in the isolation of pneumococci, *J. Bact.*, **92**:1281, 1966.

Baldwin, E.: "The Nature of Biochemistry," London, Cambridge University Press, 1967.

Bongers, L.: Energy generation and utilization in hydrogen bacteria, *J. Bact.*, **104**:145, 1970.

Breed, R. S., et al.: "Bergey's Manual of Determinative Bacteriology," Baltimore, The Williams & Wilkins Company, 1957.

Brewer, J. H.: A new petri dish cover and technique for use in the cultivation of anaerobes and microaerophiles, *Science*, **95**:587, 1942.

—— and D. L. Allgeier: Safe self-contained carbon dioxide–hydrogen anaerobic system, *Appl. Microbiol.*, **14**:986, 1968*a*.

—— and ——: A disposable anaerobic system designed for field and laboratory use, *Appl. Microbiol.*, **16**:848, 1968*b*.

—— et al.: Improved anaerobic indicator, *Appl. Microbiol.*, **14**:135, 1966.

Bryant, M. P., et al.: Hydrogen-oxidizing methane bacteria. I. Cultivation and methanogenesis, *J. Bact.*, **95**:1118, 1968.

Carroll, B. J., and B. Q. Ward: An anaerobic culturing and sampling apparatus, *Can. J. Microbiol.*, **11**:597, 1965.

Clayton, R. K.: Permeability barriers and the assay of catalase in intact cells, *Biochim. Biophys. Acta*, **36**:35, 1959.

Coffey, J. J.: Inducible synthesis of bacterial luciferase: Specificity and kinetics of induction, *J. Bact.*, **94**:1638, 1967.

Committee on Bacteriological Technic (Society of American Bacteriologists): "Manual of Microbiological Methods," New York, McGraw-Hill Book Company, 1957.

Conti, S. F., and C. R. Benedict: Carotenoids of *Rhodomicrobium vannielii*, *J. Bact.*, **83**:920, 1962.

Cormier, M. J., et al.: On the requirement and mode of action of long-chain aldehydes during bacterial bioluminescence, *Photochem. Photobiol.*, **9**:351, 1969.

Cusanovich, M. A.: Molecular weights of some cytochromes cc′, *Biochim. Biophys. Acta*, **236**:238, 1971.

DeCicco, B. T., and P. E. Stukus: Autotrophic and heterotrophic metabolism of *Hydrogenomonas*, *J. Bact.*, **95**:1469, 1968.

Deevey, E. S., Jr.: Mineral cycles, *Sci. Am.*, **223**:148, 1970.

Drucker, H., et al.: Amino acid composition, heme content, and molecular weight of cytochrome c_3 of *Desulfovibrio desulfiricans* and *Desulfovibrio vulgaris*, *Biochemistry*, **9**:1515, 1970.

Eastin, J. D., and C. B. Thorne: Carbon dioxide fixation in *Bacillus anthracis*, *J. Bact.*, **85**: 410, 1963.

von Ehlert, I.: Untersuchungen über die korrosion von eisen durch säurebildende bakterien, *Material Organismen*, **2**:297, 1967.

Eimhjellen, K. E., and S. L. Jensen: The biosynthesis of carotenoids in *Rhodopseudomonas gelatinosa*, *Biochim. Biophys. Acta*, **82**:21, 1964.

Eley, M., et al.: Bacterial bioluminescence: Comparisons of bioluminescence emission spectra, the fluorescence of luciferase reaction mixtures, and the fluorescence of flavin cations, *Biochem.*, **9**:2902, 1970.

Evans, J. M., et al.: A modification of the Brewer anaerobic jar, *J. Clin. Path.*, **18**:745, 1948.

Fildes, P.: "A System of Bacteriology," vol. 9, London, Medical Research Council, 1931.

Findley, J. E., and J. M. Akagi: Role of thiosulfate in bisulfite reduction as catalyzed by *Desulfovibrio vulgaris, J. Bact.*, **103**:741, 1970.

Friedland, J., and J. W. Hastings: Nonidentical subunits of bacterial luciferase: Their isolation and recombination to form active enzyme, *Proc. Natl. Acad. Sci.*, **58**:2336, 1967.

Galivan, J. H., and S. H. G. Allen: Methylmalonyl coenzyme A decarboxylase, *J. Biol. Chem.*, **243**:1253, 1968*a*.

—— and ——: Methylmalonyl-CoA decarboxylase: partial purification and enzymatic properties, *Arch. Biochem. Biophys.*, **126**:838, 1968*b*.

Gel'man, N. S., et al.: "Respiration and Phosphorylation of Bacteria," New York, Plenum Press, Plenum Publishing Corporation, 1967.

Goff, R. C., and R. E. Hartman: Carbon dioxide fixation by cells of *Streptococcus faecalis* var. *liquefaciens, J. Bact.*, **104**:27, 1970.

Gordon, J., et al.: Further observations on the production of hydrogen peroxide by anaerobic bacteria, *J. Path. Bact.*, **66**:527, 1953.

Gordon, R. C.: Technique of anaerobic culture eliminating reducing agents in the medium, *J. Bact.*, **85**:1441, 1963.

Gorini, L.: Effect of L-cystine on initiation of anaerobic growth of *Escherichia coli* and *Aerobacter aerogenes, J. Bact.*, **82**:305, 1961.

Gottlieb, S. F., and L. M. Pakman: Effect of high oxygen tensions on the growth of selected, aerobic, gram-negative, pathogenic bacteria, *J. Bact.*, **95**:1003, 1968.

Hadjipetrou, L. P., and A. H. Stouthamer: Energy production during nitrate respiration by *Aerobacter aerogenes, J. Gen. Microbiol.*, **38**:29, 1965.

Hammond, R. K., and D. C. White: Carotenoid formation by *Staphylococcus aureus, J. Bact.*, **103**:191, 1970.

Hartman, R. E.: Carbon dioxide fixation by extracts of *Streptococcus faecalis* var. *liquefaciens, J. Bact.*, **102**:341, 1970.

Hastings, J. W., et al.: Structurally distinct bacterial luciferases, *Biochemistry*, **8**:4681, 1969.

Hayaishi, O.: Crystalline oxygenases of pseudomonads, *Bact. Rev.*, **30**:720, 1966.

—— and M. Nozaki: Nature and mechanisms of oxygenases, *Science*, **164**:389, 1969.

Haynes, W. C., et al.: *Pseudomonas aureofaciens* Kluyver and phenazine α-carboxylic acid, its characteristic pigment, *J. Bact.*, **72**:412, 1956.

Hernandez, E., and M. J. Johnson: Energy supply and cell yield in aerobically grown microorganisms, *J. Bact.*, **94**:996, 1967.

Hertzberg, S., and S. L. Jensen: XIX. The carotenoids of *Mycobacterium phlei* strain Vera. 1. The structures of the minor carotenoids, *Acta Chem. Scand.*, **20**:1187, 1966.

Hooper, A. B.. A nitrite-reducing enzyme from *Nitrosomonas europaea, Biochim. Biophys. Acta*, **162**:49, 1968.

International Union of Biochemistry: "Enzyme Nomenclature," New York, American Elsevier Publishing Company, Inc., 1965.

Iverson, W. P.: Growth of *Desulfovibrio* on the surface of agar media, *Appl. Microbiol.,* **14**:529, 1966.

——: Disulfur monoxide: production by *Desulfovibrio, Science,* **156**:1112, 1967.

——: Corrosion of iron and formation of iron phosphide by *Desulfovibrio desulfuricans, Nature,* **217**:1265, 1968.

Jacobs, N. J., et al.: Effect of hemin and oxygen tension on growth and nitrate reduction by bacteria, *J. Bact.,* **87**:1406, 1964.

Jeris, J. S., and P. L. McCarty: The biochemistry of methane fermentation using C^{14} tracers, *J. Water Pollut. Control Fed.,* **37**:178, 1965.

Johnson, M. J.: Aerobic microbial growth at low oxygen concentrations, *J. Bact.,* **94**:101, 1967.

Kaye, D.: Effect of hyperbaric oxygen on aerobic bacteria *in vitro* and *in vivo, Proc. Soc. Exp. Biol. Med.,* **124**:1090, 1967.

Keilin, D.: "The History of Cell Respiration and Cytochrome," New York, Cambridge University Press, 1966.

Kelly, M., et al.: XXXI. C_{50}-carotenoids. 5. Carotenoids of *Halobacterium salinarium*, especially bacterioruberin, *Acta Chem. Scand.,* **24**:2169, 1970.

Khairat, O.: Modernizing Brewer and other anaerobic jars, *J. Bact.,* **87**:963, 1964.

Klein, H. A.: "Bioluminescence," Philadelphia, J. B. Lippincott Company, 1965.

Knackmuss, H.-J., et al.: The soluble blue pigment, indochrome, of *Arthrobacter polychromogenes, Europ. J. Biochem.,* **10**:90, 1969.

Kneteman, A.: A method for the cultivation of anaerobic spore forming bacteria, *J. Appl. Bact.,* **20**:101, 1957.

Knook, D. L., and R. J. Planta: Function of ubiquinone in electron transport from reduced nicotinamide adenine dinucleotide to nitrate and oxygen in *Aerobacter aerogenes, J. Bact.,* **105**:483, 1971.

Krieger, N., and J. W. Hastings: Bioluminescence: pH activity profiles of related luciferase fractions, *Science,* **161**:586, 1968.

Kuwabara, S., et al.: Crystalline bacterial luciferase from *Photobacterium fischeri, Proc. Natl. Acad. Sci. U.S.,* **53**:822, 1965.

Lachica, R. V. F.: Carbon dioxide fixation, *Enzymologia,* **34**:281, 1968.

—— and P. A. Hartman: Carbon dioxide fixation and the synthesis of aspartic acid by *S. faecium* var. *durans, Biochem. Biophys. Res. Commun.,* **32**:691, 1968.

—— and ——: Carbon dioxide fixation by cell-free extracts of group D streptococci, *Can. J. Microbiol.,* **15**:61, 1969.

Laishley, E. J., et al.: Amino acid composition of ferredoxin and rubredoxin isolated from *Desulfovibrio gigas, J. Bact.,* **98**:302, 1969.

Langenberg, K. F., et al.: Hydrogen-oxidizing methane bacteria, *J. Bact.,* **95**:1124, 1968.

la Rivière, J. W. M.: The production of surface active compounds by microorganisms and its possible significance in oil recovery. II. On the release of oil from oil-sand mixtures with the aid of sulfate reducing bacteria, *Antonie van Leeuwenhoek, J. Microbiol. Serol.*, **21**:9, 1955.

Liaaen-Jensen, S., et al.: XXVII. C_{50}-carotenoids. 3. Structure determination of dehydrogenans–P439, *Acta Chem. Scand.*, **22**:1171, 1968.

Longmuir, I. S.: Respiration rate of bacteria as a function of oxygen concentration, *Biochem. J.*, **57**:81, 1954.

MacElroy, R. D., et al.: Control of ATP-dependent CO_2 fixation in extracts of *Hydrogenomonas facilis*: NADH regulation of phosphoribulokinase, *Arch. Biochem. Biophys.*, **131**:272, 1969.

Marshall, J. H.: The production of anaerobic conditions with chromous salts, *J. Gen. Microbiol.*, **22**:645, 1960.

Mateles, R. I., and B. L. Zuber: The effect of exogenous catalase on the aerobic growth of clostridia, *Antonie van Leeuwenhoek, J. Microbiol. Serol.*, **29**:249, 1963.

Matsubara, T., and T. Mori: Studies on denitrification. IX. Nitrous oxide, its production and reduction to nitrogen, *J. Biochem.*, **64**:863, 1968.

McElroy, W. D.: Bioluminescence, *Federation Proc.*, **19**:941, 1960.

—— and H. H. Seliger: Mechanisms of bioluminescent reactions. In "Light and Life," edited by W. D. McElroy and B. Glass, Baltimore, The Johns Hopkins Press, 1961.

—— and ——: Biological luminescence, *Sci. Am.*, **207**:76, 1962.

McFadden, B. A., and C.-C. L. Tu: Regulation of autotrophic and heterotrophic carbon dioxide fixation in *Hydrogenomonas facilis, J. Bact.*, **93**:886, 1967.

—— et al.: CO_2 fixation, glutamate labeling, and the Krebs cycle in ribose-grown *Hydrogenomonas facilis, J. Bact.*, **93**:879, 1967.

Mitchell, G. W., and J. W. Hastings: Light-induced bioluminescence. Isolation and characterization of a specific protein involved in the absorption and delayed emission of light, *Biochemistry*, **9**:2699, 1970.

Miyata, M., and T. Mori: VIII. Production of nitric oxide by denitrifying reaction in the presence of tetramethyl-*p*-phenylenediamine, *J. Biochem.*, **64**:849, 1968.

Moore, M. L., et al.: An improved method for the Brewer anaerobic jar, *Am. J. Clin. Pathol.*, **41**:113, 1964.

Myoda, T., and C. H. Werkman: Carbon dioxide fixation by the genus Mycobacterium, *Iowa State J. Sci.*, **35**:73, 1960.

Nason, A.: Enzymatic pathways of nitrate, nitrite, and hydroxylamine metabolisms, *Bact. Rev.*, **26**:16, 1962.

Norgard, S., et al.: XXXII. C_{50}-carotenoids. 6. Carotenoids from *Corynebacterium poinsettiae* including four new C_{50}-diols, *Acta Chem. Scand.*, **24**:2183, 1970.

Okunuki, K., et al.: "Symposium on Structural and Chemical Aspects of Cytochromes," Tokyo, University of Tokyo Press, 1968.

Parker, C. A.: Anaerobiosis with iron wool, *Australian J. Exp. Biol. Med. Sci.*, **33**:33, 1955.

Peck, H. D., Jr.: Comparative metabolism of inorganic sulfur compounds in microorganisms, *Bact. Rev.*, **26**:67, 1962*a*.

——: The role of adenosine-5′-phosphosulfate in the reduction of sulfate to sulfite by *Desulfovibrio desulfuricans, J. Biol. Chem.*, **237**:198, 1962*b*.

Pfennig, N., et al.: 8. Isolation and characterization of a *Thiothece, Lamprocystis* and *Thiodictyon* strain and their carotenoid pigments, *Arch. Mikrobiol.*, **62**:178, 1968.

Pine, M. J.: Methane fermentation of formate by *Methanobacillus omelianskii, J. Bact.*, **75**:356, 1958.

——: and H. A. Barker: Studies on the methane fermentation. XII. The pathway of hydrogen in the acetate fermentation, *J. Bact.*, **71**:644, 1956.

—— and W. Vishniac: The methane fermentations of acetate and methanol, *J. Bact.*, **73**:736, 1957.

Powers, J. J., et al.: Anthocyanins. II. Action of anthocyanin pigments and related compounds on the growth of certain microorganisms, *Food Technol.*, **14**:626, 1960.

Pratt, D. E., et al.: Anthocyanins. I. The influence of strawberry and grape anthocyanins on the growth of certain bacteria, *Food Res.*, **25**:26, 1960.

Purkayastha, M., and R. P. Williams: Association of pigment with the cell envelope of *Serratia marcescens* (*Chromobacterium prodigiosum*), *Nature*, **187**:349, 1960.

Ramsey, H. H.: Autotrophic and heterotrophic metabolism in *Hydrogenomonas facilis, Antonie van Leeuwenhoek, J. Microbiol. Serol.*, **34**:71, 1968.

Renner, E. D., and G. E. Becker: Production of nitric oxide and nitrous oxide during denitrification by *Corynebacterium nephridii, J. Bact.*, **101**:821, 1970.

Repaske, R.: Characteristics of hydrogen bacteria, *Biotechnol. Bioeng.*, **8**:217, 1966.

Saunders, B. C., et al.: "Peroxidase," London, Butterworth & Co. (Publishers), Ltd., 1964.

Schulman, M., et al.: Total synthesis of acetate from CO_2. V., *J. Bact.*, **109**:633, 1972.

Seliger, H. H., et al.: Stereospecificity and firefly bioluminescence, a comparison of natural and synthetic luciferins, *Proc. Natl. Acad. Sci. U.S.*, **47**:1129, 1961.

Singer, T. P., ed.: "Biological Oxidations," New York, Interscience Publishers, a division of John Wiley & Sons, Inc., 1968.

Stadtman, T. C.: Methane fermentation, *Ann. Rev. Microbiol.*, **21**:121, 1967.

Starr, M. P., et al.: The intracellular blue pigment of *Pseudomonas lemonnieri, Arch. Mikrobiol.*, **59**:287, 1967.

Stephens, W. L., and M. P. Starr: Localization of carotenoid pigment in the cytoplasmic membrane of *Xanthomonas juglandis, J. Bact.*, **86**:1070, 1963.

Stukus, P. E., and B. T. DeCicco: Autotrophic and heterotrophic metabolism of *Hydrogenomonas*: Regulation of autotrophic growth by organic substrates, *J. Bact.*, **101**:339, 1970.

Suh, B., and J. M. Akagi: Formation of thiosulfate from sulfite by *Desulfovibrio vulgaris, J. Bact.*, **99**:210, 1969.

Thiele, O. W., and K. Hoffmann: Coenzyme Q_{10} in *Brucella abortus* Bang, *Naturwissenschaften*, **55**:86, 1968.

Tsuji, F. I., et al.: Luciferase-antiluciferase, *Ann. N.Y. Acad. Sci.*, **103**:715, 1963.

Tuttle, J. H., et al.: Microbial sulfate reduction and its potential utility as an acid mine water pollution abatement procedure, *Appl. Microbiol.*, **17**:297, 1969*a*.

——: Microbial dissimilatory sulfur cycle in acid mine water, *J. Bact.*, **97**:594, 1969*b*.

Ungers, G. E., and J. J. Cooney: Isolation and characterization of carotenoid pigments of *Micrococcus roseus, J. Bact.*, **96**:234, 1968.

Van 'T Riet, J., et al.: Regulation of nitrate assimilation and nitrate respiration in *Aerobacter aerogenes, J. Bact.*, **96**:1455, 1968.

Verhoeven, W.: Studies on true dissimilatory nitrate reduction. V. Nitric oxide production and consumption by microorganisms, *Antonie van Leeuwenhoek, J. Microbiol. Serol.*, **22**:385, 1956*a*.

——: Some remarks on nitrate and nitrite metabolism in microoganisms. In "Inorganic Nitrogen Metabolism," edited by W. D. McElroy and B. Glass, Baltimore, The Johns Hopkins Press, 1956*b*.

Wagner, A. F., and K. Folkers: "Vitamins and Coenzymes," New York, Interscience Publishers, a division of John Wiley & Sons, Inc., 1964.

Weeks, O. B., and A. G. Andrewes: Structure of the glycosidic carotenoid corynexanthin, *Arch. Biochem. Biophys.*, **137**:284, 1970.

—— and R. J. Garner: Biosynthesis of carotenoids in *Flavobacterium dehydrogenans* Arnaudi, *Arch. Biochem. Biophys.*, **121**:35, 1967.

—— et al.: Occurrence of C_{40} and C_{45} carotenoids in the C_{50} carotenoid system of *Flavobacterium dehydrogenans, Nature*, **224**:879, 1969.

Whitcomb, F. C., et al.: Increased carbon dioxide tension and the primary isolation of mycobacteria, *Am. Rev. Respirat. Diseases*, **86**:584, 1962.

Williams, R. P., et al.: Influence of temperature of incubation and type of growth medium on pigmentation in *Serratia marcescens, J. Bact.*, **106**:438, 1971.

——: Induction of pigmentation in nonproliferating cells of *Serratia marcescens* by addition of single amino acids, *J. Bact.*, **106**:444, 1971.

Wilson, A. T.: A simple device for the growth of anaerobic organisms in liquid media, *Proc. Soc. Exp. Biol. Med.*, **75**:515, 1950.

Wood, H. G.: The fixation of carbon dioxide and the interrelationships of the tricarboxylic acid cycle, *Physiol. Rev.*, **26**:198, 1946.

Yagi, K., ed.: "Flavins and Flavoproteins," Tokyo, University of Tokyo Press, 1968.

Zador, S.: Effect of temperature on the redox potential in bacterial cultures, *Acta Biol. Acad. Sci. Hung.*, **9**:387, 1961.

Zeikus, J. G., and R. S. Wolfe: *Methanobacterium thermoautotrophicus* sp. n., an anaerobic, autotrophic, extreme thermophile, *J. Bact.*, **109**:707, 1972.

13 Degradation, putrefaction, and synthesis of proteins by bacteria

Proteins are naturally occurring, extremely complex combinations of amino acids, being essential constituents of all living cells, both plant and animal. The name protein is derived from the Greek and means preeminence or of first importance.

Almost all soluble proteins form colloidal solutions, which indicates that they are composed of very large molecules or macromolecules (Fig. 217). One of the properties of colloids in solution is their inability to pass through certain membranes such as parchment, collodion, and animal. Most proteins form opalescent solutions, another indication that their molecules are large. They are usually amorphous, although some have been obtained in crystalline form. Colloidal substances are classed as either suspensoids or emulsoids. Suspensoids (lyophobic colloids) do not show any affinity for the dispersion medium. On the other hand, emulsoids (lyophilic colloids) show a strong affinity for the dispersion medium. Proteins form solutions of the emulsoid type. Proteins are precipitated from solution by salts of heavy metals such as mercuric chloride and silver nitrate, by acids such as tannic and phosphotungstic, and by certain dyes and detergents. Proteins are amphoteric compounds, being capable of reacting with both acids and bases to form ionizable salts. Proteins cannot be distilled; they are insoluble in organic solvents. Many are very sensitive to heat and even to the mildest reagents.

Another property of proteins is that they are hydrolyzed by (1) strong alkalies, (2) strong acids, and (3) certain groups of specific enzymes. On complete hydrolysis proteins give a mixture of compounds known as α-amino acids.

Alkalies tend to racemize the optically active products of hydrolysis and so are usually avoided. Alkaline hydrolysis also results in the partial or complete destruction of the amino acids cystine, cysteine, and arginine. Acid hydrolysis, especially in the presence of carbohydrate, results usually in the complete destruction of the amino acid tryptophan.

The amino acids are sometimes referred to as the building stones of the protein molecules. The hydrolysis of proteins to the stage of amino acids results in a complete loss of colloidal characteristics.

About 20 different amino acids have been recognized as constituents

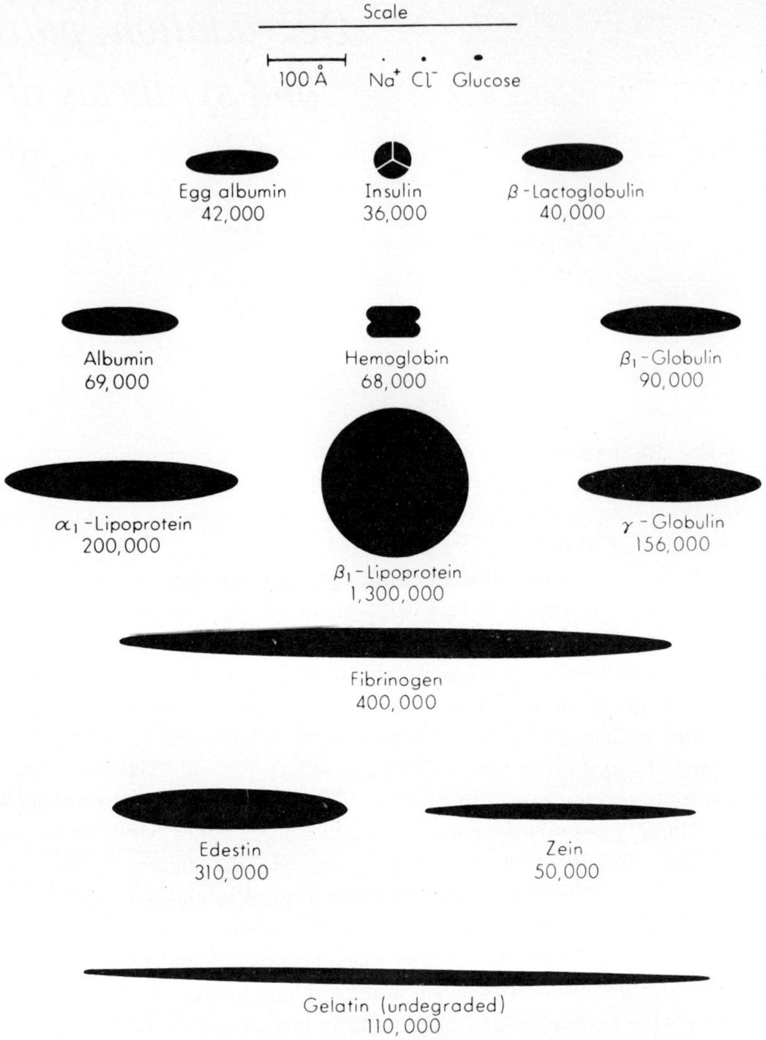

Fig. 217 Relative sizes of various protein molecules. (Courtesy of West and Todd, "Textbook of Biochemistry," The Macmillan Company.)

of proteins. With the exception of glycocoll, all are optically active. They all are soluble in water except tyrosine and cystine. Tyrosine is sparingly soluble in cold water but more soluble in hot water; cystine is sparingly soluble in both hot and cold water. They all are soluble in dilute acids and alkalies, with the exception of cystine, which dissolves with difficulty in dilute ammonia water. All the amino acids possess at least one free amino (NH_2) group and one free carboxyl (COOH) group, except proline and oxyproline, which contain an imino (NH) group in place of an amino group. This means that they are amphoteric, being capable of reacting with both acids and bases (see page 267). Most of the acids have one free amino group and free carboxyl group and are neutral in reaction; some have two amino groups to one carboxyl group and are alkaline in reaction; others have one amino to two carboxyl groups and are acid in reaction.

Structure of protein molecule From the known facts, it is agreed that the amino acids in a protein molecule are linked together principally through their α-amino and carboxyl groups to produce the so-called peptide linkage (R·CONH·R′). For example, in a tripeptide composed of glycine, alanine, and serine, the acids are joined end to end, the COOH of one acid combining with the NH_2 of the next with the loss of a molecule of H_2O:

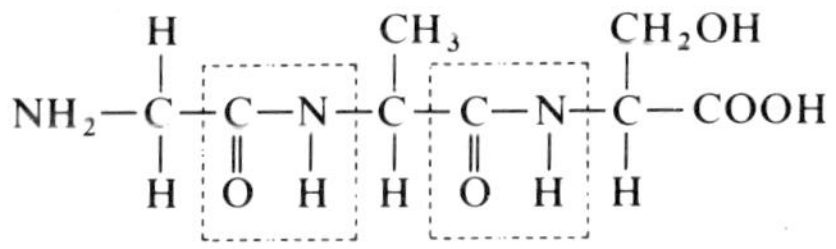

Glycyl-alanyl-serine

Under the action of acids, alkalies, or certain enzymes, peptide chains or residues break apart with the addition of water producing a hydrolysis of the peptide bond.

As stated above, protein molecules are very large and complex in structure. Insulin, a relatively small protein molecule, contains 777 atoms. Some proteins may be 50 times larger. Compare these with an inorganic substance such as sodium chloride, or common table salt, which contains only 2 atoms and you can better visualize their enormous sizes.

All enzymes are also proteins, although some work in conjunction with simpler nonprotein compounds known as coenzymes. The enzyme ribonuclease digests the cellular substance ribonucleic acid (RNA) (Fig. 218). In this two-dimensional formula by Stein and Moore (1961) are 587 atoms of carbon (C), 909 of hydrogen (H), 197 of oxygen (O), 171 of nitrogen (N), and 12 of sulfur (S), making a total of 1,876 atoms. The backbone of the chain of amino acid residues is in the darker shaded area; the side chains characteristic of the various amino acids are in the lighter shaded area. The amino acid residues are numbered from 1 (free NH_2 group, upper left) to 124 (free COOH group, upper right). Abbreviations of the amino acids are as follows: ALA = alanine; ARG = arginine; ASP = aspartic acid; ASP NH_2 = asparagine; CYS = cystine; GLU = glutamic acid; GLU NH_2 = glutamine; GLY = glycine; HIS = histidine; ILEU = isoleucine; LEU = leucine, LYS = lysine; MET = methionine; PHE = phenylalanine; PRO = proline; SER = serine; THR = threonine; TYR = tyrosine; VAL = valine.

For more information: Corrigan (1969); Fraser (1969); Kendrew (1961, 1963); Stein and Moore (1961); Witkop (1968); Dickerson (1972).

Proteolytic enzymes The hydrolysis of proteins by proteolytic enzymes results in the formation of the following fractions (including corresponding

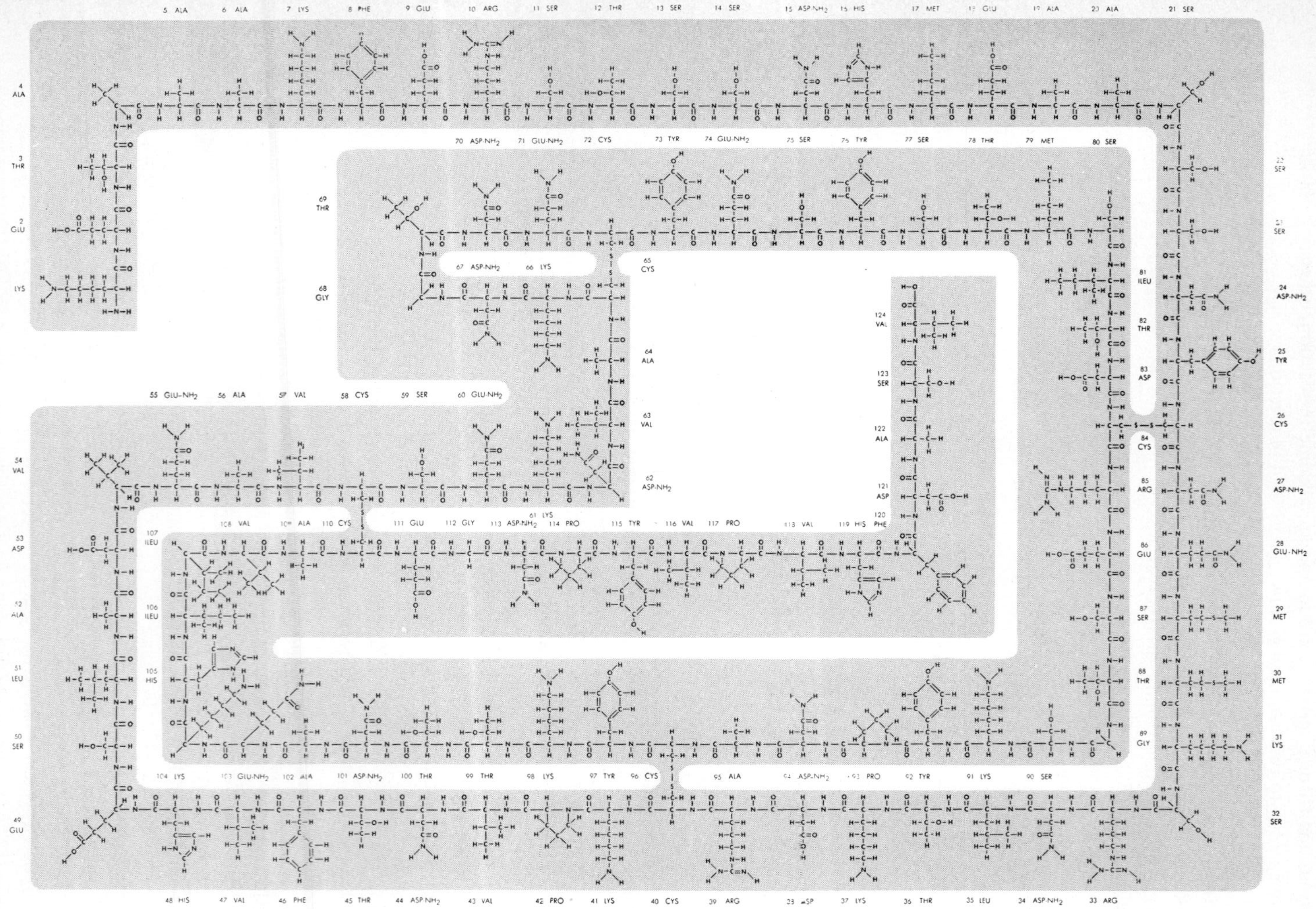

Fig. 218 A molecule of ribonuclease, an enzyme that digests ribonucleic acid. See text for description. (After Stein and Moore.)

approximate molecular weights) in the order named:

Proteins (>10,000) ⟶ proteoses (~5000) ⟶
peptones (~2000) ⟶
peptides (~1000 to 500) ⟶
dipeptides (~200) ⟶ α-amino acids (~100)

Enzymes which open up or hydrolyze peptide linkages are grouped under the hydrolases (see page 368). Some important hydrolases are subdivided as follows:[1]

III. HYDROLASES
- *C.* Acting on peptide bonds (peptide hydrolases)
  3. α-Aminopeptide aminoacidohydrolases
     - *a.* Leucine aminopeptidase. Hydrolyzes L-peptides, splitting off N-terminal residues with a free α-amino group, especially when the N-terminal residue is leucine or a related amino acid.
     - *b.* Aminopeptidase. Hydrolyzes dipeptides and tripeptides, splitting off the N-terminal residue.
     - *c.* Aminotripeptidase. Hydrolyzes tripeptides containing neutral amino acid residues, splitting off the N-terminal residue.
  4. α-Carboxypeptide aminoacidohydrolases
     - *a.* Carboxypeptidase A. Hydrolyzes peptides, splitting off the C-terminal L-amino acid residue unless this is a basic residue or proline.
     - *b.* Carboxypeptidase B. Hydrolyzes peptides having a C-terminal lysine or arginine residue, splitting off this residue.
     - *c.* Yeast carboxypeptidase. Hydrolyzes peptides having a C-terminal glycine or leucine residue, splitting off this residue.
  5. Dipeptide hydrolases
     - *a.* Glycyl-glycine hydrolase. Hydrolyzes glycyl-glycine.
     - *b.* Glycyl-leucine hydrolase. Hydrolyzes glycyl-L-leucine.
     - *c.* L-Cysteinyl-glycine hydrolase. Hydrolyzes L-cysteinyl-glycine.
     - *d.* Prolidase. Hydrolyzes dipeptides in which a proline residue is bound by its imino group.
  6. Peptidyl peptide hydrolases
     - *a.* Pepsin. Hydrolyzes peptides, including those with bonds adjacent to aromatic or dicarboxylic L-amino acid residues.
     - *b.* Rennin. Hydrolyzes peptides. Specificity may be similar to pepsin.
     - *c.* Trypsin. Hydrolyzes peptides, amides, and esters at bonds involving the carboxyl groups of L-arginine or lysine.
     - *d.* Papain. Hydrolyzes peptides, amides, and esters, especially at bonds involving basic amino acids, or leucine or glycine.

[1] This outline retains the numbering that appears in the outline on pages 376–378.

Protein hydrolysis Protein hydrolysis may be defined as the action of peptidyl peptide hydrolases on a protein resulting in the disruption of the molecule into diffusible fractions. Many peptidyl peptide hydrolases are known which have the power to break the linkage –CO NH– by the addition of water. This results in the release of free COOH and NH_2 groups.

The peptidyl peptide hydrolases of bacteria are probably extracellular enzymes. Their function is to convert indiffusible proteins into diffusible polypeptides and dipeptides which can enter the bacterial cell. The compounds produced are without offensive odor. The extracellular enzymes play no part in the respiratory activities of the cell, such function being performed by the intracellular enzymes.

The peptidyl peptide hydrolases attack both proteins and polypeptides, provided certain conditions are fulfilled. For example, trypsin attacks those bonds containing the basic arginyl or lysyl group attached to the carbonyl side of the polypeptide bond. Likewise, pepsin attacks those bonds containing the tyrosyl or phenylalanyl group attached to the imino side of the polypeptide bond as shown below:

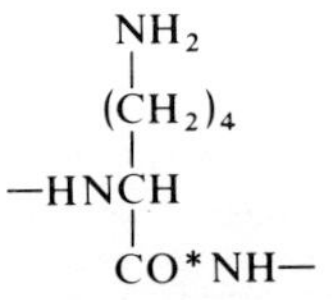

Lysyl group

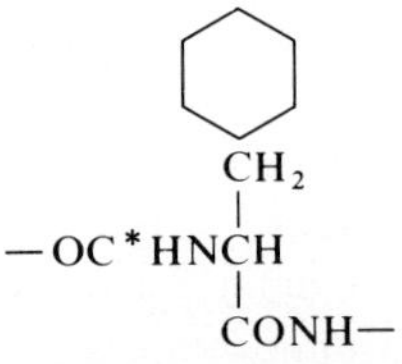

Phenylalanyl group

The structure necessary for the action of trypsin is on the left; the structure necessary for the action of pepsin is on the right. (* Denotes the peptide link hydrolyzed.)

The polypeptide and dipeptide hydrolases may be both intracellular and extracellular. They attack polypeptides and dipeptides, reducing them to mixtures of smaller units including the individual amino acids. The polypeptide and dipeptide hydrolases are also capable of attacking protein molecules, provided certain structural requirements are satisfied. For example, leucine aminopeptidase hydrolyzes L-peptides, splitting off N-terminal residues with a free α-amino group, especially when the N-terminal residue is leucine or a related amino acid. Likewise, carboxypeptidase B hydrolyzes peptides having a C-terminal lysine or arginine residue, splitting off this residue.

For more information: International Union of Biochemistry (1965); Tsuru, et al. (1970).

Putrefaction The term *putrefaction* may be defined as the anaerobic decomposition of proteins, protein split products, and nitrogenous compounds of a similar nature, resulting in the formation of incompletely oxidized, foul-smelling compounds. The reactions occur inside the cell by means of the intracellular or respiratory enzymes. The putrefactive changes are the result of the action of organisms on the amino acids. The important compounds produced include sulfides, amines, alcohols, organic acids, hydroxy acids, indole, phenol, cresol, and the gases ammonia, methane, carbon dioxide, and hydrogen. Many of the putrefactive compounds are produced only from specific amino acids; others may be produced from more than one acid.

Bacteria vary considerably in their ability to degrade proteins or protein split products. Organisms are usually designated as putrefactive or fermentative, depending upon whether they act more vigorously on proteins or on fermentable substances.

Decay Decay may be defined as the aerobic decomposition of proteins in which the products of putrefaction are completely oxidized to stable compounds having no foul odors. The process takes place in nature by aerobic or facultative organisms in the presence of air after the putrefactive changes have occurred. If the substrate is well aerated from the start, the reactions will be aerobic without the formation of foul-smelling end products.

A practical application of this principle is employed in the disposal of sewage. In one process the sewage is first digested under anaerobic conditions, resulting in the liberation of offensive odors. Then the anaerobic digestate is well aerated, after which the aerobic organisms attack the foul-smelling compounds. This results in the disappearance of the offensive odors. In another process the sewage is kept well aerated from the start, thus preventing the formation of foul-smelling, incompletely oxidized products.

Action of bacteria on proteins and amino acids

A pure native protein, when present in a medium as the only source of carbon and nitrogen, is resistant to attack even by the most proteolytic species. If, however, a small amount of peptone is added, multiplication of the organisms and degradation of the protein occur.

The results would indicate that an extracellular enzyme is necessary to convert the indiffusible protein into diffusible fractions. In the absence of an available nitrogen and carbon source, the organisms are unable to multiply and elaborate the necessary enzyme. However, peptone can enter the cell and stimulate the organisms to produce the necessary proteolytic enzyme. The enzyme is then secreted outside of the cell, where it can attack the indiffusible protein and convert it into diffusible fractions which can enter the cell.

Proteoses are intermediate products in the digestion of proteins by proteolytic enzymes and are also immune to attack by bacteria when present as the only source of nitrogen and carbon. Here again, the addition of a small amount of peptone to the medium prior to inoculation stimulates the organisms to secrete the proteolytic enzyme necessary to attack the proteose.

Action on amino acids   Bacteria may attack amino acids in a variety of ways. They may remove the NH_2 group and utilize the resulting ammonia as a source of nitrogen. They may destroy the COOH group resulting in the formation of basic compounds known as amines. These are only two of the many types of reactions produced by bacteria acting on amino acids. Other types of reactions are the following:

1. Oxidative deaminization, resulting in the formation of an α-keto acid:

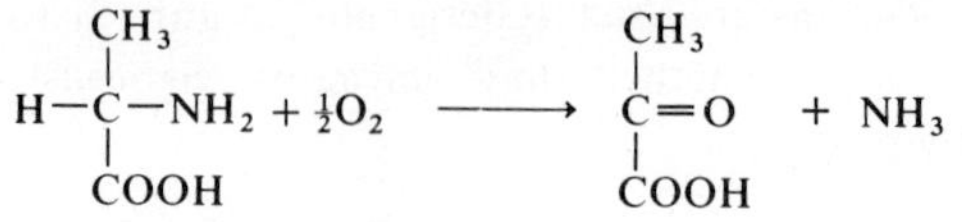

$$
\begin{array}{ccc}
CH_3 & & CH_3 \\
| & & | \\
H-C-NH_2 + \tfrac{1}{2}O_2 & \longrightarrow & C=O \quad + NH_3 \\
| & & | \\
COOH & & COOH \\
\text{Alanine} & & \text{Pyruvic acid}
\end{array}
$$

Some organisms may produce an oxidative deaminization of the dicarboxylic aspartic and glutamic acids, resulting in the formation of the corresponding α-keto acids. Others are able to metabolize the compounds to acetic acid, ammonia, and carbon dioxide.

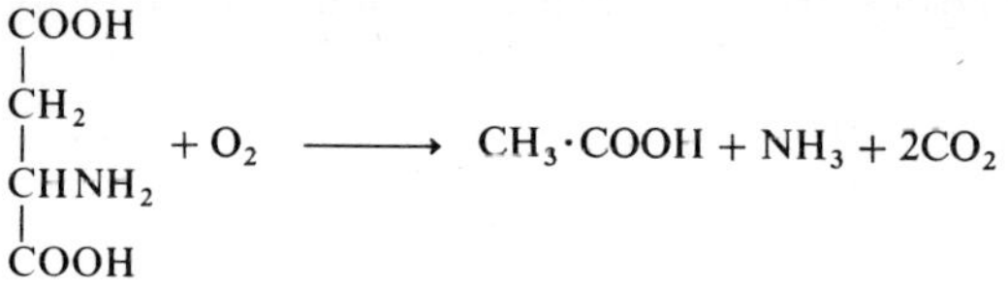

$$
\begin{array}{l}
COOH \\
| \\
CH_2 \\
| \qquad\qquad + O_2 \longrightarrow CH_3 \cdot COOH + NH_3 + 2CO_2 \\
CHNH_2 \\
| \\
COOH \\
\text{Aspartic acid} \qquad\qquad\qquad \text{Acetic acid}
\end{array}
$$

2. Reductive deaminization, resulting in the formation of a saturated acid. This reaction is generally produced by the strict or facultative anaerobes grown under anaerobic conditions:

$$
\begin{array}{ccc}
CH_3 & & CH_3 \\
| & & | \\
H-C-NH_2 + 2H & \longrightarrow & H-C-H + NH_3 \\
| & & | \\
COOH & & COOH \\
\text{Alanine} & & \text{Propionic acid}
\end{array}
$$

3. Hydrolytic deaminization, resulting in the formation of a hydroxy acid:

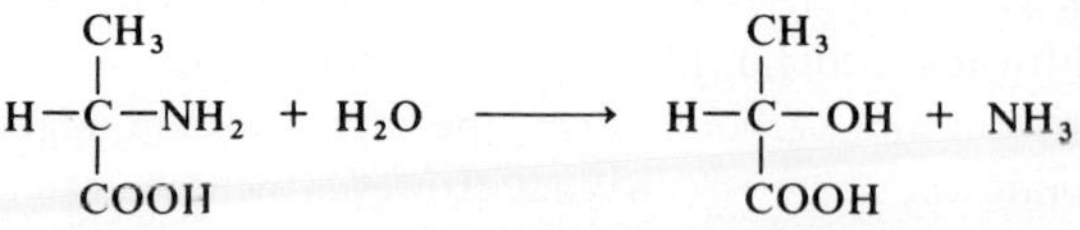

$$
\begin{array}{ccc}
CH_3 & & CH_3 \\
| & & | \\
H-C-NH_2 + H_2O & \longrightarrow & H-C-OH + NH_3 \\
| & & | \\
COOH & & COOH \\
\text{Alanine} & & \text{Lactic acid}
\end{array}
$$

Deaminization usually occurs in the bacteria grown in a neutral or slightly alkaline medium.

4. Deaminization and desaturation at the α,β-linkage, resulting in the formation of an unsaturated acid:

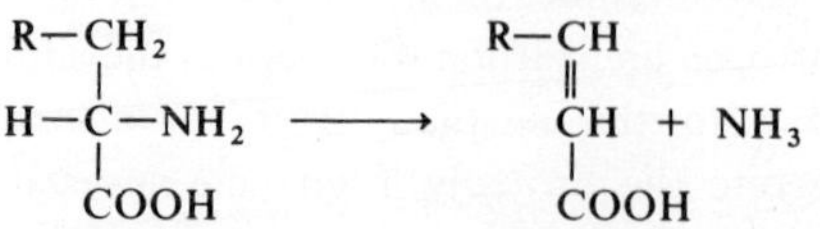

$$
\begin{array}{ccc}
R-CH_2 & & R-CH \\
| & & \| \\
H-C-NH_2 & \longrightarrow & CH + NH_3 \\
| & & | \\
COOH & & COOH \\
\text{Amino acid} & & \text{Unsaturated acid}
\end{array}
$$

5. Hydrolytic deaminization and decarboxylation, resulting in the formation of a primary alcohol with one less carbon atom:

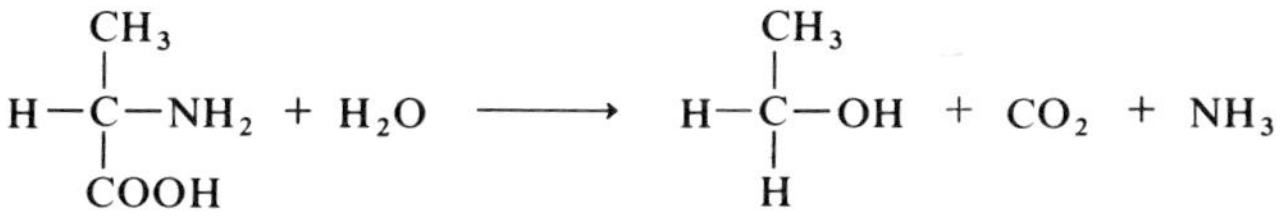

Alanine Ethyl alcohol

6. Decarboxylation or elimination of carbon dioxide, resulting in the formation of an amine with one less carbon atom:

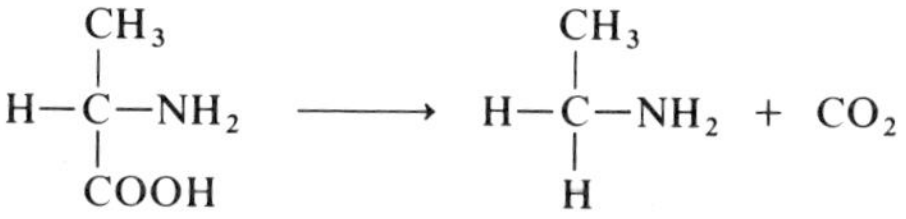

Alanine Ethylamine

Decarboxylases are generally produced by growing bacteria at a low pH. This is achieved by cultivating the organisms in a fermentable carbohydrate medium.

For more information: Haughton and King (1961).

7. Anaerobic decomposition with the liberation of hydrogen:

$$5COOH \cdot (CH_2)_2 \cdot CHNH_2 \cdot COOH + 6H_2O \longrightarrow$$

Glutamic acid

$$6CH_3 \cdot COOH + 2CH_3 \cdot (CH_2)_2 \cdot COOH + 5CO_2 + 5NH_3 + H_2$$

Acetic acid Butyric acid

8. Decomposition, resulting in the oxidation of one molecule of an amino acid and the reduction of another:

$$3CH_3 \cdot CHNH_2 \cdot COOH + 2H_2O \longrightarrow$$

Alanine

$$2CH_3 \cdot CH_2 \cdot COOH + CH_3 \cdot COOH + 3NH_3 + CO_2$$

Propionic acid Acetic acid

9. Transamination reaction in which an amino group from either aspartic or glutamic acid is transferred to the α-position in an α-keto acid:

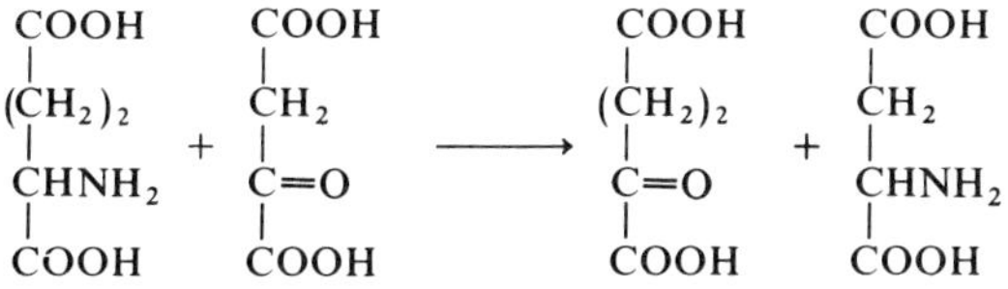

Glutamic acid Oxalacetic acid α-Keto glutaric acid Aspartic acid

10. Mutual oxidation and reduction by pairs of amino acids. Some amino acids (alanine, valine, leucine, phenylalanine) have been shown to be oxidizable by serving as hydrogen donators, whereas others (glycine, proline, hydroxyproline, ornithine, arginine) are reducible by serving as hydrogen acceptors. This reaction takes place by anaerobes cultivated under anaerobic conditions. By mixing together one amino acid from each group, i.e., a hydrogen donator and a hydrogen

acceptor, and inoculating the medium with an anaerobe, the former acid becomes oxidized and the latter reduced:

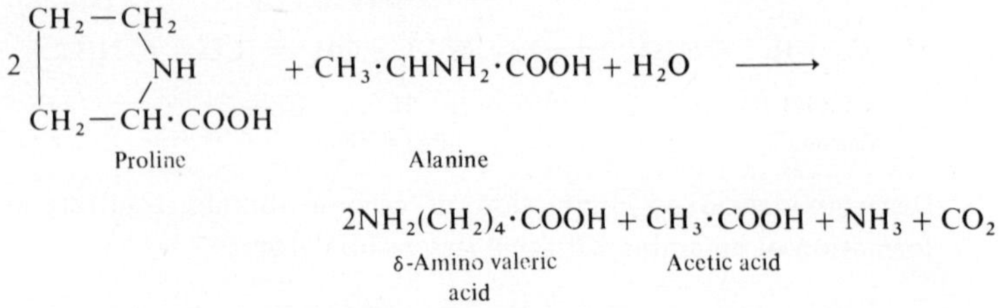

The foregoing types of chemical reactions or their combinations are able to account for nearly all the products of decomposition and putrefaction formed by bacteria from amino acids.

The names and formulas of the important amino acids present in proteins and the compounds produced from each by bacterial action are given in Table 14. The formulas of the amino acids are given at the top of each division. The table does not include all possible compounds but most of those reported to be produced by the enzymatic reactions already discussed. The numbers refer to the types of reactions given on page 454. A few compounds are not preceded by a number, indicating that the mechanism of their formation is not clearly understood.

For more information: Brysk et al. (1969); Chandra and Vining (1968); Costilow and Laycock (1968, 1969); Dyer and Costilow (1970); Fujioka and Wada (1968); Grant and Patel (1969); Gryder and Adams (1969); Kent and Gortner (1960); Kumagai et al. (1969); Morris and Koffron (1967); Raynaud and Daste (1962); Varga and Neujahr (1970).

Ptomaines Ptomaines may be defined as basic amines and diamines which are formed by the action of putrefactive bacteria on amino acids and organic bases. They are produced by decarboxylation (reaction 6, page 455). The ptomaines produced from most of the amino acids are given in Table 15. The organic base choline is an example of a nitrogenous compound that is not an amino acid.

Protein-sparing action In general, bacteria prefer fermentable carbohydrates to proteins for energy. In the presence of both types of compounds, proteins are used principally for structure. In the absence of carbohydrates, proteins are used for both structure and energy.

Rapid fermenters like *Escherichia coli* and *Proteus vulgaris* exhibit a definite sparing action on proteins. The organisms produce sufficient acid from carbohydrates to stop their growth. This occurs after a short incubation period. Under these conditions only a minimum of action occurs on the proteins of media.

In the presence of an excess of buffer, such as a mixture of primary and secondary phosphates, the limiting pH is most likely not attained. Under these conditions protein breakdown occurs simultaneously with carbohydrate utilization. This results in a negative protein-sparing action.

Slow fermenters like *Bacillus subtilis* do not produce strongly acid products and nonfermenters like *Aerobacter cloacae* produce alkaline

Table 14 Action of bacteria on the amino acids

$CH_2NH_2\cdot COOH$ Glycocoll or glycine	
(1) Glyoxylic acid, $CHO\cdot COOH$ (2) Acetic acid, $CH_3\cdot COOH$	(8) Acetic acid, $CH_3\cdot COOH + CO_2 + NH_3$ (10) Acetic acid, $CH_3\cdot COOH + NH_3$
$CH_3\cdot CHNH_2\cdot COOH$ Alanine	
(1) Pyruvic acid, $CH_3\cdot CO\cdot COOH$ (2) Propionic acid, $CH_3\cdot CH_2\cdot COOH$	(6) Ethylamine, $CH_3\cdot CH_2NH_2$ (8) Propionic acid, $CH_3\cdot CH_2\cdot COOH$, and acetic acid, $CH_3\cdot COOH$
$\begin{array}{l}CH_3\diagdown \\ \qquad CH\cdot CHNH_2\cdot COOH \\ CH_3\diagup\end{array}$ Valine	
(2) Isovaleric acid, $\begin{array}{l}CH_3\diagdown \\ \qquad CH\cdot CH_2\cdot COOH \\ CH_3\diagup\end{array}$ (5) Isobutyl alcohol, $\begin{array}{l}CH_3\diagdown \\ \qquad CH\cdot CH_2OH \\ CH_3\diagup\end{array}$	(6) Isobutylamine, $\begin{array}{l}CH_3\diagdown \\ \qquad CH\cdot CH_2NH_2 \\ CH_3\diagup\end{array}$
$\begin{array}{l}CH_3\diagdown \\ \qquad CH\cdot CH_2\cdot CHNH_2\cdot COOH \\ CH_3\diagup\end{array}$ Leucine	
(1) Keto-isocaproic acid, $\begin{array}{l}CH_3\diagdown \\ \qquad CH\cdot CH_2\cdot CO\cdot COOH \\ CH_3\diagup\end{array}$	(2) β-Isopropyl propionic acid, $\begin{array}{l}CH_3\diagdown \\ \qquad CH\cdot CH_2\cdot CH_2\cdot COOH \\ CH_3\diagup\end{array}$
$\begin{array}{l}CH_3\diagdown \\ \qquad CH\cdot CHNH_2\cdot COOH \\ C_2H_5\diagup\end{array}$ Isoleucine	
(1) β-Methylethylpyruvic acid, $\begin{array}{l}CH_3\diagdown \\ \qquad CH\cdot CO\cdot COOH \\ C_2H_5\diagup\end{array}$	(2) Methylethylpropionic acid, $\begin{array}{l}CH_3\diagdown \\ \qquad CH\cdot CH_2\cdot COOH \\ C_2H_5\diagup\end{array}$
$CH_2OH\cdot CHNH_2\cdot COOH$ Serine	
(5) Ethylene glycol, $CH_2OH\cdot CH_2OH$	(8) Propionic acid, $CH_3\cdot CH_2\cdot COOH$, and acetic acid, $CH_3\cdot COOH$

Table 14 (Continued)

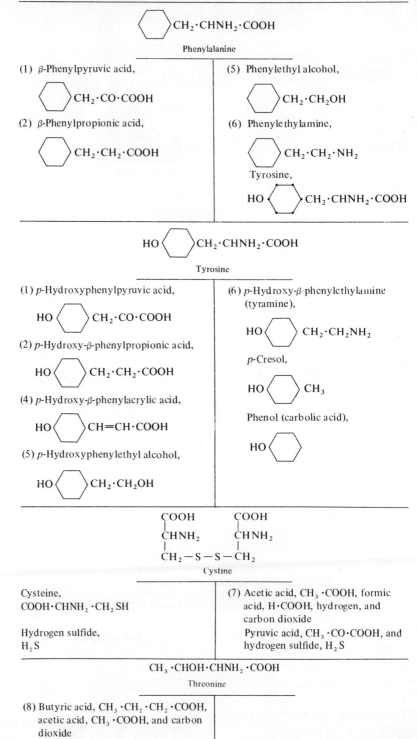

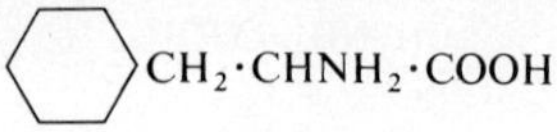

Phenylalanine

(1) β-Phenylpyruvic acid,

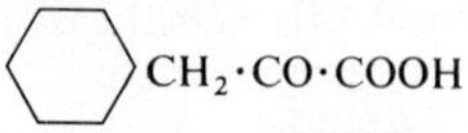

(2) β-Phenylpropionic acid,

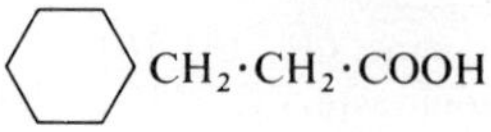

(5) Phenylethyl alcohol,

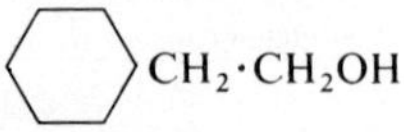

(6) Phenylethylamine,

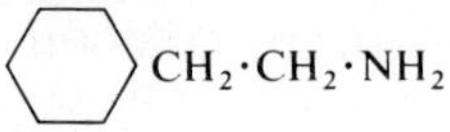

Tyrosine,

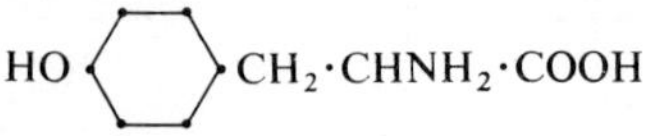

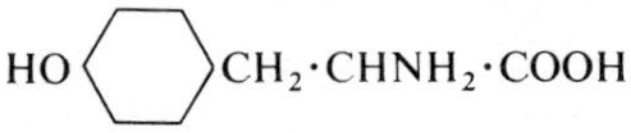

Tyrosine

(1) *p*-Hydroxyphenylpyruvic acid,

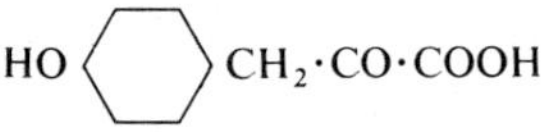

(2) *p*-Hydroxy-β-phenylpropionic acid,

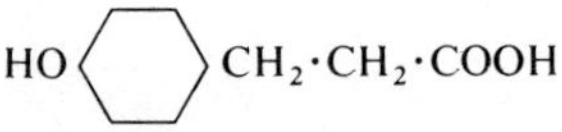

(4) *p*-Hydroxy-β-phenylacrylic acid,

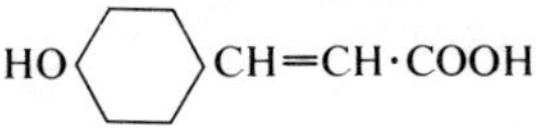

(5) *p*-Hydroxyphenylethyl alcohol,

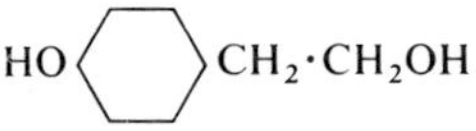

(6) *p*-Hydroxy-β-phenylethylamine (tyramine),

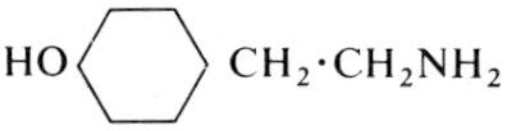

p-Cresol,

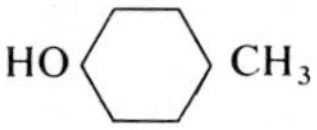

Phenol (carbolic acid),

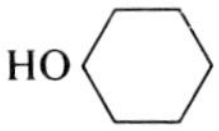

COOH COOH
| |
CHNH$_2$ CHNH$_2$
| |
CH$_2$—S—S—CH$_2$

Cystine

Cysteine,
COOH·CHNH$_2$·CH$_2$SH

Hydrogen sulfide,
H_2S

(7) Acetic acid, CH$_3$·COOH, formic acid, H·COOH, hydrogen, and carbon dioxide

Pyruvic acid, CH$_3$·CO·COOH, and hydrogen sulfide, H_2S

CH$_3$·CHOH·CHNH$_2$·COOH

Threonine

(8) Butyric acid, CH$_3$·CH$_2$·CH$_2$·COOH, acetic acid, CH$_3$·COOH, and carbon dioxide

Table 14 (Continued)

$COOH{\cdot}CH_2{\cdot}CHNH_2{\cdot}COOH$

Aspartic acid

(1) Oxalacetic acid, $COOH{\cdot}CH_2{\cdot}CO{\cdot}COOH$, and acetic acid, $CH_3{\cdot}COOH$ (2) Succinic acid, $COOH{\cdot}CH_2{\cdot}CH_2{\cdot}COOH$ (3) Malic acid, $COOH{\cdot}CH_2{\cdot}CHOH{\cdot}COOH$	(4) Fumaric acid, $COOH{\cdot}CH{=}CH{\cdot}COOH$ (6) β-Alanine, $NH_2CH_2{\cdot}CH_2{\cdot}COOH$ (7) Hydrogen and carbon dioxide (9) Oxalacetic acid, $COOH{\cdot}CH_2{\cdot}CO{\cdot}COOH$, and alanine, $CH_3{\cdot}CHNH_2{\cdot}COOH$

$COOH{\cdot}CH_2{\cdot}CH_2{\cdot}CHNH_2{\cdot}COOH$

Glutamic acid

(1) α-Ketoglutaric acid, $COOH{\cdot}CH_2{\cdot}CH_2{\cdot}CO{\cdot}COOH$ (1) Acetic acid, $CH_3{\cdot}COOH$, and carbon dioxide (2) Glutaric acid, $COOH{\cdot}CH_2{\cdot}CH_2{\cdot}CH_2{\cdot}COOH$ (5) γ-Hydroxybutyric acid, $CH_2OH{\cdot}CH_2{\cdot}CH_2{\cdot}COOH$	(6) γ-Amino butyric acid, $NH_2CH_2{\cdot}CH_2{\cdot}CH_2{\cdot}COOH$ (7) Butyric acid, $CH_3{\cdot}CH_2{\cdot}CH_2{\cdot}COOH$, acetic acid, $CH_3{\cdot}COOH$, carbon dioxide, and hydrogen (9) α-Ketoglutaric acid, $COOH{\cdot}CH_2{\cdot}CH_2{\cdot}CO{\cdot}COOH$, and aspartic acid, $COOH{\cdot}CH_2{\cdot}CHNH_2{\cdot}COOH$

$NH{:}C{\langle}^{NH_2}_{NH_2(CH_2)_3{\cdot}CHNH_2{\cdot}COOH}$

Arginine

(1) δ-Guanidine-α-keto-valeric acid, $NH{:}C{\langle}^{NH_2}_{NH(CH_2)_3{\cdot}CO{\cdot}COOH}$ (6) Agmatine, $NH{:}C{\langle}^{NH_2}_{NH(CH_2)_3{\cdot}CH_2NH_2}$	(10) δ-Guanidine valeric acid, $NH{:}C{\langle}^{NH_2}_{NH(CH_2)_4{\cdot}COOH}$ Ornithine, $NH_2(CH_2)_3{\cdot}CHNH_2{\cdot}COOH$, and urea, ${}^{NH_2}_{NH_2}{\rangle}C{=}O$

$NH_2(CH_2)_3CHNH_2{\cdot}COOH$

Ornithine

(2) δ-Aminovaleric acid (putridin), $NH_2(CH_2)_4{\cdot}COOH$ (5) γ-Aminobutyl alcohol, $NH_2(CH_2)_3CH_2OH$	(6) Tetramethylenediamine (putrescine), $NH_2(CH_2)_3CH_2NH_2$

$NH_2(CH_2)_4{\cdot}CHNH_2{\cdot}COOH$

Lysine

(2) ε-Aminocaproic acid $NH_2(CH_2)_5{\cdot}COOH$	(6) Pentamethylenediamine (cadaverine), $NH_2(CH_2)_4{\cdot}CH_2NH_2$

Table 14 (Continued)

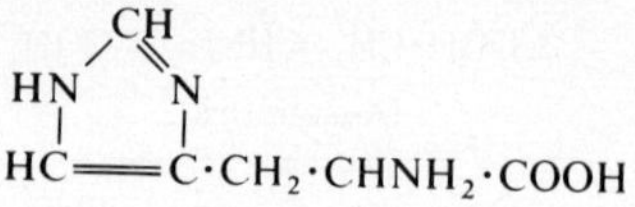

Histidine

(1) β-Imidazole pyruvic acid,

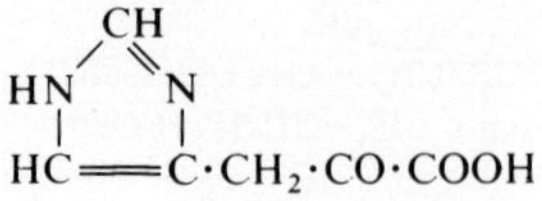

(2) β-Imidazole propionic acid,

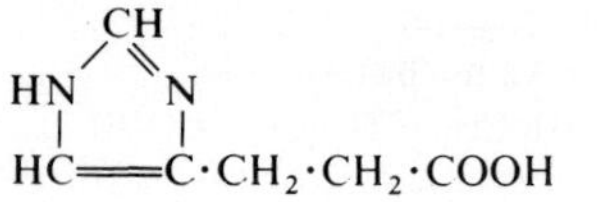

(4) β-Imidazole acrylic acid,

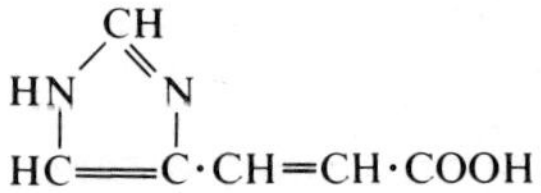

(5) β-Imidazole ethyl alcohol,

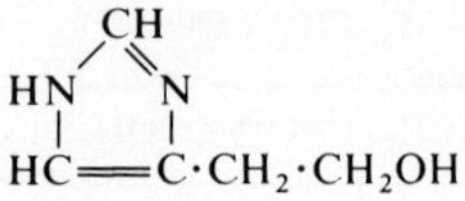

(6) β-Imidazole ethylamine (histamine),

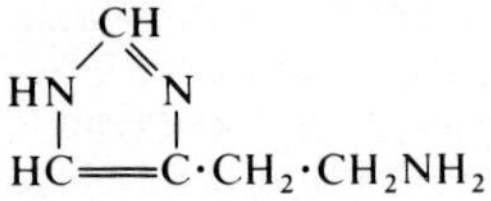

(7) Hydrogen and carbon dioxide

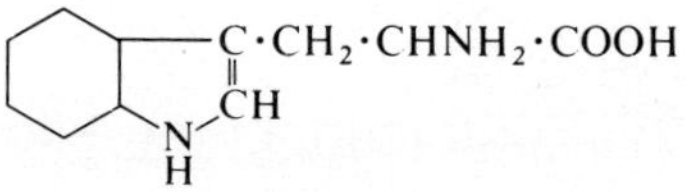

Tryptophan

(1) β-Indole pyruvic acid,

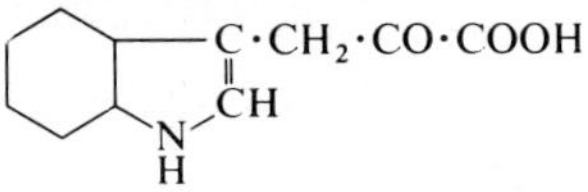

(2) β-Indole propionic acid (skatole acetic acid),

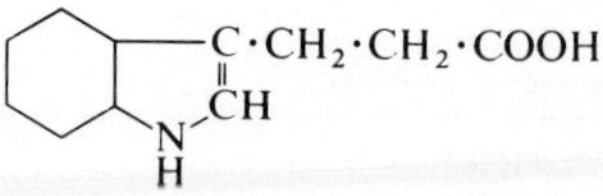

(5) β-Indole ethyl alcohol (tryptophol),

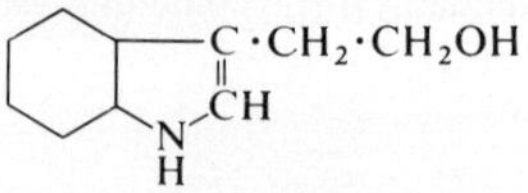

(6) β-Indole ethylamine (tryptamine),

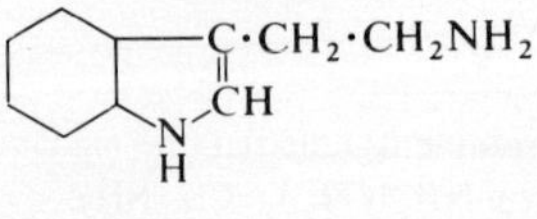

β-Indoleacetic acid,

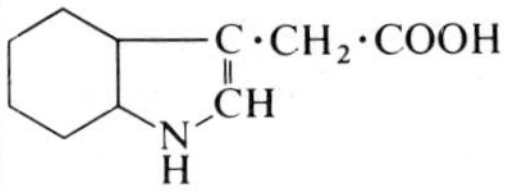

Indole,

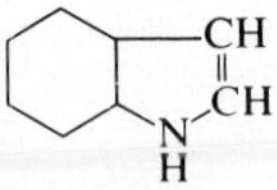

Kynurenic acid,

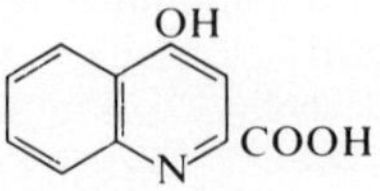

Anthranilic acid,

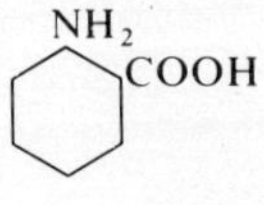

Table 14 (Continued)

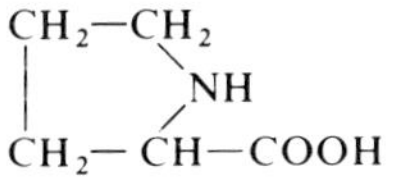

Proline

(2) δ-Aminovaleric acid, $NH_2(CH_2)_4 \cdot COOH$	(10) δ-Aminovaleric acid, $NH_2(CH_2)_4 \cdot COOH$, and carbon dioxide

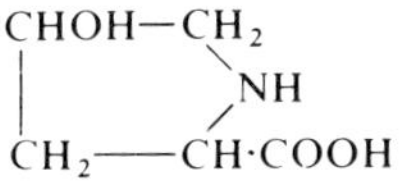

Oxyproline

(10) δ-Amino-γ-hydroxyvaleric acid, $NH_2CH_2 \cdot CHOH \cdot CH_2 \cdot CH_2 \cdot COOH$	α-Ketoglutaric acid, $COOH \cdot CH_2 \cdot CH_2 \cdot CO \cdot COOH$

products. These organisms fail to exhibit a protein-sparing action. The limiting pH is never attained, and protein breakdown is probably the same as in noncarbohydrate media.

It may be concluded that a protein-sparing action occurs only when a fermentable carbohydrate is rapidly utilized, resulting in the production of a low pH in a short time. If the acids are neutralized as formed (buffering), the inhibitory effect of carbohydrate is prevented.

Table 15 Some ptomaines formed by bacterial action

Amino acid	*Ptomaine*
Glycocoll (glycine)	Methylamine
Alanine	Ethylamine
Valine	Isobutylamine
Leucine	Isoamylamine
Phenylalanine	β-Phenylethylamine
Tyrosine	*p*-Hydroxy-β-phenylethylamine (tyramine)
Serine	Hydroxyethylamine
Cystine (cysteine)	Thioethylamine
Methionine	γ-Methylthiolpropylamine
Threonine	β-Hydroxypropylamine
Arginine	δ-Guanidine butylamine (agmatine)
Ornithine	Tetramethylenediamine (putrescine)
Lysine	Pentamethylenediamine (cadaverine)
Histidine	β-Imidazole ethylamine (histamine)
Tryptophan	β-Indole ethylamine (tryptamine)
Choline	Trimethylamine

For more information: Ginsburg (1967).

Some routine tests employed for the identification of bacteria

Many biochemical reactions are employed for the identification and classification of bacteria. Some are based on the breakdown of carbohydrates; others on certain changes produced on the nitrogenous constituents of media. The following tests are limited to the action of bacteria on proteins and their hydrolytic products.

Decomposition of tryptophan Tryptophan is decomposed by bacteria in a variety of ways, resulting in the formation of β-indole propionic acid, β-indole pyruvic acid, β-indole ethylamine, β-indole ethyl alcohol, β-indole acetic acid, indole kynurenic acid, anthranilic acid, and indigotin.

Indole Indole is a putrefactive compound produced from tryptophan by some bacteria, and tests for its presence are frequently made. Tryptophan is the only naturally occurring amino acid containing the indole ring. Some bacteria can produce indole; others cannot. The test is of value in the identification and classification of bacteria.

Tryptophan is not present in all proteins. Manufacturers select only those proteins for peptone production which are certain to yield relatively high concentrations of tryptophan. Casein, the principal protein of milk, is rich in tryptophan and is frequently used for the preparation of peptone where a strong indole test is desirable.

Some organisms are capable of oxidizing indole after it is produced. Oxidation products include isatin, kynurenic acid, formylanthranilic acid, anthranilic acid, salicylic acid, catechol, and indigotin.

To prevent the possible occurrence of a protein-sparing effect, noncarbohydrate media should be employed for the detection of indole in bacterial cultures.

For more information: Cimino and Bracci (1960).

Production of hydrogen sulfide Cystine and methionine are the two sulfur-containing amino acids present in proteins. Practically all the

Fig. 219 Production of hydrogen sulfide. Left, peptone iron agar medium inoculated with Proteus vulgaris. Hydrogen sulfide formed which reacted with the iron to give black iron sulfide. Right, same inoculated with Escherichia coli. Hydrogen sulfide not produced.

information available on the effect of bacteria on these acids has been obtained from cystine or its reduction product cysteine.

Some organisms are capable of attacking cystine with the liberation of hydrogen sulfide (H_2S); others are unable to do so. The test is of value in the identification and classification of bacteria.

Cystine does not occur in all proteins. Manufacturers select only those proteins for peptone production which are certain to contain relatively large amounts of this amino acid.

Under anaerobic conditions cystine is first reduced to two molecules of cysteine, followed by the breakdown of the cysteine to H_2S, NH_3, acetic acid, and formic acid:

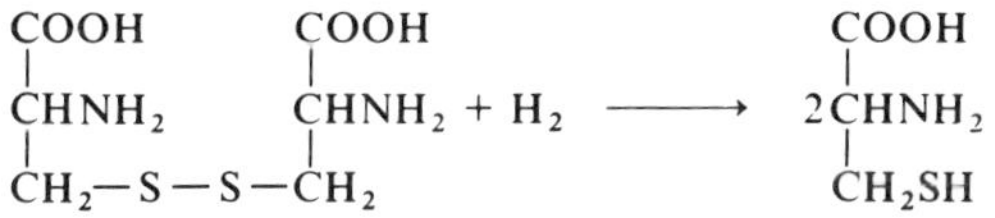

Cystine Cysteine

$$COOH{\cdot}CHNH_2{\cdot}CH_2SH + 2H_2O$$

Cysteine

$$\downarrow$$

$$H_2S + NH_3 + CH_3{\cdot}COOH + HCOOH$$

Acetic acid Formic acid

Under aerobic conditions cysteine is said to be dissimilated as follows:

$$HOOC{\cdot}CHNH_2{\cdot}CH_2SH + O \longrightarrow HOOC{\cdot}CO{\cdot}CH_2SH + NH_3$$

Cysteine

$$HOOC{\cdot}CO{\cdot}CH_2SH \longrightarrow H_2S + \text{other products}$$

H_2S reacts with heavy metals to produce colored sulfides. Metal salts which yield dark-colored sulfides are incorporated in solid media. The formation of H_2S is detected by the darkening of the media along the line of inoculation (Fig. 219).

Lead was the first metal incorporated in culture media. Since that time other metals have been employed because of their greater sensitivity to H_2S. The metals may be arranged in the following order on the basis of their sensitivities to H_2S:

Bi
Co > Fe > Pb
Ni

It may be concluded that it is not sufficient to say that an organism does or does not produce H_2S without giving the conditions under which it was investigated.

For more information: Anderson and Johansson (1963); Segal and Starkey (1969).

Liquefaction of gelatin The liquefaction of gelatin by bacteria is produced by a proteolytic enzyme generally referred to as gelatinase. The enzyme is extracellular and is concerned with the hydrolysis of the indiffusible protein prior to intracellular utilization.

The presence of the enzyme may be demonstrated by inoculating a tube of gelatin medium with the organism under investigation and incubating the culture. If the temperature of incubation is above 20°C, the gelatin will melt. The presence of gelatinase may be demonstrated by placing the culture in the refrigerator and observing whether or not hardening occurs. If the gelatin remains liquid, it indicates the organism secreted gelatinase into the medium.

Another procedure is to streak the organism over the surface of nutrient agar containing 0.4 percent gelatin. After incubation, the agar is flooded with a solution of mercuric chloride dissolved in hydrochloric acid. The reagent forms a white opaque precipitate with the unchanged gelatin, but a liquefier is surrounded by a clear zone.

Still another method is to place strips of exposed photographic film in liquid media and note the disappearance of the black gelatin layer from the film strips.

The extracellular nature of the enzyme may be demonstrated by filtering a culture and adding some of the filtrate to a gelatin medium. If enzyme is present, it will liquefy the gelatin.

The presence of a fermentable carbohydrate may result in a protein-sparing action. Under these conditions a test for gelatin liquefaction will most likely be negative even though the organism under examination is capable of attacking the protein. Therefore, noncarbohydrate media should be employed to demonstrate the ability of an organism to secrete gelatinase. The test is of value in the identification and classification of bacteria.

For more information: Le Minor and Piéchaud (1963); McDade and Weaver (1959); Smith and Goodner (1958).

Fermentation and peptonization of milk Milk contains the following major constituents: water, carbohydrate (lactose), butterfat, protein, minerals, and vitamins.

Milk protein was originally believed to be composed of three components, namely, casein, albumin, and globulin, but it is now known that there are four caseins, each with a number of genetic variants, and that the albumin and globulin are a complex group of proteins known as the whey proteins.

The use of milk as a culture medium dates back to the beginning of bacteriology. It is used as a differential medium to demonstrate a fermentation, a peptonization, or a simultaneous fermentation and peptonization.

Casein (used in the singular to represent all caseins) is present in milk entirely in colloidal suspension. Some bacteria secrete a rennin-like enzyme that is capable of hydrolyzing the protein to soluble paracasein and a peptone-like compound. The soluble paracasein then reacts with the calcium salts in solution to give a precipitate of paracasein or calcium paracaseinate. The clear liquid that separates from the curd is known as whey. This may be diagrammed as follows:

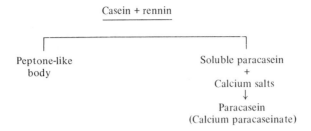

Casein + rennin

Peptone-like body

Soluble paracasein
+
Calcium salts
↓
Paracasein
(Calcium paracaseinate)

Bacteria which ferment lactose vigorously produce sufficient acid to curdle the casein. This is accompanied by the separation of a straw-colored liquid or whey. The acid stops further growth of the bacteria. Since carbohydrate spares protein, no putrefaction occurs (Fig. 220).

Bacteria which do not ferment lactose may act directly on the casein to produce a peptonization (hydrolysis of the molecules into soluble fractions), or a putrefaction. The reaction is alkaline.

Bacteria which ferment lactose very slowly fail to show a protein-sparing effect. The limiting hydrogen-ion concentration is never reached, and protein breakdown occurs simultaneously with carbohydrate utilization. The reaction may be acid or alkaline, depending upon which is dominant, fermentation or putrefaction.

The indicator litmus is frequently added to milk to detect changes in the pH of the medium. It is a feeble dye and somewhat inaccurate as an indicator. The newer and more brilliant sulfonephthalein indicators, such

Fig. 220 Fermentation and peptonization of milk. Left, litmus milk inoculated with Bacillus subtilis. The casein is completely peptonized to a clear purplish solution. Center, litmus milk inoculated with Escherichia coli. The lactose is fermented, followed by a curdling of the casein, and the indicator is completely reduced. Right, litmus milk uninoculated.

as bromocresol purple and bromothymol blue, exhibit greater sensitivity and are superior to litmus for detecting changes in reaction. However, litmus also indicates the oxidation-reduction potential of the medium, a property which is not possessed by the synthetic sulfonephthalein indicators and which often is of diagnostic importance. Some bacteria have the ability to reduce the dye to its colorless form. Because of this fact, litmus continues to enjoy popularity as an addition to milk medium.

For more information: Patton (1969); Ulrich (1944).

Production of ammonia The presence of NH_3, in bacterial cultures containing peptone, results largely from the deaminization of amino acids.

Prior to assimilation, amino acids are believed to be deaminized to NH_3 and a carbon residue. The NH_3 is utilized for structure. If a fermentable carbohydrate is absent from the medium, the organisms utilize the carbon residue for energy. Since more carbon is required for energy than nitrogen for structure, NH_3 accumulates in the medium. In the presence of a fermentable carbohydrate, a protein-sparing effect may occur, in which case the bacteria utilize amino acids principally for structure, and NH_3 does not accumulate in the medium.

In fact a medium with a fermenting carbohydrate generally shows less free NH_3 than the same medium before inoculation and incubation. A fermentable carbohydrate stimulates bacterial growth, resulting in an increased utilization of NH_3. This causes a decrease in the free NH_3 of the medium.

To prevent the possible occurrence of a protein-sparing effect, a noncarbohydrate medium should be employed for the accumulation of NH_3 in bacterial cultures.

Protein synthesis

The role of deoxyribonucleic acid (DNA) in heredity has been well established. It is the carrier of the genes. The basic function of the genes is to store and transmit information concerning the sequence of amino acids in proteins. It is generally believed that gene mutation expresses itself in a change in the sequence of amino acids of an enzyme whose formation is controlled by the gene.

Genes are located in chromosomes inside of the nucleus and cannot directly form proteins. Protein synthesis is a function of the ribonucleic acid (RNA), which is manufactured in the nucleus by DNA. The information contained in the genes is then carried to the cytoplasm by RNA. DNA is a polymer molecule in the form of a double-strand helix containing thousands of subunits. The genetic information is coded in sequences of subunits called bases.

The bacterial nucleus consists of only one molecule of DNA. By appropriate procedures it is possible to extract DNA from bacterial nuclei without rupturing the molecules, after which they may be examined intact under the electron microscope. A molecule of DNA is about 0.001 μ in diameter and about 3000 μ in length or roughly about 1000 times the length of a small bacterial cell.

RNA and ribosomes RNA is a polymer molecule containing hundreds of subunits. It transcribes from DNA the information needed to make a protein molecule and carries it to the site of protein synthesis. Because these molecules of RNA carry the genetic code to the site of protein synthesis, they are called messenger-RNA (m-RNA).

Molecules of m-RNA vary considerably in length. An average molecule has a molecular weight of about 500,000 and contains about 1500 bases. In length such a molecule measures about 3000 Å (0.33 μ).

RNA is present in both nucleus and cytoplasm but mostly in the latter. The site of protein synthesis in the cell is in the cytoplasmic particles known as ribosomes. Ribosomes coordinate the translation of the genetic information in the sequence of nucleotide bases in the m-RNA transcribed from the DNA molecule, or gene, to the sequence of amino acids in each protein manufactured by the cell.

Ribosomes are globular structures composed of about one-third protein and two-thirds RNA. The protein of ribosomes appears to be basic in nature, somewhat the same kind that binds DNA in cell nuclei. Once RNA is made, it is stable and does not break down during the life of the cell. In this respect it behaves like DNA.

Ribosomes are designated 30s, 50s, 70s, etc., depending upon their size. The size is determined by the rate, measured in Svedberg units (s), at which a particle sediments when it is centrifugated at high speed in an ultracentrifuge.

It is believed that ribosomes do not act singly but in clusters known as polyribosomes, or polysomes, working together in orderly fashion like machines on an assembly line. The number of ribosomes in a polysome may show wide variation depending upon the species. Some polysomes may contain only 3 or 4 ribosomes; others may show as many as 70 (Fig. 221).

The region of DNA that determines the amino acid sequence of a particular protein is known as the gene for that protein. Each amino acid is specified by a particular group of three bases in the DNA. The three bases adenine-adenine-thymine specify the amino acid leucine; the three bases guanine-guanine-thymine specify the amino acid proline. The sequence of the three bases in the gene corresponds to the sequence of amino acids in the protein whose structure is directed by the gene.

The coded instructions for making a protein molecule are carried to the ribosomes by the m-RNA. The m-RNA code letters are the four bases, adenine (A), guanine (G), cytosine (C), and uracil (U). A sequence of three bases is called a codon and is required to specify each of the 20 different amino acids present in proteins. When linked end to end by peptide bonds, the amino acids form polypeptide chains or proteins. Each kind of amino acid is transported to the ribosome by a form of RNA known as transfer RNA (t-RNA) which is smaller in molecular size than m-RNA. The t-RNA

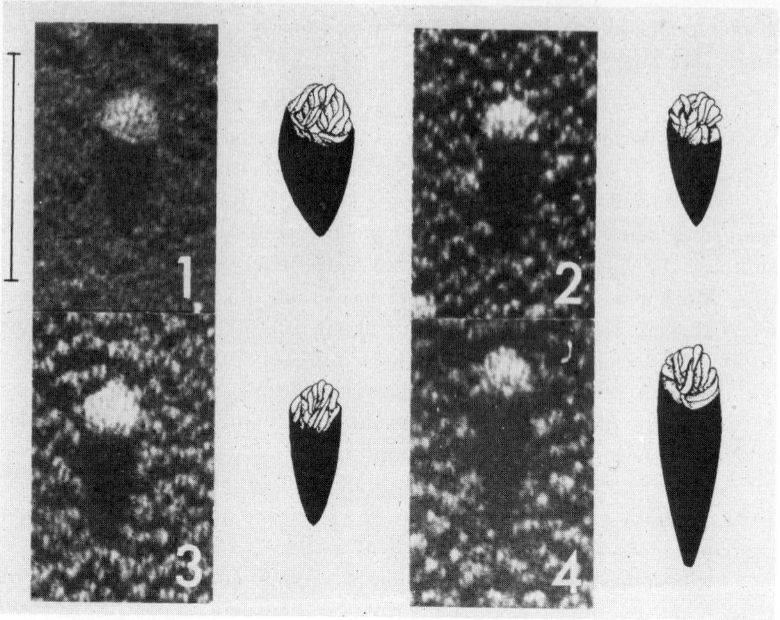

Fig. 221 Tungsten-shadowed 50S ribosomes of Escherichia coli with interpretative drawings. Bar represents 0.1 nm in all micrographs. (Courtesy of Bassel and Campbell.)

carries an anticodon that can form a temporary bond with one of the codons in m-RNA. The ribosomes move along the chain of m-RNA reading off the codons in sequence.

Rich (1963) likened the manufacture of protein to an assembly line. To quote, "In such an assembly line the product moves down the line and component parts are added to it. In the polysome assembly line, the ribosomes move down the line and each one makes a complete product. There is much evidence that the ribosomes are all alike, or at least interchangeable. They can move from one assembly line to another, making whatever protein a given line happens to call for."

The amino acid molecules reach the site of protein synthesis by being first activated by ATP, then recognized by smaller RNA molecules called transfer-RNA (t-RNA) as distinguished from m-RNA. There is believed to be a different t-RNA for each amino acid. The RNA and amino acid are joined together by the action of a specific enzyme. The t-RNA then deposits the given amino acid at a position in the polypeptide (protein) chain specified by m-RNA. The ribosome collaborates with m-RNA to link together amino acids delivered by t-RNA, thereby creating proteins.

Transfer-RNA is a single-strand polymer molecule containing about 75 base units and a molecular weight of about 25,000. It measures about 250 Å in length (Fig. 222). There are at least 20 different kinds of t-RNA. Each is characterized by a specific sequence of bases.

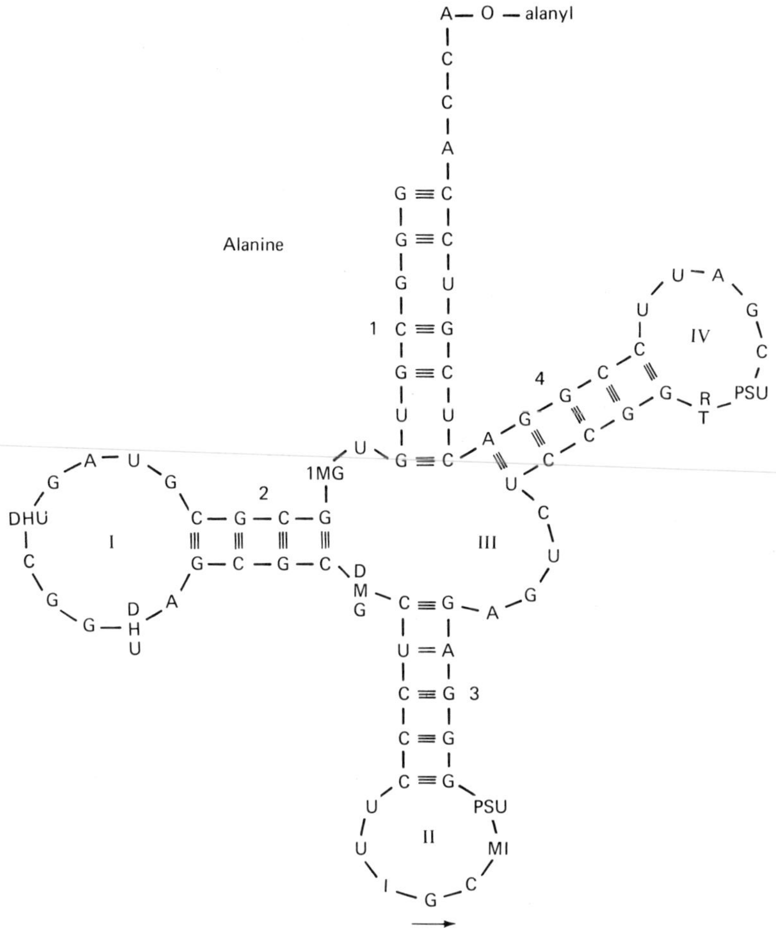

Fig. 222 Cloverleaf structure of analyl-t-RNA from yeast. Top, O-terminal adenosine is ester-linked to the carboxyl group of alanine. Bottom loop contains the anticodon -I-G-C- for alanine. Abbreviations: A = Adenosine; C = Cytidine; G = Guanosine; U = Uridine; RT = Ribothymidine; DHU = 4,5-dihydrouridine; DMG = N^2-dimethylguanosine; 1-MG = 1-methylguanosine; I = Inosine; MI = 1-methylinosine; PSU = Pseudouridine. (Courtesy of Lipmann.)

To quote further from Rich (1963): "If a ribosome is to have access to all information coded in m-RNA, it must 'read' the strand from one end to the other. As it travels it must build up a polypeptide chain, adding one amino acid after another according to instructions."

How the various components function in the synthesis of proteins is illustrated in Fig. 223. "The ribosome at the extreme left has just attached itself to the strand of m-RNA and has started synthesizing a polypeptide chain. The other four ribosomes are proportionately further along in the

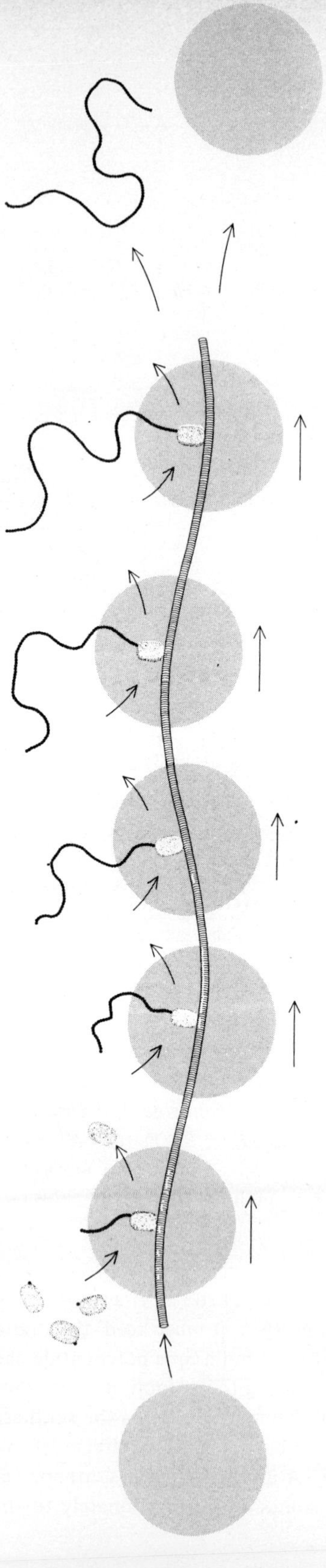

Fig. 223 Polyribosome mechanism. Ribosomes attach themselves to the long strand of messenger RNA. As each ribosome travels along the strand, it reads the information required to synthesize a complete polypeptide chain (protein). The black dots represent amino acids which are delivered by transfer RNA present in the oblong structures. See text for further information. (Redrawn from Rich.)

synthesis process and the one at the extreme right has almost completed a polypeptide chain. At each station along the way the appropriate amino acid, borne by t-RNA, is selected from the cellular milieu and added to the growing polypeptide chain. When the synthesis is complete, the ribosome liberates the polypeptide chain and itself drops off the messenger strand. At about the same time another ribosome has found its way onto the messenger strand at the other end."

The time required for one ribosome to move along the strand of m-RNA and produce a polypeptide molecule in bacteria may be as little as 10 sec.

For more information: Albrecht et al. (1970); Bassel and Campbell (1969); Connors (1969); Crick (1966): DeLey (1964); Doctor et al. (1969); Hendler (1968); Holley et al. (1965); Lipmann (1969); Nanninga (1967); Nirenberg and Leder (1964); Nomura (1969); Oppenheim et al. (1968); Petermann (1965); Schroeder (1968); Stanley et al. (1966); Traub and Nomura (1969); Umbreit (1970); Watson (1963, 1964); Young et al. (1969); Huang and Sypherd (1972).

References

Albrecht, J., et al.: An initiation factor causing dissociation of *E. coli* ribosomes, *Febs Letters,* **6**:297, 1970.

Anderson, D. S., and K. R. Johansson: Effects of glucose on the production by *Escherichia* coli of hydrogen sulfide from cysteine, *J. Gen. Microbiol.,* **30**:485, 1963.

Bassel, A., and L. L. Campbell: Surface structure of *Bacillus stearothermophilus* ribosomes, *J. Bact.,* **98**:811, 1969.

Brysk, M. M., et al.: β-Cyanoalanine formation by *Chromobacterium violaceum, J. Bact.,* **97**:322, 1969.

Chandra, P., and L. C. Vining: Conversion of phenylalanine to tyrosine by microorganisms, *Can. J. Microbiol.,* **14**:573, 1968.

Cimino, G., and R. Bracci: Analisi quantitativa della produzione di indolo in *E. coli, Riv. Biol.,* **52**:395, 1960.

Connors, P. G., et al.: Structural studies on transfer RNA: The molecular conformation in solution, *Science,* **166**:1527, 1969.

Corrigan, J. J.: D-Amino acids in animals, *Science,* **164**:142, 1969.

Costilow, R. N., and L. Laycock: Proline as an intermediate in the reductive deamination of ornithine to δ-amino-valeric acid, *J. Bact.,* **96**:1011, 1968.

—— and ——: Reactions involved in the conversion of ornithine to proline in clostridia, *J. Bact.,* **100**:662, 1969.

Crick, F. H. C.: The genetic code: III., *Sci. Am.,* **215**:55, 1966.

De Ley, J.: On the unity of bacterial ribosomes, *J. Gen. Microbiol.,* **34**:219, 1964.

Dickerson, R. E.: The structure and history of an ancient protein, *Sci. Am.,* **226**:58, 1972.

Doctor, B. P., et al.: Nucleotide sequence of *Escherichia coli* tyrosine transfer ribonucleic acid, *Science,* **163**:693, 1969.

Dyer, J. K., and R. N. Costilow: 2,4-Diaminovaleric acid: An intermediate in the anaerobic oxidation of ornithine by *Clostridium sticklandii, J. Bact.,* **101**:77, 1970.

Fraser, R. D. B.: Keratins, *Sci. Am.,* **221**:86, 1969.

Fujioka, M., and H. Wada: The bacterial oxidation of indole, *Biochim. Biophys. Acta,* **158**:70, 1968.

Ginsburg, D.: "Concerning Amines," New York, Pergamon Press, 1967.

Grant, D. J. W., and J. C. Patel: The non-oxidative decarboxylation of *p*-hydroxybenzoic acid, gentisic acid, protocatecuic acid and gallic acid by *Klebsiella aerogenes (Aerobacter aerogenes), Antonie van Leeuwenhoek, J. Microbiol. Serol.,* **35**:325, 1969.

Gryder, R. M., and E. Adams: Inducible degradation of hydroxyproline in *Pseudomonas putida*: Pathway regulation and hydroxyproline uptake, *J. Bact.,* **97**:292, 1969.

Haughton, B. G., and H. K. King: Induced formation of leucine decarboxylase in *Proteus vulgaris, Biochem. J.,* **80**:268, 1961.

Hendler, R. W.: "Protein Biosynthesis and Membrane Biochemistry," New York, John Wiley & Sons, Inc., 1968.

Holley, R. W., et al.: Structure of a ribonucleic acid, *Science,* **147**:1462, 1965.

Huang, J., and P. S. Sypherd: Comparative chemistry of 30s ribosomal proteins of *Salmonella typhimurium, J. Bact.,* **110**:300, 1972.

International Union of Biochemistry: "Enzyme Nomenclature," New York, American Elsevier Publishing Company, Inc., 1965.

Kendrew, J. C.: Three-dimensional structure of a protein, *Sci. Am.,* **205**:96, 1961.

——: Myoglobin and the structure of proteins, *Science,* **139**:1259, 1963.

Kent, M., and W. A. Gortner: Conversion of indoleacetaldehyde to indoleacetic acid by microbial contamination, *Biochem. Biophys. Res. Commun.,* **3**:319, 1960.

Kumagai, H., et al.: Synthesis of 3,4-dihydroxyphenyl-L-alanine from L-tyrosine and pyrocatechol by crystalline β-tyrosinase, *Biochem. Biophys. Res. Commun.,* **34**:266, 1969.

Le Minor, L., and M. Piéchaud: Note Technique: Une méthode rapide de recherche de la protéolyse de la gélatine, *Ann. Inst. Pasteur,* **105**:792, 1963.

Lipmann, F.: Polypeptide chain elongation in protein biosynthesis, *Science,* **164**:1024, 1969.

McDade, J. J., and R. H. Weaver: Rapid methods for the detection of gelatin hydrolysis, *J. Bact.,* **77**:60, 1959.

Morris, D. R., and K. L. Koffron: Urea production and putrescine biosynthesis by *Escherichia coli, J. Bact.,* **94**:1516, 1967.

Nanninga, N.: Fine structure observed in 50s ribosomal subunits of *Bacillus subtilis, J. Cell Biol.,* **33**:Cl, 1967.

Nirenberg, M., and P. Leder: RNA codewords and protein synthesis, *Science,* **145**:1399, 1964.

Nomura, M.: Ribosomes, *Sci. Am.,* **221**:28, 1969.

Oppenheim, J., et al.: I. Isolation, characterization and distribution of ribosomes, polyribosomes and subunits in logarithmically growing *Azotobacter*, *Biochim. Biophys. Acta,* **161**:386, 1968.

Patton, S.: Milk, *Sci. Am.,* **221**:58, 1969.

Petermann, M. L.: "The Physical and Chemical Properties of Ribosomes," New York, American Elsevier Publishing Company, Inc., 1965.

Raynaud, M., and P. Daste: Faits concernant la possibilité d'utiliser le phénol comme source de carbone par certaines bacteries retinolytiques, *Compt. Rend. Soc. Biol.,* **156**:1489, 1962.

Rich, A.: Polyribosomes, *Sci. Am.,* **209**:44, 1963.

Schroeder, W. A.: "The Primary Structure of Proteins," New York, Harper & Row, Publishers, Incorporated, 1968.

Segal, W., and R. L. Starkey: Microbial decomposition of methionine and identity of the resulting sulfur products, *J. Bact.,* **98**:908, 1969.

Smith, H. L., Jr., and K. Goodner: Detection of bacterial gelatinases by gelatin-agar plate methods, *J. Bact.,* **76**:662, 1958.

Stanley, W. M., et al.: Translation of the genetic message: Factors involved in the initiation of protein synthesis, *Proc. Natl. Acad. Sci. U.S.,* **56**:290, 1966.

Stein, W. H., and S. Moore: The structure of proteins, *Sci. Am.,* **204**:81, 1961.

Traub, P., and M. Nomura: Structure and function of *Escherichia coli* ribosomes. VI. Mechanism of assembly of 30s ribosomes studied in vitro, *J. Mol. Biol.,* **40**:391, 1969.

Tsuru, D., et al.: Studies on bacterial proteases, *Intern. J. Protein Res.,* **2**:75, 1970.

Ulrich, J. A.: New indicators to replace litmus milk, *Science,* **99**:352, 1944.

Umbreit, W. W.: Cell membranes and protein biosynthesis in microorganisms, *Wallerstein Lab. Commun.,* **33**:27, 1970.

Varga, J. M., and H. Y. Neujahr: Isolation from soil of phenol-utilizing organisms and metabolic studies on the pathways of phenol degradation, *Plant Soil,* **33**:565, 1970.

Watson, J. D.: Involvement of RNA in the synthesis of proteins, *Science,* **140**:17, 1963.

——: The synthesis of proteins upon ribosomes, *Bull. Soc. Chim. Biol.,* **46**:1399, 1964.

Witkop, B.: Chemical cleavage of proteins, *Science,* **162**:318, 1968.

Young, J. D., et al.: Structural studies on transfer RNA: Crystallization of formylmethionine and leucine transfer RNA's, *Science,* **166**:1527, 1969.

14 *Fermentation of carbohydrates and other compounds*

The term fermentation may be defined as the incomplete oxidations produced by microorganisms acting on compounds which for the most part are carbohydrates or carbohydrate-like in nature. A variety of end products are produced, many of which are of great industrial importance.

Bacterial action on carbohydrates may be either aerobic or anaerobic. In the presence of sufficient oxygen, carbohydrate breakdown may proceed to the final end products CO_2 and H_2O. In the partial or complete absence of free oxygen, incomplete combustion or anaerobic breakdown occurs.

After oxygen was discovered by Lavoisier, physiologists believed that the gas was an essential attribute for the maintenance of life, the slow combustion of organic nutrients being the source of vital energy. Pasteur showed that free oxygen was not necessary for oxidations to occur; the essential feature of fermenting agents was their ability to utilize the oxygen from organic compounds. To quote a well-known statement by Pasteur, *"La fermentation est la vie sans air."*

In anaerobic breakdown by microorganisms, oxidations are never complete. Most of the end products produced can still be further oxidized to yield more energy if free oxygen is introduced into the system. For example, glucose on complete oxidation to CO_2 and H_2O yields 674 cal, whereas in anaerobic decomposition the yield may be as low as 15 cal.

Fermentation techniques

The possible conversion reactions of microbial cells are exceedingly varied. According to Rhodes (1966),

More than 1300 single-stage transformations of extracellular substrate to a particular recoverable product are known to be brought about by microorganisms. These reactions have been classified into 14 arbitrarily selected groups. In order of decreasing frequency of occurrence, these are: Oxidation, reduction, decarboxylation, deamination, glycosylation, hydrolysis, methylation, esterification, dismutation, condensation, dehydration, amination, acetylation and amidation. . . . Detailed examination of the more than 600 examples of one of these, oxidation, makes the varied potential of the microbial cell evident:

$$a. \quad \overset{\text{H}}{\underset{|}{\overset{|}{-\text{C}}}}\text{-OH} \quad \longrightarrow \quad \overset{|}{\underset{|}{\text{C}}}=\text{O}$$

$$b. \quad \overset{\text{H}}{\underset{|}{\overset{|}{-\text{C}}}}\text{-H} \quad \longrightarrow \quad \overset{\text{H}}{\underset{|}{\overset{|}{-\text{C}}}}\text{-OH}$$

$$c. \quad \overset{\text{H}}{\underset{}{-\text{C}}}=\text{O} \quad \longrightarrow \quad -\text{COOH}$$

$$d. \quad \overset{\text{H H}}{\underset{\underset{\text{H}}{|}}{\overset{| \ |}{-\text{C}-\text{C}-}}} \quad \longrightarrow \quad \overset{\text{H}}{\overset{| \ |}{-\text{C}=\text{C}-}}$$

$$e. \quad -\text{CH}_2\text{OH} \quad \longrightarrow \quad -\text{COOH}$$

$$f. \quad \overset{\text{H}}{\underset{|}{\overset{|}{-\text{C}-}}} \quad \longrightarrow \quad \overset{\text{OH}}{\underset{|}{\overset{|}{-\text{C}-}}}$$

$$g. \quad \overset{\text{H}}{\underset{\underset{\text{H}}{|}}{\overset{|}{-\text{C}-}}} \quad \longrightarrow \quad \overset{\text{O}}{\overset{||}{-\text{C}-}}$$

$$h. \quad -\text{CH}_3 \quad \longrightarrow \quad -\text{CH}_2\text{OH}$$

$$i. \quad -\text{CH}_2\text{OH} \quad \longrightarrow \quad \overset{\text{H}}{\underset{}{-\text{C}}}=\text{O}$$

$$j. \quad \overset{}{\underset{\text{H O}}{\overset{| \ |}{-\text{C}-\text{C}-}}} \quad \longrightarrow \quad \overset{|}{\underset{}{-\text{C}}}=\text{O}$$

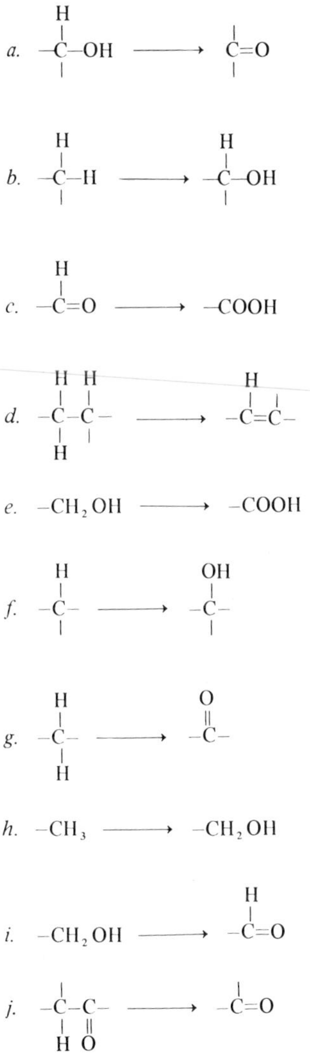

Oxidative conversions like these encompass not only all the usual substrates of biological origin—carbohydrates, fatty acids, amino acids, and steroids—but also such other materials as aliphatic and cyclic hydrocarbons. A typical fermentation involves the conversion of a carbohydrate substrate to a specific product. In its most limited form, such a fermentative conversion can be oxidation of an aldehyde to a carboxyl as in the production of gluconic acid from glucose. Correspondingly, a secondary alcohol of gluconic acid can be oxidized to a ketone to give a ketogluconate. Some organisms may bring about both transformations in sequence so that one or the other product may be alternatively obtained by selection of appropriate fermentation conditions.

The various types of fermentations which follow should give a clearer understanding of the activities of microorganisms on carbohydrates and other compounds.

Fermentation of carbohydrates

Bacteria ferment glucose in at least seven different and distinct patterns. Each pattern is associated with a particular group of organisms. The end products of the fermentations are not the same. All fermentations are believed to proceed in two stages: (1) the splitting of the glucose chain to pyruvic acid with the elimination of two pairs of hydrogen atoms; and (2) the reduction of either pyruvic acid or compounds derived from pyruvic acid by the hydrogen atoms from the first stage. Some of these fermentations are of great economic importance.

Three different pathways leading to the formation of pyruvic acid from glucose have been presented. These are known as (1) the glycolytic or Embden-Meyerhof-Parnas (EMP) pathway, after its discoverers; (2) the Entner-Doudoroff (ED) scheme found only in bacteria; and (3) the hexose monophosphate (HMP) scheme which occurs in a large group of organisms.

Ethyl alcohol (ethanol) fermentation For many years most of the industrial ethanol was produced by fermentation. At present it is produced largely from ethylene which results from petroleum refinery waste gases. Only comparatively small amounts are presently produced by fermentation.

In the fermentation process, ethanol is produced commercially by the action of yeasts on carbohydrates in molasses, grains, potatoes, whey, pineapple waste, etc., depending upon the material available. Starchy materials must first undergo hydrolysis to sugars before they can be utilized by the yeasts. This is accomplished by application of dilute acids or by the utilization of enzymes. The species of yeast generally used for this purpose is *Saccharomyces cerevisiae*, the common baker's or brewer's yeast. The organism converts from 90 to 98 percent of the sugar fermented into equimolecular quantities of ethanol and CO_2. The overall reaction may be represented by the equation

$$\underset{\text{D-Glucose}}{C_6H_{12}O_6} \rightarrow \underset{\text{Alcohol}}{2C_2H_5OH} + 2CO_2$$

Mechanism of ethanol fermentation This type of fermentation follows the EMP pathway. This scheme involves 14 known steps and at least 15 enzymes and 3 coenzyme systems. Before the enzymes and coenzymes involved were known and their reactions understood, the entire group of catalysts was given the name of zymase.

The first step involves the phosphorylation of the hexose sugar by the phosphorylating coenzyme system. This system consists of adenylic acid or adenosine monophosphate (AMP), adenosine diphosphate (ADP), and adenosine triphosphate (ATP). The first two coenzymes accept phosphate, whereas the third transfers phosphate. The formulas are as follows:

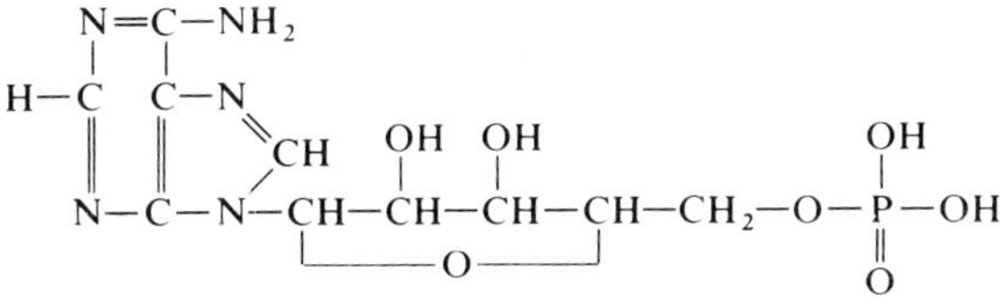

Adenosine monophosphate (AMP)

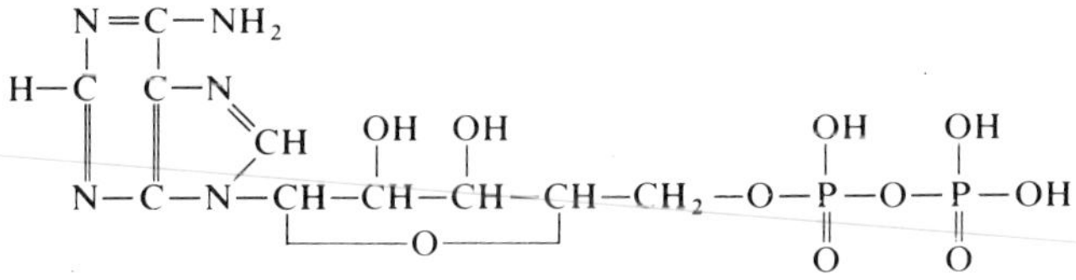

Adenosine diphosphate (ADP)

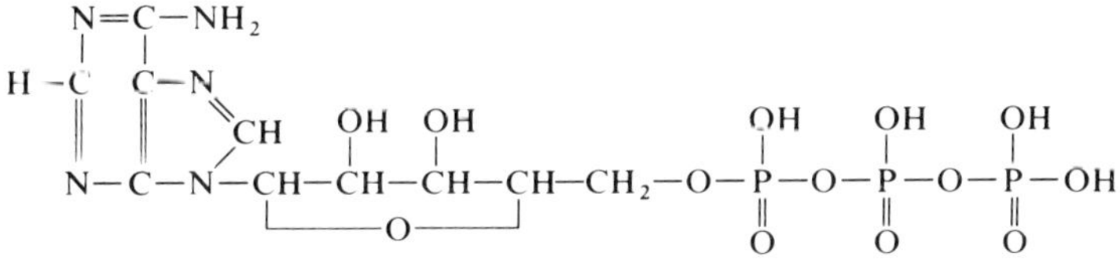

Adenosine triphosphate (ATP)

The coenzyme system is present in yeast juice and is capable of transferring phosphate to the hexose, converting the sugar first to glucose monophosphate and then to glucose diphosphate. The loss of one molecule of phosphate converts ATP to ADP. The phosphate is not utilized in the reaction. When the fermentation reaches the pyruvic acid stage, the phosphate is liberated and becomes available again for phosphorylating more sugar.

The steps in the fermentation are

1. Glucose is phosphorylated by ATP to give first D-glucose-6-phosphate, then D-fructose-6-phosphate, and finally D-fructose-1,6-diphosphate:

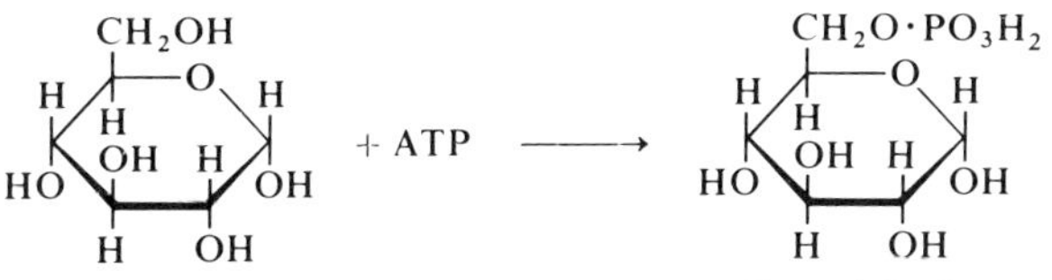

D-Glucose D-Glucose-6-phosphate

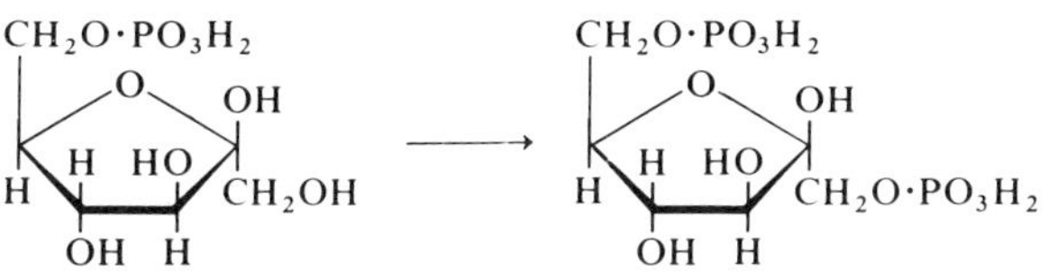

D-Fructose-6-phosphate D-Fructose-1,6-diphosphate

2. D-Fructose-1,6-diphosphate splits to form one molecule of D-glyceraldehyde-3-phosphate and one of dihydroxyacetone phosphate. The two compounds are in equilibrium with each other:

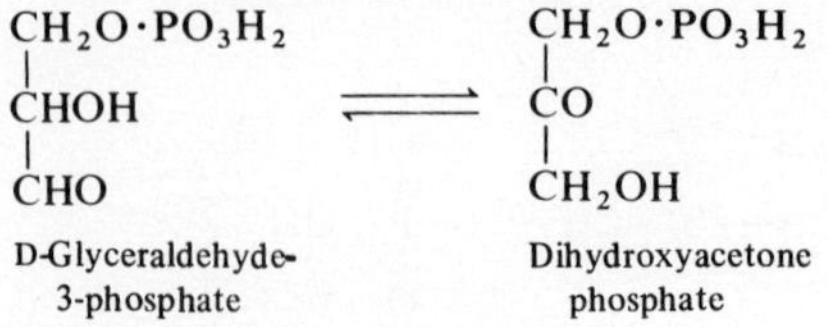

3. Dihydroxyacetone phosphate is reduced to L-glycerol 3-phosphate by $NADH_2$, then to glycerol by ADP:

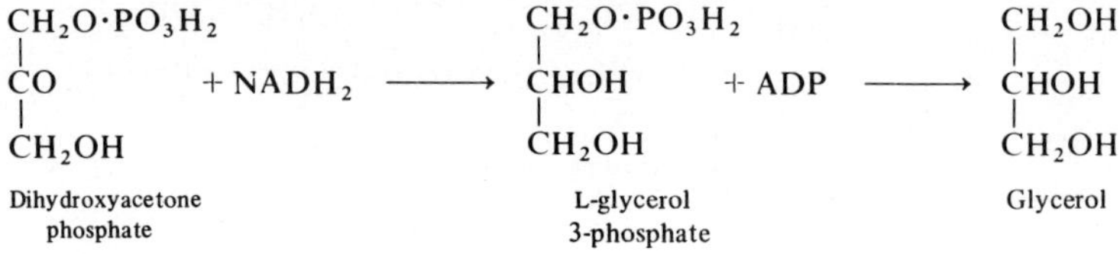

4. ATP loses one molecule of phosphate to D-glyceraldehyde-3-phosphate to yield D-1,3-diphosphoglycerate and ADP:

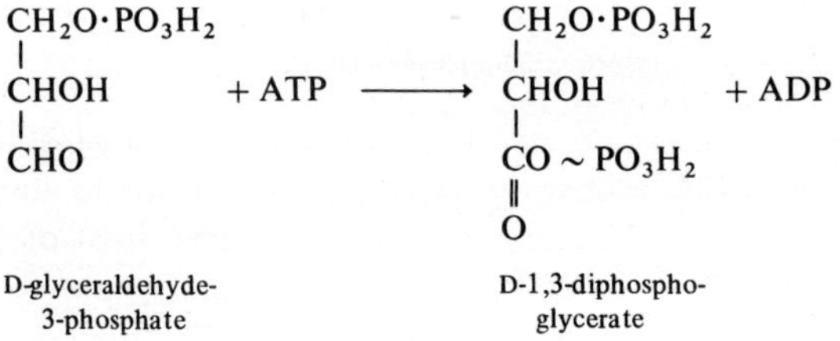

5. D-1,3-Diphosphoglycerate loses the high-energy phosphate to ADP to form D-3-phosphoglycerate and ATP:

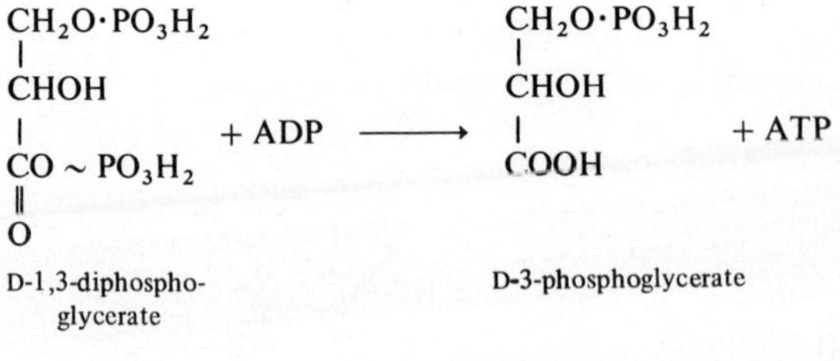

6. D-3-Phosphoglycerate is in equilibrium with D-2-phosphoglycerate:

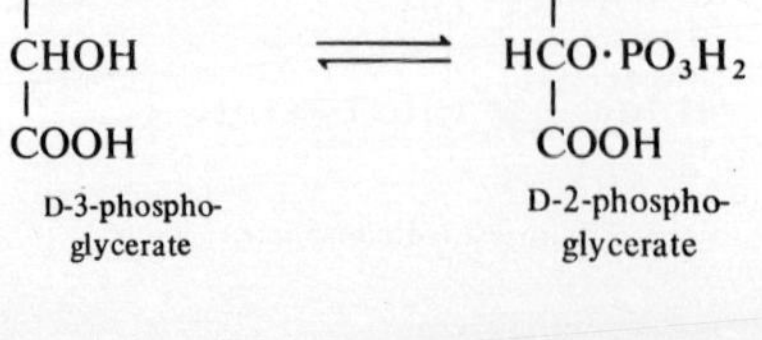

7. D-2-Phosphoglycerate loses water to yield phosphoenolpyruvate:

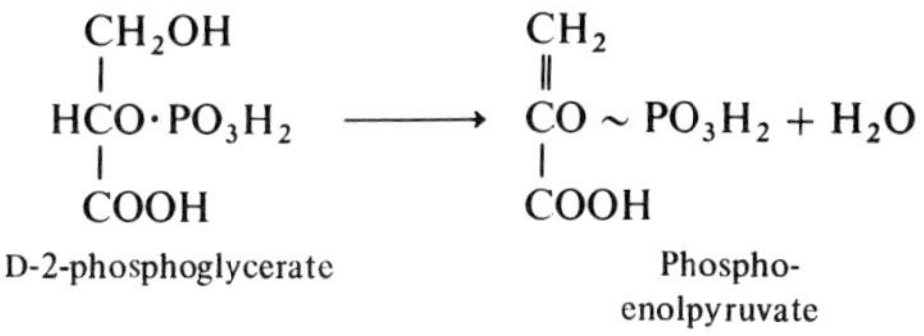

8. ADP removes the energy-rich phosphate linkage from phosphoenolpyruvate to yield pyruvate and ATP:

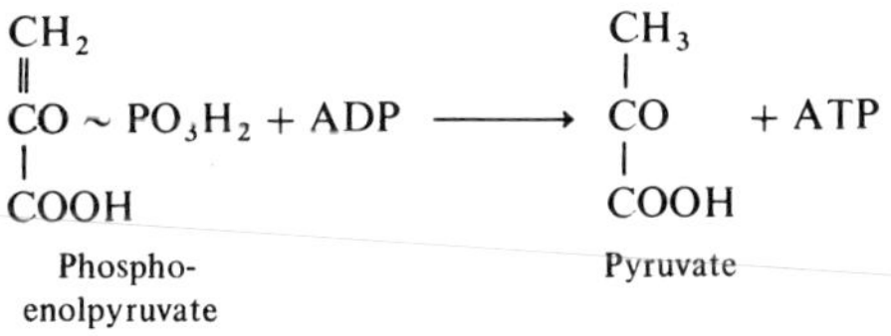

9. Pyruvate is decarboxylated to yield acetaldehyde and CO_2:

$$CH_3\cdot CO\cdot COOH \longrightarrow CH_3\cdot CHO + CO_2$$

Pyruvate — Acetaldehyde

10. Finally, acetaldehyde accepts hydrogen from $NADH_2$ to yield ethyl alcohol:

$$CH_3\cdot CHO + NADH_2 \longrightarrow CH_3\cdot CH_2OH + NAD$$

Acetaldehyde — Ethyl alcohol

The above steps are diagrammed in Fig. 224.

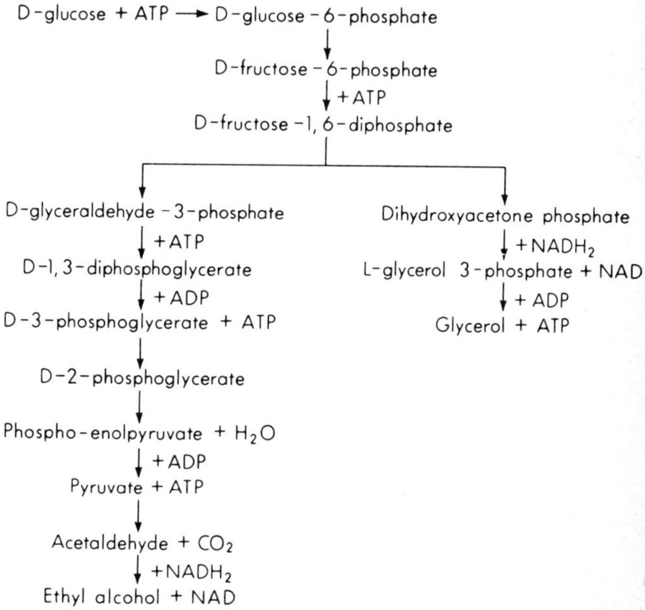

Fig. 224 Mechanism of alcoholic fermentation.

The dehydrogenation of 3-glyceraldehyde phosphate is an energy-yielding oxidation in the pathway. The hydrogen atoms are transferred to NAD to yield $NADH_2$. In the second stage of the fermentation, $NADH_2$ transfers its hydrogen to acetaldehyde, making it (NAD) available again to accept more hydrogen in the first stage.

By-products of alcoholic fermentation   In addition to alcohol and CO_2, small amounts of D-amyl alcohol, isoamyl alcohol, succinic acid, glycerol, and traces of other compounds are also produced.

A mixture of D-amyl and isoamyl alcohols, together with traces of other higher alcohols and compound ethers obtained from fermented liquors, is known as fusel oil.

D-Amyl and isoamyl alcohols are derived from the amino acids isoleucine and leucine, respectively. The alcohols are produced by deamination and decarboxylation of the amino acids:

$$\begin{array}{c} CH_3 \\ C_2H_5 \end{array}\!\!\!>\!CH \cdot CHNH_2 \cdot COOH + H_2O \longrightarrow$$

Isoleucine

$$\begin{array}{c} CH_3 \\ C_2H_5 \end{array}\!\!\!>\!CH \cdot CH_2OH + CO_2 + NH_3$$

D-Amyl alcohol

$$\begin{array}{c} CH_3 \\ CH_3 \end{array}\!\!\!>\!CH \cdot CH_2 \cdot CHNH_2 \cdot COOH + H_2O \longrightarrow$$

Leucine

$$\begin{array}{c} CH_3 \\ CH_3 \end{array}\!\!\!>\!CH \cdot CH_2 \cdot CH_2OH + CO_2 + NH_3$$

Isoamyl alcohol

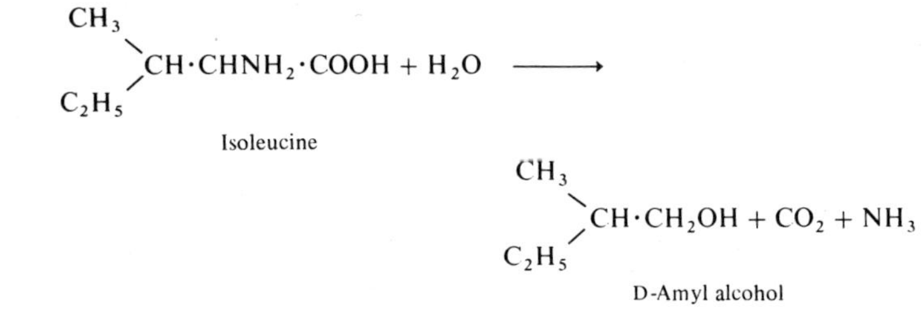

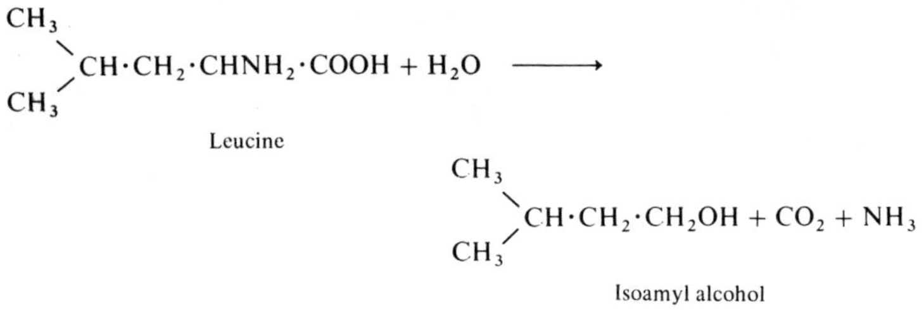

The formation of the two alcohols occurs only in the presence of a fermentable carbohydrate.

Succinic acid is produced from glutamic acid during alcoholic fermentation. As in the case of fusel oil, succinic acid formation does not occur in the absence of a fermentable carbohydrate.

Glycerol is prepared commercially chiefly as a by-product in the manufacture of soap. It is produced in alcoholic fermentation, but the yield rarely amounts to more than about 3.8 percent of the sugar fermented. However the yield can be increased by the addition of an appropriate alkali or sodium sulfite to the fermenting mixture.

As shown above, alcohol results from the reduction of acetaldehyde by hydrogen from $NADH_2$. If the hydrogen is prevented from reducing acetaldehyde, the yield of glycerol is increased. This may be accomplished by adding sodium sulfite to the fermenting mixture. The sulfite forms an addition product with the acetaldehyde, preventing it from accepting hydrogen from $NADH_2$:

$$CH_3{\cdot}CHO + Na_2SO_3 + H_2O$$

Acetaldehyde

$$\downarrow$$

CH_3
|
$CHOH + NaOH$
|
SO_3Na

Addition product

The hydrogen is believed to be accepted by dihydroxyacetone phosphate which has a strong affinity for hydrogen although less than that of acetaldehyde:

CH_2OH
|
$CO \qquad + NADH_2$
|
$CH_2O{\cdot}PO_3H_2$

Dihydroxyacetone phosphate

$$\downarrow$$

CH_2OH
|
$CHOH \qquad + H_2O$
|
$CH_2O{\cdot}PO_3H_2$

Glycerol-3-phosphate

$$\downarrow$$

CH_2OH
|
$CHOH \quad + H_3PO_4 + NAD$
|
CH_2OH

Glycerol

The yield of glycerol is proportional to the concentration of sulfite or carbonate added to the fermentation. In the sulfite process the yield may run as high as 37 percent of the sugar fermented.

The ED and HMP pathways are diagrammed in Fig. 225.

Pasteur effect Respiration in the presence of free oxygen usually results in the complete oxidation of carbohydrates to CO_2, H_2O, and cell material. Respiration may also occur in the absence of free oxygen, provided an appropriate inorganic substance such as nitrate is present as the oxidant.

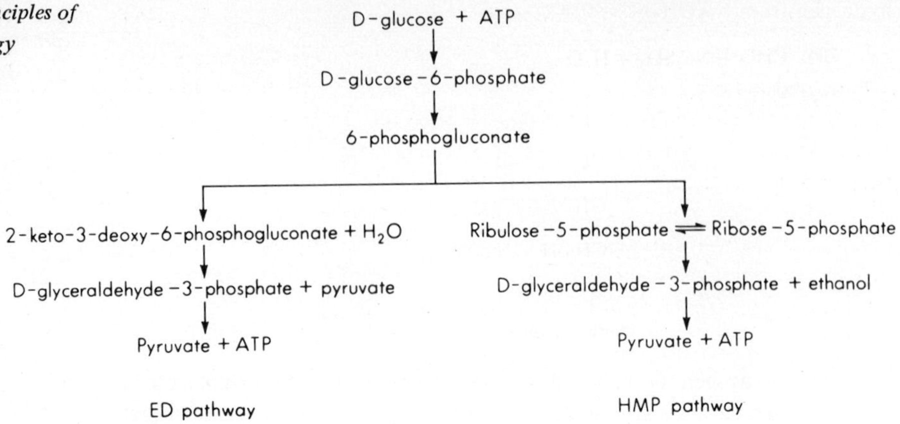

Fig. 225 The ED and HMP pathways of glucose dissimilation.

As explained in the preceding chapter, bacteria fall into three groups in their response to free oxygen: obligate aerobes, obligate anaerobes, and facultative forms.

Facultative organisms are capable of growing either in the presence of free oxygen or in its absence. They may be divided into three groups: (1) facultative bacteria which are purely fermentative and indifferent to free oxygen; (2) facultative bacteria such as the denitrifying forms; and (3) facultative bacteria which obtain energy either anaerobically (fermentation) or aerobically (respiration). The yeast organism employed in alcoholic fermentation is a good example of the third facultative type.

The denitrifying bacteria are facultative forms which can use anaerobically any substance they can utilize aerobically, provided that nitrate is present in the medium. On the other hand, facultative forms in the third group which produce fermentations do not need a substitute for free oxygen to grow under anaerobic conditions. However, not all substrates held under aerobic conditions are fermentable. An example is succinic acid, which can be utilized aerobically for respiration but cannot be fermented anaerobically and will not support growth in the absence of air.

The organisms in the second group (denitrifying) respire under both aerobic and anaerobic conditions, whereas the organisms in the third group respire only under aerobic conditions and ferment under anaerobic conditions.

The presence of oxygen suppresses fermentation. This effect was first noted by Pasteur in his observations on the growth of yeasts in alcoholic fermentation. In air (aerobic conditions) yeast growth is accelerated while sugar consumption is depressed (less alcohol formed). In the absence of air (anaerobic conditions) yeast growth is depressed while sugar consumption is increased (more alcohol formed). Pasteur showed that in the presence of air, from 4 to 10 g of glucose were required to produce 1 g of yeast cells whereas in the absence of air, from 60 to 80 g of glucose were required to produce the same mass of cells. This is sometimes referred to as the Pasteur effect.

On the other hand, Wikén and his associates (Pfennig and Wikén, 1960*a*, *b*; Scheffers, 1961; Scheffers and Wikén, 1969; Tromp and Scheffers, 1969; van Dijk and Wikén, 1968; Wikén, 1961, 1967, 1968, and Wikén et al., 1961) reported exceptions to the Pasteur effect. They found that in some yeasts, notably species of *Brettanomyces* and *Saccharomyces*, the absence of oxygen inhibited alcoholic fermentation whereas the presence of oxygen stimulated the formation of alcohol. This is the reverse of the Pasteur effect. They referred to it as a negative Pasteur effect.

For more information: Baldwin (1967); Casida (1968); Cheldelin (1961); Guymon and Cromwell (1965); Harris and Hajny (1960); Owen (1960); Rittenberg and Ponticorvo (1962); Rose (1961); Sistrom (1969); Thoukis et al. (1965); Pastan (1972).

Bacterial fermentations

The EMP scheme for alcoholic fermentation has been quite well worked out. However, the pathways of carbohydrate breakdown in some special types of bacterial fermentations are still unsettled. It is generally believed that all fermentations pass through the same steps leading to the formation of pyruvic acid, then branch off from there. Some important industrial fermentations are now discussed.

Butyric acid fermentation The organism generally employed in this type of fermentation is the anaerobic sporeformer *Clostridium butyricum*.

Cells of *C. butyricum* are straight or slightly curved rods measuring 0.7 by 5 to 7 μ, with rounded ends, occurring singly, in pairs, in short chains, and occasionally in long filaments. Spores are oval and eccentric to subterminal, swelling the cells to clostridial forms. The organism is motile and gram-positive, becoming gram-negative in old cultures.

C. butyricum and most of the other species of clostridia are able to ferment carbohydrates with the production of butyric acid but this organism gives larger yields. Because of its powerful odor, the presence of butyric acid is easily detected.

It is generally believed that the scheme from glucose to pyruvate is the same as for alcoholic fermentation and diverges at that point. The substitution of pyruvate for carbohydrate produces the same end result:

$$\underset{\text{Pyruvic acid}}{3CH_3\cdot CO\cdot COOH} + H_2O \longrightarrow \underset{\text{Butyric acid}}{CH_3\cdot CH_2\cdot CH_2\cdot COOH} + \underset{\text{Acetic acid}}{CH_3\cdot COOH} + 3CO_2 + H_2$$

Butyl alcohol and acetone fermentation The organism employed is the anaerobic sporeformer *C. acetobutylicum*.

The cells are straight rods with rounded ends, occurring singly and in pairs but not in chains. The vegetative cells measure 0.6 to 0.72 by 2.6 to 4.7 μ; the clostridia, 1.3 to 1.6 by 4.7 to 5.5 μ. Spores are ovoid, eccentric

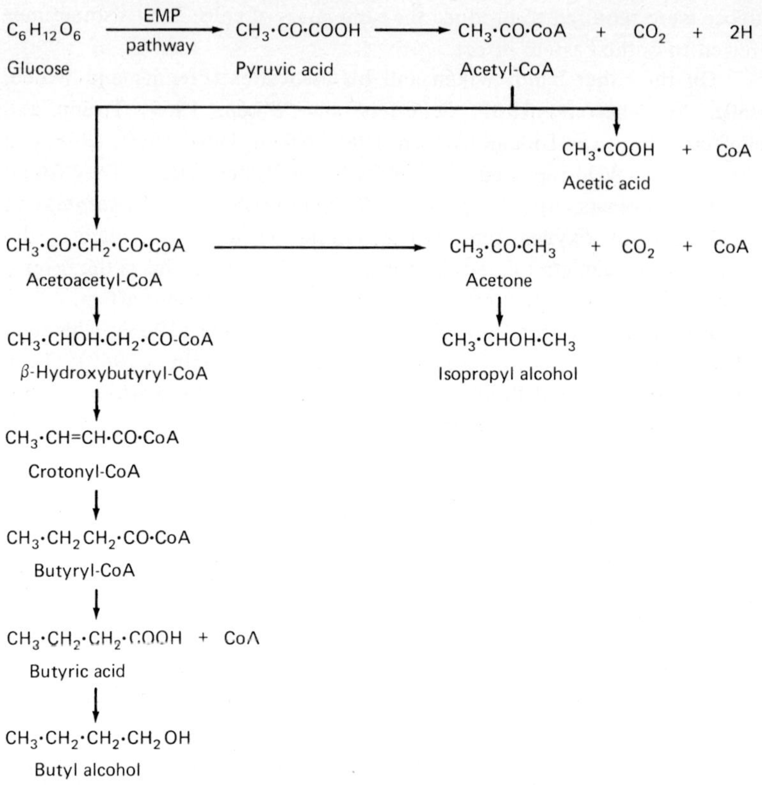

Fig. 226 Reaction scheme for the formation of butyl alcohol and acetone by Clostridium acetobutylicum.

to subterminal, swelling the cells to clostridia. Cells are not encapsulated. They are motile by means of peritrichous flagella, and they are gram-positive, becoming gram-negative.

In this fermentation, the two chief end products are butyl alcohol and acetone (Fig. 226). Other products in lesser amounts include isopropyl alcohol, butyric and acetic acids. In addition the gases CO_2 and H_2 are formed. The step in the formation of isopropyl alcohol from acetone appears to be questionable.

For more information: Casida (1968); Namba and Furusaka (1968*a*, *b*, *c*); Rose (1961, 1968).

Acetone and ethyl alcohol fermentation The organism employed in this fermentation is the aerobic sporeformer *Bacillus macerans*.

The rods measure 0.5 to 0.7 by 2.5 to 5 μ, not in chains. Cells are motile and gram-variable. Spores measure 1 to 1.5 by 1.2 to 2.5 μ, and are ellipsoidal, subterminal to terminal. Spore wall is thick and easily stained. Sporangia are swollen and clavate. Organism produces about 2 parts of ethyl alcohol to 1 part of acetone.

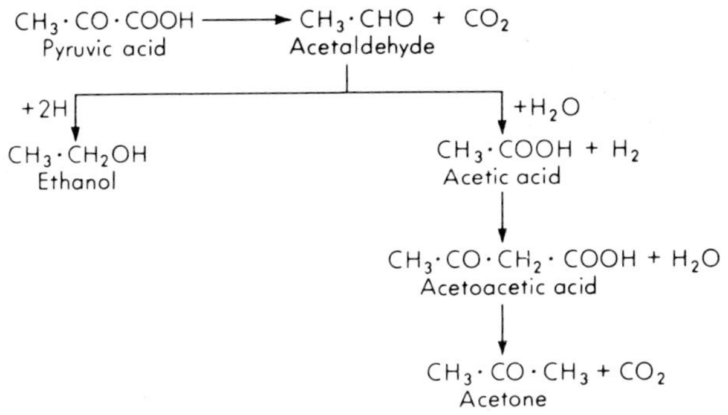

Fig. 227 Reaction scheme for the formation of acetone and ethanol by Bacillus macerans.

The reaction scheme is given in Fig. 227.

For more information: Casida (1968); Rose (1968).

Propionic acid fermentation Propionic acid is produced by a number of species of bacteria grouped in the genus *Propionibacterium.*

Propionibacterium The propionibacteria are nonmotile, non-spore-forming, gram-positive organisms which grow under anaerobic conditions in neutral media as short, diphtheroid rods, which sometimes resemble streptococci in appearance; under aerobic conditions, they grow as long, irregular, club-shaped, and branched cells. Metachromatic granules are demonstrable. They ferment carbohydrates, lactic acid, and polyhydroxy alcohols with the production of propionic acid, acetic acid, acetaldehyde, propionaldehyde, ethyl alcohol, propyl alcohol, and CO_2. In an atmosphere containing 5 percent CO_2, growth is enhanced both aerobically and anaerobically. Vitamin B requirements are relatively simple. Optimum temperature is 30°C. They are found in dairy products, especially hard cheeses.

Mechanism of propionic acid formation The propionibacteria were probably the first organisms which were definitely shown to be capable of utilizing free CO_2. This was determined by employing CO_2 containing radioactive carbon.

The scheme proposed by Allen et al. (1964) for the formation of propionic acid is given in Fig. 228.

For more information: Baldwin et al. (1962); Keenan and Bills (1968); Rose (1961, 1968); Stjernholm and Wood (1960*a, b*); Swick and Wood (1960).

Lactic acid fermentation Lactic acid was first discovered as one of the products resulting from the souring of milk. It is produced from the milk constituent lactose. It is believed to be the oldest-known acid, having been discovered by Scheele in 1780.

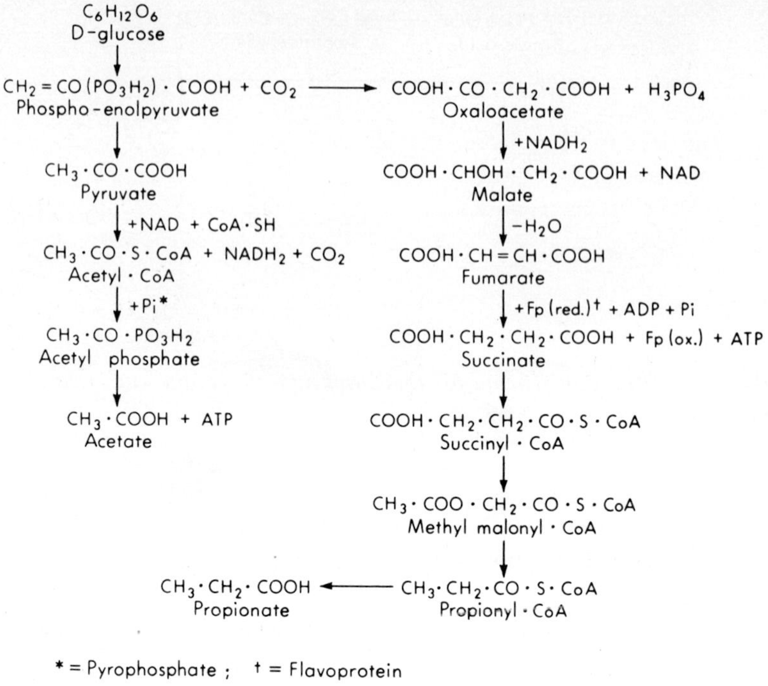

Fig. 228 Scheme for propionic acid fermentation. (After Allen et al.)

Lactic acid bacteria The designation *lactic acid bacteria* is applied to a group of organisms which ferment lactose to lactic acid as the principal, and often sole, product. The strong acid producers are found in the genera *Lactobacillus* and, to a lesser extent, *Streptococcus*, of the family *Lactobacillaceae.*

Lactobacillus Cells are rod-shaped, often long and slender. They are nonmotile and gram-positive. Pigment production is rare; when present, yellow or orange to rust or brick red. Gelatin is not liquefied. Glucose and similar aldehyde hexoses, carbohydrates which yield these simple sugars, and polyhydroxy alcohols are changed either by homofermentation to lactic acid or by heterofermentation to lactic and acetic acids, alcohol, and CO_2. Nitrates are generally not reduced. Several species grow at relatively high temperatures. Surface growth is poor because organisms are generally microaerophilic or anaerobic. They are found in fermenting dairy and plant products.

Streptococcus Cells are spherical or ovoid, rarely elongated into rods, occurring in pairs or short or long chains. They are nonmotile, except a few strains in the enterococcus group, and gram-positive. A fermentable carbohydrate or polyhydroxy alcohol is necessary for satisfactory growth in artificial media. Growth in broth culture is variable in character. Rough variants may show granular growth that tends to settle out quickly, leaving a clear supernatant. Smooth variants may show uniform turbidity with little tendency to settle out. Such variants may be noted within one species. A pellicle is never formed. Growth on agar is usually scanty. Colonies are small, usually less than 1 mm in diameter. Colony variants within a species may range from rough to smooth to mucoid. Carbohydrate fermentation is homofermentative, with D-lactic acid as the dominant end product. CO_2 is produced in very small quantities or not at all from sugar. Ethanol, acetic acid, and formic acid may be produced in appreciable quantities from glucose if allowed to ferment in alkaline media. Cytochrome systems are absent. They do not reduce nitrate, are facultative, and are fastidious in their nutrition, requiring a number of B vitamins and amino acids for growth. They are commonly found in the mouth and intestine of man and other animals, in dairy and other food products, and in fermenting plant juices.

In their action on glucose, the lactic acid organisms fall into two large groups: the homofermentative species and the heterofermentative species.

Homofermentative species The members of this group convert about 95 percent of fermentable hexoses to lactic acid:

$$\underset{\text{Glucose}}{C_6H_{12}O_6} \longrightarrow \underset{\text{Lactic acid}}{2CH_3 \cdot CHOH \cdot COOH}$$

Small amounts of volatile acids and carbon dioxide are also produced. Disaccharides are fermented in a similar manner; e.g., one mole of lactose yields four moles of lactic acid. The lactic acid may be dextrorotatory (D) or levorotatory (L), or a mixture of the two forms (DL) in equal quantities:

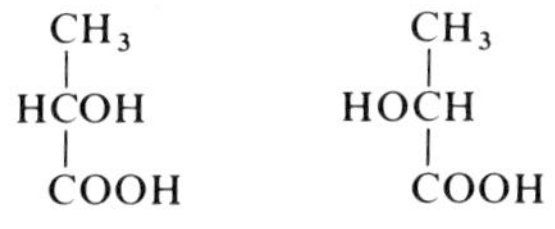

D-Lactic acid L-Lactic acid

The isomer produced is characteristic of the species.

Heterofermentative species These differ from the homofermentative species in that lactic acid is only one of several principal products formed from sugar. Other compounds include ethyl alcohol, acetic and formic acids, and carbon dioxide.

Mechanism of lactic acid formation A large number of carbohydrates, notably glucose, lactose, and sucrose, are employed. Starches of various kinds may be first hydrolyzed to sugars by treatment with mineral acids or enzymes, then fermented to lactic acid. Molasses and whey are low-priced and serve as excellent sources of carbohydrates.

Homofermentative species are the main source of industrial lactic acid. It is generally believed that the lactic fermentation follows the same pathway as in alcohol production by yeasts. As pyruvate is formed, it is reduced to lactic acid:

$$\underset{\text{Glucose}}{C_6H_{12}O_6} \longrightarrow \underset{\text{Pyruvic acid}}{CH_3 \cdot CO \cdot COOH}$$

$$\underset{\text{Pyruvic acid}}{CH_3 \cdot CO \cdot COOH} + NADH_2 \longrightarrow CH_3 \cdot CHOH \cdot COOH + NAD$$

The heterofermentative species produce a lower concentration of lactic acid accompanied by a number of other products, the most important being ethyl alcohol, acetic and formic acids, and CO_2. Some pyruvate follows the alcoholic pathway and some becomes reduced to lactic acid. The acetic and formic acids can be produced from pyruvate, according to the reaction:

$$\underset{\text{Pyruvic acid}}{CH_3 \cdot CO \cdot COOH} + H_2O \longrightarrow \underset{\text{Acetic acid}}{CH_3 \cdot COOH} + \underset{\text{Formic acid}}{HCOOH}$$

The formic acid disappears as CO_2 and H_2:

$$HCOOH \longrightarrow CO_2 + H_2 \rightleftharpoons 2H$$

For more information: Garvie (1967); Gemmell and Hodgkiss (1964); Katagiri et al. (1960); Mundt and Hammer (1968).

Silage fermentation Silage may be defined as the product resulting from the conversion of fodder, either green or mature, into succulent winter feed for livestock through processes of fermentation.

Fodder undergoes an acid fermentation which gives it an agreeable flavor and prevents spoiling. Any farm crop is satisfactory for silage production, provided it contains sufficient sugar to furnish the required amount of acid for its preservation.

Process Fodder, consisting of corn, sorghum, sugar cane, legumes, citrus pulp, potatoes, beet pulp, sunflower, etc., is cut into small pieces and stored in an airtight chamber or silo, where it is compressed to exclude air. Sufficient moisture must be present; otherwise water is added. Since the surface of plants contains the required kinds of bacteria, inoculation of the fodder is not necessary. Counts show that normally from 10,000 to 400 million bacteria per gram of material are present.

Because of the presence of carbohydrates in the plant sap, the changes which occur are fermentative rather than putrefactive. The plant cells continue to respire, and within a few hours the free oxygen is utilized and replaced with CO_2. The changes which follow are anaerobic rather than aerobic in character.

Beardsley (1956) recognized five phases in the ensiling process:

1. The continued respiration of the plant cells results in the production of CO_2, the utilization of simple carbohydrates and a flow of water from the mass due to these biochemical changes, and the mechanical compression of the crop. These events are accompanied by the evolution of heat.

2. The production of acetic acid in small amounts by organisms of the coliform group and others. This phase is of short duration and merges into phase 3.
3. The initiation of a lactic acid fermentation dependent upon the activity of lactic acid organisms (lactobacilli and streptococci) supported by adequate carbohydrate.
4. This is a stage of quiescence in the mass during which the lactic acid production passes its peak and remains constant at 1 to 1.5 percent of the fresh materials corresponding to a pH of less than 4.2.

   These four stages take place in about 3 weeks with the first three completed in about 3 days. If the material has been properly ensiled, adequate production of lactic acid has been obtained, and air is excluded, the silage should remain relatively stable and in good condition for a long time—10 years or more. If, however, the conditions within the silage are unsatisfactory and incomplete production of lactic acid is obtained, a fifth stage may set in.
5. The attack by butyric acid-producing organisms on both the residual soluble carbohydrate and the lactic acid which has already formed. In extreme cases, this is accompanied by a deaminization of amino acids with the formation of higher volatile fatty acids and ammonia; possibly a decarboxylation leading to the formation of amines and carbon dioxide.

The practical aim in the ensiling process is the production of sufficient lactic acid by certain bacteria to inhibit further breakdown of the fodder by other types of microorganisms.

Characteristics of good silage Good-quality silage has the following characteristics:

1. A pH of 4.5 or less. A lower pH is better.
2. Low volatile base content (0.5 percent or less, expressed as ammonia).
3. Lactic acid content of 3 to 5 percent or more.
4. Butyric acid content of 2 percent or less.

All characteristics are expressed on the basis of dry weight. Silage which possesses these characteristics should (1) have a mild and pleasing odor, (2) have a light to dark green color, (3) be readily accepted by cows, and (4) have a higher feeding value than silage that does not approach these standards.

Bacterial content of silage During active fermentation, each gram of silage may contain as many as 2 billion organisms. Each milliliter of the silage may show a bacterial count as high as 4 billion. If conditions within the silage are satisfactory, the microbial population should consist almost entirely of lactobacilli and streptococci. Hesser et al. (1967) in their studies on ensiled corn isolated the following species of *Lactobacillus* arranged in descending order of frequency: *L. plantarum, L. brevis, L. fermenti*, and *L. buchneri*. Other species which have been isolated from silage include *L. delbrueckii* and *L. leichmannii*. If the carbohydrate concentration is insufficient for the bacteria to produce their limiting pH, other less-acid-tolerant species will most likely be present.

For more information: Burmeister and Hartman (1966); Burmeister et al. (1966); Casida (1968); Kempton and San Clemente (1959); Langston and Bouma (1960*a*, *b*); Rydin (1961, 1963); Smith (1962).

Retting of flax and hemp Celluloses, hemicelluloses, and pectic substances are polysaccharides which form the framework of plants.

Cellulose is the most abundant cell-wall polysaccharide found in nature. It is probably the most chemically resistant of all substances elaborated by living cells. Complete hydrolysis of pure cellulose by concentrated mineral acids yields D-glucose almost quantitatively as the only constituent (see page 807).

Hemicelluloses are those cell-wall polysaccharides which may be extracted from plant tissues by treatment with dilute alkalies, either hot or cold, but not with water. They may be hydrolyzed to sugar and sugar acid units (D-xylose, D-galactose, L-arabinose, D-glucose, and uronic acids) by boiling with dilute mineral acids.

Pectic substances is a group name for those complex colloidal carbohydrate derivatives which occur in, or are prepared from, plants and contain a large proportion of anhydrogalacturonic acid units which are believed to exist in a chain-like combination. The carboxyl groups of polygalacturonic acids may be partly esterified by methyl groups and partly or completely neutralized by one or more bases. They always represent mixtures of polygalacturonic acids containing variable numbers of associated units. With the exception of pectic acid, they may also contain different proportions of methyl ester groups in the individual polygalacturonic acids. These may be unevenly distributed on the polygalacturonic acid. Acid salts may contain polymer units with different proportions of metallic ions.

Pectic compounds form the middle lamellar layer of plant cells. They have a gelatinous consistency and act as cement-like substances to bind the cells together. Their hydrolysis by bacteria is of great importance in the retting of flax and hemp.

There are two basic types of retting: (1) anaerobic and (2) aerobic.

Anaerobic retting In this method the flax or hemp stalks are immersed in water and weighted down. Water is absorbed by the plant cells, causing swelling and the extraction of water-soluble substances. The extractives amount to about 12 percent of the weight of the plants, and consist largely of sugars, glucosides, tannins, nitrogenous compounds, and coloring matter. The highly colored water now becomes a good culture medium for many kinds of organisms. The aerobes utilize the dissolved oxygen, permitting the growth of the anaerobes. The anaerobes ferment, and dissolve the pectic substances, leaving the fibers intact. Compounds produced during fermentation include acetic acid, butyric acid, lactic acid, ethyl and butyl alcohols, acetone, H_2, and CO_2. The plant material should be removed immediately after completion of the reaction to prevent overretting. The bundles are thoroughly washed to remove undesirable end products of fermentation, then spread out to air-dry. The dried material is now ready for scutching.

Several anaerobic bacterial species have been isolated from fermenting flax and hemp. The most important appear to be *C. pectinovorum*, a gram-positive sporeformer, found in soil and in naturally retting plant materials; and *C. felsineum*, also a gram-positive sporeformer but becoming gram-negative. Other anaerobic species include *C. aurantibutyricum, C. butyricum*, and *C. laniganii*.

Rosemberg (1965) in his studies on the retting of Brazilian flax isolated the following species: *Achromobacter parvulus, C. beijerinckii, C. saprogenes, C. perenne,* and *Pseudomonas aeruginosa*.

A culture of the organism selected is added to the water of the retting vat in the proportion of 1000 ml to 10,000 g of dry tissue. The vat is kept at a temperature of 37 to 38°C for 50 to 75 hr or until retting is complete.

Aerobic retting An aerobic process known as dew retting is also employed. This is not a true fermentation but is included here for convenience.

The stalks are spread out on the ground in thin layers and left exposed to the influence of climatic conditions. If necessary, the stalks are turned over once or twice. The retting materials are exposed to attack by aerobic organisms, the process being dependent upon moisture and temperature. Under these conditions the method cannot be controlled as is possible in the anaerobic process.

Molds and yeasts, rather than bacteria, are the chief retting agents in the aerobic process. The organisms which have been isolated include the molds *Cladosporium herbarum* and *Pullularia pullulans*; and the yeasts *Cryptococcus albidus, Rhodotorula glutinis*, and *R. macerans.* Some organisms are important during the winter months; others during the summer period. Since *Cladosporium* and other molds give unfavorable color to the fiber during the summer season, dew retting is better carried out in the winter months.

For more information: de Franca et al. (1969); Rose (1961).

Fermentations by coliform organisms

The coliforms and related species ferment glucose and other carbohydrates chiefly to formic, acetic, lactic, and succinic acids, ethyl alcohol, acetylmethylcarbinol (AMC), 2,3-butanediol, CO_2, and H_2.

The important products produced from the fermentation of glucose by *Escherichia coli* and *Aerobacter aerogenes* are given in Table 16.

The compounds produced by *E. coli* from glucose account for 100 percent of the carbon fermented; the same compounds produced by *A. aerogenes* account for considerably less. The discrepancy is due to the production by *A. aerogenes*, but not by *E. coli*, of the two compounds AMC and 2,3-butanediol (see page 744).

Table 16 Compounds produced in the fermentation of glucose by Escherichia coli *and* Aerobacter aerogenes

E. coli	*A. aerogenes*
Formic acid	Formic acid
Acetic acid	Acetic acid
Lactic acid	Lactic acid
Succinic acid	Succinic acid
Ethyl alcohol	Ethyl alcohol
Carbon dioxide	Carbon dioxide
Hydrogen	Hydrogen
	Acetylmethylcarbinol (acetoin)
	2,3-Butylene glycol (2,3-butanediol)

The mechanism of AMC and 2,3-butanediol formation is not clearly understood. The two compounds generally accompany each other in fermentations. Some workers believe AMC is formed first, then some of it is reduced to 2,3-butanediol; others maintain that 2,3-butanediol is produced first, then some oxidized to AMC. Perhaps the two compounds comprise an oxidation-reduction (redox) system. A low redox potential (anaerobic conditions) favors the accumulation of 2,3-butanediol; a high potential (aerobic conditions) favors the accumulation of AMC.

The scheme for the fermentation of glucose by *A. aerogenes*, according to Kluyver and van Niel (1956), is given in Fig. 229.

All organic compounds produced by living cells can be utilized by microorganisms. Many soil organisms can use AMC as a source of carbon and of energy. The breakdown can occur either aerobically or anaerobically.

For more information: Long and Patrick (1961, 1963); Paege and Gibbs (1961); Thayer and Ogg (1967).

Fig. 229 Reaction scheme for the fermentation of glucose by Aerobacter aerogenes. (After Kluyver and van Niel.)

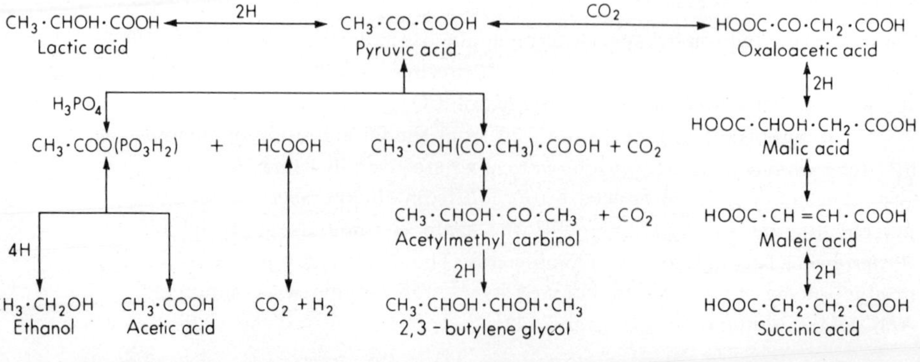

Some routine fermentation reactions employed for the identification of bacteria

Carbohydrates and compounds of a similar nature are generally added to culture media for two purposes: (1) to serve as readily available sources of energy and (2) to aid in the identification and classification of bacteria.

Carbohydrates are more readily available as sources of energy than are proteins. This means that the rate of multiplication of an organism is generally increased in the presence of a fermentable carbohydrate.

Organisms vary considerably in their ability to ferment various carbohydrates. Some bacteria are able to attack many carbohydrates; others cannot ferment any carbohydrate; all degrees of fermentability occur between these two extremes. Also, some organisms ferment carbohydrates with the production of both acid and gas, whereas others produce acid only (Fig. 230). Such information is of great value in the identification of organisms.

It is not clearly understood why an organism ferments one aldose sugar and not another having the same empirical formula. The sugars differ only in the arrangement of H atoms and OH groups around carbon atoms. There is no method of determining beforehand whether or not a particular organism is capable of fermenting a given carbohydrate. This can be determined only by making the test.

Any change in the structure of the aldose sugars results in a decrease in the frequency of fermentation of the derivatives. For example, D-gluconic, D-mannonic, and D-galactonic acids, as well as D-sorbitol, D-mannitol, and dulcitol, are fermented by fewer organisms than the corresponding aldoses. Similarly, the dicarboxylic mucic and D-saccharic

Fig. 230 Fermentation of carbohydrates. (A) Glucose; (B) lactose; (C) sucrose, all inoculated with Escherichia communior. Acid and gas produced. (D) Sucrose inoculated with E. coli. No acid or gas.

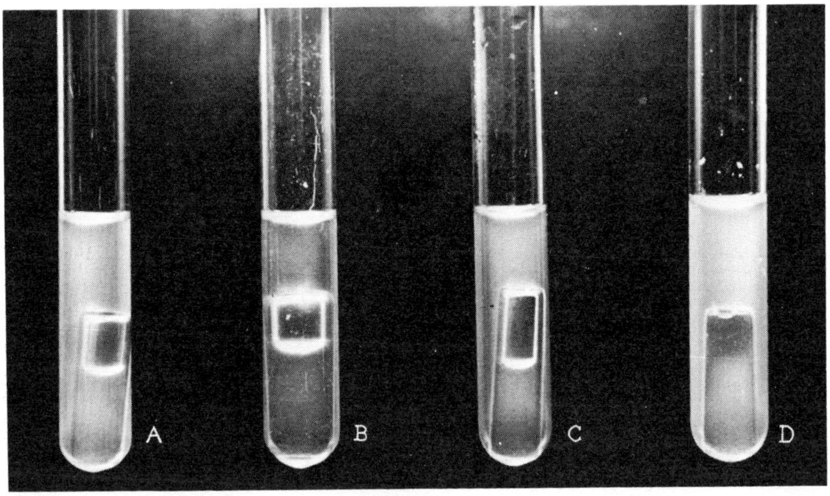

acids are attacked less frequently than the corresponding monocarboxylic D-galactonic and D-gluconic acids.

In general, fermentation becomes less frequent as the complexity of the molecule increases. This means that disaccharides are fermented less frequently than monosaccharides, trisaccharides less frequently than disaccharides, and polysaccharides less than trisaccharides. When tests on the availability of carbohydrates to microorganisms are to be made, always proceed with the monosaccharides, especially glucose. As a rule, if an organism ferments any carbohydrate, it will most likely be glucose. If glucose is fermented, then other monosaccharides should be tested before proceeding to carbohydrates of greater complexity.

Some have reported the direct fermentation of a disaccharide, such as maltose, lactose, or sucrose, without action on either of its constituent monosaccharides. Such observations are not in accord with the generally accepted view of indirect fermentation, which presupposes a hydrolysis of the disaccharide to its constituent monosaccharides before utilization takes place.

Detection of fermentation Utilization of a carbohydrate is in almost all cases accompanied by an acid reaction which may be detected by incorporating an appropriate indicator in the medium. Gas production may be detected by placing an inverted vial in the medium contained in a test tube. The evolved gas is trapped in the vial, where it rises to the top to displace the culture medium.

Importance of carbohydrate-free culture ingredients Peptones, beef extract, and yeast extract may vary considerably in composition.

In the selection of peptones and other ingredients of culture media for fermentation studies, the presence or absence of intrinsic fermentable components should be considered. Sufficient fermentable material is present in most bacteriological peptones, yeast autolysates, and meat extracts to give rise to false positive results and constitute an important cause of inaccuracy. Routine fermentation tests for identification of microorganisms may be performed with accuracy only when materials free from fermentable components are used as basic substrates.

Litmus carbohydrate media Litmus is a weakly staining dye that is employed only as an indicator. Since it is not a delicate detector of changes in reaction, it has been largely replaced by the more sensitive and brilliant sulfonephthalein indicators.

However, litmus possesses an important advantage over the newer indicators in that it is sensitive to decolorization by bacteria. It functions as an oxidation-reduction system. Like methylene blue, litmus accepts hydrogen and becomes reduced to the colorless form. The decolorization of the indicator is first noted at or near the bottom of the tube, where the dissolved oxygen is soon exhausted. In the surface layer of media exposed to air, the litmus is seldom completely decolorized because the hydrogen is transferred to oxygen instead of to indicator.

Fig. 231 Action of organisms on lactose litmus agar. Left, Bacillus subtilis; center, Staphylococcus aureus; right, Escherichia coli. B. subtilis and S. aureus produce neither acid nor gas; E. coli produces both acid and gas with the result that the agar is separated into rings and the litmus is decolorized.

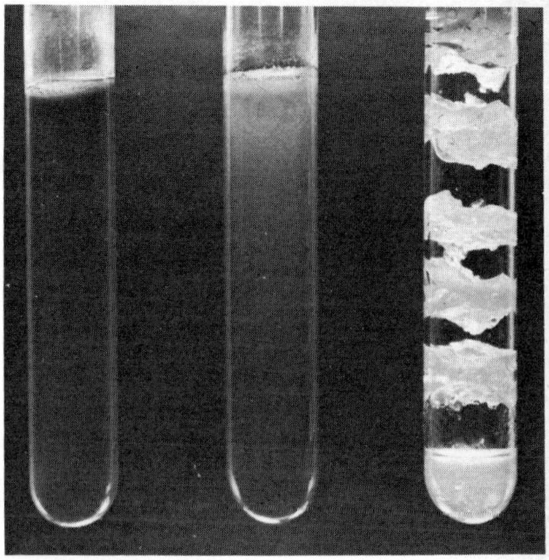

If a tube of agar with indicator is inoculated by deep stab, fermentation is indicated by a change in the color of the agar medium. Gas formation is indicated by a collection of gas bubbles in the medium or by a splitting of the agar. Under some conditions the agar is split in the form of disks with a layer of gas separating each disk (Fig. 231). Sometimes the gas pressure may be sufficient to force some of the disks and cotton stopper out of the tube.

Bromocresol purple carbohydrate media The ability of an organism to ferment a particular compound may be easily determined by streaking a loopful of the culture over the surface of nutrient agar containing the carbohydrate and indicator. Fermentation is indicated by a change in the color of the indicator (Fig. 232).

Fig. 232 Bromocresol purple lactose agar. Left, plate streaked with a culture of Bacillus subtilis. The organism produced no visible change in the color of the agar in 24 hr. Right, plate streaked with a culture of Escherichia coli. The organism produced considerable acid in 24 hr, changing the color of the agar from purple to yellow.

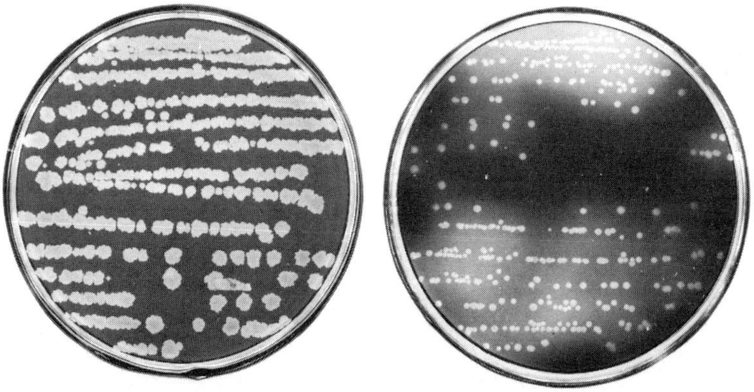

If only a few well-isolated colonies appear on the plate, acid production remains confined to areas immediately surrounding each colony. This results in a color change in the vicinity of each colony without affecting the agar free from colonies. The contrast in colors in the acid and alkaline or neutral regions of the agar becomes very striking.

In the presence of a mixed culture composed of two organisms, only one of which is capable of fermenting the carbohydrate, isolations of two different colonial forms may be successfully realized. This is possible only if the culture is highly diluted before being streaked, and the two organisms are present in approximately equal numbers.

Bromocresol purple is frequently used as the indicator in carbohydrate media. It is especially valuable for bacteriological use because of its brilliant colors at different hydrogen-ion concentrations and its resistance to decolorization by bacterial action. The sensitive range of the indicator is from pH 5.4 (yellow) to 7.0 (purple). The pK value, i.e., the point at which the dye is 50 percent dissociated, is pH 6.2. Its pH range makes it suitable for the detection of bacterial fermentations in solid carbohydrate media.

Reversal of reaction Sometimes carbohydrate agar showing an acid reaction may turn alkaline if stored in petri dishes for several days at room temperature. This has been explained on the basis of (1) insufficient carbohydrate in the medium or (2) the oxidation of the organic acids to alkaline compounds.

If a medium contains insufficient carbohydrate, an organism will be unable to produce its limiting pH. The organism continues to multiply by utilizing the nitrogenous constituents (peptones) for both structure and energy. This results in the production of alkaline products, which may be sufficient to reverse the reaction of the medium.

Reversal of the reaction of a medium may occur even if the amount of carbohydrate is sufficient for an organism to produce its limiting pH. Under these conditions simultaneous acid and alkaline reactions occur. The carbohydrate is first fermented to organic acids, followed by the oxidation of the acids to carbonates. The alkaline carbonates are responsible for the reversal of the reaction. The alkaline changes do not occur after all the carbohydrate is converted to acid, but runs simultaneously with acid production.

Fermentation of nitrogenous compounds

It is usually stated that only carbohydrates or carbohydrate-like compounds are fermented, but this is not true. It is known that some anaerobes and facultative forms are able to obtain energy by the fermentation of single amino acids, purines, and pyrimidines. The amino group of the amino acids is probably first removed as ammonia, followed by the fermentation of the carbon residues.

The following examples may be given:

1. *Clostridium tetanomorphum* ferments glutamate, cysteine, serine, histidine, aspartate, and tyrosine. Using glutamate, the products of fermentation are acetate, butyrate H_2, CO_2, and NH_3. CO_2, acetate, and butyrate originate from carbon atoms 5, 1 and 2, and 3 and 4, respectively of glutamate. Carbon 4 is found exclusively in carbon 1 of acetate, and in 1 and 3 of butyrate (Fig. 233).

2. *Peptococcus glycinophilus* ferments glycine to acetate, CO_2, and NH_3:

$$\underset{\text{Glycine}}{4CH_2NH_2 \cdot COOH} + 2H_2O \longrightarrow 4NH_3 + \underset{\text{Acetic acid}}{3CH_3 \cdot COOH} + 2CO_2$$

3. *E. coli* ferments L-lysine with the production of acetate, butyrate, and NH_3:

$$\underset{\text{L-Lysine}}{NH_2 \cdot (CH_2)_4 \cdot CHNH_2 \cdot COOH} + 2H_2O \longrightarrow$$

$$\underset{\text{Acetic acid}}{CH_3 \cdot COOH} + \underset{\text{Butyric acid}}{CH_3 \cdot CH_2 \cdot CH_2 \cdot COOH} + 2NH_3$$

4. *C. tetani* acts on aspartate, glutamate, and serine with the production of acetate, lactate, butyrate, ethyl alcohol, CO_2, and NH_3. Essentially the same products are produced from pyruvate, an intermediary in the fermentation of carbohydrates.

For more information: Deibel (1964); Dyer and Costilow (1968); Goldfine and Stadtman (1960); Hardman and Stadtman (1960); Rogosa (1969).

Acetic (vinegar) oxidation

The oxidation of alcohol to vinegar is an aerobic process and, therefore, not a true fermentation. However, the initial step, namely, the conversion of sugar to alcohol, is a true fermentation.

When a dilute alcoholic liquid is exposed to air, a film soon appears on its surface. At the same time the liquid becomes sour, owing to the oxidation of the alcohol to acetic acid. The film is composed of a viscous

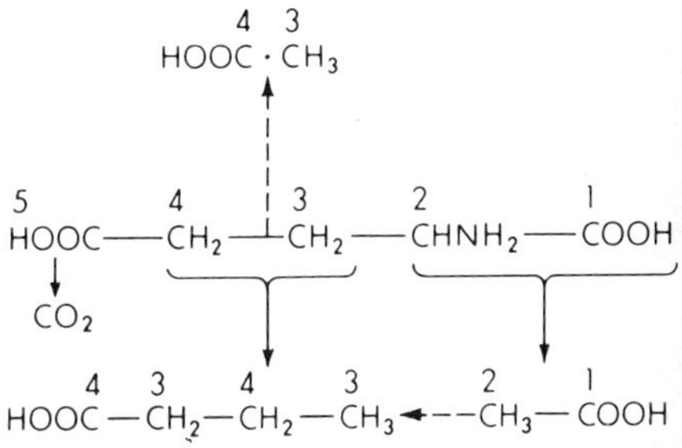

Fig. 233 Origin of products of glutamate fermentation. (After Wachsman and Barker.)

gelatinous substance, or zoogloea, in which are embedded many bacteria. It is commonly known as *mother of vinegar* because a small amount of this material is capable of acting as a starter when added to more alcoholic liquid.

Almost any alcoholic liquid may be employed in the manufacture of vinegar. The fermentation of apple cider to hard cider (alcoholic) by means of yeasts, followed by the action of the acetic bacteria, yields a product known as cider vinegar. The oxidation of wine yields a product known as wine vinegar. The alcoholic fermentation of an infusion of barley malt followed by the acetic fermentation gives rise to a preparation known as malt vinegar. Sugar vinegar results from the alcoholic fermentation of sugar, followed by the oxidation of the alcohol to acetic acid. The final product is named after the raw material used in its manufacture.

In the United States, the traditional raw material for vinegar production is apple cider. Cider vinegar is the standard to which other vinegars are compared. Within the past several years, the commercial production of wine vinegar has shown a marked increase even though it is higher-priced than other vinegars.

All vinegars are diluted to contain at least 4 percent acetic acid in the final product. In addition to acetic acid, vinegars may contain traces of other compounds such as ethyl alcohol, glycerol, esters, invert sugar, pentosans, and salts.

Acetobacter The specific bacteria present in a zoogloeal mass and capable of oxidizing dilute alcoholic solutions to vinegar are members of the genus *Acetobacter.*

Cells are ellipsoidal to rod-shaped, occurring singly, in pairs, or in short or long chains. They are motile with polar flagella, or nonmotile. Involution forms may be spherical, elongated, filamentous, club-shaped, swollen, curved, or may even appear to be branched. Young cells are gram-negative; old cells often gram-variable. They are obligate aerobes; as a rule they are strongly catalase-positive, sometimes weakly so. The cells oxidize various organic compounds to organic acids and other oxidation products which include acetic acid from ethyl alcohol, gluconic and 5-ketogluconic acids from glucose, dihydroxy-acetone from glycerol, sorbose from sorbitol, etc. Nutritional requirements vary from simple to complex. Development is generally best in yeast infusion or yeast autolysate media with added ethyl alcohol or other oxidizable substrates. Optimum temperature varies with the species. The cells are widely distributed in nature in plant materials undergoing alcoholic fermentation; they are of importance to man for their role in the completion of the carbon cycle and for the production of vinegar.

For more information: Asai (1968); Carr and Shimwell (1961); Casida (1968); McIntosh (1962); Schell and De Ley (1962).

Methods of manufacture All vinegar produced in the United States is made by acetic fermentation of an alcoholic solution by the action of an appropriate *Acetobacter* species. To ensure a successful oxidation, the alcoholic solution should be adjusted to a strength of 10 to 13 percent. This corresponds to the alcoholic content of most natural wines. If the alcoholic concentration is above 14 percent, the oxidation is incomplete; if below the minimum, much of the acetic acid may be lost by oxidation after its formation.

Three general methods are employed: (1) the Orleans process, (2) the generator process, and (3) the submerged oxidation process.

Orleans process This is the oldest commercial method known for the manufacture of vinegar. Barrels are laid horizontally and perforated near the top to permit free access of air, then filled about two-thirds full with a mixture composed of 2 parts vinegar and 3 parts wine. The wine may be raw or pasteurized, preferably the latter because it greatly reduces the percentage of abnormal reactions. Since the acetic bacteria grow better in strongly acid environments, vinegar is added to the wine to speed up the reaction and at the same time to check the growth of undesirable organisms. The temperature is kept above 70°F (21°C). At definite intervals some vinegar is drawn off and fresh wine added, making the process continuous. At best the method is too slow for large-scale commercial manufacture, but it is still practiced in the home.

Generator process This method utilizes large wooden tanks or generators filled with beechwood shavings, wood charcoal, corncobs, coke, excelsior, or other supporting materials to give greater surface (Fig. 234).

Fig. 234 Generator used in the quick vinegar process. (From Prescott and Dunn, "Industrial Microbiology.")

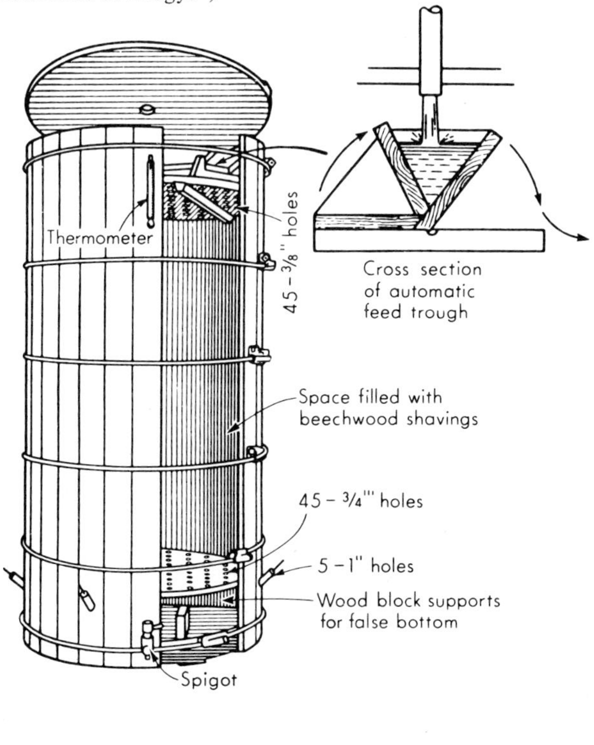

First, the supporting structure is soaked with a culture of the acetic bacteria before placing in the tank. Second, the temperature is adjusted to the optimum for the bacterial species selected. This varies from 28 to 35°C. Third, the alcoholic liquid (10 to 13 percent) is introduced at the top of the tank and allowed to trickle over the surface of the supporting material and to make contact with the bacteria. At the same time, fresh air is permitted to enter at the bottom of the tank. By the time the liquid reaches the bottom of the tank, most of the alcohol becomes oxidized to vinegar. If the oxidation is incomplete, the liquid is recirculated. This may be repeated as many times as necessary to achieve the desired acetic acid content.

A generator 10 ft in diameter and 20 ft high is capable of producing from 80 to 100 gal of vinegar per day.

Submerged oxidation process This method differs from the other two in that the bacteria are uniformly distributed throughout the alcoholic medium. The temperature of the interior is maintained at the optimum for the species of *Acetobacter* employed. Oxygen, in the form of a fine stream of air, is admitted at the bottom and the contents kept in vigorous agitation. This is the same principle employed in the production of antibiotics (see page 606). The more intimate contact of medium, bacteria, and air results in a higher rate of oxidation. Conversion of alcohol to acetic acid can be as high as 98 percent by the submerged method; 80 to 88

Fig. 235 The Cavitator generator for the production of vinegar by the submerged method. (Courtesy of Yeomans Brothers Company.)

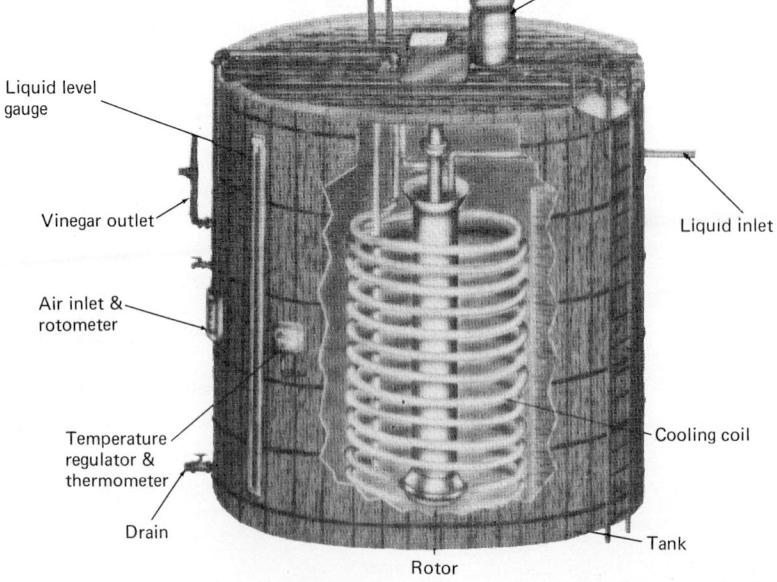

percent by the generator process; and only 70 to 80 percent by the slow Orleans method.

A generator known as the Cavitator and manufactured by Yeomans Brothers Company (1961) is available for the commercial production of vinegar by the submerged method (Fig. 235).

Commercial vinegar contains a minimum of 4 percent acetic acid. In addition to acid, traces of esters are present which are largely responsible for the pleasant odor and flavor of vinegar.

Vinegar may lose its strength on prolonged exposure to air, owing to the oxidation of the acetic acid to H_2O and CO_2 by some species of *Acetobacter*. This may be prevented by storing vinegar in well-filled, tightly stoppered containers, or by destroying the bacteria by pasteurization.

For more information: Beaman (1963); Casida (1968); Lopez et al. (1961); McNary and Dougherty (1960).

Mechanism of alcohol oxidation The aerobic oxidation of alcohol to acetic acid is believed to proceed as follows:

1. Oxygen acts as a hydrogen acceptor, converting alcohol to acetaldehyde:

$$\underset{\text{Ethyl alcohol}}{CH_3 \cdot CH_2OH + O} \longrightarrow \underset{\text{Acetaldehyde}}{CH_3 \cdot CHO + H_2O}$$

2. Acetaldehyde takes up water and becomes hydrated:

$$\underset{\text{Acetaldehyde}}{CH_3 \cdot CHO + H_2O} \longrightarrow \underset{\text{Hydrated aldehyde}}{CH_3 \cdot C(H)(OH)-OH}$$

3. Oxygen accepts two hydrogen atoms from the hydrated aldehyde to give acetic acid:

$$\underset{\text{Hydrated aldehyde}}{CH_3 \cdot C(H)(OH)-OH + O} \longrightarrow \underset{\text{Acetic acid}}{CH_3 \cdot COOH + H_2O}$$

References

Asai, T.: "Acetic Acid Bacteria," Baltimore, University Park Press, 1968.

Baldwin, E.: "The Nature of Biochemistry," London, Cambridge University Press, 1967.

Baldwin, R. L., et al.: Conversion of lactate-C^{14} to propionate by the rumen microflora, *J. Bact.*, 83:907, 1962.

Beaman, R. G.: "A Continuously Operated Submerged Fermentation Process for the Production of Vinegar," Melrose Park, Ill., Yeomans Brothers Company, 1963.

Beardsley, D. W.: Making and using silage, *Everglades Sta. Mimeo. Rep.* **56**-15, Belle Glade, Fla., May 21, 1956.

Burmeister, H. R., and P. A. Hartman: Yeasts in ensiled high-moisture corn, *Appl. Microbiol.*, **14**:35, 1966.

——— et al.: Microbiology of ensiled high-moisture corn, *Appl. Microbiol.*, **14**:31, 1966.

Carr, J. G., and J. L. Shimwell: The acetic acid bacteria, 1941–1961: A critical review, *Antonie van Leeuwenhoek, J. Microbiol. Serol.*, **27**:386, 1961.

Casida, L. E., Jr.: "Industrial Microbiology," New York, John Wiley & Sons, Inc., 1968.

Cheldelin, V. H.: "Metabolic Pathways in Microorganisms," New York, John Wiley & Sons, Inc., 1961.

de França, F. P., et al.: Retting of flax by *Aspergillus niger, Appl. Microbiol.*, **17**:7, 1969.

Deibel, R. H.: Utilization of arginine as an energy source for the growth of *Streptococcus faecalis, J. Bact.*, **87**:988, 1964.

Dyer, J. K., and R. N. Costilow: Fermentation of ornithine by *Clostridium sticklandii, J. Bact.*, **96**:1617, 1968.

Garvie, E. I.: The production of L(+) and D(–) lactic acid in cultures of some lactic acid bacteria, with a special study of *Lactobacillus acidophilus* NCDO 2, *J. Dairy Res.*, **34**:31, 1967.

Gemmell, M., and W. Hodgkiss: The physiological characters and flagellar arrangement of motile homofermentative lactobacilli, *J. Gen. Microbiol.*, **35**:519, 1964.

Goldfine, H., and E. R. Stadtman: V. The conversion of β-alanine to propionic acid by cell-free extracts of *Clostridium propionicum, J. Biol. Chem.*, **235**:2238, 1960.

Guymon, J. F., and E. A. Crowell: The formation of acetoin and diacetyl during fermentation, and the levels found in wines, *Am. J. Enol. Viticult.*, **16**:85, 1965.

Hardman, J. K., and T. C. Stadtman: I. Fermentation of γ-aminobutyric acid by *Clostridium aminobutyricum* N. sp., *J. Bact.*, **79**:544, 1960.

Harris, J. F., and G. J. Hajny: Glycerol production: A pilot-plant investigation for continuous fermentation and recovery, *J. Biochem. Microbiol. Technol. Eng.*, **2**:9, 1960.

Hesser, J. M., et al.: Lactobacilli in ensiled high-moisture corn, *Appl. Microbiol.*, **15**:49, 1967.

Katagiri, H., et al.: Studies on lactic acid fermentation. III. Mechanism of ethanol formation, *Bull. Agr. Chem. Soc. Japan*, **24**:588, 1960.

Keenan, T. W., and D. D. Bills: Volatile compounds produced by *Propionibacterium shermanii, J. Dairy Sci.*, **51**:797, 1968.

Kempton, A. G., and C. L. San Clemente: Chemistry and microbiology of forage-crop silage, *Appl. Microbiol.*, **7**:362, 1959.

Kluyver, A. J., and C. B. van Niel: "The Microbe's Contribution to Biology," Cambridge, Mass., Harvard University Press, 1956.

Langston, C. W., and C. Bouma: A study of the microorganisms from grass silage. I. The cocci, *Appl. Microbiol.,* **8**:212, 1960*a*; II. The lactobacilli, *ibid.,* **8**:223, 1960*b*.

Long, S. K., and R. Patrick: Production of 2,3-butylene glycol from citrus wastes. I. The *Aerobacter aerogenes* fermentation, *Appl. Microbiol.,* **9**:244, 1961.

—— and ——: The present status of the 2,3-butylene glycol fermentation. In "Advances in Applied Microbiology," vol. 5, New York, Academic Press, Inc., 1963.

Lopez, A., et al.: Observations on a laboratory method for submerged acetic fermentation, *Appl. Microbiol.,* **9**:425, 1961.

McIntosh, A. F.: A serological examination of some acetic acid bacteria, *Antonie van Leeuwenhoek, J. Microbiol. Serol.,* **28**:49, 1962.

McNary, R. R., and M. H. Dougherty: Citrus vinegar, *Univ. Florida Agr. Exp. Sta. Bull.,* **622**, 1960.

Mundt, J. O., and J. L. Hammer: Lactobacilli on plants, *Appl. Microbiol.,* **16**:1326, 1968.

Namba, Y., and C. Furusaka: Studies on the physiology of *Clostridium acetobutylicum*. I. Distribution of acetoacetate decarboxylase activity among strains of clostridia, *J. Agr. Chem. Soc. Japan,* **42**:311, 1968*a*; II. Acetoacetate decarboxylation under hydrogen gas phase, *ibid.,* **42**:315, 1968*b*; III. Properties of purified acetoacetyl decarboxylase, *Bull. Inst. Agr. Res. Tohoku Univ.,* **20**:1, 1968*c*.

Owen, S. P.: Industrial fermentations. In "Annual Review of Microbiology," **14**:99, 1960.

Paege, L. M., and M. Gibbs: Anaerobic dissimilation of glucose-C^{14} by *Escherichia coli, J. Bact.,* **81**:107, 1961.

Pastan, I.: Cyclic AMP, *Sci. Am.,* **227**: 97, 1972.

Pfennig, N., and T. Wikén: On a negative Pasteur effect produced in yeast by succinic acid–succinate buffer, *Pathol. Microbiol.,* **23**:359, 1960*a*.

—— and ——: Further observations on a negative Pasteur effect produced in yeast by succinic acid–succinate buffer, *Zentr. Bakteriol.,* **113**:491, 1960*b*.

Rhodes, R. A.: The fermentative capacity of microbial cells, *Chem. Eng. Progr. Symp. Ser.,* **62**:21, 1966.

Rittenberg, D., and I. Ponticorvo: On the quantitative significance of the pentose pathway in *Escherichia coli, J. Biol. Chem.,* **237**:2709, 1962.

Rogosa, M.: *Acidaminococcus* gen. n., *Acidaminococcus fermentans* sp. n., anaerobic gram-negative diplococci using amino acids as the sole energy source for growth, *J. Bact.,* **98**:756, 1969.

Rose, A. H.: "Industrial Microbiology," Washington, D.C., Butterworth, Inc., 1961.

——: "Chemical Microbiology," New York, Plenum Press, Plenum Publishing Corporation, 1968.

Rosemberg, J. A.: Bacteria responsible for the retting of Brazilian flax, *Appl. Microbiol.,* **13**:991, 1965.

Rydin, C.: X. Malt as a supplement in biological ensiling, *Arch. Mikrobiol.,* **38**:156, 1961.

—— XI: Malt-cereal mixtures and straw as supplements in biological ensiling, *Lantbrukshögskolans Ann.*, **29**:45, 1963.

Scheffers, W. A.: On the inhibition of alcoholic fermentation in *Brettanomyces* yeasts under anaerobic conditions, *Experientia,* **17**:40, 1961.

—— and T. O. Wikén: The Custers effect (negative Pasteur effect) as a diagnostic criterion for the genus *Brettanomyces, Antonie van Leeuwenhoek, J. Microbiol. Serol., Supplement,* **35**:A31, 1969.

Schell, J., and J. De Ley: Variability of acetic acid bacteria, *Antonie van Leeuwenhoek, J. Microbiol. Serol.,* **28**:445, 1962.

Sistrom, W. R.: "Microbial Life," New York, Holt, Rinehart and Winston, Inc., 1969.

Smith, L. H.: Theoretical carbohydrate requirement for alfalfa silage production, *Agron. J.,* **54**:291, 1962.

Stjernholm, R., and H. G. Wood: Trehalose and fructose as indicators of metabolism of labeled glucose by the propionic acid bacteria, *J. Biol. Chem.,* **235**:2753, 1960*a*.

—— and ——: Glycerol dissimilation and occurrence of a symmetrical three-carbon intermediate in propionic acid fermentation, *J. Biol. Chem.,* **235**:2757, 1960*b*.

Swick, R. W., and H. G. Wood: The role of transcarboxylation in propionic acid fermentation, *Proc. Natl. Acad. Sci. U.S.,* **46**:28, 1960.

Thayer, D. W., and J. E. Ogg: Aldehydes and ketones produced during fermentation of glucose by *Escherichia coli, J. Bact.,* **94**:488, 1967.

Thoukis, G., et al.: The formation of succinic acid during alcoholic fermentation, *Am. J. Enol. Viticult.,* **16**:1, 1965.

Tromp, A. G. G. M., and W. A. Scheffers: Effect of acetoin on anaerobic growth in *Brettanomyces, Antonie van Leeuwenhoek, J. Microbiol. Serol., Supplement,* **35**:H37, 1969.

van Dijk, A., and T. O. Wikén: On the free amino acid pool of yeast cells showing a negative Pasteur effect in succinic acid–succinate buffer. *Path. Microbiol.,* **32**:73, 1968.

Wikén, T.: The stimulating influence of elementary molecular oxygen and aerobically added carbohydrates on the rate of alcohol fermentation in different yeasts, *Sci. Rep. Ist. Super. Sanita,* **1**:309, 1961.

——: On "negative Pasteur effects" in yeasts. In "Aspects of Yeast Metabolism," edited by A. K. Mills and Sir H. Krebs, Oxford, Blackwell Scientific Publications, Ltd., 1968.

——: L'influence de l'oxygène sur la fermentation alcoolique par les levures, *Fermentations et Vinifications*, 2nd Symp. Internat. d'Oenologie, Bordeaux-Cognac, 1967.

—— et al.: On the existence of a negative Pasteur effect in yeasts classified in the genus *Brettanomyces* Kufferath et van Laer, *Antonie van Leeuwenhoek, J. Microbiol. Serol.,* **27**:401, 1961.

Yeomans Brothers Company: "Modern Vinegar Production," Melrose Park, Ill., 1961.

15 Differentiation and classification of bacteria

An organism can be classified only after its morphological, cultural, physiological, and pathogenic characters have been exhaustively studied.

After such a study the organism must be given a name; otherwise reference to it could not be made. Also, it is highly desirable that the name applied to the organism be understood by all workers in the field.

Two kinds of names are used: (1) common or casual names and (2) scientific or international names.

Usually the common names for the same organism will be different in each language. Because of this fact the common names are responsible for considerable confusion in bacteriology. On the other hand, scientific names are international in meaning. It is better, therefore, to refer to organisms by their scientific names which are the same in all countries.

General principles of nomenclature

The method followed in naming plants and animals was first introduced by the Swedish botanist Karl von Linné and is known as the binomial system of nomenclature. The same system is used by the zoologist and bacteriologist.

Each distinct kind of organism is called a species. Each species is given a name which consists usually of two words. The first word is the name of the genus and the second the name of the species. The genus name is taken from the Latin or Greek, or a new word compounded from Latin or Greek roots, or in rare cases from some other language. It is a noun, always capitalized, and should be used only in the singular. The species name is never capitalized.

Genera may be masculine (*Bacillus, Micrococcus, Streptococcus*), feminine (*Sarcina, Salmonella, Pasteurella*), or neuter (*Bacterium, Clostridium, Corynebacterium*). The above generic names in the plural may be written bacilli, micrococci, streptococci, sarcinae, salmonellae, pasteurellae, bacteria, clostridia, and corynebacteria, in which form they are not capitalized or italicized.

The species name may be (1) An adjective, in which case its ending should agree with the genus in gender, e.g., *Sarcina lutea, Bacillus subtilis,*

Propionibacterium arabinosum; (2) an adjective in the form of the present participle of a verb, as *Pseudomonas fluorescens, Bacillus coagulans, Clostridium perfringens*; (3) a noun in the genitive case modifying the generic name, e.g., *Clostridium muelleri, Pseudomonas lindneri, Streptococcus lactis, Brucella abortus*; (4) an explanatory noun that does not agree necessarily with the generic name in gender, as *Rhizobium phaseoli, Pseudomonas schuylkilliensis, Salmonella oslo*.

Sometimes a species is subdivided into varieties. These are also given Latin designations. An example is *Streptococcus faecalis* var. *liquefaciens*. It differs from *S. faecalis* only in being able to liquefy the acid curd produced in milk.

General principles of taxonomy

The term taxonomy may be defined as the classification of living organisms according to their natural relationships.

A satisfactory development of taxonomy is dependent upon a sound nomenclature. Regardless of whether bacteriologists will ever be able to agree on the exact classification to be employed, they should agree on some of the fundamental characteristics necessary for the development of a satisfactory bacteriological classification.

Species Each kind of organism is referred to as a species (plural, species). The term is defined in various ways, but it is generally regarded as the lowest member of a classificatory system. A bacterial species occupies a position in a classification between the genus and the variety. Since the differences between varieties are often difficult to recognize, it is the species which to the untrained observer usually represent the simplest distinct assemblages or kinds in the plant or animal kingdom.

The first described specimen of a species is spoken of as the type of the species. It is used as the type for all other species regarded as sufficiently like the type to be placed together in the same group or genus.

The term strain should never be used to indicate a biological character, but rather it should refer merely to source. An organism that is normally flagellated (motile) may be isolated from another source in which flagella are absent (nonmotile). The second form may be considered to be a nonmotile strain of the first organism but not a different species.

For more information: Buchanan (1965); Kauffmann (1963*a*); Leifson (1966).

Genus A genus (plural, genera) usually includes more than one closely related species. Some genera contain only a single species in each. These are exceptions. A genus usually contains many species.

There is no agreement as to which species should be grouped in a particular genus and probably never will be. Much of the confusion in the classification of bacteria may be attributed to this fact. However, as organisms become better known, it seems likely that much of the present confusion may be overcome and a more orderly system of classification established.

Family A family is a group of closely related genera, one of which is designated the type genus. The word is compounded from the name of the type genus by affixing the suffix *-aceae* to its root. For example, the name of the family *Micrococcaceae* is formed by combining the root of the genus *Micrococcus* with the suffix *-aceae* (*Micrococc* + *aceae*). The family names are plural.

Order An order includes a group of closely related families. It is usually named by substituting the suffix *-ales* for *-aceae* in the name of the type family. For example, the name of the order *Actinomycetales* is formed by combining the root of the family *Actinomycetaceae* with the suffix *-ales* (*Actinomycet* + *ales*).

Class A class includes closely related orders. The bacteria are placed in the class *Schizomycetes* under Division I, *Protophyta*, of the plant kingdom.

Other categories Sometimes orders are divided into suborders, families into tribes, tribes into genera, and genera into subgenera.

Citation of authors and names It is customary to place after the name of an organism the name of the discoverer, together with the citation to the literature in which the work appeared. When the name of the organism is transferred to another genus, the name of the discoverer is placed in parentheses immediately after the name of the organism, followed by the author of the new name. Citation to the literature in which the work of the new author appeared should also be included. For example, *Salmonella typhimurium* was discovered by Loeffler in 1892 and named *Bacillus typhi murium*. Later Castellani and Chalmers transferred the organism to a new genus and named it *Salmonella typhimurium*. The name of the organism is correctly designated as *Salmonella typhimurium* (Loeffler, 1892) Castellani and Chalmers, 1919. (*Bacillus typhi murium* Loeffler, *Cent. Bakt.*, 11:192, 1892; Castellani and Chalmers, "Manual of Tropical Medicine," 3d ed., page 939, 1919.)

Classification of bacteria

One purpose of a classification is to aid in the identification of an unknown organism by comparing its characteristics with others which have already been described. Unfortunately this is not always possible as the unknown may be one that has not been previously studied. Under these conditions the newly isolated organism should be described, named, and the results published so that the material becomes available to others. More than 1,500 species of bacteria have been classified but these represent probably only a small percentage of the total number of species in existence.

During the early years of bacteriology, organisms were classified entirely on the basis of morphology. Only a few species were recognized, and their classification was a comparatively simple matter. The morphological characters employed included size and shape of an organism; arrangement of the cells; presence or absence of well-defined capsules; presence or absence of spores; size, shape, and position of the spore in the cell; presence, number, and arrangement of flagella; irregular forms; presence or absence of characteristic granules; acid-fastness; gram-reaction and other differential staining procedures; and cultural and colonial characteristics.

Higher plants are differentiated almost entirely on the basis of morphology. One tree may be easily distinguished from another by differences in such characteristics as the size, shape, and color of the tree; and size, shape, and color of the leaves and seeds. In the case of bacteria, it is quite evident that the problems of classification are more difficult, owing to the fact that such minute organisms are comparatively simple spheres, rods, and spirals.

It soon became apparent that a classification based entirely on morphology was inadequate and that more characteristics were necessary. Therefore, physiological characters were introduced into the newer classifications. These include such characters as temperature relations, chromogenesis or pigment production, effect of a change in the reaction of the medium on growth, production of indole and hydrogen sulfide, reduction of nitrate, relation to oxygen, and fermentation of carbohydrates. At the present time physiological reactions are probably more important than morphological differences in the classification of bacteria.

Sometimes it becomes necessary to resort to animal inoculation and serological reactions to separate similar appearing and reacting organisms.

The classification of bacteria given here was prepared from the unabridged seventh (1957) edition of "Bergey's Manual of Determinative Bacteriology." The book is named in honor of Dr. D. H. Bergey, who was responsible for developing the first edition. The manual is administered by a committee which is known as the Board of Trustees. The actual work of revision is the responsibility of the Editorial Committee of the Board of Trustees.[1]

Position of bacteria in the plant kingdom

The plant kingdom is divided into five divisions. The bacteria are placed in that division known as the *Protophyta*. This does not mean necessarily that bacteria are more closely related to plants than to animals, although

[1] The author is greatly indebted to the Board of Trustees for permission to adapt this outline classification of the annual.

the early biologists believed that such was the case. It is quite likely that in future classifications the *Protophyta* will be separated from the plants and made a kingdom equal in rank to the plant and animal kingdoms.

The *Protophyta* is subdivided into three classes and twelve orders as follows:

DIVISION I *Protophyta*, primordial or primitive plants.

CLASS I *Schizophyceae*, the blue-green algae. Organisms which possess the photosynthetic pigment phycocyanin in addition to chlorophyll. The blue-green algae are studied with other types of algae (green, brown, red) and the higher fungi in botany. However, the blue-green algae differ structurally from all other types of *Thallophyta*. Since the blue-green algae resemble the bacteria, both groups are commonly classed in the same division of the plant kingdom.

CLASS II *Schizomycetes*. Typically unicellular plants. Cells usually small, sometimes ultra-microscopic. Frequently motile. Presence of true nuclei has been demonstrated. Individual cells may be spherical or straight, curved, or spiral rods. Cells may occur in regular or irregular masses, or even in cysts. Where they remain attached to each other after cell division, they may form chains or even definite trichomes. The latter may show some differentiation into holdfast cells and into motile or nonmotile reproductive cells. Some grow as branching mycelial threads whose diameter is not greater than that of ordinary bacterial cells, i.e., about 1 μ. Some species produce pigments. The true purple and green bacteria possess pigments much like or related to the true chlorophylls of higher plants. These pigments have photosynthetic properties. The phycocyanin found in the blue-green algae does not occur in the Schizomycetes. Multiplication is typically by cell division. Endospores are formed by some species included in *Eubacteriales*. Sporocysts are found in *Myxobacterales*. Ultramicroscopic reproductive bodies are found in *Mycoplasmatales*. The bacteria are free-living, saprophytic, parasitic, or even pathogenic. The latter types cause diseases of either plants or animals. Ten orders are recognized.

Order I *Pseudomonadales*. Straight, curved or spiral, rigid, rod-shaped bacteria. Rarely occur in pairs or chains. Cells in a few species are ellipsoidal and are frequently spoken of as being coccoid or even spherical in form. Usually about 1.0 μ in diameter, but in a few species the individual cell is larger than is normal for bacteria, reaching a size of 3.0 to 14.0 μ in diameter, and as much as 100 μ in length. Cells usually polar flagellate. When motile they sometimes bear a single flagellum; in other cases a tuft of flagella. Flagella are normally found at one or both ends of the cell, but in one genus the curved cells bear a tuft of flagella that is attached in the middle of the concave side (*Selenomonas*). Nonmotile species whose characteristics indicate that they belong to this order with closely related motile species occasionally occur. Cells are gram-negative. Cells in one suborder contain pigments that have power of photosynthesis. Cells in second suborder lack such pigments, as do all other groups of bacteria. Cells in first suborder are photoautotrophic, while chemoautotrophic species occur in the second suborder. Energy is frequently secured by oxidative processes though there are also many species that show a fermentative physiology. Cells quite frequently occur in zoogloeal masses. No endospores are found, and reproduction is by means of fission. Many species occur in coastal, swamp and pond waters, and in soil. Some are parasitic, some even pathogenic, causing diseases of fishes and other cold-blooded vertebrates. Few species cause diseases of warm-blooded mammals, including man.

Suborder I *Rhodobacteriineae.* Cells contain red, purple, brown, or green photosynthetic pigments. Sometimes also enclose granules of free sulfur.

FAMILY I *Thiorhodaceae.* Unicellular organisms often developing as cell aggregates or families of variable size and shape. Single cells have the form of spheres, ovoids, short rods, vibrios, spirals, long rods or, occasionally, chains. Occur in nature in environments containing sulfides and require light for development. Produce pigments composed of green bacteriochlorophyll and yellow and red carotenoids. Organisms anaerobic or microaerophilic with a photosynthetic metabolism in which CO_2 is reduced with the aid of special hydrogen donors without the liberation of molecular oxygen. Where these organisms are found in nature, hydrogen sulfide acts as a hydrogen donor, and sulfur accumulates as sulfur droplets in the cells. Probably all members can utilize a number of organic substances in place of hydrogen sulfide as hydrogen donors for photosynthesis. They are potentially mixotrophic.

I. Cells usually combined into aggregates.
  A. Cells grouped as regular sarcina packets.

GENUS I. *Thiosarcina.*

  B. Cells not in sarcina packets.
    1. Aggregates in the form of a flat sheet.
      a. Cells in regular arrangement, with tetrads as the common structural unit.

GENUS II. *Thiopedia.*

      aa. Cells in irregular aggregates.

GENUS III. *Thiocapsa.*

    2. Aggregates in the form of three-dimensional masses.
      a. Cells distinctly rod-shaped and arranged in a net-like structure.

GENUS IV. *Thiodictyon.*

      aa. Cells not so arranged.
        b. Cells in a common capsule, individuals rather scattered and loosely grouped.

GENUS V. *Thiothece.*

        bb. Cells in rather dense clumps.
          c. Aggregates embedded in conspicuous common slime capsule.
            d. Aggregates small, compact, often several of them enclosed together in a common capsule.

GENUS VI. *Thiocystis.*

            dd. Aggregates large and solid, later break up into small clusters.

GENUS VII. *Lamprocystis.*

          cc. Common capsule lacking or very transient.
            d. Aggregates as a whole exhibit amoeboid movements.

GENUS VIII. *Amoebobacter.*

dd. Aggregates devoid of amoeboid movements.

GENUS IX. *Thiopolycoccus.*

II. Cells usually occurring singly.

A. Cells clearly spiral-shaped.

GENUS X. *Thiospirillum.*

B. Cells not spiral-shaped.

1. Cells irregular, often swollen, distorted, or composed of long, crooked and bent rods to filaments.

GENUS XI. *Rhabdomonas.*

2. Cells regular, spherical to short rods or bean-shaped.

a. Cells spherical, as a rule nonmotile, and each one surrounded by a rather wide capsule.

GENUS XII. *Rhodothece.*

aa. Cells ellipsoidal, ovoid, short rods, or vibrios, actively motile.

GENUS XIII. *Chromatium.*

FAMILY II. *Athiorhodaceae.* Unicellular organisms of relatively small size, occurring as spheres, short rods, vibrios, long rods, and spirals. Motility due to presence of polar flagella. Gram-negative. Produce a pigment system composed of bacteriochlorophyll and one or more carotenoids, coloring the cells yellowish brown, olive-brown, dark brown or various shades of red. Color usually not observable with single cells but only with cell masses. Generally microaerophilic, although many members may grow at full atmospheric oxygen tension. Capable of development under strictly anaerobic conditions, but only in illuminated cultures by virtue of a photosynthetic metabolism. Latter dependent upon the presence of extraneous hydrogen donors, such as alcohols, fatty acids, hydroxy- and keto-acids, and does not proceed with the evolution of molecular oxygen. Members which can grow in presence of air can also be cultivated in darkness, but only under aerobic conditions.

I. Cells rod-shaped or spherical, not spiral-shaped.

GENUS I. *Rhodopseudomonas.*

II. Cells spiral-shaped.

GENUS II. *Rhodospirillum.*

FAMILY III. *Chlorobacteriaceae.* Green bacteria, usually of small size, occurring singly or in cell masses of various shapes and sizes, developing in environments containing rather high concentrations of hydrogen sulfide and exposed to light. As a rule not containing sulfur globules but frequently depositing sulfur outside the cells. Contain green pigments of a chlorophyllous nature, though not identical with the common green plant chlorophylls nor with bacteriochlorophyll. Capable of photosynthesis in the presence of hydrogen sulfide; do not liberate oxygen.

I. Free-living bacteria not intimately associated with other microbes.

A. Bacteria not united into well-defined colonies.

GENUS I. *Chlorobium.*

B. Bacteria united into characteristic aggregates.

1. Bacteria without intracellular sulfur globules (Fig. 236).

GENUS II. *Pelodictyon.*

2. Bacteria with intracellular sulfur globules.

GENUS III. *Clathrochloris.*

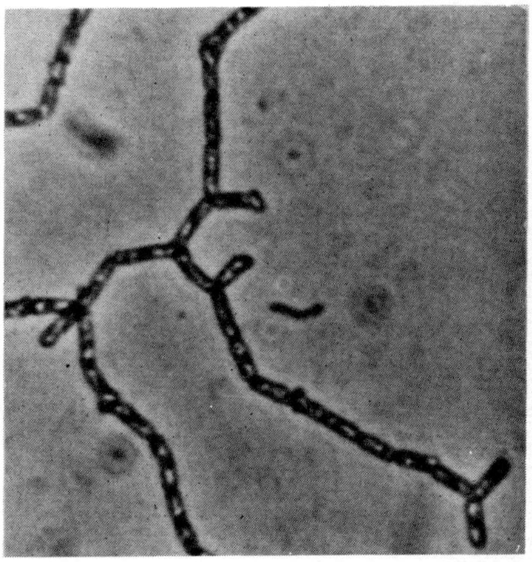

Fig. 236 Phase contrast micrograph of Pelodictyon clathratiforme. Chain of cells with several branch points. The cells are densely filled with gas vacuoles which appear as bright spots of irregular shape. x2000. (Courtesy of Pfennig and Cohen-Bazire.)

II. Green bacteria found as symbiotic aggregates with other organisms.

A. Aggregates composed of green bacteria and protozoa.

GENUS IV. *Chlorobacterium.*

B. Aggregates composed of two different types of bacteria.

1. Aggregates small, barrel-shaped, actively motile and consisting of a central, polar flagellate, rod-shaped bacterium with a covering of sulfur-green bacteria.

GENUS V. *Chlorochromatium.*

2. Aggregates large, cylindrical, nonmotile and composed of a central filamentous bacterium with a more or less extensive covering of sulfur green bacteria.

GENUS VI. *Cylindrogloea.*

Suborder II *Pseudomonadineae.* Cells that do not contain photosynthetic pigments but diffusible, water-soluble pigments, of a type not found elsewhere among bacteria, occur in many species. Also, non-water-soluble yellow or red pigments occur in some genera. Motile species are invariably polar flagellate. Some groups are strictly autotrophic, oxidizing simple inorganic compounds; others are facultatively heterotrophic or heterotrophic in their physiology. Majority of species grow well and fairly rapidly on the surfaces of ordinary culture media. Some species that attack agar or cellulose or that show other unusual types of physiology are more fastidious in their requirements. Only a few species are strictly anaerobic. Species found largely in saltwater, freshwater, or in soil. Some are parasitic and a few are pathogenic to vertebrates, including man.

FAMILY I *Nitrobacteriaceae.* Cells without endospores; rod-shaped, ellipsoidal, or even spherical, or spirillar in shape. Flagella polar, occasionally absent. Gram-negative. Organisms derive energy from oxidation of ammonia to nitrite or from nitrite to nitrate. Bacteria depend upon this oxidation for growth. Fail to grow on media containing organic matter in the absence of the specific inorganic materials used as sources of energy. Many organic compounds commonly used in standard culture media are toxic to these bacteria. Not parasitic. Commonly found in soil and freshwater.

I. Ammonia oxidized to nitrite.
   A. Zoogloeae not formed. Cells occur separately, free or in dense aggregates.
      1. Cells not spiral-shaped.
         a. Cells ellipsoidal.

GENUS I. *Nitrosomonas.*

         aa. Cells spherical.

GENUS II. *Nitrosococcus.*

      2. Cells spiral.

GENUS III. *Nitrosospira.*

   B. Zoogloeae formed.
      1. Zoogloea surrounded by a common membrane forming a cyst.

GENUS IV. *Nitrosocystis.*

      2. No common membrane surrounds the cells. The massed cells are embedded in slime.

GENUS V. *Nitrosogloea.*

II. Nitrite oxidized to nitrate.
   A. Zoogloeae not formed.

GENUS VI. *Nitrobacter.*

   B. Zoogloeae formed.

GENUS VII. *Nitrocystis.*

FAMILY II *Methanomonadaceae.* Rod-shaped organisms deriving their life energy from the oxidation of simple compounds of hydrogen or carbon. Polar flagellate when motile. Gram-negative. Found in soil and water.

I. Organisms deriving their life energy from the oxidation of simple compounds of hydrogen.
   A. Cells capable of securing growth energy by the oxidation of methane.

GENUS I. *Methanomonas.*

   B. Cells capable of securing growth energy by the oxidation of hydrogen.

GENUS II. *Hydrogenomonas.*

II. Organisms deriving their life energy from the oxidation of carbon monoxide.

GENUS III. *Carboxydomonas.*

FAMILY III *Thiobacteriaceae.* Coccoid, straight or curved, rod-shaped bacteria. Oxidize sulfur compounds, usually depositing free sulfur granules within or without the cells. Never filamentous. Colorless bacteria that are sometimes embedded in gelatinous pellicles or in gelatinous bladder-like colonies. Polar flagellate when motile. Presumably gram-negative. Found where hydrogen sulfide occurs or may oxidize free sulfur, thiosulfates, or related compounds.

I. Free sulfur granules deposited within or without the cells. Usually found in sulfurous waters or soils.
  A. Cells coccoid or straight rods.
    1. Nonmotile so far as known.

GENUS I. *Thiobacterium.*

    2. Motile by means of polar flagella so far as known.
      a. Cells rod-shaped, very large.

GENUS II. *Macromonas.*

      aa. Cells round to ovoid, large (Fig. 237).

GENUS III. *Thiovulum.*

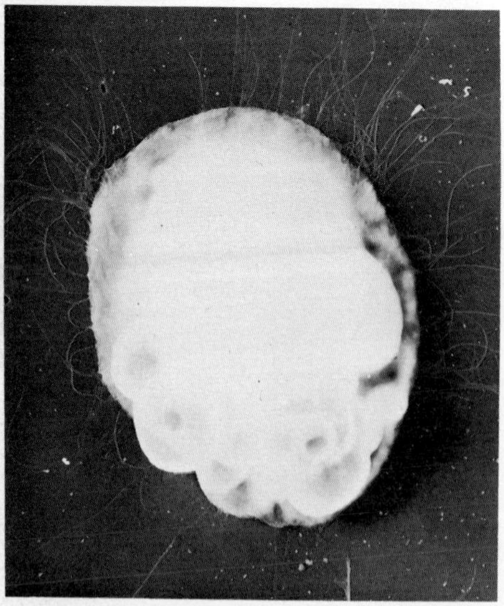

Fig. 237 Electron micrograph of Thiovulum majus showing flagella. (Courtesy of De Boer, La Rivière, and Houwink.)

B. Cells large, curved rods, somewhat pointed.

GENUS IV. *Thiospira.*

II. Oxidize free sulfur, thiosulfates, and related sulfur compounds to sulfates. Autotrophic or facultatively autotrophic.

GENUS V. *Thiobacillus.*

FAMILY IV *Pseudomonadaceae.* Cells elongate, straight rods, occasionally coccoid. Motile by means of polar flagella which are either single or in small or large tufts. A few species are nonmotile. Gram-negative. May possess either water-soluble pigments which diffuse through the medium or non-water-soluble pigments. Aerobic. Frequently oxidative in their physiology but may be fermentative. Usually found in soil or water, including seawater or even heavy brines. Many plant and a few animal pathogens.

I. Attack glucose and other sugars either oxidatively or fermentatively.
  A. Genera in which species are either known or are thought to attack glucose oxidatively.
    1. Bacteria which do not produce readily detectable acetic acid though they may oxidize ethanol. May produce a water-soluble pigment which diffuses through the medium.
      a. Cultures may or may not produce a water-soluble pigment which is bluish, greenish, or brownish in color. Rose, lilac- and yellow-colored, diffusible pigments occasionally occur (Fig. 238).

GENUS I. *Pseudomonas*

      aa. Cultures develop a yellow, non-water-soluble pigment. Cells normally monotrichous. Mostly plant pathogens which cause a necrosis.

GENUS II. *Xanthomonas.*

    2. Bacteria which produce readily detectable amounts of acetic acid by the oxidation of ethanol. The vinegar bacteria.

GENUS III. *Acetobacter.*

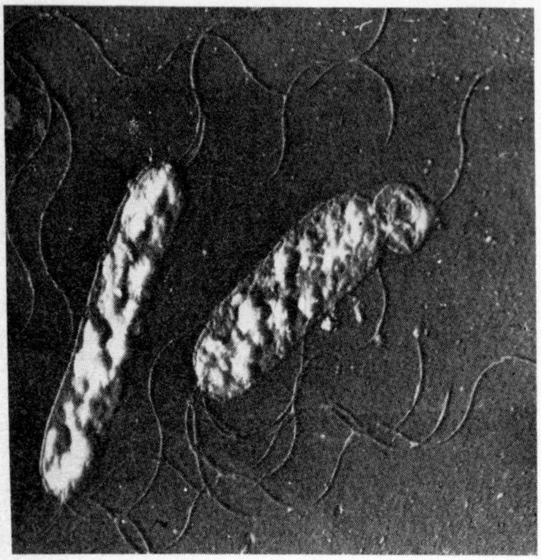

Fig. 238 Electron micrograph of Pseudomonas ovalis. (Courtesy of Wetmore and Gochenour, Jr.)

B. Genera in which the species ferment glucose, usually with the production of H_2 and CO_2.

1. Cells carry out a fermentation like that of the coliform bacteria. Usually produce acid and gas from glucose.

a. Cells not known to fix free atmospheric nitrogen.

b. Water organisms. Common species cause diseases of fishes. Also found in leeches. Not luminescent.

GENUS IV. *Aeromonas.*

bb. Luminescent bacteria commonly found on dead fishes and crustacea on seawater beaches.

GENUS V. *Photobacterium.*

aa. Cells fix free atmospheric nitrogen.

GENUS VI. *Azotomonas.*

2. Cells carry out an alcoholic fermentation similar to that of yeasts.

GENUS VII. *Zymomonas.*

II. Do not attack carbohydrates or, if so, produce only slight amounts of acid from glucose and similar sugars. Includes certain species which require at least 12 percent salt for growth.

A. Do not require salt in excess of 12 percent for growth.

1. Cells not embedded in a gelatinous matrix.
   a. Cells rod-shaped.
      b. Soil and water bacteria that are known to dissimilate alkylamines.

GENUS VIII. *Protaminobacter.*

      bb. Soil and water bacteria that are known to dissimilate alginic acid.

GENUS IX. *Alginomonas.*

   aa. Soil bacteria that are known to utilize phenol and similar aromatic compounds. Cells may be branched.

GENUS X. *Mycoplana.*

2. Cells embedded in a gelatinous matrix; this matrix may be of a branching form.

GENUS XI. *Zoogloea.*

B. Requires at least 12 percent salt before growth will take place.

GENUS XII. *Halobacterium.*

FAMILY V *Caulobacteraceae.* Nonfilamentous, rod-shaped bacteria normally attached by branching or unbranching stalks to a substrate. In one floating form the stalks are branched. Cells occur singly, in pairs, or short chains. Cells are asymmetrical in that a stalk is developed at one end of the cell or ferric hydroxide or other material is secreted from one side of the cell to form stalks. Cells are polar flagellate in the free-living state, nonmotile in the attached forms. Gram-negative. Multiply by transverse fission, the daughter cells remaining in place or swimming away as swarm cells. Typically freshwater or saltwater forms.

I. Long axis of cell coincides with axis of stalk. Stalks slender (Fig. 239).

GENUS I. *Caulobacter.*

II. Long axis of cell transverse to long axis of stalk. Stalks may be twisted and branched.
   A. Stalks are band-shaped or rounded. Contain ferric hydroxide.
      1. Stalks band-shaped and twisted. Dumbell-shaped in cross section.

GENUS II. *Gallionella.*

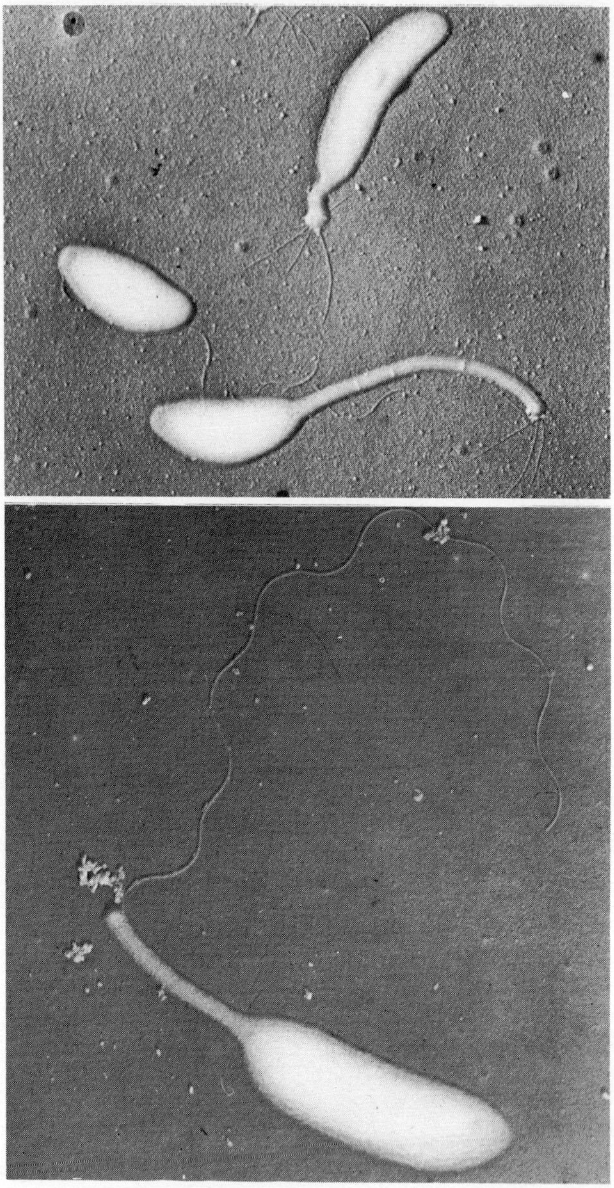

Fig. 239 Caulobacter, stalked bacteria. Upper, one cell has not yet grown a stalk; lower, young cell with one flagellum at base of stalk. (Courtesy of A. L. Houwink.)

2. Stalks horn-shaped, not twisted. Round in cross section.

GENUS III. *Siderophacus.*

B. Stalks lobose, composed of gum. Form zoogloea-like colonies. Free-floating.

GENUS IV. *Nevskia.*

FAMILY VI *Siderocapsaceae.* Cells spherical, ellipsoidal, or bacilliform. Frequently embedded in a thick, mucilaginous capsule in which iron or manganese compounds may be deposited. Motile stages, where known, are polar flagellate. Free-living in surface films or attached to the surface of submerged objects. Form deposits of iron and manganese compounds. Autotrophic, facultatively autotrophic, and heterotrophic species are included in the family. Found in freshwater.

I. Cells surrounded by capsular matter with iron compounds deposited either on the surface or throughout the capsular material.
A. Cells coccoid.
1. Cells in masses in a common capsule.

GENUS I. *Siderocapsa.*

2. Cells always in pairs in a gelatinous capsule.

GENUS II. *Siderosphaera.*

B. Cells ellipsoidal to bacilliform.
1. Cells heavily encapsulated but do not possess a torus.
a. Cells in chains in a gelatinous capsule.
b. Chains of ellipsoidal cells embedded in a gelatinous capsule, the outlines of which follow the form of the cells.

GENUS III. *Sideronema.*

bb. Rods in pairs or chains in surface films.

GENUS IV. *Ferribacterium.*

aa. Coccoid to rod-shaped cells in masses in a gelatinous capsule. Usually show an irregular arrangement of cells.

GENUS V. *Sideromonas.*

2. Cells with a thin capsule with a torus.
   a. Torus completely surrounds the cells.

GENUS VI. *Naumanniella.*

   aa. Torus open at one pole, giving the wall the appearance of a horseshoe.

GENUS VII. *Ochrobium.*

II. Nonencapsulated cells which form deposits of iron compounds in the cell wall, on the surface of the cells, or in the surrounding medium.
   A. Cells coccoid.

GENUS VIII. *Siderococcus.*

   B. Cells rod-shaped.
      1. Found in neutral or alkaline waters.

GENUS IX. *Siderobacter.*

      2. Found in acid mine wastes.

GENUS X. *Ferrobacillus.*

FAMILY VII *Spirillaceae.* Cells simple, curved or spirally twisted rods. These frequently remain attached to each other after transverse division to form chains of spirally twisted cells. Cells are rigid and usually motile by means of a single flagellum (rarely two) or a tuft of polar flagella. Gram-negative. Frequently oxidative in their physiology. Aerobic or facultatively anaerobic, although a few strict anaerobes occur among the vibrios. Largely water forms, although some are parasitic or pathogenic to higher animals and man.

I. Curved, vibrio-like rods that are rarely united into a complete ring.
   A. Cells curved; rods never united at the end into a ring-shaped cell. Usually possess a single polar flagellum.
      1. Curved rods that are not known to attack cellulose.
         a. Aerobic to anaerobic, heterotrophic vibrios.

GENUS I. *Vibrio.*

         aa. Anaerobic, facultatively autotrophic vibrios that produce hydrogen sulfide or methane.

b. Reduce sulfates to hydrogen sulfide.

GENUS II. *Desulfovibrio.*

bb. Reduce carbon dioxide to methane.

GENUS III. *Methanobacterium.*

2. Curved rods that attack cellulose.
   a. Vibrio-like cells.

GENUS IV. *Cellvibrio.*

aa. Pointed, sickle-shaped cells.

GENUS V. *Cellfalcicula.*

B. Curved rods that join ends to form a complete ring.

GENUS VI. *Microcyclus.*

II. Crescent-shaped to spiral cells that are frequently united into spiral chains of cells.
   A. Cells not embedded in zoogloeal masses.
      1. Spiral cells with polar flagellation.
         a. Possess a tuft of polar flagella.

GENUS VII. *Spirillum.*

aa. Possess a single polar flagellum.

GENUS VIII. *Paraspirillum.*

2. Crescent-shaped cells with a tuft of flagella attached to the middle of the concave side of the cell (Fig. 240).

GENUS IX. *Selenomonas.*

B. Crescent- to spiral-shaped cells embedded in a spherical mass of jelly. Found in freshwater.

GENUS X. *Myconostoc.*

Order II *Chlamydobacteriales.* Colorless, alga-like bacteria which occur in trichomes. May or may not be ensheathed. May be unbranched or may show false branching. False branching arises from a lateral displacement of the cells of the trichome within the sheath. This gives rise to a new trichome so that the sheath is branched while the trichomes are separate. Sheaths may be composed of an organic matrix impregnated with iron or manganese oxides, or they may be composed of an organic matrix free from these

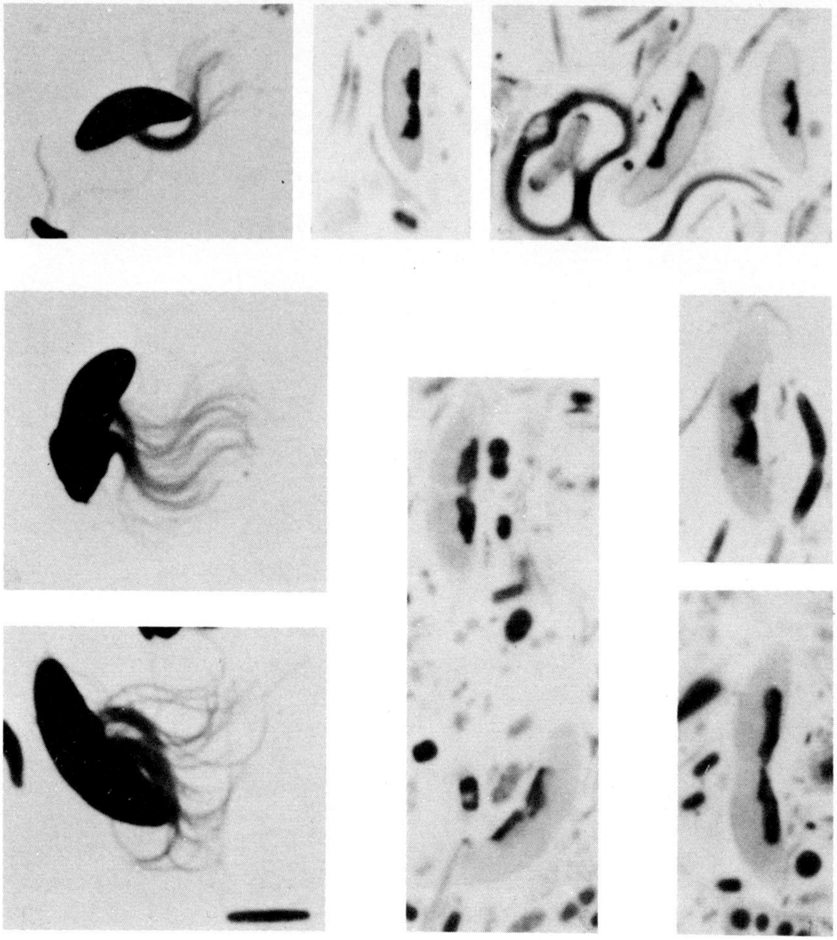

Fig. 240 Selenomonas palpitans, showing flagella arising from the middle of the concave side of the cells. Flagella are thicker at the base than at the free end. A highly refractive granule is found on the concave side at the base of the tuft of flagella. This granule stains with nuclear dyes and appears to be a true nucleus, dividing as the cell divides. (Courtesy of C. F. Robinow.)

oxides. Gram-negative. Reproduction by flagellate swarm spores or by nonmotile conidia. Endospores never developed. Freshwater and marine forms.

FAMILY I *Chlamydobacteriaceae.* Bacteria which occur in trichomes and which frequently show false branching. Sheaths, when present, may or may not be impregnated with ferric and/or manganese oxides. Cells divide transversely. Swarm cells, if developed, are usually motile by means of a tuft of flagella. Usually found in freshwater.

I. Trichomes surrounded by sheaths which are usually not impregnated with iron or manganese oxides and which do not dissolve in hydrochloric acid. Large forms, mostly sessile.

GENUS I. *Sphaerotilus.*

II. Trichomes surrounded by sheaths impregnated with oxides of iron or manganese which dissolve in strong hydrochloric acid. Free-living or sessile.

A. Individual trichomes, each with a sheath.

GENUS II. *Leptothrix.*

B. Sheaths contain more than one trichome; trichomes sometimes in a fan-like arrangement.

GENUS III. *Toxothrix.*

FAMILY II *Peloplocaceae.* Long, unbranched trichomes usually enclosed in a thin, delicate sheath. Cells within the trichomes, when in the living state, contain false vacuoles which are easily discerned by a reddish gleam of light which they emit; the cytoplasm of the cell appears bluish white. Generally nonmotile, but motile species may occur. Reproduction by transverse fission of the cell. Unattached forms found in freshwater ponds with decomposing algae.

I. Trichomes lie parallel to each other in bundles or bands.

GENUS I. *Peloploca.*

II. Trichomes occur singly.

GENUS II. *Pelonema.*

FAMILY III *Crenotrichaceae.* Trichomes attached to a firm substrate and show differentiation of base and tip. Unbranched or show false branching. Sheaths may be thin, delicate, and not encrusted with oxides of iron or manganese, or they may be plainly visible, thin, and colorless at the tip, and thick and encrusted with iron or manganese oxides at the base. Cells disk-shaped to cylindrical, dividing to produce spherical, nonmotile conidia. Individual cells may also slip out of the sheath to grow into new trichomes. Found in freshwater and saltwater.

I. Attached trichomes which are swollen at the free end.

A. Sheath thick, storing iron or manganese oxides.

GENUS I. *Crenothrix.*

B. Sheath very delicate, always colorless.

GENUS II. *Phragmidiothrix.*

II. Attached trichomes which are tapered at the free end.

GENUS III. *Clonothrix.*

Order III *Hyphomicrobiales.* Multiplication by budding or by budding and longitudinal fission. Buds may be sessile or may be borne at the tip of a slender filament which arises from the pole of a mature cell or from a filament connecting two cells. Cells may occur singly or in pairs but are found more commonly in aggregates. In some types the aggregates consist of groups of cells attached to a surface by stalks which appear to radiate from a common holdfast; in others the aggregates consist of free-floating cell groups in which the cells are attached to one another by the filament engendered in the budding process. Branching of filament may result in groups which contain several hundred cells. Cells ovoid, ellipsoidal, spherical, or pyriform. If motile, cells possess a single polar flagellum. Gram-negative so far as known. Metabolism may be heterotrophic or photosyn-

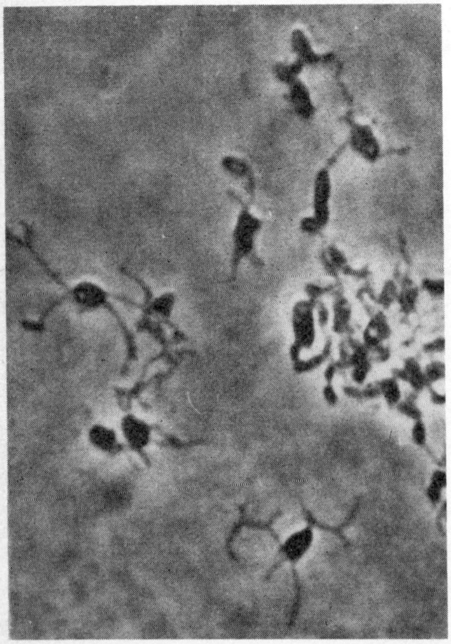

Fig. 241 Phase contrast micrograph of Hyphomicrobium sp. T37 showing varying degrees of bizarre cell shapes and branching stalks. (Courtesy of Tyler and Marshall.)

thetic. Found in mud and water of freshwater ponds and streams; also parasitic on freshwater crustacea.

FAMILY I *Hyphomicrobiaceae.* Organisms occur mainly as free-floating groups in which the cells are attached to one another by a slender, sometimes branched filament. Daughter-cell formation initiated by the outgrowth of a filament from the pole of a mature cell or from some point on a filament connecting two mature cells. Daughter cell is formed by enlargement of the tip of the filament. Gram-negative.

I. Chemoheterotrophic. Motile (Figs. 241 and 242).

GENUS I. *Hyphomicrobium.*

II. Photoheterotrophic. Nonmotile.

GENUS II. *Rhodomicrobium.*

FAMILY II *Pasteuriaceae.* Stalked bacteria with spherical or pear-shaped cells. If cells are elongated, the long axis of the cell coincides with the axis of the stalk. Stalks may be very short or absent, but when present they are usually very fine and at times arranged in whorls attached to a common holdfast. Cells multiply by longitudinal fission and/or by budding. Mostly periphytic; one species parasitic.

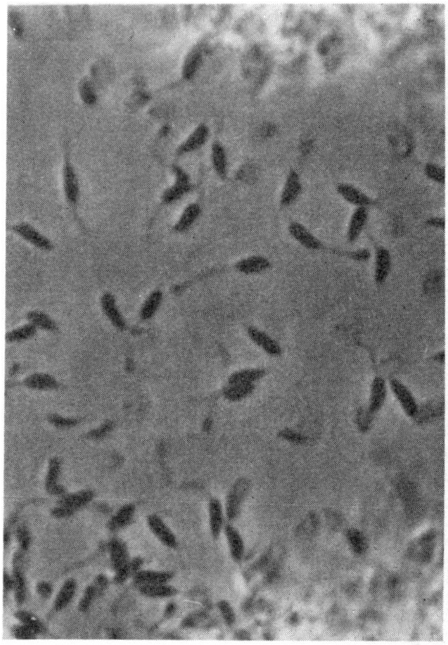

Fig. 242 Phase contrast micrograph of Hyphomicrobium sp. T37 showing classical morphology with regular buds. (Courtesy of Tyler and Marshall.)

I. Stalks lacking; cells sessile.

GENUS I. *Pasteuria.*

II. Stalks long and slender, often in whorls.

GENUS II. *Blastocaulis.*

Order IV *Eubacteriales.* Simple, undifferentiated, rigid cells which are either spherical or straight rods. Only the simplest forms of branching occur, and these only rarely. Motile and nonmotile species. Flagella are usually arranged peritrichously, but monotrichous species do occur in groups where the flagellation is normally peritrichous; such conditions appear to have been developed from ancestral peritrichrous species. Typical endospores occur in one family. All species in certain families are definitely gram-negative; in other families and groups, where the majority of species are gram-positive, at least in certain stages of growth, species occur which lose their Gram stain so readily that they are generally classed as gram-negative. Reproduction by transverse fission; occasionally cells divide in two or three planes perpendicular to each other, thereby forming tetrads or packets of eight cells. Pigments of chromogenic species commonly non-water-soluble and of a carotenoid nature; other pigments do occur however, some of which show slight powers of diffusion into agar media. Pigments nonphotosynthetic. Order includes saprophytes, parasites, and many pathogenic species; the latter cause diseases of both animals and plants. Found in saltwater, freshwater, air, soil, and in the bodies of animals and plants.

FAMILY I *Azotobacteraceae.* Relatively large rods or even cocci, sometimes almost yeast-like in appearance. Cells without endospores. Flagellation is peritrichous. Gram-negative. Obligate aerobes, usually growing in a film on the surface of culture media. Capable of fixing atmospheric nitrogen when provided with carbohydrate or other energy source. Grow best on media deficient in nitrogen. Soil and water bacteria.

GENUS I. *Azotobacter.*

FAMILY II *Rhizobiaceae.* Cells without endospores, rod-shaped, sparsely flagellated; some species nonmotile. Usually gram-negative. Grow aerobically on ordinary culture media containing glucose. Glucose and sometimes other carbohydrates are utilized without appreciable acid formation. Saprophytes, symbionts and pathogens; the latter are usually plant pathogens forming abnormal growths on roots and stems.

I. Cells capable of fixing free nitrogen when growing symbiotically on the roots of *Leguminosae.*

GENUS I. *Rhizobium.*

II. Either plant pathogens which attack roots or produce hypertrophies on stems or free-living nonchromogenic soil or water forms. Do not fix nitrogen.

GENUS II. *Agrobacterium.*

III. Usually free-living soil and water forms which produce a violet chromogenesis.

GENUS III. *Chromobacterium.*

FAMILY III *Achromobacteraceae.* Small to medium-sized rods which are usually uniform in shape. Motile by means of peritrichous flagella, or nonmotile. Gram-negative. May or may not liquefy gelatin. Growth on agar nonchromogenic to yellow, orange, brown, or even red. Pigment does not diffuse through the agar and apparently is carotenoid in nature. May produce acid but no gas from glucose and sometimes from other sugars; lactose rarely or never attacked. Certain species liquefy agar and/or attack alginates, others digest chitin. May or may not reduce nitrates. Litmus milk may be unchanged, slightly acid, or alkaline. No luminescent species are known. Generally found as saltwater, freshwater, or soil forms, less commonly found as parasites or pathogens. Some plant pathogens may belong here.

I. Do not attack agar, alginates, or chitin. Not active in the production of acid from sugars, especially lactose.

A. Nonchromogenic on ordinary agar media, although the type species of *Achromobacter* produces yellow chromogenesis on potato.

1. Litmus milk alkaline. No acid from carbohydrates.

GENUS I. *Alcaligenes.*

2. Litmus milk slightly acid (not enough to be curdled), unchanged, or alkaline. Small

amounts of acid usually produced from hexoses.

GENUS II. *Achromobacter.*

B. Yellow, orange, brown, or red chromogenesis produced on ordinary agar media; pigment non-water-soluble.

GENUS III. *Flavobacterium.*

II. Attack agar, alginates, or chitin. Slightly more active in the fermentation of sugars than is the previous group, some even attacking lactose. Non-chromogenic or chromogenic, usually with yellow or orange, always non-water-soluble pigments.

A. Attack agar and/or alginates.

GENUS IV. *Agarbacterium.*

B. Attack chitin and sometimes horny substances.

GENUS V. *Beneckea.*

FAMILY IV *Enterobacteriaceae.* Straight rods. Motile by means of peritrichous flagella, or nonmotile. Gram-negative. Grow well on artificial media. All species attack glucose producing acid or acid and gas. Some species even attack alginates or pectins. Characteristically, nitrites produced from nitrates. Antigenic composition best described as a mosaic which results in serological interrelationships among the several genera, even extending to other families. Many species live in the intestines of man and other animals, frequently causing intestinal disturbances; while others are parasitic on plants, some causing blights and soft rots; still others are saprophytic, causing decomposition of dead organic materials.

TRIBE I *Escherichieae.* Rods, either motile by means of peritrichous flagella, or occasionally nonmotile. Gelatin not liquefied, except slowly by *Aerobacter cloacae* and by *Paracolobactrum arizonae.* Ferment glucose and lactose with production of acid and gas within 24 hr at 37°C, or within 48 hr at 25 to 30°C. Some forms produce acid and gas from lactose slowly, occasionally not at all. Do not produce soft rots of vegetables.

I. Alginic acid not decomposed with production of acid and gas.

A. Lactose fermented within 48 hr.

1. Acetylmethylcarbinol not produced;

methyl red test positive; salts of citric acid may or may not be used as sole sources of carbon.

GENUS I. *Escherichia.*

2. Acetylmethylcarbinol produced; methyl red test negative; salts of citric acid used as sole sources of carbon.
   a. Usually not encapsulated; from feces, milk, dairy products, grain, and other saprophytic sources.

GENUS II. *Aerobacter.*

   aa. Usually encapsulated; from respiratory, intestinal, and urogenital tracts.

GENUS III. *Klebsiella.*

B. Lactose fermentation consistently delayed, and occasionally lactose is not fermented at all.

GENUS IV. *Paracolobactrum.*

II. Alginic acid decomposed with production of acid and gas.

GENUS V. *Alginobacter.*

TRIBE II *Erwinieae.* Motile rods which normally do not require organic nitrogen compounds for growth. Produce acid with or without gas from a variety of sugars. In some species the number of carbon compounds attacked is limited, and lactose may not be fermented. May or may not liquefy gelatin. May or may not produce nitrites from nitrates. Invade the tissues of living plants and produce dry necroses, galls, wilts, and soft rots.

GENUS VI. *Erwinia.*

TRIBE III *Serratieae.* Small, peritrichous rods. Gram-negative. Produce characteristic red pigments; white to rose-red strains that lack brilliant colors are common. Gelatin rapidly liquefied. Milk coagulated and digested. Typical species produce CO_2 and frequently H_2 from glucose and other sugars; acetic, formic, succinic, and lactic acids, acetylmethylcarbinol and 2,3-butylene glycol also produced. Coagulated blood serum is liquefied. Nitrates reduced. Aerobic. Saprophytic on decaying plant or even animal materials.

GENUS VII. *Serratia.*

TRIBE IV *Proteeae.* Straight rods. Motile by means of peritrichous flagella; generally actively motile at 25°C, but at 37°C motility may be weak or absent. Gram-negative. Two species, *Proteus vulgaris* and *P. mirabilis*, produce amoeboid colonies which show a swarming phenomenon on solid media devoid of bile salts. On moist agar, remaining species produce colonies which spread to some extent. Spreading colonies can usually be induced to swarm. Pleomorphism is characteristic only of young, actively swarming cultures. Glucose and usually various other carbohydrates, but not lactose, are fermented with the production of acid and usually gas. One species usually produces only acid. Phenylpyruvic acid is produced from phenylalanine by an oxidative deamination, and leucine rendered alkaline by an oxidative decarboxylation. Urea may or may not be decomposed. Trimethylamine oxide is reduced. Primarily from fecal matter and other putrefying materials.

GENUS VIII. *Proteus.*

TRIBE V *Salmonelleae.* Rods either motile by means of peritrichous flagella or nonmotile. Gram-negative. No spreading growth on agar. Gelatin not liquefied. Milk not peptonized. Numerous carbohydrates attacked with the production of acid or acid and gas. Lactose, sucrose, and salicin not ordinarily attacked. Acetylmethylcarbinol not produced. Urea not hydrolyzed. Found in bodies of warm-blooded animals and man, and occasionally in reptiles; frequently found in the food eaten by these animals.

I. Motile by means of peritrichous flagella (occasional strains of typhoids are nonmotile, and strains of *Salmonella gallinarum* are frequently nonmotile). Ammonium citrate normally utilized.

GENUS IX. *Salmonella.*

II. Nonmotile. Hydrogen sulfide not produced. Ammonium citrate not utilized.

GENUS X. *Shigella.*

FAMILY V *Brucellaceae.* Small, coccoid to rod-shaped cells which occur singly, in pairs, in short chains, or in groups; filamentous and pleomorphic forms occasionally found. Motile and nonmotile species occur, the motile species possessing from one to eight peritrichous flagella. May or may not be encapsulated. May or may not show bipolar staining. Gram-negative. *V* and/or *X* factors sometimes required for growth. Blood serum and similar enrichment materials may be required or may enhance growth. Increased CO_2 tension may also favor growth, especially on primary isolation. Gelatin usually not liquefied. Carbohydrates may or may not be attacked with the production of acid but no gas. Nitrites may or may not be produced from nitrates. Aerobic, facultatively anaerobic. Some invade living tissues; infection in some cases may take place by penetration of the organism through mucous membranes or through the unbroken skin. Parasites and pathogens which affect warm-blooded animals, including man, rarely cold-blooded animals.

I. Nonmotile at 37°C, but may be motile at lower temperatures.
  A. Predominantly occur singly or in masses.
    1. Cells predominantly occur singly and do not occur in masses.
      a. Grow on peptone media but may require blood serum or similar enrichment materials for growth.
        b. Show, or tend to show, bipolar staining.
          c. Attacks carbohydrates.

GENUS I. *Pasteurella.*

cc. Does not attack carbohydrates.

GENUS II. *Bordetella.*

bb. Does not show bipolar staining.

GENUS III. *Brucella.*

aa. Requires *V* and/or *X* factors for growth.

GENUS IV. *Haemophilus.*

    2. Cells predominantly occur singly and show pleomorphism and/or occur in masses.
      a. Growth occurs on ordinary media; increased CO_2 tension enhances growth, especially on primary isolation.

GENUS V. *Actinobacillus.*

aa. Growth occurs on infusion media only after growth in chick embryo.

GENUS VI. *Calymmatobacterium.*

B. Predominantly occur as diplobacilli.

GENUS VII. *Moraxella.*

II. Motile at 37°C.

A. Optimum temperature for growth, 37°C. Litmus milk becomes strongly alkaline.

GENUS II. *Bordetella.*

B. Optimum temperature for growth, between 28 and 30°C. Litmus milk unchanged.

GENUS VIII. *Noguchia.*

FAMILY VI *Bacteroidaceae.* Rods, with rounded or pointed ends, which vary in size from minute, filterable forms to long filamentous, branching forms. Marked pleomorphism may occur. Motile or nonmotile, the motile species possessing peritrichous flagella (rarely, motility has been observed without demonstrable flagella). Gram-negative. Body fluids frequently required for growth and are always stimulative. Simple carbohydrates usually fermented with the production of acid; gas may be produced in glucose or peptone media. Normally strict anaerobes, but occasionally microaerophilic species occur. Found primarily in the intestinal tracts and mucous membranes of warm-blooded animals. Sometimes pathogenic.

I. Simple, rarely pleomorphic, rod-shaped cells which are strict anaerobes.

A. Cells with diameters greater than 0.3 μ.

1. Cells with rounded ends.

GENUS I. *Bacteroides.*

2. Cells with pointed ends.

GENUS II. *Fusobacterium.*

B. Cells with diameters 0.15 μ or less.

GENUS III. *Dialister.*

II. Highly pleomorphic rods, some of which may be facultative anaerobes.

A. Strict anaerobes.

GENUS IV. *Sphaerophorus.*

B. Facultative anaerobes.

GENUS V. *Streptobacillus.*

FAMILY VII *Micrococcaceae.* Cells in free condition spherical; during division, somewhat elliptical. Endospores not produced (except in *Sarcina ureae* under special conditions). Division primarily in two or three planes; some anaerobic cells divide only in a single plane, producing chains. If cells remain in contact after division, they are frequently flattened in the plane of last division. Occur singly or in pairs, tetrads, packets, irregular masses, or even in chains. Motility rare. Gram-positive although the free-living and saprophytic species may be gram-variable or even gram-negative. Many species form a non-water-soluble, yellow, orange, pink, or red pigment. Aerobic species produce abundant growth on ordinary culture media and are capable of slight anaerobic growth. Anaerobic to aerotolerant species also occur. Heterotrophic. No visible gas is produced by the aerobic species from carbohydrates, which are frequently fermented. Anaerobic species sometimes produce gas, such as methane, carbon dioxide, and hydrogen. Gelatin often slowly liquefied. Free-living saprophytic to parasitic or even pathogenic. The typical aerobic micrococci frequently live on the skin, in skin glands, or in the skin gland secretions of *Vertebrata*; however, seawater and soil forms may occur. Anaerobic species live primarily in decomposing organic materials.

I. Aerobic to facultatively anaerobic species. Also includes some obligate anaerobes that occur in packets (*Sarcina*).

A. Cells generally found in irregular masses; occasionally they are single or in pairs.

1. Action on glucose, if any, is oxidative. Aerobic.

GENUS I. *Micrococcus.*

2. Glucose fermented anaerobically with the production of acid. Facultatively anaerobic.

GENUS II. *Staphylococcus.*

B. Cells normally occur in tetrads or packets of eight cells.

1. Parasitic species occurring in tetrads. White to pale yellow chromogenesis. Nonmotile.

GENUS III. *Gaffkya.*

2. Cells occur in packets. White, yellow,

orange, and red chromogenesis. Usually non-motile.

GENUS IV. *Sarcina.*

II. Obligate anaerobes occurring singly or in pairs, chains, or masses but never in packets; tetrads rarely formed.

A. Methane produced from various organic compounds.

GENUS V. *Methanococcus.*

B. Methane not produced.

GENUS VI. *Peptococcus.*

FAMILY VIII *Neisseriaceae.* Spherical cells occurring in pairs or in masses. Giant cells common in young cultures. Nonmotile. Gram-negative. Pigment may or may not be produced. Some species grow poorly immediately after isolation without mammalian body fluids. Aerobic, facultatively anaerobic, and anaerobic. Optimum temperature, 37°C. All known species are parasitic.

I. Cells approximately 1.0 μ in diameter, occur in pairs with adjacent sides usually flattened. Aerobic or facultatively anaerobic.

GENUS I. *Neisseria.*

II. Cells usually less than 0.5 μ in diameter, occur in pairs and masses. Anaerobic.

GENUS II. *Veillonella.*

FAMILY IX *Brevibacteriaceae.* Cells without endospores. Rod-shaped, varying from a quite short, almost coccoid form to a rather long, straight, unbranched rod. Motile or nonmotile, the motile species being peritrichous or, occasionally, monotrichous. Gram-positive. Red, reddish orange, yellow, or brown pigments may be produced. Carbohydrates may or may not be attacked. Aerobic and facultatively anaerobic species occur. Found in dairy products, soil, saltwater, freshwater, and decomposing substances of a great variety of types.

I. Short, almost coccoid, unbranched rods which do not form filaments. Acid usually produced from simple carbohydrates.

GENUS I. *Brevibacterium.*

II. Long, unbranched rods which may form filaments;

filaments may subdivide into coccoid elements. Carbohydrates are not utilized.

GENUS II. *Kurthia.*

FAMILY X *Lactobacillaceae.* Long or short rods or cocci which divide like rods in one plane only, producing chains, occasionally tetrads; filamentous as well as so-called false branching forms sometimes occur. Usually non-motile but may be motile, the motile species possessing peritrichous flagella. Gram-positive. Pigment production rare; a few species produce a yellow, orange, red, or rusty brown pigment. Gelatin liquefaction rare among the microaerophilic species but is more common among the strict anaerobes. Surface growth on all media is poor or absent. Carbohydrates are essential for good development; they are fermented to lactic acid, sometimes with volatile acids, alcohol, and carbon dioxide as by-products. Nitrites not produced from nitrates, but among the strict anaerobes there are a few species that are known to reduce nitrates, and some that have not been tested for nitrate reduction. Microaerophilic to anaerobic. Found regularly in the mouth and intestinal tract of man and other animals, in food and dairy products, and in fermenting vegetable juices; a few species are highly pathogenic.

TRIBE I *Streptococceae.* Cells spherical or elongate, dividing in one plane only, usually occurring in pairs or in chains. Gelatin rarely liquefied. None of the species grows abundantly on solid media. The microaerophilic species attack carbohydrates and polyhydroxy alcohols, producing lactic acid by homofermentation or lactic and acetic acids, alcohol, and carbon dioxide by heterofermentation; the strictly anaerobic species attack protein decomposition products, organic acids, and usually carbohydrates with the production of carbon dioxide, hydrogen, and other products. Microaerophilic to anaerobic. Catalase-negative. May or may not be pathogenic; some pathogenic species grow poorly without blood serum or other enrichment fluids. Found in various lesions and in the normal mouths and intestines of man and other animals, in food and dairy products, and in fermenting plant juices.

I. Facultatively anaerobic to microaerophilic.
   A. Homofermentative, producing only traces of end-products other than lactic acid from carbohydrates.

1. Produce dextrorotatory lactic acid from glucose.
   a. Parasites which grow poorly on artificial media. Cells usually in pairs often elongated. Bile-soluble.

   GENUS I. *Diplococcus.*

   aa. Parasites and saprophytes. Normally form short or long chains. Not soluble in bile.

   GENUS II. *Streptococcus.*

2. Produces a racemic mixture of lactic acid from glucose. Occurs singly, as tetrads, pairs, or even short chains.

   GENUS III. *Pediococcus.*

B. Heterofermentative, producing considerable amounts of carbon dioxide, ethanol, and acetic acid as well as lactic acid from carbohydrates.

GENUS IV. *Leuconostoc.*

II. Strictly anaerobic (one species becomes aerotolerant with repeated transfers).

GENUS V. *Peptostreptococcus.*

TRIBE II *Lactobacilleae.* Straight or curved rods usually occurring singly or in chains, sometimes in filaments; so-called false branching may also occur. Usually nonmotile but may be motile, the motile species possessing peritrichous flagella. Gram-positive. Gelatin may be liquefied but only by the strict anaerobes. Carbohydrates usually attacked, the end-products of fermentation including either one or a number of the following: formic, acetic, propionic, butyric, lactic, and valeric acids, alcohol, and carbon dioxide. Microaerophilic to anaerobic. Catalase-negative. May or may not be pathogenic. Found in fermenting animal and plant products; also found in the intestinal tracts and in lesions of various warm-blooded animals, including man.

I. Microaerophilic to anaerobic. Glucose fermented with the production of lactic acid or with the production of lactic and acetic acids, alcohol, and carbon dioxide.

GENUS I. *Lactobacillus.*

II. Strictly anaerobic.

A. Nonmotile.

1. Cells do not show so-called false branching.
a. Cells do not occur in long chains and/or filaments.

GENUS II. *Eubacterium.*

aa. Cells occur in long chains and/or filaments.

GENUS III. *Catenabacterium.*

2. Cells show so-called false branching.

GENUS IV. *Ramibacterium.*

B. Motile.

GENUS V. *Cillobacterium.*

FAMILY XI *Propionibacteriaceae.* Irregularly shaped rods which tend toward bending or terminal swelling (in *Butyribacterium*) or pleomorphism (in *Propionibacterium*). Nonmotile, gram-positive. Colonial development on semisolid media is slow, visible colonies seldom being discernible before 2 days. Where pigment is produced, it is brownish red. Nonproteolytic; usually saccharolytic. Ferment carbohydrates, usually lactic acid and, in some cases, polyhydroxy alcohols with the production of saturated aliphatic carboxylic acids. Anaerobic to aerotolerant, many strains of *Propionibacterium* being readily adapted to growth under aerobic conditions, with the actual utilization of oxygen. Generally catalase-positive when subjected to the usual laboratory test, but exceptions exist, particularly in *Butyribacterium*. Inhabitants of intestinal tracts of animals; also occur in materials outside the body where suitable foodstuffs are found.

I. Ferment carbohydrates and lactic acid.
A. Produce propionic and acetic acids and carbon dioxide.

GENUS I. *Propionibacterium.*

B. Produce butyric and acetic acids and carbon dioxide.

GENUS II. *Butyribacterium.*

II. Carbohydrates fermented. Glucose converted mainly to ethanol and carbon dioxide, with small amounts of acetic and other acids. Lactic acid not fermented.

GENUS III. *Zymobacterium.*

FAMILY XII *Corynebacteriaceae.* Usually nonmotile rods, frequently banded or beaded with metachromatic granules. May show marked diversity of form. Branching cells have been described in a few species but these are very uncertain. Generally gram-positive, some species being partially decolorized more easily than others. Where pigment is formed, it is grayish yellow or orange or pink. Gelatin may be liquefied. Nitrites may be produced from nitrates. Aerobic to microaerophilic; a few species are anaerobic. Animal and plant parasites and pathogens; also found in dairy products and soil.

I. Primarily pathogenic on animals and plants.
   A. Aerobic to anaerobic, pleomorphic rods that show the characteristic arrangement produced by snapping division.
      1. Animal species nonmotile but some of the plant pathogens motile.

GENUS I. *Corynebacterium.*

      2. Animal species motile by means of peritrichous flagella. Causes a monocytosis in warm-blooded animals, including man.

GENUS II. *Listeria.*

   B. Microaerophilic rods to long filaments. Nonmotile.

GENUS III. *Erysipelothrix.*

II. Live primarily on decomposing organic matter. Saprophytic.
   A. Found primarily in dairy products. Acid production weak. Lactic acid is the principal acid produced. Nonmotile.

GENUS IV. *Microbacterium.*

   B. Found primarily in soil.
      1. Decomposes cellulose. Motile and nonmotile species.

GENUS V. *Cellulomonas.*

      2. Does not decompose cellulose. Generally nonmotile. Gram-negative rods occur in

young cultures, and coccoid, gram-positive cells develop in older cultures.

GENUS VI. *Arthrobacter.*

FAMILY XIII *Bacillaceae.* Rod-shaped cells capable of producing endospores which are cylindrical, ellipsoidal, or spherical, and which are located in the center of the cell, subterminally or terminally. Sporangia do not differ from the vegetative cells except when bulged by spores larger than the cell diameter; such sporangia are spindle-shaped when spores are central and wedge- or drumstick-shaped when spores are terminal. Motile by means of peritrichous flagella or nonmotile. Usually gram-positive. Pigment formation rare. Gelatin frequently hydrolyzed. Sugars generally fermented, sometimes with production of gas. Aerobic, facultatively anaerobic; anaerobic; or anaerobic, aerotolerant. Some species capable of growth at 55°C. Mostly saprophytes, commonly found in soil; a few are animal or insect parasites or pathogens.

I. Aerobic or facultatively anaerobic. Catalase-positive.

GENUS I. *Bacillus.*

II. Anaerobic or aerotolerant; catalase not known to be produced.

GENUS II. *Clostridium.*

Order V *Actinomycetales.* Organisms forming elongated cells which have a definite tendency to branch. These hyphae do not exceed 1.5 μ and are mostly about 1.0 μ or less in diameter. In some species the cells are acid-fast. In the *Mycobacteriaceae* the mycelium is rudimentary or absent; no spores are formed. The *Actinomycetaceae*, *Streptomycetaceae*, and *Actinoplanaceae* usually produce a characteristic branching mycelium and multiply by means of special spores (oidiospores, conidia, or sporangiospores) or combinations of these spores. Special spores are formed by the fragmentation of the plasma within straight or spiral-shaped, spore-bearing hyphae; the oidiospores are formed by segmentation or by transverse division of hyphae, similar to the formation of oidia among the true fungi; the conidia are produced singly, at the end of simple or branching conidiophores; the sporangiospores are borne in spherical or variously shaped sporangia. A few species in *Nocardia* are reported to be motile. In *Actinoplanes* the sporangiospores have polar flagella and swim; in *Streptosporangium* the spores are nonmotile. Cell structure like that of the bacteria proper. Cell-wall substance is neither chitin nor cellulose. Thus it differs from the cell-wall substance of the true fungi, another indication of a closer relationship with the bacteria than with the fungi (molds). Only a few species are pathogenic. Majority found in soil or less commonly in freshwater.

FAMILY I *Mycobacteriaceae.* Cells spherical to rod-shaped; branching not evident on ordinary media. No conidia. Aerobic. Mesophilic. Gram-positive. Found in soil, dairy products, and as parasites on animals, including man.

I. Cells usually acid-fast. Rod-shaped cells that do not branch under ordinary cultural conditions.

GENUS I. *Mycobacterium.*

II. Non-acid-fast cells so far as observed. Cells generally spherical; occurring singly, in short chains, or in clumps.

GENUS II. *Mycococcus.*

FAMILY II Actinomycetaceae. Mycelium nonseptate during the early stages of growth but later may become septate and break up into short segments, rod-shaped or spherical in shape, or the mycelium may remain nonseptate and produce spores on aerial hyphae. Organisms in culture media are either colorless or produce various pigments. Some species are partially acid-fast. This family distinguished from previous one by formation of a true mycelium.

I. Obligately aerobic. Colonies are bacteria-like in nature, smooth, rough, or folded, of a soft to a dough-like consistency, sometimes compact and leathery in young stages. Most forms do not produce any aerial mycelium; a few produce a limited mycelium, the branches of which also break up into oidiospores or segmentation spores. Some species partially acid-fast.

GENUS I. *Nocardia.*

II. Anaerobic or microaerophilic; parasitic; non-acid-fast, nonproteolytic, and nondiastatic.

GENUS II. *Actinomyces.*

FAMILY III *Streptomycetaceae.* Vegetative mycelium does not fragment into bacillary or coccoid forms. Conidia borne on sporophores. Primarily soil forms, sometimes thermophilic in rotting manure. A few species are parasitic.

I. Conidia produced in aerial hyphae in chains.

GENUS I. *Streptomyces.*

II. Conidia produced terminally and singly on short conidiophores.
   A. No growth between 50 and 65°C.

GENUS II. *Micromonospora.*

   B. Growth occurs between 50 and 65°C.

GENUS III. *Thermoactinomyces.*

FAMILY IV *Actinoplanaceae.* Vegetative mycelium, usually inconspicuous, is formed in water on a variety of plant and animal parts. The aerial mycelium is lacking as a rule; it is formed in certain species and then much as in *Streptomyces*. Reproduction is by spores formed in sporangia, the spores in *Actinoplanes* possessing flagella and being motile, and those in *Streptosporangium* possessing no flagella and being nonmotile; conidia formed in many species. Culturable on a variety of artificial media and then resembling, in vegetative characters, certain species of *Nocardia, Micromonospora*, or *Streptomyces*. Widely distributed in soil and freshwater.

I. Aerial mycelium usually not formed; coiled conidiospores lacking; sporangiospores motile (Figs. 243 and 244).

GENUS I. *Actinoplanes.*

II. Aerial mycelium abundant; coiled conidiophores as well as sporangia are formed in some species; sporangiospores nonmotile.

GENUS II. *Streptosporangium.*

Order VI *Caryophanales.* Bacteria which occur as trichomes (many-celled filaments) or as shorter structures which function as hormogonia. Individual cells characterized by the presence of a central body or ring-like nucleus which frequently assumes the form of a disk; these bodies are clearly visible in the living cells. The nuclear

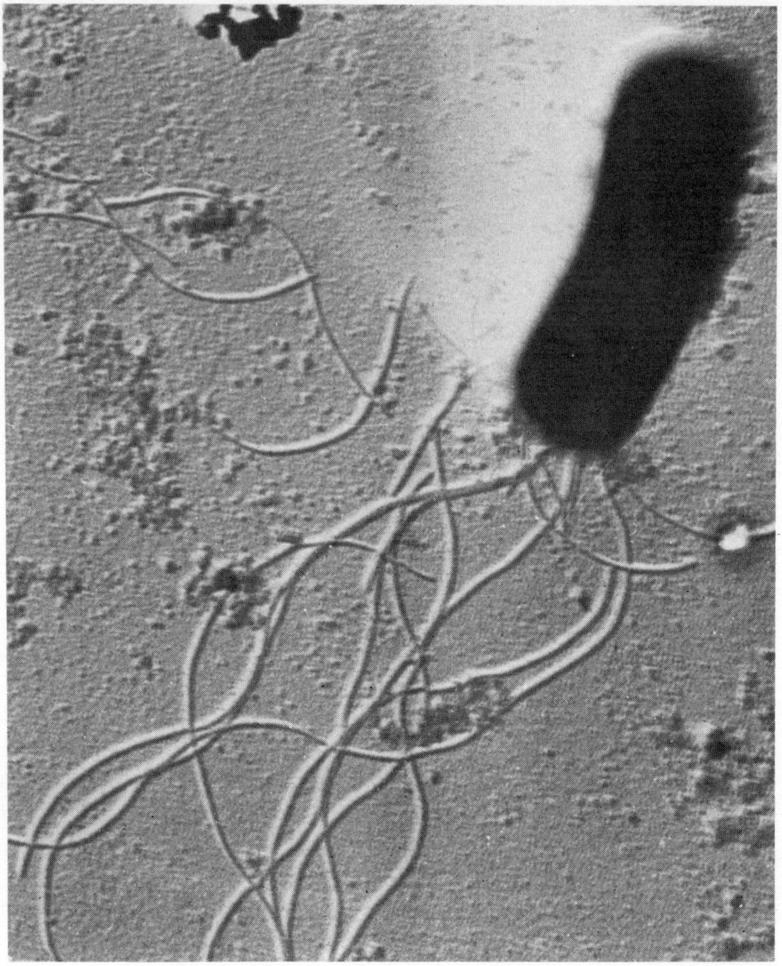

Fig. 243 Spore of Actinoplanes campanulata with flagella. (Courtesy of Higgins, Lechevalier, and Lechevalier.)

elements have a clear-cut Feulgen reaction. The trichomes are not enclosed in sheaths. Colorless. Each trichome consists of cylindrical or discoidal cells enclosed in a continuous wall. Gonidia are sometimes formed. Found in water, the intestines of arthropods and vertebrates, and in decomposing organic materials.

FAMILY I *Caryophanaceae.* Large trichomes and bacillary structures which do not form spores. Motile with peritrichous flagella or nonmotile. Organisms are found on the mucous membranes of the oral cavity of man and various other animals, in the alimentary tract of ruminants and in decomposing organic materials.

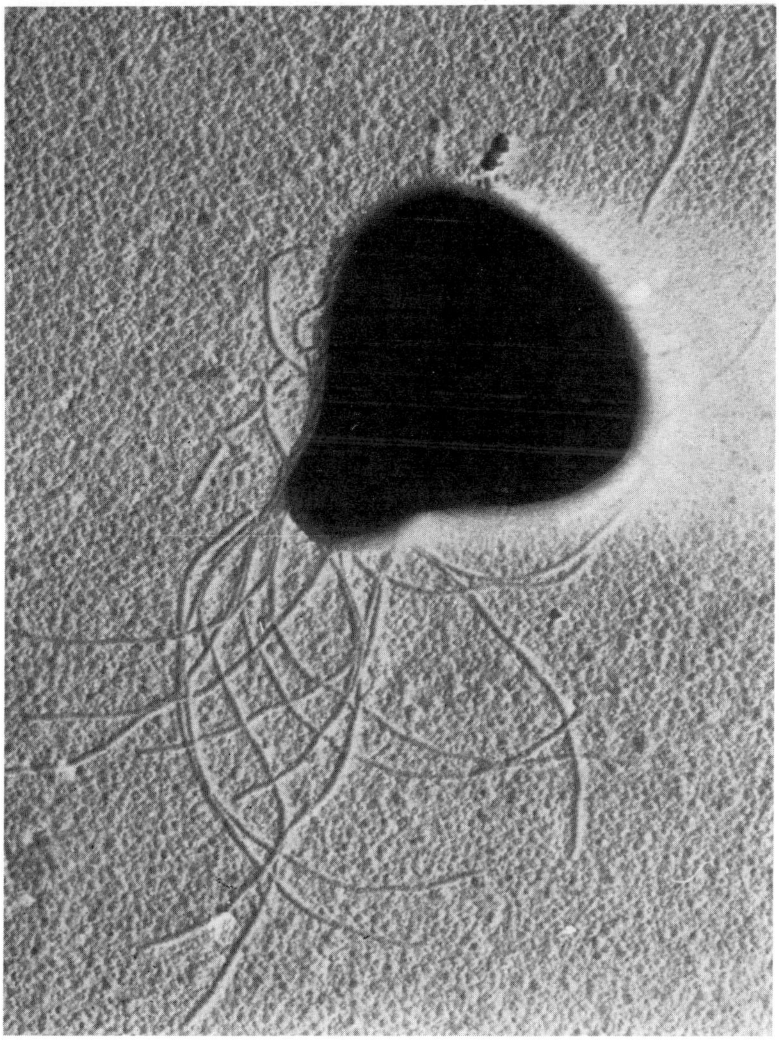

Fig. 244 Spore of Actinoplanes species with flagella; note hernia. (Courtesy of Higgins, Lechevalier, and Lechevalier.)

I. Trichomatous bacteria that are actively motile by means of peritrichous flagella.

A. Unstained trichomes show alternating light and dark bands, the dark bands being internal crosswalls.

GENUS I. *Caryophanon.*

B. Trichomes show coenocytic structure. Divide by constriction.

GENUS II. *Lineola.*

II. Nonmotile trichomes. Found in the buccal cavities of vertebrates.

GENUS III. *Simonsiella.*

FAMILY II *Oscillospiraceae.* Cells occur in trichomes of varying lengths. Trichomes are partitioned to form narrow cells, each containing a central chromatin body (disk-like nucleus); these bodies give a clear Fuelgen reaction and are embedded in hyaline protoplasm. Spores are formed by a fusion of the protoplasms of two to three neighboring cells. Actively motile by means of peritrichous flagella; nonmotile strains may occur. Parasitic in the intestinal tracts of vertebrates.

GENUS I. *Oscillospira.*

FAMILY III *Arthromitaceae.* Trichomes probably divided into cells although septa (protoplasmic?) disappear during sporulation. Disk-like nuclei alternate with thin protoplasmic segments (septa). Spores form in the distal ends of trichomes. Nonmotile. Trichomes are attached by a spherical body in groups to the intestinal walls of insects, crustaceans, and tadpoles.

GENUS I. *Arthromitus.*

GENUS II. *Coleomitus.*

Order VII *Beggiatoales.* Cells occur mostly in trichomes in three of the families and singly in the fourth family. When in contact with a substrate, the motile organisms glide over the surface or show a slow, rolling, jerky type of motion. No flagella or other organs of locomotion are known. Nonmotile trichomes may also occur. Trichomes may show bending and flexing. With respect to gliding and oscillating, the trichomes function as distinct units except in the genus *Bactoscilla*, where the trichomes show bending at the joints between the cells. Multiplication is by transverse fission throughout the entire length of the trichomes or of the singly occurring cells; gonidia occur in one family, *Leucotrichaceae*. Do not possess chlorophyll or phycocyanin. Under favorable environmental conditions, sulfur globules, sometimes in accompaniment with calcium carbonate crystals, may be found in or on the cells. Found in freshwater (with or without hydrogen sulfide) and marine habitats, in soil and in decomposing organic matter, especially algae.

FAMILY I *Beggiatoaceae.* Individual cells, generally not visible without staining, occur in trichomes; within the trichomes the cells are arranged in chains. The trichomes show a gliding motion when in contact with a substrate; they also show flexing movements. When grown in the presence of hydrogen sulfide, the trichomes contain sulfur globules. The structure of these organisms is very similar to that of the *Oscillatoriaceae*, but the cells are devoid of chlorophyll and phycocyanin. Special reproductive structures are unknown.

I. Trichomes are free and motile and are not attached to a substrate.
  A. Trichomes occur singly and are not embedded in a common slime-sheath.
    1. Trichomes straight or somewhat bent, not permanently coiled.

GENUS I. *Beggiatoa.*

    2. Trichomes coiled or spirally wound.

GENUS II. *Thiospirillopsis.*

  B. Trichomes occur in bundles and are surrounded by a slime-sheath.

GENUS III. *Thioploca.*

II. Trichomes attached to substrate at one end; apical segments, when freed, are motile until attached.

GENUS IV. *Thiothrix.*

FAMILY II *Vitreoscillaceae.* Cells occur in colorless trichomes of varying degrees of flexibility. Trichomes show a gliding motion when in contact with a substrate, the speed of movement varying inversely with the width of the trichome. One end of a trichome may become attached to a surface, the other end then becoming free-swinging. Gram-negative. The gliding habit determines the nature of growth; on agar low in nutrients, wavy, curly, or spiral colonies are produced; on rich media, drop-like colonies, resembling those of many bacteria, are formed. Do not possess chlorophyll or phycocyanin. Never contain sulfur granules. Do not hydrolyze genuine proteins. Found in dung, soil, in water with decaying plant material, and almost regularly in myxophycean scum on the surfaces of quiet waters.

I. Trichomes perceptibly septate.
  A. Trichomes divided into cells which are not

separated by empty interspaces; the trichomes may bend anywhere along their length.

GENUS I. *Vitreoscilla*.

B. Trichomes divided into cells separated by empty interspaces; the trichomes bend only at these pliable joints.

GENUS II. *Bactoscilla*.

II. Trichomes not perceptibly septate.

GENUS III. *Microscilla*.

FAMILY III *Leucotrichaceae*. Short, cylindrical cells arranged in long, colorless, unbranched, nonmotile trichomes tapering from the base to the apex. Sulfur granules may be found on the exterior of the cells under certain conditions. Trichomes commonly attached basally to solid substrates by an inconspicuous holdfast. Multiplication by means of gonidia (single, gliding cells which arise apically from the trichomes). The gonidia may aggregate to form rosettes containing up to 50 cells. The cells in the rosettes become nonmotile, develop holdfasts and elongate to form trichomes; therefore mature trichomes are characteristically arranged in the form of radial colonies, although occasionally gonidia develop singly, forming isolated trichomes. Strictly aerobic. Resemble blue-green algae in many respects but differ from them in that they do not produce photosynthetic pigments. Found in freshwater and saltwater containing decomposing algal material.

GENUS I. *Leucothrix*.

FAMILY IV *Achromatiaceae*. Large, unicellular organisms which are spherical to ovoid or shortly cylindrical with hemispherical extremities. Movements, if any, are of a slow, rolling, jerky type and are dependent upon the presence of a substrate; no special organs of locomotion are known. Division of cells is by a constriction in the middle. Do not possess photosynthetic pigments. In their natural habitat, the cells contain sulfur droplets and sometimes additional inclusions, such as large spherules of calcium carbonate. Found in freshwater and marine environments.

GENUS I. *Achromatium*.

Order VIII *Myxobacterales.* Vegetative cells are flexible rods of low refractility which exhibit gliding movement on solid surfaces and which multiply by binary, transverse fission to produce a thin, flat, rapidly extending colony. Actively motile cells at the periphery of the colony commonly occur as groups of two or three to several hundred individuals in the form of tongue-like extensions or isolated islands whose presence is virtually diagnostic of the order. The moving cells may pave the substrate with a thin layer of slime on which they rest.

FAMILY I *Cytophagaceae.* Flexible, sometimes pointed rods showing gliding motility. No fruiting bodies or resting cells (microcysts) are formed.

GENUS I. *Cytophaga.*

FAMILY II *Archangiaceae.* Resting cells are shortened rods, never enclosed in larger cysts. Fruiting bodies are irregularly swollen or twisted, or are finger-like structures.

I. Fruiting body depressed, usually irregularly delimited, the interior usually consisting of swollen or intestine-like twisted or intertwined masses, whose windings may be constricted or may jut out (project) as free ends.

GENUS I. *Archangium.*

II. Fruiting body consists of single (separate) columnar or finger-like structures arising from the substrate.

GENUS II. *Stelangium.*

FAMILY III *Sorangiaceae.* The shortened rods of the fruiting body lie in angular, usually relatively small, cysts of definite polygonal shape. Often many of these cysts are surrounded by a common membrane. The primary cyst may be differentiated from the angular or secondary cysts. No stalked forms are known.

GENUS I. *Sorangium.*

FAMILY IV *Polyangiaceae.* The resting cells are shortened and usually somewhat thickened rods which are always enclosed in cysts. Cysts may be sessile, occurring either singly or in groups and enveloped in a slime membrane, or they may be raised on stalks (cystophores) which can be either simple or branched. Cysts can occur either singly or in clusters at the tips of the stalks.

I. Cysts embedded in slime; sessile, occurring singly or as loose aggregates.

GENUS I. *Polyangium.*

II. Cysts never embedded in slime; either borne on stalks or arranged in tight clusters joined together at the base.

A. Many cysts united at base to form a large disk or rosette; either sessile or stalked.

GENUS II. *Synangium.*

B. Cysts not united at base; borne singly or in large numbers on stalks.

1. Cysts born singly on a stalk.

GENUS III. *Podangium.*

2. Numerous cysts on a stalk.

GENUS IV. *Chondromyces.*

FAMILY V *Myxococcaceae.* The rods become shortened when fruiting occurs and develop into spherical or ellipsoidal microcysts. Definite fruiting bodies are produced in three of the genera. In *Sporocytophaga* the microcysts are produced from the vegetative cells without development of fruiting bodies.

I. Definite fruiting bodies formed.

A. Microcysts not enclosed in larger cysts.

1. Fruiting bodies deliquescent.

GENUS I. *Myxococcus.*

2. Fruiting bodies firm, not deliquescent.

GENUS II. *Chondrococcus.*

B. Microcysts enclosed in larger cysts.

GENUS III. *Angiococcus.*

II. No definite fruiting bodies formed.

GENUS IV. *Sporocytophaga.*

Order IX *Spirochaetales.* Slender, flexuous bodies 6 to 500 μ in length, in the form of spirals with at least one complete turn. Some forms may show an axial filament, a lateral crista, or ridge, or transverse striations; otherwise there is no significant protoplasmic pattern. Smaller forms may have a lower refractive index than that of true bacteria, and therefore can be seen only with dark-field illumination. Some forms take aniline dyes with difficulty; Giemsa's stain is uniformly successful. Granules are formed in some species which are found in vector hosts. All forms are motile. In the true bacteria, motility is effected by flagella endowed with a lashing movement; however, no such structures exist among the spirochaetes. Terminal projections, whether derived from the periplast or from the axial filament, may assist in the movements, and it is possible that the crista has a similar function, although neither of these structures can explain the violent motion of spirochaetes. Motility consists of a rapid whirling or spinning about the long axis, which activity drives the organism forward or backward, there being no antero-posterior polarity. In addition the spirochaetes make violent, lashing movements, curling and uncurling their spirals. Multiplication by transverse fission, no sexual cycle being known. Free-living, saprophytic, and parasitic forms.

FAMILY I *Spirochaetaceae.* Coarse, spiral organisms, 30 to 500 μ in length, possessing definite protoplasmic structures. Found in stagnant freshwater or saltwater, and in the intestinal tracts of bivalve molluscs.

I. No obvious periplast membrane and no cross striations present.

GENUS I. *Spirochaeta.*

II. Periplast membrane present. Cross striations present in stained specimens.
   A. Free-living in marine ooze.

GENUS II. *Saprospira.*

   B. Parasitic on lamellibranch molluscs. Cristae are prominent.

GENUS III. *Cristispira.*

FAMILY II *Treponemataceae.* Coarse or slender spirals, 4 to 16 μ in length; longer forms are due to incomplete or delayed division. The spirals are regular or irregular and flexible or comparatively rigid. The protoplasm possesses no obvious structural features. Some cells may show terminal filaments. Some cells are visible only with dark-field illumination. Many of these organisms can be cultivated. With few exceptions, parasitic in vertebrates. Some are pathogenic.

I. Stain easily with ordinary aniline dyes.

GENUS I. *Borrelia.*

II. Stain with difficulty except with Giemsa's stain or silver impregnation.
   A. Anaerobic.

GENUS II. *Treponema.*

   B. Aerobic.

GENUS III. *Leptospira.*

Order X *Mycoplasmatales.* Highly pleomorphic organisms which possess a peculiar mode of reproduction characterized by the breaking up of filaments (with a more or less pronounced tendency to true branching) into coccoid, filterable elementary bodies. The cell bodies are soft and fragile; without special precautions they are often distorted or entirely destroyed in microscopical preparations. Nonmotile. Typical endospores are never produced. Gram-negative. Growth occurs in agar media, although most of the species have exacting nutritional requirements. Pathogenic and saprophytic species occur.

FAMILY I *Mycoplasmataceae.* Characteristics same as for the order.

GENUS I. *Mycoplasma.*

CLASS III *Microtatobiotes.* Includes the smallest of the living things. All are manifested by a dependence on other living organisms for their growth and multiplication. Parasitism is axiomatic since there is no way to determine if there are free-living forms. Most of these organisms occur intracellularly; *Rickettsia quintana* Schmincke of trench fever is an example of extracellular growth in its host, the body louse. A few of the visible forms are known to occur intranuclearly. Characteristic, intracellular inclusion bodies are often associated with the smaller agents. Hosts are represented from the highest members of the plant and animal kingdoms to the lowliest of microbial life. Some species utilize both intermediate and definitive hosts for their propagation. The largest members are the rickettsia-like organisms which are often pleomorphic, including coccoid to filamentous forms, while others show morula-like clusters of elementary bodies occurring as one or up to twenty colonies in an infected cell. Some species show larger ellipsoidal granules with a fairly compact matrix of as much as 2 μ in diameter termed initial bodies from which, in most instances, the groups of smaller elementary bodies are believed to be derived though no life cycle is postulated. At least three of these larger species, visible under the light microscope, have phases which pass through coarse or medium filters, e.g., *Coxiella burnetii* of Q fever.

The small members grade downward to filterable virus particles susceptible of measurement only by physico-chemical techniques and by special preparation under the electron microscope. Special staining procedures are required for forms visible under the light microscope and for studying characteristic pathologic reactions or associated inclusion bodies in the tissues of affected hosts. Special tissue culture techniques have been developed for the more adequate investigation of many of the species.

Order I *Rickettsiales.* Small, rod-shaped coccoid, and often pleomorphic microorganisms occurring as elementary bodies which are usually intracellular but which may occasionally be facultatively or exclusively extracellular. May also develop larger "initial bodies" as intracellular, spherical or less regular inclusions. Intracytoplasmic forms may be diffuse, compacted into colonies or morulae and may be located in special situations. Usually nonfilterable. Gram-negative. Cultivated outside the host only in living tissues, embryonated chicken eggs, or rarely in media containing body fluids. Parasitic organisms almost always intimately associated with not only reticulo-endothelial and vascular endothelial cells or erythrocytes in vertebrates, but also often in invertebrates which may act as vectors. The intracellular parasites of protozoa and other invertebrates are provisionally assigned here also. May cause diseases in man or other animals, or both. Seldom kill the invertebrate hosts.

FAMILY I *Rickettsiaceae.* Small, rod-shaped, ellipsoidal, coccoid, and diplococcus-shaped, often pleomorphic organisms which are often intimately associated with arthropod tissues, usually in an intracellular position. Gram-negative. Species pathogenic for vertebrates have not been cultivated to date in cell-free media. May be parasitic in man and other animals causing disease (typhus and related ills) that may be transmitted by invertebrate vectors (chiefly lice, fleas, ticks, and mites). Information is still inadequate for the systematic assignment of many of the species which inhabit arthropod hosts and which were originally described in this family.

TRIBE I *Rickettsieae.* Small, pleomorphic, mostly intracellular organisms adapted to existence in arthropods and pathogenic for suitable vertebrate hosts.

I. Nonfilterable; produce typhus-like rash and usually *Proteus X* agglutinins in man.

GENUS I. *Rickettsia.*

II. Filterable; produce neither rash nor Weil-Felix agglutinins in man.

GENUS II. *Coxiella.*

TRIBE II *Ehrlichieae.* Minute, rickettsia-like organisms pathogenic for certain vertebrate hosts, not including man. Adapted to existence in invertebrates, chiefly arthropods.

I. Transmitted by ticks.
   A. Transmitted transovarially; parasites of circulating monocytes of vertebrate hosts.

GENUS III. *Ehrlichia.*

   B. Not transmitted transovarially; parasites of endothelial cells of vertebrate hosts.

GENUS IV. *Cowdria.*

II. Transmitted by parasitic trematodes; pathogenic principally for canines.

GENUS V. *Neorickettsia.*

TRIBE III *Wolbachieae.* Includes many species heretofore assigned to the genus *Rickettsia* which are rickettsia-like in growth and in morphological and staining properties and which are mostly intracellular symbiotes or parasites of various species of arthropods, sometimes occupying special tissues or mycetomes. Characterization has often been not so adequate as in the preceding forms that are pathogenic for vertebrates, and differentiation has been arbitrarily assigned chiefly on the basis of presumed host-specificity, in arthropods, though differences in development and morphology are often noted.

I. No known filterability; no reported association with intracellular crystalline inclusions.
  A. Symbiotic to highly pathogenic; no mycetomes produced in hosts.

GENUS VI. *Wolbachia.*

  B. Symbiotic to the point that special mycetomes are developed for harboring the organisms, which are not pathogenic, in the host.

GENUS VII. *Symbiotes.*

II. Filterable; cause blue disease of beetle larvae; associated with intracellular crystalline inclusions; reportedly invade cell nuclei.

GENUS VIII. *Rickettsiella.*

FAMILY II *Chlamydiaceae.* Small, coccoid microorganisms with a characteristic developmental cycle. Stain with aniline dyes. Gram-negative. Have not been cultivated in cell-free media. Obligate, intracytoplasmic parasites or saprophytes. Found in various warm-blooded animals, where they are usually pathogenic.

I. Noncultivable in chicken embryonic tissues.
  A. Organisms coccoid; do not exhibit pleomorphism.

GENUS I. *Chlamydia.*

  B. Organisms usually coccoid or ellipsoidal; exhibit marked pleomorphism
    1. Pleomorphic forms small (200 nm to 2 μ). Pathogenic.
      a. Occur intracytoplasmically as prominent colonies.

GENUS II. *Colesiota.*

      aa. Occur intracytoplasmically as scattered growth.

GENUS III. *Ricolesia.*

    2. Pleomorphic forms large (2 μ). Apparently nonpathogenic; may be saprophytic.

GENUS IV. *Colettsia.*

II. Cultivable in chicken embryonic tissues.

GENUS V. *Miyagawanella.*

FAMILY III *Bartonellaceae.* Rod-shaped, coccoid, ring- or disk-shaped, filamentous and beaded microorganisms, usually less than 3 μ in greatest dimension. Parasites of the erythrocytes in man and other vertebrates. Not acid-fast. Stain lightly with many aniline dyes but distinctly with Giemsa's stain after methyl alcohol fixation; following this technique the *Bartonellaceae* are readily distinguished from the protozoa which also parasitize erythrocytes in that the former stain with no differentiation into nucleus and cytoplasm. Gram-negative. Cultivation in vitro on nonliving media has been achieved in two genera. At least one species bears a single polar flagellum in culture. Arthropod transmission has been established in the majority of genera. Cause bartonellosis in man and haemobartonellosis, grahamellosis, and eperythrozoönosis in lower animals.

I. Multiply on erythrocytes and within fixed-tissue cells. Usually possess a single, polar flagellum when cultivated in or on nonliving media. Provoke a progressive anemia or a cutaneous eruption, usually both in succession, not both coincidentally. Found in man and in *Phlebotomus* spp.

GENUS I. *Bartonella.*

II. Not known to multiply in fixed-tissue cells; parasitize erythrocytes and may multiply there. Flagella not demonstrated. Occur in mammals and possibly in other vertebrates, but not known from man.

A. Usually parasitize less than 5 percent of the total erythrocytes, rarely more. Relatively monomorphic in erythrocytes. Nonpathogenic or only slightly so. Affected little, if at all, by splenectomy. Cultivable on nonliving media. Occur within the red blood cells; epi-erythrocytic forms are problematical.

GENUS II. *Grahamella.*

B. Parasitized cells may constitute more than 90 percent of the total erythrocytes at the peak of infection. Polymorphism is marked when in or on red blood cells. May or may not be pathogenic. Marked increase in numbers following splenectomy. Cultivation on nonliving media not confirmed. Occur on the red blood cells; situation within red cells possible but not proved.

1. Extremely polymorphic; however, rods of varying sizes almost invariably occur, often in chains. Habitat predominantly epi-erythrocytic. Usually pathogenic, provoking a progressive, sometimes fatal, anemia.

GENUS III. *Haemobartonella.*

2. Fundamental morphological type is ring- or disk-shaped. Rods are one disk- or ring-diameter in length; composite rods are made of these units. Occur in great numbers in the blood plasma as well as on the erythrocytes. Usually nonpathogenic.

GENUS IV. *Eperythrozoon.*

FAMILY IV *Anaplasmataceae.* Organisms which parasitize red blood cells. There is no demonstrable multiplication in other tissues. In blood smears fixed with May-Grünwald and stained with Giemsa's stain, these organisms appear in the erythrocytes as spherical chromatic granules which stain a deep reddish-violet color. Show no differentiation into nucleus and cytoplasm. Occur naturally as parasites of ruminants. Transmitted by arthropods. Situated at or near the margin and/or at or near the center of red blood cells. Positions within the erythrocyte and/or host differences serve as bases for differentiating species. Attempts at cultivation in a variety of media have failed. Produce disease in nonsplenectomized and in splenectomized ruminants. The natural and experimental host range is fairly wide, these organisms occurring in members of the families *Bovidae* and *Camelidae*. Influenced by aureomycin and terramycin. Widely distributed throughout the world.

GENUS I. *Anaplasma.*

Order II *Virales.* Viruses are etiological agents of disease, typically of small size and capable of passing filters that retain bacteria, increasing only in the presence of living cells, giving rise to new strains by mutation. A considerable number of plant viruses have not been proved filterable; it is nevertheless customary to include these viruses with those known to be filterable because of similarities in other attributes and in the diseases induced. Some not known to be filterable are inoculable only by special techniques, as by grafting or by the use of insect vectors, and suitable methods for testing their filterability have not been developed; moreover, it is not certain that so simple a criterion as size measured in terms of filterability will prove to be an adequate indicator of the limits of the natural group. Viruses cause diseases of bacteria, plants, and animals.

Our incomplete knowledge of the entities known as viruses has made their classification, and consequently their nomenclature, a difficult matter. It is difficult to describe viruses adequately because of their small size and because they are not cultivable. Electron microscopy has enabled a determination of the size and morphology of some of the viruses. Likewise, serological methods have been developed which are proving to be useful in distinguishing between different species and types of viruses, but in many cases these methods have not been applied.

The usual characteristic that permits recognition of viruses is their capacity to produce specific diseases. Three constituent groups of viruses have come to be recognized, and to some extent named and classified, through the largely separate efforts of bacteriologists, animal pathologists, and plant pathologists. Taxonomic overlapping of the three groups, viruses affecting bacteria, viruses having human and other animal hosts, and viruses invading higher plants, can hardly be justified as yet by available evidence. Nevertheless, it has been shown that a single virus may multiply within, and cause morphological changes in, both a plant host and an insect vector. This seems to dispose of the thought that adaptation to one environment necessarily precludes the utilization of other sources for the materials needed for growth and multiplication.

For the present it seems feasible to continue with the custom, tacitly accepted in the past, of classifying bacteriophages separately as one subgroup, viruses causing diseases in higher plants as a second subgroup, and those causing diseases in man and other animals as a third subgroup. It should be recognized that this may prove to be only a temporary arrangement, necessary because we have little or no evidence to warrant taxonomic overlapping of the three groups and useful while we await critical investigations and possible development of a substitute plan capable of displaying natural relationships to better advantage. It is further possible that there may be discoveries of common physical properties which would aid in formulating an interlocking classification, for which at present we lack any substantial basis.

Because of the rapid expansion of the field by the frequent discovery of new viruses and the development of new methods for their recognition and characterization, together with some uncertainties evidenced by virologists, it does not seem appropriate, therefore, to include a formal classification of *Virales* at this time.

It may be seen that the classification of bacteria is an exceedingly difficult task. The work becomes increasingly more difficult as more species are discovered and studied. Many of the species have been moved from one genera to another and their names changed. Some have had their names changed several times since they were discovered. This is to be expected when one considers the difficulties encountered in studying such minute organisms. Because of the nature of bacteria, it is highly probable that no single classification will ever be completely acceptable to all bacteriologists. Nevertheless a classification, no matter how imperfect, is better than none at all. The one outlined here is undoubtedly the best of those which have been proposed and is in general use throughout the world.

References

Abel, K., et al.: Classification of microorganisms by analysis of chemical compositions. I. Feasibility of utilizing gas chromatography, *J. Bact.*, **85**:1039, 1963.

Ainsworth, G. C., and P. H. A. Sneath, eds.: "Microbial Classification," London, Cambridge University Press, 1962.

Allen, R. D., and P. Baumann: Structure and arrangement of flagella in species of the genus *Beneckea* and *Photobacterium fischeri*, *J. Bact.*, **107**:295, 1971.

Baird-Parker, A. C.: A classification of micrococci and staphylococci based on physiological and biochemical tests, *J. Gen. Microbiol.*, **30**: 409, 1963.

——: The classification of staphylococci and micrococci from worldwide sources, *J. Gen. Microbiol.*, **38**:363, 1965.

Baumann, L., et al.: Taxonomy of aerobic marine eubacteria, *J. Bact.*, **110**:402, 1972.

Baumann, P., et al.: Taxonomy of marine bacteria: the genus *Beneckea*, *J. Bact.*, **107**:268, 1971.

Breed, R. S., et al.: "Bergey's Manual of Determinative Bacteriology," Baltimore, The Williams & Wilkins Company, 1957.

Buchanan, R. E.: A focus on the meaning of bacterial species, *Int. Bull. Bact. Nomen. Taxon.*, **15**:25, 1965.

——— et al.: "Index Bergeyana," Baltimore, The Williams & Wilkins Company, 1966.

Cowan, S. T.: Principles and practice of bacterial taxonomy: A forward look, *J. Gen. Microbiol.*, **39**:143, 1965.

Cross, T.: Thermophilic actinomycetes, *J. Appl. Bact.*, **31**:36, 1968.

Cummins, C. S., and J. L. Johnson: Taxonomy of the clostridia: Wall composition and DNA homologies in *Clostridium butyricum* and other butyric acid-producing clostridia, *J. Gen. Microbiol.*, **67**:33, 1971.

De Boer, W. E., et al.: Observations on the morphology of *Thiovulum majus* Hinze, *Antonie van Leeuwenhoek, J. Microbiol. Serol.*, **27**:447, 1961.

Ewing, W. H.: An outline of nomenclature for the family *Enterobacteriaceae, Intern. Bull. Bact. Nomen. Taxon.*, **13**:95, 1963.

Ewing, W. H., et al.: Biochemical characterization of *Escherichia coli*, DHEW Publication No. (HSM) 72-8109, Center for Disease Control, Atlanta, Georgia 30333, 1972.

Floodgate, G. D.: Some remarks on the theoretical aspects of bacterial taxonomy, *Bact. Rev.*, **26**:277, 1962.

Focht, D. D., and W. R. Lockhart: Numerical survey of some bacterial taxa, *J. Bact.*, **90**:1314, 1965.

Frasch, C. E., and S. S. Chapman: Classification of *Neisseria meningitidis* group B into distinct serotypes. I. Serological typing by a microbactericidal method, *Infect. Immun.*, **5**:98, 1972.

Hirsch, P.: Budding nitrifying bacteria: The nomenclatural status of *Nitromicrobium germinans* Stutzer and Hartleb 1899 and *Nitrobacter winogradskyi* Winslow et al., 1917, *Intern. J. Syst. Bact.*, **20**:317, 1970.

Isenberg, H. D., and B. G. Painter: Comparison of conventional methods, the R/B system, and modified R/B system as guides to the major divisions of *Enterobacteriaceae, Appl. Microbiol.*, **22**:1126, 1971.

Kauffmann, F.: The species-definition in the family *Enterobacteriaceae, Intern. Bull. Bact. Nomen. Taxon.*, **11**:5, 1961.

———: On the species-definition, *Intern. Bull. Bact. Nomen. Taxon.*, **13**:181, 1963*a*.

———: On the classification and nomenclature of the family *Enterobacteriaceae, Intern. Bull. Bact. Nomen. Taxon.*, **13**:187, 1963*b*.

Leifson, E.: Bacterial taxonomy: A critique, *Bact. Rev.*, **30**:257, 1966.

Lewin, R. A.: A classification of flexibacteria, *J. Gen. Microbiol.*, **58**:189, 1969.

Liston, J., et al.: Quantitative approach to the study of bacterial species, *J. Bact.*, **85**:1061, 1963.

Lockhart, W. R., and P. A. Hartman: Formation of monoethic groups in quantitative bacterial taxonomy, *J. Bact.*, **85**:68, 1963.

Marmur, J., et al.: New approaches to bacterial taxonomy, *Ann. Rev. Microbiol.*, **17**:329, 1963.

Masuo, E., and T. Nakagawa: Numerical classification of bacteria IV, *Agr. Biol. Chem.*, **34**:1375, 1970.

McCurdy, H. D.: Studies on the taxonomy of the *Myxobacterales*. III. *Chondromyces* and *Stigmatella*, *Intern. J. Syst. Bact.*, **21**:40, 1971*a*; IV. *Melittangium*, *ibid.*, **21**:50, 1971*b*.

Nakagawa, T.: Numerical taxonomy of bacteria. Part 1, *Ann. Rept. Shionogi Res. Lab.*, no. 19, 1969.

Pfennig, N., and G. Cohen-Bazire: Some properties of the green bacterium *Pelodictyon clathratiforme*, *Arch. Mikrobiol.*, **59**:226, 1967.

Prévot, A. R.: "Manual for the Classification and Determination of the Anaerobic Bacteria," Philadelphia, Lea & Febiger, 1966.

Pringsheim, E. G.: Bakterien and Cyanophyceen. Übereinstimmungen und unterschiede, *Österreichischen Botanischen Z.*, **114**:324, 1967.

Skerman, V. B. D.: "The Genera of Bacteria," Baltimore, The Williams & Wilkins Company, 1967.

Steffen, D. L., et al.: Micro-complement fixation in *Klebsiella* classification, *J. Bact.*, **110**:803, 1972.

Tyler, P. A., and K. C. Marshall: Pleomorphy in stalked, budding bacteria, *J. Bact.*, **93**:1132, 1967.

Wildermuth, H., et al.: The surface structure of spores and aerial mycelium in *Streptomyces coelicolor*, *J. Ultrastruc. Res.*, **35**:168, 1971.

16 *Disinfection and disinfectants*

Various physical and chemical methods are employed for the destruction or removal of bacteria.

The usual laboratory equipment employed for the killing of bacteria by heat, or their removal by filtration, is reviewed in Chap. 8 Additional physical agents, together with the important chemical methods, are discussed here.

A number of terms are used to describe the destruction of bacteria by various processes. These terms include germicide, bactericide, antiseptic, disinfectant, viricide, fungicide, bacteriostatic agent, sanitizer, and sterilization. The terms have been and still are the cause of considerable confusion because they do not mean the same to all scientists and laymen. The definitions which follow are probably in agreement with the beliefs of the majority of microbiologists in this field.

Germicide A germicide was originally defined as an agent that killed disease organisms. It is now generally defined as any agent, physical or chemical, that kills pathogenic organisms but not necessarily their spores.

Since nonpathogenic organisms are of about the same order of susceptibility to adverse conditions as the pathogenic forms, the above definition can be applied to all bacteria.

Bactericide A bactericide is an agent that kills both pathogenic and nonpathogenic bacteria but not necessarily their spores. In practice the term is synonymous with germicide.

Antiseptic The meaning of this term has probably caused more confusion than any of the others. Some define it synonymously with germicide and bactericide; others apply the term to any agent that prevents further bacterial action whether it does so by killing the organisms or merely by preventing them from multiplying. The term antiseptic should be limited to any agent that arrests bacterial growth either by inhibiting their activity or by destroying them.

According to this definition, a germicide may also be an antiseptic, depending upon the strength of the solution, the period of action, and the nature of the test organism. A germicide in high dilution may only inhibit growth rather than kill bacteria. Also, an agent that kills bacteria in a given period of time may only inhibit growth if the exposure time is decreased. In the former case, the agent would be classed as a germicide; in the latter, an antiseptic.

Some agents are more toxic to gram-positive than to gram-negative organisms, and vice versa. Others are equally effective against both groups of organisms. This means that a substance may be a germicide against one organism and an antiseptic against another. Doubtless, other factors are also involved.

Disinfectant This term has been used rather loosely and defined in various ways but it is generally agreed that it means an agent that kills organisms but not necessarily spores.

From the above definitions, it may be concluded that the terms germicide, bactericide, and disinfectant are synonymous.

Viricide This term is applied to any agent that destroys or inactivates filterable forms known as viruses. Since viruses are of about the same order of resistance to chemical agents as bacteria, most germicides are also good viricides.

Fungicide This may be defined as an agent, usually chemical, that destroys both pathogenic and nonpathogenic molds.

Bacteriostatic agent This term was originally used in connection with the action of certain stains on bacteria, to denote a condition in which the organisms are not necessarily killed but prevented from multiplying. Some stains in appropriate concentrations do not kill bacteria but keep them in a state of suspended animation. Further dilution of the stain-bacteria mixture generally results in growth of the organisms. The stains are referred to as bacteriostatic agents and the phenomenon as bacteriostasis. Germicidal agents of mercury and silver also exhibit the same phenomenon and are referred to as bacteriostatic compounds.

Sanitizer This may be defined as any agent that reduces the bacterial count to safe levels as may be judged by public health requirements. Sanitizers are commonly applied to inanimate objects such as eating and drinking utensils and food-handling equipment.

Sterilization This is an absolute term and should always be used where reference is made to the complete destruction of all forms of life, including fungi, viruses, spores, etc.

For more information: Eisman (1961); Lawrence (1968*a*).

Dynamics of disinfection

The disinfection process does not take place at once but is a gradual operation in which the number of organisms killed in unit time is greater at the beginning and becomes less and less as the exposure period is increased. If the numbers of survivors in unit volume are plotted against time, the points fall on smooth curves. If the logarithms of the numbers of bacteria surviving in unit volume are plotted against time, a straight line is obtained.

Results of anthrax spores treated with 5 percent phenol and incubated at $33\frac{1}{3}$°C are plotted in Fig. 245. The curved line represents the numbers of anthrax spores surviving in unit volume, and the straight line the corresponding logarithms of the numbers of surviving bacteria.

The disinfection process appears to follow the mass-action law and to proceed in accordance with the monomolecular equation, or a reaction of the first order, provided the disinfectant is present in large excess:

$$-\frac{dN}{dt} = KN \quad \text{or} \quad \frac{1}{t_2 - t_1} \log \frac{N_1}{N_2} = K$$

where N_1 and N_2 represent the number of surviving bacteria in unit volume after times t_1 and t_2, respectively.

Importance of logarithmic survivor curves In the destruction of bacteria by various agents, the greater the number of cells present, the longer will be the time necessary for complete sterilization. This may be shown in the following example: A suspension contains 200,000 organisms per milliliter. If the bacteria are destroyed at the rate of 90 percent per minute, the number of survivors at the end of 7 min will be:

200,000 ⟶ 20,000 ⟶ 2000 ⟶ 200 ⟶

20 ⟶ 2 ⟶ 0.2 ⟶ 0.02

The last figure means that 2 living bacteria remain in 100 ml of the suspension. If instead of 200,000 the suspension contains only 200

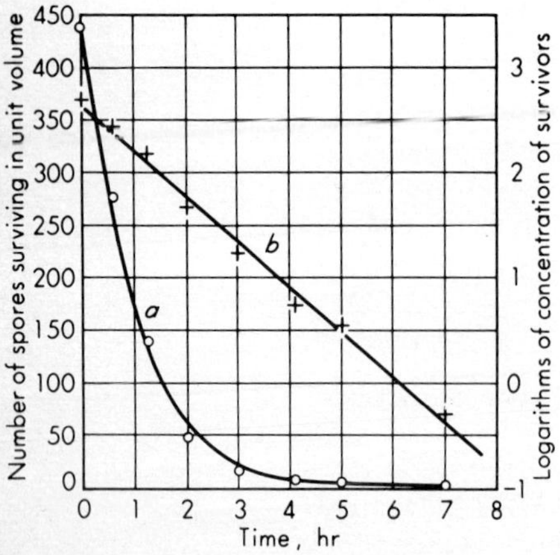

Fig. 245 Disinfection of anthrax spores with 5 percent phenol. (a) The number of survivors in unit volume are plotted against time; (b) logs of concentration of survivors are plotted against time.

organisms per milliliter, the time required to reduce the count to 2 bacteria per 100 ml will be only 4 min:

$$200 \longrightarrow 20 \longrightarrow 2 \longrightarrow 0.2 \longrightarrow 0.02$$

The relationships between bacterial numbers and the times required to produce complete sterilization are of great value in the canning and dairying industries, in bacteriology, and in surgery.

Physical agents

The physical agents which are not discussed in Chaps. 8 and 10 are included here.

Electricity Electric current at varying voltages, intensities, and periods of action has been employed for the pasteurization of milk, and for the destruction of organisms in water, sewage, etc., but the results have been too uncertain to be of practical importance. Its use has been largely abandoned, preference being given to other physical methods.

For more information: Andersen and Vad (1965).

Sonic and ultrasonic waves Sonic waves are waves of audible frequency of about 8900 Hz produced by a nickel tube vibrating in an electromagnetic field and in resonance with a 2000-V oscillating power circuit. Milk and bacterial suspensions have been treated in this manner. Reductions in viable counts up to 99 percent, after exposures of 40 to 60 min, have been reported.

Rotman (1956) treated *Escherichia coli* and *Azotobacter vinelandii* to sonic vibrations in a Raytheon 10-kc magnetostrictive oscillator for 10 min at a temperature below 4°C and a pH of 6.6, and reported structural damage to the cells (Fig. 246).

Fig. 246 Escherichia coli. Electron micrographs, uranium shadowed. Left, normal cells prior to sonic oscillation; center, cells after 10 min sonic oscillation; right, same as center but with addition of phosphate buffer, pH 7.2, to 1/3000 M final concentration. (Courtesy of Boris Rotman.)

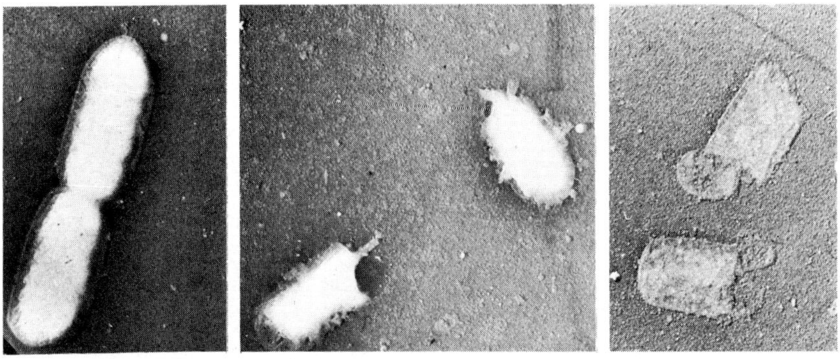

On the other hand, Huhtanen (1968) reported increases in bacterial populations after treating milk with low-frequency ultrasound, presumably by breaking up clumps of bacteria which normally occur in milk. Apparently little or no sterilizing effect occurred.

Ultrasonic waves are waves above audible frequency, of 200,000 to 1 million Hz, produced by connecting a piezoelectric crystal with a high-frequency oscillator. These waves also exert a destructive effect on organisms but there is no generally accepted theory to explain their effects.

Death occurs largely from fragmentation of the cells. It is generally accepted that a major role is played by an activity known as gaseous cavitation. This activity occurs in liquids containing dissolved gases and takes place in the presence of surfaces containing tiny cracks or crevices. Alternations of pressure in the sound field cause bubbles to grow and take part in a complex and extremely energetic motion. Under suitable conditions a tiny pocket of air expands to a size thousands of times greater than the original volume, then violently collapses to a fraction of the original size, all within a time less than that for 1 cycle of the sound. Among the effects attributed to cavity collapse are local heating and electrical discharges.

Wood and Loomis (1927) noted the fragmentation or the tearing to pieces of cells of *Spirogyra, Paramecium*, and red blood corpuscles. Harvey and Loomis (1929) found luminescent bacteria to be destroyed in 1 1/2 hr by ultrasonic treatment. Anderson et al. (1948) noted that certain large bacteriophages were sensitive to ultrasonics, whereas the small, compact viruses were relatively resistant to the shearing forces existing during cavitation of the liquid in which they were suspended.

Pisano et al. (1966) employed high-intensity airborne sonic (9.9 kc per sec) and ultrasonic (30.9 kc per sec) waves on spores of *Bacillus subtilis*. The spores were deposited on filter-paper strips, and exposed to the sound waves for periods varying from 1 to 8 hr at a temperature of 40°C and relative humidity of 40 percent. Significant reductions in viable counts were reported.

Rod-shaped bacteria, especially the large forms, appear to be easily fragmented in an ultrasonic field. On the other hand, the spherical bacteria are considerably more resistant to such treatment. The cells may be killed without being fragmented. Also, a suspending medium rich in protein (milk, peptone, gelatin, etc.) may afford considerable protection to the cells.

For more information: Anderson and Bradley (1964); Boucher and Pisano (1966); Edebo (1961*a, b*); Hughes and Nyborg (1962); Jacobs and Thornley (1954); Marr and Cota-Robles (1957); Slade and Vetter (1956); Williams and Slade (1971).

Alpha, beta, and gamma rays Radiation in a variety of forms is familiar to all of us.

To quote from Kinsman (1961):

Light is radiation we can see. Heat is radiation we can feel. Ultraviolet and X rays we neither see nor feel. None of the four can be heard or smelled. All are like light because they do not continue after the source (bulb) is turned off or removed.

Radiation from radioactive material is a stream of fast flying particles or waves which come from tiny units of matter called atoms. Atoms of a single element often consist of different kinds which behave alike chemically, yet have slightly different weights. These varieties are called "isotopes." The atoms of stable isotopes are not radioactive, but those of unstable or radioactive substances give off portions of themselves, and change into other isotopes in the process.

Man-made radiation in the form of X rays was discovered in Germany in 1895. In France, a year later, natural radioactivity was first identified with uranium. Within two years one of its main sources was isolated – the naturally radioactive element, radium. Invisible rays from this element were soon found to be of three kinds: (1) heavy particles which travel but an inch or so in air, (2) lighter particles which travel a few feet, and (3) waves similar to those of light, but too short to be seen by the human eye, which penetrate to considerable distances, even through several inches of lead. These three forms of invisible radiation are called alpha, beta and gamma rays.

Natural radioactive substances are widely distributed. They exist in minute quantities in our bodies, in water we drink, the air we breathe, the soil we cultivate, even in the materials we use for building. These tiny sources have been sending out invisible radioactive signals for millions of years. It is against this background of natural radiation that man has lived in the past and lives today. This so-called "background radiation" varies slightly from one locality to another, and also with rain or snow falls.

Within the past twenty years man has learned to take naturally stable atoms and make them artificially radioactive. It is these radioactive varieties or "radioisotopes" which are proving to be among the most useful research tools in the entire history of science. Because they are radioactive, their radiation tells where they are, even if the amount is extremely small. Their location or movement within plant and animal tissues, and in industrial and chemical processes, can, therefore, be traced by sensitive recording instruments. Used in this manner radioisotopes are spoken of as tracers.

Sulfhydryl (–SH) containing enzymes were shown by Barron and Dickman (1949) to be inactivated when solutions were irradiated with alpha rays from polonium, beta rays from strontium, and gamma rays from radium. Partial reactivation occurred on the addition of glutathione (page 418). This would indicate that the radiations inhibit such enzymes by oxidation of the –SH groups which are essential for enzyme activity.

Kempe et al. (1954) sterilized meat containing spores of *Clostridium botulinum* by application of 3.5 million reps[1] for 24 hr. Others reporting positive results with the same or different species of bacteria include Anellis et al. (1960), Darmady et al. (1961), and Wheaton and Pratt (1962).

Morgan and Reed (1954) found that spores of *C. botulinum* were more easily destroyed if given a short heat treatment prior to radiation. Licciardello and Nickerson (1962) reversed the order and found spores of *C. sporogenes* to be more sensitive to heat after preliminary exposure to radiation. Kiortsis (1968) treated *Bacillus megaterium* with radioactive zinc and reported that the combined action of zinc and γ-irradiation had a synergistic inhibitory effect on the ability of the organism to form colonies.

Radiations with high-energy electrons from a linear accelerator or gamma radiations from an isotope, such as cobalt 60, are being used commercially for sterilization of heat-sensitive products. Items such as surgical sutures, disposable plastic hypodermic syringes, catheters, animal fibers contaminated with anthrax spores, and various foods have been successfully sterilized by radiation treatment. Electron beam irradiation is a safe and effective technique for sterilizing materials in a completely sealed package, provided electrons can penetrate the product without adversely affecting the product or its package.

For more information: Bridges and Horne (1959); Deasy et al. (1968); Ginoza (1968); Gordon (1959); Hill and Phillips (1959); Hutchinson and Easter (1960); Kempe (1955); Koh et al. (1956); Pepper et al. (1956); Reitman and Tribble (1967); Smirnoff and Cantin (1967); Yamazaki (1971).

Electrons Electrons may be released by heating a tungsten filament to several thousand °C. When such electrons are acted upon by a high electric field between two metal electrodes, they are accelerated away from the negative electrode (cathode) and acquire the energy in volts which produced the field. Electrons may be acelerated to energies of 2, 3, 4 million V or more.

The penetration of high-energy electrons depends on both the electron voltage and the density of the irradiated material. Electrons accelerated to energies of 2 million V will penetrate water effectively to a depth of about 2/3 cm. The depth of penetration increases directly with voltage.

Packaged materials containing pharmaceuticals, biologicals, etc., may be effectively sterilized, provided they come within the limitations of penetration by the electrons. The rays apparently produce no reduction in potency or increase in toxicity of the irradiated materials.

The effectiveness of high-energy electrons depends upon the

[1] One rep unit is a dose of ionizing radiation capable of producing energy absorption of 93 ergs per gram of tissue.

excitation and ionization of the atoms of the organism to produce chemical changes which bring about its death.

X-rays X-rays are effective in destroying bacterial exotoxins and hemolysins. Certain compounds such as proteins, protein hydrolysates, reducing amino acids, glutathione, and thioglycollic and ascorbic acids afford protection to the rays, whereas oxidizing substances are ineffective.

The inactivation of some respiratory enzymes by x-rays is due, most likely, to oxidation of their –SH groups.

Reduction in the concentration of oxygen in bacterial suspensions results in a marked decrease in the x-ray sensitivity of *E. coli*. The cells are protected in the presence of pyruvate, formate, succinate, serine, alanine, ethanol, and hydrosulfite by virtue of their ability to remove oxygen from the system.

Powers et al. (1959) prepared a flow diagram (Fig. 247) of the distribution, release, and entrapment of electromagnetic energy pertinent to inactivation of a dry bacterial spore. X-rays produce damage in dry spores of *B. megaterium* that is not reversed by treatments during or after irradiation. This damage, called class I damage, is independent of temperature changes from 5 to 125°K, but increases from that temperature to about 30°C. Another kind of damage, called class II damage, is observed only if O_2 or NO is present during the time of irradiation. The effect is about a 25 percent increase in sensitivity and is about the same for the two gases. The quenching action of these gases could account for this effect. A third kind of damage, called class III damage, results from the formation of free radicals of long lifetimes. If these react with O_2 either during or after irradiation, they are toxic to the cell. If they are

Fig. 247 Flow diagram of the distribution, release, and entrapment of electromagnetic energy pertinent to inactivation of a dry bacterial spore. (After Powers, Webb, and Ehret.)

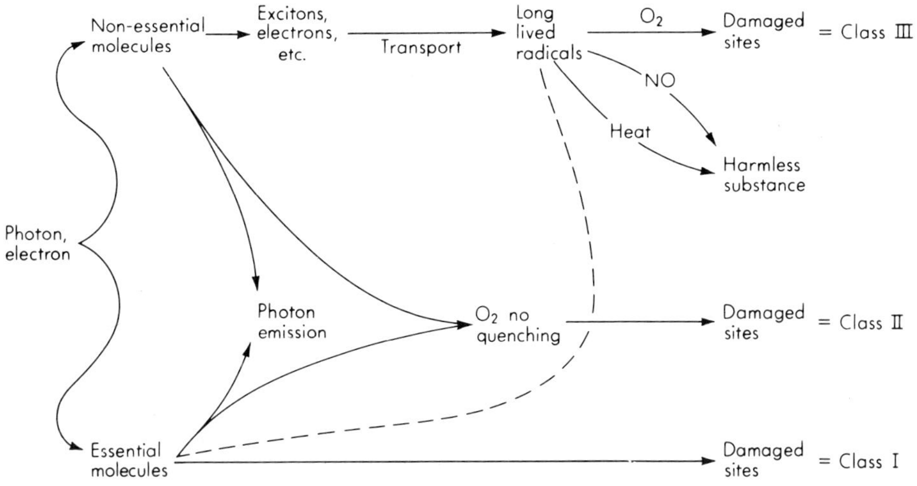

removed by heat annealment, or by treatment with radical scavengers such as NO, they are rendered harmless and apparent radiation sensitivity is reduced.

For more information: Ehret et al. (1960); Ginoza (1968); Greiff et al. (1960); Howard-Flanders and Jockey (1960); Kroeger and Kempf (1959); Powers and Kaleta (1960); Powers et al. (1960*a, b*); Sharon et al. (1963); Stapleton and Engel (1960); Stuy (1960).

Distilled water There is a difference of opinion concerning the action of distilled water on the viability of bacteria and spores. Some have reported death of vegetative cells in a few hours; others have stated that weeks were necessary. Spores resist the adverse conditions much better than vegetative forms.

Moffet and Williams (1967) reported the recovery of a number of bacterial species from stored distilled water. The organisms were limited almost exclusively to water forms, rather than to enteric bacteria or gram-positive cocci, and included the following genera: *Aerobacter, Alkaligenes, Caulobacter, Escherichia, Flavobacterium,* and *Pseudomonas.*

The volume of culture introduced into distilled water makes a great difference in the survival time. The larger the inoculum transferred into distilled water, the longer the survival period. Survival parallels the concentration of nutrients transferred. Ideally, the organisms first should be thoroughly washed before being transferred to distilled water. Other conditions causing conflicting results are the pH of the distilled water, the presence of buffers, the temperature of incubation, presence of traces of alkali dissolved from soft glass, and dissolved CO_2 and O_2 from the air.

For more information: Lee (1972).

Chemical agents

The chemical agents include acids, alkalies, salts, reducing compounds, oxidizing agents, phenols and cresols, alcohols, glycols, stains, mercury compounds, silver compounds, soaps, surface-active agents, sulfonamides, urea and its derivatives, and gaseous disinfectants.

Acids The germicidal efficiency of acids is proportional to the hydrogen-ion (H^+) concentrations of their solutions. Strong acids (HCl, H_2SO_4) are more germicidal than weak acids (acetic, lactic). Strong mineral acids adjusted to contain the same number of H^+ are of equal value as sterilizing agents.

The H^+ concentration does not necessarily account for all the disinfecting action of an acid. The organic acids are weaker in H^+ than the mineral acids, yet in equal concentrations of H^+, they are more efficient. This efficiency may be due to the additional effect of the anions or to the undissociated molecules, or to all three.

Benzoic and salicylic acids are weak acids, almost completely dissociated in neutral or alkaline solution, but almost completely undissociated in strongly acid solution. Rahn and Conn (1944) found that benzoic, salicylic, and sulfurous acids were nearly a hundred times more germicidal in strongly acid solutions than in neutral solutions. Only the undissociated molecules were germicidal; the ions exhibited only a slight toxic effect.

For more information: Bosund (1962).

Alkalies The action of alkalies as disinfectants is dependent upon the release of hydroxyl ions (OH^-) in solution. The greater the degree of dissociation, the more effective the germicidal action. Alkalies that are especially toxic to bacteria include KOH, NaOH, LiOH, and NH_4OH. Of these, KOH is the most toxic by virtue of its greater degree of dissociation, and NH_4OH is the lowest in germicidal activity because it is the least ionized.

As is true with the acids, there are exceptions to the above rule. A weak base, such as barium hydroxide [$Ba(OH)_2$] is less dissociated than KOH, yet ion for ion it is considerably more toxic. This is due to the high toxicity of the Ba^{++}. The combined action of the Ba^{++} and OH^- produces a greater germicidal effect than that exhibited by either ion acting alone.

H^+ exerts a greater toxic effect than an equal number of OH^-.

Salts Cations exert a peculiar and characteristic effect on bacterial growth. In general, low concentrations of cations stimulate growth; high concentrations are inhibitory and ultimately toxic.

Low concentrations of NaCl produce an accelerating effect on the growth of bacteria. Maximum stimulation for *E. coli* occurs at a salt concentration of 0.2 *M* and pH 7.8.

Chloride salts can be divided into two groups on the basis of their toxicity to *E. coli*. The salts in group 1 prevent growth in concentrations of 2 to 0.05 *M*; those in group 2 prevent growth in dilutions of 0.01 to 0.00001 *M*. Salts in the former group are of common occurrence in the protoplasmic environment and are considered nontoxic.

Group 1 Chlorides of Na, K, Li, NH_4, Sr, Mg, Ca, Ba, Mn, Ti, and Sn.

Group 2 Chlorides of Ni, Tl, Cu, Fe, Zn, Co, Pb, Al, Ce, Cd, and Hg.

Chlorides of Na, K, NH4, and Li produce maximum growth at a concentration of 0.25 *M* after 72 hr of incubation. Salt concentrations above or below 0.25 *M* are less stimulating.

In general, the bivalent chlorides in group 1 are more toxic than the monovalent chlorides. The concentration producing maximum growth ranges from 0.05 to 0.025 *M*.

Salts of heavy metals, particularly Hg and Ag, are toxic in relatively low concentrations. The toxicity is proportional to the concentration of the metal ions in solution. Mercury salts of the organic acids, e.g., mercury acetate, show a low degree of dissociation and are weakly germicidal.

In general, toxicity of cations for various bacteria increases with valence. The ions may be arranged in order of increasing toxicity as follows: K, Na, NH_4, Li, Sr, Mg, Ba, Ca, Mn, Ti^{3+}, Sn, Ni, Ti^{+}, Zn, Cu, Fe^{2+}, Fe^{3+}, Co, Pb, Al, Ce, Cd, Ag, Hg.

Of the anions, the chloride ion is probably the least toxic of all for *E. coli.* Appropriate amounts of Cl, I, NO_3, SO_4, PO_4, and lactate ions stimulate growth, whereas oxalate, acetate, citrate, and F ions depress growth.

Different species of bacteria vary in their susceptibility to the same ion. Organisms have been classified on the basis of their reactions to various ions.

The composition of the suspending medium is important in making a comparative study of the action of ions on bacteria. Salts exhibit a greater retarding effect in distilled water than in a protein-containing medium. This decreased retarding effect is due to the reaction of salts with proteins, resulting in a lowered concentration of salt ions in the medium.

In general, gram-positive bacteria are more sensitive to salt ions than gram-negative organisms. The same holds true for the effect of various stains on bacteria (page 577).

Mechanisms of salt action The following conclusions may be drawn on the effects of salts on bacteria:

Most salts in high dilutions produce a stimulating action; in more concentrated solutions, a toxic or germicidal effect occurs.

When the toxic range is reached, the germicidal effect is proportional to the concentration of the salt.

The toxic effect of monovalent salts can, in general, be neutralized by the addition of a bivalent salt. Likewise, the toxic effect of a bivalent salt can, in most cases, be neutralized by the addition of a monovalent salt.

In general, bivalent cations are more toxic than monovalent cations. Also, the heavier cations are usually more germicidal than the lighter cations. For example, $HgCl_2$ is more toxic than $MgCl_2$.

Salts are more germicidal in distilled water than in solutions containing protein. This is probably true for all germicidal agents. The organic matter greatly inactivates the salts, probably by combining with them and reducing the ionic concentration.

Different bacterial species vary considerably in their susceptibilities to the same salt. In general, closely allied organisms behave in a similar manner.

Gram-positive bacteria are usually more sensitive to the action of salts than gram-negative forms.

The organic acids show only slight dissociation, compared to the inorganic acids; yet some of them exhibit a strong germicidal action. The toxic effect is attributed to the action of the undissociated molecules and not to their ions. Probably the same holds true for the toxicity of the salts of organic acids.

Reducing agents Some compounds are germicidal by virtue of their powers of reduction. Sulfites (sulfurous acid, sulfur dioxide) and ferrous compounds act in this manner.

Potassium metabisulfite is used to destroy or inhibit the development of organisms found on grapes prior to the addition of starters containing pure cultures of the desired yeasts. Sulfites or metabisulfites destroy or inhibit growth of many undesirable organisms with a minimum of injury to the wine yeasts.

Oxidizing agents Some compounds which give up oxygen freely or are capable of releasing oxygen from other compounds are used for their germicidal action. Such agents destroy organisms by the process of oxidation. Among these may be mentioned peroxides, potassium permanganate, bromine, chlorine and certain compounds containing this element, such as hypochlorous acid, hypochlorites, chloramine, dichloramine, and halazone; also iodine and a large number of iodine complexes known as iodophors.

Hydrogen peroxide (H_2O_2) is an active oxidizing agent, easily decomposing into water and oxygen. The commercial solution (about 3 percent H_2O_2 in water) is capable of destroying anthrax spores in 1 hr.

A disadvantage to the use of H_2O_2 on living tissue is that it breaks down very rapidly in the presence of certain enzymes present in body fluids rendering it of limited value in vivo. Compounds such as zinc peroxide (ZnO_2) and urea-hydrogen peroxide [$CO(NH_2)_2 \cdot H_2O_2$] are more desirable in that they break down more slowly and give a more sustained action in contact with body fluids. Urea-hydrogen peroxide is employed to advantage in dental therapy and for the treatment of various forms of pharyngitis.

Potassium permanganate ($KMnO_4$) was at one time quite popular when the number of germicidal agents was limited. Its activity is increased in acid solution. A 1.1 percent solution of HCl containing 1 percent $KMnO_4$ destroys anthrax spores in 30 sec. However, the salt promptly reacts with organic matter, becoming converted into the insoluble manganese dioxide (MnO_2), a brown-staining compound. For this reason, its use as a germicide has been largely discontinued.

Sodium hypochlorite (NaClO) and calcium hypochlorite [$Ca(OCl)_2$] are important oxidizing compounds containing chlorine. When dissolved in water, they are believed to react as follows:

$$OCl^- + H_2O \rightleftharpoons HClO + OH^-$$

The active agent is free hypochlorous acid. The hydrolysis is depressed with increasing alkalinity.

The killing of bacteria by hypochlorites appears to proceed in two stages: (1) penetration of the germicide into the cell; and (2) chemical combination with the protoplasm, resulting in death of the organism.

Hypochlorites are used as household and laundry bleaches; in hygiene; for sanitizing dairy utensils and equipment; for the chlorination of water supplies, swimming pools, and sewage effluents; and for restaurant and tavern glasses and dishes.

Compounds containing active chlorine attached to a nitrogen atom are also strongly germicidal. Three such compounds are chloramine ($CH_3 \cdot C_6H_4 \cdot SO_2NNaCl \cdot 3H_2O$), dichloramine ($CH_3 \cdot C_6H_4 \cdot SO_2NCl_2$), and halazone ($COOH \cdot C_6H_4 \cdot SO_2NCl_2$). Their activity is directly proportional to the extent to which reactions of hydrolysis proceed in solution:

$$R_2{=}N{-}Cl + H_2O \longrightarrow R_2{=}N{-}H + HClO$$

$$R{-}N{=}Cl_2 + H_2O \longrightarrow R{-}N\begin{matrix} \diagup H \\ \diagdown Cl \end{matrix} + HClO$$

The active agent is HClO. Compounds of this nature release their chlorine much more slowly than the hypochlorites, giving a more sustained action.

Chlorine gas added to water reacts as follows:

$$2Cl_2 + H_2O \rightleftharpoons HCl + HClO$$

Bromine acts in a similar manner.

For more information: Baker (1969); Benarde et al. (1967); Dychdala (1968*a*); Fitzgerald and Der Vartanian (1967); Jones et al. (1968).

The mechanism of action of iodine appears to be different from that of chlorine and bromine. An aqueous solution of iodine at pH 8 or less contains chiefly two forms of iodine, namely, molecular I_2 and the tri-iodide ion, I_3^-. The I_2 is only slightly soluble in water. In the presence of an iodide, such as NaI, the solubility is increased several hundred times, the increase being in the tri-iodide form:

$$I_2 + I^- \longrightarrow I_3^-$$

Carroll (1955) found the tri-iodide ion to have negligible bactericidal activity. Since iodine for bactericidal studies is used in high dilutions, the tri-iodide ion dissociates into diatomic iodine and iodide ion unless the relative concentration of iodide (NaI, KI, etc.) is kept at a high level.

Gershenfeld and Witlin (1949) found that solutions containing free iodine displayed more effective antibacterial activity against *Staphylococcus aureus* than did chlorine or bromine, either in the presence or in the absence of organic matter.

Iodine is a suitable agent for the emergency disinfection of water supplies. Chang and Morris (1953) reported that iodine in a concentration of 5 to 10 ppm is effective against all types of waterborne pathogenic organisms within 10 min at room temperature. For this purpose iodine has the following advantages: (1) its germicidal action is less dependent on pH, temperature, and time of contact; (2) nitrogenous impurities do not impair its usefulness; and (3) side reactions leading to consumption of the germicide are less marked for iodine than for chlorinous disinfectants.

Iodophors Combinations of iodine with certain carriers to produce soluble complexes are known as iodophors. These soluble complexes slowly liberate iodine when diluted with water and give a more sustained germicidal effect.

There are many carriers capable of reacting with iodine to produce iodophors. A well-known solubilizing agent of this type is polyvinyl-pyrrolidone (PVP). Its combination with iodine produces the complex known as polyvinylpyrrolidone-iodine (PVP-I). This is known also as povidone-iodine.

It is a complex of indefinite composition. When dissolved in water, it slowly liberates most of its iodine. It is claimed to be more effective and less toxic than aqueous solutions of free iodine. It is also nonallergenic whereas some individuals are sensitive to free iodine.

For more information: Black et al. (1970*a, b*); Gershenfeld (1968); Johannesson (1959); Klebanoff (1967); Koski et al. (1966).

Phenols and cresols Phenol or carbolic acid no longer plays any significant role as a germicide but nevertheless it still finds some uses for that purpose.

The phenols and cresols are efficient germicides in fairly concentrated solutions. Phenol is soluble in water but most of the other members of the group are only slightly soluble. However, they may be held in suspension by mixing with soap, by which procedure colloidal solutions are obtained.

The emulsification of such disinfectants, which are only slightly soluble in water, results in the formation of more potent germicidal preparations. In the emulsified state the particles of germicide are adsorbed onto the surface of the emulsifying agent (soap) resulting in an increased concentration in the vicinity of the bacteria. The emulsified disinfectants are more active when freshly prepared. After a few days, the activity decreases, probably owing to a change in the colloidal state. An important commercial disinfectant of this type is the compound solution of cresol, known under the trade name of Lysol.

Emulsified germicides react with tap water to give insoluble calcium and magnesium soaps. Not only is some of the original soap destroyed but the preparations contain an undesirable turbidity. This may be avoided by preparing dilutions in distilled water. Sparkling clear solutions are produced.

Mechanism of action High concentrations of phenolics act as gross protoplasmic poisons, disrupting the cell wall, penetrating the cell, and precipitating the cellular proteins. In less concentrated solutions they inhibit protein, ribonucleic acid, and deoxyribonucleic acid synthesis. Membrane damage followed by leakage of constituents from the cellular pool is probably the primary mode of action of phenol.

For more information: Prindle and Wright (1968); Pullman and Reynolds (1965); Starr and Judis (1968).

Alcohols Ethyl alcohol is widely used as a germicide. It is germicidal rather than bacteriostatic, colorless, and much more effective against vegetative cells than spores.

In general, the maximum germicidal efficiency of alcohol is exhibited at a concentration of 70 percent by weight (77 percent by volume) in water. Smith (1947) found alcohol to be an effective germicide against *Mycobacterium tuberculosis.* The organism was killed in 15 to 30 sec by absolute, 95 percent, and even 70 percent ethyl alcohol. Ninety-five percent alcohol was found to be best for wet surfaces, 50 percent best for dry, and 70 percent best for wet or dry.

The germicidal action of aliphatic alcohols increases with the molecular weight as far as the amyl derivative (5 carbon atoms) and decreases through octyl to undecyl alcohol (11 carbon atoms). Since the alcohols decrease in solubility as the molecular weights increase, the higher members of the series are generally not employed as germicides.

Alcohols are believed to act by denaturing proteins. This occurs more readily in the presence of water than in its absence, which explains why absolute is generally less bactericidal than mixtures of alcohol and water.

Mercuric chloride dissolved in absolute alcohol is less germicidal than the same concentration in water. Mercuric chloride dissolved in 50 percent alcohol is more germicidal than a corresponding aqueous solution. The same holds true for silver nitrate. Since the toxicity of these salts is proportional to the concentration of mercury and silver ions, water is necessary for ionization to occur. On the other hand, compounds such as phenol and formaldehyde are less germicidal in the presence of even a low concentration of alcohol.

Isopropyl alcohol is slightly more efficient as a germicide than ethyl alcohol. Like ethyl alcohol, its most efficient action appears to be in a strength of about 70 percent. Since it is toxic when taken internally, it may be purchased without being subject to restrictions and heavy taxation. Isopropyl alcohol is being used more and more as a germicide and as a substitute for the ethyl compound in various manufacturing processes.

For more information: Morton (1968).

Glycols A large number of glycols, especially propylene, dipropylene, and triethylene glycols, possess certain properties which make them desirable for the disinfection of the air of enclosed spaces.

Their formulas are as follows:

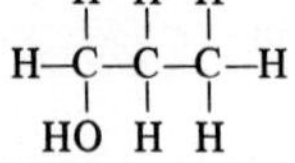

Propylene glycol

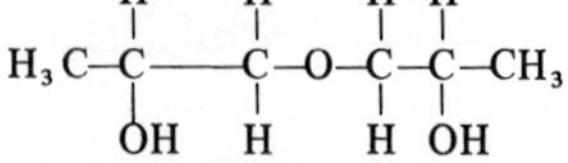

Dipropylene glycol

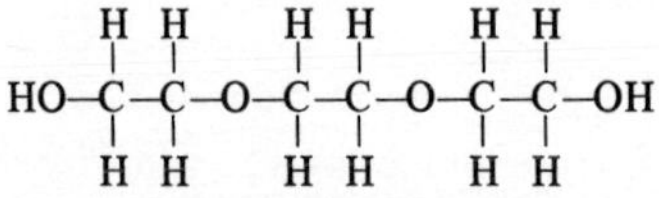

Triethylene glycol

The glycols are less active than the phenols, halogens, and detergents. Even the most highly bactericidal glycols fail to inhibit growth of bacteria in concentrations of 3 percent, and the least lethal shows an effect only in solutions above 50 percent. However, ethyl alcohol is only slightly more effective than propylene glycol, the most bactericidal of the nontoxic glycols.

In general, the higher the concentration of glycol, the more rapid the bactericidal action. In 98 percent solution both propylene and triethylene glycol killed in less than 1 min and probably in a few seconds. Propylene glycol in concentrations of 70 to 80 percent appeared to produce equally rapid killing and was found to be the most efficient of the three glycols. The action of dipropylene glycol was less regular.

It is believed that glycols sprayed into the air of enclosed spaces destroy suspended organisms by their chief physical property of hygroscopicity. The glycol molecules dissolve in the film of moisture surrounding each organism. The concentration of glycol soon builds up to a strength where it causes the withdrawal of water from the cell (plasmolysis) resulting in its death.

Stains Certain coal-tar stains, notably those of the triphenylmethane group, affect the viability of bacteria. At first this action was described as bactericidal because the organisms failed to grow. Later it was shown that the organisms were not necessarily killed but merely prevented from multiplying. The terms bacteriostasis and bacteriostatic were coined to describe this phenomenon.

In most cases, selective bacteriostatic action parallels the Gram reaction. This means that the gram-positive bacteria are more susceptible to the action of the above stains than the gram-negative forms (Fig. 248). Notable exceptions are the acid-fast bacteria (*M. tuberculosis, M. bovis,* etc.) which are gram-positive but comparatively resistant to the action of the triphenylmethane stains.

An increase in the basicity of a solution of a basic stain increases its germicidal activity; a decrease in basicity decreases its activity. Likewise, an increase in the acidity of a solution of an acid stain increases its germicidal activity; a decrease in acidity decreases its activity.

Hoffmann and Rahn (1944) found that above a certain concentration the stain acted like other germicides. The cells died in logarithmic order and in proportion to the stain concentration. The stain was more toxic to young than to old cells, and its toxicity increased slightly with an increase in pH. At lower concentrations, the stain did not give a logarithmic survivor curve and was not influenced by cell age or pH. Perhaps this effect was due to the unfavorable oxidation-reduction potential posed by the stain. In this range the cells usually overcame the action of the stain and multiplied. The stain produced an abnormally long lag period which increased with stain concentration. The length of the lag phase was inversely proportional to the log of the number of inoculated cells.

For more information: Fischer et al. (1944).

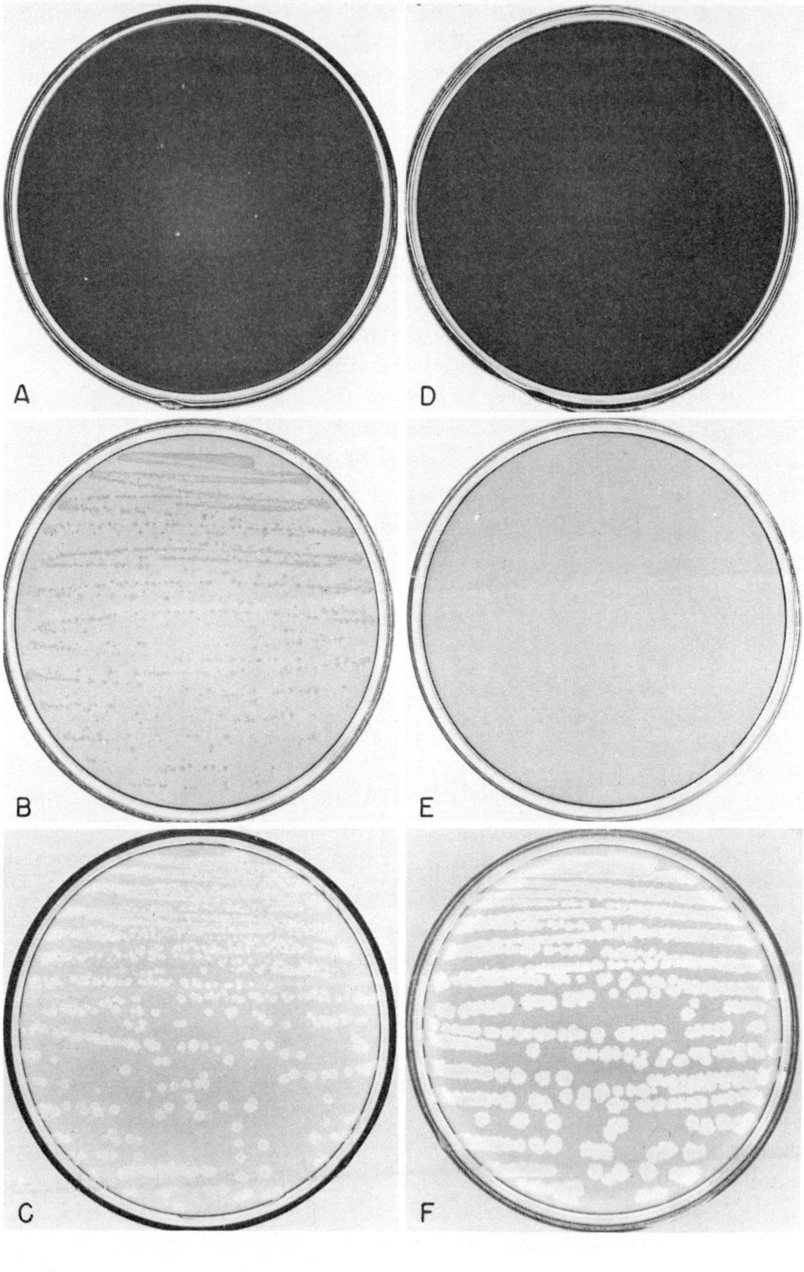

Fig. 248 Effect of crystal violet. (A, B, C) Escherichia coli streaked over the surface of nutrient agar containing, respectively, 1:5000, 1:100,000, and 1:2 million concentration of the dye. (D, E, F) same streaked with Bacillus subtilis. A 1:5000 concentration of crystal violet inhibited both organisms; a 1:100,000 concentration inhibited B. subtilis (gram-positive) but not E. coli (gram-negative); a 1:2 million concentration was not sufficient to prevent growth of B. subtilis.

Photodynamic inactivation of organisms Certain fluorescent stains, such as eosin and acridine orange, are capable of sensitizing microorganisms in liquid suspension. No action occurs in the dark but when the suspensions are exposed to sunlight, destruction of the organisms occurs. If eosin is employed, only the green rays are effective because the stain fluoresces most strongly in green light. If acridine orange is used, destruction of the organisms is most effective on exposure to violet light where the stain fluoresces most strongly. Frank and Foster (1967) reported the same stain to be effective in the destruction of Newcastle disease virus. The inactivation rate was found to be directly proportional to light intensity and to stain concentrations up to a level of approximately 100 to 200 μg per ml.

Photodynamic sensitization does not depend exclusively on fluorescent stains. Nonfluorescent stains, such as methylene blue and toluidine blue, are also effective but to a lesser degree. Bradley (1966) found methylene blue, thionin, and several other stains to be effective against an actinophage when exposed to bright light. The extent of inactivation was determined by stain concentration, duration of exposure to light, ionic strength of the suspending medium, and availability of air.

Using toluidine blue as photosensitizer, Harrison (1967) reported the following kinds of damage in bacteria: (1) loss of colony-forming ability (death); (2) damage to DNA, as evidenced by mutation; (3) damage to permeability as evidenced by leakage of 2600 Å absorbing material; and (4) damage to protein, as evidenced by inactivation of several enzymes. Because of differences in molecular structure, different photosensitizers will show differences in mode of action. For example, of the five bases in nucleic acid, methylene blue in vivo not only limits its action to guanine but apparently to the guanine of DNA. On the other hand, acridine orange has not been observed to photooxidize guanine although the stain also leads to DNA damage.

For more information: Witmer and Fraser (1971*a, b, c*).

Mercury compounds Mercurials were among the first to be used as germicides. At that time they were believed to be the most potent of the compounds available. Their effectiveness is due largely or entirely to the presence of the mercury ion.

It was later shown that mercury compounds did not necessarily kill bacteria but rather exerted a strong bacteriostatic effect which was produced by traces of mercury adhering to the walls of the cells. The bacteriostatic effect could be eliminated by the addition of ammonium sulfide to inactivate the mercury adherent to the bacteria and the small amount carried over by the inoculum into the subculturing medium. By this technique, the bacteria were usually shown to be living.

Because of their bacteriostatic property, mercury compounds with dyes or other organic radicals are employed for skin antisepsis. They liberate their mercury ions more slowly than the inorganic mercurials, giving a more prolonged effect that is less toxic and irritating to the skin, and they do not corrode instruments. Also, they retain a higher degree of antibacterial activity in the presence of organic matter than the inorganic salts of mercury.

It is generally believed that mercurials act by interfering with essential metabolites and enzymes containing the sulfhydryl (–SH) radical. Mercury combines with compounds containing this radical, producing an inactivation without demonstrable injury to the cell.

Harris et al. (1954) immersed *E. coli* in 0.01 M $HgCl_2$, then treated the cells with hydrogen sulfide. X-ray diffraction studies showed the presence of intracellular crystals of HgS. The results suggested that the site of cation adsorption was the cytoplasmic membrane of the cell.

Morton et al. (1948) reported that the organic mercurials metaphen, merthiolate, and mercurochrome possessed many shortcomings as disinfectants. *Streptococcus pyogenes*, placed in contact with these compounds, was still capable of producing fatal septicemia when introduced into animals.

The fact that bacteria, in a state of bacteriostasis, are still infectious is sufficient reason for taking precautions to eliminate the effect when testing mercurials in vitro for germicidal activity.

For more information: Brewer (1968).

Silver compounds The germicidal and antiseptic properties of the simple or inorganic silver salts are due to the presence of silver ions in solution.

Silver, like mercury, is a strongly bacteriostatic metal. The bactericidal action of the simple silver salts may be accompanied by irritation, pain, astringency, and causticity. These effects may be largely avoided or eliminated entirely by the use of colloidal organic preparations of silver.

In contrast to the inorganic salts, the colloidal preparations (silver protein strong, Argyrol, Collargol, etc.) liberate silver ions at a slow rate to give a more mild and sustained effect. They are relatively noncorrosive, nonastringent, and nonirritant. The organic silver compounds do not precipitate proteins or chlorides and retain much of their germicidal activity in the presence of these substances.

The toxicity of inorganic silver has been said to occur in two stages: (1) the immediate irritant and bactericidal effect of the silver ions; and (2) the later, milder action following re-solution and ionization of the silver-protein combinations formed in the first stage.

For more information: Romans (1968).

Soaps It has been known since the beginning of bacteriology that both soft and hard soaps are mildly antiseptic.

Soft soaps are prepared by boiling oils and fats with potassium hydroxide; hard soaps are prepared with sodium hydroxide. The soft soaps are used in preparing liquid soaps and shampoos, whereas the hard soaps are used in preparing soap powders, chips, and bars.

Soap has a number of important physical characteristics. When dissolved in water, it lowers the surface tension, forms colloidal solutions and gels, causes water to wet surfaces more rapidly, gives the solution a soapy or slippery feeling, and has the ability to emulsify and disperse oils and dirt in the solution and thus is able to cleanse.

Various chemicals such as phenols, cresols, mercuric iodide, mercuric chloride, metaphen, and chloramine have been incorporated in soaps. Most of these so-called germicidal soaps are no more useful than ordinary soaps for destroying bacteria. In fact, some compounds may lose their effectiveness in soap and may even decrease the natural antiseptic powers of the latter. For example, soaps containing cresol and phenol are less antiseptic than the cresol or phenol or soap alone when tested in equal concentrations.

A group of relatively new antibacterial agents have been shown to be effective when dissolved in soaps and are now employed commercially. These compounds in soap inhibit the development of body odors produced by bacterial activity in skin perspiration. Such preparations are marketed as deodorant toilet soaps. Consistent use of deodorant soaps permits the accumulation of the antibacterial agents on the skin, resulting in a significantly lower number of bacteria. It has also been shown that regular use of the antibacterial soaps decreases by as much as 44 percent the incidence of superficial cutaneous infections.

Probably the most important compounds in this group are:

Hexachlorophene (G-11) (no longer used)
3,4′,5–Tribromosalicylanilide (TBS)
3,4,4′–Trichlorocarbanilide (TCC)
4,4′–Dichloro-3-(trifluoromethyl) carbanilide (CF3)
3,5–Dibromosalicylanilide (3,5-DBS)

The first compound in the group is less active than the others.

Their structural formulas are as follows:

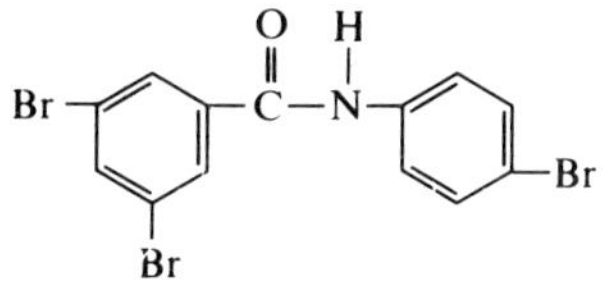

3,4′,5 Tribromosalicylanilide (TBS)

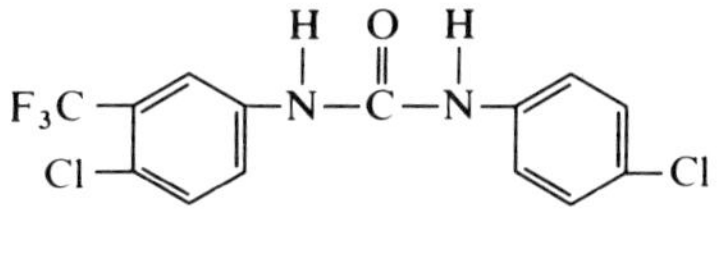

3-Trifluormethyl-4,4′-dichlorocarbanilide (CF_3)

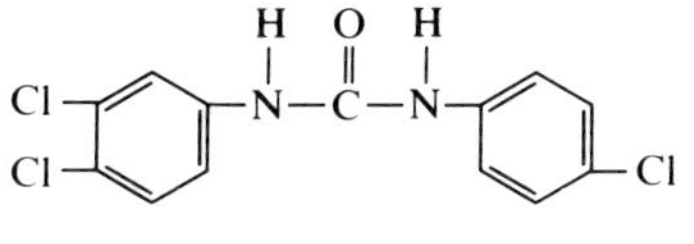

3,4,4′ Trichlorocarbanilide (TCC)

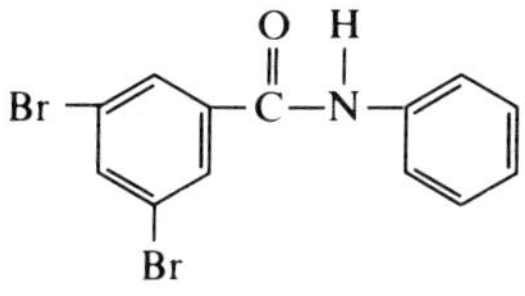

3,5-Dibromosalicylanilide (3,5-DBS)

All the above compounds are less effective in the presence of serum. Hexachlorophene loses much of its activity in soap whereas the others retain their effectiveness quite well. It has been shown to be a toxic compound and its use in soaps is no longer recommended or permitted. TBS is more effective in the presence of soap than in its absence.

For more information: Bannan and Judge (1965); Berman and Knight (1969); Ericson et al. (1968); Gump and Walter (1968); Hodes and

Stecker (1968); Jungermann et al. (1966); Leonard (1967); MacKenzie (1970); Molnar (1969); Roman and Manring (1967).

Surfactants and synthetic detergents A surface-active agent or surfactant has the property of orienting itself between two interfaces to bring them into more intimate contact. If the function of the agent is to promote wetting and penetration, it is called a surfactant. If the two interfaces are immiscible liquids, the surface-active agent lowers the interfacial tension so that emulsions are formed. Under these conditions the agent is called an emulsifier. If the surface-active agent combines both wetting and emulsifying properties to a sufficient degree, it is called a synthetic detergent.

Synthetic detergents, like soaps, consist of a hydrophobic (water-repelling) group and a hydrophilic (water-attracting) group.

The detergent and wetting class of compounds consists largely of anionic agents and possesses a negative electrical charge. They ionize in water like soaps:

Soaps	Anionic agents
$R{\cdot}COONa$	$R{\cdot}SO_3Na$
Carboxyl group	Sulfonic group

The emulsifiers for the most part are nonionic, i.e., do not ionize in water. A typical nonionic agent is glycerol monostearate, an emulsifying agent used in baking, ice cream, and cosmetics:

$$\begin{array}{l} CH_2{\cdot}OOC{\cdot}C_{17}H_{35} \\ | \\ CHOH \\ | \\ CH_2OH \end{array} \qquad \text{or} \quad RCOOR'$$

The cationic agents possess a positive electrical charge, being capable of reversing the action of soaps. They are generally substituted ammonium salts. Some well-known cationics are the quaternary ammonium compounds of the form

$$R{-}N(CH_3)_3Cl$$

Probably 75 percent of the cationic agents are quaternary ammonium salts. Most of the sanitizing and bactericidal agents belong in this group.

Surfactants have a tendency to localize in the surface layer or interface of liquids. A surfactant molecule may be diagrammatically represented by a bar for the hydrophobic (fat-soluble) group and a disk for the polar (water-soluble) group, depending upon whether the polar group is at the end or somewhere along the carbon chain:

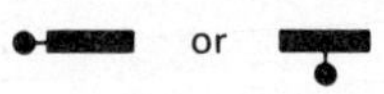

Then the surface film can be represented as follows:

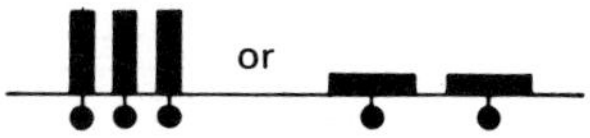

The surface of a solution containing a surfactant is actually altered, being covered with a hydrocarbon film having the thickness of one molecular layer. When an aqueous solution containing a surfactant is in contact with a lipoidal surface, the hydrophobic group of the wetter is absorbed and the polar group protrudes:

Such a surface is now capable of being wetted by water. For this reason wetting agents lower the surface tension of water. Such solutions can penetrate into openings and cracks, very small spaces, or even into the center of clumps of bacteria. The same solutions without wetters would simply bridge over openings without showing any appreciable penetration.

Surfactants are of great importance as additions to germicidal solutions intended for clinical application. They make it possible for such solutions to penetrate into infected tissues; pus, necrotic debris, bacteria, etc.

A cell surface covered with adsorbed molecules of a wetting agent may be represented as follows:

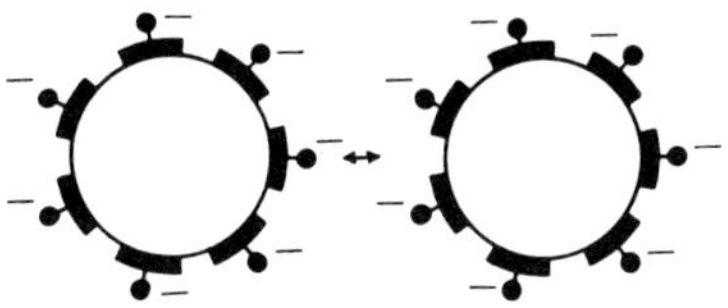

The number of negative charges on the cell surface is much higher in the presence of a surfactant than in its absence. This results in immediate dispersion and exposure of living organisms that previously were protected from the action of the germicide.

Uses of surfactants Surfactants have a wide variety of uses as manifested by their applications in detergency, solubilization, emulsification, capillary penetration, wetting, spreading, and disinfection.

The quaternary ammonium compounds are widely used as germicides. They are strongly bacteriostatic in high dilutions and germicidal in more concentrated solutions. The bacteriostatic effect may be eliminated by the addition of neutralizers such as soap, anionic detergents, and phospholipides.

The true bactericidal potency of quaternary ammonium compounds seems to be between the two extremes of very high and very low activity. A result indicating unduly high potency is obtained if the adsorbed quaternary compound is not actively removed from the treated organisms. Such inhibited cells will die eventually, but they are not killed immediately as the results seem to indicate.

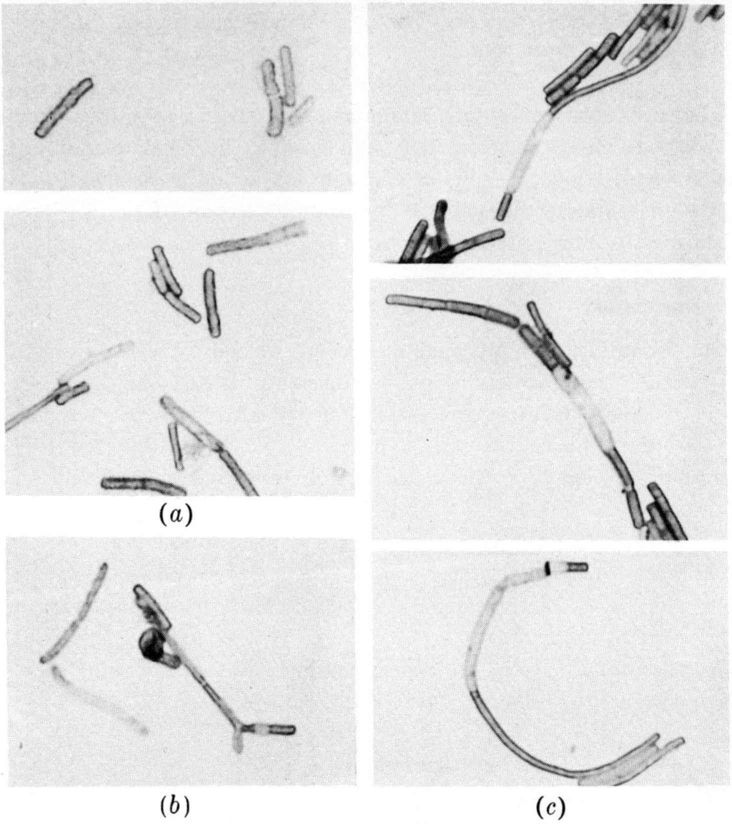

Fig. 249 Bacillus megaterium, 6-hr culture, stained with Victoria blue x2000. (a) degeneration of cytoplasmic membrane; (b) extrusion of cytoplasm; (c) empty cell walls. (After Chaplin.)

The effective quaternary ammonium compounds have many advantages over the commonly used germicides. They are noninjurious, emollient, relatively nonirritating, and active against many nonsporulating bacteria and molds. They have low surface tensions, and some possess keratolytic properties.

Chaplin (1952) treated *B. megaterium* with Roccal, a quaternary compound, and reported that three important changes occurred: (1) degeneration of the cytoplasmic membrane; (2) extrusion of the cytoplasm; and (3) a deterioration of the cell wall, in which all rigidity and tensile strength were lost (Fig. 249).

Voss (1963) reported that the germicidal effectiveness of two anionic surfactants was increased for *Staphylococcus aureus* in the presence of low concentrations of divalent cations. He believed the cations decreased the negative charge at the cell surface, causing an increased adsorption of the surfactants. This produced damage to the cytoplasmic membrane and death of the cell.

For more information: Dychdala (1968*b*); Lawrence (1968*b*); Newton (1960).

Sulfonamides The sulfonamides is a name given to a group of compounds which are derivatives of para-amino benzene sulfonamide (the amide of sulfanilic acid) commonly known as sulfanilamide:

Sulfanilic acid

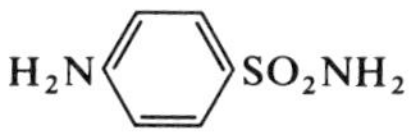

Sulfanilamide

A large number of derivatives of sulfanilamide have been prepared by substituting the hydrogen atoms of the amino radicals with other groups. Some important derivatives are the following:

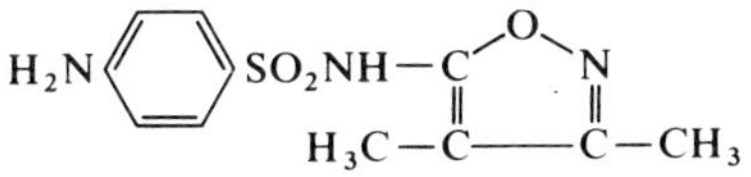

Sulfisoxazole

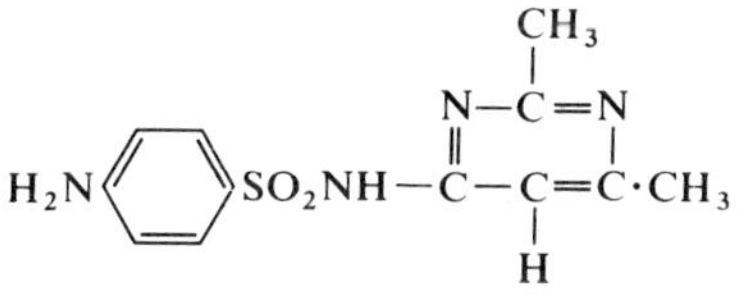

Sulfisomidine (Elkosin)

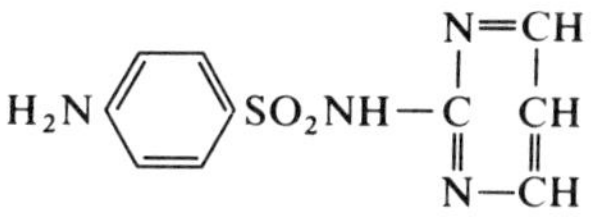

Sulfadiazine

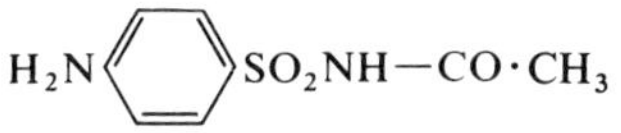

Sulfacetamide (Sulamyd)

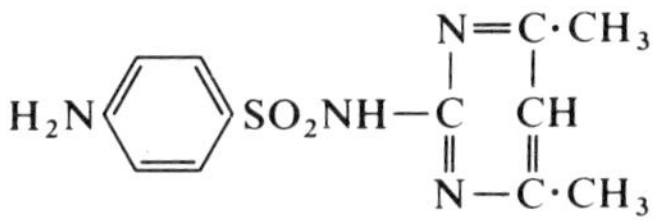

Sulfamethazine

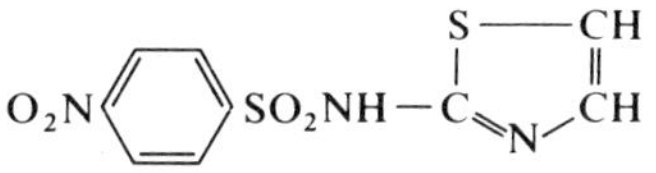

Para-nitrosulfathiazole (Nisulfazole)

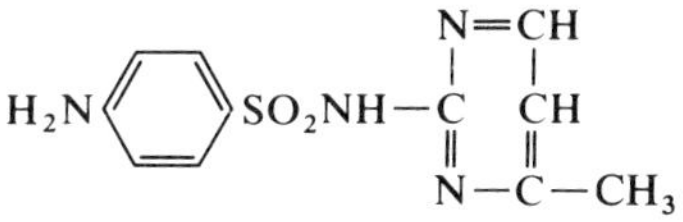

Sulfamerazine

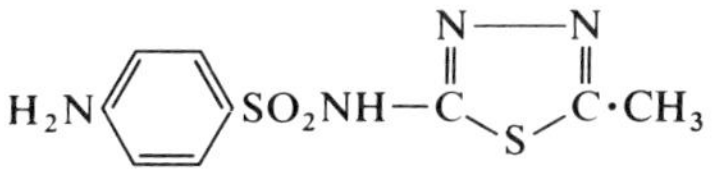

Sulfamethizole (Thiosulfil)

Clinical uses High concentrations of sulfonamides in vivo may be bactericidal but the major effect usually is bacteriostatic. Organisms may develop resistance to the drugs in the course of therapeutic use.

The choice of sulfonamide to be used in the control of infections should be made only after a bacteriological diagnosis. The derivatives vary in their usefulness to various infections. Therefore, it is necessary to select the proper derivative; otherwise the result will most likely be disappointing.

Mode of action From the many theories advanced to explain the mode of action of the sulfonamides, the one by Fildes (1940) appears to be the most logical. Woods (1940) showed that *p*-aminobenzoic acid (PABA) in high dilutions antagonized the action of sulfonamides. Fildes showed that PABA is an essential metabolite normally associated with an enzyme. The sulfonamides displace PABA from its enzyme and thereby stop this essential line of metabolism.

Organisms may develop resistance to sulfonamides in the course of therapeutic use. It has been shown that the development of resistance results from an increased synthesis of PABA by the organisms. Therefore, it is advisable to give maximum initial doses in the hope that the infection will be controlled before resistance is developed.

For more information: American Medical Association (1964); Spink et al. (1944).

Urea and its derivatives Urea and some of its derivatives, such as urethane, are highly bacteriostatic and bactericidal for many gram-negative and a few gram-positive bacteria. They also potentiate the activity of the sulfonamides, inhibit *p*-aminobenzoic acid moderately, and increase the solubility of sulfanilamide and other sulfonamides. Schlegel et al. (1961) found that urea, in concentrations present in normal concentrated urine, was bactericidal to a number of gram-negative, pathogenic, urinary-tract bacteria. Because of its ability to dissolve necrotic tissue, urea has been used in combination with the sulfonamides in the treatment of wound infections.

Urea or carbamide has the structure,

$$O=C\begin{array}{l}\diagup NH_2 \\ \diagdown NH_2\end{array}$$

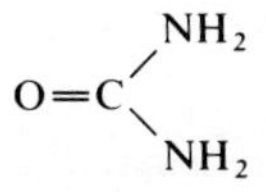

Urea

Urethane is the ethyl ester of carbamic acid:

$$O=C\begin{array}{l}\diagup NH_2 \\ \diagdown OH\end{array} \qquad O=C\begin{array}{l}\diagup NH_2 \\ \diagdown OC_2H_5\end{array}$$

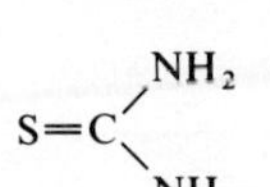

Carbamic acid   Urethane

Weinstein (1946) reported that the replacement of the $=O$ group in urea by $=S$, to give thiourea,

$$S=C\begin{array}{l}\diagup NH_2 \\ \diagdown NH_2\end{array}$$

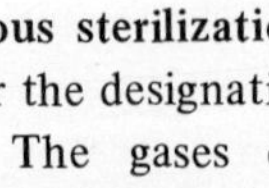

markedly increased its antibacterial activity.

For more information: Klotz and Mellody (1949).

Gaseous sterilization Gaseous sterilization has been practiced for years under the designation of fumigation.

The gases commonly used were sulfur dioxide, chlorine, and formaldehyde. These gases were widely employed for the fumigation of rooms occupied by the sick. Since sulfur dioxide and chlorine were found to be highly damaging to the materials within rooms, they were eventually replaced by formaldehyde.

Formaldehyde In order to sterilize enclosed spaces with formaldehyde, the following precautions must be observed: (1) The temperature must be high, preferably above 21°C; and (2) the relative humidity must be high, preferably near saturation. Disadvantages to its use are: (1) it is mainly a surface disinfectant, penetrating only slightly into porous materials; and (2) residual gas persists even after prolonged airing.

Formaldehyde gas for room sterilization may be generated as follows: A weighed amount of potassium permanganate crystals is placed in a deep spacious dish or bucket in the center of the room. All openings in the room, such as around windows and under doors, are sealed, except for an escape door for the operator. A measured volume of formalin solution is poured over the crystals. This results in the immediate evolution of a considerable amount of heat followed by volatilization of the formalin. The operator must leave the room immediately to avoid contact with the irritating gas. The escape door is then sealed from the outside.

Taylor et al. (1969) prepared formaldehyde gas by heating paraformaldehyde, a polymer of formaldehyde in powder form. The liberated formaldehyde gas was passed into a chamber containing the test organism. They found the range of its use to be about equal to that of formalin. The absence of any failure in the tests conducted indicated that the dry gas liberated from depolymerization of paraformaldehyde might be a more effective sterilizing agent than vaporized formalin.

Young et al. (1970) introduced vaporized formaldehyde into the sealed buildings of textile mills contaminated with anthrax spores. Pretreatment rates of surface contamination with spores were 37 percent in the initial processing area and 12.5 percent in the spinning area. Immediately after formaldehyde treatment the figures dropped to 8 percent and 1 percent respectively. Six months later the figures dropped to 1 percent and 0 percent.

The gas becomes considerably more effective if introduced into a vacuum chamber from which the contained air is almost entirely removed. In this manner the surface effect is eliminated and deep penetration of the gas is obtained, followed in most cases by complete sterilization of the contents.

The practice of fumigating rooms with formaldehyde has been largely abandoned. The most important reason for this is that fumigation as usually practiced often fails to sterilize. Also, newer gases have been introduced which are superior to formaldehyde.

However, formaldehyde is a very efficient germicide, a 5 percent solution of formalin (commercial formalin is a 37 percent solution of the gas in water) being capable of destroying anthrax spores in 1 to 2 hr.

For more information: Alder et al. (1966); Dickinson and Wagg (1967); Lloyd and Fotor (1955); Phillips (1968); Salle and Korzenovsky (1942); Spiner and Hoffman (1971); Trujillo and David (1972).

Ethylene oxide Ethylene oxide has the following structure:

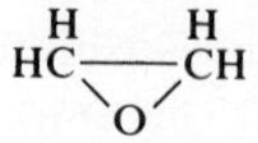

Ethylene oxide

It is a colorless gas at ordinary temperatures. Below 10.8°C it may be kept as a liquid. The gas is toxic if inhaled and causes blisters if its vapors are in contact with the skin for any period of time.

Ethylene oxide is free from many of the objections to formaldehyde. It is readily obtained and released in the pure state. It does not polymerize or condense on outer surfaces, exhibits deep penetration rather than just a surface effect, and is quickly dissipated by simple airing of treated materials.

The most important hazard of ethylene oxide is not its toxicity but its extreme flammability when mixed with air. However, this may be overcome by mixing the compound with inert gases such as carbon dioxide or the fluorinated hydrocarbons (freons). A mixture of 10 percent ethylene oxide and 90 percent carbon dioxide is marketed under the trade name of Carboxide. The carbon dioxide or the freons play no part in the activity of ethylene oxide other than to prevent flammability. If the gas is used in a vacuum chamber, inert diluents are not necessary.

Ethylene oxide is of importance in the sterilization of items which are injured by heat and excessive moisture. Advantages to its use are: few materials are damaged, it is effective at room temperature and at low humidities, it has little residual effect, it is bactericidal rather than bacteriostatic, it is effective against gram-positive and gram-negative bacteria and spores, and it offers good penetration. Disadvantages to its use are: slowness of action (hours being required rather than minutes), toxicity, costliness, flammability, and necessity of special equipment for handling.

For more information: Auerswald and Doleschel (1962); Brown and Fuerst (1963); Bruhin et al. (1968); Diding et al. (1968); Ernst and Doyle (1968); Gilbert et al. (1964); Kereluk et al. (1970*a*, *b*); Phillips (1968); Rendell-Baker and Roberts (1970); Shull (1963).

β-Propiolactone This compound has the following structure:

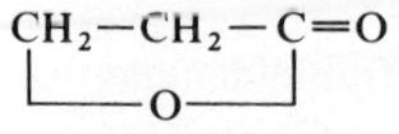

β-Propiolactone

It is a colorless, pungent liquid at ordinary temperatures. Under normal atmospheric pressure, its boiling point is 162.3°C. It may be atomized into the air by means of a special type of dispenser. Like formaldehyde, β-propiolactone requires a high relative humidity (70 percent or above) to be rapidly effective. It is not as effective as ethylene oxide in penetrating porous materials. The vapors are rapidly dissipated from enclosed spaces by simple airing.

β-Propiolactone is effective against gram-positive bacteria and spores, gram-negative organisms, molds, viruses, and bacteriophages.

The sterilizing action of β-propiolactone may be caused by its reaction with –SH and –S–S– groups of proteins, although it also reacts with –OH, $-NH_2$, and –COOH radicals.

The compound is slowly hydrolyzed in aqueous solution being converted into the inactive β-hydroxypropionic acid:

$$\underset{\beta\text{-Propiolactone}}{\underbrace{CH_2{-}CH_2{-}C{=}O}_{\text{└── O ──┘}}} + H_2O \longrightarrow \underset{\beta\text{-Hydroxypropionic acid}}{CH_2OH{-}CH_2{-}COOH}$$

For more information: Ball et al. (1961); Bruch (1961*a*, *b*); Hagiwara (1961); Hazeu and Hueck (1965); Hoffman (1968); Hoffman et al. (1966); Holme (1965); Phillips (1968).

Evaluation of germicides

The method generally employed for the evaluation of germicides is to rate them according to their phenol coefficients. In addition to this, a number of newer methods have been proposed which are designed to measure the toxicity of germicides for tissue as well as for the test bacteria.

Phenol-coefficient method The phenol coefficient may be defined as the killing power of a germicide against a test organism as compared to that of phenol under similar conditions.

The conditions which must be specified include (1) time of action of germicide, (2) temperature of test, (3) presence and amount of organic matter, (4) organism used in test, (5) age of culture, (6) composition and reaction of transfer medium, (7) proportion of disinfectant to culture, and (8) temperature and time of incubation of transfer tubes. Variations in one or more of the variables will affect the final result. Therefore, it is of utmost importance to specify the procedure followed in making the test; otherwise the final result will be of doubtful value.

The standard phenol-coefficient method in this country is that recommended by the Association of Official Agricultural Chemists (AOAC). This test in one form or another is universally employed for testing and rating disinfectants.

Time and temperature In general, germicidal action is increased with time. This means that a higher dilution may be employed with an increase in the period of action. This applies also to temperature. An increase in temperature increases the effectiveness of a germicide, making higher dilutions possible. Germicides are not affected to the same degree by an increase in time and temperature, and for this reason, no general rule can be made.

An important exception to the rule that germicidal action is increased with time is iodine. This germicide is a vigorous oxidizing agent and acts almost immediately when placed in contact with bacteria.

Organic matter Probably all germicides are reduced in activity in the presence of organic matter, although some are affected more than others. This is especially true in the presence of proteins, amino acids, and compounds of a similar nature. Results of the evaluation of germicides in aqueous solutions are quite different from those obtained when organic matter is added. The kind and amount of organic matter must be mentioned in reporting the efficiencies of germicidal substances.

Organism Germicides vary considerably in their action on different bacterial species. Some are more effective against gram-positive than against gram-negative bacteria, and vice versa. Still others display approximately the same degree of toxicity toward both groups of organisms. The name of the organism used in the test must be mentioned. The organisms generally used are *S. aureus* (gram-positive) and *Salmonella typhosa* (gram-negative).

Age of the culture In general, old organisms are more resistant to adverse environmental conditions than young cells. In practically all procedures for evaluating germicides, 24-hr cultures are specified. This precaution must be observed in order that constant and comparable results be obtained.

Composition and reaction of medium Variations in composition and pH of culture media also affect the final results. Goetchius (1950) employed beef extract from three different sources and obtained wide variations in the phenol coefficients. Klarmann and Wright (1945) obtained similar results, which led them to propose the use of semisynthetic media for more constant results in phenol-coefficient determinations. In general, an organism is more resistant to adverse conditions at its optimum pH. A change in the reaction of the medium on either side of the optimum pH increases the susceptibility of an organism to a germicide.

Proportion of disinfectant to culture A parallelism exists between the number of organisms employed in the test and the smallest amount of germicide required to destroy them. If the number of organisms is increased or decreased, the concentration of germicide required to destroy them is likewise increased or decreased.

Method The official AOAC method is as follows: A series of dilutions of phenol and germicide to be compared is prepared in sterile distilled water contained in test tubes measuring 25 mm in diameter and 150 mm in length. Each tube must contain not more than 5 ml of germicidal dilution. The tubes are placed in a rack in a water bath, previously adjusted to a temperature of 20°C, and allowed to remain for at least 5 min in order to bring the temperature of the germicidal dilutions to that of the water bath.

The test organism should be transferred daily for five successive days previous to use. A 24-hr culture must be employed in the test. The culture is shaken vigorously to break up small clumps of bacteria and is then placed in the water bath for 15 min to permit large suspended particles to settle out. One-half milliliter of culture is pipetted into each dilution of the germicide. At intervals of 5, 10, and 15 min, a 4-mm loopful of material is removed from each tube and transferred to a corresponding tube containing 10 ml of broth. The subculture tubes are incubated at 37°C for 48 hr. If the germicide is suspected of being bacteriostatic, the subculture tubes should be incubated for a longer period of time.

If mercurials, silver preparations, dyes, or other compounds exhibiting strong bacteriostatic properties are tested, it is necessary (1) to make secondary subcultures from the first subculture tubes immediately after the test has been completed, or (2) to make the first transfers to 100 ml amounts of broth contained in flasks, or (3) to make transfers to broth containing substances that combine with or destroy the germicidal agent. For example, bacteria treated with mercuric chloride contain mercury ions (Hg^{++}) adsorbed to their cell walls. In this condition, the bacteria are not necessarily killed but merely prevented from multiplying. The number of Hg^{++} may be insufficient to produce death but only a bacteriostatic effect. Sodium thioglycollate contains a sulfhydryl group that reacts with Hg^{++}. If mercury-treated organisms in a state of bacteriostasis are transferred to broth containing sodium thioglycollate, the germicide is removed from the cells by the sulfhydryl groups permitting the bacteria to multiply.

The phenol coefficient is calculated by dividing the highest dilution of germicide killing the test organism in 10 min but not in 5 min by the corresponding dilution of phenol. For example, in Table 17 the phenol coefficient would be 350/90 = 3.89. This means that germicide A is 3.89 times more effective than phenol.

If *Salmonella typhosa* was used as the test organism, the value is referred to as the *S. typhosa* phenol coefficient.

Departures from the test The phenol-coefficient test was originally designed for comparing the toxicity of phenol with phenol-like compounds. The test is now used to compare phenol with compounds which are totally unlike phenol in composition and mode of action. Phenol coefficients so obtained do not represent true comparisons of the germicides with phenol.

Table 17 Killing dilutions of phenol and germicide A for Salmonella typhosa *at different time intervals*

		Time interval, min.		
Germicide	*Dilution*	*5*	*10*	*15*
Phenol	1 : 70	–	–	–
	1 : 80	–	–	–
	1 : 90	+	–	–
	1 : 100	+	+	+
Germicide *A*	1 : 325	–	–	–
	1 : 350	+	–	–
	1 : 375	+	+	–
	1 : 400	+	+	+

Also, investigators have modified the test to favor the compounds being compared. For example, water is employed in the official test for the preparation of dilutions. Yet alcohol has been substituted for water to improve the germicidal potencies of compounds under examination. Another departure has been the use of water containing alkali for dissolving compounds which are insoluble in water.

A major criticism of the phenol-coefficient method is the difficulty of converting the phenol coefficient values into safe use-dilutions. Safe use-dilutions are of particular value where mainly surface effects are desired.

Germicides are widely used in: (1) restaurants for sterilizing dishes, utensils, cooking equipment, etc.; (2) in dairy plants and on dairy farms for treatment of milking and other equipment; (3) slaughterhouses for food-processing machinery; in breweries, wineries, bakeries, and bottling plants; etc. The object, of course, is to destroy various types of organisms likely to be present.

The efficiencies of chemicals employed in such operations depend directly upon the extent of the precleaning operations in reducing the presence of organic matter to relatively low levels. Also speed of action of the chemicals is of utmost importance.

Because of the deficiency of the phenol coefficient test, another and newer procedure, known as the use-dilution method, is employed where mainly surface disinfection on inanimate objects is desired.

For the official AOAC method of determining the use-dilution concentrations of germicides the reader is referred to the work of Bass and Stuart (1968).

For more information: Duffett et al. (1961); Shaffer and Stuart (1968); Stuart (1968).

Limitation of the phenol-coefficient test A limitation of the test is that it is of doubtful value for determining the efficiency of a germicide for clinical application. A phenol coefficient attempts to compare the toxicity of a germicide against a test organism but gives no information as to its effect on living tissue.

Tissue-toxicity test A number of methods have been proposed for comparing the effects of germicides for living tissue cells and bacteria.

Some have employed the inhibition of phagocytosis as a criterion of tissue toxicity. Others have tested the effect of germicides on the infected chorioallantoic membrane of the developing chick embryo. Still others have (1) swabbed the tails of mice with a broth suspension of the test organism and allowed the culture to dry, (2) dipped the tails in a solution of the germicide under examination, (3) allowed the solution to dry, and (4) cut off the tips of the tails and inserted them into the peritoneal cavities through small incisions. Survival of the animals indicated the effectiveness of the germicide.

Salle and Lazarus (1935), Salle and Catlin (1947), and Salle (1955) tested germicides for their effect on chick heart-tissue fragments as well as

Table 18 Toxicity of germicides for embryonic chick heart tissue and Staphylococcus aureus *at 37°C*

	Highest killing dilution for tissue and bacteria and corresponding toxicity index		
Germicide	*Tissue (A)*	*Bacteria (B)*	*Toxicity index, A/B*
Iodine, tincture	1 : 4000	1 : 20,000	0.2
Mercresin, tincture	1 : 9000	1 : 19,000	0.5
Merthiolate, tincture	1 : 15,000	1 : 4500	3.3
Metaphen, tincture	1 : 12,000	1 : 1200	10.0

on bacteria. A number known as the toxicity index was calculated from the results, which was defined as the ratio of the highest dilution of germicide required to kill the tissue fragments to the highest dilution required to kill the test organism in the same period of time. An index smaller than 1 meant that the germicide was more toxic to the bacteria than to the tissue fragments; an index larger than 1 meant that the germicide was more toxic to the tissue fragments than to the bacteria. The smaller the index, the more nearly perfect the germicide.

Several germicides tested by this technique are given in Table 18. Iodine exhibited the highest degree of efficiency, combining low tissue toxicity with high germicidal potency against *S. aureus*.

In a later communication, Salle (1961) recommended single tissue cells instead of tissue fragments for greater accuracy in the test. By this procedure each cell is in contact with the germicide, whereas in tissue fragments only the outside cells are chiefly affected, depending upon the degree of penetration of the germicide.

For more information: Ortenzio and Stuart (1968).

References

Alder, V. G., et al.: Disinfection of heat-sensitive material by low-temperature steam and formaldehyde, *J. Clin. Pathol.*, **19**:83, 1966.

American Medical Association: "New and Nonofficial Drugs," Philadelphia, J. B. Lippincott Company, 1964.

Andersen, I., and E. Vad: The influence of electric fields on bacterial growth, *Intern. J. Bioclimatol. Biometeor.*, **9**:211, 1965.

Anderson, D. L., and S. G. Bradley: The action of ultrasonic vibrations on actinophages, *J. Gen. Microbiol.*, **37**:67, 1964.

Anderson, T. F., et al.: The relative sensitivities of bacterial viruses to intense sonic vibration, *Science*, **108**:18, 1948.

Anellis, A., et al.: Resistance of *Bacillus coagulans* spores to gamma rays. Application of the multiple tube probability method, *Food Res.*, **25**:285, 1960.

Auerswald, W., and W. Doleschel: On the sterilizing effect of ethylene oxide on virus suspended in protein solutions, *Med. Exp.*, **6**:193, 1962.

Baker, R. J.: Characteristics of chlorine compounds, *J. Water Pollution Control Federation*, **41**:482, 1969.

Ball, E. L., et al.: Sterilization of regenerated collagen sutures with β-propiolactone, *Appl. Microbiol.*, **9**:269, 1961.

Bannan, E. A., and L. F. Judge: Bacteriological studies relating to handwashing. I, *Am. J. Pub. Health*, **55**:915, 1965.

Barron, E. S. G., and S. Dickman: Studies on the mechanism of action of ionizing radiations. II. Inhibition of sulfhydryl enzymes by alpha, beta, and gamma rays, *J. Gen. Physiol.*, **32**:595, 1949.

Bass, G. K., and L. S. Stuart: Methods of testing disinfectants. In "Disinfection, Sterilization, and Preservation," edited by C. A. Lawrence and S. S. Block, Philadelphia, Lea & Febiger, 1968.

Berman, R. E., and R. A. Knight: Evaluation of hand antisepsis, *Arch. Environ. Health*, **18**:781, 1969.

Bernarde, M. A., et al.: Kinetics and mechanism of bacterial disinfection by chlorine dioxide, *Appl. Microbiol.*, **15**:257, 1967.

Black, A. P., et al.: The disinfection of swimming pool waters. Part I. Comparison of iodine and chlorine as swimming pool disinfectants, *Am. J. Pub. Health*, **60**:535, 1970*a*; II. A field study of the disinfection of public swimming pools, *ibid.*, **60**:740, 1970*b*.

Bosund, I.: The action of benzoic and salicylic acids on the metabolism of microorganisms. In "Advances in Food Research," New York, Academic Press, Inc., vol. 11, 1962.

Boucher, R. M. G., and M. A. Pisano: Sterilizing effect of high intensity airborne sound & ultrasound, *Ultrasonics*, **4**:199, 1966.

Bradley, S. G.: Photosensitization of an actinophage by heteroanthracenes, *Proc. Soc. Exp. Biol. Med.*, **122**:877, 1966.

Brewer, J. H., Mercurials: Inorganic and organic. In "Disinfection, Sterilization, and Preservation," edited by C. A. Lawrence and S. S. Block, Philadelphia, Lea & Febiger, 1968.

Bridges, B. A., and T. Horne: The influence of environmental factors on the microbicidal effect of ionising radiations, *J. Appl. Bact.*, **22**:96, 1959.

Brown, B. L., and R. Fuerst: Ethylene oxide sterilization of tissue culture media, *Science*, **142**:1654, 1963.

Bruch, C. W.: Decontamination of enclosed spaces with beta-propiolactone vapor, *Am. J. Hyg.*, **73**:1, 1961*a*.

——: Gaseous sterilization. In "Annual Review of Microbiology," **15**:245, 1961*b*.

Bruhin, H., et al.: Möglichkeiten und grenzen der äthylenoxidsterilisation, *Centr. Bakt., I, Orig.*, **208**:563, 1968.

Carroll, B.: The relative germicidal activity of triiodide and diatomic iodine, *J. Bact.*, **69**:413, 1955.

Chang, S. L., and J. C. Morris: Elemental iodine as a disinfectant for drinking water, *Ind. Eng. Chem.*, **45**:1009, 1953.

Chaplin, C. E.: Bacterial resistance to quaternary ammonium disinfectants, *J. Bact.*, **63**:453, 1952.

Darmady, E. M., et al.: Radiation sterilization, *J. Clin. Pathol.*, **14**:55, 1961.

Deasy, P. B., et al.: Influence of γ-irradiation and heating in the presence of a bactericide on the inactivation of *Bacillus subtilis* spores, *Appl. Microbiol.*, **16**:810, 1968.

Dickinson, J. C., and R. E. Wagg: Use of formaldehyde for the disinfection of hospital woolen blankets in laundering, *J. Appl. Bact.*, **30**:340, 1967.

Diding, N., et al.: Ethylene oxide treatment of crude drugs. I, *Acta Pharmaceut. Suecica*, **5**:177, 1968.

Duffett, N. D., et al.: Criteria for the selection of germicides, *Am. J. Pub. Health*, **51**:1054, 1961.

Dychdala, G. R.: Chlorine and chlorine compounds. In "Disinfection, Sterilization, and Preservation," edited by C. A. Lawrence and S. S. Block, Philadelphia, Lea & Febiger, 1968*a*; Acid-anionic surfactant sanitizers, *ibid.*, 1968*b*.

Edebo, L.: Lysis of bacteria. 1. Influence of enzyme inhibitors on sonic lysis, *Acta Pathol. Microbiol. Scand.*, **52**:372, 1961*a*; The sensitivity of *Proteus mirabilis* to sonic oscillation in various phases and growth-stages, *ibid.*, **53**:202, 1961*b*.

Ehret, C. F., et al.: Thermal annealment and nitric oxide effects on free radicals in x-irradiated cells, *Science*, **132**:1768, 1960.

Eisman, P. C.: Disinfectants, antiseptics, and preservatives, *Trans. N.Y. Acad. Sci.*, **23**:709, 1961.

Ericson, C., et al.: Removal of the superficial bacterial flora of the hands: A comparison between different antibacterial preparations and soap, *Acta Chir. Scand.*, **134**:7, 1968.

Ernst, R. R., and J. E. Doyle: Sterilization with gaseous ethylene oxide: A review of chemical and physical factors, *Biotechnol. Bioengineer.*, **10**:1, 1968.

Fildes, P.: A rational approach to research in chemotherapy, *Lancet*, **238**:955, 1940.

Fischer, E., et al.: On the mechanism of bacteriostasis with triphenylmethane dyes, *J. Bact.*, **48**:439, 1944.

Fitzgerald, G. P., and M. E. Der Vartanian: Factors influencing the effectiveness of swimming pool bactericides, *Appl. Microbiol.*, **15**:504, 1967.

Frank, G. H., and J. W. Foster: Photodynamic inactivation of Newcastle disease virus with acridine orange, *Appl. Microbiol.*, **15**:1452, 1967.

Gershenfeld, L.: Iodine. In "Disinfection, Sterilization, and Preservation," edited by C. A. Lawrence and S. S. Block, Philadelphia, Lea & Febiger, 1968.

——and B. Witlin: Free halogens: A comparative study of their efficiencies as bactericidal agents, *Am. J. Pharm.*, **121**:95, 1949.

Gilbert, G. L., et al.: Effect of moisture on ethylene oxide sterilization, *Appl. Microbiol.*, **12**:496, 1964.

Ginoza, W.: Inactivation of viruses by ionizing radiation and by heat. In "Methods in Virology," New York, Academic Press, Inc., **14**:139, 1968.

Goetchius, G. R.: Testing germicides, *Soap San. Chem.*, **26**:131, 1950.

Gordon, L. E.: New horizons in sterilization. In "The Becton, Dickinson Lectures on Sterilization," Jersey City, N.J., Seton Hall College of Medicine and Dentistry, 1959.

Greiff, D., et al.: The effects of x-rays and beta rays (tritium) on the growth of *Rickettsia mooseri* and *Rickettsia akari* in embryonate eggs, *J. Exp. Med.*, **111**:841, 1960.

Gump, W. S., and G. R. Walter: The bis-phenols. In "Disinfection, Sterilization, and Preservation," edited by C. A. Lawrence and S. S. Block, Philadelphia, Lea & Febiger, 1968.

Hagiwara, S.: Inactivation of phages by β-propiolactone, *Sapporo Med. J.*, **20**:16, 1961.

Harris, J. O., et al.: A study of the location of adsorbed mercuric ions in *Escherichia coli*, *J. Bact.*, **68**:745, 1954.

Harrison, A. P., Jr.: Survival of bacteria. In "Annual Review of Microbiology," **21**:143, 1967.

Harvey, E. N., and A. L. Loomis: The destruction of luminous bacteria by high frequency sound waves, *J. Bact.*, **17**:373, 1929.

Hazeu, W., and H. J. Hueck: The use of β-propiolactone for the sterilization of heat-labile materials, *Antonie van Leeuwenhoek, J. Microbiol. Serol.*, **31**:295, 1965.

Hill, E. C., and G. O. Phillips: The inactivation of *Bacillus subtilis* spores in penicillin by gamma radiation, *J. Appl. Bact.*, **22**:8, 1959.

Hodes, L. J., and H. C. Stecker: The salicylanilides and carbanilides. In "Disinfection, Sterilization, and Preservation," edited by C. A. Lawrence and S. S. Block, Philadelphia, Lea & Febiger, 1968.

Hoffman, C. E., and O. Rahn: The bactericidal and bacteriostatic action of crystal violet, *J. Bact.*, **47**:177, 1944.

Hoffman, R. K.: Effect of bacterial cell moisture on the sporicidal activity of β-propiolactone vapor, *Appl. Microbiol.*, **16**:641, 1968.

Hoffman, R. K., et al.: β-propiolactone vapor decontamination, *Appl. Microbiol.*, **14**:989, 1966.

Holme, T.: Sterilization of microbiological media with β-propiolactone, *Biotechnol. Bioengineer.* **7**:129, 1965.

Howard-Flanders, P., and P. Jockey: Factors in the inactivation of T2 bacteriophage and mono-complex by ionizing radiations. 1. The effects of oxygen, *Intern. J. Radiation Biol.*, **2**:361, 1960.

Hughes, D. E., and W. L. Nyborg: Cell disruption by ultra-sound, *Science*, **138**:108, 1962.

Huhtanen, C. N.: Effect of low-frequency ultrasound and elevated temperatures on isolation of bacteria from raw milk, *Appl. Microbiol.,* **16**:470, 1968.

Hutchinson, F., and S. S. Easter, Jr.: A difference between biological effects of gamma rays and heavy ions, *Science,* **132**:1311, 1960.

Jacobs, S. E., and M. J. Thornley: The lethal action of ultrasonic waves on bacteria suspended in milk and other liquids, *J. Appl. Bact.,* **17**:38, 1954.

Johannesson, J. K.: Studies of the action of monobromamine on *Escherichia coli, New Zealand J. Sci.,* **2**:499, 1959.

Jones, L. A., et al.: Sporicidal activity of sodium hypochlorite at subzero temperatures, *Appl. Microbiol.,* **16**:787, 1968.

Jungermann, E., et al.: Comparative evaluation of antibacterial soaps, *J. Am. Oil Chem. Soc.,* **44**:232, 1966.

Kempe, L. L.: Combined effects of heat and radiation in food sterilization, *Appl. Microbiol.,* **3**:346, 1955.

——— et al.: Gamma ray sterilization of canned meat previously inoculated with anaerobic bacterial spores, *Appl. Microbiol.,* **2**:330, 1954.

Kereluk, K., et al.: Microbiological aspects of ethylene oxide sterilization. II. Microbial resistance to ethylene oxide, *Appl. Microbiol.,* **19**:152, 1970*a*; III. Effects of humidity and water activity on the sporicidal activity of ethylene oxide, *ibid.,* **19**:157, 1970*b*.

Kinsman, S.: Radiation simplified, *Calif. Health,* **18**:105, 1961.

Kiortsis, M.: Combined lethal effect of zinc chloride and gamma irradiation on *Bacillus megaterium, Nature,* **217**:746, 1968.

Klarmann, E. G., and E. S. Wright: Synthetic and semi-synthetic media for disinfectant testing, *Soap Sanit. Chemicals,* January 1945, p. 113.

Klebanoff, S.: Iodination of bacteria: A bactericidal mechanism, *J. Exp. Med.,* **126**:1063, 1967.

Klotz, I. M., and M. Mellody: The inhibition of growth of *Escherichia coli* by some derivatives of urea, *J. Bact.,* **57**:477, 1949.

Koh, W. Y., et al.: Relative resistances of microorganisms to cathode rays. I, *Appl. Microbiol.,* **4**:143, 1956.

Koski, T. A., et al.: Comparison of chlorine, bromine, and iodine as disinfectants for swimming pool water, *Appl. Microbiol.,* **14**:276, 1966.

Kroeger, A. V., and J. E. Kempf: Inactivation of the influenza virus by low voltage Roentgen rays, *J. Bact.,* **77**:237, 1959.

Lawrence, C. A.: Definition of terms. In "Disinfection, Sterilization, and Preservation," edited by C. A. Lawrence and S. S. Block, Philadelphia, Lea & Febiger, 1968*a*; Quaternary ammonium surface-active disinfectants, *ibid.,* 1968*b*.

Lee, J. S.: Inactivation of *Vibrio parahaemolyticus* in distilled water, *Appl. Microbiol.,* **23**:166, 1972.

Leonard, R. R.: Prevention of superficial cutaneous infections, *Arch. Dermatol.,* **95**:520, 1967.

Licciardello, J. J., and J. T. R. Nickerson: Effect of radiation environment on the thermal resistance of irradiated spores of *Clostridium sporogenes* P. A. 3679, *J. Food Sci.,* **27**:211, 1962.

Lloyd, R. S., and M. J. Foter: Efficiency of dry heat and formaldehyde in sterilizing used bedding, *Pub. Health Rep.*, **70**:810, 1955.

MacKenzie, A. R.: Effectiveness of antibacterial soaps in a healthy population, *J. Am. Med. Assoc.*, **211**:973, 1970.

Marr, A. G., and E. H. Cota-Robles: Sonic disruption of *Azotobacter vinelandii*, *J. Bact.*, **74**:79, 1957.

Moffet, H. L., and T. Williams: Bacteria recovered from distilled water and inhalation therapy equipment, *Am. J. Diseases Children*, **114**:7, 1967.

Molnar, N. M.: Antimicrobial action of soap germicide mixtures on skin, *J. Soc. Cosmetic Chemists*, **20**:103, 1969.

Morgan, B. H., and J. M. Reed: Resistance of bacterial spores to gamma irradiation, *Food Res.*, **19**:357, 1954.

Morton, H. E.: Alcohols. In "Disinfection, Sterilization, and Preservation," edited by C. A. Lawrence and S. S. Block, Philadelphia, Lea & Febiger, 1968.

——— et al.: The bacteriostatic and bactericidal actions of some mercurial compounds on hemolytic streptococci, *J. Am. Med. Assoc.*, **136**:36, 1948.

Newton, B. A.: The mechanism of the bactericidal action of surface active compounds: A summary, *J. Appl. Bact.*, **23**:345, 1960.

Ortenzio, L. F., and L. S. Stuart: Methods of testing antiseptics. In "Disinfection, Sterilization, and Preservation," edited by C. A. Lawrence and S. S. Block, Philadelphia, Lea & Febiger, 1968.

Pepper, R. E., et al.: Relative resistances of microorganisms to cathode rays, *Appl. Microbiol.*, **4**:149, 1956.

Phillips, C. R.: Gaseous sterilization. In "Disinfection, Sterilization, and Preservation," edited by C. A. Lawrence and S. S. Block, Philadelphia, Lea & Febiger, 1968.

Pisano, M. A., et al.: Sterilizing effects of high-intensity airborne sonic and ultrasonic waves, *Appl. Microbiol.*, **14**:732, 1966.

Powers, E. L., and B. F. Kaleta: Reduction of radiation sensitivity of dry bacterial spores with hydrogen sulfide, *Science*, **132**:959, 1960.

——— et al.: An oxygen effect in dry bacterial spores and its temperature dependence, *Exp. Cell. Res.*, **17**:550, 1959.

——— et al.: Storage, transfer, and utilization of energy from x-rays in dry bacterial spores, *Radiation Res., Suppl. 2*, 1960*a*.

——— et al.: Oxygen and nitric oxide as modifiers of radiation injury in spores of *Bacillus megaterium*, *Proc. Natl. Acad. Sci. U.S.*, **46**:984, 1960*b*.

Prindle, R. F., and E. S. Wright: Phenolic compounds. In "Disinfection, Sterilization, and Preservation," edited by C. A. Lawrence and S. S. Block, Philadelphia, Lea & Febiger, 1968.

Pullman, J. E., and B. L. Reynolds: Some observations on the mode of action of phenol on *Escherichia coli*, *Australian J. Pharm.*, **46**: Sept. 30, 1965.

Rahn, O., and J. E. Conn: Effect of increase in acidity on antiseptic efficiency, *Ind. Eng. Chem.*, **36**:185, 1944.

Reitman, M., and H. R. Tribble, Jr.: Inactivation of Venezuelan equine encephalomyelitis virus by γ-radiation, *Appl. Microbiol.*, **15**:1456, 1967.

Rendell-Baker, L., and R. B. Roberts: Safe use of ethylene oxide sterilization in hospitals, *J. Intern. Anesthesia Res. Soc.*, **49**:919, 1970.

Roman, D. P., and R. C. Manring: Carbanilide and salicylanilide soap bacteriostats, *Develop. Ind. Microbiol.*, **8**:21, 1967.

Romans, I. B.: Silver compounds. In "Disinfection, Sterilization, and Preservation," edited by C. A. Lawrence and S. S. Block, Philadelphia, Lea & Febiger, 1968.

Rotman, B.: On the mechanism of sonic lysis of bacteria, *J. Bact.*, **72**:827, 1956.

Salle, A. J.: An improved tissue toxicity technique for the evaluation of germicidal substances, *Appl. Microbiol.*, **3**:63, 1955.

——: The comparative toxicities of germicides for mixtures of bacteria and single tissue cells in suspension, *Arch. Mikrobiol.*, **39**:116, 1961.

—— and B. W. Catlin: Profile evaluation of germicides, *J. Am. Pharm. Assoc., Sci. Ed.*, **36**:129, 1947.

—— and M. Korzenovsky: The effect of a vacuum on the destruction of bacteria by germicides, *Proc. Soc. Exp. Biol. Med.*, **50**:12, 1942.

—— and A. S. Lazarus: A comparison of the resistance of bacteria and embryonic tissue to germicidal substances. I. Merthiolate, *Proc. Soc. Exp. Biol. Med.*, **32**:665, 1935.

Schlegel, J. U., et al.: Bactericidal effect of urea, *J. Urol.*, **86**:819, 1961.

Shaffer, C. H., Jr., and L. S. Stuart: Methods of testing sanitizers and bacteriostatic substances. In "Disinfection, Sterilization, and Preservation," edited by C. A. Lawrence and S. S. Block, Philadelphia, Lea & Febiger, 1968.

Sharon, N., et al.: Psittacosis-infected cell system for studies on radiation damage, *J. Bact.*, **86**:1139, 1963.

Shull, J. J.: Microbiological aspects of ethylene oxide sterilization, *Bull. Parenteral Drug Assoc.*, **17**:9, 1963.

Slade, H. D., and J. K. Vetter: Studies on *Streptococcus pyogenes*. I. Observations on the microscopical and biological aspects of the disintegration and solubilization of a type 6 strain by sonic oscillation, *J. Bact.*, **71**:236, 1956.

Smirnoff, W. A., and M. Cantin: Effect of gamma irradiation on the growth rate of species of *Bacillus cereus* group, *J. Invert. Pathol.*, **9**:357, 1967.

Smith, C. R.: Alcohol as a disinfectant against the tubercle bacillus, *Pub. Health Reports*, **62**:1285, 1947.

Spiner, D. R., and R. K. Hoffman: Effect of relative humidity on formaldehyde decontamination, *Appl. Microbiol.*, **22**:1138, 1971.

Spink, W. W., et al.: Para-aminobenzoic acid production by staphylococci, *J. Exp. Med.*, **79**:331, 1944.

Stapleton, G. E., and M. S. Engel: Cultural conditions as determinants of sensitivity of *Escherichia coli* to damaging agents, *J. Bact.*, **80**:544, 1960.

Starr, J. E., and J. Judis: Mechanism of action of phenolic disinfectants. VIII, *J. Pharm. Sci.*, **57**:768, 1968.

Stuart, L. S.: Introduction to antimicrobial testing methods. In "Disinfection, Sterilization, and Preservation," edited by C. A. Lawrence and S. S. Block, Philadelphia, Lea & Febiger, 1968.

Stuy, J. H.: Studies on the radiation inactivation of microorganisms. VI. X-ray-induced breakdown of deoxyribonucleic acid in *Haemophilus influenzae* and in other bacteria, *J. Bact.*, **79**:707, 1960.

Taylor, L. A., et al.: Paraformaldehyde for surface sterilization and detoxification, *Appl. Microbiol.*, **17**:614, 1969.

Trujillo, R., and T. J. David: Sporostatic and sporocidal properties of aqueous formaldehyde, *Appl. Microbiol.*, **23**:618, 1972.

Voss, J. G.: Effect of inorganic cations on bactericidal activity of anionic surfactants, *J. Bact.*, **86**:207, 1963.

Weinstein, L.: Action of urea and some of its derivatives on bacteria. V. Antibacterial activity of methyl- and thiourea, *Proc. Soc. Exp. Biol. Med.*, **63**:506, 1946.

Wheaton, E., and G. B. Pratt: Radiation survival curves of *Clostridium botulinum* spores, *J. Food Sci.*, **27**:327, 1962.

Williams, A. R., and J. S. Slade: Ultrasonic dispersal of aggregates of *Sarcina lutea, Ultrasonics,* **9**:85, 1971.

Witmer, H., and D. Fraser: Photodynamic action of proflavine on coliphage T3, I. Kinetics of inactivation, *J. Virol.*, **7**:314, 1971*a*; II. Protection by L-cysteine, *ibid.*, **7**:319, 1971*b*; III. Damages to the deoxyribonucleic acid associated with Rx1 and Rx2, *ibid.*, **7**:323, 1971*c*.

Wood, R. W., and A. L. Loomis: The physical and biological effects of high-frequency sound waves of great intensity, *Phil. Mag.*, Ser. 7, **4**:417, 1927.

Woods, D. D.: The relation of para-aminobenzoic acid to the mechanism of the action of sulfanilamide, *Brit. J. Exp. Pathol.*, **21**:74, 1940.

Young, L. S., et al.: Vaporized formaldehyde treatment of a textile mill contaminated with *Bacillus anthracis, Arch. Environ. Health,* **20**:400, 1970.

Yamazaki, K.: Studies on the radioresistance of *Bacillus* spores. Pt. I, *Agr. Biol. Chem.*, **35**:1449, 1971.

Associations of bacteria 17

Rarely are organisms found growing as pure cultures in their natural habitats. Mixed cultures of two or more species are the general rule. Because of this fact, it is sometimes erroneous to conclude from laboratory studies on pure cultures the exact changes which organisms produce when growing in nature.

Associations may exist between (1) different species of bacteria; (2) bacteria and other classes of organisms such as yeasts, molds, algae, protozoa, and plants; and (3) different classes of organisms other than bacteria. Simple mixtures of two or more bacterial species may exist in which the organisms produce no effect whatsoever on each other but such an association is rare in nature.

Symbiosis

Symbiosis may be defined as the living together of two or more species of organisms for mutual benefit. Growth and multiplication in friendly associations are generally more vigorous than in cultures of each species existing alone.

Certain soil bacteria of the genus *Rhizobium* are found growing in tumors or nodules produced on the roots of plants belonging chiefly to the family *Leguminosae*. These organisms utilize free atmospheric nitrogen and synthesize it into organic compounds. The plants are furnished available nitrogen by the bacteria, and the bacteria derive their nutrients from the plant sap. Apparently a perfect symbiotic relationship exists.

Nurmikko (1954) found that in a medium which lacked phenylalanine (required by *Lactobacillus plantarum*) and pteroylglutamic acid (required by *Streptococcus faecalis*), neither organism grew alone. However, in a mixed culture, abundant growth of both bacteria occurred. Each organism synthesized the factor required by the other; i.e., phenylalanine was synthesized by *S. faecalis*, and pteroylglutamic acid by *L. plantarum*. Both organisms were benefited by the association. Since *S. faecalis* predominated in the early stages of the association and *L. plantarum* during the latter stages, it is questionable whether this can be considered a true symbiosis.

Bachenheimer and Bennett (1961) showed that the concentration of phenylmercuric lactate required to inhibit *Pseudomonas aeruginosa* was greatly increased in the presence of *Desulfovibrio desulfuricans*. The *P. aeruginosa* furnished *D. desulfuricans* with essential metabolites, and adjusted the oxidation-reduction potential of the medium to a more favorable level. The *D. desulfuricans* produced hydrogen sulfide which protected *P. aeruginosa* from the mercurial by removing it from the environment. The relationship appeared to be syntrophic.

Chian and Mateles (1968) inoculated a mixed glucose-lactose or glucose-butyrate medium with river water, incubated under continuous culture conditions, and obtained a population composed of a pseudomonad and a coliform. The glucose was used preferentially to the other carbon source, and the utilization of the secondary carbon compound was greatly reduced at high growth rates. At high growth rates the pseudomonad predominated whereas at low and moderate growth rates the coliform was dominant. A syntrophic relationship was demonstrated by the fact that the pseudomonad could not grow alone on the glucose-butyrate medium.

Many symbioses reported in the literature are misnomers in that apparently only one of the organisms in the associations is benefited. The favorable influence of an aerobe on the growth of an anaerobe may be mentioned. The aerobe reduces the oxygen tension and creates an environment suitable for the growth of the anaerobe. The anaerobe is benefited by the association, whereas the aerobe either is not affected or is harmed. This should be regarded as an example of commensalism or of antibiosis rather than of symbiosis. True examples of bacteria growing in symbiosis with other bacterial species where both are benefited by the association are rare in nature.

For more information: Ahmadjian (1967); Bryant et al. (1967); Bungay and Krieg (1966); Henry (1966, 1967); Kreig and Pelczar (1961).

Commensalism

Commensalism may be defined as the living together of two species, one of which is benefited by the association, whereas the other is neither benefited nor harmed.

An organism may not be able to grow in a certain substrate. If, however, a second organism is present that is capable of attacking the substrate with the production of a compound that is utilized by the first organism, growth will occur. Such an association is called commensalism.

Gale (1940) reported that *Escherichia coli* could decarboxylate arginine to agmatine, and ornithine to putrescine, but could not hydrolyze arginine to ornithine. On the other hand, *S. faecalis* could hydrolyze arginine to ornithine but could not decarboxylate arginine to agmatine, or ornithine to putrescine (Fig. 250).

In a medium containing arginine, neither *E. coli* nor *S. faecalis* acting separately could produce putrescine. However, when both organisms were

Fig. 251 Bacterial synergism. Left, Escherichia coli grown in sucrose broth; center, Staphylococcus aureus grown in sucrose broth; right, a mixture of E. coli and S. aureus grown in sucrose broth with gas formation.

be obtained, namely, the failure of a gas-producing organism to form gas when grown in association with another species.

The general tendency is for one member of a pair to inhibit or outgrow the other. This usually depends on which member is present in greater numbers at the start. This tendency may be due to the elaboration by one organism of waste products detrimental to the other, to an increase in the acidity of the medium, to a higher growth rate by one of the members, etc. Therefore, the presence or absence of gas does not necessarily mean that *E. coli* is or is not present in a water sample.

Another type of synergism, namely, the cooperative action of two drugs, such that the total effect is greater than the sum of the two effects taken independently, is discussed on page 612.

Antibiosis

An organism protects itself against its enemies in various ways. It may produce metabolic waste products which change the conditions in a medium, such as pH, osmotic pressure, and surface tension, making the environment unfavorable to the growth of less tolerant organisms. It may elaborate specific toxic substances which interfere with the metabolism of other organisms to such an extent that they are either killed or prevented from multiplying. These specific toxic substances are called antibiotics, and the phenomenon, antibiosis. Antibiosis may be defined as the living together of two organisms, one of which is distinctly injurious to the other and which may result finally in the death of the latter.

On an agar plate culture of *S. aureus*, Fleming (1929) obtained a mold contaminant that produced a green pigment and prevented bacterial growth for some distance around it. He cultivated the mold in a liquid medium and found that a filtrate of the culture had the power, even when greatly diluted, to prevent growth of a number of gram-positive pathogenic bacteria. Since the mold proved to be a species of *Penicillium*, he named the antibiotic penicillin. A crude preparation of penicillin was subsequently shown to be more effective against a number of infections than the sulfa drugs. The antibiotic was found to be so nontoxic that amounts far beyond the effective curative dose could be safely administered.

Dubos (1939) isolated from soil a spore-producing bacillus that was capable of destroying living gram-positive cocci. The organism was found to be *Bacillus brevis*, a large gram-positive rod similar to *B. subtilis*. He named the antibiotic gramicidin.

Since that time thousands of antibiotics have been isolated and studied. Some are useful clinically; others are not satisfactory for clinical application but more useful for other purposes.

The field of antibiotics offers unlimited possibilities in medicine. Powerful antibiotics, such as penicillin, have proven to be of such tremendous importance that an ever-increasing search is going on in the hope that agents superior to those now in existence might be isolated.

Sources of antibiotics The soil is perhaps the richest source of microorganisms which exhibit good antibiotic action. These organisms include principally the bacteria, streptomycetes, and molds. It has been said that a teaspoonful of soil contains hundreds of millions of bacteria, millions of streptomycetes, hundreds of thousands of mold spores, tens of thousands of protozoa, and hundreds of nematode worms. In such an environment a delicate organism like *P. notatum* needs protection comparable to a tank in human warfare. One of the properties possessed by microorganisms is to excrete substances which inhibit or destroy the growth of their neighbors.

Two techniques are commonly employed in detecting and isolating organisms that give evidence of antibacterial action. The first is based on the assumption that activity against a single test species is a sufficient criterion for the selection of antagonistic organisms. In this method a sample of soil is mixed with an appropriate melted agar medium at a suitable dilution, and the mixture is poured into petri dishes. After the agar has hardened, the plate is inverted and incubated until scattered colonies appear. The plate is then flooded with a suspension of the test organism and again incubated. The test species forms a solid growth on the surface of the agar except for clear zones of inhibition. The clear zones indicate the presence of colonies from the soil sample which are capable of producing an antibiotic active against the test bacteria. The soil colonies which are surrounded by clear zones are transferred to agar slants and held for further study. By employing a series of soil agar plates, each flooded with a different test species, it is possible to isolate forms which are antagonistic to a large number of organisms.

In the second method a culture of an organism is streaked across a segment of an agar plate, then incubated until visible growth develops. Suspensions of several test organisms are then streaked at right angles to the mature culture from its edge, and the plate is again incubated until growth of all test species appears. If an antibiotic substance is formed by the organism that is being screened, the test species are inhibited at varying distances from the original culture, the lengths of the cleared spaces of inhibition being proportional to their sensitivity to the active agent diffusing from the culture producing it.

For more information: Meyers et al. (1968).

Production of antibiotics in soils As has already been stated, the majority of organisms known to produce antibiotics are soil inhabitants. There has been considerable speculation about whether such organisms actually produce antibiotics in their natural environment. It is generally accepted that if organisms do produce antibiotics in soil, they do so in localized environments where there is a favorable food supply such as in the rhizosphere of certain plants.

Wright (1956) inoculated *Trichoderma viride* into an acid soil containing pieces of wheat straw and demonstrated the presence of high concentrations of the antibiotic gliotoxin. In 1955 Wright reported the production by *P. nigricans* of the antibiotic griseofulvin.

The studies indicated that organisms capable of producing antibiotics may have different requirements and that soil conditions favoring the production of one may not be suitable for others. *T. viride* produced gliotoxin in autoclaved soil even if unsupplemented, although the yield increased if the soil was organically enriched. The beneficial effect of autoclaving the soil was believed to be due to the release of nutrients and to the removal of antagonistic organisms. On the other hand, *P. nigricans* produced griseofulvin only when the soil was both autoclaved and supplemented.

For more information: Ayers and Papavizas (1963); Jefferys (1952); Nigrelli (1962); Park (1956); Stallings (1954); Stevenson (1954, 1956*a, b*); Stout (1962).

Methods of cultivation Apparently all antibiotic-producing organisms which have been studied must have free oxygen for normal metabolic activity. Three methods of cultivation are generally employed, all based on the introduction of an ample supply of air into the environment: (1) shallow surface cultivation in which the organisms grow on the surface of liquid media and form a firm mat. This type of growth is generally characteristic of molds; (2) shallow submerged cultivation in which air penetrates the media by diffusion or as the result of mechanical agitation; and (3) deep submerged cultures which are produced by mechanical agitation combined with the introduction, under pressure, of a continuous stream of air bubbles at the bottom. This last procedure produces a more abundant growth in a shorter period of time and requires less space. It is the commercial method employed universally for antibiotic production.

Measuring antibiotic activity Three methods are commonly employed for measuring antibiotic activity: (1) Oxford cylinder-plate method, (2) agar cup-plate method, and (3) paper-disk method.

Fig. 252 Measuring antibiotic activity by the Oxford cylinder-plate method. Melted agar was mixed with Sarcina lutea and poured into a petri dish. Equal volumes of subtilin solutions ranging in concentrations from 1:1000 to 1:1 million were pipetted into the cups. The largest zone was produced by the 1:1000 dilution of subtilin.

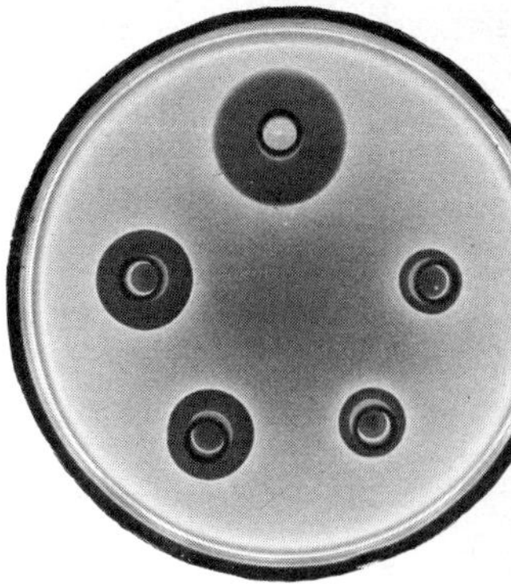

In the Oxford cylinder-plate method, melted agar is inoculated with the organism to be antagonized and poured into a petri dish. When firm, porous clay cylinders, open at both ends, are placed on the surface of the agar and filled with dilutions of the antibiotic. Around the cylinders will appear clear zones where growth has been prevented by diffusion of the antibiotic from the cylinders into the agar (Fig. 252). In the agar cup-plate method, a plate containing inoculated agar is prepared as before. One or more disks are cut in the agar and removed. The cups are filled with dilutions of the antibiotic, and the plate is incubated at the appropriate temperature. After incubation, clear zones will appear around the cups where growth has been prevented by diffusion of the antibiotic into the agar. In the paper-disk method, circles of filter paper about 10 mm in diameter are used instead of porcelain cylinders. Agar is prepared as before and poured into petri dishes. Paper disks are placed on the surface of the agar and saturated with measured volumes of antibiotic dilutions. After incubation clear zones will appear around the disks where growth has been prevented by the antibiotic (Fig. 253).

For more information: Czerkinsky et al. (1955); Simoncini et al. (1968).

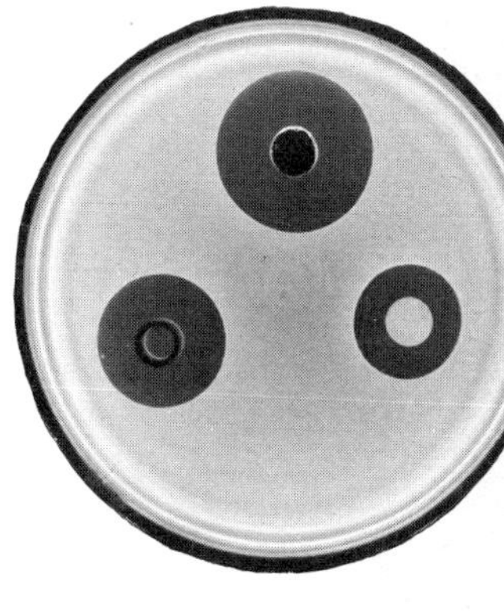

Fig. 253 Measuring antibiotic activity by various methods. Melted agar was mixed with Sarcina lutea and poured into a petri dish. A 1:1000 dilution of subtilin was used in all three tests. Top, agar cup-plate method; left, Oxford cylinder-plate method; right, paper-disk method.

Nature of antibiotics Antibiotics possess certain properties which distinguish them sharply from the usual chemical agents employed as antiseptics and germicides. An understanding of the nature, mode of action, and use of antibiotics is possible only after a consideration of their origin, chemical composition, physical properties, and biological effects in vitro and in vivo.

Antibiotics are produced by all classes of organisms such as bacteria, yeasts, molds, plants, and animals. The most useful antibiotics at present are produced by streptomycetes and other bacteria, and by molds. In most organisms the antibiotics are secreted outside of the cells and into the surrounding medium; in a few organisms the antibiotics are largely retained within the cells and must be separated by extraction.

Some antibiotics may show close similarities in their physical and chemical properties; others may differ widely. In support of the former may be mentioned the large number of penicillins which are produced by incorporating related precursors in the culture medium.

Some organisms are capable of producing more than one antibiotic. *B. brevis* produces gramicidin and tyrocidin; *Streptomyces lavendulae* produces streptothricin, lavendulin, streptolin, and chloromycetin; and *Aspergillus fumigatus* produces fumigacin, gliotoxin, and fumigatin.

Some antibiotics are produced by more than one organism. Penicillin is produced by *P. notatum, P. chrysogenum*, and *A. flavus*; streptomycin by *S. griseus* and *S. bikiniensis.*

The nature of the medium may or may not influence the effectiveness of an antibiotic. Some substrates have no appreciable effect, whereas others may reduce antibiotic activity or destroy it completely. An antibiotic may not be active in an acid medium; it may not be active in an alkaline environment; or it may be inactive in the presence of blood or blood serum.

Chemistry of some important antibiotics

Antibiotics show great variations in their chemical structures. The formulas of some important antibiotics are included here.

Penicillin The name penicillin actually covers a whole family of related compounds produced by *Penicillium* molds. The various penicillins have the same molecular core which consists of the β-lactam-thiazolidine ring to which is attached a different side-chain (R) or tail. The chemical structure of this side-chain distinguishes one kind of penicillin from another. By incorporating various precursors in the culture medium, many penicillins may be produced.

As an illustration, the use of corn-steep liquor as a culture medium caused the mold to produce penicillin G, a new product that was far more effective than any penicillin previously isolated. The substance in corn-steep liquor responsible for the appearance of penicillin G was β-phenyl-ethylamine. By adding phenylacetic acid to an appropriate medium, the same result was achieved. Manufacturers now add comparatively large amounts of this precursor to the medium to increase the yield of penicillin G.

The molecular core of penicillin has the following structure:

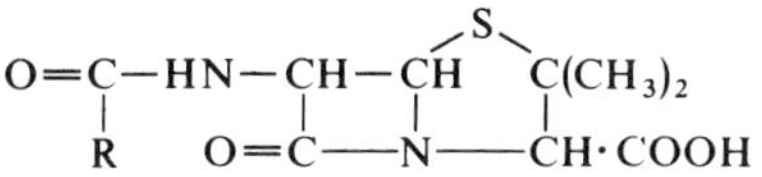

Penicillin

R may be any of the following:

Penicillin	*Side-chain R*
(G) Benzyl	
(X) *p*-Hydroxybenzyl	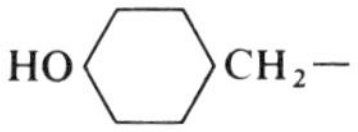
(F) 2-Pentenyl	$CH_3 \cdot CH_2 \cdot CH{=}CH \cdot CH_2-$
3-Pentenyl	$CH_3 \cdot CH{=}CH \cdot CH_2 \cdot CH_2-$
(Dihydro F) *n*-Amyl	$CH_3 \cdot CH_2 \cdot CH_2 \cdot CH_2 \cdot CH_2-$
(K) *n*-Heptyl	$CH_3 \cdot CH_2 \cdot CH_2 \cdot CH_2 \cdot CH_2 \cdot CH_2 \cdot CH_2-$

Other penicillins which have been produced either by adding various precursors to the fermentation medium or by synthesis include:

Penicillin	*Side-chain R*
α-Phenoxyethyl	$C_6H_5O(CH_3)CH-$
(V) Phenoxymethyl	$C_6H_5OCH_2-$
Phenethicillin	$C_6H_5OCH(CH_3)-$
Propicillin	$C_6H_5OCH(C_2H_5)-$
Phenbenicillin	$C_6H_5OCH(C_6H_5)-$
Rixapen	3,4-diCl$C_6H_3CH(OCH_3)-$
Methicillin	2,6-di$(CH_3O)C_6H_3-$
Nafcillin	2-$C_2H_5OC_{10}H_6-$
Oxacillin	5-CH_3-3-C_6H_5-4-$(C_3ON)-$
Cloxacillin	5-CH_3-3-(2-ClC_6H_4)-4-$(C_3ON)-$
Dicloxacillin	5-CH_3-3-(2,6-diClC_6H_3)-4-$(C_3ON)-$
Ampicillin	$C_6H_5CH(NH_2)-$

All penicillins are monocarboxylic acids which readily decarboxylate into biologically inactive derivatives. Benzyl penicillin (G) is the chief product produced by fermentation and the most widely used and marketed form of the antibiotic.

For more information: Brown et al. (1969); Childress (1967); Eykyn and Phillips (1969); Rose (1961).

Streptomycin This antibiotic is produced by *S. griseus.* It is a triacidic base having the following structure:

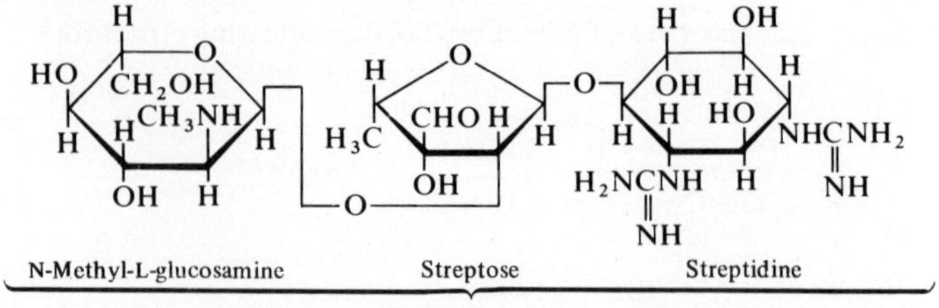

Streptomycin

Streptomycin is active against certain gram-positive and gram-negative bacteria, and the mycobacteria, especially *Mycobacterium tuberculosis.*

The antibiotic forms various salts of which the sulfate, the hydrochloride, the hydrochloride double salt with calcium chloride, and the phosphate are of commercial importance. The salts of streptomycin are readily soluble in water.

Chloramphenicol (chloromycetin) This is the first antibiotic to be synthesized on a commercial scale. The synthetic method is competitive with the fermentation process.

Chloramphenicol has the following structure:

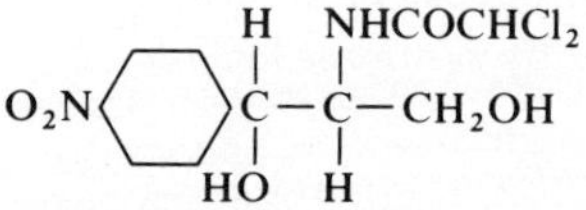

Chloramphenicol

Chloramphenicol is effective against a wide variety of microorganisms. Such a substance is known as a broad-spectrum antibiotic. It exerts strong activity against gram-positive and gram-negative cocci, bacilli, spirochetes, streptomycetes, certain rickettsiae, and the larger viruses.

The tetracyclines Chlortetracycline was the first isolated antibiotic in this group. Some are fermentation products of soil *Streptomyces*; others are prepared by chemical alteration of the basic molecule. The important analogs include Tetracycline (Tetracyn, Achromycin), Chlortetracycline (Aureomycin), Oxytetracycline (Terramycin), Methacycline (Rondomycin), Demethylchlortetracycline (Declomycin), Doxycycline (Vibramycin), and Minocycline.

They are amphoteric crystalline antibiotics which contain a common hydronaphthacene skeleton:

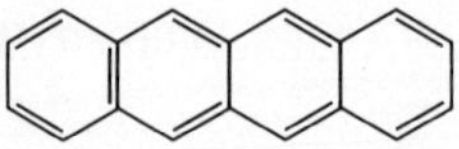

Naphthacene

The chemical structures of the three most employed analogs are as follows:

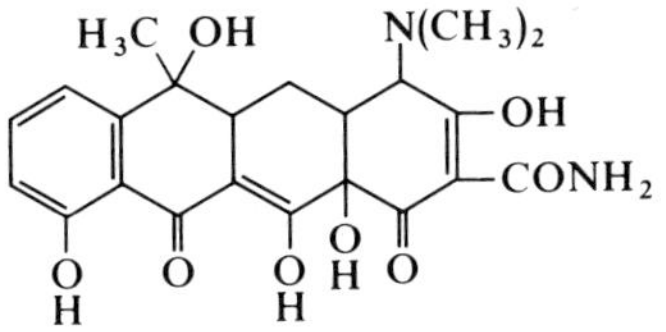

Tetracyline (Tetracyn, Achromycin)

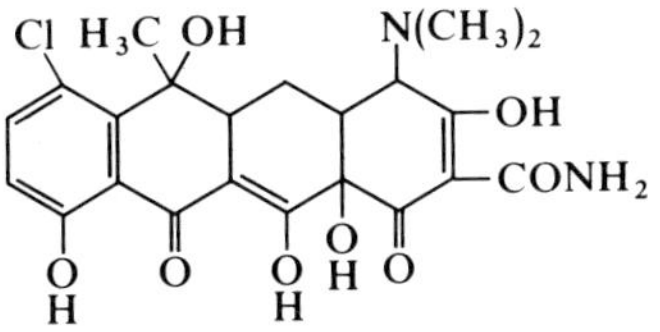

Chlortetracycline (Aureomycin)

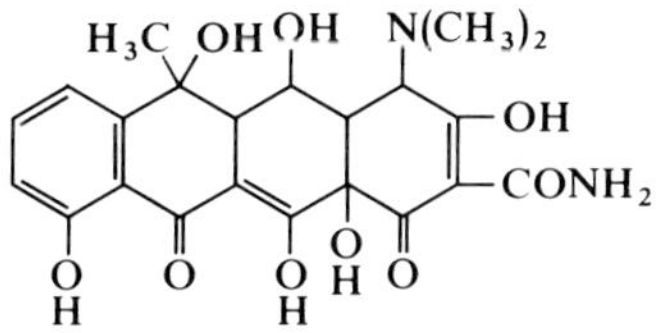

Oxytetracycline (Terramycin)

The tetracyclines are very slightly soluble in water but more soluble in dilute acid and alkali. They form salts readily with strong acids and bases. All are broad-spectrum antibiotics, being effective against a large number of gram-positive and gram-negative bacteria, rickettsiae, and the larger viruses.

For more information: Childress (1967); Steigbigel (1968).

Erythromycin and carbomycin These two antibiotics are produced by species of *Streptomyces*. They were discovered about the same time and have similar antibacterial and chemical properties. Erythromycin is also known as Ilotycin or Erythrocin; and carbomycin as Magnamycin.

Erythromycin has this structure:

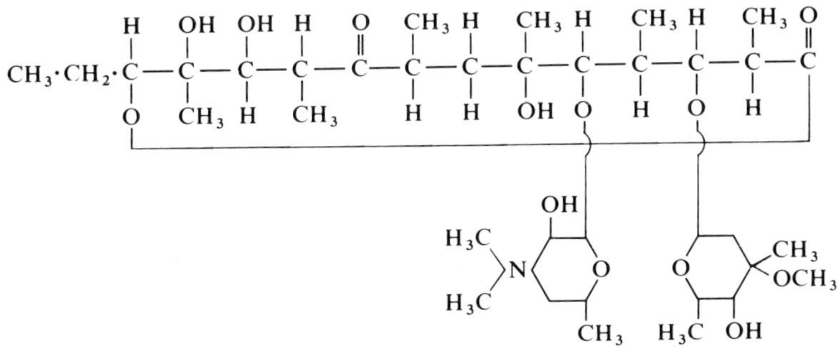

Erythromycin

Both antibiotics are active principally against gram-positive bacteria but are also effective against a few gram-negative organisms such as *Neisseria gonorrhoeae, Haemophilus influenzae*, certain larger viruses, and a few pathogenic protozoa.

Kanamycin Kanamycin is produced by strains of *S. kanamyceticus*. Kanamycin is a thermostable, water-soluble, polybasic compound having the following chemical structure:

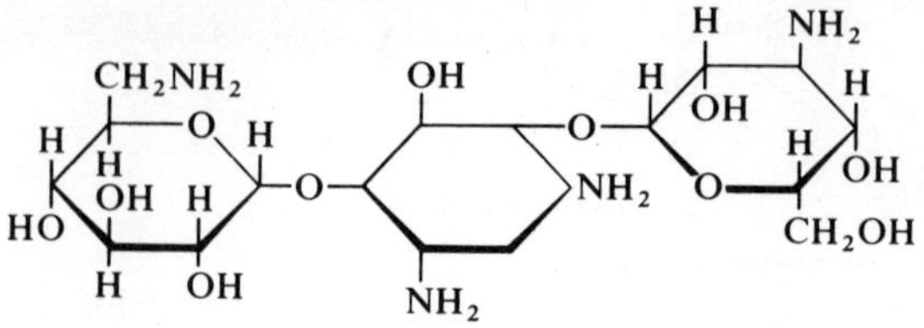

Kanamycin

The antibiotic is active against many aerobic gram-positive and gram-negative bacteria including the following: *Staphylococcus, Klebsiella, Aerobacter, Shigella, Salmonella, Neisseria, Escherichia, Proteus,* and *Mycobacterium.*

Cycloserine Cycloserine is produced by several strains of *Streptomyces.*

Chemically, the antibiotic is D-4-amino-3-isoxazolidinone having the following structure:

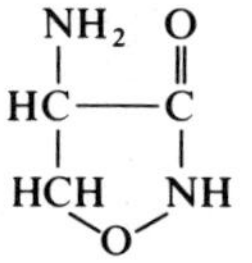

Cycloserine

Cycloserine is a wide-spectrum antibiotic, being effective against both gram-positive and gram-negative bacteria. However, its in vivo activity is less than most other antibiotics. Cycloserine is of value in those infections in which the causative organisms have become resistant to other antibiotics used in combination with other drugs in the treatment of tuberculosis.

Novobiocin Novobiocin is produced by *Streptomyces niveus* and *S. spheroides*. It is a pale yellow crystalline antibiotic having the following structure:

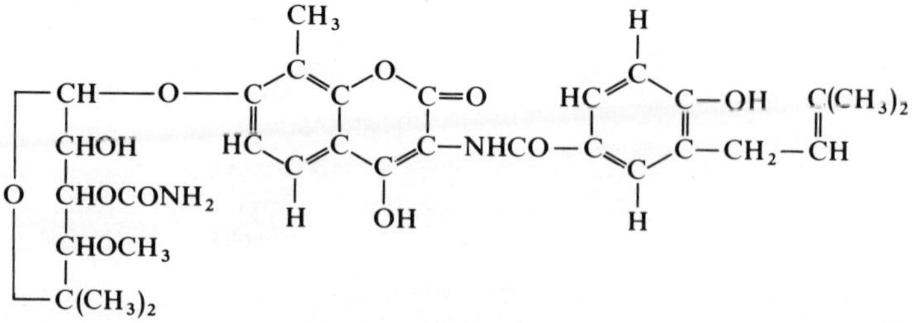

Novobiocin

Novobiocin inhibits the growth of many gram-positive organisms, being particularly active against *S. aureus*. It has little activity against most gram-negative bacteria. The antibiotic has been used in infections in which the organisms have become resistant to other antibiotics.

For more information: Kominek (1967).

Neomycin Neomycin is produced by *S. fradiae*. It is a composite antibiotic from which three components, designated Neomycin A, B, and C, have been separated. It is an amorphous base and readily forms salts with acids. The hydrochlorides of the three components are generally used.

Neomycin is a broad-spectrum antibiotic being effective against gram-positive, gram-negative, and acid-fast bacteria. It has a wider antibacterial spectrum than penicillin, streptomycin, or bacitracin. Bacteria develop resistance to Neomycin at a slower rate than to streptomycin.

Viomycin Viomycin is a strongly basic peptide antibiotic produced by *S. puniceus* var. *floridae*.

The antibiotic crystallizes as purple-colored crystals. It reacts readily with acids and is marketed chiefly as the soluble sulfate.

The antibiotic is more active against mycobacteria than other groups of bacteria. It is currently largely used as an adjunct in the treatment of various forms of tuberculosis.

For more information: Caltrider (1967).

Tyrothricin (gramicidins and tyrocidins) Tyrothricin is a mixture of polypeptide-like compounds which possess antibacterial activity. The antibiotic is composed of a family of polypeptides made up of a mixture of neutral gramicidins and basic tyrocidins.

Gramicidin consists of a mixture of four neutral polypeptide components designated A, B, C, and D. Tyrocidin is a mixture of four closely related basic polypeptide components.

Both groups are active primarily against gram-positive bacteria. Tyrothricin is ineffective when taken orally, and ineffective or even dangerous when given intravenously. It is applied locally in the treatment of ulcers and other wound infections.

Bacitracin Bacitracin is produced by strains of *B. licheniformis*. The commercial preparation is a complex polypeptide composed of one main component, Bacitracin A, and at least nine additional closely related polypeptides. Acid hydrolysis of the main component yields a mixture of L-leucine, L-isoleucine, L-cystine, L-histidine, L-lysine, D-phenylalanine, D-ornithine, D-glutamic acid, and DL-aspartic acid.

Bacitracin is antagonistic primarily for a wide variety of gram-positive bacteria. It has been used effectively in combination with other agents. The antibiotic may be administered intramuscularly, by local injection, or by topical application in the treatment of skin, eye, nose, and throat infections.

For more information: Weinberg (1967); Weinberg and Tonnis (1967).

Polymyxin (aerosporin) Polymyxin is produced by strains of *B. polymyxa*. It is a complex of at least five polypeptides designated A, B, C, D, and E. They all have similar chemical and biological properties. The free bases are only slightly soluble in water but the salts are very soluble.

The least toxic of the polymyxins is Polymyxin B. It is composed of L-leucine, L-threonine, L-phenylalanine, α-γ-diaminobutyric acid, and a lower saturated fatty acid.

Polymyxin B is marketed as the sulfate. It is active against many gram-negative bacteria. The antibiotic is administered by intramuscular injection, by the oral route for intestinal infections, and by topical application for the treatment of local infections.

For more information: American Medical Association (1964); Bodanszky and Perlman (1969); Paulus (1967); Regna (1955).

Cephalosporin C Cephalosporin C is an antibiotic closely related to penicillin. It is produced by species of *Cephalosporium*, a member of the imperfect group of molds.

It has the following structure:

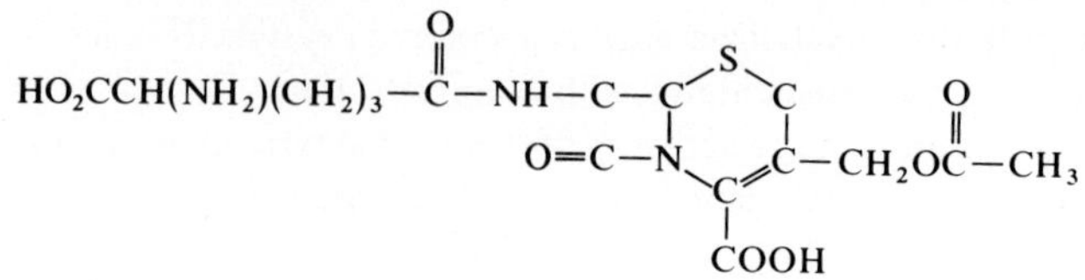

Cephalosporin C

The antibiotic is active chiefly against the gram-negative bacteria but it is also effective against some important gram-positive organisms.

For more information: Childress (1967); Flynn (1966).

Modes of action of antibiotics Antibiotics antagonize bacteria in a variety of ways. They may (1) inhibit cell-wall synthesis (penicillin, cycloserine, bacitracin); (2) inhibit protein synthesis (chloramphenicol, the tetracyclines, erythromycin, streptomycin); (3) render the cell wall and cytoplasmic membrane nonfunctional (gramicidin, tyrocidin, polymyxin). There are doubtless other mechanisms of action involved in the modes of action of antibiotics.

With the possible exception of bacitracin, no antibiotic has been shown to be active at minimal growth-inhibiting concentrations in more than one of the three categories. Bacitracin has been found to be effective in categories 1 and 2.

For more information: Gottlieb and Shaw (1967); Koike et al. (1969); Morgan et al. (1967); Reynolds (1962); Tipper and Strominger (1965); Tsukamura (1960); Weinberg (1967); Wolfe and Hahn (1964).

Antibiotics and plant diseases Attempts have been made to control plant diseases by the application of suitable antibiotics. In some instances the antibiotics are applied to the soil where they are absorbed by the roots and carried throughout the plants. In other instances the antibiotics are sprayed on plants and fruits for their surface effects.

An antibiotic may control a plant disease by (1) a direct action on a pathogen inside the plant tissue, (2) an action after transformation within the plant, and (3) a direct action on the host.

An antibiotic may be absorbed by the roots and carried in an active form throughout the plant body. Another antibiotic may not be effective as such but may be transformed within the plant to an effective compound. Still another antibiotic may be inactive against a plant pathogen in vitro but become active in the plant. The mechanism for this change is not yet clear.

For more information: Bradley (1966); Crowdy et al. (1955, 1956); Crowdy and Pramer (1955*a, b*); Dekker (1963); Krüger (1961).

Bacterial resistance to antibiotics Bacteria are able to develop resistance to antibiotics. Resistance may develop slowly or rapidly, depending upon the nature of the antibiotic. This characteristic is not limited to antibiotics but applies to other chemotherapeutic agents as well.

The classic example is the resistance acquired by staphylococci to penicillin. Resistant strains of this organism are endemic in many hospitals and are constantly threatening to break out into epidemics. Staphylococci are widely disseminated in the general population because normal and healthy people can become carriers.

When an organism acquires resistance to penicillin, it becomes necessary to switch to a derivative of the antibiotic or to an entirely different agent. This is why so many penicillins have been developed. The various penicillins have the same core but different side-chains or tails to distinguish one antibiotic from another and also to give the compounds different characteristics. An organism may be resistant to one penicillin but sensitive to another.

Resistance to penicillin is a case of genetic adaptation, or survival of the fittest. It is said that in any culture of *S. aureus* sensitive to penicillin, about one cell in 100 million is a mutant that is resistant to the antibiotic. Under normal conditions this mutant offers no survival advantages but in an environment containing penicillin the mutants are the only members which can multiply. This eventually leads to a population of staphylococci resistant to penicillin.

The mutant staphylococci are resistant to penicillin because of their ability to produce the enzyme penicillinase. This enzyme is capable of opening the β-lactam ring of penicillin to penicilloic acid, a compound that is devoid of antibiotic activity:

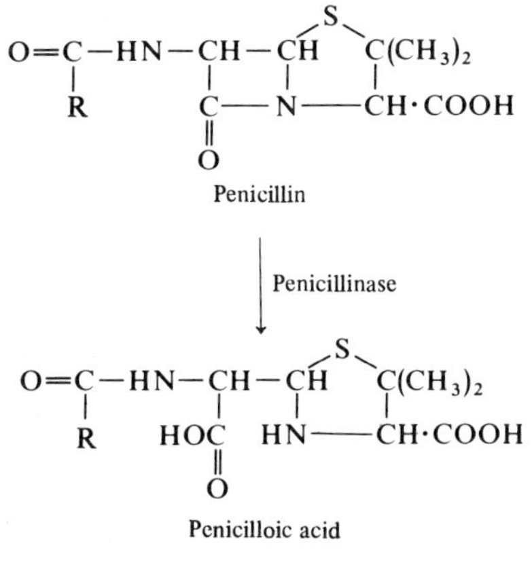

In the absence of penicillin in the medium, resistant staphylococci produce only traces of penicillinase but in the presence of the antibiotic, the enzyme concentration increases several hundred times. This explains why the mutant organisms become resistant to penicillin. This is a good example of enzyme induction or adaptation (see page 625).

Two types of resistance have been recognized: natural and acquired. Resistance is natural when it occurs among species or strains which have had no previous contact with the antibiotic under consideration. Resistance is acquired when a culture of a normally sensitive species or strain contains a few resistant forms which become predominant under the influence of the antibiotic. This gives rise to strains with acquired resistance. The ability of sensitive bacteria to give rise to antibiotic-resistant forms under the influence of the drug is a demonstration of the biochemical versatility of organisms in their struggle for existence.

The development of antibiotic-resistant forms is essentially an adaptive process since it reflects the ability of organisms to survive by adjusting themselves to adverse environmental conditions.

This form of adaptation falls into two categories:

1. Genetic adaptation, in which resistant mutants arise and overgrow the sensitive population under the influence of the antibiotic. This is called the mutation-selection theory.
2. Phenotypic adaptation, in which cytoplasmic alterations, such as adaptive enzyme formation, are induced by the antibiotic, causing at least a few of the bacteria to become more resistant without affecting the genetic apparatus. The phenotypically adapted cells eventually outnumber the sensitive cells under the selective influence of the antibiotic.

The weight of experimental evidence appears to favor the genetic adaptation concept.

Persistence of acquired resistance Acquired resistance is usually reversible. However, the ease with which resistance reverts to sensitivity depends upon a number of factors such as nature of the organism, nature of the drug, degree of resistance that has been established, and whether the resistance has been acquired by genetic or phenotypic adaptation.

When resistance has been acquired through genetic adaptation, reversion to sensitivity might occur in the absence of the antibiotic as the result of overgrowth by antibiotic-sensitive mutants. Such sensitive mutants are better fitted for survival.

If the resistance is the result of physiological or phenotypic adaptation, removal of the antibiotic should eliminate the primary stimulus for maintenance of the altered phenotype, and rapid reversion to the original phenotype would be expected to occur.

Prevention of bacterial resistance Sevag (1963), Sevag and Drabble (1962), and Drabble and Sevag (1962) reported that the two bases, spermine and spermidine, in combinations with penicillin, streptomycin, and erythromycin could completely prevent the emergence, from cultures of sensitive bacteria, of populations resistant to the antibiotics.

Table 20 *Some important antibiotics produced by bacteria and molds*

Antibiotic	*Produced by*	*Susceptible organisms*	*Property*
Bacitracin	Strain of *Bacillus subtilis*	Principally gram-positive bacteria	Topical use chiefly
Cephalosporin	Species of *Cephalosporium*	Principally gram-negative bacteria; few gram-positive cocci	Active in vivo
Chloramphenicol (Chloromycetin)	*Streptomyces venezuelae*, now chemically	Gram-positive and gram-negative bacilli and cocci, spirochetes, rickettsiae, and the larger viruses	Active in vivo
Chlortetracycline (Aureomycin)	*S. aureofaciens*	Gram-positive and gram-negative bacteria, rickettsiae, the larger viruses, some protozoa	Active in vivo
Cycloserine	Species of *Streptomyces*	Gram-positive, gram-negative, and acid-fast bacteria	Active in vivo
Erythromycin (Ilotycin)	*S. erythraeus*	Gram-positive, few gram-negative bacteria, certain larger viruses, some protozoa	Active in vivo
Kanamycin	Species of *Streptomyces*	Gram-positive, gram-negative, and acid-fast bacteria	Active in vivo
Neomycin	*S. fradiae*	Gram-positive, gram-negative, and acid-fast bacteria	Active in vivo
Novobiocin	Species of *Streptomyces*	Gram-positive, few gram-negative bacteria	Active in vivo
Oxytetracycline (Terramycin)	*S. rimosus*	Gram-positive, gram-negative bacteria, certain rickettsiae, and the larger viruses	Active in vivo
Penicillin	*Penicillium notatum* and other species	Gram-positive bacilli, gram-negative cocci, spirochetes	Active in vivo
Polymyxin (Aerosporin)	*B. polymyxa*	Principally gram-negative bacteria	Active in vivo
Streptomycin	*S. griseus*	Gram-positive, gram-negative, and acid-fast bacteria	Active in vivo
Tetracycline (Achromycin, Tetracyn)	Produced from chlortetracycline chemically	Gram-positive, gram-negative bacteria, and the larger viruses	Active in vivo
Tyrothricin (gramicidin + tyrocidin)	*B. brevis*	Principally gram-positive bacteria	Topical use chiefly
Viomycin (Viocin)	Species of *Streptomyces*	Gram-positive, gram-negative, and acid-fast bacteria	Active in vivo

S. aureus readily develops resistance to penicillin. When a sensitive culture of *S. aureus* was inoculated into a medium containing a mixture of penicillin and spermine, the organisms were completely killed. Spermine used alone had no effect on the growth of the bacteria. Unfortunately, the spermine fortified with penicillin was not effective against cells of *S. aureus* which had already built up a resistance to the antibiotic.

The mechanism of action of spermine in the prevention of drug resistance in sensitive cells is not yet clear.

For more information: Arai (1967); Bliss and Alter (1962); Doi (1968); Florey (1956); Gill and Hook (1965); Gorini (1966); Gunter and Feary (1968); Kasuga et al. (1968); Lilly Research Laboratories Staff (1957); Mitsuhashi (1967).

Antibacteriophage antibiotics Bacteriophages or bacterial viruses are submicroscopic agents which parasitize bacteria (see page 968).

Several antibiotics have been isolated which display activity against bacterial viruses. Strelitz et al. (1955*a*) isolated a colorless crystalline antibiotic that exhibited a wide range of activity against bacterial viruses. They named the compound Nybomycin. In another report (1955*b*), they announced the isolation of a second antibiotic that displayed antiphage activity.

For more information: Strelitz et al. (1956).

Importance of antibiotics Antibiotics are probably the leading chemotherapeutic agents of our time. For this reason, their number is being constantly increased. Some are useful clinically; others are either too toxic for clinical use or inactive in the presence of body fluids. Antibiotics in the latter group may be useful for other purposes. The more important antibiotics for clinical use are listed in Table 20.

For more information: Arndt and Ritts (1961); Richardson and Holt (1962); Wick et al. (1961).

References

Ahmadjian, V.: Lichens. In "Symbiosis," edited by S. M. Henry, New York, Academic Press, Inc., Vol. I, 1967.

American Medical Association: "New and Nonofficial Drugs," Philadelphia, J. B. Lippincott Company, 1964.

Arai, T.: The mechanisms of transfer of R factors, *Keio J. Med.,* **16**:55, 1967.

Arndt, W. F., and R. E. Ritts: Synergism between staphylococci and *Proteus* in mixed infection, *Proc. Soc. Exp. Biol. Med.,* **108**:166, 1961.

Ayers, W. A., and G. C. Papavizas: Violet-pigmented pseudomonads with antifungal activity from the rhizosphere of beans, *Appl. Microbiol.,* **11**:533, 1963.

Bachenheimer, A. G., and E. O. Bennett: I. The mechanism by which *Desulfovibrio desulfuricans* protects *Ps. aeruginosa* from the toxicity of mercurials, *Antonie van Leeuwenhoek, J. Microbiol. Serol.,* **27**:180, 1961.

Bliss, E. A., and B. M. Alter: Bacterial resistance to antibiotics. II. Population patterns among staphylococci, *J. Bact.,* **84**:125, 1962.

Bodanszky, M., and D. Perlman: Peptide antibiotics, *Science,* **163**:352, 1969.

Bradley, S. G.: Action of antifungal antibiotics on plants, animals, and microbes, *Develop. Ind. Microbiol.,* **7**:335, 1966.

Brown, L. D., et al.: Some active derivatives of penicillin, *Appl. Microbiol.,* **17**:339, 1969.

Bryant, M. P., et al.: *Methanobacillus omelianskii,* a symbiotic association of two species of bacteria, *Arch. Mikrobiol.,* **59**:20, 1967.

Bungay, H. R., III, and N. R. Krieg: Growth in mixed culture processes, *Chem. Eng. Progr.,* **62**:68, 1966.

Caltrider, P. G.: Viomycin. In "Antibiotics," vol. I, edited by D. Gottlieb and P. D. Shaw, Berlin, Springer-Verlag, OHG, 1967.

Chian, S. K., and R. I. Mateles: Growth of mixed cultures on mixed substrates, *Appl. Microbiol.,* **16**:1337, 1968.

Childress, S. J.: Chemical modification of antibiotics. In "Topics in Medicinal Chemistry," vol. I, edited by J. L. Rabinowitz and R. M. Myerson, New York, Interscience Publishers, a division of John Wiley & Sons, Inc., 1967.

Crowdy, S. H., et al.: The translocation of antibiotics in higher plants. I. Isolation of griseofulvin and chloramphenicol from plant tissue, *J. Exp. Botany,* **6**:371, 1955.

—— et al.: II. The movement of griseofulvin in broad bean and tomato, *J. Exp. Botany,* **7**:42, 1956.

—— and D. Pramer: The occurrence of translocated antibiotics in expressed plant sap, *Ann. Botany,* **19**:79, 1955*a*.

—— and ——: Movement of antibiotics in higher plants, *Chem. Ind.,* 1955*b*, p. 160.

Czerkinsky, G., et al.: A new method for determination of bacterial sensitivity to chemotherapeutic agents by means of paper strips, *Scand. J. Clin. Lab. Invest.,* **7**:259, 1955.

Dahiya, R. S., and M. L. Speck: Identification of stimulatory factor involved in symbiotic growth of *Streptococcus lactis* and *Streptococcus cremoris, J. Bact.,* **85**:585, 1963.

Dekker, J.: Antibiotics in the control of plant diseases. In "Annual Review of Microbiology," **17**:243, 1963.

Doi, O., et al.: Inactivation and phosphorylation of Kanamycin by drug-resistant *Staphylococcus aureus, Appl. Microbiol.,* **16**:1282, 1968.

Drabble, W. T., and M. G. Sevag: Prevention of the development of microbial resistance to drugs, *Antimicrobial Agents Chemotherapy,* 1962, p. 649.

Dubos, R. J.: Bactericidal effect of an extract of a soil bacillus on gram-positive cocci, *Proc. Soc. Exp. Biol. Med.*, **40**:311, 1939.

Eykyn, S. J., and I. Phillips: Advances in antibiotics, *Practitioner*, **203**:501, 1969.

Fleming, A.: On the antibacterial action of cultures of a *Penicillium*, with special reference to their use in the isolation of *B. influenzae, Brit. J. Exp. Pathol.*, **10**:226, 1929.

Florey, Sir H. W.: The medical aspects of the development of resistance to antibiotics, *Giorn. Microbiol.*, **2**:361, 1956.

Flynn, E. H.: Biological and chemical studies of the cephalosporins, *Antimicrobial Agents Chemotherapy*, 1966, p. 715.

Gale, E. F.: The production of amines by bacteria. 3. The production of putrescine from arginine by *Bacterium coli* in symbiosis with *Streptococcus faecalis, Biochem. J.*, **34**:853, 1940.

Gill, F. A., and E. W. Hook: Changing patterns of bacterial resistance to antimicrobial drugs, *Am. J. Med.*, **39**:780, 1965.

Gorini, L.: Antibiotics and the genetic code, *Sci. Am.*, **214**:102, 1966.

Gottlieb, D., and P. D. Shaw, eds.: "Antibiotics," vols. I, II, New York, Springer-Verlag New York, Inc., 1967.

Gunter, A. C., and T. W. Feary: Infectious drug resistance among clinically isolated *Escherichia coli, J. Bact.*, **96**:1556, 1968.

Henry, S. M., ed.: "Symbiosis," vol. I, New York, Academic Press, Inc., 1966; vol. II, 1967.

Jefferys, E. G.: The stability of antibiotics in soils, *J. Gen. Microbiol.*, **7**:295, 1952.

Kasuga, T., et al.: Drug resistance of staphylococci. VII. Genetic determinants responsible for the resistance to tetracycline, streptomycin, sulfanilamide, and penicillin, *J. Bact.*, **95**:1764, 1968.

Koike, M., et al.: Electron microscopic studies on mode of action of Polymyxin, *J. Bact.*, **97**:448, 1969.

Kominek, L. A.: Novobiocin. In "Antibiotics," vol. II, edited by D. Gottlieb and P. D. Shaw, Berlin, Springer-Verlag OHG, 1967.

Krieg, N. R., and M. J. Pelczar, Jr.: Analysis of a syntrophic growth of *Lactobacillus plantarum* and *Streptococcus faecalis, J. Gen. Microbiol.*, **25**:77, 1961.

Krüger, W.: II. Movement, stability and biological activity of antibiotics in soils and their uptake by tomato plants, *S. African J. Agri. Sci.*, **4**:301, 1961.

Lilly Research Laboratories Staff: Resistance of micro-organisms to antibiotics, **13**:22, 1957.

Meyers, E., et al.: Bioautographic technique for a rapid survey of microbial populations for the production of antibiotics and other metabolites, *Appl. Microbiol.*, **16**:10, 1968.

Mitsuhashi, S.: Epidemiological and genetical study of drug resistance in *Staphylococcus aureus, Japan J. Microbiol.*, **11**:49, 1967.

Morgan, C., et al.: Electron microscopy of chloramphenicol-treated *Escherichia coli, J. Bact.*, **93**:1987, 1967.

Nigrelli, R. F.: Antimicrobial substances from marine organisms. Introduction: The role of antibiosis in the sea, *Trans. N.Y. Acad. Sci.*, **24**:496, 1962.

Nurmikko, V.: Symbiosis experiments concerning the production and biosynthesis of certain amino acids and vitamins in associations of lactic acid bacteria, *Ann. Acad. Sci. Fennicae, Ser. A, Sec. 2*, **54**:7, 1954.

Park, D.: Effect of substrate on a microbial antagonism, with reference to soil conditions, *Trans. Brit. Mycological Soc.*, **39**:239, 1956.

Paulus, H.: Polymyxins. In "Antibiotics," vol. II, edited by D. Gottlieb and P. D. Shaw, Berlin, Springer-Verlag OHG, 1967.

Regna, P. P.: Chemistry of antibiotics of clinical importance, *Am. J. Med.*, **18**:686, 1955.

Reynolds, B. L.: The modes of action of antibiotics, *Australian J. Sci.*, **25**:243, 1962.

Richardson, M., and J. N. Holt: Synergistic action of streptomycin with other antibiotics on intracellular *Brucella abortus* in vitro, *J. Bact.*, **84**:638, 1962.

Rose, A. H.: New penicillins, *Sci. Am.*, **204**:66, 1961.

Sevag, M. G.: Prevention of microbial resistance to antibiotics, *Naval Res. Rev.*, **16**:1, 1963.

—— and W. T. Drabble: Prevention of the emergence of drug-resistant bacteria by polyamines, *Biochem. Biophys. Res. Commun.*, **8**:446, 1962.

Simoncini, F., et al.: Turbidimetric microbiological assay of antibiotics using an automatic analyzer, *Farmaco*, **23**:559, 1968.

Stallings, J. H.: Soil produced antibiotics: Plant disease and insect control, *Bact. Rev.*, **18**:131, 1954.

Steigbigel, N. H., et al.: Susceptibility of common pathogenic bacteria to seven tetracycline antibiotics in vitro, *Am. J. Med. Sci.*, **255**:179, 1968.

Stevenson, I. L.: Antibiotic production by actinomycetes in soil demonstrated by morphological changes induced in *Helminthosporium sativum*, *Nature*, **174**:598, 1954.

——: Antibiotic activity of actinomycetes in soil and their controlling effects on root rot of wheat, *J. Gen. Microbiol.*, **14**:440, 1956*a*; Antibiotic activity of actinomycetes in soil as demonstrated by direct observation techniques, *ibid.*, **15**:372, 1956*b*.

Stout, J. D.: The antibiotic relationships of some free-living bacteria, *J. Gen. Microbiol.*, **27**:209, 1962.

Strelitz, F., et al.: Nybomycin, a new antibiotic with antiphage and antibacterial properties, *Proc. Natl. Acad. Sci., U.S.*, **41**:620, 1955*a*.

——: Chrysomycin: A new antibiotic substance for bacterial viruses, *J. Bact.*, **69**:280, 1955*b*.

——: Aklavin, an antibiotic substance with antiphage activity, *J. Bact.*, **72**:90, 1956.

Tipper, D. J., and J. L. Strominger: Mechanism of action of penicillins: A

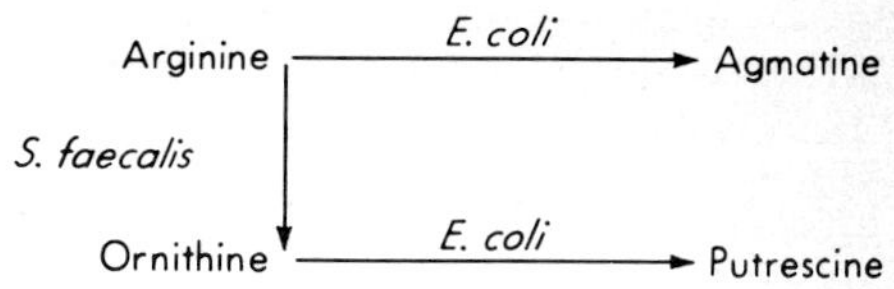

Fig. 250 Action of Escherichia coli and Streptococcus faecalis on arginine and ornithine.

inoculated into the same medium, *S. faecalis* produced ornithine, which was then acted upon by *E. coli* to give putrescine. *E. coli* was definitely benefited by the association, while *S. faecalis* probably was not affected.

Dahiya and Speck (1963) reported that *S. cremoris* and *S. lactis*, grown separately, produced lactic acid as one of the products of fermentation, and that more was manufactured from the combined growth of the two organisms. *S. lactis* produced adenine which stimulated *S. cremoris* to grow faster and to produce a greater yield of lactic acid.

Synergism

In bacteriology, synergism may be defined as the joint action of two organisms on a carbohydrate, resulting in the production of gas that is not formed by either organism cultivated separately. One of the organisms of the pair produces only acid, while the other member produces acid and gas. The acid-former degrades the carbohydrate and releases an intermediary compound that is fermented by the second organism to acid and gas.

Examples of gas-forming pairs of organisms are given in Table 19.

Results from sucrose broth inoculated with *S. aureus* and *E. coli* separately and together are shown in Fig. 251.

The phenomenon of synergism probably finds its greatest importance in the field of bacteriological water examinations. False positive presumptive tests in water analyses are sometimes caused by the associated activities of two or more bacterial species. The opposite effect might also

Table 19 Bacterial pairs that produce gas in association

Carbohydrate	*Organisms*
Lactose	Staphylococcus aureus + Salmonella schottmuelleri
	Streptococcus faecalis + S. schottmuelleri
	S. aureus + Proteus vulgaris
	S. faecalis + Salmonella paratyphi
Sucrose	S. aureus + Escherichia coli
	S. faecalis + E. coli
	Streptococcus equinus + S. schottmuelleri
	S. aureus + S. paratyphi
Mannitol	S. aureus + P. vulgaris
	Streptococcus pyogenes + P. vulgaris
	Shigella flexneri + P. vulgaris
	Salmonella typhosa + P. vulgaris

proposal based on their structural similarity to acyl-D-alanyl-D-alanine, *Proc. Natl. Acad. Sci. U.S.*, **54**:1133, 1965.

Tsukamura, M.: Mechanism of action of viomycin, *J. Biochem.*, **47**:685, 1960.

Weinberg, E. D.: Bacitracin, gramicidin, and tyrocidine. In "Antibiotics," vol. II, edited by D. Gottlieb and P. D. Shaw, Berlin, Springer-Verlag, OHG, 1967.

——— and S. M. Tonnis: Role of manganese in biosynthesis of bacitracin, *Can. J. Microbiol.*, **13**:614, 1967.

Wick, W. E., et al.: Guide to antibiotic therapy of experimental bacterial infections in mice, *J. Bact.*, **81**:233, 1961.

Wolfe, A. D., and F. E. Hahn: Erythromycin: Mode of action, *Science*, **143**:1445, 1964.

Wright, J. M.: The production of antibiotics in soil. II. Production of griseofulvin by *Penicillium nigricans*, *Ann. Appl. Biol.*, **43**:288, 1955.

———: Production of gliotoxin in soils, *Nature*, **177**:896, 1956.

Bacterial genetics[1] 18

During an early period of bacteriology, microorganisms were assumed to be capable of an almost infinite variety of morphological and physiological changes. Some of the proponents of this concept asserted that only one (or a few) species of bacteria existed and that the diverse morphological forms present in natural material were merely different developmental stages of this basic organism. These morphological changes were thought to be induced in an orderly manner by progressive alterations in the growth medium or by other environmental factors. It is now known that these descriptions of the supposed changes in the morphology of the "organism" under study were actually a series of observations on a mixed culture in which, as a response to the selectively changing growth conditions, first one and then another of the species originally present became the dominant member of the population. The objections to the doctrine of pleomorphism, as this concept was called, were pithily summarized by Brefeld, the German mycologist, who said that unless one works with pure culture, "*da kommt nur Unsinn und Penicillium glaucum heraus.*"[2]

The doctrine of pleomorphism, or unlimited morphological variation, was completely discredited after the development of simple pure culture techniques by Robert Koch. In its place the concept of monomorphism was adopted, according to which each bacterium has an absolutely fixed constancy of form and function.

Further experience revealed, however, that even when using rigidly controlled pure culture techniques, a certain amount of variation always developed among the bacteria of a given culture. Since most of the early bacteriological work was concentrated on the organisms responsible for diseases in man and animals, it was with this group that the variations were first noted.

Upon isolation and repeated subculture of many pathogens, it was

[1] This chapter was written by Dr. W. R. Romig, Department of Bacteriology, University of California, Los Angeles, Calif. The author is greatly indebted to Dr. Romig for his kindness in preparing this material for publication.

[2] "Only nonsense and *Penicillium glaucum* [a common airborne contaminant] will come of it."

found that distinct changes in colonial morphology, antigenicity, and virulence often occurred. These changes could frequently be reversed by appropriate techniques, but under ordinary conditions were quite stable. This phenomenon of reversible variation has been designated *dissociation* and was interpreted to be a natural life cycle analogous to those of some protozoan parasites; or it was thought to be due to a gradual adaptation of the culture to the changed environment within the culture tube.

While it is now accepted that bacteria have definite mechanisms for the transmission of their hereditary traits which do not differ greatly from those of other living things, this concept was not established until fairly recently. This reluctance to assign a genetic basis to the variations which demonstrably occur in bacterial cultures may be attributed to several factors.

Until Robinow published his beautiful pictures of bacterial nuclei (1944), most bacteriologists were of the opinion that bacteria possessed none of the typical structures usually associated with the storage and transmission of genetic information. Examination of thin sections of bacteria in the electron microscope has since revealed that their "nucleus" differs from nuclei of higher forms. Bacterial "nuclei" are not bounded by a nuclear membrane, nor do they contain chromosomes analogous to those of typical plants and animals. Instead the deoxyribonucleic acid (DNA) of which the nucleus is composed appears as a condensed body formed by packing into a small volume what is probably a single, very long DNA molecule.

For more information: Hayes (1968); Kellenberger (1960); Ryter and Jacob (1964).

Another barrier to the ready acceptance of a belief in the hereditary origin of a bacterial variation was the fact that bacterial properties are usually measured as a cultural response given by many millions of cells, whereas bacterial mutations occur at the level of the single cell. Since mutations occur extremely infrequently (at the rate of about 1 for every 10 million cell divisions), the mutant cell is extremely difficult to detect. Part of the reason for this difficulty is that bacteria do not have the many morphological features such as feathers, flowers, or fur in which mutational alterations may easily be detected, so that other methods of detection must be used.

Physiological and genetic modifications

Pigment production in *Serratia marcescens* is uniformly inhibited by incubation at the normal body temperature of 37°C, but when the incubation temperature is lowered, normal pigmentation again appears. Such environmentally induced changes are called *phenotypic* modifications, and they characteristically appear uniformly in all cells of a given population. Phenotypic modifications differ from genotypic changes chiefly in that they simultaneously affect most of the cells of a given culture and they are not inheritable. The *genotype* may be defined as the sum total of all the heritable cell characteristics, and sudden changes in the genotype are called *mutations.*

Thus even when pigmentation in *Serratia* is inhibited for many generations by continued incubation at 37°C, all the cells in a culture will promptly resume pigment production when the temperature is lowered. In contrast to the transient nature of phenotypic modifications, changes in the genotype of a cell are much more stable and are inherited.

Bunting (1946) has shown that mutations to the nonpigmented condition occur at fairly high rates (about 1 per 10,000 cell divisions) in normally pigmented *Serratia* and that these colorless mutants and their progeny have completely lost the ability to produce pigment, regardless of the incubation temperature or other environmental factors. Inducible enzyme systems may also contribute to the variations observed in bacterial cultures and the formation and function of some of these systems have been analyzed in detail in *Escherichia coli.* Studies of the genetic basis of the production of inducible enzymes have contributed greatly to our understanding of the mechanisms of cellular control.

For more information: Bretscher (1968); Cohn (1957); Gilbert and Muller-Hill (1966); Jacob and Monod (1961); Sheppard and Englesberg (1967).

An *inducible enzyme* is one which the cell can synthesize in quantity only when its substrate (or a chemically related compound) is present in the environment. β-galactosidase, which hydrolyzes lactose to its constituent monosaccharides, glucose and galactose, is an example of an inducible enzyme found in many bacteria. When genotypically lactose-positive (lac^+) *E. coli* are grown in media devoid of lactose, β-galactosidase cannot ordinarily be detected within the cells, and such cultures are phenotypically lactose-negative (lac^-). However, if lactose is substituted for glucose, the bacteria quickly produce large quantities of the enzyme and a *permease* to transport lactose into the cell and can now utilize lactose as their sole carbon source. It should be emphasized that the lactose has in no way altered the genotype of these bacteria. They were all genotypically capable of utilizing lactose, but the enzyme necessary for its utilization is produced only after being induced by the specific substrate. In this case only the potential for the utilization of lactose in inherited, for when lactose is removed from the environment, the cells stop producing β-galactosidase and revert to a phenotypically lac^- population.

As in the case of pigment production in *Serratia* discussed above, mutations can occur which alter the genotype of the lac^+ cells. In this event, the hereditary potential for β-galactosidase production is lost, and these *fermentation mutants* and their progeny are no longer capable of utilizing lactose, regardless of its presence or absence in the growth medium.

At this point it should be noted that mutations can usually occur at about the same rates in both directions. Cells that have lost the hereditary potential for producing a given enzyme may later *back mutate* to regain this potential. These mutational changes can be differentiated from phenotypic changes because they occur only under conditions of active multiplication, and even then much more rarely than the latter.

The mechanism of inheritance

The science of genetics is based primarily on results obtained from breeding experiments with organisms of differing genotypes. It was from a study of the segregation of characters and their reassortment during sexual reproduction that the gene concept was developed. According to this concept the gene is the basic unit of inheritance which occupies a definite position on the chromosome. Although successful breeding experiments only recently became possible using bacteria, microorganisms other than bacteria have been used as tools for genetic research ever since the pioneering work on the mechanism of inheritance in the bread mold, *Neurospora crassa*, was performed by Beadle and Tatum (Beadle, 1945).

Neurospora can grow in a minimal medium consisting of a few inorganic salts, a suitable carbon source, and biotin. From these simple compounds it is able to synthesize all the vitamins, amino acids, and other myriad cell constituents necessary for continued growth and division. After suitable mutagenic treatment, however, strains could be isolated which no longer grew unless specific supplements, such as preformed amino acids or vitamins, were added to the medium. These strains were shown to be nutritionally deficient mutants of the original strain which had lost the ability to synthesize one of the required protoplasmic constituents. By suitable breeding experiments these mutants were shown to differ from the original strain at a single gene locus.

From these experiments, Beadle and Tatum concluded that each gene which the cell inherits exerts its effect by controlling the specificity of a particular enzyme. According to their interpretation, commonly called the "one-gene-one-enzyme" theory, their mutants had lost the ability to form a particular enzyme because the gene controlling its production had been altered in some hereditary fashion.

Chemical nature of the genetic material It is now generally accepted that the genetic information which a parent cell transmits to its progeny is contained in the deoxyribonucleic acid of the chromosomes in its nucleus. It has been established that the hereditary determinants, the genes, are composed of DNA, and for this reason the chemical and physical nature of DNA has been subjected to intensive investigation in recent years.

For more information: Chargaff (1955); Davidson and Cohn (1963); Watson and Crick (1953).

Many of the known chemical and physical data on DNA were incorporated into a model of the molecular structure of this compound by Watson and Crick (for which they received the Nobel prize along with Wilkins in 1962). One of the main virtues of their model is that it immediately suggested how DNA might produce an exact copy of itself – a prime requisite for a substance assigned the role of the bearer of

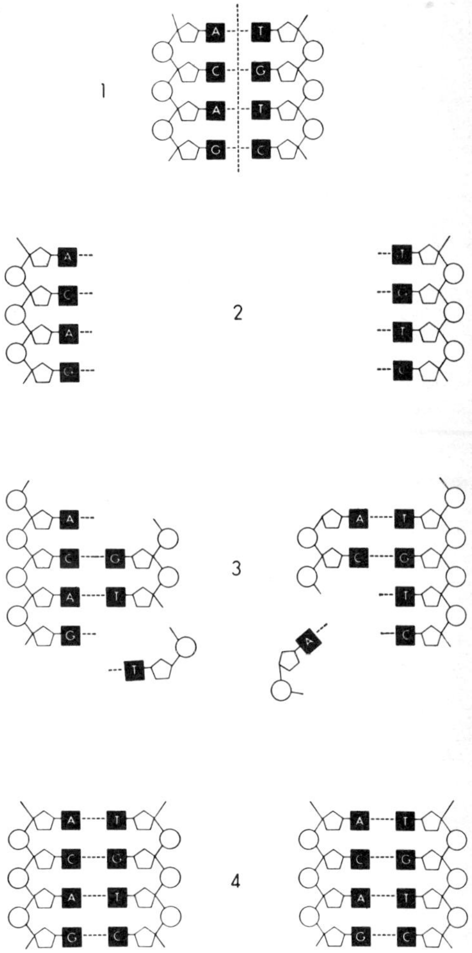

Fig. 254 Replication mechanism by which DNA might duplicate itself. A helix of two DNA chains unwinds and separates (1). Two complementary chains of DNA (2) within the cell begin to attach DNA precursor units floating loosely (3). When the proper bases are joined, two new helices will build up (4). Letters represent the bases. (Courtesy of F. H. C. Crick.)

genetic information. According to their model (Fig. 254), DNA consists of two long polynucleotide chains wound around each other in a helical fashion. Both chains are arranged so that the purine and pyrimidine bases are inside the helix and the sugar-phosphate backbones are outside. The chains are held together by hydrogen bonds formed specifically between the thymines of one chain and the adenines of the other; the guanines likewise bond to cytosines. The two chains, therefore, are exactly complementary to one another throughout their length. Preceding cell division, the chains separate and each serves as a template for the synthesis of a new complementary chain, thus forming two new helices, each of which is exactly like the original. Each daughter cell then receives one of these molecules in which the genetic information of the original cell is incorporated. This mode of duplication, in which one polynucleotide chain acts as a template to direct the synthesis of a new chain complementary to itself, is called "semiconservative," and the DNA of

E. coli has been shown to replicate in just this manner (Meselson and Stahl, 1958). DNA is a thin, enormously long molecule; in fact, the circular chromosome of *E. coli* is more than 1000 μ in circumference, yet consists of only a single piece of DNA (Cairns, 1963).

The key to the specificity of DNA lies in the order of occurrence of the purine and pyrimidine bases making up its polynucleotide chains. A gene is thought to consist of a unique "block" of about 1,000 bases along the DNA molecule which contains the information required to specify the sequence of amino acids of a particular enzyme. In this context, the purine and pyrimidine bases may be thought of as the genetic equivalent of the dots and dashes of the Morse code.

The genetic message begins at one end of the gene and is simply read off, three bases at a time, to the end. Each group of three bases (called a *codon*) specifies a particular amino acid. Since there are 4 different bases in DNA, there are 64 possible ways they can be arranged into triplets, giving more than enough codons to specify the 20 amino acids found in most proteins. As discussed elsewhere, the genetic code in the DNA is transcribed into an RNA "copy." This RNA copy, which is transcribed from only one strand of a given gene, is called messenger RNA. The DNA double helix is assumed to separate during transcription to allow one strand to serve as a template for the synthesis of messenger RNA by normal base pairing rules. Once formed, messenger RNA becomes associated with structures called ribosomes and is translated into a specific protein. During this process it serves as the template to assure that the amino acids are assembled in the correct sequence to form each specific enzyme or other protein molecule.

Mutations, or sudden heritable changes, result when the order of bases in the DNA molecule is altered. The alteration of even a single base within a gene completely changes a given codon. If the normal sequence were ATT, the change to AAT might direct an entirely different amino acid into a polypeptide chain, and cause the enzyme to completely lose its catalytic properties. Providing that such a loss of enzymatic function was not lethal to the cell, this alteration would be perpetuated in its progeny unless another rare mistake occurred to reestablish the original triplet.

For more information: Crick (1963); Nirenberg (1963).

Factors affecting mutation rates The cell machinery responsible for replication of DNA is extremely efficient, and under normal conditions only rarely is a "wrong" base incorporated into a gene. The spontaneous mutation rate can be greatly increased, however, by treating bacteria with mutagenic agents. Induced mutants occur at rates from 10 to 100,000 times more frequently than spontaneously occurring mutants among bacteria that have been exposed to mutagenic agents.

Mutagenic agents exert their effect by reacting, directly or indirectly, with DNA. Agents which have proved effective mutagens include ultraviolet light, x-rays, carcinogenic chemicals such as nitrogen and sulfur mustards, various peroxides and epoxides, and purine and pyrimidine analogs. Some mutagens such as nitrous acid and hydroxylamine have been shown to cause mutations in vitro by allowing them to react with purified transforming DNA, and then introducing the altered DNA into recipient bacteria. Bacteria have proved quite useful for screening various compounds or agents for possible mutagenic activity, partly because large populations of them are easily obtained and scored, and also because the problem of chemical diffusion is greatly minimized in bacteria.

For more information: the excellent review about chemical mutagenesis published by Krieg (1963); Zamenhof (1959).

The mutational origin of bacterial resistance

The mode of acquisition of resistance by sensitive bacteria has both practical and theoretical interest and has been intensively investigated for many years. The phenomenon of acquired resistance by bacteria was recognized by Ehrlich soon after his discovery of successful chemotherapeutic drugs, but the mechanisms whereby resistance to various toxic agents is acquired were discovered only recently. The experiments which first proved the true nature of the phenomenon were performed by Luria and Delbrück (1943), and are generally conceded to have formed the basis for modern work in the field of bacterial genetics. For this reason their approach will be discussed in some detail.

At the time Luria and Delbrück performed their experiments, two explanations concerning the nature of acquired drug resistance were prevalent. According to one of these, the *direct adaptation* theory, sensitive bacteria become resistant as a result of an interaction between the toxic agent and some members of the sensitive population. According to this theory, resistance occurs as a direct adaptive response of some bacteria which is induced by placing them in contact with the toxic agent; the theory is identical with that of Lamarck who believed that environmental effects caused changes in structure, and that such acquired characters could be hereditarily transmitted to the offspring.

As an alternative explanation, the *spontaneous mutation* theory suggested that undirected spontaneous mutations occur at a constant rate during normal growth of all bacterial cultures. According to this view, some of these mutations cause the cells receiving them to become resistant to one or another of a variety of toxic agents. Such cells become resistant in the absence of the agent, and exposure to one of these agents merely serves as a selective force by eliminating the sensitive parents from the culture.

Unfortunately, direct experimental tests were not available for discriminating between these two conflicting hypotheses. Spontaneous mutations occur at such low frequencies – on the order of 1 per 10 million cell divisions – that it is impossible to isolate directly and test a suspected resistant mutant from a normally growing culture. The only method for detecting the presence of a few resistant cells in a predominantly sensitive population consisted of placing the culture in a growth medium containing the toxic agent. The limitations imposed by this method resulted in a paradoxical situation in which the postulated mutants could be detected only if the agent was added; while by the other hypothesis, addition of the agent evoked resistance by a directed adaptive process.

The fluctuation test This seeming paradox was resolved by an ingenious experiment called the fluctuation test that may be performed in the following manner:

A young bacterial suspension containing about 500 *E. coli* per milliliter is divided into two cultures, A and B, of 10 ml each. Culture A is then further subdivided into a series of 50 tubes, each containing 0.2 ml. Both cultures are incubated until a predetermined population density is attained. Under these conditions, the cell concentration in the two cultures will be approximately equal. The number of streptomycin-resistant colonies which develop when the contents of each of the small tubes are poured onto the surface of streptomycin-containing agar plates is then compared with the number which develop when a similar series of 0.2-ml samples from culture B are pipetted onto the surface of streptomycin-containing agar (Fig. 255).

Luria and Delbrück reasoned that if bacterial resistance developed by direct adaptation to the streptomycin, there should be no significant

Fig. 255 Fluctuation test. See text for details.

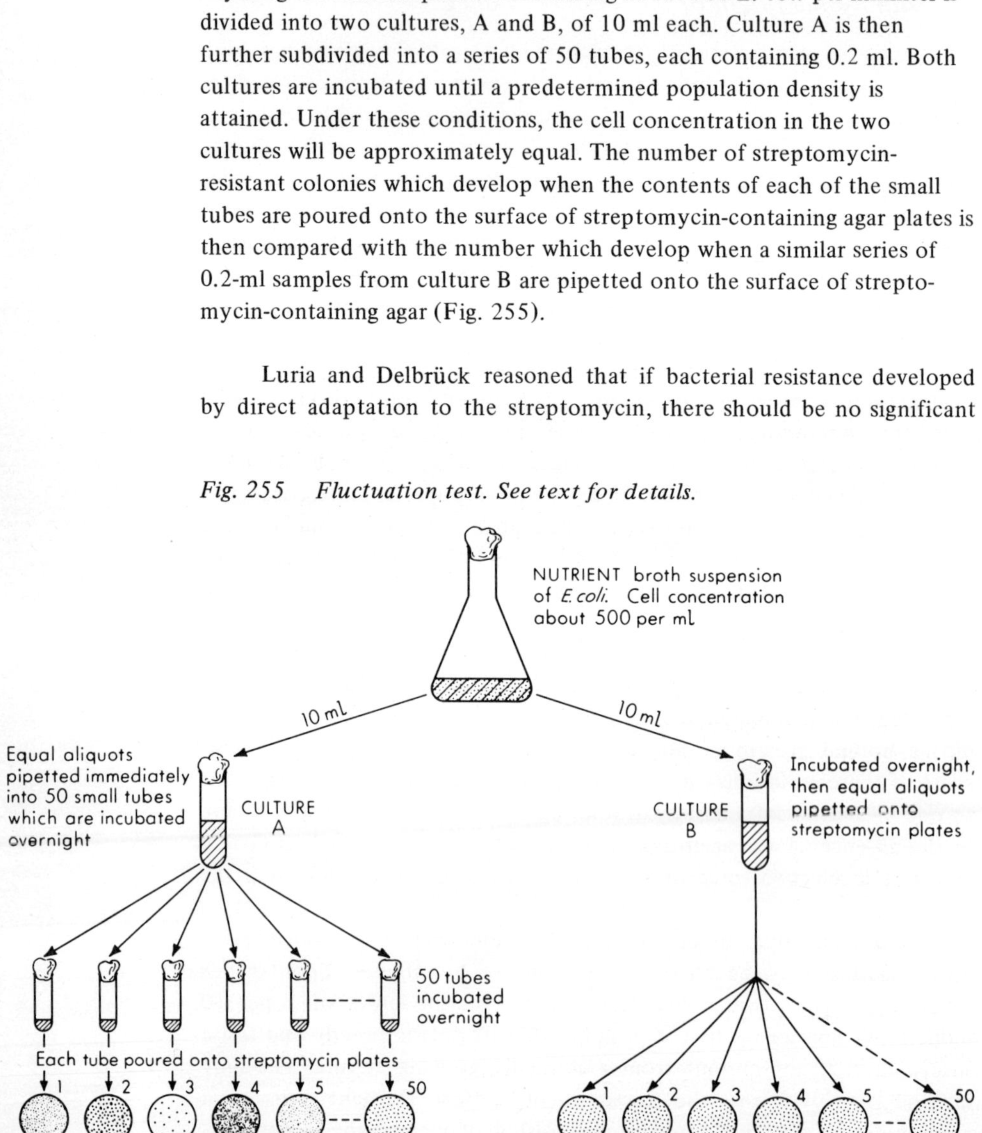

differences in the numbers of resistant colonies on any of the plates derived from the two cultures. By this theory it should make no difference whether the cultures were incubated in a series of separate tubes or were all grown together in one large tube, since they are all presumed to be sensitive until brought into contact with streptomycin.

If, on the other hand, streptomycin resistance resulted from spontaneous mutations during growth in streptomycin-free medium, there should be a large fluctuation between the numbers of resistant colonies which developed from each of the 50 small tubes. This fluctuation in the numbers of resistant bacteria in a series of parallel cultures is a consequence of the randomness in time of the occurrence of spontaneous mutations and of the clonal nature of bacterial multiplication. If the mutation to resistance occurred early during cultural growth, the resistant mutant would pass through a large number of cell divisions, and there would be many resistant bacteria present at the time of testing. If the mutation to resistance occurred just before the cells were added to streptomycin, only one resistant colony should appear on the plate.

The similar series of samples all taken from the same container (culture B) serve as a control on the experimental technique. By either hypothesis, random samples taken from it should contain the same number of resistant bacteria.

The results obtained from the fluctuation test unequivocally support the spontaneous mutation theory. A large variance was observed in numbers of resistant colonies derived from the series of small tubes, while on the plates all taken from the single large culture, the variance was about equal to the mean and was no larger than would be expected from ordinary sampling errors.

Indirect selection Several other experimental methods have since been devised which more directly support the concept of the spontaneous origin of bacterial resistance. The most unambiguous of these is the method of indirect selection by replica plating developed by the Lederbergs (1952). By this procedure, samples from all the colonies on a plate may be transferred simultaneously to another plate by means of a velveteen-covered stamp pad. The outstanding advantage of this method is that it allows the spatial relationships between the colonies on the "master" plate and the "copy" to be maintained (Fig. 256).

The replica plating technique was used to verify the spontaneous origin of bacterial resistance in the following manner: Several million streptomycin-sensitive bacteria were spread onto the surface of a streptomycin-free agar plate and incubated a few hours to allow microcolonies consisting of the descendants of the original cells to be formed. The sterile velveteen stamp was then pressed onto the plate containing these microcolonies so that a few of the bacteria from each of them adhered to the fibers. Next the pad was pressed against the surface of a streptomycin-containing plate, and the bacteria transferred to it were incubated until resistant colonies appeared. These colonies were then superimposed over the original plate, and the bacterial growth in the area corresponding to

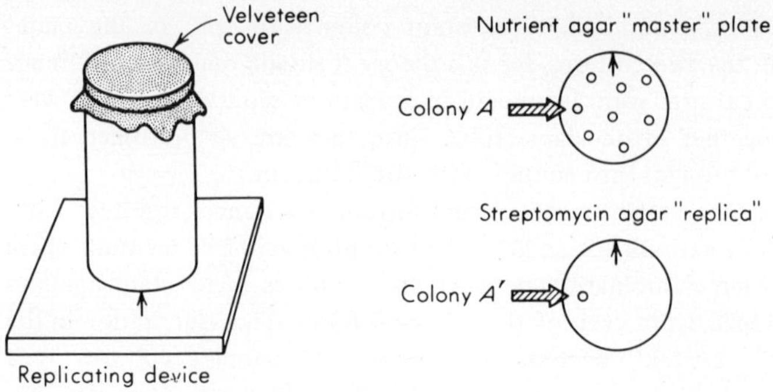

Fig. 256 Replica plating. Colonies on the nutrient agar plate are pressed onto the velveteen stamp pad, and the adhering bacteria are then transferred to a sterile streptomycin agar plate. After incubation, colony A on the nutrient agar plate corresponding to colony A′ on the streptomycin plate is located by superimposing the plates over one another. Colony A on nutrient agar may be picked and proved to be resistant to the drug, even though it has never come into direct contact with it. See text for further detail.

the resistant colonies was picked into fresh nutrient broth. If this step is carefully performed, there will be almost a hundredfold increase in the proportion of resistant mutants after incubation of this fresh culture.

Following overnight incubation of this new culture, diluted aliquots of it were spread onto the surface of fresh, drug-free agar plates, and the same procedure of incubation and replication was repeated as in the previous step. This time more resistant colonies were found on the drug-containing plates than in the first cycle. After one or two more cycles of enrichment, a colony was located on the plain agar plate which exactly corresponded to a colony on the drug-containing replica. The colony on the plain agar was picked and proved to consist entirely of resistant cells. The important point in this procedure is that the resistant colony on the plain agar arose without ever coming directly into contact with the streptomycin, so the drug could not possibly have interacted with it to induce resistance.

As a result of these and similar experiments (Demerec, 1948; Newcombe, 1949; Cavalli-Sforza, 1957), it is now generally held that mutation followed by selection is the main source of variation found among bacteria of the same species. For a discussion of the opposing viewpoint, however, see Dean and Hinshelwood (1957).

Therapeutic significance of acquired resistance With the advent of widespread, and often ill-advised, antibiotic therapy, the study of mutants resistant to these compounds has assumed great importance. Most organisms normally sensitive to a given antibiotic spontaneously mutate to drug resistance at a low but constant rate even when growing in their natural habitats in the absence of drugs. These spontaneously appearing resistant mutants normally make up only a minute fraction of a bacterial population, but when the sensitive majority are eliminated by the drug, they rapidly establish themselves as the dominant flora. Fortunately, mutation to resistance to one antibiotic does not necessarily confer resistance to other antibiotics. Therefore, if two antibiotics are applied simultaneously, the chance that a given sensitive bacterium will have mutated to resistance to both is extremely small. For this reason, dual antibiotic therapy is often preferred to treatment with any single antibiotic, especially in treating tuberculosis.

It has also been shown that resistance to many antibiotics arises as a series of individual mutational steps, each mutation conferring a higher degree of resistance to the organism than the one preceding it. It is therefore recommended that these drugs be administered at the highest tolerated level to minimize the chances of selecting mutants resistant to the lower levels of the antibiotic. For with continued antibiotic therapy, these low-level resistant mutants may mutate to progressively more resistant states and become completely indifferent to the presence of the antibiotic.

For more information: Braun (1965); Brock (1966); Hayes (1968).

Nutritional mutants Bacterial mutants with inherited nutritional deficiencies similar to those studied in *Neurospora* by Beadle and Tatum are easily isolated and have been used extensively in biochemical and genetic studies. Nutritionally deficient mutants of bacteria are called *auxotrophs*, which means that they can be grown only upon minimal medium that has been supplemented with growth factors not required by the parental strain. The parental strain is called a *prototroph*, which signifies that the bacteria can grow on unsupplemented minimal medium. Studies with many auxotrophic mutants have revealed that, as in the case of *Neurospora* mutants, they have lost the ability to form one of the enzymes required for the biosynthesis of an essential protoplasmic constituent.

By isolating many mutants which require the same compound for growth, it is usually possible to obtain mutants blocked at different steps in the biosynthetic pathway of this compound. By analyzing the growth responses of these mutants when tested with possible biosynthetic intermediates of this pathway, it is often possible to determine the steps used for the synthesis of this compound by normal bacteria.

As an example, a number of mutants were isolated which cannot grow in minimal medium unless it is supplemented with the amino acid arginine. These mutants may be classified biochemically by determining which compounds can still support normal growth when they are substituted for arginine. Group I mutants may grow only if arginine is included in the medium; group II may grow normally if either arginine or citrulline-supplemented minimal medium is used; and group III may respond to supplementation with arginine, or citrulline, or ornithine. From these observations it may be deduced that the normal biosynthetic sequence for arginine is:

$$X \xrightarrow{III} \text{ornithine} \xrightarrow{II} \text{citrulline} \xrightarrow{I} \text{arginine}$$

Mutants of group I are blocked in the enzymatic conversion of citrulline to arginine; mutants of group II are blocked in the conversion of ornithine to citrulline, but if provided with citrulline can convert it to arginine; while mutants of group III must be blocked in the formation of ornithine, but if supplied with either ornithine or citrulline can enzymatically convert them on to arginine.

Mutants blocked in the normal biosynthetic pathway may accumulate precursors behind the mutant block. In the example above, mutants of group I excrete citrulline when deprived of arginine and mutants of group II excrete ornithine.

The utility of the technique of mutant analysis for determining biosynthetic pathways was greatly facilitated by simple methods developed for isolating auxotrophic mutants. In most cases the number of mutants in a normal population is increased by first treating the bacteria with a mutagen such as ultraviolet light or nitrosoguanidine. These agents increase the proportion of mutants in the population, but most of the bacteria in such a treated culture usually remain nonmutant. However, these numerically superior prototrophs can be eliminated from the culture by the penicillin method developed independently by Lederberg and Zinder (1948) and by Davis (1948). Their technique depends upon the fact that penicillin exerts its bactericidal effect only against growing cells. This occurs because penicillin specifically inhibits one of the steps used by bacteria in making their cell walls. Prototrophs exposed to penicillin in minimal medium continue normal growth, but since new cell-wall synthesis is inhibited, they burst the weakened cell wall and are destroyed. However, auxotrophs exposed to penicillin in minimal medium are not killed because they cannot grow under these conditions and hence do not burst their cell walls. After the prototrophs have been killed, the penicillin is removed, and the surviving auxotrophs can be recovered by plating them onto appropriately supplemented media.

For further details: Davis (1955); Hartman and Suskind (1965); Wagner and Mitchell (1964).

Nutritional deficiencies have also been extensively used by the bacterial geneticist as specific markers in various genetic studies. The rate at which auxotrophic mutants back mutate to nutritional independence is fairly stable and is easily measured. To detect revertants, all that is required is to plate a large, known number of auxotrophs onto minimal medium and to count the number of nutritionally independent colonies that arise. This spontaneous back mutation is greatly increased by various mutagenic agents and has been used for screening the effectiveness of many mutagens.

Mutations which affect virulence Virulence is defined as the capacity of an organism to produce disease. However, it is understood to depend not only on properties inherent in the organism, but also on the host defense mechanisms. Therefore, it is not, strictly speaking, possible to measure virulence without considering the state of the host. At the risk of oversimplification, however, the following discussion will be confined solely to bacterial properties influencing this relationship.

The ability of a given pathogen to produce capsular material is one of the more obvious properties influencing its degree of virulence. Freshly isolated pathogens are often thickly encapsulated, but upon continued cultivation in vitro, may lose the ability to form capsules. These nonencapsulated mutants are usually much less virulent than the original isolates. The capsule probably increases virulence by protecting the pathogen from the macrophages of the host, and possibly from certain of the circulating antibodies and antibacterial drugs.

Some bacteria are virulent solely because of the toxins that they produce. In these organisms, mutations affecting the type or quantity of toxin produced have profound effects on their degree of virulence.

Genetic exchange in bacteria

Genetic recombination may be defined as the occurrence of offspring whose combinations of genes differ from those that were present in the parents. In "higher" plants and animals recombination can result from the independent assortment of chromosomes or by the process of crossing over which occurs between homologous chromosomes. In bacteria and viruses, recombination occurs when a new, recombinant chromosome is formed from DNA contributed by two different organisms. Three different methods have been described by means of which DNA may be transferred from "donor" to "recipient" bacteria and which may result in the appearance of recombinant forms.

In *bacterial transformation*, the recipient cell in which genetic recombination occurs is capable of taking up "naked," soluble DNA from its environment. In *transduction*, bacteriophages act as vectors to carry the bacterial DNA from the donor bacteria in which the phages grew to the recipient bacteria. In *bacterial conjugation*, direct contact occurs between two sexually differentiated cells, and one of these, the donor, directly transfers chromosomal DNA to the other, the recipient. These three phenomena will be discussed in the order of their discovery.

Bacterial transformation Bacterial transformation refers to the hereditary alteration in the properties of one bacterium mediated by soluble DNA obtained from a different bacterium. The phenomenon of transformation was discovered by Griffith (1928) in *Diplococcus pneumoniae,* and has since been reported in several genera.

Virulent pneumococci characteristically produce appreciable quantities of polysaccharide capsular material, and consequently form smooth, glistening colonies on artificial medium (Fig. 257). Each strain of pneumoccoci produces a highly specific kind of polysaccharide which differs chemically and antigenically from the kinds of polysaccharide

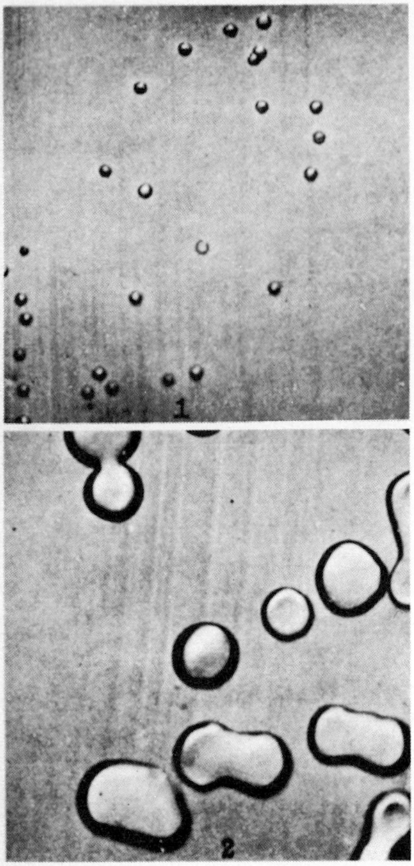

Fig. 257 (1) Pneumococcus Type II (R) colonies; (2) pneumococcus Type II (R) variant grown on agar medium with the addition of active transforming principle isolated from Type III pneumococci. The smooth, glistening, mucoid colonies shown are characteristic of Type III pneumococcus and readily distinguishable from the small rough colonies of the parent R strain. (After Avery, MacLeod, and McCarty.)

produced by all other strains. On the basis of antigenic differences between the polysaccharide capsules they produce, almost 100 different strains, or types, of pneumococci have been reported. After repeated subculture on laboratory medium, some of the smooth organisms may lose by mutation the ability to form polysaccharide capsular material. The colonies formed by such mutants assume a rough granular appearance. These nonencapsulated, "rough" strains are usually avirulent and cannot be typed serologically. While back mutations from the nonencapsulated rough state to the smooth, encapsulated state can sometimes occur, no cases of mutation from one type to another have been found.

While studying a possible correlation between pneumococcal type specificity and pathogenicity, Griffith injected a mouse with a mixture containing a few living rough pneumococci derived from a Type II culture (Type II-*R*) and a large number of heat-killed Type III smooth cells (Type III-*S*). The mouse subsequently died, and living Type III-*S* pneumococci could be isolated from it. As was later found, this type transformation resulted from the uptake by the living Type II-*R* bacteria of DNA released from the heat-killed Type III-*S* pneumococci.

This phenomenon was confirmed in other laboratories, and it was later found possible to obtain type transformation in vitro by exposing living Type II-*R* pneumoccoci to cell-free extracts of the Type III-*S* cultures (Dawson and Sia, 1931; Alloway, 1933). These newly formed Type III-*S* bacteria could, in turn, be extracted to obtain more transforming principle, and be used to transform other Type II-*R* cultures into Type III-*S*. The transforming principle was thus shown to be capable of self-duplication because the quantities extracted from the newly transformed cells greatly exceeded the quantity required to transform them.

Avery et al. (1944) demonstrated that the transforming principle found in the cell-free extracts consisted principally, if not exclusively, of highly polymerized DNA. Their discovery did much to focus the attention of biologists on the importance of nucleic acids. Prior to their report, most biologists were of the opinion that the protein portion of the chromosomes determined genetic specificity, while the DNA portion acted merely as an accessory substance.

Several traits other than type specificity in pneumococci have since been shown amenable to transformation. These include transformation of fermentation markers and transformation from the antibiotic-sensitive state to antibiotic resistance and vice versa. Transformation of traits such as streptomycin resistance to recipient bacteria sensitive to this drug have greatly aided quantitative studies on transformation in pneumococci. This is because even a few resistant transformants among a large number of sensitive recipients that have been exposed to "resistant DNA" are easily detected by plating the culture onto drug-containing agar.

Studies on transformation have revealed that conditions suitable for uptake of donor DNA into recipient bacteria occur only transiently during cultural growth. During a relatively short interval, transformable bacteria become *competent* and during this period of competency they can efficiently take up and incorporate donor DNA added to their growth medium. In most cases it is not known what factors provoke the acquisition of competence in a bacterial population; nor is it completely understood why, upon continued growth, the recipient population again become unable to incorporate added DNA. It is clear that the uptake of DNA by competent bacteria is a relatively specific process because even highly competent bacteria take up added RNA or protein very poorly, if at all, nor do they incorporate denatured DNA or highly fragmented DNA to any appreciable extent. In some species of transformable bacteria competence may be provoked by addition to non-competent populations of a protein "competence factor" produced by other competent populations. It is also known that the uptake of DNA by competent cells is an energy-requiring process, and any agent that interferes with energy metabolism of competent bacteria abolishes their ability to incorporate DNA.

After DNA has entered the recipient bacteria, it is physically integrated by a series of enzymatic steps into a homologous section of the DNA of the recipient's chromosome. Although the sequence and exact nature of these enzymatic reactions are not known, they result in replacement of an existing section of the recipient's DNA with a corresponding section of the previously incorporated donor DNA.

The ability to be transformed has been reported in a number of bacterial species including *Haemophilus, Neisseria, Streptococcus, Rhizobium*, and various species of bacilli. Extensive studies have been made of transformation in *B. subtilis* because it can be grown and transformed in simple chemically defined media and it has been possible to easily obtain and transform many nutritional markers. In addition, genetic studies on sporulation have become possible using *B. subtilis* and it has also been possible to easily infect recipient cultures of this organism with purified DNA extracted from bacteriophage.

For more information: Hayes (1968); Schaeffer (1964); Stent (1971).

Bacterial conjugation Sexual conjugation in bacteria was first demonstrated by Lederberg and Tatum in 1946 using mutants of *Escherichia coli* K-12 especially derived for this purpose. These workers were aware that several unsuccessful attempts had been made to demonstrate genetic exchange in *E. coli* and therefore designed their experiments so that even very rare recombinants could be detected. For this purpose they constructed polyauxotrophic mutants from the parental prototrophic *E. coli* K-12. The normal prototrophic strain can synthesize all its necessary protoplasmic constituents from a simple minimal medium containing an inorganic nitrogen source, a few minerals and glucose. As previously described for *Neurospora*, mutants can be obtained which lack the ability to synthesize one or more of the required cellular components such as amino acids, nucleic acid bases, or vitamins. Polyauxotrophic strains were constructed by first isolating an auxotrophic mutant which could not make, for instance, its own amino acid, methionine. This mutant was grown in minimal medium supplemented with methionine, and a diauxotrophic mutant was isolated from it which could not grow unless an additional preformed organic compound such as biotin was also added to the medium. By continuing this process, polyauxotrophs were obtained which were blocked in as many as 3 or 4 different genes, and hence required 3 or 4 different nutritional supplements for growth. Such polyauxotrophs were deemed essential for these experiments because it is almost impossible for them to completely mutate to the prototrophic state. This is because back mutation from nutritional independence for any given requirement occurs at the low frequency of about once per 10^{6-7} cell divisions. Since each back mutation is an independent event, the likelihood that even two back mutations will simultaneously occur in a given organism is the product of each individual frequency, or about once in 10^{12} to 10^{14} cell divisions.

In the experiments to detect genetic exchange, Lederberg and Tatum used two of these polyauxotrophic mutants whose genotypes can be written as follows:

Polyauxotroph 1. Met^+ Bio^+ Thr^- Leu^- Thi^-

Polyauxotroph 2. Met^- Bio^- Thr^+ Leu^+ Thi^+

These symbols stand for methionine, biotin, threonine, leucine, and thiamine, respectively. The plus (+) superscript represents the capacity of the organism to synthesize the compound and the minus (−) superscript denotes its lack of ability to synthesize the compound.

These polyauxotrophic mutants were tested for their ability to recombine by first growing each culture in a "complete" medium to which the growth factors threonine, leucine, and thiamine in the first case and methionine and biotin in the second case had been added. Both cultures were then washed and samples of them were mixed together and plated into a minimal medium free of any growth factors. As controls, appropriate samples of each culture were plated separately into minimal medium. Only prototrophs, or cells with the genotype Met^+ Bio^+ Thr^+ Leu^+ Thi^+ are able to grow in minimal medium. Colonies of this type appeared at the rate of about one for every million auxotrophic parents in the mixture, but no colonies at all appeared on the control plates. Since the appearance of prototrophs from the mixed culture could not be due to back mutation of either parent, their formation was attributed to genetic exchange between the two parental strains.

It was assumed that the polyauxotrophic mutants conjugated to form a fusion zygote containing the nuclei from both parents, and that during reduction division from the diploid to the haploid state, portions of their chromosomes were exchanged by the process of crossing over. We can represent the fusion zygote as:

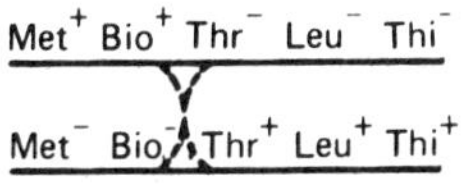

It may be seen that a single crossover between Bio and Thr results in a recombinant chromosome containing the Met^+ Bio^+ genes from parent #1 and the Thr^+ Leu^+ Thi^+ genes from the other, and that such a recombinant should be prototrophic for all the genes under consideration. Of course, other crossovers should also be possible, but only recombinants with the constitution as outlined above would be able to form colonies on unsupplemented minimal medium. The plating conditions used here are called *selective*, which means that only recombinant classes receiving the nonmutant allele can multiply.

Since the prototrophs could conceivably arise by a transformation reaction similar to that found in the pneumococci, several experiments were devised to test this possibility. One of the most straightforward of these was performed by Davis (1950). He constructed a culture device consisting of a U tube in which the two arms were separated by a sintered-glass filter whose pore size was small enough to hold back bacteria, but which allowed free passage of the medium and any soluble cellular products that might be present. He inoculated each arm of the U tube with one each of the two auxotrophic mutants, and during incubation flushed the medium through the filter from one side of the tube to the other. Prototrophs were never recovered from bacterial populations under these conditions. Results of these and similar experiments indicated that bacterial recombination of *E. coli* requires cell-to-cell contact and strengthened the concept that it results from a true sexual fusion.

Lederberg also designed experiments to determine whether genetic linkage in *E. coli* is similar to that of other organisms. For this purpose he used polyauxotrophs 1 and 2 as parents in his mating mixtures, but each contained alternative forms of several *nonselective* mutant genes in addition to their normal chromosome a small circular piece of DNA called nonselective markers included resistance or sensitivity to various agents such as bacteriophages or antibiotics as well as ability or lack of ability to ferment various carbohydrates. Nonselective markers do not have to be present in the normal form to enable the recombinants to form colonies on minimal medium. Therefore either form can be inherited along with the selective marker from one or the other parent, thus revealing their tendencies to be inherited together, or degree of linkage. Results from these analyses revealed that genes of *E. coli* may be linked and can be ordered on a linear map similar to the genetic maps of higher plants and animals. It was therefore inferred that bacteria possess genetic structures analogous to the chromosomes of higher organisms previously studied by geneticists.

Further experiments revealed that *E. coli* may be sexually differentiated into male, donor bacteria called F^+ and female, recipient bacteria called F^- (Fig. 258). It has since been shown that donor F^+ cells possess in addition to their normal chromosome, a small circular piece of DNA called the sex-, or F-factor. Sex factor DNA replicates independently of the donor's normal chromosome, but under most conditions F^+ bacteria contain only one copy of sex factor DNA per nucleus. The sex factor is about 1/100 as large as the bacterial chromosome, and thus contains enough DNA to specify about 40 genes. Some of these genes are known to control sex factor replication and others specify the synthesis of structures called sex pili. The sex pili are thin, hollow appendages which protrude from the surface of donor bacteria. They must be present for conjugation to occur, and it seems likely that the sex pili make up the "conjugation tube" through which DNA is transmitted from the donor to the recipient bacteria (Brinton, 1964).

Copies of the sex factor are synthesized by the F^+ cell and are transmitted very efficiently during bacterial conjugation to the F^- recipients. Female cells which acquire sex factors are thereby converted to the F^+ state and thereafter have all the properties of donor cells. The sex factor can be selectively eliminated from F^+ cells by growing them in the presence of acridine orange and other inhibitory agents. When this occurs, the cells lose all donor characteristics and revert to the F^- condition. They

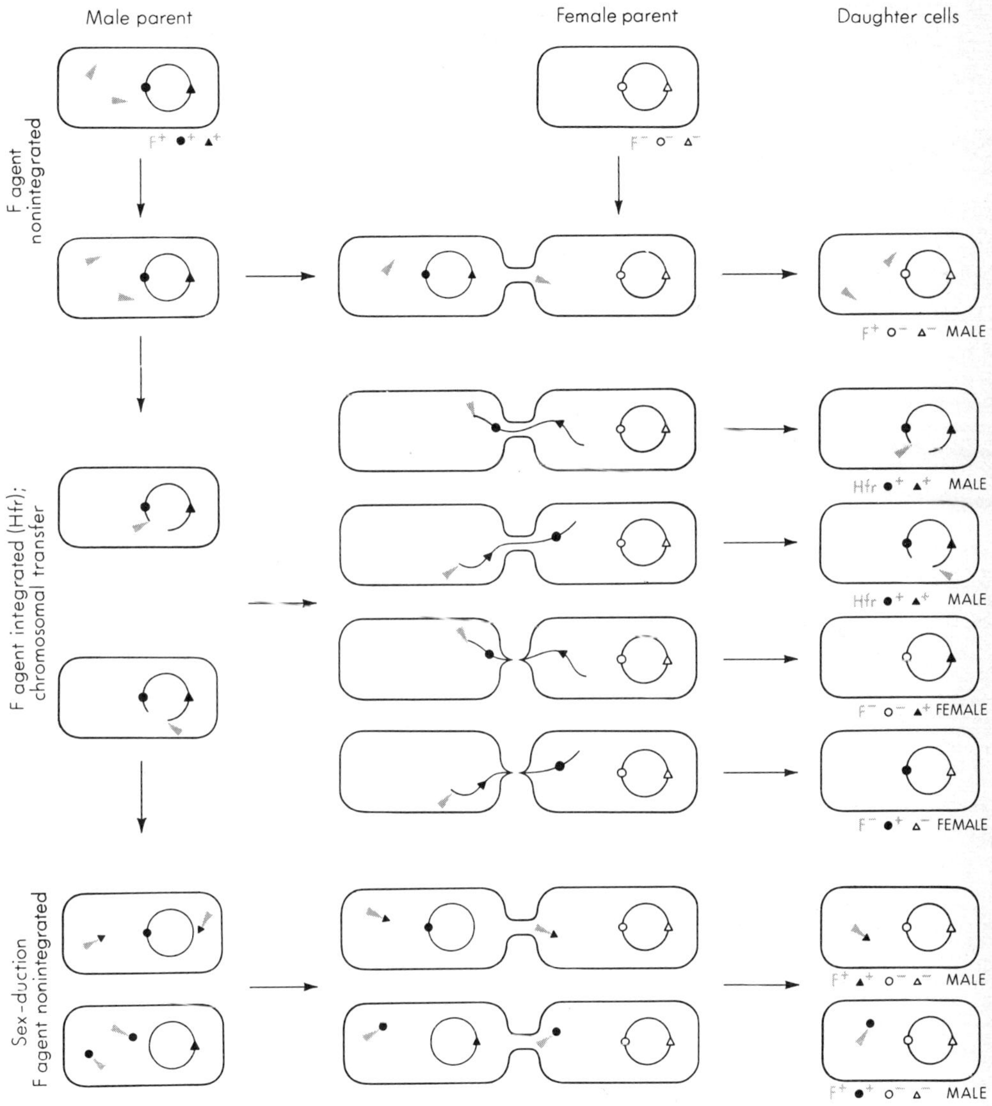

Fig. 258 F, or sex, agent, indicated by toned wedge, is a versatile and busy "broker" in genes. It can be attached to the bacterial chromosomes (integrated) or unattached (nonintegrated) and can alternate between the two states. When nonintegrated, it usually transmits only itself when bacteria conjugate (top sequence). When integrated, it opens chromosome ring and is the last marker transferred in conjugation (middle sequence). Daughters may inherit markers in combinations other than those shown. When F agent leaves integrated state (bottom), it may remove a marker and transfer it (sexduction).

can then be reconverted to the donor state only if they again receive a sex factor by conjugation with fresh F^+ donors. Even though the sex factor is transferred very efficiently in $F^+ \times F^-$ crosses, transfer of chromosomal genes from male to female strains under these conditions is a very rare event.

Soon after the discovery of the sex factor had been made, mutants of the original F^+ strains were isolated that transmitted genetic markers to the F^- strains at a much higher frequency than their parents. These mutants were called "*h*igh-*f*requency *re*combinants" (Hfr) because they transfer chromosomal genes to the female F^- strains during conjugation at a frequency about 20,000 times higher than their parental F^+ strains. However, Hfr strains are not able to efficiently transmit the sex factor to female bacteria, and unlike their F^+ parents, their donor ability is not eliminated by growing them in acridine orange.

Detailed insight into the mechanism of genetic transfer was obtained from the ingenious experiments using these Hfr mutants which were performed by Wollman et al. (1956). They mixed Hfr and F^- strains together and at timed intervals after mixing, samples were removed from the conjugation mixture and violently agitated to interrupt conjugation. Each agitated suspension, which contained the separated mating pairs, was then plated onto the proper medium to detect those genes which had been transferred from the male to the female bacteria prior to agitation. The agitation served to break the Hfr chromosome at the cytoplasmic bridge. Therefore only those Hfr genes which had already been transferred to the F^- recipient before agitation were available to participate in recombination. They found that genes were transferred into the F^- strain in a definite order and at a fixed rate, which showed that these particular Hfr donors all transmitted their chromosomes to the F^- recipients in the same sequence.

From one particular F^+ strain, Jacob and Wollman (1961) next isolated a series of different Hfr "mutants." Upon examining these new Hfr isolates by interrupted mating, they found that although a given Hfr always transferred its chromosome in the same fixed sequence, this sequence differed from one Hfr to another. They postulated that an Hfr arises from an F^+ parent as a consequence of the integration of the sex factor into the bacterial chromosome. Different Hfr donors are the result of the sex factor's capacity to be integrated into different regions of the donor's chromosome. Their analyses of gene transfer by these Hfr donors revealed that the genes of *E. coli* are arranged on a closed circular chromosome. During conjugation, the bacterial chromosome is transmitted from the Hfr donor to the F^- recipient in the order determined by the site of integration of the sex factor into the chromosomal DNA.

It was shown that F^+ donors as such probably do not transmit any chromosomal genes to recipients, but can only transmit the sex factor. When the sex factor, which is "free" in the F^+ bacteria, becomes integrated into the chromosome to form the Hfr donor, the donor chromosome is broken during conjugation and the end opposite the sex factor is always inserted first into the recipient and the attached sex factor is the last chromosomal element to be transferred. In most experiments, even without agitation, chromosomal transfer from Hfr to F^- recipients is usually incomplete, so that F^- recombinants rarely receive the sex factor. In the rare cases where an entire chromosome, including the attached sex factor, is transferred, the F^- recombinant is converted to an Hfr.

Hfr donors can revert to the F^+ condition and when this occurs, the sex factor is released from the bacterial chromosome and resumes its independent replication. Occasionally upon reversion from the Hfr to the F^+ state, the sex factor is not cleanly released from the chromosome and carries with it a few bacterial genes. Consequently the resulting sex factor contains not only the genes required for normal sex factor activity, but also those genes picked up from the bacterial chromosome. When cultures of these bacteria are mixed with F^- recipients, the sex factor is transmitted very efficiently to the recipients and carries along with it its added bacterial genes. These genes may express themselves in the resulting F^+ cells and may be detected by normal selective techniques. This process whereby bacterial genes are transmitted from donor to recipient as part of the sex factor has been called *sexduction* (Jacob and Wollman, 1961).

Resistance factors Self-replicating genetic factors other than the F^- factor have since been reported in various genera of bacteria. Increasing attention has been directed to a particular class of these factors which confer the property of transmissible drug resistance to cells which contain them. These resistance transfer factors (R factors) were first reported in Japan in the 1950s following an outbreak of bacillary dysentery. The bacteria recovered from patients suffering from this disease were discovered to be simultaneously resistant to four different antibiotics normally used to treat this disease. Investigation of these isolates revealed them to be capable of transmitting their multiple drug resistance properties to sensitive bacteria and that transfer occurred as a result of bacterial conjugation. As in the case of F transfer previously discussed, recipients receiving the R factor are likewise converted to donors and can transmit the factor to additional sensitive recipients.

Analyses have revealed that R factors contain genes for directing the synthesis of sex pili similar to the pili specified by F; they contain genes for independent replication and transfer of the R factor DNA; and in addition they contain genes which confer high degrees of antibiotic resistance to bacteria which harbor them. Antibiotic resistance associated with the R factors may result from the enzymatic destruction of the drug or it may be a consequence of a decreased permeability which limits the passage of the drug into the cell. R factors have been reported which confer resistance to penicillins, chloramphenicol, kanamycin, tetracyclines, cephalosporin, streptomycin, sulfanilamide, and other related drugs. They have been found in, or shown capable of being transferred to, *Shigella, Escherichia, Klebsiella, Salmonella, Vibrio, Proteus*, and others, and so are widespread in nature and of obvious concern in the epidemiology of these organisms.

For more information: Hayes (1968); Mitsuhashi (1971); Watanabe (1967).

Transduction (Fig. 259) Still another type of unidirectional genetic exchange has been demonstrated in bacteria, one in which a temperate virus (bacteriophage) serves as the vector. This phenomenon, called *transduction*, was discovered by Zinder and Lederberg (1952) while investigating *Salmonella* for evidences of sexual recombination. Two different auxotrophic *Salmonella* were mixed together and were then plated into minimal medium on which only prototrophs can grow. It was found that many more prototrophs than could be accounted for on the basis of back mutation were formed under these conditions. Unlike the experience with recombining *E. coli* K-12 strains described above, when the Davis U-tube experiment was performed with these organisms, prototrophs occurred very rapidly in one arm of the tube. This, of course, suggested that some soluble products from a donor strain were traversing the sintered-glass filter and transforming the recipient auxotroph to prototrophy.

Further investigation of this phenomenon revealed, however, that a bacteriophage was released from one strain, a *lysogenic* culture, which carried the phage in a hereditary, latent state and only rarely released some

Fig. 259 Transduction is similar to sexduction and was discovered earlier. In transduction the agent for transferring bacterial genes is a virus particle rather than an F agent. The virus injects its genes (color) into bacterial cell A, and the genes create new copies of the virus. Occasionally the new virus particles so formed enclose a few genes from the chromosome of the bacterial host along with a few viral genes. These imperfect viruses are able to inject their contents into another cell (cell B) but are unable to destroy it. In this way, genes (solid black shapes) can be transferred from cell A to the daughters of cell B.

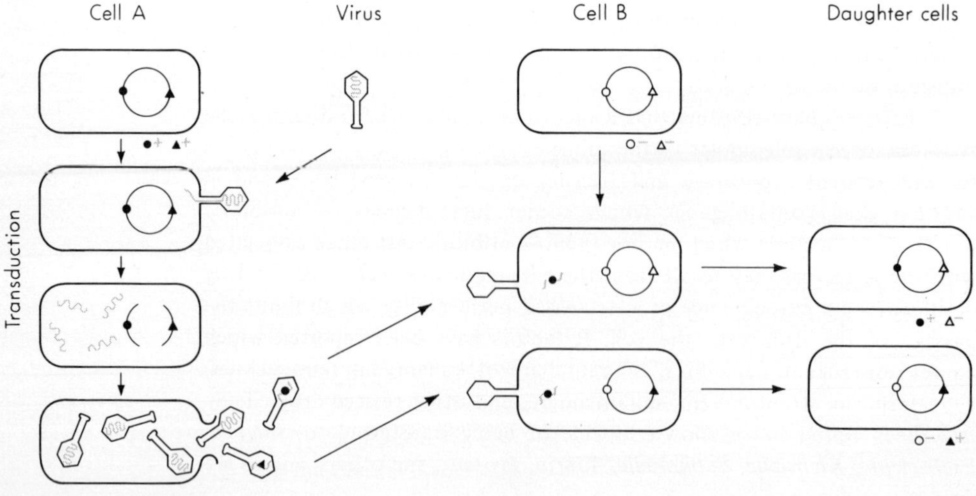

of them into the medium. The phage, which can pass through bacterial filters, was then able to infect the other strain (the donor) and lyse it. During intracellular replication of the phage in the second organism, parts of the bacterial chromosome were adventitiously included within the phage, so that when it again traversed the glass filter, it carried back to the recipient strain part of the donor's genetic information.

The genetic information that temperate bacteriophage can transmit from one bacterium to another is quite small, usually only one gene at a time. In some cases, however, when the genes are located very close together on the bacterial chromosome, both of them are included within one phage and are transferred as a unit to the recipient bacterium.

Transduction in *Salmonella* is called *generalized* transduction to signify that almost any bacterial gene can be transmitted from donor to recipient, although at a low frequency. In *E. coli* it was later found that phage *lambda* could transduce, but unlike the phage of *Salmonella*, its activity was limited solely to genes controlling the ability to ferment galactose. This kind of transduction is called *specialized* transduction (Morse et al., 1956).

In transduction the phage acts as a vector for the transmission of genetic information, the specificity of which is determined by the genotype of the host on which the phage was grown. A different phenomenon called *phage conversion* was discovered in *Corynebacterium diphtheriae* (Groman, 1955). All toxigenic diphtheria organisms seem to be infected with a temperate phage, and if these organisms are "cured" of this virus, they are rendered nontoxigenic. These "cured" strains can be readily rendered toxigenic again by reinfecting them with the same temperate phage. In this system the ability to induce toxin production seems to be a function of the phage itself, rather than to depend on the genes of the host bacterium.

For more information: Bertani (1958); Hayes (1968); Lwoff (1959); Stent (1971).

References

Alloway, J. L.: Further observations on the use of pneumococcus extracts in effecting transformation of types in vitro, *J. Exp. Med.*, **57**:265, 1933.

Avery, O. T., et al.: Studies on the chemical nature of the substance inducing transformation of pneumococcal types: Induction of transformation by a desoxyribonucleic acid fraction isolated from pneumococcus Type III, *J. Exp. Med.*, **79**:137, 1944.

Beadle, G. W.: Biochemical genetics, *Chem. Rev.*, **37**:15, 1954.

Bertani, G.: Lysogeny, *Advan. Virus Res.*, **5**:151, 1958.

Braun, W.: "Bacterial Genetics," Philadelphia, W. B. Saunders Company, 1965.

Bretscher, M. S.: How repressor molecules function, *Nature (London)*, **217**:509, 1968.

Brinton, C. C.: The structure, formation, synthesis and genetic control of bacterial pili and a molecular model for DNA and RNA transport in gram negative bacteria, *Trans. N.Y. Acad. Sci. (Ser. II)*, **27**:1003, 1964.

Brock, T. D., Streptomycin, *Symp. Soc. Gen. Microbiol.:* **16**:131, 1966.

Bryson, V., and W. Szybalski: Microbial drug resistance, *Advan. Genet.*, **7**:1, 1955.

Bunting, M. I.: The inheritance of color in bacteria, with special reference to *Serratia marcescens, Cold Spring Harbor Symp. Quant. Biol.*, **11**:25, 1946.

Cairns, J.: The bacterial chromosome and its manner of replication as seen by autoradiography, *J. Mol. Biol.*, **6**:208, 1963.

Cavalli-Sforza, L. L.: Indirect selection and origin of resistance. In "Drug Resistance in Micro-organisms," edited by G. E. W. Wolstenholme and C. M. O'Conner, London, J. and A. Churchill, Ltd., 1957.

Chargaff, E.: Isolation and composition of the deoxypentose nucleic acids and of the corresponding nucleoproteins. In "The Nucleic Acids," vol. I, New York, Academic Press, Inc., 1955, pp. 307-358.

Clark, A. J., and E. A. Adelberg: Bacterial conjugation, *Ann. Rev. Microbiol.*, **16**:289, 1962.

Cohn, M.: Contributions of studies on the beta-galactosidase of *Escherichia coli* to our understanding of enzyme synthesis, *Bact. Rev.*, **21**:140, 1957.

Crick, F. H. C.: The structure of the hereditary material, *Sci. Am.*, **191**: 54, 1954.

——The recent excitement in the coding problem. In "Progress in Nucleic Acid Research," vol. I, edited by J. N. Davidson and W. E. Cohn, New York, Academic Press, Inc., 1963.

Davis, B. D.: Isolation of biochemically deficient mutants of bacteria by penicillin, *J. Am. Chem. Soc.*, **70**:4267, 1948.

——: Nonfilterability of the agents of genetic recombination in *Escherichia coli, J. Bact.*, **60**:507, 1950.

——: Nutritional and enzymatic studies on microbial mutants. In "Perspectives and Horizons in Microbiology," edited by S. Waksman, New Brunswick, N.J., Rutgers University Press, 1955.

Dawson, M. H., and R. H. P. Sia: In vitro transformation of pneumococcal types, *J. Exp. Med.*, **54**:681, 1931.

Dean, A. C. R., and C. Hinshelwood: Aspects of the problem of drug resistance in bacteria. In "Drug Resistance in Microorganisms," edited by G. E. W. Wolstenholme and C. M. O'Conner, London, J. and A. Churchill, Ltd., 1957.

Demerec, M.: Origin of bacterial resistance to antibiotics, *J. Bact.*, **56**:63, 1948.

Driskell-Zamenhof, P.: Bacterial episomes. In "The Bacteria," vol. V, edited by I. C. Gunsalus and R. Y. Stanier, New York, Academic Press, Inc., 1964.

Gilbert, W., and B. Muller-Hill: Isolation of the *lac* repressor, *Proc. Natl. Acad. Sci., Wash.*, **56**:1891, 1966.

Griffith, F.: The significance of pneumococcal types, *J. Hyg.*, **27**:113, 1928.

Groman, N. B.: Evidence for the active role of bacteriophage in the conversion of nontoxigenic *Corynebacterium diphtheriae* to toxin production, *J. Bact.*, **69**:9, 1955.

Hartman, P. E.: Transduction: A comparative review. In "The Chemical Basis of Heredity," edited by W. D. McElroy and B. Glass, Baltimore, The Johns Hopkins Press, 1957.

—— and S. R. Suskind: "Gene Action," Englewood Cliffs, N.J., Prentice-Hall, Inc., 1965.

Hayes, W.: "The Genetics of Bacteria and Their Viruses," 2d ed., Oxford, Blackwell Scientific Publications, Ltd., 1968.

——: The structure and function of the bacterial chromosome, *Symp. Soc. Gen. Microbiol.*, **15**:294, 1965.

Jacob, F., and E. L. Wollman: "Sexuality and the Genetics of Bacteria," New York, Academic Press, Inc., 1961.

—— and J. Monod: Genetic regulatory mechanisms in the synthesis of proteins, *J. Mol. Biol.*, **3**:318, 1961.

Kellenberger, E.: The physical state of the bacterial nucleus, *Symp. Soc. Gen. Microbiol.*, **10**:39, 1960.

Krieg, D. R.: Specificity of chemical mutagenesis. In "Progress in Nucleic Acid Research," vol. II, edited by J. N. Davidson and W. E. Cohn, New York, Academic Press, Inc., 1963.

Lederberg, J.: Viruses, genes and cells, *Bact. Rev.*, **21**:133, 1957.

—— and E. Lederberg: Replica plating and indirect selection of bacterial mutants, *J. Bact.*, **63**:399, 1952.

—— and E. L. Tatum: Gene recombination in *Escherichia coli*, *Nature*, **158**:588, 1946.

—— and N. Zinder: Concentration of biochemical mutants of bacteria with penicillin, *J. Am. Chem. Soc.*, **70**:4267, 1948.

Luria, S. E., and M. Delbrück: Mutations of bacteria from virus sensitivity to virus resistance, *Genetics*, **28**:491, 1943.

Lwoff, A.: Bacteriophage as a model of host-virus relationship. In "The Viruses," edited by F. M. Burnet and W. M. Stanley, New York, Academic Press, Inc., 1959.

Meselson, M., and F. W. Stahl: The replication of DNA in *Escherichia coli*, *Proc. Natl. Acad. Sci. U.S.*, **44**:671, 1958.

Mitsuhashi, S.: "Transferable Drug Resistance Factor R," Baltimore, University Park Press, 1971.

Morse, M. L., et al.: Transduction in *Escherichia coli*, *Genetics*, **41**:142, 1956.

Newcombe, H. B.: Origin of bacterial variants, *Nature*, **164**:150, 1949.

Nirenberg, M. W.: The genetic code: II, *Sci. Am.*, **208**:80, 1963.

Robinow, C. F.: Cytological observations of *Bact. coli*, *Proteus vulgaris* and various aerobic spore-forming bacteria, with special reference to their nuclear structure, *J. Hyg.*, **43**:413, 1944.

——: Morphology of the bacterial nucleus, *Brit. Med. Rev.*, **18**:31, 1962.

Ryter, A., and F. Jacob: Étude au microscope électronique de la liaison entre noyau et mésosome chez *Bacillus subtilis, Ann. Inst. Pasteur*, **107**:384, 1964.

Schaeffer, P.: Transformation. In "The Bacteria," vol. V, edited by I. C. Gunsalus and R. Y. Stanier, New York, Academic Press, Inc., 1964.

Sheppard, D., and E. Englesberg: Positive control in the L-arabinose gene-enzyme complex of *Escherichia coli* B/r as exhibited with stable merodiploids, *Cold Spring Harbor Symp. Quant. Biol.*, **31**:345, 1966.

Stent, G. S.: "Molecular Genetics, An Introductory Narrative," San Francisco, W. H. Freeman and Company, 1971.

Wagner, R. P., and H. K. Mitchell: "Genetics and Metabolism," 2d ed., New York, John Wiley & Sons, Inc., 1964.

Watanabe, T.: Infectious drug resistance, *Sci. Am.*, **217**:19, 1967.

Watson, J. D., and F. H. C. Crick: The structure of DNA, *Cold Spring Harbor Symp. Quant. Biol.*, **18**:123, 1953.

Wollman, E. L., et al.: Conjugation and genetic recombination in *Escherichia coli* K-12, *Cold Spring Harbor Symp. Quant. Biol.*, **21**:141, 1956.

Zamenhof, S.: "The Chemistry of Heredity," Springfield, Ill., Charles C Thomas, Publisher, 1959.

Zinder, N. D., and J. Lederberg: Genetic exchange in *Salmonella, J. Bact.*, **64**:679, 1952.

Bacteriology of air 19

The atmosphere consists of a mixture of permanent gases and variable quantities of water and solid particles. Its gaseous content, vapor pressure, and suspended matter are not constant in composition.

According to Landsberg (1951), air has the following composition:

Element	*Volume, percent*	*Range*
Nitrogen	78.03	
Oxygen	20.99	
Argon	0.94	
Carbon dioxide	0.03	
Hydrogen	0.01	
Neon	0.0012	
Krypton	0.0010	
Helium	0.0004	
Xenon	0.0001	
Ozone	Very variable	2 to 20 x 10^{-7}
Water vapor	Very variable	0 to 4
Dust	Very variable	0 to millions of particles per ml

The composition shows slight variations with latitude and to a lesser extent with altitude. The ozone owes its existence in the atmosphere to photosynthesis from oxygen under the influence of solar ultraviolet radiation. This process takes place at heights of 15 to 22 miles, where the ozone concentration is much greater than at the surface. Most of the water vapor is concentrated in the lower atmosphere. The amount is usually about 1.2 percent by volume, but in cold weather this quantity may fall almost to zero. Almost any sample contains suspended matter consisting of dust, bacteria, yeasts, molds, pollen grains, etc. Unlike the gaseous content, the suspended matter shows considerable variation.

Air is not a natural environment for the growth and reproduction of microorganisms. It does not contain the necessary amount of moisture and kinds of nutrients in a form utilizable by bacteria and other microscopic organisms. Therefore, air does not possess a flora. Yet organisms are found in air, and their presence is of considerable importance economically and to public health.

Microorganisms are introduced into the air by various forces, the principal source being from dust particles containing dry vegetative cells and spores. The organisms for the most part are saprophytes, i.e., forms that live on dead organic matter. They are of great importance to the canner, in sugar refineries, dairies, biological laboratories, etc. In short, they are the organisms responsible for contaminations from the air.

The species vary somewhat, depending upon the locality. However, certain forms are quite uniformly present. Molds and yeasts are commonly found in the air and in some localities even outnumber the bacteria. These organisms produce spores which are capable of resisting unfavorable conditions for long periods of time. The aerobic spore-forming bacilli from the soil are found quite frequently in the air. The best-known member of this group is *Bacillus subtilis*. It is known as the hay bacillus and is one of the most common bacterial organisms found in nature. Its natural habitat is in the soil and on vegetation. Since it is a spore-forming organism, it is very resistant to drying and other unfavorable environmental conditions. Sarcinae and micrococci also are found in air. The spherical, saprophytic, chromogenic organisms found in air usually belong to these two genera.

Number and kinds of organisms in air

The number of organisms in the air is dependent upon the activity in the environment and upon the amount of dust stirred up. An active environment shows a higher bacterial count than a less active one. The numbers in dirty, untidy rooms are greater than in clean rooms.

A rich, fertile, cultivated soil shows a higher viable count than a sandy, or clay, uncultivated soil. It follows that the air above the fertile, cultivated soil will contain more organisms than the air above the poor soil. Likewise, the air above a bare surface contains more organisms than the air above land covered with vegetation. This means that where the earth is bare the organisms can be blown more easily into the air because the earth is not protected from air currents.

Marine air also contains microorganisms, although to a lesser extent than terrestrial air. Kelly and Layne (1957*a, b*) collected air samples over the Atlantic Ocean at altitudes of 8000 to 9000 ft. Species of the following genera were isolated: *Achromobacter, Bacillus, Corynebacterium, Flavobacterium, Micrococcus, Sarcina,* and *Staphylococcus.* The investigators concluded that the organisms from marine air were similar in kinds to those found over land. Tropical air showed higher percentages of gram-negative rods and gram-positive pleomorphic rods than micrococci and sporeformers, whereas arctic air showed the opposite.

Microorganisms remain in air for varying periods of time, depending upon the speed of the air current, the size of the particles on which they are attached, and the humidity of the atmosphere. Organisms in the free state are slightly heavier than air and settle out slowly in a quiet atmosphere. However, a gentle air current is capable of keeping them in suspension almost indefinitely. Organisms attached to dust particles or in droplets of moisture settle out at a much faster rate.

A damp or humid atmosphere contains fewer organisms than a dry one, owing to the fact that the organisms are carried down by the droplets of moisture. The air of a refrigerator is usually free from all organisms. Therefore, air during the dry summer months contains many more organisms than during the wet winter months. Gently expired air from the lungs is sterile. The moist passages of the upper respiratory tract remove the bacteria from the air. Cotton stoppers in pipettes are not necessary as far as contamination of the contents is concerned. They are inserted as a protection against aspirating infectious or other material into the mouth.

Altitudes attained by microorganisms

Organisms have been recovered at altitudes as high as 20,000 ft or more through the chance action of air currents. This is of particular significance as it suggests the almost limitless possibilities of travel in a horizontal direction. The survival of such forms despite the many influences which are unfavorable to their existence is also significant in view of the length of time they may remain viable.

Pollen grains and dust also have been recovered at high altitudes and indicates the importance of air as a vehicle for the transmission of wind-borne particles over a wide area.

For more information: Fulton (1966); Fulton and Mitchell (1966); Timmons et al. (1966).

Enumeration of bacteria in air

Biological contaminants suspended in air are referred to as aerosols. Each particle of an aerosol may vary in size from less than 1 μ to 50 μ, or possibly more. The particles may consist of a single organism in the free state or a clump composed of many organisms. Sometimes an organism is not free-floating but adhering to a dust particle. Since conditions in the atmosphere are not favorable to the existence of vegetative cells, most of the viable organisms found in air are spores. Some vegetative cells such as staphylococci, streptococci, and mycobacteria are better able to resist drying. They may survive for relatively long periods of time and be carried considerable distances by air currents. If they are permitted to settle in a quiet atmosphere, they may again become airborne during activities such as sweeping and bedmaking.

The number and variety of methods which have been proposed for the enumeration of organisms in air are so great that it is beyond the scope of this book to review more than a few of them. For a more complete coverage, the reader is referred to the references given at the end of this section.

On the basis of their mode of action, aerosol samplers may be classified as follows: impingement in liquids, impaction on solid surfaces,

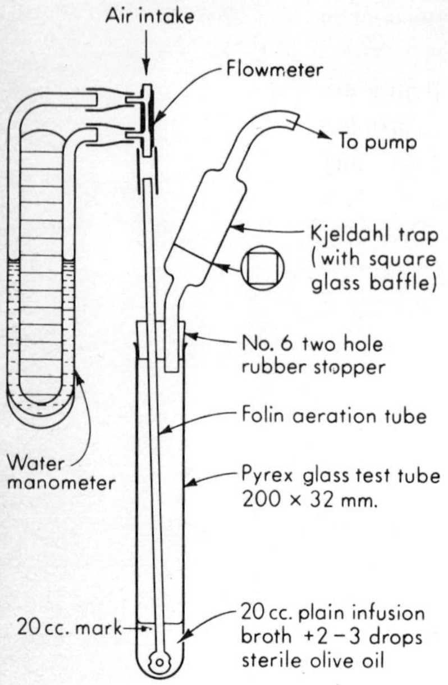

Fig. 260 *Device for the collection of airborne bacteria.* (After Lemon.)

filtration, sedimentation, centrifugation, electrostatic precipitation, and thermal precipitation.

Impingement in liquids In samplers of this type, the air is drawn through very small openings or a capillary tube and bubbled through the liquid. The airborne particles are wetted and entrapped in the collecting medium. Since the air is bubbled through the liquid, fragmentation of clumps of organisms will occur.

The Lemon (1943) sampler consists of a glass Folin aeration tube with a bulb at one end perforated by six holes (Fig. 260). The tube is passed through a two-hole rubber stopper, and the bulb centered near the bottom of the containing test tube. The Kjeldahl trap with square glass baffle is shortened at both ends for convenience, and a slight bend made in the intake so that this may be inserted into the remaining hole of the stopper. A small flowmeter measures the rate of air flow entering the upper open end of the Folin tube. An air pump is attached to the exhaust end of the Kjeldahl trap.

The entire bubbler should be sterilized by autoclaving or by rinsing with 70 percent alcohol and drying. Air drawn in at the rate of 25 to 30 liters per min is dispersed through 20 ml of broth containing two or three drops of olive oil to prevent foaming. Room-air samples generally require about 300 liters of air for satisfactory results.

Kluyver and Visser (1950) found the capillary impinger type of sampler to be the most efficient and satisfactory (Fig. 261). The air is brought in contact with water through a capillary tube to break up any clumps of bacteria. An aliquot part of the water is mixed with melted nutrient agar in a petri dish. After incubation, the colonies are counted. The

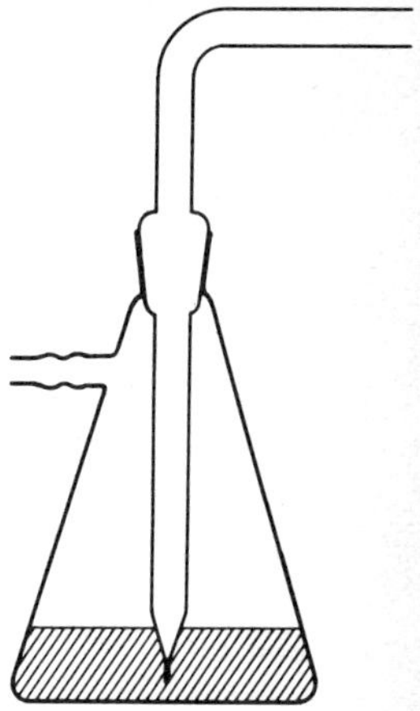

Fig. 261 Capillary impinger. See text for description. (After Kluyver and Visser.)

device recovers at least 99 percent of the spores of *B. cereus* from artificially contaminated air.

For more information: Buchanan et al. (1968).

Impaction on solid surfaces Samplers of this type deposit suspended particles directly on solid surfaces.

The organisms generally are deposited on agar medium. The agar is incubated until visible colonies appear, and then counts are made.

The Hollaender and Dalla Valle (1939) sampler consists of a brass container with a removable bottom (Fig. 262). The container is fitted with an inverted 3 1/4-in glass funnel which sits approximately 1/2 in from the bottom of a petri dish. The latter is placed in the lower portion of the container before use, then screwed tightly against the gasket. The inside of the funnel and rim are swabbed with alcohol before use. The air sample passes through the funnel stem, and the airborne organisms are impinged upon the agar medium. The air is drawn by means of a pump in series with a flowmeter. A sampling rate of 28 liters per min was found to be the most effective. The method is simple, portable, and efficient, and all testing is carried out in one operation.

Another device of this type is the Andersen (1958) air sampler (Fig. 263).

Fig. 262 Funnel device used for sampling airborne bacteria. (After Hollaender and Dalla Valle.)

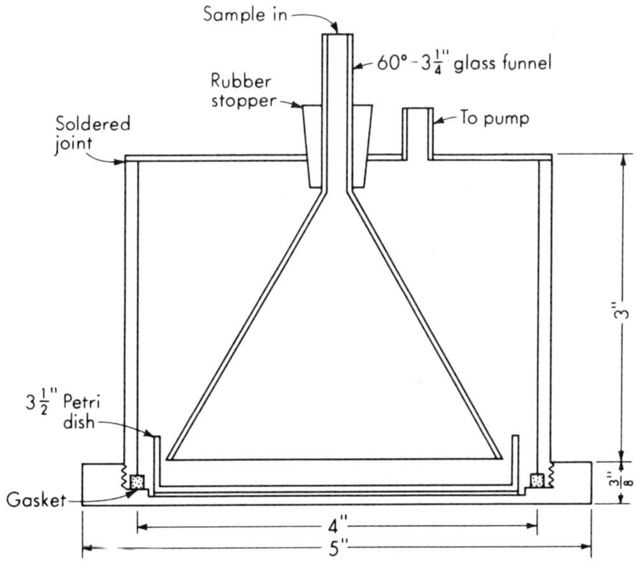

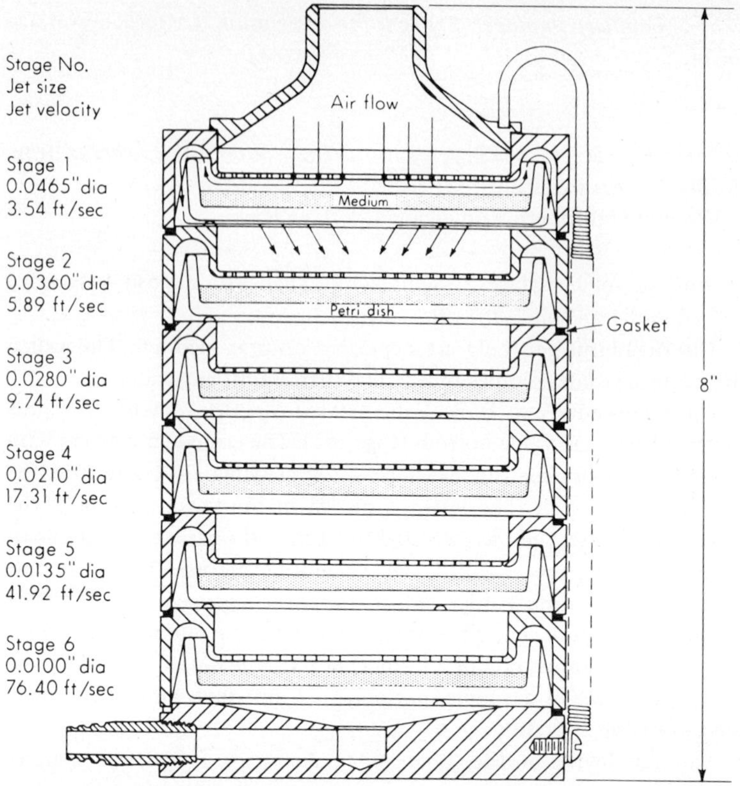

Fig. 263 Andersen air sampler. See text for description. (Courtesy of Andersen.)

The sampler automatically separates air particles into six aerodynamic sizes which indicate respiratory tract penetration and hazard. The viable organisms are collected on agar plates, incubated, and counted.

The aerodynamic size of airborne particles is important in studies concerned with lung penetration.

The Andersen sampler has 400 drilled small jets in each stage instead of one large jet. For this reason the device is extremely sensitive and efficient.

For more information: Martin (1967).

Filtration Filters of the type of Polypore and Millipore membranes are used for the removal of organisms from air. These are discussed in Chap. 8, page 262.

The diameters of the collected particles are larger than the pores of the membranes and are retained on or within a few microns of the filter surface. After filtration, the membrane is agitated in a suitable liquid to disperse the particles; then aliquots of the suspension are mixed with melted agar and incubated, and counts are determined. The counts usually are higher than the actual number of viable particles because clumps of organisms may be fragmented during agitation of the sample.

For more information: Roelants et al. (1968).

Sedimentation Airborne organisms may be collected by exposing agar plates for definite periods of time. Then the covers are replaced, plates incubated, and counts determined (Fig. 264).

This is one of the simplest procedures used for air examination but is of no value from a quantitative standpoint. It does not indicate the number of organisms present in a given volume of air. However, the method does give relative results and is commonly employed for that purpose.

Air movements may influence the results by causing the deposition of the larger-sized particles. This may give an erroneous picture of the actual distribution of organisms in air. On the other hand, when aerosols are collected in a closed area free from air movements, a quantitative measure of particle size distribution may be obtained by microscopic examination.

Centrifugation Samplers of this type employ centrifugal force for propulsion of air particles to the collecting surface. When air moves in a circular direction at high speed, the suspended aerosols are impacted on the collecting surface by a force proportional to the particle velocity and mass.

In one type of centrifuge, the sampler remains stationary while the aerosol travels in a circular path, the larger particles being collected at the

Fig. 264 Organisms in air. Left, colonies developing on an agar plate exposed to a quiet atmosphere for 10 min; right, colonies developing on an agar plate exposed to an active atmosphere for the same period of time.

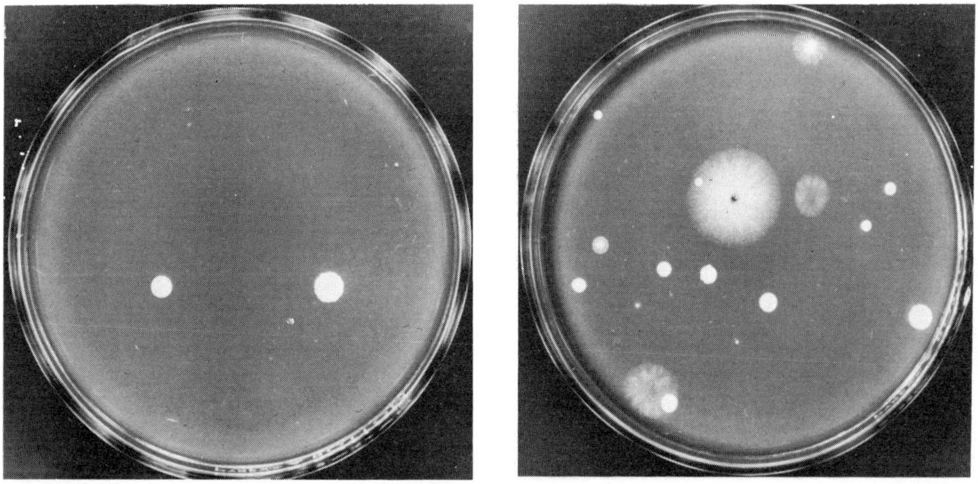

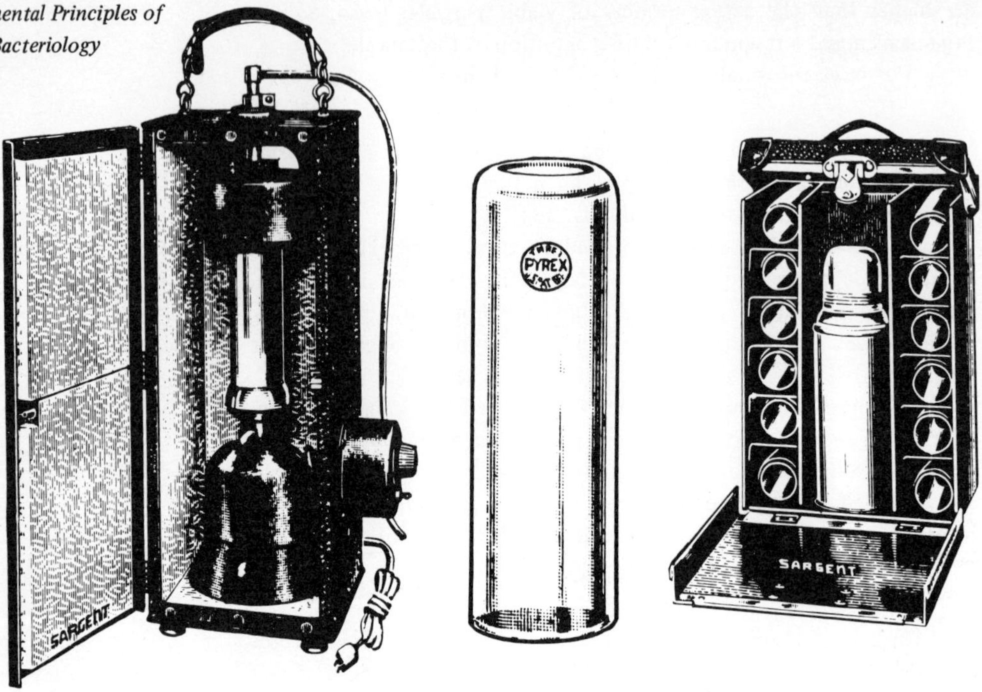

Fig. 265 Wells air centrifuge. Left, centrifuge; center, sample tube; right, auxiliary case containing sample tubes and melted agar. See text for description.

bottom. In another type, the collecting vessel and aerosol rotate at high velocity resulting in the impaction of particles on the walls.

Wells (1933) utilized the principle of the latter type for the separation of particles from air (Fig. 265). The centrifuge consists of a head assembly with exhaust fan, air inlet tube, and chamber to contain the sample tube. The variable-speed motor is mounted in an aluminum housing and equipped with a 300-step rheostat, providing a range of speeds from 2000 to 4500 rpm. Air flow is measured by a manometer tube with reservoir bottle and rubber connecting tube. The head assembly fan and sample chamber are driven by the motor through a hub on which rests the sample tube. The air inlet tube, which is suspended from the case top by means of a rubber disk and clamp, serves as the upper shaft.

An auxiliary case is provided for carrying 12 sample tubes and melted agar.

In use, the culture medium in the fluid state is poured into the sample tube. The tube is inserted in the centrifuge and the current switched on. The incoming air is mixed with the agar and the medium is spread as a thin film on the wall of the sample tube by the centrifugal force of rotation. The organisms are deposited in the agar film and, after incubation, may be counted.

Fig. 266 Cylindrical sampler. The concave surface (B) at the bottom is necessary to avoid breaking of the solid medium during centrifugation. See text for description. (After Borzani and Vairo.)

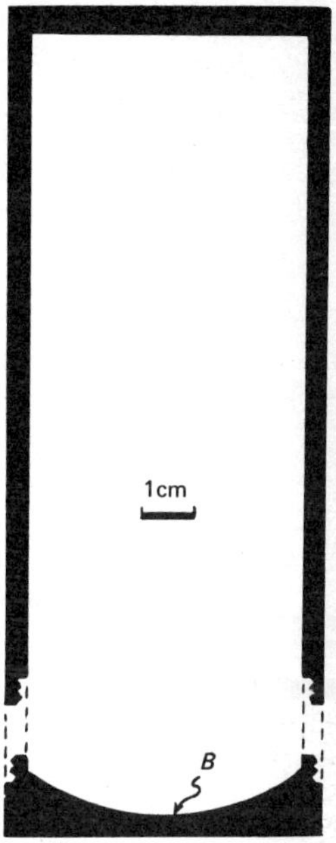

The Wells centrifuge is simple to operate, quickly manipulated, and portable. All testing is carried out in one step.

Borzani and Vairo (1967) recommended a sampler composed of aluminum alloy in cylinder form (Fig. 266). The cylinder contains a certain volume of air and a solid culture medium at the bottom. It is centrifugated, the time and rate being predetermined. The suspended cells are deposited on the surface of the medium, and after incubation the colonies are counted and the concentration of organisms calculated.

The solid medium can be substituted by a suitable liquid, such as 5 percent ethanol solution. In this case aliquots of the solution are plated in melted agar and counts of the entire volume determined.

Electrostatic precipitation Samplers using this method collect particles by drawing air over an electrically charged surface.

Luckiesh et al. (1946) devised a sampler holding two petri dishes in separate plastic units with removable covers (Fig. 267). A small electrically operated blower draws air at equal rates through the two units. One unit has the lower electrode negative and the upper electrode, a fairly flat metal cone, positive. In the other unit, the electrical conditions are reversed.

Both positively and negatively charged organisms exist in the air simultaneously. Those having a positive charge are collected on the petri

Fig. 267 A vertical section of the duplex electrostatic air sampler. (After Luckiesh, Taylor, and Holladay.)

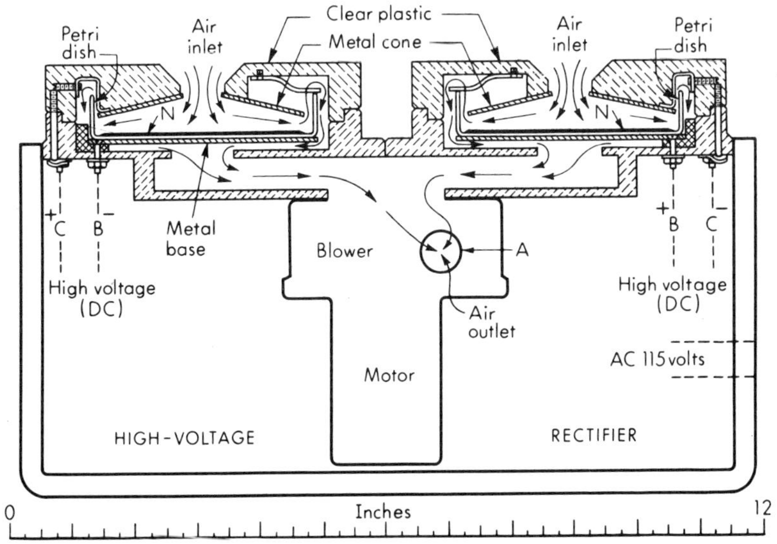

dish placed over the negative electrode; those negatively charged are collected on the petri dish placed over the positive electrode.

Thermal precipitation Samplers of this type collect particles on surfaces by means of thermal gradients. Airborne particles are repelled by hot surfaces and deposited on colder surfaces by forces proportional to the temperature gradient. Since such samplers require careful adjustment and the air sampling rates are quite low, they are not commonly used.

For more information: Andersen and Andersen (1962); Artenstein et al. (1967*a*, 1968); Errington and Powell (1969); Gregory and Monteith (1967); Harstad et al. (1967); Technical Development Laboratories (1959); Warner et al. (1964).

Importance of state of suspension

Organisms suspended in air are seldom found in the free state. They are usually attached to floating particles such as dust, saliva, and carbon. It is necessary to distinguish between ubiquitous saprophytic soil organisms raised as a dust and those from body tissues introduced into the air during coughing, sneezing, talking, and singing. The former have no pathogenic significance; the latter do.

Newer techniques developed during the past few years have stimulated a renewed interest in air microbiology. Formerly it was believed that air played an insignificant role in the transfer of infections from one person to another. This view is no longer tenable since it has been definitely established that air is capable of transmitting infections, especially those of the respiratory tract, and that infections transferred in this manner may be of frequent occurrence. This is especially true in closed spaces such as in rooms, offices, theaters, barracks, and halls.

The state of suspension plays a very important role on the settling velocity of organisms in air. Organisms in the free state are slightly heavier than air and settle out very slowly in a quiet atmosphere. A gentle current is capable of keeping them in suspension almost indefinitely. Dust particles laden with organisms settle out rapidly and remain in a quiet atmosphere for a relatively short period of time. Droplets expelled into the air during coughing and sneezing do not necessarily fall immediately to the ground within a short distance from their source. As droplets decrease in size by evaporation, the surface exposed to air resistance increases. This means that droplets remain suspended longer as they decrease in size.

Some droplets are of such size that complete evaporation occurs in falling the height of a man. This droplet size has been estimated to be about 0.1 mm in diameter. The residues of droplets of this size will float or drift with the slightest air current and become in effect a part of the atmosphere.

Air infections are said to be produced by two types of droplets, depending upon their sizes. Droplet infection proper applies to droplets larger than 0.1 mm in diameter, which rapidly settle out a short distance from their source and before drying occurs. The other type may be called airborne infection and applies to the dried residues of droplets (droplet nuclei) derived from droplets less than 0.1 mm in diameter. The time they remain suspended in air depends upon the activity of the atmosphere. Droplet infection may become airborne infection when large droplets evaporate in settling to the floor, and then are lifted into the air as dust. Droplet infection remains localized and concentrated, whereas airborne infection may be carried long distances and is dilute.

Airborne pathogenic organisms

Dust in hospital wards, army barracks, and other places where people congregate becomes highly contaminated with certain viruses and pathogenic bacteria associated with diseases of the respiratory tract, particularly streptococci. The dispersion of these organisms into the air from floors, bedclothes, and clothes of the room occupants at the time of floor sweeping, bedmaking, and dressing results in a general contamination of the environment, providing many opportunities for the spread of disease by direct or indirect contact with the dust.

Virulent β-streptococci have been repeatedly isolated from floor sweepings of public places, particularly schools. Such isolated strains were found to survive for more than 5 days when inoculated into sterile dry dust, and for 31 days when spread over the surface of floorboards at room temperature.

Other infections which have been transmitted by aerosols include brucellosis, tuberculosis, tularemia, pneumonia, pneumonic plague, anthrax, influenza, mumps, diphtheria, scarlet fever, meningitis, chicken pox, Q fever, typhus, coccidiomycosis, and those in which staphylococci are involved.

For more information: Artenstein et al. (1967*b*); de Jong and Winkler (1968); Favero et al. (1968); Kenny and Sabel (1968); Kethley et al. (1963); Lidwell and Towers (1969); McDermott (1961); May and Druett (1968); Riley et al. (1962); Sciple et al. (1967); Walter et al. (1969); Williams (1967).

Effect of atmospheric temperature and humidity

In general, aerosols of microorganisms decay or lose viability very rapidly. Webb (1959, 1960*a, b*) showed that decay occurred in two stages: (1) a rapid death in the first few seconds of cloud age; and (2) a slower rate thereafter. The death of airborne cells decreased as humidity increased and temperature decreased. Low and high relative humidity hastened the decay rate for most organisms. The optimal range occurred between 40 and 80 percent humidity. Webb correlated the rate of death of air organisms with the amount of water bonded to the cellular proteins. Cells suspended in solutions of various compounds contained less water. Inositol produced the smallest change in the water content and afforded a large measure of stability to the airborne cells. He suggested that compounds enhance survival by replacing water molecules in protein structures during desiccation.

Wright et al. (1968*a*) aerosolized cultures of *Mycoplasma laidlawii* into enclosed air spaces and reported maximal survival at relative humidities (RH) of 25 percent or lower and at 75 percent or higher. The number of organisms decreased rapidly when exposed to midrange humidities, e.g., at 40 percent only 1 percent of the organisms initially present was recovered. In a later report (1968*b*) they found that when *M. pneumoniae* was used, the most lethal RH levels were at 60 and 80 percent.

Ehrlich et al. (1970) reported that airborne survival of *Serratia marcescens* and *Escherichia coli* was closely related to temperature. An increase in temperature from −18° to 49°C resulted in a progressive increase of the death rate. An increase in temperature from 24 to 49°C resulted in significantly reduced aerosol recoveries of the two organisms. At −40°C the aerosol recovery was consistently lower than at −18 to 24°C.

Aerosols of viruses behave in a manner similar to the bacteria. Harper (1961) found that virus particles of influenza, poliomyelitis, and vaccinia survived better in air at low temperatures (7 to 24°C). All viruses showed better survival at relative humidities of 17 to 25 percent except poliomyelitis which survived better at humidities of 80 to 81 percent.

The results would indicate some variation in RH depending upon the organism or virus aerosolized.

Wright et al. (1969) prepared eight aerosols of *M. pneumoniae* at RH between 0 and 85 percent and five temperatures between 10 and 43°C. Survival was found to be a function of both RH and temperature. At all temperatures, survival was best at the extremes of RH. Irrespective of RH, an increase in temperature resulted in a decreased airborne survival time.

For more information: Buckland and Tyrrell (1962); Cox et al. (1970); Hatch and Dimmick (1966); Webb (1961, 1963*a, b, c*); Webb et al. (1963); Wilkinson (1966); Barlow (1972); Donaldson (1972).

Potential hazards of laboratory techniques

Infections have occurred among laboratory workers handling disease organisms. Some of the organisms involved are *Brucella abortus, B. melitensis, Pasteurella tularensis,* and *Salmonella typhosa*. In most cases the source of these infections has been obscure and has not been related to known accidents in handling cultures. Many laboratory techniques have been accepted with little consideration of the role they may play in the dissemination of organisms into the air.

The addition of the finest safety device to a laboratory does not ensure against the possibility of laboratory infections. Equipment is not a substitute for careful laboratory technique. All laboratory workers should be instructed to follow safe procedures to guard against the possible dissemination of disease organisms.

Johansson and Ferris (1946) showed that certain accepted laboratory operations, such as pipetting, pouring, and vigorous agitation of dilution blanks, often produced bacterial contaminations of the surrounding air and environment. Slightly more than one-half of the laboratory operations revealed droplet aerosols formed by blowing the last drop from pipettes or removing the stoppers from dilution blanks after vigorous agitation. The dangers in handling some of the highly infectious agents with the commonly used laboratory equipment were apparent.

Barbeito et al. (1961) reported that 20 petri dish cultures of *P. tularensis* were dropped in the hallway of a laboratory building. At the time of the accident or during the cleanup, 14 persons entered the hallway from adjoining rooms. Of these, 5 became infected, 2 being as far as 70 ft from the site of the accident.

During the past several years, the preservation of bacteria and viruses by drying (lyophilization) has increased to such an extent that almost every laboratory has occasion to treat organisms by this technique. Organisms in the desiccated state are easily disseminated into the air. Reitman et al. (1954*a, b, c*) showed that the lyophilizer became heavily contaminated during its operation. Danger to the operator appeared to lie in the contamination of the hands from the manifold outlets. Heavy contamination of the apparatus precluded disassembling before sterilization.

For more information: Ederer (1965); Pike et al. (1965); Sulkin (1961); Wedum (1961, 1964).

Air sanitation

Methods for the continuous disinfection of air of enclosed spaces have been employed and found to be practicable. These may be classified as follows:

1. Suppression of dust, particularly through application of oil emulsions to floors, blankets, and bed linen.
2. Inactivation of organisms by mists and vapors.
3. Inactivation of organisms by ultraviolet rays.

The bacterial content of dust found in homes, schools, factories, offices, and hospitals varies with the different environments. Saprophytic organisms usually predominate but parasitic and pathogenic agents may be found in large numbers. Both healthy and ill individuals more or less continually extrude bacteria into their environments.

Dust has been studied particularly in relation to respiratory tract infections, skin infections, and secondary infections of burns and wounds. Large numbers of hemolytic streptococci, staphylococci, pneumococci, diphtheria bacilli, and tubercle bacilli have been demonstrated in floor dust of hospital wards. These organisms survive in the environment for long periods of time. Little is known concerning the survival of viruses in dust. Influenza A virus has been shown to survive in floor dust up to 10 days without loss of its ability to produce infection in susceptible animals.

The great proportion of organisms expelled from the respiratory tract in droplets and droplet nuclei eventually settle to form a part of the bacterial component of dust.

Suppression of dust Oiling floors, bedclothes, and other textiles is a highly effective method for the control of dust, lint, and dust-borne bacteria. The action of the oil is a mechanical one only. Methods are available for the treatment of surfaces and fabrics with oil which fall within the range of practicability with respect to simplicity of application and cost. Although the most important environmental reservoirs of pathogens found in hospital wards are the floors and bedclothes, all surfaces, as well as textiles, should be oiled to bring about maximum dust control. Studies thus far indicate that oiling of floors, bedclothes, and other textiles can effect a significant reduction in the incidence of respiratory-tract infections of streptococcal etiology in hospitals and army barracks.

Litsky and Litsky (1968) recommended dust removal by the use of dry-vacuum pickup followed by the application of an appropriate disinfectant-detergent solution.

For more information: Coriell et al. (1968); Dixon et al. (1966); Loosli (1948); Shechmeister and Greenspan (1947); Westwood et al. (1971); Shreve et al. (1970).

Effect of mists and sprays Hypochlorous acid or a hypochlorite in a final concentration of 1:2 million is sufficient to produce 90 to 99 percent reduction of influenza virus suspended in air. In general, the same concentration applies to bacteria.

The effectiveness of hypochlorous acid and hypochlorite aerosols for airborne organisms depends upon the moisture content. The degree of inactivation of streptococci and staphylococci is slight if the relative humidity falls below 50 percent. If the humidity is maintained between 70 and 90 percent, the organisms are rapidly killed.

Propylene glycol atomized into air in a concentration of 1:4 million produces an immediate and complete sterilization of suspended aerosols of streptococci, staphylococci, pneumococci, *H. influenzae*, and influenza virus. The killing process is most effective at a temperature of 27°C and a relative humidity of 45 to 70 percent. The same holds true for triethylene glycol aerosols.

Glycols are believed to exert their bactericidal action by the chief physical property of hygroscopicity. The glycol molecules dissolve in the film of moisture surrounding each organism. When the glycol reaches a certain concentration, moisture is drawn out of the cell, resulting in its death.

Friedman et al. (1968) spray-fogged hospital rooms with a commercially available quaternary ammonium surface-active disinfectant and found it to be an effective means of reducing the number of detectable airborne and surface bacteria. Rooms vacated by patients infected with staphylococci, streptococci, pseudomonads, and salmonellae were found to be effectively decontaminated of most of the detectable organisms by the fogging procedure.

For more information: Dunham (1968); Lawrence (1968); Lester et al. (1949).

Effect of ultraviolet light Ultraviolet rays centering around 2537 angstroms (Å) emitted by special lamps exert a pronounced destructive action on bacteria, viruses, and other microorganisms suspended in air (see page 303). However, the highly germicidal range is between 2000 and 2967 Å with a maximum at 2650 Å (Fig. 268). Rays emitted by the sun do not extend below the region of 2950 Å because of the earth's protective atmosphere. Although the high intensity of sunlight makes up somewhat for the lack of energy in the lower wavelengths, the germicidal effectiveness varies enormously from hour to hour during the day and with the seasons. Germicidal lamps, on the other hand, are available at all times for production of rays at their predominant wavelength of 2537 Å.

Radiant disinfection of air depends on type of contamination, state of suspension, humidity of the atmosphere, volume of space, quality of the radiation, strength of ray, length of ray, total exposure, uniformity of exposure, and air motion.

The germicidal lamp is a low-pressure mercury lamp designed so that as much energy as possible is generated by the mercury vapor at its predominant wavelength of 2537 Å. This amounts to about 60 percent of input energy, or watts, transformed into this single wavelength. The special glass transmits about 74 percent of this energy, or close to half of the total energy input. Only 2 percent of the input energy is transmitted into visible light; the remainder (48 percent) is transformed into heat.

Rays from ultraviolet lamps are capable of destroying at least 99 percent of influenza virus particles suspended in air in a matter of seconds.

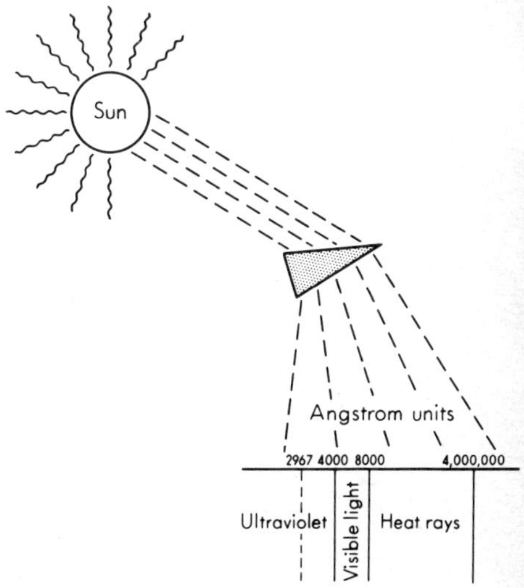

Fig. 268 Radiant energy from the sun may be divided into three broad groups: long-wave energy, such as heat, which is invisible; visible energy which produces light and color; and short-wave energy such as invisible ultraviolet. It is invisible ultraviolet that produces the germicidal effect. (Courtesy of General Electric Company.)

Such lamps are particularly valuable in preventing the spread of virus infections of the upper respiratory tract. Ultraviolet radiations applied to the floors and upper air in barracks housing recruits produced a marked reduction in viral respiratory infections. The same was true of throat infections caused by streptococci.

Room sanitation Outdoor air practically never contains hazardous numbers of human disease-producing organisms. Air of outdoor purity is an essential part of man's environment, and the need for it is well recognized. Air sanitation is inherent in the dispersion and dilution provided by a large volume of air. The amount of air needed indoors depends upon the volume of the room, and the number of normal and sick individuals present.

The development of the germicidal tube has made it possible to obtain air sanitation indoors which is equivalent to that of outdoor air. To provide outdoor benefits to indoor air, it would be necessary to change the air in a room about 60 times per hour. Germicidal tubes properly installed

Fig. 269 Three basic methods for installing germicidal lamps. Upper, side wall; center, ceiling; lower, above floor. (Courtesy of General Electric Company.)

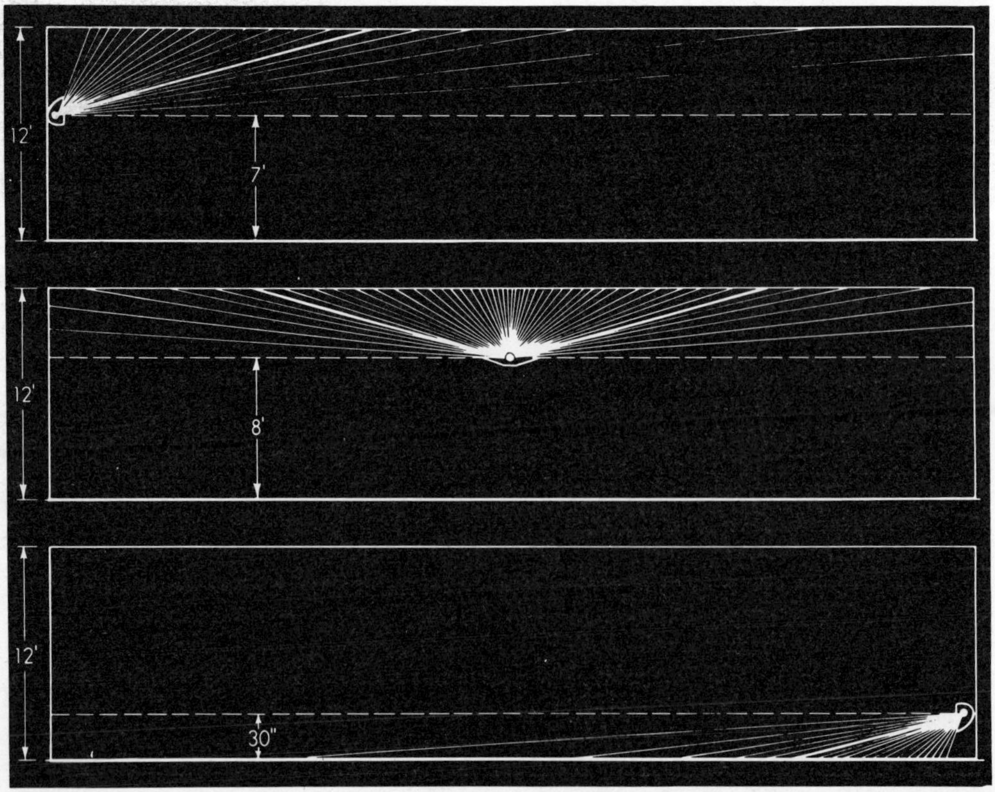

make it possible to achieve the equivalent of 100 or more air changes per hour.

Intensities of germicidal ultraviolet irradiation sufficient to kill airborne organisms before they have had time to travel from one person to another would be very irritating to the face and eyes. For this reason the rays must be limited to the upper and lower parts of rooms where the faces of people would not be exposed for more than a few minutes per day. These limitations lead to three basic methods of installing germicidal lamps: (1) on the side wall or (2) from the ceiling of a room to irradiate the air above the 7-ft level, and (3) on the side wall to irradiate the floor and air below the 30-in level (Fig. 269).

Air warmed by heat sources such as radiators and floor lamps rises upward to the ceiling (Fig. 270). This forces cooler air down along the cool outside walls to the floor, where it is again warmed and rises. This normal air circulation is utilized for diluting contaminated air with disinfected air. Disinfected air from above the head level and below the 30-in level is as good as outdoor air for room sanitation by ventilating dilution.

Operating-room sanitation The problem in surgery is chiefly that of protecting tissues exposed during the operation. Debilitated tissues exposed during long operations may become contaminated with pus-forming and fever-producing organisms not ordinarily a hazard. This is especially serious during brain surgery and thoracoplasty operations, during which the tissues may be exposed for hours.

Fig. 270 How air circulates. Air warmed by radiators, floor lamps, etc., rises toward the ceiling. This forces cooler air down along the cool outside walls to the floor, where it is again warmed and rises. Germicidal units are mounted on the walls so that the radiation from the tubes disinfects the room's upper air. (Courtesy of General Electric Company.)

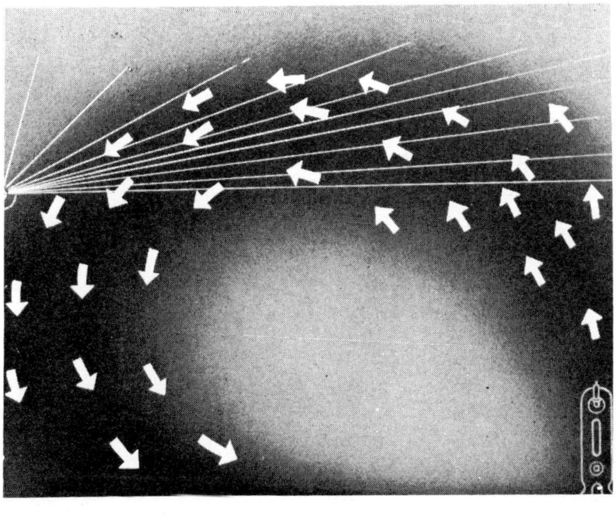

Fig. 271 Ultraviolet air sanitation in the operating room. See text for description. (Courtesy of General Electric Company.)

Although the aseptic techniques in the operating room are superior to those in other parts of the hospital, the practice can be further improved by the application of floor and air irradiation. In order to protect the hands, arms, and faces of the operating personnel, the air should be irradiated above the 7-ft level and below the 2 1/2- to 3-ft level (Fig. 271).

Product sanitation Product sanitation includes the application of ultraviolet rays for the protection of foods and fruit juices; treatment of pharmaceuticals and biologicals; sterilization of instruments and towels used by the barber and beautician; washing and sterilization of dishes, drinking glasses, silverware, and cutlery used in restaurants and other public places; treatment of foods in domestic refrigerators to reduce odors and provide some protection by the disinfection of the enclosed air; and the treatment of meats in storage to reduce spoilage and make it less necessary to use lower temperatures (Figs. 272 and 273).

To prevent bacteria and mold from forming on meat in coolers and holding rooms, the temperature must be kept at 1.1 to 2.2°C. Under such temperatures, the humidity drops to about 60 percent with the result that the meat dries out. Weight is reduced, and some of the meat juices are lost. As meat is dehydrated, the surface darkens in color and must be trimmed. Losses due to trimming, drying out, bacterial slime, and mold may run as high as 15 percent for an average retail meat dealer.

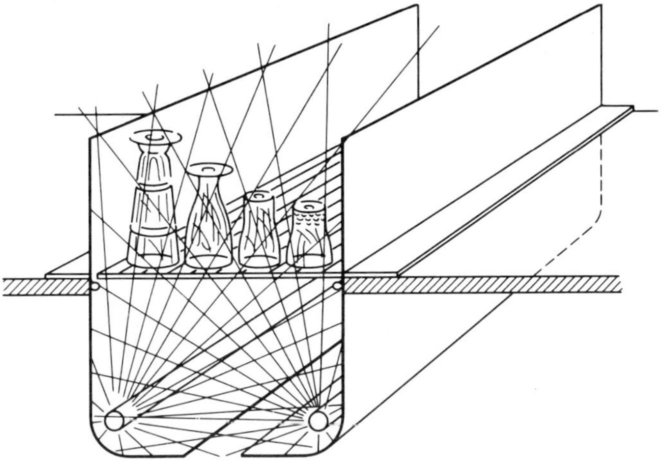

Fig. 272 Sanitary storage of drinking glasses combined with germicidal ultraviolet treatment. (Courtesy of General Electric Company.)

Fig. 273 Simplified drawings of the various techniques of air sanitation and product protection. Clear areas represent effective sterilization of air by proper installation of germicidal lamps. (Courtesy of General Electric Company.)

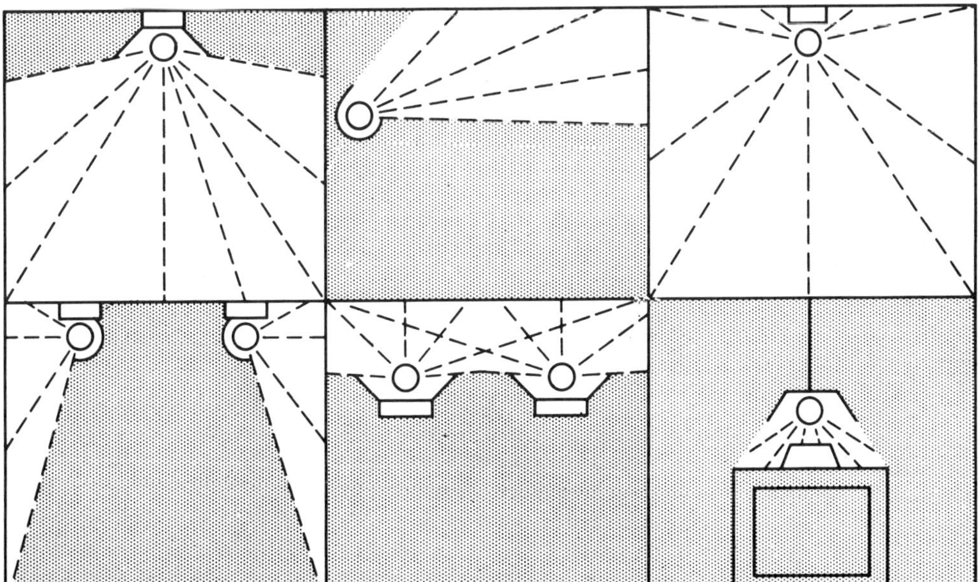

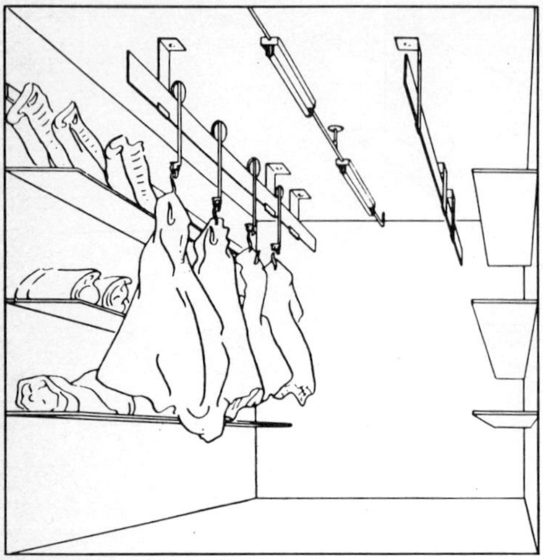

Fig. 274 Sanitary storage of meat by installation of germicidal lamps in coolers and holding rooms. (Courtesy of General Electric Company.)

When germicidal tubes are used in coolers, temperatures can be increased to 7.2°C (Fig. 274). This permits higher humidities, which reduce dehydration and discoloration of the product. Sufficient amounts of ultraviolet light can prevent mold from forming, and reduce trimming waste. Meats may be held in storage longer, and certain freshening processes (rinsing) can be eliminated. Odors are minimized. Mold growth can be prevented on walls, floors, and ceilings.

For more information: Riley and Kaufman (1971); Riley et al. (1971); Shechmeister (1968).

References

Andersen, A. A.: New sampler for the collection, sizing, and enumeration of viable airborne particles, *J. Bact.*, **76**:471, 1958.

Andersen, A. A., and M. R. Andersen: A monitor for airborne bacteria, *Appl. Microbiol.*, **10**:181, 1962.

Artenstein, M. S., et al.: Large-volume air sampling of human respiratory disease pathogens, *Am. J. Epidemiol.*, **85**:479, 1967*a*.

——: Acute respiratory disease and meningococcal infection in army recruits, *J. Am. Med. Assoc.*, **201**:1004, 1967*b*.

——: Large-volume air sampling for meningococci and adenoviruses, *Am. J. Epidemiol.*, **87**:567, 1968.

Barbeito, M. S., et al.: Infectious bacterial aerosol from dropped petri dish cultures, *Am. J. Med. Technol.*, Nov.-Dec., 1961, p. 318.

Barlow, D. F.: The aerosol stability of a strain of foot-and-mouth disease virus and the effects on stability of precipitation with ammonium sulfate, methanol or polyethylene glycol, *J. Gen. Virol.*, **15**:17, 1972.

Borzani, W., and M. L. R. Vairo: Counting of microorganisms in air by centrifugation, *Biotechnol. Bioengineer.*, **9**:439, 1967.

Buchanan, L. M., et al.: Novel multi-slit large-volume air sampler, *Appl. Microbiol.,* **16**:1120, 1968.

Buckland, F. E., and D. A. J. Tyrrell: Loss of infectivity on drying various viruses, *Nature,* **195**:1063, 1962.

Coriell, L. L., et al.: Medical applications of dust-free rooms, *J. Am. Med. Assoc.,* **203**:1038, 1968.

Cox, C. S., et al.: Experimental technique for studying aerosols of lyophilized bacteria, *Appl. Microbiol.,* **20**:927, 1970.

de Jong, J. C., and K. C. Winkler: The inactivation of poliovirus in aerosols, *J. Hyg.,* **66**:557, 1968.

Dixon, G. J., et al.: Persistence of poliomyelitis virus on cotton and wool fabrics, *Appl. Microbiol.,* **14**:183, 1966.

Donaldson, A. I.: The influence of relative humidity on the aerosol stability of different strains of foot-and-mouth disease virus suspended in saliva, *J. Gen. Virol.,* **15**:25, 1972.

Dunham, W. B.: Virucidal agents. In "Disinfection, Sterilization, and Preservation," edited by C. A. Lawrence and S. S. Block, Philadelphia, Lea & Febiger, 1968.

Ederer, G. M.: Dissemination of bacteria by laboratory personnel, *Am. J. Med. Technol.,* March-April 1965, p. 108.

Ehrlich, R., et al.: Relationship between atmospheric temperature and survival of airborne bacteria, *Appl. Microbiol.,* **19**:245, 1970.

Errington, F. P., and E. O. Powell: A cyclone separator for aerosol sampling in the field, *J. Hyg.,* **67**:387, 1969.

Favero, M. S., et al.: Comparison of microbial contamination levels among hospital operating rooms and industrial clean rooms, *Appl. Microbiol.,* **16**:480, 1968.

Friedman, H., et al.: Terminal disinfection in hospitals with quaternary ammonium compounds by use of a spray-fog technique, *Appl. Microbiol.,* **16**:223, 1968.

Fulton, J. D.: Microorganisms of the upper atmosphere. III. Relationship between altitude and micropopulation, *Appl. Microbiol.,* **14**:237, 1966.

—— and R. B. Mitchell: II. Microorganisms in two types of air masses at 690 meters over a city, *Appl. Microbiol.,* **14**:232, 1966.

Gregory, P. H., and J. L. Monteith, eds.: "Airborne Microbes," London, Cambridge University Press, 1967.

Harper, G. J.: Airborne microorganisms: survival tests for four viruses, *J. Hyg.,* **59**:479, 1961.

Harstad, J. B., et al.: Air filtration of submicron virus aerosols, *Am. J. Pub. Health,* **57**:2186, 1967.

Hatch, M. T., and R. L. Dimmick: Physiological responses of airborne bacteria to shifts in relative humidity, *Bact. Rev.,* **30**:597, 1966.

Hollaender, A., and J. M. Dalla Valle: A simple device for sampling air-borne bacteria, *Pub. Health Reports,* **54**:574, 1939.

Johansson, K. R., and D. H. Ferris: Photography of air-borne particles during bacteriological plating operations, *J. Infectious Diseases,* **78**:238, 1946.

Kelly, C. D., and S. Layne: Bacteria found in the air over Canada and the American Arctic, *Can. J. Microbiol.,* **3**:447, 1957*a*; Bacteria found in the air over the Atlantic Ocean, *ibid.,* **3**:457, 1957*b*.

Kenny, M. T., and F. L. Sabel: Particle size distribution of *Serratia marcescens* aerosols created during common laboratory procedures and simulated laboratory accidents, *Appl. Microbiol.,* **16**:1146, 1968.

Kethley, T. W., et al.: Adequate expression for average particle size of microbiological aerosols, *Appl. Microbiol.,* **11**:188, 1963.

Kluyver, A. J., and J. Visser: The determination of microorganisms in air, *Antonie van Leeuwenhoek, J. Microbiol. Serol.,* **16**:299, 1950.

Landsberg, H.: Origin and physics of the atmosphere, *Trans. N.Y. Acad. Sci.,* **13**:154, 1951.

Lawrence, C. A.: Quaternary ammonium surface-active disinfectants. In "Disinfection, Sterilization, and Preservation," edited by C. A. Lawrence and S. S. Block, Philadelphia, Lea & Febiger, 1968.

Lemon, H. M.: A method for collection of bacteria from air and textiles, *Proc. Soc. Exp. Biol. Med.,* **54**:298, 1943.

Lester, W., Jr., et al.: The rate of bactericidal action of triethylene glycol vapor on microorganisms dispersed into the air in small droplets, *Am. J. Hyg.,* **50**:175, 1949.

Lidwell, O. M., and A. G. Towers: Protection from microbial contamination in a room ventilated by a uni-directional air flow, *J. Hyg.,* **67**:95, 1969.

Litsky, B. Y., and W. Litsky: Investigations on decontamination of hospital surfaces by the use of disinfectant-detergents, *Am. J. Pub. Health,* **58**:534, 1968.

Loosli, C. G.: Problem of dust control for the disinfection of air, *Am. J. Pub. Health,* **38**:409, 1948.

Luckiesh, M., et al.: Sampling devices for air-borne bacteria, *J. Bact.,* **52**:55, 1946.

McDermott, W., ed.: "Conference on Air-borne Infection," Baltimore, The Williams & Wilkins Company, 1961.

Martin, A., Jr.: Air sampling for bacteria, *Proc. Penn. Acad. Sci.,* **41**: 1967.

May, K. R., and H. A. Druett: A microthread technique for studying the viability of microbes in a simulated air-borne state, *J. Gen. Microbiol.,* **51**:353, 1968.

Pike, R. M., et al.: Continuing importance of laboratory-acquired infections, *Am. J. Pub. Health,* **55**:190, 1965.

Reitman, M. R., et al.: Potential infectious hazards of laboratory techniques. III. Viral techniques, *J. Bact.*, **68**:549, 1954*a*; I. Lyophilization, *ibid.,* **68**:541, 1954*b*; II. The handling of lyophilized cultures, *ibid.,* **68**:545, 1954*c*.

Riley, R. L., and J. E. Kaufman: Air disinfection in corridors by upper air irradiation with ultraviolet, *Arch. Environ. Health,* **22**:551, 1971.

——— et al.: Infectiousness of air from a tuberculosis ward, *Am. Rev. Respirat. Diseases,* **85**:511, 1962.

——— et al.: Room air disinfection by ultraviolet irradiation of upper air, *Arch. Environ. Health,* **23**:35, 1971.

Roelants, P., et al.: Evaluation of a commercial air filter for removal of viruses from the air, *Appl. Microbiol.,* **16**:1465, 1968.

Sciple, G. W., et al.: Recovery of microorganisms shed by humans into a sterilized environment, *Appl. Microbiol.,* **15**:1388, 1967.

Shechmeister, I. L.: Sterilization by ultraviolet radiations. In "Disinfection, Sterilization, and Preservation," edited by C. A. Lawrence and S. S. Block, Philadelphia, Lea & Febiger, 1968.

——— and F. S. Greenspan: The relation of the oil treatment of floors and bedding to the control of respiratory diseases among naval personnel, *Am. J. Hyg.,* **46**:376, 1947.

Shreve, W. B., et al.: Air cleaning devices for reduction in number of airborne bacteria, *J. Dent. Res.,* **49**:1078, 1970.

Sulkin, S. E.: Laboratory-acquired infections. In "Conference on Airborne Infection," edited by W. McDermott, Baltimore, The Williams & Wilkins Company, 1961.

Technical Development Laboratories: Sampling microbiological aerosols, *U.S. Pub. Health Serv. Pub. Health Monograph 60,* 1959.

Timmons, D. E., et al.: Microorganisms of the upper atmosphere. I. Instrumentation for isokinetic air sampling at altitude, *Appl. Microbiol.,* **14**:229, 1966.

Walter, W. G., et al.: Studies of airborne bacteria in a partially carpeted hospital, *Health Lab. Sci.,* **6**:140, 1969.

Warner, P., et al.: Convenient sampling of air bacteria in operating rooms, *Can. Med. Assoc. J.,* **90**:537, 1964.

Webb, S. J.: Factors affecting the viability of air-borne bacteria. I. Bacteria aerosolized from distilled water, *Can. J. Microbiol.,* **5**:649, 1959; II. The effect of chemical additives on the behavior of air-borne cells, *ibid.,* **6**:71, 1960*a*; III. The role of bonded water and protein structure in the death of air-borne cells, *ibid.,* **6**:89, 1960*b*; V. The effect of desiccation on some metabolic systems of *Escherichia coli, ibid.,* **7**:621, 1961; VI. The action of inositol on lactose oxidation by desiccated *Escherichia coli, Can. J. Biochem. Physiol.,* **41**:455, 1963*a*; VII. The relationship between the structure of chemical additives and their action on air-borne cells, *ibid.,* **41**:867, 1963*b*.

———: The effect of relative humidity and light on air-dried organisms, *J. Appl. Bact.,* **26**:307, 1963*c*.

——— et al.: The effect of relative humidity and inositol on air-borne viruses, *Can. J. Microbiol.,* **9**:87, 1963.

Wedum, A. G.: Control of laboratory air-borne infection. In "Conference on Air-borne Infection," edited by W. McDermott, Baltimore, The Williams & Wilkins Company, 1961.

———: II. Airborne infection in the laboratory, *Am. J. Pub. Health,* **54**:1669, 1964.

Wells, W. F.: Apparatus for study of the bacterial behavior of air, *Am. J. Public Health,* **25**:58, 1933.

Westwood, J. C. N., et al.: Hospital sanitation: the massive bacterial contamination of the wet mop, *Appl. Microbiol.*, **21**:693, 1971.

Wilkinson, T. R.: Survival of bacteria on metal surfaces, *Appl. Microbiol.*, **14**:303, 1966.

Williams, R. E. O.: Spread of airborne bacteria pathogenic for man, *Symp. Soc. Gen. Microbiol.*, no. 18, 1967.

Wright, D. N., et al.: Effect of temperature on survival of air-borne *Mycoplasma pneumoniae, J. Bact.*, **99**:491, 1969.

——: Survival of air-borne *Mycoplasma* as affected by relative humidity, *J. Bact.*, **95**:251, 1968*a*; Role of relative humidity in the survival of air-borne *Mycoplasma pneumoniae, ibid.*, **96**:970, 1968*b*.

Bacteriology of the sea[1] 20

The sea has been described as a gigantic laboratory or chemostat in which a great diversity of organisms compete for nutrients. The kinds of organisms present, their abundance, and their physiological activities are influenced by each other and by environmental conditions. As contrasted with other major habitats, the marine environment is distinctive in its vast area and volume, its low content of organic matter, its relatively high salinity, preponderantly low temperature, and high hydrostatic pressures at great depths.

Approximately 71 percent of the earth's surface is covered with water, nearly 99 percent of which is seawater. Except in highly localized areas in coastal regions, the salt content of seawater ranges from 3.2 to 3.6 percent, the average being 3.433 percent, usually expressed by oceanographers as 34.33 per mille. The average depth of the sea is about 3,800 m. The greatest known depth, the Challenger Deep in the Mariana Trench, is slightly more than 11,000 m, or nearly 36,200 feet. At this depth the hydrostatic pressure is about 1,160 atm or 17,050 psi. Nearly 90 percent of the sea is deeper than 1,000 m, meaning that nearly 90 percent of the marine environment (or about 56 percent of the earth's biosphere by area) has pressures ranging from 100 to 1,160 atm. At depths exceeding 1,000 m, the temperature of seawater and the sea floor is nearly always colder than 3°C. More than 90 percent of the marine environment is never warmer than 5°C. In Arctic and Antarctic areas, surface water is close to −1.86°C, the freezing point of normal seawater.

Neritic (shallow, near-shore) areas, which make up less than 1 percent of the marine environment, are transitional zones in which salinity, turbidity, organic content, flora, fauna, and other conditions differ markedly from oceanic or pelagic areas.

[1] This chapter was written by Dr. Claude E. ZoBell of the Scripps Institution of Oceanography, University of California, La Jolla, Calif. The author is greatly indebted to Dr. ZoBell for his kindness in preparing this material for publication.

Table 21 Percent of several hundred water samples in which the stated numbers of bacteria were found in water samples collected from depths ranging from 1,000 to 11,000 m (ZoBell, 1968)

Bacteria per milliliter	< 1	1-9	10-99	10^2	10^3	10^4
Percent of samples	13.1	58.1	17.2	7.4	3.3	0.6

Occurrence of bacteria in the sea

Bacteria have been found at all depths and at all latitudes in seawater in abundances ranging from only a few per liter to as many as 10^8 per ml. The topmost layers of bottom sediments generally contain from 10^3 to 10^9 bacteria per g (wet weight). Bacteria are generally most abundant near shore, particularly in polluted areas, and sparsest at great depths in the open ocean. Although very sporadically distributed in both time and space, in the water column bacteria tend to decrease in abundance with depth. Ordinarily one finds from 10 to 10^5 bacteria per ml in the topmost 1,000 m of water and usually fewer than 10^4 per ml at depths exceeding 1,000 m (Table 21).

At depths exceeding a few hundred meters, the abundance of bacteria in bottom deposits does not appear to be a function of water depth, although the rate of bacterial activity is influenced by hydrostatic pressure which increases with depth (ZoBell, 1964*a*, 1970). At 20 different Mid-Pacific Expedition stations, Morita and ZoBell (1955) found from 10 to 10^4 bacteria per gram of red clay and globigerina ooze collected from depths of 1,700 to 5,940 m. A few orders of magnitude more bacteria occurred in the topmost layers of nearly all cores than in strata 50 to 750 cm below the mud-water interface. Table 22 illustrates the typical vertical distribution of bacteria in some long cores collected by ZoBell (1942) from the floor of the Pacific Ocean off the coast of southern California.

Table 22 Number of bacteria per gram of marine mud from different core depths

	Station number		
Core depth, cm	*14:37*	*14:45*	*14:53*
0–2.5	38,000,000	7,500,000	840,000
2.5–5	940,000	250,000	102,000
10–12.5	88,000	160,000	63,000
22.5–25	36,000	23,000	19,000
35–37.5	2,400	8,700	1,500
47.5–50	400	2,100	2,200
97.5–100	330	200	190
147.5–150	600	430	60
197.5–200	130	100	140
247.5–250	290	150	140

From thousands of determinations made in different parts of the sea by various investigators and arbitrarily taking 10^{-12} g per cell as the mean wet weight of marine bacteria, it is calculated that the standing stock of bacteria in the world oceans is between 10^5 and 10^6 metric tons. The generation time of these bacteria ranges from less than 1 hour to many months, or maybe much longer at the lowest temperatures and highest pressures occurring in the sea. The shortest generation time reported (Eagon, 1962) is 9.8 min for *Pseudomonas natriegens* at 37°C. Observations (Kriss and Lambina, 1955) on the reproduction of bacteria on glass slides submerged at different depths in the Arctic Ocean indicated generation times of 33 to 200 hours at −1.8 to 2°C. This approximates the findings of Jannasch (1969), who employed chemostats.

Characteristics of marine bacteria

Whether there are specific marine bacteria is more a matter of semantics than of science. If marine is defined as being synonymous with oceanic, a good case can be made for the specificity of marine bacteria. They differ from bacteria in other environmental domains primarily in their capacity to survive and grow in the sea.

In coastal regions, where there is an interchange of microflora between land and sea via running water, wind, or otherwise, there is no sharp demarcation between soil or freshwater forms and truly marine species. Excepting spores, most freshwater forms perish within a few hours or days under oceanic conditions. However, certain bacterial endospores and fungus spores seem to survive in a dormant state for prolonged periods, particularly in deep-sea sediments where the temperature is low and high pressure also has a preservative effect (ZoBell, 1968).

In the open ocean, one finds relatively more gram-negative rods and vibrios than in soil or other freshwater habitats, including raw sewage, milk, etc. Likewise, a larger percentage of oceanic bacteria are motile and more produce pigments. Cocci and actinomycetes are of much more limited occurrence in the sea than in terrestrial environments.

Most oceanic bacteria grow more slowly and form smaller colonies on submerged slides or on nutrient agar than bacteria in soil, sewage, or freshwater. Oceanic bacteria tend to be somewhat more proteolytic and not so commonly saccharolytic as soil or freshwater forms. Facultative aerobes predominate in the sea, there being relatively few obligate aerobes or obligate anaerobes.

Although only a small percentage of oceanic species digest agar, most agar liquefying bacteria come from the sea. Stanier (1941), Humm (1946), Yaphe (1957), and Shinano (1965) have made extensive studies on marine agar digesters. Most bioluminescent bacteria also come from the sea (Harvey, 1952; Hendrie et al., 1970), but only a small percentage of marine species are luminous. Saprophytic, symbiotic, and commensal species of luminous bacteria are widely distributed in the sea (Haneda, 1955).

Aquatic bacteria in general and particularly those found in the open ocean can subsist on extremely low concentrations of nutrients. Some are effective at substrate concentrations below 20 μg per liter (Hobbie and Wright, 1966). The content of organic matter in seawater in the open ocean is generally less than 1 mg per liter.

Temperature tolerance A large percentage of the bacteria found in the open ocean grow at temperatures in the neighborhood of 0°C (Morita, 1966), although the optimum for most species is between 12 and 25°C. Relatively few grow in nutrient media at temperatures higher than 30°C (ZoBell and Conn, 1940). Heating deep-sea samples for 10 min at 30 or 40°C has been shown to kill 20 to 80 percent of the bacteria (ZoBell, 1959).

Pressure tolerance Deep-sea pressures influence the distribution and physiological reaction rates of organisms in the sea. The growth of all microbial species, marine as well as terrestrial, is retarded by hydrostatic pressures 100 to 400 atm higher than their natural habitat (ZoBell and Johnson, 1949; ZoBell and Oppenheimer, 1950). Except for bacterial spores and baroduric species isolated from high-pressure environments, most bacteria are killed by prolonged (a few hours to several days) pressurization at 400 to 600 atm. Pressure tolerance is influenced by temperature, chemical composition of the medium, pH, gas tension, species peculiarities, and other factors (ZoBell, 1964*a*, 1970). Barophilic bacteria which grow at pressures exceeding 1,000 atm have been demonstrated in samples collected from depths exceeding 10,000 m. Among the many mechanisms whereby increased pressure affects physiological reaction rates and microbial survival, the inactivation of essential enzymes appears to be one of the most important (ZoBell and Kim, 1972).

Hydrogen-ion concentration requirements Most marine bacteria grow best in nutrient media at pH 7.2 to 7.6. This is somewhat less alkaline than seawater, which ranges from pH 7.5 to 8.5. Bacteria grow in marine sediments throughout the range of pH 6.4 to 9.4.

Salt requirements The majority of the bacteria isolated from the open ocean require seawater or its equivalent for optimal growth. Marine bacteria have a highly specific need for Na^+ (MacLeod, 1965, 1968). Their need for Mg^{++} or for the proper combination of Mg^{++} and Ca^{++} exceeds such needs for most terrestrial species. Probably due to the lack of essential trace elements and differences in the combinations of minerals, neither artificial seawater nor isotonic salt solutions have proved to be as good as natural seawater for the cultivation of certain marine microorganisms. Twelve elements, including oxygen (which occurs in sulfate, bicarbonate, borate, etc.), make up more than 99.99 percent of the solids dissolved in seawater (Table 23).

Most marine bacteria tolerate hypotonicity (with respect to normal

Table 23 Major constituents in seawater having a total salt content of 35 g per kg or about 3.5 percent

Constituent	*g/kg*	*Constituent*	*g/kg*
Chloride	19.253	Bicarbonate	0.142
Sodium	10.700	Bromide	0.067
Sulfate	2.710	Borate	0.023
Magnesium	1.284	Strontium	0.008
Calcium	0.413	Fluoride	0.0013
Potassium	0.387	Trace elements	0.0027

seawater) much better than hypertonicity. Table 24 gives average colony counts (as percent of maximum values) obtained by plating various samples in nutrient agar prepared with different concentrations of sea salts. These data show that more than 75 percent of the bacteria in seawater and marine mud samples differed in their salinity requirements from bacteria in certain other habitats. Note that bacteria in normal seawater and marine mud are only slightly halophilic (Larsen, 1967) as compared with extreme halophiles found in marine salterns and Great Salt Lake.

Kriss et al. (1967) tested 2,500 cultures obtained from the high seas for salt tolerance. All except 5 grew well in nutrient agar prepared with normal seawater, 36 percent grew almost equally well on nutrient agar prepared with either normal seawater or freshwater fortified with 0.85 percent NaCl and other essential minerals, and 63.4 percent of the cultures grew preferentially or exclusively on nutrient agar prepared with normal seawater. Rheinheimer (1968) observed a relationship between salinity and the characteristics of bacteria occurring in the Baltic Sea, in which the salinity ranges from less than 0.3 to 1.6 percent.

Attachment propensities A high percentage of marine bacteria have a tendency to grow attached to solid surfaces (ZoBell, 1943). This is the

Table 24 Average percent of bacteria in samples from various sources which formed colonies in nutrient agar prepared with different concentrations of sea salts

	Salt concentration of nutrient agar					
Kind of sample	*0.5%*	*2.0%*	*3.5%*	*7.0%*	*15%*	*30%*
Seawater	19.2	71.7	*100.0*	38.3	7.4	0.0
Marine mud	23.9	74.2	*100.0*	43.5	6.1	0.2
Marine saltern	4.3	12.6	19.6	30.4	84.9	*100.0*
Great Salt Lake	1.7	8.0	14.1	26.3	92.2	*100.0*
Garden soil	*100.0*	56.1	25.2	9.8	4.6	0.4
Sewage	*100.0*	44.8	13.5	5.7	2.7	0.0
River water	*100.0*	47.3	22.8	6.2	1.9	0.0

basis of the submerged slide technique used extensively to study the occurrence and growth rates of bacteria in the sea (Kriss, 1963; Rodina, 1972). Numerous species occur attached to plankton or to larger organisms. According to Waksman and Vartiovaara (1938), most of the bacteria in marine bottom deposits occur adsorbed on or attached to solid particles. Surfaces of sand grains are well-defined habitats for marine microbes (Meadows and Anderson, 1968). Corpe (1970) has reviewed the adhesive properties of marine bacteria and their relationship to the biofouling of submerged surfaces both natural and man-made.

Importance of bacteria in the sea

Bacteria have far-reaching effects on biological, geochemical, and physicochemical conditions in the sea. The majority are beneficial to the economy of the sea and to the welfare of man. Their most obvious harmful activity is causing infectious diseases in marine organisms. Relatively few species are known to be pathogenic (Snieszko, 1970; Sindermann, 1970). With few exceptions, the sea is inhospitable for most human pathogens.

Because such a high proportion of marine bacteria thrive at the temperature of ice water, the spoilage of fish and other food products from the sea presents special problems (Shewan, 1962; Wood, 1967). Spoilage of marine fish stored at $-3^{\circ}C$ has been reported by Shaw and Shewan (1968). Fish caught in polluted waters are likely to carry food-poisoning staphylococci and salmonellae (Appleman et al., 1964). *Clostridium botulinum* is quite commonly present in fish, clams, oysters, and crabs from coastal waters (Craig *et al.,* 1968).

Certain bacteria and allied microorganisms promote the fouling of submerged surfaces, such as ships' bottoms and water conduits (Skerman, 1956). Other harmful bacteria bring about the deterioration of various man-made structures such as fishnets, ropes, sailcloth, wooden structures, concrete, etc. (Wood, 1967).

In localized areas, particularly in polluted bays, estuaries, or stagnant basins, bacteria may vitiate water or bottom deposits by depleting dissolved oxygen, by producing hydrogen sulfide, or by forming toxic amines. On the whole, though, bacterial activity is more likely to minimize the adverse effects of pollution.

Geochemical activities The most important activity of bacteria in the sea is the mineralization, modification, and synthesis of organic matter (ZoBell, 1964*b*). Virtually all kinds of organic compounds, including carbohydrates, lipids, proteins, hydrocarbons, and chitins, are oxidized by marine bacteria. Approximately half of the carbon may be converted into bacterial cell substance, the remainder being released as carbon dioxide, methane, and other metabolic products. Sulfate, hydrogen sulfide, phosphate, and ammonium are other common products resulting from the microbial decomposition of plant and animal remains. Under aerobic conditions, much of the ammonium is oxidized to nitrite or nitrate.

Both chemosynthetic and photosynthetic autotrophs occur in the sea, where they contribute to the synthesis of bacterial biomass. Photosynthetic bacteria are confined largely to shallow bottoms, brackish water, and masses of decomposing vegetation in which they may be more important in the transformation of sulfur than in the synthesis of organic matter. Chemosynthetic bacteria appear to be the only primary producers of organic matter below the euphotic zone, generally less than 200 m. Rather widely distributed at all depths in the sea are bacteria which fix carbon dioxide while obtaining their energy from the oxidation of such substances as ammonium, nitrite, molecular hydrogen, hydrogen sulfide, sulfur, methane, or carbon monoxide. Bacteria in the Black Sea were observed by Sorokin (1965) to fix from 0.03 to 10 g CO_2 per sq m per day.

ZoBell (1963) has reviewed some of the many ways in which bacteria transform sulfur and its compounds, including the fractionation of isotopes. Besides contributing to the origin of oil and the deposition of free sulfur, one of the most important functions of sulfate-reducing bacteria is making fossil oxygen in sulfate available for the oxidation of organic matter in bottom deposits. In reviewing the microbial geochemistry of oxygen, ZoBell (1972) discusses mechanisms whereby this element is captured by or released from various sulfur, carbon, nitrogen, phosphorus, iron, and silicon compounds.

Application of the acetylene reduction technique (Stewart et al., 1967) has greatly simplified the detection of nitrogen-fixing microorganisms in nature:

$$HC{\equiv}CH + H_2 \xrightarrow{\text{nitrogenase}} H_2CH_2$$

$$N{\equiv}N + 3H_2 \xrightarrow{\text{nitrogenase}} 2NH_3$$

More than 100 species, representing 16 genera of bacteria, 14 genera of blue-green algae, and several fungi have been shown to fix nitrogen (Stewart, 1967; Postgate, 1971). Many of these have been found in seawater or bottom deposits (Pshenin, 1966).

Some of the foregoing and several other geochemical activities of bacteria are discussed in much greater detail by Kuznetsov et al. (1963) and Zajic (1969).

Physicochemical conditions Certain chemical transformations caused by bacteria tend to affect the pH. The pH may be increased by (1) formation of ammonia as in deaminization reactions, (2) reduction of nitrate, nitrite, or sulfate, (3) oxidation or decarboxylation of organic acids, (4) reduction or assimilation of carbon dioxide, and by numerous other reactions. On the other hand, conditions may be made more acidic by the (1) bacterial oxidation of ammonia to nitrite or nitrate, (2) oxidation of sulfur or hydrogen sulfide to sulfate, (3) formation of organic acids or carbon dioxide as from carbohydrates, (4) liberation of phosphate from phosphoproteins or phospholipids, etc. By and large, such acidifying or alkalinizing processes tend to neutralize each other in the sea, except in localized environments where one process or the other may predominate. In microenvironments, particularly in bottom sediments, bacterial activities may be responsible for pH conditions differing from the mean by 0.5 to 4 units. The pH has pronounced effects on the state or solubility of iron, manganese, calcium carbonate, phosphates, and many other minerals.

The Eh or redox potential also affects the state of iron, manganese, and other electronically unstable compounds (Jacob, 1970). ZoBell (1946) has summarized some of the many ways in which bacterial activities influence the redox potential of marine sediments. Reducing conditions are believed to be conducive to the formation and preservation of petroleum hydrocarbons.

Origin of oil It is generally agreed by geologists that petroleum has been formed in marine sediments from the organic remains of plants and animals. Bacteria are believed to promote the transformations in many ways (ZoBell, 1952).

In highly reducing (anaerobic) environments, bacteria tend to convert organic matter partly into carbon compounds which are relatively richer in hydrogen and poorer in oxygen, nitrogen, sulfur, and phosphorus. This results in the formation of carbon compounds that are more petroleum-like than their predecessors. Petroleum consists largely of gaseous, liquid, and solid hydrocarbons.

Appreciable quantities of methane are produced by bacteria in soil, swamps, sewage, and marine sediments. Minute quantities of various liquid and solid hydrocarbons occur in the cells of many microbial species. While it is not to be concluded that bacteria produce petroleum, they seem to contribute to the origin of oil and its modification in many ways (Davis, 1967).

Oxidation of oil Under favorable aerobic conditions, virtually all kinds of hydrocarbons are susceptible to microbial oxidation (Foster, 1962; Treccani, 1962; ZoBell, 1950). Davis (1967) discusses some of the many ways in which industry is exploiting the ability of bacteria, yeasts, and fungi to metabolize hydrocarbons. Microbes are believed to bring about the degradation of mineral oils polluting marine environments (ZoBell, 1969, 1971).

Bacteriology of marine fish

Escherichia coli and allied coliform bacteria are not normal inhabitants of the intestines of marine fish. The presence of coliform bacteria or *Streptococcus faecalis* indicates that the fish has been taken from polluted water or else contamination has occurred during handling, transportation, or marketing (Shewan, 1961; Appleman et al., 1964). The normal microflora of the intestine and viscera depends largely on the feeding habits of the fish and its food intake. Whereas the intestinal contents of feeding fish may contain up to 10^8 bacteria per gram, the intestinal tract of nonfeeding fish is often almost entirely free of bacteria.

The flesh and body fluids of newly caught, healthy fish are generally free of bacteria, but the slime on the surface of fish and their gills carry heavy bacterial loads, anywhere from 10^3 to 10^6 bacteria per sq cm of surface (Shewan, 1962). Up to 95 percent of the bacteria isolated from cold-water environments, i.e., colder than 10 or 15°C, are psychrophilic, growing at temperatures below 5°C but not at temperatures much above 25°C. The majority of the bacteria on marine fish grow best in media prepared with 50 to 75 percent seawater or with 1.5 to 2.5 percent NaCl (Simidu and Hasuo, 1968).

The aerobic microflora of fish consist predominantly of members of the genera *Pseudomonas, Achromobacter, Vibrio,* and *Flavobacterium* (Shewan, 1962). Less common are various micrococci and coryneforms. Anaerobes are generally absent in the surface slime and gills of freshly caught fish, but anaerobes always occur in the gut of feeding fish. Most commonly detected are *C. welchii, C. sporogenes, C. tertium, C. tetani,* and *C. botulinum*.

Though normally free of human pathogens, contaminated fish and shellfish may carry toxigenic *C. botulinum* (Cann et al., 1965), various pathogenic or food-poisoning *Salmonella, Vibrio parahaemolyticus*, another common cause of food poisoning (Baross and Liston, 1970), ciguatera disease toxins (Hashimoto et al., 1969), and paralytic toxins produced by the marine dinoflagellate *Gonyaulax* (Prakash, 1966).

Virus-induced hepatitis has been traced to contaminated oysters (Roos, 1956). Various enteroviruses have been found in oysters collected from seawater having a coliform index of less than 70 per ml (Metcalf and Stiles, 1968).

Bacteriology of other marine animals

Except for a few dozen species of fish, shellfish, and marine mammals, mostly of commercial importance, only a few desultory observations have been made on the microflora of other kinds of animals inhabiting the sea – about 185,000 species representing 31 of the 32 commonly recognized phyla. How marine animals are helped, harmed, or inhabited by bacteria remains an unsolved problem of great significance. From fragmentary studies, it appears that the integuments and digestive tracts of most aquatic animals are veritable bacterial gardens.

Bacteria may be beneficial by aiding animals in the digestion of food such as chitin, polysaccharides, and other complex carbon products. Bacteria may aid animal nutrition by providing vitamins or other accessory growth factors which the animal is unable to synthesize. Being highly nutritious and generally readily digestible, bacteria constitute an important part of the diet of many animal species (Hall, 1962; Fenchel, 1968; Seki and Kennedy, 1969). This function of bacteria may be of the greatest importance in the deep sea where both organotrophic and chemolithotropic forms may produce biomass.

Animals which ingest particulate organic matter, particularly bacteria and other microorganisms, are said to be *phagotrophic* (Wood, 1965, 1967, 1968). Certain marine ecologists (Fenchel, 1968) call bacteria-eating animals *bacteriophages*. This term is not to be confused with bacterial viruses (Kott and Ben Ari, 1968) or *Bdellovibrio bacteriovorus* (Paoletti, 1970), both of which occur in the sea.

References

Appleman, M. D., N. Bain, and J. M. Shewan: A study of some organisms of public health significance from fish and fishery products, *J. Appl. Bact.*, **27**:69-77, 1964.

Baross, J., and Liston, J.: Occurrence of *Vibrio parahaemolyticus* and related hemolytic vibrios in marine environments of Washington State, *Appl. Microbiol.*, **20**:179-186, 1970.

Cann, D. C., B. B. Wilson, and C. Hobbs: Incidence of *Clostridium botulinum* in bottom deposits in British coastal waters, *J. Appl. Bact.*, **31**:511-514, 1968.

Corpe, W. A.: Attachment of marine bacteria to solid surfaces. In "Adhesion in Biological Systems," edited by R. S. Manly, New York, Academic Press, Inc., 1970, pp. 73-87.

Craig, J. M., S. Hayes, and K. S. Pilcher: Incidence of *Clostridium botulinum* Type E in salmon and other marine fish in the Pacific Northwest, *Appl. Microbiol.*, **16**:553-557, 1968.

Davis, J. B.: "Petroleum Microbiology," New York, American Elsevier Publishing Company, Inc., 1967.

Eagon, R. G.: *Pseudomonas natriegens*, a marine bacterium with a generation time of less than 10 minutes, *J. Bact.*, **83**:736-737, 1962.

Fenchel, T.: The ecology of marine microbenthos. II. The food of marine benthic ciliates, *Ophelia*, **5**:73-121, 1968.

Foster, J. W.: Hydrocarbons as substrates for microorganisms, *Antonie van Leeuwenhoek, J. Microbiol. Serol.*, **28**:241-274, 1962.

Hall, H. H.: Applied microbiology in animal nutrition, *Advan. Appl. Microbiol.*, **4**:77-99, 1962.

Haneda, Y.: Luminous organisms of Japan and the Far East. In "The Luminescence of Biological Systems," edited by F. H. Johnson, Washington, D.C., American Association for the Advancement of Science, pp. 335-385, 1955.

Harvey, E. N.: "Bioluminescence," New York, Academic Press, Inc., 1952.

Hashimoto, Y., S. Konosu, T. Yasumoto, and H. Kamiya: Studies on marine toxins. XXI. Occurrence of ciguatoxin and ciguaterin in ciguatoxic fishes in the Ryukyu and Amami islands, *Bull. Japan Soc. Sci. Fisheries*, **35**:316-326, 1969.

Hendrie, M. S., W. Hodgkiss, and J. M. Shewan: The identification, taxonomy and classification of luminous bacteria, *J. Gen. Microbiol.*, **64**:151-169, 1970.

Hobbie, J. E., and R. T. Wright: Competition between planktonic bacteria and algae for organic solutes. In "Primary Productivity in Aquatic

Environments," edited by C. R. Goldman, Berkeley, University of California Press, 1966, pp. 175-185.

Humm, H. J.: Marine agar-digesting bacteria of the South Atlantic coast, *Duke Univ. Marine Sta. Bull.*, **3**:45-75, 1946.

Jacob, H.-E.: Redox potential. In "Methods in Microbiology," vol. 2, edited by J. R. Norris and D. W. Ribbon, New York, Academic Press, Inc., 1970, pp. 91-123.

Jannasch, H. W.: Estimation of bacterial growth rates in natural waters, *J. Bacteriol.*, **99**:156-160, 1969.

Kott, Y., and Ben Ari, H.: Bacteriophages as marine pollution indicators, *Rev. Intern. Oceanog. Med.*, **9**:207-217, 1968.

Kriss, A. E.: "Marine Microbiology (Deep Sea)," London, Oliver & Boyd Ltd., 1963.

—— and Lambina, V. A.: Rapid increase of microorganisms in the ocean in the region of the North Pole, *Usp. Sovrem. Biol.*, **39**:366-373, 1955 (in Russian).

——, I. E. Mishustina, N. Mitskevich, and E. V. Zemtsova: "Microbial Population of Oceans and Seas," New York, St. Martin's Press, Inc., 1967.

Kuznetsov, S. I., M. V. Ivanov, and N. N. Lyalikova: "Introduction to Geological Microbiology," New York, McGraw-Hill Book Company, 1963.

Larsen, H.: Biochemical aspects of extreme halophilism. In "Advances in Microbial Physiology," edited by A. R. Rose and J. F. Wilkinson, New York, Academic Press, Inc., 1967, pp. 97-132.

MacLeod, R. A.: The question of the existence of specific marine bacteria, *Bact. Rev.*, **29**:9-23, 1965.

—— : On the role of inorganic ions in the physiology of marine bacteria. In "Advances in Microbiology of the Sea," vol. 1, edited by M. R. Droop and E. J. F. Wood, New York, Academic Press, Inc., 1968, pp. 95-126.

Meadows, P. A., and Anderson, J. G.: Micro-organisms attached to marine sand grains, *J. Marine Biol. Assoc. U.K.*, **48**:161-175, 1968.

Metcalf, T. G., and Stiles, W. C.: Enterovirus within an estuarine environment, *Am. J. Epidemiol.*, **88**:379-388, 1968.

Morita, R. Y.: Marine psychrophilic bacteria, *Oceanog. Marine Biol. Ann. Rev.*, **4**:105-121, 1966.

—— and ZoBell, C. E.: Occurrence of bacteria in pelagic sediments collected during the mid-Pacific expedition, *Deep-Sea Res.*, **3**:66-73, 1955.

Paoletti, A.: Facteurs biologiques d'autoepuration des eaux de mer: Points clairs et points obscurs d'une question discutee, *Rev. Intern. Oceanog. Med.*, **18**:33-68, 1970.

Postgate, J.: Relevant aspects of the physiological chemistry of nitrogen fixation. In "Microbes and Biological Productivity," edited by D. E. Hughes and A. H. Rose, London, Cambridge University Press, 1971, pp. 287-307.

Prakash, A.: A "red water" bloom of *Gonyaulax catenella* in the Strait of Georgia and its relation to paralytic shellfish toxicity, *J. Fisheries Res. Board Can.*, **23**:1265-1270, 1966.

Pshenin, L. N.: "Biology of Marine Nitrogen Fixers," Kiev, Naukova Dumka, 1966 (in Russian).

Rheinheimer, G.: Beobachtungen über den Einfluss von Salzgehaltsschwankungen auf die Bakterienflora der westlichen Ostsee, *Sarsia*, **34**:253-262, 1968.

Rodina, A. G.: "Methods in Aquatic Microbiology," Baltimore, University Park Press, 1972.

Roos, B.: Hepatitis epidemic conveyed by oysters, *Svenska Läkartidn.*, **53**:989-1003, 1956.

Seki, H., and Kennedy, O. D.: Marine bacteria and other heterotrophs as food for zooplankton in the Strait of Georgia during winter, *J. Fisheries Res. Board Can.*, **26**:3165-3173, 1969.

Shaw, B. G., and Shewan, J. M.: Psychrophilic spoilage bacteria of fish, *J. Appl. Bact.*, **31**:89-96, 1968.

Shewan, J. M.: The microbiology of sea-water fish. In "Fish as Food," edited by G. Borgstrom, New York, Academic Press, Inc., 1961, pp. 487-560.

———: The bacteriology of fresh and spoiling fish and some related chemical changes. In "Recent Advances in Food Science," edited by J. Hawthorn and J. Leitch, London, Butterworth & Co. (Publishers), Ltd., 1962, pp. 167-193.

Shinano, H.: Agar softening bacteria isolated from sea water, *Bull. Japan Soc. Sci. Fisheries*, **31**:840-847, 1965.

Simidu, U., and Hasuo, K.: Salt dependency of the bacterial flora of marine fish, *J. Gen. Microbiol.*, **52**:347-354, 1968.

Sindermann, C. J.: "Principal Diseases of Marine Fish and Shellfish," New York, Academic Press, Inc., 1970.

Skerman, T. M.: The nature and development of primary films on surfaces submerged in the sea, *New Zealand J. Sci. Technol.*, **38**:44-57, 1956.

Snieszko, S. F.: "A Symposium on Diseases of Fishes and Shellfishes," Washington, D.C., American Fisheries Society, 1970.

Sorokin, J. I.: On the trophic role of chemosynthesis and bacterial biosynthesis in water bodies, *Mem. Ist. Ital. Idrobiol.*, **18**(Suppl.):187-205, 1965.

Stanier, R. Y.: Studies on marine agar-digesting bacteria, *J. Bact.*, **42**:527-559, 1941.

Stewart, W. D. P.: Nitrogen-fixing plants, *Science*, **158**:1426-1432, 1967.

———, G. P. Fitzgerald, and R. H. Burris: *In situ* studies on N_2 fixation using the acetylene reduction technique, *Proc. Natl. Acad. Sci.*, **58** (Part II):2071-2078, 1967.

Treccani, V.: Microbial degradation of hydrocarbons, *Progr. Indust. Microbiol.*, **4**:3-33, 1962.

Waksman, S. A., and Vartiovaara, U.: The adsorption of bacteria by marine bottom, *Biol. Bull.*, **74**:56-63, 1938.

Wood, E. J. F.: "Marine Microbial Ecology," New York, Reinhold Publishing Corporation, 1965.

——: "Microbiology of Oceans and Estuaries," New York, Elsevier Publishing Company, Inc., 1967.

——: Perspectives in marine microbiology. In "Advances in Microbiology of the Sea," vol. 1, edited by M. R. Droop and E. J. F. Wood, New York, Academic Press, Inc., 1968, pp. 1-22.

Yaphe, W.: The use of agarase from *Pseudomonas atlantica* in the identification of agar in marine algae (Rhodophyceae), *Can. J. Microbiol.*, **3**:987-993, 1957.

Zajic, J. E.: "Microbial Biogeochemistry," New York, Academic Press, Inc., 1969.

ZoBell, C. E.: Changes produced by microorganisms in sediments after deposition, *J. Sediment. Petrology*, **12**:127-136, 1942.

——: The effect of solid surfaces upon bacterial activity, *J. Bact.*, 46:39-56, 1943.

——: Studies on redox potential of marine sediments, *Bull. Am. Assoc. Petrol. Geol.*, **30**:477-513, 1946.

——: Assimilation of hydrocarbons by microorganisms, *Advan. Enzymol.*, **10**:443-486, 1950.

——: Part played by bacteria in petroleum formation, *J. Sediment. Petrol.*, 22:42-49, 1952.

——: Introduction to marine microbiology, *New Zealand Oceanog. Inst. Mem.*, 3:7-23, 1959.

——: Organic geochemistry of sulfur. In "Organic Geochemistry," edited by I. A. Breger, Oxford, Pergamon Press, 1963, pp. 543-578.

——: Hydrostatic pressure as a factor affecting the activities of marine microbes. In "Recent Researches in the Fields of Hydrosphere, Atmosphere and Nuclear Geochemistry," edited by Y. Miyake and T. Koyama, Tokyo, Maruzen Co., Ltd., 1964*a*, pp. 83-116.

——: Geochemical aspects of the microbial modification of carbon compounds. In "Advances in Organic Geochemistry," edited by U. Colombo and G. Hobson, Oxford, Pergamon Press, pp. 339-356, 1964*b*.

——: Bacterial life in the deep sea, *Bull. Misaki Marine Biol. Inst., Kyoto Univ.*, no. 12, pp. 77-96, 1968.

——: Microbial modification of crude oil in the sea. In "Proceedings API/FWPCA Conference on Prevention and Control of Oil Spills," publ. 4040, Washington, D.C., American Petroleum Institute, 1969, pp. 317-326.

——: Pressure effects on morphology and life processes of bacteria. In "High Pressure Effects on Cellular Processes," edited by A. M. Zimmerman, New York, Academic Press, Inc., 1970, pp. 85-130.

——: Sources and biodegradation of carcinogenic hydrocarbons. In "API/EPA/USCG Conference, Prevention and Control of Oil Spills," Washington, D.C., American Petroleum Institute, 1971, pp. 441-451.

ZoBell, C. E.: Microbial biogeochemistry of oxygen, *Akad. Nauk SSSR Izvestia, Ser. Biol.*, no. 1:23-42, 1972.

——and Conn, J. E.: Studies on the thermal sensitivity of marine bacteria, *J. Bact.*, **40**:223-238, 1940.

——and Johnson, F. H.: The influence of hydrostatic pressure on the growth and viability of terrestrial and marine bacteria, *J. Bact.*, **57**:179-189, 1949.

——and Kim, J.: Effects of deep-sea pressures on microbial enzyme systems. In "The Effects of Pressure on Organisms," edited by M. A. Sleigh and A. G. Macdonald, London, Cambridge University Press, pp. 125-146, 1972.

——and Oppenheimer, C. H.: Some effects of hydrostatic pressure on the multiplication and morphology of marine bacteria, *J. Bact.*, **60**: 771-781, 1950.

Bacteriology of water 21

Water receives its bacterial content from air, soil, sewage, organic wastes, dead plants and animals, etc. This means that at times almost any organism may be found in water. Most of the bacteria find conditions unfavorable and soon die. Those species which survive and are constantly present constitute the natural flora of water.

The great majority of the bacteria found in nature live on dead or decaying organic matter. They are called saprophytes (sapro, rotten; phyte, plant) and belong to the so-called metatrophic group of organisms. Saprophytes are commonly present in large numbers in humus, a brown or black material found in the surface layer of soil, formed by the partial decomposition of vegetable or animal matter.

Natural waters are commonly grouped into four well-marked classes: (1) atmospheric waters, (2) surface waters, (3) stored waters, and (4) ground waters.

Atmospheric waters Rain and snow are included under the atmospheric waters. Sometimes these may contain considerable numbers of bacteria, owing chiefly to the high content of dust in the air. After snow or heavy rain the atmosphere is washed nearly free of organisms so that many sterile plates, each inoculated with 1 ml of water, may be obtained.

Surface waters As soon as the raindrops and snowflakes touch the earth, they become contaminated by the microorganisms in the soil. These are then known as surface waters. The extent of the contamination is dependent upon the numbers of organisms in the soil and, also, upon the kinds and quantities of food materials dissolved out of the soil by the water. The bacterial counts of surface waters are apt to show great variations. This is particularly true in the fall and spring, the seasons of heavy rains and melting snows. The washoff from the soil may upset the existing equilibrium in the surface water, resulting in considerable variation in the flora and bacterial content.

The first result of a mild rain is to greatly increase the bacterial contamination of a body of water. A prolonged rain exerts an opposite effect, owing to the fact that after the main impurities have been removed from the upper layers of the soil, the subsequent rainfall acts merely as a diluent of the body of water. Rivers usually show their highest count during the rainy period.

Stored waters The effect of storage is to greatly decrease the numbers of organisms in water. The forces which tend to produce bacterial self-purification now come into play. These are sedimentation, activities of other organisms, ultraviolet light, temperature, food supply, and perhaps osmotic effects.

Bacteria have a specific gravity slightly greater than that of distilled water, which means that they will slowly settle in a still body of water. However, the greatest factor responsible for the sedimentation of bacteria is their attachment to suspended particles. The suspended particles in settling mechanically remove the organisms from the upper layers of the water.

Predatory protozoa present in waters play an important role in decreasing the number of bacteria. Protozoa require living or dead bacteria for food and easily engulf large numbers of these organisms, provided the water contains sufficient dissolved oxygen. In the absence of bacteria and dissolved oxygen, the protozoa gradually disappear.

Direct sunlight is toxic to vegetative bacterial cells and even to spores if the action is sufficiently prolonged. Diffuse light is less effective as a sterilizing agent. In a water supply, the toxicity of ultraviolet rays is inversely proportional to its turbidity. In other words, the light rays are practically without action in a turbid water. In a clear water ultraviolet rays may be effective to a depth of 1 to 3 m.

Gameson and Saxon (1967) immersed mixtures of seawater and fresh macerated sewage in glass bottles at depths as low as 4 m below the sea surface and reported a pronounced effect of daylight on the mortality of coliform bacteria. The rate of death at any period of the year was approximately proportional to the intensity of short-wave radiations received by the sample.

Increasing temperatures exert a harmful effect upon the survival of some organisms in water, especially those capable of producing disease. On the other hand, multiplication of certain soil and intestinal forms may actually occur when the temperature of the water is increased. It has been shown that *Escherichia coli* is capable of multiplying when inoculated in autoclaved water and incubated at 37°C. Cultures in autoclaved water stored at 22°C show higher counts and remain viable longer than cultures kept at 37°C. Raw waters stored at 22°C also show increases in bacterial numbers, but not as great as in autoclaved waters.

An increase in the food supply usually results in an increase in bacterial numbers. On the other hand, certain toxic substances such as acids and bases produce marked reductions in the numbers of viable organisms. Various dissolved gases such as carbon dioxide and hydrogen also show a toxic effect. Environmental factors generally produce marked fluctuations in the bacterial counts. Apparently this is not due to any one factor but to a group of factors acting as a whole.

Obviously, all the factors which operate to decrease the numbers of bacteria in water will be more effective with an increase in time. This may be represented mathematically as follows:

$$\log \frac{N_1}{N_2} = kt$$

where N_1 = number of organisms at the beginning
N_2 = number of organisms at the end
t = time
k = a constant that varies with temperature and other environmental factors

For more information: Cody et al. (1961); Potter (1960).

Groundwaters Groundwaters are, in general, relatively free from bacteria because of the filtering action of the earth through which the waters have penetrated. This filtering action removes not only most of the bacteria but also any suspended organic food particles. Deep wells usually contain fewer organisms than water from shallow wells, owing to the deeper layers of filtering material. The distance that bacteria travel through soil depends upon a number of factors, including the permeability of the soil, the hydraulic gradient of the groundwater, and the climatic conditions.

The majority of bacteria found in water belong to the following groups: (1) fluorescent bacteria (*Pseudomonas, Alginomonas,* etc.); (2) chromogenic rods (*Xanthomonas*, etc.); (3) coliform group (*Escherichia, Aerobacter*, etc.); (4) *Proteus* group; (5) non-gas-forming, nonchromogenic, non-spore-forming rods; (6) sporeformers of the genus *Bacillus*; and (7) pigmented and nonpigmented cocci (*Micrococcus*).

Quantitative bacteriological examination of water

Experience indicates that the determination of the approximate numbers of bacteria in water multiplying at 35 and 20°C may yield useful information concerning the quality of the water and may provide supporting data regarding the significance of the results of the coliform tests (see below). The standard plate count also may be useful in judging the efficiency of the various water-treatment methods in use.

Collection of sample The water sample is collected in a clean, sterile 100-ml glass-stoppered or screw-cap bottle. The neck and top of the bottle are covered with a parchment paper cap and tied in place with string. The bottle is then sterilized in an autoclave at 15-lb pressure (121°C) for 30 min. The purpose of the covering is to keep the neck and stopper of the bottle free from contaminating bacteria.

Bottles for the collection of water containing residual chlorine should contain a dechlorinating agent such as sodium thiosulfate. The amount added should be 10 mg per 100-ml bottle.

To obtain a representative sample from a tap, the water should be run for 2 or 3 min to wash away any contaminating organisms likely to be present around the opening of the faucet. Also, changes in bacterial content may occur in small pipes; some species tend to die, others to multiply.

The bottle is grasped with the right hand. The stopper is removed with the left hand, holding it by the paper covering. The bottle is filled to 2 cm from the stopper, to permit mixing of the contents. After collection, the stopper with covering is carefully replaced to avoid contamination, and the paper tied in place with string. The fingers must not touch the inside of the neck or stopper; otherwise contamination of the contents may occur and lead to an erroneous result.

In sampling a still body of water, the cap is first removed with the left hand. The bottle is plunged mouth downward to a depth of about 30 cm, then inverted. When almost filled, it is removed and stoppered. If any current exists, the mouth of the bottle should be directed against it to avoid bacterial contamination from the fingers.

After a sample is collected and stored, a rapid change in the bacterial content takes place. The number of organisms usually shows a marked increase. In some cases, the increase is gradual; in others, it is very rapid. The increase in numbers is caused by the multiplication of the typical water bacteria. Disease and other organisms whose natural habitat is the intestinal tract of man and animals tend to die rapidly.

An increase in bacterial numbers is accelerated by an increase in temperature. Because of the rapid changes which may take place in bottled waters on standing, all samples should be examined as quickly as possible. If samples cannot be processed within 1 hr after collection, they should be iced for shipment to the laboratory. The samples should be held below 10°C during a maximum transport period of 6 hr. On arrival at the laboratory the samples should be processed within 2 hr. In no case should the elapsed time between collection and examination exceed 30 hr.

Method The sample bottle should be shaken vigorously at least 25 times to obtain a uniform distribution of organisms. Then measured amounts are transferred to petri dishes and mixed thoroughly with liquefied agar previously cooled to 45°C. After the agar has solidified, the plates are inverted and incubated at 35 ± 0.5°C for 24 ± 2 hr, or at 20 ± 0.5°C for 48 ± 3 hr.

Only plates showing from 30 to 300 colonies should be counted. If more than 300 organisms are present, many will fail to develop owing to the inhibitory action of the waste products secreted by those organisms which develop first.

The counts are expressed in numbers per milliliter of sample and are designated "standard plate count at 35°C" or "standard plate count at 20°C."

Various factors influencing the numbers of colonies developing on agar include (1) composition and reaction of the medium, (2) temperature and period of incubation, (3) presence of an abundant supply of oxygen, and (4) moisture. Unless such factors are controlled, the counts may show wide variation.

The above method and its many modifications give only a fraction of the total count. The nitrifying and other autotrophic bacteria are cultivated on synthetic media composed of inorganic salts. They are unable to grow on the usual nonsynthetic types of media. The obligate anaerobic bacteria fail to multiply in the presence of oxygen. Certain parasitic forms do not grow unless special media are used. Sulfate-reducing bacteria require the presence of sulfates for growth. Cellulose-dissolving bacteria grow very poorly or not at all unless cellulose is present in the medium. The result is that most bacteria found in water escape detection. Twenty to seventy times more organisms have been enumerated by the direct microscopic method than by the agar-plate procedure.

No great error is introduced by failure to obtain the total bacterial count of water. The sanitary bacteriologist is not interested in such organisms as the anaerobic spore formers, the nitrifying and other autotrophic forms, and the pathogenic organisms, but in a group as rapidly growing, rich-food-loving bacteria found in sewage. Most of these organisms are members of the coliform group.

Waterborne disease

The most important bacterial diseases transmitted by water are typhoid, dysentery, and cholera. Since they are intestinal diseases, the causative agents are found in feces. Therefore, the presence of sewage in a water supply means that one or more of these disease organisms may be present and that the water is potentially dangerous for human consumption.

Theoretically it would be better to examine water for the presence of disease organisms to determine its sanitary quality but such a procedure cannot be recommended as a routine practice. The techniques available at present are so complicated and time-consuming that the result, when obtained, would lose much of its significance and might even be confusing in a particular study of pollution.

Therefore the methods which follow do not provide for the detection, isolation, or enumeration of pathogenic bacteria in water. They are intended to indicate the degree of contamination of the water with wastes from human or animal sources. As stated above, attempts to isolate pathogenic bacteria from water on a routine basis have not proven fruitful and in general have ended in complete failure to obtain satisfactory results.

Coliform organisms in sewage The coliform group is defined to include all aerobic and facultative anaerobic, gram-negative, non-spore-forming, rod-shaped species which ferment lactose with the production of acid and gas within 48 hr at 35°C. Probably the most important members found in sewage-polluted waters and relatively easy to isolate are *E. coli, E. freundii*, and *A. aerogenes*.

Some coliform species or varieties have been designated fecal because they are commonly found in feces; others have been called nonfecal because they are believed to be normal inhabitants of soil. However, in the tests which follow, no attempt is made to differentiate between fecal and nonfecal types. Such a differentiation has been shown to be of limited value in determining the suitability of water for human consumption, as contamination with either type renders the water potentially dangerous and unsafe from a sanitary standpoint.

For more information: Gallagher and Spino (1968); Geldreich et al. (1964); Kabler et al. (1964); Klein and Casida (1967); Van Donsel et al. (1967).

STANDARD MULTIPLE TUBE FERMENTATION TECHNIQUE

The routine standard tests are the (1) presumptive test, (2) confirmed test, and (3) completed test.

Presumptive test Graduated amounts of water are transferred to a series of fermentation tubes containing lactose broth or lauryl tryptose broth of proper strength. It is the usual practice to inoculate five fermentation tubes each with 10 ml water, one tube with 1 ml water, and another with 0.1 ml water.

The tubes are incubated at $35 \pm 0.5°C$ and examined at the end of 24 ± 2 hr. If no gas has formed, the tubes are reincubated and examined at the end of 48 ± 3 hr. The formation of gas in any of the tubes within 48 ± 3 hr, regardless of the amount, constitutes a positive presumptive test (Fig. 275); the absence of gas formation within that period constitutes a negative test, and no further tests need be performed.

All tubes showing gas at the end of 24 or 48 hr of incubation shall be subjected to the confirmed test.

Conditions necessary for gas formation Chambers (1950) showed that 40 to 390 million coliforms per milliliter were required to produce

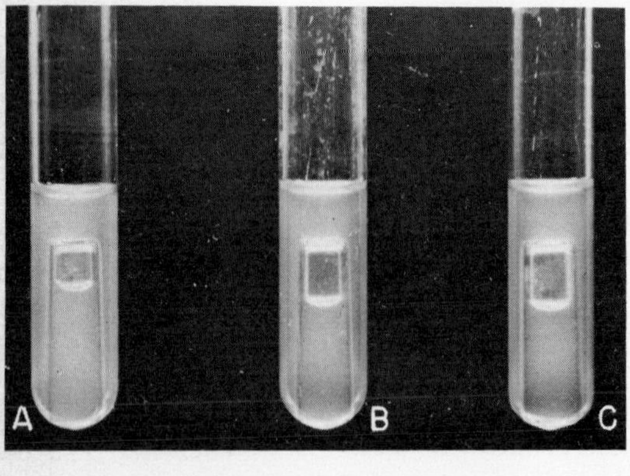

Fig. 275 Lactose broth fermentation tubes. (A) 1-ml water sample; (B) 0.5-ml water sample; (C) 0.1-ml water sample. All tubes show fermentation of the lactose with the production of acid and gas.

visible gas in fermentation broth. The average number was 170 million per ml. Different coliforms varied in the population densities required to produce gas. In most cases, coliform counts of 75 million or more per ml were required to produce visible gas. The number of coliform bacteria originally inoculated into fermentation broth had little effect on the population density required to produce gas but did have a marked effect on the time necessary for gas formation. In some cases high ratios of noncoliforms to coliforms prevented the formation of gas even though coliforms were present in considerable numbers in the original water. This probably explains why coliforms are isolated occasionally from fermentation tubes which do not demonstrate gas formation.

False positive presumptive tests A positive test does not mean necessarily that coliform organisms are present. False positive presumptive tests may be produced by other organisms capable of fermenting lactose with the formation of acid and gas. Also, positive presumptive tests may result from a type of association known as synergism (page 603). Synergism may be defined as the joint action of two organisms on a carbohydrate, resulting in the production of gas that is not formed by either species when grown separately.

Elimination of false presumptive tests Probably the most important method for the elimination of false presumptive tests is to incorporate an appropriate amount of a suitable triphenylmethane stain in the fermentation medium. Synergism is frequently caused by a gram-positive and a gram-negative organism growing together. A concentration of stain just sufficient to inhibit growth of the gram-positive species should have no effect on the gram-negative bacteria. This will eliminate a synergistic effect. Also, the method will eliminate false presumptive tests produced by gas-forming gram-positive aerobes and anaerobes.

Confirmed test All fermentation tubes showing gas within 48 hr at 35°C shall be utilized in the confirmed test.

Eosin methylene blue agar, Endo agar, or brilliant green lactose bile broth fermentation tubes may be used in the test.

Eosin methylene blue agar This medium is prepared by adding definite quantities of the two stains eosin and methylene blue to a melted lactose agar base and pouring about 15 ml into each petri dish.

A loopful of culture from each positive fermentation tube is streaked over the surface of E.M.B. agar. The plates are inverted and incubated at 35 ± 0.5°C for 24 ± 2 hr.

Three types of colonies develop on the medium:

1. Typical – nucleated, with or without metallic sheen
2. Atypical – opaque, nonnucleated mucoid after 24-hr incubation, pink
3. Negative – all others

If typical coliform colonies appear on the plates after 24 ± 2 hr, the confirmed test may be considered positive (Fig. 276).

If only atypical colonies appear, the confirmed test cannot be

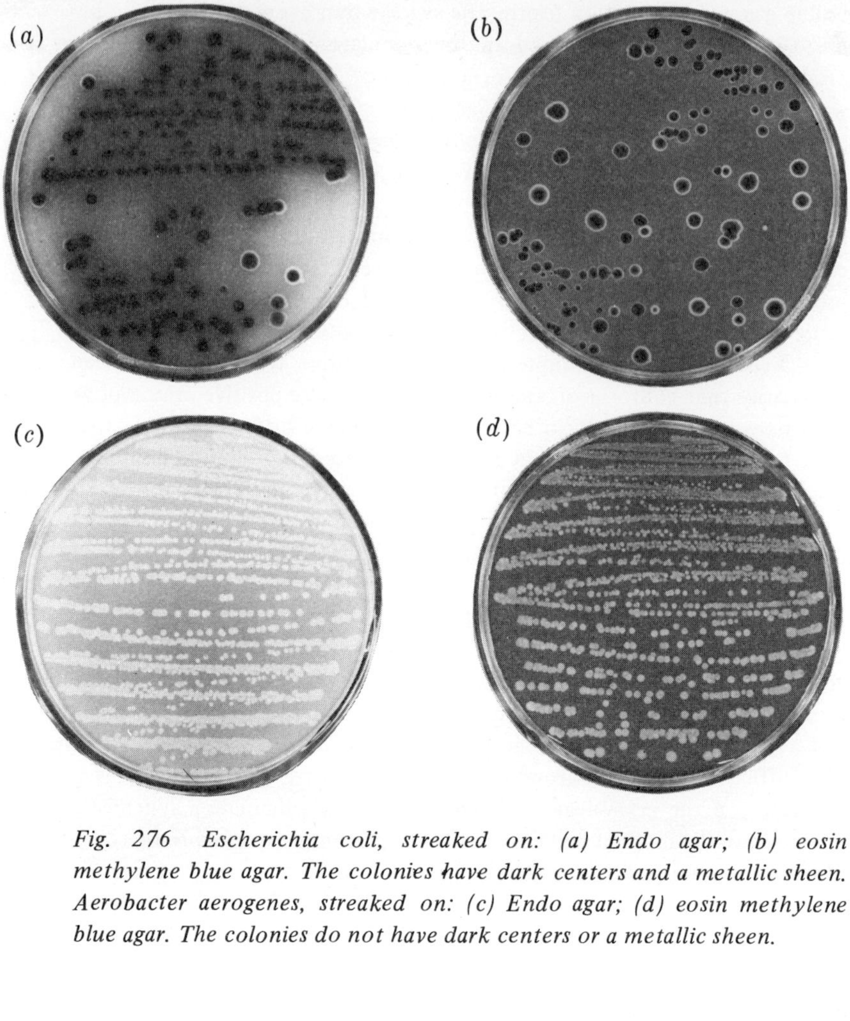

Fig. 276 Escherichia coli, streaked on: (a) Endo agar; (b) eosin methylene blue agar. The colonies have dark centers and a metallic sheen. Aerobacter aerogenes, streaked on: (c) Endo agar; (d) eosin methylene blue agar. The colonies do not have dark centers or a metallic sheen.

considered negative, since some coliforms fail to produce typical colonies on this medium, or the colonies develop slowly. Regardless of whether typical or atypical colonies appear, it is necessary to complete the test as directed below.

If no colonies, or noncoliform colonies develop within 24 ± 2 hr, the confirmed test may be considered negative.

Wynne et al. (1942) found the color of coliforms on this medium depended on two factors: (1) the reaction of eosin (an acidic stain) with methylene blue (a basic stain) to form a compound of either acidic or neutral nature and (2) the formation by lactose-fermenting organisms of sufficient acid to cause this stain compound to be taken up by individual cells of a colony. The non-lactose-fermenting organisms were not colored because the stain compound was not taken up in basic solution.

The medium is relatively stable. Prepared plates may be kept in the refrigerator for at least 1 week before use. The plates should not be exposed unnecessarily to the light; otherwise toxic substances may develop in the medium (see page 309).

Endo agar This medium is prepared by adding basic fuchsin, previously decolorized with sodium sulfite, to a melted lactose agar base, and pouring about 15 ml into each petri dish.

The basic fuchsin of commerce is a mixture of pararosaniline, rosaniline, and magenta II. Using pararosaniline as an example, the decolorization of the stain by sodium sulfite is as follows:

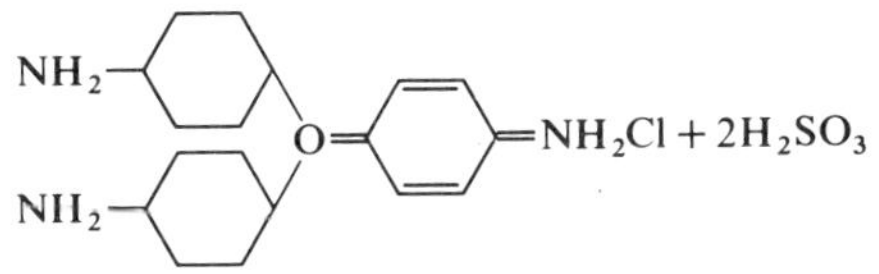

Quinoid salt (colored)

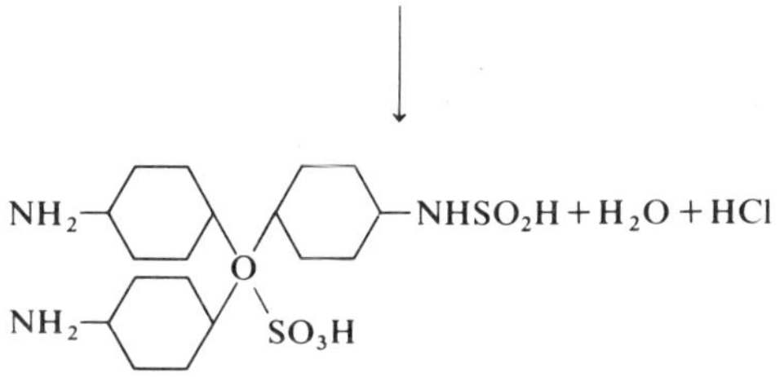

Nonquinoid compound (colorless)

A loopful of culture from each positive fermentation tube is streaked over the surface of Endo agar. The plates are inverted and incubated at 35 ± 0.5°C for 24 ± 2 hr.

The remarks made in the previous section with respect to the types of colonies developing on E.M.B. agar also apply here.

It is believed that Endo agar acts as a trapping medium for acetaldehyde, which is responsible for the reaction of typical colonies. Acetaldehyde is an intermediary compound produced in the fermentation of lactose.

As the acetaldehyde is produced, it reacts with the sulfite to form an addition compound. This releases the basic fuchsin from the combination to give the agar a deep red color. The metallic gold-like sheen imparted to the surface of the typical colonies is due to the precipitation of the liberated stain. However, the restored stain is different chemically from the original compound, being more purple in color. The reaction appears to be more complicated than just the removal of sulfite accompanied by the liberation of basic fuchsin.

Brilliant green lactose bile broth A loopful of culture from each positive presumptive tube is transferred to brilliant green lactose bile broth fermentation tubes. The primary tubes are gently shaken before making transfers. The tubes are incubated at 35 ± 0.5°C for 48 ± 3 hr.

A positive test is indicated by the presence of gas in any amount in the inverted vial within the incubation period.

Completed test This test may be performed on the typical or atypical colonies developing on E.M.B. or Endo agar, or on the brilliant green lactose bile broth fermentation tubes showing gas in the confirmed test. If the latter medium is used in the confirmed test, an E.M.B. or Endo agar plate is streaked from each fermentation tube showing gas. The plates are incubated at 35 ± 0.5°C for 24 ± 2 hr.

The purpose of the completed test is to determine if (1) the colonies developing on E.M.B. or Endo agar are again capable of fermenting lactose with the formation of acid and gas, and (2) the organisms transferred to agar slants show the morphological and tinctorial picture of members of the coliform group.

At least one typical colony or, if no typical colonies are present, at least two atypical colonies considered likely to be members of the coliform group are each transferred to lactose fermentation tubes and to nutrient agar slants. All tubes are incubated at 35 ± 0.5°C for a period not to exceed 48 ± 3 hr. Gram stains are prepared from the agar slant cultures.

The formation of gas in any amount in the fermentation tubes and the demonstration of gram-negative, non-spore-forming rods on the agar slants shall constitute a positive completed test for members of the coliform group. The absence of gas or failure to show the presence of rods answering to the above description in a gas-forming culture shall constitute a negative completed test.

For more information: American Public Health Association (1971).

Computing the most probable numbers of coliform organisms

Table 25 is based on the general formula of Hoskins (1934) for calculating the numbers of coliform organisms present in 100 ml of water. The figures are based on the employment of five tubes, each containing 10 ml of water

Table 25 Most probable numbers of coliform organisms present in 100 ml of a water sample

Number of positive lactose broth tubes			Number of coliform organisms	Number of positive lactose broth tubes			Number of coliform organisms
10 ml	1 ml	0.1 ml		10 ml	1 ml	0.1 ml	
0	0	0	0	3	0	0	8.8
0	1	0	2	3	0	1	11
0	1	1	4	3	1	0	12
1	0	0	2.2	4	0	0	15
1	0	1	4	4	0	1	20
1	1	0	4.4	4	1	0	21
2	0	0	5	5	0	0	38
2	0	1	7	5	0	1	96
2	1	0	7.6	5	1	0	240

sample, one tube containing 1 ml of water sample, and one tube containing 0.1 ml of sample.

For more information: Levine et al. (1955).

STANDARD MEMBRANE FILTER TECHNIQUE

Goetz and Tsuneishi (1951) described a new method for the enumeration of coliform bacteria in water which they named the molecular filter technique. It is now generally referred to as the membrane filter technique.

The technique has now been designated a standard method for the examination of water, sewage, and related problems. It may be used for determining the sanitary quality of a water supply after parallel testing has demonstrated that it yields information equal to that from the fermentation broth procedure already discussed.

Method A sterile membrane filter is placed in the holding apparatus shown in Fig. 277. The bacteria are collected by passing the water sample through the membrane with the aid of a partial vacuum. The membrane and side of the funnel are rinsed with a small volume of sterile distilled water. The membrane is then removed aseptically and placed on an absorbent disk, previously saturated with culture medium and contained in a petri dish. The medium passes through the membrane and nourishes the bacteria present thereon. After sufficient incubation, each organism will

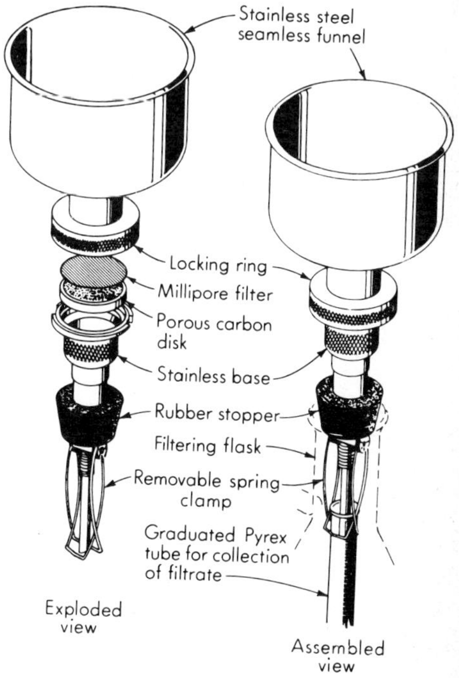

Fig. 277 Millipore filter holder. The unit consists of a funnel of stainless steel clamped to a base containing a molecular filter. The stem of the base is inserted into a filter flask through a rubber stopper.

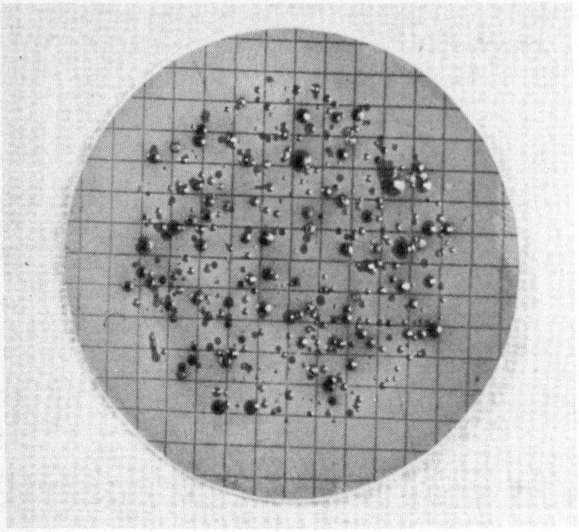

Fig. 278 Culture of Escherichia coli on molecular filter. (Courtesy of Millipore Filter Corporation, Bedford, Massachusetts.)

grow and produce a visible colony on the membrane (Fig. 278). The incubation time and the temperature depend on the organism to be grown and the medium used.

The medium generally used is a modified Endo (M-Endo) medium (page 695). On this medium all organisms that produce a dark (purplish-green) colony with a metallic sheen in 20 $\pm$ 2 hr at 35 $\pm$ 0.5°C are considered members of the coliform group. The sheen may appear in the central area or cover the entire colony. The coliform group as defined here is roughly equivalent in sanitary significance to the coliform group as previously defined.

Sometimes coliform organisms may fail to produce typical colonies. Under these conditions, the presence of large numbers of similarly appearing atypical colonies should be identified by subculture study.

Advantages to membrane technique The membrane filter possesses the following advantages in water examinations:

1. It permits the concentration of small numbers of organisms from large quantities of water, thereby increasing the accuracy and reliability of such determinations.
2. It minimizes spreading of colonies and allows the combination of any number of bacteria, from a few to 5000 at one time. This has the advantage of reducing the number of laboratory dilutions and duplicate incubations.
3. It permits separation of the organisms from their nutrient at any time for the purpose of either changing the medium or partially or totally inhibiting further development.
4. It gives a direct count instead of a determination of the most probable number from positive fermentation tubes (Table 25).
5. It saves time in comparison with the multiple tube fermentation test.
6. It permits better and faster differentiation of bacteria.
7. It gives a permanent record in the form of the preserved filter disks.

8. It permits field filtration of the sample and shipment of the filters to the laboratory on a preservative medium.
9. The method is extremely useful in emergencies.

Disadvantages to membrane technique The following disadvantages have been mentioned:

1. In turbid waters heavily contaminated with algae or other material, clogging of the filter will prevent testing of sufficient sample to provide a reliable indication of coliform density and may interfere with the development of characteristic coliform colonies during incubation.
2. In waters with a high density of noncoliforms, the ratio of membrane filter density estimates to M.P.N. estimates will be excessively low.

For more information: American Public Health Association (1971); Buras and Kott (1969); Ehrlich (1960); Gärtner and Schyma (1961); Guthrie and Reeder (1969).

TENTATIVE FECAL STREPTOCOCCI GROUP TECHNIQUES

In some areas of the world, fecal streptococci (Lancefield's group D *Streptococcus*), rather than coliforms, are used as indicators of fecal pollution of water. The fecal streptococci include the following species and varieties: *S. faecalis, S. faecalis* var. *liquefaciens, S. faecalis* var. *zymogenes*; *S. faecium, S. faecium* var. *durans*; *S. bovis,* and *S. equinus.*

The normal habitat of these organisms is the intestinal tracts of man and animals, and they are not known to be found in nature in the absence of fecal contamination from the above sources. Fecal streptococci die out rapidly in soil. On the other hand, some believe that the coliforms do not always represent fecal pollution because the organisms may persist in soil and water for long periods of time and occasionally multiply outside the animal body.

Tests for the presence of fecal streptococci are not yet proposed as substitutes for coliform determinations. However, such tests may be used to confirm the source of coliforms found to be present in a sample. Therefore, tests for fecal streptococci may be of value where there is a question concerning the sanitary significance of coliform bacteria isolated from a sample.

In natural waters the most frequent proportion of coliforms to fecal streptococci ranges from 10:1 to 1:2.

The multiple tube fermentation technique and the membrane filter method may be utilized to determine the presence of fecal streptococci. Both determinations are tentative.

For more information: American Public Health Association (1971); Breed et al. (1957); Burman (1961); Clark (1969); Hartman et al. (1966); Hashimoto (1961); Horie (1959*a*, *b*, 1960); Kenner et al. (1960); Kjellander (1960); Medrek and Litsky (1960).

Differentiation of fecal from nonfecal coliform organisms

The two standard procedures already discussed make no distinction between so-called fecal and nonfecal types. The American Public Health Association feels that any attempt to evaluate a water supply on the basis of a distinction between the two types is unwarranted. Contamination with either type of waste renders the water potentially unsatisfactory and unsafe from a sanitary standpoint. However, differentiation may be advantageous under some conditions where the identity of a specific member of the group present may indicate the source of the pollution.

The coliform group Studies of recent years have emphasized the complexity of the coliform group. The general practice is to classify the members into fecal *E. coli*, and nonfecal *E. freundii* and *Aerobacter aerogenes*. The classification is based on the results of four tests, namely, indole, methyl red, Voges-Proskauer, and sodium citrate.

For convenience, the mnemonic IMViC has been coined to designate these four reactions. The letter I is for indole, M for methyl red, V for Voges-Proskauer, and C for citrate. The letter i between the V and C is added solely for euphony. On the basis of these four tests, the coliforms may be classified as shown in Table 26.

Probably all types of coliform organisms may be found in feces. *E. coli* will most likely be found in fresh pollution derived from several sources, but other types, without *E. coli*, may sometimes be present in a fresh pollution from a single source.

There is little evidence that any of the coliform organisms multiply on grasses, on grains, or in soils. Therefore, it is questionable whether grasses, grains, and soils can be considered normal habitats of the members of the coliform group.

However, there is evidence that coliforms do multiply on such organic materials as leather washers, wood, swimming-pool ropes, jute packing, and

Table 26 A simplified reaction classification of members of the coliform group

Organism	*Indole*	*Methyl red*	*Voges-Proskauer*	*Sodium citrate*	*Commonly designated source*
Escherichia coli					
Variety I	+	+	–	–	fecal
Variety II	–	+	–	–	fecal
E. freundii (Intermediates)					
Variety I	–	+	–	±	nonfecal
Variety II	+	+	–	+	nonfecal
Aerobacter aerogenes					
Variety I	–	–	+	±	nonfecal
Variety II	±	–	+	+	nonfecal

even in water pipes. The presence of large numbers of coliforms of a single type in water obtained from wells, springs, or a single distribution system is suggestive of multiplication on the above materials.

Members of the coliform group vary in the amount of gas (a mixture of carbon dioxide and hydrogen) produced in lactose fermentation broth. Typical strains of *E. coli* rarely produce more than 25 percent of gas in the inverted vial, whereas typical strains of *A. aerogenes* produce from 75 to 100 percent of gas. This indicates differences in the carbohydrate metabolism of the two subgroups.

Typical *E. coli* produce carbon dioxide and hydrogen in approximately equal volumes and have been termed the low-ratio organisms. Typical *A. aerogenes* produce about twice as much carbon dioxide as hydrogen and have been called the high-ratio organisms.

The high ratio *A. aerogenes* as well as *E. freundii* are only occasionally found in the intestinal contents of man and animals (about 6 percent). They are normally found in the soil. For this reason some believe that their presence in water is of doubtful sanitary significance. On the other hand, the low-ratio *E. coli* are rarely found in soil but constitute the predominating organisms in the intestinal contents of man and animals. They are only occasionally found in localities not showing recent fecal pollutions.

For more information: Geldreich et al. (1962*b*); Tennant and Reid (1961); Thomas et al. (1960).

IMViC and elevated temperature tests

The procedures which follow make an attempt to distinguish between fecal and nonfecal types of coliforms. The tests are employed in some laboratories but should be regarded as tentative rather than as standard.

Methyl red test Cultures of *Escherichia* rapidly ferment lactose with the formation of acids until the pH drops to about 5.0. This acidity is sufficient to prevent further growth. The same principle applies to any strongly fermentative organism, although the final pH values may show wide variation.

From a given amount of a fermentable sugar, *Aerobacter* produces less acid than *Escherichia*. This is due largely to the fact that *Aerobacter* converts some of the sugar to 2:3 butanediol and acetylmethylcarbinol (acetoin), whereas *Escherichia* does not. The amount of carbohydrate that is just sufficient for *Escherichia* to produce its maximum acidity is inadequate for *Aerobacter* to produce its limiting hydrogen-ion concentration. The result is that *Escherichia* will be stopped in its growth. On the contrary, *Aerobacter* will exhaust the sugar without producing its limiting pH and continue to grow, utilizing the peptone of the medium for both structure and energy. The medium becomes progressively less acid (more alkaline) in reaction.

The amount of buffer present influences the final pH attained by an organism when grown in the presence of a fermentable substance. As the buffer content is increased, the final acidity is decreased. This means that as fermentation is prolonged, metabolic waste products other than acids accumulate to produce an increased toxic effect on the organisms.

The medium used for the test contains 0.5 percent glucose and is buffered with dipotassium phosphate and peptone to give a limiting pH of about 5.0 when inoculated with typical *Escherichia*. Typical *Aerobacter* give a higher final pH. The methyl red indicator used in the test is turned red by typical *Escherichia* and yellow by typical *Aerobacter*.

Voges-Proskauer test Voges and Proskauer observed that cultures of some bacteria, when made alkaline with potassium hydroxide, developed a red color on standing for 24 hr. The development of a red color constitutes a positive V-P reaction.

Distinct differences exist in the carbohydrate metabolism of typical *Escherichia* and *Aerobacter*. The fermentation of glucose by the two groups yields the products shown in Table 27.

The figures show that 87 percent of the carbon is accounted for in the case of *E. coli* and 69 percent in cultures of *A. aerogenes*. The discrepancy is due to the formation by *A. aerogenes* of 2:3-butanediol ($CH_3 \cdot CHOH \cdot CHOH \cdot CH_3$) and acetylmethylcarbinol ($CH_3 \cdot CO \cdot CHOH \cdot CH_3$) but not by *E. coli*. The acetylmethylcarbinol in the presence of potassium hydroxide and air is further oxidized to diacetyl ($CH_3 \cdot CO \cdot OC \cdot CH_3$) which, in the presence of peptone, gives a red color. The constituent of peptone responsible for the red color is the guanidine nucleus [$NH{:}C(NH_2)NH \cdot R$] of the amino acid arginine.

The presence of acetylmethylcarbinol (acetoin) may be easily detected in bacterial cultures. The organism being studied is heavily inoculated into a tube of glucose broth and incubated at 30°C for 24 to 48 hr. Then 0.6 ml of α-naphthol (5 g α-naphthol in 100 ml 95 percent alcohol) is added, followed by 0.2 ml of 40 percent potassium hydroxide solution. It is important to shake for about 5 sec after addition of each reagent. A positive test is indicated by the appearance of a red color in from 2 to 4 hr. Results should be read not later than 4 hr after addition of the reagents.

Table 27

	Percent by weight of glucose fermented	
Product	*Aerobacter aerogenes*	*Escherichia coli*
Alcohol	17.10	12.85
Acetic acid	5.10	18.84
Succinic acid	2.40	5.20
Formic acid	1.00	0.00
Lactic acid	5.50	31.90
Carbon dioxide	38.00	18.10
Total	69.10	86.89
Ratio, CO_2/H_2	2.40	0.83

Juni (1952) explained the reaction by demonstrating the presence of two enzymes in extracts of *A. aerogenes*, one producing α-acetolactic acid from pyruvic acid (an intermediary compound), and the other capable of decarboxylating α-acetolactic acid to acetoin but having no action on pyruvic acid:

$$\underset{\text{Pyruvic acid}}{2CH_3\cdot CO\cdot COOH} \longrightarrow \underset{\alpha\text{-acetolactic acid}}{CH_3\cdot CO\cdot COH\,(COOH)\cdot CH_3} + CO_2$$

$$\xrightarrow{\alpha\text{-acetolactic dehydrogenase}} \underset{\text{Acetoin}}{CH_3\cdot CO\cdot CHOH\cdot CH_3} + CO_2$$

With some bacteria, the V-P reaction may weaken on prolonged incubation; with others it may even become negative. This is probably caused by partial or total reduction of acetoin to 2:3-butanediol which does not give the reaction, or by oxidation to diacetyl followed by further metabolism to other compounds. Some strains of V-P positive and also some V-P negative bacteria are capable of destroying acetoin. The compound may be utilized as a source of carbon. This behavior, no doubt, explains why some transiently positive reactions become weaker as incubation is prolonged. Therefore, tests for the presence of acetoin should not be unnecessarily delayed.

The V-P test distinguishes to a high degree between fecal and nonfecal types of coliforms.

For more information: Barry and Feeney (1967); Eddy (1961).

Sodium citrate test Koser (1923, 1924) showed that coliforms could be separated into two groups on the basis of their action on sodium citrate. Typical fecal *Escherichia* was unable to utilize citrate as the only source of carbon; typical nonfecal *Aerobacter* utilized citrate readily.

Lara and Stokes (1952) observed that living cells of typical *Escherichia* did not oxidize citrate, whereas dried cells did. This would indicate that perhaps impermeability of living cells of *Escherichia* to citrate might be responsible for their inability to grow in the medium.

Citrate is said to be dissimilated by *Aerobacter* to acetate, pyruvate, and carbon dioxide as follows:

$$\underset{\text{Sodium citrate}}{COONaCH_2\cdot C(OH)(COONa)\cdot CH_2COONa} + H_2O \longrightarrow$$

$$\underset{\text{Sodium oxaloacetate}}{COONa\cdot CH_2\cdot CO\cdot COONa} + \underset{\text{Sodium acetate}}{CH_3\cdot COONa} + H_2O$$

$$\underset{\text{Sodium oxaloacetate}}{COONa\cdot CH_2\cdot CO\cdot COONa} + H_2O \longrightarrow \underset{\text{Sodium pyruvate}}{CH_3\cdot CO\cdot COONa} + NaOH + CO_2$$

The reaction occurs in an anaerobic environment. The enzyme is adaptive and suppressed by growth in the presence of air.

O'Brien and Stern (1969) showed that anaerobic growth of *Aerobacter* on citrate as a carbon source required the presence of NA^+. The growth rate increased with increasing Na^+ concentration and was optimal at a molar concentration of 0.1 Na^+. The Na^+ requirement was specific and could not be replaced by other cations.

The sodium citrate test appears to correlate more closely with the source of the coliforms than any of the other IMViC tests.

For more information: Ferlin and Karabinos (1954).

Elevated temperature tests The IMViC tests are considered by many as controversial procedures and do not satisfactorily differentiate the fecal from the nonfecal types of coliforms.

Eijkman (1904) proposed a test that he believed differentiated between the two groups. The method was based on the assumption that only fecal types of coliforms from warm-blooded animals could grow at 46°C with the production of acid and gas.

Stuart et al. (1942) found that *Aerobacter* and intermediates seldom produced gas from lactose at 45.5°C, whereas *Escherichia* seldom failed to do so. In their Eijkman characteristics, the intermediates were more closely related to *Aerobacter* than to *Escherichia*.

Taylor (1945) reported that 97 percent of cultures of typical or type I *E. coli* (Table 28) examined fermented lactose with the production of acid and gas between 40 and 44°C. The number was not appreciably reduced at 45°C but was markedly reduced at 46°C. On the other hand, 15 percent of cultures of typical or type I *Aerobacter* were found to be positive at 44°C but only 2 percent at 45°C.

A buffered tryptose lactose broth medium may be used with good results. The medium is placed in fermentation tubes, inoculated with pure cultures of coliforms, and incubated in a water-jacketed incubator at a temperature of 45.5 ± 0.2°C for 48 hr. Gas production constitutes a positive test for fecal coliforms.

Another elevated temperature test that is frequently performed is based on the use of a buffered boric acid lactose broth (BALB) medium. The broth selectively inhibits growth and gas production of *Aerobacter* and intermediate members of the coliform group, whereas practically all fecal *Escherichia* cultures produce gas in 48 hr.

Table 28 Production of acid and gas in lactose broth by various types of coliform bacteria at different temperatures

Type	*No. of cultures tested*	*Percentage of cultures forming acid and gas at* 37°C	40°C	42°C	43°C	44°C	45°C	46°C
Escherichia coli, type I	96	100	97	97	97	97	93	52
E. coli, type II	50	100	64	36	32	28		
Intermediate type I	78	100	78	56	23	0		
Intermediate type II	53	100	55	4	2	0		
Aerobacter aerogenes,								
type I (or	80	100	83	53	35	15	2	
A. cloacae)	52	100	–	–	–	15		
A. aerogenes, type II	25	100	80	12	0	0		

The fermentation medium is first warmed to 37°C, then inoculated with a loopful of a broth culture of the organism under study. The tube is incubated at 43 ± 0.5°C for 48 hr. Gas production constitutes a positive step for fecal coliforms.

The two elevated temperature tests show a positive correlation of at least 95 percent with coliforms from fecal sources, which is much higher than with the IMViC tests, and are generally preferred in water examinations.

For more information: American Public Health Association (1971); Amsterdam and Wolfe (1968); Geldreich et al. (1962*a*); Hendricks (1970); von Graevenitz (1971); Wolfe and Amsterdam (1968).

Paracolon bacteria

The paracolon bacteria occupy a position intermediate between the coliforms (*Escherichia* – intermediates – *Aerobacter*) and the *Salmonella*. Borman et al. (1944) proposed the genus name *Paracolobactrum* for this group.

The *Paracolobactrum* has some of the cultural characteristics of the coliforms and some of the pathogenicity of the *Salmonella*. The cells are short rods, and gram-negative. Fermentation of lactose is consistently delayed; occasionally lactose is not fermented. Certain forms attack carbohydrates characteristically at 20 to 30°C but not at 37°C. The production of acetylmethylcarbinol may likewise be influenced by incubation temperature. Antigenic relationships to other genera in the family (see page 530) are common, even with respect to major antigens. The organisms are found in surface water, in soil, on grains, and in the intestinal tracts of animals and man.

The paracolons, like the coliform bacteria, can be divided into three subgroups: paracolon *Escherichia*, paracolon intermediates, and paracolon *Aerobacter*, according to their IMViC reactions. A frequent property of the paracolons is that they either do not ferment lactose or attack it very slowly. For this reason plate colonies on differential media are often mistaken for pathogenic nonlactose fermenters. The organisms are frequently the cause of food poisoning and have been mistaken for members of the *Salmonella*.

Purification of water

Sewage organisms, especially the pathogenic forms, do not live long in water. The temperature and food conditions are not conducive to survival for any length of time. Also many protozoa are present in water at times, and these depend largely on bacteria as a source of food.

However, a municipal water supply must usually pass through a series of purification steps to make it safe for human consumption. In some cases the water may be sufficiently pure to require only the addition of a disinfectant. In other instances the condition of the water is such as to require three stages of purification, namely, sedimentation, filtration, and disinfection.

Sedimentation In this stage the water is run into large tanks where it is allowed to stand for the suspended matter to settle out. Since this is a very slow process, it may be hastened by the addition of a chemical coagulant, such as aluminum sulfate, which reacts with the alkaline salts to produce insoluble gelatinous aluminum hydroxide. The insoluble material rapidly settles out, carrying with it bacteria and other particulate matter in suspension.

Filtration In the second stage the water is passed through sand filters. These are of two types: the slow sand filter and the rapid sand filter. When the slow filter is employed, it is usually not necessary to add a chemical coagulant. A slimy gelatinous film soon forms on the surface of the sand; the film acts as a filtering medium, removing bacteria and other suspended matter. The filtering surface eventually becomes too thick and must be removed.

In the rapid method the coagulant forms a mat or floc on the surface of the sand which acts as a filtering medium. It should remove bacteria and particles of similar size. Since this type of filter can handle more water per surface area than the slow type, it is more extensively used.

Disinfection Since not all bacteria are necessarily removed by the sand filters, those that remain can be destroyed by the addition of a suitable disinfectant. Chlorine is effective in minute amounts and is generally used for this purpose. Automatic machines are available which feed chlorine gas directly into the water. Usually sufficient chlorine is added to provide a residual concentration of 1 part of available chlorine per 1 million parts of water.

For more information: Chang and Morris (1953); Geldreich and Clarke (1966); Kabler (1959); Kabler et al. (1961); Okun (1962).

References

American Public Health Association: "Standard Methods for the Examination of Water and Wastewater," New York, 1971.

Amsterdam, D., and M. W. Wolfe: Comparison of reagent-impregnated paper strips and conventional tests for distinguishing *Escherichia* from *Aerobacter*: Correlation with colonial morphology, *Appl. Microbiol.,* **16**:1460, 1968.

Barry, A. L., and K. L. Feeney: Two quick methods for Voges-Proskauer test, *Appl. Microbiol.,* **15**:1138, 1967.

Borman, E. K., et al.: Taxonomy of the family *Enterobacteriaceae, J. Bact.,* **48**:351, 1944.

Breed, R. S., et al.: "Bergey's Manual of Determinative Bacteriology," Baltimore, The Williams & Wilkins Company, 1957.

Buras, N., and Y. Kott: A simple test for differentiation between *E. coli* and *A. aerogenes, Water Res.,* **3**:973, 1969.

Burman, N. P.: Some observations on coli-aerogenes bacteria and streptococci in water, *J. Appl. Bact.,* **24**:368, 1961.

Chambers, C. W.: Relationships of coliform bacteria to gas production in media containing lactose, *Public Health Rept.,* **65**:619, 1950.

Chang, S. L., and J. C. Morris: Elemental iodine as a disinfectant for drinking water, *Ind. Eng. Chem.,* **45**:1009, 1953.

Clark, J. A.: The detection of various bacteria indicative of water pollution by a presence-absence (P-A) procedure, *Can. J. Microbiol.,* **15**:771, 1969.

Cody, R. M., et al.: Coliform population in stored sewage, *J. Water Pollution Control Federation,* **33**:164, 1961.

Eddy, B. P.: The Voges-Proskauer reaction and its significance: A review, *J. Appl. Bact.,* **24**:27, 1961.

Ehrlich, R.: Application of membrane filters. In "Advances in Applied Microbiology," vol. II, New York, Academic Press, Inc., 1960.

Eijkman, C.: Die Gärungsprobe bei 46°C. als Hilfsmittel bei der Trinkwasseruntersuchung, *Centr. Bakt., I, Orig.,* **37**: 742, 1904.

Ewing, W. H., and B. R. Davis: Biochemical characterization of *Citrobacter diversus* (Burkey) Werkman and Gillen and designation of the neotype strain, *Intern. J. Systematic Bact.,* 22:12, 1972.

Ferlin, H. J., and J. V. Karabinos: Differential media for *Escherichia coli* and *Aerobacter aerogenes, J. Wash. Acad. Sci.,* **44**:303, 1954.

Gallagher, T. P., and D. F. Spino: The significance of numbers of coliform bacteria as an indicator of enteric pathogens, *Water Res.,* 2:169, 1968.

Gameson, A. L. H., and J. R. Saxon: Field studies on effect of daylight on mortality of coliform bacteria, *Water Res.,* **1**:279, 1967.

Gärtner, H., and D. Schyma: Über den Nachweis von Coli-Bakterien und Salmonellen aus dem Wasser mit einem löslichen Ultra-filter, *Arch. Hyg. Bakteriol.,* **145**:81, 1961.

Geldreich, E. E., and N. A. Clarke: Bacterial pollution indicators in the intestinal tract of fresh-water fish, *Appl. Microbiol.,* **14**:429, 1966.

——— et al.: Type distribution of coliform bacteria in the feces of warm-blooded animals, *J. Water Pollution Control Federation,* **34**:295, 1962*a*.

———: The faecal coli-aerogenes flora of soils from various geographical areas, *J. Appl. Bact.,* **25**:87, 1962*b*.

——— et al.: Occurrence of coliforms, fecal coliforms, and streptococci on vegetation and insects, *Appl. Microbiol.,* **12**:63, 1964.

Goetz, A., and N. Tsuneishi: Application of molecular filter membranes to the bacteriological analysis of water, *J. Am. Water Works Assoc.,* **43**:943, 1951.

Guthrie, R. K., and D. J. Reeder: Membrane filter-fluorescent-antibody method for detection and enumeration of bacteria in water, *Appl. Microbiol.,* **17**:399, 1969.

Hartman, P. A., et al.: Indicator organisms: A review. I. Taxonomy of the fecal streptococci, *Intern. J. Systematic Bact.*, **16**:197, 1966.

Hashimoto, H.: Hygienic studies on Enterococcus group in milk, meat and their products, especially on significance as an indicator of pollution, *J. Fac. Fisheries Animal Husbandry, Hiroshima Univ.*, **3**:503, 1961.

Hendricks, C. W.: Formic hydrogenlyase induction as a basis for the Eikjman fecal coliform concept, *Appl. Microbiol.*, **19**:441, 1970.

Horie, S.: Studies on enterococci as pollution indices of food and drink. I. Some observations on the distribution of enterococci, *Bull. Japan Soc. Sci. Fisheries*, **25**:294, 1959*a*; II. Subdivision of enterococci on the basis of the fermentation tests, *ibid.*, **25**:488, 1959*b*; IX. Viability of enterococci in waters and foodstuffs, *ibid.*, **26**:614, 1960.

Hoskins, J. K.: Most probable numbers for evaluation of coli-aerogenes tests by fermentation tube method, *Public Health Rept.*, **49**:393, 1934.

Juni, E.: Mechanisms of formation of acetoin by bacteria, *J. Biol. Chem.*, **195**:715, 1952.

Kabler, P. W.: Removal of pathogenic microorganisms by sewage treatment processes, *Sewage Ind. Wastes*, **31**:1373, 1959.

—— et al.: Viricidal efficiency of disinfectants in water, *Public Health Rept.*, **76**:565, 1961.

—— et al.: Sanitary significance of coliform and fecal coliform organisms in surface water, *Public Health Rept.*, **79**:58, 1964.

Kenner, B. A., et al.: Fecal streptococci. II. Quantification of streptococci in feces, *Am. J. Public Health*, **50**:1553, 1960.

Kjellander, J.: Enteric streptococci as indicators of fecal contamination of water, *Acta Pathol. Microbiol. Scand., Suppl.* 136, vol. 48, 1960.

Klein, D. A., and L. E. Casida, Jr.: *Escherichia coli* die-out from normal soil as related to nutrient availability and the indigenous microflora, *Can. J. Microbiol.*, **13**:1461, 1967.

Koser, S. A.: Utilization of the salts of organic acids by the colon-aerogenes group, *J. Bact.*, **8**:493, 1923.

——: Correlation of citrate utilization by members of the colonaerogenes group with other differential characteristics and with habitat, *J. Bact.*, **9**:59, 1924.

Lara, F. J. S., and J. L. Stokes: Oxidation of citrate by *Escherichia coli*, *J. Bact.*, **62**:415, 1952.

Levine, M., et al.: Simultaneous determination of coliform and *Escherichia coli* indices, *Appl. Microbiol.*, **3**:310, 1955.

Medrek, T. F., and W. Litsky: Comparative incidence of coliform bacteria and enterococci in undisturbed soil, *Appl. Microbiol.*, **8**:60, 1960.

O'Brien, R. W., and J. R. Stern: Requirement for sodium in the anaerobic growth of *Aerobacter aerogenes* on citrate, *J. Bact.*, **98**:388, 1969.

Okun, D. A.: Wastewater treatment in Europe, *J. Water Pollution Control Federation*, **34**:704, 1962.

Potter, L. F.: The effect of pH on the development of bacteria in water stored in glass containers, *Can. J. Microbiol.,* **6**:257, 1960.

Sen, R., and S. N. Ghosh: A comparison of the enterococci index with coliform index as indication of faecal pollution of water, *Indian J. Med. Res.,* **58**:1168, 1970.

Stuart, C. A., et al.: Eijkman relationships of the coliform and related bacteria, *J. Bact.,* **43**: 557, 1942.

Taylor, C. B.: The effect of temperature of incubation on the results of tests for differentiating species of coliform bacteria, *J. Hyg.,* **44**:109, 1945.

Tennant, A. D., and J. E. Reid: Coliform bacteria in sea water and shellfish. I. Lactose fermentation at 35.5° and 44°C., *Can. J. Microbiol.,* **7**:725, 1961.

Thomas, S. B., et al.: An ecological study of the coli-aerogenes bacteria of surface soil, *J. Appl. Bact.,* **23**:169, 1960.

Van Donsel, D. J., et al.: Seasonal variations in survival of indicator bacteria in soil and their contribution to storm-water pollution. *Appl., Microbiol.,* **15**:1362, 1967.

von Graevenitz, A.: Practical substitution for the indole, methyl red, Voges-Proskauer, citrate system, *Appl. Microbiol.,* **21**:1107, 1971.

Wolfe, M. W., and D. Amsterdam: New diagnostic system for the identification of lactose-fermenting Gram-negative rods, *Appl. Microbiol.,* **16**:1528, 1968.

Wynne, E. S., et al.: Mechanism of the selective action of eosin-methylene-blue agar on the enteric group, *Stain Technol.,* **17**:11, 1942.

22 *Bacteriology of milk and milk products*

Milk is considered to be the most satisfactory single food substance elaborated by nature. It contains about 3.8 percent fat, 3.2 percent protein, 4.8 percent carbohydrate, 0.7 percent minerals, and 87.5 percent water. In addition to the major constituents, milk contains a large number of substances that occur in minute amounts. Among them are fatty acids, amino acids, sugars other than lactose, sugar phosphates, proteoses, peptones, nitrogenous bases, enzymes, and vitamins.

The U.S. Public Health Service (1967) defines milk and its products as follows:

Milk Milk is the lacteal secretion, practically free from colostrum, obtained by the complete milking of one or more healthy cows, which contains not less than 8 1/4 percent milk solids-not-fat and not less than 3 1/4 percent milkfat.

The food value of milk depends upon its milkfat and its solids-not-fat content. If either of these is reduced below the range for normal market milk, the proteins, carbohydrates, minerals, and certain vitamins are also reduced. Practical experience shows that 3 1/4 percent fat and 8 1/4 percent solids-not-fat are reasonable minima for mixed-herd milk.

Milk produced within 15 days before calving or 5 days after calving is likely to contain colostrum. Colostrum tends to produce intestinal disturbance in children.

Goat milk Goat milk is the lacteal secretion, practically free from colostrum, obtained by the complete milking of healthy goats. The word "milk" shall be interpreted to include goat milk.

Cream Cream is the sweet, fatty liquid separated from milk, with or without the addition of milk or skim milk, which contains not less than 18 percent milkfat.

Light cream, coffee cream, or table cream Light cream, coffee cream, or table cream is cream which contains not less than 18 percent but less than 30 percent milkfat.

Whipping cream Whipping cream is cream which contains not less than 30 percent milkfat.

Light whipping cream Light whipping cream is cream which contains not less than 30 percent but less than 36 percent milkfat.

Heavy cream or heavy whipping cream Heavy cream or heavy whipping cream is cream which contains not less than 36 percent milkfat.

Whipped cream Whipped cream is whipping cream into which air or gas has been incorporated.

Whipped light cream, coffee cream, or table cream Whipped light cream, coffee cream, or table cream is light cream, coffee cream, or table cream into which air or gas has been incorporated.

Sour cream or cultured sour cream Sour cream or cultured sour cream is a fluid or semifluid cream resulting from the souring, by lactic acid–producing bacteria or similar culture, of pasteurized cream, which contains not less than 0.20 percent acidity expressed as lactic acid.

Half-and-half Half-and-half is a product consisting of a mixture of milk and cream which contains not less than 10.5 percent milkfat.

Sour half-and-half or cultured half-and-half Sour half-and-half or cultured half-and-half is fluid or semifluid half-and-half derived from the souring, by lactic acid producing bacteria or similar culture, of pasteurized half-and-half, which contains not less than 0.2 percent acidity expressed as lactic acid.

Reconstituted or recombined milk and milk products Reconstituted or recombined milk and/or milk products shall mean milk or milk products defined in this section which result from the recombining of milk constituents with potable water.

Concentrated milk Concentrated milk is a fluid product, unsterilized and unsweetened, resulting from the removal of a considerable portion of the water from milk, which, when combined with potable water, results in a product conforming with the standards of milkfat and solids-not-fat of milk as defined above.

Concentrated milk products Concentrated milk products shall be taken to mean and to include homogenized concentrated milk, vitamin D concentrated milk, concentrated skim milk, fortified concentrated skim milk, concentrated lowfat milk, fortified concentrated lowfat milk, concentrated flavored milk, concentrated flavored milk products, and similar concentrated products made from concentrated milk or concentrated skim milk, and which, when combined with potable water in accordance with instructions printed on the container, conform with the definitions of the corresponding milk products in this section.

Frozen milk concentrate Frozen milk concentrate is a frozen milk product with a composition of milkfat and milk solids-not-fat in such proportions that when a given volume of concentrate is mixed with a given volume of water the reconstituted product conforms to the milkfat and milk solids-not-fat requirements of whole milk. In the manufacturing process, water may be used to adjust the primary concentrate to the final desired concentration. The adjusted primary concentrate is pasteurized, packaged, and immediately frozen. This product is stored, transported, and sold in the frozen state.

Skim milk or skimmed milk Skim milk or skimmed milk is milk from which sufficient milkfat has been removed to reduce its milkfat content to less than 0.50 percent.

Low-fat milk Low-fat milk is milk from which a sufficient portion of milkfat has been removed to reduce its milkfat content to not less than 0.50 percent and not more than 2.0 percent.

Vitamin D milk and milk products Vitamin D milk and milk products are milk and milk products, the vitamin D content of which has been increased by an approved method to at least 400 U.S.P. units per quart.

Fortified milk and milk products Fortified milk and milk products are milk and milk products other than vitamin D milk and milk products, the vitamin and/or mineral content of which have been increased by a method and in the amount approved by the health authority.

Homogenized milk Homogenized milk is milk which has been treated to ensure breakup of the fat globules to such an extent that, after 48 hours of quiescent storage at 45°F, no visible cream separation occurs on the milk, and the fat percentage of the top 100 ml of milk in a quart, or of proportionate volumes in containers of other sizes, does not differ by more than 10 percent from the fat percentage of the remaining milk as determined after thorough mixing. The word "milk" shall be interpreted to include homogenized milk.

Flavored milk or milk products Flavored milk or milk products shall mean milk and milk products as defined in this Ordinance to which has been added a flavor and/or sweetener.

Eggnog-flavored milk Eggnog-flavored milk is a milk product consisting of a mixture of at least 3.25 percent butterfat, at least 0.5 percent egg yolk solids, sweetener, and flavoring. Emulsifier and a maximum of 0.5 percent stabilizer may be added.

Eggnog Eggnog is a milk product consisting of a mixture of milk or milk product of at least 6.0 percent butterfat, at least 1.0 percent egg yolk solids, sweetener, and flavoring. Emulsifier and not over 0.5 percent stabilizer may be added.

Buttermilk Buttermilk is a fluid product resulting from the manufacture of butter from milk or cream. It contains not less than 8 1/4 percent of milk solids-not-fat.

Cultured buttermilk Cultured buttermilk is a fluid product resulting from the souring, by lactic acid producing bacteria or similar culture, of pasteurized skim milk or pasteurized low-fat milk.

Cultured milk or cultured whole milk buttermilk Cultured milk or cultured whole milk buttermilk is a fluid product resulting from the souring, by lactic acid producing bacteria or similar culture, of pasteurized milk.

Acidified milk and milk products Acidified milk and milk products are milk and milk products obtained by the addition of food grade acids to pasteurized cream, half-and-half, milk, low-fat milk, or skim milk, resulting in a product acidity of not less than 0.20 percent expressed as lactic acid.

Milk products Milk products include cream, light cream, coffee cream, table cream, whipping cream, light whipping cream, heavy cream, heavy whipping cream, whipped cream, whipped light cream, whipped coffee cream, whipped table cream, sour cream, cultured sour cream, half-and-half, sour half-and-half, cultured half-and-half, reconstituted or recombined milk and milk products, concentrated milk, concentrated milk products, skim milk, skimmed milk, low-fat milk, fortified milk and milk products, vitamin D milk and milk products, homogenized milk, flavored milk or milk products, eggnog, eggnog-flavored milk, buttermilk, cultured buttermilk, cultured milk, cultured whole milk buttermilk, and acidified milk and milk products.

This definition is not intended to include such products as sterilized milk and milk products hermetically sealed in a container and so processed, either before or after sealing, as to prevent microbial spoilage, or evaporated milk, condensed milk, ice cream and other frozen desserts, butter, dry milk products, or cheese except when they are combined with other substances to produce any pasteurized milk or milk product defined herein.

For more information: McKenzie (1971); Patton (1969).

Color of milk The color of milk is due largely to the presence of carotene. Carotene exists in at least three isomeric forms: α-carotene, β-carotene, and γ-carotene. Another pigment closely related to carotene is cryptoxanthin, which occurs in yellow corn. All these pigments are precursors of vitamin A. One molecule of β-carotene is capable of yielding two molecules of vitamin A; one molecule of each of the others yields only one molecule of vitamin A.

Carotene is found in hay, grass, green leaves, some fruits, carrots, etc. The carotene content of cow's milk is dependent upon the carotene content of the ration. Not all the carotene of the ration is converted into vitamin A. When cows consume carotene-containing foods, some of the pigment is converted into vitamin A and some is found unchanged in the milk. Vitamin A imparts no color to milk, whereas carotene gives milk a yellow color.

Milk also contains several other vitamins. Bottled milk exposed to sunshine may lose all of its ascorbic acid (vitamin C) in 30 min and all of its riboflavin (vitamin B_2) after 2 hr. Therefore, milk should not be allowed to stand in strong sunlight for any appreciable length of time.

Normal souring of milk

Reaction of fresh milk Milk when freshly drawn may show considerable variation in reaction. As a general rule, the pH is slightly acid, ranging from 6.3 to 7.2, with an average of about 6.75. The pH fluctuates at different stages of the milking operation. The fore milk is usually the lowest in acidity, the middle milk the highest, and the strippings between the two.

Changes in the reaction of milk On standing, unsterilized milk rapidly ferments, with the production chiefly of lactic acid from the lactose of the medium. The first stage is believed to be a hydrolysis of the lactose to one molecule of glucose and one of galactose. In the second stage, the hexoses are fermented to lactic acid.

It is generally stated that acidity in milk is first detected by taste when the pH drops to about 6.0. As the acid concentration continues to increase, it eventually causes a precipitation of the casein. This is said to occur when the pH reaches 4.78 to 4.64. Boiling produces a curdling of milk at a higher pH (lower acidity). The acidity continues to rise until the concentration is sufficient to prevent growth of the bacteria producing the fermentation. The lactic acid produced in the milk prevents the growth of most bacterial types likely to be present and thus acts as a preservative.

Molds and yeasts may contaminate milk. They are capable of growing in much higher acid environments than bacteria likely to be found in milk. They utilize the organic acids (chiefly lactic) and make the milk more alkaline in reaction. Conditions now become favorable for the rapid decomposition of the milk proteins by the growth of putrefactive bacteria. As a rule, several weeks are required for putrefactive changes to occur. The utilization of the acid occurs at a faster rate if the milk is placed in shallow, well-aerated layers. This is the general cycle of changes that occur in raw milk when allowed to stand at ordinary temperature.

Lactic acid bacteria A number of organisms are concerned in the production of lactic acid from the lactose sugar in milk. Some produce relatively large amounts of lactic acid; others form smaller amounts together with other products.

The principal lactic acid-producing bacteria include (1) *Streptococcus lactis* and related species and (2) *Lactobacillus* species.

Streptococcus lactis This organism is responsible for the normal souring of milk. It is of widespread occurrence in dairy products. Several varieties of the organism are recognized which show differences in flavor produced, character of the fermented milk, rate of acid formation, rate of litmus reduction, and in other ways. Proof of the similarities of the different varieties or strains has been established by means of specific grouping immune sera.

Cells of *S. lactis* are characterized as being ovoid in shape and elongated in the direction of the chain; measuring 0.5 to 1.0 μ in diameter; occurring mostly in pairs or short chains, sometimes in long chains; and are gram-positive (Fig. 279).

Fig. 279 Streptococcus lactis, the cause of normal souring of milk.

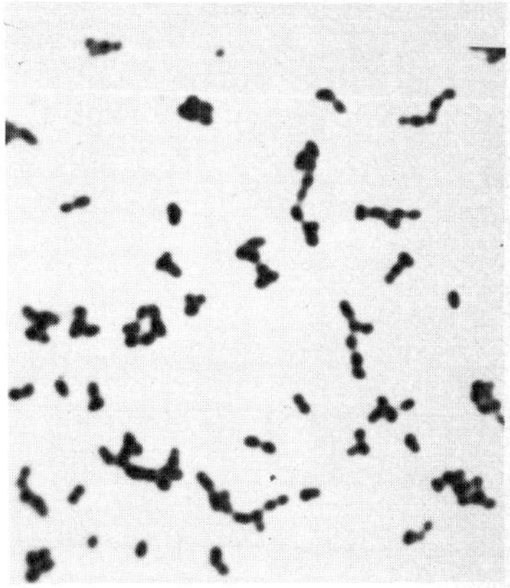

Distinctive characteristics *S. lactis* is homofermentative, utilizing a number of carbohydrates with the production chiefly of D-lactic acid; gas is not formed; growth occurs at 10°C or below and at 40°C, but not at 45°C; growth occurs in 4 percent but not in 6.5 percent sodium chloride; growth occurs at pH 9.2 but not at pH 9.6; reduction of litmus is rapid and complete before curdling of milk.

S. lactis is a common contaminant of milk and milk products. It seldom occurs in the udders of cows. Hay and grains are believed to be the natural habitat of the organism. Since these are consumed by cows, the organisms are found in manure. *S. lactis* can be obtained from the coat of the cow. It is generally believed that the organism finds its way into milk from manurial contamination.

S. cremoris The morphological characteristics are similar to *S. lactis* (Fig. 280).

Distinctive characteristics *S. cremoris* ferments several carbohydrates with the production of acid but no gas; growth occurs at 10°C and below but not at 40°C, with optimum temperature below 30°C; litmus is completely reduced before curdling of milk; no growth occurs in 4 percent sodium chloride; it is commonly employed in commercial dairy starters; it is readily distinguished from *S. lactis* by its inability to grow at 40°C, in 4 percent NaCl broth, or at pH 9.2.

S. cremoris is commonly found in raw milk and milk products, and in commercial starters for butter and cheese manufacture; it is probably of plant origin.

For more information: Claydon and Fryer (1960).

Lactobacillus species These are characterized as rod-shaped cells, often long and slender. They are nonmotile and gram-positive. They can ferment a large number of carbohydrates to acid. The homofermentative species yield almost entirely lactic acid; the heterofermentative species produce other end products besides lactic acid. Several species grow at relatively high (50 to 65°C) temperatures. Surface growth may be poor because species are generally microaerophilic or anaerobic.

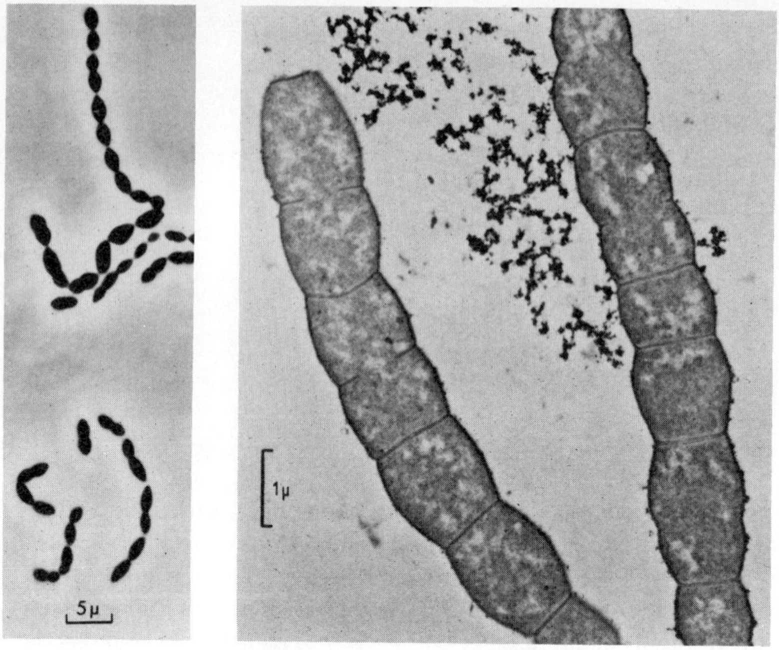

Fig. 280 Streptococcus cremoris, filamentous form. Left, phase contrast microscopy. Scale mark = 5 μ; right, same under electron microscope. Scale mark = 1 μ. (Courtesy of McDonald.)

Lactobacilli are widely distributed in nature being found in milk, butter, cheese, sour and aromatic foods, cattle feed, normal saliva, gastric juice, fermented milk, soil, feces, etc.

The important species are homofermentative and are found in dairy products. They include *L. casei, L. acidophilus, L. bulgaricus, L. helveticus,* and *L. plantarum.*

L. casei *L. casei* is isolated from milk and cheese. It utilizes casein and is important in cheese ripening. It is isolated from compressed yeast and fermenting milk. Found in dairy and plant products.

L. acidophilus *L. acidophilus* is isolated from feces of milk-fed infants. It is also found in the feces of older persons on diets rich in milk, lactose, or dextrin.

L. bulgaricus *L. bulgaricus* was originally isolated from yoghurt. Its optimum temperature is between 45 and 50°C. It is probably present in many milk products if they are held at high temperature.

L. helveticus *L. helveticus* is isolated from sour milk and cheese and is widely distributed in dairy products.

L. plantarum *L. plantarum* is isolated from cheese, butter, kefir, fermenting potatoes, beets, corn, sauerkraut, cucumber pickles, spoiled tomato products, feces, etc. It is widely distributed in nature, particularly in fermenting plant and animal products.

For more information: Brown and Van Demark (1968).

Quantitative examination of milk

Normal udders of cows are probably never sterile. Organisms are present in abundance in freshly drawn milk. The first milk drawn shows the highest count, the middle milk shows a smaller count, and the strippings the least of all. The comparatively high count of the fore milk is the result of the washing out of the easily removable organisms present in the milk passages. The numbers washed out become less and less during the milking process. It is the general practice to discard the first few streams of milk which contain the highest count. This may decrease the bacterial count by as much as 5 percent.

A high bacterial count means that the milk has come from diseased udders, has been collected under insanitary conditions, or has been kept warm enough to permit bacterial growth. It would seem, therefore, that the contamination of milk by bacteria is largely preventable.

The methods commonly employed for enumerating the bacterial population of milk include (1) the agar plate method, (2) the direct microscopic method, (3) the methylene blue reduction method, and (4) the resazurin method.

Agar plate method The agar plate method is used more than any other for the enumeration of the number of bacteria in milk.

The method consists of preparing a series of dilutions of the milk sample, transferring 1-ml amounts to petri dishes, mixing with 10 to 12 ml of melted agar cooled to 45°C before pouring, incubating the plates at 32 ± 1°C for 48 hr ± 3 hr, and counting visible colonies developing on the agar. The result is expressed as the number of bacteria per milliliter of undiluted milk by the standard plate count.

The factors which influence plate counts include temperature of incubation, period of incubation, composition of plating medium, amount of oxygen, etc. Unless a standard procedure is followed, the results from different laboratories may show wide variation.

Advantages of the plate method The agar plate method is especially suited to determinations where bacterial numbers are low. The method is generally used for testing pasteurized milk and the higher-grade raw milk. It is the only method approved by the American Association of Medical Milk Commissions for examining samples of certified milk.

Objections to the plate method Objections to the method are numerous. The most important are probably the following: Pathogenic bacteria usually are not detected. If they grow, they cannot be distinguished from nonpathogenic forms by appearance. The number of colonies appearing on agar do not represent all organisms present in the milk. Many of the organisms fail to develop. Anaerobic organisms do not find conditions favorable for growth. This means that no single medium or given set of conditions is capable of giving growth of all viable organisms likely to be found in milk. A temperature of 32 ± 1°C is not the optimum for the growth of all organisms. Shaking the sample does not break up all clumps or groups of bacteria. Chains of streptococci usually remain intact and record as only one colony. The colony counts represent only a fraction of the total bacterial content of milk. Agar counts should be regarded as estimates rather than exact numbers.

Because of the long incubation period, the milk is usually consumed before information on the number of bacteria present is available. However, the quality of a milk supply cannot be determined on the basis of one bacterial count. A series of counts is necessary to establish the quality of the milk source.

For more information: American Public Health Association (1967); Huhtanen (1968).

Direct microscopic method This method consists of spreading 0.01 ml of milk sample uniformly over a ruled area of 1 sq cm on a glass slide, allowing the film to dry, removing the fat, staining the film with methylene blue, and examining the sample under the oil-immersion objective previously calibrated to determine the area of the field. The individual cells are counted and the results are expressed as numbers per milliliter of milk.

Advantages to the microscopic method The method possesses a number of advantages over the plate procedure. Results can be obtained quickly, usually in about 10 or 15 min. Since less work is required, more samples can be examined by this procedure than by the plate method. The amount of equipment necessary is much less. The slides may be preserved as a permanent record and examined whenever occasion arises, whereas plates must be examined and discarded. Some idea of the morphological types can be obtained from slide preparations. This is frequently of great value in determining the cause of the bacterial count. Microscopic examination reveals the presence of leucocytes and other body cells in milk. An excessive number of leucocytes indicates a diseased condition of the udder. The slide method gives a better quantitative determination than the agar plate method.

Some believe that the method is less useful for pasteurized milk because most of the bacteria are dead. This is not necessarily so. It has been shown that bacteria killed by heat treatment usually disintegrate and gradually lose their ability to stain within a few hours after pasteurization. Even if the bacteria remain intact, they are usually distinguished from living forms by their inability to stain intensely.

Objections to the microscopic method Unless the milk sample contains a high count, the microscopic method may be the source of considerable error. A large factor is used for converting the number of bacteria per field to the number per milliliter of milk. Significant errors in the average number of organisms per field are not likely to be of great importance when bacteria are numerous in milk. However, in low-count milk a considerable error may be introduced. This is especially true where many fields show no bacteria and some fields show a chain or cluster of organisms to produce great variation in the count per field. Other objections involve inaccuracies in measurement of 0.01-ml quantities, faulty preparation and staining of slides, failure of bacteria to stain, and errors in observation and in calculation.

For more information: American Public Health Association (1967); Moats (1961).

Comparison of counts by the two methods The microscopic method gives much higher individual cell counts than the agar method. The differences between the counts by the two methods are considerably greater on samples showing low bacterial counts than on those showing high counts. The organisms in low-count milk generally represent external contaminants that fail to develop on agar, whereas those organisms in high-count milk are forms which have developed in the milk. Also, low-count milk usually shows a greater percentage of clumps than high-count milk.

Bacterial standards are generally based on microscopic clump counts or agar plate counts, the standards being the same by either method. However, if individual cells are counted, the slide method gives higher results, the ratio being 3.33:1. Therefore total cell counts by the slide method are approximately 3.33 times greater than counts by the plate method. The word clumps refers to isolated cells as well as to groups of bacteria.

Methylene blue reduction method The method depends upon the ability of bacteria in milk to grow and to consume the dissolved oxygen, which reduces the oxidation-reduction potential in the medium. Certain enzymes present in bacteria, known as oxidoreductases, are then able to produce oxidations by the removal of hydrogen. The methylene blue accepts the hydrogen and becomes reduced to the colorless or leuco compound. The speed of decoloration is an indication of the rate at which oxidation takes place.

Methylene blue is of value in making a rapid survey of the quality of raw milk. The rate of decolorization depends upon the number of organisms present. The test can be employed to determine, in a rough way, the bacterial population of a milk sample. The procedure is quickly and easily carried out and with a minimum of expense. It is particularly valuable in making rapid inspections of large numbers of samples to determine whether the milk received by companies answers the requirements prescribed by law.

The test is expressed as the time required for a known concentration (about 1:250,000) of methylene blue thiocyanate to lose its color when the milk is incubated at 35 to 37°C to hasten oxygen consumption and to shorten the period of observation. Under some conditions the blue color does not disappear uniformly. In such cases the end point is taken as the time required for the milk to show no blue color after it is mixed.

There is not always good agreement between the methylene blue reduction time and the agar plate count because (1) some organisms fail to grow on nutrient agar; (2) a clump of organisms records as only one colony, whereas the rate of decolorization is due to the combined effect of each member of the mass; (3) the rate of decolorization of the stain is not the same for all organisms; and (4) the test becomes less accurate as the reduction time is increased, freshly drawn milk requiring at least 10 hr to decolorize methylene blue.

Milk may be classified on the basis of the methylene blue reduction time as follows:

Class 1. Excellent milk, not decolorized in 8 hr.

Class 2. Good milk, decolorized in less than 8 hr, but not less than 6 hr.

Class 3. Fair milk, decolorized in less than 6 hr, but not less than 2 hr.

Class 4. Poor milk, decolorized in less than 2 hr.

The reduction time limit for grade A milk to be pasteurized is 5½ hr; for milk to be consumed raw, 8 hr; for raw cream to be pasteurized, 5 hr; and for cream to be consumed raw, 7 hr.

Resazurin method As milk for pasteurization improves in quality, the methylene blue reduction time increases. This requires incubation for inconveniently long periods. To provide comparable information in less time, the resazurin test is recommended as a substitute for the methylene blue reduction test.

Resazurin (diazoresorcinol) is an oxidation-reduction indicator having a pH range of 6.5 to 3.8. At pH 6.5 and above, the indicator is purple in color; at pH 3.8 and below, it is pink. A gradual color change occurs between these two extremes.

On complete reduction of resazurin to a pink color, the compound resorufin is formed in milk. This reaction is irreversible. In the second stage, the resorufin is further reduced to dihydroresorufin, followed by a gradual disappearance of the pink color. This reaction is reversible.

The indicator is capable of being reduced by bacteria. The greater the bacterial count, the faster the stain is reduced. Because of this fact, it may be used to estimate the number of bacteria in milk. The test provides results in less time than by the methylene blue reduction method. Since resazurin is also sensitive to the reducing action of leucocytes, it reveals the presence of the udder disease known as mastitis (page 739).

The test is performed as follows: The milk sample is thoroughly mixed and 10 ml pipetted into a 5/8- by 6-in. test tube. One ml of a 1:20,000 dilution of resazurin is added to the milk and the tube is inverted three times slowly to mix well. The tube is placed in a 35 to 37°C water bath and examined at the end of each of three successive hourly intervals. The color of the milk should be compared with a series of standard disks prepared for the purpose. If the color remains blue or lilac (disks 6 or 5) for 1 hr, the milk is normal in bacterial content. If mauve or mauve-pink (disk less than 5), the bacterial content is high and the milk should be regarded as abnormal. If a disk number less than 3 is obtained in 1/2 hr, the milk is grossly abnormal. If the stain is completely reduced (disk O) in 1 hr or less, the milk contains many pus cells and is probably teeming with mastitis organisms.

For more information: American Public Health Association (1967); U.S. Public Health Service (1967).

Grading of milk

According to the Milk Ordinance and Code of the U.S. Public Health Service (1967), milk and certain milk products are graded on the basis of their bacterial counts for the following reasons:

It is widely accepted that the bacterial count of milk and certain milk products is an index of the sanitary quality. A high count does not necessarily mean that disease organisms are present, and a low count does not necessarily mean that disease organisms are absent; but a high bacterial count does mean that the milk has come from diseased udders, or has been milked or handled under undesirable conditions, or has been kept warm enough to permit bacterial growth. This means, in the first two cases, that the chances of infection have been increased, and, in the last case, that any bacterial contamination which may have reached the milk has been permitted to increase to more dangerous proportions. In general, therefore, a high count means a greater likelihood of disease transmission. On the other hand, a wrong interpretation of the significance of low bacterial counts should be avoided, since low-count milk can be obtained from cows with brucellosis or tuberculosis, or can have been handled by typhoid carriers or under unclean conditions.

When coliform organisms are present in pasteurized milk, they usually indicate that the milk has been contaminated after pasteurization. The phosphatase test is an index of the efficiency of pasteurization.

Collection of samples Samples for bacteriological examinations are collected by inspectors or other officials. They may be taken at random, and should be representative of the milk or milk products to be tested. In order to yield significant results, samples shall be taken while in the possession of the producer or of the plant, and must be collected in such a manner that they will fairly represent the condition of the milk as received at the plant intake or as finally delivered to the consumer.

At least 10 ml of well-agitated milk is collected and placed in a sterile sample bottle, which should be of such size that only about two-thirds of it is filled. This provides sufficient air space for vigorous agitation to ensure a uniform suspension of organisms before plating the milk. The sample, when taken, must be kept below 45°F; however, it may be kept above 45 but below 50°F when it is to be plated within 4 hr. It shall be protected against freezing.

Experience indicates that four or more samples should be examined before attempting to grade the supply. The samples should be taken from each supply during each 6-month grading period. The average bacterial plate count is expressed as the logarithmic average of the plate counts of the last four consecutive samples taken on separate days throughout the 6-month grading period.

Milk standards The number of organisms permissible in different grades of milk vary somewhat, depending upon standards set up by local public health authorities.

The highest grade of milk is known as certified milk, which is safeguarded at every step in its production, collection, and distribution. Milk collected under conditions not so carefully controlled is graded as A, B, or C. The ratings are based upon the bacterial count of milk and also upon the hygienic conditions under which it was produced. The standards of the various grades reported here are those set up by the U.S. Public Health Service milk ordinance (1967). They are as follows:

Certified milk – raw This is raw milk which conforms with the latest requirements of the American Association of Medical Milk Commissions in force at the time of adoption of this ordinance, and which is produced under the supervision of a medical milk commission reporting monthly to the health officer, and of the State health authority or the municipal or county health officer. The usual standard is that the count must not go above 10,000 bacteria per milliliter.

All milk having counts in excess of this number must be placed in the following grades:

Grade A raw milk for pasteurization This is raw milk from producer dairies conforming with the following items of sanitation. The bacterial plate count or the direct microscopic clump count of the milk, as delivered from the farm, shall not exceed 100,000 per milliliter from an individual producer prior to commingling with other producer milk, and not exceeding 300,000 per ml. as commingled milk prior to pasteurization.

All milk for pasteurization shall be from herds which are located in a modified accredited tuberculosis-free area, as determined by the U.S. Department of Agriculture, and which have been tested for tuberculosis not more than 6 years prior to the adoption of this ordinance and at least every 6 years after such test.

The health of the cow is a very important consideration because a number of diseases of cattle, including tuberculosis, brucellosis, Q fever, salmonellosis, staphylococcic infection, and streptococcic infection, may be transmitted to man through the medium of milk. The organisms of most of these diseases may get into the milk either directly from the udder, or indirectly through infected body discharges which may drop, splash, or be blown into the milk.

Grade B raw milk for pasteurization This is raw milk which does not meet the bacterial standard for Grade A raw milk for pasteurization, but which conforms with all other requirements. The bacterial plate count or the direct microscopic clump count of the milk, as delivered from the farm, shall not exceed 1 million per milliliter.

Grade C raw milk for pasteurization This is raw milk which does not meet the requirements for Grade B raw milk for pasteurization.

Certified milk – pasteurized This is certified milk (raw) which has been pasteurized, cooled, and bottled in a milk plant which conforms with the requirements for Grade A pasteurized milk.

Grade A pasteurized milk This is Grade A raw milk for pasteurization which has been pasteurized, cooled, and placed in the final container in a milk plant which conforms with certain items of sanitation. In all cases the milk shall show efficient pasteurization as evidenced by satisfactory phosphatase test, and at no time after pasteurization and before delivery shall the milk have a bacterial plate count exceeding 20,000 per milliliter, or a coliform count exceeding 10 per milliliter: Provided that the raw milk at no time between dumping and pasteurization shall have a bacterial plate count or a direct microscopic clump count exceeding 400,000 per milliliter.

Grade B pasteurized milk This is pasteurized milk which does not meet the bacterial-count standard for Grade A pasteurized milk, but which conforms with all other requirements for Grade A pasteurized milk. It has been made from raw milk for pasteurization of not less than Grade B quality, and has a bacterial plate count after pasteurization and before delivery not exceeding 50,000 per milliliter.

Grade C pasteurized milk This is pasteurized milk which does not meet the requirements for Grade B pasteurized milk.

For more information: Nilsson (1959); U.S. Public Health Service (1967).

Pasteurization

The destruction of all organisms in milk is called sterilization. The high temperature required to achieve this result would impart a cooked flavor to the milk. Such milk is objectionable for two reasons: (1) The cooked flavor is not so pleasant as that of unheated milk, and (2) heating to such a high temperature might result in a decrease in the vitamin content. These objections are largely overcome by heating milk to temperatures lower than that required to sterilize completely but sufficiently high to destroy all disease organisms likely to be present.

The destruction of disease and most other organisms in milk without attempting complete sterilization is called pasteurization.

According to the U.S. Public Health Service (1967):

The terms "pasteurization," "pasteurized," and similar terms shall mean the process of heating every particle of milk or milk product to at least 145°F., and holding it continuously at or above this temperature for at least 30 min., or to at least 161°F., and holding it continuously at or above this temperature for at least 15 sec., in equipment which is properly operated and approved by the health authority: *Provided*, That milk products which have a higher milkfat content than milk and/or contain added sweeteners shall be heated to at least 150°F., and held continuously at or above this temperature for at least 30 min., or to at least 166°F., and held continuously at or above this temperature for at least 15 sec.: *Provided further*, That nothing in this definition shall be construed as barring any other pasteurization process which has been recognized by the United States Public Health Service to be equally efficient and which is approved by the State health authority.

Public-health reason The public-health value of pasteurization is unanimously agreed upon by health officials. Long experience conclusively shows its value in the prevention of diseases which may be transmitted through milk. Pasteurization is the only practical, commercial measure which, if properly applied to all milk, will destroy all milk-borne disease organisms. Examination of cows and milk handlers, while desirable and of great value, can be done only at intervals and, therefore, it is possible for pathogenic bacteria to enter the milk for varying periods before the disease condition is discovered. Disease bacteria may also enter milk accidentally from other sources, such as flies and contaminated water and utensils. It has been demonstrated that the time-temperature combinations of 143°F for 30 min, and 161°F for 15 sec, if applied to every particle of milk, will devitalize all milk-borne pathogens. Compilations of outbreaks of milk-borne disease by the U.S. Public Health Service, over many years, indicate that the risk of contracting disease from raw milk is approximately 50 times as great as from milk labeled "pasteurized."

A note of caution is in order. Although pasteurization devitalizes the organisms, it does not destroy the toxins that may be formed in milk when certain staphylococci are present (as from udder infections), and when the milk is not properly refrigerated before pasteurization. Such toxins may cause severe illness.

Numerous studies and observations clearly prove that the food value of milk is not significantly impaired by pasteurization.

The pasteurization process reduces the bacterial count 99 to 100 percent (Fig. 281), depending upon the number and kinds present.

For more information: Deane et al. (1967); Overcast and Skean (1959); Pietermaat and Van Dyck (1959); U.S. Public Health Service (1967).

Influence of temperature on the keeping quality of milk

The number of organisms in milk at the outset depends upon the degree of care exercised in its collection, handling, and storing. For a period after collection, the count tends to remain constant or decrease somewhat. Then the number increases rapidly unless milk is stored at a low temperature. The storage temperature determines to a large extent the bacterial count and microflora of milk.

Germicidal property of milk Freshly drawn milk contains substances which are capable of exerting a bactericidal action. These substances are destroyed by heat, but the temperature required varies for different organisms.

Morris (1945) heated raw milk to temperatures of both 52 and 53°C

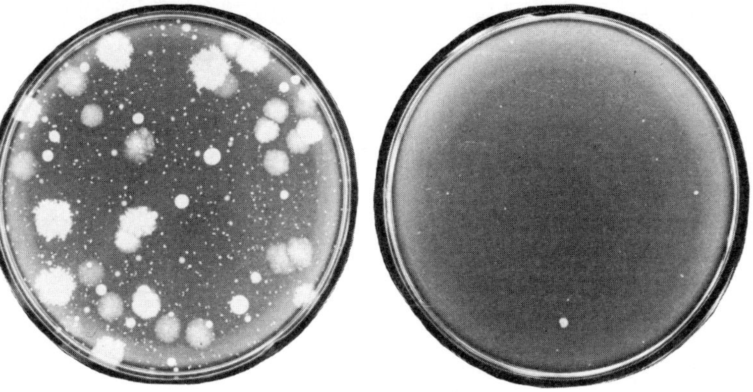

Fig. 281 Pasteurization of milk. Left, 1 ml of a 1:10 dilution of milk before heating; right, 1 ml of a 1.10 dilution after heating.

for 30 min, then inoculated each of the samples with a young culture of coliform organisms. Plate counts were made immediately after inoculation and after holding the milk for 4 hr at 37°C. From the results (Table 29), it would appear that the bactericidal substances are completely destroyed by heating the milk at 53°C for 30 min and that this destruction is critical to within 1°C.

For more information: Randolph and Gould (1966).

The bactericidal property is of little practical importance unless milk is stored at a low temperature immediately after collection. Milk cooled promptly to 4°C and held at this temperature for 10 hr retained to a large extent its bactericidal action. On the other hand, the germicidal property was almost completely lost when a temperature of 16°C was used.

Milk held below freezing point When milk is frozen to −0.55°C and held at that temperature, no multiplication of organisms occurs. The methods by which bacteria obtain their nutrients preclude growth as long as the milk is frozen throughout. In fact, milk treated in this manner may show a decrease in bacterial numbers. When frozen milk is thawed, the

Table 29 Effect of heat on the germicidal property of milk

Temperature to which milk was heated for 30 min before inoculation	*Count per ml immediately*	*Count per ml after 4 hr at 37°C*
Culture no. 1:		
52°C	1,328,000	1,000
53°C	848,000	40,000,000
Culture no. 2:		
52°C	316,000	31,000
53°C	640,000	28,000,000

bacteria present start multiplying again, the temperature determining the rate of multiplication and the type of change occurring.

Milk held just above freezing If raw milk, or milk pasteurized at temperatures below 70°C, is kept at 0 to 5°C for 24 hr, the plate count decreases. After about 1 week there is an increase over the original plate count. At the same time the number of organisms capable of liquefying gelatin increases. This continues until enormous numbers are present. Some of the organisms are acid formers; others are neutral types; still others are strongly proteolytic forms.

This is followed by decomposition and putrefaction of the casein. Under these conditions, toxic waste products may be present, rendering the milk unfit or even dangerous for human consumption. Milk is generally stored at 0°C but the time it may be kept at this temperature should not be over 10 days, for the above reasons.

The bacterial flora of milk kept at different temperatures is as follows:

0 to 5°C: *Pseudomonas* species predominate
5 to 10°C: *Pseudomonas* species, *Proteus vulgaris, Micrococcus* species, *Alcaligenes viscolactis, A. marshallii*
10 to 15°C: *S. acidominimus, S. agalactiae, S. cremoris, S. durans, S. dysgalactiae, S. faecalis, S. lactis, S. uberis*
15 to 30°C: Streptococci, especially *S. lactis*
30 to 40°C: *A. aerogenes, Escherichia coli*, lactic acid-forming rods including *L. brevis, L. bulgaricus, L. casei, L. caucasicus, L. fermenti, L. helveticus, L. lactis, L. leichmannii, L. plantarum, L. thermophilus*, and a few streptococci
40 to 50°C: Lactic acid-forming rods including *L. bulgaricus, L. caucasicus, L. fermenti, L. helveticus, L. lactis, L. thermophilus, S. faecalis, S. thermophilus*, and yeasts

The smaller the initial plate count, the longer will be the time required to sour the milk. As already stated, the fluorescent bacteria (*Pseudomonas*) found in soil are able to multiply at a temperature as low as 3°C. If milk is to be kept for any length of time, it should be frozen. At somewhat higher temperatures, species of *Proteus* develop, with the result that putrefactive compounds accumulate in the milk.

Coagulation seldom takes place in milk stored below 10°C. Above this temperature, a coagulum forms in a few days owing to the combined action of rennin and acid-producing organisms. At a temperature of 20°C, the bacterial flora are composed of about 90 percent streptococci. This results in a rapid coagulation of the milk. The acidity produced is sufficient to inhibit the growth of most other species of organisms likely to be present. Above 30°C, rod-shaped bacteria predominate, which are capable of producing still higher concentrations of lactic acid. Also, this is the most favorable temperature range for the growth of the butyric acid-producing anaerobes.

The aerobic organisms grow best near the surface of milk, where there is an abundant supply of dissolved oxygen. The organisms predominating near the bottom include the anaerobes and *S. lactis*. This means that the spontaneous curdling of milk usually starts at the bottom.

For more information: Stadhouders (1968); Yotis and Teodoro (1957).

Undesirable organisms in milk

Great care must be exercised during the milking process; otherwise a great variety of additional bacteria may enter the milk. Since particles of manure and soil are always found adhering to the skin of the cow, they may be an important source of contamination. Additional organisms may enter the milk from the milker, the air of the barn, improperly cleaned utensils, milking machines, etc.

Probably the most important undesirable organisms isolated from milk are the following:

Coliform organisms Isolated instances are on record in which coliform organisms have been recovered from milk taken directly from the udder. However, this is unusual; for all practical purposes, normal milk as it comes from the udder is entirely free from coliform bacteria.

The presence of coliforms in milk usually indicates contamination from manure. As milk leaves the farm, it almost always contains members of the coliform group regardless of the care exercised in its production. Under careful conditions raw milk usually contains less than 100 coliforms per ml. Under careless conditions of production the number may run as high as 2000 per ml. Regardless of the conditions of production, raw milk probably always contains coliforms in varying numbers.

The examination of water for the presence of coliforms is a standard procedure for the determination of the healthfulness of a water supply. On the other hand, it is controversial as to whether the coliform test should be used for the determination of the healthfulness of milk and milk products. In the former case, the presence of coliforms indicates fecal pollution and the possible presence of intestinal pathogens. In the latter case, coliforms may indicate manurial pollution, which is not subject to the intestinal infections common to man.

Coliforms are, in general, completely destroyed in the pasteurization process. Only occasionally are strains encountered which are capable of resisting pasteurization. Therefore, the presence of coliforms in milk immediately after pasteurization indicates that (1) the milk was not properly processed or (2) it became contaminated after pasteurization. Under these conditions, the coliform test should prove valuable for control purposes.

The procedures followed for the identification of coliforms in milk are similar to those employed for the examination of water (page 692).

Importance in dairy products Coliform organisms are undesirable in dairy products. In addition to acid and gas production, they may produce objectionable odors and flavors. When such products are used in the manufacture of butter and cheese, the objectionable qualities are commonly carried over, producing off-flavors. Also, the gas produced may give cheese a blown appearance.

For more information: Blankenagel and Okello-Uma (1969); Read et al. (1961); Thomas (1955); Jones (1971).

Ropy or slimy milk A number of organisms are capable of producing a condition in milk known as ropiness. The milk becomes ropy or slimy and may be pulled out into long threads. Sometimes the change is very slight; at other times the ropy consistency may be so pronounced the milk may be drawn out into threads 3 ft or more in length.

Generally the ropy condition does not manifest itself immediately when milk is produced or processed but only after 24 or 48 hr of incubation. A milk dealer does not know he has a ropy-milk problem until after the milk is delivered to the home.

Several organisms are capable of producing this condition in milk. Probably the most important species is *A. viscolactis*. The organism produces its maximum amount of ropiness at a temperature of about 25°C.

A. viscolactis is a small rod measuring 0.6 to 1.0 by 0.8 to 2.6 μ, occurring singly, in pairs, or in short chains. It is frequently found as almost spherical cells. It is nonmotile and gram-negative. The organism produces both ropiness and a pellicle in milk. The pellicle is the result of the aerobic character of the organism. This explains why the ropy condition is frequently noted only in the cream layer. The organism produces an alkaline reaction in milk with no coagulation of the casein.

Jones (1954*a, b, c*) found *A. viscolactis* to be a pleomorphic gram-variable rod that produced ropiness and an alkaline reaction in milk. Loss of smoothness or mucoid colony characteristic was accompanied by a decrease or loss of the ability to produce ropiness in milk and most likely occurred during prolonged storage of cultures without transfer.

Ropiness was found to result from attack on the albumin of the milk. No evidence of polysaccharide synthesis was obtained. Therefore the slime is neither capsular material nor carbohydrate gum.

Some members of the coliform group have been responsible for producing the ropy condition of milk. Most strains are unable to do so, but occasionally a strong ropy producer is encountered. *A. aerogenes* is of more frequent occurrence than *E. coli*.

Organisms causing ropiness are commonly found in pools, wells, and streams. Water from such sources contaminates the dairy utensils and equipment. The organisms must be removed as quickly as possible; otherwise great economic losses may result. All utensils and equipment coming in contact with the milk should be sterilized. The barns should be thoroughly cleaned and disinfected. The flanks of the cows should be washed with an appropriate disinfectant. The organisms causing ropiness should be destroyed by pasteurization, but outbreaks sometimes occur in pasteurized milk. They are the result of recontamination from the plant equipment after pasteurization.

Slime-producing lactic acid organisms have been used in the manufacture of cheese, but such practice has been largely discontinued, owing to the fact that it is difficult to separate the whey from the cheese. Also, the presence of the organisms in cream results in a poor yield of butter. The property of producing slime appears to be lost by growing the organisms at higher temperatures. Conversely, some organisms that ordinarily do not produce slime can be made to do so if cultivated at lower temperatures.

For more information: Marth et al. (1964).

Stormy fermentation of milk The organism *Clostridium perfringens* produces a condition in milk known as stormy fermentation (Fig. 282). It is a strongly saccharolytic organism, rapidly fermenting the lactose of milk with the production of acid and gas. The curd becomes torn to shreds by the vigorous fermentation and gas production. The acid prevents further growth of the organism with the result that the milk proteins are not attacked.

C. perfringens is a strictly anaerobic, spore-forming organism of widespread occurrence in nature. It is found in the feces of animals and man, in sewage, and in soil. The organism is generally considered to be the most important etiological agent of gas gangrene.

C. perfringens is a short, thick rod, measuring 1.0 to 1.5 by 4.0 to 8.0 μ, occurring singly and in pairs, less frequently in short chains. Spores are ovoid, central to eccentric, not swelling the cells. The spores are not destroyed during pasteurization. Rods are encapsulated. The cells are nonmotile and gram-positive.

Since the organism is present in the intestinal contents of cows, its presence in milk generally indicates manurial contamination.

Fig. 282 Stormy fermentation of milk. Left, fermentation of a sample of milk free from Clostridium perfringens; right, fermentation of a sample of milk containing C. perfringens. The vaspar seal is pushed up by the gas pressure, and the curd is riddled with bubbles.

Colored milk Several organisms have been isolated from milk which are capable of imparting brilliant colors. These changes are of occasional occurrence and of minor importance economically.

Many organisms isolated from milk are capable of producing colored colonies, but they should not be confused with those species which elaborate brilliant pigments in milk. Because the organisms in the latter group are strongly aerobic, pigment formation is first observed in the surface layer of milk or cream.

Blue milk Blue milk is produced by the growth of *Pseudomonas syncyanea*. It is a rod-shaped organism with rounded ends, occurring singly, occasionally in chains, and measuring 0.7 by 2.0 to 4.0 μ. It is motile with two to four polar flagella, and gram-negative. Optimum growth temperature for the organisms is 25°C.

Red milk Red milk is the result of the growth of *Serratia marcescens*, a small rod, sometimes almost spherical, measuring 0.5 by 0.5 to 1.0 μ, occurring singly and occasionally in chains of five or six cells. It is motile by means of four peritrichous flagella, and gram-negative.

The organism produces an acid reaction accompanied by a soft coagulum. A red surface growth develops. Little or no digestion of the casein occurs. Its optimum growth temperature is about 25°C.

Yellow milk Yellow milk is produced by the growth of *P. synxantha*, a rod-shaped organism occurring singly, in pairs, and measuring 0.5 to 0.6 by 1.3 to 2.2 μ. It is motile with polar flagella, and gram-negative. Its optimum growth temperature is 20°C.

Acid is produced in milk with coagulation. Casein is slowly digested. The organism produces an intense diffusible yellow or orange pigment in cream or in the cream layer of milk.

These and other pigmented forms are found in soil and water from which they gain entrance to milk. They are easily destroyed in the approved pasteurization processes. Their presence in milk and cream in viable form would indicate faulty pasteurization or contamination after such treatment.

For more information: Seitz et al. (1961).

Psychrophilic bacteria

The psychrophiles are cold-loving bacteria, being capable of multiplying at 5°C and below, regardless of their optimum temperatures. They are, therefore, capable of growing at refrigeration temperatures and are primarily responsible for limiting the keeping quality of milk and many milk products in which they may produce a wide variety of spoilage defects. The defects may result in the production of many off-flavors which are characterized as fruity, stale, musty, bitter, rancid, and even putrid.

The psychrophilic bacteria most commonly encountered are members of the genera *Achromobacter, Aerobacter, Alcaligenes,*

Escherichia, Flavobacterium, Pseudomonas, and *Vibrio*. The general consensus is that *Pseudomonas* is the most commonly encountered, and this is true not only for milk products, but also for meat, fish, poultry, and eggs. The various species are killed in the pasteurization process but they are sometimes found in pasteurized milk. It is believed that contamination takes place after pasteurization from equipment, cans, bottles, and water.

The above genera are characterized as follows:

Achromobacter These organisms are non-pigment-forming rods; they are motile by means of peritrichous flagella, or nonmotile; litmus milk is faintly acid to alkaline, or unchanged; they are gram-negative, and occur in saltwater, freshwater, and soil.

Aerobacter These organisms are short rods; they are motile by means of peritrichous flagella, or nonmotile; they grow readily on ordinary media, are strongly fermentative, aerobic or facultatively anaerobic, and are widely distributed in nature.

Alcaligenes These are rods, either motile by means of peritrichous flagella, or nonmotile. Litmus milk is turned alkaline with or without peptonization; carbohydrates are not utilized; they are gram-negative; and they are generally found in intestinal tracts of vertebrates and in dairy products.

Escherichia These are short rods, motile or nonmotile; they are strongly fermentative, and gram-negative; they are found in feces; occasionally they are pathogenic to man (enteritis, peritonitis, cystitis, etc.); and they are widely distributed in nature.

Flavobacterium These organisms are rods, motile by means of peritrichous flagella, or nonmotile; they produce yellow, orange, red, or yellow-brown pigment on appropriate media; the cells are commonly proteolytic; their nutritional requirements are usually not complex; they are aerobic to facultatively anaerobic, and gram-negative; and they are found in freshwater, saltwater, and soil.

Pseudomonas These are rods, monotrichous, lophotrichous, or nonmotile; they frequently develop fluorescent, diffusible pigments; sometimes their pigments are indiffusible; usually the cells are inactive on lactose, some species hydrolyze fats, and attack hydrocarbons; they are gram-negative, and found in soil, freshwater, seawater, and heavy brines; many species are plant pathogens; a few are animal pathogens.

Vibrio These are short, curved rods, appearing singly or united into spirals; they are motile by means of a single, polar flagellum which is usually relatively short; and rarely two or three flagella in one tuft. Cells grow well and rapidly on standard culture media. They are heterotrophic, varying greatly in their nutritional requirements. Species are aerobic, facultative anaerobic, and anaerobic. Organisms are widely distributed in nature being found in saltwater, freshwater, and soil. Some are parasitic and pathogenic.

For more information: Druce and Thomas (1970); Elliker et al. (1964); Gelpi (1963); Morita (1968); Panes and Thomas (1959); Schultze and Olson (1960*a*, *b*); Thomas (1960); Witter (1961).

Thermoduric bacteria

Bacteria growing best at temperatures of 25 to 40°C are termed mesophilic organisms.

Some mesophilic forms are capable of surviving pasteurization although they do not necessarily grow at such temperatures. These resistant species are called thermoduric bacteria (thermo, heat; durans, to endure).

The most common thermoduric bacteria are found in the genera *Bacillus, Corynebacterium, Microbacterium, Micrococcus,* and *Streptococcus*. The important species include:

Bacillus *B. brevis, B. cereus, B. coagulans, B. licheniformis, B. polymyxa, B. pumilus,* and *B. subtilis*. All members are aerobic or facultative anaerobic spore-forming rods.

Corynebacterium *C. bovis*, and other species.

Microbacterium *M. flavum, M. lacticum*. Members are among the most heat-resistant non-spore-forming bacteria known, surviving a temperature of 72°C for 15 to 30 min in skim milk.

Micrococcus *M. caseolyticus, M. conglomeratus, M. flavus, M. freudenreichii, M. luteus, M. varians*. Optimum growth temperature of all species ranges from 20 to 25°C.

Streptococcus *S. bovis, S. cremoris, S. durans, S. faecalis, S. faecalis* var. *liquefaciens, S. faecalis* var. *zymogenes, S. lactis, S. thermophilus, S. uberis.*

Source of thermoduric organisms The major factor in high counts of thermoduric bacteria in pasteurized milk is the presence of such organisms in the raw product. Sanitary conditions on the farm and in the processing plant must be high to prevent the entrance of such organisms in milk.

Thermodurics are sometimes found in infected udders and find their way into pails, cans, milking machines, and other equipment used for handling milk. Unless pieces of equipment are properly sterilized after each use, they will continue to contaminate each new lot of milk.

Excessive numbers of thermoduric bacteria in milk make it difficult to meet grading standards. Their presence in milk is generally indicative of insanitary conditions on the farm.

For more information: Mantere-Alhonen (1969); Thomas et al. (1963); Thomas et al. (1967); Yano et al. (1968).

Thermophilic bacteria

Some organisms present in milk are able to survive pasteurization and can grow at such temperatures. They have an optimum range of 50 to 55°C. These high-temperature forms are referred to as thermophilic bacteria (thermo, heat; phile, to like).

Thermophilic bacteria grow well at the temperatures used in pasteurization, especially when the low-temperature method is followed. For the most part the thermophilic forms encountered in milk are spore-

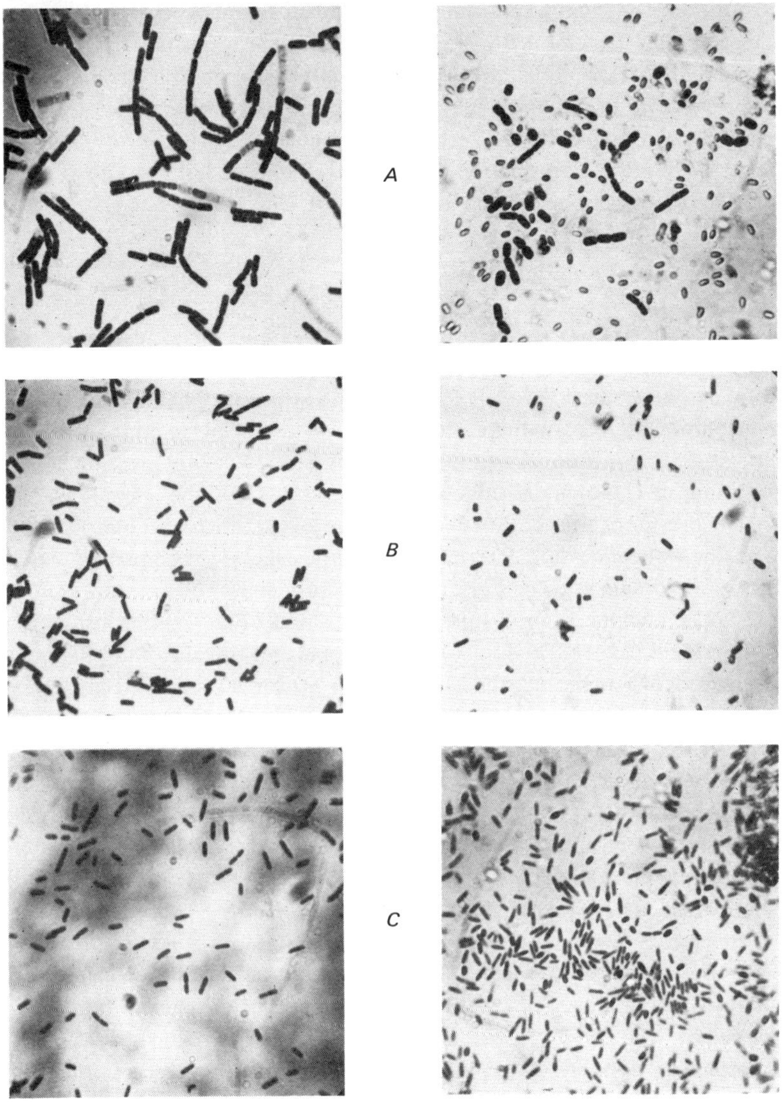

Fig. 283 Thermophilic bacteria from raw milk. (A) Bacillus cereus; (B) Bacillus subtilis; (C) Bacillus brevis. (Courtesy of Yano, Irie, Morichi, and Kenbo.)

formers. They are found in two genera, the aerobic *Bacillus* (Fig. 283) and the anaerobic *Clostridium*. In addition, there are some thermophilic species which are not spore producers.

The thermophiles are found in soil, air, manure, improperly cleaned milking machines, and on utensils and pasteurizing equipment, and undoubtedly get into milk from these sources. If the utensils and equipment are not thoroughly scrubbed daily, the organisms collect in the milk films and act as foci for seeding new lots.

Significance of thermoduric and thermophilic bacteria Thermoduric and thermophilic bacteria are not pathogenic and are of no importance from the standpoint of public health. The organisms indicate, for the most part, improper care in cleaning milking utensils and pasteurizers. The organisms may be controlled by proper care of all equipment.

For more information: Fabian (1946); Galesloot and Labots (1959*a, b*); Higginbottom and Taylor (1960); Yano et al. (1968).

Phosphatase test

The phosphatase test is employed to determine the efficiency of the pasteurization process. The test is based on the property of the heat-labile enzyme phosphatase, present in raw milk, to liberate phenol from phenyl phosphoric ester (disodium phenyl phosphate). The amount of phenol liberated is proportional to the amount of active enzyme present. On the addition of 2,6-dichloroquinonechloroimide (CQC), the phenol is converted to indophenol blue. The blue color is extracted with butyl alcohol and the concentration determined quantitatively by comparing with a series of permanent color standards in a colorimeter.

The enzyme is always present in raw milk. When milk is heated to a temperature of 143°F for 30 min, 96 percent of the phosphatase is destroyed; when heated above 145°F (63°C) for 30 min, all the enzyme present is destroyed.

The presence of the enzyme indicates either faulty pasteurization or subsequent addition of raw milk. The amount of phosphatase present may be easily and quickly measured colorimetrically. Disease organisms likely to be present in milk are killed at a temperature lower than that required to destroy the enzyme. Therefore, a heat treatment adequate to destroy the phosphatase should ensure a milk that is free from the common pathogenic bacteria.

Several organisms have been shown to be capable of elaborating phosphatase in milk. The organisms are members of the genera *Aerobacter, Klebsiella*, and *Micrococcus*. This observation may explain some positive phosphatase tests attributed to manual and mechanical defects in the dairy when in reality the false tests resulted from phosphatase produced by bacteria.

For more information: American Public Health Association (1967); U.S. Public Health Service (1967).

Antibiotics in milk

Among potential growth inhibitors found in milk and cream, probably the most important are the antibiotics.

Antibiotics have been used chiefly to combat mastitis infections of the udders of cows. These include penicillin, streptomycin, and aureomycin. The antibiotic selected for treatment depends upon the type of organism producing the infection.

Probably most of the antibiotic is excreted in the first milking after administration of the drug. Smaller and smaller quantities are excreted in subsequent milkings unless more injections of the drug are given.

It is advisable to discard the milk from several milkings after completion of the antibiotic treatment; otherwise the drug may interfere with the lactic fermentation in the preparation of butter, cheese, and fermented milk. For example, penicillin is effective chiefly against gram-positive bacteria. The concentration excreted in milk may be sufficient to interfere with the growth of *S. lactis* but not certain gram-negative forms which may produce some undesirable changes.

Galesloot (1956), in his studies on yoghurt (a fermented milk prepared from a mixture of *S. thermophilus* and *L. bulgaricus*), found penicillin to eliminate the cocci selectively without exhibiting any appreciable effect on the lactobacilli (Fig. 284). This resulted in a modified type of fermentation.

The presence of antibiotics in milk is undesirable for other reasons. Many disease-producing bacteria may develop resistance to antibiotics. The frequent presence of such drugs in milk may result in the development of resistant forms of potential disease producers in the human body. Under these conditions the antibiotics would not be effective, if, at a later date, such organisms produced an infection in the host. It would then be necessary to switch to a different antibiotic not present in milk, or to resort to some other type of treatment. Also the milk may be contaminated with sufficient penicillin or other antibiotic to cause reactions in persons who are unusually sensitive to the drug. For these reasons it is generally recommended that milk taken from dairy animals within 72 hr after the latest treatment for mastitis must not be used for human consumption.

The addition of antibiotics to milk as preservatives has not been approved by the U.S. Public Health Service.

For more information: American Public Health Association (1967); Blackburn (1956); Hargrove and Plowman (1959); Overby (1967); Palmer and Kosikowski (1967); Rosanove (1960); Snyder et al. (1961); Welch et al. (1956).

Milk-borne infection

Milk is an excellent culture medium for the growth of a great variety of organisms. Pathogenic as well as saprophytic forms may remain viable and even multiply in milk. For this reason it is difficult to obtain pure milk and to keep it pure. The prevention of milk-borne disease is one of the most important problems of public health.

Pathogenic organisms of both bovine and human origin have been isolated from milk. Many serious epidemics were caused by the consumption of such products before this fact was clearly recognized. This is to be expected when one takes into consideration the enormous quantities of milk and its products that are consumed daily. Even today, epidemics are

bacteria looks and tastes normal and gives no warning to the consumer. The disease organisms present in milk may be derived from (1) diseased cows or (2) persons collecting and handling milk.

Diseases of bovine origin The health of the cow is a very important consideration, because a number of diseases of cattle, including tuberculosis, brucellosis, Q fever, salmonellosis, staphylococcic infection, streptococcic infection, and foot-and-mouth disease, may be transmitted to man through the medium of milk. The organisms of most of these diseases may get into the milk either directly from the udder or indirectly through infected body discharges which may drop, splash, or be blown into the milk.

Tuberculosis Tuberculosis of cattle is produced by the bovine tubercle bacillus *Mycobacterium bovis*. The disease has been largely eradicated in this country but is still common in some parts of the world. The organism is very similar in appearance to the human type, being usually shorter and plumper.

Children, especially those under five years of age, are highly susceptible to the organism and may become infected by drinking milk from tuberculous cows. If the udders of cows are infected, contamination of the milk cannot be avoided.

If cows are suffering from tuberculosis of the lungs, the sputum is swallowed, with the result that the organisms appear in the feces. Since most milk may contain some excreta, it is likely to show the presence of such organisms. It is doubtful that the organism multiplies in milk, but it can live and may retain its virulence for a considerable time.

The great reduction in the incidence of bovine tuberculosis in man in recent years indicates that the practice of good sanitation in animal husbandry, the testing of cattle and removal of the reactors from the herds, and the pasteurization of milk have been effective in the control of this disease. The reservoir of bovine tuberculosis still exists, however; hence, constant vigilance against this disease must be continued by industry and health agencies.

For more information: Chapman et al. (1965). See also page 927.

Johne's disease The etiological agent is a small acid-fast rod known as *M. paratuberculosis*. The organism has been isolated from the intestinal mucosa of cattle suffering from chronic diarrhea. It is probably identical with the organism isolated from a similar disease in sheep.

M. paratuberculosis is an obligate parasite. It may be cultivated on artificial culture media only in the presence of dead acid-fast bacilli.

For more information: Smith (1960).

Brucellosis Another disease organism found in cow's milk is *Brucella abortus*, which produces contagious abortion in cows. The organism may produce the same disease in mares, sheep, rabbits, and guinea pigs. Other organisms producing similar results are *B. melitensis* from goats and *B. suis* from hogs. *B. melitensis* may also infect cows and hogs and be excreted in milk. *B. suis* produces abortion in swine and frequently attacks horses, dogs, cows, monkeys, and laboratory animals.

The three organisms are pathogenic for man, producing the disease known as undulant fever or brucellosis (after Bruce, who first isolated *B. melitensis* from the spleen of patients who had died of the disease on the island of Malta).

Brucellosis may be contracted by drinking raw milk or, less frequently, certified milk. Pasteurized milk should be safe, since the organism is destroyed in the heat process. Because of this fact, many public health authorities believe that all milk should be pasteurized before it reaches the consumer.

The incidence of brucellosis in man is increasing at the present time, and a greater effort is required to reduce the extent of infection in cattle and its transmission to man through milk.

For more information: Barrow et al. (1968); Manthei (1968); U.S. Public Health Service (1967). See also page 897.

Q fever This is a newly recognized disease of cattle which may be transmitted to man through the consumption of milk. Natural infections occur among sheep, cattle, and goats. The etiological agent is the rickettsial organism *Coxiella burnetii*.

Within the last few years, Q fever has reached endemic proportions in man in some parts of this country. A vaccine has been developed which protects cattle and probably laboratory personnel from infection. Pasteurization of milk is the most practical safeguard against its transmission to man through milk.

For more information: See page 1053.

Salmonellosis Members of the genus *Salmonella* produce a variety of infections in man and animals. In man the organisms may give rise to (1) enteric fever, (2) gastroenteritis, and (3) septicemia.

Some species or strains are natural pathogens for cattle and other domestic warmblooded animals, being present in the intestinal contents. Outbreaks of salmonellosis have occurred from drinking contaminated milk.

For more information: Marth (1969); McDonough and Hargrove (1968).

Toxigenic staphylococci Some strains of *Staphylococcus aureus* elaborate a potent exotoxin. The consumption of milk containing a toxin-producing strain may result in severe gastroenteritis.

Most enterotoxigenic strains of staphylococci are members of the coagulase-positive group (page 948). Therefore, only coagulase-positive strains are generally considered potentially enterotoxigenic, although there have been many exceptions.

Clark and Nelson (1961) recovered coagulase-positive staphylococci from all samples of raw milk examined. The initial counts ranged from 25 to 3300 per ml. The organisms did not multiply in milk at 4°C but did increase in numbers at 10°C.

For more information: Clark et al. (1961); Donnelly et al. (1968).

Mastitis Mastitis is one of the costliest diseases confronting the American dairyman. It is estimated that the disease occurs in about one cow out of every four.

Mastitis is an inflammatory and, generally, highly communicable disease of the bovine udder. An infection may be acute, having a short but severe course which may lead to the loss of one or more quarters of the udder, or even death; or the infection may be chronic, flaring up at intervals, resulting in the eventual loss of part or the entire udder.

A number of organisms are involved, the most important being *Staphylococcus aureus, Streptococcus agalactiae*, and *E. coli*. Other organisms which have been isolated include *Corynebacterium pyogenes, Mycoplasma agalactiae* and possibly other species, *Pseudomonas aeruginosa, Streptococcus dysgalactiae* and *S. uberis*. The organism most commonly associated with the disease is *S. aureus*.

Mastitis is harmful not only to the cow but also to man, since milk from an infected animal can be a means of spreading such organisms to man. If the mastitis is severe, pus and blood may appear in the milk. Milk containing appreciable numbers of the organisms and blood cells must be regarded as unfit for human consumption.

All organisms associated with mastitis are killed by pasteurization.

For more information: Barnum (1963); Glawischnig et al. (1968); Kästli (1967); Kehoe et al. (1967); Kroger and Jasper (1967); McDonald et al. (1970); Mitchell and Schultze (1962); Nurmi and Koiranen (1967), Olson et al. (1970); Raghavan et al. (1968); Luedecke et al. (1972).

Foot-and-mouth disease This is a highly contagious viral disease of domestic animals (Fig. 285). The virus produces fever, digestive disturbances, and a vesicular eruption on the mucous membranes of the mouth, on the skin between the toes, and on the udder and teats of the cow. From the vesicles, the virus may gain entrance to saliva, urine, feces, and milk. The virus also has a consistent affinity to all areas of bovine skin even though gross cutaneous lesions usually are found only in pedal area.

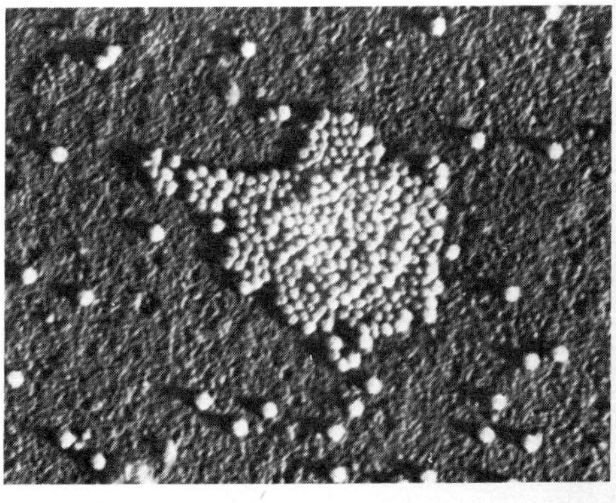

Fig. 285 Electron micrograph of foot-and-mouth disease virus particles. x79,000. The particles are approximately 22 nm in diameter. (Courtesy of Bachrach and Breese.)

The infection may be transmitted by feeding and drinking troughs, stalls, cattle cars, etc. The virus produces a high death rate among cattle; in man the disease runs a mild course.

Vaccines are now available for increasing the immunity in susceptible animals. One such vaccine is prepared by infecting monolayers of bovine kidney cells growing in flasks.

For more information: Ahl (1968); Argentina–United States Joint Commission on Foot-and-Mouth Disease (1966); Bachrach and Breese (1958); Bachrach et al. (1964); Breese and Graves (1966); DeBoer and Bachrach (1961); Fellowes (1965, 1966); Gailiunas and Cottral (1966, 1967); Sellers (1968); Smith and Hugh-Jones (1969); Wild et al. (1969).

Diseases of human origin Some of the important diseases of human origin that have been disseminated by milk are (1) typhoid fever, (2) scarlet fever, (3) diphtheria, (4) septic sore throat, (5) infantile diarrhea, and (6) poliomyelitis. The organisms may be transferred to milk by contaminated hands of the workers; by droplets expelled during coughing, sneezing, and talking; by moistening the hands with saliva during wet milking; and in other ways.

Typhoid fever Many typhoid epidemics have resulted from drinking milk contaminated with the organism *Salmonella typhosa*. Further investigations revealed that in each outbreak usually only one dairy supplying the milk was responsible for the epidemic. The organisms were introduced into the milk by typhoid carriers or unrecognized cases of typhoid fever among the workers at the dairies. The isolation of such individuals resulted in the disappearance of new typhoid cases in the communities. Epidemics of typhoid fever have been traced to a lesser extent to the consumption of ice cream, cheese, and butter. Because of the improved sanitary conditions of production and handling of milk and milk products in this country, such epidemics are a rare occurrence today.

According to the U.S. Public Health Service:

> Investigations of milkborne disease outbreaks have shown that body discharges of infected milk handlers are the most frequent source of contamination responsible for such outbreaks. The diseases so transmitted through milk include typhoid fever, dysentery, diphtheria, septic sore throat, scarlet fever, and tuberculosis.
>
> The health officer, or a physician authorized by him, shall examine and take a careful morbidity history of each person connected with a pasteurization plant, or about to be employed by one, whose work will bring him into contact with the processing, handling, storage, or transportation of milk, milk products, containers, or equipment. If such examination or history should suggest that such person may be a carrier of, or infected with, the organisms of typhoid or paratyphoid fever, or any other communicable disease likely to be transmitted through milk, he shall secure appropriate specimens of body discharges and cause them to be examined in a laboratory approved by him or by the State health authori-

ties for such examinations, and, if the results justify, such persons shall be barred from such employment.

For more information: See page 939.

Scarlet fever and septic sore throat Both diseases are produced by *Streptococcus pyogenes*, a pus-producing organism. Epidemics have been caused by the consumption of milk containing this organism. The milk may become contaminated by handlers or by infected udders. Usually a milker suffering from scarlet fever or sore throat infects the udders with the organisms by contaminated hands. The organisms rapidly multiply in the udders. Abscesses form and the milk becomes heavily contaminated. Such milk may produce septic sore throat or scarlet fever in persons consuming the product.

For more information: See page 952.

Diphtheria The causative agent is *Corynebacterium diphtheriae*. The cells are straight or slightly curved rods and frequently swollen at one or both ends. They are nonmotile and gram-positive.

Diphtheria is spread principally by direct contact with human sources through either active cases of the disease or carriers. Although epidemics of milk-borne diphtheria are comparatively rare, a few outbreaks have been traced to infected milk handlers.

For more information: See page 906.

Fermented milk

Milk is probably the most important article of food among many pastoral tribes of Europe and Asia. Because of the primitive sanitary conditions under which the people live, milk is usually allowed to ferment before it is consumed. The high acidity produced preserves the milk for an indefinite period.

The consumption of sour milk preparations is widespread because of their supposedly therapeutic value. The fact that they appear under various names does not mean necessarily that each product is fermented with a different organism. The names identify the country or region where they are produced. Some preparations contain only one organism; others result from the combined action of two or more organisms. The latter furnish good examples of microbial associations.

Some well-known fermented milk preparations are the following:

Yoghurt This is a semisolid, cheese-like or sometimes thickly fluid or jelly-like, preparation made from milk partly evaporated and then fermented by *Lactobacillus bulgaricus*. In Bulgaria, where the preparation is very popular, the fermentation is initiated with a starter, saved from a previous lot. Yoghurt is nonalcoholic. The organism produces from 1.5 to 2.5 percent lactic acid.

Matzoon This is the sour milk preparation of Armenia, being similar to yoghurt in flavor and flora.

Gioddu Gioddu is the sour milk product prepared on the island of Sardinia. It is also similar to yoghurt in its method of preparation.

Leban This is an Egyptian drink prepared by the action of lactic acid bacteria and yeasts on cow's, goat's, or buffalo's milk. The bacteria produce lactic acid while the yeasts manufacture alcohol and carbon dioxide. It is an intoxicating, effervescent, fermented milk preparation.

Kumiss Kumiss is a Russian product prepared by the fermentation of milk by yeasts, lactobacilli, and lactic streptococci. The yeasts produce alcohol and carbon dioxide; the bacteria chiefly lactic acid. It is an intoxicating, fermented liquor originally made by the Tartars from mare's or camel's milk.

Kefir Kefir is prepared by inoculating milk with kefir grains, a symbiotic association of yeasts and lactic acid bacteria. Kefir grains resemble minute cauliflowers in appearance. If the grains are transferred daily to fresh milk, their weight doubles in 7 to 10 days. According to La Rivière (1969), kefir grains are composed of the yeasts *Saccharomyces delbrückii* and *Torulopsis holmii*, and the bacterium *L. brevis*. The fermentation yields alcohol, carbon dioxide, and lactic acid. It is used as a food and as a medicine in the Caucasus.

Curds This is the fermented milk preparation of Ceylon. It is manufactured from cow's or buffalo's milk. The milk is boiled, cooled, and while still warm, inoculated with some curd from the previous lot. The milk is allowed to ferment for about 36 hr. The organisms involved include yeasts, *Streptococcus lactis*, and a *Lactobacillus*. The preparation is similar to kefir and kumiss in taste and composition.

Bulgarian and acidophilus milk

Yoghurt is consumed in large quantities as an article of diet by the people of Bulgaria. Metschnikoff (1908) noted that centenarians were more numerous in Bulgaria, in proportion to population, than in other countries. He believed that the increase in the life span was due to the ingestion of large quantities of sour milk, produced by the action of the rod-shaped, gram-positive organism *L. bulgaricus*. Because of this fact, Metschnikoff advocated the consumption of Bulgarian milk for the prolongation of life.

According to Metschnikoff, growth of *L. bulgaricus* in the intestinal tract produced a high concentration of lactic acid, which inhibited the growth of the putrefactive bacteria. Disorders that were supposedly associated with autointoxication (absorption of putrefactive waste products from the intestinal tract) would not occur.

Later Hull and Rettger (1915) found the conclusions of Metschnikoff to be incorrect. They reported that *L. bulgaricus* was not a normal inhabitant of the intestinal tract of man and, therefore, did not become acclimated to the new environment. They showed that the consumption of

Bulgarian milk stimulated the growth of *L. acidophilus*, an organism normally present in the intestines of adults.

L. acidophilus was first isolated from the feces of breast-fed infants. It is present in the intestinal contents of adults. On a mixed diet, the numbers are small. If the diet is supplemented with large quantities of milk or carbohydrates, such as lactose or dextrin, the numbers are greatly increased. The organisms ferment the carbohydrate with the production of a high concentration of lactic acid (about 3 percent), which is sufficient to inhibit the growth of the putrefactive types (*E. coli*, etc.). In the absence of a high carbohydrate diet, the flora again become predominantly putrefactive in character. The numbers are also increased by the ingestion of milk fermented by *L. acidophilus*, especially when taken with lactose or dextrin to increase the fermentable constituents in the intestines.

For more information: Huhtanen and Williams (1963).

Butter

Butter consists of about 80.5 percent butterfat, 16.5 percent water, 2 percent table salt, 1 percent curd, and traces of lactose. The amounts of casein and lactose present depend upon the extent to which butter is washed during manufacture. Since the salt is completely dissolved in the water, the liquid portion of butter consists of about a 30 percent saline solution.

Butter is prepared by churning fresh, sweet cream, either raw or pasteurized, to separate the fat globules from the other constituents. This method necessitates churning daily while the milk or cream is still fresh. A more popular method is to allow the cream to sour first, after which the butterfat may be more easily separated from the casein.

The cream may be soured naturally or by the addition of a culture of organisms known as a starter. Many starters are used, depending upon the type of organism desired. The advantages of first souring the cream are: (1) the yield of butter is increased, owing to a better separation of the butterfat, and (2) the aroma and flavor may be greatly improved, especially with the use of starters, by proper selection of organisms.

Before a starter is added, it is necessary to destroy practically all the bacteria already in the milk or cream. For this purpose a pasteurization temperature of 71°C is applied for a period of 30 min. Such treatment results in the destruction of at least 99.9 percent of the organisms present.

Butter cultures (starters) Butter cultures consist of a mixture of two types of organisms: (1) those producing a high acidity (lactic acid, chiefly) and (2) those imparting the characteristic aroma and flavor to butter.

Lactic acid types The lactic acid types consist generally of *Streptococcus lactis*, the organism responsible for the normal souring of milk. Another organism of this type is *S. cremoris*. These organisms produce fairly large amounts of lactic acid from the lactose of milk together with small amounts of secondary products. Growth of the organisms in milk does not result in the product having a butter-culture flavor. However, the compounds formed by the lactic bacteria greatly influence the action of the flavor organisms, resulting in a product having a more pronounced aroma and flavor.

The organisms produce from 0.7 to 1.2 percent lactic acid in milk. Those figures correspond to a pH range of 4.6 to 4.0 approximately.

Characteristics of *S. lactis* are given on page 714.

Cells of *S. cremoris* are spheres or ovoid-shaped and elongated in the direction of the chain; measuring 0.6 to 1.0 μ in diameter; usually form long chains in milk but some cultures occur predominantly as pairs. The cells are gram-positive. Optimum growth temperature is below 30°C. They may survive 60°C for 30 min. Organisms fail to grow in 4 percent salt solution. They are commonly employed in commercial starters for butter and cheese manufacture. Natural habitat is probably on plants (Fig. 280).

Aroma and flavor type Organisms responsible for the aroma and flavor of butter consist generally of a mixture of *Leuconostoc dextranicum* and *L. citrovorum*.

Cells are normally spheres, 0.9 to 1.2 μ in diameter, occurring in pairs and in short or long chains. They are gram-positive. Optimum growth temperature is 20 to 25°C. The organisms are isolated from dairy products. Their natural habitat is on plant materials and in milk and milk products.

The aroma and flavor of butter are dependent upon the citric acid content of milk. Milk normally contains about 0.18 percent citric acid. *L. citrovorum* is capable of attacking the citric acid with the formation of acetic acid, possibly some formic acid and propionic acids, carbon dioxide, and acetylmethylcarbinol ($CH_3 \cdot CO \cdot CHOH \cdot CH_3$). This compound is also known as acetoin. The addition of more citric acid to milk results in a proportionate increase in the content of acetoin. Under conditions of high acidity, some acetoin is oxidized to diacetyl ($CH_3 \cdot CO \cdot OC \cdot CH_3$). Under conditions of low acidity and suitable temperature, some acetoin is reduced to 2,3-butylene glycol or more correctly known as 2,3-butanediol ($CH_3 \cdot CHOH \cdot CHOH \cdot CH_3$).

The citric fermentation, according to Keenan and Bills (1968), is shown in Fig. 286.

Diacetyl in high dilution suggests the odor of butter. Acetylmethylcarbinol in pure form is odorless; in the impure state, it gives off an odor not unlike that of diacetyl. Cultures having a satisfactory aroma and flavor contain relatively large amounts of these two compounds. Milk is rarely curdled by the flavor and aroma organisms.

Preparation and use of butter culture The butter culture is generally prepared as follows: The milk or cream is pasteurized at a temperature of 70 to 85°C for 30 min to diminish foreign bacteria, then inoculated with the desired organisms. After the starter is prepared, it should be handled with great care to prevent entrance of organisms likely to produce undesirable changes. Butter cultures are commonly ripened at a temperature of 21.1 to 22.2°C for maximum development of aroma and

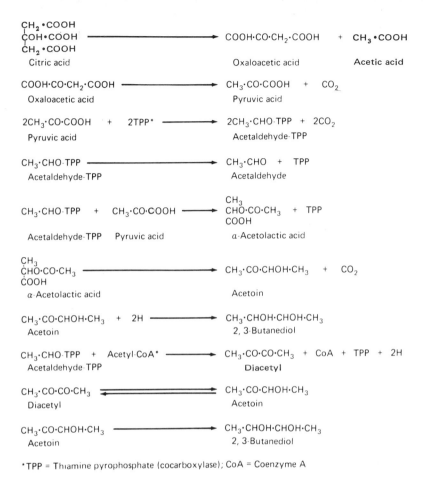

$$\begin{matrix} CH_2\cdot COOH \\ COH\cdot COOH \\ CH_2\cdot COOH \end{matrix} \longrightarrow COOH\cdot CO\cdot CH_2\cdot COOH + CH_3\cdot COOH$$

Citric acid → Oxaloacetic acid + Acetic acid

$$COOH\cdot CO\cdot CH_2\cdot COOH \longrightarrow CH_3\cdot CO\cdot COOH + CO_2$$

Oxaloacetic acid → Pyruvic acid

$$2CH_3\cdot CO\cdot COOH + 2TPP^* \longrightarrow 2CH_3\cdot CHO\text{-}TPP + 2CO_2$$

Pyruvic acid → Acetaldehyde-TPP

$$CH_3\cdot CHO\text{-}TPP \longrightarrow CH_3\cdot CHO + TPP$$

Acetaldehyde-TPP → Acetaldehyde

$$CH_3\cdot CHO\text{-}TPP + CH_3\cdot CO\cdot COOH \longrightarrow \begin{matrix} CH_3 \\ CHO\cdot CO\cdot CH_3 \\ COOH \end{matrix} + TPP$$

Acetaldehyde-TPP + Pyruvic acid → α-Acetolactic acid

$$\begin{matrix} CH_3 \\ CHO\cdot CO\cdot CH_3 \\ COOH \end{matrix} \longrightarrow CH_3\cdot CO\cdot CHOH\cdot CH_3 + CO_2$$

α-Acetolactic acid → Acetoin

$$CH_3\cdot CO\cdot CHOH\cdot CH_3 + 2H \longrightarrow CH_3\cdot CHOH\cdot CHOH\cdot CH_3$$

Acetoin → 2, 3-Butanediol

$$CH_3\cdot CHO\text{-}TPP + Acetyl\text{-}CoA^* \longrightarrow CH_3\cdot CO\cdot CO\cdot CH_3 + CoA + TPP + 2H$$

Acetaldehyde-TPP → Diacetyl

$$CH_3\cdot CO\cdot CO\cdot CH_3 \rightleftharpoons CH_3\cdot CO\cdot CHOH\cdot CH_3$$

Diacetyl ⇌ Acetoin

$$CH_3\cdot CO\cdot CHOH\cdot CH_3 \longrightarrow CH_3\cdot CHOH\cdot CHOH\cdot CH_3$$

Acetoin → 2, 3-Butanediol

*TPP = Thiamine pyrophosphate (cocarboxylase); CoA = Coenzyme A

Fig. 286 The citric fermentation as proposed by Keenan and Bills.

flavor. The flavor compounds are not produced in significant amounts until the butter culture has an acidity of about 0.8 percent. Usually 0.25 to 1 percent butter culture is added to cream that is to be ripened for the preparation of butter.

Elliker (1945) showed that the loss of aroma of butter, butter substitutes, and other food products was caused by the action of certain organisms on the diacetyl content. Some members of the genus *Pseudomonas* (*P. fluorescens, P. fragi, P. putrefaciens*, etc.) were found to be active in destroying diacetyl with accompanying loss of flavor. The results emphasized the importance of keeping butter and other products containing diacetyl free from contaminating organisms.

For more information: Harvey and Collins (1961); Keenan (1968); Lee et al. (1970); Peebles et al. (1969); Speckman and Collins (1968); Stadhouders et al. (1969, 1971); Branen and Keenan (1971).

Undesirable changes in butter Butter prepared from sweet unpasteurized cream contains the same microflora as the cream from which it was prepared. Also the bacterial changes which take place during storage are the same as those which occur in the milk or cream under the same conditions. Butter prepared from cream previously pasteurized at high temperatures, and then inoculated, generally contains only those organisms which were added to promote ripening. Some molds and yeasts may be present which result from air contamination. Since molds are generally aerobic, they grow chiefly on the surface of butter.

Undesirable changes which take place after butter has been manufactured are produced largely by the growth of organisms. Many of the organisms are present as a result of contamination of butter after its manufacture. Therefore, the same precautions followed in handling milk and cream also apply to butter. The extent of recontamination is roughly an indication of the care exercised in handling and storing butter.

Some butter defects and their causes are the following:

Surface taint This is caused by *P. putrefaciens*, commonly found in raw milk and cream, butter, water, soil, and creamery equipment.

Rancidity The first stage is believed to be a hydrolysis of the butterfat into glycerol and fatty acids. During this stage butter acquires a strong odor of butyric and caproic acids. A lipase capable of doing this is normally present in raw milk or is secreted by various contaminating organisms such as *P. fragi* and *P. fluorescens*. Since the enzyme is destroyed in the pasteurization process, the defect can be controlled by heat treatment.

Malty flavor This is produced by a strain or variety of *Streptococcus lactis*. The organism has been isolated from raw milk. It is destroyed when milk or cream is pasteurized at 70 to 85°C for 30 min.

Skunk-like odor This is produced by *P. mephitica*. The organism is presumably derived from the rinse water.

Fruity flavor Fruity aromas may be imparted to dairy products by the growth of *P. fragi* in the milk or cream used in their manufacture.

Black discoloration This is a defect produced by *P. nigrifaciens*. The organism produces a black to reddish-brown discoloration of butter. It is widely distributed in nature.

Surface discoloration This is produced by molds of the genera *Cladosporium, Alternaria, Aspergillus, Mucor*, and *Rhizopus.*

Yeasty flavor This is caused by the yeasts *Torula cremoris* and *T. sphaerica.*

Metallic flavor This is caused by various organisms, including a strain of *S. lactis.*

Tallowiness in butter A tallow-like odor is produced by oxidation. This may result from the action of ultraviolet rays of sunlight or from oxidases naturally present in milk. Certain mold enzymes are capable of producing a similar effect. Changes may be prevented by high-temperature pasteurization, which destroys the oxidases.

For more information: Downes (1959); Morgan (1970*a, b*); Sheldon et al. (1971).

Quantitative bacteriological examination of butter Butter is not a favorable medium for bacterial growth. Multiplication usually occurs only in the minute droplets of water containing salt and traces of casein and lactose. Since the high salt content makes this an unfavorable medium, butter never shows bacterial counts as high as those of milk. The count is highest in freshly prepared butter and becomes less and less as butter ages.

Yeasts and molds, when numerous, are objectionable in butter. Their presence indicates either improper pasteurization of the milk or cream, or recontamination after pasteurization.

The bacterial flora of the surface of butter differ from that of the interior, owing to contamination from the air and differences in the oxygen concentration. For these reasons it is difficult to obtain representative samples for examination. A sample is usually obtained by removing a cylinder of butter from a cake by means of a sterile sampler. The butter is then melted on a water bath at a temperature not to exceed 40°C, and dilutions are prepared in sterile dilution blanks previously heated to the same temperature.

Ice cream

Ice cream is a frozen dairy product composed of cream and other milk derivatives, stabilizers, emulsifiers, sugar, flavoring materials, and coloring.

The milk derivatives consist of sweetened condensed skim milk, nonfat dry milk, butter oil, etc.

Stabilizers include sodium alginate, Irish moss (*Chondrus*), locust bean gum, cellulose gum, karaya gum, gelatin, etc. They are added to give ice cream a smooth texture by preventing the formation of large ice crystals.

Emulsifiers are added to produce dryness in ice cream. Dryness eliminates dripping and reduces cleanup. A stiff, dry ice cream extrudes easily, forms readily, cuts and molds into novelty shapes.

The emulsifier functions by causing the dispersed fat globules to bunch together like grapes. This "bunches of grapes" effect or proper agglomeration of dispersed fat globules determines the degree of dryness of ice cream.

Glyceryl monostearate esters have been used for this purpose. The newer and better emulsifiers are created by combining conventional glyceryl monostearates with surfactants or solubilizers, such as Tween 80 or Tween 65, to achieve the optimum in whipping and dryness.

Unlike butter and some cheeses, bacteria play no part in the preparation of ice cream.

Bacterial counts of ice cream The bacterial content of ice cream depends largely upon (1) the number present in the milk or cream at the time of preparation and (2) the number present in the various ingredients employed in its manufacture.

High total counts usually indicate neglect and unsanitary conditions. These may result from (1) poor quality of ingredients used, (2) improper pasteurization, (3) contamination after pasteurization, (4) improper aging, (5) unsanitary equipment, and (6) negligent and untrained personnel.

Organisms which may be found in ice cream include coliform bacteria, micrococci, streptococci, spore-forming rods, yeasts, and molds. Some are of dairy origin; others are from the ingredients used in its manufacture.

Ice cream is stored at temperatures of -17.8 to $-28.9°C$. At these temperatures, there is a gradual but slow decrease in the bacterial population. The lactic acid organisms, i.e., those responsible for the souring of cream, fail to multiply. The presence of disease-producing organisms is generally the result of using contaminated cream in its manufacture. Cold is not a sterilizing agent and should not be depended upon to destroy all disease organisms in ice cream.

Cheese

According to the standards of the Food and Drug Administration of the U.S. Department of Agriculture, cheese is defined as "... the product made from the separated curd obtained by coagulating the casein of milk, skimmed milk, or milk enriched with cream." The coagulation is accomplished by means of rennet or other suitable enzyme, lactic fermentation, or by a combination of the two. The curd may be modified by heat, pressure, ripening ferments, special molds, or suitable seasoning.

The name "cheese" unqualified means Cheddar cheese (American cheese, American Cheddar cheese).

Two general processes are used for the preparation of curd. One method depends upon the addition of an appropriate enzyme; the other, on the activities of organisms. Since the changes produced by organisms are dependent upon the presence of enzymes, the latter is also an enzymatic process. The cheeses in the first group are known as rennin-curd cheeses; those in the second group as acid-curd cheeses.

Steps in cheese making Five steps are followed in the preparation of cheese. These are (1) inoculating pasteurized milk with lactic acid bacteria, (2) curdling the milk, (3) draining the curd and pressing it into desired shapes, (4) adding salt, and (5) ripening.

Starter cultures may consist of one or a combination of two or more of the following organisms: *Streptococcus lactis, S. cremoris, S. thermophilus, Lactobacillus bulgaricus, L. helveticus, L. lactis*, etc. These organisms are capable of fermenting lactose with the production of sufficient acid to curdle the milk.

Casein is dispersed in milk as calcium caseinate. Lactic acid, as it is formed, reacts with the calcium to give calcium lactate. When the calcium content is reduced to a certain low point, the casein precipitates. The clear straw-colored liquid that separates from the curd is known as whey.

Curd may also be prepared by adding the enzyme rennin to milk. The enzyme reacts with the calcium caseinate to form calcium paracaseinate. Then the calcium paracaseinate reacts with free calcium ions to give a curd. The insoluble curd is produced at the pH of milk.

The curd is separated from the whey by draining, with or without the use of pressure. Draining without pressure results in the production of soft cheeses; draining with pressure results in the formation of hard cheeses.

The solid curd is molded into various shapes according to the variety of cheese being manufactured. Sodium chloride is added by (1) floating the cheese in a strong brine solution or (2) rubbing the surface with dry salt. The salt slowly diffuses throughout the cheese.

Freshly prepared and molded curd is known as green cheese. It has a bland and slightly sour taste. The curd is tough when chewed and some varieties are rubbery in consistency. To be made satisfactory for consumption, it must be set aside to ripen. Certain conditions, such as temperature and moisture, are carefully controlled during the ripening process. The cheese changes considerably during this stage. The insoluble casein is rendered soluble, and the digestibility is greatly improved. The consistency changes to a softer product. Also the flavor, characteristic of the finished product, develops during the ripening period.

As has already been said, the changes which occur during ripening are largely enzymatic. However, this does not explain all the changes which occur during the long aging period. The flavors which develop are not the result of enzymatic action but of the associated activities of bacteria, yeasts, and molds. The flavors can be attributed to complex associations of chemical compounds produced during the manufacturing and ripening processes by the degradation of protein, fat, and lactose. The enzymes improve the consistency and digestibility of cheeses but play no part in improving the aromas and flavors. The latter are dependent upon the kinds and numbers of organisms present.

The typical flavor of Cheddar cheese is due to a complex mixture of volatile compounds which are present in exceedingly low concentrations. Tests have revealed the presence of the following compounds: acetaldehyde, acetoin, acetone, 2-butanone, diacetyl, formaldehyde, glyoxal, 2-heptanone, α-keto isocaproic acid, methional, 3-methylbutanal, 2-nonanone, oxalsuccinic acid, 2-pentanone, propionaldehyde, pyruvic acid, 2-tridecanone, and 2-undecanone. Doubtless other compounds are present which have not been identified.

Hard cheeses Hard-curd cheeses are prepared from curd subjected to heavy pressure to remove as much of the whey as possible. This gives a very hard, tough curd that does not become softened to any extent during the ripening period. Since the curd is very compact and tough, the ripening stage requires a considerable period of time to produce a satisfactory product. Enzymatic and bacterial changes proceed simultaneously.

Examples of hard-curd cheeses include Cheddar, Edam, Parmesan, Provolone, Romano, and Swiss.

Soft cheeses Soft cheeses are prepared by allowing the whey to drain from the curd without the application of pressure. Cheeses prepared in this manner contain more moisture than the hard-curd cheeses, which results in a much softer finished product.

Examples of soft-curd cheeses include brick, Brie, Camembert, Limburger, and Roquefort.

Camembert cheese is made by shaping the curd into the desired size and form. The surface is inoculated with spores of the mold *Penicillium camemberti*. Enzymes secreted by the mold growth act upon the surface of the cheese to produce a slow liquefaction of the casein. The enzymes gradually penetrate the product until the entire curd is affected. The result is the formation of a soft, creamy mass at the completion of the ripening period. The enzymes are responsible for the consistency of the product, while the growth of the mold contributes to the flavor and aroma.

In cheeses of the Roquefort type, spores of the blue-green mold *P. roqueforti* are inoculated into the curd. Since the mold is aerobic, holes are punched into the curd to facilitate development of the mold throughout the curd. The enzymes elaborated by the organism soften the casein, while certain metabolic products impart the characteristic aroma and flavor to the cheese.

Desirable organisms Many organisms are responsible for the aromas, flavors, and characteristics of the various types of cheeses. Each type has its own characteristic flora. Some important organisms which have been isolated from cheeses are: (1) *S. cremoris* (page 744); (2) *S. faecalis*, ovoid cells elongated in direction of chain, occurring mostly in pairs or short chains, gram-positive, with some strains actively motile; some strains fermenting citric acid; organism survives 60°C for 30 min; (3) *S. lactis* (page 714); (4) *S. thermophilus*, spherical or ovoid cells, occurring in pairs to long chains, gram-positive, optimum temperature 40 to 45°C, and survives 65°C for 30 min; (5) *L. bulgaricus*, slender rods with rounded ends, often in chains, nonmotile, gram-positive with optimum temperature 50 to 62.8°C; (6) *L. casei*, short or long rods, occurring in short or long chains, nonmotile, gram-positive, optimum temperature 36°C; (7) *L. helveticus*, large rods, occurring singly and in chains, nonmotile, gram-positive, with optimum temperature 40 to 42°C; (8) *L. lactis*, rods appearing as long forms with a tendency to grow into threads, often strongly curling, occurring singly or in pairs in young vigorous cultures, gram-positive, with optimum temperature 40°C; (9) *L. plantarum*, large rods, occurring singly or in short chains, with rounded ends, cells becoming longer in acid media, gram-positive, with optimum temperature

30°C; (10) *Leuconostoc citrovorum* (page 744); (11) *L. dextranicum* (page 744); (12) *Propionibacterium shermanii*, small, spherical cells, occurring mostly in pairs and short chains, nonmotile, gram-positive, and responsible for characteristics of Swiss cheese.

Undesirable organisms The presence of undesirable organisms is responsible for numerous types of faulty cheeses. Some faults affect the taste; others are concerned with the appearance of the finished product. The milk becomes contaminated through carelessness in its collection and handling. Considerable losses are experienced at times by cheese manufacturers. For this reason it is generally advisable to use milk previously pasteurized and then inoculated with a desirable organism rather than to start with raw milk.

A common fault is swollen or blown cheese, which results from the fermentation of the lactose to acid and gas. The gas bubbles cause the cheese to swell until it may actually burst. Unpleasant flavors are also produced. Coliform species and certain yeasts may be involved.

The presence of putrefactive organisms may be responsible for putrid odors and flavors. The bacteria grow and become active when the acidity of the cheese is reduced during the ripening period.

The presence of pigment-producing organisms may be responsible for discolorations in cheese. These may be caused also by chemical reactions with metals, such as copper and iron from the utensils used in handling the raw materials. Red and brown spots in Emmentaler cheese are produced by the growth of chromogenic *Propionibacterium*. *Lactobacillus brevis* is responsible for the appearance of rusty spots in Cheddar cheese. Surface discolorations are produced by many molds such as *P. casei*, *Cladosporium herbarum*, *Monilia niger*, and *Oöspora crustacea*. Red and yellow torulae (false yeasts) play some part in the process.

For more information: Babel (1953); Badings et al. (1968); Harwalkar and Seitz (1971); Lubert and Frazier (1955).

Cheese and its relation to disease Many epidemics have been traced to the consumption of cheese contaminated with disease organisms. The organisms most commonly associated with cheese-borne infections are *Brucella melitensis*, *Clostridium botulinum*, *Staphylococcus aureus*, *Salmonella choleraesuis*, *S. schottmuelleri*, *S. typhimurium*, and *S. typhosa*.

Gilman and Marquardt (1951) recovered *B. abortus* from five out of six Italian cheese curds made from raw milk. After pasteurization, they were unable to recover the organisms from either the milk or the curd prepared from the same lot.

Thatcher and Ross (1960) and Donnelly et al. (1964) reported that Cheddar cheese frequently contained large numbers of coagulase-positive *Staphylococcus aureus* (page 951). Substantial contamination could result from infected bovine udders, inadequate overnight cooling, presence of antibiotic residues, and contamination with strains of *S. aureus* resistant to antibiotics present in milk.

Multiplication of *S. aureus* could occur during the period of heat treatment after coagulation of the casein. The presence of low levels of penicillin in the milk allowed multiplication of *S. aureus* and inhibition of the lactic starter culture.

Two public health hazards were envisaged: (1) *S. aureus* food poisoning; and (2) dissemination of large populations of virulent strains of *S. aureus*.

From available evidence, it appears that if cheese prepared from raw milk is allowed to ripen for at least 90 days, all pathogenic organisms present should be either dead or attenuated. A ripening period of 120 days should eliminate all viable disease bacteria. However, a combination of pasteurization and a 90-day aging period should be more nearly ideal, as well as economically sound, in the preparation of a safe and mature cheese.

For more information: Davis (1965); Keenan et al. (1968); Edelsten et al. (1969); Franklin and Sharpe (1963); Gehrig and Knight (1963); Goepfert et al. (1968); Mabbitt (1961); Marth (1963); McDonough et al. (1967); Perry (1961); Read et al. (1965); Reiter et al. (1964); Reiter and Møller-Madsen (1963); Sternberg (1971); Vedamuthu et al. (1966*a, b*).

References

Ahl, R.: Untersuchungen zur thermoinaktivierung des Maul-und Klauenseuche-virus, *Arch. Ges. Virusforsch.*, **24**:361, 1968.

American Public Health Association: "Standard Methods for the Examination of Dairy Products," New York, 1967.

Argentina–United States Joint Commission on Foot-and-Mouth Disease: Studies on foot-and-mouth disease, *Nat. Acad. Sci.–Nat. Res. Council Publ. no. 1343*, 1966.

Babel, F. J.: The role of fungi in cheese ripening, *Econ. Botany*, **7**:27, 1953.

Bachrach, H. L., and S. S. Breese, Jr.: Purification and electron microscopy of foot-and-mouth disease virus, *Proc. Soc. Exp. Biol. Med.*, **97**:659, 1958.

——— et al.: Chemical and physical properties of virtually pure foot-and-mouth disease virus, *Am. J. Vet. Res.*, **25**:333, 1964.

Badings, H. T., et al.: Phenolic flavor in cheese, *J. Dairy Sci.*, **51**:31, 1968.

Barnum, D. A.: Bovine mastitis and its relation to milk quality, *Can. J. Public Health*, **54**:129, 1963.

Barrow, G. I., et al.: *Brucella abortus* in fresh cream and cream products, *Brit. Med. J.*, **2**:596, 1968.

Blackburn, P. S.: Antibiotic treatment of mastitis and its effect on the cell content of the milk, *J. Dairy Res.*, **23**:225, 1956.

Blankenagel, G., and I. Okello-Uma: Gram-negative bacteria in raw milk, *Can. Inst. Food Technol. J.*, **2**:69, 1969.

Branen, A. L., and T. W. Keenan: Diacetyl and acetoin production by *Lactobacillus casei*, *Appl. Microbiol.*, **22**:517, 1971.

Breese, S. S., Jr., and J. H. Graves: Electron microscopic observation of crystalline arrays of foot-and-mouth disease virus, *J. Bact.*, **92**:1835, 1966.

Brown, J. P., and P. J. Van Demark: Respiration of *Lactobacillus casei*, *Can. J. Microbiol.*, **14**:829, 1968.

Chapman, J. S., et al.: Isolation of mycobacteria from raw milk, *Am. Rev. Respirat. Diseases*, **91**:351, 1965.

Clark, W. S., Jr., and F. E. Nelson: Multiplication of coagulase-positive staphylococci in grade A raw milk samples, *J. Dairy Sci.*, **44**:232, 1961.

——— et al.: Characterization of coagulase-positive staphylococci isolated from raw milk, *Appl. Microbiol.*, **9**:195, 1961.

Claydon, T. J., and H. C. Fryer: Effect of raw milk storage and bacterial development on subsequent lactic culture activity, *Appl. Microbiol.*, **8**:278, 1960.

Davis, J. G.: "Cheese," vols. I and II, New York, American Elsevier Publishing Company, Inc., 1965.

Deane, D. D., et al.: Pasteurization treatment and consumer acceptance, *J. Dairy Sci.*, **50**:1216, 1967.

DeBoer, C. J., and H. L. Bachrach: The multiplication of foot-and-mouth disease virus in trypsinized calf kidney and tongue cells and its use as immunizing and complement-fixing antigens, *J. Immunol.*, **86**:282, 1961.

Donnelly, C. B., et al.: Occurrence of coagulase-positive staphylococci in cheddar cheese, *Appl. Microbiol.*, **12**:311, 1964.

——— et al.: Production of enterotoxin A in milk, *Appl. Microbiol.*, **16**:917, 1968.

Downes, T. E. H.: The lipolytic deterioration of butter by microorganisms, *S. African J. Agr. Sci.*, **2**:527, 1959.

Druce, R. G., and S. B. Thomas: An ecological study of the psychrotrophic bacteria of soil, water, grass and hay, *J. Appl. Bact.*, **33**:420, 1970.

Edelsten, D., et al.: A study of microbial milk coagulating enzymes (microbial rennets), Copenhagen, Denmark, *Royal Veterinary and Agricultural University Yearbook*, page 201, 1969.

Elliker, P. R.: Effect of various bacteria on diacetyl content and flavor of butter, *J. Dairy Sci.*, **28**:93, 1945.

——— et al.: Psychrophilic bacteria and keeping quality of pasteurized dairy products, *J. Milk Food Technol.*, **27**:69, 1964.

Fabian, F. W.: Significance of thermoduric and thermophilic bacteria in milk and their control, *Milk Technol.*, **9**:273, 1946.

spread through milk, but they are of rare occurrence compared to the number reported during the early years of public health.

The abnormal changes that occur in milk are usually easily detected by appearance, taste, and smell. However, the presence of disease organisms cannot be detected in that manner. Milk containing disease

Fig. 284 Effect of penicillin on Streptococcus thermophilus and Lactobacillus bulgaricus in yoghurt. Upper left, no penicillin added; upper right, 0.01 μg penicillin added; lower left, same as in preceding; lower right, 0.03 μg penicillin added. The streptococci gradually dissappear, leaving the more resistant lactobacilli. This gives a modified type of fermentation. (Courtesy of Galesloot.)

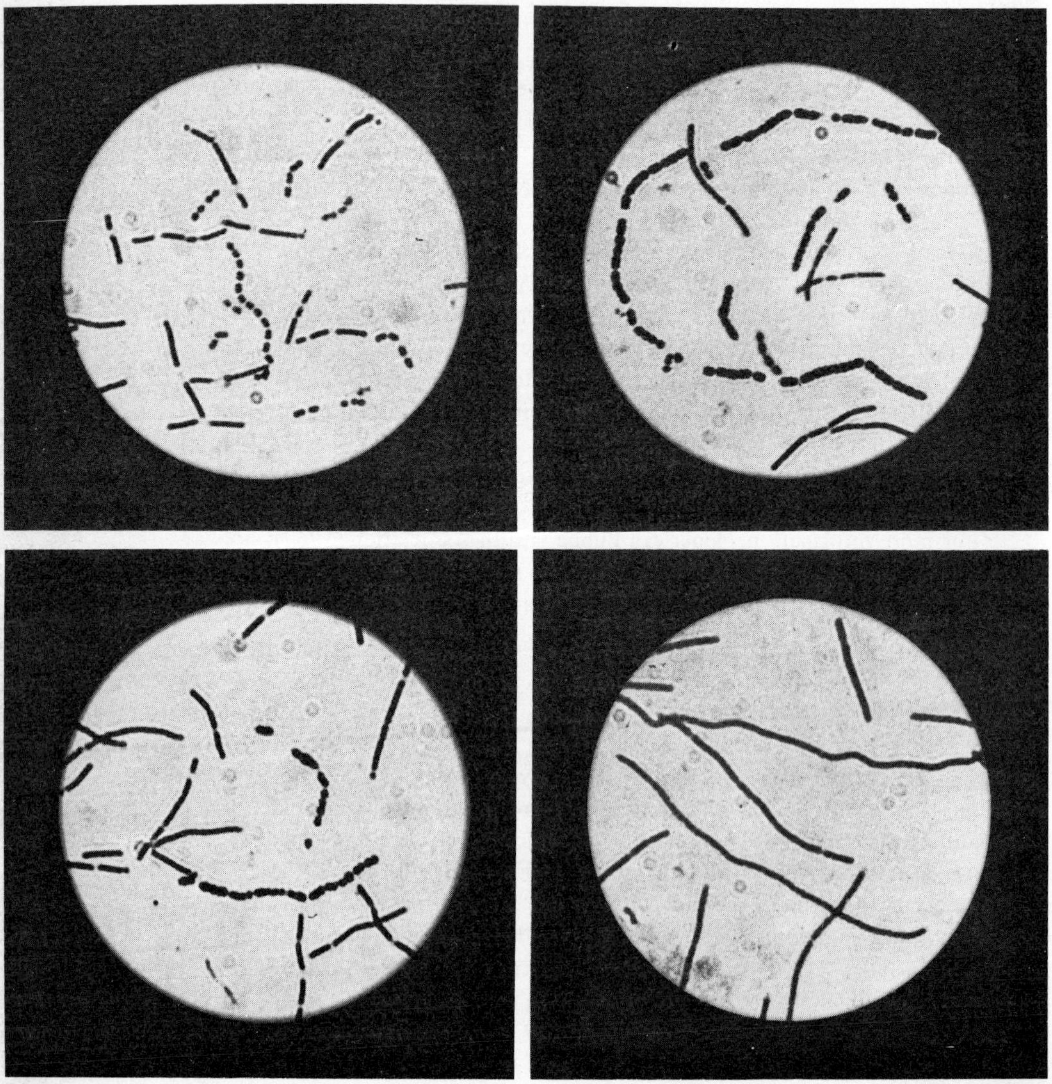

Fellowes, O. N.: Freeze-drying of foot-and-mouth disease virus and storage stability of the infectivity of dried virus at 4°C, *Appl. Microbiol.*, **13**:496, 1965.

——: Inactivation of foot-and-mouth disease virus by interaction of dye and visible light, *Appl. Microbiol.*, **14**:86, 1966.

Franklin, J. G., and M. E. Sharpe: The incidence of bacteria in cheese milk and Cheddar cheese and their association with flavour, *J. Dairy Res.*, **30**:87, 1963.

Gailiunas, P., and G. E. Cottral: Presence and persistence of foot-and-mouth disease virus in bovine skin, *J. Bact.*, **91**:2333, 1966.

—— and ——: Survival of foot-and-mouth disease virus in bovine hides, *Am. J. Vet. Res.*, **28**:1079, 1967.

Galesloot, T. E.: Invloed van penicilline op de morphologie van yoghurt-enzuursel-bacterien, *Ned. Melk Zuiveltijdschr.*, **10**:64, 1956.

—— and H. Labots: Over de bacteriologie van gesteriliseerde melk & melk produkten, *Ned. Melk Zuiveltijdschr.*, **13**:16, 1959*a*.

—— and ——: Thermofiele sporevormers in melk, vooral met betrekking tot de bereiding van gesteriliseerde melk en chocolade-melk, *Ned. Melk Zuiveltijdschr.*, **13**:155, 1959*b*.

Gehrig, R. F., and S. G. Knight: Fatty acid oxidation by spores of *Penicillium roqueforti*, *Appl. Microbiol.*, **11**:166, 1963.

Gelpi, A. J., Jr.: Psychrophilic bacteria causing rancidity and related off-flavors in pasteurized milk, *Southern Dairy Prod. J.*, p. 40, April 1963.

Gilman, H. L., and J. C. Marquardt: The occurrence and survival of *Brucella abortus* in Italian cheese curd made from raw and pasteurized milk, *J. Milk Food Technol.*, **14**:55, 1951.

Glawischnig, E., et al.: Die staphylokokkenmastitiden in Österreich, *Wien. Tieraerztl. Monatsschr.*, **12**:792, 1968.

Goepfert, J. M., et al.: Behavior of *Salmonella typhimurium* during manufacture and curing of Cheddar cheese, *Appl. Microbiol.*, **16**:862, 1968.

Hargrove, R. E., and R. D. Plowman: Indirect detection of antibiotics in milk, *Proc. XV Intern. Dairy Cong.*, **3**:1411, 1959.

Harvey, R. J., and E. B. Collins: Role of citritase in acetoin formation by *Streptococcus diacetilactis* and *Leuconostoc citrovorum*, *J. Bact.*, **82**:954, 1961.

Harwalkar, V. R., and E. W. Seitz: Production of bitter flavor components by lactic cultures, *J. Dairy Sci.*, **54**:12, 1971.

Higginbottom, C., and M. M. Taylor: The growth of aerobic spore-forming bacilli in sterilized milk, *J. Dairy Res.*, **27**:221, 1960.

Huhtanen, C. N.: Incubation temperatures and raw milk bacterial counts, *J. Milk Food Technol.*, **31**:154, 1968.

—— and W. L. Williams: Factors which increase acid production in milk by lactobacilli, *Appl. Microbiol.*, **11**:20, 1963.

Hull, T. G., and L. F. Rettger: The influence of milk and carbohydrate feeding on the intestinal flora of white rats, *Zentr. Bakteriol., Abt. 1, Origin.*, **75**:219, 1915.

Jones, A. M.: *Escherichia coli* in retail samples of milk and their resistance to antibiotics, *The Lancet*, August 14, 1971, p. 347.

Jones, D. L.: Ropy milk: the biology of rope production, *Food Res.*, **19**:246, 1954*a*; II. The influence of physicochemical environment on the formation of ropy substance, *ibid.*, **19**:250, 1954*b*; III. Chemical studies of ropy substance, *ibid.*, **19**:254, 1954*c*.

Kästli, P.: Das verhalten von staphylokokken im kuheuter, *Pathol. Microbiol.*, **30**:849, 1967.

Keenan, T. W.: Production of acetic acid and other volatile compounds by *Leuconostoc citrovorum* and *Leuconostoc dextranicum*, *Appl. Microbiol.*, **16**:1881, 1968.

—— and D. D. Bills: Metabolism of volatile compounds by lactic starter culture microorganisms. A review, *J. Dairy Sci.*, **51**:1561, 1968.

—— et al.: Metabolism of volatile compounds of *Pediococcus cerevisiae* and their occurrence in Cheddar cheese, *J. Dairy Sci.*, **51**:1737, 1968.

Kehoe, J. M., et al.: Studies of bovine mycoplasma mastitis, *J. Infect. Diseases*, **117**:171, 1967.

Kroger, D., and D. E. Jasper: Relationships between Wisconsin mastitis test scores and cell counts in milk, *J. Dairy Sci.*, **50**:1226, 1967.

La Rivière, J. W. M.: Ecology of yeasts in the kefir grain, *Antonie van Leeuwenhoek, J. Microbiol. Serol.*, **35**:D15, 1969.

Lee, S. Y., et al.: Diacetyl production by *Propionibacterium shermanii* in milk cultures, *Can. J. Microbiol.*, **16**:1231, 1970.

Lubert, D. J., and W. C. Frazier: Microbiology of the surface ripening of brick cheese, *J. Dairy Sci.*, **38**:981, 1955.

Luedecke, L. O., et al.: Effect of freezing and storage at −20 C on survival of mastitis pathogens, *J. Dairy Sci.*, **55**:417, 1972.

Mabbitt, L. A.: Reviews of the progress of dairy science, *J. Dairy Res.*, **28**:303, 1961.

Mantere-Alhonen, S.: Über mikrobakterien-arten in zweimal pasteurisierter milch, *Milchwissenschaft*, **24**:488, 1969.

Manthei, C. A.: Application of research to bovine brucellosis control and eradication programs, *J. Dairy Sci.*, **51**:1115, 1968.

Marth, E. H.: Microbiological and chemical aspects of Cheddar cheese ripening. A review, *J. Dairy Sci.*, **46**:869, 1963.

——: Salmonellae and salmonellosis associated with milk and milk products: A review, *J. Dairy Sci.*, **52**:283, 1969.

—— et al.: Ropiness in milk caused by a strain of *Escherichia* intermedia, *J. Dairy Sci.*, **47**:1265, 1964.

McDonald, I. J.: Filamentous forms of *Streptococcus cremoris* and *Streptococcus lactis*. Observations on structure and susceptibility to lysis, *Can. J. Microbiol.*, **17**:897, 1971.

McDonald, T. J., et al.: Aerobic gram-negative rods isolated from bovine udder infections, *Am. J. Vet. Res.*, **31**:1937, 1970.

McDonough, F. E., and R. E. Hargrove: Heat resistance of *Salmonella* in dried milk, *J. Dairy Sci.*, **51**:1587, 1968.

——— et al.: The fate of salmonellae in the manufacture of cottage cheese, *J. Milk Food Technol.*, **30**:354, 1967.

McKenzie, H. A.: "Milk Proteins. Chemistry and Molecular Biology," vols. I and II, New York, Academic Press, Inc., 1971.

Mitchell, R. G., and W. D. Schultze: Mastitis, *West Va. Univ. Agr. Exp. Sta. Bull.*, no. 467, 1962.

Moats, W. A.: Chemical changes in bacteria heated in milk as related to loss of stainability, *J. Dairy Sci.*, **44**:1431, 1961.

Morgan, M. E.: Microbial flavor defects in dairy products and methods for their simulation. I. Malty flavor, *J. Dairy Sci.*, **53**:270, 1970*a*; II. Fruity flavor, *ibid.*, **53**:273, 1970*b*.

Morita, R. Y.: The basic nature of marine psychrophilic bacteria, *Bull. Misaki Marine Biol. Inst. Kyoto Univ.*, February 1968.

Morris, C. S.: Presence in raw cow's milk of a bactericidal substance specific for certain strains of coliform organisms, *Nature*, **155**:22, 1945.

Nilsson, G.: Reducing properties of normal and abnormal milk and their importance in bacteriological grading of milk, *Bact. Rev.*, **23**:41, 1959.

Nurmi, E. V., and L. Koiranen: Bacteriological review of milk samples from clinical mastitis, *Nord. Veterinarmed.*, **19**:36, 1967.

Olson, J. C., et al.: Enterotoxigenicity of *Staphylococcus aureus* cultures isolated from acute cases of bovine mastitis, *Appl. Microbiol.*, **20**:605, 1970.

Overby, A. J.: Elimination of leocillin with dihydrostreptomycin after intramammary treatment, *Milchwissenschaft*, **22**:20, 1967.

Overcast, W. W., and J. D. Skean: Growth of certain lipolytic microorganisms at 4°C and their influence on free fat acidity and flavor of pasteurized milk, *J. Dairy Sci.*, **42**:1479, 1959.

Palmer, J. M. A., and F. V. Kosikowski: Simple ultrasensitive test for detecting penicillin in milk, *J. Dairy Sci.*, **50**:1390, 1967.

Panes, J. J., and S. B. Thomas: The multiplication of coli-aerogenes bacteria in milk stored at 3-5°C, *J. Appl. Bact.*, **22**:272, 1959.

Patton, S.: Milk, *Sci. Am.*, **221**:58, 1969.

Peebles, M. M., et al.: Preparation of concentrated lactic *Streptococcus* starters, *Appl. Microbiol.*, **17**:805, 1969.

Perry, K. D.: A comparison of the influence of *Streptococcus lactis* and *Str. cremoris* starters in the flavour of Cheddar cheese, *J. Dairy Res.*, **28**:221, 1961.

Pietermaat, F., and W. Van Dyck: La pasteurisation du lait à haute fréquence, *Neth. Milk Dairy J.*, **13**:45, 1959.

Raghavan, R., et al.: Bovine mastitis due to *Pseudomonas pyocyanea*, *Mysore J. Agr. Sci.*, **2**:214, 1968.

Randolph, H. E., and I. A. Gould: Effect of the inherent properties of milk on production of acid by selected lactic cultures, *J. Dairy Sci.*, **49**:254, 1966.

Read, R. B., Jr., et al.: Studies on thermal destruction of *Escherichia coli* in milk and milk products, *Appl. Microbiol.*, **9**:415, 1961.

—— et al.: Assay of staphylococcal enterotoxin from cheese, *J. Dairy Sci.*, **48**:420, 1965.

Reiter, B., and A. Møller-Madsen: Review of the progress of dairy science, *J. Dairy Res.*, **30**:419, 1963.

—— et al.: Factors affecting the multiplication and survival of coagulase positive staphylococci in Cheddar cheese, *J. Dairy Res.*, **31**:261, 1964.

Rosanove, R.: Contamination of milk with penicillin, *Minn. Med.*, **43**:306, 1960.

Schultze, W. D., and J. C. Olson, Jr.: Studies on psychrophilic bacteria. I. Distribution in stored commercial dairy products, *J. Dairy Sci.*, **43**:346, 1960*a*; II. Psychrophilic coliform bacteria in stored commercial dairy products, *ibid.*, **43**:351, 1960*b*.

Seitz, E. W., et al.: A pigment-producing spoilage bacterium responsible for violet discoloration of refrigerated market milk and cream, *Appl. Microbiol.*, **9**:287, 1961.

Sellers, R. F.: The inactivation of foot-and-mouth disease virus by chemicals and disinfectants, *Vet. Record*, **83**:504, 1968.

Sheldon, R. M., et al.: Chemical nature of malty flavor and aroma produced by *Streptococcus lactis* var. *maltigenes*, *Appl. Microbiol.*, **22**:263, 1971.

Smith, H. W.: The examination of milk for the presence of *Mycobacterium johnei*, *J. Pathol. Bact.*, **80**:440, 1960.

Smith, L. P., and M. E. Hugh-Jones: The weather factor in foot-and-mouth disease epidemics, *Nature*, **223**:712, 1969.

Snyder, W. W., et al.: Residues in milk, blood, and urine resulting from various types of antibiotic administration, *Quart. Bull. Mich. Agr. Exp. Sta.*, **43**:1, 1961.

Speckman, R. A., and E. B. Collins: Diacetyl biosynthesis in *Streptococcus diacetilactis* and *Leuconostoc citrovorum*, *J. Bact.*, **95**:174, 1968.

Stadhouders, J.: Cooling of raw milk immediately after production as a main factor for controlling the bacterial growth during storage at 4°C, *Neth. Milk Dairy J.*, **22**:173, 1968.

—— et al.: A study of the optimum conditions of freezing and storing concentrated mesophilic starters, *Neth. Milk and Dairy J.*, **25**:229, 1971.

—— et al.: Preservation of starters and mass production of starter bacteria, *Neth. Milk Dairy J.*, **23**:182, 1969.

Sternberg, M. Z.: Crystalline milk-clotting protease from *Mucor miehei*, *J. Dairy Sci.*, **54**:159, 1971.

Thatcher, F. S., and D. Ross: Multiplication of staphylococci during cheese making, *Can. J. Public Health*, **51**:226, 1960.

Thomas, S. B.: Coli-aerogenes bacteria in raw milk, *J. Appl. Bact.*, **18**:331, 1955.

Thomas, S. B.: Psychrophilic bacteria in refrigerated raw milk, *Dairy Eng.*, **77**:5, 1960.

—— et al.: Incidence and significance of thermoduric bacteria in farm milk supplies: A reappraisal and review, *J. Appl. Bact.*, **30**:265, 1967.

Thomas, W. R., et al.: Effect of temperature and time of plate incubation on the enumeration of pasteurization-resistant bacteria in milk, *J. Milk Food Technol.*, **26**:357, 1963.

U.S. Public Health Service: "Grade A Pasteurized Milk Ordinance," pub. no. 229, U.S. Department of Health, Education, and Welfare, 1967.

Vedamuthu, E. R., et al.: Flavor and texture in Cheddar cheese. I. Role of mixed strain lactic starter cultures, *J. Dairy Sci.*, **49**:144, 1966*a*; II. Carbonyl compounds produced by mixed strain lactic starter cultures, *ibid.*, **49**:151, 1966*b*.

Welch, H., et al.: Antibiotics in fluid market milk, *Antibiot. Chemotherapy*, **6**:369, 1956.

Wild, T. F., et al.: Surface structure of foot-and-mouth disease virus, *J. Gen. Virol.*, **4**:313, 1969.

Witter, L. D.: Psychrophilic bacteria: A review, *J. Dairy Sci.*, **44**:983, 1961.

Yano, N., et al.: Mesophilic bacilli in raw milk, *Bull. Natl. Inst. Animal Ind.*, **17**:21, 1968.

Yotis, W., and R. Teodoro: The influence of temperature on the generation time of bacteria commonly found in milk, *J. Dairy Res.*, **24**:27, 1957.

Bacteriology of food 23

Spoilage involves changes which render foods unattractive, unsalable, and unwholesome. According to Oser (1946), spoiled food may be non-injurious even though unfit for human consumption. The stock terms used to describe spoiled foods are filthy, putrid, and decomposed. All such foods are not necessarily inedible. For example, Limburger cheese owes its popularity to the process of putrefaction by which it is made. Soy sauce likewise is manufactured by a process of enzymatic decomposition. Yet these are neither inedible nor uneaten despite these characteristics. In many instances selected microorganisms are inoculated into food products in order to produce certain specific types of decomposition.

The causative factors involved in food spoilage are physical, chemical, and biological. These rarely proceed as single processes but usually go on simultaneously or in sequence so that any case of spoilage may be of an extremely complex nature.

Spoilage by radiation One of the most important physical factors responsible for food spoilage is radiation. This may involve visible, ultraviolet, or infrared rays.

Visible light may cause discoloration or undesirable flavors in foods. Many liquid foods are protected from light by being placed in dark bottles. Ultraviolet light is effective in destroying bacteria, yeasts, and molds but, in so doing, may induce other types of spoilage. For example, fats may be oxidized to rancid-smelling compounds, off-colors may be imparted to foods, and milk may lose much of its riboflavin content.

Infrared rays produce an increase in the temperature. Excessive heat may result in dehydration of foods; loss of volatile constituents; alteration of proteins; and changes in weight, volume, texture, and general appearance.

Spoilage by pressure Pressure is often responsible for food spoilage. Fruits and vegetables may become unsalable by crushing, pressing, or bruising. Even though damage is not complete, loss of weight and nutrients may occur because of expression of liquids.

Spoilage by freezing Treatment of foods by freezing is a most effective method of preservation. The effect is to retard physical, chemical, and biological changes. Unless the process is carefully controlled, undesirable changes in flavor, texture, and keeping qualities may occur.

In slow freezing, large ice crystals may form which cause a disruption of cell walls. On the other hand, quick freezing produces very small ice crystals which do not cause a disintegration of cell walls. Foods so treated may be satisfactorily preserved without spoilage.

Spoilage by enzymes Enzymes continue to function even though fruits and vegetables are picked and animals are slaughtered. Unless the enzymes are destroyed, they continue to act during processing and storage of foods, with the result that considerable spoilage may occur. Oxidizing enzymes catalyze the destruction of ascorbic acid (vitamin C) and may produce a deterioration in flavor. Proteolytic enzymes produce autolysis and putrefaction of meats. Amylolytic enzymes hydrolyze carbohydrates to smaller units. Lipolytic enzymes produce a breakdown of fats to glycerol and fatty acids. The fatty acids may be oxidized to peroxides and aldehydes, which are associated with rancidity.

Proper pH control is of considerable importance in the processing of foods. Some natural fruit and vegetable pigments are indicators, and their colors are dependent upon the reactions of the products. Proteins, carbohydrates, and fats may undergo undesirable changes if the reactions are unfavorable. Vitamins are particularly sensitive to change in pH. Vitamin A is stable in alkaline media; pantothenic acid is stable only in neutral environments; thiamine and ascorbic acid are stable only in acid media. Because of their different characteristics, it is difficult or impossible to preserve all vitamins in a single product.

Spoilage by organisms Of all agents involved in the spoilage of foods, the activities of living organisms are undoubtedly the most important. Spoilage is caused principally by bacteria, yeasts, and molds. The organisms may be pathogenic or nonpathogenic; spore-forming or non-spore-forming; psychrophilic, mesophilic, or thermophilic; aerobic or anaerobic; each requiring special means of prevention and control.

Different kinds of organisms produce different types of changes in food. The decomposition of foods rich in carbohydrates results usually in various types of fermentations. The action of organisms on high-protein foods results in putrefactions. The products of the former are usually harmless, whereas those of the latter are objectionable and even dangerous.

Bacteria are more exacting in their requirements than either the yeasts or the molds. This means that yeasts and molds can multiply under conditions unfavorable to the growth of bacteria. Bacteria require relatively large amounts of moisture, hydrogen-ion concentrations usually near the neutral point, and relatively low osmotic pressures for growth and multiplication. Yeasts can tolerate less moisture, are less exacting in their pH requirements, and can multiply in solutions having higher osmotic pressures. Molds are the least exacting of the fungi. They can withstand relatively high acidities, require far less moisture, even grow on substances which are almost dry, and can tolerate extremely high osmotic pressures.

Methods employed for the preservation of food

Various methods have been employed for the preservation of food. Modifications in the methods were made from time to time as their modes of action became known. The methods commonly employed involve the use of (1) heat, (2) cold, (3) dehydration, (4) preservatives, (5) high osmotic pressures, (6) antibiotics, and (7) radiations.

HEAT

The use of heat is the method commonly employed in home and commercial treatment of fruits, vegetables, meats, and fishes. Heat is used to effect either a complete sterilization or a reduction in the numbers of organisms present. In the latter case, the organisms which are not killed are prevented from multiplying. Heat is effective in destroying all forms of life. It is desirable not to greatly exceed the minimum temperature required to sterilize a product; otherwise alterations may occur in its appearance, flavor, and composition. Since not necessarily all organisms are killed, the term "processing" is generally used in referring to heat-treated foods.

The diverse details of canning procedures necessarily vary with the nature of the product to be preserved. However, certain important operations common to all procedures are (1) cleansing, (2) blanching, (3) exhausting or preheating, (4) sealing the container, (5) heat-processing the container, and (6) cooling the container after thermal processing.

Cleansing The first and one of the most important steps is thorough cleansing of the food materials. Cleansing serves two purposes: (1) it makes a better-looking product and (2) it reduces substantially the load of spoilage organisms that may place a heavy burden on the heat process.

Cleansing may be effected by various types of washers. The raw materials are subjected to high-pressure sprays or strong-flowing streams of water while passing along a moving belt or while being dropped in agitating or revolving screens. With certain food materials, dirt and other large adhering particles are mechanically removed by means of revolving or agitating screens or by strong blasts of air.

Blanching The blanch involves the immersion of raw food materials (fruits and vegetables) into warm or hot water, or exposure to live steam. This is practiced for several reasons. Bianching may serve only as a hot-water wash where adhering materials cannot be removed with cold water. It may soften fibrous plant tissue so that it will either contract (lose water) or expand (take up water). This ensures proper filling of the container. During the blanching operation, respiratory gases are expelled.

This prevents strain on the can during processing and favors the development of a higher vacuum in the finished product. Blanching inhibits the action of respiratory enzymes, especially those of oxidation, to give a product of superior quality and nutritive value. Lastly, blanching fixes the natural color of certain products and makes them more attractive in appearance.

Exhausting or preheating All canning procedures provide for the exclusion of most of the oxygen (air) for two reasons: (1) it may react with the food materials and the interior of the container to affect the quality and nutritive value of the contents; and (2) it may cause undue strain on the container during the processing operation.

The procedure followed in removing gases consists of passing the open can, containing the raw food, through an exhaust box in which hot water or steam is used to expand the food and expel air and other gases from the contents and the head-space area of the can. After the gases are expelled, the can is immediately sealed, heat-processed, and cooled. During the cooling, the contents of the can contract, creating a vacuum. This is accepted as evidence of soundness of the canned product.

With some products, the same effect is produced by preheating the food in kettles, filling into cans while still hot, and immediately sealing the containers. With other products, an exhausting effect is produced by adding boiling water, sirup, or brine to the food in the tin. With still other products, exhausting is accomplished by mechanical means rather than by the use of heat. Special machines are used for withdrawing the air from the cans and sealing at the same time. This process is known as vacuum packing.

Sealing the container Each container must be properly sealed before being subjected to the heating process. The heat destroys the organisms present in the food materials, and the seal prevents recontamination of the contents. The sealing operation is one of the most important steps in the canning procedure.

Heat-processing the container The processing operation usually involves the application of steam under pressure (autoclave). This destroys all contaminating organisms including spores, in one operation, preventing spoilage of the contents in storage.

The time required for processing canned foods depends upon various factors, such as character and composition of the food, types and numbers of organisms likely to be present, and pH of the food. Heat penetrates to the center of containers by conduction and convection. Heat penetration of solid foods takes place by conduction and is relatively slow. Penetration of liquid foods takes place by conduction and convection, and is relatively fast. The size of the food particles also influences the speed of penetration by heat. There is an inverse ratio between size of particles and speed of penetration.

Bacteria are generally more easily killed in an acid or an alkaline environment than in a neutral one. Therefore, fruits and vegetables are more easily processed than meats and fish. Also, heat penetrates the former more readily than the latter. The temperature and time of processing must be determined for each kind of food. In general, non-spore-forming organisms in a liquid food are destroyed at a temperature of 60°C in 1 hr, or at 70 to 80°C in a few minutes. Spores are more resistant, requiring temperatures of 115°C for 30 min or 120°C for 15 min.

In the processing of foods, an excessive period of heating is avoided to prevent injury to the product. A long exposure at a relatively low temperature is usually preferable to a shorter exposure at a higher temperature. This applies especially to canned fruits.

Cooling the container The last operation in the commercial process involves rapid cooling of the sealed containers. This is necessary to check the action of the heat and prevent undue softening or change in color of the contents. The containers may be cooled by air or water.

Air cooling is accomplished in well-ventilated, specially designed storage rooms where the containers are stacked in rows with ample space for efficient circulation of air.

Water cooling is accomplished by allowing water to run into the autoclave in which the containers are processed, or the containers may be removed from the sterilizer and conveyed through tanks of cold water or through cold-water showers.

For more information: Angelotti et al. (1961); Ball (1947); Goldblith et al. (1961); Hill et al. (1967); Simonsen (1968); Sognefest et al. (1948).

COLD

Two methods are employed in the preservation of foods by cold temperatures: (1) chilling and (2) freezing.

In the chilling method, the temperature is kept just above the freezing point. This is the condition encountered in the ice or electric refrigerator in the home. The physical state of the food is unaltered. Chilling retards but does not prevent multiplication of all organisms. Therefore, chilled foods cannot be kept for many weeks.

Unless foods are completely frozen, spoilage is likely to occur. Some organisms grow slowly in the presence of free water, and their enzymes continue to function even though the cells are dead.

The freezing method is considered the simplest, safest, and sanest method for the preservation of food, provided the following five points are considered: (1) careful selection, (2) proper packaging, (3) freezing at −18°C or lower, (4) storage at −18°C with a minimum fluctuation, and (5) avoidance of too long storage.

Careful selection is the first step to successful preservation. Blemishes and decayed spots are preserved with the good unless eliminated before freezing. Any mold or slime on meat must be carefully trimmed before processing. Freezing will not overcome a product that is damaged from the start.

A number of high-class packaging containers and wrappers have been developed and should be used for proper food preservation. The characteristics of a good packaging material are the following: (1) clean and sanitary; (2) odorless, flavorless, and impervious to odors; (3) mechanically practical; (4) attractive in appearance; (5) protective against desiccation; (6) resistant to oxidation; (7) tough and not brittle at low temperatures; (8) greaseproof and stainproof; and (9) resistant to moisture, vapor, and oxygen (air).

Rapid freezing gives the best results in food preservation. Slow freezing refers to temperatures above $-18^\circ C$. Sharp freezing applies to temperatures between -18 and $-29^\circ C$. Quick freezing refers to still lower temperatures. Foods should be frozen rapidly and at temperatures below $-18^\circ C$. Packages should be completely frozen before packing tightly in a storage cabinet.

The vitamin content of foods is rapidly lost at temperatures above $-18^\circ C$. Also, rancidity in foods develops more rapidly at higher temperatures. After foods are frozen, the storage temperature should be $-18^\circ C$ with a minimum fluctuation. Fluctuation promotes dehydration. Snow and ice on the inside of a package usually indicate a fluctuation in the temperature in the storage room.

Bacterial growth can occur on meat held at a temperature of $-3^\circ C$, and yeasts and molds can multiply at $-7^\circ C$. Bacteria die more rapidly at -1 to $-5^\circ C$ than at lower temperatures. Flocculation of bacterial protein is more pronounced at temperatures slightly below the freezing point.

The bacteria chiefly responsible for growth and spoilage at low temperatures are the psychrophilic forms (see page 730). These organisms can multiply at defrost temperatures near $0^\circ C$. They liberate proteolytic and other enzymes at temperatures which are considered adequate for prevention of microbial growth. Enzymes are responsible for off-flavors and odors in advance of the development of large numbers of psychrophiles. Enzymatic activity can be slowed by decreasing the temperature but cannot be eliminated entirely.

Contrary to popular belief, frozen foods cannot be kept indefinitely. Foods slowly deteriorate in the frozen state. It is best to plan on a normal turnover with the season. As examples, beef and lamb may be safely preserved for 1 year; fresh pork should not be stored over 9 months; and ground meats may be safely kept for 6 months.

A large number of pathogenic and nonpathogenic species have been isolated from frozen foods. These include: *Achromobacter* species, *Bacillus* species, *Escherichia coli* and other coliforms, *Pseudomonas* species, *Salmonella* species and serotypes, *Shigella dysenteriae, Staphylococcus aureus, Streptococcus faecalis* and *S. faecalis* var. *liquefaciens,* etc.

Microbial survival and metabolism at low temperatures are influenced by many factors. Some species survive best at a temperature of about

−25°C; others are best preserved at temperatures around 10°C; still others require optimum temperatures near 50°C. As would be expected, the temperatures vary for the same organism in different foods. There is no rule that can be followed.

For more information: Brady et al. (1949); Elliott and Michener (1961, 1965); Georgala and Hurst (1963); Mackintosh et al. (1949); Michener et al. (1960); Ott et al. (1961); Peterson and Gunderson (1960); Raj and Liston (1961); Splittstoesser et al. (1961*a, b*); Surkiewicz (1966); Yates and Aref (1969).

DEHYDRATION

The preservation of foods by dehydration is of ancient origin. The method is now of great industrial importance. Practically every type of food is prepared in dehydrated form, including nuts, vegetables, fruits, meats, fish, eggs, milk, and soups.

Dehydrated foods are probably never sterile. In this respect they differ from most canned foods. Therefore, it is of the utmost importance to prevent entrance of organisms which are capable of producing food poisoning into such foods. This applies especially to the toxigenic forms of *Clostridium botulinum* and certain strains of *Staphylococcus aureus*. Dehydrated foods should also be free from certain intestinal forms likely to be pathogenic to man when taken by mouth, e.g., members of the genera *Salmonella* and *Shigella*. The bacterial count of dried foods should be reasonably low to avoid decomposition or development of undesirable flavors during the period of reconstitution.

To prevent the formation of bacterial toxins or the development of organisms pathogenic to man, the food product should be dried at a temperature where significant growth is unlikely to occur. A temperature of about 50°C is probably the minimum. Where some heating below this temperature is unavoidable, because of loss of quality in the product, the period of dehydration should not exceed 4 hr. Bacterial growth generally does not occur in foods containing less than 15 percent water.

Dehydration of foods is a valuable procedure for several important reasons. Dried foods may be easily preserved for future use. This means that certain foods may be utilized over longer periods of time rather than for only a short season of the year. Dehydration greatly reduces the bulk of a product, conserves space, and facilitates handling. This is a decided advantage from the standpoint of transportation costs. Most of the dehydrated products, if properly prepared, are very good substitutes for fresh foods, being detected from the normal product with difficulty. Dehydrated foods do not require sterilization or the maintenance of sterile conditions during preparation. Since no waste is involved, they are more economical to use. Only that amount necessary for use each time need be prepared.

The use of dehydrated foods also presents several decided disadvantages. Dried products require a long soaking period to restore the water lost by evaporation. The period required for rehydration varies with different foods. If this is not carefully done, the results are likely to be unsatisfactory. Sometimes dehydrated foods become infested with insects owing to improper packaging or handling. Sometimes dried foods become moistened, creating favorable conditions for the growth of bacteria, yeasts, and molds. This applies especially to the hygroscopic foods, i.e., those which readily absorb moisture from the air.

PRESERVATIVES

Some foods are preserved by the addition of chemicals. These act either by killing the organisms or preventing them from multiplying.

An ideal antiseptic is one that affects microorganisms adversely without producing any harmful physiological effect on the consumer. Apparently such a compound is not yet known. All the commonly used preservatives exert some physiological action on the human body and, unless employed in minute amounts, may produce harmful effects.

The inorganic chemicals commonly employed include boric acid and borates, nitric acid and nitrates, nitrous acid and nitrites, sulfurous acid and sulfites.

Boric acid is a weak antiseptic, saturated solutions being unable to kill bacteria. However, it does prevent the growth of most bacteria and has been used to preserve butter.

The color of fresh, unheated muscle tissue is due to the presence of a red pigment known as myohemoglobin or myoglobin. This pigment is an integral part of tissue that does not circulate in the blood.

Sodium nitrate and sodium nitrite are usually added to brines for the pickling of meats. Usually some bacterial species on meat are able to reduce the nitrate to nitrite. In the event the nitrate is not reduced, a small amount of nitrite is added as a precautionary measure. In an acid brine the nitrite is changed to nitrous acid which reacts with the myohemoglobin to give nitric oxide myohemoglobin, according to the reaction:

$$\text{NO} + \text{myohemoglobin} \longrightarrow \text{nitric oxide myohemoglobin}$$

This gives the cured meat a bright red color, making it more attractive in appearance. The nitrite plays no part in producing the cured flavor. Nitric oxide myohemoglobin, when heated to temperatures which coagulate proteins, becomes converted to nitric oxide myochromogen. This compound is red also, and is one of the objectives gained in curing meats. In addition to the development of color, nitrate and nitrite produce an inhibitory effect on bacterial growth, nitrite being more effective in this respect than nitrate.

Sulfurous acid and sulfites are added to wines as preservatives. The addition of sulfite to discolored meat restores the original red color to the product. It enjoyed great popularity as an addition to hamburger prepared from old scrap meat, but its use for this purpose is now prohibited by law.

The organic chemicals added to foods include benzoic acid and benzoates, salicylic acid and salicylates, formaldehyde, creosote, and sorbic acid.

Benzoic acid and benzoates are used for the preservation of vegetables. A small amount of sodium benzoate is sometimes added to tomato catsup. Salicylic acid and salicylates are used as preservatives of fruits and vegetables. Formaldehyde was formerly used as a preservative of milk, but its use for this purpose is forbidden by law. The value of wood smoke in the curing of meats is due to the presence of a small amount of creosote furnished by the burning wood. Sorbic acid is an effective inhibitor of gaseous fermentation by yeasts, which are responsible for large economic losses in brine-cured cucumbers due to bloater or hollow cucumber formation. It inhibits growth of yeasts, molds, and some bacteria but does not interfere with the activities of the organisms involved in the lactic fermentation.

For more information: Costilow (1957); Costilow et al. (1957); Etchells et al. (1961, 1968); Johnson et al. (1969); Mahoney (1961); McBean (1967); Patterson (1963); Reith and Szakály (1967*a, b*); Robinson and Hills (1959).

HIGH OSMOTIC PRESSURES

Some foods are protected from microbial attack by the presence of high concentrations of salt or sugar. These compounds act largely by osmosis or the withdrawal of water from the cell. This causes shrinkage of the protoplasmic contents followed by death of the cell.

The preserving action of common salt may involve more than its ability to dehydrate. Magnesium sulfate has a greater dehydrating action on proteins than common salt, yet it is less efficient in preventing growth of *Staphylococcus aureus.*

Bacteria capable of resisting high salt concentrations are called halophilic (salt-loving) organisms. With the exception of the halophiles, practically no multiplication of organisms occurs in salt concentrations of 25 percent. A strength of 10 percent markedly inhibits the growth of the great majority of bacterial species. The pathogenic or disease-producing bacteria are less resistant to strong saline solutions than the saprophytic forms.

Cane sugar in a concentration of 60 to 70 percent usually prevents growth of all types of microorganisms. Occasionally molds may be seen growing on the surface of a closed jar of jelly or bottle of syrup. This is caused by the evaporation of water which, not being able to escape, condenses back on the surface of the jelly or syrup to produce a layer of less concentrated sugar solution. Some molds can multiply in this weak sugar solution. If the surface is kept dry, growth would not occur.

Bacteria, yeasts, and molds are not as sensitive to osmotic changes as higher plant and animal cells. For this reason, solutions having extremely high osmotic pressures must be used either to kill them or prevent their multiplication.

For more information: Gochnauer and Kushner (1969).

ANTIBIOTICS

Food preservation in canning is achieved by killing the microbial contaminants with heat. The extent of the heating depends largely upon the pH of the product. Acid foods having a pH below 4.5 are conventionally processed in boiling water under atmospheric pressure. The heat-resistant spores likely to be present are, in general, destroyed during the heat treatment or, if not killed, do not germinate under such strongly acid conditions.

However, with meats, vegetables, and other low-acid foods, the problem is different. Foods having a pH above 4.5 are conventionally processed under high steam pressures and temperatures. Even then the procedure is not always successful. Spores associated with some foods are among the most heat-resistant bodies known and make absolute sterility almost impossible to achieve without impairing the quality of the product or without rendering it unfit for human consumption.

Because of the limitations of the thermal processing procedure, antibiotics capable of destroying or lowering the thermal resistance of spores have been incorporated in nonacid and low-acid foods to eliminate the need for excessive heat treatment. The antibiotics are destroyed by the heat during the cooking stage. The most commonly used antibiotics for this purpose include chlortetracycline (aureomycin), nisin, nystatin, oxytetracycline (terramycin), subtilin, and tylosin. Of these, chlortetracycline (CTC) has been employed more successfully than any of the others.

Poultry Chlortetracycline is extensively used for the preservation of poultry. It may be applied by (1) dipping freshly dressed and eviscerated birds in a chill tank containing ice and water plus 20 ppm of CTC, (2) feeding a mixture containing 1000 g CTC per ton of feed for several days prior to slaughtering, or (3) adding 500 g CTC per ton of drinking water for one day prior to slaughtering. The birds are stored at 3°C or lower. By such treatment, the shelf life of poultry is markedly increased over the controls.

Meat Antibiotics may be applied to the surface of meat as with poultry, or they may be injected into the animals shortly before slaughtering. The latter procedure permits circulation of the antibiotics to all tissues of the animals, affording better preservation. The meat may be stored at chilling temperatures or in cold storage.

Fish For optimum effectiveness in retarding spoilage, antibiotics must be applied to fish as soon as possible after capture. CTC has been used more than any other antibiotic for this purpose. It may be incorporated in the ice that is used for icing fish, or in chilled seawater in which they are to be transported. By such treatment, the keeping quality of commercial eviscerated fish is considerably improved.

For more information: Abbey et al. (1960); Buttiaux and Catsaras

(1967); Cory and Byrnes (1963); Deatherage (1962); Elliott (1968); Hawley (1962); Hobbs et al. (1960); Novak et al. (1960); Parks et al. (1960); Phillips et al. (1961); Segmiller et al. (1965); Southcott and Boyd (1965); Tarr (1959); Thatcher and Loit (1961); Thomson et al. (1967); Thornley et al. (1960); Volkova (1961); Walker et al. (1961); Wells et al. (1963); Wheaton and Hays (1964); Yuzawa (1968).

RADIATIONS

Numerous and extensive investigations have been conducted on the preservation of foods by various ionizing radiations. These include alpha, beta, and gamma rays; ultraviolet light; protons; and neutrons. Gamma rays have been used more than any of the others for the sterilization of foods. They have good penetration, being effective to a depth of about 15 cm in most foods. The depth is dependent upon the period of exposure.

The radiations function by causing ionization of molecules of the absorbing materials and destroy contaminating organisms without producing an appreciable rise in temperature. Destruction of organisms by ionizing radiations is sometimes called cold sterilization.

Experimental irradiations have been conducted at various facilities such as the (1) Materials Testing Reactor Gamma Irradiation Facilities, Idaho Falls, Idaho; (2) Atomic Energy of Canada, Limited; (3) High Level Gamma Irradiation Facility, Argonne National Laboratory, Lemont, Illinois; (4) Dugway Proving Grounds of the Army, Dugway, Utah; (5) Midwest Irradiation Center, Rockford, Illinois; and (6) General Electric Company, Milwaukee, Wisconsin.

Since the use of ionizing radiations for the preservation of foods is still in the experimental stage, the reader is referred to the following publications for a clearer understanding of the subject: American Chemical Society (1957); Anellis and Werkowski (1968); Anellis et al. (1965, 1967); Beraha et al. (1961); Brown et al. (1960); Comer et al. (1963); Drake et al. (1960); Duggan et al. (1963*a, b*); Dyer et al. (1966); Erdman et al. (1961*a, b*); Ijichi et al. (1964); Ingram and Thornley (1961); Kazanas and Emerson (1968); Kempe and Graikoski (1964); Lerke and Farber (1960); Miyauchi (1960); Ostovar et al. (1967); Solberg and Riha (1969); Stavrić et al. (1968*a, b*); Watanabe (1960).

Microbiology of meat

The presence of living organisms in the tissues and blood of healthy animals is still a controversial issue. Some believe viable organisms may be isolated from living tissues; others are of the opinion that their presence is agonal and post mortem rather than ante mortem.

Autolysis of tissues Frozen or cold-storage meats can be kept for long periods without showing any signs of spoilage. On the other hand, meats kept at higher temperatures (chilled) show spoilage in much shorter periods. The changes result from the action of enzymes normally present in meat and those elaborated by the contaminating organisms. Proteins are first hydrolyzed to amino acids, then putrefied with the liberation of bad odors. A short action of the proteolytic enzymes is beneficial in tenderizing meats; a prolonged action results in excessive decomposition accompanied by putrefaction.

Ripening or aging of meats This refers to the practice of holding meat at a temperature of 1 to 3°C to increase the tenderness and improve the flavor. Good grades of beef may be ripened for 14 days; lamb for 7 to 8 days but not longer. Ripening is accomplished by the action of the proteolytic enzymes present in the tissues and in the contaminating organisms.

Bacteria growing at 1 to 3°C are chiefly the psychrophilic forms, i.e., those which like cold temperatures. The psychrophiles are found chiefly in the genera *Achromobacter, Aerobacter, Aeromonas, Alcaligenes, Chromobacterium, Escherichia, Flavobacterium, Micrococcus,* and *Pseudomonas.* Members of the last genus are probably the most commonly encountered.

Ordinary beef is held at 2.2 to 3.3°C for 5 days after slaughter before it is released. If the meat is to be ripened to increase tenderness and improve flavor, it is stored at low temperatures. Beef held at the following temperatures and times shows the same degree of tenderness: 21 days at 1.1°C, 8 days at 4.4°C, 5 days at 8.3°C and 3 days at 15.6°C. Rey et al. (1970) recommended a holding temperature of 16°C for a short period, followed by storage at 5°C.

Meats are also tenderized by the use of enzyme preparations. Many proteolytic enzymes are employed, including ficin from figs, bromelin from fresh pineapple juice, papain from papaya, etc. The enzymes are allowed to act until the meat shows the proper degree of tenderness. The meat is then heated to destroy the enzyme; otherwise it tends to become mushy and butyrous in texture during culinary heating. Meat in this condition is organoleptically undesirable.

For more information: Jay (1966); McIntosh (1967).

Sanitary quality of meats The bacterial counts cannot be used in judging the sanitary quality of meats. There appears to be no correlation between bacterial numbers and sanitary quality. It is not necessarily the numbers but the kinds of organisms that determine the quality of the product. Samples showing high counts of saprophytic organisms may produce no harmful effects when ingested. On the other hand, samples showing low counts may produce harmful effects. Apparently the best criteria for judging quality are appearance, feel, and smell. Such an examination is sometimes referred to as the organoleptic test.

Bacterial flora of meats Both aerobic and anaerobic species are concerned in the spoilage of meats. First the aerobes utilize the free oxygen and create conditions favorable to the growth of the anaerobes.

Then both aerobes and anaerobes, growing anaerobically, can attack the proteins and liberate foul-smelling compounds. The process of putrefaction in nature involves the action of both aerobes and anaerobes, but the changes produced are anaerobic in character.

The organisms concerned in meat spoilage may be grouped as (1) gram-positive, aerobic, spore-bearing rods; (2) gram-negative, aerobic, non-spore-forming rods; (3) cocci; (4) anaerobes; and (5) molds and yeasts.

The gram-positive, aerobic, spore-bearing rods include *Bacillus cereus, B. megaterium, B. pumilus, B. subtilis,* etc. These organisms are saprophytes and liquefy gelatin rapidly.

The gram-negative, aerobic, non-spore-forming rods include *Aerobacter cloacae, E. coli, Proteus mirabilis, P. vulgaris, Pseudomonas aeruginosa, P. fluorescens,* and *P. putrefaciens.* The *Proteus* species are strongly proteolytic and putrefactive, resembling the clostridia in their characteristics.

The cocci isolated from meats include *M. candidus, M. caseolyticus, M. conglomeratus, M. cryophilus, M. flavus, M. freudenreichii, M. varians, Sarcina aurantiaca, Staphylococcus aureus,* and *Streptococcus faecalis.* All are gram-positive.

The most pronounced changes in meats are produced by the anaerobic, spore-bearing rods, which include *Clostridium aerofoetidum, C. bifermentans, C. histolyticum, C. lentoputrescens, C. perfringens,* and *C. sporogenes.* These organisms liberate foul-smelling odors in proteinaceous foods.

The molds which have been isolated from meat include species of *Alternaria, Aspergillus, Cladosporium, Monilia, Mucor, Penicillium,* and *Sporotrichium.* Those belonging to the genera *Mucor* and *Penicillium* have been isolated with greater frequency than any of the others. Spores of molds are commonly present in air. Several species of yeasts have been isolated from meat kept under refrigeration.

Molds are generally aerobic, growing on the surface of meat. They may produce pigments which impart discolorations to meat. Molds may be removed by wiping or trimming the surface layer. If molds are allowed to grow without being checked, unpleasant odors and flavors may be developed in the meat.

Dried beef Beef is dried so that it may be preserved over long periods.

The beef is cured by soaking in a brine solution. It must be well cured; otherwise decomposition may occur during drying and smoking. A typical brine has the following composition: sodium chloride, 25 g; sodium nitrate, 0.05 g; sodium nitrite, 0.10 g; sugar (sucrose), 4 g; water, to make 100 g.

After the meat is cured, it is soaked to remove an excessive amount of salt. The soaking period depends upon the kind and nature of meat, and the length of time it was cured in the brine.

Next, the meat is dried in a room heated to 57°C and provided with good circulation to remove moist air. The drying is completed in 5 to 9 days. Sometimes both drying and smoking are practiced. Smoke is produced by burning hardwoods such as hickory and maple, which preserves the meat and gives it a desirable flavor.

Finally, the dried beef is chilled to 1°C to facilitate slicing. The sliced product is now ready for packaging and distribution.

Greening of cured meat Cured meats and meat products are subject to a type of bacterial spoilage in which the meat pigment is oxidized to a greenish color. According to Niven et al. (1954), the organisms involved are a group of heterofermentative lactobacilli which grow at low temperatures and produce hydrogen peroxide. The peroxide reacts with the meat pigment to give a green color.

For more information: Ayres (1960, 1963); Clark and Lentz (1969); Goldman et al. (1963); Jay (1966*a*, *b*, 1967); Niven and Evans (1957); Steinkraus and Ayres (1964); Stringer et al. (1969); Turner and Campbell (1962).

Microbiology of fish

Fish spoilage is largely a bacteriological problem. A knowledge of the kinds of bacteria and their characteristics is of importance in improving methods of handling and preserving fish and fish products.

The flesh and internal organs of healthy fish are generally believed to be sterile. In contrast, the slime, gills, and intestines usually carry heavy bacterial loads. Fish may show up to 10 million viable organisms per square centimeter of surface, or per gram of gill tissue, or per milliliter of intestinal fluid. Since counts may show wide seasonal variations, conclusions should not be too hastily drawn.

The organisms encountered most frequently on the external surfaces of fish are members of the genera *Achromobacter, Corynebacterium, Flavobacterium, Micrococcus,* and *Pseudomonas;* and to a lesser extent the genera *Aeromonas, Alcaligenes, Bacillus, Brevibacterium, Escherichia, Proteus, Sarcina, Serratia,* and *Vibrio.*

Almost all bacteria associated with fish are facultative psychrophilic forms having a temperature range from −7.5 to 30°C. Of 71 organisms isolated by one investigator, 10 grew at −7.5°C, 22 at −5°C, and 65 at 0°C. Fish spoilage proceeds about twice as fast at 2.8°C as at −0.3°C. One of the best methods of retarding fish spoilage is to store it at a temperature as close as possible to the freezing point of the muscle (−1.1°C). Undoubtedly one of the reasons why fish usually spoils more rapidly than meat under storage conditions is that the former is richly contaminated with psychrophiles, whereas the latter is more likely to be contaminated with a higher proportion of mesophiles.

As is true for meat, bacterial counts are valueless as a measure of the degree of spoilage of fish. It is not so much the numbers as it is the kinds of bacteria that are important in determining sanitary quality. A fish sample giving a very high bacterial count may show less spoilage than one having a comparatively low bacterial population.

Reddening of salted fish Members of the genus *Halobacterium* are obligate, halophilic (salt-loving), rod-shaped bacteria which are highly pleomorphic. They require at least 12 percent salt for growth and will even live in saturated brines (about 35 percent salt). Motile species have polar flagella; some species are nonmotile. They are gram-negative. Organisms are generally chromogenic, producing non-water-soluble carotenoid pigments which vary in shade from colorless to orange or even brilliant red. Most species produce a reddening of salted fish where untreated solar salt is used. They are abundant in tidal pools along shores of tropical seas. They redden the water in the pools where solar salt is produced soon after the brine is concentrated to 18 percent salt. The organisms are commonly found in untreated solar salt.

For more information: Bramstedt and Auerbach (1961); Chai et al. (1968); Colwell and Liston (1960); Craig et al. (1968); Evelyn and McDermott (1961); Fieger and Novak (1961); Geldreich and Clarke (1966); Hayward and MacCallum (1969); Liston (1960); Potter and Baker (1961); Shewan (1961); Shewan et al. (1960); Simidu and Hasuo (1968); Simidu et al. (1969); Tarr (1961); Varga and Anderson (1968).

Microbiology of poultry

Nagel et al. (1960) made a study of organisms associated with spoilage of chilled fryers. The chickens were purchased in the open market and stored at 4.4°C for 7 days before sampling. Of 103 isolates, 88 were classified as *Pseudomonas*, 2 as *Aeromonas*, and 13 as *Achromobacter-Alcaligenes*. No attempt was made to classify the latter into either genus.

Kraft et al. (1963) froze turkeys by brine immersion and reported marked decreases in bacterial numbers on the skin surfaces. Experimental inoculation preliminary to freezing resulted in a 98 to 99 percent reduction in the total surface flora. Freezing was more destructive to coliforms than to the enterococci. The former were eliminated after 90 days, whereas 2 to 3 percent of the latter were still viable after 3 to 5 months.

For more information: Kraft and Ayres (1964, 1965); Wilkerson et al. (1961).

Microbiology of eggs

An eggshell is composed largely of calcite, a crystalline form of calcium carbonate. It is a porous structure, the pores being large enough to permit passage of gases and microscopic solid particles. The pores are the means by which air passes through the shell to furnish oxygen to the developing chick embryo.

Microbial contamination of eggs Normally, the oviduct of the hen is sterile and, therefore, the shell and internal contents of the egg are also sterile. As the egg leaves the oviduct at a temperature of 107°F (41.7°C),

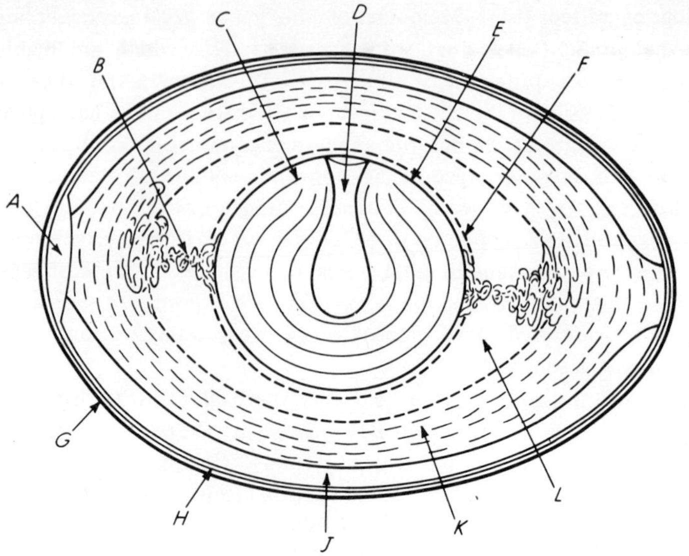

Fig. 287 The structure of the hen's egg, as shown by a section through the long axis. (A) air cell; (B) chalaza; (C) yolk; (D) germinal disk and white yolk; (E) vitelline membrane; (F) film of mucin; (G) shell; (H) shell membranes; (J) outer thin white; (K) thick white; (L) inner thin white. (After Brooks.)

the shell is completely filled with yolk, white, and shell membranes (Fig. 287). During cooling, the contents shrink and air is drawn through the 6000 to 8000 shell pores to form the air cell. Further transpiration of gas occurs as the egg ages.

The shell may be soiled with blood, manure, feathers, nest material, or broken eggs. These substances are most likely contaminated with bacteria, which may be drawn through the shell as the egg cools. Studies have shown that from 5 to 12 percent of eggs are internally contaminated within a few hours after they are laid. Whether they eventually decompose from bacterial causes depends upon the kinds and concentration of organisms, the natural bactericidal properties of the egg, environment, and time.

Shell membranes Two membranes cover the eggshell, one on the outside and another on the inside. They act as physical barriers to entrance of bacteria into the egg. This is probably their only function, since they exert no known bactericidal activity against several types of bacteria that cause spoilage in eggs. Brown et al. (1965) found the inner shell membrane to be a more effective barrier to bacteria than either the outer shell membrane or shell.

Rizk et al. (1966) reported that of the disinfectants used for the sterilization of egg shells none of them could eliminate *Salmonella* from the inner shell membrane after the organisms had penetrated the shell.

Bactericidal property of egg white Egg white is not a favorable medium for the growth of bacteria. It contains several antibacterial substances, namely, lysozyme, avidin, conalbumin, and ovomucoid, and it has a high pH.

Gram-positive organisms are particularly sensitive to the lytic action of lysozyme. Gram-negative forms are less sensitive and only these have been isolated from spoiled eggs.

Organisms which require the growth factor biotin preformed are unable to grow in egg white. The protein avidin in egg white forms a complex with the vitamin, making it unavailable to bacteria.

Conalbumin also is capable of forming a complex with the iron of egg white rendering the metal unavailable to the invading bacteria. However, the addition to egg white of sufficient iron to saturate the conalbumin permits previously suppressed iron-requiring bacteria to grow.

Fresh egg white has a pH of 7.6 to 7.8. On standing it loses CO_2, and after a few days the pH may go as high as 9.3. Such a high pH is beyond the growth range of most bacteria.

Ovomucoid, another protein in egg white, has the ability to combine with the proteolytic enzyme trypsin, thereby inhibiting its action. However, the protein has no effect on rate or extent of growth of gram-negative bacteria.

Organisms in eggs As stated above, the oviduct of the hen is normally sterile, which means that the shell and contents of the egg are free from bacteria. Sometimes the ovaries and oviduct become infected and the bacteria are deposited inside the egg. More frequently, however, the egg becomes contaminated after it is laid. The organisms are deposited on the shell from the intestinal tract of the hen and from other sources in the environment.

The most important disease bacteria isolated from contaminated eggs are members of the genus *Salmonella.* This genus includes species causing typhoid fever and food poisoning in man.

Of the nondisease producers, a large number have been isolated from spoiled eggs. These are too numerous to mention, but some important genera are *Bacillus, Escherichia, Proteus,* and *Pseudomonas*. Also included is an extensive list of molds.

For more information: Adler (1965); Akashi (1968); Banwart and Kreitzer (1969); Banwart et al. (1968); Board (1964, 1965); Board et al. (1964); Brooks (1960); Garibaldi (1960); Garibaldi and Bayne (1962); Graves and MacLaury (1962); Hartung and Stadelman (1965); Lifshitz et al. (1964); Sauter and Petersen (1969); Simmons et al. (1970); Taylor (1970); Vadehra et al. (1970); Williams and Whittemore (1967).

Washing of eggs From 10 to 25 percent of eggs produced under average farm conditions are soiled at the time of gathering. Under poor management, the percentage is much higher. Even under good conditions, the number may run from 5 to 10 percent. Soiled eggs have a lower market value than clean eggs. Therefore, the washing of soiled eggs is a well-established practice.

During washing, bacteria are carried through the shells and membranes into the egg. Thus, the problem of producing and marketing clean, uncontaminated eggs is primarily one of sanitation.

Wet-cleaning operations are more desirable than dry methods. The former result in labor-saving time, less breakage, more thorough cleaning, and less spoilage if sanitizers are used.

The types of compounds which have been used in egg-washing solutions include alcohol, acids, alkalies, detergents, detergent-sanitizers, formalin, hypochlorites, salts, soaps, and water glass. Bierer et al. (1961*a*, *b*, 1962) found a 1 percent solution of zinc sulfate to be superior to all others for washing and sterilizing egg shells. The zinc sulfate had no detrimental effect on the hatchability of the eggs.

It is obvious that the washing method cannot affect the incidence of eggs contaminated before they reach that stage, but the use of a good cleaning method should reduce the proportion of eggs which become contaminated during the washing process.

It is generally believed that considerable numbers of eggs are contaminated during improper washing operations. Bacteria and molds contained in the filth removed from the dirty eggs are dispersed in the washing liquid. If the solution is cooler than the internal temperature of the egg, negative pressure is obtained and organisms enter the egg through the pores. Contamination may also occur later if the eggs are transferred from a cold room to a warm or humid atmosphere. Condensation forms on the shell, making it possible for motile cells to enter through the pores.

For more information: Bierer and Barnett (1965); Sauter (1966); Östlund (1971*a*, *b*).

Preservation of eggs A number of methods are employed for the preservation of eggs: (1) cold storage, (2) freezing, (3) drying, (4) immersion in sodium silicate solution (water glass), (5) packing in brine or sawdust, (6) coating with petroleum jelly, (7) wrapping in oiled paper, (8) coating with paraffin, (9) immersion in lime water (solution of calcium hydroxide), and (10) dipping in light paraffin oil.

In the cold-storage process the eggs are kept at a temperature of about -6°C. If the temperature goes below this point, a nonreversible change takes place in the albumin, preventing the egg from being restored to its normal transparent condition.

Frozen eggs are kept at a temperature of about -18°C or lower until needed. They are used in large quantities by candymakers, bakers, and egg-noodle and macaroni manufacturers, and in other industries. The eggs are removed from the shell, placed in large metal containers, and frozen. Often the whites and yolks are separated because they may be used for different purposes. They are usually cheaper than fresh eggs because they are prepared during the periods of high production.

Egg white is now being pasteurized to destroy contaminating

organisms which may be present. This applies especially to salmonellae which are frequently present on egg shells and are one of the groups of organisms involved in food poisoning. Pasteurization prevents or reduces spoilage of the egg white. The high pH value (9.0) of the albumin and the presence of a number of bacterial inhibitors, including lysozyme, make it a much more selective environment than that of whole egg.

Eggs may be dried unseparated or first separated into whites and yolks and then dried. Dried yolks, whites, and unseparated eggs are used to a considerable extent in prepared cake and doughnut flours, ice creams, macaroni and noodles, and bakery products such as meringue and marshmallows. Dried eggs have good keeping qualities if kept dry and cool. However, the presence of a small amount of moisture and a warm atmosphere may be sufficient to permit the growth of organisms and cause spoilage in a short time. It is desirable, therefore, to store dried eggs in a cool place, protected from an excessive amount of moisture.

The other methods used for preserving eggs are designed to prevent the passage of air (oxygen) through the pores of the shell and into the eggs. Regardless of which one of these procedures is followed, the eggs must be kept under cold conditions to prevent decomposition by the enzymes normally present in the egg.

For more information: Barnes and Corry (1969); Cotterill (1968); Garibaldi et al. (1969*a, b*).

Microbiology of sweetened condensed milk

Sweetened condensed milk is prepared (1) by pasteurizing milk at a temperature of 80 to 90°C for 1 min to destroy enzymes and bacteria which may cause undesirable physical and chemical changes, (2) by adding about 13 percent sucrose, and (3) by heating under reduced pressure to decrease the volume to about one-third of the original. The final product contains about 40 percent sucrose. It is transferred to sterile containers and capped.

Sweetened condensed milk is not sterilized in the containers. The high osmotic pressure produced by the added sugar is sufficient to prevent growth of almost all organisms.

Organisms in milk Organisms isolated from milk include anaerobes, *B. subtilis, E. coli,* micrococci, streptococci, thermophiles, and yeasts. The yeasts are believed to be the most common group producing spoilage. They ferment the sugar with the formation of acid and gas which gives the cans a blown appearance.

The sources of yeasts are (1) contamination of the original fresh milk, (2) contamination of the sucrose, and (3) contamination of the air of the plant. The yeasts are probably not harmful. Fermented milk is objectionable but most likely free from any toxic substances.

For more information: Bartram (1967); Borgstrom (1968); Frazier (1967); Thatcher and Clark (1968).

Food poisoning

Food poisoning refers to the ingestion of food contaminated either with harmful bacteria or with certain soluble excretory products known as toxins. It does not include the toxications which follow the consumption of noxious plants (mushrooms), poisonous fish (mussels), or decomposed foods containing chemical poisons (arsenic, lead, fluorides), or idiosyncrasies associated with certain plant and animal poisons.

At one time food poisoning was believed to be caused by the consumption of decomposed food containing certain chemical compounds known as ptomaines. The term is taken from the Greek and means a dead body. Ptomaines are produced in putrefied meat and other proteinaceous foods. They are basic substances and belong to the group of compounds known as amines. They result chiefly from the decarboxylation of amino acids. A typical reaction is the following:

$$\underset{\text{Alanine}}{CH_3 \cdot CHNH_2 \cdot COOH} \longrightarrow \underset{\text{Ethylamine}}{CH_3 \cdot CH_2NH_2} + CO_2$$

The amino acid alanine loses carbon dioxide and is converted into ethylamine, a ptomaine. This reaction occurs only at a pH below 7.0 and when putrefaction is in an advanced stage. Ptomaines are poisonous when injected into the tissues, but there appears to be very little evidence that they produce any toxicity when taken by mouth. The ptomaine theory of intoxication is a misconception.

Presence of bacteria in foods The fact that bacteria are present in foods does not mean necessarily that they are harmful. Many saprophytic forms can attack proteins and release ptomaines during the later stages of decomposition. However, most bacteria which are capable of putrefying proteins are harmless when taken by mouth. Facultative and anaerobic species normally present in the intestinal tract of man, and capable of producting putrefactions, are harmless when ingested with food.

Food-poisoning organisms There are at least three genera of bacteria responsible for true food poisoning: (1) *Staphylococcus (S. aureus)*; (2) *Salmonella (S. enteritidis, S. typhimurium*, etc.); and (3) *Clostridium (C. botulinum*, several types, all nonovolytic, *C. parabotulinum*, several types, all ovolytic, *C. perfringens*). In addition, *E. coli* and *P. vulgaris* have been involved in occasional outbreaks of food poisoning. *Streptococcus*, α type *(S. faecalis, S. faecalis* var. *liquefaciens, S. faecalis* var. *zymogenes)*, has been reported as being responsible for some cases of food poisoning but the evidence is questionable.

Staphylococcus aureus Certain strains of *S. aureus*, under favorable conditions, produce a potent enterotoxin. The ingestion of foods containing the enterotoxin gives rise to an intoxication, but the symptoms generally are mild. Death rarely occurs.

S. aureus is widely distributed in nature (Fig. 288). It is found on

Fig. 288 Staphylococcus aureus, an enterotoxin-producing strain.

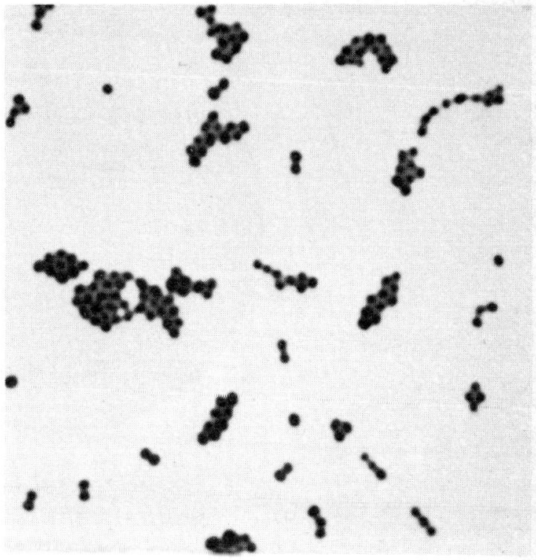

the nasal mucous membrane and skin (hair follicle); in boils, carbuncles, pyemia, osteomyelitis, infected cuts or breaks in the skin; and in foods. The human nose is the main reservoir of the organism.

Foods most commonly involved include those which are eaten cold, e.g., hams, pressed meats, salads, raw milk and dairy products, and pastries. Such foods are readily contaminated by secretions from the nose, throat, and lesions on the skin.

Symptoms appear in about 3 hr after ingestion of foods containing the enterotoxin, but the range is from 0.5 to 7 hr, depending upon dosage and susceptibility. The symptoms are nausea, vomiting, abdominal cramps, and diarrhea. In more severe cases, debility is great. The patient is weak and prostrate and may go into shock. Blood may be passed in the vomitus or stools. Symptoms begin to disappear usually in a few hours, and recovery is complete in a few days. Body temperature may be subnormal.

In pure cultures, toxin production occurs in 4 to 6 hr. Toxin is produced in the temperature range of 15.5 to 49°C. Foods held at temperatures below 15°C do not develop toxin; hence the importance of keeping susceptible foods refrigerated at all times.

Not all strains of *S. aureus* produce enterotoxin. Therefore, the presence of *S. aureus* in foods does not mean necessarily that an intoxication will result from its consumption. Those strains that do produce enterotoxin are also capable of coagulating citrated human or rabbit plasma. They are referred to as coagulase-positive strains.

Some strains of *S. aureus* liquefy gelatin; others do not. Stone (1935) believed that there was a correlation between gelatin liquefaction and enterotoxin production. A nutrient gelatin-agar plate was streaked with a loopful of a broth culture of the strain under examination and incubated at 37°C for 24 hr. Then a strong solution of ammonium sulfate was poured over the surface of the plate. The ammonium sulfate precipitates protein (gelatin), giving the agar an opaque appearance. The

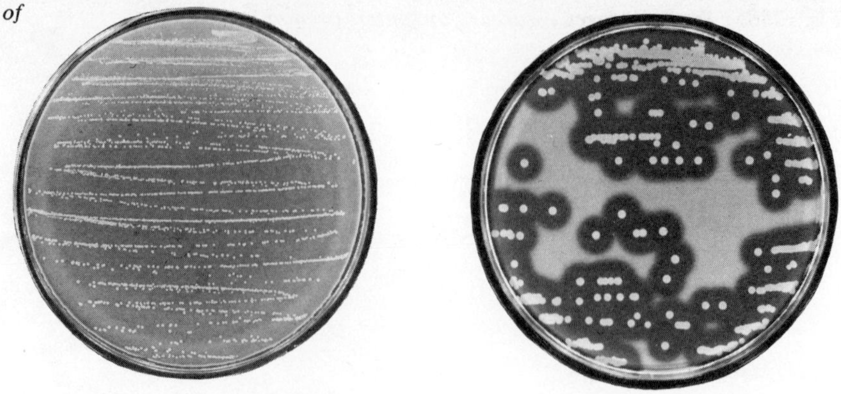

Fig. 289 Left, strain of Staphylococcus aureus not capable of liquefying gelatin; right, strain capable of liquefying gelatin. The gelatin agar plates were flooded with a solution of ammonium sulfate to precipitate the gelatin. Clear zones around the colonies indicate digestion of the gelatin by a gelatinase.

gelatin in the area surrounding each colony is digested, giving a clear agar (Fig. 289). A typical reaction shows clear zones around each colony with the remainder of the agar appearing opaque. The test does not always show perfect correlation between toxigenicity and gelatin liquefaction.

For more information: Bryan (1968*a*); Casman et al. (1963); Cockburn et al. (1962); Crisley et al. (1964); McCoy and Faber (1966); McLean et al. (1968); Munch-Petersen (1963); Thatcher and Robinson (1962); Zehren and Zehren (1968).

Salmonella The salmonellae are widely distributed in nature, being found in the intestinal contents of warm-blooded animals, including man; also in birds and occasionally in reptiles. They are excreted in feces during acute illness and by convalescent carriers, and are disseminated by water, foods, insects, and other living creatures coming in contact with contaminated feces (see page 940).

All members of the genus *Salmonella* are pathogenic for man, for animals, or for both. With the exception of *S. typhosa,* and *S. paratyphi* A, B, and C, they are considered zoonoses, diseases or infections transmitted between vertebrate animals and man. Such transmission is, however, seldom direct, but usually results from eating food of animal origin.

Members of this genus produce chiefly endotoxins. The organisms are ingested with contaminated food and reach the intestinal tract. Those which are not destroyed continue to multiply. The endotoxins are liberated after death and digestion of the bacteria (Fig. 290).

Symptoms appear in from 6 to 12 hr. Infection is characterized by continued fever, acute gastroenteritis, involvement of the lymphoid tissue of the intestines, enlargement of the spleen, and sometimes rose spots on the trunk. Infection is usually accompanied by diarrhea, and frequently by

Fig. 290 Salmonella enteritidis, an organism that multiplies in the intestinal tract and produces food poisoning by the elaboration of an endotoxin.

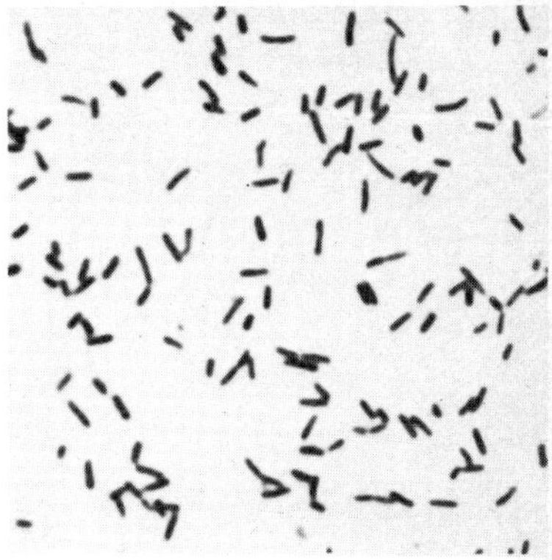

headache, nausea, and vomiting. Organisms may be present in the feces, urine, and blood. Recovery is usually complete within 2 to 4 days.

Salmonellae are disseminated in various ways. Rats and mice are generally believed to be involved in spreading the organisms. The bacteria have been recovered from a number of foods. They are found on the shell, and in the contents of clean fresh eggs. After processing to egg powder, a high percentage of samples show the presence of salmonellae. Also, under faulty operating conditions, the organisms may multiply on the processing equipment.

Salmonella has been isolated from desiccated coconut. The coconut is believed to be contaminated by handlers during processing under primitive conditions. The organisms are particularly resistant to desiccation.

Poultry as it is processed is generally heavily exposed to *Salmonella* contamination. Infections may result from consumption of insufficiently or improperly cooked fowl.

Pork products purchased in the open market showed the presence of *Salmonella.* Evidence indicates that this is not the result of infected hogs on the farm but rather of dissemination of the organisms in the abattoirs. This appears to apply equally well to cattle. It has been shown that the tools used in the preparation of meat and meat products may become contaminated, from which source the organisms may be easily disseminated into other products.

Of the many possible modes of dissemination of the salmonellae the distribution of these potential pathogens by the egg, poultry, and meat-processing industries appears to be of high importance.

The cycle of transmission, according to Bryan (1968*b*), is shown in Fig. 291.

For more information: Bryan et al. (1968); Carroll and Ward (1967); Cockburn et al. (1962); Granville et al. (1969); Insalata et al. (1968); Morris et al. (1970); Morris and Wells (1970); Morris and Ayres (1960); Read et al. (1968); Sadler and Corstvet (1965); Schaffner et al. (1967);

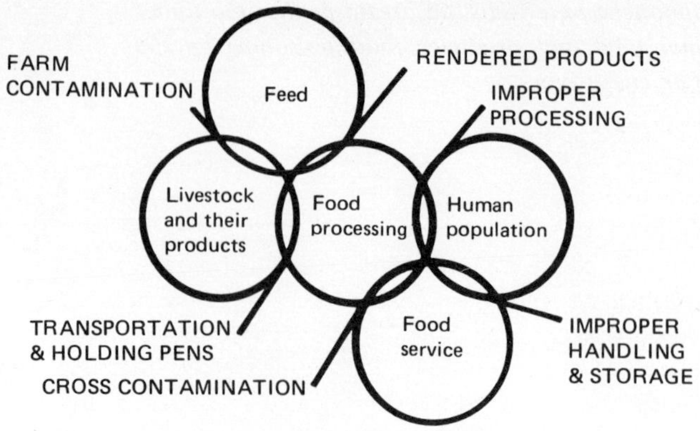

Fig. 291 Cycle of transmission of salmonellae. (After Bryan.)

Scott (1968); Van Schothorst and Kampelmacher (1967); Weissman and Carpenter (1969); Williams and Newell (1968); Woodburn and Stadelman (1968); Zindel and Bennett (1968); Dega et al. (1972).

Clostridium Members causing food poisoning include *C. botulinum* (types B, C, D, E), *C. parabotulinum* (types A, B), and *C. perfringens* (types A and B).

C. botulinum and *C. parabotulinum* produce their intoxications by the secretion of powerful exotoxins. Spores of the organisms are commonly found in the soil and the intestinal tracts of man and animals but they do not produce infections under natural conditions.

The spores are found in the soil and are likely to contaminate many kinds of foods. Since the organisms are anaerobic, the spores will not germinate unless the free oxygen in the environment is greatly reduced. The spores are very resistant to heat. Unless they are destroyed during thermal processing, conditions are likely to be created which favor their germination (Fig. 292). The vegetative cells then secrete powerful exotoxins into the food. The mere tasting of such food may cause death. The toxins can pass through the stomach wall and intestines unchanged, differing in this respect from most bacterial toxins.

Symptoms usually develop suddenly and consist of gastrointestinal pain, headache, diarrhea or constipation, and prostration. These may be followed by several types of paralyses of the central nervous system. Death from botulism occurs by cardiac or respiratory paralysis, usually in 3 to 7 days. Severity of symptoms depends upon the amount of toxin ingested.

Since the toxins are destroyed by heat, freshly cooked foods are safe. Nonsterile canned foods and certain meat products which are consumed cold may contain toxin. Commercial processing of foods in this country is generally adequate. Practically all cases of botulism result from the consumption of home-canned, insufficiently processed foods.

Fig. 292 Bacilli and spores of Clostridium botulinum.

Spoiled canned foods or those which are suspected of being spoiled should never be tasted. If the container is contaminated with *Clostridium*, sufficient toxin may be present in a minute portion of the food to cause death. The suspected food should not be fed to animals or chickens, as they may be highly susceptible. Such foods are best disposed of by thorough cooking to destroy the toxin.

C. perfringens (types A and B) is responsible for an ever-increasing number of outbreaks of food poisoning. The organism has been extensively studied in England where it is more commonly known as *C. welchii.*

The organism is widely distributed, being found in soil, the normal intestinal tract of man and animals, and in many foods. Raw meat is often contaminated with the spores before reaching the consumer. Cooked meats and meat products are chiefly the reported vehicles of infection. Heat labile spores have been isolated from cooked meat which has caused outbreaks. This would imply that the food was contaminated after cooking. Prevention must be placed on proper temperature control of cooked meat, that is to say, keeping it hot enough (above 60°C) prior to serving to destroy germinating spores, or cold enough (below 7°C) to retard germination and growth of the vegetative cells.

Symptoms include diarrhea and abdominal pain with little or no fever, nausea or vomiting. Symptoms occur 8 to 24 hr after ingestion with an average of 8 to 12 hr. Rapid recovery usually follows within 24 hr after onset. Death rarely occurs in healthy persons. The symptoms resemble both an infection and an intoxication. There appears to be some doubt as to which is the predominating cause of the symptoms.

For more information: Abrahamsson et al. (1965); Beers and Reich (1969); Cockburn et al. (1962); Crisley (1967); Duda and Slack (1969); Fantasia and Duran (1969); Foster (1968); Foster and Sugiyama (1967); Hall et al. (1963); Hauschild and Hilsheimer (1968); Hauschild and Thatcher (1967); Kautter (1964); Licciardello et al. (1969); Nakamura and Kelly (1968); Pace et al. (1967); Petty (1965); Reed (1964); Simonetti and D'Arca (1967); Stefanye et al. (1967); Sutton and Hobbs (1968); Wheatley (1969); Johnston and Loynes (1971).

References

Abbey, A., et al.: The effectiveness of acronized chlortetracycline in poultry preservation following long term commercial use, *Food Technol.*, **14**:609, 1960.

Abrahamsson, K., et al.: Toxin production by *Clostridium botulinum*, type E, in vacuum-packed, irradiated fresh fish in relation to changes of the associated microflora, *Can. J. Microbiol.*, **11**:523, 1965.

Adler, H. E.: *Salmonella* in eggs: An appraisal, *Food Technol.*, **19**:191, 1965.

Akashi, A.: Lytic action of egg white lysozyme on the food contaminating microorganisms. Lytic action on *Bacillus subtilis*, *J. Food Hyg. Soc. Japan*, **9**:97, 1968.

American Chemical Society: "Radiation Preservation of Foods," Washington, American Chemical Society, 1967.

Anellis, A., and S. Werkowski: Estimation of radiation resistance values of microorganisms in food products, *Appl. Microbiol.*, **16**:1300, 1968.

——— et al.: Radiation sterilization of bacon for military feeding, *Appl. Microbiol.*, **13**:37, 1965.

——— et al.: Radiation sterilization of prototype military foods. II. Cured ham, *Appl. Microbiol.*, **15**:166, 1967.

Angelotti, R., et al.: Time-temperature effects on salmonellae and staphylococci in foods. III. Thermal death time studies, *Appl. Microbiol.*, **9**:308, 1961.

Ayres, J. C.: The relationship of organisms of the genus *Pseudomonas* to the spoilage of meat, poultry and eggs, *J. Appl. Bact.*, **23**:471, 1960.

———: Low temperature organisms as indexes of quality of fresh meat. In "Microbiological Quality of Foods," New York, Academic Press, Inc., 1963.

Ball, C. O.: Processing for sterilization, *Food Ind.*, **19**:44, 174, 338, 1947.

Banwart, G. J., and M. Kreitzer: Further studies on the screening technique for determining *Salmonella*-negative samples of pasteurized dried egg, *Poultry Sci.*, **48**:237, 1969.

——— et al.: Screening method for determining *Salmonella*-negative samples of pasteurized dried whole egg, *Poultry Sci.*, **47**:598, 1968.

Barnes, E. M., and J. E. L. Corry: Microbial flora of raw and pasteurized egg albumen, *J. Appl. Bact.*, **32**:193, 1969.

Bartram, M. T.: International microbiological standards for foods, *J. Milk Food Technol.*, **30**:349, 1967.

Beers, W. H., and E. Reich: Isolation and characterization of *Clostridium botulinum* type B toxin, *J. Biol. Chem.*, **244**:4473, 1969.

Beraha, L., et al.: Gamma radiation in the control of decay in strawberries, grapes, and apples, *Food Technol.*, **15**:94, 1961.

Bierer, B. W., and B. D. Barnett: Killing *Salmonella* on egg-shells with disinfectants, *J. Am. Vet. Med. Assoc.*, **140**:159, 1962.

——— and ———: Killing *Salmonella* on egg-shells by increasing wash water temperature, *J. Am. Vet. Med. Assoc.*, **146**:735, 1965.

——— et al.: Experimentally killing *Salmonella typhimurium* on egg shells by washing, *Poultry Sci.*, **40**:1009, 1961*a*.

——— et al.: Germicidal efficiency of egg washing compounds on eggs artificially contaminated with *Salmonella typhimurium*, *Poultry Sci.*, **40**:148, 1961*b*.

Board, R. G.: The growth of gram-negative bacteria in the hen's egg, *J. Appl. Bact.*, **27**:350, 1964.

———: Bacterial growth on and penetration of the shell membranes of the hen's egg, *J. Appl. Bact.*, **28**:197, 1965.

——— et al.: The microbiological contamination of egg shells and egg packing materials, *Poultry Sci.*, **43**:584, 1964.

Borgstrom, G.: "Principles of Food Science," New York, The Macmillan Company, 1968.

Brady, D. E., et al.: Storage of frozen meats, poultry, eggs, fruits, and vegetables, *Missouri Univ. Agr. Exp. Sta. Res. Bull.*, no. 440, 1949.

Bramstedt, F., and M. Auerbach: The spoilage of fresh-water fish. In "Fish as Food," edited by G. Borgstrom, New York, Academic Press, Inc., 1961.

Brooks, J.: Mechanism of the multiplication of *Pseudomonas* in the hen's egg, *J. Appl. Bact.*, **23**:499, 1960.

Brown, W. E., et al.: The role of the inner shell membrane in bacterial penetration of chicken eggs, *Poultry Sci.*, **44**:1323, 1965.

Brown, W. L., et al.: Radiation resistance of the natural bacterial flora of cured ham, *Food Technol.*, **14**:622, 1960.

Bryan, F. L.: What the sanitarian should know about staphylococci and salmonellae in non-dairy products. I. Staphylococci, *J. Milk Food Technol.*, **31**:110, 1968*a*; II. Salmonellae, *ibid.*, **31**:131, 1968*b*.

——— et al.: Salmonellae associated with further-processed turkey products, *Appl. Microbiol.*, **16**:1, 1968.

Buttiaux, R., and M. Catsaras: Les bacteries psychrotrophes des viandes entreposees en chambre froid, *Société Française de Microbiologie*, Jan. 5, 1967.

Carroll, B. J., and B. Q. Ward. Control of salmonellae in fish meal, *Fishery Ind. Res.*, **4**:29, 1967.

Casman, E. P., et al.: Staphylococcal growth and enterotoxin production in meat, *Appl. Microbiol.*, **11**:498, 1963.

Chai, T., et al.: II. Relative incidence of *Pseudomonas putrefaciens* and fluorescent pseudomonads on haddock fillets, *Appl. Microbiol.*, **16**:1738, 1968.

Clark, D. S., and C. P. Lentz: The effect of carbon dioxide on the growth of slime producing bacteria on fresh beef, *Can. Inst. Food. Tech. J.*, **2**:72, 1969.

Cockburn, W. C., et al.: "Food Poisoning," London, The Royal Society of Health, 1962.

Colwell, R. R., and J. Liston: Microbiology of shellfish, *Appl. Microbiol.*, **8**:104, 1960.

Comer, A. G., et al.: Gamma irradiation of *Salmonella* species in frozen whole egg, *Can. J. Microbiol.*, **9**:321, 1963.

Cory, R. R., and J. M. Byrnes: Oxytetracycline-resistant coliforms in commercial poultry products, *Appl. Microbiol.*, **11**:481, 1963.

Costilow, R. N.: Sorbic acid as a selective agent for cucumber fermentations. III, *Food Technol.*, **11**:591, 1957.

——— et al.: II. Effect of sorbic acid on the yeast and lactic fermentations in brined cucumbers, *Appl. Microbiol.*, **5**:373, 1957.

Cotterill, O. J.: Equivalent pasteurization temperatures to kill salmonellae in liquid egg white at various pH levels, *Poultry Sci.*, **47**:354, 1968.

Craig, J. M., et al.: Incidence of *Clostridium botulinum* type E in salmon and other marine fish in the Pacific Northwest, *Appl. Microbiol.*, **16**:553, 1968.

Crisley, F. D.: Considerations of hand sanitation in the food handling environment. In "Developments in Industrial Microbiology," vol. 8, Washington, American Institute of Biological Sciences, 1967.

Crisley, F. D., et al.: Multiplication of *Staphylococcus aureus* in synthetic cream fillings and pies, *Public Health Rept.*, **79**:369, 1964.

Deatherage, F. E.: Antibiotics in the preservation of meat. In "Antibiotics in Agriculture," edited by M. Woodbine, London, Butterworth & Co. (Publishers), Ltd., 1962.

Dega, C. A., et al.: Heat resistance of salmonellae in concentrated milk, *Appl. Microbiol.*, **23**:415, 1972.

Drake, S. D., et al.: The effect of heat and irradiation on the microflora of canned hams, *Food Res.*, **25**:270, 1960.

Duda, J. J., and J. M. Slack: Toxin production in *Clostridium botulinum* as demonstrated by electron microscopy, *J. Bact.*, **97**:900, 1969.

Duggan, D. E., et al.: Inactivation of the radiation-resistant spoilage bacterium *Micrococcus radiodurans*, I, *Appl. Microbiol.*, **11**:398, 1963*a*; II, *ibid.*, **11**:413, 1963*b*.

Dyer, J. K., et al.: Radiation survival of food pathogens in complex media, *Appl. Microbiol.*, **14**:92, 1966.

Elliott, R. P.: Bacterial contamination and growth in modern food preserving systems, *J. Environmental Health*, **30**:510, 1968.

—— and H. D. Michener: Microbiological standards and handling codes for chilled and frozen foods: A review, *Appl. Microbiol.*, **9**:452, 1961.

—— and ——: Psychrophilic micro-organisms in foods, *U.S. Dept. Agr. Tech. Bull.* no. 1320, 1965.

Erdman, I. E., et al.: Studies on the irradiation of micro-organisms in relation to food preservation. I. The comparative sensitivities of specific bacteria of public health significance, *Can. J. Microbiol.*, **7**:199, 1961*a*; II. Irradiation resistant mutants, *ibid.*, **7**:207, 1961*b*.

Etchells, J. L., et al.: Influence of sorbic acid on populations and species of yeasts occurring in cucumber fermentations, *Appl. Microbiol.*, **9**:139, 1961.

—— et al.: Bloater formation by gas-forming lactic acid bacteria in cucumber fermentations, *Appl. Microbiol.*, **16**:1029, 1968.

Evelyn, T. P. T., and L. A. McDermott: Bacteriological studies of freshwater fish. I. Isolation of aerobic bacteria from several species of Ontario fish, *Can. J. Microbiol.*, **7**:375, 1961.

Fantasia, L. D., and A. P. Duran: Incidence of *Clostridium botulinum* type E in commercially and laboratory dressed whitefish chubs, *Food Technol.*, **23**:85, 1969.

Fieger, E. A., and A. F. Novak: Microbiology of shellfish deterioration. In "Fish as Food," edited by G. Borgstrom, New York, Academic Press, Inc., 1961.

Foster, E. M.: Microbial problems in today's foods, *J. Am. Dietet. Assoc.*, **52**:485, 1968.

—— and H. Sugiyama: Recent developments in botulism research, *Health Lab. Sci.*, **4**:193, 1967.

Frazier, W. C.: "Food Microbiology," 2d ed., New York, McGraw-Hill Book Company, 1967.

Garibaldi, J. A.: Factors in egg white which control growth of bacteria, *Food Res.*, **25**:337, 1960.

—— and H. G. Bayne: Iron and the bacterial spoilage of shell eggs, *J. Food Sci.*, **27**:57, 1962.

—— et al.: Effect of pH and chelating agents on the heat resistance and viability of *Salmonella typhimurium* Tm-1 and *Salmonella senftenberg* 775W in egg white, *Appl. Microbiol.*, **18**:318, 1969*a*.

—— et al.: Number of salmonellae in commercially broken eggs before pasteurization, *Poultry Sci.*, **48**:1096, 1969*b*.

Geldreich, E. E., and N. A. Clarke: Bacterial pollution indicators in the intestinal tract of freshwater fish, *Appl. Microbiol.*, **14**:429, 1966.

Georgala, D. L., and A. Hurst: The survival of food poisoning bacteria in frozen foods, *J. Appl. Bact.*, **26**:346, 1963.

Gochnauer, M. B., and D. J. Kushner: Growth and nutrition of extremely halophilic bacteria, *Can. J. Microbiol.*, **15**:1157, 1969.

Goldblith, S. A., et al.: "Introduction to Thermal Processing of Foods," Westport, Conn., The Avi Publishing Co., Inc., 1961.

Goldman, M., et al.: Interrelationship between temperature and sodium chloride on growth of lactic acid bacteria isolated from meat-curing brines, *J. Bact.*, **85**:1017, 1963.

Granville, A., et al.: Ouvriers de l'industrie de la viande porteurs de *Salmonella*, *Ann. Méd. Vét.*, **4**:224, 1969.

Graves, R. C., and D. W. MacLaury: The effects of temperature, vapor pressure and absolute humidity on bacterial contamination of shell eggs, *Poultry Sci.*, **41**:1219, 1962.

Hall, H. E., et al.: Characteristics of *Clostridium perfringens* strains associated with food and food-borne disease, *J. Bact.*, **85**:1094, 1963.

Hartung, T. E., and W. J. Stadelman: *Pseudomonas fluorescens* penetration of egg shell membranes as influenced by shell porosity, age of egg and degree of bacterial challenge, *Poultry Sci.*, **42**:147, 1963.

Hauschild, A. H. W., and R. Hilsheimer: Heterogeneity of *Clostridium botulinum* type A toxin, *Can. J. Microbiol.*, **14**:805, 1968.

—— and F. S. Thatcher: Experimental food poisoning with heat-susceptible *Clostridium perfringens* type A, *J. Food Sci.*, **32**:467, 1967.

Hawley, H. B.: The uses of antibiotics in canning. In "Antibiotics in Agriculture," edited by M. Woodbine, London, Butterworth & Co. (Publishers), Ltd., 1962.

Hayward, M. J., and W. A. MacCallum: Bacteria counts on cod and flounder fillets produced commercially from fish frozen at sea and thawed in water, *J. Fisheries Res. Board Can.*, **26**:3217, 1969.

Hill, J. J., et al.: Thermal conductivity of various meats, *Food Technol.*, **21**:91, 1967.

Hobbs, B. C., et al.: Antibiotic treatment of poultry in relation to

Salmonella typhimurium, Bull. Ministry, Health Public Health Lab. Service, **19**:178, 1960.

Ijichi, K., et al.: Effects of ultraviolet irradiation of egg liquids on *Salmonella* destruction and performance quality with emphasis on egg white, *Food Technol.,* **18**:124, 1964.

Ingram, M., and M. J. Thornley: The effect of low temperatures on the inactivation by ionizing radiations of *Clostridium botulinum* spores in meat, *J. Appl. Bact.,* **24**:94, 1961.

Insalata, N. F., et al.: *Salmonella* and the food industry: A proposed protection plan. In "Developments in Industrial Microbiology," vol. 9, Washington, D.C., American Institute of Biological Sciences, 1968.

Jay, J. M.: Response of the extract-release volume and water-holding capacity phenomena to microbiologically spoiled beef and aged beef, *Appl. Microbiol.,* **14**:492, 1966.

——: Influence of postmortem conditions on muscle microbiology. In "The Physiology and Biochemistry of Muscle as a Food," edited by E. J. Briskey, R. G. Cassens, and B. B. Marsh, Madison, The University of Wisconsin Press, 1970.

——: Nature, characteristics, and proteolytic properties of beef spoilage bacteria at low and high temperatures, *Appl. Microbiol.,* **15**:943, 1967.

Johnston, M. A., and R. Loynes: Inhibition of *Clostridium botulinum* by sodium nitrite as affected by bacteriological media and meat suspensions, *Can. Inst. Food Technol. J.,* **4**:179, 1971.

—— et al.: Inhibition of *Clostridium botulinum* by sodium nitrite in a bacteriological medium and in meat, *Can. Inst. Food Technol. J.,* **2**:52, 1969.

Kautter, D. A.: *Clostridium botulinum* type E in smoked fish, *J. Food Sci.,* **29**:843, 1964.

Kazanas, N., and J. A. Emerson: Effect of γ irradiation on the microflora of freshwater fish, *Appl. Microbiol.,* **16**:242, 1968.

Kempe, L. L., and J. T. Graikoski: Effect of gamma irradiation on the spoilage of canned pork luncheon meat, *Food Technol.,* **18**:134, 1964.

Kraft, A. A., and J. C. Ayres: Development of microorganisms and fluorescence on poultry dipped in water containing iron, *J. Food Sci.,* **29**:218, 1964.

—— and ——: Development of microorganisms and fluorescence of poultry chilled in water containing iron or magnesium, *J. Food Sci.,* **30**:154, 1965.

—— et al.: Effect of method of freezing on survival of microorganisms on turkey, *Poultry Sci.,* **42**:128, 1963.

Lerke, P. A., and L. Farber: Effect of electron beam irradiation on the microbial content of spices and teas, *Food Technol.,* **14**:266, 1960.

Licciardello, J. J., et al.: Effect of irradiation temperature on inactivation of *Clostridium botulinum* toxin type E by gamma rays, *J. Appl. Bact.*, **32**:476, 1969.

Lifshitz, A., and R. C. Baker: Some physical properties of the egg shell membranes in relation to their resistance to bacterial penetration, *Poultry Sci.*, **43**:527, 1964.

—— et al.: The relative importance of chick egg exterior structures in resisting bacterial penetration, *J. Food Sci.*, **29**:94, 1964.

Liston, J.: The bacterial flora of fish caught in the Pacific, *J. Appl. Bact.*, **23**:469, 1960.

Mackintosh, D. L., et al.: Preserving foods by freezing, *Kansas Agr. Exp. Sta.* circ. 249, 1949.

Mahoney, J. F.: The function of food additives, *Am. J. Public Health*, **51**:1101, 1961.

McBean, D. M.: Levels of free and combined sulfur dioxide in fruits during sulfuring and drying, *Food Technol.*, **21**:112, 1967.

McCoy, D. W., and J. E. Faber: Influence of food microorganisms on staphylococcal growth and enterotoxin production in meat, *Appl. Microbiol.*, **14**:372, 1966.

McIntosh, E. N.: Effect of post-mortem aging and enzyme tenderizers on mucoprotein of bovine skeletal muscle, *J. Food Sci.*, **32**:210, 1967.

McLean, R. A., et al.: Effects of meat-curing salts and temperature on production of staphylococcal enterotoxin B, *J. Bact.*, **95**:1207, 1968.

Michener, H. D., et al.: Time-temperature tolerance of frozen foods, 22. Relationship of bacterial population to temperature, *Food Technol.*, **14**:290, 1960.

Miyauchi, D. T.: Irradiation preservation of Pacific Northwest fish. I. Cod fillets, *Food Technol.*, **14**:379, 1960.

Morris, G. K., et al.: Salmonellae in fish meal plants: relative amounts of contamination at various stages of processing and a method of control, *Appl. Microbiol.*, **19**:401, 1970.

—— and J. G. Wells: *Salmonella* contamination in a poultry-processing plant, *Appl. Microbiol.*, **19**:795, 1970.

Morris, T. G., and J. C. Ayres: Incidence of salmonellae on commercially processed poultry, *Poultry Sci.*, **39**:1131, 1960.

Munch-Petersen, E.: Staphylococci in food and food intoxication, *J. Food Sci.*, **28**:692, 1963.

Nagel, C. W., et al.: Microorganisms associated with spoilage of refrigerated poultry, *Food Technol.*, **14**:21, 1960.

Nakamura, M., and K. D. Kelly: Incidence of *Clostridium perfringens* in fish and fish products, *Health Lab. Sci.*, **5**:84, 1968.

Niven, C. F., Jr., and J. B. Evans: *Lactobacillus viridescens* nov. species, a heterofermentative species that produces a green discoloration of cured meat pigments, *J. Bact.*, **73**:758, 1957.

—— et al.: Thermal tolerance studies on the heterofermentative lactobacilli that cause greening of cured meat products, *Appl. Microbiol.*, **2**:26, 1954.

Novak, A. F., et al.: In vitro effects of chlortetracycline on bacteria indigenous to gulf shrimp and oysters, *Food Technol.*, **14**:585, 1960.

Oser, B. L.: How foods spoil, *Food Ind.*, **18**:1683, 1946.

Östlund, K.: Bacteriology of washed and unwashed eggs, I. Bacterial counts of egg contents after storage, *Acta Vet. Scand.*, **12**:467, 1971*a*; II. Penetration of *Salmonella* bacteria through the eggshell, *ibid.*, **12**:479, 1971*b*.

Ostovar, K., et al.: Preservation of fresh whitefish with gamma radiation, *J. Fisheries Res. Bd. Can.*, **24**:9, 1967.

Ott, T. M., et al.: Thermal destruction of *Streptococcus faecalis* in prepared frozen foods, *Food Sci.*, **26**:1, 1961.

Pace, P. J., et al.: Demonstration and isolation of *Clostridium botulinum* types from whitefish chubs collected at fish smoking plants of the Milwaukee area, *Appl. Microbiol.*, **15**:877, 1967.

Parks, T. R., et al.: Thermal inactivation of chlortetracycline in various meat menstrua, *Appl. Microbiol.*, **8**:305, 1960.

Patterson, J. T.: Salt tolerance and nitrate reduction by micrococci from fresh pork, curing pickles and bacon, *J. Appl. Bact.*, **26**:80, 1963.

Peterson, A. C., and M. F. Gunderson: Role of psychrophilic bacteria in frozen food spoilage, *Food Technol.*, **14**:413, 1960.

Petty, C. S.: Botulism: The disease and the toxin, *Am. J. Med. Sci.*, **249**:133, 1965.

Phillips, A. W., et al.: Experimental preservation of fresh beef with antibiotics and radiation, *Food Technol.*, **15**:13, 1961.

Potter, L. F., and G. E. Baker: The role of fish as conveyors of microorganisms in aquatic environments, *Can. J. Microbiol.*, **7**:595, 1961.

Raj, H., and J. Liston: Survival of bacteria of public health significance in frozen sea foods, *Food Technol.*, **15**:429, 1961.

Read, R. B., Jr., et al.: Thermal resistance of salmonellae isolated from dry milk, *Appl. Microbiol.*, **16**:998, 1968.

Reed, J. R.: A review of botulism in the United States, *J. Environmental Health*, **27**:632, 1964.

Reith, J. F., and M. Szakály: Formation and stability of nitric oxide myoglobin. I. Studies with model systems, *J. Food Sci.*, **32**:188, 1967*a*; II. Studies on meat, *ibid.*, **32**:194, 1967*b*.

Rey, C. R., et al.: Microbial changes in meat during aging at elevated temperature and later refrigerated storage, *Food Technol.*, **24**:67, 1970.

Riemann, H., ed.: "Food-Borne Infections and Intoxications," New York, Academic Press, Inc., 1969.

Rizk, S. S., et al.: Disinfection of eggs artificially inoculated with salmonellae. I. Application of several disinfectants, *Poultry Sci.*, **45**:764, 1966.

Robinson, J. F., and C. H. Hills: Preservation of fruit products by sodium sorbate and mild heat, *Food Technol.*, **13**:251, 1959.

Sadler, W. W., and R. E. Corstvet: Second survey of market poultry for *Salmonella* infection, *Appl. Microbiol.*, **13**:348, 1965.

Sauter, E. A.: The effect of lye on the incidence of fluorescent spoilage in washed eggs, *Poultry Sci.*, **45**:131, 1966.

—— and C. F. Petersen: The effect of egg shell quality on penetration by *Pseudomonas fluorescens, Poultry Sci.*, **48**:1525, 1969.

Schaffner, C. P., et al.: Coconut and *Salmonella* infection, *Appl. Microbiol.*, **15**:471, 1967.

Scott, R. G.: Retailing of barbecued chickens: A survey in British Columbia, *Can. J. Public Health,* **59**:385, 1968.

Segmiller, J. L., et al.: The efficacy of nisin and tylosin lactate in selected heat-sterilized food products, *J. Food Sci.*, **30**:166, 1965.

Shewan, J. M.: The microbiology of sea-water fish. In "Fish as Food," edited by G. Borgstrom, New York, Academic Press, Inc., 1961.

—— et al.: The *Pseudomonas* and *Achromobacter* groups of bacteria in the spoilage of marine white fish, *J. Appl. Bact.*, **23**:463, 1960.

Simidu, U., and K. Hasuo: Salt dependency of the bacterial flora of marine fish, *J. Gen. Microbiol.*, **52**:347, 1968.

—— et al.: Microflora of fresh and stored flatfish, *Kareius bicoloratus, Bull. Japan. Soc. Sci. Fisheries,* **35**:77, 1969.

Simmons, E. R., et al.: Effect of moisture and temperature on ability of salmonellae to infect shell eggs, *Poultry Sci.*, **49**:761, 1970.

Simonetti, A. D'A., and S. U. D'Arca: *Cl. perfringens* e tossinfezioni alimentari, *Nuovi Ann. Igiene Microbiologia,* **28**:129, 1967.

Simonsen, B.: Aerobic spore-forming bacteria that "blow" canned meats, *Nord. Vet.-Med.*, **20**:121, 1968.

Sognefest, P., et al.: Effect of pH on thermal process requirements of canned foods, *Food Res.*, **13**:400, 1948.

Solberg, M., and W. E. Riha, Jr.: Microbial control using ultraviolet radiations. I, *Food Technol.*, **23**:83, 1969.

Southcott, B. A., and J. W. Boyd: Tetracycline antibiotics in shrimp preservation, *J. Fisheries Res. Bd. Can.*, **22**:117, 1965.

Splittstoesser, D. F., et al.: Control of microorganisms during preparation of vegetables for freezing. I. Green beans, *Food Technol.*, **15**:329, 1961*a*; II. Peas and corn, *ibid.*, **15**:332, 1961*b*.

Stavrić, S., et al.: Effects of γ-irradiation on *Escherichia coli* wild type and its radiation-resistant mutants. I. Post-irradiation synthesis of DNA, *Intern. J. Radiation Biol.*, **14**:403, 1968*a*; II. Post-irradiation degradation of DNA, *ibid.*, **14**:411, 1968*b*.

Stefanye, D., et al.: Amino acid composition of crystalline botulinum toxin type A, *J. Bact.*, **94**:277, 1967.

Steinkraus, K. H., and J. C. Ayres: Incidence of putrefactive anaerobic spores in meat, *J. Food Sci.*, **29**:87, 1964.

Stone, R. V.: A cultural method for classifying staphylococci as of the "food poisoning" type, *Proc. Soc. Exp. Biol. Med.*, **33**:185, 1935.

Stringer, W. C., et al.: Microbial profiles of fresh beef, *Food Technol.*, **23**:97, 1969.

Surkiewicz, B. F.: Bacteriological survey of the frozen prepared foods industry. I, *Appl. Microbiol.*, **14**:21, 1966.

Sutton, R. G. A., and B. C. Hobbs: Food poisoning caused by heat-sensitive *Clostridium welchii*. A report of five recent outbreaks, *J. Hyg.*, **66**:135, 1968.

Tarr, H. L. A.: Antibiotics as a preservative measure, National Conference, Canadian Institute of Food Technologists, Toronto, 1959.

——: Chemical control of microbiological deterioration. In "Fish as Food," edited by G. Borgstrom, New York, Academic Press, Inc., 1961.

Taylor, T. G.: How an eggshell is made, *Sci. Am.*, **222**:88, 1970.

Thatcher, F. S., and D. S. Clark, eds.: "Microorganisms in Foods," Toronto, University of Toronto Press, 1968.

—— and A. Loit: Comparative microflora of chlortetracycline-treated and nontreated poultry with special reference to public health aspects, *Appl. Microbiol.*, **9**:39, 1961.

—— and J. Robinson: Food poisoning: An analysis of staphylococcal toxins, *J. Appl. Microbiol.*, **25**:378, 1962.

Thomson, J. E., et al.: Effect of chlorine antibiotics, β-propiolactone, acids, and washing on *Salmonella typhimurium* on eviscerated fryer chickens, *Poultry Sci.*, **46**:146, 1967.

Thornley, M. J., et al.: The effects of antibiotics and irradiation on the *Pseudomonas-Achromobacter* flora of chilled poultry, *J. Appl. Bact.*, **23**:487, 1960.

Turner, R. J., and N. E. R. Campbell: A bacteriological survey of certain processed meats, *Can. J. Public Health*, **53**:382, 1962.

Vadehra, D. V., et al.: Infection routes of bacteria into chicken eggs, *J. Food Sci.*, **35**:61, 1970.

Van Schothorst, M., and E. H. Kampelmacher: *Salmonella* in meat imported from South American countries, *J. Hyg.*, **65**:321, 1967.

Varga, S., and G. W. Anderson: Significance of coliforms and enterococci in fish products, *Appl. Microbiol.*, **16**:193, 1968.

Volkova, L. P.: On the use of nistatin for combating meat mould, *Mikrobiologiya*, **30**:158, 1961.

Walker, E. A., et al.: The effect of chlortetracycline on eviscerated turkeys commercially processed in Great Britain, *J. Appl. Bact.*, **24**:104, 1961.

Watanabe, W.: Effect of gamma-ray upon food microorganisms. V. Studies on survival of *E. coli* irradiated with gamma-ray, *Bull. Agr. Chem. Soc. Japan*, **24**:84, 1960.

Weissman, M. A., and J. A. Carpenter: Incidence of salmonellae in meat and meat products, *Appl. Microbiol.*, **17**:899, 1969.

Wells, F. E., et al.: Growth of psychrophiles. II. Growth of poultry meat spoilage bacteria and some effects of chlortetracycline, *Food Sci.*, **28**:254, 1963.

Wheatley, W.: *Clostridium perfringens.* Food borne illness, *J. Environmental Health*, **31**:568, 1969.

Wheaton, E., and G. L. Hays: Antibiotics and the control of spoilage in canned foods, *Food Technol.*, **18**:147, 1964.

Wilkerson, W. B., et al.: Occurrence of enterococci and coliform organisms on fresh and stored poultry, *Food Technol.*, **15**:286, 1961.

Williams, J. E., and A. D. Whittemore: A method for studying microbial penetration through the outer structures of the avian egg, *Avian Diseases*, **11**:467, 1967.

Williams, L. P., and K. W. Newell: Sources of salmonellas in market swine, *J. Hyg.*, **66**:281, 1968.

Woodburn, M., and W. J. Stadelman: Salmonellae contamination of production and processing facilities for broilers and ducklings, *Poultry Sci.*, **47**:777, 1968.

Yates, A. R., and M. M. Aref: Comparison of liquid nitrogen and conventional freezing methods on bacterial count in frozen vegetables, *Can. Inst. Food Technol. J.*, **2**:1, 1969.

Yuzawa, S.: Evaluation of tylosin as a food preservative. (1) Antimicrobial action of tylosin, *J. Med. Soc. Toho Univ.*, **15**:498, 1968.

Zehren, V. L., and V. F. Zehren: Relation of acid development during cheese-making to development of staphylococcal enterotoxin A, *J. Dairy Sci.*, **51**:645, 1968.

Zindel, H. C., and M. V. Bennett: Salmonellae in poultry feeds, *Poultry Sci.*, **47**:1925, 1968.

24 *Bacteriology of soil*

A soil is the product of weathered rock. However, a mass of weathered rock does not constitute a soil without the intervention of living processes. Soil has been subjected to the action of air and water, which have altered and removed some of the original components so that the proportions of the various substances are not the same as in the parent rock. The mineral particles constitute the basis or foundation of soil but not the whole of it (Fig. 293).

Each kind of soil has its own individuality. The distinctive feature of this individuality is the soil profile, which consists of a series of layers different from one another in color, texture, structure, and other ways. Each layer is called a horizon, and the succession of layers down to the weathered rock forms the complete soil profile. The profile is best studied in soil that has not been disturbed by cultivation, because horizons near the surface become destroyed by mixing, which occurs during cultivation.

Numbers and kinds of organisms in soil

Soils are excellent culture media for the growth of many kinds of organisms. The microscopic life of soils includes bacteria, yeasts, molds, algae, diatoms, and protozoa. The latter includes amoebae, flagellates, ciliates, and rotifers. In addition, there are present various nematodes, insects, etc.

A spoonful of fertile soil is said to contain more microorganisms than there are people in this world. Multiply this by the number of spoonfuls of soil in an acre, and you have figures which are astronomical. The bacteria in an acre of soil of average fertility would weigh as much as a medium-sized dairy cow. This seething mass of microorganisms constitutes a crop of 3 to 5 tons per acre-foot of soil that the farmer grows beneath the surface, in addition to the crop he grows above ground. If this crop of organisms beneath the surface is not fed adequately, the crop above ground may suffer from competition, disease, or other adverse effects of the microbes. Without the microorganic life, the soil would not be the dynamic, perpetual system that sustains all plants and indirectly all animal life.

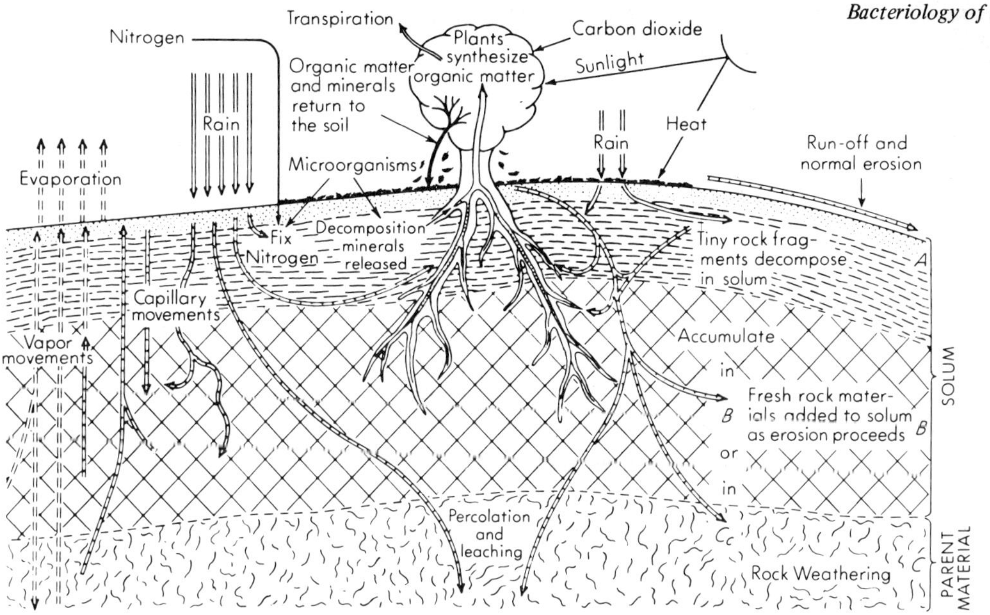

Fig. 293 Diagram showing the interlocking cycles leading to the conversion of rock into soil, making for the upkeep of a cover of vegetation on the ground and the maintenance of the soil. (From Williamson.)

Most soil organisms are found in the surface layers. The numbers decrease with depth. A well-aerated soil contains more organisms than one lacking in an abundance of oxygen. The numbers and kinds of organisms found in soil depend upon the nature of the soil, depth, season of the year, state of cultivation, reaction, amount of organic matter, temperature, moisture, etc.

Anaerobic conditions commonly exist in the soil. Carbon sources, as plant and animal residues, introduced into the soil are attacked by the microbial population resulting in a decrease of available O_2 accompanied by an increase of CO_2. The composition of the soil atmosphere, therefore, fluctuates between extremes of high and low tensions of O_2 and CO_2. Stotzky and Goos (1965) prepared plates from four samples of tropical soils to determine the effects on growth of CO_2, O_2, and mixtures of the two gases. Incubation of plates in atmospheres containing more than 90 percent CO_2 reduced the number of developing organisms. Inhibition by high CO_2 tensions was the result primarily of CO_2 concentration and not absence of O_2. A higher percentage of organisms developed in an atmosphere of 100 percent N_2 than in CO_2 tensions less than 100 percent.

Methods are available for counting the organisms in soil as well as for isolating the various species in pure culture. Since the organisms may vary considerably in their growth requirements, many types of culture media must be employed. The organisms may be aerobic, anaerobic, or facultative types.

Many of the species of organisms present in soil grow in association with others. The phenomena of symbiosis, synergism, commensalism, and antagonism are believed to be of common occurrence. Two species growing together frequently elaborate metabolic waste products not produced by either organism when grown in pure culture. A product of metabolism of one organism may serve as a nutrient for another species. Antagonistic organisms are usually present and serve to combat other species, especially certain important plant pathogens. This explains why it is exceedingly difficult to determine from laboratory studies on pure cultures what actually takes place in the soil. The various types of associations are discussed in Chap. 17, page 601.

In general, the same media used for the cultivation of heterotrophic bacteria are employed for the isolation of most soil species. However, the soil contains some species which do not grow on the usual culture media. Special media and methods must be employed for their cultivation. These include symbiotic nitrogen-fixing bacteria, nonsymbiotic nitrogen-fixing bacteria, sulfur-oxidizing forms, sulfate-reducing species, urea-decomposing bacteria, cellulose-decomposing forms, and ammonia-oxidizing species.

Functions of organisms in soil

One of the important functions of soil organisms is to decompose various kinds of organic matter of plant and animal origin. This includes stable manures, green manures, plant stubble, plant roots, organic fertilizers, and other products. The decomposition of such compounds is the result of the activities of bacteria, molds, protozoa, worms, and other organisms present in the soil. Each group selects certain constituents of the organic matter suitable for synthesizing its own characteristic protoplasm.

The organic compounds added to soil as a result of biological action include various sugars, pentosans, celluloses, lignins, proteins, amino acids, fats, waxes, tannins, and pigments. These are decomposed further, resulting in the liberation of soluble organic and inorganic compounds. The inorganic compounds, notably NH_3 and its salts, may be utilized by plants as a source of nitrogen.

Organic materials, especially stable and green manures, are said to produce four distinct effects upon soil processes and upon plant growth:

1. They supply inorganic nutrients to plants, especially nitrogen and phosphorus.
2. They affect the physical conditions of the soil, especially the moisture-holding and buffering capacities.
3. They supply certain specific elements that may be limiting factors for the growth of some plants.
4. They favor the development of organisms that secrete substances antagonistic to the growth of certain specific forms responsible for plant diseases.

Higher plants are indispensable to human welfare, and the activities of microbes in the soil are intimately related to plant growth. If it were not for the growth of organisms in soil, all animals, including man, and plants would soon perish. Therefore, life could not exist in the complete absence of soil microorganisms.

Biology of soil

Soil structure has a significant influence on erosion, water intake, and crop growth. A stable granulated soil will permit rapid water uptake, drainage, aeration, and beneficial microbic activity, whereas a dispersed or compact soil has a low infiltration rate. Crops grow more luxuriantly in fertilized, well-aerated, drained, and granulated soils. The most important factors in maintaining or improving the granulation of soils are the presence of microorganisms and their decomposition products (Fig. 294).

A good soil is made up of about 5 percent organic and 95 percent inorganic matter. The influence the organic matter exerts is out of all proportion to its weight.

Organisms are active in the formation and stabilization of soil structure. A good soil structure is essential for the smooth integration and

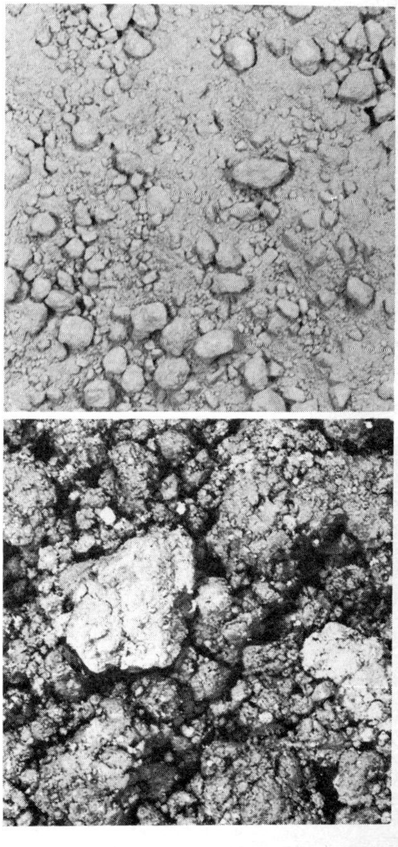

Fig. 294 Soil structure developed by fungi. Top, untreated Peorian loess; bottom, Peorian loess in which the growth of fungi has been greatly stimulated. (After McCalla. Courtesy of Soil Conservation Service.)

performance of the multigroups of organisms in soil. A deterioration of soil structure suppresses the activities of the beneficial groups and promotes the growth of organisms that render nutrients unavailable. An unfavorable plant root medium soon develops.

Good soil structure (1) increases water intake and drainage; (2) has a direct effect on plant growth; (3) influences biological activity such as nitrogen fixation, nitrification, and decay of organic matter; (4) holds to a minimum the activities of anaerobic organisms which reduce sulfates and nitrates to products unavailable or toxic to plants.

A desirable soil structure can be maintained by following a rotation including a sod crop, and frequent use of manures and crop residues.

For more information: Bollen and Glennie (1961); Chesters et al. (1957); Martin and Richards (1963); McCalla (1958); Aspiras et al. (1971).

Nitrogen cycle

Nitrogen is the cornerstone of the structural requirements of all living cells. It is absolutely necessary for the growth of all organisms. Without an available supply of this element, life could not endure. The biological fixation of nitrogen far exceeds all other cycles in magnitude.

A continuous transformation of nitrogen takes place in the soil by various groups of organisms. In the breakdown of protein, the first step is a hydrolysis of the molecules to their constituent building stones, or amino acids. The amino acids are then deaminized with the liberation of ammonia:

1. Proteins + $H_2O \rightarrow$ amino acids.
2. Amino acids + $H_2O \rightarrow$ ammonia + carbon residue.

After ammonia has been liberated from various nitrogenous compounds, it may be (1) assimilated by soil organisms and again synthesized into proteins, (2) used by higher plant life as a source of nitrogen, (3) absorbed by the colloidal substances in soil and bound as ammonia, and (4) acted upon by other soil forms and oxidized first to nitrites and then to nitrates. The organisms responsible for this last set of reactions belong to the autotrophic group of bacteria and are incapable of utilizing organic compounds for structural or energy purposes.

The nitrates may be utilized by various microorganisms and by higher plants and synthesized into proteins, or they may be reduced first to nitrites and finally to free nitrogen. The free nitrogen is lost as far as being available to plant and most microscopic soil life. However, certain bacteria found in the soil have the ability to utilize free nitrogen and make it available to plant life.

Two groups of organisms are responsible for nitrogen fixation. The organisms in one group are nonsymbiotic nitrogen fixers; those in the other group fix nitrogen only when growing in symbiosis on the roots of certain plants. The free nitrogen is transformed into organic compounds which are made available to plants and soil organisms. The nitrogen

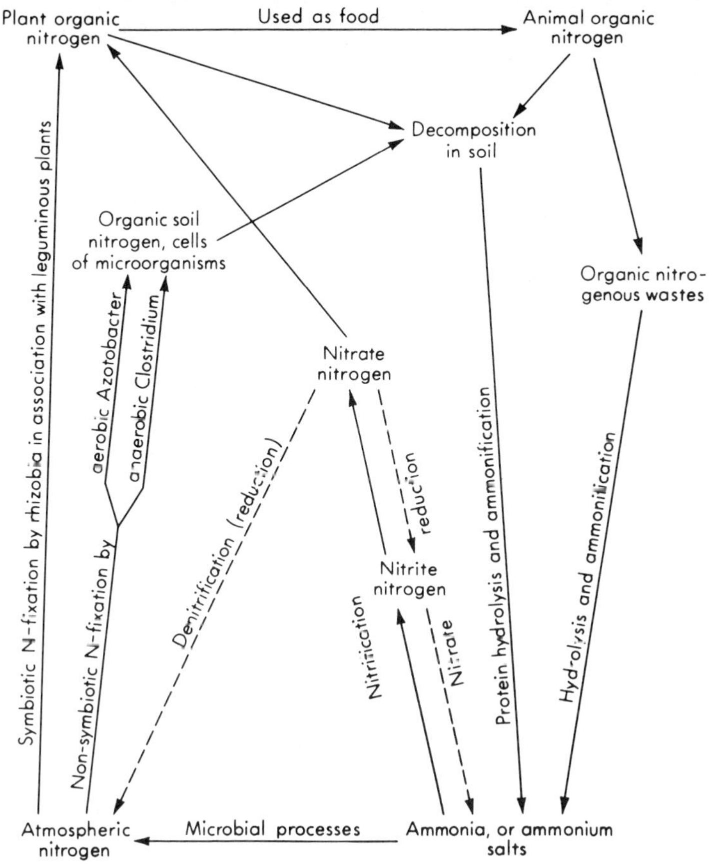

Fig. 295 *The nitrogen cycle.* (After Allen.)

compounds are then synthesized into proteins. The nitrogen cycle is diagrammed in Fig. 295.

The amount of energy derived by bacteria from the various reactions in the nitrogen cycle, according to Delwiche (1970), is shown in Table 30. The most profitable are the denitrification reactions whereas the least amount of energy is released in the nitrification reactions. Only nitrogen fixation (last step) calls for an input of energy.

For more information: Delwiche (1970).

Quantitative examination of soil

Soil is the natural habitat of a considerable number of bacterial species, varying from strict autotrophic to heterotrophic forms. Because of this fact, many types of culture media are required to obtain an accurate appraisal of the numbers and kinds of organisms present. Such a procedure would be out of the question where a rapid determination of the approximate numbers present in a soil sample is desired.

Table 30 Energy yields from various reactions of the nitrogen cycle.

Reaction	*Energy yield (kilocalories)*
Denitrification	
1. $C_6H_{12}O_6 + 6KNO_3 \longrightarrow 6CO_2 + 3H_2O + 6KOH + 3N_2O$ (Glucose; Potassium nitrate; Potassium hydroxide; Nitrous oxide)	545
2. $5C_6H_{12}O_6 + 24KNO_3 \longrightarrow 30CO_2 + 18H_2O + 24KOH + 12N_2$	570 per mole of glucose
3. $5S + 6KNO_3 + 2CaCO_3 \longrightarrow 3K_2SO_4 + 2CaSO_4 + 2CO_2 + 3N_2$ (Sulfur; Calcium carbonate; Potassium sulfate; Calcium sulfate)	132 per mole of sulfur
Respiration	
4. $C_6H_{12}O_6 + 6O_2 \longrightarrow 6CO_2 + 6H_2O$	686
Ammonification	
5. $CH_2NH_2COOH + 1\frac{1}{2}O_2 \longrightarrow 2CO_2 + H_2O + NH_3$ (Glycocoll; Oxygen; Ammonia)	176
Nitrification	
6. $NH_3 + 1\frac{1}{2}O_2 \longrightarrow HNO_2 + H_2O$ (Nitrous acid)	66
7. $KNO_2 + \frac{1}{2}O_2 \longrightarrow KNO_3$ (Potassium nitrite)	17.5
Nitrogen fixation	
8. $N_2 \longrightarrow 2N$ (activation of nitrogen)	–160
9. $2N + 3H_2 \longrightarrow 2NH_3$	12.8

Soil colloids Bacterial organisms are found chiefly in the layer of colloidal material surrounding the inorganic particles of soil. Conn and Conn (1940), McCalla (1940), and others noted that bacteria grew better in the presence of colloidal clay than in its absence. McCalla attributed the stimulation of bacterial growth to the catalytic effect of the clay in speeding up biochemical reactions, either by providing a more efficient utilization of nutritive material or by decreasing the toxic effects of waste products by adsorbing them.

Bacteria in contact with soil colloids adsorb cations. Under normal conditions, the solid material in the soil constitutes about 80 percent and the water content about 20 percent. The water is present around and

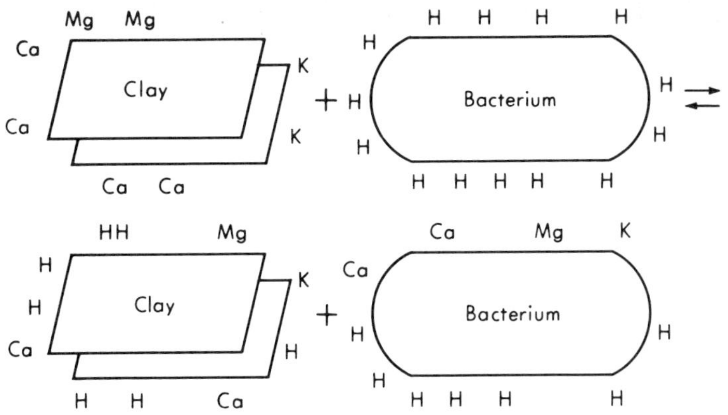

Fig. 296 Mechanism of ion exchange between soil bacteria and colloidal clay. (After McCalla.)

between the particles of soil. Normally, the bacteria probably live in the water films that adhere to the surface of the colloid particles containing adsorbed ions. On the basis of the displacement of adsorbed methylene blue, the ions would be expected to be adsorbed by the bacteria from the soil colloids in the following series:

$$H > Al > Fe > Mn > Ba > Ca > Mg > K > NH_4 > Na$$

McCalla suggested that, in the adsorption of nutrients from soil by bacteria, an exchange of adsorbed bases occurred between bacteria and soil colloid as shown in Fig. 296. In bacterial metabolism H^+ are produced which may be adsorbed at the cell's surface. When a colloidal clay particle, saturated with adsorbed bases, contacts a bacterium saturated with H^+, an exchange of ions takes place until an equilibrium is reached. As this equilibrium is upset by the more complete utilization of the adsorbed basic ions in bacterial metabolism, a further exchange may occur, the colloidal clay functioning as a constant reservoir for basic ions utilized in the growth of the bacteria.

It is difficult to remove or separate the organisms from this colloidal layer. This means that the number of colonies appearing on an agar plate is not an accurate index of the population of a soil sample. Also, no single culture medium is satisfactory for the growth of all species present in soil.

Two methods are generally employed for estimating the numbers of microorganisms in soil: (1) the agar plate method and (2) the direct microscopic method.

Agar plate method A weighed sample of soil is mixed with a known volume of sterile water contained in a screw-cap bottle. The sample is shaken vigorously to separate as many organisms as possible from the colloidal material surrounding the soil particles. After the coarse particles have settled, a series of dilutions is prepared from the suspension. Aliquot portions from each dilution are transferred to petri dishes and mixed with

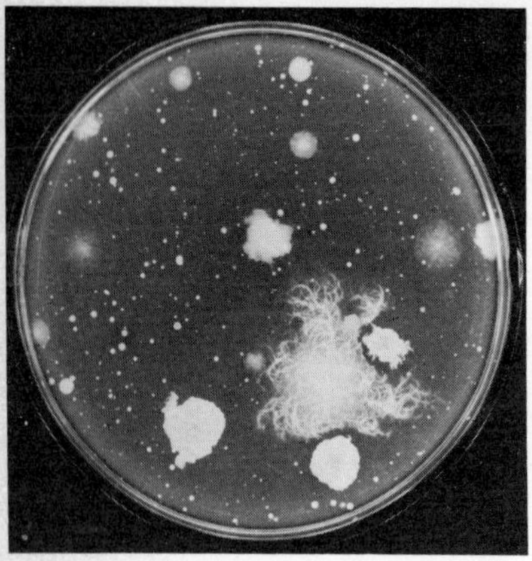

Fig. 297 Quantitative examination of soil by the agar plate method.

melted agar. The plates are inverted and incubated at 25°C for 2 to 14 days. The colonies are counted and the results expressed as the number of organisms per gram of soil (Fig. 297).

Numerous media have been used for isolating soil bacteria. A medium containing a small amount of organic matter is more satisfactory than one rich in this constituent. Peptone is particularly unsatisfactory as it permits the overgrowth of spore formers and proteolytic *Pseudomonas*, which prevent the development of the slow-growing species. Results are also unsatisfactory if the temperature of incubation is higher than 25°C.

Objections to the agar plate method The enumeration of the soil population by this method presents several serious errors. Obligate anaerobes do not grow in the presence of oxygen; autotrophic bacteria fail to multiply in organic media; nonsymbiotic nitrogen fixers grow to a limited extent only; many cellulose-decomposing organisms do not grow on the commonly used media; sulfate reducers grow only in a medium containing sulfate; etc. The counts represent only a fraction of the total bacterial population of soil. However, the method does give an estimate of the number of organisms in soil capable of growing on a nutrient agar medium.

Molds are commonly present in soil, although the numbers appearing on agar plates represent only a small percentage of the total counts. Yeasts are less prevalent than molds, increasing in numbers in acid soils and in soils of orchards and vineyards.

The counts by the agar plate method vary from 200,000 to 100 million per gram of soil. The colonies appearing on plates consist of about 10 to 40 percent actinomycetes, 50 to 80 percent non-spore-formers, and 3 to 10 percent sporebearers.

For more information: Egdell et al. (1960).

Direct microscopic method One part of soil is suspended in 10 times its weight of a 0.015 percent aqueous solution of agar. The purpose of the agar is to fix the organisms to the slide. An area of 4 sq cm is ruled off on a microscope slide by means of a china-marking pencil. One-tenth milliliter of the soil suspension is transferred to the ruled area and spread out uniformly. The slide is dried on a flat surface over a water bath, and then covered with a solution consisting of 1 percent rose bengal dissolved in a 5 percent aqueous solution of phenol.

In a good preparation, the bacteria take a deep pink or red color, whereas the mineral constituents do not stain. Some of the dead organic matter appears light pink, but most of it stains either yellow or not at all. If the bacteria appear faintly stained or if everything is colored pink, a new preparation should be made. The former condition generally means that the slide has been washed too long; the latter indicates that the staining solution is too old.

The slide is examined under a calibrated oil-immersion objective and the number of organisms per field counted. At least 25 fields are counted and an average taken. Knowing the area and count per field, it is a simple matter to calculate the number of organisms present in a gram of soil.

The direct counts are from 5 to 20 times greater than by the agar plate method. Much of the discrepancy is due to bacteria which fail to grow on the plates rather than to clumps of organisms which do not break up in the plating method.

Objections to the direct method The method records dead organisms which do not develop on agar plates. The organisms must be evenly distributed over the slide. It is advisable to prepare several slides and to take an average of the counts. A disturbing factor is the difficulty of recognizing the bacteria. Many soil forms are too small to be easily distinguished from soil particles. Also many soil particles resemble bacteria. Therefore, considerable experience is necessary in making accurate determinations of the soil population by the direct method.

For more information: Casida (1962); Santoro and Stotzky (1967); Skinner et al. (1952).

Variations in soil counts The numbers of organisms in soil are not uniform even over a very small area. In order to increase the accuracy of the determination, several samples from the same plot should be collected and an average taken of all determinations for computing the final count. A single determination may be considered as valueless for calculating the soil population.

It has been reported that if soil is partially sterilized with steam, or with a volatile antiseptic such as toluene, the bacterial count first decreases, followed by a sharp increase in numbers and activity. If a little untreated soil is added, the bacterial count again decreases. The partial sterilization of the soil destroys the protozoa but not the bacteria and is followed by a sharp increase in the latter. The addition of an untreated soil reintroduces protozoa and results again in a decrease in bacterial numbers. The important protozoa responsible for the daily variations in bacterial numbers include ciliates, flagellates, and rhizopods. These organisms are widely distributed in soils. Therefore, if soil determinations are to have any value, samples should be taken from different areas of the same plot and at frequent intervals over a long period.

Soil bacteria differ in their edibility by protozoa. Certain groups, such as the root-nodule bacteria, generally are resistant to attack by micropredators, whereas some strains of *Aerobacter* are attacked by all soil protozoa which have been studied.

Qualitative studies of soil organisms

Lochhead (1940) showed that the qualitative nature of the soil microflora was markedly influenced by the growing plant. In the rhizosphere (the zone influenced by root excretions), the gram-negative rods were increased in numbers, whereas the gram-positive rods, cocci, and spore bearers were less abundant.

Holding (1960) reported that the average proportion of gram-negative bacteria in plant-free soil was 7 percent, whereas in the rhizosphere it was 20 percent. *Pseudomonas* species made up a large proportion of the gram-negative flora. Other species present were members of the genera *Xanthomonas, Chromobacterium, Agrobacterium, Flavobacterium, Achromobacter, Alcaligenes, Aeromonas, Aerobacter, Bacterium,* and *Cytophaga.*

Chan and Katznelson (1961) set up a model system using a species of *Pseudomonas* as a representative of the rhizosphere, and *Arthrobacter globiformis* (gram-positive) as a representative of the plant-free soil flora. *A. globiformis* was strongly suppressed in the presence of the *Pseudomonas* in a medium containing root extracts of mature plants. The *Pseudomonas* was unaffected in the association.

Most rhizosphere bacteria are saprophytes. Some live on the root surface; others penetrate the roots. Some are restricted to the cortical cells, whereas others go deeper, passing between the cells and invading them. The bacteria may be harmless, or they may exert a favorable or a harmful effect on the development of the host.

Owens et al. (1969) exposed soil to volatile compounds that emanated from alfalfa and other plant residues, and reported an immediate rise in the respiration rate of the soil microflora, which was followed by an increase in the numbers of bacteria and molds. Of the 20 or more volatile compounds isolated, acetaldehyde, isobutyraldehyde, isovaleraldehyde, 2-methylbutanal, and/or valeraldehyde, methanol, and ethanol were identified as the active compounds. Acetaldehyde alone accounted for almost one-half of the increase in soil respiration and caused preferential increases in numbers of soil molds.

Bacteria in the rhizosphere differ from those beyond this zone in being physiologically more active. Also they show (1) a greater proportion of motile forms, (2) a pronounced increase in chromogenic bacteria, and (3) a higher incidence of gelatin liquefiers and glucose fermenters.

This increased growth is unquestionably related to the supply of inorganic and organic nutrients at the root-soil interface. Plant excretion of inorganic substances, growth factors, amino nitrogen, and a variety of other substances have been generally accepted. Also the food materials supplied by sloughed-off root caps, root hairs, and cortical and epidermal cells should not be overlooked, since they are rapidly attacked by soil organisms and supply considerable quantities of organic and inorganic nutrients.

Plant roots affect microbial growth, and the plant in turn is affected by the increased activity of the microbial population of the rhizosphere. The information available is inadequate to indicate whether plant development is increased or impaired by the presence of the rhizosphere organisms.

For more information: Babieva and Belianin (1966); Canada Department of Agriculture (1938–1957); Chan et al. (1963); Dart and Mercer (1964); Iswaran and Sen (1961); Lochhead (1958*a*, *b*); Rouatt and Katznelson (1961); Rovira (1965); Starkey (1958); Vágnerová and Vančura (1962); Trolldenier (1971).

Classification of soil bacteria

Winogradsky placed all soil bacteria into either of two groups, (1) the zymogenous or (2) the autochthonous types.

Zymogenous bacteria The zymogenous types consist of actively fermenting organisms requiring for their activity ingredients which are quickly exhausted. They are involved in processes in which organic matter is made available to plants. The added organic matter is rapidly attacked in successive stages, each of which involves a specific group of organisms. The organisms increase rapidly whenever furnished with the special nutrients to which they are adapted. When the special nutrients are exhausted the organisms return again to low numbers until another occasion for active growth occurs. Some of these organisms require special media for laboratory cultivation; others grow well on ordinary media.

Organisms in this group include the nitrifiers, nitrogen fixers, cellulose-hydrolyzing bacteria, sulfur-oxidizing forms, acid-fast species, molds, streptomycetes, sporeforming *Bacillus* species, and non-sporeforming *Pseudomonas* or closely related species.

Autochthonous bacteria The autochthonous types are more familiarly known as the indigenous or native bacteria. They are found in soil in fairly high and quite constant numbers. They do not change in numbers by the presence or absence of specific nutrients. The exact function of these organisms in soil is not clearly understood. They appear to utilize the same nutrients (salts) as higher plants and display a low level of activity in soil. They utilize the same sources of nitrogen as plants and may be looked upon as rivals. Since bacteria are short-lived and are readily autolyzed after death, their rivalry with plants cannot be considered serious. The autochthonous bacteria produce weak fermentation reactions which make their classification difficult. Because of this fact, they are classified almost entirely on morphology.

The group includes some of the streptomycetes, which are more appropriately classified here rather than under the zymogenous bacteria.

A large number of closely related soil species have characteristics intermediate between the true bacteria and the streptomycetes. The organisms are somewhat pleomorphic. In young cells they appear as rods which may vary in size and shape from straight to bent, curved, swollen, or club-shaped forms. Short filament formation with rudimentary budding may occur. These organisms are placed in the genus *Arthrobacter*. They resemble streptomycetes only in their ability to produce occasional short mycelia.

A number of small, nonpleomorphic, nonsporing, motile rods are found in soil; they differ from the *Arthrobacter* in showing no tendency toward filament formation or the appearance of coccoid forms. These organisms are placed in the genus *Agrobacterium*. Most of these species are plant pathogens.

For more information: Bowie et al (1969); Skyring and Quadling (1969).

Production of ammonia

Most species of bacteria are capable of decomposing proteins and protein-split products with the liberation of ammonia as one of the compounds. The ammonia is released, prior to the utilization of the carbon residue, chiefly for energy. The production of ammonia is an essential stage in the formation of nitrate in soil. Most plant crops are largely dependent on soil nitrates for structure and growth.

The decomposition of nitrogenous compounds for energy takes place only in the absence of a rapidly utilizable carbohydrate. In the presence of a rapidly fermentable carbohydrate, bacteria derive their energy from this source, utilizing the nitrogenous compounds for structure only. This results in a greatly lowered production of ammonia and nitrate. Therefore, maximum ammonia production takes place only in the absence of a rapidly fermentable compound.

Urea-hydrolyzing bacteria Compounds other than proteins and their degradation products are also capable of yielding ammonia by bacterial action.

Stable and barnyard manures are often used as fertilizers because of their nitrogen content. They help to replenish the nitrogen supply of the soil. A high content of urine is often present in such waste material. The most important nitrogen compound present in urine is urea. Many organisms have the power of converting the urea to ammonium carbonate and finally to free ammonia and carbon dioxide, according to the equations

$$O{=}C\begin{matrix}\diagup NH_2 \\ \diagdown NH_2\end{matrix} + 2H_2O \longrightarrow (NH_4)_2CO_3$$

$$(NH_4)_2CO_3 \longrightarrow 2NH_3 + CO_2 + H_2O$$

The presence of urea-hydrolyzing organisms in soil is demonstrated by incorporating urea in a special medium. The liberation of ammonia from the medium indicates the presence of organisms capable of attacking the compound. The important urea-decomposing bacteria include (1) *B. pasteurii*, a gram-variable spore-forming rod with rounded ends, usually not in chains, producing round, terminal to subterminal spores which cause a bulging of the sporangia, optimum growth temperature 28 to 35°C; (2) *Micrococcus ureae*, a gram-variable, nonmotile, aerobic coccus, appearing singly, in pairs, and in clumps, optimum temperature 25°C; (3) *Sarcina ureae*, a gram-positive coccus, occurring singly, in pairs, and in packets, motile by means of a single flagellum, endospores of an unusual type, located centrally, resist heating to 80°C for 10 min, optimum growth temperature 20°C.

The urea bacteria thrive best in media containing urea, especially when made alkaline with ammonium carbonate. The organisms are capable of converting urea rapidly to ammonium carbonate. They are commonly found in air, water, soil, and manure. About 2 percent of the organisms present in surface soil and about 10 percent in manure are capable of decomposing urea.

In addition to the above, some members of the genus *Proteus* may utilize urea. This applies also to many coliform and paracolon cultures, especially to the *Aerobacter*.

For more information: Bornside and Kallio (1956*a*, *b*).

Digestion of cellulose

The greater part of the organic matter in soil is decomposed by bacteria in the process of acquiring energy. The simpler carbohydrates and some of the polysaccharides are attacked by a large number of soil bacteria. The addition of such compounds to soil causes a rapid increase in the numbers of the zymogenous species.

Cellulose constitutes the chief part of the solid framework (cell walls) of plants. It is one of the most important constituents added to soil. Under normal conditions of temperature and moisture, cellulose disappears almost completely and quite rapidly.

The cellulose molecule is built up of units of β-glucose. Two molecules of β-glucose are combined through a 1,4-linkage to give β-cellobiose. The cellulose molecule is a simple linear polymer of 1000 to

10,000 units of cellobiose linked end to end through 1,4-β-glucosidic linkages:

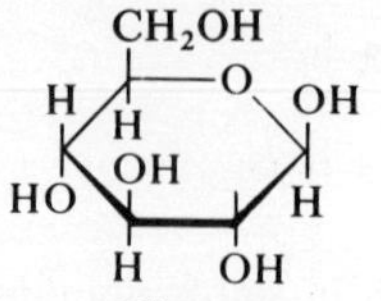

β-Glucose

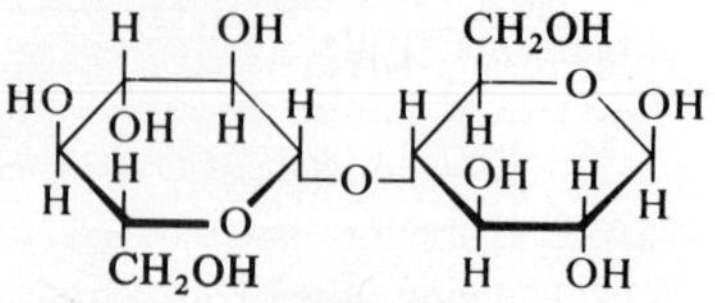

β-Cellobiose

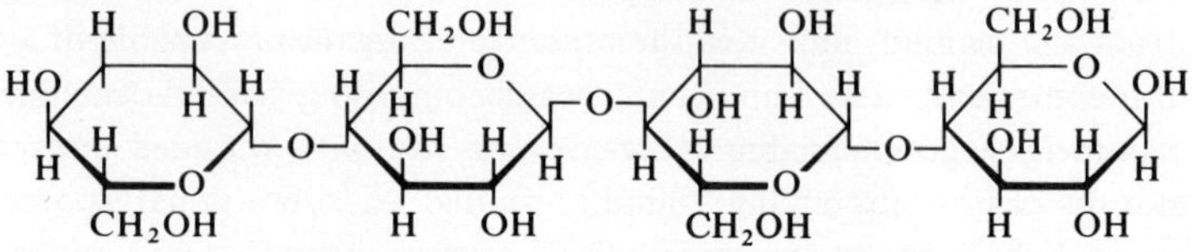

Cellulose chain

Mark et al. (1969) considered the basic protofibril of native cellulose as a helical spring of diameter $2R = 35$ Å; pitch $h = 40$ Å; and rectangular cross section of $a = 7.86$ Å, and $b = 35$ Å (Fig. 298).

Two enzymes are believed to be involved in the hydrolysis of cellulose. The first enzyme β-1,4-glucan glucanohydrolase, hydrolyzes cellulose to cellobiose. The second enzyme, β-D-glucoside glucohydrolase, hydrolyzes cellobiose to two molecules of β-glucose. The enzymes are of tremendous importance in the dissolution of insoluble cellulose in the soil. Since animals do not elaborate β-1,4-glucan glucanohydrolase, the presence of cellulose-hydrolyzing bacteria in the intestines of herbivorous animals is responsible for the hydrolysis of some cellulose.

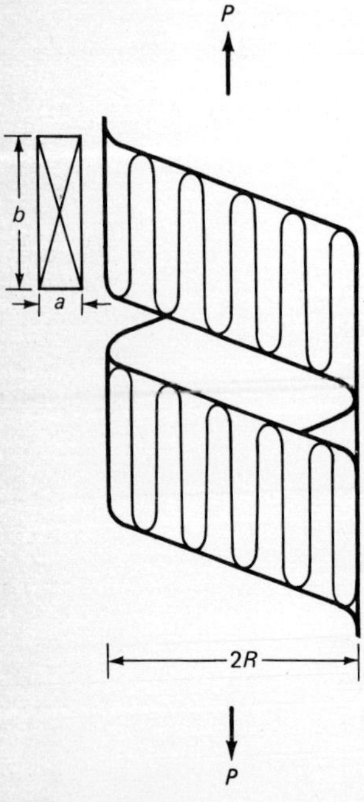

Fig. 298 The basic protofibril of native cellulose. See text for description. (After Mark, Kaloni, Tang, and Gillis.)

Organisms involved Cellulose is attacked by a wide variety of soil bacteria, including (1) aerobic mesophilic species, (2) aerobic thermophilic forms, (3) anaerobic mesophils, and (4) anaerobic thermophilic types. The organisms utilize cellulose as a source of energy. In addition to the bacteria, a number of molds and protozoa are able to attack cellulose.

Aerobic species Almost all the aerobic mesophilic cellulolytic bacteria so far isolated have been placed in the following genera: (1) *Cellfalcicula*, short rods or spindles with pointed ends, gram-negative; (2) *Cellulomonas*, small pleomorphic rods, gram-negative; (3) *Cellvibrio*, long, slightly curved rods with rounded ends, gram-negative; (4) *Cytophaga*, flexible, sometimes pointed rods, gram-negative; (5) *Pseudomonas*, rods, sometimes coccoid, gram-negative; (6) *Sporocytophaga*, vegetative cells, rod-shaped, microcysts spherical; (7) *Vibrio*, short, curved rods, gram-negative. With the exception of *P. erythra*, all are capable of growing well in the presence of a variety of carbohydrates.

The presence of aerobic cellulolytic bacteria can be easily demonstrated by adding soil to an inorganic medium containing cellulose (filter paper, cotton) and incubating for several days at the appropriate temperature. The cellulose is slowly attacked and solubilized.

Anaerobic species The anaerobic cellulolytic bacteria may be classified as follows: (1) mesophilic non-spore-forming rods and cocci, (2) mesophilic spore formers, (3) thermophilic spore producers, and (4) streptomycetes. Most of the anaerobic forms encountered in nature are members of the genus *Clostridium*.

Some anaerobic species are capable of dissolving cellulose at temperatures of 60 to 65°C. These organisms are widely distributed in nature, being found in soil, river mud, and the intestinal contents of animals. They may be isolated by inoculating an inorganic medium containing cellulose with an infusion of rapidly decomposing manure and incubating anaerobically at 65°C. If the organisms are present, the cellulose will be dissolved in 6 to 8 days.

The end products of bacterial action on cellulose consist chiefly of acetic, butyric, formic, lactic, propionic, and succinic acids; alcohol; carbon dioxide; and hydrogen. Cellulolytic organisms show considerable variation in the kinds and quantities of end products produced.

Molds A considerable number of molds and mold-like organisms have been shown to be capable of utilizing cellulose. The species which have been most extensively investigated are members of the genera *Alternaria, Aspergillus, Chaetomium, Cladosporium, Curvularia, Fusarium, Merulius, Monilia, Mucor, Myrothecium, Paxillus, Penicillium, Polyporus, Rhizoctonia, Rhizopus, Stachybotrys, Streptomyces,* and *Trichoderma*.

Molds are commonly found growing on old paper, especially if kept in a damp or humid atmosphere. They have been responsible for the destruction of many valuable papers and books.

For more information: Fergus (1969); Kadota (1956); King (1961); Kistner and Gouws (1964); McBee (1950); Simpson and Marsh (1964); Went and de Jong (1966); Leatherwood and Sharma (1972).

Biological nitrogen fixation

Relatively few species of organisms are able to take free nitrogen gas from the atmosphere and convert it into utilizable compounds. Some of these species are free-living and capable of fixing nitrogen without the aid of other living forms. They are referred to as nonsymbiotic nitrogen-fixing organisms. Others fix free nitrogen only when growing in symbiosis with certain plants. These are called symbiotic nitrogen-fixing organisms.

Nonsymbiotic nitrogen fixation Bacterial species which are capable of fixing nitrogen nonsymbiotically in the soil have been reported from the following genera: (1) *Achromobacter*, (2) *Azotobacter*, (3) *Bacillus*, (4) *Chlorobium*, (5) *Chromatium*, (6) *Clostridium*, (7) *Derxia*, (8) *Pseudomonas*, (9) *Rhodomicrobium*, (10) *Rhodopseudomonas*, (11) *Rhodospirillum*, and (12) *Streptomyces*. Of these, the most important species are members of the genus *Azotobacter*.

Azotobacter The cells are relatively large rods or even cocci, sometimes almost yeast-like in appearance. Endospores are not produced. However, the presence of cysts has been recognized. These structures fall in between vegetative cells and endospores in their resistance to harmful agents. Flagellation is typically peritrichous. They are gram-negative, obligate aerobes, usually growing in a film on the surface of media. They are capable of fixing atmospheric nitrogen when provided with carbohydrate or another energy source. Cells grow best on media deficient in nitrogen. They are soil and water bacteria (Fig. 299).

In the presence of nitrogen compounds such as ammonium salts, nitrate, urea, aspartic acid, and asparagine, the organisms fail to fix nitrogen. However, the relative availability and probable occurrence of fixed nitrogen compounds in soils are such that they would be unlikely markedly to retard nitrogen fixation by *Azotobacter* under most conditions.

Isolation of Azotobacter The presence of *Azotobacter* may be demonstrated by adding a small amount of fertile soil to a mineral medium containing mannitol. The culture is incubated at 25°C. After several days, a pellicle begins to form on the surface of the medium. A smear prepared from the pellicle, then stained, reveals the presence of many typical cells of *Azotobacter*. Depending upon the species, the cells may or may not be surrounded by slime. New cultures are prepared by transferring some of the pellicle to fresh medium. After several transfers, the culture becomes heavily enriched in *Azotobacter*. A loopful of the pellicle is now streaked over the surface of the same medium solidified with agar. The plate is incubated at 25°C until colonies appear. A pure culture is obtained by transferring a typical colony to fresh medium.

The organisms are widely distributed in soil. Their absence from some soils is probably due to an unfavorable pH. *Azotobacter* cannot develop if the reaction is more acid than pH 6.

The size and shape of cells are subject to great variation, being dependent upon a number of environmental factors, such as composition

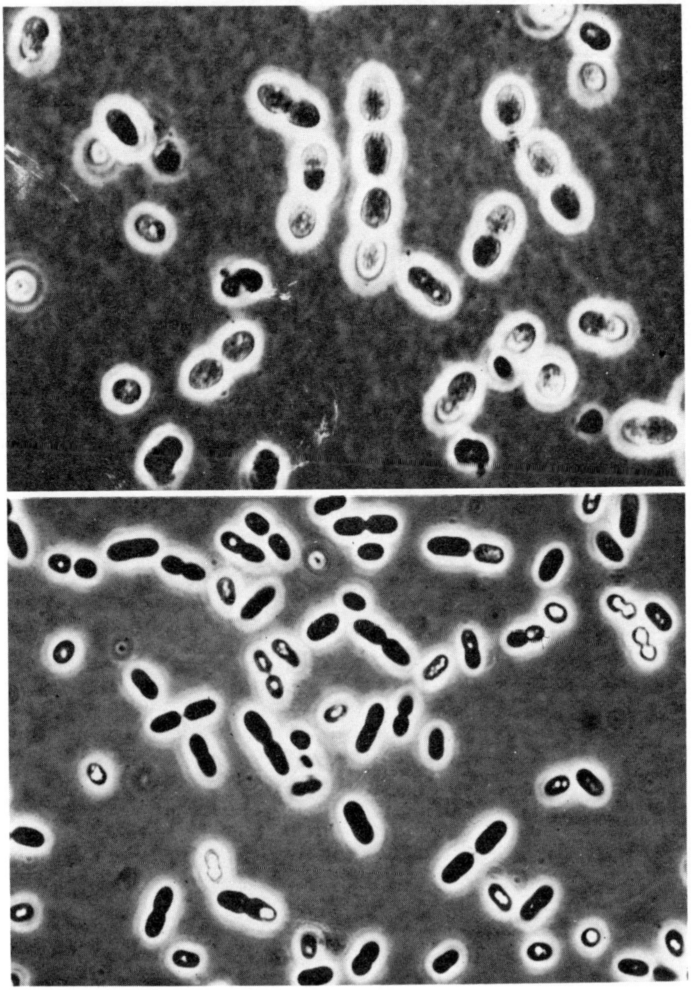

Fig. 299 Azotobacter, phase-contrast micrographs. Top, A. agilis; bottom, A. chroococcum. (Courtesy of H. L. Jensen.)

of the medium, presence of essential minerals, presence of oxygen, temperature of incubation, and reaction of the medium.

Azotobacter cysts Spores and cysts are dormant forms of some bacteria. They are highly resistant to adverse environmental conditions although cysts appear to fall in between the vegetative cells and spores in this respect.

Cysts have been described as the dormant stages in the life cycles of some members of the *Azotobacter* (Fig. 300). Resistance appears to reside in the cyst coat although there is some evidence to indicate the resistances to desiccation, ultraviolet radiation, and ultrasonic treatment may not be due entirely to the cyst wall structure. Removal of the outer coat by appropriate means makes the surviving central body as vulnerable to most harmful agents as the vegetative form.

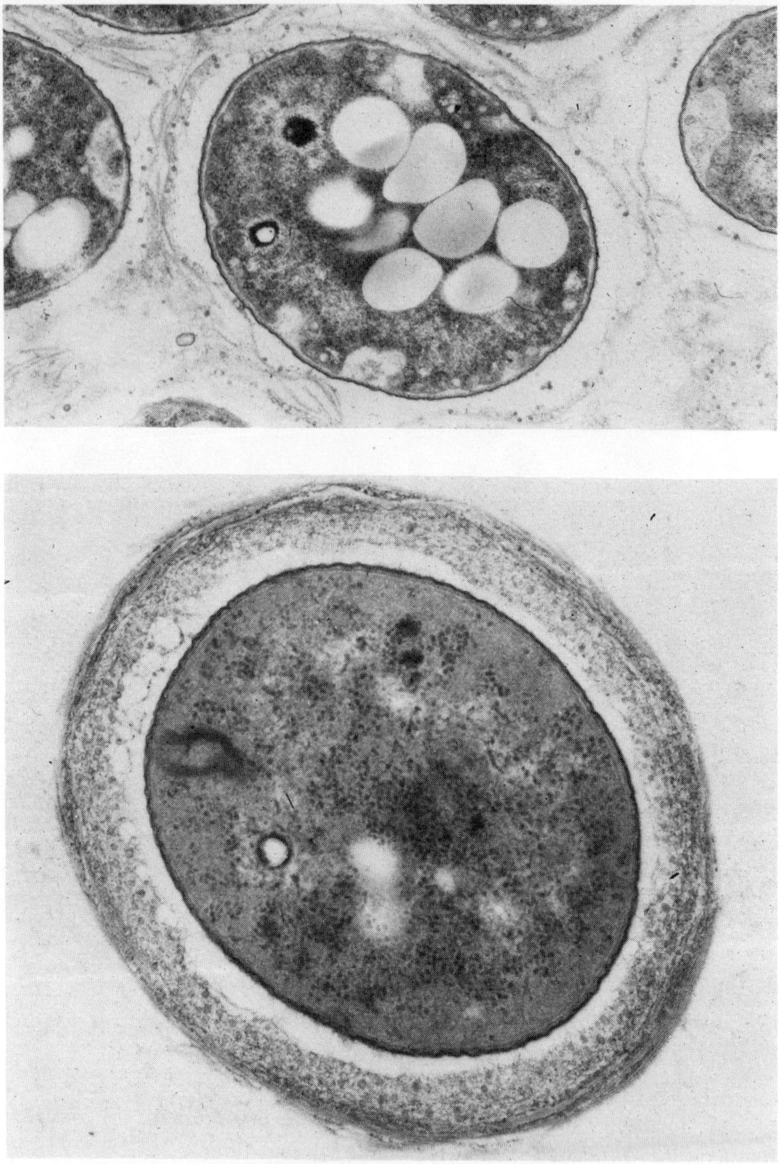

Fig. 300 Azotobacter vinelandii cysts. Upper, young, two-day-old encysting cell with fibrous material and vesicles surrounding the cell. X43,500. Lower, mature cyst. Identifiable structures external to the central body include the exine and the intine containing vesicles. X70,000. (Courtesy of Pope and Wyss.)

For more information: Bortels and Henkel (1968); Bortels and Olivares (1967); Bulen (1965); Hitchins and Sadoff (1970); Koo et al. (1969); Kramer and Socolofsky (1970); Lin and Sadoff (1969); Pope and Wyss (1970); Sadoff et al. (1971); Stevenson and Socolofsky (1966); Vela and Cagle (1969); Vela et al. (1970); Wyss et al. (1961); Vela and Cagle (1972).

Ferredoxin Ferredoxin, an electron-transferring protein that links hydrogenase with electron donors or acceptors, was isolated from *Clostridium pasteurianum* by Mortenson et al. (1962). It is composed of five iron sulfide groups attached to an acidic protein at its active site.

Ferredoxin protein varies somewhat in composition depending upon its source. *C. pasteurianum* ferredoxin has been shown to be composed of 55 amino acid molecules arranged in the sequence shown in Fig. 301.

Mechanism of action of ferredoxin Considerable progress toward an understanding of the mechanism of action of ferredoxin has been made in recent years due largely to the successful preparation of a cell-free extract capable of fixing atmospheric nitrogen.

Ferredoxin is present in all clostridia tested and in other obligate anaerobes. It functions as a carrier of electrons from low-potential oxidoreductase systems to hydrogenase and after combining them with protons releases them as molecular hydrogen. It also functions in reductase systems when H_2 is the electron donor and can transfer electrons directly from low-potential oxidoreductase systems to cellular electron carriers other than hydrogenase.

According to Fottrell (1968),

Pyruvate functions both as electron donor and energy source in non-symbiotic nitrogen fixation (Fig. 302). In the phosphoroclastic reaction,

Fig. 301 The amino acid sequence of Clostridium pasteurianum ferredoxin. (After Bayer, Jung, and Hagenmaier.)

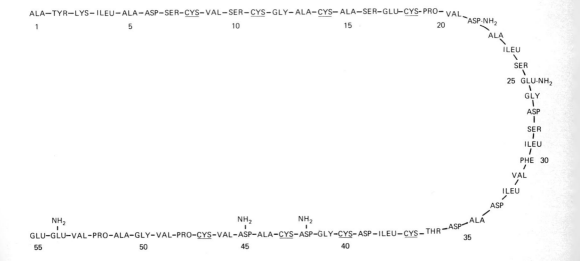

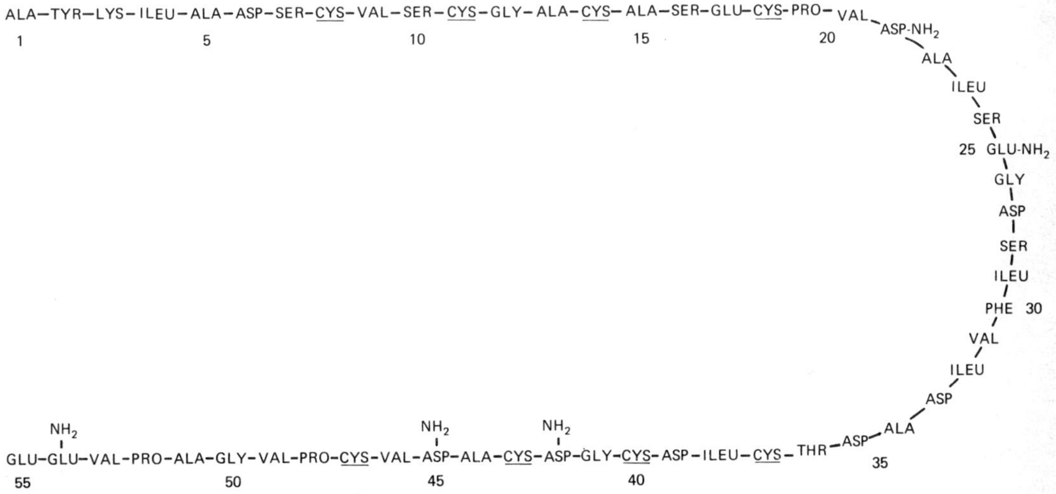

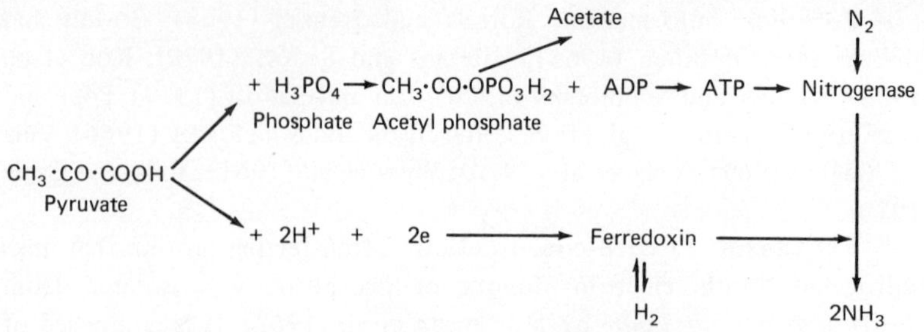

Fig. 302 The role of pyruvate in cell-free nitrogen fixation (After Fottrell.)

pyruvate forms acetyl phosphate which in the presence of ADP gives rise to ATP, the immediate energy source for nitrogen fixation.

The nitrogenase enzyme system catalyzes the adsorption and reduction of N_2. Nitrogenase is active only in cells fixing nitrogen and is repressed by a source of combined nitrogen such as ammonium salts in the medium. The hydrogenase enzyme system catalyzes the transfer of electrons from pyruvate or H_2 to ferredoxin.

Nitrogen fixation by extracts of *C. pasteurianum* is linked with pyruvic acid metabolism where pyruvate utilization occurs primarily by the phosphoroclastic reaction. Fixation has been observed to occur only while pyruvate is being actively consumed. The ratio of pyruvate consumed to nitrogen fixed is approximately 100:1.

Biological inoculation of soil Attempts have been made to increase the nitrogen content of soil by inoculation with pure cultures of *Azotobacter*. In most experiments no effect was noted on the growth of nonlegume plants, even though high numbers of bacteria became well established in soil. In general, the numbers of bacteria steadily decreased following establishment of growth. In some experiments a detrimental influence was observed. Failures have been attributed to (1) absence of a favorable environment, such as temperature, moisture, oxygen concentration, nutrients, and soil pH; (2) absence of a carbon source; and (3) injurious effects of end products liberated in the fermentation of added carbohydrate.

For more information: Arnon (1965); Bayer et al. (1968); Benemann et al. (1969); Buchanan et al. (1969); Eck and Dayhoff (1966); Kennedy et al. (1968); Mortenson (1963, 1964*a*, *b*, 1965); Mortenson et al. (1963); Rall et al. (1969); Rao and Matsubara (1970); Rao et al. (1969); Witz et al. (1967); Yoch et al. (1969); Shanmugam and Arnon (1972); Shanmugam et al. (1972).

Symbiotic nitrogen fixation It was shown several generations ago that the growth of certain plants in soil increased its fertility. This stimulated the growth of the succeeding plant crop. The plants responsible for this stimulation were members of the family *Leguminosae*.

Hellriegal and Wilfarth (1888) demonstrated that the stimulation was due to an increase in the nitrogen supply of the soil, which resulted from the presence of small tumor-like growths or nodules on the roots of the plants. In the absence of nodules, no stimulation of growth of the succeeding plant crop occurred.

The formation of nodules on roots is the result of the associated growth of plant and bacteria. The bacteria are members of the genus *Rhizobium* and are commonly referred to as the root-nodule bacteria. The organisms live in the cells of the plant roots, where their growth and metabolic activities cause a swelling or nodule to form on the root. The organisms utilize the free nitrogen of the atmosphere and synthesize it into nitrogen compounds. The plant obtains its nitrogen from the synthetic activities of the bacteria, while the organisms derive their food from the plant. The plant and bacteria live together for mutual benefit. Such an association is known as symbiosis.

Formation of the nodule The rhizobia live in the soil in the free state. Recent work indicates that they fix nitrogen only when growing in association with the plant nodule. They gain entrance to the plant through the root hairs or other epidermal cells. The bacteria multiply rapidly, forming long filaments in the root hairs and in the parenchyma of the root. The organisms cause a rapid proliferation of the surrounding tissue in the innermost cells of the root cortex, which results in the formation of a young nodule. The young nodule pushes out the overlying parenchyma and epidermis and produces a swelling on the side of the root. A nodule consists of a mass of thin-walled parenchyma cells, which are usually almost filled with the bacteria (Fig. 303). A corky layer and branches of a vascular system are also present. This system provides the bacteria with

Fig. 303 Early nodulation of a pea seedling resulting from seed inoculation. (Courtesy of O. N. Allen.)

their nutrients, and the plant in turn takes away the nitrogen compounds synthesized by the bacteria.

Nodules excised from the plant are also capable of fixing nitrogen, even though for only a limited period of time, indicating that attachment to the plant is not necessary for the reaction to occur. Nitrogen fixation does not occur in crushed nodules. However, nitrogen fixation has been achieved in extracts of root nodules.

Nodules vary in shape and size according to the plants on which they occur. Nodules on clover are round or oval; those on pea are commonly round, elongated, and frequently clustered; on the bean and soybean plants nodules are relatively large, round, and firmly attached to the root; those on alfalfa are usually long, finger-like growths.

Plants involved Nodule formation occurs probably exclusively on plants of the family Leguminosae. Reports of the association of rhizobia with species of plants from other families are without confirmation. On the other hand, not all leguminous plants are attacked by the rhizobia.

There are an estimated 10,000 species in the family Leguminosae. According to Allen and Allen (1961), a total of only 1285 species have been examined and of these, 1112 or about 86 percent showed nodulation.

Rhizobium The presence of rhizobia may be demonstrated by crushing a nodule between two glass slides, fixing the smear, and staining the organisms by the usual technique. If the nodule is washed in water, sterilized by immersion in a disinfectant, washed again, and the prepared material streaked over the surface of a satisfactory medium, such as Ashby's mannitol phosphate agar, colonies should appear on the plate in 5 to 10 days at an incubation temperature of 25°C.

The rhizobia are rod-shaped cells, measuring 0.5 to 0.9 by 1.2 to 3 μ. They are motile when young, commonly changing to bacteroidal forms (1) upon artificial media containing alkaloids or glucosides, or in which aciditv

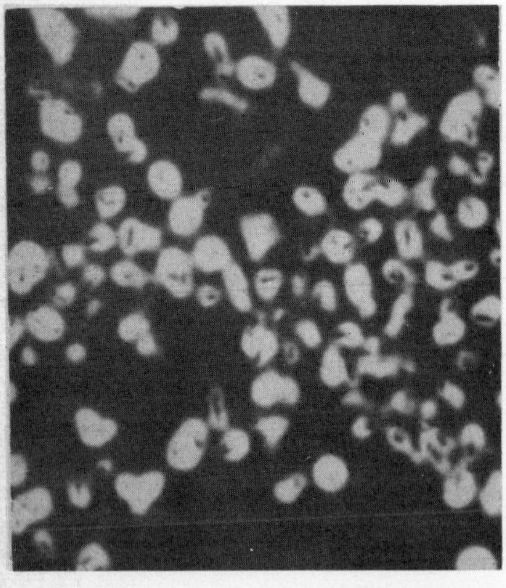

Fig. 304 Rhizobium trifolii, capsulated strain. (Courtesy of Dudman.)

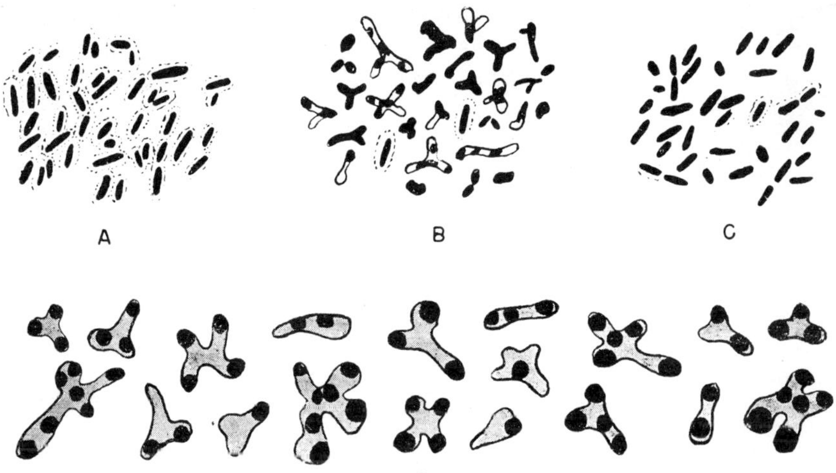

Fig. 305 Rods and bacteroids from root nodules. (A) rods from a white ineffective nodule; (B) bacteroids from a red effective nodule; (C) rods from a green ineffective nodule; (D) various forms of bacteroids. (After Virtanen.)

is increased; or (2) during symbiosis within the nodule. They are gram-negative and aerobic. Dudman (1968) reported the presence of capsules in some strains of *R. trifolii* (Fig. 304).

Bacteroids may be irregular with X-, Y-, star-, and club-shaped forms, or rods may be swollen, vacuolated, and with or without branching (Fig. 305). Spherical cells or cocci may also be present. Cocci arise from the swollen or vacuolated cells, which later return to the rod form. There is considerable variation among the different species. Some have considered the various bizarre forms as orderly stages in the life history of the organisms.

The genus includes six species: *R. leguminosarum, R. phaseoli, R. trifolii, R. lupini, R. japonicum*, and *R. meliloti*. A specific name has not been given to the organism that produces nodules on plants of the so-called "cowpea" group. Walker and Brown (1935) proposed a consolidation of the soybean and cowpea groups as being inoculated by a single species, *R. japonicum*. Reid and Baldwin (1937) would include the lupine group also.

The colonial characteristics of the organisms show some variation, depending upon the plants from which they are isolated. Some species show raised, glistening, semitranslucent, white, slimy, and occasionally viscous colonies with considerable gum formation. Other species show small, slightly raised, glistening, opaque, white, and butyrous colonies with little gum formation.

Although bacteroids predominate in the nodule, they are usually absent on artificial cultivation (Figs. 306 and 307). After death and

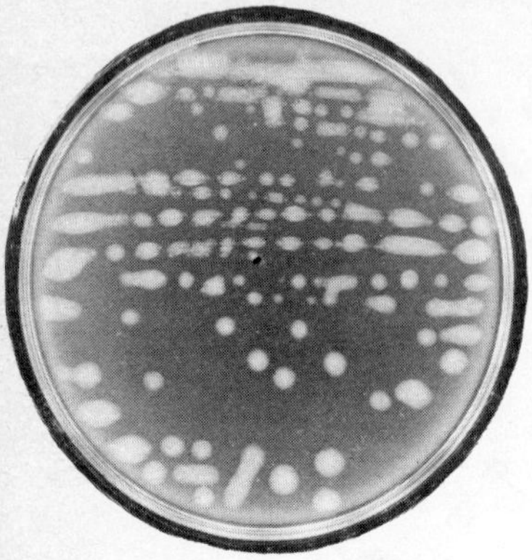

Fig. 306 Agar plate culture of Rhizobium leguminosarum incubated 6 days at room temperature. Note the mucoid appearance of the colonies, showing considerable tendency to run together.

dissolution of the nodule, the organisms return to the soil as rod-shaped or ovoid cells which may become flagellated. In this stage, the organisms swarm in the soil moisture, and it is probable that by such means they reach the roots of their host plants.

Cross-inoculation groups There are probably as many strains of root-nodule bacteria as susceptible plants. Nitrogen fixers isolated from the nodules on one legume are not necessarily capable of producing nodules on another legume. The clover plant grows best when inoculated with nodule bacteria from clover; peas grow best with nodule bacteria from peas; etc. The right pair must be brought together for successful results. The plant and bacteria must be highly compatible. This high selectivity was responsible for the recognition of cross-inoculation groups of leguminous plants. Any plant within such a group is inoculable with a culture of the proper organism, usually comprising several strains known to inoculate effectively all legumes in that particular group.

Cross-inoculation groups have been open to criticism because some strains of bacteria are effective for more than one group. But for all practical purposes these groups offer a convenient and workable plan for the preparation of inoculant cultures. Farmers have been accustomed to

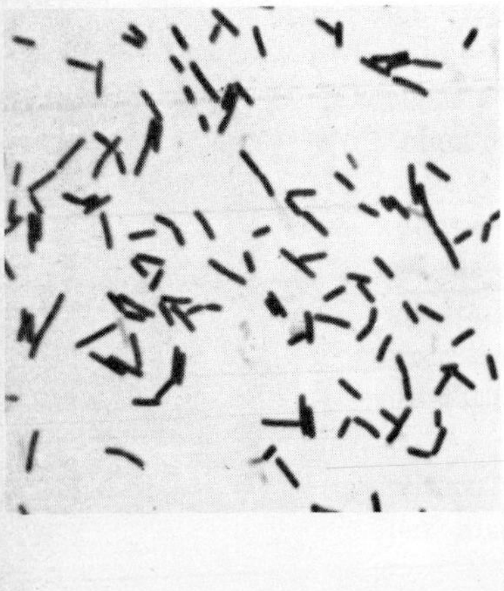

Fig. 307 Rhizobium leguminosarum from culture on mannitol agar.

ordering legume cultures according to group designations. Seven groups are generally recognized: alfalfa, bean, clover, cowpea, lupine, pea and vetch, and soybean. Wilson (1948) distinguished 22 cross-inoculation groups among the legume bacteria. The causes of this host-plant specificity are unknown.

Some of the root-nodule bacteria show a strong tendency to produce large quantities of gum when cultivated on carbohydrate media. Gum formation may be so pronounced as to render the medium strongly viscid. Organisms in different cross-inoculation groups vary considerably in their capacity to produce gum. The gum is nitrogen-free, water-soluble, and precipitated from solution by alcohol or acetone. On heating with a mineral acid, the gum is hydrolyzed to a reducing sugar.

Variation in nodules Virtanen (1947) observed two types of nodules on pea plants. One type effectively fixed nitrogen, whereas the other did not. The plants containing the ineffective nodules showed no better growth than the uninoculated controls. Vertanen observed that the ineffective nodules contained only rods surrounded by a slimy layer; the effective nodules contained noticeably swollen bacteroidal forms (Fig. 305). He believed that the slimy layer around the ineffective organisms interfered with their nutrition and oxygen uptake.

The root nodules of leguminous plants that actively fix nitrogen contain a red pigment, whereas the ineffective nodules are devoid of pigment. A positive correlation appears to exist between the concentration of red pigment and the ability of the nodules to fix nitrogen. The pigment changes to green when nitrogen fixation ceases in annual plants at the end of vegetative growth or when the plants are placed in the dark for a few days. The red pigment may be concentrated by treating a suspension of crushed nodules with ammonium sulfate. The pigment is capable of storing and transporting oxygen. Chemically it is a hemoprotein having absorption bands similar to hemoglobin. Virtanen named the pigment leghemoglobin. The green pigment may also be isolated from the green nodules in the same manner as leghemoglobin from the red nodules. It is also a chromoprotein. However, the absorption bands of the green and red pigments are not the same.

More recently it has been shown that leghemoglobin is not directly involved in nitrogen fixation. Fixation occurred even after removal of leghemoglobin from cell-free nodule extracts. It has been suggested that the possible function of leghemoglobin is to permit O_2 to penetrate to the bacteria at very low levels of free O_2 but at a rate sufficiently fast to allow nitrogen fixation to proceed. Leghemoglobin, therefore, appears to function by preserving the low oxygen tension necessary for the activity of the nitrogen-fixing enzymes, and also, in allowing oxygen, which is necessary for nitrogen fixation, to diffuse quickly to the bacteria.

Mechanism of symbiotic nitrogen fixation The requirements for nitrogen fixation in extracts of rhizobia are similar to those that have been established for extracts from nonsymbiotic nitrogen-fixing bacteria.

In both systems NH_3 was shown to be the first stable intermediate that could be isolated. The NH_3 was incorporated into a number of compounds, mainly glutamic and aspartic acids, alanine, and glutamine. From available evidence so far, some of the properties of the enzymes which have been isolated from both the free-living (nonsymbiotic) bacteria and the rhizobia (symbiotic) appear to be quite similar.

From 50 to 80 percent of the fixed nitrogen is excreted from the nodule into the soil. In marked contrast, *Azotobacter* excretes only 10 to 25 percent. Most of the fixed nitrogen is used by the bacteria to synthesize cell material. However, it makes little difference in the end as the organisms soon die and autolyze, releasing their nitrogen to the soil.

Artificial inoculation of plants The efficiency of symbiotic nitrogen fixation has been greatly increased by (1) choice of the proper legumes, (2) development of new varieties of leguminous plants, (3) artificial inoculation of seeds with pure cultures of root-nodule bacteria, and (4) adjustment of the environment to optimum conditions.

Cultures of the rhizobia are frequently added to soil to increase the nitrogen supply. This is generally practiced by soaking the seeds of a legume in a culture of the appropriate species. This procedure has become a well-established practice, especially when a legume plant is seeded for the first time.

For successful results it is extremely important that the strain of organism selected be one that is known to produce effective nodules on the legume. There appears to be no rule to follow in the selection except the actual testing of various strains on the plant to be affected. Greenhouse tests should be supplemented with actual field tests to make sure of the proper selection for the mass production of commercial cultures.

The addition of an abundant supply of nitrate to soil results in a luxuriant plant growth. Under such conditions nodules are probably not formed even though the seeds have been soaked in a culture of the organisms. The suppression of nodule formation is due to a change in the metabolism of the plants. On the other hand, the addition of calcium and phosphorus to soil results in a stimulation of nodule production. Other factors influencing nodule formation include reaction, temperature, moisture, and oxygen content of soil.

For more information: Date and Decker (1965); De Ley and Rassel (1965); Dilworth and Parker (1969); Fottrell (1968); Gibson (1967*a, b*); Holland (1966); Jordan (1962); Loneragan et al. (1961); Means and Erdman (1963); Nutman (1967).

Strictly autotrophic bacteria

Green plants are capable of effecting a synthesis of their own organic compounds. They combine CO_2 with H_2O in the presence of chlorophyll and sunlight as a source of energy to form carbohydrate. Nitrogen as nitrate, or other inorganic nitrogen compounds, is absorbed from the soil and synthesized into protein. The other required elements in the form of inorganic compounds are also taken from the soil.

Some bacteria cannot synthesize their complex protoplasm from simple inorganic materials but must have organic compounds such as proteins, peptones, amino acids, and vitamins for growth. These are classified as strictly heterotrophic organisms.

Others are able to grow in media devoid of organic matter. They are capable of synthesizing all cellular substances from simple inorganic compounds and CO_2 as the only and irreplaceable source of carbon. These are the strictly autotrophic bacteria.

However, most bacteria so far isolated are able to utilize both inorganic and organic compounds for structure and growth. These are the facultative forms, and they constitute the largest group of bacteria found in nature.

The number of strictly autotrophic species is comparatively small. A greater number are facultatively autotrophic, being capable of existing both autotrophically and heterotrophically. The strict forms include the nitrifying bacteria and some of the sulfur and iron organisms. The facultative forms can obtain their energy from (1) the oxidation of inorganic substances and carbon from CO_2 as the sole source or (2) purely organic compounds. The hydrogen bacteria, and some of the sulfur and iron forms, are facultatively autotrophic.

For more information: Bisset and Grace (1954); Lees (1962*a, b*); Umbreit (1962); Woods and Lascelles (1954).

Nitrifying organisms The strict autotrophic nitrifying bacteria obtain their energy from the oxidation of ammonia to nitrite, and from nitrite to nitrate. They depend upon this oxidation for growth. The nitrifiers fail to grow unless the specific nitrogen compounds are available for energy. Also, they must have CO_2 as the sole source of carbon.

Nitrosification The oxidation of ammonia to nitrite is called nitrosification:

$$2NH_3 + 3O_2 \longrightarrow 2HNO_2 + 2H_2O + 79 \text{ Cal}$$

The organisms involved are members of the genera *Nitrosococcus, Nitrosocystis, Nitrosogloea, Nitrosomonas,* and *Nitrosospira*.

Clark and Schmidt (1967*a, b*) reported that several amino acids supplied exogenously at low levels (1 μg per ml or less) were readily taken up and incorporated into cellular fractions by resting cells of *Nitrosomonas europaea*. Although growing cells of *N. europaea* were able to incorporate amino acids into cellular material, Hooper (1969) found that they were unable to grow heterotrophically. Williams and Watson (1968) reported that resting cells of *Nitrosocystis oceanus*, incubated with glucose and glutamate, and cells grown in the presence of glucose, pyruvate, and the amino acids glutamate and methionine, incorporated these compounds into cellular material but at a level too low to provide the cells with their major needs of carbon and energy.

Cells of the above genera are rod-shaped, ellipsoidal, spherical, or spirillar; aerobic; nonsporulating; and grow best at about 25°C. Organisms fail to grow on media containing organic matter unless compounds of ammonia are present.

Isolation of organisms. Nitrite formation takes place best in neutral or slightly alkaline media. Since the organisms are strongly aerobic, cultures should be exposed in shallow layers, preferably in Erlenmeyer flasks. The ammonia is oxidized to nitrous acid. As acid is formed, it becomes neutralized to magnesium nitrite by the insoluble magnesium carbonate in the medium. The reaction continues until all ammonia is oxidized to nitrite. The addition of more ammonium salt makes the process continuous, provided there is an excess of carbonate to neutralize the nitrous acid as it is formed. When all magnesium carbonate disappears, the organisms should be transferred to fresh medium. Colonies may be obtained by streaking the organisms over the surface of the same medium solidified with silicic acid. Inorganic silicic acid is superior to agar for the cultivation of nitrifying bacteria.

For more information: Funk and Krulwich (1964); Ghosh (1967); Morrill and Dawson (1967); Pramer (1957); Sommers and Harris (1968).

Nitrification The oxidation of nitrite to nitrate is called nitrification:

$$HNO_2 + \tfrac{1}{2}O_2 \longrightarrow HNO_3 + 21.6\ \text{Cal}$$

Some autotrophic nitrifiers are unable to oxidize ammonia to nitrite but can oxidize nitrite to nitrate. These organisms are placed in the genera *Nitrobacter* and *Nitrocystis.* The bacteria are aerobic, and grow best at about 25°C.

Isolation of organisms. The medium employed is similar to that for the cultivation of the preceding group, except that sodium nitrite is substituted for the ammonium salt. Since these organisms also are strongly aerobic, the medium is best dispensed in shallow layers. After inoculation with the soil sample, the culture is incubated at 25°C until growth appears. Several transfers to fresh medium are necessary to obtain a culture rich in nitrifying bacteria.

The nitrifiers are peculiarly sensitive to alkalinity, oxidizing little, if any, nitrite at values just above pH 7.7. Just below this valuc, the oxidation of nitrite to nitrate proceeds so rapidly that the purely biological nature of the reaction is open to question.

Since nitrite is seldom found in soils, it may be assumed that the two groups of nitrifiers almost always accompany each other. They are widely distributed in nature, being found in practically all neutral and alkaline soils, and in water. They are the agents primarily responsible for the appearance of nitrate in soils which in turn is utilized by higher plants for the synthesis of cell material.

For more information: Lees (1954); Silver (1961); Tsien et al. (1968); Van Gool and Laudelout (1967).

Sulfur-oxidizing bacteria Sulfur is widely distributed in nature, being among the 13 most abundant elements making up the earth's crust.

Sulfur is found in both the combined and the free state. In the combined form it exists in both inorganic and organic combination. Inorganic sulfates and sulfides are probably the most important compounds modified by bacterial action. Sulfates are utilized in the synthesis of sulfur-containing amino acids. They are also important as sources of oxygen for the oxidation of organic matter under anaerobic conditions. Sulfides may be oxidized by some bacteria for energy purposes.

Gases emitted from volcanoes contain sulfur dioxide and hydrogen sulfide. Sulfur is present in large quantities in sulfur springs. Volcanic sulfur usually occurs as a sublimate on the walls of vents, probably as the result of action between hydrogen sulfide and sulfur dioxide. The commercial source of sulfur is chiefly as crude brimstone obtained from the sides of volcanoes or mined in certain parts of the world. Extensive deposits are found on the island of Sicily and in Louisiana. Commercially it occurs as brimstone, or as flowers of sulfur, which is prepared from the crude brimstone. Sulfur finds its way into the soil from the decomposition of native rock, from organic manures, and from rainwater.

Sulfur is one of the elements absolutely essential to living organisms. It enters into the compositions of all plant and animal cells (see Fig. 308).

The addition of sulfur to a soil low in the element results in a marked stimulation of growth. As the plants and animals die and decompose, the sulfur finds its way into the atmosphere. The gases are dissolved by rainwater and again returned to the soil. Some soil organisms are able to convert the sulfur-containing gases to sulfates. Some of the sulfates are utilized by growing plants, and some are leached out by waters and carried off to the ocean. The sulfates in the ocean may be reduced to sulfides and then precipitated by iron as iron sulfide, or they may be converted into insoluble calcium sulfate (gypsum). Deposits of gypsum are believed to have been formed in this manner. Deposits of sulfur probably resulted from the reduction of sulfate to sulfite and then to free sulfur, or from the oxidation of hydrogen sulfide. The sulfur cycle may be diagrammed as shown in Fig. 309.

Thiobacillus This genus contains both obligately autotrophic and facultatively autotrophic species.

The cells are small, rod-shaped, and gram-negative. They are nonmotile or motile, usually by means of a single polar flagellum. Energy is derived principally from the oxidation of elemental sulfur or thiosulfate but in some cases also from sulfide, sulfite, and polythionates. The principal product of oxidation is sulfate but sulfur is sometimes formed. One species derives energy from the oxidation of inorganic ferrous iron compounds to ferric hydroxide. The organisms become heavily encrusted with the reddish-colored compound. They grow under acid or alkaline conditions and derive carbon from CO_2 or from bicarbonates in solution. Some species are anaerobic in the presence of nitrate. They are found in soil, mine wastewaters, sewage effluents, and related sources.

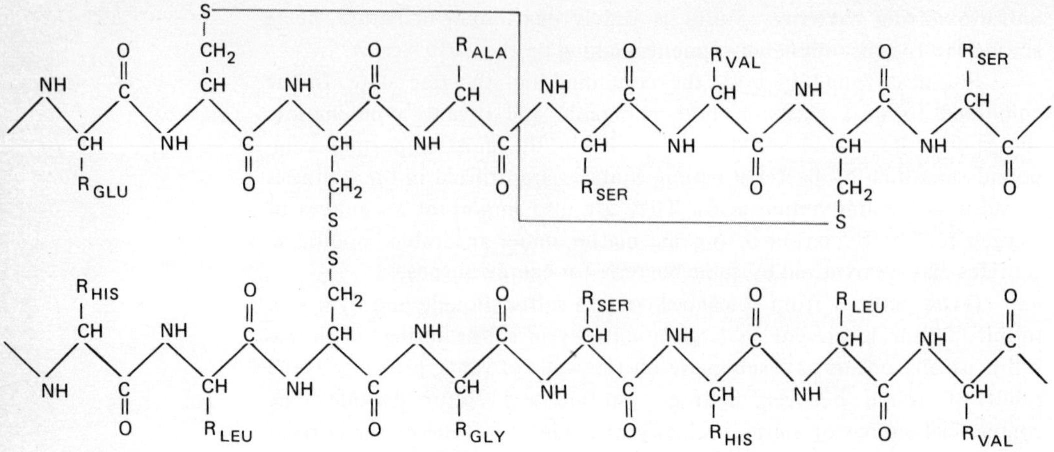

Fig. 308 According to Deevey (1970), "Basic function of sulfur in living matter appears to be to provide a linkage between the polypeptide chains in a protein molecule. These linkages help the protein maintain its three-dimensional shape so that it can perform its function. In this segment of a bovine insulin molecule, disulfide bonds (color) are formed between sulfur atoms, which are present in the amino acid cystine. Cystine is a subunit of both polypeptide chains. Because the molecule is displayed in two dimensions it is flattened. Therefore the bond between the top and bottom cystine groups on the upper chain appears broken. In the normal three-dimensional state, however, this chain is twisted and folded because of the disulfide bond in a way indicated by the colored line that joins the two sulfur atoms. (The other bond is one of two that links the chains.) The shape of the insulin molecule maintained by these bonds enables it to control the metabolism of sugar. The other amino acids in this molecular segment, whose side chains are indicated by the letter R, are glutamic acid (GLU), alanine (ALA), serine (SER), valine (VAL), histidine (HIS), leucine (LEU) and glycine (GLY)." (Courtesy of E. S. Deevey, Jr.)

Fig. 309 The sulfur cycle.

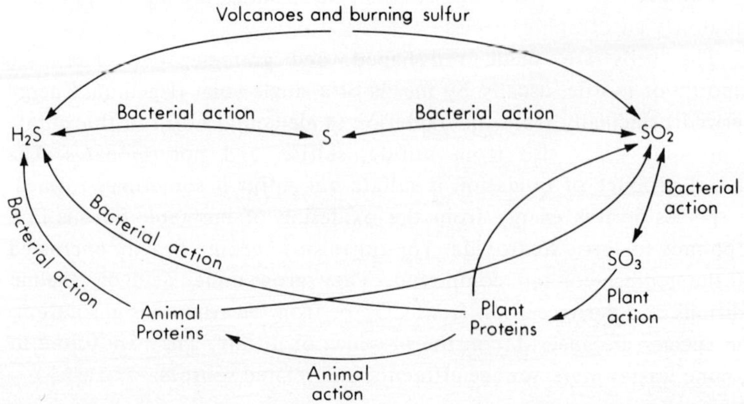

Oxidation of sulfite The following pathway has been proposed for the oxidation of sulfite by species of *Thiobacillus*:

$$2SO_3^{2-} + 2\text{ adenosine monophosphate (AMP)} + \text{adenosine-5'-phosphosulfate reductase (APS reductase)} \longrightarrow 2APS + 4e$$

$$2APS + 2\text{ pyrophosphate (Pi)} + \text{adenosine diphosphate-sulfurylase (ADP-sulfurylase)} \longrightarrow 2ADP + 2SO_4^{2-}$$

AMP is regenerated by the action of adenylate kinase:

$$2ADP + \text{adenylate kinase} \longrightarrow AMP + \text{adenosine triphosphate (ATP)}$$

Oxidation of thiosulfate Thiosulfate is oxidized to sulfate by all species of thiobacilli. Charles and Suzuki (1966) proposed the following reactions for its oxidation by whole cells and cell-free extracts of *T. novellus*:

$$S_2O_3^{2-} + \text{thiosulfate cleaving enzyme} \longrightarrow S + SO_3^{2-}$$

$$S + O_2 + H_2O + \text{sulfur oxidizing enzyme} \longrightarrow SO_3^{2-} + 2H^+$$

$$2SO_3^{2-} + \underset{\text{Ferricytochrome c}}{4\text{cyt.c } Fe^{3+}} + 2H_2O + \text{sulfite:oxygen oxidoreductase} \longrightarrow 2SO_4^{2-} + \underset{\text{Ferrocytochrome c}}{4\text{cyt.c } Fe^{2+}} + 4H^+$$

$$4\text{cyt.c } Fe^{2+} + O_2 + 4H^+ + \text{cyt.c:oxygen oxidoreductase} \longrightarrow 4\text{cyt.c } Fe^{3+} + 2H_2O$$

Overall reaction,

$$S_2O_3^{2-} + 2O_2 + H_2O \longrightarrow 2SO_4^{2-} + 2H^+$$

Thiobacillus thiooxidans This species may be cultivated in an acid mineral medium containing free sulfur. Since it is strictly aerobic, growth is hastened if the medium is exposed in shallow layers. The organism is of special interest in that it produces more acid from the oxidation of sulfur and continues to live in a more acid medium than any other microbe yet reported. The pH of the medium may drop to 0.6 or less. The organism actually lives and grows in 5 to 7 percent sulfuric acid.

The oxidation of sulfur to sulfuric acid may be represented by the following reaction:

$$2S + 3O_2 + 2H_2O \longrightarrow 2H_2SO_4$$

Corrosion of concrete and sealing mixtures The thiobacilli are known to cause serious corrosion of concrete. Sufficient sulfuric acid is produced from the oxidation of sulfur or hydrogen sulfide to dissolve the concrete. The species principally involved are *T. thioparus, T. concretivorus,* and *T. thiooxidans*.

Materials containing sulfur are used to seal the joints of cast-iron pipes. The molten mixture is poured into the joints, where it forms an impervious cement. Ordinarily this sealing material remains intact for years, but occasionally the sulfur cement breaks down under biological action. The principal organism producing this breakdown is *T. thiooxidans*, although other species are doubtless involved. The sulfur is first oxidized to sulfate, followed by the reduction of the sulfate to sulfide by sulfate-reducing bacteria (page 430).

Thiobacillus ferrooxidans This species differs from the others in the genus in that it can derive energy from the oxidation of inorganic ferrous iron as well as from thiosulfates. Elemental sulfur is not appreciably utilized. Ferrous iron becomes quickly and easily oxidized to ferric ions in acid media where ferric iron is not otherwise produced in quantity. Cultures maintained on ferrous media lose the ability to oxidize thiosulfates but colonies maintained on thiosulfate media retain their capacity to oxidize iron.

For more information: Baalsrud (1954); Beck and Brown (1968); Butler and Umbreit (1966); Duncan et al. (1967); Johnson and Abraham (1969); Lazaroff (1963); Matin and Rittenberg (1970*a, b*); Moriarty and Nicholas (1969, 1970); Peck and Stulberg (1962); Peeters and Aleem (1970); Ross et al. (1968); Shively et al. (1970); Tilton et al. (1967*a, b*); Trudinger (1967); ZoBell (1963).

Facultatively autotrophic sulfur bacteria

A number of sulfur bacteria appear to be facultatively autotrophic rather than strict autotrophs. They are treated here for convenience.

The organisms are commonly present in water containing dissolved hydrogen sulfide. The gas results from the decomposition of organic matter by saprophytic organisms. The bacteria may be present in such organic materials as decomposing seaweed, rock pools containing dead algae and other lower forms of plant life, stagnant woodland pools, and sewage. Sulfur bacteria have been found in sulfur hot springs and in sulfur mines. Some species are able to grow in water pipes and cause serious obstructions. Foul odors and tastes are produced after death and decomposition of the organisms.

The important facultative sulfur bacteria are found in the families *Achromatiaceae, Beggiatoaceae, Chlorobacteriaceae, Thiobacteriaceae*, and *Thiorhodaceae*.

Achromatiaceae These are large unicellular organisms which are spherical to ovoid or shortly cylindrical with hemispherical extremities. Movements, if any, are of a slow, rolling, jerky type and are dependent upon the presence of a substrate. No special organs of locomotion are known. Division of cells is by a constriction in the middle. They do not possess photosynthetic pigments. In their natural habitat, the cells contain sulfur droplets and sometimes additional inclusions, such as large spherules of calcium carbonate. They are found in freshwater and marine environments; apparently require hydrogen sulfide.

Beggiatoaceae The individual cells are generally not visible unless stained. The cells are arranged in chains within trichomes. The trichomes show a gliding motion when in contact with a substrate; they also show flexing movements. The trichomes contain sulfur globules when cultivated in the presence of hydrogen sulfide. The cells do not contain chlorophyll and phycocyanin. They are found in both freshwater and marine environments containing hydrogen sulfide.

For more information: Burton et al. (1966); Faust and Wolfe (1961); Pringsheim (1964); Scotten and Stokes (1962).

Chlorobacteriaceae These are green bacteria, usually of small size, occurring singly or in cell masses of various shapes and sizes, developing in environments containing rather high concentrations of hydrogen sulfide and exposed to light. As a rule, they do not contain sulfur globules but frequently deposit elemental sulfur outside the cells. They contain chlorophyll-like pigments and are capable of photosynthesis in the presence of hydrogen sulfide; do not liberate oxygen.

Thiobacteriaceae The cells are coccoid, straight, or curved rods. They oxidize sulfur compounds, usually depositing free sulfur granules within or without the cells. They are never filamentous. They are colorless sulfur bacteria which are sometimes embedded in gelatinous pellicles or in gelatinous bladder-like colonies. The organisms are polar flagellate when motile. They are found where hydrogen sulfide occurs or may oxidize free sulfur, thiosulfates, or related compounds.

Thiorhodaceae These are unicellular organisms, often developing as cell aggregates or families of variable size and shape. Single cells are spheres, ovoids, short rods, vibrios, spirals, long rods, or occasionally chains. They occur in nature in environments containing sulfides and require light for development. They produce a pigment system composed of green bacteriochlorophyll and yellow and red carotenoids. They appear as bluish violet, pale purple, brownish to deep red cell masses. Single cells usually appear to be unpigmented. The organisms are anaerobic to microaerophilic with a photosynthetic metabolism in which carbon dioxide is reduced with the aid of special hydrogen donors without the liberation of molecular oxygen. Where these organisms are found in nature, hydrogen sulfide acts as a hydrogen donor, and sulfur accumulates as sulfur droplets in the cells. Probably all members of the group can utilize a number of organic substances in place of hydrogen sulfide as hydrogen donors for photosynthesis.

For more information: Cusanovich and Kamen (1968*a*, *b*); Cusanovich et al. (1968); Hurlbert (1967).

Iron-oxidizing bacteria

Iron is found in all living cells. In some species only minute amounts are present but regardless of the quantitative aspects, the element is absolutely necessary for growth and well-being of all plants and animals.

Magnesium rather than iron is present in chlorophyll, the green coloring matter of plants. However, in the absence of iron, leaves do not become green. Also, if iron is withheld from a plant in which the chlorophyll is well developed, the color gradually fades to yellow.

Certain organisms found in water and soil are capable of taking up iron and accumulating it on the surfaces of their cells where it is quickly oxidized to ferric hydroxide. The important iron bacteria are found in the orders *Pseudomonadales* and *Chlamydobacteriales*.

Facultatively autotrophic iron bacteria

Ferrobacillus ferrooxidans is a small bacillus occurring singly, in pairs, rarely in chains; the cells are not united to form colonies. It obtains its carbon from carbon dioxide and energy from the oxidation of ferrous iron to ferric sulfate or insoluble ferric hydroxide in acid environments. The organism has been cultivated on organic matter in the absence of an oxidizable iron source and is, therefore, not to be regarded as a strict but rather a facultative autotroph.

Margalith et al. (1966) showed that the organism can derive its energy from the oxidation of elemental sulfur as well as from ferrous iron. The iron-oxidizing enzyme was found to be constitutive, since the ferrous ion was oxidized as rapidly by sulfur-grown cells as by cells cultivated on iron. Sulfur-grown cells had a better capacity for oxidizing sulfur than did iron-grown cells; however, no oxidation lag was observed in either case. The sulfur-oxidizing system was not inducible.

The oxidation of ferrous sulfate to ferric sulfate in an acid environment is as follows:

$$4Fe^{2+} + O_2 + 4H^+ \longrightarrow 4Fe^{3+} + 2H_2O$$

The optimum pH for growth is 3.5. The organism is indigenous to bituminous coal regions.

For more information: Blaylock and Nason (1963); Din et al. (1967); Dugan and Lundgren (1965); Kinsel (1960); Lees et al. (1969); Shafia and Wilkinson (1969); Short et al. (1969); Wang and Lundgren (1968, 1969).

Nonautotrophic iron bacteria

A number of species of nonautotrophic iron bacteria have been isolated from iron-bearing waters. These organisms are capable of utilizing both inorganic and organic matter. They are placed in the following genera:

Gallionella Cells are kidney-shaped or rounded; placed at the end of the stalk with the long axis of the cell transverse to the long axis of the stalk. Stalks secreted by the cells are slender and twisted. They branch dichotomously or in the form of umbels. Stalks are more or less dumbbell- or biscuit-shaped in cross section, composed of ferric hydroxide, completely dissolving in weak acids. Two polar flagella are present when the cells are motile. They are gram-negative. Multiplication is by fission of the cells, the daughter cells remaining at first at the end of the stalk; later they may be liberated as swarm cells. Organisms grow only in iron-bearing waters and are found in both freshwater and saltwater (Fig. 310).

Fig. 310 Iron bacteria. (1) Leptothrix ochracea; (2, 3, 4, 5, 6) Gallionella ferruginea. (After Ellis.)

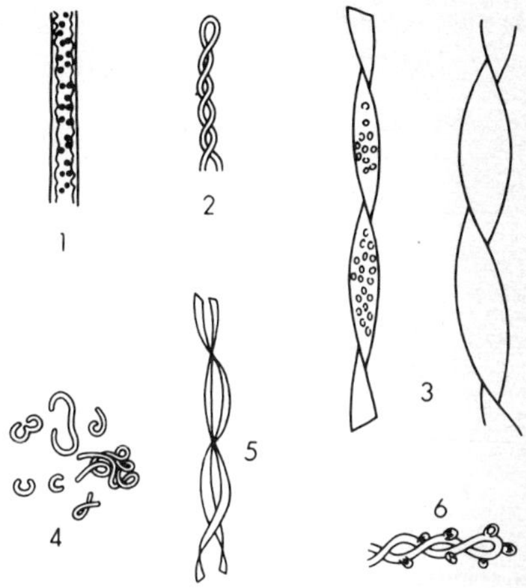

Organisms attach themselves to pipes and cause extensive deposits of iron which interfere with the flow of water. For this reason, they are sometimes called water-pest bacteria.

For more information: Hanert (1968); Sharpley (1961).

Sphaerotilus The cells are attached or free-floating, colorless trichomes showing false branching, though this may be rare in some species. Sheath shows a homogeneous structure; may become yellowish or brown with deposition of iron oxide. Deposition of iron is dependent on environmental factors, not on physiological ability to store iron. Trichomes consist of rod-shaped or ellipsoidal cells surrounded by a firm sheath (Fig. 311). Multiplication occurs both by nonmotile conidia and by motile swarm cells, the latter with a subpolar tuft of flagella. They are found in freshwater.

For more information: Bisset and Brown (1969); Dias et al. (1968); Mulder (1964); Romano and Gleason (1964); Stokes (1954).

Leptothrix Trichomes are of cylindrical, colorless cells with a sheath at first thin and colorless, later thicker, yellow or brown, encrusted with iron or manganese oxide. Oxides may be dissolved by dilute acid, exposing the inner cells. If sheath contains manganese oxide, it does not dissolve completely in weak acids. Under the electron microscope, the sheath shows an alveolar structure. Multiplication is by cell division with individual cells occasionally slipping out of the sheath as reproductive cells. These are sometimes motile with a tuft of flagella. False branching may occur. They are usually found in freshwater. Best-known species, *Leptothrix ochracea*, is worldwide in distribution (Fig. 310).

Toxothrix Trichomes are composed of cylindrical, colorless cells with a thin primary sheath; latter soon becomes impregnated with iron oxide. Trichomes lie loosely, longitudinally together, in slightly spirally twisted

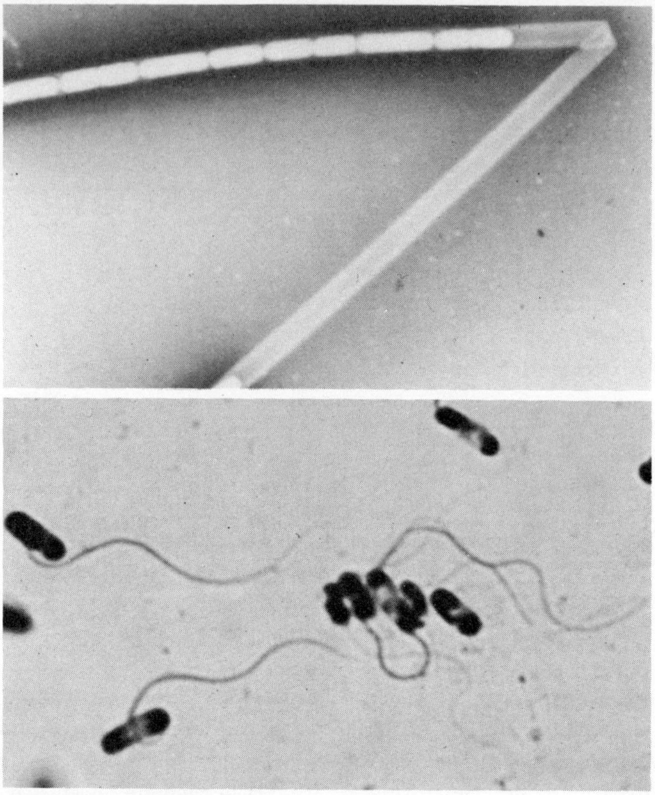

Fig. 311 Photomicrographs of Sphaerotilus natans. Upper, cells surrounded by a sheath; lower, individual cells with tufts of subpolar flagella. (Courtesy of Stokes.)

rolls. Continued repetition of this process leads to the development of a thick, secondary sheath from which parallel bundles may separate. False branching may occur. Sheaths do not completely dissolve in weak acids. Cells may slip out of the sheath and become motile swarm spores. Found in iron-bearing freshwaters.

Crenothrix Trichomes are attached to a firm substrate and swollen at the free end. They are unbranched or show false branching. Sheaths surrounding trichomes are plainly visible, thin and colorless at the tip, and encrusted with iron or manganese oxides at the base. Cells are disk-shaped to cylindrical, dividing to produce spherical, nonmotile conidia of two types: micro- and macroconidia. Individual cells may slip out of sheath and form new trichomes. Organisms are found in stagnant and running waters which contain organic matter and iron salts. *Crenothrix polyspora* is the only species recognized.

Clonothrix Attached trichomes show false branching. Sheaths are organic, encrusted with iron or manganese, broader at the base and tapering toward the tip. Cells are colorless and cylindrical. Reproduction is by spherical conidia formed in chains by transverse fission of cells; conidia formation is acropetal, limited to short branches of the young portions of the trichomes. Only species, *Clonothrix putealis*, is widely distributed in rivers and streams with gravelly, manganese-bearing bottoms; also found in waterworks and pipelines, where it may cause technical difficulties.

Ferruginous waters usually show the presence of a yellowish- or reddish-colored slime on the stream bottom. The color is due to the deposition of iron in the outer sheaths of the filaments. The accumulation of iron and its oxidation to ferric hydroxide result in the formation of a hard and inelastic membrane, which eventually leads to the death of the organisms. Old filaments show a higher iron content than young filaments. In some cases, young cells are completely lacking in a deposition of iron in their sheaths. The iron oxide may be removed by application of dilute hydrochloric acid, after which the outer membrane becomes visible.

For more information: Breed et al. (1957); Pringsheim (1949).

Bacterial photosynthesis

Photosynthesis is the process by which green plants manufacture carbohydrate from CO_2 and H_2O in chlorophyll-containing tissues with the aid of radiant energy from the sun. The first and last steps of the process may be represented by the equation

$$6CO_2 + 6H_2O \xrightarrow{\text{Light}} \underset{\text{Glucose}}{C_6H_{12}O_6} + 6O_2$$

The CO_2 is reduced to carbohydrate and the H_2O is decomposed with the liberation of oxygen. The reaction is strictly aerobic.

Certain bacteria classified in the families *Thiorhodaceae, Chlorobacteriaceae*, and *Athiorhodaceae* contain a pigment system consisting of bacteriochlorophyll and various carotenoids and are also capable of photosynthesis. The Thiorhodaceae includes the autotrophic purple sulfur bacteria. They contain sulfur globules in the presence of hydrogen sulfide. The Chlorobacteriaceae embraces the autotrophic green sulfur bacteria. The chlorophyllous compound differs from the chlorophyll in the Thiorhodaceae. The Athiorhodaceae includes the purple and brown nonsulfur bacteria. They do not contain sulfur globules even in the presence of hydrogen sulfide. The first two are considered primarily autotrophic; the latter is heterotrophic.

The characteristics of the first two families have already been given. A description of the Athiorhodaceae is as follows:

They are unicellular bacteria, of relatively small size, occurring as spheres, short rods, vibrios, long rods, and spirals. They are motile by means of polar flagella and are gram-negative. The cells produce a pigment system

Table 31 Distinctions between photosynthetic bacteria and green plants

Organism	*Chlorophyll type*	*Substrate*	*Relation to oxygen*
Green sulfur bacteria (*Chlorobium*)	Chlorobium chlorophyll	Inorganic sulfur compounds	Strictly anaerobic
Purple sulfur bacteria (*Chromatium*)	Bacteriochlorophyll	Inorganic or organic compounds	Strictly anaerobic
Purple to red or brown nonsulfur bacteria (*Rhodospirillum, Rhodopseudomonas*)	Bacteriochlorophyll	Organic compounds + vitamins	Strictly anaerobic or microaerophilic
Green plants	Chlorophyll *a*	H_2O	Aerobic

composed of bacteriochlorophyll and one or more carotenoids, coloring the cells yellowish brown, olive brown, dark brown, or various shades of red. Color is usually not observable with single cells but only with cell masses. The organisms are generally microaerophilic, although many representatives may grow at full atmospheric oxygen tension. They are capable of development under strictly anaerobic conditions, but only in illuminated cultures by virtue of a photosynthetic metabolism. The latter is dependent upon the presence of extraneous hydrogen donors, such as alcohols, fatty acids, hydroxy and keto acids, and does not proceed with the evolution of molecular oxygen. Those members which can grow in the presence of air can also be cultivated in darkness, but only under aerobic conditions.

Photosynthesis in green and purple sulfur bacteria The green and some of the purple sulfur bacteria can develop in inorganic media containing hydrogen sulfide, in the presence of light, and under strictly anaerobic conditions. They use CO_2 as the sole source of carbon for cellular synthesis. The CO_2 is reduced not by H_2O as occurs in green plants but by hydrogen sulfide or other inorganic compounds. The hydrogen sulfide is oxidized to free sulfur which is deposited outside the cells. Since the reaction is anaerobic, free oxygen is not released as occurs in green plants.

The purple sulfur bacteria can utilize simple organic compounds as the major carbon source for cellular synthesis. However, in order to have photosynthesis, light and anaerobic conditions must be maintained.

The distinctions among photosynthesis in the green sulfur bacteria, purple sulfur bacteria, purple to red or brown nonsulfur bacteria and green plants are given in Table 31.

The various chlorophylls which have been isolated from photosynthetic bacteria are not the same. Jensen et al. (1964) proposed a new nomenclature for the bacterial chlorophylls (Table 32).

The oxidation of hydrogen sulfide, accompanied by a reduction of CO_2, by the green and purple sulfur bacteria is believed to proceed as follows:

$$\textbf{1.}\quad CO_2 + 2H_2S \xrightarrow{\text{Light}} (CH_2O) + H_2O + 2S$$

Table 32 Current and proposed nomenclature of bacterial chlorophylls

Current designation	*Proposed designation*
Bacteriochlorophyll	Bacteriochlorophyll a
Bacteriochlorophyll b[5]	Bacteriochlorophyll b
Chlorobium chlorophyll-660	Bacteriochlorophylls c
Chlorobium chlorophyll-650	Bacteriochlorophylls d

2. $$3CO_2 + 2S + 5H_2O \xrightarrow{\text{Light}} 3(CH_2O) + 2H_2SO_4$$

Overall $$2CO_2 + H_2S + 2H_2O \xrightarrow{\text{Light}} 2(CH_2O) + H_2SO_4$$

The brown and red (sometimes called purple) nonsulfur bacteria of the family Athiorhodaceae also contain photosynthetic pigments and are light-sensitive under anaerobic conditions but show the following dissimilarities to the green and purple sulfur bacteria.

1. They do not contain sulfur granules.
2. They reduce CO_2 in the presence of extraneous hydrogen donors, including alcohols, fatty acids, hydroxy and keto acids, without the evolution of O_2. In the presence of butyric acid the reaction becomes

$$CO_2 + C_3H_7 \cdot COOH + H_2O \xrightarrow{\text{Light}} 5(CH_2O)$$

3. They may produce H_2 and CO_2 in the presence of (*a*) dicarboxylic acid, (*b*) glutamic or aspartic acid as a source of nitrogen, and (*c*) biotin as a growth factor. The net formation of both CO_2 and H_2 is unusual in view of the fact these organisms can utilize both H_2 and CO_2 according to the reaction

$$2H_2 + CO_2 \xrightarrow{\text{Light}} \text{“cell material”}$$

The green and purple sulfur bacteria can be adapted to grow in media containing simple organic compounds in place of H_2S. Likewise, the brown and red nonsulfur bacteria can be adapted to grow in media containing H_2S in place of certain simple organic compounds. Therefore, the following equation may be used to express the photosynthetic reaction:

$$CO_2 + 2H_2A \xrightarrow{\text{Light}} (CH_2O) + H_2O + 2A$$

The photosynthetic process in bacteria is very similar to plant photosynthesis (Fig. 312). According to Vernon (1964), the bacteriochlorophyll system absorbs light to produce both an oxidizing component and a reducing component. The reducing component can be used for cellular reductions, including the reduction of CO_2 to carbohydrate. Both

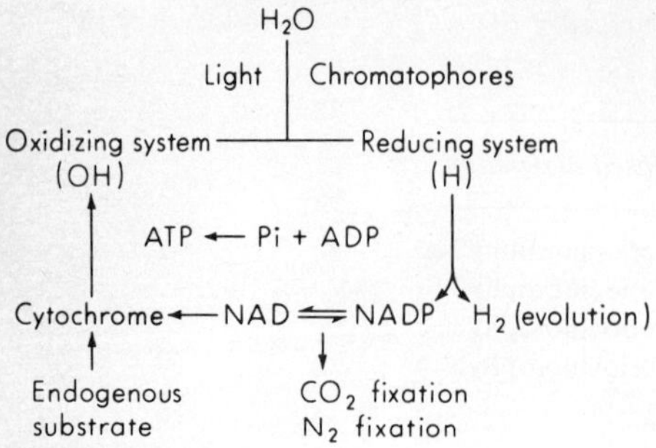

Fig. 312 Schematic representation of bacterial photosynthesis. See text for description. (After Vernon.)

components can be linked through a series of enzymatic reactions to yield ATP by the process of photophosphorylation. In these reactions the bacteria are similar to the plants but differ in their use of the oxidizing component. Plants evolve O_2; bacteria do not. Consequently the oxidizing power generated photochemically partially goes to accomplish oxidation of cellular components. This represents the main difference between plant and bacterial photosynthesis.

For more information: Arnon (1961); Clayton and Sistrom (1964); Fuller and Conti (1963); Gest et al. (1963); Kamen (1963); Levine (1968, 1969); Pfennig (1967); Pfennig and Cohen-Bazire (1967); Stanier (1961); Thiele (1968); Vernon (1964, 1968); Whittenbury and McLee (1967).

Phosphorus cycle

Phosphorus does not occur in the free state in nature. It is found in the form of phosphates in the minerals chloroapatite [calcium phosphate chloride, $3Ca_3(PO_4)_2 \cdot CaCl_2$], fluorapatite [calcium phosphate fluoride, $3Ca_3(PO_4)_2 \cdot CaF_2$], vivianite (ferrous phosphate octahydrate, $3FeO \cdot P_2O_5 \cdot 8H_2O$), and wavellite (aluminum phosphate tridecahydrate, $3Al_2O_3 \cdot 2P_2O_5 \cdot 13H_2O$). Small quantities occur in granite rocks; also in fertile soil. On the surface of an alkaline and oxidized earth, it tends to be immobilized as calcium phosphate or ferric phosphate.

Fig. 313 The phosphorus cycle. (After Lotka.)

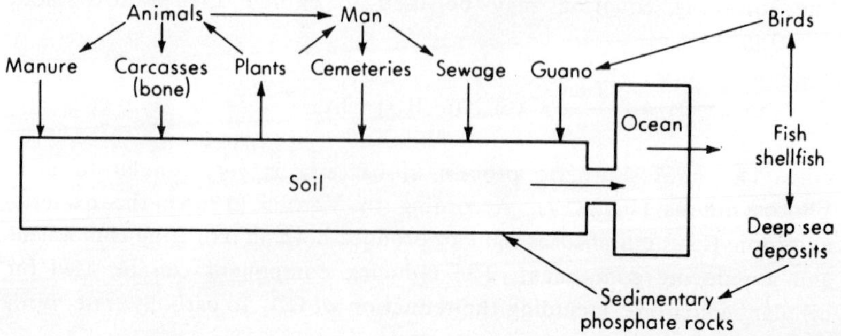

Phosphorus is an essential constituent of protoplasm, nervous tissue, and bones. The framework of man and animals consists largely of calcium phosphate. The element is found also in many organic compounds including ribonucleic acid (RNA) and deoxyribonucleic acid (DNA).

In undisturbed nature, phosphorus is absent or present in insufficient amounts. Soil infertility is generally interpreted to mean a shortage or absence of the element.

The phosphorus cycle is diagrammed in Fig. 313.

References

Allen, E. K., and O. N. Allen: The scope of nodulation in the *Leguminosae*. In "Recent Advances in Botany," Toronto, University of Toronto Press, 1961.

Arnon, D. I.: Cell-free photosynthesis and the energy conversion process. In "Light and Life," edited by W. D. McElroy and Bentley Glass, Baltimore, The Johns Hopkins Press, 1961.

Aspiras, R. B., et al.: The role of microorganisms in the stabilization of soil aggregates, *Soil Biol. Biochem.*, **3**:347, 1971.

——: Ferredoxin and photosynthesis, *Science*, **149**:1460, 1965.

Baalsrud, K.: Some aspects of the physiology of thiobacilli. In "Autotrophic Microorganisms," edited by B. A. Fry and J. L. Peel, New York, Cambridge University Press, 1954.

Babieva, I. P., and A. I. Belianin: Yeast in rhizosphere, *Mikrobiologiya,* **35**:712, 1966.

Bayer, E., et al.: Untersuchungen zur totalsynthese des ferredoxins: I. Synthese der aminosauresequenz von *C. pasteurianum* ferredoxin, *Tetrahedron,* **24**:4853, 1968.

Beck, J. V., and D. G. Brown: Direct sulfide oxidation in the solubilization of sulfide ores by *Thiobacillus ferrooxidans, J. Bact.,* **96**:1433, 1968.

Benemann, J. R., et al.: The electron transport system in nitrogen fixation by *Azotobacter*, I. Azotoflavin as an electron carrier, *Proc. Natl. Acad. Sci.,* **64**:1079, 1969.

Bisset, K. A., and D . Brown: Some electron microscope observations on the morphology of *Sphaerotilus natans, Giorn. Microbiol.,* **17**:97, 1969.

——and J. B. Grace: The nature and relationships of autotrophic bacteria. In "Autotrophic Microorganisms," edited by B. A. Fry and J. L. Peel, New York, Cambridge University Press, 1954.

Blaylock, B. A., and A. Nason: Electron transport systems of the chemoautotroph *Ferrobacillus ferrooxidans, J. Biol. Chem.,* **238**:3453, 1963.

Bollen, W. B., and D. W. Glennie: Sawdust, bark, and other wood wastes for soil conditioning and mulching, *Forest Prod. J.,* **11**:38, 1961.

Bornside, G. H., and R. E. Kallio: Urea-hydrolyzing bacilli. I. A physiological approach to identification, *J. Bact.,* **71**:627, 1956*a*; II. Nutritional profiles, *ibid.,* **71**:655, 1956*b*.

Bortels, H., and H. G. Henkel: Über die farbstoffbildung durch *Azotobacter chroococcum.* II. Das chromogen, *Arch. Mikrobiol.,* **60**:99, 1968.

—— and J. Olivares: Über die farbstoffbildung durch *Azotobacter chroococcum, Arch. Mikrobiol.,* **58**:6, 1967.

Bowie, I. S., et al.: Identification of aerobic heterotrophic soil bacteria to generic level by using multipoint inoculation techniques, *Can. J. Microbiol.,* **15**:297, 1969.

Breed, R. S., et al., "Bergey's Manual of Determinative Bacteriology," Baltimore, The Williams & Wilkins Company, 1957.

Buchanan, B., et al.: Ferredoxin from the photosynthetic bacterium, *Chlorobium thiosulfatophilum.* A link to ferredoxins from nonphotosynthetic bacteria, *Biochim. Biophys. Acta,* **189**:46, 1969.

Bulen, W. A.: Biological nitrogen fixation, *Science,* **147**:310, 1965.

Burton, S. D., et al.: Utilization of acetate by *Beggiatoa, J. Bact.,* **91**:1192, 1966.

Butler, R. G., and W. W. Umbreit: Absorption and utilization of organic matter by the strict autotroph, *Thiobacillus thiooxidans,* with special reference to aspartic acid, *J. Bact.,* **91**:661, 1966.

Canada Department of Agriculture: Qualitative studies of soil microorganisms, Ottawa, I-XV, 1938–1957.

Casida, L. E., Jr.: On the isolation and growth of individual microbial cells from soil, *Can. J. Microbiol.,* **8**:115, 1962.

Chan, E. C. S., and H. Katznelson: Growth interactions of *Arthrobacter globiformis* and *Pseudomonas* sp. in relation to the rhizosphere effect, *Can. J. Microbiol.,* **7**:759, 1961.

—— et al.: The influence of soil and root extracts on the associative growth of selected soil bacteria, *Can. J. Microbiol.,* **9**:187, 1963.

Charles, A. M., and I. Suzuki: Mechanism of thiosulfate oxidation by *Thiobacillus novellus, Biochim. Biophys. Acta,* **128**:510, 1966.

Chesters, G., et al.: Soil aggregation in relation to various soil constituents, *Soil Sci. Soc. Am. Proc.,* **21**:272, 1957.

Clark, C., and E. L. Schmidt: Growth response of *Nitrosomonas europaea* to amino acids, *J. Bact.,* **93**:1302, 1967*a*.

—— and ——: Uptake and utilization of amino acids by resting cells of *Nitrosomonas europaea, J. Bact.,* **93**:1309, 1967*b*.

Clayton, R. K., and W. R. Sistrom: The importance of reaction centers for the photochemistry of photosynthesis, *Proc. Natl. Acad. Sci. U.S.,* **52**:67, 1964.

Conn, H. J., and J. E. Conn: The stimulating effect of colloids upon the growth of certain bacteria, *J. Bact.,* **39**:99, 1940.

Cusanovich, M. A., and M. D. Kamen: Light-induced electron transport in *Chromatium* strain D. I. Isolation and characterization of *Chromatium* chromatophores, *Biochim. Biophys. Acta,* **153**:376, 1968*a*; III. Photophosphorylation by *Chromatium* chromatophores, *ibid.,* **153**:418, 1968*b*.

—— et al.: II. Light-induced absorbance changes in *Chromatium* chromatophores, *Biochim. Biophys. Acta,* **153**:397, 1968.

Dart, P. J., and F. V. Mercer: The legume rhizosphere, *Arch. Mikrobiol.*, **47**:344, 1964.

Date, R. A., and A. M. Decker: Minimal antigenic constitution of 28 strains of *Rhizobium japonicum, Can. J. Microbiol.*, **11**:1, 1965.

Deevey, E. S., Jr.: Mineral cycles, *Sci. Am.*, **223**:148, 1970.

De Ley, J., and A. Rassel: DNA base composition, flagellation and taxonomy of the genus *Rhizobium, J. Gen. Microbiol.*, **41**:85, 1965.

Delwiche, C. C.: The nitrogen cycle, *Sci. Am.*, **223**:136, 1970.

Dias, F. F., et al.: Calcium nutrition of *Sphaerotilus* growing in a continuous-flow apparatus, *Appl. Microbiol.*, **16**:1364, 1968.

Dilworth, M. J., and C. A. Parker: Development of the nitrogen-fixing system in legumes, *J. Theoret. Biol.*, **25**:208, 1969.

Din, G. A., et al.: Ferrous iron oxidation by *Ferrobacillus ferrooxidans, Can. J. Biochem.*, **45**:1523, 1967.

Dudman, W. F.: Capsulation in *Rhizobium* species, *J. Bact.*, **95**:1200, 1968.

Dugan, P. R., and D. G. Lundgren: Energy supply for chemoautotroph *Ferrobacillus ferrooxidans, J. Bact.*, **89**:825, 1965.

Duncan, D. W., et al.: Role of *Thiobacillus ferrooxidans* in the oxidation of sulfide minerals, *Can. J. Microbiol.*, **13**:397, 1967.

Eck, R. V., and M. O. Dayhoff: Evolution of the structure of ferredoxin based on living relics of primitive amino acid sequences, *Science*, **152**:363, 1966.

Egdell, J. W., et al.: Some studies of the colony count technique for soil bacteria, *J. Appl. Bact.*, **23**:69, 1960.

Faust, L., and R. S. Wolfe: Enrichment and cultivation of *Beggiatoa alba, J. Bact.*, **81**:99, 1961.

Fergus, C. L.: The cellulolytic activity of thermophilic fungi and actinomycetes, *Mycologia*, **61**:120, 1969.

Fottrell, P. F.: Recent advances in biological nitrogen fixation, *Sci. Progr. (Oxford)*, **56**:541, 1968.

Fuller, R. C., and S. F. Conti: The microbial photosynthetic apparatus. In "Studies on Microalgae and Photosynthetic Bacteria," Tokyo, Institute of Applied Microbiology, University of Tokyo, 1963.

Funk, H. B., and T. A. Krulwich: Preparation of clear silica gels that can be streaked, *J. Bact.*, **88**:1200, 1964.

Gest, H., et al.: "Bacterial Photosynthesis," Yellow Springs, Ohio, The Antioch Press, 1963.

Ghosh, H. K.: A simplified and inexpensive method of making silica gel for growing bacteria, *Indian J. Microbiol.*, **7**:151, 1967.

Gibson, A. H.: Physical environment and symbiotic nitrogen fixation. IV. Factors affecting the early stages of nodulation, *Australian J. Biol. Sci.*, **20**:1087, 1967*a*; V. Effect of time of exposure to unfavorable root temperatures, *ibid.*, **20**:1105, 1967*b*.

Hanert, H.: Untersuchungen zur isolierung, stoffwechselphysiologie und morphologie von *Gallionella ferruginea* Ehrenberg, *Arch. Mikrobiol.*, **60**:348, 1968.

Hitchins, V. M., and H. L. Sadoff: Morphogenesis of cysts in *Azotobacter vinelandii, J. Bact.,* **104**:492, 1970.

Holding, A. J.: The properties and classification of the predominant gram-negative bacteria occurring in soil, *J. Appl. Bact.,* **23**:515, 1960.

Holland, A. A.: Serologic characteristics of certain root-nodule bacteria of legumes, *Antonie van Leeuwenhoek, J. Microbiol. Serol.,* **32**:410, 1966.

Hooper, A. B.: Biochemical basis of obligate autotrophy in *Nitrosomonas europaea, J. Bact.,* **97**:776, 1969.

Hurlbert, R. E.: Effect of oxygen on viability and substrate utilization in *Chromatium, J. Bact.,* **93**:1346, 1967.

Iswaran, V., and A. Sen: *Azotobacter* spp. in the rhizosphere of some agricultural crops, *Proc. Indian Acad. Sci.,* **53**:182, 1961.

Jensen, A., et al.: Chlorophylls of photosynthetic bacteria, *Biochim. Biophys. Acta,* **88**:466, 1964.

Johnson, E. J., and S. Abraham: Assimilation and metabolism of exogenous organic compounds by the strict autotrophs *Thiobacillus thioparus* and *Thiobacillus neapolitanus, J. Bact.,* **97**:1198, 1969.

Jordan, D. C.: The bacteroids of the genus *Rhizobium, Bact. Rev.,* **26**:119, 1962.

Kadota, H.: A study on the marine aerobic cellulose-decomposing bacteria, *Mem. Coll. Agr., Kyoto Univ.,* no. 74, 1956.

Kamen, M. D.: "Primary Processes in Photosynthesis," New York, Academic Press, Inc., 1963.

Kennedy, I. R., et al.: N_2 fixation by purified components of the N_2-fixing system of *Clostridium pasteurianum, Biochim. Biophys. Acta,* **153**:777, 1968.

King, K. W.: Microbial degradation of cellulose, *Tech. Bull. 154, Va. Polytech. Inst.,* 1961.

Kinsel, N. A.: New sulfur oxidizing iron bacterium: *Ferrobacillus sulfooxidans* sp. n., *J. Bact.,* **80**:628, 1960.

Kistner, A., and L. Gouws: Cellulolytic cocci occurring in the rumen of sheep conditioned to Lucerne hay, *J. Gen. Microbiol.,* **34**:447, 1964.

Koo, V. M., et al.: Surface structure of *Azotobacter vinelandii* cysts as revealed by freeze cleaving, *J. Bact.,* **100**:1105, 1969.

Kramer, M. J., and M. D. Socolofsky: Characterization of the central body of the *Azotobacter* cyst, *Antonie van Leeuwenhoek, J. Microbiol. Serol.,* **36**:119, 1970.

Lazaroff, N.: Sulfate requirement for iron oxidation by *Thiobacillus ferrooxidans, J. Bact.,* **85**:78, 1963.

Leatherwood, J. M., and M. P. Sharma: Novel anaerobic cellulolytic bacterium, *J. Bact.,* **110**:751, 1972.

Lees, H.: The biochemistry of the nitrifying bacteria. In "Autotrophic Microorganisms," edited by B. A. Fry and J. L. Peel, New York, Cambridge University Press, 1954.

——: IV. Some thoughts on the energetics of chemosynthesis, *Bact. Rev.*, **26**:165, 1962*a*.

——: The unremarkable autotrophs, College Park, University of Maryland, Department of Microbiology, 1962*b*.

—— et al.: The thermodynamics of iron oxidation by the ferrobacilli, *Can. J. Microbiol.*, **15**:43, 1969.

Levine, R. P.: Genetic dissection of photosynthesis, *Science*, **162**:768, 1968.

——: The mechanism of photosynthesis, *Sci. Am.*, **221**:58, 1969.

Lin, L. P., and H. L. Sadoff: Chemical composition of *Azotobacter vinelandii* cysts, *J. Bact.*, **100**:480, 1969.

Lochhead, A. G.: Qualitative studies of soil microorganisms. III. Influence of plant growth on the character of the bacterial flora, *Can. J. Res.*, **18**:42, 1940.

——: The soil microflora, the plant, and the root pathogen, *Trans. Roy. Soc. Can.*, **52**:17, 1958*a*; soil bacteria and growth-promoting substances, *Bact. Rev.*, **22**:145, 1958*b*.

Loneragan, J. F., et al.: Survival of *Rhizobium* on clover seeds, *Nature*, **192**:526, 1961.

Margalith, P., et al.: Sulfur oxidation by the iron bacterium *Ferrobacillus ferrooxidans*, *J. Bact.*, **92**:1706, 1966.

Mark, R. E., et al.: Cellulose: refutation of a folded-chain structure, *Science*, **164**:72, 1969.

Martin, J. P., and S. J. Richards: Decomposition and binding action of a polysaccharide from *Chromobacterium violaceum* in soil, *J. Bact.*, **85**:1288, 1963.

Matin, A., and S. C. Rittenberg: Utilization of glucose in heterotrophic media by *Thiobacillus intermedius*, *J. Bact.*, **104**:234, 1970*a*.

—— and ——: Regulation of glucose metabolism in *Thiobacillus intermedius*, *J. Bact.*, **104**:239, 1970*b*.

McBee, R. H.: The anaerobic thermophilic cellulolytic bacteria, *Bact. Rev.*, **14**:51, 1950.

McCalla, T. M.: Physico-chemical behavior of soil bacteria in relation to the soil colloid, *J. Bact.*, **40**:33, 1940.

——: Microbial and related studies of stubble mulching, *J. Soil Water Conserv.*, **13**:255, 1958.

Means, U. M., and L. W. Erdman: Longevity and efficiency of rhizobial cultures, *Soil Sci. Soc. Am. Proc.*, **27**:305, 1963.

Moriarty, D. J. W., and D. J. D. Nicholas: Enzymic sulphide oxidation by *Thiobacillus concretivorus*, *Biochim. Biophys. Acta*, **184**:114, 1969.

—— and ——: Products of sulphide oxidation in extracts of *Thiobacillus concretivorus*, *Biochim. Biophys. Acta*, **197**:143, 1970.

Morrill, L. G., and J. E. Dawson: Patterns observed for the oxidation of ammonium to nitrate by soil organisms, *Soil Sci. Soc. Am. Proc.*, **31**:757, 1967.

Mortenson, L. E.: Nitrogen fixation: Role of ferredoxin in anaerobic metabolism. In "Annual Review of Microbiology," **17**:115, 1963.

——: Purification and analysis of ferredoxin from *Clostridium pasteurianum, Biochim. Biophys. Acta,* **81**:71, 1964*a*.

——: Ferredoxin requirement for nitrogen fixation by extracts of *Clostridium pasteurianum, Biochim. Biophys. Acta.,* **81**:473, 1964*b*.

——: Nitrogen fixation in extracts of *Clostridium pasteurianum*. In "Non-Heme Iron Proteins: Role in Energy Conversion," edited by A. San Pietro, Yellow Springs, Ohio, The Antioch Press, 1965.

—— et al.: III. Nitrogen fixation by enzyme preparations, *Bact. Rev.,* **26**:42, 1962.

—— et al.: Ferredoxin in the phosphoroclastic reaction of pyruvic acid and its relation to nitrogen fixation in *Clostridium pasteurianum, J. Biol. Chem.,* **238**:794, 1963.

Mulder, E. G.: Iron bacteria, particularly those of the *Sphaerotilus-Leptothrix* group, and industrial problems, *J. Appl. Bact.,* **27**:151, 1964.

Nutman, P. S.: Varietal differences in the nodulation of subterranean clover, *Australian J. Agr. Res.,* **18**:381, 1967.

Owens, L. D., et al.: Identification of plant volatiles that stimulate microbial respiration and growth in soil, *Phytopathology,* **59**:1468, 1969.

Peck, H. D., Jr., and M. P. Stulberg: O^{18} studies on the mechanism of sulfate formation and phosphorylation in extracts of *Thiobacillus thioparus, J. Biol. Chem.,* **237**:1648, 1962.

Peeters, T., and M. I. H. Aleem: Oxidation of sulfur compounds and electron transport in *Thiobacillus denitrificans, Arch. Mikrobiol.,* **71**:319, 1970.

Pfennig, N.: Photosynthetic bacteria. In "Annual Review of Microbiology," **21**:285, 1967.

—— and G. Cohen-Bazire: Some properties of the green bacterium *Pelodictyon clathratiforme, Arch. Mikrobiol.,* **59**:226, 1967.

Pope, L. M., and O. Wyss: Outer layers of the *Azotobacter vinelandii* cyst, *J. Bact.,* **102**:234, 1970.

Pramer, D.: The influence of physical and chemical factors on the preparation of silica gel media, *Appl. Microbiol.,* **5**:392, 1957.

Pringsheim, E. G.: Iron bacteria, *Biol. Rev.,* **24**:200, 1949.

——: Heterotrophism and species concepts in *Beggiatoa, Am. J. Botany,* **51**:898, 1964.

Rall, S. C., et al.: The amino acid sequence of ferredoxin from *Clostridium acidi-urici, Biochemistry,* **8**:2486, 1969.

Rao, K. K., and H. Matsubara: The amino acid sequence of taro ferredoxin, *Biochem. Biophys. Res. Commun.,* **38**:500, 1970.

—— et al.: Amino acid composition and terminal sequences of ferredoxins from two photosynthetic green bacteria, *J. Bact.,* **100**:1411, 1969.

Romano, A. H., and D. J. Gleason: Pattern of sheath synthesis in *Sphaerotilus natans, J. Bact.,* **88**:1145, 1964.

Ross, A. J., et al.: Electron transport and coupled phosphorylation in the

chemoautotroph *Thiobacillus neapolitanus, Biochem. Biophys. Res. Commun.*, **32**:301, 1968.
Rouatt, J. W., and H. Katznelson: A study of the bacteria on the root surface and in the rhizosphere soil of crop plants, *J. Appl. Bact.*, **24**:164, 1961.
Rovira, A. D.: Interactions between plant roots and soil microorganisms. In "Annual Review of Microbiology," **19**:241, 1965.
Sadoff, H. L., et al.: Physiological studies of encystment in *Azotobacter vinelandii, J. Bact.*, **105**:185, 1971.
Santoro, T., and G. Stotzky: Effect of electrolyte composition and pH on the particle size distribution of microorganisms and clay minerals as determined by the electrical sensing zone method, *Arch. Biochem. Biophys.*, **122**:664, 1967.
Scotten, H. L., and J. L. Stokes: Isolation and properties of *Beggiatoa, Arch. Mikrobiol.*, **42**:353, 1962.
Shafia, F., and R. F. Wilkinson, Jr.: Growth of *Ferrobacillus ferrooxidans* on organic matter, *J. Bact.*, **97**:256, 1969.
Shanmugam, K. T., and D. I. Arnon: Effect of ferredoxin on bacterial photophosphorylation, *Biochim. Biophys. Acta*, **256**:487, 1972.
—— et al.: Ferredoxins in light- and dark-grown photosynthetic cells with special reference to *Rhodospirillum rubrum, Biochim. Biophys. Acta*, **256**:477, 1972.
Sharpley, J. M.: The occurrence of *Gallionella* in salt water, *Appl. Microbiol.*, **9**:380, 1961.
Shively, J. M., et al.: Comparative ultrastructure of the thiobacilli, *J. Bact.*, **101**:618, 1970.
Short, S. A., et al.: Phospholipid metabolism in *Ferrobacillus ferrooxidans, J. Bact.*, **99**:142, 1969.
Silver, W. S.: Studies on nitrite oxidizing micro-organisms: I. Nitrite oxidation by *Nitrobacter, Soil Sci. Soc. Am. Proc.*, **25**:197, 1961.
Simpson, M. E., and P. B. Marsh: Cellulose decomposition by the aspergilli, *U.S. Dept. Agr. Tech. Bull.* no. 1303, 1964.
Skinner, F. A., et al.: A comparison of a direct- and a plate-counting technique for the quantitative estimation of soil micro-organisms, *J. Gen. Microbiol.*, **6**:261, 1952.
Skyring, G. W., and C. Quadling: Soil bacteria: principal component analysis of descriptions of named cultures, *Can. J. Microbiol.*, **15**:141, 1969.
Sommers, L. E., and R. F. Harris: Routine preparation of silica gel media using silicate solutions of varying pH, *J. Bact.*, **95**:1174, 1968.
Stanier, R. Y.: Photosynthetic mechanisms in bacteria and plants: Development of a unitary concept, *Bact. Rev.*, **25**:1, 1961.
Starkey, R. L.: Interrelations between micro-organisms and plant roots in the rhizosphere, *Bact. Rev.*, **22**:154, 1958.
Stevenson, L. H., and M. D. Socolofsky: Cyst formation and poly-β-hydroxybutyric acid accumulation in *Azotobacter, J. Bact.*, **91**:304, 1966.

Stokes, J. L.: Studies on the filamentous sheathed iron bacterium *Sphaerotilus natans, J. Bact.*, **67**:278, 1954.

Stotzky, G., and R. D. Goos: Effect of high CO_2 and low O_2 tensions on the soil microbiota, *Can. J. Microbiol.*, **11**:853, 1965.

Thiele, H. H.: Die verwertung einfacher organischer substrate durch *Thiorhodaceae, Arch. Mikrobiol.*, **60**:124, 1968.

Tilton, R. C., et al.: Marine thiobacilli. I. Isolation and distribution, *Can. J. Microbiol.*, **13**:1521, 1967*a*; II. Culture and ultrastructure, *ibid.*, **13**:1529, 1967*b*.

Trolldenier, G.: Influence of nitrogen and potassium nutrition of wheat and oxygen supply of roots on bacterial numbers, respiration of roots and denitrification in the rhizosphere, *Zentr. Bakt.*, Abt. II, **126**:130, 1971.

Trudinger, P. A.: The metabolism of inorganic sulphur compounds by thiobacilli, *Rev. Pure Appl. Chem.*, **17**:1, 1967.

Tsien, H.-C., et al.: Fine structure and the localization of the nitrite oxidizing system in *Nitrobacter winogradskyi, Antonie van Leeuwenhoek, J. Microbiol. Serol.*, **34**:483, 1968.

Umbreit, W. W.: II. The comparative physiology of autotrophic bacteria, *Bact. Rev.*, **26**:145, 1962.

Vágnerová, K., and V. Vančura: Production and utilization of amino acids by various species of rhizosphere bacteria, *Folia Microbiol.*, **7**:55, 1962.

Van Gool, A., and H. Laudelout: Spectrophotometric and kinetic study of nitrite and formate oxidation in *Nitrobacter winogradskyi, J. Bact.*, **93**:215, 1967.

Vela, G. R., and G. D. Cagle: Aggregates of *Azotobacter vinelandii* cysts, *Can. J. Microbiol.*, **18**:371, 1972.

——— and ———: Formation of fragile cysts by a strain of *Azotobacter chroococcum, J. Gen. Microbiol.*, **57**:365, 1969.

——— et al.: Ultrastructure of *Azotobacter vinelandii, J. Bact.*, **104**:933, 1970.

Vernon L. P.: Bacterial photosynthesis. In "Annual Review of Plant Physiology," **15**:73, 1964.

———: Photochemical and electron transport reactions of bacterial photosynthesis, *Bact. Rev.*, **32**:243, 1968.

Virtanen, A. I.: The biology and chemistry of nitrogen fixation by legume bacteria, *Biol. Rev.*, **22**:239, 1947.

Walker, R. H., and P. E. Brown: The nomenclature of the cowpea group of root-nodule bacteria, *Soil Sci.*, **39**:221, 1935.

Wang, W. S., and D. G. Lundgren: Peptidoglycan of a chemolithotrophic bacterium, *Ferrobacillus ferrooxidans, J. Bact.*, **95**:1851, 1968.

——— and ———: Poly-β-hydroxybutyrate in the chemolithotrophic bacterium *Ferrobacillus ferrooxidans, J. Bact.*, **97**:947, 1969.

Went, J. C., and F. de Jong: Decomposition of cellulose in soils, *Antonie van Leeuwenhoek, J. Microbiol. Serol.*, **32**:39, 1966.

Whittenbury, R., and A. G. McLee: *Rhodopseudomonas palustris* and *Rh. viridis*: Photosynthetic budding bacteria, *Arch. Mikrobiol.,* **59**:324, 1967.

Williams, P. J. L., and S. W. Watson: Autotrophy in *Nitrosocystis oceanus, J. Bact.,* **96**:1640, 1968.

Wilson, J. K.: Symbiotic segregation of strains of the root nodule bacteria by leguminous plants, *Cornell Univ. Agr. Exp. Sta. Mem.* no. 279, 1948.

Witz, D. F., et al.: Nitrogen fixation by growing cells and cell-free extracts of the *Bacillaceae, Arch. Mikrobiol.,* **55**:369, 1967.

Woods, D. D., and J. Lascelles: The no-man's land between the autotrophic and heterotrophic ways of life. In "Autotrophic Microorganisms," edited by B. A. Fry and J. L. Peel, New York, Cambridge University Press, 1954.

Wyss, O., et al.: Development and germination of the *Azotobacter* cyst, *J. Biophys. Biochem. Cytol.,* **10**:555, 1961.

Yoch, D. C., et al.: The electron transport system in nitrogen fixation by *Azotobacter*. II. Isolation and function of a new type of ferredoxin, *Proc. Natl. Acad. Sci.,* **64**:1404, 1969.

ZoBell, C. E.: Organic geochemistry of sulfur. In "Organic Geochemistry," New York, Pergamon Press, 1963.

25 *Nucleoproteins and nucleic acids of bacteria*

Nucleoproteins Nucleoproteins are compounds of nucleic acid and protein. They are called nucleoproteins because of their association with cell nuclei. Nucleoproteins are widely distributed in nature, being found in both the nuclei and cytoplasmic contents of living cells. Some virus particles consist entirely of nucleoprotein; other viruses contain additional compounds making them more complex in structure. Different proteins and nucleic acids have been isolated, indicating that many kinds of nucleoproteins occur in nature.

The proteins present are basic in character, being members of the protamines and the histones. Their basicity is due to the presence of relatively large amounts of the diamino monocarboxylic acids, namely arginine, lysine, and histidine. Pure nucleoproteins have been crystallized. However the more complicated viruses containing other compounds in addition to nucleoproteins have not been crystallized.

Nucleoproteins give an acid reaction and are insoluble in water. They are soluble in weak alkali but are precipitated from solution on the addition of acid. Nucleoproteins are very complex in composition and unstable chemically. The nucleic acid may be separated from the protein moiety by salts or by cautious addition of acids or alkalies. Some enzymes or weak acids split off some protein, transforming the compound into a mixture of protein and nuclein. The nuclein still contains some protein. More prolonged enzymatic action or treatment with acid removes the remainder of the protein, setting free nucleic acid.

Nucleic acids Nucleic acids are polymers of nucleotide residues (consisting of phosphoric acid-sugar-base) joined together by sugar phosphate esterifications. Because of their high phosphoric acid content, they are acid in reaction.

There are two types: (1) ribonucleic acid (RNA) and (2) deoxyribonucleic acid (DNA). RNA occurs largely in the cytoplasm of living cells with smaller amounts in the nucleus; DNA appears to exist entirely or almost entirely within the nucleus.

Molecules of RNA and DNA both contain normally only four kinds of nucleotides. However the two molecules yield different nucleotides.

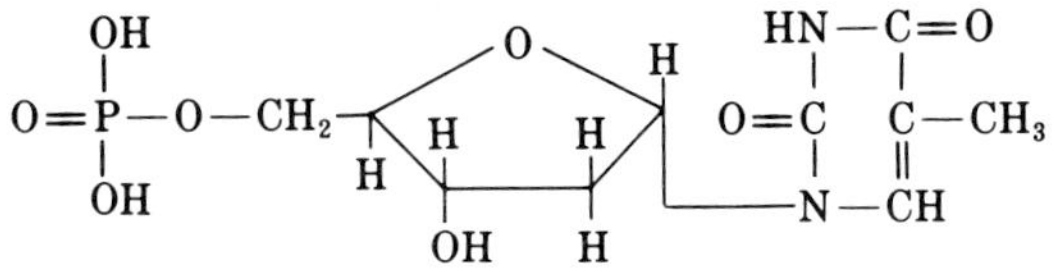

Thymidylic acid

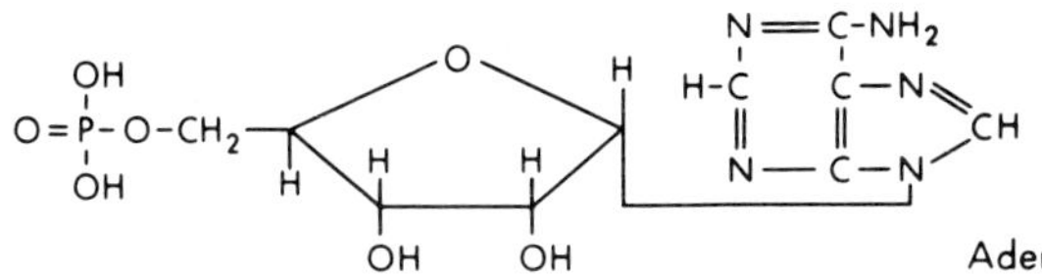

Adenylic acid

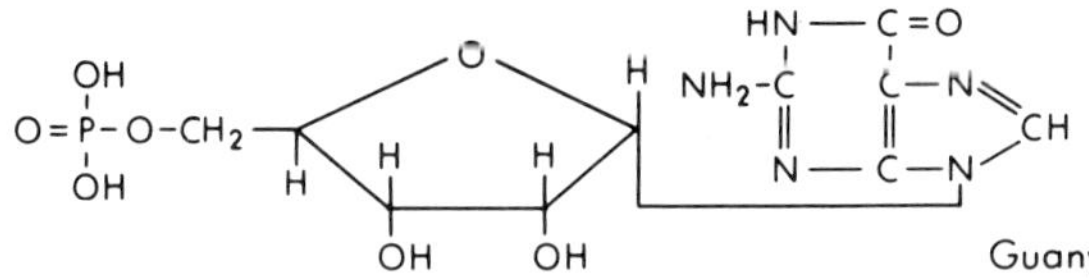

Guanylic acid

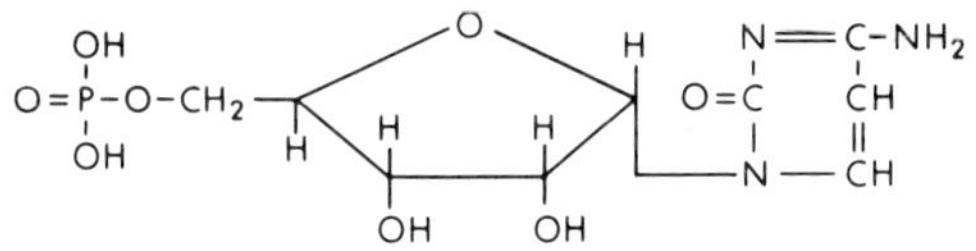

Cytidylic acid

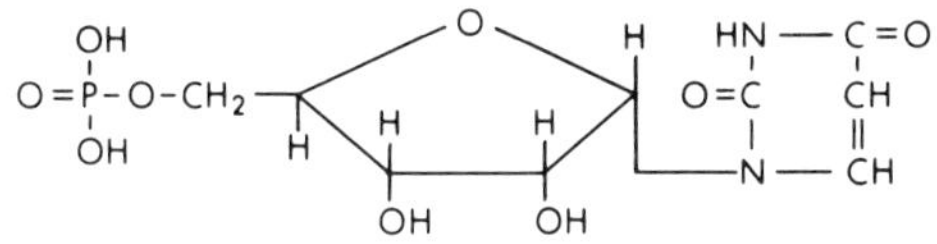

Uridylic acid

Fig. 314 Structures of the four nucleotides from RNA.

Even though each molecule contains only four nucleotides, a large variety of different nucleic acids is possible by varying the number and sequence of the few nucleotides that are available.

The structures of the four nucleotides from RNA are shown in Fig. 314.

It may be seen that all four nucleotides contain the pentose sugar ribose with phosphoric acid attached to carbon 5. The four bases (adenine, guanine, cytosine, uracil) are attached to carbon 1 of the sugar. Some RNA molecules are unusual and give bases of slightly different structure. A nucleotide takes its name from the purine or pyrimidine base present. For example, a nucleotide containing guanine is called guanylic acid; one containing cytosine is called cytidylic acid; etc. The nucleotides are joined together by an ester linkage involving the phosphoric acid of one and the ribose of another through carbon 3 to produce chains known collectively as polynucleotides (Fig. 315). Therefore a molecule of RNA consists of a

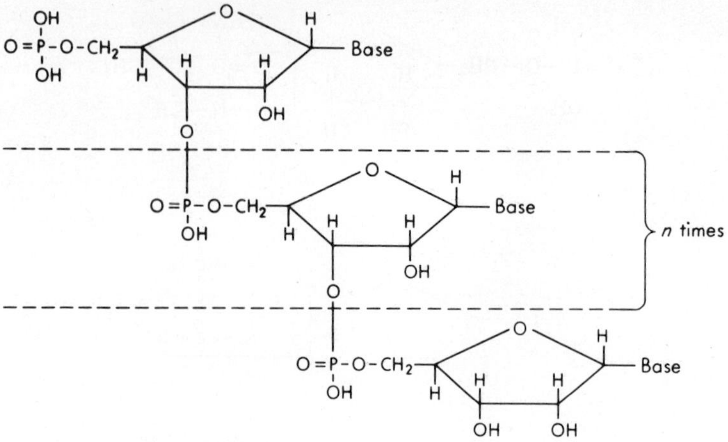

Fig. 315 Structure of RNA.

chain of alternating residues of ribose and phosphoric acid with one of the four bases linked to carbon 1 of each ribose molecule.

The four nucleotides normally found in DNA closely resemble those from RNA but differ in the following respects: (1) The sugar has no oxygen atom on carbon 2 and is called D-2-deoxyribose; (2) the base thymine is substituted for uracil. The structure of the nucleotide thymidylic acid containing thymine and D-2-deoxyribose is shown at the top of page 845. In a few instances DNA has yielded nucleotides in which cytosine contains a methyl or hydroxymethyl group attached to carbon 5. The nucleotides in DNA are linked together as in RNA. Also as in RNA, a molecule of DNA consists of a chain of alternating residues of D-2-deoxyribose and phosphoric acid with one of the four bases normally linked to carbon 1 of each pentose molecule.

For many years a nucleic acid molecule was believed to consist of the four mononucleotides joined together to give a tetranucleotide. Information now available refutes this belief. Nucleic acid molecules have now been shown to be very long, thin, thread-like filaments consisting of large numbers of nucleotide units joined together as shown in Fig. 316. The D-ribose (or D-2-deoxyribose) phosphate grouping is repeated many times to form the main chain of the molecule, leaving the bases protruding on one side.

DNA By means of x-ray information, it has been established that the -pentose-phosphate- chain in the DNA molecules does not lie on a straight line but is arranged in the form of a helix. Also, two such chains are involved forming a double-stranded spiral (Fig. 179).

The bases form a core that lies within the two spiral structures and are held together in pairs by means of hydrogen bonds. Each complete turn of the spiral contains ten hydrogen-bonded pairs. The hydrogen bonds hold the two helices and their bases together to give cohesion and strength to the entire structure.

Fig. 316 Double-stranded ribonucleic acid from cytoplasmic polyhedrosis virus of the silkworm. (Courtesy of Miura, Fujii, Sakaki, Fuke, and Kawase.)

The diameter of the double-stranded molecule is very constant throughout. The purine base adenine of one chain always lies opposite to the pyrimidine base thymine of the other chain; and in like fashion the purine base guanine always lies opposite to the pyrimidine base cytosine. The DNA molecule normally contains equivalent quantities of adenine and thymine.

The spiral may be likened to a staircase. Also, the molecules of DNA and RNA can differ from one another by the length and order of the nucleotides in a chain. Each nucleotide is not repeated along a chain at constant intervals. In this manner, the number of different molecules of DNA and RNA is almost unlimited. It should be mentioned that both single-stranded and double-stranded DNA and RNA have been isolated from bacteria and viruses. In general DNA occurs more often as double-stranded and RNA as single-stranded. Exceptions have been noted in both acids, especially among the viruses.

DNA is the genetic material of all animals, plants, microorganisms, and some viruses. In other viruses, RNA is the genetic material. These are

Fig. 317 Superhelical circular deoxyribonucleic acid. Upper left, twisted DNA molecule; lower and right, relatively untwisted DNA molecule. (Courtesy of Bode and MacHattie.)

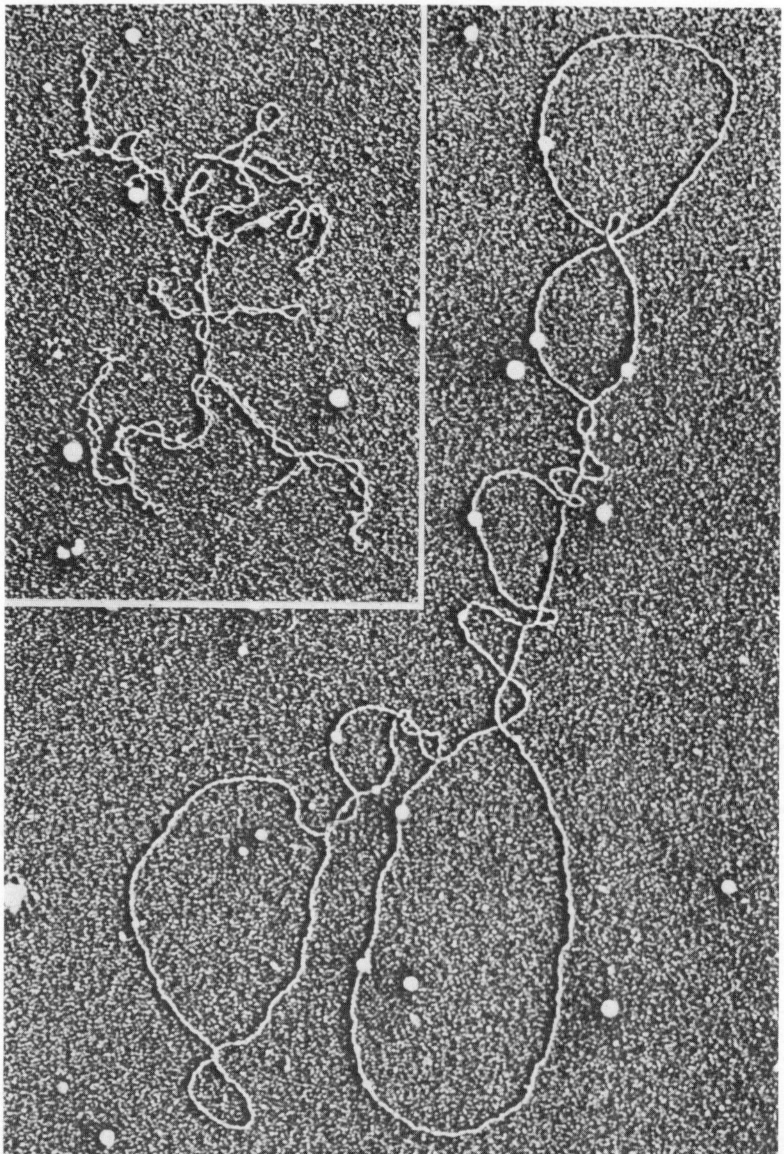

the only compounds present in cells which can carry genetic information. They make up the so-called genes of cells.

Electron micrographs of filaments of DNA and RNA molecules are shown in Figs. 316 and 317.

Chromosomes It is well established that chromosomes consist chiefly of DNA. Before a cell divides, its chromosomes divide into pairs with one of each pair passing into the new cell, followed by separation of the daughter cell from the mother cell. The new chromosomal material is an exact duplication of the old which means that the new cell is a copy of the old. Barring modification from outside influences, chromosomes reproduce themselves with 100 percent accuracy.

DNA has been established as constituting the chemical basis of heredity. The genes which control heredity are strung out along the lengths of the chromosomes. Each gene corresponds to a segment of the DNA molecule. All the various characteristics of a new organism are the result of the inheritance of the genes from the mother cell.

For more information: Baldwin (1967); Bendet et al. (1962); Bernstein (1970); Bode and MacHattie (1968); Britten and Kohne (1968, 1970); Cairns (1961); Cairns and Davern (1967); Cold Spring Harbor Lab. Quant. Biol. (1963); Freifelder (1969); Fuller (1961); Hanawalt and Haynes (1967); Hayashi et al. (1963); Holley (1966); Langridge and Gomatos (1963); Miura et al. (1968); Nisioka et al. (1969); Polanyi (1968); Rush et al. (1969); Takahashi et al. (1969); Thomas (1967); Vasquez and Kleinschmidt (1968); Westmoreland et al. (1969); Watson (1968); Wilkins (1963); Williams (1963); Davidson (1972); Dressler (1970); Mendelson (1972); Ohe and Weissman (1971).

References

Baldwin, E.: "The Nature of Biochemistry," Cambridge, Cambridge University Press, 1967.

Bendet, I., et al.: The size of T3 DNA, *J. Mol. Biol.*, 5:76, 1962.

Bernstein, C.: A comparison of the number of nucleotides per unit length in *Escherichia coli* and T4 chromosomes, *Biophys. J.*, **10**:1154, 1970.

Bode, V. C., and L. A. MacHattie: Electron microscopy of superhelical circular λ DNA, *J. Mol. Biol.*, **32**:673, 1968.

Britten, R. J., and D. E. Kohne: Repeated sequences in DNA, *Science*, **161**:529, 1968.

—— and ——: Repeated segments of DNA, *Sci. Am.*, **222**:24, 1970.

Cairns, J.: An estimate of the length of the DNA molecule of T2 bacteriophage by autoradiography, *J. Mol. Biol.*, **3**:756, 1961.

—— and C. I. Davern: The mechanics of DNA replication in bacteria, *J. Cell Physiol.*, **70**:65, 1967.

Cold Spring Harbor Lab. Quant. Biol.: "Synthesis and Structure of Macromolecules," Cold Spring Harbor, N.Y., 1963.

Davidson, J. N.: "The Biochemistry of the Nucleic Acids," London, Methuen & Co., Ltd., 1972.

Dressler, D.: The rolling circle for ϕX DNA replication, II. Synthesis of single-stranded circles, *Proc. Natl. Acad. Sci.,* **67**:1934, 1970.

Freifelder, D.: Single-strand breaks in bacterial DNA associated with thymine starvation, *J. Mol. Biol.,* **45**:1, 1969.

Fuller, W.: Two-stranded helical configurations for ribonucleic acid, *J. Mol. Biol.,* **3**:175, 1961.

Hanawalt, P. C., and R. H. Haynes: The repair of DNA, *Sci. Am.,* **216**:36, 1967.

Hayashi, M., et al.: Replicating form of a single-stranded DNA virus: Isolation and properties, *Science,* **140**:1313, 1963.

Holley, R. W.: The nucleotide sequence of a nucleic acid, *Sci. Am.,* **214**:30, 1966.

Langridge, R., and P. J. Gomatos: The structure of RNA, *Science,* **141**:694, 1963.

Mendelson, N. H.: Deoxyribonucleic acid distribution in *Bacillus subtilis* independent of cell elongation, *J. Bact.,* **111**:156, 1972.

Miura, K., et al.: Double-stranded ribonucleic acid from cytoplasmic polyhedrosis virus of the silkworm, *J. Virology,* **2**:1211, 1968.

Nisioka, T., et al.: Composite circular forms of R factor deoxyribonucleic acid molecules, *J. Bact.,* **97**:376, 1969.

Ohe, K., and S. M. Weissman: The nucleotide sequence of a low molecular weight ribonucleic acid from cells infected with adenovirus 2, *J. Biol. Chem.,* **246**:6991, 1971.

Polanyi, M.: Life's irreducible structure, *Science,* **160**:1308, 1968.

Rush, M. G., et al.: Circular deoxyribonucleic acid from *Shigella dysenteriae* Y6R, *J. Bact.,* **100**:803, 1969.

Takahashi, M., et al.: Estimation of relative molecular length of DNA by electrophoresis in agarose gel, *Biochim. Biophys. Acta,* **174**:183, 1969.

Thomas, C. A., Jr.: The rule of the ring, *J. Cell. Physiol.,* **70**:13, 1967.

Vasquez, C., and A. K. Kleinschmidt: Electron microscopy of RNA strands released from individual reovirus particles, *J. Mol. Biol.,* **34**:137, 1968.

Watson, J. D.: "The Double Helix," New York, Atheneum Publishers, 1968.

Westmoreland, B. C., et al.: Mapping of deletions and substitutions in heteroduplex DNA molecules of bacteriophage lambda by electron microscopy, *Science,* **163**:1343, 1969.

Wilkins, M. H. F.: Molecular configuration of nucleic acids, *Science,* **140**:941, 1963.

Williams, R. C.: Macromolecules, *Science,* **141**:934, 1963.

Immunity and the immune response[1] 26

Immunology is the study of a specific response of the higher organisms directed against foreign materials. The term "immunity" is best defined as the altered capability of a host cell or organism to react to a particular stimulus. The stimulating material is termed an antigen. The science of immunology has a broad biological scope, involving concepts of recognition, specificity, and biologic memory. Immunity, however, was historically applied to that state of nonsusceptibility following contact with a very narrow group of antigens from disease-producing microorganisms.

There were two fortuitous occurrences which provoked the most rapid early advances in the field of immunology, long before the concept of an antigen or antibody molecule was even suggested. First was the very obvious nature of smallpox infection. Voltaire wrote in his letters of 1733 (Humphrey and White, 1964) that "three score persons in every hundred have the smallpox. Of these three score, twenty die of it in the most favorable season of life, and as many more wear the disagreeable remains of it on their faces so long as they live." The loathsome sign of recovery from smallpox served as a constant reminder that infection was ever present and served to stimulate medical scientists to work toward its eradication. The first such measure was "variolation" or the injection of material from a smallpox pustule into young children with the view toward protecting them from smallpox in later life. This was at best a perilous procedure, and fatalities occasionally resulted.

Edward Jenner, an English country physician and naturalist, who had been subject to the rigors of variolation as a child and remembered it in nightmares, was successful in proving that the old wives' tales about milkmaids never succumbing to smallpox infection had a factual basis. The cowpox virus, which caused the scars on the girls' hands, was closely enough related to the smallpox virus to induce immunity of smallpox. These observations served as a basis for the science of immunology.

[1] This chapter was written by Dr. Eli E. Sercarz, Department of Bacteriology, University of California, Los Angeles, Calif. The author is greatly indebted to Dr. Sercarz for his kindness in preparing this material for publication.

Natural immunity

Resistance to infection Before we discuss the nature of the specific immune response, let us consider the genetic factors underlying natural resistance to infection. The terms resistance and natural immunity have often been used to describe the natural insusceptibility of a host to an infectious agent and/or its products; this is in contradistinction to the specific immunity which results from exposure to the agent. The term natural immunity is a misnomer and should not be used for resistance, since resistance relates to a genetically acquired, and not an environmentally acquired, phenomenon (Chap. 18). The existence of naturally occurring differences in resistance among animals of similar genetic composition is well recognized. European sheep breeds are less resistant to anthrax than are the Algerian breeds. By selective breeding, mice have been obtained which were completely resistant to yellow fever. Webster (1933) also demonstrated that resistance to typhoid fever in mice is controlled by dominant genes. Additionally, it was shown that no significant change in host resistance could be determined by challenge with a number of other infectious agents. The inherited genotype responsible for resistance to disease is subject to phenotypic modulation, and expression of resistance would be demonstrated by exposure of the host to the infectious agent.

The success of a mutant is dependent on its ability to survive in a selective environment and to pass this characteristic of resistance on to succeeding generations. This might then be the basis of resistance of the Semitic peoples and the great susceptibility of the Negro races to tuberculosis. The Semites have lived in parts of the world where tuberculosis is an important disease and a selective environmental factor for many centuries. Those individuals who, as a result of randomly occurring mutation, possessed a genetic resistance to disease, survived to childbearing age, and genetically transmitted this resistance to their offspring. In many instances the resistance is not absolute. For example, a survey conducted among Polish Jews showed a normal tuberculosis mortality of 71 cases per 100,000 in 1938. Under the effects of the stress of Nazi persecution, starvation, and crowding, the rates reached 205 cases per 100,000 in 1940 and eventually 601 per 100,000 in 1942 (Dubos and Dubos, 1952).

Environmental factors, the host, and the parasite all interact, and the result of this interaction may be disease. Theobald Smith expressed these relationships as: $D = NV/R$, where D is disease, N is numbers of organisms, V is virulence of the parasite, and R is "resistance" of the host, which may be defined to include the aforementioned genetic factors plus a specific acquired immunity factor.

Some microorganisms show a specific affinity or tropism for certain hosts or host tissues. The reasons for this have long been questioned, and several experiments have shown that one factor might be a competition between host and parasite for nutrients. Bacon et al. (1951) have shown, for example, that mutants of *Salmonella typhimurium*, unable to synthesize their own purines, were far less virulent for mice than wild-type organisms. The mouse species has a marked lack of free purines in body tissues, so unless purines were injected with the purine-deficient mutants, the microorganism could not cause disease.

Plasmodium berghei, causative agent of malaria, requires *p*-amino-benzoic acid (PABA) for growth. Milk is deficient in PABA, and if rats are fed a milk diet, they are resistant to the organism. If the milk diet is supplemented with PABA, the rats again become susceptible (Hawking, 1954).

In addition to the presence or absence of specific metabolic intermediates, other factors participate in the nonspecific defense or resistance of the host. Foremost of these would be the physical barriers of the skin and mucous membranes, substances such as sweat, fatty acids of the skin, acids in the stomach and vagina, lysozyme in tears, enzymes of the digestive tract, complement, interferon, and possibly also properdin. The nonimmune destruction of microorganisms takes place in phagocytes which contain destructive enzymes, e.g., deoxyribonuclease, ribonuclease, cathepsins. These phagocytes engulf and destroy invading organisms much more rapidly, however, in the presence of a specific antibody. Temperature, diet, and hormone level have also been shown to exert an effect on the host-parasite relationship.

Complement Changes in resistance may be due to alteration in the level of complement components. The complement system is a group of serum proteins (C1-C9) which occur in all normal vertebrate animals, cooperating with the immunoglobulins and cellular elements in preserving the integrity of the body. The individual components act as enzymes either singly or in combination with one another. They exist as inactive precursors and become activated in the course of their sequential reactions, yielding products which manifest immune lysis of bacteria and red cells, inflammation, and enhancement of phagocytosis. Primary activation of C1 depends upon steric modification of antibody by antigen; complement will not react with uncombined antigen or antibody. Some complement components are heat-labile (56°C), and its total activity is quite labile.

Individuals have been found with specific deficiencies in certain complement components and these situations may serve to define the function of portions of the complement hierarchy.

For further information on complement: Müller-Eberhard (1969); and the International Symposium on the Biological Activities of Complement. Dubos (1954) and Elberg (1956) should be consulted for general views on resistance.

Acquired immunity

The nature of the immune response Let us consider the fate of foreign material, such as poliomyelitis virus, after injection into the rabbit. If one follows the numbers of viruses left in the bloodstream using a suitable method of enumeration, they will be seen to decrease slowly over the first few days. Then, dramatically, the viruses will disappear completely from the circulation. After this has taken place, free antibody molecules may be found in the serum; it was the antibody molecules which combined with the viruses in circulation, causing them to be picked up and destroyed by the macrophages. The level of free antipoliomyelitis antibody will rise to a peak some time between the first and second week after virus injection; then it starts to decline. Such a response, termed a "primary response," follows the first contact of the animal with a foreign material. A similar sort of response in the rabbit would follow upon the injection of sheep erythrocytes, bovine serum albumin, a suspension of pneumococci, or the type-specific capsular polysaccharide from the pneumococcus. In each case, the antiserum would have the unique quality of reacting specifically with the foreign material which stimulated its synthesis.

The memory feature of the immune response is demonstrable several weeks to several years following the first exposure to the foregoing material. Upon reintroduction of the material, such as the poliomyelitis virus, the level (or titer) of antibody in the serum rises, is apparent earlier, and reaches a peak level many times higher. This is called the "secondary" or anamnestic response. For a comparison between the primary and secondary responses, see Fig. 318.

Immune tolerance or paralysis An antigenic molecule contains chemical groupings or antigenic determinants which can call forth an antibody response; however, the body itself is composed of many molecules which, if injected into another animal, would cause an antibody response. The body must distinguish between "self" and "foreign"; otherwise antibodies directed against its own antigens would destroy it. It can be said that an organism has a "tolerance" to its own antigens in that they do not provoke an immune response. The importance of this dictum was recognized by Paul Ehrlich in the early years of this century when he coined the term "horror autotoxicus." It was reemphasized by Sir Macfarlane Burnet in the forties. The prediction was made by Burnet (1949) that if an antigenic substance could be presented to the organism early enough in its development, it might be treated as "self" and throughout its lifetime it would be incapable of eliciting an immunological response. This prediction was verified by Medawar and his colleagues in a brilliant series of skin-grafting experiments. When newborn mice of strain CBA were injected with live cells from strain-A mice, the recipients upon reaching adulthood were unable to react against skin grafts from strain-A mice. Thus the grafts either remained permanently in place or were rejected only after an extended period.

This acquired tolerance of foreign skin grafts was shown to be because of a central failure of the immune machinery to react to the antigen. The maintenance of the tolerant state required the continued presence of antigen, which in this situation was provided by the continued

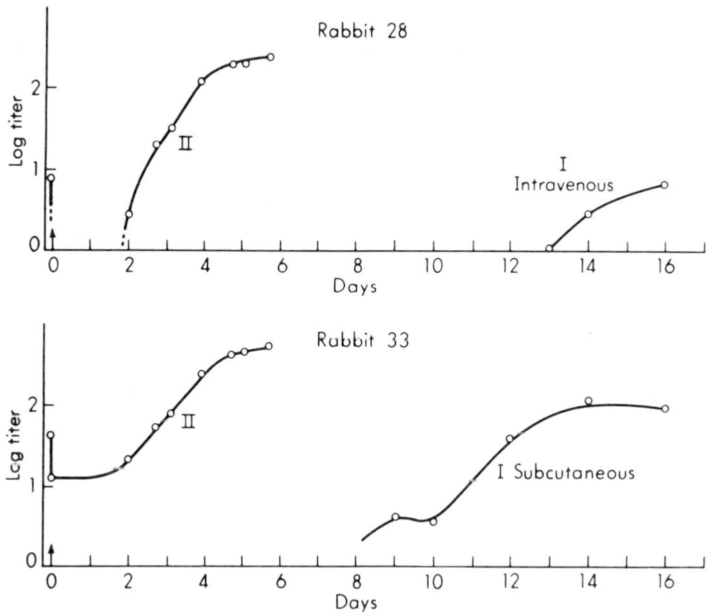

Fig. 318 The primary and secondary antitoxic responses of two rabbits to staphylococcal toxoid. Both primary and secondary injections were given at the point 0. Curve I is the slowly appearing primary response; curve II the rapid secondary response to intravenous injection made 18 days after the primary stimulus. (Courtesy of Burnet and Fenner, 1949.)

persistence of the A-line cells transferred on the day of birth, making the recipient a CBA/A chimera.

Other workers have shown that simpler antigen complexes than cells, for example bovine serum albumin, can similarly induce tolerance. Prior to the allograft tolerance experiments, Felton (1955) had demonstrated inhibition of the immune response to pneumococcal polysaccharide in adult mice given too large a dose of the immunogen. He called this effect "immunological paralysis"; other workers also have noted a similar effect with nonpolysaccharide antigens (Gowland, 1965; Smith, 1961). It is unlikely that there is a significant difference in mechanism between neonatal tolerance and adult paralysis (Dresser & Mitchison, 1968).

Thus, it appears that too much antigen is not desirable if one wishes to induce an immune response. Very small numbers of particles, for example 10^6 ϕX-174 bacteriophages (Uhr et al., 1962), can induce an immune response in guinea pigs. With increasing doses, up to a plateau at roughly 10^{11} phage particles, there is a direct correlation between antigen dose and the immune response. In the case of serum protein antigens such as the albumins and globulins, where one can administer as many as 10^{17} molecules conveniently, although a dose-response relationship is also found, a plateau is reached beyond which excess antigen can lower subsequent responsiveness. Various explanations for this dual nature of antigen will be discussed later in this chapter.

Antibodies Antibodies are proteins found in the blood serum and related fluids, produced in response to antigens. Antibody will react specifically with the inducing antigen, but its presence may or may not be related to the immune status (re: infectious disease) of the host. In many instances the antibody may be harmful rather than protective.

Antibodies may be separated from other serum proteins by specific methods involving reaction with the antigen responsible for their induction, or by nonspecific methods based on the physical and chemical properties of the antibody molecule. Early workers characterized proteins with regard to their solubility in salt solutions, but it was Tiselius who first separated proteins on the basis of their migration in an electric field. He postulated that, since each protein is composed of different amino acids, the total net charge on the proteins would be different and this would specifically determine their rate of migration in an electric field. Tiselius was able to define four major plasma protein groups by this method. The fastest-moving protein was albumin, followed by three globulin groups which he named alpha, beta, and gamma, respectively. The globulins which showed antibody activity, immunoglobulins, were found to be primarily in the gamma fraction and secondarily in the slower-moving beta globulins. As would be expected, an animal which has been extensively immunized will often have an increased content of gamma globulin in its serum. The entire immunoglobulin array of molecules is sometimes absent in human patients with the disease hypogammaglobulinemia; in other diseases, only one of the immunoglobulins may be absent.

The immunoglobulins have been subjected to extensive study and analysis of structure and function. The major findings include explanations for the vast heterogeneity of these molecules. Of primary importance was the realization that the basic unit of all antibody molecules is a four-polypeptide chain array, throughout the diverse immunoglobulin families. For example, in the human, five major families are found: IgG, IgM, IgA, IgD, and IgE. An antiserum may contain antibodies of similar specificity in each of these classes.

IgG, the major immunoglobulin found in serum, consists of two identical heavy (*H*) chains (molecular weight 55,000) and two identical light (*L*) chains (molecular weight 22,500), the four chains being held together by three disulfide bonds plus noncovalent forces (see Fig. 319).

The two antigen-binding sites make up a small proportion of its total surface. Both *H* and *L* chains contribute to the active site. This can be appreciated from Fig. 320, where the combining sites are depicted at the upper tips of the Y-shaped molecule. Porter had first shown that the enzyme papain could split the IgG molecule at 2 vulnerable spots, into three pieces. Two of these bear the combining sites (Fab), and include all the light chain and one half of the heavy chain; the remaining Fc piece does not combine with antigen but is essential for certain biological

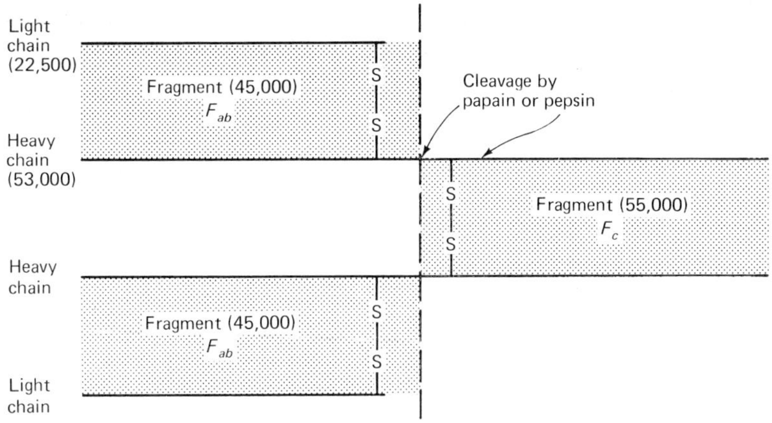

Fig. 319 Line representation of the IgG molecule, emphasizing disulfide bonds interconnecting the four chains of the molecule and the molecular weights of the different units. The hatched areas indicate the three fragments which are produced after treatment by papain or pepsin. (Adapted from Nisonoff and Thorbecke, 1964.)

activities of γ-globulin, such as fixation to skin cells, complement binding, and crossing of the placenta.

Despite the physicochemical similarities among molecules of the same antibody class, they display a wide array of specificities. It is a fascinating recent discovery that these different specificities are derived from unique amino acid sequences in both the light and heavy chains. The possibility of analyzing the exact position of the binding site variability arises from the availability of unique molecular species of antibody in mice and men bearing plasma cell tumors or myelomas. The immunoglobulin of such an individual is composed almost exclusively of a myeloma protein which is actually a homogeneous antibody molecule; its great importance for chemical analysis is obvious. The specificity of some myeloma proteins is known.

The first analysis of amino acid sequences of several myeloma protein light chains was performed by Hilschmann and Craig (1965) and sequence work has been carried out extensively ever since. The results show that all human kappa light chains are essentially identical throughout the carboxy-terminal half of the molecule (110 amino acids). *All* variant regions occur within the amino-terminal 110 amino acids. Likewise, heavy chains also contain a variable region within the amino-terminal quarter of the molecule. Heavy chains have three times as much invariant region as light chains.

Genetic differences in the invariant regions do occur. For example, some human kappa chains have leucine at position 19 (Inv 1) whereas others have valine (Inv 3). These differences are termed *allotypic,* and homozygous Inv 1 individuals can make antibody to Inv 3 kappa chains.

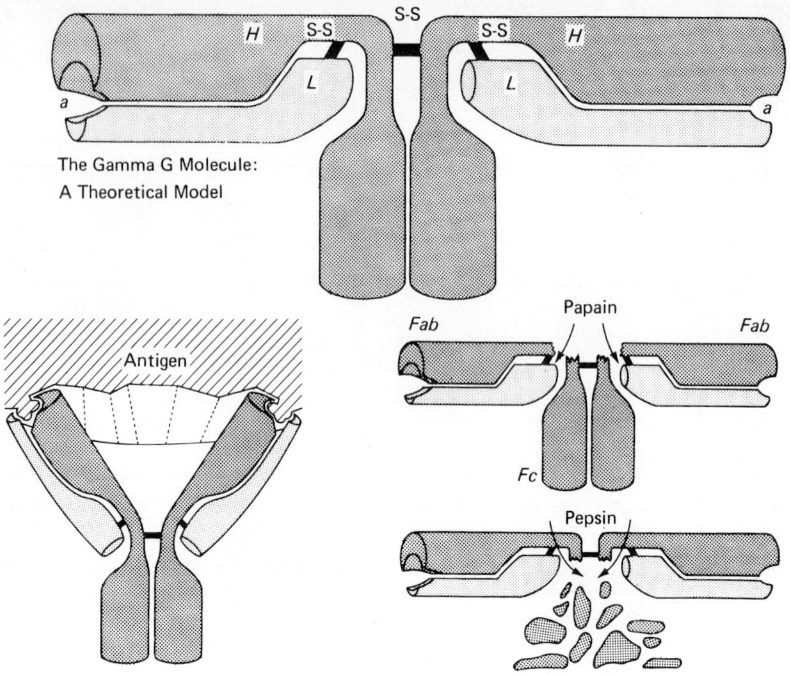

Fig. 320 In model of IgG (top) both heavy (H) and light (L) chains contribute to the two antibody combining sites (a). Three interchain disulfide bonds (S–S) are seen. Flexibility of chains permits bivalent combination with a single antigen (lower left). When antibody is treated with papain, it splits into three parts. One (Fc) has two half H chains, is crystallizable (homogeneous) and inactive. The others (Fab) each contain a combining site. It is believed that these are associated with the variable portion of the H and L chains and therefore with immunologic specificity. Pepsin treatment leaves the two Fab units united in a bivalent fragment including both combining sites, while "chewing up" the Fc fragment. A single disulfide bond joins the heavy chains in rabbit IgG (as shown above); there are two to five such bonds in human IgG. (Courtesy of Nisonoff.)

Actually, current thought assumes that two genes can determine a single immunoglobulin chain; the variable 110 amino-terminal part of light and heavy chains are determined by separate "v" genes from the common or invariant remainder of the molecules which are determined by "c" genes (see Haber, "Two genes: One polypeptide chain," 1972).

In Table 33 are listed some of the characteristics of the different immunoglobulin classes, using the human example. A few special points should be underlined. IgG is the most prevalent molecule in the serum, but because of its slow catabolism, this does not necessarily indicate that it is *synthesized* daily in the greatest amount.

IgA synthesis throughout the body is vigorous: it is dominant in all fluids bathing the organs and systems which are exposed to the "external"

Table 33 *Properties of immunoglobulins*

	IgG	*IgM*	*Secretory IgA*	*IgD*	*IgE*
Synonym	γG	Macroglobulin γM	γA	γD	Reagin γE
Approximate Svedberg sedimentation constant	7*S*	19*S*	11.4*S*	6*S*	7.5*S*
Component light chain	*K* or λ	*K* or λ	*K* or λ	*K* or λ	*K* or λ
Heavy chain subclasses	$\gamma 1, \gamma 2, \gamma 3, \gamma 4$	μ	$\alpha 1, \alpha 2$	δ	ϵ
Number of chains	4	20	8 + *J* chain + secretory comp.	4	4
Approximate molecular weight	160,000	900,000	400,000	150,000	200,000
Valence	2	10	?	?	?
Complement fixation	+	+	–	–	–
Placental transfer	+	–	–	–	–
Concentration in serum (mg/ml)	12	3.5	1.8 Serum IgA	0.030	0.0005

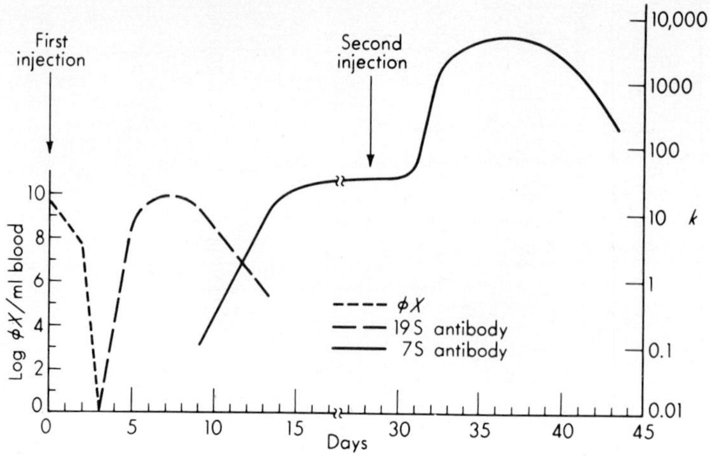

Fig. 321 Antibody response to ϕX phage in the guinea pig after two intravenous injections of 10^{11} ϕX particles administered 1 month apart. See text for further details. (From Uhr, 1964.)

environment – gastrointestinal secretions, respiratory mucus, tears, saliva, breast milk and colostrum, genitourinary secretions. The example of IgA in Table 33 is the *secretory* IgA. Serum IgA is a four-chain molecule of molecular weight 170,000, while secretory IgA has two such units and an additional *J* chain of about 25,000 molecular weight (Mestecky et al., 1971) and a glycoprotein secretory component (SC) (Halpern and Koshland, 1970).

IgM is found on the surface of lymphoid cells and is the first of the antibodies to appear phylogenetically in evolution, ontogenetically in the development of the individual, and also in the response to antigen injection. Figure 321 shows the relationship between IgM (19S) synthesis and IgG (7S) synthesis following the first and second injections of a bacterial virus *ϕX*-174. After intravenous injection of *ϕX*, there is a short phase of "nonimmune" elimination, followed by an accelerated "immune" elimination due to antibody, and total disappearance of the virus occurring at 3 days. The initial exponential phase of IgM synthesis lasts until 5 days; a similar exponential phase of IgG synthesis is not detected for 7 days. Whereas IgM synthesis soon ceases, nonexponential IgG synthesis continues for months or even years. After the second injection, IgG is primarily produced. Note especially in the figure that a logarithmic scale is used to plot the antibody titers (*K* = rate constant for neutralization).

Evidence is available to indicate that the same cell can produce IgM as well as IgG antibody (Nossal et al., 1964; Mellors and Korngold, 1963). This observation lends support to theories postulating a switchover within the cell from IgM to IgG synthesis.

No function for IgD is known. IgE antibody will be discussed below, in the section on allergy.

Allergy due to antibody

After priming with an antigen, an individual may develop hypersensitivity or allergy (after Von Pirquet, 1906) to that "allergen." On subsequent contact, dramatic results may become evident within minutes or hours: this type of allergy has been termed "immediate-type hypersensitivity." It is caused by the contact of antigen with sensitized cells and the release of pharmacological mediators from these cells that have effects on a variety of target organs. We will analyze these reactions by considering (1) sensitization, (2) latent period, (3) type of antibody, (4) mediator released, (5) effects produced.

1. The amount of antigen needed to sensitize varies considerably with the allergen and the animal. For example, egg albumin sensitizes in doses of 0.0001 to 1.0 mg in the guinea pig, in a single injection. Mice become sensitized only upon incorporation of the antigen in an adjuvant, a term referring to materials which enhance immunogenicity.
2. A latent period of several weeks must then follow sensitization. During this time, "skin-sensitizing" antibodies are produced and attach to mast cells and basophilic leucocytes by a site found in the Fc region of the antibody molecule. Cells throughout the body become sensitized and the reactivity of lungs, mesentery, uterus, etc., can be demonstrated in vitro. Such antibody has been called "reaginic" or "atopic" antibody.
3. In man, the reaginic antibody belongs to a special class IgE, with a unique heavy chain ϵ bearing the cell-attachment site. Some evidence in support of this statement is presented in Fig. 322.

   Patients have been found with IgE myelomas allowing chemical characterization of IgE. The epsilon heavy chains are unusually long with a molecular weight of 75,000. IgE has been demonstrated in nasal washings and sputum from asthmatic patients, and IgE-forming cells have been found in tonsils, adenoid tissues, bronchial and peritoneal lymph nodes, but very few in the spleen or subcutaneous lymph nodes. Other animals also have reaginic antibodies that sensitize the same species for allergic reactions: monkey, rabbit, mouse, rat, and dog.
4. As a result of allergen combining with cell-bound IgE antibody, a variety of vasoactive amines are released into the circulation. Histamine is a major mediator that is released from mast cells. It induces an increase in vascular permeability, constriction of bronchiolar and other smooth muscle and the elicitation of itching and pain. Another material is "slow-reacting substance (SRS)" which produces a lower and more prolonged contraction of the guinea pig ileum than histamine. An eosinophil chemotactic factor has been demonstrated in guinea pig lung. Other mediators have also been described.
5. Anaphylactic shock, and subsequent death, can occur in a sensitized animal if enough antigen simultaneously encounters tissue-fixed antibody to cause the cell damage and release of pharmacological mediators, all in one burst. This is usually only caused by parenteral injection, after which antigen reaches the bloodstream rapidly in high

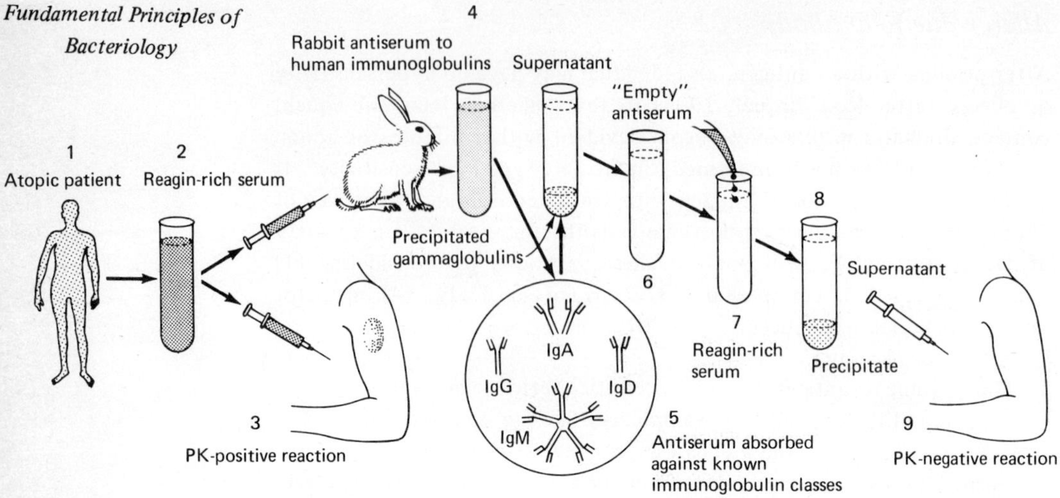

Fig. 322 Scheme above outlines the critical experiments that led to conclusion that reaginic antibodies belong to a distinct immunoglobulin class – IgE. Patient with atopic allergy (1) acts as source of reagin-rich serum (2). Injected into arm of a nonatopic individual it gives positive PK (passive transfer) (3). The reagin-rich serum is injected into a rabbit to make an antiserum against human immunoglobulins (4). This antiserum is then absorbed against known immunoglobulin classes (5) so that the resulting supernatant is devoid of anti-gamma globulin antibodies, or "empty" (6). Next, a small quantity of reagin-rich serum is added to the so-called empty antiserum (7) and another precipitate produced by ultracentrifugation (8). The supernatant is now injected into a nonatopic individual and there is no PK reaction, indicating that the antibodies possessing reaginic activity, introduced after other antibody classes had been removed, are now themselves precipitated out and the antiserum is truly empty in terms of immunoglobulin activity. (Courtesy of Ishizaka.)

concentration. Intracutaneous injection of eliciting antigen gives rise to local anaphylaxis, characterized as the "wheal and flare reaction" which is due to local dilatation of capillaries.

In diseases such as serum sickness which follow the administration of large amounts of serum, antibodies are produced to the foreign serum components leading to the presence of immune complexes in the circulation. With the increased vascular permeability following the release of mediators, large immune complexes become entrapped in vessel walls and cause injury.

Ways exist to block the anaphylactic response. Blocking antibodies of other classes may tie up the antigen. Purified Fc portion of IgE can block reagin uptake. Identification of the enzyme system releasing histamine might lead to a specific antagonist; until then, antihistamines can mitigate symptoms.

It is of interest to speculate on the biological significance of reaginic antibody. Its beneficial aspects, leading to evolutionary retention, are still unclear.

For more information: Austen and Becker (1968); Ishizaka and Ishizaka (1971); Schild (1971).

Cells and organs involved in the synthesis of antibodies

Very early in the history of immunology, it was found that there was a higher concentration of antibody in the spleen than in the blood as early as 2 days after antigen stimulation. This indicated that the spleen might be an antibody-producing organ. Many decades later it was shown that the lymph node draining the site of injection of an antigen was a major site for the formation and storage of antibody. The spleen and lymph nodes are the *peripheral* organs of antibody formation which can be regarded as sites of interaction of the chief responsive cells in the system: macrophages, lymphocytes and plasma cells (see Fig. 323). There are also *central* lymphoid organs – the thymus and the bone marrow. In these organs, the lymphoid cells originate and start their differentiation but do not fully develop their functional potential (see Fig. 324).

The "reticuloendothelial" system is one which engulfs and ingests foreign materials and includes a variety of phagocytic and antigen-trapping cells. There are macrophages in the peritoneal cavity and lining the sinuses of peripheral lymphoid tissue. They are specialized for the phagocytosis and digestion of antigens and have an extensive library of hydrolytic enzymes for this purpose. This catabolic or processing function may be necessary in preparing effective stimulatory fragments from erythrocyte antigens, for example (Shortman and Palmer, 1971). RNA-associated antigen derived from macrophages may be especially effective (Askonas and Rhodes, 1965). Extensively degraded antigen is soon excreted.

Fig. 323 Schematic representation of thymic-dependent and bursa-dependent areas in a peripheral lymph node. (Courtesy of Meuwissen, Stutman, and Good.)

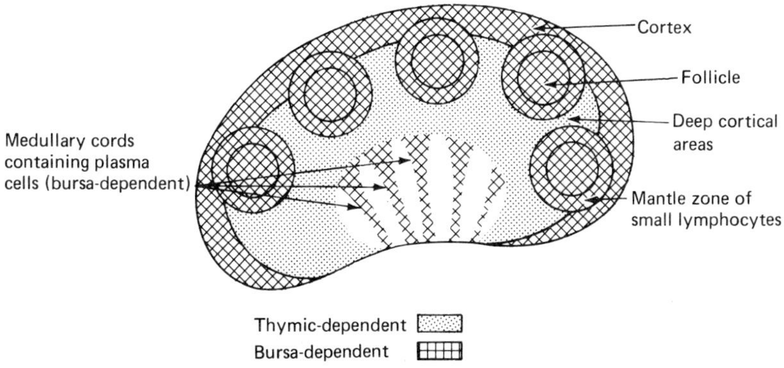

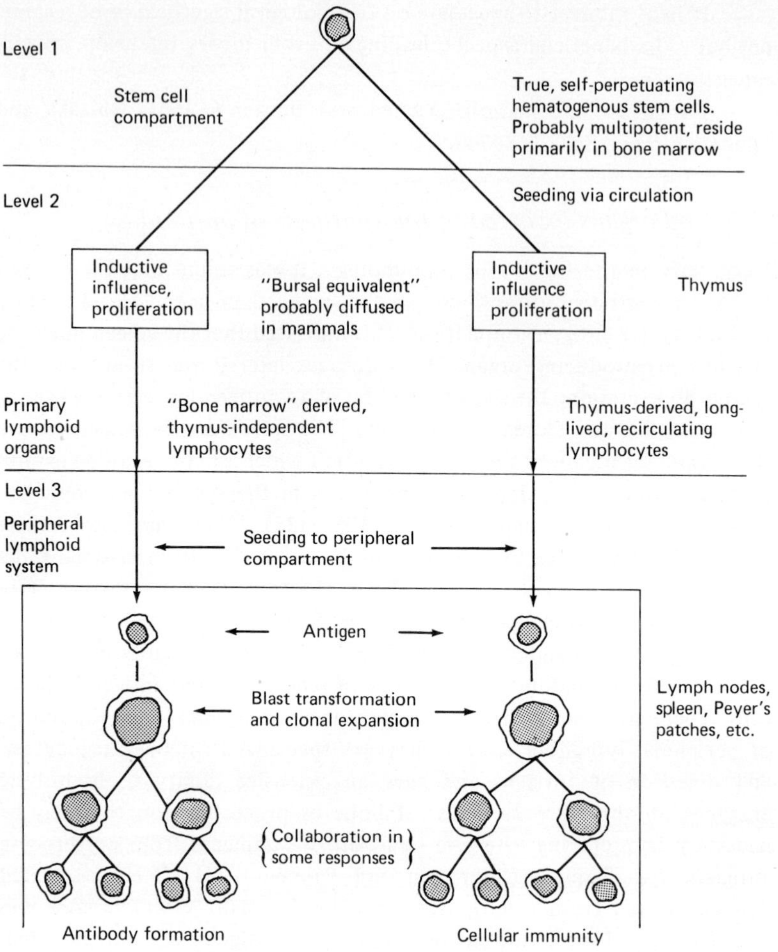

Fig. 324 A schematic overview of the three main functional compartments of the lymphoid system. (Courtesy of Nossal and Ada.)

However, more relevant to the immune response may be the small proportion of antigen retained intact on the plasma membrane of cells (Unanue and Askonas, 1968). Furthermore, dendritic cells with a profusion of cytoplasmic processes bear antigen on their surface and are found interdigitating with lymphocytes in the lymphoid follicles of the spleen and lymph node. It is presumed that antigen in this location can readily stimulate lymphocyte receptors.

During the time since the publication of the last edition of this text, there has been a revolution in thinking about the cellular elements in the immune response. Two populations of lymphocytic cells are involved in establishing the immune response: one, the T population, is concerned with the cellular responses that do not seem to be dependent on free antibody and has passed through the thymus (after originating in the bone marrow); the other, the B population, is concerned with secreting antibody and is marrow-derived. In the chicken, the functional division within the lymphoid system is made more obvious by the presence of the bursa of Fabricius, a primary organ for the B cell equivalent to the thymus in generating T cells.

The antibody response of mice to most antigens is depressed by thymectomy, because the mouse becomes deficient in thymus-derived or T lymphocytes. Likewise, in the chicken, bursectomy depresses the humoral response to antigens, but leaves the cellular T responses unaffected. Both T and B cells have antigen-specific receptors; both are capable of storing immune memory. The T cells do not synthesize antibody but somehow help the marrow-derived B lymphocyte to do so. The exact nature of the cooperation is unknown: one idea is that the cell-bound receptor of the T lymphocytes serves to concentrate antigens at critical sites where it can more easily stimulate B lymphocytes. In addition, there is some evidence that a nonspecific local stimulator of B cells may be secreted by activated T cells.

In a group of carefully observed "experiments of nature," Dr. Robert Good and his colleagues have studied T and B cell function in humans. Some babies with congenital "Swiss agammaglobulinemia" seem to lack both a T and a B system and are highly susceptible to infection. Other patients retain the T system intact but lack a functional B system: they can display delayed hypersensitivity reactions with all their manifestations and reject skin grafts. Similarly, there are patients who produce normal amounts of all the immunoglobulins but fail to produce cellular immunities: they lack the thymic-dependent system of lymphocytes and may be able to be cured by thymus transplantation.

We may look upon the T and B cell systems as two parallel mechanisms which have evolved to contend with a variety of antigenic challenges. The T system responds to very small quantities of antigen and the activated T cell produces a variety of factors which affect nearby cells – chemotactic factors for macrophages, migration inhibitory factor (MIF), cytotoxic factors, mitogenic factors, etc. It is thought to constitute a surveillance system which rids the body of errant new cells (cancer cells) continuously. The B system seems to be especially suited for response to larger amounts of antigen and the effector cells secrete antibody molecules of a wide variety of types. These antibody molecules can inactivate toxins, can prepare bacteria for phagocytosis, and can aggregate soluble antigens to prepare them for phagocytosis.

At times, the T and B systems antagonize each other. It is known that activated T cells, capable of destroying a specific tumor, can be prevented from doing so in the presence of antibody ("blocking") also directed against the tumor. Removal of the antibody allows T cells to destroy the tumor. The relevance to cancer prevention hardly needs underlining.

The B precursor cell, after being stimulated to divide by antigen (Dutton and Mishell, 1967), differentiates during several generations to a plasma cell, which produces antibody at a rapid rate and then disappears. It is remarkable that before the primary or the secondary injection of antigen, the plasma cell, which is going to synthesize the largest part of the antibody, is not even present. This fact was an indication of its short lifetime, and pointed to a precursor of different appearance. Plasma cells were first implicated as the synthetic cells by Fagraeus (1948) who showed that the peak of antibody synthesis correlated with the peak of appearance of plasma cells in spleen cultures. Leduc et al. (1955) later demonstrated, with the fluorescent antibody technique, that it was the plasma cells which showed antibody localization when lymph nodes were examined at the height of the antibody response.

Cell-mediated immunity (CMI) Many changes in the dynamic equilibrium between host and parasite are mediated by antibodies synthesized by B cells and transported in the blood or tissue fluids.

Persons with agammaglobulinemia, however, are still able to maintain protection against a large variety of microbial agents. Humoral antibodies are not usually effective against those agents which are able to survive intracellularly and proliferate. In such cases T cell hypersensitivity mechanisms are the major participants in host defense.

It was Koch who first described the local reaction which occurred after tubercle bacilli were injected into the skin of previously infected guinea pigs. This reaction reaches a maximum in one day and is characterized by a massive infiltration of mononuclear cells at the site of antigen administration. This "delayed reaction" is characteristic of a type of reactivity which cannot be transferred to a recipient by serum, but only by the injection of cells from a sensitized donor. No pharmacological mediators such as histamine are released. Allograft and tumor immunity are considered to be manifestations of CMI; so is contact sensitivity to poison ivy.

Specifically committed T lymphocytes can sometimes destroy target cells without assistance, by some kind of direct interaction. There is also an indirect mechanism in which the T cell function is amplified by means of soluble factors generated by interaction of T cells with the specific antigen. For example, sensitized mouse T spleen cells, when cultured with the specific antigen, release a migration inhibitory factor which inhibits the migration of macrophages. It is an acidic glycoprotein with a molecular weight between 35,000 and 55,000. MIF can be demonstrated visually in vitro when peritoneal exudate cells (sensitized lymphocytes plus macrophages) are prevented from migrating out of a capillary tube in the presence of specific antigen. MIF is just one of a large group of CMI-related soluble factors which can damage certain cells or cause others to transform or divide. The macrophages and blood-borne monocytes which arrive at the skin injection site or a natural local depot of antigen are secondary participants in the immunologic reaction. Their display of increased phagocytic and microbicidal capacity is not specific and can be directed against antigenically unrelated organisms.

The order of the events occurring in CMI is of interest: the reaction of antigen with cell-bound antibody occurs within 30 min; MIF production takes a matter of hours; the eventual effect of MIF and other killing materials on target cells may take several days.

The "antibody" of CMI or delayed hypersensitivity is unknown. However, it is probable that studies on the antigen binding receptors of T cells with erythrocyte or protein antigens will arrive at a description sooner than with the more poorly defined cellular antigens.

The allograft reaction If tissue is removed from one part of a person's body and surgically grafted to another part of his body, the graft is called an autograft and the tissue will remain and grow normally. Tissue removed from one person and grafted to another's body is called an allograft because the transfer is done within the same species. This graft will grow for a few days and then will be rejected by the host, die, and slough off. This sloughing reaction is called an allograft (allo, other) reaction. The only exception to this rule is tissue exchanges between identical twins, which will grow like autografts. During World War II, P. B. Medawar studied the basic mechanism responsible for graft rejection, and in 1944 he published the following observations: (1) Allografts, then called homografts, survive for a short period of time, during which time they cannot be distinguished from autografts, but then are sloughed; (2) small homografts survive longer than large ones; and (3) reactions at the periphery of rejected grafts looked quite similar to the lesions of delayed hypersensitivity. The fact that the larger grafts were rejected faster than smaller ones indicated a dose-response effect. This, combined with the resemblance of the cellular picture to an immune process, led him to postulate that the allograft reaction was indeed an immune response and that an anamnestic response could be demonstrated by measuring the survival time of the primary and secondary grafts. One group of inbred mice was grafted with skin from a different inbred strain, and the time necessary to slough each graft was measured. The recipients were then regrafted with skin from the same donors and, as expected, the second grafts were rejected much faster. Hence, by this criterion, the allograft reaction is an immune response. Highly inbred mice will accept each other's tissue grafts and reject grafts from different inbred lines. This shows that the antigenic structure of tissues depends upon the genetic constitution, which differs from person to person, or mouse to mouse.

Graft versus host reactions It has been shown that the host can reject the graft; however, if the graft includes cells which can respond immunologically, the graft can also reject the host. This is best evidenced in studies on immune tolerance. A neonatal host is injected with spleen cells from an adult recipient. These cells find their way to the neonatal host's lymphoid tissues, making the host a chimera; he is incapable of recognizing these spleen cells as foreign tissue and cannot produce antibody against them. They are not tolerant to his tissue, however, and may react against the host tissue surrounding them. This can lead to stunted growth, called "runt disease" (Billingham, 1959),

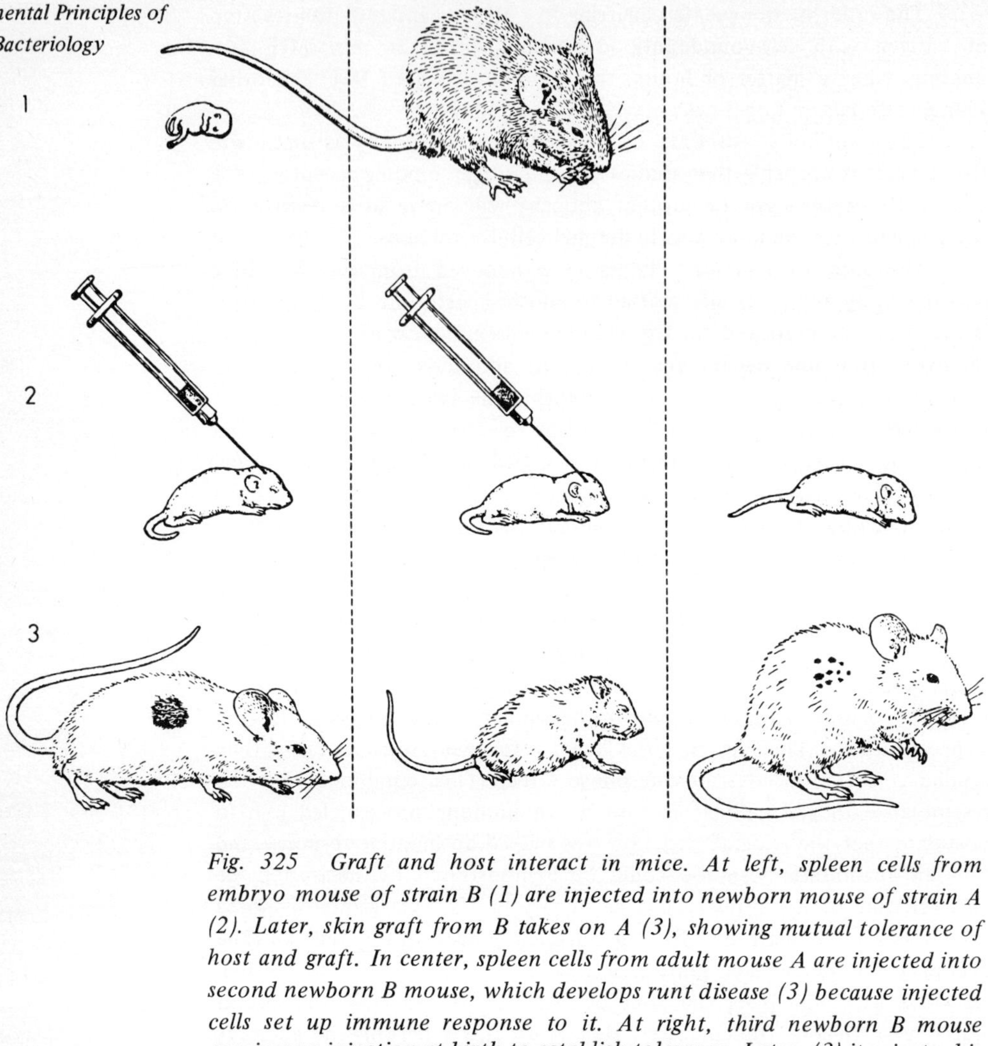

Fig. 325 Graft and host interact in mice. At left, spleen cells from embryo mouse of strain B (1) are injected into newborn mouse of strain A (2). Later, skin graft from B takes on A (3), showing mutual tolerance of host and graft. In center, spleen cells from adult mouse A are injected into second newborn B mouse, which develops runt disease (3) because injected cells set up immune response to it. At right, third newborn B mouse receives no injection at birth to establish tolerance. Later, (3) it rejects skin graft from strain A. (From Burnet, 1961.)

and eventual death. Aspects of graft-host interactions are illustrated in Fig. 325.

A pregnant woman provides another instance of foreign tissues in intimate contact with host tissues. Part of the fetal antigens are inherited from the father, and they are foreign to the mother. The maternal and fetal bloodstreams never mix, and therefore, in most instances there is no antibody reaction against the fetus. However, it is necessary for the maternal and fetal circulation to run very closely alongside each other to allow nutrients to diffuse from the maternal bloodstream through a membrane to the fetal blood. This membrane is not permeable to cells, but in the long course of a pregnancy, fetal red-cell antigens can penetrate this barrier. In most cases this is harmless, but if the fetal red cells contain antigens inherited from the father which the mother does not possess, her body will produce antibodies to the antigens. Some of these antibodies can recross the placental barrier, react with the fetal cells, and cause disease and possible death.

The allograft reaction is classically defined as one which is cell-mediated and closely related to the delayed-type hypersensitivities, as opposed to mediation by circulating antibodies. If, however, pregnancy in mammals is included in our definition of the allograft reaction, the role of circulating antibody must also be indicated since it by itself has been demonstrated to cause hemolytic disease of the newborn.

The cellular mediation of transplantation reactions is best demonstrated by explaining an "adoptive-transfer" experiment. If mouse *A* is x-irradiated, it can make no immune response to allografts from mouse *B*; a skin graft from strain *B* will be accepted. However, if immune lymphoid tissue from an *A* mouse which has been immunized against *B* determinants is now "adoptively" transferred to the x-irradiated *A* mouse bearing a *B* graft, the graft will be sloughed. Serum from the immune *A* donor mouse, used in place of the lymphoid cells, has no effect.

It now seems almost within our grasp to save lives through organ transplants, and experiments are underway in many laboratories to overcome the obstacles to this goal. These experiments include careful matching on the basis of histocompatibility between human donor and recipients, the use of immuno-suppressive drugs in the recipient, and the possible use of cadaver donor material.

Antigens

A complete antigen is capable of inducing an immune response in a suitable host, and is specifically reactive with the antibody produced. If both criteria are met, the material is said to be *immunogenic*; if the second condition only is true, the material is merely *antigenic.*

A minimum molecular weight of several thousand is generally necessary for immunogenicity. The reason is not known; possibly small molecules cannot fold into a rigid, stable pattern which would then be recognized as foreign by the cell. Small molecules may, however, be immunogenic if they are combined with some large protein. This is called a carrier protein, and may even be the body proteins of the host. Hapten-carrier conjugates evoke antibodies which will react with the hapten as well as the carrier. Sensitization of the animal to carrier determinants is a prerequisite for a successful response to the hapten. It is thought that any determinant can function either as a carrier or haptenic determinant. In each individual case of a T and B cell collaboration, the immunogen binds to a T cell receptor by one of its carrier determinants and presents its haptenic determinants to neighboring B cells.

At times, a "hapten" may be much larger than its "carrier." Azobenzenearsonate-tyrosine (MW 409) induces CMI with no antibody formation. Poly-γ-D-glutamic acid (MW 35,000) induces anti-PGA antibody but no CMI. Conjugates of ABA-tyrosine and PGA give rise to anti-PGA antibody *and* cellular immunity to the ABA-tyrosine; however, no cellular immunity is elicited to PGA (Maurer and Gill, 1971). Thus, ABA-tyrosine acts as a "carrier" and PGA as a "hapten" in this case.

A primary requirement for a good immunogen is "foreignness." In order for a material to be immunogenic, it must be sufficiently different from self to be recognized as foreign. Sheep insulin is not immunogenic when reinjected into the donor sheep. Beef insulin is almost identical to sheep insulin, differing only in a single amino acid. As would be expected, it is a very weak antigen when injected into a sheep. Similarly, chicken proteins are weak antigens in other birds, but rather good antigens in mammals.

A third requirement is a degree of molecular rigidity allowing the formation of a stable pattern or determinant. A continuously flexible molecule might present no discrete determinant to the antibody-producing mechanism for a sufficiently long period of time. Often conditions which make a molecule more rigid enhance its immunogenicity. An aromatic ring system at the determinant sites of an antigen will confer this added rigidity and cause increased immunogenicity, and for many years it was thought that this aromatic ring was an essential requirement for a good antigen. Sela and Arnon (1960) have disproved this by demonstrating that nonaromatic ring systems also enhance immunogenicity. The fact still remains that rigidity is an essential ingredient, no matter how it is conferred.

A fourth requirement is availability of the determinant. It has been shown that if a polypeptide antigen is artificially constructed where potentially immunogenic determinants are buried within the molecular complex, antibody is made only to the exposed regions of the antigen (Sela, 1962). For example, in Fig. 326, antibody is only produced to the tyrosyl (tyr) determinants when they are in the position shown in the diagram at the left.

Charged amino acid side-chains have been implicated in antigen – antibody site interactions. However, it has been shown that charged groups are not necessary for immunogenicity (Sela and Fuchs, 1963).

Bacteria, red blood cells, or protein molecules are commonly used antigens, but it must be realized that the antibodies produced to these large entities are not directed against the whole, but only to restricted portions of them. A molecule of human serum albumin can be split into three parts, and different groups of antibodies in such an antialbumin serum will react with each of the three parts (Lapresle et al., 1959). Hence human serum albumin has at least three, and probably more, determinants.

A more obvious example is provided by *S. typhosa*. This bacterium has flagellar antigens as well as cell-wall antigens. The flagella can be sheared from the bacterium and shown to react with anti-*Salmonella* serum. The antiflagellar antibody can be removed from this serum by adding flagella, allowing the antibody-antigen reaction to take place, and

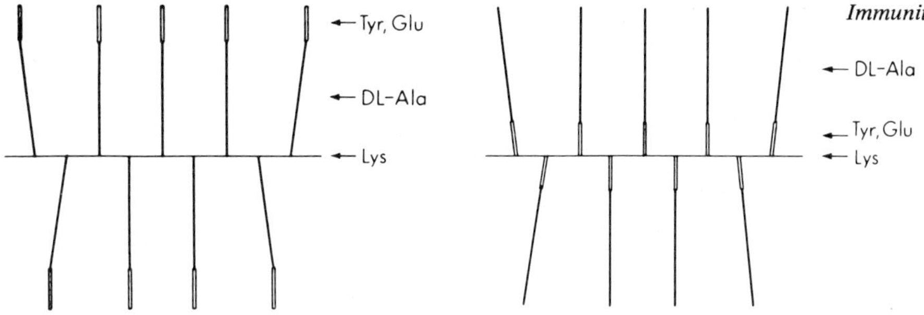

Fig. 326 Schematic diagram showing importance of availability of antigenic determinants. In (T,G)–A L, the tyrosyl-glutamyl side-chains are at the ends of artificial polypeptides and are immunogenic. In the molecule A–(T,G)–L, the tyrosyl-glutamyl determinants are covered with poly–DL–alanine and no antibody is formed. (From Sela, 1962.)

then removing the complexes (absorption); the serum remaining will still be capable of reacting with *Salmonella* cell walls.

Antibodies produced to cell-wall antigens of a particular *Salmonella* will sometimes react with the cell wall of a different strain of *Salmonella*. This is not evidence of nonspecificity, but rather of the presence of an identical antigenic determinant, and is called a "cross-reaction." The careful study of cross-reactions in the *Salmonella* has led to an extensive classification system (Kauffmann-White scheme) of the cell-wall and flagellar antigens. At the present time, chemical information can indicate the exact nature of some of the determinants referred to by number and letter in this classification. If a particular organism has the cell-wall (*O* antigen) designation Group E, 3, 10, 15, the antiserum to this organism will react with Group E, 1, 3, 9 *Salmonella* because of their chemical similarity (determinant 3). To remove this cross-reacting antibody, the serum could be absorbed with any organism containing determinant 3 in its cell wall, leaving only antibodies to 10 and 15 in the antiserum. A schematic representation of the basis for cross-reactions is shown in Fig. 327.

Cross-reactions in other systems, for example that of the type-specific pneumococcal polysaccharides, have permitted Heidelberger (1956) to make predictions about the chemical constitution of the polysaccharides before any detailed chemical knowledge was available.

Blood group antigens An analysis of the antigenic constitution of human red cells and the antibodies directed against them will review the concepts we have discussed. The surface envelope of the red blood cells is composed of large molecules which are primarily polysaccharides closely bound to a polypeptide moiety. Both of these groups are necessary for serological specificity. A red-cell envelope is a mosaic containing many determinants. The ABO system of blood types is only the most prominent

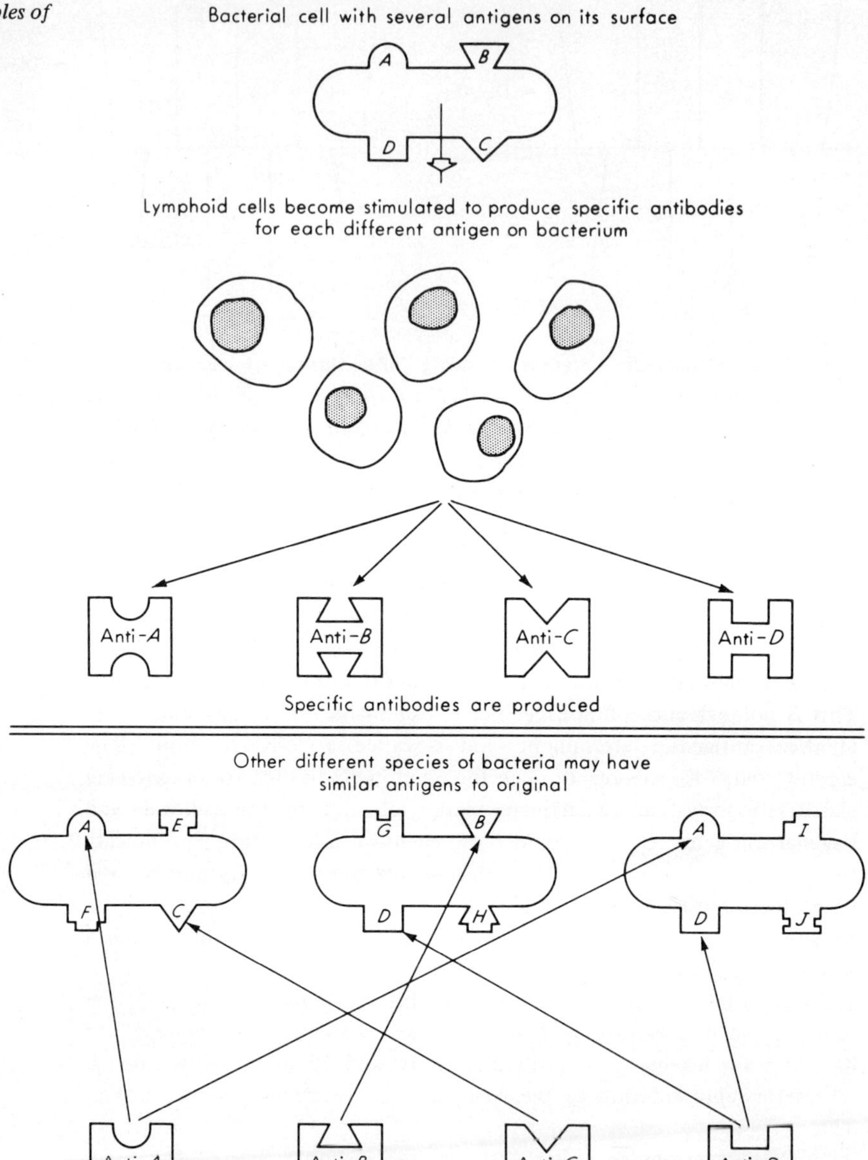
Bacterial cell with several antigens on its surface
A
B
D
C
Lymphoid cells become stimulated to produce specific antibodies for each different antigen on bacterium
Anti-A
Anti-B
Anti-C
Anti-D
Specific antibodies are produced
Other different species of bacteria may have similar antigens to original
A
E
F
C
G
B
D
H
A
I
D
J
Anti A
Anti-B
Anti-C
Anti-D

Fig. 327 Schematic basis for cross-reactions to bacterial surface antigens.

of approximately a dozen groups, each with its gallery of specific determinants.

Landsteiner (1901) found the serum of some people agglutinated the red blood cells of others. It was soon determined that humans could be classified into four groups on the basis of three antigenic determinants, A and B and O (H). People with determinant A on their red cells have anti-B antibody in their serum; conversely, those with B antigen on their cells have anti-A antibody in their serum. People with both A and B antigen on their red cells have no antibodies to the A or B determinants of any human red cells. These patterns would be expected according to the rules concerning autoantigens. Type O individuals, with neither A nor B determinants, will make antibodies early in life to either or both of these polysaccharide components. A and B substances are widespread in nature, and recurrent contact with these determinants initiates and maintains the level of antibody.

These are the main groups, but there are so many other antigenic determinants on red cells that the blood of any individual is unique. This interesting biological fact has its medico-legal importance in cases of disputed paternity. The study of blood group composition has also clarified the nature of transfusion reactions and blood incompatibilities.

One most notable case was the discovery of the Rh blood group. Levine and Stetson (1939) found an antibody in a woman who had just given birth to a deformed stillborn fetus, that could agglutinate the red cells of her husband as well as 85 percent of ABO compatible donors. It was postulated that the blood of the fetus contained a dominant factor, inherited from the father but lacking from the mother, which was responsible for the isoimmunization. This woman's serum recognized what is now known as the Rh_0 antigen; Landsteiner and Wiener (1940) showed that by injecting the blood from a rhesus monkey into rabbits, a similar serum was obtained which agglutinated the cells of the same 85 percent of donors. The Rh blood system is quite complex and made up of many individual antigenic determinants. The present status of the subject is reviewed in Wiener and Wexler (1963), and Race and Sanger (1964).

Antigen-antibody reactions

The reaction between antigens and their specific antibodies may be thought of as occurring in two steps. The primary reaction is association, and is not visible. There are methods of detecting this, but they will not be discussed here. The manifestation of the reaction is evident later, and may take several forms, depending on the nature of the antigen. Soluble antigens will precipitate from solution while large-particulate antigens, such as cells, will agglutinate to form large clumps which then settle out.

The precipitin reaction This is a specific chemical reaction between complementary regions on the two reagents, antigen and antibody. The antigen and antibody molecules can combine in a number of different proportions, so the precipitate in different portions of the precipitin curve (Fig. 328) is not of identical molecular composition. One obtains such a precipitin curve by adding increasing amounts of antigen to tubes containing a fixed concentration of antibody. Molecules of IgG antibody have a valence of two. Antigen molecules such as bovine serum albumin can combine with five or six antibody molecules, while tobacco mosaic virus (TMV) has a valence close to one thousand. The opportunity

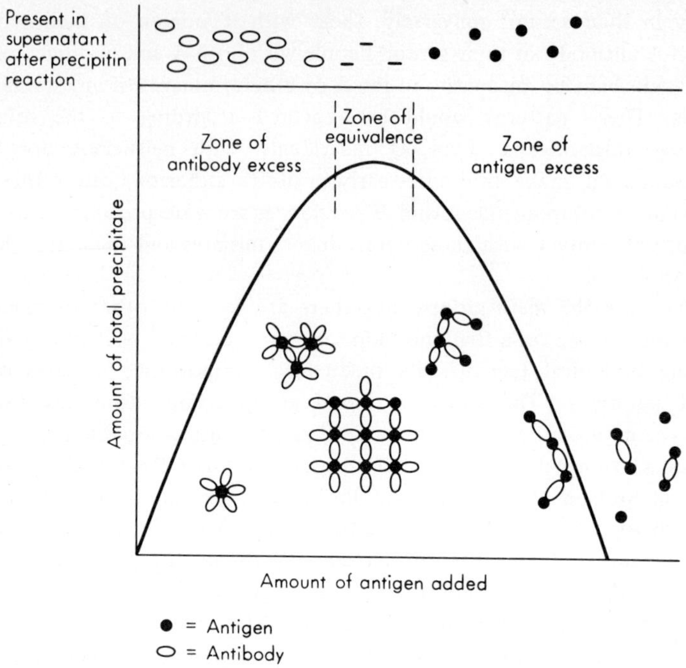

Fig. 328 Course of the quantitative precipitin reaction. The curve represents the amount of antigen-antibody precipitate throughout the three zones of reaction. The zones listed at the top are defined on the basis of the supernatant test (e.g., zone of equivalence is that zone in which neither antigen nor antibody remain in the supernatant after precipitation). The drawings give a general idea about the molecular ratio in different regions of the curve.

exists for specific lattice formation (Marrack, 1934) between alternate molecules of antigen and antibody, linking them to form large aggregates which are the precipitates visible in most regions of the precipitin curve. In the region of antigen excess, no precipitate forms because the valences of antibody are saturated; the complexes have the formula Ag_2-Ab.

Although not as sensitive as many other immunological tests, the relative sensitivity of precipitin tests has been very useful in detecting small quantities of contaminants in biological materials. Diffusion in agar gel can be combined with the precipitin reaction to yield a valuable tool for detection of various antigenic substances. These materials and their antibodies will diffuse out from wells punched in the agar in concentric circles; where a specific antigen-antibody system meets, a precipitate will form in the gel. This will result in an arc at a distance from the well which is dependent on the molecular size and concentration of the reactants. An example is shown in Fig. 329, which pictures the precipitation of diphtheria toxin and its antitoxin. This dangerous protein exotoxin is produced under specially defined conditions of growth of *Coryne-*

Fig. 329 Immunodiffusion study of the precipitation between crude and purified diphtheria toxin, and crude antitoxin. See text for further details. (From Boyd, 1956.)

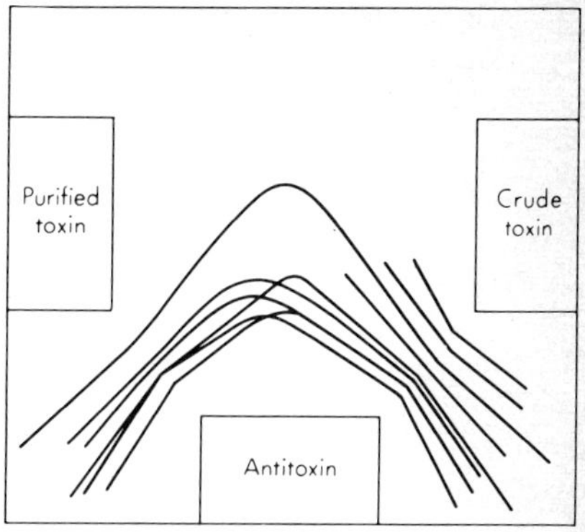

bacterium diphtheriae. When a crude preparation of the toxin is reacted against its antitoxin, careful study of the figure will indicate at least nine different precipitating antigen-antibody systems at the right. Note especially that the so-called purified toxin still has at least six molecular components. It appears from the immunodiffusion reaction that only three components were removed during the purification; these were in the arcs which only appear on the right side of the figure.

Another recent use has been an immunodiffusion pregnancy test, in which the chorionic gonadotropin hormone in the urine of a pregnant woman is detected by means of a specific antihormone antibody placed in the agar gel.

Detection of "incomplete" nonagglutinating antibodies We have already spoken of agglutination with regard to bacteria and red blood cells. Sometimes human antibody to Rh antigens, for example, will combine with but not agglutinate the Rh-positive red cells; it is called an "incomplete" antibody. Therefore, a special method is needed to detect this antibody, which is shown in Fig. 330. The anti-Rh antibody is allowed to react with the red cells, then antihuman gamma globulin made in a rabbit is added, and the complexes will be agglutinated. Another method for detection is called an inhibition reaction. The anti-Rh antibody is allowed to combine with the red cells, and then a standard amount of complete anti-red-cell antibody is added to the system. The visible reaction between the complete antibody and the red cell will be diminished, because most of the surface of the red cells is already covered with the incomplete antibody.

Complement fixation Another very sensitive method which tests for the presence of antibody or antigen makes use of the quality of complement to add, or "fix" to, antigen-antibody aggregates. As mentioned earlier, complement will lyse red blood cells which have combined with anti-red-cell antibody. To test for the presence of antihuman serum albumin in a serum, place human serum albumin (HSA) in dilutions of the serum, incubate, then add a small, measured amount of complement. If anti-HSA were present, an antigen-antibody complex would have formed and the complement would have combined with it. If red blood cells

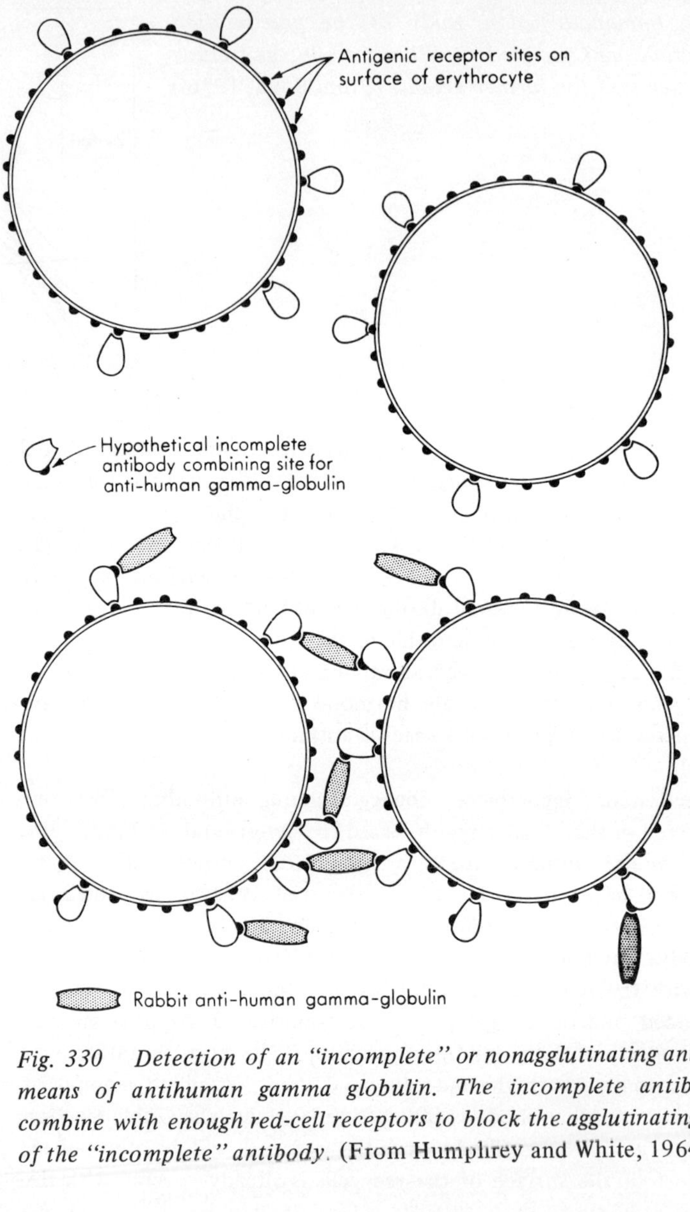

Fig. 330 Detection of an "incomplete" or nonagglutinating antibody by means of antihuman gamma globulin. The incomplete antibody may combine with enough red-cell receptors to block the agglutinating reaction of the "incomplete" antibody. (From Humphrey and White, 1964.)

which have been incubated with anti-red-cell antibody are then added, there will be no available complement to combine with this new antigen-antibody complex; the result will be no red cell lysis. However, if there was no anti-HSA in the original serum, the complement will remain free to combine with the new system, lysing the red cells. An example of a complement-fixation test with gonococcal antigen is presented in Fig. 331.

It is easy to see that if one starts with known antibody and is trying to detect antigen, the complement-fixation test is also applicable. Further advantages of the complement-fixation test are that it is useful in relatively

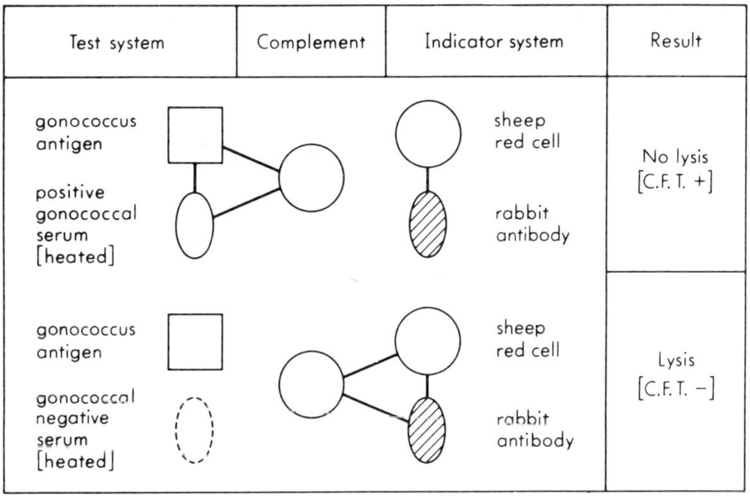

Fig. 331 The complement-fixation reaction, used here to demonstrate the presence of antibody to the gonococcus. C.F.T. = Complement-fixation titer. For further details, please see text. (From Humphrey and White, 1964.)

impure systems where the impurities would inhibit precipitation, and with soluble or particulate antigens. One of the prevalent applications of the complement-fixation test is in the Wassermann test for the diagnosis of syphilis.

Immunofluorescence Some ingenious tricks have been used to take advantage of the great specificity of immunological reactions in diagnosis, as well as in localizing the site of formation of proteins in the body. One outstanding example is the use of fluorescent antibodies which are visualized through a fluorescence microscope. The elements of this assay are shown schematically in Fig. 332.

Coons et al. (1955) and Leduc et al. (1955) have shown that an antibody can be labeled with the fluorescein dye molecule without losing its immunochemical specificity. An antibody such as fluorescein-labeled

Fig. 332 Direct immunofluorescence test. Antigen, perhaps in a tissue section, is detected by combination with the fluorescein-labeled antibody. The dye molecule is represented as an electric light bulb.

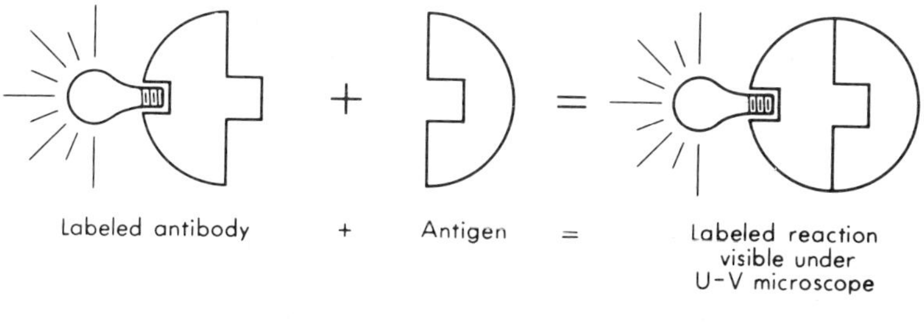

anti-*Shigella* may be used to detect *Shigella* in a smear containing many other microorganisms. Similarly, fluorescent antiinsulin could be used to detect the presence of insulin in tissue sections of pancreas.

For more information: Nairn (1970).

Theories of antibody formation

The nature of the events underlying immune competence has been the subject of speculation throughout the years of this century. Paul Ehrlich, in his 1900 Croonian lecture, set forth his side-chain theory. He visualized normal cells covered with receptors which resembled side-chains of the cell, which would unite specifically with chemical materials bathing the cell, just as a key fits a lock. Hence, if a bacterial toxin were injected and reached the cell, antitoxin formation would result since "the antitoxins represent nothing more than the side-chains produced in excess during regeneration and therefore pushed off from the protoplasm to exist in a free state."

Thus, in Ehrlich's concept, the ability of an antigen to stimulate antibody formation did not represent any new instruction to the relevant cells; they were normally engaged in producing these side-chains.

"Instructive" theories of antibody formation, nevertheless, were most popular up until the mid-fifties. Instructive theories postulate that the antigenic determinant is present and affects the final configuration of the antibody. Presumably this might have occurred at several stages during synthesis: at the DNA level (Schweet and Owen, 1957; Burnet, 1949), at the RNA level (Haurowitz, 1965), or finally the antigen may act as a template and stamp its imprint into a "nonspecific" gamma globulin, if this exists, and convert it to antibody (Pauling, 1940). Pauling assumed that "all antibody molecules contain the same polypeptide chain as normal globulin, and differ from normal globulin only in the configuration of the chain; that is, in the way that the chain is coiled in the molecule."

The time was ripe for the reintroduction by Jerne (1955) and Burnet (1959) of the early ideas of Ehrlich, when it became probable that all antibody patterns were already present at the time of antigen administration, and the antigen merely chooses the unit that fits it best because of chemical complementarity. The cornerstone of Burnet's clonal selection theory is that one cell is genetically capable of producing only one antibody. Therefore, if there are 10,000 different chemical configurations to which antibody can be made, there are initially about 10,000 different types of immune cells throughout the body. When antigen is injected, it contacts that cell which will produce antibody specific for it. This contact not only causes the antibody to be made, but also results in the selective proliferation of the cell into a clone.

A special feature of this formulation is its ease in explaining immune tolerance. During embryonic life, those clones which can react with antigenic determinants in the body are destroyed, probably in the thymus, and the organism is tolerant to "self" antigens. Thus, the concept of a

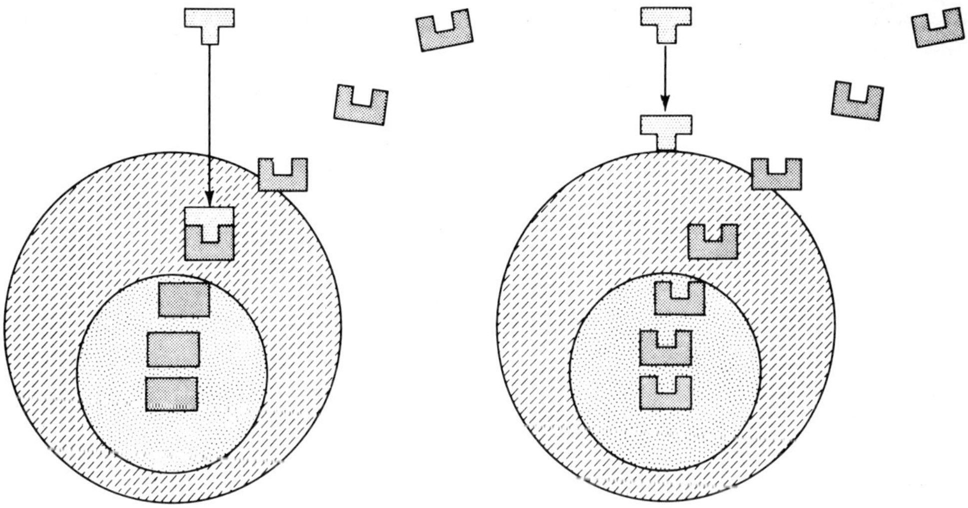

Fig. 333 Two theories of immunity are depicted. Left, according to the classical "instructive" hypothesis, antigen enters a plasma cell and forms a template from which a complementary antibody is produced. Right, the "clonal selection" theory suggests that the mere contact of a given antigen and a given plasma cell signals DNA in the cell nucleus to start directing production of antibody. (From Burnet, 1961.)

"tolerant cell" is not permitted under the clonal-selection theory. The clonal-selection theory and one of the instructive theories are compared in Fig. 333.

Results indicating that a single allotype, a single class, and only one antibody specificity could be produced by a single plasma cell, or a single myeloma cell, gave some experimental support to the idea of clonal selection. However, the critical question is whether the *precursor* cells are unispecific or multispecific. Recent evidence (Miller et al., 1971) supports the notion that precursor cells are multispecific and, furthermore, can express several classes of antibody and both allotypic specificities on their surface. Current work must now explain how restriction of this multipotentiality is achieved.

The most interesting controversies in immunologic theory concern the origin of diversity in the system. Multiple gene theory suggests that the variable portion of each gene differs in at least one base pair and that thousands of such genes are inherited. The other theories state that a limited number of different genes are inherited and that the diversity is generated during somatic development by mutation or recombination. At the level of models explaining cellular activation, and self-nonself discrimination, discussion revolves around the nature of the signal to receptors on antigen-sensitive cells.

However, it is still true that the phenomena underlying the basis of memory or tolerance, the exact determination of specificity and the true interrelationship of the cells and organs in the immune response, remain to some degree unsolved. Understanding at the molecular level seems close.

Infection, disease, and immunity

The response of a mammalian host to foreign antigenic invasion is a marvelously coordinated one, which makes use of nonspecific as well as specific elements. There are the migrating T cells and macrophages, capable of recognizing and processing the penetrating antigen with buildup of responding cells through localized proliferation and recruitment from the circulation. Antigen can thus be localized and destroyed. Specific cells also make their way to the lymph nodes and spleen leading to an amplification of the local response. This cellular monitoring system affords early antigen recognition.

In the lymphoid organs, there is close apposition among the macrophages, and B and T lymphocytes, allowing for a cooperative response. The macrophages may be necessary for the processing of antigen for either T or B cell stimulation. Once triggered, there is a clonal expansion of these cell lines and the products of progeny effector cells ultimately neutralize the antigen. In some cases, it appears that cytotoxic lymphocytes alone may be responsible for target-cell killing. Resistance to some virus infections, on the other hand, may be entirely due to activated mononuclear phagocytes.

Nevertheless, it is the *combination* of these specific elements and several nonspecific agents such as interferon and lysozyme which constitute the complete defense system. Derepression of the interferon cistron is accomplished by a wide variety of viruses, microorganisms, microbial extracts, and synthetic polymers. Furthermore, interaction of viral as well as nonviral antigens with sensitized lymphocytes releases interferon which then interferes with the translation steps in viral replication.

Lysozyme is widely dispersed in the animal and plant kingdoms and is found in the skin, macrophages, and in all the secretory fluids of mammals. There, together with IgA antibody and complement, a lytic system is formed which is effective against the various microorganisms for which the IgA is specific. Without lysozyme, the system doesn't work.

The presence of a battery of defense mechanisms doesn't guarantee the elimination of the offending microorganism. Infection with a microbe is not necessarily coupled with disease. One of the earliest recognized and best-known examples of a symptomless carrier of a disease agent was Typhoid Mary, a domestic who worked for eight different families over a 10-year period about the time of World War I. Despite showing no symptoms of disease, she carried live, virulent typhoid organisms and succeeded in causing at least seven epidemics which involved more than 200 people. Typhoid is strictly a human infection, and it is known that the disease persists in human populations for long periods of time with no evidence of clinical symptoms. Studies have shown the symptomless carrier, usually a postmenopausal female, to be the most important reservoir of infection.

Numerous other examples of host-parasite relationships involving a healthy host and potentially pathogenic infectious agents may be cited. A few of the more widely found carrier states in human populations would include the streptococcus, pneumococcus, meningococcus, and diphtheria carriers; and certainly the staphylococcus carried by hospital personnel should be included here. Most, if not all, of these latent infections are the consequence of recovery from disease. Since the host was originally susceptible and now has an altered response to the infectious agent, he can be considered immune, although the mechanism of the immunity is not well understood.

Certain viral examples are also of interest. The suppression of vaccinia virus replication within peritoneal exudate cells occurs *only* in the immune animal. Yet vaccinia may persist in macrophages and be responsible for the continued humoral antibody response seen following infection.

Another fascinating case is that of the LCM (lymphocytic choriomeningitis virus). Mice infected in utero with LCM have circulating virus-antibody complexes by the first day of birth. Those strains of mice making the greatest antiviral response have earlier and more marked disease. Actually, the macromolecular aggregates of virus and antibody might themselves constitute a pathogenic agent causing disease characteristic of immune complexes.

These examples indicate that the ways of the immune system are not completely fathomed at the level of protection of the individual. Nevertheless, the value of active and if necessary, passive, immunization to a number of microbial agents is undoubted. Sometimes, as in the case of syphilis, the need is great for a protective immunogen. With cancer, it is believed that the immune system continually acts as a surveillance mechanism, ridding the body of unwanted precancerous variant cells. When cancer does strike, it has been shown (Hellstrom, 1971) that activated T cells do exist, but they are prevented from attacking the tumor by "blocking antibody." Future research should bring new insights regarding experimental manipulation of the immune response.

References

Askonas, B. A., and J. M. Rhodes: Immunogenicity of antigen-containing ribonucleic acid preparations from macrophages, *Nature*, **205**:470, 1965.

Attardi, G., M. Cohn, K. Horibata, and E. S. Lennox: Antibody formation by rabbit lymph node cells. I. II. III, *J. Immunol.*, **92**:335, 1964.

Austen, K. F., and E. L. Becker, eds.: "Biochemistry of the Acute Allergic Reactions," Philadelphia, F. A. Davis Company, 1968.

Bacon, G. A., T. W. Burrows, and M. Yates: The effect of biochemical mutants on virulence of *Bacterium typhosum*: The loss of virulence of certain mutants, *Brit. J. Exp. Pathol.*, **32**:85, 1951.

Billingham, R. E.: Reactions of grafts against their hosts, *Science*, **130**:947, 1959.

——, L. Brent, and P. B. Medawar: Actively acquired tolerance of foreign cells, *Nature*, **172**:603, 1953.

Boyd, W. C.: "Fundamentals of Immunology," 4th ed., New York, Interscience Publishers, a division of John Wiley & Sons, Inc., 1966.

Burnet, F. M.: "The Clonal Selection Theory of Acquired Immunity," Nashville, Tenn., Vanderbilt University Press, 1959.

——: The mechanism of immunity, *Sci. Am.*, **204**:58, 1961.

——: "The Integrity of the Body," Cambridge, Mass., Harvard University Press, 1962.

—— and F. Fenner: "The Production of Antibodies," 2d ed., Melbourne, Macmillan and Co., 1949.

Carpenter, P.: "Immunology and Serology," Philadelphia, W. B. Saunders Company, 1956.

Coombs, R. R. A., A. E. Mourant, and R. R. Race: A new test for the detection of weak and "incomplete" Rh, *Brit. J. Exp. Pathol.*, **26**:255, 1945.

Coons, A. H., E. H. Leduc, and J. M. Connolly: Studies on antibody production. I. A method for the histochemical demonstration of specific antibody and its application to a study of the hyperimmune rabbit, *J. Exp. Med.*, **102**:49, 1955.

Dresser, D. W., and N. A. Mitchison: Mechanism of immunological paralysis, *Advan. Immunol.*, **8**:129, 1968.

Dubos, R. J.: "Biochemical Determinants of Microbial Diseases," Cambridge, Mass., Harvard University Press, 1954.

Dutton, R. W., and R. I. Mishell: Cellular events in the immune response: The *in vitro* response of normal spleen cells to erythrocyte antigens, *Symp. Quant. Biol.*, **32**:407, 1967.

Edelman, G. M., and J. A. Gally: A model for the 7 S antibody molecule, *Proc. Nat. Acad. Sci. U.S.*, **51**:846, 1964.

Ehrlich, P.: On immunity with special reference to cell life, *Proc. Roy. Soc. (London)*, **66**:424, 1900.

Eisen, H. N., J. R. Little, C. K. Osterland, and E. S. Simms: A myeloma protein with antibody activity, *Symp. Quant. Biol.*, **32**:75, 1967.

Elberg, S. S.: Factors affecting resistance to infection, *Ann. Rev. Microbiol.*, **10**:1, 1956.

Fagraeus, A.: Antibody production in relation to the development of plasma cells, *Acta Med. Scand.*, **130**:3, 1948 (Suppl. 204).

Felton, L. D.: Studies on the mechanism of the immunological paralysis induced in mice by pneumococcal polysaccharides, *J. Immunol.*, **74**:17, 1955.

Gowland, G.: Induction of transplantation tolerance in adult animals, *Brit. Med. Bull.*, **21**:123, 1965.

Haber, E., convener: Symposium on "Two Genes, One Polypeptide Chain," *Federation Proceedings,* **31**:176 *et seq.,* 1972.

Halpern, M. S., and Koshland, M. E.: Secretory component of IgA, *Nature,* **228**:1276, 1970.

Haurowitz, F.: Antibody formation and the coding problem, *Nature,* **205**:847, 1965.

Hawking, F.: Milk, *p*-aminobenzoate and malaria of rats and monkeys, *Brit. Med. J.,* **1**:425, 1954.

Heidelberger, M.: "Lectures in Immunochemistry," New York, Academic Press, Inc., 1956.

Hellstrom, K. E., et al.: Cell mediated immunity to human tumor antigens, *Progr. Immunol.,* **1**:940, 1971.

Hilschmann, N., and L. Craig: Amino-acid sequence studies with Bence-Jones proteins, *Proc. Natl. Acad. Sci. U.S.,* **53**:1403, 1965.

Humphrey, J. H., and R. G. White: "Immunology for Students of Medicine," Oxford, Blackwell Scientific Publications, Ltd., 1964, p.2.

"International Symposium on the Biological Activities of Complement," Basel, Switzerland, Karger Publisher, 1971.

Ishizaka, K., and Ishizaka, T.: Immunoglobulin E and homocytotropic properties, *Progr. Immunol.,* **1**:859, 1971.

Jerne, N. K.: The natural selection theory of antibody diversity, *Proc. Natl. Acad. Sci.,* **41**:849, 1955.

Landsteiner, K.: Über Agglutinatienserscheinungen Normalen Menschlichen Blutes, *Wien. Klin. Wochschr.,* **14**:1132, 1901.

——: "The Specificity of Serological Reactions," Cambridge, Mass., Harvard University Press, 1945.

—— and A. S. Wiener: An agglutinable factor in human blood recognized by immune sera for rhesus blood, *Proc. Soc. Exp. Biol. Med.,* **43**:233, 1940.

Lapresle, C., M. Kaminski, and C. E. Tanner: Immunochemical study of the enzymatic degradation of human serum albumin: An analysis of the antigenic structure of a protein molecule, *J. Immunol.,* **82**:94, 1959.

Lederberg, J.: Genes and antibodies, *Science,* **129**:1649, 1959.

Levine, P., and R. E. Stetson: An unusual case of intra-group agglutination, *J. Am. Med. Assoc.,* **113**:126, 1939.

Marrack, J. R.: "The Chemistry of Antigens and Antibodies," Oxford, England; printed by John Johnson at the University Press, 1934.

Maurer, P. H., and T. J. Gill: Structural requirements for immunogenicity, *Progr. Immunol.,* **1**:1175, 1971.

Medawar, P. B.: The behaviour and fate of skin autografts and skin homografts in rabbits, *J. Anat. Soc. India.,* **78**:176, 1944.

Mellors, R. C., and L. Korngold: The cellular origins of human immunoglobulins, *J. Exp. Med.,* **118**:387, 1963.

Mestecky, J., J. Zikar, and W. T. Butler: The nature of secretory IgA, *Science,* **171**:1163, 1971.

Miller, A., D. DeLuca, J. Decker, R. Ezzell, and E. Sercarz: Specific binding of antigen to lymphocytes, *Am. J. Pathol.*, **65**(no. 2):451, November 1971.

Miller, J. F. A. P., et al.: The immunological significance of the thymus, *Advan. Immunol.*, 2:111, 1962.

Müller-Eberhard, H. J.: Complement, *Ann. Rev. Biochem.*, **38**:389, 1969.

Nairn, R. C.: "Fluorescent Protein Tracing," 2d ed., Edinburgh, E. & S. Livingstone Ltd., 1970.

Nisonoff, A., and G. J. Thorbecke: Immunochemistry, *Ann. Rev. Biochem.*, **33**:355, 1964.

Nossal, G. J. V., and O. Mäkelä: Elaboration of antibodies by single cells, *Ann. Rev. Microbiol.*, **16**:53, 1962.

Nossal, G. J. V., A. Szenberg, G. L. Ada, and C. M. Austin. Single cell studies on 19 S antibody production, *J. Exp. Med.*, **119**:485, 1964.

Osoba, D., and J. F. A. P. Miller: The lymphoid tissues and immune responses of neonatally thymectomized mice bearing thymus tissue in millipore diffusion chambers, *J. Exp. Med.*, **113**:177, 1965.

Pauling, L.: A theory of the structure and process of formation of antibodies, *J. Am. Chem. Soc.*, **62**:2643, 1940.

Race, R. R., and R. Sanger: "Blood Groups in Man," 4th ed., Oxford, Basil Blackwell & Mott, Ltd., 1964.

Raffel, S.: "Immunity," 2d ed., New York, Appleton-Century-Crofts, Inc., 1961.

Schild, H. O.: In "Second International Symposium on the Biochemistry of the Acute Allergic Reactions," edited by K. F. Austen and E. L. Becker, Oxford, Basil Blackwell & Mott, Ltd., 1972.

Schweet, R. S., and R. D. Owen: Concepts of protein synthesis in relation to antibody formation, *J. Cellular Comp. Physiol.*, **50** (Suppl. 1):199, 1957.

Sela, M.: Some contributions of the study of synthetic polypeptides to the understanding of the chemical basis of antigenicity. Edited by Stahmann, M. A. In "International Symposium of Polyamino Acids, Polypeptides, and Proteins," 1962.

——— and R. Arnon: Studies on the chemical basis of the antigenicity of proteins. 3. The role of rigidity in the antigenicity of polypeptidyl gelatins, *Biochem. J.*, **77**:394, 1960.

——— and S. Fuchs: A synthetic polypeptide devoid of charge, *Biochim. Biophys. Acta*, **74**:796, 1963.

Shortman, K., and J. Palmer: The requirement for macrophages in the in vitro immune response. In "Cellular Immunology," vol. 2, no. 5, 1971.

Smith, R. T.: Immunological tolerance of nonliving antigens. In "Advances in Immunology," edited by W. H. Taliaferro and J. H. Humphrey, New York, Academic Press, Inc., **1**:67, 1961.

Smith, T.: "Parasitism and Disease," Princeton, N.J., Princeton University Press, 1934.

Stetson, C. A.: The role of humoral antibody in the homograft reaction. In "Advances in Immunology," vol. 3, edited by F. J. Dixon, Jr., and J. H. Humphrey, New York, Academic Press, Inc., 1963, p. 397.

Thorbecke, G. J., and B. Benacerraf: The reticulo-endothelial system and immunological phenomena, *Progr. Allergy,* **6**:559, 1962.

Uhr, J. W.: The heterogeneity of the immune response, *Science,* **145**:457, 1964.

——, M. S. Finkelstein, and J. H. Baumann: Antibody formation. III. The primary and secondary antibody response to bacteriophage ϕX174 in guinea pigs, *J. Exp. Med.,* **115**:655, 1962.

Unanue, E. R., and Askonas, B. A.: Persistence of immunogenicity of antigen after uptake by macrophages, *J. Exp. Med.,* **127**:915, 1968.

Von Pirquet, C. F.: Allergie, *Munch. Med. Wochschr.,* **53**:1457, 1906.

Webster, L. T.. Inherited and acquired factors in resistance to infection. I. Development of resistant and susceptible lines of mice through selective breeding, *J. Exp. Med.,* **57**:793, 1933.

Wiener, A. S., and I. B. Wexler: "An Rh-Hr Syllabus: The Types and Their Applications," New York, Grune & Stratton, Inc., 1963.

Wolstenholme, G. E. W., ed.: "The Immunologically Competent Cell," *Ciba Found. Study Group No. 16*, London, Churchill, 1963.

27 *Bacterial diseases of man*

Diseases of man are caused by various classes of organisms such as bacteria, yeasts, molds, protozoa, viruses, and rickettsiae.

Bacteria Thousands of bacterial species have been isolated and studied, but only a small number are capable of producing infections in man. Some bacteria are quite specific, attacking only one host; others are less specific, being capable of naturally infecting more than one host. Scarlet fever occurs naturally only in man. On the other hand the anthrax bacillus attacks cattle, sheep, and horses as well as man.

Yeasts Some yeasts or yeast-like organisms are parasitic for man. *Candida albicans* produces generally an ulcerative condition of the mouth and throat which may later become localized in some internal organ; or it may produce a generalized infection. When the infection remains confined to the mucous membranes of the mouth and throat, it is known as *thrush.*

North American systemic blastomycosis is a chronic granulomatous mycosis primarily of the lungs. It begins with fever and upper respiratory symptoms resembling influenza. Infection is accompanied with cough, purulent sputum, and abscesses in the subcutaneous tissues, bones, central nervous system, and visceral organs. If unchecked, the disease is highly fatal. The disease is produced by *Blastomyces dermatitidis.*

Coccidioidal granuloma is produced by the yeast-like organism *Coccidioides immitis.* The disease manifests itself in so many forms that no general description can be given. Bronchial or pulmonary lesions are almost always present. The skin and subcutaneous tissues are usually involved. The lesions consist of firm or soft nodules, abscesses, ulcers, sinus infections, etc. The bones and joints may also be involved. The organisms appear in the tissues as large, round or spherical cells with thick cell walls. They are sometimes spoken of as double-contoured bodies. Spores appear in the larger cells. On maturity, the spore-filled cells rupture, releasing the spores. Each spore increases to full size and then repeats the cycle.

Molds Molds and mold-like organisms produce some important infections.

Probably the best-known parasitic molds are those producing dermatophytosis in man. Several species of the genera *Microsporum* and *Trichophyton* are involved. These organisms produce superficial infections of the keratinized epidermis, hair, hair sheaths, nails, and skin. Such infections are generally referred to as *ringworm*. When the infection is confined to the feet, it is called *athlete's foot*.

Protozoa A number of important protozoal diseases of man include malaria, which is produced by several species of *Plasmodium*; African sleeping sickness, produced by two species of *Trypanosoma*; amoebic dysentery, produced by *Entamoeba histolytica*; kala-azar, by *Leishmania donovani*; and schistosomiasis, produced by at least three species of *Schistosoma*.

Zoonoses

The term zoonoses has been defined as infectious diseases transmitted naturally between vertebrate animals and man. Any list of zoonoses should include only those infections where there is either proof or strong circumstantial evidence that there is transmission between animals and man. More than 150 zoonoses are now recognized.

Man has known for centuries that epidemics occur among animals, but it was sometime later before he learned that animals may also transmit diseases to man (Fig. 334).

Epizootics are epidemics of animal diseases. They often threaten the health of animals and may lead to enormous economic losses. Epizootics may also lead to human sickness, and are a threat to public health.

For more information: Francis (1967); World Health Organization (1967*a*).

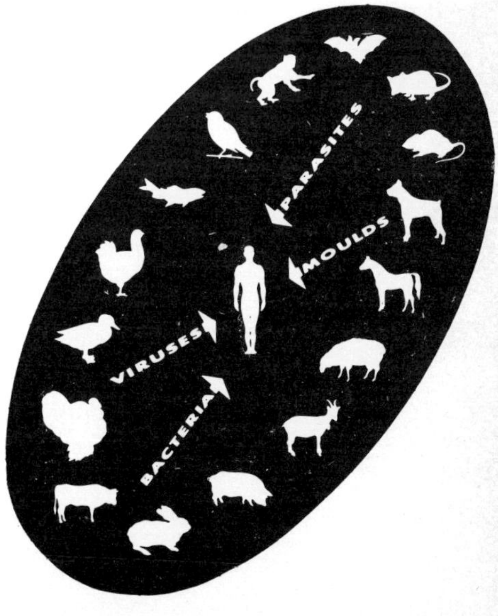

Fig. 334 Schematic representation of disease organisms transmitted from animals to man. (Courtesy of World Health Organization.)

Bacterial diseases of man

The number of organisms producing diseases in various hosts is so great, it is beyond the scope of a textbook on fundamentals to give a detailed description of each etiological agent. For this reason, only a brief outline of the most important pathogenic bacteria for man is included here for convenient reference. Several excellent textbooks and articles on disease organisms are listed at the end of this chapter and may be consulted where additional information is desired.

Actinomyces

The *Actinomyces* are members of the order *Actinomycetales,* which have characteristics intermediate between the true bacteria and the molds. They produce a true mycelium. The vegetative mycelium fragments into elements of irregular size and may exhibit angular branching. Conidia are not produced. The organisms are not acid-fast. They are anaerobic to microaerophilic.

A. israelii Erect aerial hyphae are produced under reduced oxygen tension. The hyphae are occasionally septate but no spores are formed; they measure 1 μ or more in diameter. Large club-shaped forms greater than 5 μ in diameter can be seen in morbid tissues. Substrate mycelium is initially unicellular, and branches may extend into the medium in long filaments, or may exhibit fragmentation and characteristic angular branching. They are nonmotile, are not acid-fast, and are gram-positive (Fig. 335).

Disease produced The organism produces a chronic granulomatous process, generally localized in the jaw, lungs, or abdomen, which is characterized by swellings, at first firm but later breaking down to form multiple draining sinuses which penetrate to the surface. Discharges from sinuses contain "sulfur granules" which are colonies of the organisms. The course of the disease is long and recovery seldom occurs.

Source of infection *A. israelii* is the usual pathogen of man and *A. bovis* of animals only. In man the source of the infection is the oral cavity where the organism lives around normal carious teeth and in tonsillar crypts without producing an apparent infection.

Mode of transmission It is believed that the organism passes from man to man by contact. From the mouth, the organism may be swallowed, inhaled, or introduced into jaw tissues by injury.

Incubation period The incubation period is unknown.

Susceptibility and immunity Natural susceptibility is low. There is no immunity following an attack.

Prevalence The disease occurs infrequently in man (all races may be affected). It is more common in males than in females. It is primarily a disease of domestic animals.

Prevention and control The disease can be prevented by: hygiene of oral cavity; inspection of meat and condemnation of infected carcasses;

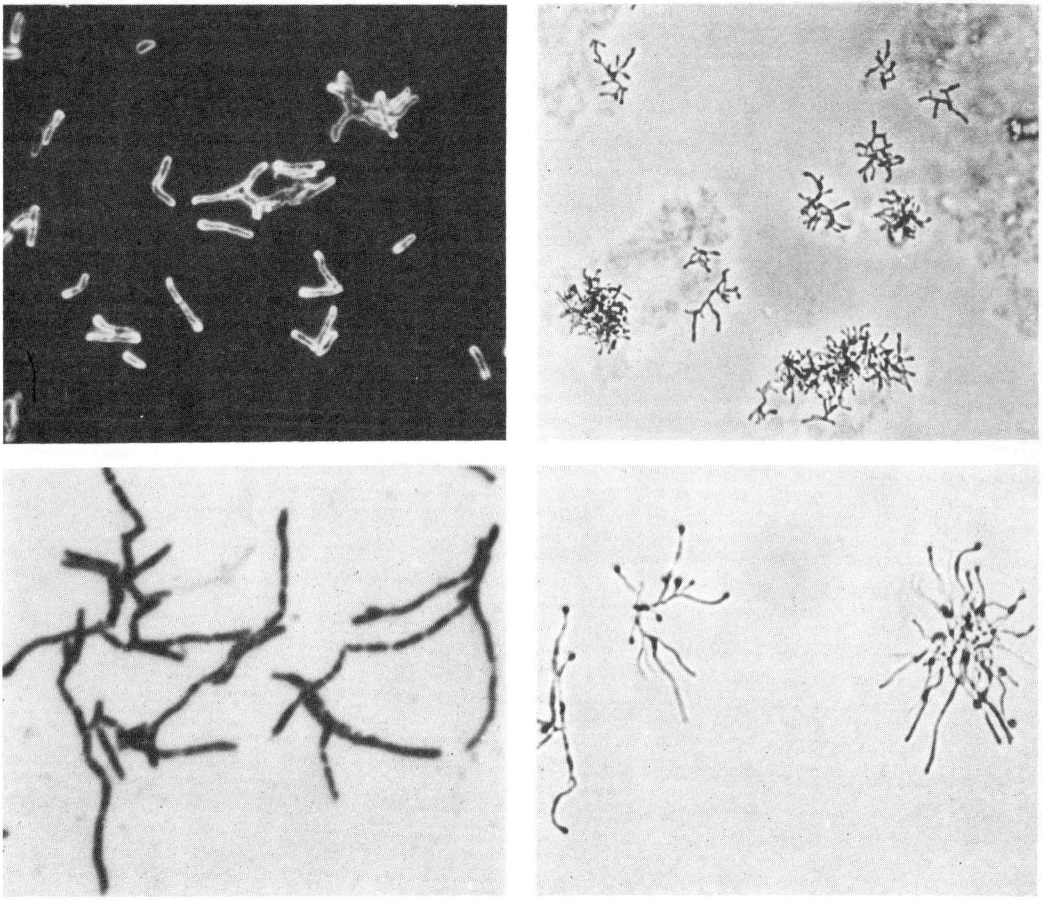

Fig. 335 Actinomyces israelii. Upper left, dark-field micrograph from 72-hr growth in thiogylcollate broth, showing V and Y forms; lower left, Gram stain from 72-hr growth in thioglycollate broth, showing elongated filaments, branching, and some irregular staining; upper right, unstained 24-hr microcolonies growing on beef heart infusion agar, showing filaments with multiple short angular branches; lower right, unstained 24-hr microcolonies growing on beef heart infusion agar, showing branching, filamentous or "spider" colonies with no distinct center. (Courtesy of Slack, Landfried, and Gerencser.)

and destruction of sources of infection. Discharges from lesions and articles soiled by lesions should be disinfected. Care should be taken to prevent contact with lesions.

Treatment Prolonged administration of sulfonamides, penicillin, chloramphenicol, or chlortetracycline is usually effective. There is no spontaneous recovery.

For more information: Georg and Coleman (1968); Pine et al. (1960); Slack et al. (1969).

Bacillus

The anthrax organism is classified in this genus, the members being capable of producing heat-resistant spores. Spores may be cylindrical, ellipsoidal, or spherical, and located centrally, subterminally, or terminally. Almost all species are gram-positive. With the exception of the anthrax bacillus, all members are saprophytic and usually not pathogenic. *Bacillus subtilis* has been known to become pathogenic at times, but this is the exception rather than the rule. Members are typically aerobic but some are facultative.

Spores are not produced under anaerobic conditions. On the other hand, the anaerobic spore-producing species do form spores under anaerobic conditions. This fact offers a means for the separation of the aerobic from the anaerobic spore formers. By the application of heat to a mixed culture, the vegetative cells are destroyed and leave only the anaerobic spores, which are capable of germinating into vegetative cells under favorable conditions.

Members of the genus *Bacillus* are universally distributed in soil and water. Spores and vegetative cells of such species are easily carried into the air by gentle air currents. This explains why viable spores of such organisms are universally present in air and are responsible for many laboratory contaminations of culture media and cultures.

B. anthracis The cells are rod-shaped, measuring 1 to 1.3 by 3 to 10 μ, with square or concave ends, occurring in long chains. They are nonmotile and gram-positive.

The spores are ellipsoidal to cylindrical, measuring 0.8 to 1.0 by 1.3 to 1.5 μ, central or paracentral, often in chains. Germination is polar.

Disease produced *B. anthracis* causes anthrax, an acute specific disease of cattle, sheep, and swine, sometimes occurring in workers handling wool and hides of animals affected with the disease (Fig. 336). It usually occurs as a febrile disease of animals that runs a rapid course and terminates in a septicemia. Mortality rate may run as high as 80 percent. The infection causes a marked enlargement of the spleen, in which may be found enormous numbers of bacilli.

Two forms occur in man: cutaneous (malignant pustule) and internal anthrax. Cutaneous anthrax is produced by direct inoculation through a cut or abrasion in the skin. This type occurs most frequently in persons working with livestock. It is characterized by the appearance of a small furuncle within 12 to 24 hr after entrance of the organisms. The furuncle ulcerates and discharges a seropurulent exudate, which may heal and disappear, or gangrene may set in followed by a septicemia. This usually terminates fatally in about 5 days.

The internal or pulmonary type is contracted by inhalation or by swallowing spores of *B. anthracis.* The disease is characterized by a pneumonia that generally terminates fatally. Before death, it is possible to isolate the organism from the sputum. The organism may also be recovered from the blood and spinal fluid.

Fig. 336 Bacillus anthracis. Left, smear from the liver of an experimentally inoculated guinea pig. Note the absence of spores. Right, smear from a 72-hr agar slant culture. (From Muir, "Bacteriological Atlas," E. and S. Livingstone, Edinburgh, Scotland.)

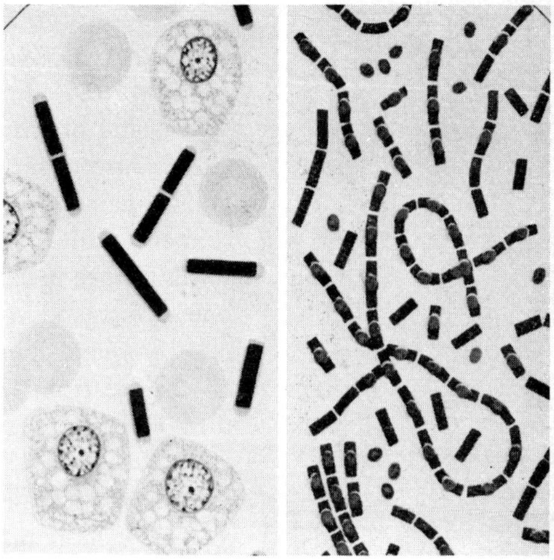

Diagnosis In the skin type, smears may be prepared from the seropurulent exudate and stained by Gram's method. The presence of large, gram-positive, encapsulated organisms without spores is strong evidence for the presence of *B. anthracis.* The organism may be confirmed by guinea-pig inoculation. The animals usually die in 12 hr to 3 days with a septicemia.

In the pneumonic type, sputum and blood are examined by the Gram technique. Cultures may be prepared by inoculating blood into broth and examining for characteristic organisms after an incubation period of 24 hr. A confirmation test may be made by guinea-pig inoculation as given above.

Source of infection Sources of infection are hair, hides, wool, flesh, and feces of infected animals and their manufactured products.

Mode of transmission The organisms are transmitted by inhalation of spores, ingestion of insufficiently cooked food, flies, and accidental inoculation by wounds or scratch.

Incubation period The incubation period is not over 7 days, usually less than 4 days. In pulmonary cases may be within 24 hr.

Susceptibility and immunity Man is less susceptible to the disease than the herbivora but more so than the carnivora. Immunity may develop after recovery from the disease. Active artificial immunity produced in animals by the use of a vaccine.

Cell-free vaccine effective in preventing cutaneous and probably inhalation anthrax. Recommended for veterinarians and others handling potentially contaminated raw materials.

Prevalence The disease is rare in humans and is mostly associated with occurrence of disease in animals, or with handling hides, hair, and other products from infected animals. Epizootics occur in cattle and sheep.

Prevention and control Animals known to have the disease should be destroyed. Exposed animals should be immunized with vaccine. Milk from infected animals should not be used. Disinfection of discharges from lesions and of articles soiled by such discharges. All hair, wool, and bristles from sources not known to be free of anthrax should be disinfected. Human beings handling hides, wool, and hair should report immediately any skin abrasion. Infection has occurred from the use of shaving brushes and toothbrushes made from unsterilized bristles. The spores are very resistant, being destroyed in the autoclave at temperatures above 120°C.

Infected human beings should be isolated until lesions have healed. Discharges from lesions and articles soiled by such discharges should be disinfected.

Treatment Penicillin is the drug of choice. Tetracyclines or other broad-spectrum antibiotics may be used.

For more information: Albrink (1961); Brachman et al. (1966); Fish and Lincoln (1968); Fubra (1966); Lincoln et al. (1964); Ward et al. (1965); Dalldorf (1971).

Bordetella

Members of genus are minute coccobacilli. Motile and nonmotile species occur. On primary isolation, some species dependent upon complex media; all are hemolytic. Carbohydrates not fermented. A dermonecrotic toxin is produced. All are parasitic.

B. pertussis These are minute coccobacilli, measuring 0.2 to 0.3 by 1 μ, occurring singly, in pairs, and occasionally in short chains. Capsules may be demonstrated. They show a tendency to bipolar staining and are nonmotile and gram-negative.

The blood medium is excellent for isolation and maintenance. Charcoal may be used instead of blood in certain agar media.

Various forms of the organism have been isolated, including smooth, rough, and intermediate types. The smooth forms are pathogenic, whereas the rough and intermediate forms are not. The organisms may exist in four phases on the basis of serological reactions. Freshly isolated or phase I strains are encapsulated, virulent for laboratory animals, hemolytic, and require the presence of the X and V factors (page 231). The phase I properties are lost on artificial cultivation, the organisms changing to phase II, III, or IV. Only virulent organisms in phase I are suitable for the production of vaccines.

Disease produced The bacillus is believed to be the cause of whooping cough. The organism is sometimes referred to as the Bordet-Gengou bacillus after the names of its discoverers. Whooping cough is an acute, specific, infectious disease of the trachea and bronchi. It is characterized by a cough typical of the disease and lasts 1 to 2 months. The disease starts as a catarrhal condition followed by an irritating cough. The cough becomes paroxysmal after a period of 1 to 2 weeks. The paroxysms consist of a repeated series of violent coughs often followed by a characteristic long-drawn whoop during inhalation. Paroxysms are sometimes followed by vomiting. The period of communicability probably does not last longer than 3 weeks after the cough appears.

Whooping cough shows its greatest incidence in children under five years of age, and the death rate is highest in those under one year of age. Children suffering from the disease show a predisposition to infections by micrococci, streptococci, pneumococci, and tubercle bacilli.

There has been a marked decline in case development and mortality during the past 20 years chiefly in communities practicing active immunization and having good medical care. In many less developed countries incidence continues at a high rate and the disease is among the most lethal of the common communicable diseases of childhood.

Diagnosis The organism may be recovered by the cough-plate method. This consists of exposing a petri dish, containing an appropriate medium, before a patient's mouth during a cough in the early paroxysmal stage of the infection. The plate is then incubated, and characteristic colonies are isolated.

Source of infection Sources of infection are discharges from mucous membranes of larynx and bronchi of infected persons.

Mode of transmission The disease is transmitted by direct contact with an infected person or with the discharges from an infected person. It is easily spread among children by personal contact. There is no evidence of a carrier state.

Incubation period The incubation period is usually 7 days, is almost always within 10 days, and does not exceed 21 days.

Susceptibility and immunity Susceptibility to the disease is general; there is no natural immunity. Children under seven are most susceptible to infection: children under two are most susceptible to fatal attack. One attack confers a definite immunity but not for life; second attacks are known to occur.

Prevalence The disease is common among children everywhere regardless of race or climate. About 15 percent of cases occur in children under two years of age.

Prevention and control All children under five years of age should be vaccinated. This is especially advisable in infants two months old.

Infected individuals should be isolated, especially from children. There should be disinfection of discharges from nose and throat of patient and articles soiled with such discharges.

Young unvaccinated children exposed to the infection should be given one of the tetracycline antibiotics on first appearance of symptoms. The tetracyclines and chloramphenicol given early shorten the period of communicability, but unless given in the early stages of the disease, do not modify the clinical manifestations.

For more information: Lane (1968*a*, *b*); Litkenhous and Liu (1967); Turner (1961).

Borrelia

Genus is placed in the order *Spirochaetales.* Members have characteristics intermediate between true bacteria and protozoa. Cells measure 8 to 16 μ in length, with coarse, shallow, irregular spirals, a few of which may be obtuse-angled. They generally taper terminally into fine filaments. They are parasitic upon many forms of animal life. Some are pathogenic for man, other mammals, or birds. Organisms are generally hematophytic or are found on mucous membranes. Some are transmitted by the bites of arthropods.

B. recurrentis They are cylindrical or slightly flattened cells, measuring 0.35 to 0.5 by 8 to 16 μ, with pointed ends. The spirals are large, wavy, and inconstant; they number about five. Spiral amplitude is 1.5 μ. Terminal, finely spiral filaments are present. Motility is by active corkscrew motion without polarity. Lashing movements are common in drawn blood. Organism is gram-negative.

Disease produced *Borrelia recurrentis* causes epidemic relapsing fever (Fig. 337), which is a systemic spirochetal disease with short febrile paroxysms lasting 2 to 9 days, alternating with afebrile periods of 2 to 4 days. Relapses average 1 to 10, usually not more than 2. General eruption occurs on the body. The duration of the louse-borne disease is 13 to 16 days; the tick-borne disease usually lasts longer.

The disease is epidemic when louse-borne (*Pediculus humanus*), and endemic when carried by ticks (principally *Ornithodorus turicata and O. hermsi* in United States; *O. rudis* and *O. talaje* in Central and South America; *O. moubata* in Africa; *O. tholozani* in Near, Middle, and Far East). Louse-borne disease occurs in Asia; East, North, and Central Africa; and South America. Tick-borne disease is widespread throughout tropical Africa, Spain, Saudi Arabia, Iran, India, Central Asia, and North and South America.

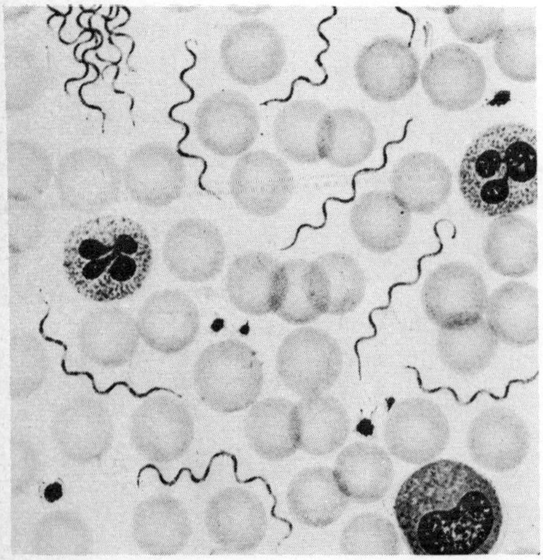

Fig. 337 Borrelia recurrentis. Blood smear from a case of European relapsing fever. (From Muir, "Bacteriological Atlas," E. and S. Livingstone, Edinburgh, Scotland.)

Diagnosis Diagnosis is made by demonstrating the organisms in dark-field preparations of fresh blood or by inoculating rats or mice with 1 ml of patient's blood taken during the pyrexial period and before crisis.

Source of infection The reservoir of louse-borne disease is man; the reservoir of tick-borne is wild rodents. In the United States the reservoir is principally ground squirrels and prairie dogs; also ticks through transovarian transmission.

Mode of transmission The disease is transmitted by the bite of a louse and the rubbing of its feces into the abrasion in the skin, or by the bite of an infected tick. The louse remains infective for life; the infected tick may remain infective for years. In Africa, certain ticks transmit the organisms directly from man to man.

Incubation period This is usually 8 days and can be as long as 12 days.

Susceptibility and immunity Susceptibility is general. Active immunity is produced during course of disease. Duration of immunity after recovery is probably not more than 2 years.

Prevention and control The disease can be prevented or controlled by application of insecticide at appropriate intervals to people living under poor, unhygienic conditions and improvement of living conditions (this should include frequent bathing and washing of clothing). Also helpful are delousing of patient's clothing and bedroom, application of insecticide to patient's clothing and body, and application of insecticide to all persons in contact with infected individuals.

Penicillin G in adequate dosage is effective; also effective are the tetracyclines and chloramphenicol. Arsenical therapy is widely used.

B. vincentii The organisms measure 0.3 by 8 to 12 μ, with three to eight irregular shallow spirals. They are motile with a rapid, progressive, vibratory motion, and gram-negative.

Cells are cultivated under anaerobic conditions. Cultures may show long forms with only a writhing motion (Fig. 338).

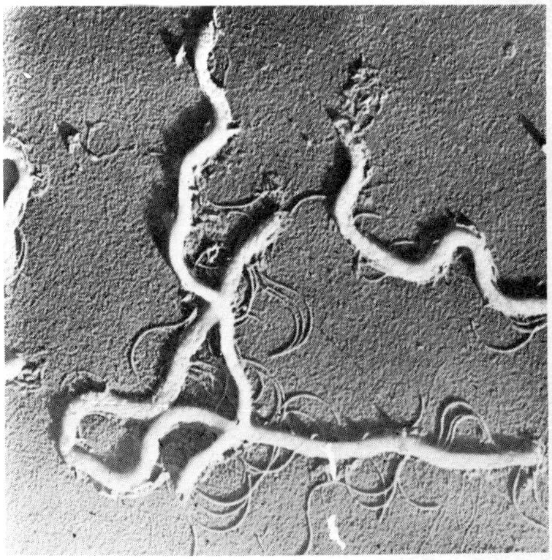

Fig. 338 Electron micrograph of Borrelia vincentii. X11,000. (After Hampp, Scott, and Wyckoff.)

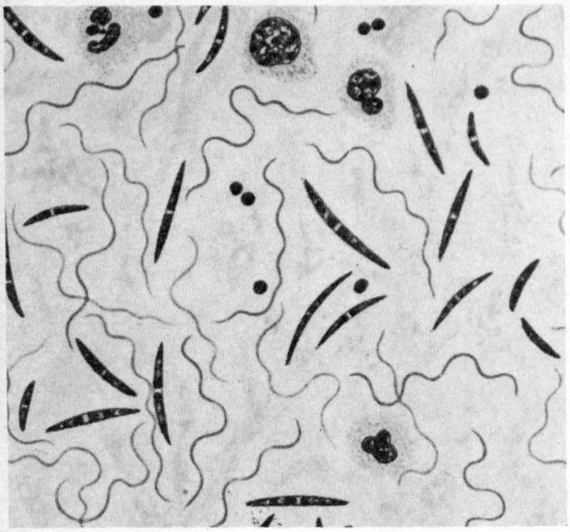

Fig. 339 Borrelia vincentii and Fusobacterium fusiforme growing in association. Smear prepared from a throat swab, taken from a case of Vincent's angina. (From Muir, "Bacteriological Atlas," E. and S. Livingstone, Edinburgh, Scotland.)

Disease produced *B. vincentii* occurs in association with *Fusobacterium fusiforme* in Vincent's angina, and in acute infection of the tonsils or neighboring parts, and is characterized by the appearance of a pseudomembranous inflammation followed by ulceration. The disease is sometimes called *trench mouth* (Fig. 339). The lungs may also become involved. When this occurs, the patient may present the clinical and anatomical picture of pulmonary gangrene, pulmonary abscess, or bronchiectasis.

Cells of *F. fusiforme* are straight or slightly curved rods, 0.5 to 1 by 8 to 16 μ, occurring in pairs with blunt ends together and outer ends pointed, sometimes in short, curved chains or long spirillum-like threads. Granules are present. They are nonmotile, anaerobic, and gram-negative.

Diagnosis The disease is diagnosed by preparing smears direct from the deeper ulcerated areas, staining by Gram technique, and examining under oil-immersion objective. A characteristic smear shows presence of spirochetes and bacilli in large numbers.

Source of infection Infection is from deposit on teeth and in the oral cavity.

Mode of transmission The disease is not ordinarily communicable. Under unusual conditions of crowding, such as may prevail among soldiers, the infection may become transmissible. The disease appears to be associated with a state of lowered resistance. The tonsillar ulceration occurs often in individuals whose resistance has been lowered by such diseases as measles, tuberculosis, diabetes, and scarlet fever.

Susceptibility and immunity Susceptibility to infection is general. Acquired immunity does not follow recovery from the disease.

Prevention and control Local lesions can usually be controlled by treatment with arsenicals or penicillin. Neoarsphenamine, sulfarsphenamine, or bismarsen is effective in curing the pulmonary infection if administered during the first few days of the disease before the beginning of necrosis.

For more information: Dean and Singleton (1945).

Brucella

Genus includes three pathogenic species which produce abortion in animals, Malta or undulant fever in man, and a wasting disease of chickens. Organisms invade animal tissue, producing infections of the genital organs, the mammary gland, and the intestinal tract. The genus is named after Bruce, who was the first to isolate the organism of undulant fever from the spleens of persons who had died of the disease on the island of Malta in the Mediterranean.

B. abortus Cells are short ellipsoidal rods, measuring 0.3 to 0.4 μ in length, occurring singly, in pairs, and rarely in short chains. The organism requires 10 percent CO_2 for isolation; it becomes aerobic after several transfers. It is nonmotile and gram-negative.

Disease produced *Brucella abortus* is the cause of infectious abortion in cattle (the same effect is produced in mares, sheep, rabbits, and guinea pigs). It causes undulant fever or brucellosis in man.

The disease is a generalized infection with gradual or insidious onset, and is characterized by irregular fever usually of prolonged duration, headache, sweating, chills, pain in the joints and muscles. The usual case follows a long febrile period, returning to normal by lysis. In the United States, febrile lapses are uncommon. Recovery is usual but disability may be pronounced.

Diagnosis Since brucellosis is generally accompanied by a septicemia, a diagnosis can usually be made on the basis of a blood culture.

Agglutinins and complement-fixing antibodies are present in the serum of patients suffering from the disease. In the event the blood culture is negative, it is generally desirable to test the patient's serum for the presence of agglutinins. An agglutination titer of 1:100 or above is positive evidence of undulant fever.

Source of infection Sources of infection are tissues, blood, milk, and urine of infected animals, especially cattle, sheep, swine, goats, horses, reindeer, and Alaskan caribou. Laboratory infections are quite common.

Mode of transmission The disease is transmitted by ingestion of milk and dairy products from infected animals; also by direct contact with tissues, blood, urine, aborted fetuses, and placentas of such animals.

Incubation period The incubation period is highly variable; it is usually 5 to 21 days, but occasionally is considerably longer.

Susceptibility and immunity Most persons have some natural immunity or have acquired partial immunity by ingestion of small doses of the organism. Duration of immunity not known. One attack of undulant fever usually protects against a second attack.

Prevalence The disease has a worldwide occurrence, but is especially prevalent in the Mediterranean countries of Europe and North Africa. It affects persons of all races; it is more prevalent in males than in females because of occupational risks. Outbreaks occur among consumers of raw milk or milk products from cows, sheep, and goats. *B. suis* infections have a greater rate of occurrence in the Middle West (United States).

Prevention and control The source of infection should be ascertained. All milk supplies from cows, goats, and sheep should be pasteurized, although boiling of milk may be more practicable. Animals should be tested by agglutination technique, and the positive reactors should be segregated or slaughtered. The general public and meat handlers should be made aware of the nature of the disease and the mode of transmission. There should be vaccination of calves. Care should be taken in handling and disposal of discharges and fetus from an aborting animal. There should be inspection of all meats, especially pork and pork products.

The disease should be diagnosed early and body discharges should be disinfected. The tetracycline antibiotics generally produce rapid subsidence of fever and symptoms within several days. A relapse may occur. More satisfactory results are obtained from a combination of Aureomycin and dihydrostreptomycin over a period of at least 3 weeks. Blood transfusion may be necessary where acute anemia exists. The disease is seldom fatal.

B. melitensis Cells are short ellipsoidal rods, measuring 0.3 to 0.4 μ in length, occurring singly, in pairs, and rarely in short chains (Fig. 340). They are nonmotile and gram-negative.

Disease produced *B. melitensis* is the cause of infectious abortion in goats and undulant fever (brucellosis) in man. It may infect cows and hogs and be excreted in their milk. The disease is infectious for all domestic animals.

The disease is a generalized infection with gradual or insidious onset, and is characterized by irregular fever, usually of prolonged duration, headache, sweating, chills, and pain in the joints and muscles. The usual case follows a long febrile period, returning to normal by lysis. In the

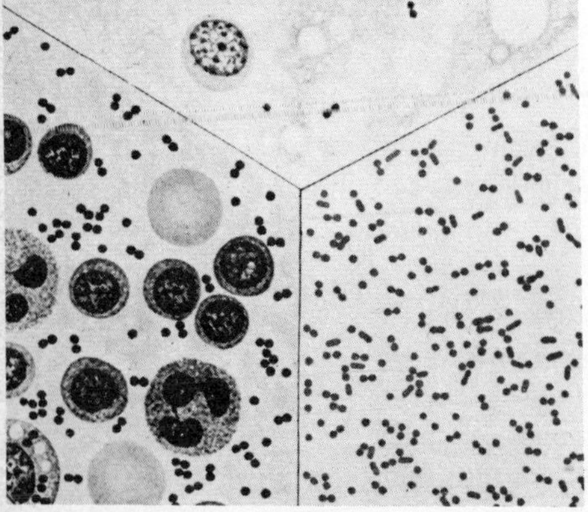

Fig. 340 Brucella melitensis. Upper, smear of milk from an infected goat; lower left, spleen smear from a case of Malta fever; lower right, smear prepared from a young culture. (From Muir, "Bacteriological Atlas," E. and S. Livingstone, Edinburgh, Scotland.)

United States, febrile lapses are uncommon. Recovery is usual, but disability may be pronounced.

Diagnosis Since brucellosis is generally accompanied by a septicemia, a diagnosis usually can be made on the basis of a blood culture. Blood cultures are positive in about 80 percent of cases after the second day, and the infection may continue in the septicemic form for a number of months.

Agglutinins and complement-fixing antibodies are present in the serum of patients suffering from the disease. In the event the blood culture is negative, it is generally desirable to test the patient's serum for the presence of agglutinins. Agglutinins generally occur in the serum about the tenth day of the fever. An agglutination titer of 1:100 or above is positive evidence of undulant fever. Since antiserum for *B. melitensis* will cross-agglutinate with *B. abortus* and *B. suis,* agglutinin-absorption tests are necessary for diagnosis.

Source of infection Sources of infection are tissues, blood, milk, and urine of infected goats. The disease is spread to man through the milk of such animals. The disease in both goats and man is a septicemia. Laboratory infections are quite common.

Mode of transmission The disease is transmitted by ingestion of milk and milk products from infected animals; also by direct contact with tissues, blood, urine, aborted fetuses, and placentas of such animals.

Incubation period The incubation period is highly variable, usually 5 to 21 days, and is occasionally considerably longer.

Susceptibility and immunity Most persons have some natural immunity or have acquired partial immunity by ingestion of small doses of the organism. The duration of immunity is not known. One attack of undulant fever usually protects against a second attack.

Prevalence The disease has a worldwide occurrence, but is especially prevalent in the Mediterranean countries of Europe and North Africa. It affects persons of all races; it is more prevalent in males than in females because of occupational risks. Outbreaks occur among consumers of raw milk or milk products from cows, sheep, and goats. *B. suis* infections have greater rate of occurrence in the Middle West (United States).

Prevention and control The source of infection should be ascertained. All milk supplies from cows and goats should be pasteurized, although boiling of milk may be more practicable. Animals should be tested by agglutination technique, and positive reactors should be segregated or slaughtered. The general public and meat handlers should be made aware of the nature of disease and the mode of transmission. There should be vaccination of animals. Care should be taken in handling and disposal of discharges and fetus from an aborting animal. There should be inspection of all meats, especially pork and pork products.

The disease should be diagnosed early and body discharges should be disinfected. The tetracycline antibiotics generally produce rapid subsidence of fever and symptoms within several days. A relapse may occur.

More satisfactory results are obtained from a combination of Aureomycin and dihydrostreptomycin over a period of at least 3 weeks. Blood transfusion may be necessary where acute anemia exists. The disease is seldom fatal.

B. suis Cells are short ellipsoidal rods, measuring 0.3 to 0.4 μ in length, occurring singly, in pairs, and rarely in short chains. The organism is nonmotile and gram-negative.

Disease produced *B. suis* is the cause of undulant fever (brucellosis) in man and abortion in hogs. It may infect cows and goats and be excreted in their milk; it is infectious for all domestic animals.

The disease is a generalized infection with gradual or insidious onset, and is characterized by irregular fever usually of prolonged duration, headache, sweating, chills, pain in the joints and muscles. The usual case follows a long febrile period, returning to normal by lysis. In the United States, febrile lapses are uncommon. Recovery is usual but disability may be pronounced.

Diagnosis Since brucellosis is generally accompanied by a septicemia, a diagnosis can usually be made on the basis of a blood culture. Blood cultures are positive in about 80 percent of the cases after the second day, and the infection may continue in the septicemic form for a number of months.

Agglutinins and complement-fixing antibodies are present in the serum of patients suffering from the disease. In the event the blood culture is negative, it is generally desirable to test the patient's serum for the presence of agglutinins. Agglutinins generally occur in the serum about the tenth day of the fever. An agglutination titer of 1:100 or above is positive evidence of undulant fever. Since antiserum for *B. suis* will cross-agglutinate with *B. melitensis* and *B. abortus,* agglutinin-absorption tests are necessary for diagnosis.

Source of infection Sources of infection are tissues, blood, milk, and urine of infected animals. The disease is spread to man through the milk of such animals. The disease in both hogs and man is a septicemia. Laboratory infections are quite common.

Mode of transmission The disease is transmitted by ingestion of milk and milk products from infected animals; also by direct contact with tissues, blood, urine, aborted fetuses, and placentas of such animals.

Incubation period The incubation period is highly variable, usually 5 to 21 days, but occasionally considerably longer.

Susceptibility and immunity Most persons have some natural immunity or have acquired partial immunity by ingestion of small doses of the organism. The duration of immunity is not known. One attack of undulant fever usually protects against a second attack.

Prevalence The disease has a worldwide occurrence, but is especially prevalent in the Mediterranean countries of Europe and North Africa. It affects persons of all races; it is more prevalent in males than in females because of occupational risks. Outbreaks occur among consumers of raw milk or milk products from cows, sheep, and goats. *B. suis* infections have a greater rate of occurrence in the Middle West of the United States.

Prevention and control The source of infection should be ascertained. All milk supplies should be pasteurized, although boiling of milk may be more practicable. Animals should be tested by agglutination technique, and positive reactors should be segregated or slaughtered. The general public and meat handlers should be made aware of the nature of the disease and the mode of transmission. There should be vaccination of animals. Care should be taken in handling and disposal of discharges and fetus from an aborting animal. There should be inspection of all meats, especially pork and pork products.

The disease should be diagnosed early and body discharges should be disinfected. The tetracycline antibiotics generally produce rapid subsidence of fever and symptoms within several days. A relapse may occur. More satisfactory results are obtained from a combination of Aureomycin and dihydrostreptomycin over a period of at least 3 weeks. Blood transfusion may be necessary where acute anemia exists. The disease is seldom fatal.

For more information: Agius (1965); De Petris et al. (1964); Ellwood et al. (1967); Karlsbad et al. (1964); Spink and Bradley (1960).

Clostridium

Clostridia are rod-shaped cells, often swollen at sporulation, producing clostridial, plectridial, clavate, or navicular forms. They are motile by means of peritrichous flagella, although occasionally nonmotile; they are generally gram-positive. Many species are saccharolytic and fermentative, producing various acids, gases (CO_2, H_2, CH_4), and variable amounts of neutral products, i.e., alcohols and acetone. Other species are proteolytic, some attacking proteins with putrefaction or more complete proteolysis. Organisms are strictly anaerobic or anaerobic, aerotolerant. They are catalase-lacking except in small amounts in certain aerotolerant forms. Few species are obligately thermophilic. Exotoxins are sometimes produced. Cells are commonly found in soil and in the intestinal tracts of man and other animals.

Cl. botulinum, types A, B, C, D, E, F *Cl. botulinum* comprises a large number of toxic species conveniently divided into a nonovolytic (*Cl. botulinum*) and an ovolytic (*Cl. parabotulinum)* group. Authorities are not yet in agreement on fermentation and on variant subtypes, and the present groups are only tentative and subject to revision.

Cells are rod-shaped, measuring 0.5 to 0.8 by 3 to 8 μ, with rounded ends, occurring singly, in pairs, and in short to occasionally long chains. The organism is motile by means of peritrichous flagella and gram-positive. The spores are ovoid, central, subterminal, and terminal at maturation, swelling the cells.

Disease produced *Cl. botulinum* causes a highly fatal afebrile bacterial intoxication, not an infection. The organism is the cause of botulism in man and limberneck in chickens. It produces a powerful exotoxin that is neurotoxic both on injection and on feeding. Animals susceptible to the toxin include monkeys, rabbits, guinea pigs, and cats.

Symptoms develop suddenly with gastrointestinal pain, headache, dizziness, weakness, constipation or diarrhea, prostration, and several types of paralyses of the central nervous system which are produced by the extracellular neurotropic toxin. Death occurs by cardiac or respiratory paralysis, usually in 3 to 7 days. The severity of symptoms depends upon the amount of toxin ingested in relation to body weight.

Most outbreaks are caused by type A, B, or E toxins of *Cl. botulinum*; a few by types C and F. Type E outbreaks are usually related to fish.

Diagnosis Biological or toxicological tests may confirm the presence of the organism or its toxin in suspected food.

Source of infection Contaminated smoked, pickled, or canned foods improperly processed are the sources of infection (page 782). Toxin is produced only under anaerobic conditions. Toxin is easily destroyed by boiling, but spores are inactivated only at the temperature of an autoclave. Ordinary refrigeration does not necessarily prevent toxin production.

Mode of transmission The disease is transmitted through the consumption of smoked, pickled, or canned foods containing the exotoxin from containers improperly processed. In the United States most intoxications are caused by consumption of home-canned vegetables and fruits or by fish; in Europe most poisonings are caused by the consumption of sausages and by smoked or preserved meats or fish.

Incubation period Symptoms usually occur within 18 hr after partaking of contaminated food, possibly longer, depending upon amount of food consumed and its toxin content. In general, the shorter the incubation period the more severe the illness and the higher the fatality rate.

Susceptibility and immunity Susceptibility to the toxin is general. Passive immunity with specific antitoxin is of value before symptoms have appeared.

Prevalence Sporadic cases occur in all countries, usually from home-canned food.

Prevention and control The disease can be prevented or controlled by inspection of commercial processing of canned and preserved foods and education of housewives in methods for the safe processing of home-canned foods. Since the toxin is destroyed on boiling, all home-canned foods should be boiled for 3 min before serving.

Specific polyvalent antitoxin will neutralize the toxin and should be administered preferably before symptoms have developed.

Specific treatment consists of intravenous and intramuscular injections of antitoxin.

For more information: Herrero et al. (1967); Licciardello et al. (1967); Lynt et al. (1967).

Cl. histolyticum Cells are rod-shaped, measuring 0.5 to 0.7 by 3 to 5 μ, occurring singly and in pairs. The organism is motile by means of peritrichous flagella and gram-positive. Spores are ovoid and subterminal, swelling the cells. They are anaerobic and aerotolerant, but scanty growth is obtained under aerobic conditions.

Disease produced *Cl. histolyticum* was originally isolated from war wounds; it is apparently widely but sparsely dispersed in soil. The organism produces at least three antigenic components in toxic culture filtrates: (1) *alpha,* lethal and necrotizing toxin; (2) *beta,* collagenase; and (3) cysteine-activated proteinase which attacks altered collagen but not native collagen. Intramuscular injection of a small amount of culture into a guinea pig produces rapid digestion of muscle tissue. It is not toxic on feeding, and is pathogenic for small laboratory animals.

Source of infection The organism is found in soil.

Mode of transmission The disease is produced by entrance of the organism into the broken skin, where it multiplies and produces an extracellular cytolytic toxin.

Susceptibility and immunity Susceptibility to toxin is general. The exotoxin may be neutralized by injection of homologous antitoxin.

For more information: MacLennan (1962).

Cl. perfringens Cells are short, thick rods, measuring 1 to 1.5 by 4 to 8 μ, occurring singly and in pairs, less frequently in short chains. The organism is encapsulated, nonmotile, and gram-positive. Spores are ovoid, central to eccentric, not swelling the cells.

At least six types have been established on the basis of variety and nature of toxins present in culture filtrates. Type A is classic human gas gangrene organism; type F causes hemolytic enteritis and enteritis necroticans of humans. Antitoxin from any one type will neutralize the toxin from the other types.

Disease produced *Cl. perfringens* causes an intestinal disorder with sudden onset of abdominal colic followed by diarrhea. Nausea is common but vomiting is usually absent. It is generally a mild illness of short duration and rarely fatal in healthy persons.

The organism is the most frequent cause of gas gangrene in man (Fig. 341). It produces extensive necrosis and considerable gas in tissues. Gas bubbles cause an expansion in the tissues accompanied by pressure, which results in cutting off the blood supply. This causes the affected tissues to die. Organisms may be recovered from liver and heart blood.

Cl. perfringens produces a powerful exotoxin which may be obtained in crude form by filtering a culture of the organisms. The exotoxin aids in weakening the patient. The injection of animals with culture filtrates results in the development of a potent antitoxin.

Source of infection Sources of infection are soil, street dust, milk, and human and animal feces.

Mode of transmission Ingestion of food contaminated with fecal material and soil. Most outbreaks are associated with consumption of raw

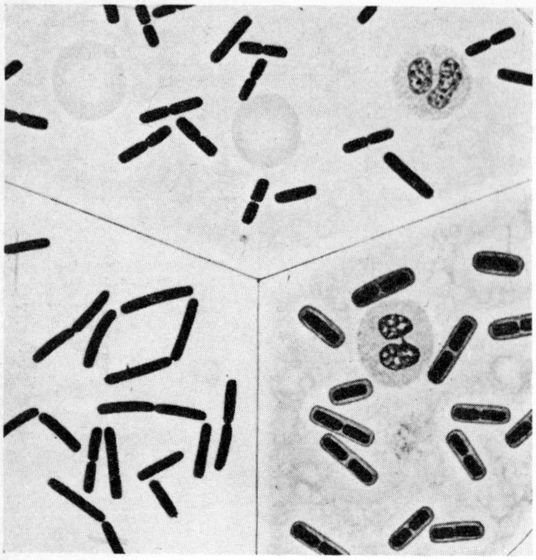

Fig. 341 Clostridium perfringens. Upper, smear of the exudate from a case of human gas gangrene; lower left, smear from a young agar culture; lower right, another smear from a case of gas gangrene. (From Muir, "Bacteriological Atlas," E. and S. Livingstone, Edinburgh, Scotland.)

or undercooked meats, especially meat pies, stews, turkey, and chicken. Outbreaks have been traced to catering firms, restaurants, and cafeterias.

Gas gangrene is produced by entrance of organisms in the broken skin or wound where it multiplies and secretes a potent exotoxin.

Incubation period In food poisoning the incubation period is from 8 to 22 hr; usually 10 to 12 hr.

Susceptibility and immunity Susceptibility to infection is general. Immunity following recovery has not been observed.

Prevention and control Meat dishes should be thoroughly heated and served hot; otherwise refrigerated until serving time. Reheating should be rapid. large pieces of meat should be reduced to smaller pieces to permit penetration of heat throughout. Food handlers should be instructed in importance of proper cooking, especially of meat dishes.

Antitoxic serums have been used both for prophylaxis and for therapeutic administration in cases of gas gangrene and beneficial results have been reported.

For more information: MacLennan (1962).

Cl. tetani Cells are rod-shaped, measuring 0.4 to 0.6 by 4 to 8 μ, with rounded ends, occurring singly, in pairs, and often in long chains and filaments. They are motile by means of peritrichous flagella and are gram-positive. Spores are spherical, terminal, swelling the cells (Fig. 342).

Disease produced *Cl. tetani* is the cause of tetanus or lockjaw in man. It is an acute disease produced by toxin of the organism growing anaerobically at site of injury, and characterized by headache, difficulty in swallowing and opening the mouth, owing to spasms of the masseter and neck muscles. This is accompanied by a slight stiffness of the neck and spasm of the cheek muscles. Spasms may spread to trunk and back. Swallowing becomes increasingly more difficult. Organisms rarely invade tissues but remain localized in the wound, where they secrete a powerful

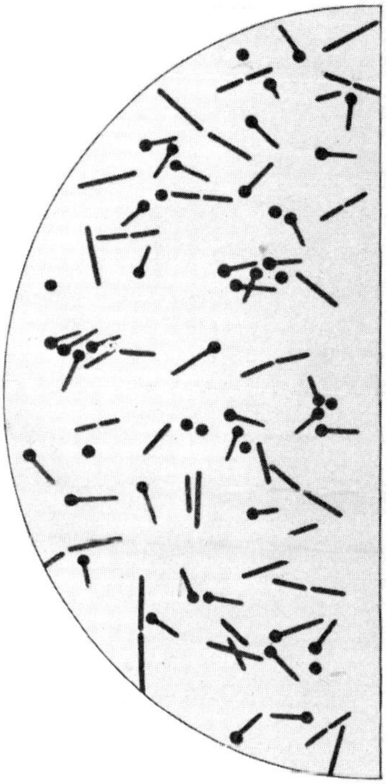

Fig. 342 Clostridium tetani. Smear from a 72-hr glucose agar stab culture. The spores are round and situated at one end of the rod. (From Muir, "Bacteriological Atlas," E. and S. Livingstone, Edinburgh, Scotland.)

toxin. Inoculation of horses and other animals with culture filtrates results in development of a powerful antitoxin.

Diagnosis The disease is diagnosed by preparing smears from infected materials, staining by Gram's method, and examining under the microscope for typical cells with spores.

Source of infection Sources of infection are soil, street dust, and human and animal feces, especially the latter where the organism is a normal and harmless inhabitant.

Mode of transmission The disease is produced by entrance of spore into the broken skin or wound where it germinates, multiplies and secretes a powerful exotoxin. Toxin is intensely toxic on injection but not on feeding.

Incubation period Symptoms develop after the toxin reaches the central nervous system. This may vary from 4 days to 3 weeks, depending upon the character, extent, and location of wound.

Susceptibility and immunity Susceptibility is general. Active artificial immunity may be produced by use of tetanus toxoid. Three doses of 1 ml each at intervals of 3 weeks are generally given and are considered capable of producing a high concentration of antitoxic antibodies. Immunity is said to last for about 5 years. Toxoid precipitated by alum and given in two injections of 1 ml each at intervals of 2 to 4 weeks has been reported to produce a greater and more rapid immunity than toxoid.

Prevalence The disease is worldwide in distribution, but is most frequent in North America among young males and in summer months. It is prevalent especially in wounds contaminated with fertile or manured soil.

Prevention and control Active immunization with tetanus toxoid is desirable for those likely to be exposed to infection. Active immunization with toxoid is advisable in infancy or early childhood.

The use of tetanus antitoxin is an effective protection against the disease. The injection of about 1500 U.S.A. units of antitoxin in slight injuries and about 2000 to 3000 U.S.A. units in more severe injuries will prevent tetanus. Since antitoxin tends to disappear, an additional injection should be given within 10 days.

After disease has developed, large doses of antitoxin are injected by spinal puncture. At the same time, an intravenous injection of about 10,000 units should be administered. Also penicillin in large doses should be given intramuscularly. The intraspinal injection should be repeated every 24 hr until three doses have been given. If intraspinal doses cannot be given, larger intravenous injections should be administered.

For more information: Altemeier and Hummel (1966); Bizzini et al. (1970); Bytchenko (1966); Holmes and Ryan (1971); Steigman (1968); Zacharias and Björklund (1968).

Corynebacterium

The corynebacteria are straight to slightly curved rods with irregularly stained segments, sometimes granules. They frequently show club-shaped swellings. Snapping division produces angular and palisade (picket-fence) arrangements of cells. They are gram-positive to variable, sometimes young cells and sometimes old cells being gram-negative. Granules are invariably gram-positive. Generally cells are quite aerobic, but microaerophilic or even anaerobic species occur. Some pathogenic species produce a powerful exotoxin. They are widely distributed in nature. Best-known species are parasites and pathogens on man and domestic animals. Several species are well-known plant pathogens; still other common species are found in dairy products, water, and soil.

C. diphtheriae Cells are rod-shaped, varying greatly in size from 0.3 to 0.8 by 1 to 8 μ and occurring singly. Rods are straight or slightly curved, frequently club-shaped at one or both ends. As a rule, rods do not stain uniformly with methylene blue but show alternate bands of stained and unstained material, and one or more metachromatic granules which are best shown by special stains. The organism is nonmotile and gram-positive (Figs. 343 and 344).

Disease produced *C. diphtheriae* is the cause of diphtheria in man. It is pathogenic for guinea pigs, kittens, and rabbits. The organism localizes in the throat, where it produces a powerful extracellular toxin which is absorbed into the blood stream and may cause death unless neutralized by antibodies known as antitoxin.

Diphtheria is an acute, febrile infection, generally of the nose,

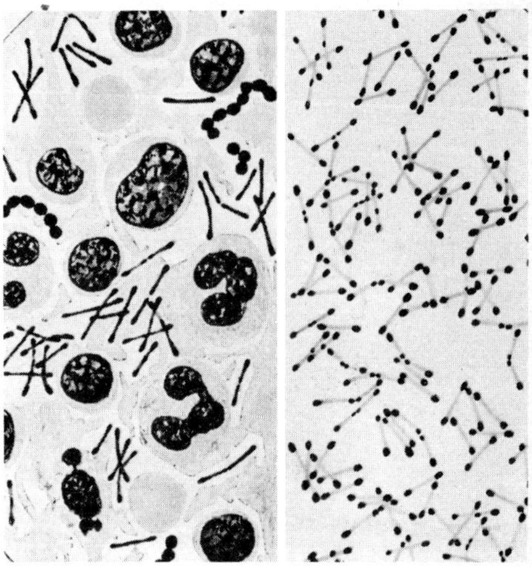

Fig. 343 Corynebacterium diphtheriae. Left, smear prepared from a throat swab taken from a case of diphtheria; right, smear from a 12-hr culture on Loeffler's blood serum medium. (From Muir, "Bacteriological Atlas," E. and S. Livingstone, Edinburgh, Scotland.)

throat, and tonsils. The throat becomes considerably inflamed, especially the fauces, where a grayish false membrane is formed. The membrane may eventually spread to the entire respiratory tract. The toxic action of the organism results in the destruction of the superficial layer of cells. This is followed by the exudation of a plasma-like fluid which clots and covers the surface of the injured mucous membrane with a tough elastic network of fibrin in which are embedded dead cells and bacteria. The toxin elaborated by the organisms produces an injurious action on the kidneys, and muscles of the heart. Injury to the heart is probably the most important action of the toxin.

Carriers Convalescents usually harbor the organisms for 3 to 4 weeks, after which the bacteria gradually disappear. Three negative throat examinations are generally required before a convalescent is released from quarantine. However, studies on the bacterial flora of normal throats have revealed the presence of a surprisingly high percentage of persons who habitually have the diphtheria organisms in their throats. These are chronic carriers and of no importance unless the organisms are of the virulent type.

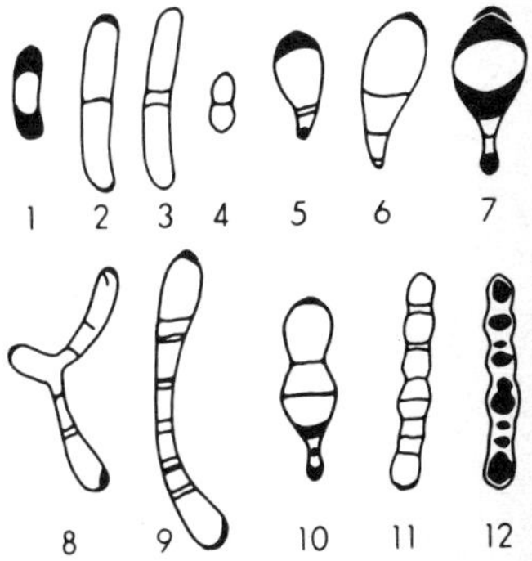

Fig. 344 Typical cell forms of Corynebacterium diphtheriae. (1, 2, 3, 4) short and long rod-shaped forms; (5, 6, 7) forms club-shaped at one end; (8) branched form; (9) cell club-shaped at both ends; (11) form with cell walls; (12) same cell with nuclear material in the cell compartments. (After Hewitt.)

Whether they are or not can be determined in the laboratory by performing virulence tests on animals.

Diagnosis A sterile swab is rubbed over the tonsillar region of the throat or the pharynx and then streaked over the surface of coagulated serum known as Loeffler's medium. The swab is also rubbed over the surface of a glass slide and stained with methylene blue. The culture is incubated at 37°C for 24 hr, and the slide is examined under the microscope for the presence of characteristic barred and granular rods. The slide examination gives a preliminary idea of what to expect from the culture.

Since the culture on blood serum will show the presence of many kinds of organisms, it is necessary to streak some of the mixed growth over the surface of a solid medium contained in a petri dish to obtain *C. diphtheriae* in pure form. An excellent preparation for this purpose is known as cystine-tellurite medium. Typical colonies of *C. diphtheriae* on this medium are opaque and dark gray or black in color. The organisms are capable of absorbing the potassium tellurite and reducing it to the colored metal. Characteristic colonies are transferred to tubes of Loeffler's coagulated serum medium, incubated at 37°C for 24 hr, and then tested for virulence.

Virulence test The method generally employed is the intracutaneous test, performed as follows: Two guinea pigs are used, one being the control. The abdomens of the animals are shaved. The control pig is given 250 units of diphtheria antitoxin intraperitoneally. The growth from a 24-hr culture of the organism to be tested is emulsified in about 20 ml of salt solution and 0.15 ml injected intracutaneously into each pig. If the culture contains virulent diphtheria bacilli, it will produce toxin which will have no effect on the immunized pig but will produce in the other pig a definite local inflammatory lesion in 24 hr which becomes necrotic in 48 to 72 hr.

Schick test This test is employed to determine the susceptibility of an individual to diphtheria. The test is performed as follows: Diphtheria toxin is diluted so that 0.1 ml contains one-fiftieth of the minimum lethal dose (MLD) required to kill a 250-g guinea pig in 96 hr. One-tenth milliliter of toxin so diluted is injected intracutaneously in the arm. The same amount of heated toxin is injected in the other arm as a control. The injected areas are examined daily for several days. Usually the fourth day gives the most reliable readings. The reactions may be recorded as follows:

1. A positive test is indicated by the presence of a slightly raised area of redness 1 to 2 cm in diameter which appears in 24 to 36 hr and reaches a maximum in 48 to 72 hr. The reaction persists for about 1 week, then gradually fades, and finally disappears. The control arm shows no reaction.
2. A negative test does not show an area of redness in the test or control arm.

Source of infection Sources of infection are discharges from diphtheritic lesions of the pharynx, larynx, trachea, nose, conjunctiva, and vagina; secretions from the healthy pharynx and nose of carriers may also contain the organisms.

Mode of transmission Disease is spread from person to person by fingers; by articles such as eating utensils, toys, pencils, handkerchiefs containing nasal discharges and saliva; and by inhalation of droplets expelled from the throat during coughing and sneezing.

Incubation period The period is usually 2 to 5 days, sometimes longer.

Susceptibility and immunity One attack of diphtheria usually confers immunity for life. Resistance to disease increases with age. This is believed to be caused by continued exposure to the disease. Infants born of immune mothers are relatively immune, a passive protection usually lost before the sixth month. Passive temporary immunity for a few days or weeks can be conferred, and active immunity of prolonged duration can be artificially induced.

Prevalence The disease is endemic and epidemic, and occurs more frequently in temperate zones and during fall and winter months. One-half of deaths occur in children under five years of age before they have had an opportunity to develop immunity to disease.

Prevention and control Suspected cases of diphtheria should be isolated. Persons showing a positive Schick test may be susceptible to the disease. All children should be vaccinated against the disease. Active immunization of susceptible individuals may be practiced by use of two types of preparations: (1) toxoid and (2) alum-precipitated toxoid.

Toxoid Toxoid is prepared by treating toxin with formaldehyde to destroy its toxic properties without affecting its ability to stimulate the production of antitoxin. Usually three doses at intervals of 1 month are necessary to give a negative Schick test.

Alum-precipitated toxoid Toxoid precipitated with alum is superior to ordinary toxoid as an immunizing agent. The precipitate is insoluble and remains in the tissue for a long period of time before it is completely absorbed. This affords a more prolonged antigenic response. Usually two injections are necessary to render Schick-positive individuals negative.

Milk should be pasteurized to render it free of possible contamination.

All articles which have been in contact with patient and all articles soiled by discharges of patient should be disinfected.

Therapy Passive immunization with diphtheria antitoxin is necessary in clinical cases. The antitoxin is administered as early in the disease as possible. The dosage is about 10,000 units in mild cases and 30,000 to 50,000 units in severe cases. The antitoxin neutralizes the damaging effect of the toxin (page 873). Passive immunization lasts usually 2 to 4 weeks. If infection still persists, the dose should be repeated. Penicillin or erythromycin may be used in conjunction with antitoxin, but it is not a substitute for the latter.

For more information: Beattie (1949); Bojlén and Scheibel (1955); Stainer et al. (1968); Uchida and Yoneda (1967); Michel (1972).

Diplococcus

The pneumococcus (*D. pneumoniae*) is the causative agent of lobar pneumonia. The disease is nearly always caused by this organism, although other bacteria are occasionally involved.

D. pneumoniae Cells are oval or spherical, measuring 0.5 to 1.25 μ, typically in pairs, occasionally singly or in short chains. Distal ends of each pair of cells tend to be pointed or lancet-shaped. The organism is encapsulated and nonmotile (Figs. 345 and 346). Young cells are gram-positive.

Whole bile or 10 percent solutions of sodium taurocholate or sodium glycocholate added to actively growing broth cultures will dissolve the organisms.

The pneumococci are generally classified according to types. The cells contain two types of antigens: (1) the so-called "somatic antigen" and (2) the "polysaccharide hapten" or soluble specific substances (SSS). The somatic antigen is probably a nucleoprotein and is found to react with all pneumococci regardless of types. The carbohydrate hapten is type specific and serves to differentiate the various types. In other words, by means of immunological reactions, the various types of pneumococci can be distinguished from one another by the composition of the polysaccharide comprising the capsular material. These polysaccharides belong to a group of substances called haptens, or partial antigens. They are not antigenic in themselves but may become so when combined with protein.

At least 32 types of *D. pneumoniae* are recognized on the basis of serological reactions, chiefly the Quellung phenomenon as induced by type-specific immune rabbit serums.

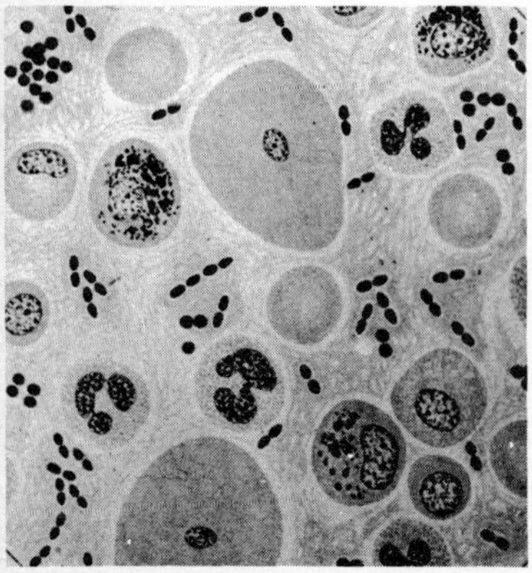

Fig. 345 Diplococcus pneumoniae. Smear of sputum from a case of lobar pneumonia. (From Muir, "Bacteriological Atlas," E. and S. Livingstone, Edinburgh, Scotland.)

Fig. 346 Diplococcus pneumoniae. Smear prepared from the heart blood of an infected mouse. (From Muir, "Bacteriological Atlas," E. and S. Livingstone, Edinburgh, Scotland.)

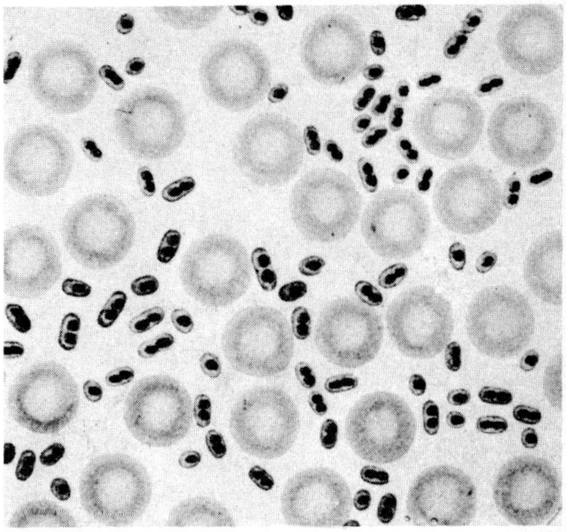

Typing of pneumococci A number of methods are employed for the determination of pneumococcal types:

1. In the precipitin test, a sample of sputum (about 1 ml) from a suspected case of pneumonia is injected into the peritoneal cavity of a mouse. The animal will appear sick in about 8 hr. The mouse is usually killed after this period of time, the peritoneal cavity opened, and the exudate washed into a petri dish with the aid of about 2 ml of saline solution. The washings are centrifugated, and the supernatant liquid is used for the precipitin test.
2. In the microscopic agglutination test, the peritoneal exudate is prepared as given under 1, and small drops of the washings are placed on a slide. The various type serums are added to the drops and spread out in thin films. The films are allowed to dry, then stained and examined under the microscope for the presence of clumps. The type of serum producing clumps indicates the type of organism present in the peritoneal exudate.
3. The Quellung test is based on the observation that the capsules of pneumococci become swollen when placed in contact with specific immune serum. A loopful of undiluted immune serum is placed on a slide and mixed with a fleck of sputum. A loopful of dilute methylene blue solution is then added, and the mixture is examined under the microscope. If the mixture is homologous, the capsules will appear swollen with very distinct outlines. The test is made with many types of antiserums and as many flecks of sputum.

Pneumococci, regardless of serological type, manifest three chief culture phases or stages: mucoid, smooth, and rough. The mucoid form represents the typical phase of the species. The most frequently observed dissociative trend is $M \rightarrow S \rightarrow R$. Serological types are recognizable only in the mucoid form, owing to the presence of type-specific polysaccharides in the capsular material. Both smooth and rough forms are devoid of capsular material, but possess species-specific antigens common to all members of the species. Smooth and rough forms are nonpathogenic, possess distinctive growth characteristics, and require special technique for accurate observations.

Disease produced *D. pneumoniae* is the commonest cause of lobar pneumonia. The disease is characterized by sudden onset with chill followed by fever, often pain in the chest, accompanied by cough and leucocytosis. Incidence may be as high as 95 percent. Organism is present in the alveoli and bronchioles of the lung, in the lymph channels, and sometimes in the blood. Organism may also produce pericarditis, arthritis, meningitis, otitis media, mastoiditis, endocarditis, rhinitis, tonsillitis, conjunctivitis, septicemia, osteomyelitis, and peritonitis.

Diagnosis Organisms may be detected by appropriate laboratory examinations of sputum discharges of the respiratory tract.

Source of infection Sources of infection are sputum, blood, and exudates in pneumonia; cerebrospinal fluid in meningitis and saliva from respiratory tract of normal individuals can also be sources.

Mode of transmission The disease is transmitted by direct contact with infected person or carrier and by inhalation of droplets expelled from the throat during coughing and sneezing. Also indirectly through articles freshly soiled with discharges of nose and throat of patient or carrier.

Incubation period This period is generally believed to be 1 to 3 days.

Susceptibility and immunity Resistance is generally high but may be lowered by exposure to wet and cold, by physical and mental fatigue, and by alcoholism. Acquired immunity may follow an attack of pneumonia. Immunity to the homologous type usually follows an attack and may last for months or years.

Prevalence The disease affects a large proportion of the population, with no race, color, or sex exempt; it occurs in all climates, but most often in winter and spring and in colder regions. Epidemics occur under conditions of crowding.

Prevention and control Crowding in living and sleeping quarters should be avoided. General resistance should be maintained by adequate nourishing food, sufficient sleep, fresh air, and personal hygiene.

Suspected cases of pneumonia should be isolated. Discharges from nose and throat, and articles soiled by such discharges should be disinfected.

Treatment consists of intramuscular or oral administration of the appropriate form of penicillin. Tetracycline antibiotics are equally effective. Sulfonamides and erythromycin are usually effective.

Mortality rate has been reduced to an insignificant figure since the employment of antibiotics.

For more information: Austrian (1960); Roy et al. (1970).

Escherichia

The escherichiae are short rods, motile or nonmotile. Glucose and lactose are fermented with production of acid and gas. Acetylmethylcarbinol is not produced. Methyl red test is positive. Carbon dioxide and hydrogen are produced in approximately equal volumes from glucose. Generally they are not able to utilize uric acid as sole source of nitrogen. Cells are found in feces; occasionally they are pathogenic to man, producing enteritis, peritonitis, cystitis, etc. They are gram-negative, and widely distributed in nature.

Disease produced *E. coli* is a normal inhabitant of the intestinal tract of man and other vertebrate animals. It is generally nonpathogenic; in certain instances some strains have been found to overcome the defense mechanisms of the body to produce septicemia, peritonitis, inflammation of the liver and gall bladder, cystitis, meningitis, pneumonia, and other infections. The more virulent strains resist phagocytosis by macrophages in vivo and by polymorphonuclear leucocytes in vitro. Since the organism is found in the intestinal contents, its presence in water and foods generally means contamination with fecal material.

For more information: Lovell and Rees (1960); Medearis and Kenny (1968); Tillotson and Lerner (1967).

Gaffkya

The members of this genus occur in the animal body and in special media as tetrads, whereas in ordinary culture media they occur in pairs and irregular masses. They are aerobic to anaerobic. The most important species is *G. tetragena.*

G. tetragena Cells are spherical, measuring 0.6 to 0.8 μ, with pseudocapsule (in body fluids) surrounding four of the cells, forming typical tetrads. The organism is gram-positive.

Disease produced *G. tetragena* is generally believed to be a normal inhabitant of sputum or saliva. It is found in tuberculous sputum, in the blood in cases of septicemia, in the pus of abscesses, and in the spinal fluid in meningitis. It appears to be a secondary invader of low virulence, invading the tissues only when weakened by some other infectious organism. It is pathogenic for Japanese mice (Fig. 347).

For more information: Mageau and Roberson (1969).

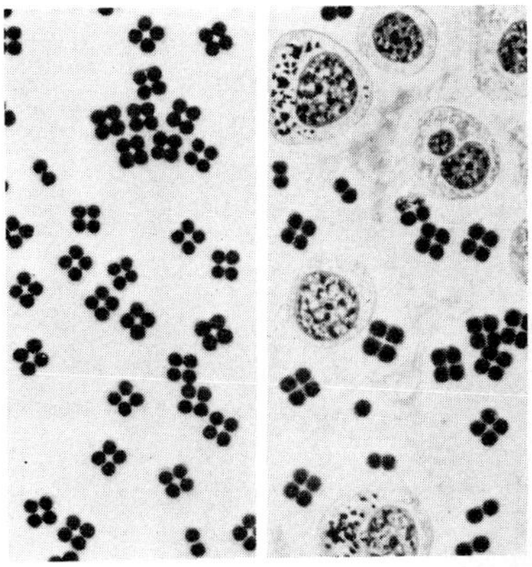

Fig. 347 Gaffkya tetragena. Left, smear from a 24-hr agar slant culture; right, smear of sputum from a case of chronic tuberculosis. (From Muir, "Bacteriological Atlas," E. and S. Livingstone, Edinburgh, Scotland.)

Haemophilus

This genus contains several important disease-producing species. Cells are minute, rod-shaped, sometimes thread-forming, and pleomorphic; they are nonmotile and gram-negative. They are strict parasites, growing only in presence of certain growth accessory substances. Organisms are found in various lesions and secretions, as well as in normal respiratory tracts of vertebrates.

H. influenzae Cells are rod-shaped, measuring 0.2 to 0.3 by 0.5 to 2 μ, occurring singly and in pairs, occasionally in short chains, and even in long threads. Frequently a marked tendency to bipolar staining is shown. The organisms are nonmotile and gram-negative.

H. influenzae is an obligate parasite growing only in the presence of hemoglobin and other body fluids. It will not grow in the absence of the X and V factors present in blood (page 231). On the basis of precipitin reaction, six types (A to F) are recognized. Majority of strains from respiratory tract not type specific.

Disease produced H. influenzae at one time was believed to be the cause of influenza, but the disease is now known to be produced by a filterable virus (page 1018). Organism commonly present in normal nose and throat and has been found to be a secondary invader in some bacterial and viral infections, including scarlet fever, measles, chickenpox, and whooping cough. It is also believed to be responsible for cases of endocarditis, sinusitis, meningitis, bronchopneumonia, and acute infectious conjunctivitis or "pinkeye" (Figs. 348 and 349). Type B organisms were found to be the cause of severe throat infections in children.

Source of infection Sources of infection are discharges from nose and throat or articles soiled by such discharges.

Mode of transmission The disease is transmitted by contact with

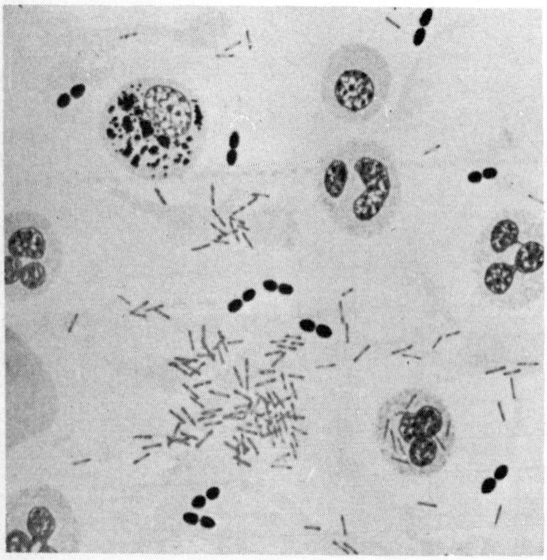

Fig. 348 Haemophilus influenzae. Smear of sputum from a case of influenzal pneumonia. The oval-shaped cells are Diplococcus pneumoniae. (From Muir, "Bacteriological Atlas," E. and S. Livingstone, Edinburgh, Scotland.)

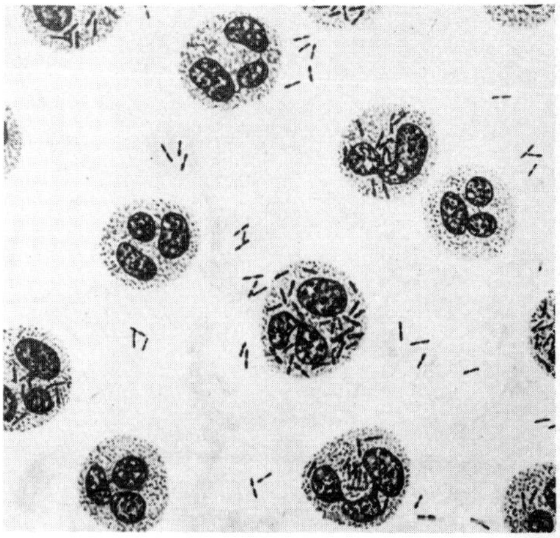

Fig. 349 Haemophilus influenzae. Smear of pus from a case of acute conjunctivitis. (From Muir, "Bacteriological Atlas," E. and S. Livingstone, Edinburgh, Scotland.)

an infected person, use of towels or other freshly contaminated articles, and inhalation of droplets expelled during coughing, sneezing, and talking.

Incubation period The period is usually 24 to 72 hr.

Susceptibility and immunity Children under five years are most often affected; incidence decreases with age. Immunity after attack is low-grade and variable.

Prevalence The organism may be isolated from pharynges of almost all normal individuals, and virulent encapsulated strains are found in chronic infections of the nasal sinuses and in the pharynges following viral colds.

Prevention and control The disease may be prevented and controlled by isolation of infected individuals, avoidance of overcrowding, and disinfection of articles soiled by discharges from nose and throat.

The antibiotics Aureomycin, Terramycin, and streptomycin have been found useful in controlling infections.

For more information: White (1963).

H. ducreyi Cells are small rods, measuring 0.5 by 1.5 to 2 μ, with rounded ends, occurring singly and in short chains. The organism is nonmotile and gram-negative; it requires the X factor for growth.

Disease produced *H. ducreyi* is the cause of soft chancre or chancroid, an acute inflammatory lesion that occurs upon the genitals or, less frequently, the skin surrounding the genitals. Lesion starts as a small pustule which eventually ruptures to form an open ulcer. Infection easily spreads to other areas. Genital lesions frequently accompanied by painful inflammatory swelling and suppuration of regional lymph nodes.

Source of infection Discharges from ulcerated lesions are the source of infection (Fig. 350).

Mode of transmission The disease is transmitted chiefly by sexual intercourse or by articles soiled with discharges from ulcerated lesions. Organism quickly loses its viability outside of body and soon dies.

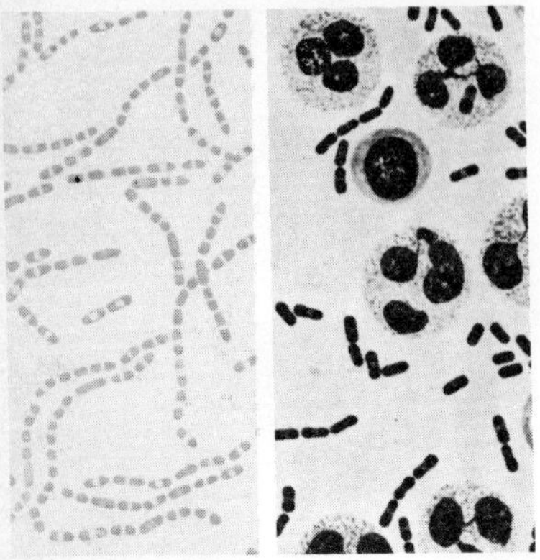

Fig. 350 Haemophilus ducreyi, the cause of soft chancre or chancroid. Left, smear from blood culture; right, smear from a soft chancre. (From Muir, "Bacteriological Atlas," E. and S. Livingstone, Edinburgh, Scotland.)

Incubation period The incubation period is from 3 to 5 days, occasionally longer.

Susceptibility and immunity Susceptibility to disease is general; there is probably no natural immunity. One attack does not confer protection against subsequent infection.

Prevalence The disease is widespread in distribution, and is particularly common where sexual promiscuity occurs.

Prevention and control The disease is spread largely by sexual contact; it can be prevented by prophylactic measures before, during, and following exposure to disease.

Avoid sexual contact until lesions are healed. Sulfonamides are specific for treatment. Streptomycin, tetracyclines, or chloramphenicol may be used if organisms have developed resistance to sulfonamides. It is not a serious disease and yields readily to treatment.

For more information: Deacon et al. (1956).

H. aegyptius Cells are small rods, measuring 0.25 to 0.5 by 1 to 2.5 μ, occurring singly, occasionally in short chains, and at times in the form of threads. The organisms show bipolar staining and are nonmotile and gram-negative. Both V and X factors are required for growth.

Disease produced *H. aegyptius* is the cause of acute bacterial conjunctivitis, commonly known as sore eye or pinkeye. Infection produces lacrimation, irritation, and vascular injection of conjunctivae of one or both eyes, followed by edema of the lids, photophobia, pain, and mucopurulent discharge. It is a nonfatal disease lasting 2 to 3 weeks. In most cases, the disease runs a mild course.

Diagnosis Bacteriological culture or microscopic examination of smear of exudate. Differentiation should be made from viral conjunctivitis (trachoma) and idiopathic conjunctivitis (Reiter's syndrome).

Source of infection Discharges from conjunctiva or upper respiratory tract of infected persons are sources of infection.

Modes of transmission The disease is transmitted by contact with infected persons through contaminated fingers, or articles soiled with discharges.

Incubation period The period is generally 24 to 72 hr.

Susceptibility and immunity Children under five most often affected; incidence decreases with age. Immunity is of low grade and variable.

Prevalence Disease is widespread throughout the world, particularly in warmer climates. It is largely confined to southern states and California in summer and early autumn.

Prevention and control The disease can be prevented and controlled by personal cleanliness and treatment of affected eyes. Discharges and articles soiled by such discharges should be disinfected. Children should not attend school during acute stage.

Specific treatment consists of local application of tetracycline antibiotics, or a sulfonamide such as sodium sulfacetamide.

Klebsiella

Genus includes short, somewhat plump rods, with rounded ends, usually occurring singly. The cells are encapsulated in the mucoid phase, and are nonmotile and gram-negative. They are aerobic, growing well on ordinary culture media; they are encountered frequently in respiratory, intestinal, and urogenital tracts of man. They may also be isolated from a variety of animals.

K. pneumoniae Cells are rods, measuring 0.3 to 0.5 by 5 μ, often four to five times as long as broad, with rounded ends, occurring singly and in pairs. They are encapsulated, nonmotile, and gram-negative (Fig. 351). Organism was first isolated by Friedländer and is usually referred to as Friedländer's bacillus. It is found frequently associated with upper respiratory infections in man. In most instances, organism is present as secondary invader. Less than 1 percent of pneumonias are caused by this organism.

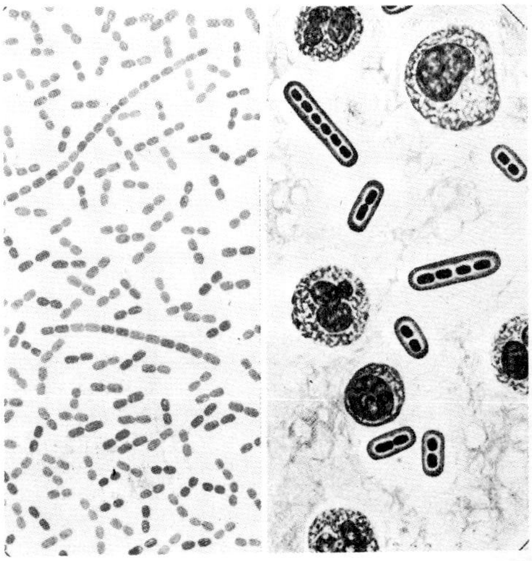

Fig. 351 Klebsiella pneumoniae (Friedländer's bacillus). Left, smear from a culture; right , smear of sputum. (From Muir, "Bacteriological Atlas," E. and S. Livingstone, Edinburgh, Scotland.)

Disease produced The organism is associated with infections of the respiratory, intestinal, and genitourinary tracts of man. It can be isolated from the lungs in lobar pneumonia, and is associated with pneumonia and other inflammations of the respiratory tract. *K. pneumoniae* may also produce otitis media, empyema, pericarditis, meningitis, and septicemia.

Source of infection Buccal and nasal discharges of infected persons or carriers and articles contaminated with such discharges are the sources of infection.

Mode of transmission The disease is transmitted by droplet spread, by direct contact with infected person or carrier, and by articles soiled with discharges from nose and throat of such person.

Susceptibility and immunity Organism is carried in nasopharynx of 1 percent of normal individuals. Susceptibility is of low grade, highest in infants and young children, and in the aged. Immunity is relatively slight and of short duration.

Prevention and control The disease may be prevented and controlled by isolation of infected persons and concurrent disinfection of discharges from mouth, nose, and contaminated articles. Treatment consists of the use of streptomycin in combination with tetracyclines or chloramphenicol.

For more information: Ørskov (1957).

Leptospira

The leptospirae are finely coiled organisms 6 to 20 μ in length. Spirals are 0.3 μ in depth and 0.4 to 0.5 μ in amplitude. In liquid medium one or both ends are bent into a semicircular hook, each involving one-tenth to one-eighth of the organism. Spinning movements occur in liquid and vermiform in semisolid agar, forward or backward. The organisms stain with difficulty except with Giemsa stain and silver impregnation. Cells require oxygen for growth. They are chiefly saprophytic organisms, being found in water, and sometimes in normal mouth. Best-known member producing disease is *L. icterohaemorrhagiae*, the cause of infectious jaundice or Weil's disease (Figs. 352 and 353). Other species or serotypes include *L. autumnalis, L. canicola,* and *L. pomona.* At present 18 serogroups and more than 100 serotypes are recognized from various parts of the world.

L. icterohaemorrhagiae, L. autumnalis, L. canicola, L. pomona These organisms produce a group of acute infectious diseases with fever, headache, chills, severe malaise, vomiting, muscular pains, conjunctivitis, infrequently jaundice, renal insufficiency, hemolytic anemia, and hemorrhage in skin and mucous membranes. Clinical illness lasts 1 to 3 weeks; relapses may occur. Fatality is low, increasing with advancing age. Disease is known under various names: Weil's disease, canicola fever, hemorrhagic jaundice, and Fort Bragg fever.

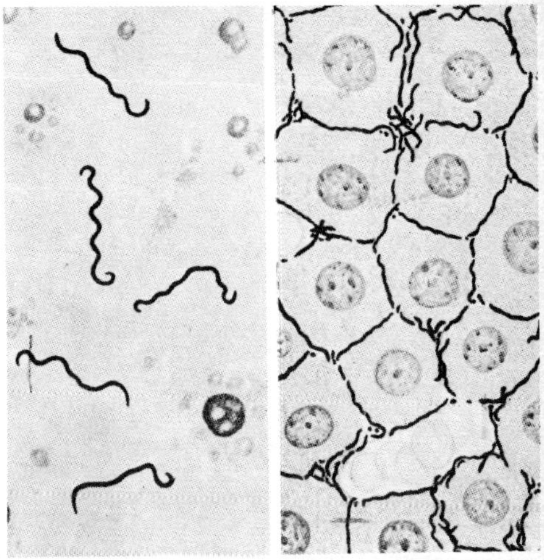

Fig. 352 Leptospira icterohaemorrhagiae, the cause of Weil's disease or infectious jaundice in man. Left, smear of urine; right, section of liver from an infected rat. (From Muir, "Bacteriological Atlas," E. and S. Livingstone, Edinburgh, Scotland.)

Differentiation of serotypes is based on antigenic structure which can be determined by agglutination.

Diagnosis Disease is diagnosed by inoculating guinea pigs or hamsters with blood taken early in course of disease.

Source of infection Urine and feces of rats, dogs, cats, mice, deer, foxes, skunks, raccoons, and opossums are sources of infection. Water and soil become contaminated with discharges of infected animals.

Mode of transmission Infection in man probably occurs through rubbing contaminated soil into the skin, eyes, and nose, or from swallowing contaminated water. Disease shows selection for such trades as fish dealers, abattoir workers, sewer workers, miners, veterinarians, and agriculturists. Occasionally infections result from handling dogs, cattle, swine, and other animals. Infection results from penetration of abraded skin or mucous membrane, and possibly by ingestion.

Incubation period The period is from 4 to 19 days, usually 9 to 10 days.

Susceptibility and immunity Susceptibility is general. A refractory state develops following recovery. Immune bodies may be demonstrated for a considerable period after recovery. Urine may show the presence of the organism for months after convalescence.

Prevalence Disease is of worldwide distribution in rats. Sporadic cases in humans are reported throughout the United States.

Prevention and control Swimming in potentially contaminated waters should be avoided. Rodent control in human habitations should be practiced.

Urine and feces of patients should be disinfected.

Penicillin, streptomycin, and the tetracyclines are leptospirocidal in vitro but of undemonstrated value in vivo.

For more information: Burgdorfer (1956); Coghlan et al. (1967); Diesch et al. (1967); Doherty (1967*a*, *b*); Ellinghausen and McCullough

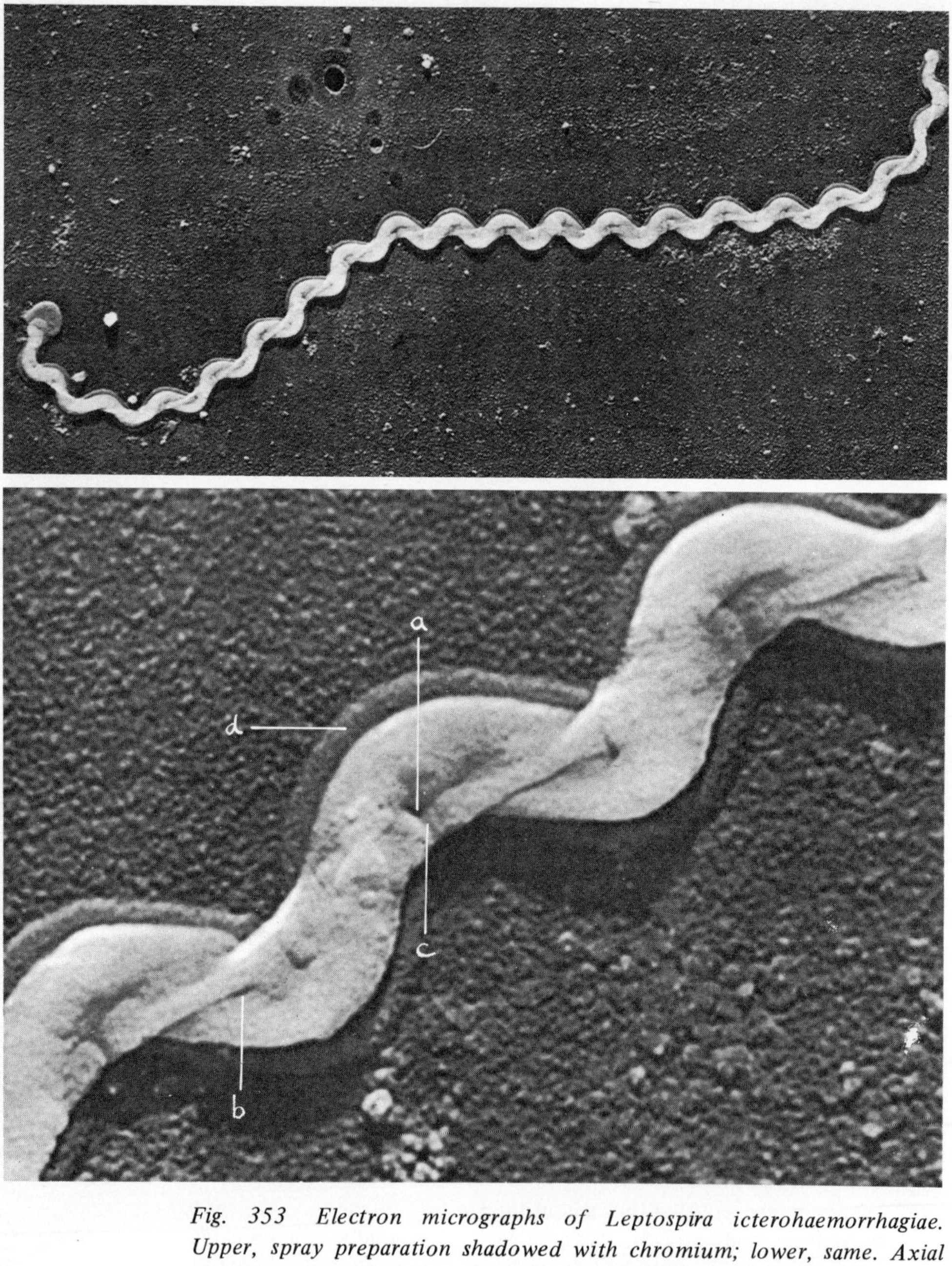

Fig. 353 Electron micrographs of Leptospira icterohaemorrhagiae. Upper, spray preparation shadowed with chromium; lower, same. Axial filament is separated from the protoplasmic spiral by a narrow space at (a) and is more closely approximated at (b). Distortion of protoplasmic spiral is seen at (c). Sheath is seen at (d). (Courtesy of Simpson and White.)

(1965); Ritchie and Ellinghausen (1965); Simpson and White (1964); World Health Organization (1967*b*); Yanagihara and Mifuchi (1968).

Listeria

Genus includes one species of small rods, motile by means of peritrichous flagella. The cells are gram-positive and grow freely on ordinary media. They are parasitic in warm-blooded animals and pathogenic for man.

L. monocytogenes Cells are small rods, measuring 0.4 to 0.5 by 0.5 to 2 μ, with rounded ends, slightly curved in some culture media, occurring singly and in V-shaped or parallel pairs. They are motile and gram-positive.

Disease produced Listeriosis is an acute meningitis with or without septicemia. Onset of meningitis is usually sudden, with fever, headache, nausea, vomiting, and signs of meningeal irritation. May be followed by delirium, coma, sometimes collapse, and shock. Endocarditis, localized internal or external abscesses, and pustular or papular cutaneous lesions may occur. Septicemic infection is an acute, mild, febrile illness, sometimes with influenza-like symptoms, which involves the fetuses of pregnant women. Infant may be stillborn, born with massive septicemia, or develop meningitis in the neonatal period. Case fatality is about 50 percent in newborn, approaching 100 percent when onset occurs in first 4 days.

Diagnosis The disease is diagnosed by isolation of organisms from spinal fluid, blood, or lesions.

Source of infection Infected domestic and wild mammals, fowl, and man are sources of infection. Organism frequently found to be free-living in water, mud, and ensilage.

Mode of transmission This is not known except for infections transferred from mother to fetus, and those from direct contact with infected material.

Incubation period The period is probably 4 days to 3 weeks. Fetus is usually infected within several days after disease appears in mother.

Susceptibility and immunity Unborn and newborn are highly susceptible. Children and young adults are generally resistant, but adults over 40 are less resistant. There is little evidence of acquired immunity following recovery.

Prevalence The disease is sporadic, rarely epidemic; it occurs in all seasons, and is more prevalent in males than in females. Abortion may occur as early as second month of pregnancy, but usually in fifth or sixth month. Incidence in United States is low.

Prevention and control Precautions by veterinarians in handling aborted fetuses can prevent and control the disease.

Infected persons should be isolated until bacteria are no longer present in body discharges. Discharges and materials soiled by such discharges should be disinfected.

Tetracyclines, penicillin, and chloramphenicol are effective for treatment.

For more information: Armstrong and Sword (1964); Coppel and Youmans (1969*a*); Gray and Killinger (1966); Holder and Sword (1969); McCallum and Sword (1970); Sword (1966).

Moraxella

Cells are small, short rods, which occur as diplobacilli and are sometimes described as diplococci; they occasionally occur singly. They are non-motile and gram-negative. They do not require V or X factor for growth. Organisms are aerobic. They are found as parasites and pathogens in warm-blooded animals, especially in association with diseases of the eye.

M. lacunata Cells are short rods, measuring 0.4 to 0.5 by 2 μ, occurring singly, in pairs, and short chains. Ends are rounded or square in the chains. The organism is nonmotile and gram-negative.

Disease produced *M. lacunata* is the cause of subacute infectious conjunctivitis, or angular conjunctivitis which usually attacks both eyes (Fig. 354). There is rarely much swelling or ulceration of the conjunctiva. The condition runs a subacute or chronic course.

Diagnosis The diagnosis is made by smear preparations of the pus, which is especially abundant during the night.

Fig. 354 Moraxella lacunata. Smear of lacrimal secretion from a case of chronic conjunctivitis. (From Muir, "Bacteriological Atlas," E. and S. Livingstone, Edinburgh, Scotland.)

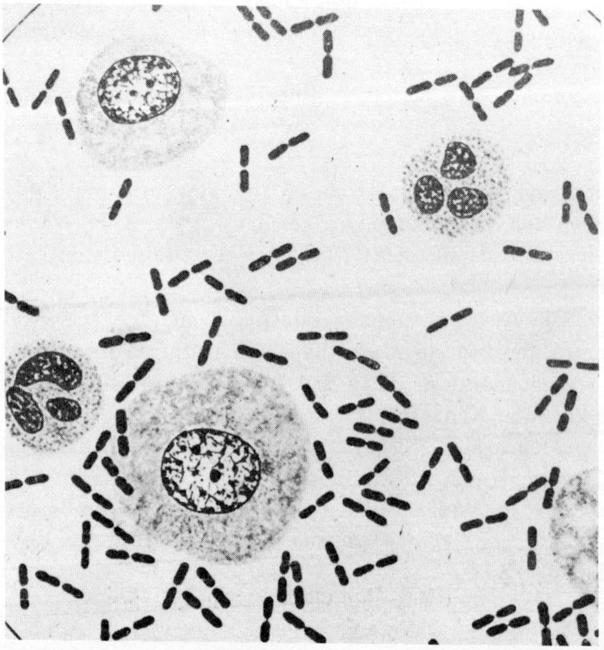

Source of infection Discharges from conjunctivas are the source of infections.

Mode of transmission The disease is transmitted by contact with infected person or with articles freshly soiled with discharges from such person.

Incubation period The period is usually 24 to 72 hr.

Susceptibility and immunity Children under five years of age are most often affected. Immunity is low-grade and variable.

Prevalence The disease is widespread and common throughout world, particularly in warm areas; it is sometimes epidemic.

Prevention and control Conjunctival discharges and articles soiled by such discharges should be disinfected.

Tetracycline antibiotics or sulfonamides are useful in treatment of infection.

Mycobacterium

Mycobacteria differ from most other species in containing a higher percentage of fatty substances. The fatty material is stained with difficulty, but once stained it resists decolorization with acid. Because of this fact, they are called acid-fast bacteria. Pathogenic species produce diseases characterized by the presence of nodules or tubercles in various parts of body.

Cells are slender rods, straight or slightly curved, occasionally slender filaments but branching forms rarely occur. They are nonmotile, aerobic, and non-spore-forming. Two obligate parasites have not been cultivated on culture media.

M. tuberculosis Cells are rod-shaped, measuring 0.3 to 0.6 by 0.5 to 4 μ, straight or slightly curved, occurring singly and in occasional threads. They are sometimes swollen, clavate, or even branched. The organisms may stain uniformly or irregularly, showing banded or beaded forms. They are acid-fast and gram-positive. Growth in all media is slow, requiring several days or weeks for development. The cell contains mycolic acid to which it owes its acid-fatness.

Disease produced *M. tuberculosis* is the cause of tuberculosis in man, monkey, dog, and parrot. It is experimentally very pathogenic for guinea pigs but not for rabbits, cats, goats, oxen, or domestic fowl (Figs. 355 and 356).

Disease is characterized by insidious onset with parenchymal pulmonary infiltration, recognizable by x-ray examination before constitutional symptoms or physical signs appear. Pleurisy is almost always first symptom. Advanced disease is accompanied by cough, fever, fatigue, and loss in weight.

Pulmonary tuberculosis generally arises as reactivation of a lateral focus and runs a chronic variable, and often asymptomatic, course with exacerbations and remissions. It is capable of arrest or relapse at any stage.

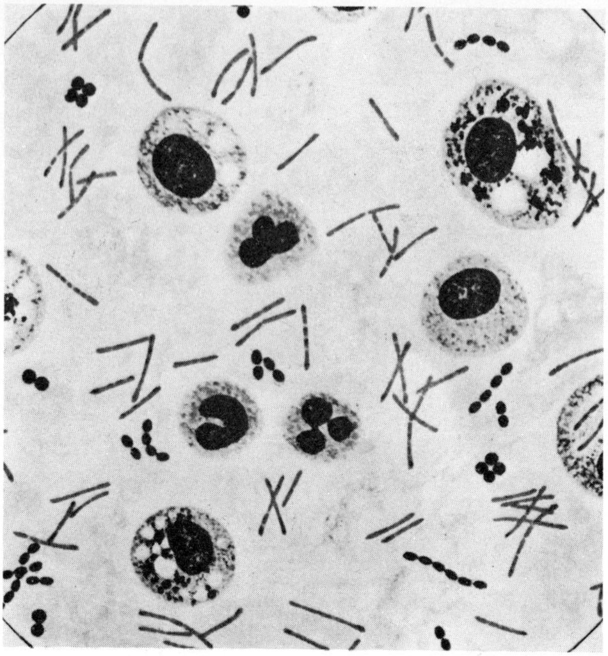

Fig. 355 Mycobacterium tuberculosis. Smear of sputum from a case of pulmonary tuberculosis. (From Muir, "Bacteriological Atlas," E. and S. Livingstone, Edinburgh, Scotland.)

Diagnosis Suspected material such as sputum, urine, feces, cerebrospinal fluid, or stomach contents is examined for the presence of tubercle bacilli.

Sputum may be smeared on a slide and stained by the Ziehl-Neelsen acid-fast technique. The presence of typical organisms is usually indicative of infection with the tubercle bacillus.

It is usually advisable to concentrate the tubercle bacilli before making laboratory tests. This may be performed as follows: The infected material is treated with 3 percent sodium hydroxide to digest the sputum, pus, or other material with which the organisms are mixed. The digested material is then neutralized with acid and centrifugated. Only the sediment is retained. The sediment may be used for (1) preparation of smears for staining and direct microscopic examination, (2) injection into animals, or (3) inoculation of culture media.

Animal inoculation A guinea pig is inoculated in the groin or muscle of the thigh with some of the sediment. Enlargement of the regional lymphatic glands occurs in 2 or 3 weeks, and the animal usually dies in about 6 to 8 weeks. On autopsy, the animal shows necrotic areas in the liver and spleen and enlarged lymph nodes filled with caseous material. The lungs and kidneys are rarely attacked.

Inoculation of culture media Some of the sediment may be used for the inoculation of appropriate culture media. A variety of media may

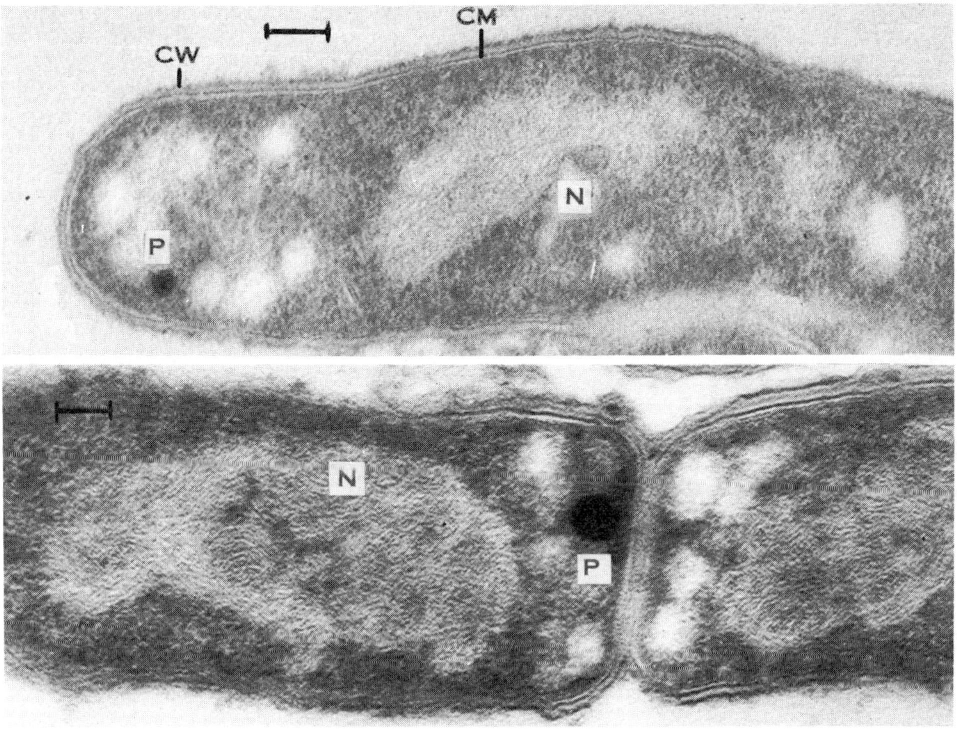

Fig. 356 Electron micrographs of ultrathin sections of Mycobacterium tuberculosis. Upper, cell wall (CW), cytoplasmic membrane (CM), nuclear apparatus (N), and polyphosphate granule (P) may be seen. The CW and CM are composed of two dense inner and outer layers and a less dense interspace. Lower, two cells still joined together. Fine fibrils in parallel arrangement fill the less dense nuclear region. (Courtesy of Koike and Takeya.)

be used, consisting largely of egg, glycerin, and some dye to kill or inhibit the growth of contaminating organisms. The media are tubed, slanted, and then sterilized by heat. During the heating process, the egg albumin is coagulated, producing solid media.

Use of tuberculins A large number of tuberculins are available. They are prepared in different ways but consist of filtrates of liquid cultures of the tubercle bacillus. They contain certain products liberated after the death and disintegration of the organisms.

The first tuberculin preparation, known as "Koch's old tuberculin," is prepared as follows: The organisms are cultivated in a slightly alkaline 5 percent glycerin peptone broth for 6 to 8 weeks. The culture is concentrated in a water bath, heated at 80°C until reduced to one-tenth of its original volume. The culture is then filtered to remove bacterial debris. The clear filtrate contains the tuberculin.

Tuberculins are used to test the sensitiveness of persons or animals to proteins of the tubercle bacillus. By injecting tuberculin intradermally, a positive test appears in 6 to 8 hr, reaches a maximum in 24 to 48 hr, and generally subsides in 6 to 10 days. It is characterized by a reddening of the skin about 1 cm in diameter. The reaction is positive in those having active or healed lesions. Since most individuals have healed tuberculous lesions, the test is of limited value.

Source of infection The source of infection is primarily from man, although it sometimes is from cattle. Other sources are respiratory secretions of persons with pulmonary tuberculosis, and occasionally milk of tuberculous cows. The human type causes almost all pulmonary cases; the bovine type causes most extrapulmonary cases.

Prevalence Tuberculosis is one of the most common communicable diseases of man; it is endemic in practically all populations and races. Incidence and mortality rates are declining. Mortality is highest among infants and adults beyond middle age.

Bovine type in man is rare in the United States, Canada, Scandinavia, and the Netherlands; it is rapidly disappearing in France, Great Britain, Germany, and Switzerland.

Mode of transmission The disease is transmitted through discharges of the respiratory tract, and less frequently through discharges of the digestive tract; by inhaling droplets expelled during coughing, sneezing, talking, or singing; by kissing; by use of contaminated eating and drinking utensils; by contaminated dust, flies, etc. Infection usually results from continued contact with an infected person.

Bovine type results from consumption of raw milk or dairy products from infected cows, from airborne contamination in barns, and from handling contaminated animal products.

Incubation period The incubation period is variable, depending upon type of disease, age, etc. The time from infection to demonstrable primary lesion is about 4 to 6 weeks. It may take years for the disease to become progressive pulmonary tuberculosis.

Susceptibility and immunity Susceptibility to infection is general; it is higher in children under three years of age, lowest from three to twelve years, and then relatively high for the remainder of life. Susceptibility is greater in undernourished, fatigued, and neglected persons than in well-fed and well-cared-for persons. Relapses of long latent infections account for large proportion of active cases.

Natural immunity is generally negative.

Prevention and control The public should be made aware of the danger of tuberculosis, its mode of spread, and methods of control. Overcrowding should be avoided. Working and living conditions should be improved. Isolation and treatment of infected individuals is necessary. Milk and milk products should be pasteurized. Babies and tuberculous mothers should be separated at birth. Tuberculous cattle must be slaughtered. Tuberculous patients should be prohibited from handling foods for public consumption.

Preventive treatment of those individuals in household contact with active cases consists of administration of isoniazid. Isoniazid also useful in reducing risk of relapse in arrested cases not previously treated with effective antibacterial drugs.

Sputum and articles soiled with it, such as handkerchiefs, towels, paper, and eating and drinking utensils should be disinfected, as should rooms previously occupied by tuberculous patients. Vaccination of uninfected individuals with BCG (Bacillus Calmette-Guérin) may be useful. The cure is based largely on rest, good wholesome food, fresh air, sunshine, and freedom from worry. Surgical, chemotherapeutic, and antibiotic therapy are all supplementary and cannot be substituted for the basic rest treatment.

Combination of streptomycin and para-aminosalicylic acid (PAS) is commonly employed in adults for 6 months to a year or more. Isoniazid either alone or in combination with PAS or streptomycin, or both, is also useful.

For more information: Coppel and Youmans (1969*b*); Dannenberg (1968); Dubos (1949); Frappier and Guy (1949, 1950); Koike and Takeya (1961); Lincoln (1967); Markovits and Vilkas (1969); Mitchell (1967); Panisset et al. (1960); Springett (1969); Wolinsky and Rynearson (1968); Xalabarder (1969); Youmans and Youmans (1969).

M. bovis Rods are shorter and plumper than those of human species, measuring 1 to 1.5 μ. Very short forms are frequently intermixed with somewhat larger forms. They stain regularly or irregularly, and are acid-fast and gram-positive.

The organism is less easily cultivated than the human species.

Disease produced *M. bovis* is the cause of tuberculosis in cattle. The disease is transmissible to man and domestic animals. It is pathogenic for monkey, goat, sheep, pig, cat, parrot, cockatoo, and other birds; it is experimentally highly pathogenic for rabbit and guinea pig, but more highly pathogenic for animals than human species (Fig. 357).

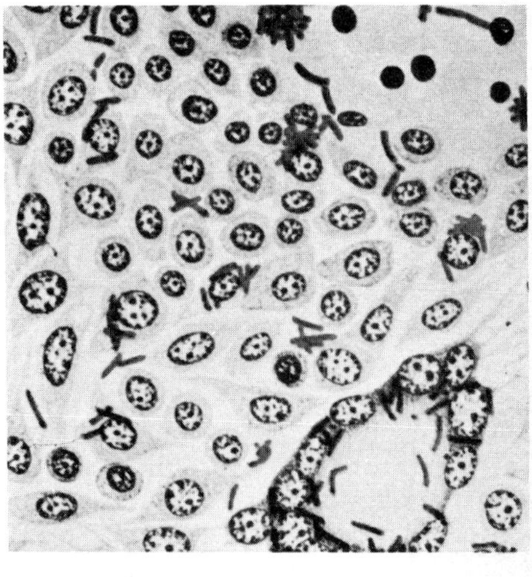

Fig. 357 Mycobacterium bovis. Section of udder of cow suffering from chronic tuberculosis. (From Muir, "Bacteriological Atlas," E. and S. Livingstone, Edinburgh, Scotland.)

Source of infection Tuberculous cows may eliminate organisms in feces, urine, and milk. Milk from infected udders may contain enormous numbers of the organisms.

Mode of transmission Infection is spread to healthy cows through milk, urine, and feces of diseased cows. Children, particularly those under five years of age, may become infected by drinking contaminated milk from diseased cows.

Susceptibility and immunity Susceptibility to infection is general; it is higher in children under three years of age, lowest from three to twelve years.

Natural immunity to disease is generally negative.

Prevention and control The public should be made aware of the danger of tuberculosis, its mode of spread, and methods of control. Isolation and treatment of infected individuals is necessary. Milk and milk products should be pasteurized. Tuberculous cattle should be slaughtered. Tuberculous patients should be prohibited from handling foods for public consumption.

Discharges and articles soiled with such discharges should be disinfected, as should rooms previously occupied by tuberculous patients. Vaccination of uninfected individuals with BCG may be useful. This is a bovine strain rendered avirulent by cultivating for many generations on a bile-glycerol-potato medium. This change appears to be permanent, the loss of virulence not being restored when the organisms are again transferred to the usual culture media. The cure is based largely on rest, good wholesome food, fresh air, sunshine, and freedom from worry. Surgical, chemotherapeutic, and antibiotic therapy are all supplementary and cannot be substituted for the basic rest treatment.

A combination of streptomycin and para-aminosalicylic acid is commonly employed in adults for 6 months to a year or more. Isoniazid either alone or in combination with PAS or streptomycin, or both, is also useful.

M. leprae Cells are rods, measuring 0.3 to 0.5 by 1 to 8 μ, with parallel sides and rounded ends, staining evenly or at times beaded. When numerous, as from lepromatous cases, they are generally arranged in clumps, rounded masses, or in groups of bacilli side by side. They are strongly acid-fast and gram-positive.

Disease produced *M. leprae* causes Hansen's disease, or leprosy in man. It is a chronic, mildly communicable disease that occurs in three forms: (1) lepromatous (muscle) type; (2) anesthetic (nerve) type; and (3) mixed type or combination of (1) and (2). In the lepromatous type there are diffuse skin lesions and invasion of mucous membranes of the upper respiratory tract as well as some viscera. Skin lesions may ulcerate. In the anesthetic type there is relatively early nerve involvement. Progress of the disease is slow; death is rarely due to leprosy but to other causes.

Diagnosis The bacilli occur in enormous numbers in nodular (lepromatous) cases of the disease and sparsely in the neural forms. The disease may be recognized by lesions of the skin and mucous membranes and by neurological manifestations. The present bacteriological means of identification depend on (1) acid-fast staining and (2) failure of the organism to grow on bacteriological media or in laboratory animals. Heated suspensions of the bacilli recovered from nodules produce a positive lepromin reaction in 75 to 97 percent of normal persons and of neural cases of leprosy, but usually produce no reaction in lepromatous individuals.

Source of infection Discharges from lesions are the source of infection.

Mode of transmission The disease is transmitted by intimate and prolonged contact with lepers. Nodules on skin may liquefy, ulcerate, and discharge great masses of the bacilli. Patients showing the presence of acid-fast organisms in smears, even though ulcers are not present, are potentially open cases. It is generally believed that the disease is the result of living under filthy, unsanitary conditions, but there is very little evidence to substantiate this statement. Children contract the disease more easily than adults. Babies of lepers rarely, if ever, become infected if separated from them at birth. Workers in leper colonies seldom contract the disease.

Incubation period The period is not known, although some claim that it is 1 to 7 years.

Susceptibility and immunity Susceptibility is uncertain; there is no racial immunity.

Prevalence The disease occurs mostly in tropics and subtropics. Highest reported rates are in equatorial Africa, West Africa, and Congo. The disease is also found in temperate climates like China and Korea. It is endemic in low form in Greece, Portugal, Spain, Hawaii, Canal Zone, Puerto Rico, Virgin Islands, Southeast Texas, Louisiana, Florida, and southern California.

Prevention and control Leprosy may be contracted in adult life but is usually acquired in childhood. Infants born of leprous patients should be separated from them at birth.

Positive cases should be isolated in leprosaria until disease has been arrested. Usually three negative bacteriological tests at intervals of 6 months are required before a patient is released. Paroled individuals should be reexamined every 6 months thereafter. Discharges should be disinfected. Living premises of patient should be thoroughly cleaned. Treatment is based chiefly on rest, good food, fresh air, sunshine, and freedom from worry. Surgical, chemotherapeutic, and antibiotic therapy are all supplementary and cannot be substituted for the basic rest treatment.

Specific treatment consists of administration of sulfones, diasone and diaminodiphenylsulfone (DDS) orally, and promin intravenously. When sulfones are not tolerated, thiourea derivatives and sulfonamides may be substituted. Penicillin of value in control of secondary infections.

For more information: Chang and Andersen (1969); DeSouza-Araujo (1958); Hanks (1966); Kirchheimer (1964); Leiker (1966); Pedley

(1967, 1970); Rees (1964); Sato and Imi (1968); Shepard (1963); Trautman (1965); Levy (1971).

Neisseria

Genus includes several well-known species. Cells are spherical, occurring in pairs with adjacent sides flattened. Growth on nonenriched media may be poor. The cells are gram-negative, and aerobic or facultatively anaerobic. They are parasites of animals so far as known.

N. gonorrhoeae Cells are spherical, measuring 0.6 to 1 μ, occurring singly and in pairs, the sides flattened where they are in contact, and are usually described as coffee-bean-shaped.

Disease produced *N. gonorrhoeae* is the cause of gonorrhea or gonococcal urethritis in man (Fig. 358). Organism attacks chiefly the columnar and transitional epithelium in both sexes, producing an acute catarrhal condition. In males, a thick, yellow, purulent discharge from the anterior urethra appears in 3 to 9 days after exposure to the organisms. There is a marked tendency for the infection to spread, producing in the male epididymitis, prostatitis, cystitis, and other inflammatory conditions. In females, a mild urethritis occurs a few days after exposure. Next the pelvis becomes invaded at the first, second, or third menstrual period. Finally a chronic stage follows. The organism may also invade the blood stream and be carried to various parts of the body.

The organism shows a predilection for the synovial membranes of the joints producing gonorrheal rheumatism, and for the heart valves, causing endocarditis. Organism may persist for many years and is probably never completely eliminated. Death is rare, but early and late manifestations are common and seriously incapacitating.

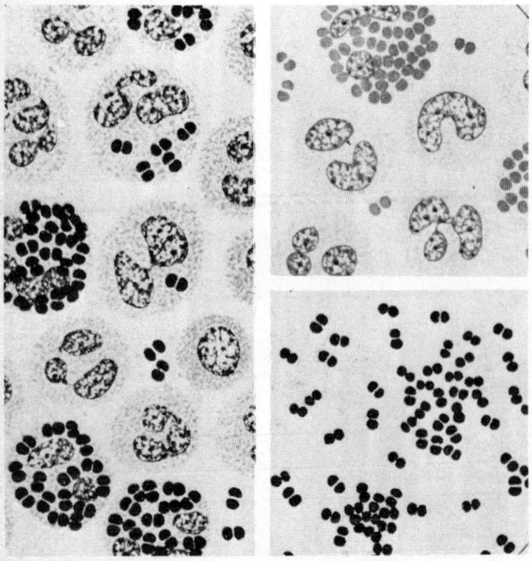

Fig. 358 Neisseria gonorrhoeae, the cause of gonorrhea. Left, a smear of pus showing the presence of the organism within the polymorphonuclear leucocytes; upper right, same but stained differently; lower right, smear from a pure culture. (From Muir, "Bacteriological Atlas," E. and S. Livingstone, Edinburgh, Scotland.)

Diagnosis Direct smears are prepared from urethral discharges and stained by Gram method. Smears are examined for presence of gram-negative intracellular diplococci slightly flattened at adjacent surfaces. Organisms may be cultivated by streaking some of urethral pus over the surface of a blood agar plate and incubating culture in an atmosphere containing 10 percent CO_2. This may be approximated by placing plate in a screw-cap jar with a lighted candle and replacing cover.

Oxidase test The gonococcus produces an oxidase which causes a 1 percent solution of dimethyl-paraphenylenediamine to turn first pink, then rose, magenta, and finally black. The test is made by flooding a 24-hr agar plate culture with the reagent and noting the development of the above series of colors.

Source of infection The sources of infection are purulent venereal discharges from both males and females.

Mode of transmission Infection in almost every case is transmitted by direct contact from person to person (sexual intercourse), rarely by direct contact with articles freshly soiled with discharges of such persons.

Incubation period Period is from 1 to 14 days, usually 3 to 5 days.

Susceptibility and immunity Susceptibility is general. An attack of gonorrhea produces very little, if any, immunity.

Prevalence Disease is common throughout the world, particularly among people of low economic status. It affects both sexes, and at all ages, especially period of greatest sexual activity. In recent years incidence of disease has increased worldwide.

Prevention and control Adequate diagnostic facilities for identification of the gonococcus are necessary. Public facilities for prompt and adequate treatment of infected persons should be available. Public should be educated on mode of transmission and how to avoid infection. Houses of prostitution should be controlled or eliminated. Personal hygiene should be practiced before or immediately after promiscuous sexual intercourse.

Sexual intercourse should be avoided during period of infection.

Discharges from lesions and articles soiled with such discharges should be disinfected.

Specific treatment consists of administration of procaine penicillin or other penicillins in one intramuscular injection. Females require much larger doses than males. Patients sensitive to penicillin may be given streptomycin, tetracyclines, chloramphenicol, or oleandomycin.

Ophthalmia neonatorum This is an acute infection of the newborn characterized by redness and swelling of the conjunctiva of one or both eyes, with mucopurulent or purulent discharge in which the gonococcus may be identified by microscopic and cultural examination. The disease is a consequence of maternal infection. If neglected this is probably the most common cause of corneal ulcer and blindness.

Approximately 10 percent of all cases of blindness are due to gonococcal infections. This is easily prevented by instillation of one or two drops of 2 percent silver nitrate in each eye of the newborn.

Incubation period Usually 36 to 48 hr.

Susceptibility and immunity Susceptibility is general. No known immunity following an attack.

Prevention and control Care in disposal of conjunctival discharges and articles soiled by such discharges.

Prompt treatment consisting of parenteral penicillin; a 1 percent tetracycline solution, or an oily suspension may be applied locally as an adjunct and as a safeguard against penicillin sensitization.

For more information: Brookes and Hedén (1967); Kellogg et al. (1963).

N. meningitidis Cells are spherical, measuring 0.6 to 1 μ, occasionally larger, occurring singly, in pairs with adjacent sides flattened, or occasionally in tetrads. The organism is gram-negative.

Meningococci may be differentiated into four main groups on the basis of agglutination reactions with immune serum. Groups are designated A, B, C, and D. Most strains fall into one or another of these four groups. As is true with pneumococci, presence of capsular substances is believed to be responsible for the immunological specificities of the various meningococcal types.

Disease produced An acute bacterial disease is produced, characterized by sudden onset, fever, headache, nausea, often vomiting, with signs of meningeal irritation, and frequently a petechial rash, often followed by delirium and coma.

Organism attacks the base and cortex of brain and the surfaces of spinal cord. Organism is present in spinal fluid; if growth is heavy, fluid may be turbid. Organisms may appear both free and within certain leucocytes. Meningococci are frequently present in blood stream of patients. Organism has been isolated from persons with arthritis and pericarditis, and from nasopharynx of those with rhinopharyngitis (Fig. 359).

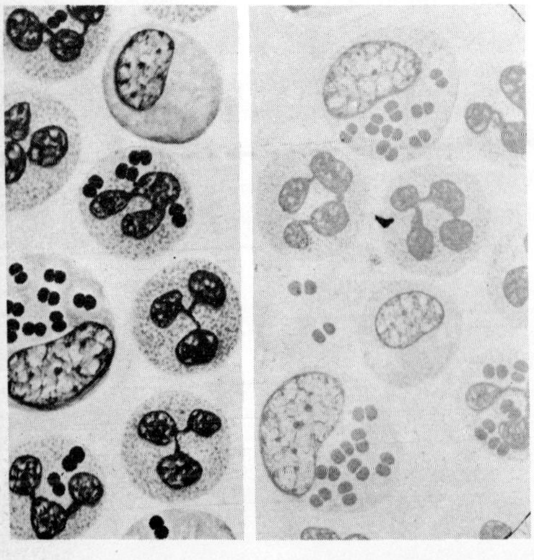

Fig. 359 Neisseria meningitidis, the cause of meningococcus meningitis. Left, smear of sediment from a centrifugalized specimen of spinal fluid; right, same but stained differently. (From Muir, "Bacteriological Atlas," E. and S. Livingstone, Edinburgh, Scotland.)

Diagnosis The meningococcus is present in large numbers in the spinal fluid of persons with the disease and may be diagnosed by centrifugating the spinal fluid and examining stained smears of the sediment. The typical picture is the presence of gram-negative intracellular diplococci. Typical smears of the meningococcus and gonococcus are indistinguishable. They may be distinguished by their cultural reactions.

N. meningitidis may be cultivated by streaking some sediment from spinal fluid over the surface of a blood agar or heated (chocolate) blood agar plate. The plate is then incubated in an atmosphere containing about 10 percent CO_2.

As in the case of gonococcus, this organism also produces an oxidase that is capable of oxidizing dimethyl-paraphenylenediamine to a black color. The test is made by flooding a 24-hr plate culture with a 1 percent solution of the reagent and noting the development of a black color in colonies of the meningococcus.

Source of infection The nasopharynx, blood, cerebrospinal fluid, conjunctiva, pus from joints, etc., of persons suffering from disease are sources of infection. Carriers rather than cases are usually the source; the carrier rate may be as high as 25 percent without epidemics and 50 percent or more during epidemics.

Mode of transmission The disease is disseminated by direct contact and by droplet infection during coughing and sneezing. The organism is frequently found in the nasopharynx of healthy persons. Such individuals are referred to as carriers because they are able to spread the disease to others. Organisms are very easily killed when outside the body and probably never reach a new individual except by direct contact. Epidemics generally develop during periods of overcrowding such as occur in army camps.

Incubation period The incubation period is from 2 to 10 days, commonly 3 to 4.

Susceptibility and immunity Susceptibility to disease is slight. Younger age groups are more susceptible, but disease may occur at any age. Agglutinins against the organisms are demonstrable in the blood stream after an infection but the duration of immunity is uncertain.

Prevalence The disease is endemic and epidemic, and is widely distributed throughout the world. It occurs most frequently during winter and spring. An epidemic wave may last 2 to 3 years. It is mainly a disease of children and young adults, males more than females, and occurs in crowded living conditions such as in army barracks.

Prevention and control Overcrowding should be avoided. Infected persons must be isolated. Discharges from nose and throat and articles soiled with such discharges should be disinfected. Specific treatment consists of administration of penicillin intravenously for 24 hr, then orally. If outbreak is caused by sulfonamide-sensitive strains, sulfadiazine may be given intravenously.

For more information: Branham (1956).

Nocardia

Organisms grow as slender filaments or rods, frequently swollen, occasionally branched, forming mycelium which, after attaining a certain size, assumes appearance of bacterium-like growths. Shorter rods and coccoid forms are present in older cultures. Conidia are not formed. Cells are nonmotile; they form no endospores; they are aerobic. Organisms are gram-positive.

In early stages organisms form typical mycelium, with hyphae showing true branching, and mycelium are nonseptate. Filaments soon form transverse walls and mycelium breaks up into cylindrical short cells, and then into coccoid forms. On culture media, coccoid forms germinate into mycelia. Cycle continues for 2 to 7 days. Chlamydospores are present in older cultures and coccoid forms are changed into resistant cells.

Multiplication is by fission and budding.

N. asteroides This organism causes nocardiosis, a chronic mycotic disease often initiated in the lungs, spreading to produce peritonitis, meningitis, brain abscess, and other pyogenic infections. It is highly fatal.

Diagnosis Microscopic examination of stained smears of sputum, pus, or spinal fluid reveals branching hyphae, partially acid-fast.

The micrographs shown in Fig. 360 by Beaman and Shankel (1969) are of a nonpathogenic species of *Nocardia* but they show the type of growth that is characteristic of the genus.

Source of infection Soil and dust are the sources of infection.

Mode of transmission The disease is transmitted by direct contact with contaminated soil through minor traumatic wounds and abrasions. Pulmonary infection by inhalation of contaminated dust is a second mode of transmission.

Incubation period The incubation period is unknown.

Susceptibility and immunity These factors are unknown.

Prevalence The disease has occasional sporadic occurrence in man and animals throughout the world. No information on differences in age, sex, or race is available.

Prevention and control Discharges and articles soiled by such discharges should be disinfected.

Specific treatment of systemic infections consists of administration of sulfonamides in high doses. The drugs must be given early and for prolonged periods.

Pasteurella

The pasteurellae are small, ellipsoidal to elongated rods which show bipolar staining by special methods. They are gram-negative. Cells are facultatively anaerobic. They may require low oxidation-reduction potential on primary isolation. Organisms are parasitic on man, other animals, and birds.

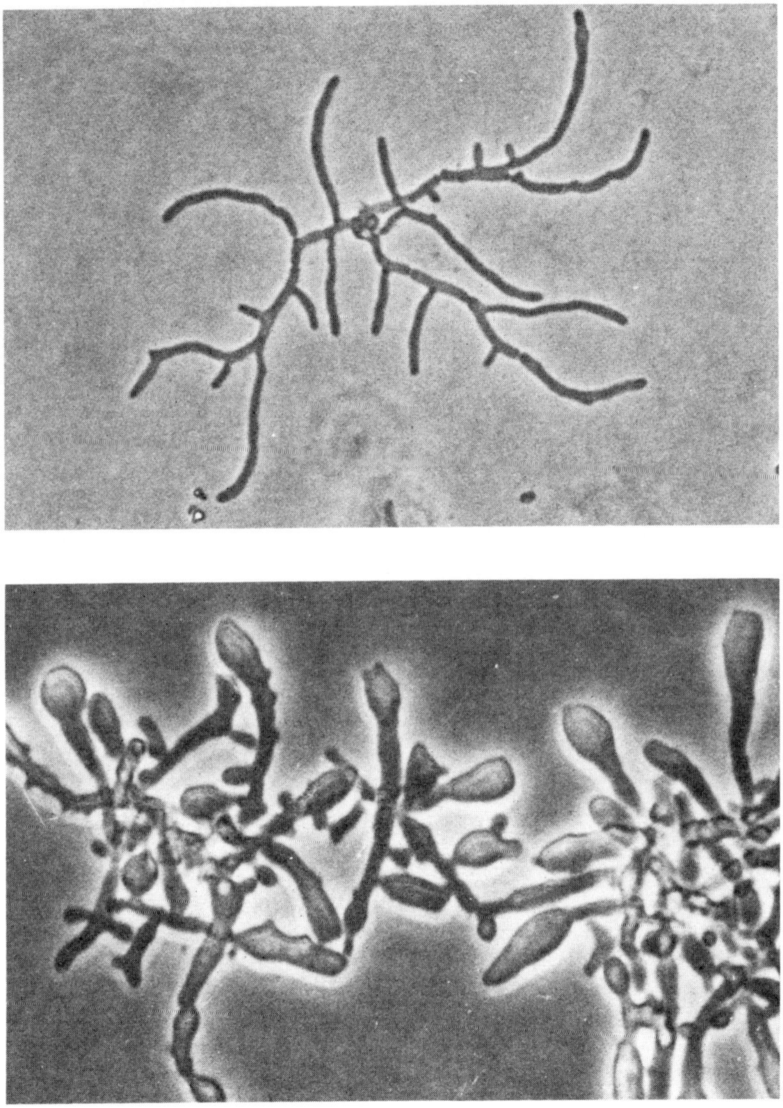

Fig. 360 Phase contrast micrographs of a nonpathogenic species of Nocardia. Upper, grown on tryptone agar for 12 hr at 30° C; lower, grown on brain heart infusion agar for 12 hr at 30° C. (Courtesy of Beaman and Shankel.)

P. pestis Cells are rods, measuring 1 by 2 μ, and occurring singly. They show characteristic bladder, safety pin, and ring involution forms. The organisms are polar staining, nonmotile, and gram-negative.

Disease produced *P. pestis* is the cause of plague in man, rats, ground squirrels, and other rodents. Infectious for mice, guinea pigs, and rabbits.

Human plague may be of three types: (1) bubonic, (2) pneumonic, and (3) septicemic.

Bubonic plague is a severe and highly fatal disease, characterized by high fever, weakness, buboes, sometimes pustules and subcutaneous hemorrhages. The bacteria are carried through the blood and lymph vessels to the lymph glands in the groin, armpits, neck, etc. The bacteria multiply, produce pus, and cause an enlargement of the glands. The glands may ulcerate and discharge their contents. The enlarged glands are referred to as buboes, and the infection is known as bubonic plague.

Pneumonic plague gives the picture of a virulent septic pneumonia. The lungs become engorged, and hemorrhages appear under the pleura. Bacteria are found in large numbers in the peribronchial lymph spaces and in the adjoining alveoli.

The disease may be mild, or it may take an acute septicemic form which generally produces rapid death.

Diagnosis Blood cultures prove positive in about 30 percent of cases. Direct smears can be prepared from open buboes, or such material can be inoculated into culture media. If the material is contaminated, it can be purified by inoculating a guinea pig and isolating the plague bacillus from the heart's blood. In the pneumonic type, the organisms are usually present in large numbers in sputum and may be recognized by direct smear.

Source of infection Organisms are found in buboes, blood, pleural effusion, spleen, and liver of infected persons and rodents. In addition, organisms are present in the sputum in cases of pneumonic plague (Fig. 361). The infection usually reaches man through contact with diseased rats.

Mode of transmission The disease is transmitted from rat to rat and from rat to man by infected rat fleas, the most important of which are *Xenopsylla cheopis* and *Ceratophyllus fasciatus*. The flea becomes infected by feeding on a diseased rat. The flea next feeds on a person and at the

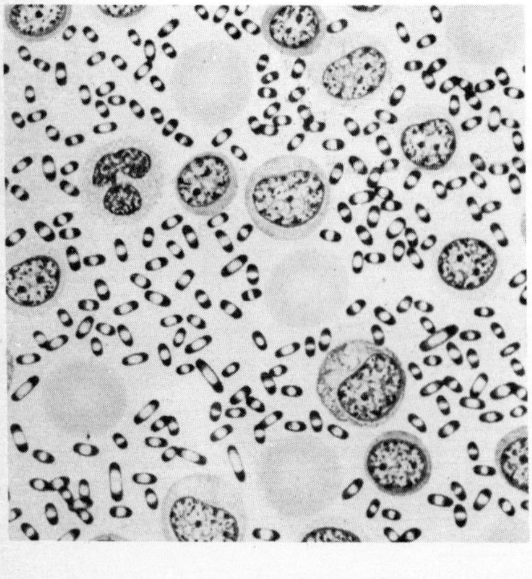

Fig 361 Pasteurella pestis, the cause of bubonic, pneumonic, and septicemic plague in man. Smear prepared from a bubo. (From Muir, "Bacteriological Atlas," E. and S. Livingstone, Edinburgh, Scotland.)

same time deposits feces and possibly some regurgitated blood. The bacilli are then rubbed into the skin by scratching.

Also pneumonic plague and pharyngeal plague are spread by the airborne route or by contact with exhaled droplets, or sputum from patients with pneumonic plague, or patients with bubonic plague who develop the pneumonic type.

Incubation period The period is usually 3 to 6 days.

Susceptibility and immunity Susceptibility to disease is general. Immunity after recovery is temporary and relative. Active immunization with a plague bacterin may confer protection for some months. Repeated stimulating injections are necessary.

Prevalence The disease rarely occurs in North America. Sporadic cases occur where persons are exposed to infected wild rodents. Disease is distributed in various parts of the world.

Prevention and control Rat extermination, ratproofing, and other necessary measures help to control the disease. Ratproofing of ships is also advisable.

Infected individuals should be isolated. Sputum and articles soiled with sputum in pneumonic type should be disinfected. Terminal disinfection of walls, floors, and furniture of rooms occupied by patients with plague serves to control the disease.

Streptomycin and the tetracyclines are highly effective for all forms of plague when used early. Recurrence of fever during streptomycin therapy may indicate pneumonia caused by gram-positive cocci. Under such conditions penicillin should be used with streptomycin. Sometimes sulfadiazine used if antibiotics not available.

For more information: Janssen and Surgalla (1969); Kadis et al. (1969); Smith and Packman (1966); Stanley and Smith (1967); Warren et al. (1955).

P. tularensis The name of this organism has now been changed to *Francisella tularensis*. There are equal numbers of cocci and rods, measuring 0.2 by 0.2 to 0.7 μ, occurring singly. The organism is extremely pleomorphic and may show bipolar staining. It is nonmotile, gram-negative, aerobic, and filterable through Berkefeld filters.

Disease produced *F. tularensis* is the cause of tularemia in man. Onset of disease is sudden with chills and fever present. Patient is usually prostrated and confined to bed. Ulcer usually appears at site of original infection. Fever may last 3 to 4 weeks, followed by slow convalescence. Lymph glands may become swollen and tender, and suppurate in 50 percent of infections. Organism is infectious for rabbits, guinea pigs, rats, muskrats, water rats, gray mice, ground squirrels, beavers, and lemmings.

Diagnosis Disease is diagnosed by (1) animal inoculation, (2) isolation of organism from lesions or discharges, and (3) agglutination test with serum from patient.

Source of infection Organism was originally isolated from California ground squirrels and later from more than 30 other forms of wildlife in the United States and elsewhere. Also from rabbits, hares, deerfly (*Chrysops discalis*), wood tick (*Dermacentor andersoni* and *D. variabilis*), and possibly other biting insects. Hunters, cooks, and butchers may contract the disease during rabbit-hunting season.

Mode of transmission The disease is transmitted by bites of infected flies and ticks and by inoculation through handling of diseased animals, especially in skinning rabbits during the hunting season. Ingestion of insufficiently cooked meat from a diseased animal or contaminated drinking water can cause the disease.

Incubation period Period is from 1 to 10 days, usually about 3 days.

Susceptibility and immunity All ages are susceptible. Recovery from an attack is followed by permanent immunity. An immune person exposed to the disease may develop a local infection through a break in the skin, but this does not cause any constitutional disturbance. Disease is most prevalent during the hunting season.

Prevention and control One should avoid bites from, or handling of, flies and ticks when working in infected areas during bloodsucking season. Rubber gloves should be used in the handling and dressing of rabbits or in performing autopsies on animals likely to be infected. Meat from wild rabbits and other susceptible animals should be thoroughly cooked before eating. Water from infected areas should not be used unless first boiled or disinfected.

Discharges from ulcer, lymph glands, or conjunctiva of infected person should be disinfected.

For specific treatment, streptomycin, the tetracyclines, and chloramphenicol are effective when continued for 4 to 5 days after temperature becomes normal.

For more information: Eigelsbach et al. (1962); Hornick and Eigelsbach (1966); Landay et al. (1968); Owen et al. (1964); Saslaw et al. (1961*a, b*); Shepard et al. (1955).

Salmonella

Rods are motile by means of peritrichous flagella or nonmotile. They are found in bodies of warm-blooded animals, including man, and occasionally in reptiles; they are frequently found in food eaten by these animals. Cells are gram-negative. They are aerobic, facultatively anaerobic.

Within the limits of the genus, serological relationships are the chief means of identifying new strains. Over 400 serotypes have been recognized and more are being isolated almost daily. There is general dissatisfaction with granting of species rank to each of the rapidly mounting number of serotypes.

As the morphology, staining properties, and physiology of the bacteria belonging to the various types are practically identical, only the antigenic structure, source, and habitat have been recorded for the majority of the types listed.

The nomenclature adopted is in accordance with the view that the recognition of similar antigenic structures really identifies serotypes rather than species. In a way, serotypes are varieties in a taxonomic sense, though like horticultural varieties in higher plants, they do not exactly correspond with varieties as usually defined by taxonomists. Where cultural differences rather than antigenic structure have been used to subdivide species, those subdivisions are designated as varieties. However, species rank has been accorded to those organisms (10 in number) which are easily identified because they are commonly encountered and/or cause rather well-established syndromes.

For more information: Breed et al. (1957); Cherubin et al. (1969); Edwards (1962); Kauffmann (1964); Kauffmann and Edwards (1957); Prost and Riemann (1967); Roantree (1967); Schaffner et al. (1967); Steele (1966, 1969).

S. typhosa Cells are rods, measuring 0.6 to 0.7 by 2 to 3 μ, occurring singly, in pairs, and occasionally in short chains. The organism is motile by means of peritrichous flagella, but is sometimes nonmotile. It is gram-negative (Fig. 362).

Disease produced *S. typhosa* is the cause of typhoid fever in man. It is a systemic disease characterized by fever, malaise, slow pulse, involvement of lymphoid tissues, enlargement of spleen, rose spots on trunk, irritation of walls of intestinal tract with formation of ulcers, and constipation more commonly than diarrhea. During first 2 weeks the organism may be recovered from blood; after second week, from urine and feces. Widal reaction becomes positive during second week. Disease may be transferred to laboratory animals by inoculation.

Carriers Typhoid fever may be transferred from person to person by individuals known as carriers. A carrier is one who has recovered from the disease but still continues to discharge the bacilli in the intestinal contents. These organisms are no longer pathogenic to the carrier but are capable of producing typhoid fever when they reach the intestinal tracts of other persons. Carriers generally harbor the organisms in their gall bladder, which is believed to be the reservoir of the bacilli. Removal of the gall bladder appears to be the best method for the treatment of some carriers.

Diagnosis Diagnosis is made from examination of antigenic structure: 9, 12 (*Vi*) *d* (see Breed et al., 1957). Motile species contain two antigenic components: (1) flagellar or *H* antigen and (2) somatic or *O* antigen (page 870). Agglutinins may be produced against both components. During infection, both kinds of agglutinins are present in blood serum. Typhoid fever may be diagnosed by testing for presence of agglutinins in blood stream.

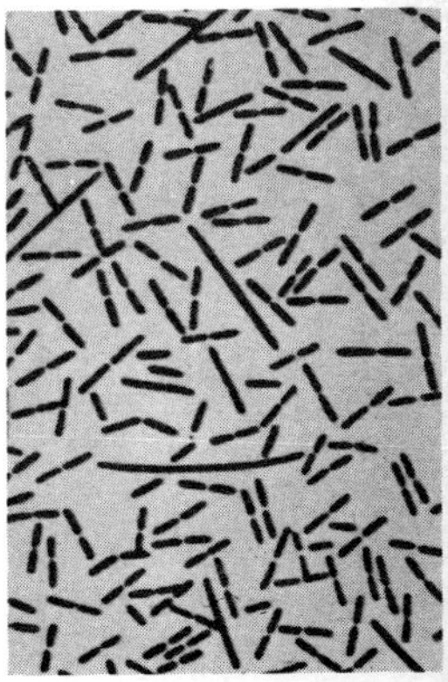

Fig. 362 Salmonella typhosa. Smear from a 24-hr agar slant culture. (From Muir, "Bacteriological Atlas," E. and S. Livingstone, Edinburgh, Scotland.)

Widal reaction This reaction is a specific agglutination test for diagnosis of typhoid fever. The test is performed by mixing gradually increasing dilutions of patient's serum with a suspension of typhoid bacilli and observing for presence of agglutination of the organisms. Since reaction of patient's serum does not become positive until during second week of infection, test of no value in early days of disease.

During first few days of disease, it is better to make a diagnosis by isolating the organism from feces, preparing a suspension, and testing for agglutination against a specific immune serum.

Source of infection Feces and urine of infected persons or carriers are the sources of infection.

Mode of transmission Disease is transmitted by direct or indirect contact with feces or urine of patient or carrier. Foods contaminated by fingers of typhoid patients or carriers often transmit the organisms. Most common source of typhoid outbreaks is through milk contaminated by a dairy worker. Oysters and shellfish, grown in sewage-polluted waters, may harbor the organism. Waterborne epidemics due to sewage contamination sometimes occur.

Incubation period The period is from 3 to 35 days, usually 7 to 21 days.

Susceptibility and immunity Susceptibility to disease is general. Natural immunity exists to some extent in adults. Permanent acquired immunity usually follows recovery from disease. Active artificial immunity of about 2 years' duration is developed by inoculation with typhoid bacterin.

Bacterin is prepared by growing a freshly isolated smooth strain on a solid medium, suspending growth in saline, standardizing suspension to contain about 1 billion cells per milliliter, killing organisms at 53°C for 1 hr, and then preserving with 0.25 percent tricresol.

Three injections of bacterin in doses of 500 million, 1 billion, and 1 billion at intervals of 7 to 10 days are generally sufficient for establishing a satisfactory active immunity.

Prevalence The disease is widely distributed throughout the world. It is endemic in some rural areas in United States but occurring now commonly as sporadic cases and small carrier epidemics. Incidence is steadily falling owing to protection of water, food, and milk supplies.

Prevention and control Preventive and controlling measures include: protection and purification of water supplies; sanitary disposal of human sewage; pasteurization of milk and milk products; sanitary control of foods and shellfish; fly control and protection of foods against fly contamination; periodic examination of individuals who handle foods for public consumption; immunization of population by use of a bacterin; and education of public on sources of infection and modes of transmission.

Infected individuals should be isolated. All bowel and urinary discharges and articles contaminated with such discharges should be disinfected. Vaccination of susceptible members in family or household of patient is helpful.

The administration of chloramphenicol in large oral dose is followed by oral doses every 6 hr until temperature is normal, and then smaller doses are given for total of 2 weeks. If chloramphenicol is contraindicated or ineffective, a tetracycline antibiotic may be substituted.

For more information: Anderson et al. (1961); Hejfec et al. (1968); Hobbs (1961); Merselis et al. (1964); Woodward (1970).

S. schottmuelleri Cells are rods, measuring 0.6 to 0.7 by 2 to 3 μ, occurring singly and in pairs. They are usually motile by means of peritrichous flagella, and are gram-negative.

Disease produced *S. schottmuelleri* is the cause of paratyphoid fever in man. It is also responsible for cases of food poisoning (page 778). The organism is not a natural pathogen of other animals.

Infection is characterized by fever, involvement of the lymphoid tissues of the intestines, enlargement of the spleen, and sometimes rose spots on the trunk; it is usually accompanied by a diarrheal condition. Organism may be present in feces, urine, and blood, and may be identified by fermentation and serological reactions.

Source of infection Feces and urine of infected persons or carriers and water or foods contaminated with discharges of infected persons or healthy carriers are the sources of infection.

Mode of transmission The disease is transmitted by direct contact with infected persons or by articles soiled with discharges of infected persons; also, through water, food, and milk contaminated with discharges of infected persons or carriers; and by insects.

Incubation period The period is from 1 to 10 days.

Susceptibility and immunity Susceptibility is general. Natural immunity is believed to exist in some persons. Acquired immunity is usually permanent after recovery from disease. Active artificial immunity of about 2 years' duration is developed after inoculation with a bacterin.

Prevalence The disease occurs sporadically or in limited outbreaks from contact with infected persons or from contaminated water, milk, and other foods.

Prevention and control Preventive and controlling measures include: protection and purification of water supplies; sanitary disposal of human sewage; pasteurization of milk and milk products; sanitary control of foods and shellfish; fly control and protection of foods against fly contamination; periodic examination of individuals who handle foods for public consumption; immunization of population by use of a bacterin; education of public on sources of infection and modes of transmission.

Infected individuals should be isolated. All bowel and urinary discharges and articles contaminated with such discharges should be disinfected. Vaccination of susceptible members in family or household of patient is helpful.

Administration of chloramphenicol in large oral dose is followed by oral doses every 6 hr until temperature is normal, and then smaller doses are given for total period of 2 weeks. If chloramphenicol is contraindicated or ineffective, a tetracycline antibiotic may be substituted.

S. paratyphi Cells are rods, measuring 0.6 by 3 to 4 μ, occurring singly. Organism is usually motile by means of peritrichous flagella, and is gram-negative.

Disease produced *S. paratyphi* is the cause of paratyphoid fever in man. It is also responsible for cases of food poisoning (page 778). It is not known to be a natural pathogen of other animals.

Disease is characterized by sudden onset, with continued fever, involvement of lymphoid mesenteric tissues and of intestines, enlargement of spleen, sometimes rose spots on trunk, and usually diarrhea. Fatality is lower than for typhoid.

Diagnosis Organism may be present in feces, urine, and blood, and may be identified by fermentation and serological reactions.

Source of infection Feces and urine of infected persons or carriers and water or foods contaminated with discharges of infected persons or healthy carriers are the sources of infection.

Mode of transmission The disease is transmitted by direct contact with infected persons or by articles soiled with discharges of infected persons; through water, food, and milk contaminated with discharges of infected persons or carriers; and by insects.

Incubation period The period is from 1 to 10 days.

Susceptibility and immunity Susceptibility is general. Natural immunity is believed to exist in some persons. Acquired immunity is usually permanent after recovery from disease. Active artificial immunity of about 2 years' duration is developed after inoculation with a bacterin.

Prevalence The disease occurs sporadically or in limited outbreaks from contact with infected persons or from contaminated water, milk, and other foods.

Prevention and control Preventive and controlling measures include: protection and purification of water supplies; sanitary disposal of human sewage; pasteurization of milk and milk products; sanitary control of foods and shellfish; fly control and protection of foods against fly contamination; periodic examination of individuals who handle foods for public consumption; immunization of population by use of a bacterin; education of public on sources of infection and modes of transmission.

Infected individuals should be isolated. All bowel and urinary discharges and articles contaminated with such discharges should be disinfected. Vaccination of susceptible members in the family or household of patient is a helpful preventive.

Chloramphenicol is administered in large oral dose every 6 hr until temperature is normal, and then smaller doses are given for total period of 2 weeks. If chloramphenicol is contraindicated or ineffective, a tetracycline antibiotic may be substituted.

S. typhimurium Cells are rods, measuring 0.5 by 1 to 1.5 μ, occurring singly. The organism is motile by means of peritrichous flagella and is gram-negative.

Disease produced *S. typhimurium* is a cause of food poisoning in man. It is a natural pathogen for all warm-blooded animals.

An acute gastroenteritis is acquired from ingestion of contaminated food. Onset is sudden, with abdominal pain, diarrhea, and frequent vomiting, fever is generally present.

Organism may localize in any tissue of body, producing abscesses, and causing arthritis, cholecystitis, endocarditis, meningitis, pericarditis, pneumonia, or pyelonephritis.

Source of infection Feces and urine of infected persons or carriers and water or foods contaminated with discharges of infected persons or healthy carriers are the sources of infection.

Mode of transmission The disease is transmitted by direct contact with infected persons, or by articles soiled with discharges of infected persons or carriers; through water, food, and milk contaminated with discharges of infected persons or carriers; and by flies.

Incubation period The period is from 6 to 48 hr, and is usually 12 to 24 hr.

Susceptibility and immunity Susceptibility is general. Natural immunity is believed to exist in some persons. Acquired immunity is usually permanent after recovery from disease. Active artificial immunity of about 2 years' duration is developed after inoculation with a bacterin.

Prevention and control Preventive and controlling measures include: protection and purification of water supplies; sanitary disposal of human sewage; pasteurization of milk and milk products; sanitary control of foods and shellfish; fly control and protection of foods against fly contamination; periodic examination of individuals who handle foods for public consumption; immunization of population by use of a bacterin; education of public on sources of infection and modes of transmission.

Infected individuals should be isolated. All bowel and urinary discharges and articles contaminated with such discharges should be disinfected. Susceptible members in the family or household of patient should be vaccinated.

Chloramphenicol and tetracyclines have limited and irregular effect.

For more information: Badakhsh and Herzberg (1969).

S. enteritidis Cells are rods, measuring 0.6 to 0.7 by 2 to 3 μ, occurring singly, in pairs, and occasionally in short chains. The organism is motile by means of peritrichous flagella and is gram-negative.

Disease produced *S. enteritidis* is a cause of food poisoning in man (page 778). It is also found in domestic and wild animals, particularly rodents.

An acute gastroenteritis is acquired from consumption of contaminated food. Onset is sudden, with abdominal pain, diarrhea, and frequent vomiting; fever is generally present.

Organism may localize in any tissue of body, producing abscesses, and causing arthritis, cholecystitis, endocarditis, meningitis, pericarditis, pneumonia, or pyelonephritis.

Organism may be recovered from feces or from site of localized infection during acute illness.

Source of infection Feces and urine of infected persons or carriers and water or foods contaminated with discharges of infected persons or healthy carriers are the sources of infection.

Mode of transmission The disease is transmitted by direct contact with infected persons, or by articles soiled with discharges of infected persons; through water, food, and milk contaminated with discharges of infected persons or carriers; and by insects.

Incubation period The period is from 6 to 48 hr, and is usually 12 to 24 hr.

Susceptibility and immunity Susceptibility is general. Natural immunity is believed to exist in some persons. Acquired immunity is usually permanent after recovery from disease. Active artificial immunity of about 2 years' duration is developed after inoculation with a bacterin.

Prevalence The disease occurs sporadically or in limited outbreaks from contact with infected persons or from contaminated water, milk, and other foods.

Prevention and control Preventive and controlling measures include: protection and purification of water supplies; sanitary disposal of human sewage; pasteurization of milk and milk products; sanitary control of foods and shellfish; fly control and protection of foods against fly contamination; periodic examination of individuals who handle foods for public consumption; immunization of population by use of a bacterin; education of public on sources of infection and modes of transmission.

Infected individuals should be isolated. All bowel and urinary discharges and articles contaminated with such discharges should be disinfected. Susceptible members in the family or household of patient should be vaccinated.

Chloramphenicol and tetracyclines have limited and irregular effect.

Shigella

The shigellae are rod-shaped organisms, nonmotile, aerobic, and gram-negative. Lactose is ordinarily not fermented, but some species may attack the sugar very slowly. The species possess distinctive antigenic structures. They form a fairly homogeneous group. There are currently over 30 serotypes which are divided into 4 main groups: Group A includes *S. dysenteriae*; group B, *S. flexneri*; group C, *S. boydii*; and group D, *S. sonnei*. Some species are pathogenic causing dysenteries. All species live in the bodies of warm-blooded animals. The organisms are found in polluted water supplies and in flies.

S. dysenteriae Cells are rods, measuring 0.4 to 0.6 by 1 to 3 μ, occurring singly. They are nonmotile and gram-negative.

Disease produced *S. dysenteriae* is a cause of bacillary dysentery in man and monkeys (Fig. 363). This disease is also called shigellosis. It is

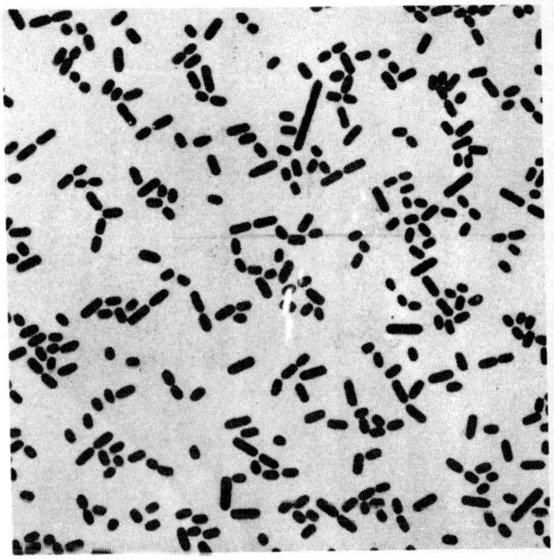

Fig. 363 Shigella dysenteriae. Smear from a 24-hr culture. (From Muir, "Bacteriological Atlas," E. and S. Livingstone, Edinburgh, Scotland.)

an acute intestinal disease characterized by diarrhea, fever, often vomiting, cramps, and tenesmus. In severe cases, stools may contain blood, mucus, and pus. Severe infections are most frequent in infants and in elderly, debilitated persons.

Organism produces an active exotoxin which can be changed to toxoid by formalin and heat. It is believed that the lesions produced in the gastrointestinal tract of persons suffering from disease are caused by the toxin rather than the direct action of the bacteria.

Diagnosis Diagnosis is made through isolation of organisms from feces or preferably from rectal swabs.

Carriers Dysentery may be transferred from person to person by carriers. A carrier is one who has recovered from the disease but still continues to discharge the bacilli in the intestinal contents. These organisms are no longer pathogenic to the carrier but are capable of producing dysentery when they reach the intestinal tracts of other persons.

Source of infection Bowel discharges of infected persons and carriers are the sources of infection. Healthy carriers are common.

Mode of transmission The disease is transmitted by direct contact, by eating contaminated foods, by articles soiled with discharges from infected persons or carriers, by drinking contaminated water, and by flies. Disease is most prevalent in the summer months.

Incubation period The period is from 1 to 7 days, usually less than 4 days.

Susceptibility and immunity Susceptibility to disease is general. Disease is more common and symptoms are more severe in children than in adults. Slight acquired immunity is of relatively short duration after recovery from the disease.

Prevalence The disease may be endemic, epidemic, or sporadic. Reduction occurs wherever water supplies are rendered safe, milk is pasteurized, and sewage is disposed of in a hygienic manner. The disease is more common in summer months.

Prevention and control Preventive and controlling measures include: purification of public water supplies; pasteurization of public milk supplies; sanitary disposal of sewage; hygienic preparation and handling of public food supplies; periodic examination of individuals who handle foods for public consumption; extermination of flies.

Infected persons should be isolated during period of communicability.

Aureomycin, Terramycin, streptomycin, and chloramphenicol given parenterally are very effective in relieving symptoms and greatly reducing numbers of organisms in from 24 to 48 hr. Sulfadiazine may be used alone where antibiotics are not available. Fluid and electrolyte replacement is an important consideration.

For more information: Christie (1968); Ewing et al. (1971).

S. flexneri Cells are rods, measuring 0.5 by 1 to 1.5 μ, occurring singly, often filamentous and irregularly shaped in old cultures. They are nonmotile and gram-negative.

Disease produced *S. flexneri* is the most common cause of dysentery epidemics, sometimes of infantile gastroenteritis. It is found in feces of sick and convalescents and of carriers of dysentery bacilli.

S. flexneri produces an acute intestinal disease characterized by diarrhea, fever, tenesmus, and frequent stools containing blood, mucus, and pus. Symptoms generally are milder than in *S. dysenteriae* infections.

Carriers Disease may be transferred from person to person by carriers. A carrier is one who has recovered from the disease but still continues to discharge the bacilli in the intestinal contents. Organisms are no longer pathogenic to the carrier but may produce dysentery when they reach the intestinal tracts of other persons.

Source of infection Bowel discharges of infected persons and carriers are the sources of infection. Healthy carriers are common.

Mode of transmission The disease is transmitted by direct contact, by eating contaminated foods, by articles soiled with discharges from infected persons or carriers, by drinking contaminated water, and by flies.

Incubation period The period is from 1 to 7 days, usually less than 4 days.

Susceptibility and immunity Susceptibility to disease is general. Disease is more common and symptoms are more severe in children than in adults. Slight acquired immunity is of relatively short duration after recovery from the disease.

Prevalence The disease may be endemic, epidemic, or sporadic. Reduction occurs wherever water supplies are rendered safe, milk is pasteurized, and sewage is disposed of in a hygienic manner. The disease is more common in summer months.

Prevention and control Preventive and controlling measures include: purification of public water supplies; pasteurization of public milk supplies; sanitary disposal of sewage; hygienic preparation and handling of public food supplies; periodic examination of individuals who handle foods for public consumption; extermination of flies.

Infected persons should be isolated during period of communicability.

Aureomycin, Terramycin, streptomycin, and chloramphenicol given parenterally are very effective in relieving symptoms and greatly reducing numbers of organisms in from 24 to 48 hr. Sulfadiazine may be used alone where antibiotics are not available. Fluid and electrolyte replacement in severe cases is an important consideration.

For more information: Ewing et al. (1971); Rowiński (1967*a*, *b*).

S. sonnei Cells are rods. Organism is nonmotile and gram-negative.

Disease produced *S. sonnei* is a cause of mild dysentery in man and of infantile gastroenteritis. This is an acute intestinal disease characterized by diarrhea, fever, tenesmus, and frequent stools containing blood, mucus, and pus. Symptoms generally are milder than in *S. dysenteriae* infections.

Carriers Disease may be transferred from person to person by carriers (page 946).

Source of infection Bowel discharges of infected persons and carriers are the sources of infection. Healthy carriers are common.

Mode of transmission The disease is transmitted by direct contact, by eating contaminated foods, by articles soiled with discharges from infected persons or carriers, by drinking contaminated water, and by flies.

Incubation period The period is from 1 to 7 days, usually less than 4 days.

Susceptibility and immunity Susceptibility to disease is general. Disease is more common and symptoms are more severe in children than in adults. Slight acquired immunity is of relatively short duration after recovery from the disease.

Prevalence The disease's occurrence may be endemic, epidemic, or sporadic. Reduction occurs wherever water supplies are rendered safe, milk is pasteurized, and sewage is disposed of in a hygienic manner. The disease is more common in summer months.

Prevention and control Preventive and controlling measures include: purification of public water supplies; pasteurization of public milk supplies; sanitary disposal of sewage; hygienic preparation and handling of public food supplies; periodic examination of individuals who handle foods for public consumption; extermination of flies.

Infected persons should be isolated during period of communicability.

Aureomycin, Terramycin, streptomycin, and chloramphenicol given parenterally are very effective in relieving symptoms and greatly reducing numbers of organisms in from 24 to 48 hr. Sulfadiazine may be used alone where antibiotics are not available. Fluid and electrolyte replacement in severe cases is an important consideration.

For more information: Ewing et al. (1971).

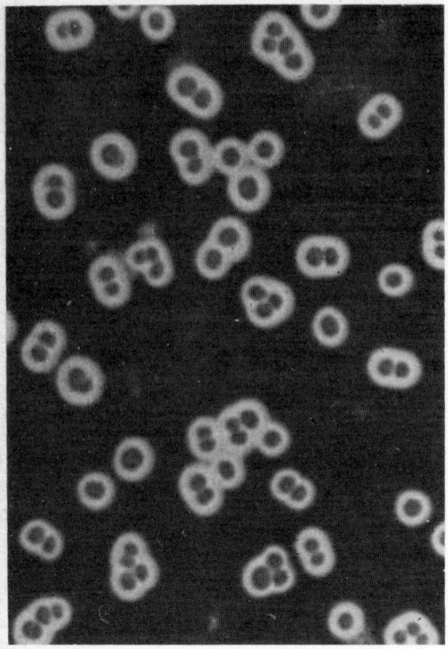

Fig. 364 Staphylococcus aureus, a capsulated strain. (Courtesy of Scott.)

Staphylococcus

Cells are spherical, occurring singly, in pairs, in tetrads, and in irregular clusters. They are nonmotile and gram-positive. Many strains produce an orange or yellow pigment, particularly on media containing high levels of NaCl. Cells are facultative, growing very well anaerobically in the presence of a fermentable carbohydrate but growing even better aerobically. Coagulase-positive strains produce a variety of toxins and are potentially pathogenic and may cause food-poisoning. Frequently cells are found on skin, in skin glands, on the nasal and other mucous membranes of warm-blooded animals, and in a variety of food products. They are facultative parasites and saprophytes.

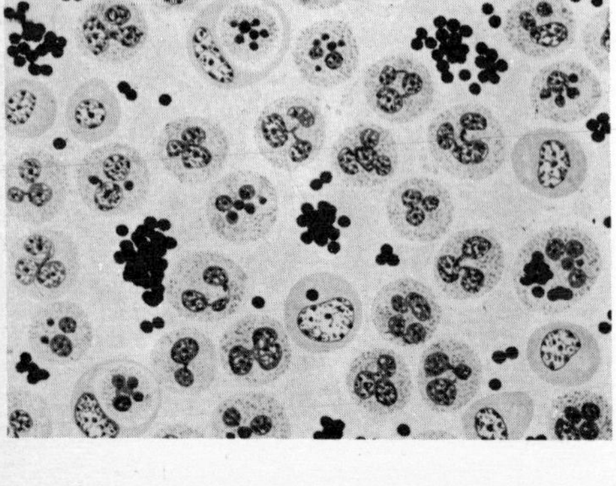

Fig. 365 Staphylococcus aureus. Smear of pus showing bacteria and white blood cells. (From Muir, "Bacteriological Atlas," E. and S. Livingstone, Edinburgh, Scotland.)

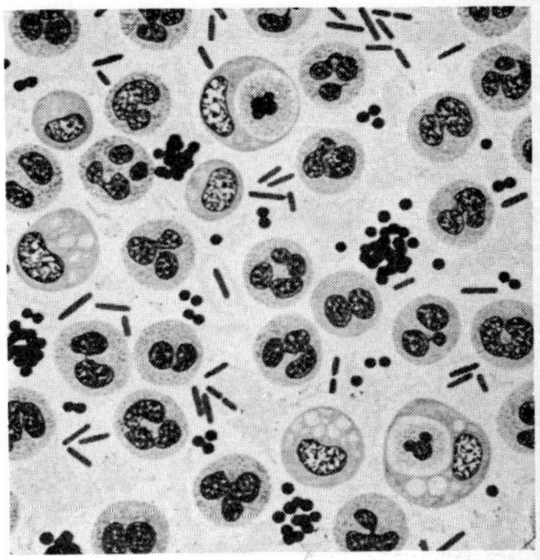

Fig. 366 Staphylococcus and Pseudomonas aeruginosa. Smear of pus showing bacteria and white blood cells. The two organisms are frequently found together in pyogenic infections. (From Muir, "Bacteriological Atlas," E. and S. Livingstone, Edinburgh, Scotland.)

S. aureus Cells are spherical, measuring 0.8 to 1 μ in diameter, occurring singly, in pairs, in short chains, and in irregular clumps. They are nonmotile and gram-positive.

Some strains develop a golden-yellow, water-insoluble pigment; others show a lemon-colored, water-insoluble pigment; still others are nonpigmented. Some strains are capsulated (Fig. 364).

Disease produced The organisms are normally present on skin, and their entrance into cut or scratch may lead to infection. *S. aureus* is the cause of boils, furuncles, abscesses, and suppuration in wounds (Figs. 365 and 366). Pus consists largely of an accumulation of bacteria and polymorphonuclear leucocytes in the infected area. Organism rarely produces septicemia but may be a secondary invader in peritonitis, pyemia, cystitis, osteomyelitis, meningitis, lung abscess, and brain abscess. Constitutional symptoms rarely occur. If lesions are widespread, fever, malaise, headache and anorexia may be present.

Certain strains under favorable conditions produce not only exotoxins (hemotoxin, dermatoxin, lethal toxin, etc.) but also a potent enterotoxin which is a significant cause of food poisoning. Normally also capable of coagulating citrated human or rabbit plasma (Fig. 367). Many strains produce an enzyme capable of dissolving such clots 8 to 12 hr after incubation.

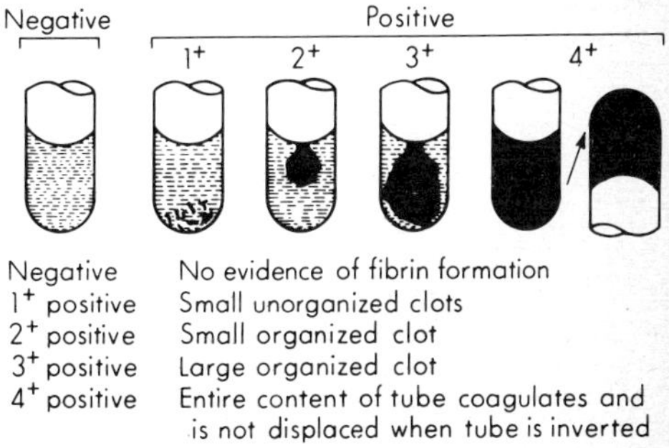

Fig. 367 Types of coagulase test reactions. Negative, no evidence of fibrin formation; 1+ positive, small unorganized clots; 2+ positive, small organized clot; 3+ positive, large organized clot; 4+ positive, entire content of tube coagulates and is not displaced when tube is inverted. (Courtesy of Warner-Chilcott Laboratories.)

Source of infection The sources of infection are pus, skin, air, contaminated clothing, food, water, etc. From 30 to 40 percent of normal persons may carry coagulase-positive staphylococci. Draining sinuses or purulent discharges are notorious causes of epidemics.

Mode of transmission The infection is transmitted by entrance of organism into a cut or break in the skin, or by contact with a person who has a purulent lesion or who is an asymptomatic nasal carrier.

Incubation period The period is variable and indefinite; it is usually 4 to 10 days.

Susceptibility and immunity Susceptibility to infections is general. Stock polyvalent bacterins, autogenous bacterins, and other heat-killed preparations have been used but very little immunity is produced.

Prevention and control Public health education in personal hygiene should be available. Discharges and articles soiled by such discharges should be disinfected.

For systemic infections, oral oxacillin or parenteral methicillin may be used for those strains which are resistant to penicillin. Penicillins and other antibiotics, such as cephalosporin and vancomycin, have revolutionized the treament of staphylococcal infections.

For more information: Bergdoll (1967); Bernheimer et al. (1968); Blair (1962, 1965); Casman et al. (1967); Drummond and Tager (1963); Eickhoff (1967); Ekstedt and Yoshida (1969); Foster (1967); Gow and Robinson (1969); Haque and Baldwin (1969); Maverakis and Wiley (1969); Oeding (1960); Scott (1969); Shay and Clendenin (1963); Smith (1968); Walker et al. (1969); World Health Organization (1968); Zolli and San Clemente (1963); Yoshida and Naito (1972).

Streptococcus

Cells are spherical or ovoid, rarely elongated into rods, occurring in pairs or short or long chains. They are nonmotile and gram-positive. Capsules are not regularly formed but may become conspicuous with some species under certain conditions. Growth on artificial media is slight. Agar colonies are very small, and bile-insoluble. Latter property is used to differentiate streptococci from pneumococci, which are bile-soluble. The organisms are found wherever organic matter containing sugars accumulates. Cells are regularly found in mouth and intestine of man and other animals, dairy products, and fermenting plant juices.

Organisms are found associated with a variety of pathological conditions including erysipelas, scarlet fever, streptococcal sore throat, puerperal fever, cellulitis, otitis media, lymphadenitis, mastoiditis, osteomyelitis, peritonitis, septicemia, impetigo contagiosa, and various skin and wound infections.

Classification of streptococci The streptococci are among the most difficult groups of bacteria to classify. One of the earliest classifications is that proposed by Brown (1919), who divided the organisms into three groups according to their effect on blood agar:

1. Alpha streptococci produce a greenish coloration (methemoglobin formation) of the medium and partial hemolysis in the immediate vicinity of the colonies.
2. Beta streptococci produce completely hemolyzed clear, colorless zones around the colonies.
3. Gamma streptococci have no effect on blood agar.

The most important contribution to methods for the classification of the streptococci is the serological technique (precipitin test) proposed by Lancefield (1933). On the basis of this method, the streptococci may be placed into the following groups:

Group A *S. pyogenes*. Under this species are placed those organisms causing scarlet fever, erysipelas, tonsillitis, puerperal fever, septicemia, and sore throat. They are hemolytic, liquefy fibrin, and do not curdle milk or hydrolyze sodium hippurate.

Group B *S. agalactiae*. This species has been isolated from mastitis in cows and occasionally from human sources. It curdles milk, hydrolyzes sodium hippurate, and does not liquefy fibrin. Most strains are hemolytic.

Group C This group includes three rather clearly defined biochemical groups: (1) *S. equi*, the cause of "strangles" in horses; (2) the "animal pyogenes" *Streptococcus*; and (3) the "human C" *Streptococcus*. Some of these have been isolated from animals; others are of human origin.

Group D This group includes both hemolytic and nonhemolytic types. The most important member is *S. faecalis* var. *zymogenes*. Other members are *S. faecalis* and *S. faecalis* var. *liquefaciens*.

Group E This group includes nonpathogenic streptococci isolated from milk. They are hemolytic, and do not liquefy fibrin or hydrolyze sodium hippurate.

Group F This organism is generally present in normal throats and is sometimes referred to as the "minute hemolytic *Streptococcus*." On blood agar plates, the organism produces extremely small pinpoint colonies, frequently barely visible, but surrounded by a zone of true hemolysis.

Group G This is a heterogeneous group of hemolytic streptococci which have been isolated from the normal human throat and nose, vagina, skin, and feces. They are not believed to be of any importance in producing disease in humans.

S. pyogenes Cells are spherical or ovoid, measuring 0.6 to 1 μ in diameter in cultures, usually spherical in blood and inflammatory exudates, occurring in pairs or chains; in broth culture, they usually occur in long chains. The organisms are gram-positive.

S. pyogenes is placed in Lancefield's group A. This group may be subdivided into serological types by precipitin technique on basis of capsular protein *M.* antigen. The antigen is associated with virulence, and antibodies to which it gives rise are primarily concerned with the specific protective action of immune sera. At least 40 types have been identified.

Culture filtrates of typical strains are capable of hemolyzing red blood cells. Soluble toxin is called a hemolysin. Two types of hemolysin can be elaborated: one being oxygen-sensitive (streptolysin O), and the other oxygen-stable (streptolysin S). On blood agar, organism produces a type of hemolysis referred to as β-hemolysis that possesses considerable diagnostic importance.

Disease produced *S. pyogenes* is found in human mouth, throat, respiratory tract, and inflammatory exudates. It produces septic sore throat, septicemia, erysipelas, scarlet fever, puerperal fever, cellulitis, mastoiditis, osteomyelitis, otitis media, peritonitis, and various skin and wound infections (Fig. 368).

Scarlet fever Scarlet fever is an acute febrile disease of the throat accompanied by an exudative tonsillitis or pharyngitis, leucocytosis, adenopathy, strawberry tongue, and rash. Invasion of other parts of the body may occur, resulting in infections of the middle ear, kidneys, etc.

The scarlet rash is caused by an extracellular erythrogenic toxin.

Fig. 368 Streptococcus pyogenes. Smear from human pus showing the typical appearance of hemolytic streptococci together with polymorphonuclear and mononuclear leucocytes. (From Muir, "Bacteriological Atlas," E. and S. Livingstone, Edinburgh, Scotland.)

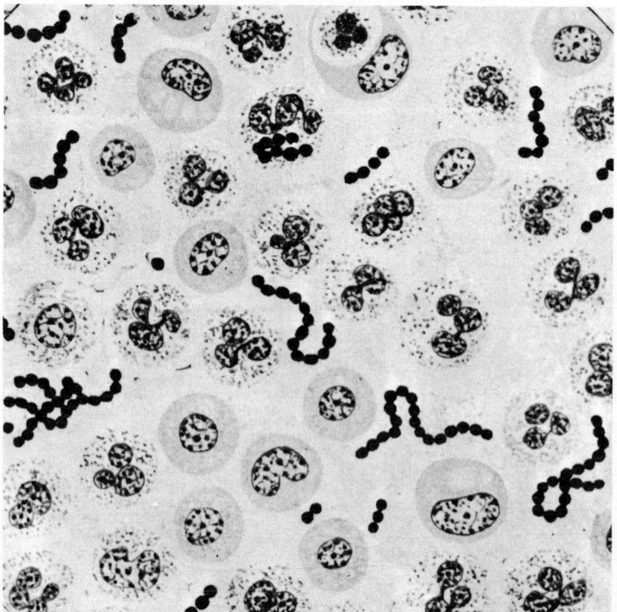

Immunity to the disease is an immunity to the toxin rather than to the organism. If the organism is not a good toxin producer or if the patient is immune to the toxin, streptococcal sore throat results.

Scarlet fever is diagnosed by its clinical symptoms and by the isolation of the specific organisms from the throat. For susceptibility to the disease, the Dick test may be used. This test consists of the intradermal injection of 0.1 ml of a known strength of toxin; the reaction is read after 24 hr. A positive test manifests itself as a bright red area 1.5 to 3 cm or more in diameter, with swelling and tenderness of the skin.

An antitoxin may be prepared by immunizing animals against culture filtrates of the scarlet fever strain of *S. pyogenes.* Administration of the antitoxin in cases of scarlet fever produces a favorable result on the outcome of the infection. The antitoxin neutralizes the damaging effect of the toxin and, in so doing, decreases the duration of the rash, changes the character and extent of desquamation, and reduces the number of complications.

Erysipelas Erysipelas is an acute infection characterized by fever, leucocytosis, constitutional symptoms, and a red, tender edematous spreading lesion of the skin, frequently with a definite raised border. Face and legs are common areas. Disease may be accompanied by a bacteremia in patients suffering from a debilitating disease.

Puerperal fever Puerperal fever is an acute streptococcal infection usually with fever, and accompanied by local and general symptoms, and signs of invasion of the genital tract. Fatality is low when properly treated. Disease also caused by other bacteria including *Staphylococcus aureus, E. coli, C. perfringens*, etc.

Source of infection Acutely ill or convalescent patients and carriers are sources of infection. Discharges from nose, throat, purulent lesions, objects contaminated with such discharges, contaminated milk and other foods are also sources.

Mode of transmission The infection is transmitted by direct contact with patient or carrier, indirectly from contaminated objects, inhalation of droplets from an infected person, and inhalation of contaminated dust. Epidemics are caused by contaminated milk and other foods.

Incubation period The period is usually 1 to 3 days, rarely longer.

Susceptibility and immunity Immunity against types of one group does not protect against infection by types from other groups. Bacterial immunity slight and of temporary duration. Exception is immunity to scarlet fever. This organism secretes an exotoxin against which an antitoxin is produced. Immunity against toxin and to rash develops in about 1 week and is usually permanent. Repeated attacks of sore throat or other streptococcal disease caused by different type are relatively frequent.

One attack of erysipelas may predispose person to subsequent attacks.

Prevalence These diseases are most prevalent in temperate zones, and are less common in semitropical and tropical climates. Highest incidence of scarlet fever and streptococcal sore throat occurs during late winter and spring.

Prevention and control Pasteurization of milk supplies, exclusion of infected individuals or carriers from handling foods, and care in treating cuts and abrasions are some preventive and controlling measures. Infected individuals should be isolated. Dressings, discharges, and clothing from infected persons should be disinfected; floors, tabletops, and contaminated objects should similarly be disinfected.

Specific treatment consists of administration of various penicillins, such as penicillin G, procaine penicillin G in aluminum monostearate in oil, procaine penicillin G in aqueous suspension, and phenoxymethyl penicillin orally. Erythromycin may be used by patients sensitive to penicillin.

For more information: Basiliere et al. (1968); Nowlan and Deibel (1967*a, b*); World Health Organization (1968).

Treponema

Cells measure 3 to 18 μ in length, with acute, regular or irregular spirals; longer forms are due to incomplete division. Terminal filament may be present. Organisms are weakly refractive by dark-field illumination in

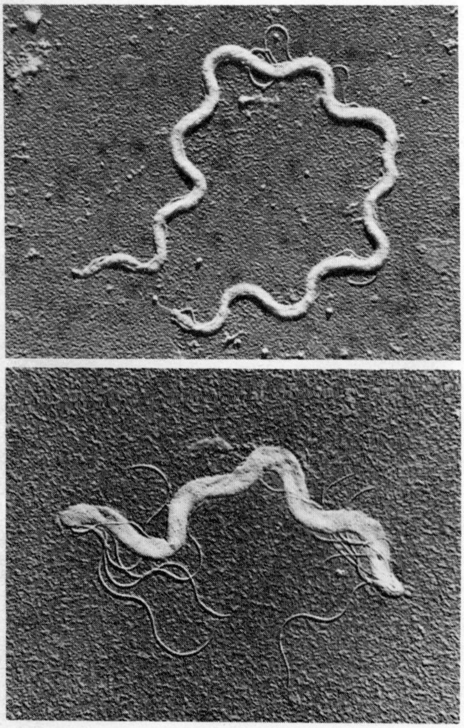

Fig. 369 Electron micrographs of Treponema pallidum. Upper, Nichols strain grown in Nelson and Mayer medium for 20 hr at 35°C. Cells have not separated. X15,600. Lower, Reiter strain grown in Brewer's thioglycollate medium for 24 hr. X12,500. (Courtesy of D'Alessandro and Zaffiro.)

living preparations. They are cultivated under strictly anaerobic conditions. Some are pathogenic and parasitic for man and other animals. They generally produce local lesions in tissues.

T. pallidum Cells occur as very fine protoplasmic spirals, 0.25 to 0.3 by 6 to 14 μ. Spiral amplitude is 1 μ, regular, and fixed; spiral depth is 0.5 to 1 μ. Terminal spiral filament is present. Organisms are weakly refractive in living state by dark-field illumination. They are motile by means of a sluggish, drifting motion; they are stiffly flexible, rarely rotating. The organisms appear black with silver impregnation methods. They are cultivated with difficulty under strict anaerobiosis in ascitic fluid with addition of fresh rabbit kidney (Fig. 369).

Disease produced *T. pallidum* is the cause of syphilis in man. Syphilis is acquired almost entirely by sexual contact. Disease manifests itself as a primary lesion, which starts as a papule at the site of infection, increases in size, and finally ulcerates. Ulcer is generally referred to as a chancre. Invasion of blood precedes initial lesion. Infection without chancre is fairly frequent. This is followed by constitutional symptoms and lesions of the skin and mucous membranes. Secondary lesions eventually heal and may reappear during the first 5 years after infection. Later manifestations may include disturbances of cardiovascular and central nervous systems. In congenital syphilis, only secondary and late manifestations are observed. Early acquired syphilis does not result in death but late manifestations shorten life, impair health, and limit occupational efficiency.

Diagnosis Disease in primary stage may be diagnosed by examining the serous exudate from a chancre under dark-ground illumination (Fig. 370). Presence of spirochetes indicates a syphilitic infection. Disease in later stages may be diagnosed by serological reactions (page 875).

Source of infection The main sources of infection are discharges from lesions of the skin and mucous membranes and blood of infected

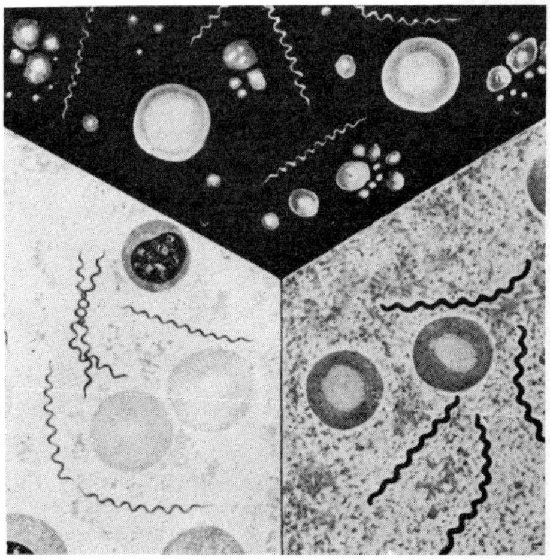

Fig. 370 Treponema pallidum, the cause of human syphilis. Upper, exudate from a primary sore, viewed by dark-ground illumination; lower left, smear of material from a chancre; lower right, same but stained by a different method. (From Muir, "Bacteriological Atlas," E. and S. Livingstone, Edinburgh, Scotland.)

individuals. Only rarely do articles freshly soiled with discharges serve as the source of infection.

Mode of transmission Syphilis is transmitted by direct personal contact with syphilitic individual, chiefly by sexual intercourse; occasionally it is transmitted by kissing or by dental instruments; only rarely is infection through articles freshly soiled with discharges. Transmission by syphilitic mother to offspring through placenta results in congenital syphilis.

Incubation period Incubation period is from 10 to 90 days; the average is about 21 days.

Susceptibility and immunity Susceptibility to disease is universal. Recovery is said to confer some immunity, although reinfections do occur.

Prevalence Disease is worldwide in distribution, varying with age, sex, and race. Occurs most frequently between ages of eighteen and thirty. Since 1957 the incidence of syphilis has shown a marked increase throughout the world.

Prevention and control Health and sex education, preparation for marriage, and premarital and prenatal examinations as part of general physical examination are some preventive measures.

Adequate treatment facilities, including free distribution of antibiotics to physicians, should be available for treatment of all cases. Legislation making examinations before marriage compulsory should be enacted. Houses of prostitution should be controlled or eliminated. Public should be educated on nature, characteristics, prevalence, mode of transmission, how to avoid infection, and how to secure prompt treatment in case of infection. Personal hygiene before or immediately after sexual intercourse with those exposed to infection should be practiced. Infected individuals should be isolated in the communicable stage. Discharges from open lesions and of articles soiled with such discharges should be disinfected.

Specific treatment consists of the use of large doses of long-acting penicillin, such as benzathine penicillin. Persons sensitive to penicillin may be treated with Erythromycin, the tetracyclines, or chloramphenicol. Follow-up checks of serological reactions should be required.

For more information: Bredt (1968); D'Alessandro and Zaffiro (1961); Drusin et al. (1969); Jackson and Black (1971); Jepsen et al. (1968); Nicholas (1967); Nicholas and Beerman (1965); Willcox and Guthe (1966); Cox (1972); Miller (1971).

Vibrio

Cells are short, curved rods, single or united into spirals. They are motile by means of single, polar flagellum which is usually relatively short; rarely there are two or three flagella in one tuft. Organisms are heterotrophic, varying greatly in nutritional requirements. There are aerobic, facultative anaerobic, and anaerobic species. They are widely distributed as saprophytic forms in saltwater, freshwater, and in soil; they also occur as parasites and pathogens.

V. cholerae Cells are slightly curved rods, measuring 0.3 to 0.6 by 1 to 5 μ occurring singly and in spiral chains. Cells may be long, thin, and delicate, or short and thick. They may lose their curved form on artificial cultivation. Organisms are motile by means of a single polar flagellum and are gram-negative.

Organism tolerates high alkalinity. pH 7.6 to 8 is optimum; for primary isolation, pH 9 to 9.6 should be used.

Vibrios have been classified into six groups on the basis of their protein and polysaccharide structures.

Disease produced *V. cholerae* is the cause of cholera in man. It is a serious acute intestinal disease characterized by sudden onset, vomiting, watery stools, rapid dehydration, acidosis and collapse. Mild cases may show only a diarrhea; severe infections may cause death within a few hours after onset.

Diagnosis The disease is diagnosed through isolation of organisms from feces or rectal swabs and confirmation by appropriate tests.

Carriers Patients convalescing from the disease usually continue to eliminate the organisms in the feces for about 7 to 14 days after recovery. Healthy carriers may also be found who excrete the cholera vibrio without exhibiting any signs of the disease. Both convalescent and healthy carriers play an important role in the dissemination of the disease.

Source of infection Intestinal contents and vomitus of infected persons and feces of convalescent or healthy carriers are sources of infection. Food and water also are sources.

Mode of transmission The disease is transmitted by water and foods, by contact with infected persons or carriers, or articles soiled with discharges from such persons, and by flies.

Incubation period The period is from a few hours to 5 days, usually 3 days.

Susceptibility and immunity Susceptibility to disease is general. Natural resistance to infection varies. Clinical attack confers a temporary immunity which may last for several years. Active artificial immunity for about 6 to 12 months may be produced by the use of vaccines.

Prevalence Epidemics of the disease occur in India and adjacent areas in Southeastern Asia. More recently epidemics have occurred in Iran, Iraq, Southern USSR, Afghanistan, and the Western Pacific. The disease does not occur in the Western Hemisphere.

Prevention and control Preventive and controlling measures include: sanitary disposal of human excreta; protection of water by boiling; pasteurization of milk and dairy products; sanitary preparation and handling of food in public places; control or destruction of houseflies; education of public in personal cleanliness. Persons suffering from the disease should be isolated. Stools and vomitus and articles soiled by such discharges should be disinfected. Food left by patient should be destroyed by burning. Room occupied by patient should be thoroughly cleaned and disinfected. Carriers should be isolated. Cholera vaccine should be given in exposed population groups.

Specific treatment consists of prompt and adequate replacement of fluid and electrolyte. Tetracycline antibiotics may reduce volume and duration of diarrhea, and they may shorten the period of excretion of organisms.

For more information: Burrows (1962); Felsenfeld (1967); Gallut and Nicolle (1965); Hirschhorn and Greenough (1971); World Health Organization (1967*c*).

For more general information on the pathogenic bacteria: Austrian (1968); Benenson (1970); Breed et al. (1957); Burnet (1962); Burrows (1968); Christie (1969); Gillies and Dodds (1968); Hanks (1966); Hare (1967); Henneberg (1969); Litsky and Litsky (1968); Neter (1969); Parish (1968); Parry (1969); Potter et al. (1968); Prevot et al. (1967); Smith et al. (1968); H. Smith (1968); I. M. Smith (1968); Smith and Holdeman (1968); Stewart (1968); Whitby (1968); Wilson and Miles (1964).

References

Agius, E.: The incidence of human brucellosis in Malta, *Arch. Inst. Pasteur Tunis,* **42**:31, 1965.

Albrink, W. S.: Pathogenesis of inhalation anthrax, *Bact. Rev.,* **25**:268, 1961.

Altemeier, W. A., and R. P. Hummel: Treatment of tetanus, *Surgery,* **60**:495, 1966.

Anderson, E. S., et al.: The detection of the typhoid carrier state, *J. Hyg.,* **59**:231, 1961.

Armstrong, A. S., and C. P. Sword: Cellular resistance in listeriosis, *J. Infect. Diseases,* **114**:258, 1964.

Austrian, R.: The Gram stain and the etiology of lobar pneumonia, an historical note, *Bact. Rev.,* **24**:261, 1960.

——: The bacterial flora of the respiratory tract. Some knowns and unknowns, *Yale J. Biol. Med.,* **40**:400, 1968.

Badakhsh, F. F., and M. Herzberg: Deoxycholate-treated, nontoxic, whole-cell vaccine protective against experimental salmonellosis of mice, *J. Bact.,* **100**:738, 1969.

Basiliere, J. L.: Streptococcal pneumonia, *Am. J. Med.,* **44**:580, 1968.

Beaman, B. L., and D. M. Shankel: Ultrastructure of *Nocardia* cell growth and development on defined and complex agar media, *J. Bact.,* **99**:876, 1969.

Beattie, M. I.: Occurrence and distribution of types of *C. diphtheriae, Am. J. Public Health,* **39**:1458, 1949.

Benenson, A. S., ed.: "Control of Communicable Diseases in Man," New York, American Public Health Association, 1970.

Bergdoll, M. S.: The staphylococcal enterotoxins. In "Biochemistry of

Some Foodborne Microbial Toxins," edited by R. I. Matelas and G. N. Wogan, Cambridge, Mass., The M.I.T. Press, 1967.

Bernheimer, A. W., et al.: Lytic effects of staphylococcal α-toxin and δ-hemolysin, *J. Bact.*, **96**:487, 1968.

Bizzini, B., et al.: Chemical characterization of tetanus toxin and toxoid. Amino acid composition, number of SH and S-S groups and N-terminal amino acid, *European J. Biochem.*, **17**:100, 1970.

Blair, J. E.: What is a *Staphylococcus*? *Bact. Rev.*, **26**:375, 1962.

——: Host-parasite relationships: A summation, *Ann. N.Y. Acad. Sci.*, **128**:451, 1965.

Bojlén, K., and I. Scheibel: The duration of immunity following diphtheria vaccination, *Danish Med. Bull.*, **2**:70, 1955.

Brachman, P. S., et al.: Industrial inhalation anthrax, *Bact. Rev.*, **30**:646, 1966.

Branham, S. E.: Milestones in the history of the meningococcus, *Can. J. Microbiol.*, **2**:175, 1956.

Bredt, W.: Terminalfäden bei Reiter-treponemen, *Centr. Bakt., I, Orig.*, **206**:123, 1968.

Breed, R. S., et al.: "Bergey's Manual of Determinative Bacteriology," Baltimore, The Williams & Wilkins Company, 1957.

Brookes, R., and C.-G. Hedén: Dense cultures of *Neisseria gonorrhoeae* in liquid medium, *Appl. Microbiol.*, **15**:219, 1967.

Brown, J. H.: "The Use of Blood Agar for the Study of Streptococci," New York, Rockefeller Institute for Medical Research, Monograph 9, 1919.

Burgdorfer, W.: I. Transmission of *Leptospira pomona* by the argasid tick, *Ornithodoros turicata*, and the persistence of this organism in its tissues, *Exp. Parasitol.*, **5**:571, 1956.

Burnet, F. M.: "Natural History of Infectious Disease," New York, Cambridge University Press, 1962.

Burrows, W.: Effective immunity to enteric infection with special reference to cholera, College Park, University of Maryland, Department of Microbiology, 1962.

——: "Textbook of Microbiology," Philadelphia, W. B. Saunders Company, 1968.

Bytchenko, B.: Geographical distribution of tetanus in the world, 1951–60, *Bull. World Health Organ.*, **34**:71, 1966.

Casman, E. P., et al.: Identification of a fourth staphylococcal enterotoxin, enterotoxin D, *J. Bact.*, **94**:1875, 1967.

Chang, Y. T., and R. N. Andersen: Morphological changes of *Mycobacterium lepraemurium* grown in cultures of mouse peritoneal macrophages, *J. Bact.*, **99**:867, 1969.

Cherubin, C. E., et al.: Symptoms, septicemia and death in salmonellosis, *Am. J. Epidemiol.*, **90**:285, 1969.

Christie, A. B.: Bacillary dysentery, *British Med. J.*, **2**:285, 1968.

——: "Infectious Diseases," Edinburgh, E. and S. Livingstone Ltd., 1969.

Coghlan, J. D., et al.: Low-temperature preservation of *Leptospira,* preliminary communication, *J. Hyg.,* **65**:373, 1967.

Coppel, S., and G. P. Youmans: Specificity of acquired resistance produced by immunization with *Listeria monocytogenes, J. Bact.,* **97**:121, 1969*a*.

—— and —— : Specificity of acquired resistance produced by immunization with mycobacterial cells and mycobacterial fractions, *J. Bact.,* **97**:114, 1969*b*.

Cox, C. D.: Shape of *Treponema pallidum, J. Bact.,* 109: **943**, 1972.

D'Alessandro, G., and P. Zaffiro: Sull'esistenza di un involucro similcapsulare nel *T. pallidum, Giorn. Microbiol.,* **9**:151, 1961.

Dalldorf, F. G., et al.: Woolsorters' disease, *Arch. Pathol.,* **92**:418, 1971.

Dannenberg, A. M., Jr.: Cellular hypersensitivity and cellular immunity in the pathogenesis of tuberculosis: Specificity, systemic and local nature, and associated macrophage enzymes, *Bact. Rev.,* **32**:85, 1968.

Deacon, W. E., et al.: I. A simple procedure for the isolation and identification of *Hemophilus ducreyi, J. Invest. Dermatol.,* **26**:399, 1956.

Dean, H. T., and D. E. Singleton: Vincent's infection: A wartime disease, *Am. J. Public Health,* **35**:433, 1945.

De Petris, S., et al.: The ultrastructure of S and R variants of *Brucella abortus* grown on a lifeless medium, *J. Gen. Microbiol.,* **35**:373, 1964.

DeSouza-Araujo, H. C.: Experiments in immunology of leprosy by means of inoculation of patients with living and dead suspensions of acid-fast bacilli cultures, Tokyo, *VII Intern. Cong. Leprol.,* 1958.

Diesch, S. L., et al.: Human leptospirosis acquired from squirrels, *New Engl. J. Med.,* **276**:838, 1967.

Doherty, P. C.: Streptomycin treatment of bovine carriers of *Leptospira pomona, Australian Vet. J.,* **43**:138, 1967*a*.

——: Bovine *Leptospira pomona* infection: Environmental contamination and the spread of the disease in a susceptible herd, *Queensland J. Agr. Animal Sci.,* **24**:329, 1967*b*.

Drummond, M. C., and M. Tager: Fibrinogen clotting and fibrino-peptide formation by staphylocoagulase and the coagulase-reacting factor, *J. Bact.,* **85**:628, 1963.

Drusin, L. M., et al.: Electron microscopy of *Treponema pallidum* occurring in a human primary lesion, *J. Bact.,* **97**:951, 1969.

Dubos, R. J.: Immunological aspects of BCG vaccination, *Am. Rev. Tuberc.,* **60**:670, 1949.

Edwards, P. R.: "Serologic Examination of *Salmonella* Cultures for Epidemiologic Purposes," Atlanta, Ga. Public Health Service, Communicable Disease Center, 1962.

Eickhoff, T. C.: Hospital-acquired staphylococcal infection. In "Preventive Medicine," edited by D. W. Clark and B. MacMahon, New York, Little, Brown and Company, 1967.

Eigelsbach, H. T., et al.: I. Host-parasite relationship in monkeys vaccinated intracutaneously or aerogenically, *J. Bact.*, **84**:1020, 1962.

Ekstedt, R. D., and K. Yoshida: Immunity to staphylococcal infection in mice: Effect of living versus killed vaccine, role of circulating antibody, and induction of protection-inducing antigen(s) in vitro, *J. Bact.*, **100**:745, 1969.

Ellinghausen, H. C., Jr., and W. G. McCullough: Nutrition of *Leptospira pomona* and growth of 13 other serotypes: Fractionation of oleic albumin complex and a medium of bovine albumin and polysorbate 80, *Am. J. Vet. Res.*, **26**:45, 1965.

Ellwood, D. C., et al.: The chemical basis of the virulence of *Brucella abortus*. VIII. The identity of purified immunogenic material from culture filtrate and from the cell-wall of *Brucella abortus* grown in vitro, *British J. Exp. Pathol.*, **48**:28, 1967.

Ewing, W. H., and colleagues: "Biochemical Reactions of *Shigella*," U.S. Department of Health, Education, and Welfare, Public Health Service, publ. no. (HSM) 72-8081, 1971.

Felsenfeld, O.: "The Cholera Problem," St. Louis, Mo., Green, 1967.

Fish, D. C., and R. E. Lincoln: In vivo-produced anthrax toxin, *J. Bact.*, **95**:919, 1968.

Foster, E. A.: Tissue injury by toxins in experimental staphylococcal infections, *Am. J. Pathol.*, **51**:913, 1967.

Francis, J.: Zoonoses and public health, *Australian Vet. J.*, **43**:311, 1967.

Frappier, A., and R. Guy: The use of BCG, *Can. Med. Assoc. J.*, **61**:18, 1949.

——— and ———: A new and practical B.C.G. skin test (the B.C.G. scarification test) for the detection of the total tuberculous allergy, *Can. J. Public Health*, **41**:72, 1950.

Fubra, E. S.: Nonproteolytic, avirulent *Bacillus anthracis* as a live vaccine, *J. Bact.*, **91**:930, 1966.

Gallut, J., and P. Nicolle: Évolution récente, épidémiologique et bactériologique, du choléra: Étude de quelques souches récemment isolées, *Bull. Soc. Pathol. Exotique*, **58**:728, 1965.

Georg, L. K., and R. M. Coleman: Comparative pathogenicity of various *Actinomyces* species. In "The Actinomycetales," *Jena Intern. Symp. Taxon.*, September 1968.

Gillies, R. R., and T. C. Dodds: "Bacteriology Illustrated," Edinburgh, E. and S. Livingstone Ltd., 1968.

Gow, J. A., and J. Robinson: Properties of purified staphylococcal β-hemolysin, *J. Bact.*, **97**:1026, 1969.

Gray, M. L., and A. H. Killinger: *Listeria monocytogenes* and listeric infections, *Bact. Rev.*, **30**:309, 1966.

Hanks, J. H.: Host-dependent microbes, *Bact. Rev.*, **30**:114, 1966.

Haque, R.-U. and J. N. Baldwin: Purification and properties of staphylococcal beta hemolysin. II. Purification of beta hemolysin, *J. Bact.*, **100**:751, 1969.

Hare, R.: "An Outline of Bacteriology and Immunity," London, Longmans, Green & Co., Ltd., 1967.

Hejfec, L. B., et al.: Controlled field trials of paratyphoid B vaccine and evaluation of the effectiveness of a single administration of typhoid vaccine, *Bull. World Health Organ.*, **38**:907, 1968.

Henneberg, G.: "Pictorial Atlas of Pathogenic Microorganisms," vol. III, Stuttgart, Gustav Fischer Verlag, 1969.

Herrero, B. A., et al.: Experimental botulism in monkeys: A clinical pathological study, *Exp. Mol. Pathol.*, **6**:84, 1967.

Hirschhorn, N., and W. B. Greenough III: Cholera, *Sci. Am.*, **225**:15, 1971.

Hobbs, B. C.: Public health significance of *Salmonella* carriers in livestock and birds, *J. Appl. Bact.*, **24**:340, 1961.

Holder, I. A., and C. P. Sword: Characterization and biological activity of the monocytosis-producing agent of *Listeria monocytogenes*, *J. Bact.*, **97**:603, 1969.

Holmes, M. J., and W. L. Ryan: Amino acid analysis and molecular weight determination of tetanus toxin, *Infection Immun.*, **3**:133, 1971.

Hornick, R. B., and H. T. Eigelsbach: Aerogenic immunization of man with live tularemia vaccine, *Bact. Rev.*, **30**:532, 1966.

Jackson, S., and S. H. Black: Ultrastructure of *Treponema pallidum* Nichols following lysis by physical and chemical methods. I. Envelope, wall, membrane and fibrils, *Arch. Mikrobiol.*, **76**:308, 1971.

Janssen, W. A., and M. J. Surgalla: Plague bacillus: Survival within host phagocytes, *Science*, **163**:950, 1969.

Jepsen, O. B., et al.: Electron microscopy of *Treponema pallidum* Nichols, *Acta Pathol. Microbiol. Scand.*, **74**:241, 1968.

Kadis, S., et al.: Plague toxin, *Sci. Am.*, **220**:93, 1969.

Karlsbad, G., et al.: Electron microscope observations of *Brucella abortus* grown within monocytes in vitro, *J. Gen. Microbiol.*, **35**:383, 1964.

Kauffmann, F.: Supplement zum Kauffmann-White-Schema (VII), *Acta Pathol. Microbiol. Scand.*, **61**:583, 1964.

—— and P. R. Edwards: A revised, simplified Kauffmann-White schema, *Acta Pathol. Microbiol. Scand.*, **41**:242, 1957.

Kellogg, D. S., et al.: *Neisseria gonorrhoeae*. I. Virulence genetically linked to clonal variation, *J. Bact.*, **85**:1274, 1963.

Kirchheimer, W. F.: Survey of recent leprosy research, *Public Health Rept.*, **79**:481, 1964.

Koike, M., and K . Takeya: Fine structures of intracytoplasmic organelles of mycobacteria, *J. Biophys. Biochem. Cytol.*, **9**:597, 1961.

Lancefield, R. C.: A serological differentiation of human and other groups of hemolytic streptococci, *J. Exp. Med.*, **57**:571, 1933.

Landay, M. E., et al.: Toxicity of *Pasteurella tularensis* killed by ionizing radiation, *J. Bact.*, **96**:804, 1968.

Lane, A. G.: Detoxification of liquid cultures of *Bordetella pertussis* by forced aeration at high pH, *Appl. Microbiol.*, **16**:1211, 1968*a*.

——: Appearance of mouse-lethal toxin in liquid cultures of *Bordetella pertussis, Appl. Microbiol.*, **16**:1400, 1968*b*.

Leiker, D. L.: Classification of leprosy, *Leprosy Rev.*, **37**:7, 1966.

Levy, L.: The effect of freezing and storage at −60°C on the viability of *Mycobacterium leprae, Cryobiology*, **8**:574, 1971.

Licciardello, J. J., et al.: Thermal inactivation of type E botulinum toxin, *Appl. Microbiol.*, **15**:249, 1967.

Lincoln, E. M.: Epidemics of tuberculosis, *Arch. Environ. Health*, **14**:473, 1967.

Lincoln, R. E., et al.: Anthrax. In "Advances in Veterinary Science," New York, Academic Press, Inc., 1964.

Litkenhous, C., and P. V. Liu: Bacteriocin produced by *Bordetella pertussis, J. Bact.*, **93**:1484, 1967.

Litsky, B. Y., and W. Litsky: Oral hygiene, a factor in the control of microbial contamination of the hospital environment, *J. Environ. Health*, **31**:36, 1968.

Lovell, R., and T. A. Rees: A filterable haemolysin from *Escherichia coli, Nature*, **188**:755, 1960.

Lynt, R. K., Jr., et al.: Serological studies of *Clostridium botulinum* type E and related organisms, *J. Bact.*, **93**:27, 1967.

MacLennan, J. D.: The histotoxic clostridial infections of man, *Bact. Rev.*, **26**:177, 1962.

Mageau, R. P., and B. S. Roberson: Association of toxic capsule and cell wall mucopeptide with virulence in *Gaffkya tetragena, J. Bact.*, **97**:16, 1969.

Markovits, J., and E. Vilkas: Etude des cires D d'une souche humaine virulente de *Mycobacterium tuberculosis, Biochim. Biophys. Acta*, **192**:49, 1969.

Maverakis, N. H., and B. B. Wiley: Evidence for a multiplicity of capsular types among *Staphylococcus aureus* strains, *J. Bact.*, **99**:472, 1969.

McCallum, R. E., and C. P. Sword: Mechanisms of pathogenesis in *Listeria monocytogenes* infection. IV. Hepatic carbohydrate metabolism and function in experimental listeriosis, *Infection Immun.*, **1**:183, 1970.

Medearis, D. N., Jr., and J. F. Kenny: Observations concerning the pathogenesis of *E. coli* infections in mice, *J. Immunol.*, **101**:534, 1968.

Merselis, J. G., Jr., et al.: Quantitative bacteriology of the typhoid carrier state, *Am. J. Trop. Med. Hyg.*, **13**:425, 1964.

Michel, A., et al.: Partial characterization of diphtheria toxin and its subunits, *Biochim. Biophys. Acta*, **257**:249, 1972.

Miller, J. N., ed.: "Spirochetes in Body Fluids and Tissues," Springfield, Ill., Charles C Thomas, 1971.

Mitchell, R. S.: Control of tuberculosis, *New Engl. J. Med.*, **276**:842, 905, 1967.

Neter, E.: "Medical Microbiology," Philadelphia, F. A. Davis Company, 1969.

Nicholas, L.: Serodiagnosis of syphilis, *Arch. Dermatol.*, **96**:324, 1967.

——— and H. Beerman: Present day serodiagnosis of syphilis. A review of some of the recent literature, *Am. J. Med. Sci.,* **249**:466, 1965.

Nowlan, S. S., and R. H. Deibel: Group Q streptococci. I. Ecology, serology, physiology, and relationship to established enterococci, *J. Bact.,* **94**:291, 1967*a*; II. Nutritional characteristics and growth relationship to thymine, folate, and folinate, *ibid.,* **94**:297, 1967*b*.

Oeding, P.: Antigenic properties of *Staphylococcus aureus, Bact. Rev.,* **24**:374, 1960.

Ørskov, I.: Biochemical types in the *Klebsiella* group, *Acta Pathol. Microbiol. Scand.,* **40**:155, 1957.

Owen, C. R., et al.: Comparative studies of *Francisella tularensis* and *Francisella novicida, J. Bact.,* **87**:676, 1964.

Panisset, M., et al.: II. Comparaison de vaccins BCG lyophilisés, préparés a partir de cultures de trois ages différents, *Ann. Inst. Pasteur,* **99**:496, 1960.

Parish, H. J.: "Victory with Vaccines," Edinburgh, E. & S. Livingstone Ltd., 1968.

Parry, W. H.: "Infectious Diseases," London, English Universities Press, Ltd., 1969.

Pedley, J. C.: The presence of *M. leprae* in human milk, *Leprosy Rev.,* **38**:239, 1967.

———: Composite skin contact smears: A method of demonstrating the non-emergence of *Mycobacterium leprae* from intact lepromatous skin, *Leprosy Rev.,* **41**:31, 1970.

Pine, L., et al.: Studies of the morphological, physiological, and biochemical characters of *Actinomyces bovis, J. Gen. Microbiol.,* **23**:403, 1960.

Potter, C. W., et al.: "Introduction to Medical Microbiology," London, Butterworth & Co. (Publishers), Ltd., 1968.

Prevot, A. R., et al.: "Les Bactéries Anaérobies," Paris, Dunod, 1967.

Prost, E., and H. Riemann: Food-borne salmonellosis. In "Annual Review of Microbiology," **21**:495, 1967.

Rees, R. J. W.: Limited multiplication of acid-fast bacilli in the footpads of mice inoculated with *Mycobacterium leprae, Brit. J. Exp. Pathol.,* **45**:207, 1964.

Ritchie, A. E., and H. C. Ellinghausen: Electron microscopy of leptospires. I. Anatomical features of *Leptospira pomona, J. Bact.,* **89**:223, 1965.

Roantree, R. J.: *Salmonella O* antigens and virulence. In "Annual Review of Microbiology," **21**:443, 1967.

Rowiński, S.: *S-R* dissociation in *Shigella flexneri* populations. I. The genetic mechanism of development of rough forms, *Arch. Immunol. Therap. Exp.,* **15**:640, 1967*a*; II. Selection of *R* forms under certain environmental conditions, *ibid.,* **15**:652, 1967*b*.

Roy, N., et al.: The specific substance from *Diplococcus pneumoniae* type 31, *Carbohydrate Res.,* **12**:89, 1970.

Saslaw, S., et al.: Tularemia vaccine study. I. Intracutaneous challenge, *Arch. Internal Med.*, **107**:689, 1961*a*; II. Respiratory challenge, *ibid.*, **107**:702, 1961*b*.

Sato, S., and M. Imi: The surface structure of *M. leprae, Intern. J. Leprosy*, **36**:303, 1968.

Schaffner, C. P., et al.: Coconut and *Salmonella* infection, *Appl. Microbiol.*, **15**:471, 1967.

Scott, A. C.: A capsulate *Staphylococcus aureus, J. Med. Microbiol.*, 2:253, 1969.

Shay, D. E., and G. G. Clendenin: Incidence of coagulase-positive staphylococci in the upper respiratory tract of dental students and a study of their transmission during a routine dental prophylaxis, *J. Dental Res.*, **42**:110, 1963.

Shepard, C. C.: Leprosy germs grown in mouse foot pads, *Sci. News Letter*, **83**.88, 1963.

—— et al.: Electron microscopically revealed structural elements of *Bacterium tularense* and their in vitro and in vivo role in immunologic reactions, *J. Immunol.*, **75**:7, 1955.

Simpson, C. F., and F. H. White: Ultrastructural variations between hooked and nonhooked leptospires, *J. Infect. Diseases*, **114**:69, 1964.

Slack, J. M., et al.: Morphological, biochemical, and serological studies on 64 strains of *Actinomyces israelii, J. Bact.*, **97**:873, 1969.

Smith, D. T., et al.: "Zinsser Microbiology," New York, Appleton-Century-Crofts, Inc., 1968.

Smith, H.: Biochemical challenge of microbial pathogenicity, *Bact. Rev.*, **32**:164, 1968.

—— and L. P. Packman: A filtered non-toxic plague vaccine which protects guinea-pigs and mice, *Brit. J. Exp. Pathol.*, **47**:25, 1966.

Smith, I. M.: Death from staphylococci, *Sci. Am.*, **218**:84, 1968.

Smith, L. De S., and L. V. Holdeman: "The Pathogenic Anaerobic Bacteria," Springfield, Ill., Charles C Thomas, Publisher, 1968.

Spink, W. W., and G. M. Bradley: Persistent parasitism in experimental brucellosis: Attempts to eliminate brucellae with long-term tetracycline therapy, *J. Lab. Clin. Med.*, **55**:535, 1960.

Springett, V. H.: B.C.G. vaccination in Birmingham: An assessment of its contribution to tuberculosis control in 1956–67, *Tubercle (London)*, **50**:159, 1969.

Stainer, D. W., et al.: Preparation and properties of diphtheria toxoids in submerged culture. III. Development of a new semisynthetic medium, *Can. J. Microbiol.*, **14**:1155, 1968.

Stanley, J. L., and H. Smith: The chemical basis of the virulence of *Pasteurella pestis*. IV. The components of the guinea-pig toxin, *Brit. J. Exp. Pathol.*, **48**:124, 1967.

Steele, J. H.: Salmonellosis, *Proc. U.S. Livestock Sanit. Assoc.*, p. 457, October 1966.

——: Salmonellosis, *Arch. Environ. Health,* **19**:871, 1969.

Steigman, A. J.: Abuse of tetanus toxoid, *J. Pediat.,* **72**:753, 1968.

Stewart, G. T.: Limitations of the germ theory, *Lancet*, p. 1077, May 18, 1968.

Sword, C. P.: Mechanisms of pathogenesis in *Listeria monocytogenes* infection. I. Influence of iron, *J. Bact.,* **92**:536, 1966.

Tillotson, J. R., and A. M. Lerner: Characteristics of pneumonias caused by *Escherichia coli, New Engl. J. Med.,* **277**:115, 1967.

Trautman, J. R.: The management of leprosy and its complications, *New Engl. J. Med.,* **273**:756, 1965.

Turner, G. C.: Cultivation of *Bordetella pertussis* on agar media, *J. Pathol. Bacteriol.,* **81**:15, 1961.

Uchida, T., and M. Yoneda: Estimation of the time required for the process of diphtheria toxin formation, *Biken's J.,* **10**:121, 1967.

Walker, W. S., et al.: Isolation and partial characterization of a staphylococcal leucocyte cytotoxin, *J. Bact.,* **97**:1005, 1969.

Ward, M. K., et al.: Studies on anthrax infections in immunized guinea pigs, *J. Infect. Diseases,* **115**:59, 1965.

Warren, J., et al.: II. Immunological properties of purified *Pasteurella pestis* toxin, *J. Bact.,* **70**:170, 1955.

Whitby, L. E. H.: "Medical Bacteriology," London, Churchill, 1968.

White, D. C.: Respiratory systems in the hemin-requiring *Haemophilus* species, *J. Bact.,* **85**:84, 1963.

Willcox, R. R., and T. Guthe: *Treponema pallidum, Suppl. Bull. World Health Organ.,* **35**:1, 1966.

Wilson, G. S., and A. A. Miles: "Principles of Bacteriology and Immunity," London, Edward Arnold (Publishers), Ltd., 1964.

Wolinsky, E., and T. K. Rynearson: Mycobacteria in soil and their relation to disease-associated strains, *Am. Rev. Respirat. Diseases,* **97**:1032, 1968.

Woodward, T. E.: The unmasking of typhoid fever, *S. Am. Med. J.,* **44**:99, 1970.

World Health Organization: Joint FAO/WHO expert committee on zoonoses, *World Health Organ. Tech. Rept. Ser.* no. 378, 1967*a*.

——: Current problems in leptospirosis research, *World Health Organ. Tech. Rept. Ser.* no. 380, 1967*b*.

——: WHO expert committee on cholera, *World Health Organ. Tech. Rept. Ser.* no. 352, 1967*c*.

——: Streptococcal and staphylococcal infections, *World Health Organ. Tech. Rept. Ser.* no. 394, 1968.

Xalabarder, P.: Cien mil (100,000) reacciones de fijación del complemento en tuberculosis, *Publ. Inst. Antituberculoso,* **18**:41, 1969.

Yanagihara, Y., and I. Mifuchi: Microfibers present in surface structure of *Leptospira, J. Bact.,* **95**:2403, 1968.

Yoshida, K., and Y. Naito: Comparison of capsular types of *Staphylococcus aureus* strains, *Infection and Immunity,* **5**:143, 1972.

Youmans, G. P., and A. S. Youmans: Immunizing capacity of viable and killed attenuated mycobacterial cells against experimental tuberculosis infection, *J. Bact.*, **97**:107, 1969.

Zacharias, B., and M. Björklund: Continuous production of *Clostridium tetani* toxin, *Appl. Microbiol.*, **16**:69, 1968.

Zolli, Z., Jr., and C. L. San Clemente: Purification and characterization of staphylocoagulase, *J. Bact.*, **86**:527, 1963.

28 *Bacterial viruses*[1]

Bacteria are susceptible to infection by submicroscopic entities called *bacteriophages*, or more simply *phages*. Phages are viruses whose parasitic activity, so far as known, is limited strictly to bacteria. The following definition for viruses has been proposed. They are "strictly intracellular and potentially pathogenic entities with an infectious phase: (1) possessing only one type of nucleic acid; (2) reproduced from their genetic material; (3) unable to grow and to undergo binary fission; and (4) devoid of a Lipmann system" (Lwoff, 1959).

What are now called bacteriophages were first reported by Twort (1915) who noticed that after prolonged incubation some micrococcal colonies underwent a "glassy transformation" and could no longer be subcultured onto fresh medium. When these glassy areas were transferred with an inoculating loop to normal micrococcal colonies, they too became transparent and were no longer viable. He also found that the glassy material could dissolve, or lyse, a broth culture of the same bacteria even after filtration through Berkefeld filters, and that the lytic principle was transmissible in series.

Independently of Twort, D'Hérelle (1917) also discovered a transmissible disease of bacteria, and it was he who named the lytic principle bacteriophage, which literally means bacteria-eating agent. D'Hérelle believed that phages were submicroscopic, filterable microbes obligately parasitic on bacteria. He strongly believed that phages would be therapeutically useful because of their ability to destroy disease-causing bacteria. The novel *Arrowsmith* by Sinclair Lewis fairly accurately describes some of the early studies of this facet of phage research.

Morphology and composition of bacteriophages The most intensively studied group of phages belong to the T series, numbered from 1 through 7, that attack the nonmotile strain B of *Escherichia coli*. All seven of these phages are composed almost exclusively of deoxyribonucleic acid (DNA) and protein, which are present in roughly equal amounts. The DNA

[1] This chapter was written by Dr. W. R. Romig, Department of Bacteriology, University of California, Los Angeles, Calif. The author is greatly indebted to Dr. Romig for his kindness in preparing this material for publication.

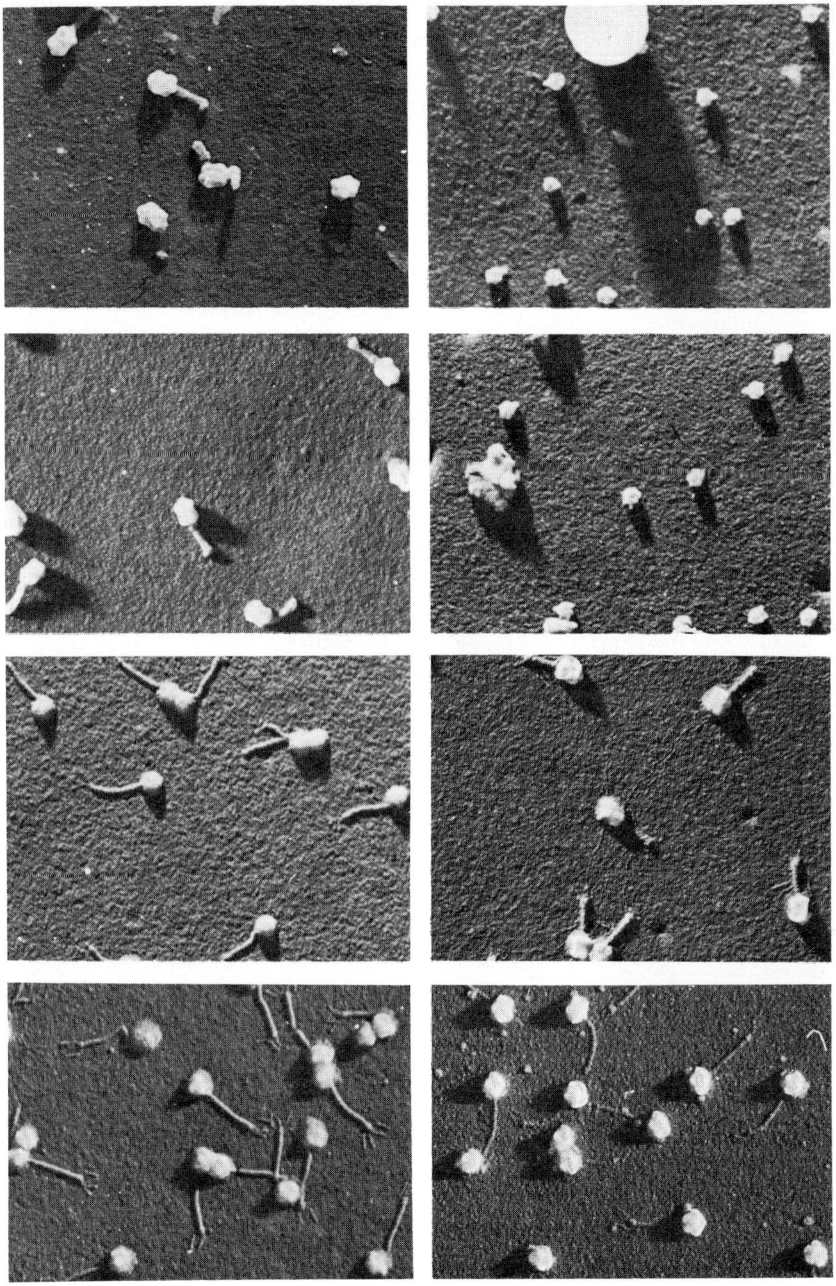

Fig. 371 The seven T phages. From left to right, top row, phages T2 and T3, frozen-dried; second row, phages T6 and T7, frozen-dried; third row, phages T1 and T4, air-dried; bottom row, phage T5, air-dried, deposited in gelatin, and same, air-dried, fixed in formalin. (Courtesy of Williams and Fraser, Virus Laboratory, University of California, Berkeley, Calif.)

apparently consists of one continuous molecule, which may be almost 50 μ long, tightly condensed in the phage head, and covered with a layer of protein. All the T phages, except T3 and T7, have a tadpole shape with hexagonal heads and long tails. T3 and T7 have very short tails. The size of the T phages varies from about 200 to 65 nm in length and 70 to 50 nm in width (see Fig. 371).

Other bacteriophages for *E. coli* have been studied whose morphology and composition differ drastically from the T phages. Loeb and Zinder (1961) discovered "male specific" phages capable of infecting only F^+ or Hfr bacteria. These phages have no visible tail appendages at all and contain ribonucleic acid (RNA) rather than DNA within their protein shell. This phage, called f2, is much smaller than the T phages, and bacteria infected with it release about 10,000 mature phage particles per infected cell (see Fig. 372).

Sinsheimer (1959) studied a small (20-nm diameter) DNA-containing phage, φX174, and showed that the nucleic acid of the mature particle contains only one polynucleotide strand. After infecting a susceptible *E. coli*, the single-stranded DNA is converted to a double-stranded *replicative form*, but only one strand is finally enclosed in a protein coat and released by the bacterium as a mature, infective particle. Guthrie and Sinsheimer (1960) and Sekiguchi et al. (1960) showed that the purified DNA extracted from mature φX174 phages could infect protoplasts of *E. coli* and that the mature virus particles obtained from the infected protoplasts were identical in all respects to the viruses used to prepare infectious DNA. Other studies have shown that both the single-stranded DNA of the mature particle and the double-stranded replicative form are closed circles, about 1.7 μ long.

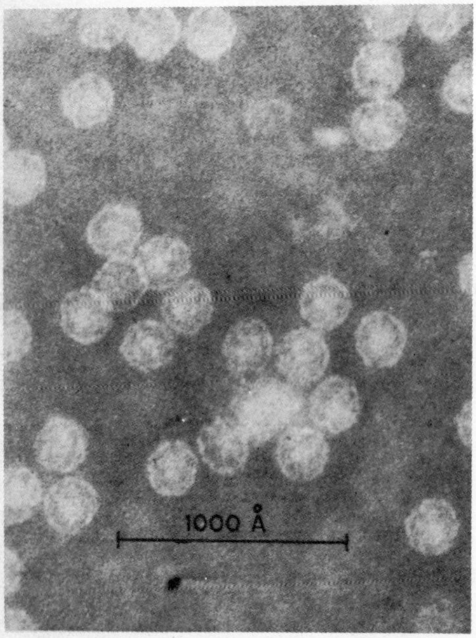

Fig. 372 An electron micrograph of f2. The phages were negatively stained by embedding in neutral phosphotungstate. (Courtesy of Loeb and Zinder.)

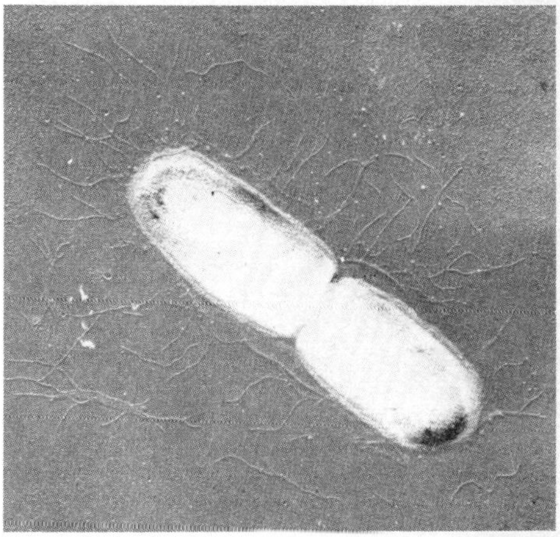

Fig. 373 Thread-like structures, typical of M13 phages, surround the bacteria. (Courtesy of Hofschneider and Preuss.)

For more information: Freifelder et al. (1964); Hayashi et al. (1963); Kleinschmidt et al. (1963).

A thin, rod-shaped bacteriophage, M13, has been described by Hofschneider and Preuss (1963) that does not lyse the infected *E. coli* cells during its release and whose bacterial host continues to multiply while liberating phage. Details of the infective process of this phage have not been clarified (see Fig. 373).

Growth of virulent bacteriophage Most of our knowledge concerning the details of bacteriophage reproduction has come from studies on the virulent T-even (T2, T4, T6) phages, which is summarized below.

The nonmotile phage particle is brought into contact with a bacterium by brownian motion and attaches by the tip of its tail to a specific receptor site on the bacterial cell wall. The tail *fibers* are probably the phage organelles responsible for specific fixation, and help to anchor the particle to the sensitive bacterium. After adsorption has occurred, a phage enzyme, probably lysozyme present in the tail fibers, digests a hole in the bacterial cell wall. The tail *sheath* then contracts and forces the tail *core* into the cell. The tail core is a hollow tube through which the DNA contained in the head is forced into the bacterial cell. After the DNA has been injected into the bacterium, the protein part of the phage is dispensable, and can be stripped from the bacterial surface without affecting the outcome of the infective process (Hershey and Chase, 1952). The protein of the phage acts, then, merely to protect the DNA in its extracellular sojourn and to deliver it into a suitable host bacterium (see Figs. 374 and 375).

After the DNA has penetrated the susceptible cell, it is referred to as *vegetative* phage. In this state the phage is ordinarily noninfective, and no phage activity can be detected in broken bacteria immediately after infection.

The bacterial nucleus is broken down soon after phage adsorption (Luria and Human, 1950), and the cell ceases therewith to synthesize

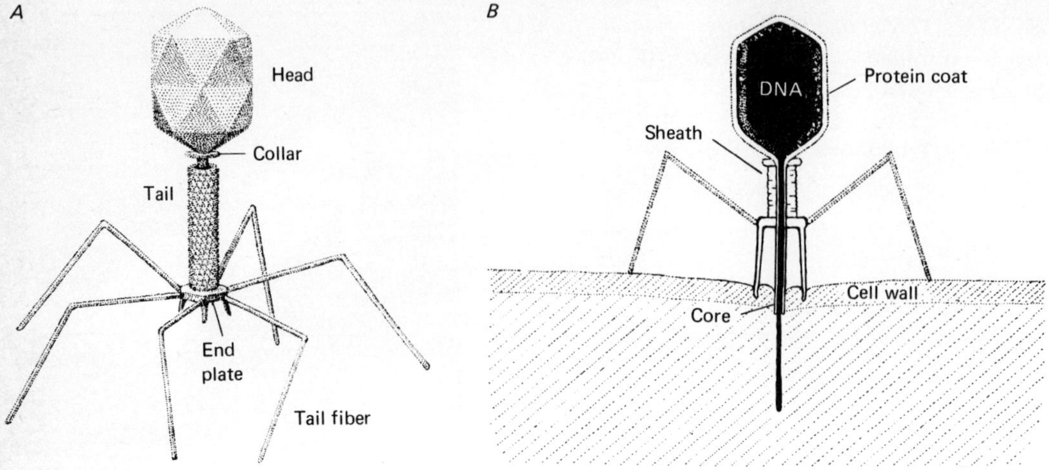

Fig. 374 *T4 bacterial virus is an assembly of protein components (A). The head is a protein membrane, shaped like a kind of prolate icosahedron with 30 facets and filled with deoxyribonucleic acid. It is attached by a neck to a tail consisting of a hollow core surrounded by a contractile sheath and based on a spiked end plate to which six fibers are attached. The spikes and fibers affix the virus to a bacterial cell wall (B). The sheath contracts, driving the core through the wall, and viral DNA enters the cell.*

bacterial components. The injected phage DNA contains all the information required for synthesis of new phage DNA and protein and for the assembly of these components into mature, infective particles. The new enzymes required for manufacture of phage DNA and protein are synthesized by preexisting bacterial machinery, using as a template the hereditary information encoded in the phage DNA. The phage DNA is replicated manyfold and at about the same time protein components of the head, sheath, core, etc., are also manufactured. The phage DNA condenses within the bacterial cytoplasm and is enclosed by the head protein, which, in turn, is joined to the phage tail.

About 25 min after the initial particle is adsorbed, from 100 to 200 mature particles are synthesized. A phage-induced lysozyme then destroys the bacterial cell wall to release the mature phages that are now able to initiate a new infective cycle.

Fig. 375 *Morphogenetic pathway has three principal branches leading independently to the formation of heads, tails and tail fibers, which then combine to form complete virus particles. The numbers refer to the gene product or products involved at each step. The solid portions of the arrows indicate the steps that have been shown to occur in extracts.*

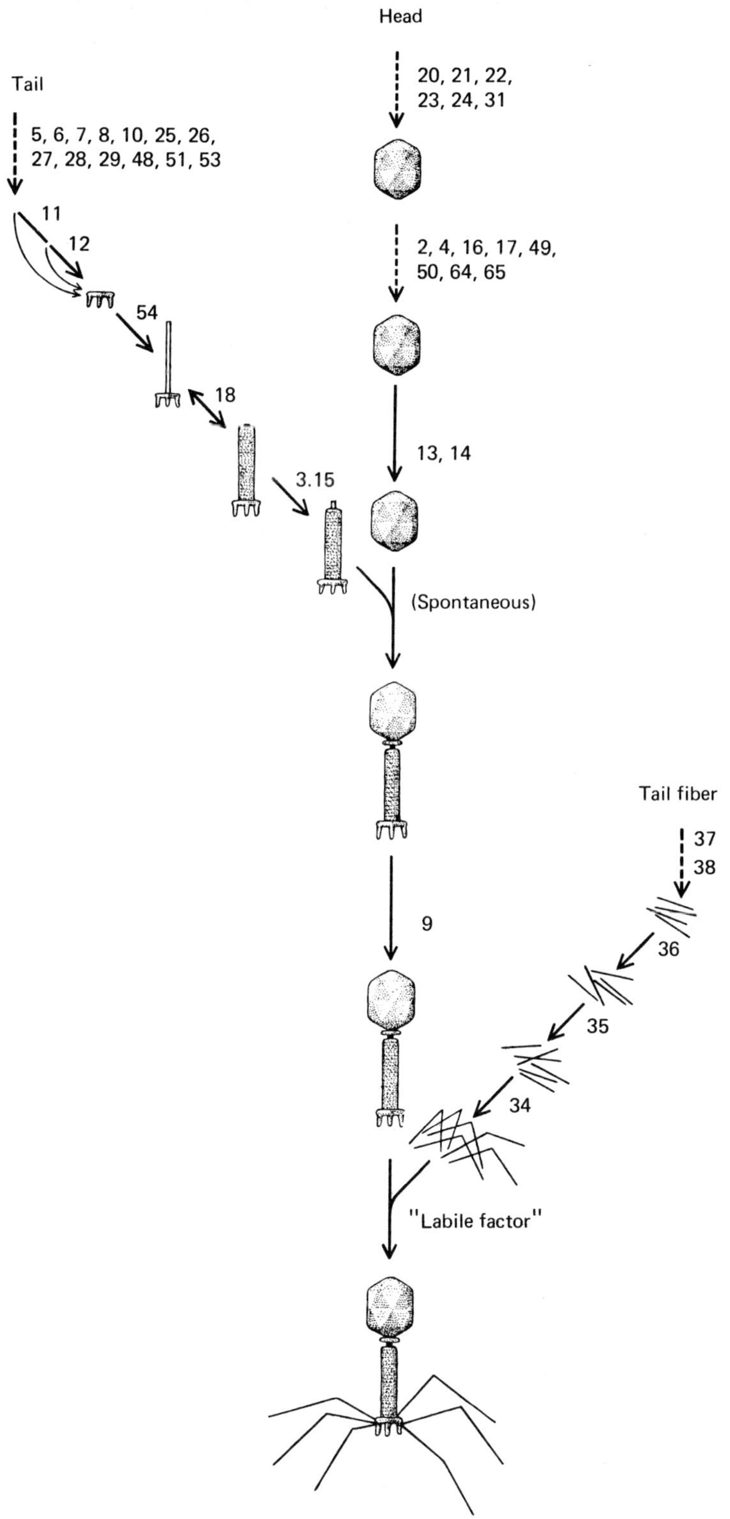
Head
20, 21, 22,
23, 24, 31
Tail
5, 6, 7, 8, 10, 25, 26,
27, 28, 29, 48, 51, 53
11
12
2, 4, 16, 17, 49,
50, 64, 65
54
18
13, 14
3.15
(Spontaneous)
Tail fiber
37
38
9
36
35
34
"Labile factor"

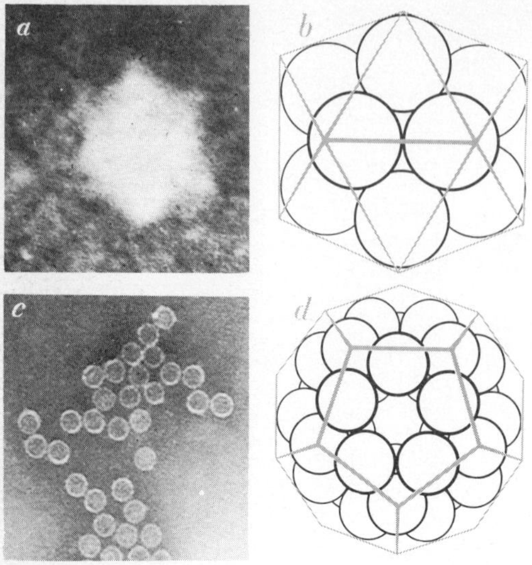

Fig. 376 Bacteriophage ϕX174, magnified 750,000 diameters in (a), appears to consist of 12 capsomeres arranged in icosahedral symmetry as shown in (b). In other micrographs (c) smaller subunits seem to be arranged in ring-like structures. Each capsomere might actually be formed from five subunits as shown in d. Thirty such subunits would form a dodecahedron. (Courtesy of Horne.)

Other phages are illustrated in Figs. 376 to 379.

Detailed knowledge of the sequence of events and their genetic controls which occur during phage replication in infected bacteria was gained by skillful exploitation of *conditionally lethal* phage mutants. These mutations can occur in almost all phage genes and they result in the synthesis of an aberrant protein specified by the mutant genes. Ordinarily the aberrant proteins are nonfunctional and normally phages with such

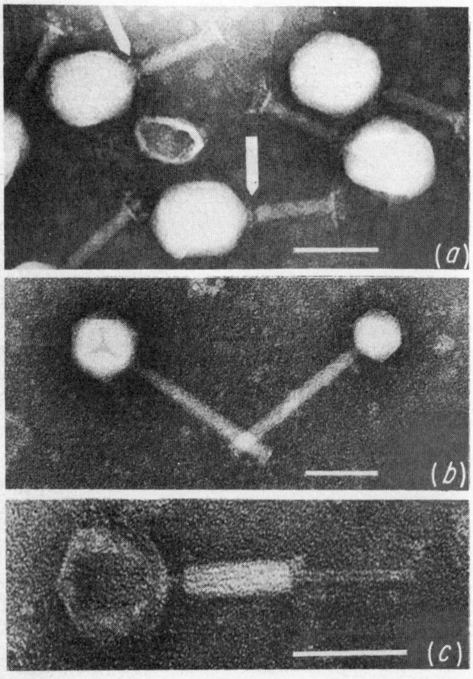

Fig. 377 (a) T4 bacteriophage showing "collars" at arrows; (b) P1 bacteriophage showing massive tail and heads of two sizes; and (c) contracted tail sheath. (Courtesy of Anderson.)

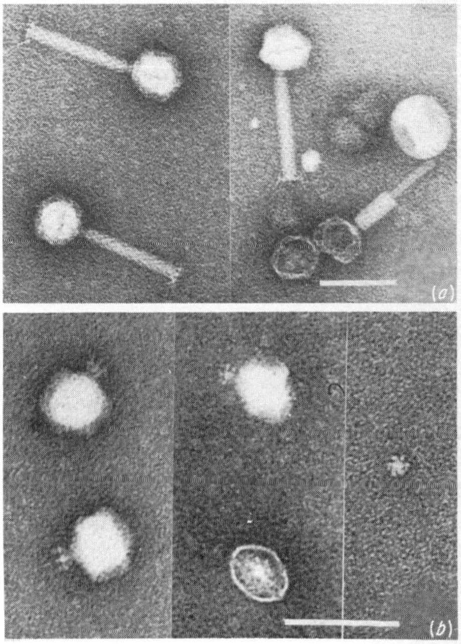

Fig. 378 *(a) P22 bacteriophage after 10 min in 3 percent* H_2O_2 *and 30 percent* C_2H_5OH*; (b) normal particles and base plates seen end on.* (Courtesy of Anderson.)

mutational defects would fail to replicate and hence could not be propagated and studied. Edgar and Epstein (1965) discovered conditionally lethal mutations in phage genes in which the mutant protein is produced in either a functional or a nonfunctional form depending on

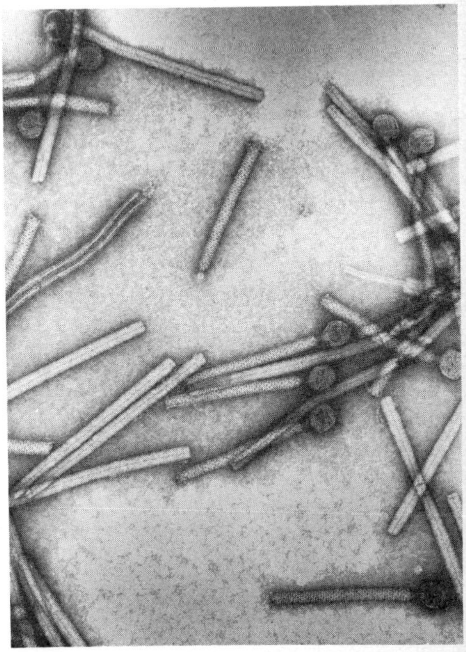

Fig. 379 *SPα bacteriophage found in lysogenic Bacillus subtilis 168 ultraviolet induction. The long rods are tobacco mosaic viruses added for comparison.* (Eiserling and Romig, unpublished.)

growth conditions. Under *permissive* conditions, the protein is functional and infection proceeds almost normally; under alternative *nonpermissive* or *restrictive* conditions, the mutation is expressed as an altered protein and infection proceeds only until the protein is required and then stops at that point. By determining the step at which phage development is blocked under restricting conditions, it is possible to deduce the normal developmental sequence. By use of conditionally lethal mutants, it has been possible to identify about 80 genes of phage T4, and to determine their normal function and relative positions on the T4 genetic map.

Phage-infected bacteria maintained under restrictive conditions have been examined by chemical means and by observing them in the electron microscope. It has thus been possible to determine the site of the mutational defect. For instance, some mutations interfere with phage DNA synthesis, while other mutations may result in failure of the phage to make functional heads. Phage mutations in products formed late in infection generally do not affect the formation of other phage products. Therefore, a mutant which cannot make normal heads may be observed to effectively direct the synthesis of many phage tails, but of course, no infective particles will result (Fig. 380).

Further information on the biosynthetic pathways of phage synthesis have been obtained following the discovery that, under proper conditions, certain steps in phage formation can occur in cell-free extracts in the test tube. If complete heads and complete tails are mixed together, intact, infectious particles can be formed by the association of these two components. By using other combinations of the structural units which make up a completed phage particle, it has been possible to deduce the normal pathway of biosynthesis of different phage components and to determine the genes that control them (Fig. 375; Wood and Edgar, 1967).

For more information: Benzer (1962); Burnet and Stanley (1959); Hayes (1964); Hershey (1957); and Stent (1960, 1963).

Lysogeny The outcome of infection of a susceptible bacterium with a *temperate* phage can either closely resemble that of a virulent phage by going through the lytic cycle with the production of new phage particles, or alternatively, it can take an entirely different course that culminates in the establishment of a lysogenic bacterium. Which of these alternatives actually occurs depends on the genetic constitution of the temperate phage, the physiological state of the sensitive bacterium, and various environmental factors. In establishing the lysogenic state, the injected phage DNA does not multiply vegetatively, but instead is reduced to the *prophage* state.

In a lysogenic bacterium the prophage (the phage genetic material) is generally inserted at a specific site into the bacterial chromosome. As such it behaves like a bacterial gene and is replicated in synchrony with the bacterial chromosome. Each descendant of a lysogenic bacterium receives a copy of the phage genome along with its copy of the bacterial

Fig. 380 Two assembly reactions occur in the experiment illustrated above: union of heads and tails and attachment of fibers. One phage (left), with a defective tail gene, produces heads and fibers. The other (right), with a mutation in a head gene, produces tails and fibers. When the two extracts are mixed and incubated, the parts assemble to produce infectious virus.

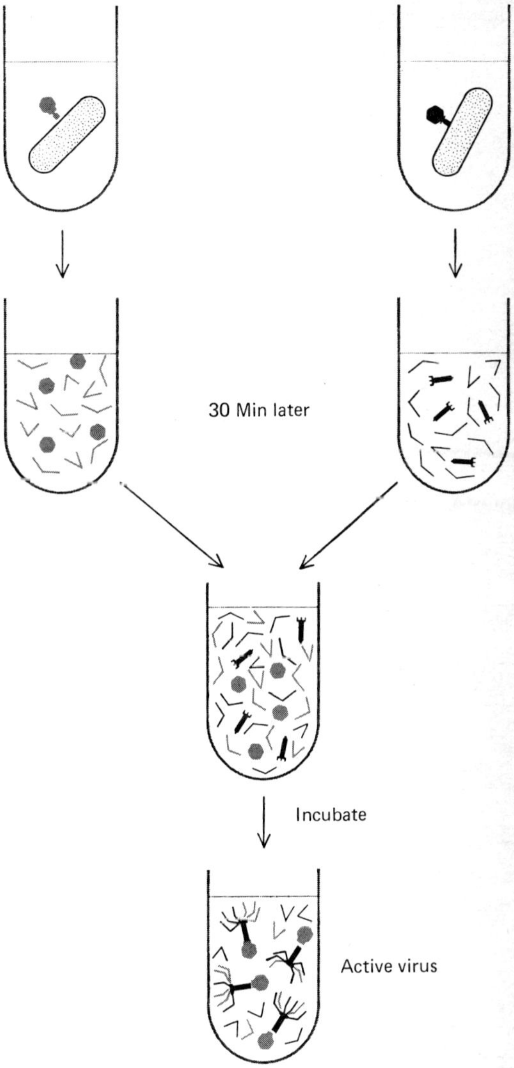

chromosome. As long as the prophage is retained in its chromosomal location, it has no adverse effects on the bacterium that harbors it. Occasionally, however, the prophage escapes the control of the bacterial chromosome and initiates vegetative replication. When this occurs, the cell lyses and releases mature phage into the medium. This process, called spontaneous induction, can be greatly accelerated by treating some lysogenic bacteria with inducing agents such as ultraviolet light or carcinogenic chemicals. These inducing agents cause almost every cell of a lysogenic culture to produce mature, infective phage particles.

The interactions of temperate phage with its bacterial host are shown diagrammatically in Fig. 381.

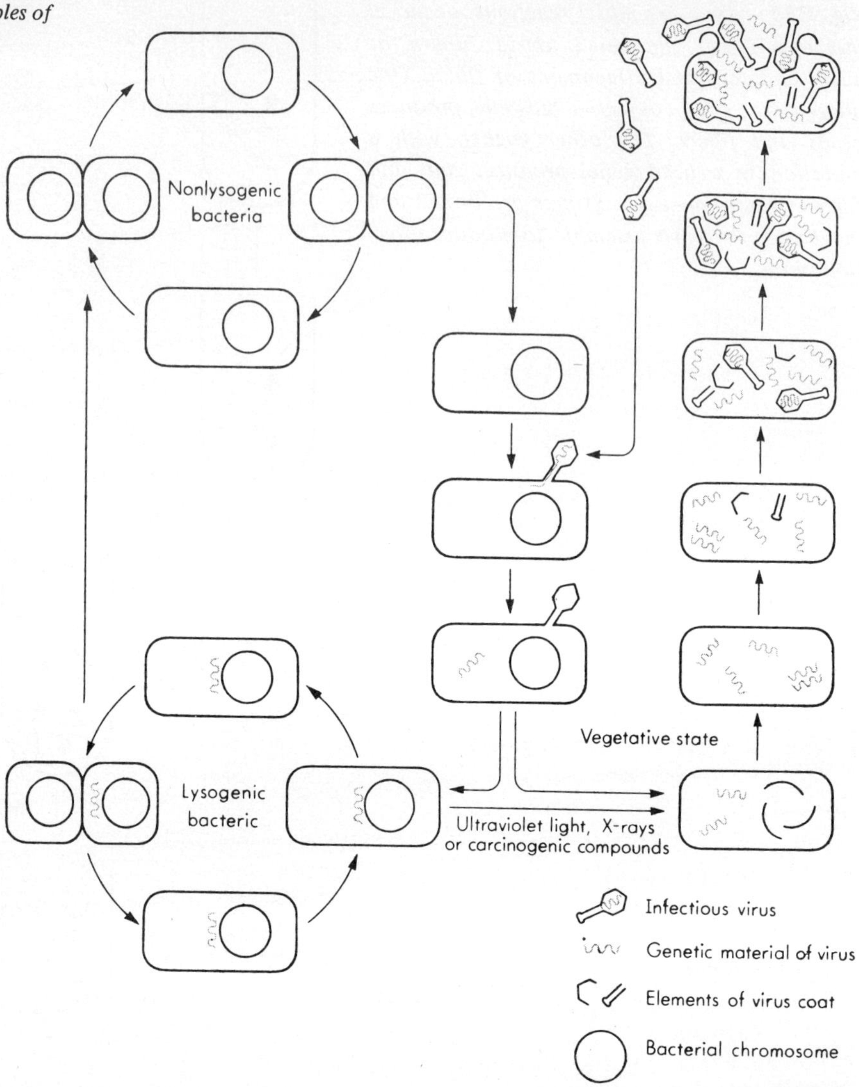

Fig. 381 Life cycle of bacterial virus shows that, for the bacterium attacked, infection and death are not inevitable. After the genes of the virus enter a cell descended from a completely healthy line (top left), the cell may take either of two paths. One (far right) leads to destruction as the virus enters the vegetative state, makes complete copies of its infective self and bursts open the cell, a process called lysis. The other path leads to the so-called lysogenic state, in which the viral genes attach themselves to the bacterial chromosome and become a provirus; the cell lives. Exposure to ultraviolet light, however, can dislodge the provirus and induce the vegetative state. The provirus is sometimes lost during cell division, running the cell to the nonlysogenic state.

Fig. 382 Bacteriophage plaques (colonies). A mixture of about 1×10^8 *bacteria and 100 phage particles was plated on appropriate medium in a petri dish. The bacteria formed a solid growth over the surface except for clear circular spots that represented phage plaques, each containing about* 10^{11} *phage particles.* (Courtesy of Demerec.)

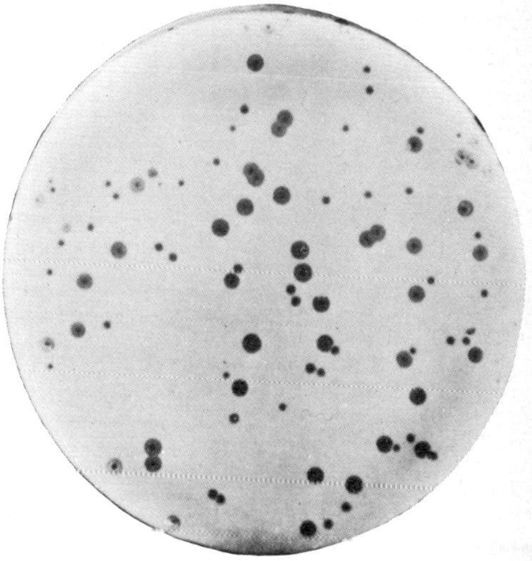

Enumeration of phage particles The number of infective phages in a suspension can be determined by counting the number of plaques produced by a suitably diluted phage sample on a bacterial lawn. In this method a few phage particles are mixed with a large excess of bacteria in soft (0.85 percent), molten agar and poured onto the surface of a plate of ordinary agar medium. The uninfected bacteria form a confluent layer of growth in the soft agar, except for the cleared areas (plaques) which represent sites of phage multiplication (Fig. 382). Particles may also be enumerated by quantitative electron microscopy. It has been found that from 0.5 to 0.9 of the visible particles form plaques when assayed as above.

Phage taxonomy Phages can be classified by several criteria, but it has not proved possible to determine their natural relationships to one another. Criteria used to classify phages include host specificity, that is, the kinds of bacteria they can infect; serological relationships; size and morphology, generally as determined with the electron microscope; chemical composition; susceptibility to inactivation; and genetic exchange. Of these, the first two are most easily accomplished. Antiserum produced by injecting purified phages into laboratory animals neutralizes the phages and prevents them from forming plaques when mixed with sensitive bacteria. These antisera are specific for the phage used to produce them, but will also neutralize closely related phages; they are without effect on nonrelated species.

For more information on phages: Adams (1959); Bertani (1953); Jacob and Wollman (1957); Lwoff (1953).

References

Adams, M. H.: "Bacteriophages," New York, Interscience Publishers, Inc., 1959.

Benzer, S.: The fine structure of a gene, *Sci. Am.*, **206**:70, 1962.

Bertani, G.: Lysogeny. In "Advances in Virus Research," vol. 5, New York, Academic Press, Inc., 1958, p. 151.

Bertani, G.: Infections bacteriophagiques secondaires des bacteries lysogenes, *Ann. Institut Pasteur,* **84**:1, 1953.

Burnet, F. M., and W. M. Stanley: "The Viruses," vols. I, II, and III, New York, Academic Press, Inc., 1959. See chapters by Lwoff, Garen, and Kozloff; Stent, Levinthal, Jacob and Wollman; Stahl, Luria, Evan, and Schachman; and Williams.

Edgar, R. S., and R. H. Epstein: The genetics of a bacterial virus, *Sci. Am.*, February 1965.

Freifelder, D., A. K. Kleinschmidt, and R. L. Sinsheimer: Electron microscopy of single-stranded DNA: Circularity of DNA of bacteriophage ϕX174, *Science,* **146**:254, 1964.

Guthrie, S. D., and R. L. Sinsheimer: Infection of protoplasts of *Escherichia coli* by subviral particles of bacteriophage ϕX174, *J. Mol. Biol.,* **2**:297, 1960.

Hayashi, M., M. N. Hayashi, and S. Spiegelman: Replicating form of a single-stranded DNA virus: Isolation and properties, *Science,* **140**:1313, 1963.

Hayes, W.: "The Genetics of Bacteria and Their Viruses," Oxford, Blackwell Scientific Publications, Ltd., 1964.

D'Hérelle, F.: Sur un microbe invisible antagoniste des bacilles dysentériques, *C. R. Acad. Sci. Paris,* **165**:373, 1917. See also, "The bacteriophage," *Sci. News,* no. 14, p. 44, Harmondsworth, Penguin Press, 1949.

Hershey, A. D.: Bacteriophages as genetic and biochemical systems. In "Advances in Virus Research," vol. 4, New York, Academic Press, Inc., 1957, p. 25.

——— and M. Chase: Independent functions of viral protein and nucleic acid in growth of bacteriophage, *J. Gen. Physiol.,* **36**:39, 1952.

Hofschneider, P. H., and A. Preuss: M13 bacteriophage liberation from intact bacteria as revealed by electron microscopy, *J. Mol. Biol.,* 7:450, 1963.

Jacob, F., and E. L. Wollman: Genetic aspects of lysogeny. In "Symposium on the Chemical Basis of Heredity," edited by W. D. McElroy and B. Glass, Baltimore, The Johns Hopkins Press, 1957.

Kleinschmidt, A. K., A. Burton, and R. L. Sinsheimer: Electron microscopy of the replicative form of the DNA of bacteriophage ϕX174, *Science,* **142**:961, 1963.

Loeb, T., and N. D. Zinder: A bacteriophage containing RNA, *Proc. Natl. Acad. Sci. U.S.*, **47**:282, 1961.

———: Lysogeny, *Bact. Rev.,* **17**:269, 1953.

Luria, S. E., and M. L. Human: Chromatin staining of bacteria during bacteriophage infection, *J. Bact.,* **59**:551, 1950.

Lwoff, A.: Bacteriophage as a model of host-virus relationship. In "The Viruses," vol. II, edited by F. M. Burnet and W. M. Stanley, New York, Academic Press, Inc., 1959.

Lwoff, A.: Lysogeny, *Bact. Rev.,* **17**:269, 1953.

Sekiguchi, M., A. Taketo, and Y. Takagi: An infective deoxyribonucleic acid from bacteriophage ϕX174, *Biochim. Biophys. Acta,* **45**:199, 1960.

Sinsheimer, R. L.: A single-stranded deoxyribonucleic acid from bacteriophage ϕX174, *J. Mol. Biol.,* **1**:43, 1959.

——, B. Starman, C. Nagler, and S. Guthrie: The process of infection with bacteriophage ϕX174. I. Evidence for a replicative form, *J. Mol. Biol.,* **4**:142, 1962.

Stent, G. S.: "Papers on Bacterial Viruses," Boston, Little, Brown and Company, 1960.

——: "Molecular Biology of Bacterial Viruses," San Francisco, W. H. Freeman and Company, 1963.

Twort, F. W.: An investigation on the nature of the ultramicroscopic viruses, *Lancet* (2), **189**:1241, 1915.

Wood, W. B., and R. S. Edgar: Building a bacterial virus, *Sci. Am.,* July 1967.

29 *Bacterial and viral diseases of plants*

Bacteria

The first recorded observation on a bacterial disease of plants dates back to the work of Burrill (1881), who discovered the causative organism of pear blight. This work was confirmed by Waite (1891), who isolated the etiological agent and proved its pathogenicity. Since then a large number of bacterial plant pathogens have been isolated and described. It is safe to assume that there are as many bacterial diseases of plants as of man and animals.

Before an organism can be stated definitely to be the causative agent of a plant disease, it must be isolated from the plant tissue and its pathogenicity proved beyond doubt. Koch (1883) postulated certain requirements that should be met before an organism could be said to be the cause of a specific disease. These requirements have been generally accepted by both plant and animal pathologists. Koch's postulates are as follows:

1. An organism must be consistently associated with the disease in question.
2. The organism must be isolated in pure culture and accurately described.
3. The organism in pure culture, when inoculated into healthy plants, must be capable of reproducing the disease.
4. The organism must be reisolated from the diseased plant tissue and shown to be identical with the original species.

Toxins in plant disease

Owens (1969) reported that bacterial plant pathogens may cause disease by (1) producing one or more toxic substances; (2) excreting enzymes which degrade cell walls, causing tissue disorganization; (3) destroying plant hormones causing abnormal growth of the host; (4) physically blocking the water-conducting vessels of the host plant by their prolific growth or by the production of viscous polysaccharides; or (5) disrupting the metabolism of the host by producing antimetabolites (see page 233).

The toxins of plant pathogens may be defined as nonenzymatic substances which injure plant cells or interfere with their metabolism. Such toxins are smaller in molecular size than the toxins elaborated by animal pathogens and they do not induce antibody formation when injected into animals.

Bacterial diseases of plants

The bacterial diseases of plants may be placed into five groups on the basis of the location and character of the lesions produced: (1) soft rots; (2) vascular diseases or wilts; (3) blights; (4) intumescence diseases; and (5) local lesions or spots.

Soft rots Organisms responsible for soft rots reduce the plant tissue to a soft, very moist, pulpy mass. The condition may be better recognized as a state of rottenness. The attack may or may not be due to a specific organism.

The organisms producing soft rots differ from other forms found in soil in that they attack healthy plant tissue by the secretion of an extracellular protopectinase. The enzyme dissolves the pectin or cement-like material that binds the plant cells. The action is probably hydrolytic, resulting in the liberation of soluble sugars which are utilized by the bacteria for food. The plant becomes reduced to a mass of separate cells, which become converted later into a slimy, pulpy material.

In most cases, the specific organism is accompanied or closely followed by many saprophytic soil bacteria and fungi. These organisms find a favorable environment in the exposed cells and produce relatively large quantities of ammonia by the deaminization of the amino acids present in the proteins of dead plant tissue. The ammonia produces a destruction of the neighboring plant cells and rapidly reduces the plant to a slimy, pulpy, foul-smelling mass. The unpleasant odor is due to the secondary invaders. Plants decayed by pure cultures of the specific disease organisms do not give off an objectionable odor.

Organisms producing soft rots are placed in the genus *Erwinia*. It is characterized as follows: Motile rods which normally do not require organic nitrogen compounds for growth. Produce acid with or without visible gas from a number of sugars. In some species the number of carbon compounds attacked is limited, and lactose may not be fermented. May or may not liquefy gelatin. May or may not reduce nitrate to nitrite. Invade the tissues of living plants and produce dry necroses, galls, wilts, and soft rots.

Some important species of *Erwinia* producing soft rots include: *E. ananas*, the cause of brown rot of the fruitlets of pineapple; *E. aroideae*, the cause of soft rot of calla, potato, eggplant, cauliflower, radish, cucumber, cabbage, parsnip, turnip, and tomato; *E. atroseptica*, responsible for black rot of stem and tuber of potato and other vegetables; *E. carotovora*, the cause of black rot of carrot, cabbage, celery, cucumber, eggplant, muskmelon, onion, parsnip, pepper, potato, radish, tomato, turnip, iris, hyacinth, and other plants. *E. chrysanthemi*, responsible for a soft rot on many fleshy vegetables; *E. dissolvens*, isolated from rotting corn stalks; and *E. rhapontici*, the cause of crown rot of rhubarb.

For more information: Dye (1969*a*); Grula et al. (1968*a, b*); Starr and Moran (1962).

Vascular diseases or wilts The bacterial wilts constitute a group of important destructive diseases of plants. The infecting organisms multiply and accumulate in large numbers in the vascular system, causing an interruption in the flow of sap in the plant. A complete interruption in the flow of sap results in a rapid wilting of the plant. A partial interruption results in the growth of a sickly plant which makes poor headway and finally dies. In some cases death is due to the action of secondary invaders.

Organisms causing vascular diseases are found in the genera *Corynebacterium, Erwinia, Pseudomonas,* and *Xanthomonas.*

The *Corynebacterium* consists of straight or slightly curved rods with irregularly stained segments, sometimes granules. Frequently show club-shaped swellings. Snapping division produces angular and palisade arrangements of cells. Gram-positive, sometimes young cells and sometimes old cells losing the stain easily. Granules invariably gram-positive. May or may not liquefy gelatin. May or may not ferment sugars.

The genus *Pseudomonas* is characterized as follows: cells monotrichous, lophotrichous, or nonmotile. Gram-negative. Frequently develop fluorescent, diffusible pigments of many colors. Sometimes the pigments are bright red or yellow and nondiffusible. Many species fail to develop any pigmentation. Frequently reduce nitrate to either nitrite, or to ammonia, or to free nitrogen.

In the genus *Xanthomonas*, the cells are usually monotrichous. A yellow water-insoluble pigment is produced on agar. Proteins usually readily digested. Hydrogen sulfide is produced. Some species liquefy pectin, others do not.

Some important organisms causing wilt diseases are: *X. stewartii*, the cause of wilt disease of corn; *C. insidiosum*, the agent of vascular disease of alfalfa; *C. michiganense*, the cause of canker of tomato; *E. tracheiphila*, the etiological agent of wilt of cucumber, cantaloupe, muskmelon, pumpkin, and squash; *P. solanacearum*, the cause of brown rot of potato, tobacco, and tomato; and *X. vasculorum*, the causative organism of gummosis of sugar cane.

For more information: Dye (1969*b*).

Blights Organisms producing blight diseases are capable of penetrating considerable distances between cells, leaving the neighboring tissue intact. The bacteria grow in the plant juices without producing any digestion of the tissues. The rods produce usually a discoloration of the leaves and branches. Death is due probably to an interference with the flow of plant sap.

Organisms producing plant blights are found in the genera *Erwinia*,

Pseudomonas, and *Xanthomonas*. Some important species are: *E. amylovora*, the agent of fire blight or pear blight, isolated from the blossoms, leaves, and twigs of diseased pear and apple trees (Fig. 383); *P. medicaginis*, the etiological agent of stem blight of alfalfa, isolated from brown lesions on the leaves and stems; *P. mori*, the cause of mulberry blight, isolated from blighted shoots; *P. pisi*, the agent of stem blight of field and garden peas, isolated from water-soaked lesions on stems and petioles; *X. juglandis*, the cause of walnut blight, isolated from black spots on the leaves and nuts of English walnuts; and *X. phaseoli*, the cause of blight of bean, hyacinth bean, lupine, and other plants.

For more information: Dye (1968); Huang and Goodman (1970).

Intumescence diseases Some bacteria produce galls or tumors on plants. These excrescences or abnormal growths are produced by the action of the organisms on the meristematic tissue of the plants. Tissues infected in this manner are grouped under the intumescence diseases.

In some infections the galls remain small; in others they may assume large proportions. Sugar beets have been known to carry tumors larger than the original plants. The bacteria are believed to elaborate some irritating metabolic product that causes rapid division of the neighboring plant cells. Intracellular organisms are not necessary for the development of characteristic lesions.

Organisms producing intumescence diseases are classified in the genera *Agrobacterium, Corynebacterium, Pseudomonas,* and *Xanthomonas.*

The genus *Agrobacterium* is characterized as follows: Small, short rods which are typically motile by means of 1 to 4 peritrichous flagella. Ordinarily gram-negative. Gelatin is either slowly liquefied or not at all. Free nitrogen is not fixed but other inorganic forms of nitrogen can ordinarily be utilized.

Some important species producing intumescence diseases are: *A. gypsophilae*, the cause of galls on *Gypsophila paniculata* and related plants; *A. rhizogenes*, the agent producing hairy-root of apple and other plants; *A. rubi*, the cause of cane gall of raspberries and blackberries; *A. tumefaciens*, the etiological agent of galls on Paris daisy (Fig. 384); *A. pseudotsugae*, the cause of galls on Douglas fir in California; *C. fascians*, the agent of fasciation on sweet pea, chrysanthemum, geranium, petunia, tobacco, etc.; *P. savastanoi*, the cause of olive galls; *P. tonelliana*, the agent of oleander galls; and *X. beticola*, the cause of galls on sugar beets and garden beets.

For more information: Boyd et al. (1970); El Khalifa and Lippincott (1968); Heberlein and Lippincott (1967); Lippincott and Lippincott (1969, 1970*a, b*); Mani (1963); Stonier (1956); Fujiwara and Fukui (1972).

Local lesions or spots In many plant diseases the attack is restricted to a small area around the point of entry. These diseases are grouped under local lesions or spots.

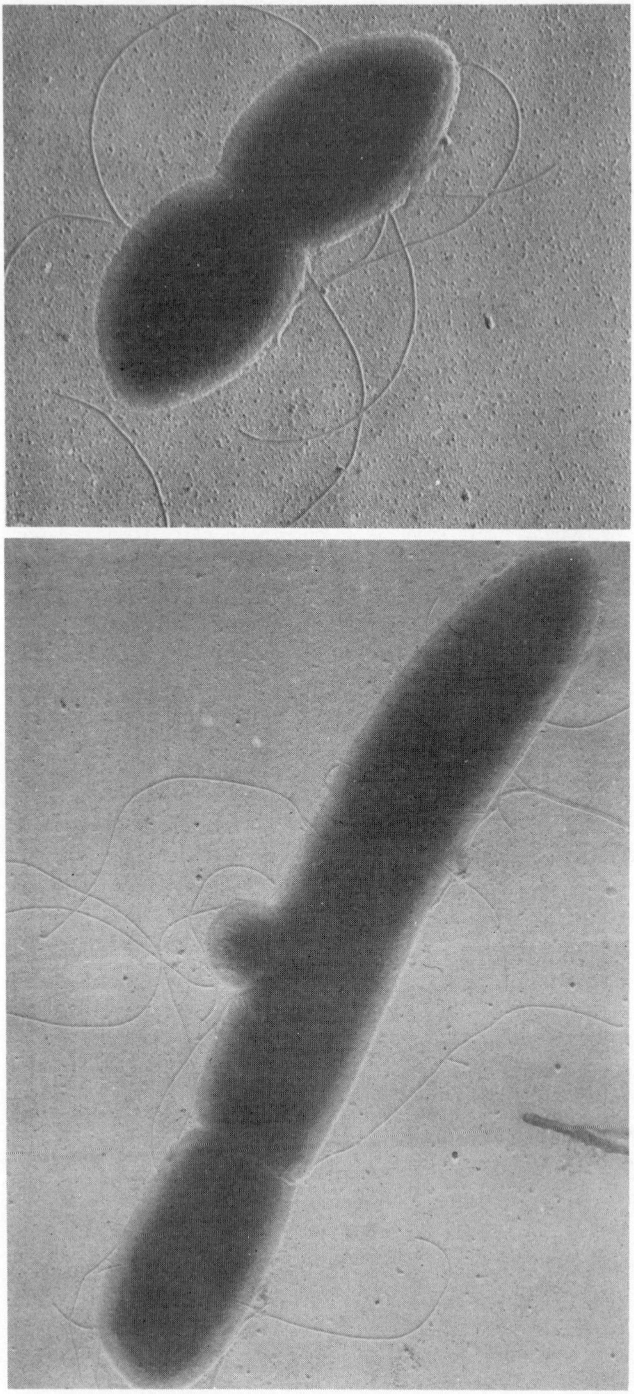

Fig. 383 Electron micrographs of Erwinia amylovora. Upper, dividing, normal rod-shaped cell; lower, filamentous cell with minicell. (Courtesy of Huang and Goodman.)

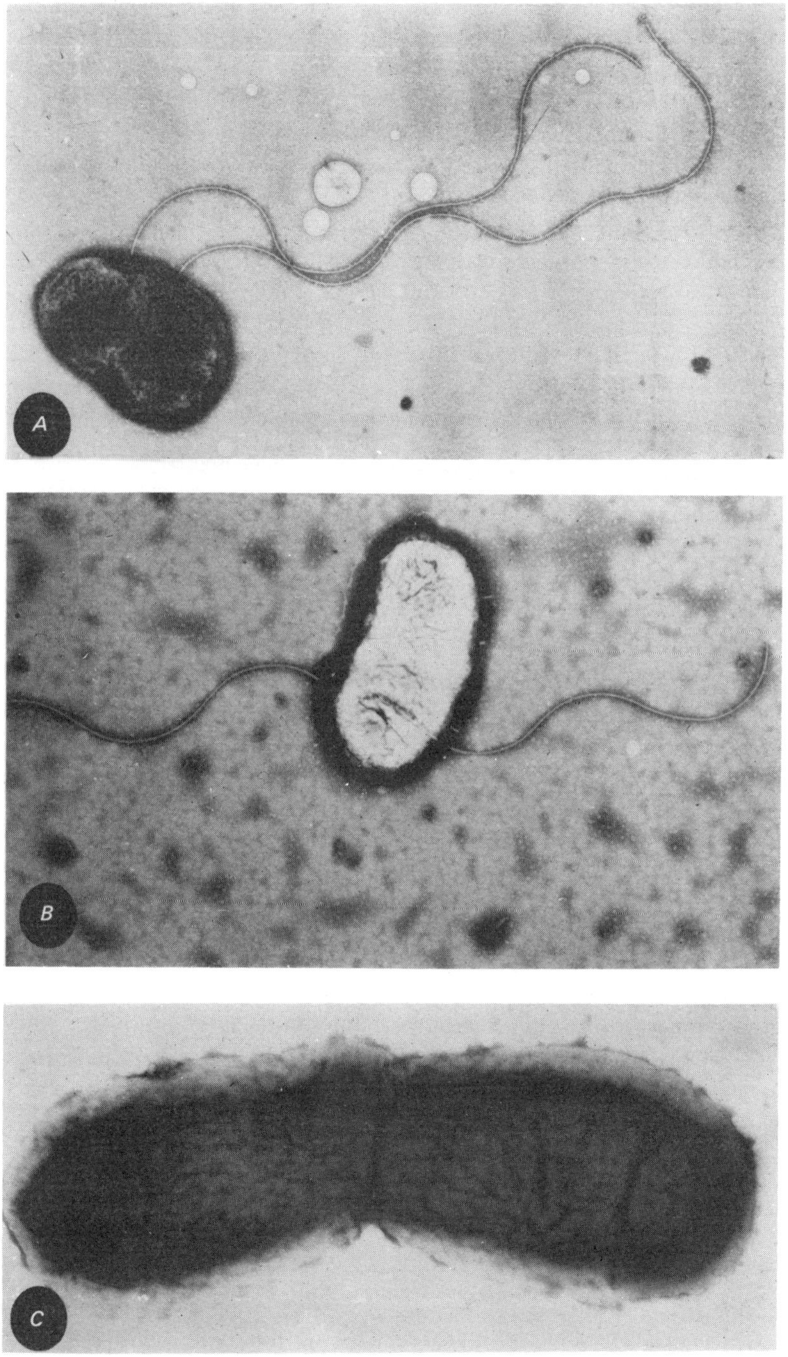

Fig. 384 Agrobacterium tumefaciens. Strain A, with flagella; strain B, with attached and broken flagella; strain C (same strain as B), two cells held together by capsular slime. (Courtesy of Boyd, Hildebrandt, and Allen.)

The organisms responsible for leaf-spot diseases produce a vigorous attack on the plant, with the result that the cells become heavily infected and strongly discolored. The discolored areas dry up and frequently fall out, leaving holes in the leaves.

Some organisms causing leaf-spot diseases are *P. angulata*, the agent of angular leaf spot of tobacco; *P. maculicola*, the cause of cauliflower spot; *P. mellea*, the etiological agent of leaf spot of tobacco; *X. cucurbitae*, the cause of leaf spot of squash and related plants; *X. malvacearum*, the agent of angular leaf spot of cotton; and *X. ricinicola*, the cause of leaf spot of castor bean.

Mode of infection The mode of entry of bacteria into the plant is usually through wounds. Roots, leaves, and stems are easily injured mechanically by agricultural implements, by animals, etc. Plants become easily infected following injury to the roots, whereas sound plants remain free from bacterial attack. Hailstones are known to produce injury to plants and make them vulnerable to infection. However, the usual cause of plant injury is through the bite of various insects. Sometimes the insects carry the etiological agent on their mouth parts, making it possible to injure and infect the plant in one operation.

In many of the leaf and fruit infections, the organisms gain entry through natural openings known as stomata. The organisms pass from the stomata into the intercellular spaces. The bacteria greatly reduce the resistance of the cells by suffocation or poisoning and make it possible for the etiological agent to enter the affected plant cells.

Bacteria may enter plants by way of the hydathodes or organs for the excretion of water. An excessive elimination of water results in the collection of considerable moisture on the plant surface. Bacteria readily collect in the water droplets, making it possible for some to gain entrance to the plant.

Lenticels are also unprotected openings, which may offer bacteria a path for invasion of the plant. These organs are cortical pores in the stems of woody plants through which air penetrates to the interior.

Many insects are responsible for plant infections. Their proboscises or legs act as carriers of bacteria that are capable of attacking the plant. This is especially true of those plants which produce nectars designed to attract bees and other insects for the fertilization of flowers.

For more information: Carter (1962); Pirone et al. (1960); Sands et al. (1970); Van der Plank (1963); Wheeler and Luke (1963).

Plant viruses

A virus is an infectious agent generally too small to be seen with the usual light microscope. Its name is taken from the Latin and means a poison.

Viruses show the following characteristics:

1. They are generally below visibility under the light microscope, but may be seen with the electron microscope.

2. They have the ability to produce disease in almost all living things.
3. They reproduce themselves only within functioning cells.
4. They produce typical and similar disease in suitable hosts in unbroken series.
5. They are antigenic, being capable of stimulating antibody production when introduced into an animal body.
6. They show great capacity for variation.

Iwanowski (1892) was probably the first to report the existence of ultramicroscopic particles capable of producing disease. He showed that the agent causing tobacco mosaic disease passed through a filter that retained all the bacteria then known. Since that time many filter-passing agents causing diseases of plants, animals, and bacteria have been discovered.

There are probably more plant diseases caused by viruses than by bacteria. They are the causes of some of the most destructive diseases of agricultural crops. It is safe to say that almost all cultivated plants are affected by at least 1 virus. It is not uncommon to encounter plants affected by 2 or more viruses. For example, the potato is susceptible to at least 25 viruses, and the tobacco plant to at least 12 virus infections. Because of these multiple infections, it is often very difficult, if not impossible, to identify a virus by the symptoms produced.

Shapes and sizes of viruses Viruses show considerable variation in shapes and sizes. The shapes of plant viruses fall into two general classes: the anisometric and the isometric viruses.

The anisometric viruses are rod-shaped, and are the most common. They may be rigid or long flexuous rods. An example of the former is the tobacco mosaic virus (TMV); an example of the latter is the potato X virus. Some rod-shaped viruses are shown in Figs. 385 to 388. The

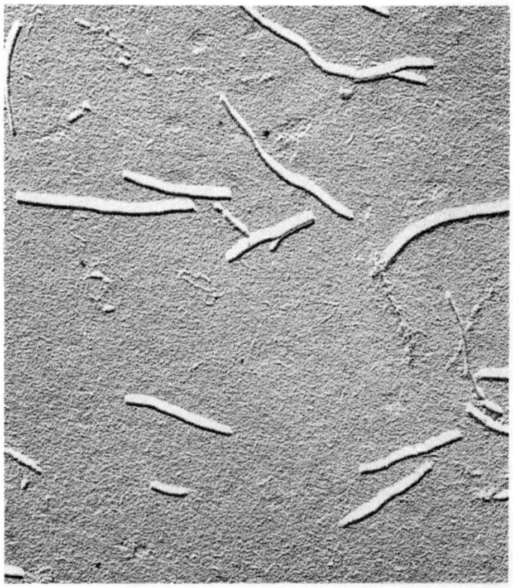

Fig. 385 Clover mosaic virus. The particles are flexuous rods. (Courtesy of Fry, Grogan, and Lyttleton.)

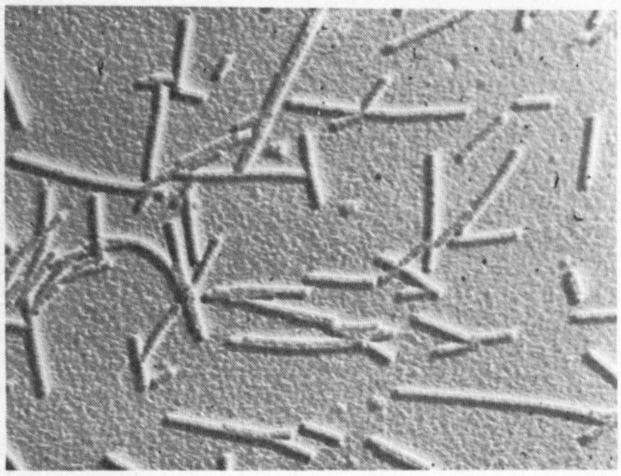

Fig. 386 Sweet-potato mosaic virus. (Courtesy of Sill, Lal, and del Rosaria.)

isometric viruses appear to be spherical but are actually polyhedral. Most appear to be icosahedral, i.e., having 20 sides. Electron micrographs of a number of isometric viruses are shown in Figs. 389 to 395.

Viruses are nonmotile and do not multiply by fission as is characteristic of bacteria.

Bacteria are measured in microns (μ). A micron is one-thousandth of a millimeter or one twenty-five thousandths of an inch. Viruses are much smaller than bacteria and are measured in nanometers (nm). A nanometer is one-thousandth of a micron.

Viruses cover the range from 16 to 300 nm. An individual TMV

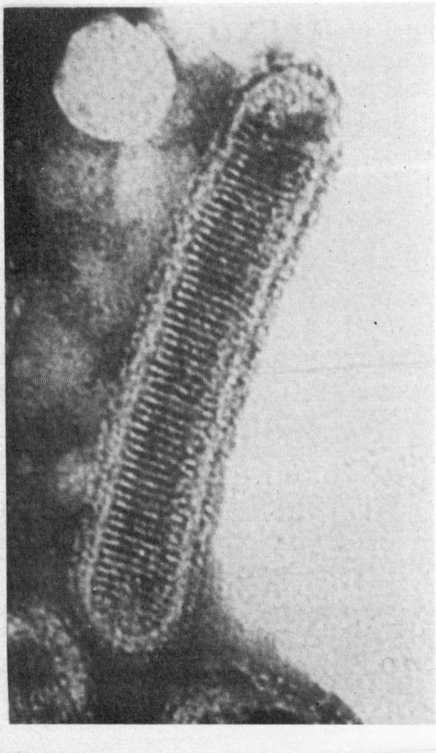

Fir. 387 Electron micrograph of a highly purified preparation of lettuce necrotic yellows virus. The bacilliform particle consists of an outer envelope covered with uniformly arranged projections and an internal component consisting of a long filament coiled in a regular low helix. (Courtesy of Wolanski, Francki, and Chambers.)

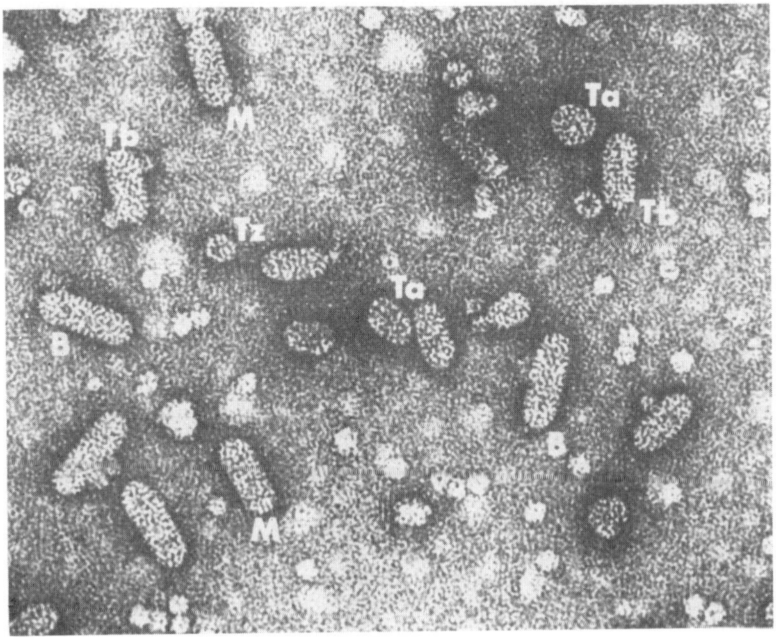

Fig. 388 Alfalfa mosaic virus. The small material is normal plant protein. (Courtesy of Hull, Hills, and Markham.)

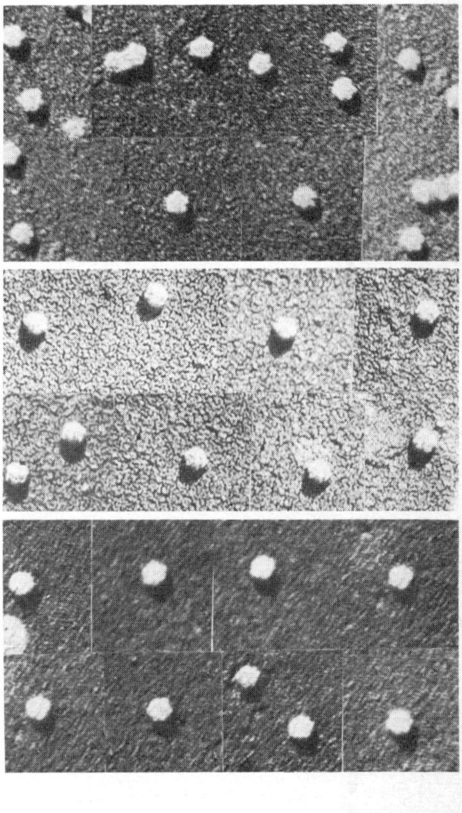

Fig. 389 Upper, tomato black ring virus; center, arabis mosaic virus; lower, raspberry ring spot virus. All X172,000. (Courtesy of Harrison and Nixon.)

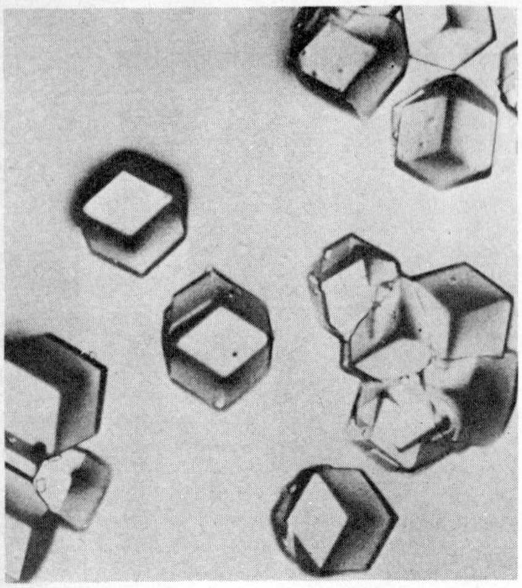

Fig. 390 *Crystalline bushy stunt virus.* (After Stanley.)

particle measures 18 x 300 nm. Some viruses are smaller than the largest protein molecules; a few are larger than the smallest bacteria.

For more information: Finch and Klug (1966); Klug et al. (1966); van Kammen et al. (1966).

Composition of viruses Stanley (1935) was the first to isolate, from Turkish tobacco plants diseased with TMV, a high-molecular-weight compound in crystalline form that possessed the properties of TMV. Repeated crystallizations failed to change the infectivity of the crystals (Fig. 396). Similar crystals have been obtained from unrelated plants diseased with TMV. The reports of Stanley on TMV have stimulated work on other agents causing plant diseases, and a number of different viruses have been obtained in crystalline form.

Viruses are usually found scattered among the molecules of other substances in cells. Only occasionally are they obtained in the form of pure crystals. To date about 20 viruses have been obtained in pure form.

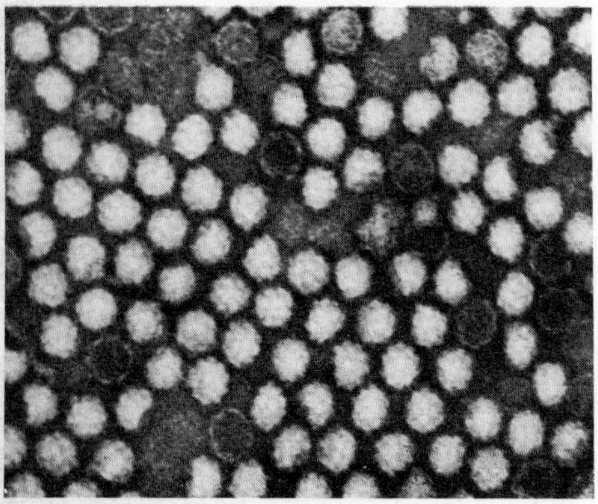

Fig. 391 *Turnip yellow mosaic virus.* (Courtesy of Huxley.)

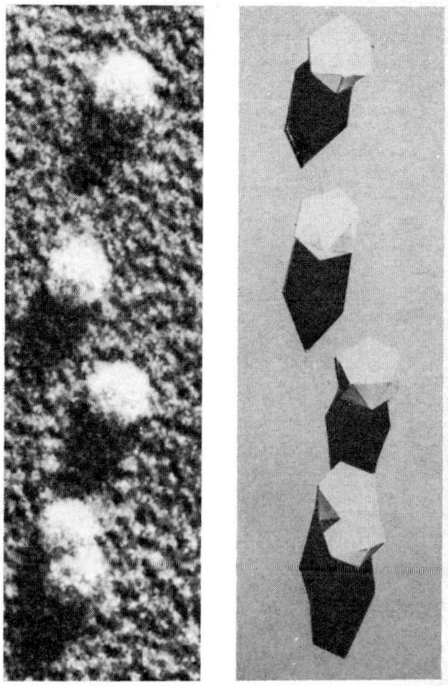

Fig. 392 Left, brome grass mosaic virus; right, icosahedral models of the particles shown on the left. (Courtesy of Kaesberg.)

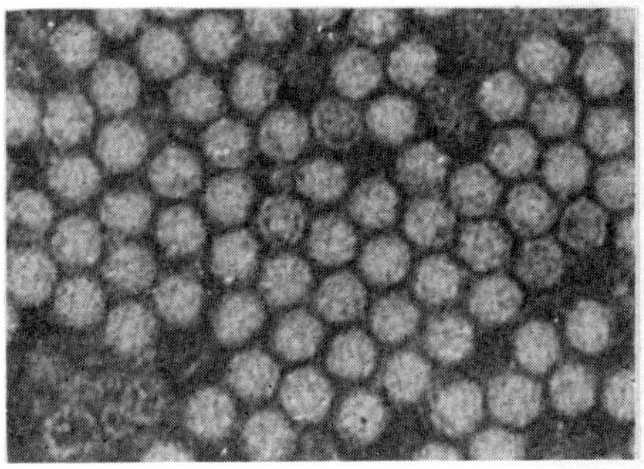

Fig. 393 Broad bean wilt virus. (Courtesy of Taylor, Smith, Reinganum, and Gibbs.)

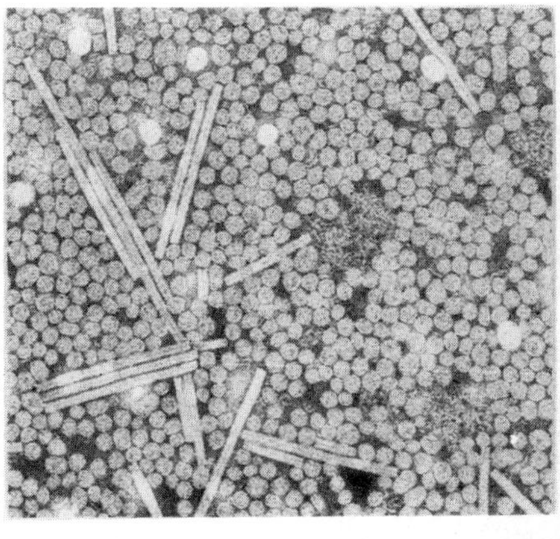

Fig. 394 Rose mosaic virus. Tobacco mosaic virus was added as a size standard. (Courtesy of Fulton.)

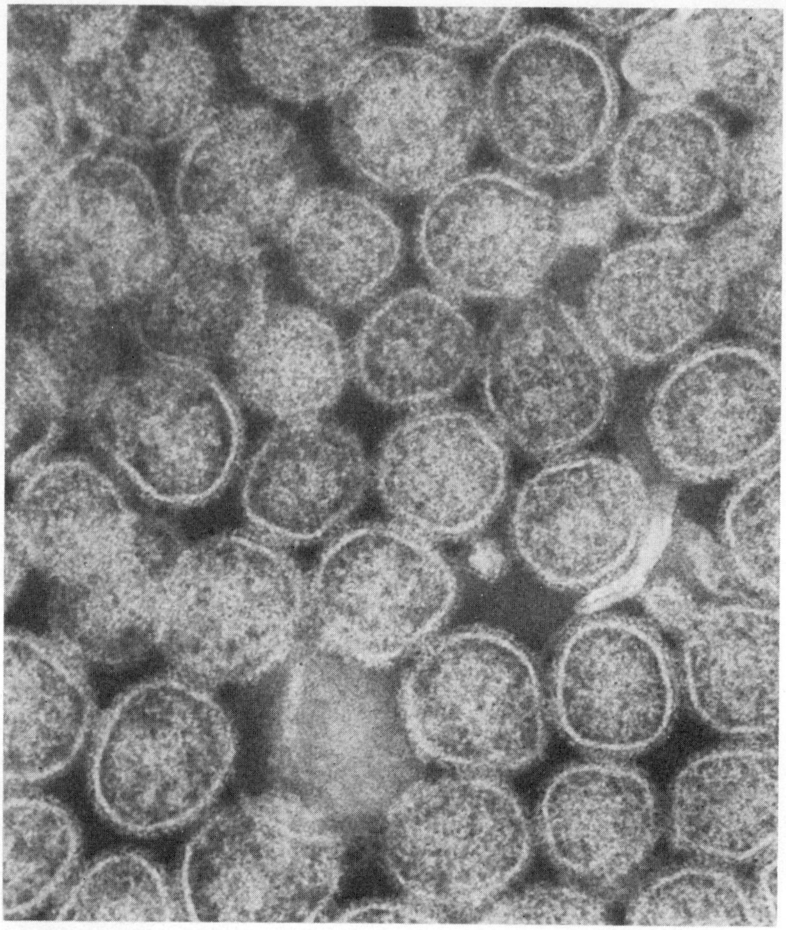

Fig. 395 Tomato spotted wilt virus. The membranes around the particles are clearly shown. Some of the particles have a tail or stalk. (Courtesy of van Kammen, Henstra, and Ie.)

They are stable and inactive in the pure form when stored away from living cells they normally infect.

A simple virus, such as TMV, is a molecule composed of two substances, nucleic acid and protein. The protein is combined with the nucleic acid and forms a protective coating around the acid. Chemically it is a nucleoprotein. Since nucleic acid is an acid, it is combined with a basic protein, i.e., one rich in the diamino-monocarboxylic acids.

In TMV about 95 percent of the molecule is protein and 5 percent nucleic acid. Viruses may differ in the ratio of protein to nucleic acid and also in the kind of nucleic acid.

Some of the larger viruses consist of more than just nucleoprotein. They may contain enzymes, phospholipides, and other compounds found in bacterial cells. Such viruses have not been crystallized.

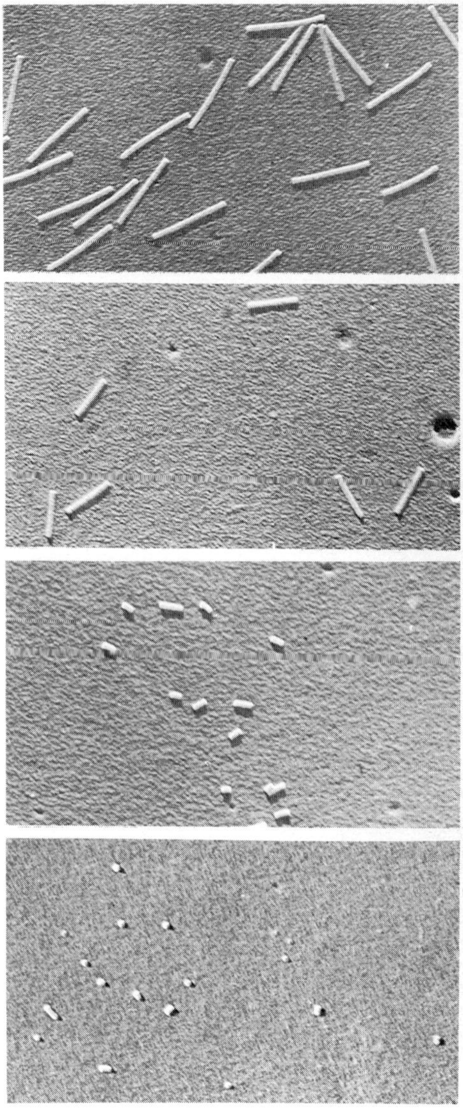

Fig. 396 Tobacco mosaic virus crystals. Only crystals measuring approximately 300 nm in length (top) are infective. (Courtesy of Steere.)

Two kinds of nucleic acids are found in living cells: Ribonucleic acid (RNA) and deoxyribonucleic acid (DNA). Almost all plant viruses contain only RNA; a few have been reported to contain only DNA. No plant virus has been isolated that contains both RNA and DNA. Cauliflower mosaic virus has been reported as containing double-stranded DNA. In TMV, and in most other plant viruses, the RNA is single-stranded, which means that it exists as a single nucleotide chain (see page 844).

For more information: Jackson et al. (1971); Russell et al. (1971).

Existence of viruses The only way to recognize the existence of viruses is to examine tissues damaged by such agents. Disease is the only satisfactory evidence of the presence of an infective virus.

Viruses may be cultivated but only in living cells. This distinguishes them from bacteria and most other organisms which may be propagated in nonliving materials.

Some viruses attack only plant cells; others only animal cells; still others only bacteria and other microorganisms. This is one method used for separating closely appearing viruses. They may also be separated by the kinds of symptoms produced in identical hosts.

Nature of viruses The exact nature of plant viruses is not clearly understood.

Living cells are associated with organization, movement, irritability, growth, and reproduction. Probably the most important of these are growth and reproduction. Inanimate objects may possess some of the above characteristics, but they are unable to reproduce themselves. Without growth and reproduction there could be no mutation. The capacity to reproduce and mutate is characteristic only of living cells.

Viruses are able to grow, reproduce, and mutate, but they do not possess the other properties of living matter. They are lifeless and quite simple in comparison to living cells. They exhibit some of the characteristics of living cells, some of the characteristics of inert chemical molecules, and they differ among themselves.

Viruses are giant molecules possessing the characteristics of other giant molecules. But once inside susceptible cells, they reproduce and behave as though they were part of the living cells. They appear to fall in between chemical molecules and living cells.

Bacterial viruses (bacteriophages) are more complex in structure. They have heads and tails and all, or most of them, have special tail fibers by which they attach themselves to bacteria.

Some of the animal viruses, such as vaccinia, are still more complex in structure. Vaccinia has an outer membrane which makes it appear more like a single-celled organism. Also, its huge size in comparison to the molecular viruses makes it visible under the light microscope.

Structure of TMV As explained above, TMV is composed of protein and RNA. The protein is combined with the RNA as ribonucleoprotein. It is arranged around the particle as a helical structure and forms a protective covering for the RNA. There is a hollow central region along the rod axis with a radius of 20 Å, and the RNA forming the axial core is located at a radius of 40 Å. The RNA is therefore deeply embedded in the outer protein component, and its structural arrangement and chain direction are closely related to the protein helical array.

TMV contains about 2200 protein subunits. Each subunit is composed of 158 amino acids arranged in definite sequence. Each nucleic acid core contains 6500 units. The virus molecule contains 5.25 million atoms. It can be seen from this that the molecule is a giant in size, with a molecular weight in millions, compared with most molecules.

Many strains of TMV have been recognized in nature. The nucleoprotein molecules isolated from various sources are similar in physical and chemical properties, yet different from each other and from ordinary TMV. A virus may become modified after cultivation in an unnatural host. Such a modified virus is most likely accompanied by a change in the properties of the specific nucleoprotein. Typical symptoms of a plant virus disease are produced only by a nonmodified (nonmutated) strain inoculated into a natural host plant.

Viruses are not only capable of reproduction, but are also subject to mutation. This ability to change makes it difficult to combat virus infections. A mutated virus is different antigenically from the original nonmutated strain. Mutation is a form of survival and viruses are forever changing themselves as a protection against adverse conditions. The nucleic acid, not the protein, is the carrier of the genetic information that is necessary in the reproduction of viruses. It is intimately involved in the process of mutation.

For information on how viruses kill cells, see page 971.

Chemistry of TMV TMV is a ribonucleoprotein and can be broken down into RNA and protein. The protein is composed of some 2200 identical subunits. Each subunit can be hydrolyzed to its constituent amino acids. The RNA consists of a long single-stranded chain of about 6500 nucleotides held together by phosphodiester linkages between molecules of the pentose sugar ribose. Each nucleotide consists of, in addition to phosphate and ribose, one of the four bases adenine, guanine, cytosine, or uracil. In contrast to cellular RNA, unusual bases or nucleotides have not been found in viral RNA.

Chromosomes of living cells are also composed of nucleoproteins. The virus nucleoproteins and chromosomes are able to control their duplication and to manufacture exact replicas of themselves. Information obtained on the structure and function of viruses aids us in understanding what actually causes or induces mutations.

For more information: Agrawal (1964); Frist et al. (1965); Fry et al. (1960); Fulton (1967); Harrison and Crowley (1965); Harrison et al. (1965); Harrison (1969); Hart (1961); Hitchborn and Hills (1967); Horne and Wildy (1961); Hull et al. (1969*a, b*); Huxley and Zubay (1960); Kleczkowski (1963); Markham et al. (1964); Mattern (1962); Milne (1970); Nilsson-Tillgren (1969); Nixon and Gibbs (1960); Nozu and Okada (1968); Schiffer and Edmundson (1968); Steere (1963); Symington and Commoner (1967); Taylor et al. (1968).

Dissociation and reconstruction of viruses Fraenkel-Conrat and Williams (1955) separated RNA from the protein of TMV. The free RNA was found to be infectious, whereas the protein was not. However, the free RNA was not as infectious as the complete virus with its heavy protein covering.

RNA in the complete TMV was about 1000 times more infectious than the same amount of RNA in the free form.

RNA is present in the core of TMV as a long string-like molecule. It is easily broken, damaged, or attacked by enzymes, unless protected from outside forces. If any part of this thread-like structure is broken, it loses its infectivity in part or completely. The outer protein coat most likely acts as a protection against damage to the RNA.

The same workers found that if pure protein units and pure RNA were brought together, a recombination occurred to form the original virus molecule. The reconstituted virus infected tobacco plants and reproduced itself just as well as the unaltered TMV (Fig. 397). Recombination also occurred between proteins and RNA isolated from different strains of TMV. Virus particles reconstituted from proteins and RNA from different strains yielded very active preparations, being equal to or higher in infectivity than the parent strain from which the proteins and RNA were obtained.

The nature of the disease provoked by mixed virus preparations resembled in each case that characteristic of the virus supplying the RNA. In other words, it is the RNA and not the protein that is the genetically active material. In contrast to these properties, the serological characteristics of mixed virus preparations were those of the virus supplying the protein. Therefore, immunological specificity is primarily an attribute of the protein, and infectivity an attribute of the RNA.

The chemistry and properties of RNA and DNA are given in Chap. 25, page 844.

For more information: Bosch et al. (1967); Fraenkel-Conrat and

Fig. 397 Fragmented and reconstituted virus. Left, tobacco mosaic virus protein; right, particles of reconstituted virus. Morphology identical with undegraded TMV. (Courtesy of Fraenkel-Conrat.)

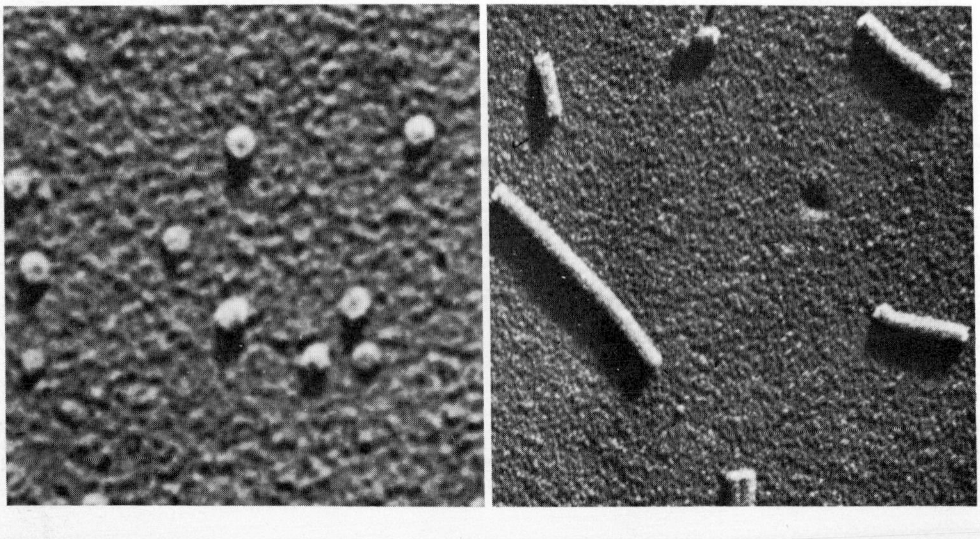

Singer (1959, 1964); Kaper (1969); Semancik and Reynolds (1969); Wagner and Bancroft (1968); Goto and Taniguchi (1971); Okada and Ohno (1972).

Preservation of viruses Plant virus culture collections maintained in stock plants have a number of disadvantages: (1) danger of contamination and mutation; (2) loss of old cultures; and (3) necessity of large storage (glasshouse) space.

Probably the best method for preservation is by the rapid removal of water from frozen virus suspensions by vacuum sublimation. This is better known as lyophilization (see page 321). Hollings and Stone (1970) lyophilized 74 plant viruses in infective sap, followed by storage in vacuum-sealed ampoules at room temperature. Fifty-seven viruses survived at least 1 year and 19 of them over 10 years. They concluded that lyophilization offered many advantages for maintaining a culture collection of sap-transmissible viruses.

Agents responsible for the dissemination of viruses

Probably all viruses causing plant diseases are disseminated by insects. Other methods, of less importance, include (1) wind, (2) water, (3) soil, (4) seed, and (5) pollen.

Wind Wind plays a minor role in the dissemination of plant viruses. However, there may be exceptions. For example, TMV is very resistant to desiccation and may be windborne in the form of dried, crumbled plant tissue. The virus is capable of readily infecting healthy plants through slight wounds.

Water Water appears to be of little importance in the spread of plant viruses. Here again there may be exceptions. It is possible to infect healthy plants with tobacco necrosis by bathing the roots in water containing the virus. The virus can infect healthy plants without artificial wounding.

Soil The soil itself is not an agent for the transmission of plant viruses. Infection may occur through roots and other underground parts of plants by soil water or by insects working in the soil. Since it is very difficult to observe underground parts of plants, the mechanism of infection by this route remains obscure. However, it has been definitely shown that wheat mosaic virus may be transmitted underground through roots or the crown, or both. The virus is capable of surviving in the soil for some time and is difficult to remove by thorough washing. The mechanism for the entrance of the virus into plants is not known.

Seed Seed transmission of plant diseases does occur for some viruses. Bean mosaic virus is transmitted in this manner in about 50 percent of plants under experimental conditions. The results are usually inconsistent.

A plant may show both healthy and infected seeds. Plants of the family *Leguminosae* appear to be more susceptible to infection by this route than plants of other families.

Pollen Virus infections may be transmitted to seeds by pollen from infected plants. In the plant Jimson weed, up to 79 percent of the seeds may become infected. Bean seeds may also be infected in this manner.

Insect transmission The most important agents for transmitting plant viruses are insects. Some insects transmit the virus mechanically, others biologically. The former method occurs usually in those insects which have chewing mouth parts. The latter method occurs only in the sucking insects, but transmission by these insects is not always biological.

According to Leach (1940), biological transmission of plant viruses by insects has usually one or more of the following attributes:

1. An apparent multiplication or increase of the virus in the insect's body.
2. An incubation period in the body of the insect, i.e., a necessary period after feeding on infected plants before the insect becomes infective or viruliferous.
3. A degree of specificity between the insect and the virus that it transmits.
4. An obligatory relationship.
5. A relation between the age or life stage of the insect and its ability to transmit the virus.
6. Congenital transmission of the virus from one generation to the next.

For more information: Hewitt and Grogan (1967); Hildebrand (1959); Lojek and Orlob (1969).

For general information on the plant viruses: Bawden (1964); Beemster and Dijkstra (1966); Esau (1968); Matthews (1970); Smith (1962, 1968); Stanley and Valens (1961).

References

Agrawal, H. O.: Electron microscope observations on the structure of the protein shell of turnip yellow mosaic virus, *Neth. J. Plant Pathol.*, **70**:175, 1964.

Bawden, F. C.: "Plant Viruses and Virus Diseases," New York, The Ronald Press Company, 1964.

Beemster, A. B. R., and J. Dijkstra: "Viruses of Plants," Amsterdam, North-Holland Publishing Company, 1966.

Bosch, L., et al.: *In situ* breakage of turnip yellow mosaic virus RNA and *in situ* aggregation of the fragments, *Virology*, **31**:453, 1967.

Boyd, R. J., et al.: Electron microscopy of phages for *Agrobacterium tumefaciens, Arch. Mikrobiol.*, **73**:47, 1970.

Burrill, T. J.: Anthrax of fruit trees; or the so-called fire-blight of pear, and twig blight of apple trees, *Proc. Am. Assoc. Advan. Sci.*, **29**:583, 1881.

Carter, W.: "Insects in Relation to Plant Disease," New York, Interscience Publishers, a division of John Wiley & Sons, Inc., 1962.

Dye, D. W.: A taxonomic study of the genus *Erwinia*. I. The "amylovora" group, *New Zealand J. Sci.,* **11**:590, 1968; II. The "carotovora" group, *ibid.,* **12**:81, 1969*a*; III. The "herbicola" group, *ibid.,* **12**:223, 1969*b*.

El Khalifa, M. D., and J. A. Lippincott: The influence of plant-growth factors on the initiation and growth of crown-gall tumours on primary pinto bean leaves, *J. Exp. Botany,* **19**:749, 1968.

Esau, K.: "Viruses in Plant Hosts," Madison, The University of Wisconsin Press, 1968.

Finch, J. T., and A. Klug: Arrangement of protein subunits and the distribution of nucleic acid in turnip yellow mosaic virus, II. Electron microscopic studies, *J. Mol. Biol.,* **15**:344, 1966.

Fraenkel-Conrat, H., and B. Singer: Virus constitution. II. Combination of protein and nucleic acid from different strains, *Biochim. Biophys. Acta,* **24**:540, 1959.

—— and ——: Reconstitution of tobacco mosaic virus. IV. Inhibition by enzymes and other proteins, and use of polynucleotides, *Virology,* **23**:354, 1964.

—— and R. C. Williams: Reconstitution of active tobacco mosaic virus from its inactive protein and nucleic acid components, *Proc. Natl. Acad. Sci. U.S.,* **41**:690, 1955.

Frist, R. H., et al.: The protein subunit of cucumber virus 4: Degradation of viruses by succinylation, *Virology,* **26**:558, 1965.

Fry, P. R., et al.: Physical and chemical properties of clover mosaic virus, *Phytopathology,* **50**:175, 1960.

Fujiwara, T., and S. Fukui: Isolation of morphological mutants of *Agrobacterium tumefaciens, J. Bact.,* **110**:743, 1972.

Fulton, R. W.: Purification and serology of rose mosaic virus, *Phytopathology,* **57**:1197, 1967.

Goto, T., and T. Taniguchi: Some characters of viruses reconstituted with components from different tobacco mosaic virus strains, *Japan. J. Microbiol.,* **15**:443, 1971.

Grula, E. A., et al.: Cell division in a species of *Erwinia.* X. Morphology of the nuclear body in filaments produced by growth in the presence of D-serine, *Can. J. Microbiol.,* **14**:293, 1968*a*.

——: Cell division in *Erwinia*: Inhibition of nuclear body division in filaments grown in penicillin or mitomycin C, *Science,* **161**:164, 1968*b*.

Harrison, B. D., and N. C. Crowley: Properties and structure of lettuce necrotic yellows virus, *Virology,* **26**:297, 1965.

—— et al.: Lengths and structure of particles of barley stripe mosaic virus, *Virology,* **26**:284, 1965.

Harrison, S. C.: Structure of tomato bushy stunt virus. I. The spherically averaged electron density, *J. Mol. Biol.,* **42**:457, 1969.

Hart, R. G.: Surface structure of tobacco mosaic virus, *J. Mol. Biol.*, 3:701, 1961.

Heberlein, G. T., and J. A. Lippincott: Ultraviolet-induced changes in the infectivity of *Agrobacterium tumefaciens, J. Bact.*, **93**:1246, 1967.

Hewitt, W. B., and R. G. Grogan: Unusual vectors of plant viruses, *Ann. Rev. Microbiol.*, **21**:205, 1967.

Hildebrand, E. M.: Importance of microscopic openings in vector transmission of plant viruses and bacteria, *Plant Disease Rept.*, **43**:715, 1959.

Hitchborn, J. H., and G. J. Hills: Tubular structures associated with turnip yellow mosaic virus in vivo, *Science*, **157**:705, 1967.

Hollings, M., and O. M. Stone: The long-term survival of some plant viruses preserved by lyophilization, *Ann. Appl. Biol.*, **65**:411, 1970.

Horne, R. W., and P. Wildy: Symmetry in virus architecture, *Virology*, **15**:348, 1961.

Huang, P.-Y., and R. N. Goodman: Morphology and ultra-structure of normal rod-shaped and filamentous forms of *Erwinia amylovora*, *J. Bact.*, **102**:862, 1970.

Hull, R., et al.: Studies on alfalfa mosaic virus. II. The structure of the virus components, *Virology*, **37**:416, 1969*a*.

—— et al.: I. The protein and nucleic acid, *Virology*, **37**:404, 1969*b*.

Huxley, H. E., and G. Zubay: The structure of the protein shell of turnip yellow mosaic virus, *J. Mol. Biol.*, **2**:189, 1960.

Iwanowski, D.: Ueber die Mosaikkrankheit der Tabakspflanze, *Centr. Bakt. Abt. II*, **5**:250, 1892.

Jackson, A. O., et al.: Replication of tobacco mosaic virus. I. Isolation and characterization of double-stranded forms of ribonucleic acid, *Virology*, **45**:182, 1971.

Kaper, J. M.: Nucleic acid-protein interactions in turnip yellow mosaic virus, *Science*, **166**:248, 1969.

Klug, A., et al.: Arrangement of protein subunits and the distribution of nucleic acid in turnip yellow mosaic virus. I. X-ray diffraction studies, *J. Mol. Biol.*, **15**:315, 1966.

Kleczkowski, A.: Protein of tobacco mosaic virus, *Biol. Rev.*, **38**:364, 1963.

Leach, J. G.: "Insect Transmission of Plant Diseases," New York, McGraw-Hill Book Company, 1940.

Lippincott, J. A., and B. B. Lippincott: Tumour-initiating ability and nutrition in the genus *Agrobacterium*, *J. Gen. Microbiol.*, **59**:57, 1969.

—— and ——: Enhanced tumour initiation by mixtures of tumorigenic and nontumorigenic strains of *Agrobacterium*, *Infection Immun.*, **2**:623, 1970*a*.

—— and ——: Lysopine and octopine promote crown-gall tumor growth in vivo, *Science*, **170**:176, 1970*b*.

Lojek, J. S., and G. B. Orlob: Aphid transmission of tobacco mosaic virus, *Science*, **164**:1407, 1969.

Mani, M. S.: "The Ecology of Plant Galls," The Hague, Dr. W. Junk Publishers, 1963.

Markham, R., et al.: The anatomy of the tobacco mosaic virus, *Virology*, **22**:342, 1964.

Mattern, C. F. T.: Electron microscopic observations of tobacco mosaic virus structure, *Virology*, **17**:76, 1962.

Matthews, R. E. F.: "Plant Virology," New York, Academic Press, Inc., 1970.

Milne, R. G.: An electron microscope study of tomato spotted wilt virus in sections of infected cells and in negative stain preparations, *J. Gen. Virol.*, **6**:267, 1970.

Nilsson-Tillgren, T.: Studies on the biosynthesis of TMV. II. On the RNA synthesis of infected cells, *Mol. Gen. Genet.*, **105**:191, 1969.

Nixon, H. L., and A. J. Gibbs: Electron microscope observations on the structure of turnip yellow mosaic virus, *J. Mol. Biol.*, **2**:197, 1960.

Nozu, Y., and Y. Okada: Amino acid sequence of a common Japanese strain of tobacco mosaic virus, *J. Mol. Biol.*, **35**:643, 1968.

Okada, Y., and T. Ohno: Assembly mechanism of tobacco mosaic virus particle from its ribonucleic acid and protein, *Mol. Gen. Genet.*, **114**:205, 1972.

Owens, L. D.: Toxins in plant disease: Structure and mode of action, *Science*, **165**:18, 1969.

Pirone, P. P., et al.: "Diseases and Pests of Ornamental Plants," New York, The Ronald Press Company, 1960.

Russell, G. J., et al.: The double-stranded DNA of cauliflower mosaic virus, *J. Gen. Virol.*, **11**:129, 1971.

Sands, D. C., et al.: Taxonomy of phytopathogenic pseudomonads, *J. Bact.*, **101**:9, 1970.

Schiffer, M., and A. B. Edmundson: Correlation of amino acid sequence and conformation in tobacco mosaic virus, *Biophys. J.*, **8**:29, 1968.

Semancik, J. S., and D. A. Reynolds: Assembly of protein and nucleoprotein particles from extracted tobacco rattle virus protein and RNA, *Science*, **164**:559, 1969.

Smith, K. M.: "Viruses," London, Cambridge University Press, 1962.

——: "Plant Viruses," London, Methuen & Co., Ltd., 1968.

Stanley, W. M.: Isolation of a crystalline protein possessing the properties of tobacco mosaic virus, *Science*, **81**:644, 1935.

—— and E. G. Valens: "Viruses and the Nature of Life," New York, E. P. Dutton & Co., Inc., 1961.

Starr, M. P., and F. Moran: Eliminative split of pectic substances by phytopathogenic soft-rot bacteria, *Science*, **135**:920, 1962.

Steere, R. L.: Tobacco mosaic virus: Purifying and sorting associated particles according to length, *Science*, **140**:1089, 1963.

Stonier, T.: Radioautographic evidence for the intercellular location of crown gall bacteria, *Am. J. Botany*, **43**:647, 1956.

Symington, J., and B. Commoner: Rod length in protein-stripped tobacco mosaic virus, *Virology*, **32**:1, 1967.

Taylor, R. H., et al.: Purification and properties of broad bean wilt virus, *Australian J. Biol. Sci.*, **21**:929, 1968.

Van der Plank, J.: "Plant Diseases," New York, Academic Press, Inc., 1963.

van Kammen, A., et al.: Morphology of tomato spotted wilt virus, *Virology*, **30**:574, 1966.

Wagner, G. W., and J. B. Bancroft: The self-assembly of spherical viruses with mixed coat proteins, *Virology*, **34**:748, 1968.

Wheeler, H., and H. H. Luke: Microbial toxins in plant disease. In "Annual Review of Microbiology," **17**:223, 1963.

Viral diseases of man 30

The etiological agents of disease discussed in Chap. 27 can be seen with the aid of a light microscope. With few exceptions the animal viruses cannot be seen with a light microscope. However, their presence may be demonstrated by examination under an electron microscope, or by the inoculation of virus-containing materials into susceptible animals.

Animal viruses

The animal viruses, like the plant viruses, are also composed of molecules of nucleoprotein, the nucleic acid being either ribonucleic acid (RNA) or deoxyribonucleic acid (DNA). Exceptions include a few animal viruses which are very much larger in size and composed of other compounds in addition to nucleoprotein. They appear to be like bacteria in this respect. As far as known, animal viruses do not attack plants or bacteria, and vice versa. The characteristics of animal viruses are discussed here.

Nature of viruses Viruses show no evidence of respiration or biochemical activity. Studied in vitro, they do not exhibit any of the characteristics of living material. Yet, when inoculated into a susceptible host, they are able to invade the cells and multiply intracellularly. All evidence points to the fact that they are obligate intracellular parasites and do not multiply outside living cells.

The rickettsiae occupy a position intermediate between the viruses and the bacteria. They possess some of the properties of the viruses and some of the bacteria. Like the viruses, they multiply only in the presence of living tissue. Like the bacteria, they show the more complex morphological structures, and multiply by binary fission. Their growth is inhibited by the same drugs which antagonize bacteria (Table 34).

Viruses show considerable variation in size. They range from about 300 nm for the large pox viruses to about 20 nm for the small foot-and-mouth disease virus. The genetic material of the viruses is either RNA or DNA, never the two in combination. Some of the larger viruses can be seen stained or unstained under the ordinary light microscope. The smaller viruses, such as the poliomyelitis virus, cannot be seen under the

Table 34 Some properties of bacteria, rickettsiae, and viruses

Properties	*Bacteria*	*Rickettsiae*	*Viruses*
Cultivation	Lifeless media	Living cells	Living cells
Size	500 nm	500 nm	250 nm
Visibility	Light microscope	Light microscope	Electron microscope
Filterability	Nonfilterable	Nonfilterable	Filterable
Multiplication	Binary fission	Binary fission	Complex
Nucleic acids	DNA and RNA	DNA and RNA	DNA or RNA
Antibiotics	Sensitive	Sensitive	Resistant

light microscope but become visible in unstained preparations under the electron microscope.

Inclusion bodies Individual virus particles are sometimes referred to as elementary bodies. Masses of viruses or "colonies" are sometimes called intracellular inclusions.

The infectious property of viruses resides in the elementary bodies. These bodies bear the same relationship to virus infections as the pneumococcus does to pneumonia.

The larger elementary bodies, i.e., those over 200 nm, can be seen under the light microscope in stained preparations. The smaller elementary bodies can be seen only under the electron microscope.

Intracellular inclusions may be present in the nucleus or the cytoplasm of parasitized cells. Nuclear inclusions are eosinophilic. Inclusions of this type are found in chickenpox, yellow fever, herpes simplex, and poliomyelitis. Cytoplasmic inclusions may be eosinophilic or basophilic. Inclusions of the former type are found in fowlpox, rabies, and vaccinia; the latter type are present in trachoma, psittacosis, and pneumonitis.

Structure of viruses Viruses are the smallest biological structures that embody all the information necessary for their own reproduction.

The complete viral particle is known as the virion and consists of a protein and a nucleic acid. The protein (capsid) forms a protective coating or shell around a core of nucleic acid. The nucleic acid is either RNA or DNA. Only the nucleic acid enters the cell that the virus is capable of attacking. The nucleic acid directs the cell to produce more of the complete virus particles, after which the cell ruptures and the new particles are ready to attack more intact cells.

The smaller viruses possess an extremely high degree of structural symmetry. The subunits (capsomeres) visible under an electron microscope are most likely individual protein molecules packed together to form a simple geometric structure. The larger viruses are more complex in structure and a certain degree of flexibility begins to appear.

Most small viruses are polyhedrons, having from 4 to 20 sides. The envelopes of such viruses probably consist of a number of identical protein subunits packed together in a symmetrical pattern (Fig. 398).

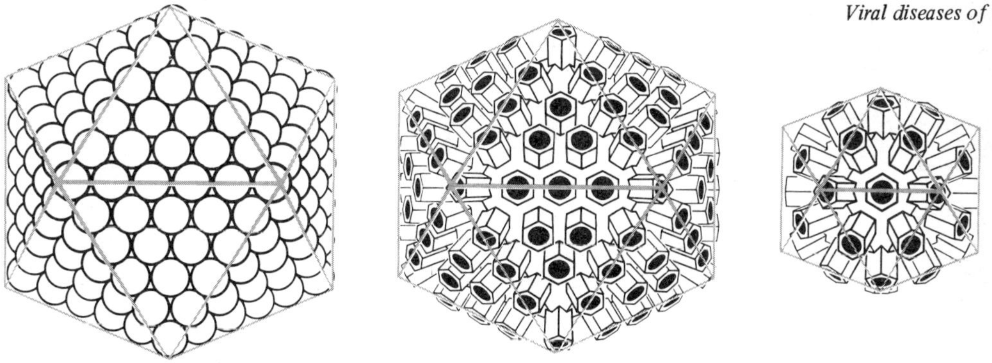

Fig. 398 Structure of virus particles. Left, drawing shows how the 252 surface subunits or capsomeres of an adenovirus particle are arranged with icosahedral symmetry. There are 12 capsomeres on corners and 240 on faces. Center, drawing shows how the 162 capsomeres of a herpes virus particle are arranged with icosahedral symmetry. Right, drawing shows how the 42 capsomeres of polyoma virus are arranged with icosahedral symmetry. (After Horne.)

Other viruses show a characteristic helical symmetry. Such viruses are tobacco mosaic (TMV), and the myxoviruses (influenza, mumps, Newcastle disease, etc.). TMV has 2130 elongated capsomeres, consisting of protein molecules arranged around a hollow core (Fig. 399). The helical coil embedded in the capsomeres represents viral RNA.

The capsid of influenza virus carries surface projections that contain a protein known as hemagglutinin (Fig. 399). It is so named because it agglutinates red blood cells.

The larger viruses are more complex in structure and symmetry. Examples of viruses in this group are variola, vaccinia, and cowpox. These viruses are large enough to be seen under a light microscope.

For more information: Finch and Klug (1966); Horne (1963); Jordan and Mayor (1962); Klug et al. (1966); Melnick et al. (1968); Richter (1963); Schäfer (1963).

Size and shape of virus particles The size of a virus may be determined by (1) filtration through collodion and millipore membranes; (2) gradient centrifugation in colloidal silica; (3) diffusion; and (4) measurement under a light microscope if the virus particles are large, or by electron microscopy if the particles are beyond the range of visibility of an ordinary light microscope. The most accurate results are obtained by means of an electron microscope.

Viruses show great variation in size and shape, ranging from about 10 to 300 nm (Table 35). Some viruses are smaller than the largest protein molecule; others are larger than the smallest bacteria. In between these two extremes, other viruses form an almost continuous spectrum. Some

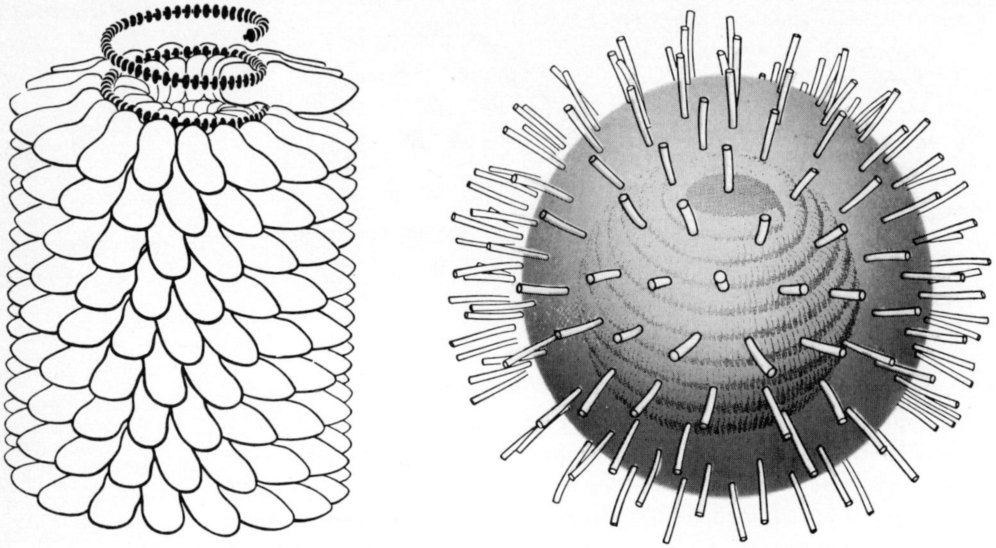

Fig. 399 Structure of virus particles. Left, drawing shows a rod-like particle of TMV with 2130 elongated capsomeres, consisting of protein molecules, arranged around a hollow core. The helical coil embedded in the capsomeres represents viral RNA. Right, drawing shows a coiled structure (RNA) in the core of an influenza virus particle. (After Horne.)

viruses are spherical, some are ovoid, others are cubes or minute parallelepipeds, still others are rod-shaped.

For more information: Pertoft et al. (1967); Ver et al. (1968); Wallis and Melnick (1967).

Cultivation of viruses Living cells of animal tissue are required for the successful propagation of animal viruses. The viruses grow in the intracellular position and are liberated only after rupture of the cell. Although viruses multiply only in living cells, it does not follow that all viruses can be cultivated in this manner. Many viruses have not been successfully cultivated. Some animal viruses may be propagated in experimental animals; in minced embryo tissue; on the chorioallantoic membrane of the developing chick; in living tissue fragments embedded in plasma; and in trypsinized tissue suspensions (single cells) placed in test tubes and rotated in the horizontal position to encourage growth in monolayers on the glass walls.

Most viruses growing in living tissue produce little or no change in cells that can be detected by microscopic examination. With certain other viruses, such as herpes and cowpox, inclusions may be formed, or cytopathogenic degeneration may occur.

For more information: Oroszlan and Rich (1964); Scherp et al. (1963).

Table 35 Size of some viruses and reference materials determined by electron microscopy

Particle	*Diameter or width x length nm*
Red blood corpuscle	7500
Serratia marcescens	750
Rickettsia prowazekii (epidemic typhus)	500 x 1100
Rickettsia typhi (endemic typhus)	450 x 1000
Chlamydia psittaci (psittacosis)	455
Lymphogranuloma venereum	330
Canary pox	265 x 311
Smallpox	275
Fowlpox	264 x 322
Mumps	230
Chickenpox	175
Tobacco mosaic	15 x 280
Herpes simplex	150
Measles	140
Rabies	125
Influenza B (Lee)	123
Newcastle disease	115
Influenza A (PR8)	100
Staphylococcus bacteriophage	100
Adenovirus	90
Escherichia coli bacteriophage T1	Head 50, tail 12 x 120
Equine encephalitis (Eastern)	42
Equine encephalitis (Western)	40
Tomato bushy stunt	26
Poliomyelitis (Lansing)	25
Coxsackie	24
Yellow fever	22
Hemocyanin molecule	22
Foot-and-mouth disease	10
Hemoglobin molecule (horse)	3 x 15
Egg albumin	2.5 x 10

Resistance of viruses Viruses are heat-labile and sensitive to the same agents that affect bacteria, but they are, in general, more resistant. They are readily destroyed at 60°C for 30 min.

Viruses survive for long periods when kept in the frozen state at subzero temperatures. The best temperature for prolonged storage of influenza virus PR8 strain, with maximum retention of infectivity, was found to be −70°C or below; for vaccinia and herpes viruses, the storage temperature for maximum retention of infectivity was reported to be −80°C or below. Infectivity is rapidly lost when viruses are stored in the usual home refrigerator (about 4°C).

Some viruses are resistant to desiccation; most are not. Infectious smallpox virus has been recovered from dry pox crusts stored at room temperature for at least 1 year.

The addition of 0.5 percent phenol to vaccine virus does not destroy its infectivity. Glycerin in a concentration of 50 percent gradually destroys bacteria but has no appreciable effect on viruses. On the other hand, viruses are susceptible to oxidative destruction. Jordan et al. (1969) reported that a washing additive containing 200 ppm of available chlorine effectively inactivated type II poliomyelitis virus. On the other hand, a soap containing a germicide or a synthetic detergent containing surfactants, carbonates, phosphates, and a quaternary germicide did not appear to influence the effectiveness of the virus.

Virus infections do not respond to treatment with sulfa drugs, penicillins, and other antibiotics. This resistance is generally believed to be associated with the method of multiplication of viruses, which are dependent upon the parasitized cell and have no metabolic process that can be blocked by drugs. Exceptions are the large basophilic viruses, such as psittacosis, trachoma, and pneumonitis, which are sensitive to a number of antibiotics. The antibiotics are not considered virucidal but are effective by virtue of their ability to interfere with the process of multiplication.

For more information: Gard (1960); Hurrell (1967); Pollard (1960); Stasny et al. (1968).

Mutation of viruses Viruses show great capacity of variation or mutation. Many types of mutants have been recognized on the basis of host range, plaque size, morphology, resistance to heat, drugs, and other inhibitors. The ability of a particle to mutate is a property not ordinarily ascribed to molecules but to living organisms. Once a virus becomes modified, it generally persists indefinitely.

Perhaps the most practically useful mutation of a virus is one that involves the loss of invasiveness or virulence. For example, influenza virus loses its virulence for mice on passage through embryonated eggs. The virulence may be restored by several passages of this avirulent strain through mice. Virulence is increased for mice but there is no increase in virulence for the chick embryo.

Another example would be the decrease in virulence of smallpox virus for man. Vaccine virus is derived from smallpox by repeated passage through calves. Antigenically, smallpox is indistinguishable from vaccinia, differing only in its ability to produce the disease smallpox.

Immunity in virus diseases Recovery from a virus disease usually confers strong, almost lasting immunity. Second attacks of such virus diseases as smallpox, mumps, measles, poliomyelitis, and chickenpox rarely occur. Various immune substances have been identified in the circulating blood including agglutinins, precipitins, and complement-fixing and neutralizing antibodies. This does not differ in any way from specific immunity induced in bacterial infections. On the other hand, with some virus infections such as influenza, the common cold, fever blisters, etc., the immunity developed is of short duration.

It is well known that some species are resistant to viral infections whereas others are not. For example humans are highly susceptible to influenza and poliomyelitis; most animal species are not susceptible to the human strains.

Recovery from a virus disease does not mean necessarily that the etiological agent is entirely eliminated from the host. The virus may persist in the recovered individual for life. The agent is believed to be stored in certain living cells where it cannot be spread to others, or come in contact with circulating antibodies. Consequently, if the host cells are not destroyed by the virus, the two can live together without the infecting agent ever coming in contact with the humoral antibodies.

Interferons Virus-infected cells elaborate a substance that is capable of protecting normal cells against some but not all viral infections. It is produced by cells, does not directly inactivate virus, inhibits replication of virus and infectious nucleic acid intracellularly, is relatively species-specific, is not inactivated by antibodies produced against the virus, and is not necessarily produced by a virus or a component of a virus. This substance is a small nondialyzable protein molecule having a molecular weight of 20,000 to 34,000 and has been given the name interferon.

Since the protective substance varies in composition, depending upon the infecting virus, the term is used in the plural. Interferons are produced in vivo before antibodies may be demonstrated and diminish as antibodies are formed. Interferons are capable of protecting the infected cells and may be produced in large enough amounts to protect neighboring cells or even cells at more distant sites. Interferons act by inhibiting some stage in the intracellular synthesis of virus, probably by synthesis of viral nucleic acid.

For more information: Bedson et al. (1967); Glasgow (1965); Hilleman and Tytell (1971); Levine and Nichol (1970); Murthy and Anders (1970); Rita (1968); Wolstenholme and O'Connor (1968); Bodo (1971).

Infectious nucleic acids Naked or free viral nucleic acids are infectious, whereas the protein shells are not. Nucleic acids are not affected by antibodies prepared against the whole virus. In other words, immunological specificity is primarily an attribute of the protein, and infectivity an attribute of the nucleic acid. However, the efficiency of infection is lower in the nucleic acid preparations than in the intact viruses (nucleic acid + protein).

According to Herriott (1961),

> The release from infected tissues of even a small proportion of total virus as free nucleic acid could, in an otherwise immune individual, lead to a low level of infection which would, perhaps, explain permanent immunity. If the proportion of nucleic acid is higher, or if the nucleic acid is resistant to nucleases because of an inert envelope, such conditions as "carrier" states – that is, viremias with or without antibodies – are possible.

Classification of viruses The number of known viruses increases almost daily. Virtually all living creatures, even bacteria, have transmissible viruses. Acute respiratory disease is caused not by one but by a host of viruses, each contributing only a small segment of the symptoms. Also, poliomyelitis is known to be caused by three distinct viruses, all members of the enterovirus family. In addition, there are at least 60 other enteroviruses closely related to poliomyelitis.

Some important viral infections of man are classified in Table 36.

Table 36 Viral infections of man

I. RNA (riboviruses)

A. Picornaviruses.

1. Polioviruses. Three distinct serotypes.
Produce poliomyelitis, pharyngitis and fever.

2. ECHO viruses (enteric cytopathic human orphan viruses). Over 30 distinct serotypes are known.
Produce aseptic meningitis, meningo-encephalitis with rash, diarrhea neonatorum, common colds.

3. Coxsackie viruses.
Type A. Usually do not grow well in tissue culture. Over 24 types are known.
Produce herpangina, aseptic meningitis, viral pharyngitis.
Type B. Grow well in tissue culture. Six serotypes are known.
Produce aseptic meningitis, myocarditis neonatorum, acute febrile respiratory illness.

4. Rhinoviruses. Found in nose and pharynx.
Type M. Grow in primary monkey kidney and human embryo kidney. Four or five serotypes are known.
Type H. Grow only in human embryo kidney cultures. At least 80 serotypes are known.
Produce the common cold.

B. REO viruses (respiratory enteric orphan viruses). Grow within the cytoplasm of many types of tissue cells giving rise to characteristic inclusions. Three types are known.
Produce steatorrheic enteritis, mild respiratory illness.

C. Arboviruses, arthropod-borne. Many grow well in tissue cultures of mammalian and avian tissue.
Group A. Differentiated from following group (B) by physicochemical properties of their hemagglutinins. Fifteen strains are recognized.
Produce Eastern, Western, Venezuelan, and African encephalomyelitis.
Group B. Contains 33 types.
Produce yellow fever, dengue fever, Omsk hemorrhagic fever, Japanese B encephalitis, St. Louis encephalitis, Murray Valley encephalitis, West Nile encephalitis, louping-ill.
Group C. Differentiated from A and B by characteristics of their hemagglutinins, and serologically.
Produce unnamed febrile illness.
Miscellaneous, ungrouped.
Produce Rift Valley fever, Sicilian phlebotomus fever, Colorado tick fever.

D. Myxoviruses. Group is subdivided by differences in biological characteristics.
Subgroup I. True influenza viruses, human strains A, A_1 and A_2 of type A, type B, and type C.

Produce classical influenza, acute bronchitis, viral pneumonia, croup, common colds, rabies.

Subgroup II. Mumps virus, Newcastle disease virus, and parainfluenza viruses, types 1, 2, 3, 4.

Produce mumps (parotitis), orchitis, mumps encephalitis, acute febrile respiratory illness, acute bronchitis, viral pneumonia, croup, common colds, possibly measles (rubeola), and German measles (rubella).

II. DNA (deoxyviruses)

A. Adenoviruses. Do not grow in chick embryos but grow well in tissue cultures within the cell nuclei. Most strains show hemagglutination. 31 different types.

Produce acute respiratory disease (ARD), pharyngo-conjunctival fever, epidemic keratoconjunctivitis (EKC), acute follicular conjunctivitis, viral pneumonia, acute febrile respiratory illness of children, common colds.

B. Herpes viruses. Grow on the allantoic membrane of the developing chick embryo and in tissue culture, initially within cell nuclei.

Produce herpes labialis, eczema herpeticum, herpetic conjunctivitis, herpetic encephalitis, cytomegalic inclusion disease, varicella (chickenpox), herpes zoster.

C. Pox viruses. Most grow well in tissue cultures and produce pocks on the allantoic membrane of the developing chick embryo. Virus development is entirely cytoplasmic.

Produce variola (smallpox), alastrim, vaccinia, cowpox.

For more information: Bedson et al. (1967); Cohen (1969); Huebner (1962); Rhodes and Van Rooyen (1968); World Health Organization (1967); Wilner (1969).

A brief description of some of the important viral diseases classified in Table 36 follows.

Poliomyelitis (infantile paralysis) Virus one of smallest known, measuring about 27 nm in diameter. It is a ribonucleoprotein and one strain (MEF-1) has been prepared in crystalline form. The RNA may be separated from the protein and is capable of producing infections although the severity of the symptoms is lower than with the intact nucleoprotein.

The virus is one of the most stable known. It resists inactivation by alcohol, phenol, formalin, chlorine, antibiotics, repeated freezing and thawing, ultraviolet light, lyophilization, and heating to 65°C for 1 hr.

Disease produced Poliomyelitis is widely prevalent. In only a small proportion of infected persons is the disease clinically recognizable. In its recognizable form poliomyelitis is an acute, systemic infectious disease that involves the central nervous system. Disease characterized usually by fever, headache, vomiting, constipation, drowsiness alternating with irritability, almost always stiffness of neck and spine, tremor, and exaggeration of muscular reflexes. In about half of such cases paralysis may develop in the first few days, which shows a marked tendency for improvement after it has reached its height. Diagnosis depends upon detection of a flaccid paralysis characteristically irregular in its involvement of muscular tissue.

All strains fall into three immunologically distinct types designated types 1, 2, and 3 (Fig. 400). Type 1 is most commonly associated with paralytic polio; type 3 less frequently; and type 2 uncommonly.

Diagnosis Virus can be isolated by tissue culture from throat secretions, or feces, early in infection. Disease usually recognized on clinical grounds.

Source of infection Nose and throat discharges of infected persons; also from those not suffering from clinically recognized attack of disease. Virus also present in feces.

Virus recovered from throat swabs and throat washings of poliomyelitis patients collected 3 to 13 days after onset of disease.

A study of four households attacked by poliomyelitis provided evidence of widespread distribution of virus in the members of these units. Of 20 members in these households, 16 had polio virus in their intestinal discharges, 7 had virus in the oropharynx.

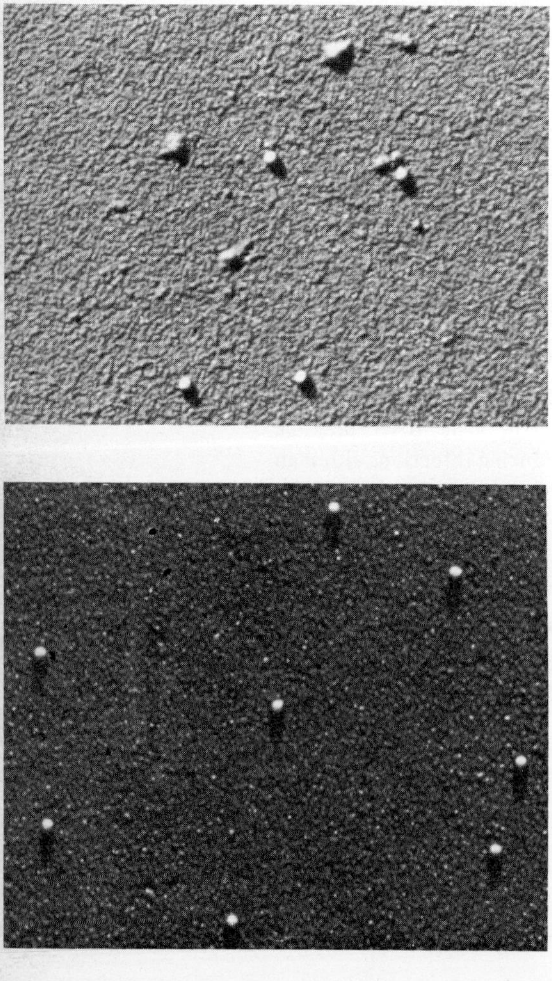

Fig. 400 Poliomyelitis virus particles. Upper, Leon strain, type 3; lower, Mahoney strain, type 1. (Courtesy of Sabin, Hennessen, and Warren.)

Mode of transmission Believed virus enters by way of nose and mouth, either from a carrier or person with a subclinical infection. Some evidence disease may be spread by milk, water, swimming pools, sewage, and insects. Flies have been shown to be contaminated with virus.

Strain of human polio virus in stool, after addition to river water, survived for 188 days and retained property of producing paralysis in monkeys.

In samples of natural waters having a pH range of 7.9 to 8.3, virus was inactivated within 10 min in presence of 0.05 ppm free chlorine. At a pH range of 10 to 11.25, 0.1 to 0.15 ppm free chlorine was necessary to achieve same results.

Incubation period Usually 7 to 14 days; may be as early as 3 and as late as 21 days. Period of greatest communicability may be from latter part of incubation period to first week of the acute illness.

Susceptibility and immunity Susceptibility to infection is general. Children are believed to be more susceptible than adults, although definite proof is still lacking. Active immunity produced after recovery from the disease. The duration of immunity is unknown, but second attacks are rare.

Prevalence Infection widespread throughout the world. Paralytic cases more prevalent in temperate climates. Cases more numerous in summer and early fall. Children one to sixteen years of age appear to be more susceptible than adults.

Prevention and control Active immunization of all persons against three types of poliovirus.

Attenuated oral vaccine is administered by mouth in two ways: a separate dose of each of the three types at intervals of about 8 weeks; or two doses of the three types mixed together at intervals of 4 to 6 weeks, with a third dose 6 to 12 months later.

Formalin-inactivated virus (dead) vaccine also provides protection but is less effective. The vaccine may be given in four injections: the first three at intervals of 6 weeks, and the fourth injection 6 months after the third.

Isolation of infected persons for 7 days. Disinfection of throat discharges and feces and articles soiled by such discharges.

Isolation of children with fever pending outcome of diagnosis. Protection of children from contact with others during an epidemic. Avoidance of nose and throat operations on children during an epidemic.

Vaccination with trivalent preparation of those exposed to disease.

No specific treatment known once symptoms appear.

For more information: Benenson (1970); Bishop and Koch (1967); Cockburn and Drozdov (1970); Dane et al. (1961*a, b*); Dick et al. (1961); Hill et al. (1967); Horne and Nagington (1959); Koch et al. (1960); Ludwig and Smull (1963); Lund (1961); McGregor and Mayor (1968); Meenan and Hillary (1963); Schaffer et al. (1960); Wiley et al. (1962); Berge et al. (1971).

Arthropod-borne viral diseases

This group includes a large number of arboviruses known to produce disease in man. The number is growing almost daily. Some important members are briefly discussed under three subgroups: (1) viral encephalitides; (2) viral fevers; and (3) viral hemorrhagic fevers.

1. **Viral encephalitides (mosquito-borne)** Some important members are Eastern equine, Western equine, Japanese B, Murray Valley, and St. Louis encephalitis viruses. Each disease is caused by a specific virus.

Acute inflammatory diseases of short duration, involving parts of brain, spinal cords, and meninges. Mild cases often occur as aseptic meningitis. Severe cases usually have acute onset, fever, stupor, disorientation, coma, tremors, sometimes convulsions in infants, and spastic paralysis. Death rate from 5 to 60 percent.

Source of infection True reservoir possibly bird, rodent, bat, reptile, amphibian, or surviving adult mosquito. Source for man is infective mosquito.

Mode of transmission Bite of mosquitoes. Mosquitoes acquire infection from birds, less frequently pigs and horses. Most important vectors are: for Eastern equine, probably *Culiseta melanura*; Western equine, *Culex tarsalis*; Japanese B, *Culex tritaeniorhynchus* and *Culex gelidus;* Murray Valley, *Culex annulirostris*; St. Louis, *Culex tarsalis, Culex pipiens-quinquefasciatus, Culex nigripalpus.*

Incubation period Usually 5 to 15 days.

Susceptibility and immunity Susceptibility generally highest in infancy and old age. Recovery or exposure to disease gives lasting immunity.

Prevalence Eastern equine recognized in Eastern and North Central United States, adjacent Canada, parts of Central and South America, Caribbean islands; Western equine in Western United States, Canada, Florida, South America; Japanese B in Western Pacific islands, Japan, Asia, Korea, Singapore, India; Murray Valley in Australia, New Guinea; St. Louis in United States, Caribbean, Panama, Brazil. Occurs during periods of high temperature and many mosquitoes.

Prevention and control Destruction of larvae and elimination of breeding places. Killing mosquitoes by spraying. Screening sleeping quarters. Avoid exposure to mosquitoes during biting hours.

No specific treatment is known.

2. **Viral fevers (mosquito-borne)** Important member of the group is dengue fever; also known as breakbone fever.

Disease produced An acute febrile disease of sharp onset, fever lasting about 5 days, accompanied by intense headache, postorbital pains, joint and muscle pains, and eruption. Eruption appears usually 3 to 4 days after onset of fever. Recovery associated with prolonged fatigue and depression. Fatality exceedingly low.

Source of infection From blood of infected person. Immediate source of infection is an infective mosquito.

Mode of transmission By bite of mosquito, *Aëdes aegypti, A. albopictus,* and *A. scutellaris*. Mosquito becomes infective 8 to 11 days after feeding on infected person and remains so for life.

Incubation period From 3 to 15 days, usually 5 to 6 days.

Susceptibility and immunity Susceptibility to disease universal. Homologous immunity of long duration; heterologous immunity of short duration.

Prevalence May occur wherever specific mosquitoes exist. Disease found mainly in the tropics and subtropics.

Prevention and control Elimination of mosquitoes and breeding places. Use of repellents.

Isolation of patients in screened rooms for at least 5 days after onset.

No specific treatment is known.

3. **Viral hemorrhagic fevers (mosquito-borne)** Important member of group is yellow fever.

Particles smaller than most viruses, measuring only 22 nm in diameter. Virus readily passes through Seitz, Berkefeld, and Chamberland filters; is readily inactivated by heat and germicides; may be preserved in 50 percent glycerol for several months; retains its activity in the frozen state for a long time; resists desiccation from frozen state and may remain viable for many months; is inactivated by dilution in physiologic salt solution; and multiplies in tissue culture.

Disease produced An acute, specific virus disease characterized by sudden onset, fever, chills, prostration, headache, muscular pain, some destruction of red blood cells, congestion of mucous membranes, black vomit, mild albuminuria, and jaundice. Leucopenia is the rule. Disease of short duration. Fatality among indigenous populations of endemic regions less than 5 percent; for others may be as high as 40 percent.

Source of infection Immediate source is an infective mosquito. Man plays no part in transmitting yellow fever, nor in maintaining the virus.

Mode of transmission By bite of mosquito *A. aegypti*. In South America by bite of several species of forest mosquitoes of genus *Haemagogus,* and *A. leucocelaenus*. In tropical Africa, *A. africanus, A. simpsoni*, etc.

Incubation period From 3 to 6 days.

Susceptibility and immunity Susceptibility to disease is general. Permanent acquired immunity follows recovery from disease. Antibodies appear in blood within first week. Second attacks unknown.

Vaccine prepared by inoculating chick embryos with virus and incubating at 37°C for 4 days. Embryos harvested and reduced to a pulp in a blender. Juice measured into ampoules and desiccated from the frozen state. Ampoules filled with nitrogen gas before sealing. Vaccine stored in refrigerator. For use, vaccine reconstituted with saline and injected subcutaneously.

Prevalence Jungle form present from Mexico south to Central and South America except in Salvador, Uruguay, and Chile. Endemic among humans and some animals chiefly in Western and Central Africa. Still endemic in jungle form in Panama.

Prevention and control Eradication of mosquitoes involved. Vaccination of persons living in yellow fever areas and those planning to visit such areas. Antibodies appear 7 to 10 days after vaccination and persist for at least 17 years.

Protection of patients in screened rooms or rooms sprayed with insecticide. Vaccination of persons exposed to disease.

No specific treatment is known.

For more information: Benenson (1970); Kissling (1960).

Influenza The causative agent of influenza is classified with the myxoviruses.

The disease was formerly believed to be caused by the bacterium *Haemophilus influenzae* but it is now known to be produced by several strains of a virus. In typical severe cases, *H. influenzae* is also present but as a secondary invader only, the virus being the primary cause of the disease (see page 914).

The virus is capable of passing through most filters; resists freezing for about 2 weeks; and retains its potency in 50 percent glycerol for the same length of time. When dried from the frozen state (lyophilized), the virus retains its potency for at least 6 weeks in the refrigerator. Beardmore et al. (1968) lyophilized influenza virus in allantoic fluid of embryonated eggs and found the virus to be much more stable than fluid virus preparations, retaining infectivity up to 94 days when stored in the refrigerator (4°C). Virus is inactivated by mercurials which would indicate the presence of $-SH$ groups (cysteine). The inactive virus is reactivated by sodium thioglycollate. Lee strain (type B) virus is inactivated at 50 to 54°C in 15 min; and by as little as 0.05 percent formalin in 18 hr. The virus is inactivated by intense ultraviolet irradiation.

The virus is sensitive to strong oxidizing agents such as iodine, salts of heavy metals, Mercurochrome, and the wetters Phemerol, Roccal, and sodium dodecyl sulfate. It is only slightly affected by reducing agents, sulfathiazole, dilute phenol solutions, glucose, ammonium sulfate, calcium chloride, and sodium thiosulfate.

Disease produced An acute viral infection characterized by sudden onset, fever of 1 to 7 days' duration, catarrh of respiratory tract (sometimes alimentary tract), pains in head and muscles, coryza, sore throat, bronchitis, marked prostration, and tendency to pneumonic complications. Deaths are concentrated in the elderly and those suffering from debilitating disease.

Two distinct types of influenza virus, designated as types A and B, have been long identified. Type A is the older and more widely distributed

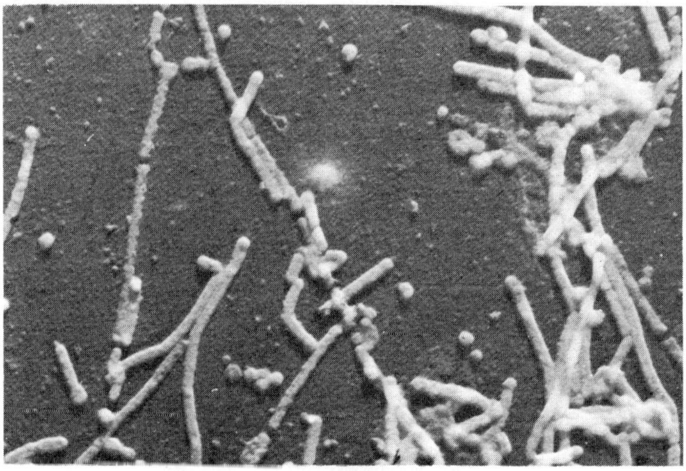

Fig. 401 Electron micrograph of influenza A virus (PR8 strain). Spherical and filamentous forms are seen in close association. X25,300. (After Murphy, Karzon, and Bang.)

(Fig. 401); type B has usually been found in smaller and more localized outbreaks. Some outbreaks are not caused by either A or B but by a third type designated C.

Two subtypes of A have been isolated from man and designated A1 and A2. Type A is now considered to be a family of viruses. At least three subtypes of B have been recognized. C appears to be antigenically homogeneous.

Apostolov and Flewett (1969) compared the fine structures of influenza viruses A and C as seen in thin sections (Fig. 402). In the walls of virus A the structures encountered from the surface inward were: (1) a layer of hollow spikes open at both ends; (2) an electron-transparent layer of uniform width; and (3) a finely granulated layer. The electron-transparent layer of virus C was much less prominent. The spikes were irregularly placed and often absent. Sections of C virus filaments displayed a hexagonal pattern. The internal component of virus A measured about 6 nm in diameter and sometimes appeared to be hollow. The internal component of virus C measured about 9 nm across and appeared to be helical.

Diagnosis Recovery of virus from throat washings or demonstration of rise in serum antibodies obtained during acute and convalescent stages.

Source of infection Discharges from nose and throat of infected persons or from articles freshly soiled with discharges from such individuals.

Mode of transmission By direct contact with infected persons, by droplet infection, or by articles freshly soiled with discharges from nose and throat of infected individuals.

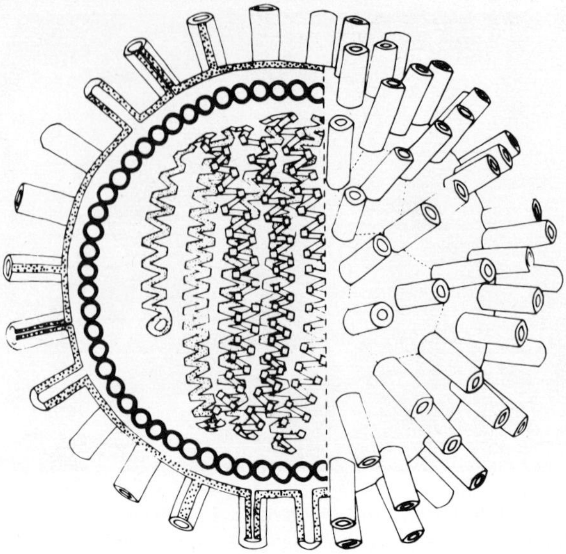

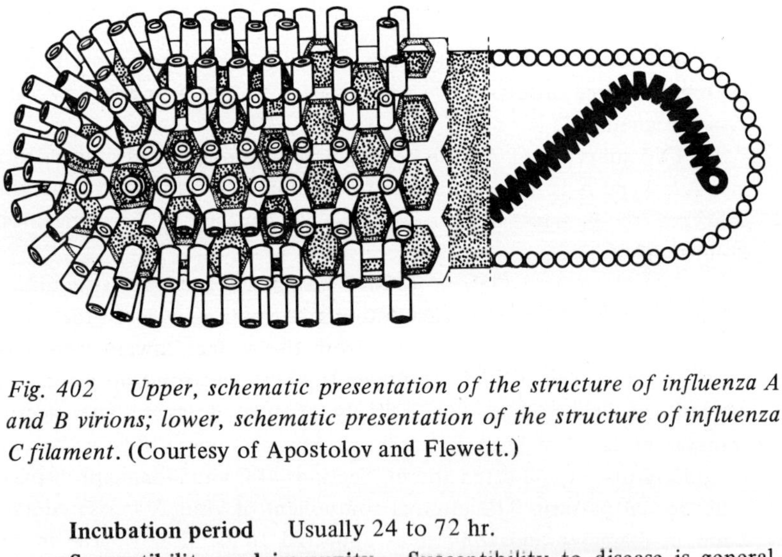

Fig. 402 Upper, schematic presentation of the structure of influenza A and B virions; lower, schematic presentation of the structure of influenza C filament. (Courtesy of Apostolov and Flewett.)

Incubation period Usually 24 to 72 hr.

Susceptibility and immunity Susceptibility to disease is general, although some have natural immunity. Acquired immunity of short duration, possibly as long as 1 year, follows recovery and is effective only against the particular type which caused the infection.

Prevalence Epidemics may affect up to 50 percent of population within 4 to 6 weeks. Pandemics occur at irregular intervals.

Prevention and control Education of public to the dangers of droplet infection from spitting, sneezing, and coughing in the presence of others. Use of common eating and drinking utensils, towels, etc., should be avoided. Use of disposable tissues and napkins should be encouraged.

Vaccines recommended for persons highly susceptible to disease, and for those engaged in essential community services. Vaccines should be administered in advance of infection.

Antibiotics or sulfa drugs are not effective but may be used to combat secondary invaders.

For more information: Benenson (1970); Duesberg (1969); Hollós (1960); Hoyle et al. (1961); Morgan and Rose (1968); Nermut and Frank (1971); Payne (1967); Schulze (1970); Pons (1972).

Acute viral respiratory disease

Under this title are grouped a number of acute viral respiratory illnesses. The group may be subdivided into: (1) the more severe febrile illnesses, including pneumonias and croups and (2) the less severe nonfebrile common colds. Diseases in both subgroups are undifferentiated in the sense that neither is regularly identified with a specific virus.

1. **Acute febrile respiratory disease** Viral diseases in this subgroup are characterized by fever, chills or chilliness, headache or general aching, malaise, anorexia, and in infants occasional gastrointestinal disturbances. Symptoms may be accompanied by nasal obstruction or nasal discharge with rhinitis, pharyngitis or tonsillitis, laryngitis, bronchitis or pneumonitis, or pneumonia. Symptoms usually subside in 2 to 5 days without complications. Examples of clinical syndromes are pharyngoconjunctival fever, croup, primary atypical pneumonia, acute respiratory disease of recruits (ARD), severe cold, and upper respiratory infection (URI).

At least 20 serotypes of the myxovirus, adenovirus, and picorna virus families are involved in undifferentiated acute febrile illnesses.

Source of infection Discharges from nose and throat of infected persons. Possibly intestinal tract with adenoviruses. Adenoviruses may remain latent in tonsils and adenoids.

Mode of transmission Direct contact or by droplet spread. Indirectly by articles freshly soiled with discharges from an infected person.

Incubation period From few days to a week or more; commonly 5 days for adenovirus; 3 to 6 days for parainfluenza virus; and 4 to 5 days for respiratory syncytial virus.

Susceptibility and immunity Susceptibility universal. Illness more frequent and severe in childhood. Antibodies produced but of short duration.

Prevalence Endemic in most parts of world. More prevalent during winter. Annual incidence high, especially in children. Under special conditions, three-fourths of population may be disabled.

Prevention and control Avoid crowding in living and sleeping quarters. Education of public in personal hygiene such as covering the mouth and nose during coughing and sneezing. Sanitary disposal of discharges from nose and throat. Adequate and good food, rest, fresh air, and exercise.

Patients should avoid exposure to others. Rest in bed during acute stage. Sanitary disposal of contaminated articles.

Antibiotics and other drugs are of little or no value. Indiscriminate use of antibiotics should be discouraged. Such agents should be reserved for the treatment of identified bacterial infections against which they are known to be effective.

2. **Common cold** Produced by many viruses, although at one time numerous bacteria were believed to be responsible for the disease.

Disease produced An acute catarrhal condition of the upper respiratory tract characterized by coryza, lacrimation, chilliness, malaise, irritated nasopharynx, lasting 2 to 7 days. Fever rarely in adults. Probably never fatal. Catarrhal sinusitis, otitis media, laryngitis, tracheitis, and bronchitis frequent secondary bacterial complications.

Agents involved include over 30 ECHO viruses, 30 types of Coxsackie viruses, at least 80 serotypes of rhinoviruses, some myxoviruses, and at least 31 types of adenoviruses. Viruses in other families also have been involved.

Source of infection Discharges from nose and throat of infected persons or from articles freshly soiled with discharges from such persons.

Mode of transmission Usually directly by droplets of infected saliva sprayed into the air during coughing, sneezing, and talking, or indirectly from articles freshly soiled with such discharges.

Incubation period Probably 12 to 72 hr. Communicability limited to early stages of disease, although the virus remains in discharges for an undetermined period.

Susceptibility and immunity Susceptibility in disease is universal. Temporary active immunity of approximately 1 month follows recovery from the disease.

Prevalence Virus colds are of world-wide distribution. Most individuals contract one or more colds each year. In temperate zones incidence rises in fall, winter, and spring. Both sexes about equally susceptible. Incidence higher in children under five years of age and becomes less after twenty years.

Prevention and control Infected persons should avoid contact with others. Rest in bed during the acute stage is advisable. Nasal and mouth discharges should be kept away from others and disposed of, preferably, by burning. Disinfection of eating and drinking utensils.

For more information: Adams (1967); Benenson (1970); Kingston et al. (1962); Melnick et al. (1963); Tyrrell (1962); Verlinde (1962).

Rabies (hydrophobia, canine madness) Virus particles are large and spherical, measuring 125 to 150 nm in diameter (Fig. 403). Particles capable of passing through Berkefeld and the coarser Chamberland filters, but not through Seitz filter pads. Infected tissues may be stored in undiluted glycerol for several weeks at room temperature, for several months in the refrigerator, and for 1 to 2 years at subzero temperatures.

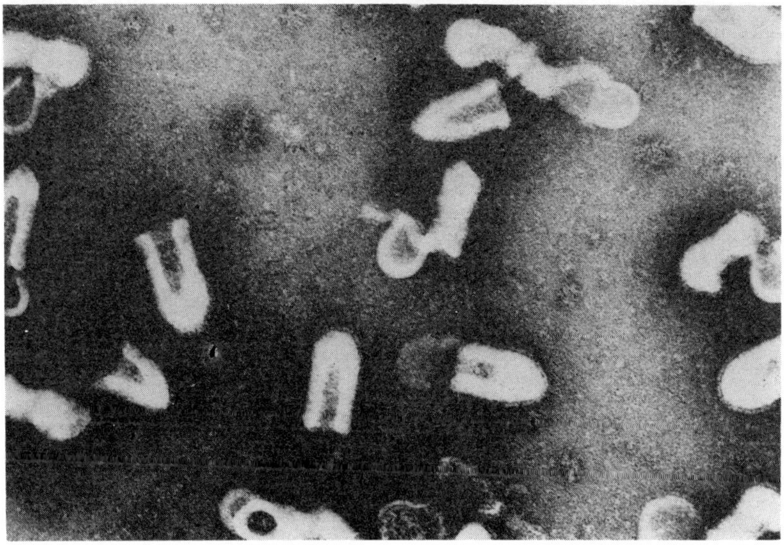

Fig. 403 Purified extracellular rabies virus. (Courtesy of Sokol, Kuwert, Wiktor, Hummeler, and Koprowski.)

The virus is reduced in virulence (attenuated) by (1) drying infected spinal cords of rabbits suspended over pellets of potassium hydroxide as a dehydrating agent, (2) exposure to 1 percent phenol, and (3) heating to 54 to 56°C for 1 hr or less. Virus best preserved by drying from the frozen state followed by storage at refrigerator temperature. Repeated freezing and thawing of virus suspensions results in loss of infectivity.

Disease produced An acute encephalitis caused by a neurotropic virus acquired from the bite of a rabid animal, usually the dog. Disease characterized by depression, itching at site of primary infection, and fever. Patient becomes uneasy, swallowing becomes difficult, salivation marked, followed by attacks of delirium. Paralysis of the face muscles, eyes, and tongue appears, gradually spreading to the trunk and limbs. Spasms of muscles of deglutition in attempts to swallow may lead to fear of water, hence the name hydrophobia. Death due to respiratory paralysis. Duration from 2 to 6 days, sometimes longer.

Cause of death determined by demonstrating presence of Negri bodies in nerve cells of brain or spinal cord, or by emulsifying small portion of the hippocampus in sterile saline and inoculating subdurally into guinea pigs or rabbits. Death occurs in about 16 days, and Negri bodies can be demonstrated in the brain tissue.

Source of infection Infected animals, chiefly dogs; vampire bats also involved. Source of infection in rabies of man is saliva of rabid animals.

Mode of transmission Virus present in saliva and usually transmitted by bite of rabid animal. Infections have occurred by contact of saliva with a scratch or break in skin. Airborne spread from bats to man in caves where bats are roosting.

Incubation period Usually 4 to 6 weeks, sometimes shorter or longer, depending upon site of wound in relation to richness of nerve supply and distance of nerve path to brain.

Susceptibility and immunity Susceptibility to disease is general. Natural immunity in man and animals not known to exist. Active artificial immunity may be developed by use of vaccine.

Prevalence Occurs throughout the world except in Australia, New Zealand, Japan, Hawaii, and other islands of the Pacific, some islands of the West Indies, Great Britain, Ireland, Norway, and Sweden. Rabies primarily a disease of dogs. Incidence in man is low.

Prevention and control A dog or other animal that has bitten a person should be isolated and observed for a proper period of time. If rabies is suspected, animal should be killed and brain examined for presence of Negri bodies. If examination is positive, person bitten should be given antirabic vaccination immediately before symptoms appear.

Wound caused by bite or scratch of a suspected animal should be thoroughly cleaned and irrigated with a solution of tincture of green soap or other satisfactory antiseptic. Dogs over six months of age should be vaccinated annually.

Vaccine consists of an emulsion of infected rabbit brain in saline containing 0.25 percent phenol, and incubated at 37°C to kill the virus. Recommended treatment consists of daily subcutaneous injections for 14 days. Treatment is generally sufficient to produce an active artificial immunity. A vaccine prepared from virus cultivated in nonnervous tissue, such as duck embryo, and killed, is preferred to the rabbit brain vaccine. Passive immunization with hyperimmune serum often supplements vaccination. Vaccines are useless after symptoms appear.

Use of vaccines has been instrumental in greatly reducing the mortality rate. In persons so treated, the death rate has dropped to 1 percent.

Isolation of infected persons for duration of illness. Disinfection of saliva and articles soiled with saliva.

No specific treatment other than administration of vaccine prior to appearance of symptoms.

For more information: Campbell et al. (1968); Hummeler et al. (1967, 1968); Humphrey (1967); Kaplan (1969); Morecki and Zimmerman (1969); Sokol et al. (1968); Walker (1969); Neurath et al. (1972).

Psittacosis (ornithosis, parrot fever) Psittacosis is produced by *Chlamydia psittaci.* It is not a true virus but is placed here for convenience.

Virus particles relatively large, spherical, measuring 455 nm in diameter, and approaching the size of small bacteria. Particles are visible under a light microscope. Virus is filterable through membranes which retain bacteria. Virus stains gram-negatively, is susceptible to sulfonamides, penicillin, tetracyclines, and other chemotherapeutic agents which antagonize bacteria. However, in its mode of multiplication intracellularly and in its relationship to the metabolism of animal cells, it behaves more like the true viruses than do other classes of microorganisms.

Virus is easily cultivated on the chorioallantoic membrane of the developing chick embryo, and in tissue cells (Fig. 404).

Disease produced Psittacosis is a contagious disease of parrots, parakeets, lovebirds, canaries, and other birds. It resembles influenza and is transmissible to man. In man disease characterized by high fever, headache, backache, thirst, changes in tongue and pharynx, stupor or depression, pulse usually slow in relation to temperature, diarrhea or constipation, enlargement of spleen, symptoms of atypical pneumonia or of a typhoidal state, with rales and cardiac dullness. Sputum mucopurulent, not copious. White blood count is normal or slightly increased early, followed by leucopenia. Human infections may be severe but are most often mild in character. Disease rarely produces death.

Source of infection Canaries, pigeons, parrots, parakeets, lovebirds, and other birds; occasionally man. Birds that appear well occasionally transmit the infection. Source of infection is excreta of birds and sputum of man.

Mode of transmission Virus present in blood, saliva, and feces of infected birds. Transmitted by contact with such birds or their recent surroundings. Laboratory infections are common.

Incubation period From 4 to 15 days, usually 10 days.

Susceptibility and immunity. All ages susceptible. Disease more severe in higher age groups. Recovery does not always confer lasting immunity.

Prevalence Disease worldwide in distribution. Outbreaks usually sudden and caused by exposure to sick birds. Occupational disease of persons associated with pet shops, aviaries, poultry farms, processing and rendering plants. Deaths usually confined to adults over 30 years of age. Mild cases may result from exposure to birds not necessarily sick (carriers).

Prevention and control Strict regulation of traffic in birds of the parrot family. Education of public to dangers of birds of the parrot family, particularly to those freshly imported. Homes and pet shops harboring infected birds should be quarantined.

Psittacine birds offered in commerce should be raised under psittacosis-free conditions. Tetracycline-impregnated bird seed has proved very effective in controlling the infection in birds.

Disease may be diagnosed by presence of virus in saliva and blood during the first week of the infection. Serum contains complement-fixing antibodies. Infected persons should be isolated during the febrile and acute clinical stage. Masks should be worn when handling patients with coughs. Disinfection of discharges and articles soiled with such discharges. Infected birds should be disposed of by burning. Buildings in which infected birds were housed should be thoroughly cleaned and disinfected.

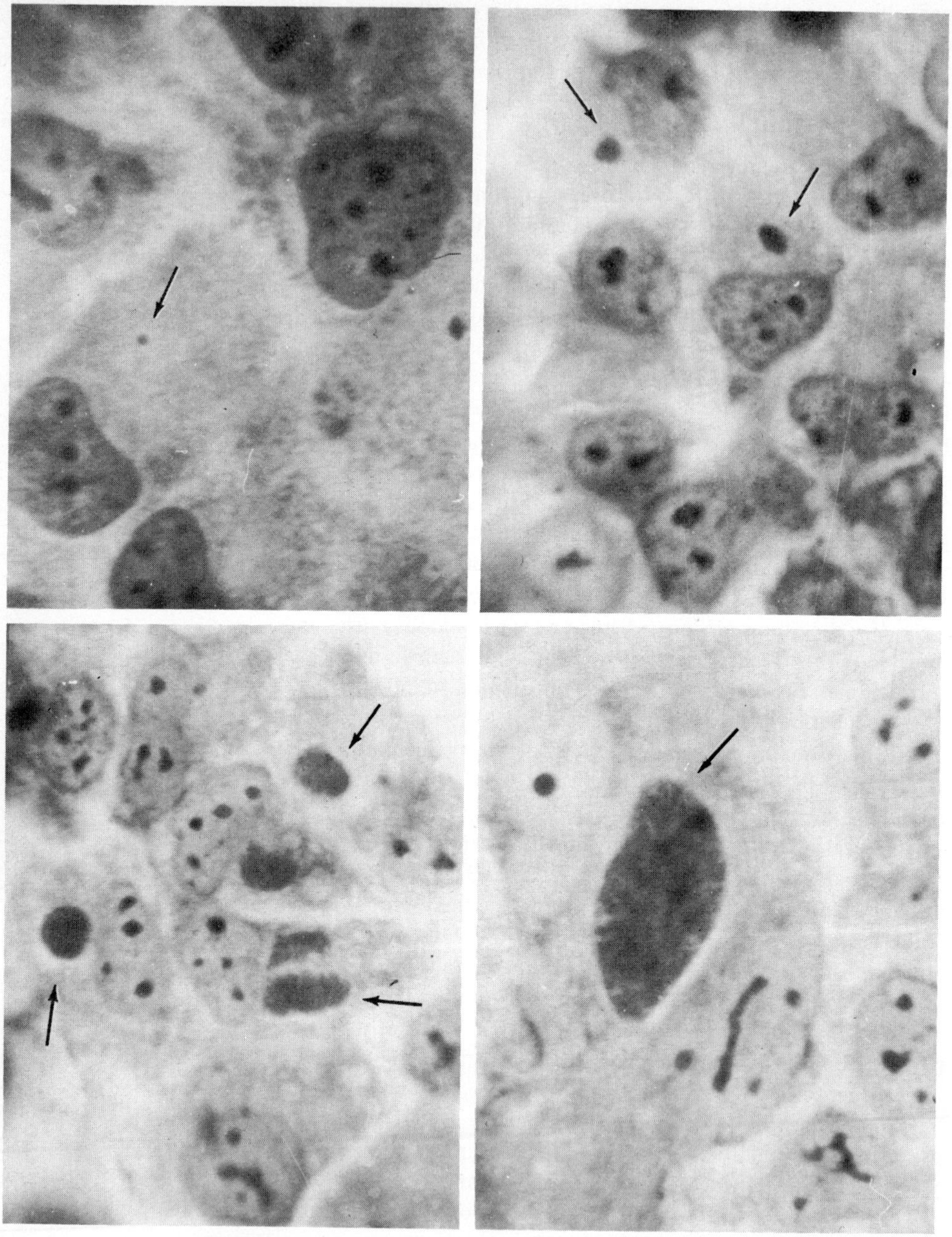

Fig. 404 Morphological development of psittacosis virus in human liver cells. Culture shows virus inclusions after following hours of incubation: Upper left, 12 hr; upper right, 16 hr; lower left, 24 hr; lower right, 36 hr. (Courtesy of Officer and Brown.)

Specific treatment consists of administration of the tetracycline antibiotics or chloramphenicol until patient afebrile. Repeat if relapse occurs.

For more information: Benenson (1970); Dean et al. (1964); Officer and Brown (1960); Tamura et al. (1971).

Mumps (infectious parotitis) Mumps virus is a member of the myxovirus group. It has been cultivated in the yolk sac, amniotic sac, and allantoic sac of the developing chick embryo. Duc-Nguyen and Rosenblum (1967) reported that mumps virus was roughly spherical in shape and measured from 100 to 800 nm. Weil et al. (1948) found the virus particles to measure from 106 to 282 nm with an average of 190 nm (Fig. 405).

Disease produced Mumps is an acute, specific, contagious disease characterized by fever and inflammation of the salivary glands. The parotid, submaxillary, and sublingual glands may be infected, although the parotid is most frequently involved. Pancreatitis, neuritis, arthritis, pericarditis, orchitis, and meningoencephalitis may occur. About one-third of exposed susceptible persons may be infected without showing any symptoms. Mumps is more serious in young adults. Death is exceedingly rare.

Source of infection Secretions of the mouth and possibly the nose.

Mode of transmission By direct contact with infected persons or by articles freshly soiled with discharges from mouth and nose of such individuals.

Incubation period From 12 to 26 days, usually 18 days.

Susceptibility and immunity Susceptibility to disease is general.

Fig. 405 Electron micrograph of mumps virus shadowed with chromium. X28,800. (After Weil, Beard, Sharp, and Beard.)

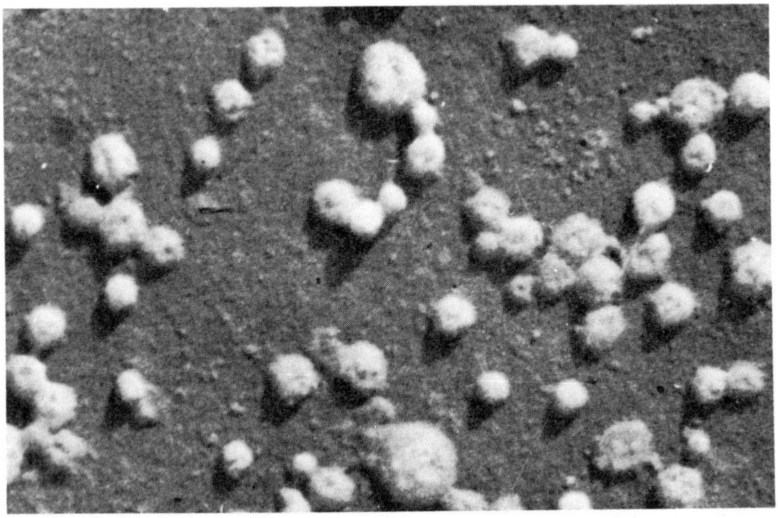

Recovery from infection usually confers permanent active immunity. Complement-fixing antibodies regularly appear or increase in concentration in the sera of persons during an attack of mumps or during convalescence.

Prevalence Mumps less prevalent than other diseases of childhood. Occurs more frequently during winter and spring. Occurrence of disease sporadic and epidemic. In large cities, it is endemic. Outbreaks frequently occur during periods of overcrowding.

Prevention and control Infected persons should be isolated until 9 days after swelling of salivary glands has subsided.

Disinfection of eating and drinking utensils, and articles soiled with secretions from nose and throat.

Live attenuated vaccine prepared by passage of the virus in embryonated hens' eggs is available and should be used prior to occurrence of a natural infection. The neutralizing antibodies induced by the vaccine persist for at least 2 years without substantial decline and the pattern closely resembles that for antibody persistence after naturally acquired mumps.

No specific treatment is known.

For more information: Benenson (1970); Davidson et al. (1967); Hilleman et al. (1967, 1968).

Measles (rubeola, morbilli) A filterable virus capable of passing through Berkefeld N and Seitz filters. In the presence of serum, virus can be preserved in refrigerator at $-72^{\circ}C$ for at least 1 year with little drop in titer. Virus relatively unstable at $25^{\circ}C$ and at $37^{\circ}C$, rapid inactivation. At $25^{\circ}C$ virus most stable at pH 7 and 8. Virus may be lyophilized from frozen state without loss of titer. Virus inactivated by β- and γ-radiations. Particles withstand 10 percent ether for 40 min. Particle size about 140 nm in diameter.

Disease produced A specific, highly contagious disease characterized by fever; catarrhal symptoms of the eyes, nose, and throat; Koplik spots on the buccal mucosa; a cutaneous rash followed by desquamation during convalescence. Disease is more severe in adults. Death from uncomplicated measles is rare. Pneumonia may follow measles, especially in children under two years of age.

Source of infection Secretions from nose and throat of infected persons.

Mode of transmission Directly from person to person; by droplets of infected saliva sprayed into air during coughing, sneezing, and talking; by articles freshly soiled with discharges from an infected person. One of most readily transmitted communicable diseases.

Incubation period After exposure, fever appears in about 10 days; rash in 13 to 15 days, occasionally shorter or longer.

Susceptibility and immunity Susceptibility to disease is general. Disease occurs most commonly in children between five and fourteen years of age.

Permanent acquired immunity usually follows recovery from disease.

Prevalence Universal; almost all persons (up to 90 percent) have had an attack at some time during life. Disease common in childhood.

Prevention and control Live attenuated and inactivated vaccines are available. Single injection of live attenuated vaccine induces active immunity in 95 percent of susceptible children for at least 7 years. Adults are nearly all immune and vaccination is rarely indicated. Use of inactivated virus vaccine not recommended because of short-lived protection. Also reactions are unusual and severe when followed by natural infection.

Immune globulin may be used for passive immunization of children under three years of age in families or institutions where measles occurs and for whom vaccine is contraindicated.

Isolation of infected persons during periods of communicability to protect them from possibility of reinfection and as protection to others. Isolation period usually 7 days from first appearance of rash. Disinfection of articles soiled with fresh discharges from nose and throat of infected persons.

Passive immunity may be transferred to healthy individuals before symptoms of measles appear by the injection of convalescent serum or serum from a person who has recovered from the disease. Such passive immunity may persist for about 4 weeks. During an epidemic, convalescent serum may either prevent the disease or modify the severity of the attack. In the latter instance a mild case of measles is usually sufficient to produce a lasting immunity.

Antibiotics or sulfa drugs not effective but may be used to combat complications due to secondary invaders.

For more information: Black and Sheridan (1960); Enders et al. (1960); Haggerty et al. (1960); Hilleman et al. (1962); Katz et al. (1960*a*, *b*); Kempe and Fulginiti (1965); Lepow et al. (1960); Warren and Cutchins (1962).

Rubella (German measles) Rubella virus particles are relatively very small in size. Best et al. (1967) found the virus particles to measure from 50 to 75 nm in diameter. Smith and Hobbins (1969) reported similar results with an average particle size of 55 nm in diameter (Fig. 406). Some particles showed spicule surface subunits. Its small size and the absence of 180A ribonucleoprotein in disrupted preparations suggested that rubella virus should not be included with the myxoviruses.

Disease produced A specific mild viral infection characterized by fever, a cutaneous eruption sometimes resembling that of measles, or scarlet fever, or both. Children exhibit few or no constitutional symptoms but adults may experience low-grade fever, headache, malaise, mild coryza, and conjunctivitis. As many as 20 to 50 percent of infections may occur without rash. Disease almost always accompanied by enlargement of the postauricular, suboccipital, and cervical lymph nodes. Mild catarrhal symptoms may be present. Arthritis may accompany some infections in adults, especially in females.

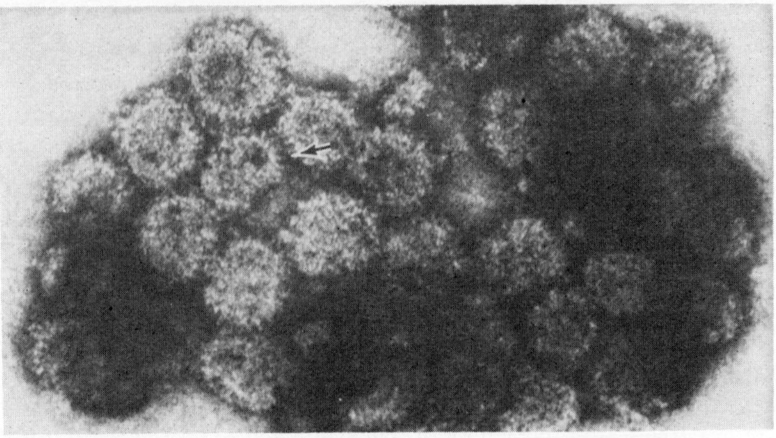

Fig. 406 Rubella (German measles) virus particles associated with visible band, infectivity, and hemagglutination activity. Arrow points to surface spicules seen occasionally on some particles. (Courtesy of Smith and Hobbins.)

A series of congenital defects occurs in about 25 percent of infants born to mothers who had acquired rubella during the first 3 months of pregnancy. Such defects include cataracts, microphthalmia, mental retardation, deafness, cardiac defects, jaundice, and distinctive bone defects.

Source of infection Nasopharyngeal secretions of infected person.

Mode of transmission Directly from person to person; by droplets of infected saliva sprayed into the air during coughing, sneezing, and talking; by articles freshly soiled with discharges of an infected individual.

Incubation period From 14 to 21 days, usually 18 days.

Susceptibility and immunity Susceptibility to disease is general in children. Permanent acquired immunity usually follows recovery from disease. Disease more prevalent in adults than measles.

Prevalence Occurs most commonly in children. More prevalent in winter and spring. Worldwide in distribution.

Prevention and control Efforts to control disease prompted by hazard of congenital defects in offspring of mothers who acquire disease during pregnancy.

Attenuated live virus vaccine is available which protects 90 to 95 percent of susceptible individuals against natural exposure. Vaccine should not be administered to a pregnant woman because of the danger of producing a transient viremia. All children over 1 year of age should be vaccinated with the hope of increasing immunity to the point where virus transmission cannot take place.

No specific treatment is known.

For more information: Bellanti et al. (1965); Buynak et al. (1968); Chagnon and Pavilanis (1970); Dowdle et al. (1970); Hilleman (1968);

Ingalls et al. (1967); Meyer et al. (1968); Parkman et al. (1966); Sever (1967); Vaheri et al. (1969).

Herpes simplex This is one of the DNA-containing viruses. Virus particles are spherical, measuring about 150 nm in diameter. An electron micrograph (Fig. 407) by Nii et al. (1968) shows a typical intranuclear crystal composed of capsids and cores.

Disease produced Herpes simplex is a viral infection characterized by latency and repeated, recurrent local lesions. About 10 percent of primary infections may appear as a mild or severe illness, accompanied by fever and malaise lasting about one week. Infection is associated with a gingivostomatitis accompanied by vesicular lesions in the oropharynx, or a severe keratoconjunctivitis, a vulvovaginitis, a cutaneous eruption, or a meningoencephalitis. Reactivation of a latent infection commonly results in the appearance of fever blisters or cold sores on the face and lips which tend to disappear in a few days. Five to 7 percent of cases of meningoencephalitis are caused by this virus.

Diagnosis Disease may be confirmed by isolation of virus from lesions or spinal fluid, or by a rise in specific neutralizing antibodies. Frequently virus can also be isolated from healthy persons.

Source of infection From saliva and hands of infected individuals or from articles soiled with discharges from lesions.

Mode of transmission By contact with virus in saliva of carriers is probably the most important method of spread. Virus also spread by direct contact with contaminated hands of attending physicians or nurses. Transmission to nonimmune adults often occurs by sexual contact.

Incubation period Up to 2 weeks.

Susceptibility and immunity Man is believed to be universally susceptible. An attack may be followed by a considerable degree of

Fig. 407 Herpes simplex virus. Typical intranuclear crystal composed of capsids and cores. Note the small proportion of dense cores. (Courtesy of Nii, Morgan, and Rose.)

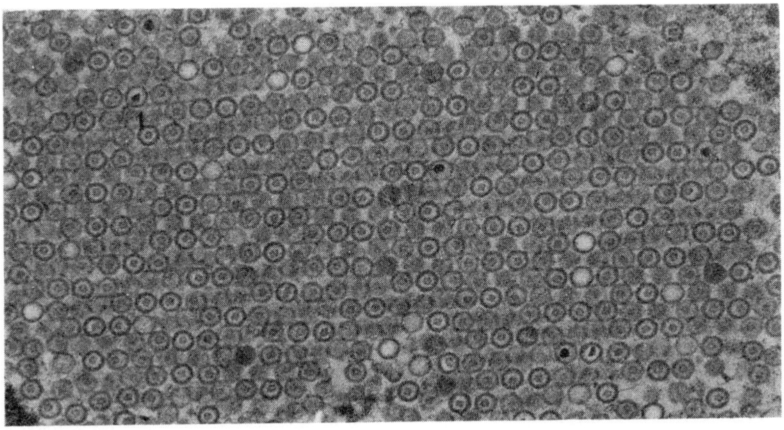

immunity. Humoral antibodies may be demonstrated. Since herpes may recur in persons whose blood contains neutralizing antibodies, it is assumed that the virus persists in the body tissues, and that the immunity is to a considerable extent nonspecific. Herpes constitutes one of the most striking exceptions to the general rule that virus infections usually leave behind a solid and lasting immunity.

Prevalence Disease worldwide in distribution. From 70 to 90 percent of adults possess circulating antibodies. Infection most prevalent before the fifth year.

Prevention and control Personal hygiene and health education to minimize the transfer of contaminated material. No vaccine is available.

No specific treatment is known.

For more information: Benenson (1970); Hampar et al. (1971).

Chickenpox (varicella) Virus particles contain DNA, are spherical, and measure about 175 nm in diameter (Fig. 408).

Disease produced An acute, extremely contagious viral disease characterized by sudden onset, fever, mild constitutional symptoms, a cutaneous eruption involving the superficial layers of the skin, lasting 3 to 4 days, and leaving a granular scab. Vesicles tend to be more abundant on the covered parts of the body. Sometimes vesicles may be so few as to escape observation. Lesions may appear on scalp and mucous membranes

Fig. 408 Electron micrographs of varicella virus from human cases of chickenpox, gold-shadowed. X24,800. (After Nagler and Rake.)

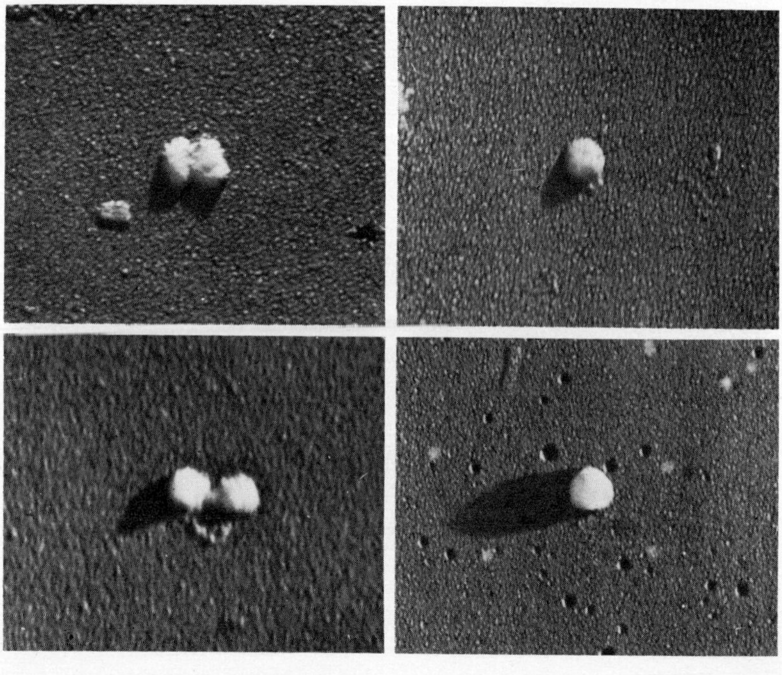

of upper respiratory tract. Rarely fatal. Primary viral pneumonia commonest cause of death in adults, and septic complications and encephalitis in children.

Herpes zoster or shingles is caused by same virus. Vesicles with an erythematous base are produced on skin areas supplied by sensory nerves of a single or associated group of dorsal root ganglia. Severe pain and paresthesia are common. Disease occurs mainly in elderly people.

Source of infection Secretions from respiratory tract of infected persons. Chickenpox may be contracted from patients with herpes zoster.

Mode of transmission From person to person by direct contact or by droplet or airborne spread of secretions from respiratory tract of infected persons; by articles freshly soiled with discharges from infected persons. Disease may be communicable before eruption is in evidence.

Incubation period From 2 to 3 weeks, usually 14 to 16 days.

Disease communicable up to 5 days before and not more than 6 days after appearance of vesicles. Especially communicable in early stages of eruption.

Susceptibility and immunity Susceptibility to the disease is universal among those who have not had an attack. About 70 percent of persons have had the disease by the time they are fifteen years of age. Disease ordinarily more severe in adults than in children. Recovery usually confers permanent active immunity. Infection remains latent and may reappear years later as herpes zoster, usually in adults but sometimes in children.

Prevention and control Isolation of infected persons for period of communicability. Disinfection of discharges from nose and throat and articles soiled by such discharges.

No specific treatment is known.

For more information: Benenson (1970).

Smallpox (variola) Virus capable of passing through most filters; resistant to low temperatures, to 50 percent glycerol, and to 0.5 percent phenol; sensitive to heat, being destroyed at 55°C or over. Cytoplasmic inclusions are characteristic of the infection. Inclusions may be demonstrated in various tissues, but most characteristic in epithelial cells (Fig. 409).

Disease produced An acute, specific, systemic viral disease characterized by sudden onset, fever, malaise, headache, severe backache, occasionally abdominal pain, and prostration. Temperature drops after 2 to 4 days followed by a deep-seated rash, which passes through successive stages of macules, papules, vesicles, pustules, and finally scabs. The scabs drop off at the end of about the third week. Lesions are more profuse on irritated areas, prominences, and extensor surfaces than on protected areas; most abundant and earliest on face, next on forearms and wrists, and favoring limbs more than trunk; more abundant on shoulders and chest than on loins and abdomen. Disease sometimes confused with varicella (chickenpox).

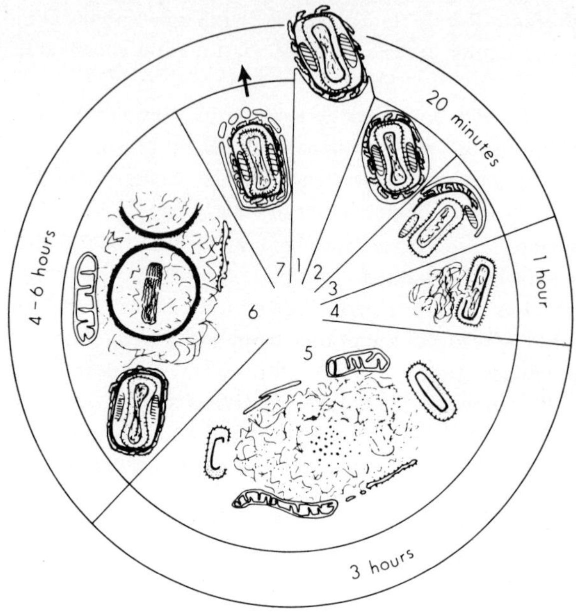

Fig. 409 Diagram illustrating the sequence of vaccinia uptake and replication in tissue cells cultivated in vitro. (Courtesy of Dales.)

Three types of smallpox are recognized: (1) variola major or classical smallpox; (2) variola minor or alastrim; and (3) vaccinia.

Alstrim and vaccinia are more mild forms of the virus, presumably derived from the classical form but have become altered in virulence. Vaccinia virus is employed for the preparation of commercial vaccines for immunization of persons prior to coming in contact with the more virulent variola virus.

Diagnosis Isolation of the virus from blood during febrile pre-eruptive stage or from fluid, scrapings, or crusts of skin lesions, and cultivation on the chorioallantoic membrane of the developing chick embryo, or in tissue culture. Also material from cutaneous lesions may be examined by electron microscopy.

Source of infection Respiratory discharges of patients and lesions of skin and mucous membrane, or articles soiled with such materials. Separated scabs can remain infectious for several years.

Mode of transmission Contact with diseased persons; by articles soiled with discharges from such persons.

Incubation period From 7 to 16 days; usually 9 to 12 days to onset of illness and 3 to 4 days more to onset of rash.

Susceptibility and immunity Susceptibility to infection is universal, but not every individual exposed to virus contracts the disease. Permanent active immunity usually follows recovery from the disease. Artificial active immunity may be effective for as long as 20 years or for less than 2 years. Complete protection is afforded by revaccination every 3 to 5 years.

Vaccine-immune serum contains agglutinating, precipitating, complement-fixing, and neutralizing antibodies.

Prevalence Distribution sporadic or endemic and occurs almost everywhere. Frequency greatest in winter, and least in summer. Disease repeatedly introduced into countries free of smallpox by international travelers.

Prevention and control Vaccination in early childhood before exposure to disease; revaccination every 3 to 5 years.

Vaccine most generally employed is prepared from calf lymph, obtained by rubbing the virus into the scarified abdomen of calves six months old. After 5 days, the scarified areas are scraped under aseptic conditions. The harvested pulp is mixed with twice its weight of water and passed through a sieve. The emulsion is preserved by the addition of glycerin and phenol to give a final concentration of 50 percent of the former and 0.5 percent of the latter.

Vaccine maintains its potency for about 6 months if stored below freezing at all times before use. At a temperature below 10°C, it should maintain its potency for about 14 days.

Freeze-dried vaccine, to be reconstituted at time of use, is generally available. Its use assures a potency not often possible under field conditions in endemic or tropical areas.

Isolation of infected persons in screened wards. Disinfection of mouth and nasal discharges and articles soiled by such discharges. Terminal sterilization of mattresses and all bedding materials.

Vaccination of all persons in contact with infected individuals, or quarantine for at least 16 days.

No specific treatment known.

For more information: Joklik (1962, 1966); Kaplan and Micklem (1961); McCrea et al. (1962); McCrea and Lipman (1967); Noyes (1962); Pfau and McCrea (1962); Sarov and Becker (1967); Sidwell et al. (1966); Woodson (1968); Zwartouw (1964); Easterbrook and Rozee (1971).

For general information on the viruses: Andrewes (1967); Barry and Mahy (1970); Bedson et al. (1967); Burnet (1960); Cohen (1969); Crawford (1968); Fenner and White (1970); Fraenkel-Conrat (1969); Hilleman (1969); Höglund (1968); Horsfall and Tamm (1965); Prier (1960); Rhodes and Van Rooyen (1968); Waterson (1968).

References

Adams, J. M.: "Viruses and Colds," New York, American Elsevier Publishing Company, Inc., 1967.

Andrewes, C. H.: "The Natural History of Viruses," London, Weidenfeld and Nicolson, 1967.

Apostolov, K., and T. H. Flewett: Further observations on the structure of influenza viruses A and C, *J. Gen. Virol.*, **4**:365, 1969.

Barry, R. D., and B. W. J. Mahy, eds.: "The Biology of Large RNA Viruses," New York, Academic Press, Inc., 1970.

Beardmore, W. B., et al.: Preservation of influenza virus infectivity by lyophilization, *Appl. Microbiol.*, **16**:362, 1968.

Bedson, S., et al.: "Virus and Rickettsial Diseases of Man," London, Edward Arnold (Publishers) Ltd., 1967.

Bellanti, J. A., et al.: Congenital rubella, *Am. J. Diseases Children,* **110**:464, 1965.

Benenson, A. S., ed.: "Control of Communicable Diseases in Man," New York, American Public Health Association, 1970.

Berge, T. O., et al.: Preservation of enteroviruses by freeze-drying, *Appl. Microbiol.,* **22**:850, 1971.

Best, J. M., et al.: Morphological characteristics of rubella virus, *Lancet,* July 29, 1967, p. 237.

Bishop, J. M., and G. Koch: Purification and characterization of poliovirus-induced infectious double-stranded ribonucleic acid, *J. Biol. Chem.,* **242**:1736, 1967.

Black, F. L., and S. R. Sheridan: Studies on an attenuated measles-virus vaccine. IV, *New Engl. J. Med.,* **263**:165, 1960.

Bodo, G.: Interferone. Zelluläre proteine mit antiviraler wirksamkeit, *Naturwissenschaften,* **58**:425, 1971.

Burnet, F. M.: "Principles of Animal Virology," New York, Academic Press, Inc., 1960.

Buynak, E. B., et al.: Live attenuated rubella virus vaccines prepared in duck embryo cell culture, *J. Am. Med. Assoc.,* **204**:195, 1968.

Campbell, J. B., et al.: Present trends and the future in rabies research, *Bull. World Health Organ.,* **38**:373, 1968.

Chagnon, A., and V. Pavilanis: Epidemiological studies on rubella, *Can. Med. Assoc. J.,* **102**:933, 1970.

Cockburn, W. C., and S. G. Drozdov: Poliomyelitis in the world, *Bull. World Health Organ.,* **42**:405, 1970.

Cohen, A.: "Textbook of Medical Virology," Oxford, Blackwell Scientific Publications, Ltd., 1969.

Crawford, L. V.: The structure and function of viral nucleic acid, *Symp. Soc. Gen. Microbiol.,* no. 18, The Molecular Biology of Viruses, 1968.

Dales, S., and R. Kajioka: The cycle of multiplication of vaccinia virus in Earle's strain L cells. I. Uptake and penetration, *Virology,* **24**:278, 1964.

Dane, D. S., et al.: Vaccination against poliomyelitis with live virus vaccines, 6, *Brit. Med. J.,* **2**:259, 1961*a*; 8, *ibid.,* **2**:269, 1961*b*.

Davidson, W. L., et al.: Vaccination of adults with live attenuated mumps virus vaccine, *J. Am. Med. Assoc.,* **201**:995, 1967.

Dean, D. J., et al.: Psittacosis in man and birds, *Public Health Rept.,* **79**:101, 1964.

Dick, G. W. A., et al.: Vaccination against poliomyelitis with live virus vaccines, 7, *Brit. Med. J.,* **2**:266, 1961.

Dowdle, W. R., et al.: WHO collaborative study of the sero-epidemiology of rubella in Caribbean and Middle and South American populations in 1968, *Bull. World Health Organ.*, **42**:419, 1970.

Duc-Nguyen, H., and E. N. Rosenblum: Immunoelectron microscopy of the morphogenesis of mumps virus, *J. Virol.*, **1**:415, 1967.

Duesberg, P. H.: Distinct subunits of the ribonucleoprotein of influenza virus, *J. Mol. Biol.*, **42**:485, 1969.

Enders, J. F., et al.: Studies on an attenuated measles-virus vaccine. I, *New Engl. J. Med.*, **263**:153, 1960.

Easterbrook, K. B., and K. R. Rozee: The intracellular development of vaccinia virus as observed in freeze-etched preparations, *Can. J. Microbiol.*, **17**:753, 1971.

Fenner, F., and D. O. White: "Medical Virology," New York, Academic Press, Inc., 1970.

Finch, J. T., and A. Klug: Arrangement of protein subunits and the distribution of nucleic acid in turnip yellow mosaic virus. II. Electron microscopic studies, *J. Mol. Biol.*, **15**:344, 1966.

Fraenkel-Conrat, H.: "The Chemistry and Biology of Viruses," New York, Academic Press, Inc., 1969.

Gard, S.: Theoretical considerations in the inactivation of viruses by chemical means, *Ann. N.Y. Acad. Sci.*, **83**:638, 1960.

Glasgow, L. A.: Interferon: A review, *J. Pediat.*, **67**:104, 1965.

Haggerty, R. J., et al.: Studies on an attenuated measles-virus vaccine. VII, *New Engl. J. Med.*, **263**:178, 1960.

Hampar, B., et al.: Serologic classification of herpes simplex viruses, *J. Immunol.*, **106**:580, 1971.

Herriott, R. M.: Infectious nucleic acids, a new dimension in virology, *Science*, **134**:256, 1961.

Hill, W. F., Jr., et al.: Survival of poliovirus in flowing turbid seawater treated with ultraviolet light, *Appl. Microbiol.*, **15**:533, 1967.

Hilleman, M. R.: The control of viral diseases, with special reference to mumps, rubella, and interferon, *Clin. Pharmacol. Therap.*, **9**:517, 1968.

——: Toward control of viral infections of man, *Science*, **164**:506, 1969.

—— and A. A. Tytell: The induction of interferon, *Sci. Am.*, **225**:26, 1971.

—— et al.: Studies of live attenuated measles virus vaccine in man. II. Appraisal of efficacy, *Am. J. Public Health*, **52**:44, 1962.

—— et al.: Live, attenuated mumps-virus vaccine. IV. Protective efficacy as measured in a field evaluation, *New Engl. J. Med.*, **276**:252, 1967.

—— et al.: Live, attenuated mumps-virus vaccine, *New Engl. J. Med.*, **278**:227, 1968.

Höglund, S.: Electron microscopic studies on some virus and immunoglobulin components, *Arkiv Kemi*, **28**:505, 1968.

Hollós, I.: Electron microscopic examination of complete influenza viruses. III, *Acta Microbiol.*, Budapest, 1960.

Horne, R. W.: The structure of viruses, *Sci. Am.*, **208**:48, 1963.

—— and J. Nagington: Electron microscope studies of the development and structure of poliomyelitis virus, *J. Mol. Biol.*, **1**:333, 1959.

Horsfall, F. L., Jr., and I. Tamm, eds.: "Viral and Rickettsial Infections of Man," Philadelphia, J. B. Lippincott Company, 1965.

Hoyle, L., et al.: The structure and composition of the myxoviruses. II, *Virology*, **13**:448, 1961.

Huebner, R. J.: New "frontiers" for virus disease research, *J. Hyg. Epidemiol. Microbiol. Immunol.*, **6**:34, 1962.

Hummeler, K., et al.: Structure and development of rabies virus in tissue culture, *J. Virol.*, **1**:152, 1967.

——: Morphology of the nucleoprotein component of rabies virus, *J. Virol.*, **2**:1191, 1968.

Humphrey, G. L.: Rabies. Suggested indications for treatment of exposed persons, *Calif. Med.*, **107**:363, 1967.

Hurrell, J. M. W.: Methods of storing viruses at low temperatures with particular reference to the myxovirus group, *J. Med. Lab. Tech.*, **24**:30, 1967.

Ingalls, T. H., et al.: Rubella: Epidemiology, virology, and immunology, *Am. J. Med. Sci.*, **253**:349, 1967.

Joklik, W. K.: The multiplication of poxvirus DNA, *Cold Spring Harbor Symp. Quant. Biol.*, **27**:199, 1962.

——: The poxviruses, *Bact. Rev.*, **30**:33, 1966.

Jordan, L. E., and H. D. Mayor: The fine structure of reovirus, a new member of the icosahedral series, *Virology*, **17**:597, 1962.

Jordan, W. E., et al.: Antiviral effectiveness of chlorine bleach in household laundry use, *Am. J. Diseases Children*, **117**:313, 1969.

Kaplan, C., and L. R. Micklem: A method for preparing smallpox vaccine on a large scale in cultured cells, *J. Hyg.*, **59**:171, 1961.

Kaplan, M. M.: Epidemiology of rabies, *Nature*, **221**:421, 1969.

Katz, S. L., et al.: Studies on an attenuated measles-virus vaccine. II, *New Engl. J. Med.*, **263**:159, 1960*a*.

——: Studies on an attenuated measles-virus vaccine. VIII, *New Engl. J. Med.*, **263**:180, 1960*b*.

Kempe, C. H., and V. A. Fulginiti: The pathogenesis of measles virus infection, *Arch. Ges. Virusforsch.*, **16**:103, 1965.

Kingston, D., et al.: The epidemiology of the common cold. III, *J. Hyg.*, **60**:341, 1962.

Kissling, R. E.: The arthropod-borne viruses of man and other animals, *Ann. Rev. Microbiol.*, **14**:261, 1960.

Klug, A., et al.: Arrangement of protein subunits and the distribution of nucleic acid in turnip yellow mosaic virus, *J. Mol. Biol.*, **15**:315, 1966.

Koch, G., et al.: Quantitative studies on the infectivity of ribonucleic acid from partially purified and highly purified poliovirus preparations, *Virology*, **10**:329, 1960.

Lepow, M. L., et al.: Studies on an attenuated measles-virus vaccine. V, *New Engl. J. Med.*, **263**:170, 1960.

Levine, S., and F. R. Nichol: Interferon inducers, *BioScience,* **20**:696, 1970.

Ludwig, E. H., and C. E. Smull: Infectivity of histone-poliovirus ribonucleic acid preparations, *J. Bact.,* **85**:1334, 1963.

Lund, E.: Inactivation of poliomyelitis virus by chlorination at different oxidation potentials, *Arch. Ges. Virusforsch.,* **11**:330, 1961.

McCrea, J. F., et al.: Surface structures of chemically treated vaccinia virus revealed by negative staining, *Virology,* **17**:208, 1962.

—— and M. B. Lipman: Strand-length measurements of normal and 5-iodo-2′-deoxyuridine-treated vaccinia virus deoxyribonucleic acid released by the Kleinschmidt Method, *J. Virol.,* **1**:1037, 1967.

McGregor, S., and H. D. Mayor: Biophysical studies on rhinovirus and poliovirus. I. Morphology of viral ribonucleoprotein, *J. Virol.,* **2**:149, 1968.

Meenan, P. N., and I. B. Hillary: Poliovirus in the upper respiratory tract of household contacts, *Lancet,* p. 907, Nov. 2, 1963.

Melnick, J. L., et al.: Rhinoviruses: A description, *Science,* **141**:152, 1963.

——: Intracellular herpesvirus aggregate in the form of a pentagonal dipyramidal crystal-like structure, *J. Virol.,* **2**:78, 1968.

Meyer, H. M., et al.: Clinical studies with experimental live rubella virus vaccine (strain HPV-77), *Am. J. Diseases Children,* **115**:648, 1968.

Morecki, R., and H. M. Zimmerman: Human rabies encephalitis, *Arch. Neurol.,* **20**:599, 1969.

Morgan, C., and H. M. Rose: Structure and development of viruses as observed in the electron microscope, *J. Virol.,* **2**:925, 1968.

Murthy, Y. K. S., and H.-P. Anders: Interferons, *Angew. Chem.,* **9**:480, 1970.

Nermut, M. V., and H. Frank: Fine structure of influenza A_2 (Singapore) as revealed by negative staining, freeze-drying and freeze-etching, *J. Gen. Virol.,* **10**:37, 1971.

Neurath, A. R., et al.: Characterization of subviral components resulting from treatment of rabies virus with tri(n-butyl) phosphate, *J. Gen. Virol.,* **14**:33, 1972.

Nii, S., et al.: Electron microscopy of herpes simplex virus. II. Sequence of development, *J. Virol.,* **2**:517, 1968.

Noyes, W. F.: The surface fine structure of vaccinia virus, *Virology,* **17**:282, 1962.

Officer, J. E., and A. Brown: Growth of psittacosis virus in tissue culture, *J. Infect. Diseases,* **107**:283, 1960.

Oroszlan, S., and M. A. Rich: Human wart virus: In vitro cultivation, *Science,* **146**:531, 1964.

Parkman, P. D., et al.: Attenuated rubella virus. I. Development and laboratory characterization, *New Engl. J. Med.,* **275**:569, 1966.

Payne, A. M.-M.: Acute infectious respiratory diseases, *Arch. Environ. Health,* **14**:730, 1967.

Pertoft, H., et al.: Gradient centrifugation of viruses in colloidal silica, *Virology,* **33**:185, 1967.

Pfau, C. J., and J. F. McCrea: Some unusual properties of vaccinia virus deoxyribonucleic acid, *Biochim. Biophys. Acta,* **55**:271, 1962.

Pollard, E. C.: Theory of the physical means of the inactivation of viruses, *Ann. N.Y. Acad. Sci.,* **83**:654, 1960.

Pons, M. W.: Studies on the replication of influenza virus RNA, *Virology,* **47**:823, 1972.

Prier, J. E.: Live virus immunizing agents, *J. Am. Vet. Med. Assoc.,* **137**:577, 1960.

Rhodes, A. J., and C. E. Van Rooyen: "Textbook of Virology," Baltimore, The Williams & Wilkins Company, 1968.

Richter, A.: Structure of viral nucleoproteins, *Ann. Rev. Microbiol.,* **17**:415, 1963.

Rita, G., ed.: "The Interferons," New York, Academic Press, Inc., 1968.

Sarov, I., and Y. Becker: Studies on vaccinia virus DNA, *Virology,* **33**:369, 1967.

Schäfer, W.: Structure of some animal viruses and significance of their components, *Bact. Rev.,* **27**:1, 1963.

Schaffer, F. L., et al.: Base composition of the ribonucleic acids of the three types of poliovirus, *Virology,* **10**:530, 1960.

Scherp, H. W., et al.: Continuously cultured tissue cells and viral vaccines, *Science,* **139**:15, 1963.

Schulze, I. T.: The structure of influenza virus. I. The polypeptides of the virion, *Virology,* **42**:890, 1970.

Sever, J. L.: Epidemiology of rubella, *Arch. Opthal.,* **77**:427, 1967.

Sidwell, R. W., et al.: I. Persistence of vaccinia virus on cotton and wool fabrics, *Appl. Microbiol.,* **14**:55, 1966.

Smith, K. O., and T. E. Hobbins: Physical characteristics of rubella virus, *J. Immunol.,* **102**:1016, 1969.

Sokol, F., et al.: Purification of rabies virus grown in tissue culture, *J. Virol.,* **2**:836, 1968.

Stasny, J. T., et al.: Effect of formamide on the capsid morphology of adenovirus types 4 and 7, *J. Virol.,* **2**:1429, 1968.

Tamura, A., et al.: Electron microscopic observations on the structure of the envelopes of mature elementary bodies and developmental reticulate forms of *Chlamydia psittaci, J. Bact.,* **105**:355, 1971.

Tyrrell, D. A. J.: Some viruses isolated from common colds, *J. Hyg. Epidemiol. Microbiol. Immunol.,* **6**:136, 1962.

Vaheri, A., et al.: Purification of rubella virus particles, *J. Gen. Virol.,* **5**:39, 1969.

Ver, B. A., et al.: Efficient filtration and sizing of viruses with membrane filters, *J. Virol.,* **2**:21, 1968.

Verlinde, J. D.: ECHO viruses: General and pathogenic properties, *Arch. Ges. Virusforsch.,* **13**:173, 1962.

Walker, V. C. R.: Rabies today: Man and animals, *Can. Vet. J.,* **10**:11, 1969.

Wallis, C., and J. L. Melnick: Concentration of viruses from sewage by adsorption on Millipore membranes, *Bull. World Health Organ.,* **36**:219, 1967.

Warren, J., and E. C. Cutchins: Immunization of man against measles: Potential vaccines and problems, *Am. J. Public Health,* **52**:80, 1962.

Waterson, A. P.: "Introduction to Animal Virology," Cambridge, England, Cambridge University Press, 1968.

Weil, M. L., et al.: Purification and sedimentation and electron micrographic characters of the mumps virus, *Proc. Soc. Exp. Biol. Med.,* **68**:309, 1948.

Wiley, J. S., et al.: Enterovirus in sewage during a poliomyelitis epidemic, *J. Water Pollution Control Federation,* **34**:168, 1962.

Wilner, B. I.: "A Classification of the Major Groups of Human and other Animal Viruses," Minneapolis, Burgess Publishing Company, 1969.

Wolstenholme, G. E. W., and M. O'Connor, eds.: "Interferon," London, Churchill, 1968.

Woodson, B.: Recent progress in poxvirus research, *Bact. Rev.,* **32**:127, 1968.

World Health Organization: Arboviruses and human disease, *World Health Organ. Tech. Rept.* ser. 369, 1967.

Zwartouw, H. T.: The chemical composition of vaccinia virus, *J. Gen. Microbiol.,* **34**:115, 1964.

31 *Rickettsial diseases of man*

The rickettsial diseases or rickettsioses of man may be defined as specific infections induced by small, often pleomorphic, rod-shaped to coccoid organisms which usually occur intracytoplasmically in lice, fleas, ticks, and mites. Occasionally they occur extracellularly in gut lumen. They bear a morphologic resemblance to bacteria but are biologically related to viruses. The organisms measure from 0.2 to 0.5 μ in diameter. When multiplying they pass through a regular sequence of developmental stages. The cytoplasmic colonies they produce are basophilic, differing in this respect from the viruses which are acidophilic. The rickettsiae contain both types of nucleic acids, deoxyribonucleic acid (DNA), and ribonucleic acid (RNA), as well as enzymes which function in energy metabolism, whereas viruses contain either DNA or RNA but not both. This is another distinguishing characteristic between rickettsiae and viruses. The rickettsiae are readily visible in microscopic preparations when examined with a light microscope. Like the bacteria, they multiply by binary fission; like the viruses they have not been cultivated on lifeless media. The cell membrane contains muramic acid, also found in the cell walls of bacteria. They are sensitive to sulfonamides and to some antibiotics whereas viruses are not. The pathologic lesions occur chiefly in the blood vessels, being caused by the presence of the organisms and not by their toxins.

Organisms The rickettsiae are so named in honor of Howard Taylor Ricketts, who was the first to give a description of the organisms in connection with his studies on Rocky Mountain spotted fever and later on typhus fever. He succeeded in isolating from the blood of typhus fever patients very short bacillus-like rods measuring about 0.3 μ in diameter and 2 μ in length. The organisms were stained readily by Giemsa stain and possessed a faintly stained bar through the middle, giving each organism the appearance of a diplobacillus. They are nonmotile and gram-negative. Ricketts' observations were later confirmed by Da Rocha-Lima (1916).

Following Ricketts' discovery, Hegler and Prowazek (1913) reported the presence of similar organisms in the blood of patients with typhus fever and in lice that had fed on infected persons.

The rickettsiae are found typically in arthropods. The species pathogenic for man occur intracytoplasmically only, or both intracytoplasmically and intranuclearly. The organisms have not been cultivated on the usual laboratory media. However, they can be cultivated in the various tissue culture media. From the standpoint of size, the rickettsiae occupy a position intermediate between the bacteria and the viruses. With the exception of one pathogenic species, they do not pass through filters that retain bacteria.

Classification Depending upon the specific arthropod vector, the important rickettsial diseases may be classified as follows:

I. Typhus fever
  A. Louse-borne (epidemic or classical typhus)
  B. Flea-borne (murine or endemic typhus)
II. Mite-borne Tsutsugamushi fever (scrub typhus)
III. Spotted fever group
  A. Tick-borne typhus (Rocky Mountain spotted fever)
  B. Rickettsialpox
IV. Q fever

I. Typhus fever This disease is generally differentiated into (1) epidemic or classical typhus (louse-borne), and (2) murine or endemic typhus (flea-borne).

A. Epidemic or classical typhus The etiological agent is *Rickettsia prowazekii*. Organisms are minute, coccobacillary, sometimes ellipsoidal or long, rod-shaped cells which are occasionally filamentous. Often occur in pairs and occasionally in chains. In infected lice the minute coccoid and paired coccoid forms predominate over the short and long rods and over the filamentous forms, which are up to 40 μ in length. Single cells from yolk sacs measure 0.3 to 0.7 by 0.5 to 2.0 μ (Figs. 410 and 411). Organisms occur intracytoplasmically in vascular endothelial cells and in serosal cells. When stained by the Giemsa method, the two individuals of a pair are connected by a zone of faintly blue-stained material. Nonmotile. Gram-negative.

Organisms have been cultivated in plasma tissue cultures of mammalian cells, in the louse intestine, and in the chorioallantoic membrane and yolk sac of the developing chick embryo. Cells readily inactivated at 50°C in 15 to 30 min by 0.5 percent phenol and 0.1 percent formalin.

Disease produced Onset of disease variable, often sudden, and characterized by headache, chills, fever, prostration, and general pains, followed by a macular eruption on the fifth or sixth day, and toxemia. The disease terminates by rapid lysis after about 14 days of fever. Organisms may be found in the blood vessels of the skin, kidneys, muscles, brain, and testes. Mortality varies from 10 to 40 percent, depending upon epidemics and age. Recrudescence of infection (Brill's disease) many years after an initial typhus episode without intervention of lice has been confirmed. Pathogenic for apes, monkeys, guinea pigs, cotton rats, gerbils, and lice.

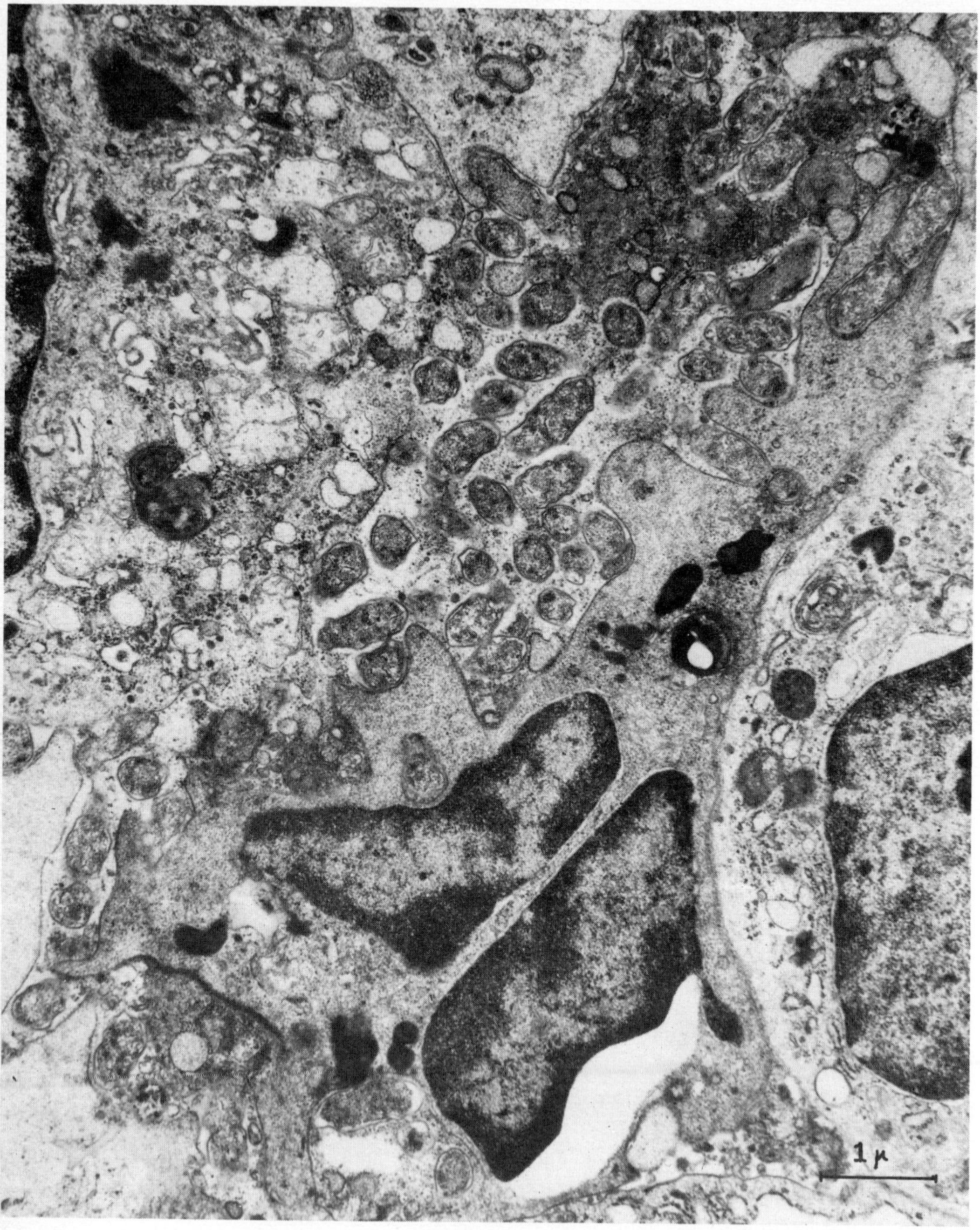

Fig. 410 Intracellular and extracellular Rickettsia prowazekii. A granulocyte may be seen in the course of phagocytizing some of the rickettsiae. (Courtesy of Jadin, Creemers, Jadin, and Giroud.)

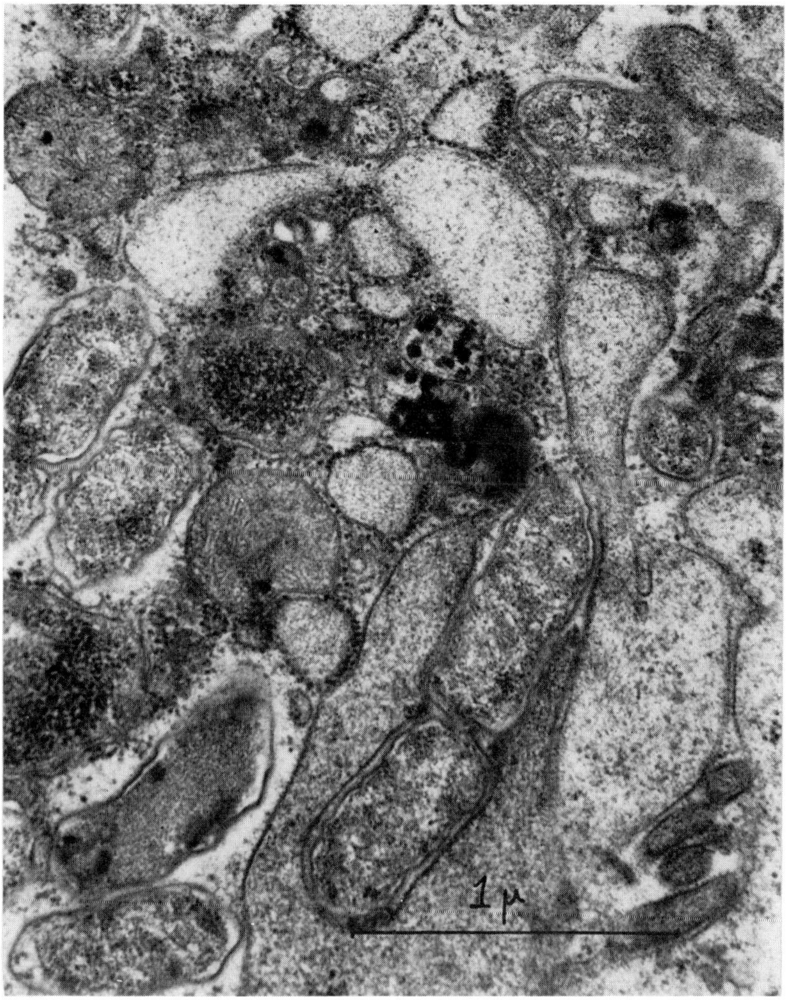

Fig. 411 Upper right-hand corner of Fig. 410 at greater magnification. One rickettsia already engulfed by the phagocyte is undergoing division. (Courtesy of Jadin, Creemers, Jadin, and Giroud.)

Serum from patients with epidemic typhus agglutinates *Proteus* OX_{19} and OX_2. This test is referred to as the Weil-Felix reaction. Former strain more commonly agglutinated and is strain customarily employed. Reaction usually becomes positive during second week of disease, reaches its height about time of convalescence, then disappears rather rapidly.

Source of infection Man is reservoir. Lice infected by feeding on blood of man with febrile disease. Patients with recrudescent typhus can infect lice and probably serve as foci for new outbreaks.

Mode of transmission Organism transmitted from man to man by the louse *Pediculus humanus* that has fed on an infected person. Organisms present in feces of louse. Cells inoculated by scratching louse feces into the wound produced by the bite or into other superficial breaks in the skin. Dirty clothing contaminated with louse feces may disseminate organisms into air, from where they may reach the respiratory tract.

Incubation period From 6 to 15 days, most often 12 days.

Susceptibility and immunity Susceptibility to disease is general. Acquired immunity follows recovery from disease, but it is not always permanent.

Prevalence Disease prevalent in most colder regions of world where appreciable groups of people live under unhygienic conditions and are louse-infested. Until recently, epidemics were frequent among military and refugee populations, and in famine areas.

Prevention and control Immunization of susceptible persons entering typhus areas. Vaccine prepared by growing rickettsiae in yolk sac of developing chick embryo. Suspension after purification is inactivated by formalin. Vaccine generally given in two injections, and confers considerable protection. Reimmunization should be practiced every few months where danger of disease is present. Vaccination reduces risk of infection, modifies course of disease, and lowers mortality rate.

Application of DDT or 1 percent lindane to clothing of persons living under conditions favoring lousiness. Improvement of living conditions. Frequent bathing and washing of clothes.

Delousing of infected persons. Disinfection of clothing and bedding of patient and contacts. Treatment of hair for louse eggs. Persons exposed to typhus should be quarantined for 15 days.

Fig. 412 Electron micrograph of Rickettsia typhi, the etiological agent of endemic or murine typhus (flea-borne). X20,000. (After van Rooyen and Scott.)

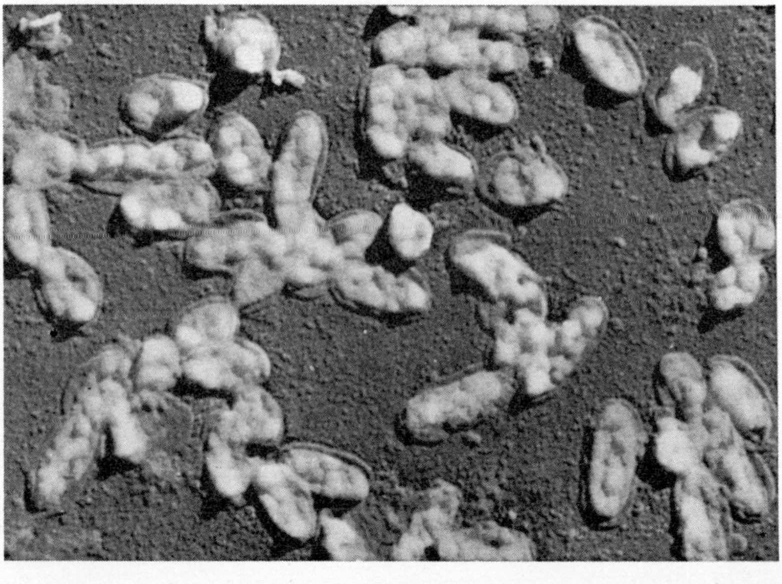

Treatment consists of application of tetracyclines or chloramphenicol orally daily until patient becomes afebrile, then for one additional day.

For more information: Benenson (1970); Bird et al. (1967); Jadin et al. (1968); Snyder (1965); Stuart-Harris (1967); Wisseman (1968).

B. Murine or endemic typhus The etiological agent is *R. typhi* (*R. mooseri*). Cells resemble *R. prowazekii* in morphology and staining properties (Figs. 412 and 413). Nonmotile. Gram-negative.

Organisms may be cultivated in plasma tissue cultures of mammalian cells; in fleas and lice; in the peritoneal cavities of x-irradiated rats; in the lungs of white mice and white rats following intranasal inoculation; in the lungs of rabbits following intratracheal inoculation; and in the chorioallantoic membrane and yolk sac of the developing chick embryo.

Disease produced Clinical picture resembles that of the epidemic or louse-borne type, except that symptoms are generally much less severe. Rash usually does not appear before the fifth day and may comprise only a few macules which tend to disappear after a day or so. Death rate for all ages is about 2 percent.

Distinguishable from rickettsiae of Rocky Mountain spotted fever, Q fever, and Tsutsugamushi disease by complement-fixation, agglutination, and precipitation tests, less readily from *R. prowazekii* by these tests. Serum from patients agglutinates *Proteus* OX_{19}.

Source of infection Fleas infected from rats, especially *Rattus norvegicus*, which are the reservoir.

Mode of transmission Organism transmitted from rodent to man by the flea *Xenopsylla cheopis*. Organisms present in feces of flea. Cells

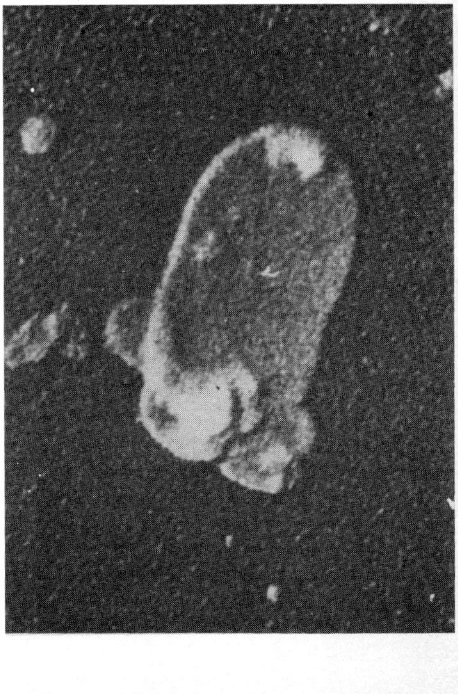

Fig. 413 *Cell wall of Rickettsia typhi.* (Courtesy Wood and Wisseman.)

inoculated by scratching louse feces into wound produced by bite of insect. Dirty clothing contaminated with dried louse feces may disseminate organisms into the air from where they may be inhaled.

Incubation period From 6 to 14 days, usually 12 days.

Susceptibility and immunity Susceptibility to disease is general. Acquired immunity follows recovery from disease but is not always permanent.

Prevalence Widely distributed in tropical, semitropical, and temperate climates where people and rats occupy same buildings. Increased rat populations on farms makes disease primarily rural instead of urban as once prevailed. Transmission occurs throughout year but more frequently during warmer months.

Prevention and control Application of insecticides with residual action to rat runs, burrows, and rat shelters to reduce flea populations. Control of rat population by trapping, poisoning, and ratproofing. Trapping and poisoning must be continuous to be of any practical value. Ratproofing only method that may be considered of permanent value.

Vaccine prepared against endemic typhus in same manner as for epidemic type and considered equally effective.

Treatment consists of application of tetracycline antibiotics or chloramphenicol orally daily until patient becomes afebrile, then for one additional day.

For more information: Benenson (1970); Myers et al. (1967); Snyder (1965); Stuart-Harris (1967); Wood and Wisseman (1967).

II. Tsutsugamushi disease (mite-borne, scrub typhus) The etiological agent is *R. tsutsugamushi* (*R. orientalis*) (Fig. 414). Small, pleomorphic bacterium-like organisms, usually wider and less sharply defined than cells of *R. prowazekii* and *R. typhi*. Ellipsoidal or rod-shaped, often appearing as a diplococcus or as a short bacillus with bipolar staining resembling that of the plague bacillus. Diffusely distributed in the cytoplasm of the cell. Measure 0.3 to 0.5 by 0.8 to 2.0 μ. Nonmotile. Gram-negative.

Cultivated in plasma tissue cultures of mammalian cells, on the chorioallantoic membrane and in the yolk sac of the chick embryo.

Cells readily inactivated by heat and chemicals. Destroyed at 50°C in 10 min, and by 0.5 percent phenol and 0.1 percent formalin.

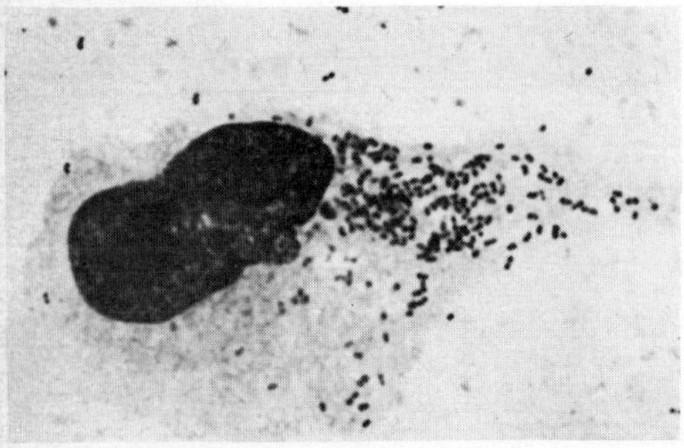

Fig. 414 Rickettsia tsutsugamushi in infected animal tissue cell. (After Philip.)

Disease produced An acute, specific infection of man prevalent in Japan and characterized by sudden onset, fever, malaise, chills, and headache. Primary sores develop at site of insect bite, caused by secretions of the insect and accompanied by adenitis of the regional lymph nodes. A skin eruption appears on the trunk, may spread to the arms and legs, and ordinarily disappears after several days. Cough and x-ray evidence of pneumonitis are common. The spleen and liver are congested and there may be cloudy swelling in the parenchymatous organs, and cellular necrosis. Convalescence usually occurs in 2 to 3 weeks.

Diagnosis Isolation of organism by inoculating patients' blood into mice. Complement-fixation and fluorescent antibody tests supplement the Weil-Felix reaction, using *Proteus OXK*, in serological diagnosis.

Source of infection Infected larval mites of *Leptotrombidium akamushi, L. deliensis* and related species depending upon locality. The infection is passed from generation to generation in mites, and maintained by feeding upon diseased wild rodents, especially mice and rats.

Mode of transmission By the bite of infected larval mites. The mites become infected in the larval stage by feeding on infected rodents (Fig. 415).

Incubation period Usually 10 to 12 days; may be as long as 21 days.

Fig. 415 The "cycle" of Rickettsia tsutsugamushi in nature through two generations in mites. (After Philip.)

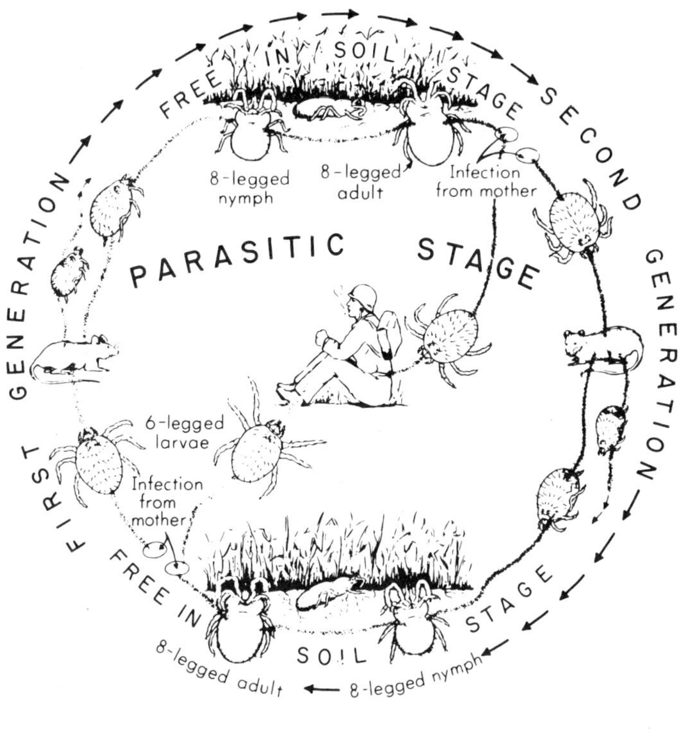

Susceptibility and immunity Susceptibility to disease is general. Recovery from an attack confers long-lasting immunity against the homologous strain. However, immunity is only transient against heterologous strains. Second and even third attacks are not uncommon among persons who spend their lives in endemic areas.

Serum from patient usually agglutinates *Proteus OXK*. Agglutination reaction usually becomes positive toward end of second week, reaches its peak during convalescence, then disappears rather rapidly.

Prevalence Widely distributed. Found in eastern and southeastern Asia, northern Australia and adjacent islands, India, Japan, Taiwan, and East Indies. Mainly restricted to adult workers who frequent scrub or overgrown terrain. Epidemics may occur when susceptible persons brought into endemic areas.

Prevention and control Wearing of miteproof clothing recommended in areas infested with mites. Grass and scrub on localities for camp sites should be cut level with the ground and burned. Camp area should also be burned, preferably with a flame thrower. Cots should be used to keep bedding from contact with the ground. Sleeping on ground should be avoided. Body should be kept clean by thorough soaping and scrubbing of skin. Impregnation of clothing with antimite fluids and powders.

Specific treatment consists of daily doses of one of the tetracycline antibiotics or chloramphenicol until patient is afebrile. A second course of treatments should be given after one week to prevent relapse.

For more information: Benenson (1970); Philip (1948, 1949); Smadel and Elisberg (1965); Stuart-Harris (1967).

III. Spotted fever group

A. Rocky Mountain spotted fever (tick-borne typhus) Etiological agent is *R. rickettsii*. Minute paired organisms surrounded by a narrow clear zone or halo; often lanceolate, resembling in appearance a minute pair of pneumococci. Average 0.6 by 1.2 μ. Nonmotile (Fig. 417). In the tick three forms may be recognized by Giemsa stain: (1) pale blue

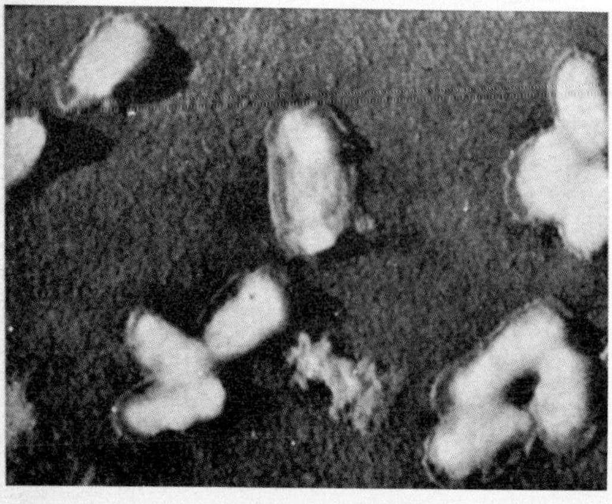

Fig. 416 Electron micrograph of Rickettsia akari, the etiological agent of rickettsialpox. X20,000. (After van Rooyen and Scott.)

bacillary forms, curved and club-shaped; (2) smaller, bluish rods with deeply staining chromatoid granules; and (3) more deeply staining, purplish, lanceolate forms. Occurs in the cytoplasm and nucleus in all types of cells in the tick; also occurs in mammals in the vascular endothelium, in macrophages, in the serosal cells of the peritoneal cavity, and in smooth muscle cells of arteriolar walls.

May be cultivated in plasma tissue cultures of mammalian cells, on the chorioallantoic membrane, in the yolk sac of the chick embryo, and in ticks.

White blood cell count usually shows a slight increase over the normal. Serum from patients with spotted fever agglutinates *Proteus* OX_{19} and usually OX_2. Distinguishable from *R. prowazekii* and *R. typhi* by complement fixation and by agglutination with specific antigens. Agglutination reaction usually becomes positive toward end of second week, reaches its peak during convalescence, then disappears rather rapidly.

Cells inactivated at 50°C in 10 min, and by 0.5 percent phenol and 0.1 percent formalin. Destroyed by desiccation in about 10 hr.

Disease produced A specific infectious disease characterized by sudden onset, fever, headache, conjunctival injection, a maculopapular rash on the extremities on about the third day, spreading over most of the body including palms and soles before becoming petechial. Also enlargement of the spleen, and catarrh of the respiratory tract. Patients show a history of either tick bite or exposure to ticks. Death rate varies with age and locality; in the United States for all ages, it is about 20 percent.

Source of infection Infected ticks. In the eastern and southern United States, the common vector is the dog tick *Dermacentor variabilis*; in the northwestern States it is the wood tick *D. andersoni*; in the southwestern United States, it may occasionally be the lone star tick *Amblyomma americanum*. In Brazil, the common vector is *A. cajennense*. Infection passed from generation to generation in ticks. Probably maintained in ticks by larvae feeding on susceptible wild rodents.

Mode of transmission Disease transmitted by bite of infected tick or contact with tick material, such as blood or feces on the unbroken skin.

Incubation period From 3 to 10 days.

Susceptibility and immunity Susceptibility to disease is general. Acquired immunity follows recovery from disease, but it is not always permanent.

Prevalence Occurs throughout the United States, western Canada, western and central Mexico, Colombia, and Brazil. Most prevalent in the South Atlantic states, less prevalent in the Rocky Mountain region. Disease occurs more frequently in spring and early summer, corresponding to time of appearance of adult ticks.

Prevention and control Avoidance of areas known to be infested with ticks. Removal of ticks from body as promptly as possible without crushing. Destruction of ticks in infested areas by clearing and burning vegetation, by destruction of small animals known to be infested with ticks.

Destruction of all ticks on patients. Specific treatment consists of daily oral doses of tetracycline antibiotics or chloramphenicol until patient is afebrile, then for one or two additional days.

For more information: Benenson (1970); Burgdorfer and Lackman (1960); Burgdorfer et al. (1968); Commission on Rickettsial Diseases (1960); Stuart-Harris (1967); Weiss et al. (1967); Woodward and Jackson (1965).

B. **Rickettsialpox** Etiological agent is *R. akari*. Minute diplobacilli and bipolar stained rods. Resemble typical rickettsiae morphologically, measure 0.6 by 0.9 to 1.4 μ, occur intracytoplasmically. Nonmotile. Gram-negative (Fig. 416).

Fig. 417 Ultrathin section of Rickettsia rickettsii in nucleus of epithelial cell of the wood tick Dermacentor andersoni. (Courtesy of Burgdorfer, Anacker, Bird, and Bertram.)

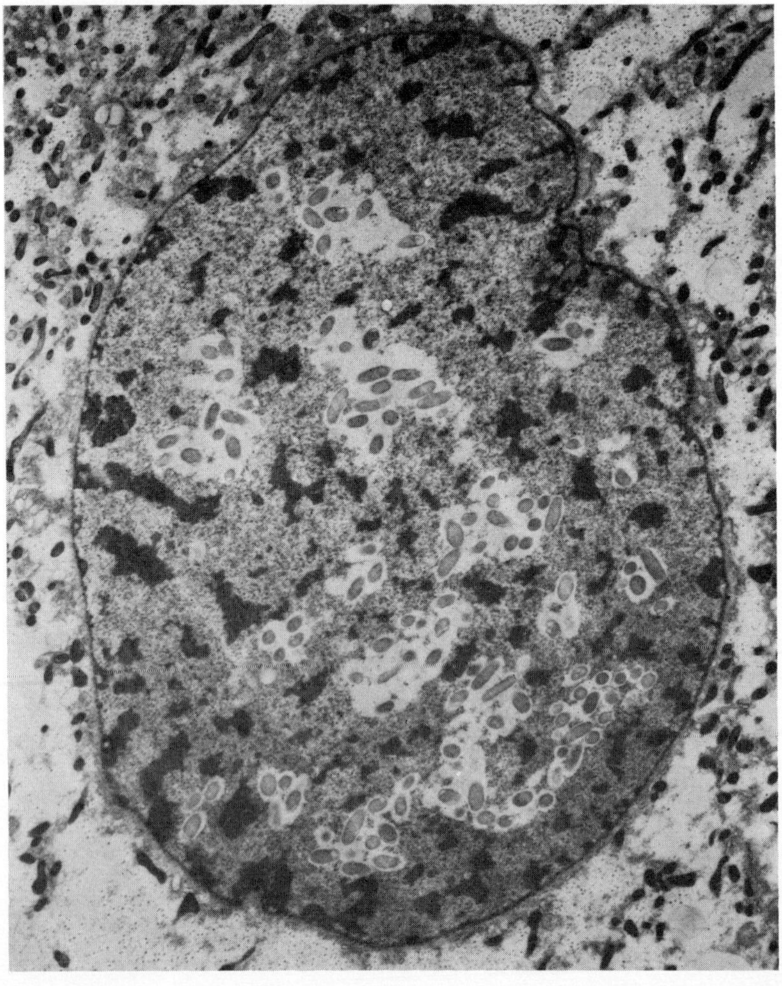

May be cultivated in yolk sac of chick embryo and in intrarectally injected body lice. No growth on artificial media.

Disease produced Disease characterized by initial skin lesion, chills, fever, rash, and mild to severe course. Initial lesion is a firm red papule and appears in about one week in advance of fever; becomes vesicular, then covered by a scab, and after 3 weeks leaves a small pigmented scar. Headache, muscular pain, and general malaise are frequent. A secondary rash appears 3 to 4 days after onset of fever without any characteristic distribution; seldom on palms or soles; progresses through papular and papulovesicular stages, lasting less than a week and leaving no scars.

Diagnosis Complement-fixation test becomes positive between second and third week of disease.

Source of infection Reservoir is infected house mouse (*Mus musculus*). Source of infection is infective mite.

Mode of transmission Organism transmitted from mouse to mouse, probably from mouse to man by a rodent mite (*Allodermanyssus sanguineus*).

Incubation period Probably 10 to 24 days.

Susceptibility and immunity Susceptibility to disease is general. Immunity follows an attack but duration unknown.

Prevention and control Elimination of rodents.

Specific treatment consists of use of one of the tetracycline antibiotics or chloramphenicol.

For more information: Benenson (1970); Breed et al. (1957); Fuller (1954); Horsfall and Tamm (1965); Stuart-Harris (1967); Woodward and Jackson (1965).

IV. Q fever Etiological agent is *Coxiella burnetii* (Fig. 418). Small, bacterium-like, pleomorphic organisms varying in size from coccoid forms to well-marked rods (Fig. 418). Occur as intracellular microcolonies with diffuse or compact distribution of organisms through the cytoplasm. Also seen extracellularly, where they appear as small, lanceolate rods, diplobacilli, and occasionally segmented filamentous forms. Small lanceolate rods, 0.25 by 0.4 to 0.5 μ, bipolar forms 0.25 by 1 μ, diplobacilli 0.25 by 1.5 μ. Nonmotile. Gram-negative.

May be cultivated in plasma tissue cultures, in the yolk sacs of chick embryos, and by injection into meal worms and certain other arthropods.

Agent readily passes through Berkefeld N filters, which are impermeable to ordinary bacteria, and W filters, which are impermeable to typhus and Rocky Mountain spotted fever rickettsiae. Resists a temperature of 60°C for 1 hr, and 0.5 percent formalin and 1 percent phenol for 24 hr when tested in fertile eggs.

Disease produced Q fever characterized by sudden onset, fever, chilly sensations, headache, weakness, malaise, and severe sweats. Virus-like or atypical pneumonia occurs in majority of cases, with mild cough,

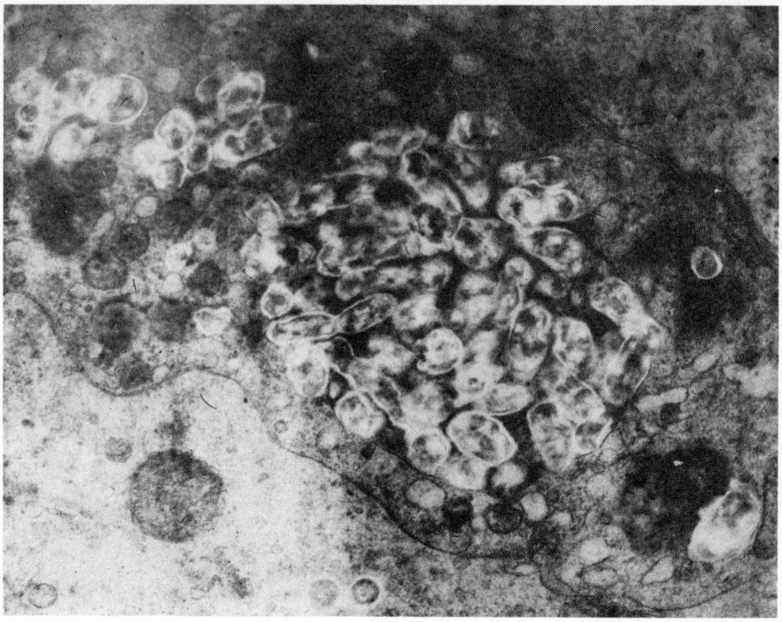

Fig. 418 Coxiella burnetii, both free in the cytoplasm and within a vacuole in guinea pig liver. (Courtesy of Handley, Paretsky, and Stueckemann.)

scanty expectoration, and chest pain. Upper respiratory tract essentially normal. Death rate negligible.

Diagnosis By complement-fixation or agglutination tests, or by recovery of organisms from blood of patient, a hazardous procedure for laboratory workers.

Source of infection Patient frequently shows history of contact with cattle, sheep, goat, or tick. Agent pathogenic for man, guinea pig, and white mouse. Mild infections occur in monkey, dog, white rat, rabbit, and cow. Etiologic agent abundant in mammary gland and milk of infected cows; also present in milk of sheep and goats. Common source of human infection is dust contaminated with products of domestic animals, especially placental tissues, birth fluids, and raw milk.

Mode of transmission Usually by airborne dissemination of organisms. Contact with infected animals, such as cows, sheep, goats, has accounted for a large proportion of cases reported. Agent may or may not be destroyed in pasteurization. Milk important medium for dissemination of organisms. Infected ticks may also be involved in transmission.

Incubation period Usually 2 or 3 weeks.

Susceptibility and immunity Susceptibility is general. An attack confers immunity for an indefinite period. Vaccination increases resistance to disease.

Prevalence Disease widespread. Endemic in California and several other states, affecting veterinarians, dairy workers, and farmers. Epidemics have occurred among laboratory workers, in stockyards, meat-packing plants, and wool-processing factories.

Prevention and control Immunization with vaccine found useful for laboratory workers, and others handling animals. Pasteurization or boiling of milk from cows, sheep, and goats. Control of importation of domestic animals.

Disinfection of sputum, excreta, and blood of infected individual and articles soiled by such materials.

Specific treatment consists of administration of tetracycline antibiotics or chloramphenicol orally and continued for several days after patient is afebrile. Repeat treatment if relapse occurs.

For more information: Bedson et al. (1967); Benenson (1970); Berman et al. (1961); Handley et al. (1967); Horsfall and Tamm (1965); Jones and Paretsky (1967); Linde and Urbach (1963); Luoto (1960); Nermut et al. (1968); Ormsbee (1965); Paretsky (1968); Stoenner and Lackman (!960); Stuart-Harris (1967); Tigertt et al. (1961).

References

Bedson, S., et al.: "Virus and Rickettsial Diseases of Man," London, Edward Arnold (Publishers) Ltd., 1967.

Benenson, A. S., ed.: "Control of Communicable Diseases in Man," New York, American Public Health Association, 1970.

Berman, S., et al.: Method for the production of a purified dry Q fever vaccine, *J. Bact.*, **81**:794, 1961.

Bird, R. G., et al.: Fine structure of *Rickettsia prowazeki* in the haemocytes of ticks *Hyalomma dromedarii, Acta Virol.*, **11**:60, 1967.

Breed, R. S., et al.: "Bergey's Manual of Determinative Bacteriology," Baltimore, The William & Wilkins Company, 1957.

Burgdorfer, W., and D. Lackman: Identification of *Rickettsia rickettsii* in the wood tick, *Dermacentor andersoni* by means of fluorescent antibody, *J. Infect. Diseases,* **107**:241, 1960.

—— et al.: Intranuclear growth of *Rickettsia rickettsii, J. Bact.*, **96**:1415, 1968.

Commission on Rickettsial Diseases: "Symposium on the Spotted Fever Groups of Rickettsiae," Washington, D.C., Walter Reed Army Institute of Research, 1960.

da Rocha-Lima, H.: Zur aetiologie des fleckfiebers, *Zentr. für Allgemein Path. und Path. Anat.*, **27**:45, 1916.

Fuller, H. S.: Studies of rickettsialpox. III. Life cycle of the mite vector, *Allodermanyssus sanguineus, Am. J. Hyg.*, **59**:236, 1954.

Handley, J., et al.: Electron microscopic observations of *Coxiella burnetii* in the guinea pig, *J. Bact.*, **94**:263, 1967.

Horsfall, F. L., Jr., and I. Tamm, eds.: "Viral and Rickettsial Infections of Man," Philadelphia, J. B. Lippincott Company, 1965.

Jadin, J., et al.: Ultrastructure of *Rickettsia prowazeki, Acta Virol.*, **12**:7, 1968.

Jones, F., Jr., and D. Paretsky: Physiology of rickettsiae. VI. Host-independent synthesis of polyribonucleotides by *Coxiella burnetii, J. Bact.*, **93**:1063, 1967.

Linde, K., and H. Urbach: Complement-fixing *Coxiella burnetii* antigen prepared from infective yolk sacs by trypsin treatment, *Acta Virol.*, **7**:90, 1963.

Luoto, L.: Report on the nationwide occurrence of Q fever infections in cattle, *Public Health Rept.*, **75**:135, 1960.

Myers, W. F., et al.: Permeability properties of *Rickettsia mooseri, J. Bact.*, **93**:950, 1967.

Nermut, M. V., et al.: Electron microscopy of *Coxiella burneti* phase I and II, *Acta Virol.*, **12**:446, 1968.

Ormsbee, R. A.: Q fever *Rickettsia*. In "Viral and Rickettsial Infections of Man," edited by F. L. Horsfall, Jr., and I. Tamm, Philadelphia, J. B. Lippincott Company, 1965.

Paretsky, D.: Biochemistry of rickettsiae and their infected hosts, with special reference to *Coxiella burneti, Centr. Bakt., I, Orig.*, **206**: 283, 1968.

Philip, C. B.: Tsutsugamushi disease (scrub typhus) in World War II, *J. Parasitol.*, **34**:169, 1948.

——: Scrub typhus, or tsutsugamushi disease, *Sci. Monthly*, **69**:281, 1949.

Smadel, J. E., and B. L. Elisberg: Scrub typhus *Rickettsia*. In "Viral and Rickettsial Infections of Man," edited by F. L. Horsfall, Jr., and I. Tamm, Philadelphia, J. B. Lippincott Company, 1965.

Snyder, J. C.: Typhus fever rickettsiae. In "Viral and Rickettsial Infections of Man," edited by F. L. Horsfall, Jr., and I. Tamm, Philadelphia, J. B. Lippincott Company, 1965.

Stoenner, H. G., and D. B. Lackman: The biologic properties of *Coxiella burnetii* isolated from rodents collected in Utah, *Am. J. Hyg.*, **71**:45, 1960.

Stuart-Harris, C. H.: The rickettsial diseases. In "Virus and Rickettsial Diseases of Man," edited by S. Bedson et al., London, Edward Arnold (Publishers) Ltd., 1967.

Tigertt, W. D., et al.: Airborne Q fever, *Bact. Rev.*, **25**:285, 1961.

Weiss, E., et al.: Metabolic activity of purified suspensions of *Rickettsia rickettsi, Nature*, **213**:1020, 1967.

Wisseman, C. L., Jr.: Some biological properties of rickettsiae pathogenic for man, *Centr. Bakt., I, Orig.*, **206**:299, 1968.

Wood, W. H., Jr., and C. L. Wisseman, Jr.: The cell wall of *Rickettsia mooseri.* I. Morphology and chemical composition, *J. Bact.*, **93**:1113, 1967.

Woodward, T. E., and E. B. Jackson: Spotted fever rickettsiae. In "Viral and Rickettsial Infections of Man," edited by F. L. Horsfall, Jr., and I. Tamm, Philadelphia, J. B. Lippincott Company, 1965.

Index

Index